# SCHAUM'S SOLVED PROBLEMS SERIES

## 2500 SOLVED PROBLEMS IN

# FLUID MECHANICS AND HYDRAULICS

by

**Jack B. Evett, Ph.D**

**Cheng Liu, M.S.**

The University of North Carolina at Charlotte

## McGRAW-HILL, INC.

New York   St. Louis   San Francisco   Auckland   Bogotá   Caracas   Lisbon
London   Madrid   Mexico City   Milan   Montreal   New Delhi
San Juan   Singapore   Sydney   Tokyo   Toronto

Jack B. Evett, Ph.D., *Professor of Civil Engineering,* and Cheng Liu, M.S., *Associate Professor of Civil Engineering Technology,* both at the *University of North Carolina at Charlotte.*

Both authors have extensive teaching experience in the domain of fluid mechanics and hydraulics. They are coauthors of a textbook in fluid mechanics for the McGraw-Hill College Division.

 This book is printed on recycled paper containing a minimum of 50% total recycled fiber with 10% postconsumer de-inked fiber. Soybean based inks are used on the cover and text.

Project supervision by The Total Book.
Cover design by Wanda Siedlecka.
Index by Hugh C. Maddocks, Ph.D.

**Library of Congress Cataloging-in-Publication Data**

Evett, Jack B.
    2500 solved problems in fluid mechanics and hydraulics / by Jack B. Evett, Cheng Liu.
        p.    cm. — (Schaum's solved problems series)
    ISBN 0-07-019783-0
    1. Fluid mechanics—Problems, exercises, etc. 2. Hydraulics—Problems, exercises, etc. I. Liu, Cheng.    II. Title. III. Title: Twenty-five hundred solved problems in fluid mechanics and hydraulics. IV. Series.
TA357.3.E84    1988
620.1′06′076—dc 19                                                  88-13373
                                                                                        CIP


    4 5 6 7 8 9 0 SHP/SHP     9 3


ISBN 0-07-019784-9 (Formerly published under ISBN 0-07-019783-0)

# CONTENTS

# To the Student

This book contains precisely 2500 completely solved problems in the areas of fluid mechanics and hydraulics. Virtually all types of problems ordinarily encountered in study and practice in these areas are covered. Not only you, but teachers, practitioners, and graduates reviewing for engineering licensing examinations should find these problems valuable.

To acquaint you with our "approach," particular steps taken in presenting the problems and their solutions are itemized below.

- First and most important of all, each problem and its solution are essentially independent and self-contained. That is to say, each contains all the data, equations, and computations necessary to find the answers. Thus, you should be able to pick a problem anywhere and follow its solution without having to review whatever precedes it. The exception to this is the occasional problem that specifically refers to, and carries over information from, a previous problem.
- In the solutions, our objective has been to present any needed equation first and then clearly to evaluate each term in the equation in order to find the answer. The terms may be evaluated separately or within the equation itself. For example, when solving an equation that has the parameter "area" as one of its terms, the area term ($A$) may be evaluated separately and its value substituted into the equation [as in Prob. 14.209], or it may be evaluated within the equation itself [as in Prob. 14.94].
- Virtually every number appearing in a solution is either "given" information (appearing as data in the statement of the problem or on an accompanying illustration), a previously computed value within the problem, a conversion factor (obtainable from the List of Conversion Factors), or a physical property (obtainable from a table or illustration in the Appendix). For example, in Prob. 1.77, the number 1.49, which does not appear elsewhere in the problem, is the dynamic viscosity ($\mu$) of glycerin; it was obtained from Table A-3 in the Appendix.
- We have tried to include all but the most familiar items in the List of Abbreviations and Symbols. Hence, when an unknown sign is encountered in a problem or its solution, a scan of that list should prove helpful. Thus, the infrequently used symbol $\psi$ is encountered in Prob. 25.6. According to the list, $\psi$ represents the stream function, and you are quickly on your way to a solution.

Every problem solution in this book has been checked, but, with 2500 in all, it is inevitable that some mistakes will slip through. We would appreciate it if you would take the time to communicate any mistakes you find to us, so that they may be corrected in future printings. We wish to thank Bill Langley, of The University of North Carolina at Charlotte, who assisted us with some of the problem selection and preparation.

# Abbreviations and Symbols

| | |
|---|---|
| $a$ | acceleration or area |
| $A$ | area |
| abs | absolute |
| $\alpha$ (alpha) | angle between absolute velocity of fluid in hydraulic machine and linear velocity of a point on a rotating body or coefficient of thermal expansion or dimensionless ratio of similitude |
| atm | atmosphere |
| atmos | atmospheric |
| $\beta$ (beta) | angle between relative velocity in hydraulic machines and linear velocity of a point on a rotating body or coefficient of compressibility or ratio of obstruction diameter to duct diameter |
| $b$ | surface width or other width |
| $B$ | surface width or other width |
| bhp | brake horsepower |
| bp | brake power |
| Btu | British thermal unit |
| $c$ | speed of sound or wave speed (celerity) |
| $C$ | Celsius or discharge coefficient or speed of propagation |
| cal | calorie |
| c.b. or CB | center of buoyancy |
| $C_c$ | coefficient of contraction |
| $C_d$ | coefficient of discharge |
| $C_D$ | drag coefficient |
| $C_f$ | friction-drag coefficient |
| $C_F$ | force coefficient |
| cfs | cubic foot per second |
| c.g. or CG | center of gravity |
| $C_I$ | Pitot tube coefficient |
| $C_L$ | lift coefficient |
| cm | centimeter ($10^{-2}$ m) |
| cP | centipoise |
| c.p. | center of pressure |
| $c_P$ | specific heat at constant pressure |
| $c_v$ | specific heat at constant volume |
| $C_v$ | coefficient of velocity |
| $C_w$ | weir coefficient |
| $d$ | depth or diameter |
| $D$ | depth or diameter or drag force |
| $\delta$ (delta) | thickness of boundary layer |
| $\delta_1$ (delta) | thickness of the viscous sublayer |
| $\Delta$ (Delta) | change in (or difference between) |
| $d_c$ | critical depth |
| $D_{\text{eff}}$ | effective diameter |
| $D_h$ | hydraulic diameter |
| $d_m$ | mean depth |
| $d_n$ | normal depth |
| $d_N$ | normal depth |
| $E$ | modulus of elasticity or specific energy or velocity approach factor |
| $e_h$ | hydraulic efficiency |
| el | elevation |
| $\eta$ (eta) | pump or turbine efficiency |
| $\epsilon$ (epsilon) | height or surface roughness |
| $E_p$ | pump energy |
| $E_t$ | turbine energy |
| exp | exponential |
| $f$ | frequency of oscillation (cycles per second) or friction factor |

| | |
|---|---|
| F | Fahrenheit or force |
| $F_b$ | buoyant force |
| $F_D$ | drag force |
| $F_H$ | horizontal force |
| $F_L$ | lift force |
| fps | foot per second |
| F.S. | factor of safety |
| ft | foot |
| $F_U$ | uplift force on a dam |
| $F_V$ | vertical force |
| g | acceleration due to gravity or gage height or gram |
| G | weight flow rate |
| gal | gallon |
| $\gamma$ (gamma) | specific (or unit) weight |
| $\Gamma$ (Gamma) | circulation |
| GN | giganewton ($10^9$ N) |
| GPa | gigapascal ($10^9$ Pa) |
| gpm | gallons per minute |
| h | enthalpy per unit mass or height or depth or pressure head or hour |
| $\bar{h}$ | average height or depth or head |
| $\hat{h}$ | enthalpy per unit weight |
| H | energy head or total energy head |
| $h_1$ | unit head loss |
| $h_{cg}$ | vertical depth to center of gravity |
| $h_{cp}$ | vertical depth to center of pressure |
| $h_f$ | head loss due to friction |
| Hg | mercury |
| HGL | hydraulic grade line |
| $h_L$ | total head loss |
| $h_m$ | head loss due to minor losses |
| hp | horsepower |
| Hz | hertz (cycles per second) |
| I | inflow or moment of inertia |
| ID | inside diameter |
| in | inch |
| $\infty$ (infinity) | sometimes used as a subscript to indicate upstream |
| J | joule |
| K | bulk modulus of elasticity or Kelvin or minor loss coefficient |
| k | specific heat ratio |
| kcal | kilocalorie ($10^3$ cal) |
| kg | kilogram ($10^3$ g) |
| kJ | kilojoule ($10^3$ J) |
| km | kilometer ($10^3$ m) |
| kN | kilonewton ($10^3$ N) |
| kPa | kilopascal ($10^3$ Pa) |
| kW | kilowatt ($10^3$ W) |
| L | length or lift force or liter |
| $\lambda$ (lambda) | model ratio or wave length |
| lb | pound |
| $lb_m$ | pound mass |
| $L_e$ | equivalent length |
| $L_m$ | linear dimension in model |
| $L_p$ | linear dimension in prototype |
| m | mass or meter |
| $\dot{m}$ | mass flow rate |
| $\underline{M}$ | mass flow rate or molecular weight or moment or torque |
| $\overline{MB}$ | distance from center of buoyancy to metacenter |
| mbar | millibar ($10^{-3}$ bar) |
| mc | metacenter |
| mgd | million gallons per day |

| | |
|---|---|
| ml | milliliter ($10^{-3}$ L) |
| min | minute |
| mm | millimeter ($10^{-3}$ meter) |
| MN | meganewton ($10^6$ N) |
| MPa | megapascal ($10^6$ Pa) |
| mph | mile per hour |
| MR | manometer reading |
| $\mu$ (mu) | absolute or dynamic viscosity |
| MW | megawatt ($10^6$ W) |
| $n$ | Manning roughness coefficient or number of moles |
| $N$ | newton or rotational speed |
| $N_B$ | Brinkman number |
| $N_F$ | Froude number |
| $N_M$ | Mach number |
| NPSH | net positive suction head |
| $N_R$ | Reynolds number |
| $N_s$ | specific speed of pump or turbine |
| $v$ (nu) | kinematic viscosity |
| $N_W$ | Weber number |
| O | outflow |
| OD | outside diameter |
| $\Omega$ (ohm) | rotational rate |
| $\omega$ (omega) | angular velocity |
| $p$ | pressure or poise |
| $P$ | force (usually resulting from an applied pressure) or power |
| Pa | pascal |
| $\phi$ (phi) | peripheral-velocity factor |
| $\pi$ (pi) | constant $= 3.14159265$ |
| $\Pi$ (pi) | dimensionless parameter |
| $P_r$ | power ratio |
| $p_s$ | stagnation pressure |
| psi | pound per square inch |
| $\psi$ (psi) | stream function |
| psia | pound per square inch absolute |
| psig | pound per square inch gage |
| $p^{*\prime}$ | pressure for condition at $N_M = 1/\sqrt{k}$ |
| $p_v$ | vapor pressure |
| $p_w$ | wetted perimeter |
| $q$ | flow rate per unit width or heat per unit mass |
| $Q$ | discharge or heat or volume flow rate |
| $Q_H$ | heat transferred per unit weight of fluid |
| $Q/w$ | volume flow rate per unit width of channel |
| qt | quart |
| $r$ | radius |
| $R$ | gas constant or Rankine or resultant force or hydraulic radius |
| $R'$ | manometer reading |
| rad | radian |
| $R_c$ | critical hydraulic radius |
| $R_h$ | hydraulic radius |
| $\rho$ (rho) | mass density |
| $r_i$ | inside radius |
| $r_o$ | outside radius |
| rpm | revolutions per minute |
| $R_u$ | universal gas constant |
| $s$ | entropy of a substance or second or slope |
| $S$ | slope or storage |
| $s_c$ | critical slope |
| s.g. | specific gravity |
| s.g.$_M$ | specific gravity of manometer fluid |
| s.g.$_F$ | specific gravity of flowing fluid |

| | |
|---|---|
| $\sigma$ (sigma) | pump cavitation parameter or stress or surface tension |
| $\sigma'$ | cavitation index |
| $\Sigma$ (sigma) | summation |
| $S$ | specific gravity of flowing fluid |
| $S_0$ | specific gravity of manometer fluid |
| $t$ | thickness or time |
| $T$ | surface width or temperature or torque or tension |
| $\tau$ (tau) | shear stress |
| $\tau_0$ (tau) | shear stress at the wall |
| $T_s$ | stagnation temperature |
| $u$ | velocity |
| $u_{C_L}$ | centerline velocity |
| $U$ | velocity |
| $v$ | velocity |
| $v_c$ | critical velocity |
| $V$ | velocity or volume |
| $v_{av}$ | average velocity |
| $V_{C_L}$ | centerline velocity |
| $V_d$ | volume of fluid displaced |
| $V_m$ | velocity in model |
| $V_p$ | velocity in prototype |
| $V_s$ | specific volume |
| $v_*$ | shear velocity |
| $v_t$ | tangential velocity |
| $v_T$ | terminal velocity |
| $w$ | width |
| $W$ | watt or weight or weight flow rate or work |
| $x_{cp}$ | distance from center of gravity to center of pressure in $x$ direction |
| $\xi$ (xi) | vorticity |
| $y$ | depth |
| $y_c$ | critical depth |
| $y_{cp}$ | distance from center of gravity to center of pressure in $y$ direction |
| $y_n$ | normal depth |
| $y_N$ | normal depth |
| $z_{cg}$ | inclined distance from liquid surface to center of gravity |
| $z_{cp}$ | inclined distance from liquid surface to center of pressure |

# Conversion Factors

$0.00001667 \text{ m}^3/\text{s} = 1 \text{ L/min}$
$0.002228 \text{ ft}^3/\text{s} = 1 \text{ gal/min}$
$0.0145 \text{ lb/in}^2 = 1 \text{ mbar}$
$0.3048 \text{ m} = 1 \text{ ft}$
$2.54 \text{ cm} = 1 \text{ in}$
$3.281 \text{ ft} = 1 \text{ m}$
$4 \text{ qt} = 1 \text{ gal}$
$4.184 \text{ kJ} = 1 \text{ kcal}$
$4.448 \text{ N} = 1 \text{ lb}$
$6.894 \text{ kN/m}^2 = 1 \text{ lb/in}^2$.
$7.48 \text{ gal} = 1 \text{ ft}^3$
$12 \text{ in} = 1 \text{ ft}$
$14.59 \text{ kg} = 1 \text{ slug}$
$25.4 \text{ mm} = 1 \text{ in}$
$60 \text{ min} = 1 \text{ h}$
$60 \text{ s} = 1 \text{ min}$
$100 \text{ cm} = 1 \text{ m}$
$100 \text{ kPa} = 1 \text{ bar}$

$101.3 \text{ kPa} = 1 \text{ atm}$
$144 \text{ in}^2 = 1 \text{ ft}^2$
$550 \text{ ft-lb/s} = 1 \text{ hp}$
$778 \text{ ft-lb} = 1 \text{ Btu}$
$1000 \text{ N} = 1 \text{ kN}$
$1000 \text{ L} = 1 \text{ m}^3$
$1000 \text{ mm} = 1 \text{ m}$
$1000 \text{ Pa} = 1 \text{ kPa}$
$1728 \text{ in}^3 = 1 \text{ ft}^3$
$2000 \text{ lb} = 1 \text{ ton}$
$3600 \text{ s} = 1 \text{ h}$
$4184 \text{ J} = 1 \text{ kcal}$
$5280 \text{ ft} = 1 \text{ mile}$
$86\,400 \text{ s} = 1 \text{ day}$
$1\,000\,000 \text{ N} = 1 \text{ MN}$
$1\,000\,000 \text{ Pa} = 1 \text{ MPa}$
$1\,000\,000\,000 \text{ N} = 1 \text{ GN}$
$1\,000\,000\,000 \text{ Pa} = 1 \text{ GPa}$

**Note:** For many problems in this chapter, values of various physical properties of fluids are obtained from Tables A-1 through A-8 in the Appendix.

**1.1** A reservoir of glycerin (glyc) has a mass of 1200 kg and a volume of 0.952 m³. Find the glycerin's weight ($W$), mass density ($\rho$), specific weight ($\gamma$), and specific gravity (s.g.).

$$F = W = ma = (1200)(9.81) = 11\,770\,\text{N} \quad \text{or} \quad 11.77\,\text{kN}$$
$$\rho = m/V = 1200/0.952 = 1261\,\text{kg/m}^3$$
$$\gamma = W/V = 11.77/0.952 = 12.36\,\text{kN/m}^3$$
$$\text{s.g.} = \gamma_{\text{glyc}}/\gamma_{\text{H}_2\text{O at 4 °C}} = 12.36/9.81 = 1.26$$

**1.2** A body requires a force of 100 N to accelerate it at a rate of 0.20 m/s². Determine the mass of the body in kilograms and in slugs.

$$F = ma$$
$$100 = (m)(0.20)$$
$$m = 500\,\text{kg} = 500/14.59 = 34.3\,\text{slugs}$$

**1.3** A reservoir of carbon tetrachloride ($CCl_4$) has a mass of 500 kg and a volume of 0.315 m³. Find the carbon tetrachloride's weight, mass density, specific weight, and specific gravity.

$$F = W = ma = (500)(9.81) = 4905\,\text{N} \quad \text{or} \quad 4.905\,\text{kN}$$
$$\rho = m/V = 500/0.315 = 1587\,\text{kg/m}^3$$
$$\gamma = W/V = 4.905/0.315 = 15.57\,\text{kN/m}^3$$
$$\text{s.g.} = \gamma_{CCl_4}/\gamma_{\text{H}_2\text{O at 4 °C}} = 15.57/9.81 = 1.59$$

**1.4** The weight of a body is 100 lb. Determine (a) its weight in newtons, (b) its mass in kilograms, and (c) the rate of acceleration [in both feet per second per second (ft/s²) and meters per second per second (m/s²)] if a net force of 50 lb is applied to the body.

(a)
$$W = (100)(4.448) = 444.8\,\text{N}$$

(b)
$$F = W = ma \quad 444.8 = (m)(9.81) \quad m = 45.34\,\text{kg}$$

(c)
$$m = 45.34/14.59 = 3.108\,\text{slugs}$$
$$F = ma \quad 50 = 3.108a \quad a = 16.09\,\text{ft/s}^2 = (16.09)(0.3048) = 4.904\,\text{m/s}^2$$

**1.5** The specific gravity of ethyl alcohol is 0.79. Calculate its specific weight (in both pounds per cubic foot and kilonewtons per cubic meter) and mass density (in both slugs per cubic foot and kilograms per cubic meter).

$$\gamma = (0.79)(62.4) = 49.3\,\text{lb/ft}^3 \qquad \gamma = (0.79)(9.79) = 7.73\,\text{kN/m}^3$$
$$\rho = (0.79)(1.94) = 1.53\,\text{slugs/ft}^3 \qquad \rho = (0.79)(1000) = 790\,\text{kg/m}^3$$

**1.6** A quart of water weights about 2.08 lb. Compute its mass in slugs and in kilograms.

$$F = W = ma \quad 2.08 = (m)(32.2)$$
$$m = 0.0646\,\text{slug} \quad m = (0.0646)(14.59) = 0.943\,\text{kg}$$

**1.7** One cubic foot of glycerin has a mass of 2.44 slugs. Find its specific weight in both pounds per cubic foot and kilonewtons per cubic meter.

$F = W = ma = (2.44)(32.2) = 78.6\,\text{lb}$. Since the glycerin's volume is 1 ft³, $\gamma = 78.6\,\text{lb/ft}^3 =$ $(78.6)(4.448)/(0.3048)^3 = 12\,350\,\text{N/m}^3$, or $12.35\,\text{kN/m}^3$.

**1.8**    A quart of SAE 30 oil at 68 °F weighs about 1.85 lb. Calculate the oil's specific weight, mass density, and specific gravity.

$$V = 1/[(4)(7.48)] = 0.03342 \text{ ft}^3$$
$$\gamma = W/V = 1.85/0.03342 = 55.4 \text{ lb/ft}^3$$
$$\rho = \gamma/g = 55.4/32.2 = 1.72 \text{ slugs/ft}^3$$
$$\text{s.g.} = \gamma_{\text{oil}}/\gamma_{\text{H}_2\text{O at 4°C}} = 55.4/62.4 = 0.888$$

**1.9**    The volume of a rock is found to be 0.00015 m³. If the rock's specific gravity is 2.60, what is its weight?

$$\gamma_{\text{rock}} = (2.60)(9.79) = 25.5 \text{ kN/m}^3 \qquad W_{\text{rock}} = (25.5)(0.00015) = 0.00382 \text{ kN} \quad \text{or} \quad 3.82 \text{ N}$$

**1.10**    A certain gasoline weighs 46.5 lb/ft³. What are its mass density, specific volume, and specific gravity?

$$\rho = \gamma/g = 46.5/32.2 = 1.44 \text{ slugs/ft}^3 \qquad V_s = 1/\rho = 1/1.44 = 0.694 \text{ ft}^3/\text{slug}$$
$$\text{s.g.} = 1.44/1.94 = 0.742$$

**1.11**    If the specific weight of a substance is 8.2 kN/m³, what is its mass density?

$$\rho = \gamma/g = 8200/9.81 = 836 \text{ kg/m}^3$$

**1.12**    An object at a certain location has a mass of 2.0 kg and weighs 19.0 N on a spring balance. What is the acceleration due to gravity at this location?

$$F = W = ma \qquad 19.0 = 2.0a \qquad a = 9.50 \text{ m/s}^2$$

**1.13**    If an object has a mass of 2.0 slugs at sea level, what would its mass be at a location where the acceleration due to gravity is 30.00 ft/s²?

■  Since the mass of an object does not change, its mass will be 2.0 slugs at that location.

**1.14**    What would be the weight of a 3-kg mass on a planet where the acceleration due to gravity is 10.00 m/s²?

$$F = W = ma = (3)(10.00) = 30.00 \text{ N}$$

**1.15**    Determine the weight of a 5-slug boulder at a place where the acceleration due to gravity is 31.7 ft/s².

$$F = W = ma = (5)(31.7) = 158 \text{ lb}$$

**1.16**    If 200 ft³ of oil weighs 10 520 lb, calculate its specific weight, density, and specific gravity.

$$\gamma = W/V = 10\,520/200 = 52.6 \text{ lb/ft}^3 \qquad \rho = \gamma/g = 52.6/32.2 = 1.63 \text{ slugs/ft}^3$$
$$\text{s.g.} = \gamma_{\text{oil}}/\gamma_{\text{H}_2\text{O at 4°C}} = 52.6/62.4 = 0.843$$

**1.17**    Find the height of the free surface if 0.8 ft³ of water is poured into a conical tank (Fig. 1-1) 20 in high with a base radius of 10 in. How much additional water is required to fill the tank?

$$V_{\text{cone}} = \pi r^2 h/3 = \pi(10)^2(20)/3 = 2094 \text{ in}^3 \qquad V_{\text{H}_2\text{O}} = 0.8 \text{ ft}^3 = 1382 \text{ in}^3$$

Additional water needed = 2095 − 1382 = 713 in³. From Fig. 1-1, $r_o/10 = h_o/20$, or $r_o = h_o/2.0$;
$V_{\text{empty (top) cone}} = \pi(h_o/2.0)^2 h_o/3 = 713$; $h_o = 13.96$ in. Free surface will be 20 − 13.96, or 6.04 in above base of tank.

r = r_o

h_o

20 in

r = 10 in

**Fig. 1-1**

**1.18** If the tank of Prob. 1.17 holds 30.5 kg of salad oil, what is the density of the oil?

$$V_{cone} = 2094 \text{ in}^3 \quad \text{(from Prob. 1.17)}$$
$$= \tfrac{2094}{1728}(0.3048)^3 = 0.03431 \text{ m}^3$$
$$\rho = m/V = 30.5/0.03431 = 889 \text{ kg/m}^3$$

**1.19** Under standard conditions a certain gas weighs 0.14 lb/ft³. Calculate its density, specific volume, and specific gravity relative to air weighing 0.075 lb/ft³.

$$\rho = \gamma/g = 0.14/32.2 = 0.00435 \text{ slug/ft}^3 \qquad V_s = 1/\rho = 1/0.00435 = 230 \text{ ft}^3/\text{slug}$$
$$\text{s.g.} = 0.14/0.075 = 1.87$$

**1.20** If the specific volume of a gas is 360 ft³/slug, what is its specific weight?

$$\rho = 1/V_s = \tfrac{1}{360} = 0.002778 \text{ slug/ft}^3 \qquad \gamma = \rho g = (0.002778)(32.2) = 0.0895 \text{ lb/ft}^3$$

**1.21** A vertical glass cylinder contains 900.00 mL of water at 10 °C; the height of the water column is 90.00 cm. The water and its container are heated to 80 °C. Assuming no evaporation, what will be the height of the water if the coefficient of thermal expansion ($\alpha$) for the glass is $3.6 \times 10^{-6}$ °C$^{-1}$?

$$\text{Mass of water} = \rho V = \rho_{10}V_{10} = \rho_{80}V_{80} \quad (1000)(900 \times 10^{-6}) = 971V_{80} \qquad V_{80} = 926.9 \times 10^{-6} \text{ m}^3 = 926.9 \text{ cm}^3$$
$$A_{10} = V_{10}/h_{10} = 900.00/90.00 = 10.000 \text{ cm}^2$$
$$A_{10} = \pi r_{10}^2 \qquad 10.000 = \pi r_{10}^2 \qquad r_{10} = 1.7841 \text{ cm}$$
$$r_{80} = r_{10}[1 + (\Delta T)(\alpha)] = (1.7841)[1 + (80 - 10)(3.6 \times 10^{-6})] = 1.7845 \text{ cm}$$
$$A_{80} = \pi r_{80}^2 = \pi(1.7845)^2 = 10.004 \text{ cm}^2 \qquad h_{80} = V_{80}/A_{80} = 926.9/10.004 = 92.65 \text{ cm}$$

**1.22** If a vessel that contains 3.500 ft³ of water at 50 °F and atmospheric pressure is heated to 160 °F, what will be the percentage change in its volume? What weight of water must be removed to maintain the original volume?

$$\text{Weight of water} = \gamma V = \gamma_{50}V_{50} = \gamma_{160}V_{160} \qquad (62.4)(3.500) = 61.0V_{160} \qquad V_{160} = 3.5803 \text{ ft}^3$$

Change in volume = $(3.5803 - 3.500)/3.000 = 0.027$, or 2.7% (increase). Must remove $(3.5803 - 3.500)(61.0)$, or 4.90 lb.

**1.23** A vertical, cylindrical tank with a diameter of 12.00 m and a depth of 4.00 m is filled to the top with water at 20 °C. If the water is heated to 50 °C, how much water will spill over?

$$V_{tank} = (V_{H_2O})_{20} = \pi(12.00/2)^2(4.00) = 452.4 \text{ m}^3$$
$$W_{H_2O} = (9.79)(452.4) = 4429 \text{ kN} \qquad (V_{H_2O})_{50} = 4429/9.69 = 457.1 \text{ m}^3$$
$$\text{Volume of water spilled} = 457.1 - 452.4 = 4.7 \text{ m}^3$$

**1.24** A thick, closed, steel chamber is filled with water at 50 °F and atmospheric pressure. If the temperature of water and chamber is raised to 100 °F, find the new pressure of the water. The coefficient of thermal expansion of steel is $6.5 \times 10^{-6}$ per °F.

The volume of water would attempt to increase as the cube of the linear dimension; hence, $V_{90} = V_{50}[1 + (100 - 50)(6.5 \times 10^{-6})]^3 = 1.000975V_{50}$; weight of water = $\gamma V = \gamma_{50}V_{50} = \gamma_{90}V_{90}$, $62.4V_{50} = \gamma_{90}(1.000975V_{50})$, $\gamma_{90} = 62.34$ lb/ft³. From Fig. A-3, $p_{90} = 1300$ psia (approximately).

**1.25** A liquid compressed in a cylinder has a volume of 1000 cm³ at 1 MN/m² and a volume of 995 cm³ at 2 MN/m². What is its bulk modulus of elasticity ($K$)?

$$K = -\frac{\Delta p}{\Delta V/V} = -\frac{2 - 1}{(995 - 1000)/1000} = 200 \text{ MPa}$$

**1.26** Find the bulk modulus of elasticity of a liquid if a pressure of 150 psi applied to 10 ft³ of the liquid causes a volume reduction of 0.02 ft³.

$$K = -\frac{\Delta p}{\Delta V/V} = -\frac{(150 - 0)(144)}{-0.02/10} = 10\,800\,000 \text{ lb/ft}^2 \quad \text{or} \quad 75\,000 \text{ psi}$$

**1.27** If $K = 2.2$ GPa is the bulk modulus of elasticity for water, what pressure is required to reduce a volume by 0.6 percent?

$$K = -\frac{\Delta p}{\Delta V/V} \qquad 2.2 = -\frac{p_2 - 0}{-0.006} \qquad p_2 = 0.0132 \text{ GPa} \quad \text{or} \quad 13.2 \text{ MPa}$$

**1.28** Find the change in volume of $1.00000$ ft$^3$ of water at 80 °F when subjected to a pressure increase of 300 psi. Water's bulk modulus of elasticity at this temperature is 325 000 psi.

$$K = -\frac{\Delta p}{\Delta V/V} \qquad 325\,000 = -\frac{300 - 0}{\Delta V/1.00000} \qquad \Delta V = -0.00092 \text{ ft}^3$$

**1.29** From the following test data, determine the bulk modulus of elasticity of water: at 500 psi the volume was $1.000$ ft$^3$, and at 3500 psi the volume was $0.990$ ft$^3$.

$$K = -\frac{\Delta p}{\Delta V/V} = -\frac{500 - 3500}{(1.000 - 0.990)/1.000} = 300\,000 \text{ psi}$$

**1.30** A rigid steel container is partially filled with a liquid at 15 atm. The volume of the liquid is $1.23200$ L. At a pressure of 30 atm, the volume of the liquid is $1.23100$ L. Find the average bulk modulus of elasticity of the liquid over the given range of pressure if the temperature after compression is allowed to return to its initial value. What is the coefficient of compressibility ($\beta$)?

$$K = -\frac{\Delta p}{\Delta V/V} = -\frac{(30 - 15)(101.3)}{(1.23100 - 1.23200)/1.23200} = 1.872 \times 10^6 \text{ kPa} \quad \text{or} \quad 1.872 \text{ GPa}$$

$$\beta = 1/K = 1/1.872 = 0.534 \text{ GPa}^{-1}$$

**1.31** A heavy tank contains oil ($A$) and water ($B$) subject to variable air pressure; the dimensions shown in Fig. 1-2 correspond to 1 atm. If air is slowly added from a pump to bring pressure $p$ up to 1 MPa gage, what will be the total downward movement of the free surface of oil and air? Take average values of bulk moduli of elasticity of the liquids as 2050 MPa for oil and 2075 MPa for water. Assume the container does not change volume. Neglect hydrostatic pressures.

$$K = -\frac{\Delta p}{\Delta V/V} \qquad 2050 = -\frac{1 - 0}{\Delta V_{\text{oil}}/[600\pi(300)^2/4]} \qquad \Delta V_{\text{oil}} = -20\,690 \text{ mm}^3$$

$$2075 = -\frac{1 - 0}{\Delta V_{\text{H}_2\text{O}}/[700\pi(300)^2/4]} \qquad \Delta V_{\text{H}_2\text{O}} = -23\,850 \text{ mm}^3$$

$$\Delta V_{\text{total}} = -44\,540 \text{ mm}^3$$

Let $x$ = distance the upper free surface moves. $-44\,540 = -[\pi(300)^2/4]x$, $x = 0.630$ mm.

**Fig. 1-2**

**1.32** A thin-walled spherical tank is filled with water at a pressure of 4666 psig; the tank's volume is then $805.407$ in$^3$. If the water is released from the tank, how many pounds will be collected at atmospheric pressure? $805.4069$ in$^3$

when the pressure is 4666 psig. Use 305 000 psi as an average value of the bulk modulus of elasticity.

$$K = -\frac{\Delta p}{\Delta V/V} \qquad 305\,000 = -\frac{0 - 4666}{(V_2 - 805.407)/805.407} \qquad V_2 = 817.73 \text{ in}^3$$

$$W = (62.4)(817.73/1728) = 29.5 \text{ lb}$$

**1.33** Water in a hydraulic press, initially at 20 psia, is subjected to a pressure of 17 000 psia at 68 °F. Determine the percentage decrease in specific volume if the average bulk modulus of elasticity is 365 000 psi.

$$K = -\frac{\Delta p}{\Delta V/V} \qquad 365\,000 = -\frac{17\,000 - 20}{\Delta V/V_1} \qquad \frac{\Delta V}{V_1} = -0.0465 \quad \text{or} \quad 4.65\% \text{ decrease}$$

**1.34** At a depth of 7 km in the ocean, the pressure is 71.6 MPa. Assume a specific weight at the surface of 10.05 kN/m³ and an average bulk modulus of elasticity of 2.34 GPa for that pressure range. Find (*a*) the change in specific volume between the surface and 7 km; (*b*) the specific volume at 7 km; (*c*) the specific weight at 7 km.

(*a*) $$(V_s)_1 = 1/\rho_1 = g/\gamma_1 = 9.81/10\,050 = 0.0009761 \text{ m}^3/\text{kg}$$

$$K = -\frac{\Delta p}{\Delta V_s/V_s} \qquad 2.34 \times 10^9 = -\frac{71.6 \times 10^6 - 0}{\Delta V_s/0.0009761} \qquad \Delta V_s = -0.0000299 \text{ m}^3/\text{kg}$$

(*b*) $$(V_s)_2 = (V_s)_1 + \Delta V_s = 0.0009761 - 0.0000299 = 0.000946 \text{ m}^3/\text{kg}$$
(*c*) $$\gamma_2 = g/V_2 = 9.81/0.000946 = 10\,370 \text{ N/m}^3$$

**1.35** Approximately what pressure must be applied to water at 60 °F to reduce its volume 2.5 percent?

$$K = -\frac{\Delta p}{\Delta V/V} \qquad 311\,000 = -\frac{p_2 - 0}{0.025} \qquad p_2 = 7775 \text{ psi}$$

**1.36** A gas at 20 °C and 0.21 MPa abs has a volume of 41 L and a gas constant ($R$) of 210 m · N/(kg · K). Determine the density and mass of the gas.

$$\rho = p/RT = 0.21 \times 10^6/[(210)(20 + 273)] = 3.41 \text{ kg/m}^3 \qquad m = \rho V = (3.41)(0.041) = 0.140 \text{ kg}$$

**1.37** What is the specific weight of air at 70 psia and 70 °F?

$\gamma = p/RT$. From Table A-6, $R = 53.3 \text{ ft/°R}$; $\gamma = (70)(144)/[(53.3)(70 + 460)] = 0.357 \text{ lb/ft}^3$.

**Note:** $p/RT$ gives $\rho$ (Prob. 1.36) or $\gamma$ (Prob. 1.37), depending on the value of $R$ used. Corresponding values of $R$ in Table A-6 differ by a factor of $g$.

**1.38** Calculate the density of water vapor at 350 kPa abs and 20 °C if its gas constant ($R$) is 0.462 kPa · m³/kg · K.

$$\rho = p/RT = 350/[(0.462)(20 + 273)] = 2.59 \text{ kg/m}^3$$

**1.39** Nitrogen gas (molecular weight 28) occupies a volume of 4.0 ft³ at 2500 lb/ft² abs and 750 °R. What are its specific volume and specific weight?

$$R = R_u/M = 49\,709/28 = 1775 \text{ ft} \cdot \text{lb/(slug} \cdot \text{°R)}$$

[where $R_u$, the universal gas constant, $= 49\,709 \text{ ft} \cdot \text{lb/(slug} \cdot \text{°R)}$]

$$\rho = 1/V_s = p/RT = 2500/[(1775)(750)] \qquad V_s = 532.5 \text{ ft}^3/\text{slug}$$

$$\gamma = \rho g = (1/V_s)(g) = (1/532.5)(32.2) = 0.0605 \text{ lb/ft}^3$$

**1.40** One kilogram of hydrogen is confined in a volume of 200 L at −45 °C. What is the pressure if $R$ is 4.115 kJ/kg · K?

$$p = \rho RT = (m/V)RT = (1/0.200)(4115)(-45 + 273) = 4.691 \times 10^6 \text{ Pa} \quad \text{or} \quad 4.691 \text{ MPa abs}$$

**1.41** What is the specific weight of air at a temperature of 30 °C and a pressure of 470 kPa abs?

$$\gamma = p/RT = 470/[(29.3)(30 + 273)] = 0.0529 \text{ kN/m}^3$$

**1.42**  Find the mass density of helium at a temperature of 39 °F and a pressure of 26.9 psig, if atmospheric pressure is 14.9 psia.

$$\rho = p/RT = (14.9 + 26.9)(144)/[(12\,420)(39 + 460)]$$
$$= 0.000971\,\text{lb}\cdot\text{s}^2/\text{ft}^4 \quad\text{or}\quad 0.000971\,\text{slug/ft}^3$$

**1.43**  The temperature and pressure of nitrogen in a tank are 28 °C and 600 kPa abs, respectively. Determine the specific weight of the nitrogen.

$$\gamma = p/RT = 600/[(30.3)(28 + 273)] = 0.0658\,\text{kN/m}^3$$

**1.44**  The temperature and pressure of oxygen in a container are 60 °F and 20.0 psig, respectively. Determine the oxygen's mass density if atmospheric pressure is 14.7 psia.

$$\rho = p/RT = (20.0 + 14.7)(144)/[(1552)(60 + 460)] = 0.00619\,\text{slug/ft}^3$$

**1.45**  Calculate the specific weight and density of methane at 100 °F and 120 psia.

$$\gamma = p/RT = (120)(144)/[(96.2)(100 + 460)] = 0.321\,\text{lb/ft}^3$$
$$\rho = \gamma/g = 0.321/32.2 = 0.00997\,\text{slug/ft}^3$$

**1.46**  At 90 °F and 30.0 psia, the specific weight of a certain gas was 0.0877 lb/ft$^3$. Determine the gas constant and density of this gas.

$$\gamma = p/RT \quad 0.0877 = (30.0)(144)/[(R)(90 + 460)] \quad R = 89.6\,\text{ft/°R}$$
$$\rho = \gamma/g = 0.0877/32.2 = 0.00272\,\text{slug/ft}^3$$

**1.47**  A cylinder contains 12.5 ft$^3$ of air at 120 °F and 40 psia. The air is then compressed to 2.50 ft$^3$. (*a*) Assuming isothermal conditions, what are the pressure at the new volume and the bulk modulus of elasticity? (*b*) Assuming adiabatic conditions, what are the final pressure and temperature and the bulk modulus of elasticity?

(*a*)
$$p_1 V_1 = p_2 V_2 \quad\text{(for isothermal conditions)}$$
$$(40)(12.5) = (p_2')(2.50)$$
$$p_2' = 200\,\text{psia}$$
$$K = -\frac{\Delta p}{\Delta V/V} = -\frac{40 - 200}{(12.5 - 2.5)/12.5} = 200\,\text{psi}$$

(*b*) $p_1 V_1^k = p_2 V_2^k$ (for adiabatic conditions). From Table A-6, $k = 1.40$. $(40)(12.5)^{1.40} = (p_2')(2.50)^{1.40}$, $p_2' = 381$ psia; $T_2/T_1 = (p_2/p_1)^{(k-1)/k}$, $T_2/(120 + 460) = (\frac{381}{40})^{(1.40-1)/1.40}$, $T_2 = 1104$ °R, or 644 °F; $K = kp' = (1.40)(381) = 533$ psi.

**1.48**  Air is kept at a pressure of 200 kPa and a temperature of 30 °C in a 500-L container. What is the mass of the air?

$$\rho = p/RT = [(200)(1000)]/[(287)(30 + 273)] = 2.300\,\text{kg/m}^3 \qquad m = (2.300)(\tfrac{500}{1000}) = 1.15\,\text{kg}$$

**1.49**  An ideal gas has its pressure doubled and its specific volume decreased by two-thirds. If the initial temperature is 80 °F, what is the final temperature?

$$\rho = 1/V_s = p/RT \qquad pV_s = RT \qquad p_1(V_s)_1 = RT_1 \qquad p_2(V_s)_2 = RT_2$$
$$(p_2/p_1)[(V_s)_2/(V_s)_1] = (R/R)(T_2/T_1) \qquad (2)(\tfrac{1}{3}) = T_2/(80 + 460) \qquad T_2 = 360\,\text{°R} \quad\text{or}\quad -100\,\text{°F}$$

**1.50** The tank of a leaky air compressor originally holds 90 L of air at 33 °C and 225 kPa. During a compression process, 4 grams of air is lost; the remaining air occupies 42 L at 550 kPa. What is the temperature of the remaining air?

■ $\rho_1 = p_1/RT_1 = (225 \times 10^3)/[(287)(33 + 273)] = 2.562 \text{ kg/m}^3$     $m = (2.562)(0.090) = 0.2306 \text{ kg}$

$\rho_2 = p_2/RT_2$     $(0.2306 - 0.004)/0.042 = (550 \times 10^3)/(287T_2)$     $T_2 = 355 \text{ K}$   or   $82 °C$

**1.51** In a piston-and-cylinder apparatus the initial volume of air is 90 L at a pressure of 130 kPa and temperature of 26 °C. If the pressure is doubled while the volume is decreased to 56 L, compute the final temperature and density of the air.

■ $\rho_1 = p_1/RT_1 = (130 \times 10^3)/[(287)(26 + 273)] = 1.515 \text{ kg/m}^3$     $m = (1.515)(0.090) = 0.1364 \text{ kg}$

$\rho_2 = p_2/RT_2$     $0.1364/\frac{56}{1000} = (2)(130 \times 10^3)/(287T_2)$     $T_2 = 372 \text{ K}$   or   $99 °C$

$\rho = 0.1364/(0.056) = 2.44 \text{ kg/m}^3$

**1.52** For 2 lb mol of air with a molecular weight of 29, a temperature of 90 °F, and a pressure of 2.5 atm, what is the volume?

■ $pV/nM = RT$     $[(2.5)(14.7)(144)]\{V/[(2)(29)]\} = (53.3)(90 + 460)$     $V = 321 \text{ ft}^3$

**1.53** If nitrogen has a molecular weight of 28, what is its density according to the perfect gas law when $p = 0.290 \text{ MPa}$ and $T = 30 °C$?

■ $R = R_u/M = 8312/28 = 297 \text{ J/(kg} \cdot \text{K)}$     [where $R_u = 8312 \text{ J/(kg} \cdot \text{K)}$]

$\rho = p/RT = 290\,000/[(297)(30 + 273)] = 3.22 \text{ kg/m}^3$

**1.54** If a gas occupies 1 m³ at 1 atm pressure, what pressure is required to reduce the volume of the gas by 2 percent under isothermal conditions if the fluid is (a) air, (b) argon, and (c) hydrogen?

■ $pV = nRT = $ constant for isothermal conditions. Therefore, if $V$ drops to $0.98V_o$, $p$ must rise to $(1/0.98)p_o$, or $1.020p_o$. This is true for any perfect gas.

**1.55** (a) Calculate the density, specific weight, and specific volume of oxygen at 100 °F and 15 psia. (b) What would be the temperature and pressure of this gas if it were compressed isentropically to 40 percent of its original volume? (c) If the process described in (b) had been isothermal, what would the temperature and pressure have been?

■ (a)     $\rho = p/RT = (15)(144)/[(1552)(100 + 460)] = 0.00248 \text{ slug/ft}^3$

$\gamma = \rho g = (0.00248)(32.2) = 0.0799 \text{ lb/ft}^3$     $V_s = 1/\rho = 1/0.00248 = 403 \text{ ft}^3/\text{slug}$

(b)     $p_1(V_s)_1^k = p_2(V_s)_2^k$     $[(15)(144)](403)^{1.40} = [(p_2)(144)][(0.40)(403)]^{1.40}$     $p_2 = 54.1 \text{ psia}$

$p_2 = \rho_2 RT_2$     $(54.1)(144) = (0.00248/0.40)(1552)(T_2 + 460)$     $T_2 = 350 °F$

(c) If isothermal, $T_2 = T_1 = 100 °F$ and $pV = $ constant.

$[(15)(144)](403) = [(p_2)(144)][(0.40)(403)]$     $p_2 = 37.5 \text{ psia}$

**1.56** Calculate the density, specific weight, and volume of chloride gas at 25 °C and pressure of 600 000 N/m² abs.

■     $\rho = p/RT = 600\,000/[(118)(25 + 273)] = 17.1 \text{ kg/m}^3$

$\gamma = \rho g = (17.1)(9.81) = 168 \text{ N/m}^3$     $V_s = 1/\rho = 1/17.1 = 0.0585 \text{ m}^3/\text{kg}$

**1.57** If methane gas has a specific gravity of 0.55 relative to air at 14.7 psia and 68 °F, what are its specific weight and specific volume at that same pressure and temperature? What is the value of $R$ for the gas?

■     $\gamma_{air} = p/RT = (14.7)(144)/[(53.3)(68 + 460)] = 0.07522 \text{ lb/ft}^3$

$\gamma_{gas} = (0.55)(0.07522) = 0.0414 \text{ lb/ft}^3$

$V_s = 1/\rho = g/\gamma$     $(V_s)_{gas} = 32.2/0.0414 = 778 \text{ ft}^3/\text{slug}$

Since $R$ varies inversely with density for fixed pressure and temperature, $R_{gas} = 53.3/0.55 = 96.9 \text{ ft/°R}$.

**1.58** A gas at 40 °C under a pressure of 21.868 bar abs has a unit weight of 362 N/m³. What is the value of $R$ for this gas? What gas might this be?

▌ $$\gamma = p/RT \qquad 362 = (21.868 \times 10^5)/[(R)(40+273)] \qquad R = 19.3 \text{ m/K}$$

This gas might be carbon dioxide, since its gas constant is 19.3 m/K (from Table A-6).

**1.59** If water vapor ($R = 85.7$ ft/°R) in the atmosphere has a partial pressure of 0.60 psia and the temperature is 80 °F, what is its specific weight?

▌ $$\gamma = p/RT = (0.60)(144)/[(85.7)(80+460)] = 0.00187 \text{ lb/ft}^3$$

**1.60** Refer to Prob. 1.59. If the barometer reads 14.60 psia, calculate the partial pressure of the air, its specific weight, and the specific weight of the atmosphere (air plus water vapor).

▌ $$p_{air} = 14.60 - 0.60 = 14.00 \text{ psia} \qquad \gamma = p/RT$$

$$\gamma_{air} = (14.00)(144)/[(53.3)(80+460)] = 0.0700 \text{ lb/ft}^3 \qquad \gamma_{atm} = \gamma_{air} + \gamma_{H_2O(vap)}$$

$$\gamma_{H_2O(vap)} = 0.00187 \text{ lb/ft}^3 \qquad \text{(from Prob. 1.59)} \qquad \gamma_{atm} = 0.0700 + 0.00187 = 0.0719 \text{ lb/ft}^3$$

**1.61** (*a*) Calculate the density, specific weight, and specific volume of oxygen at 20 °C and 40 kPa abs. (*b*) If the oxygen is enclosed in a rigid container, what will be the pressure if the temperature is reduced to −100 °C?

▌ (*a*) $$\rho = p/RT = (40)(1000)/[(260)(20+273)] = 0.525 \text{ kg/m}^3$$

$$\gamma = \rho g = (0.525)(9.81) = 5.15 \text{ N/m}^3 \qquad V_s = 1/\rho = 1/0.525 = 1.90 \text{ m}^3/\text{kg}$$

(*b*) $\rho = 1/V_s = p/RT$. Since $V_s$ and $R$ are constants, $V_s/R = T/p = \text{constant}$, $(20+273)/40 = (-100+273)/p_2$, $p_2 = 23.6$ kPa.

**1.62** Helium at 149 kPa abs and 10 °C is isentropically compressed to one-fourth of its original volume. What is its final pressure?

▌ $$p_1 V_1^k = p_2 V_2^k \qquad 149 V_1^{1.66} = (p_2)(V_1/4)^{1.66} \qquad p_2 = 1488 \text{ kPa abs}$$

**1.63** (*a*) If 9 ft³ of an ideal gas at 75 °F and 22 psia is compressed isothermally to 2 ft³, what is the resulting pressure? (*b*) What would the pressure and temperature have been if the process had been isentropic?

▌ (*a*) $$p_1 V_1 = p_2 V_2 \qquad (22)(9) = (p_2)(2) \qquad p_2 = 99 \text{ psia}$$

(*b*) $$p_1 V_1^k = p_2 V_2^k \qquad (22)(9)^{1.30} = (p_2)(2)^{1.30} \qquad p_2 = 155 \text{ psia}$$

$$T_2/T_1 = (p_2/p_1)^{(k-1)/k} \qquad T_2/(75+460) = (\tfrac{155}{22})^{(1.30-1)/1.30} \qquad T_2 = 840 \text{ °R} \quad \text{or} \quad 380 \text{ °F}$$

**1.64** (*a*) If 12 m³ of nitrogen at 30 °C and 125 kPa abs is permitted to expand isothermally to 30 m³, what is the resulting pressure? (*b*) What would the pressure and temperature have been if the process had been isentropic?

▌ (*a*) $$p_1 V_1 = p_2 V_2 \qquad (125)(12) = (p_2)(30) \qquad p_2 = 50.0 \text{ kPa abs}$$

(*b*) $$p_1 V_1^k = p_2 V_2^k \qquad (125)(12)^{1.40} = (p_2)(30)^{1.40} \qquad p_2 = 34.7 \text{ kPa abs}$$

$$T_2/T_1 = (p_2/p_1)^{(k-1)/k} \qquad T_2/(30+273) = (34.7/125)^{(1.40-1)/1.40} \qquad T_2 = 210 \text{ K} \quad \text{or} \quad -63 \text{ °C}$$

**1.65** If the viscosity of water at 68 °F is 0.01008 poise, compute its absolute viscosity ($\mu$) in pound-seconds per square foot. If the specific gravity at 68 °F is 0.998, compute its kinematic viscosity ($\nu$) in square feet per second.

▌ The poise is measured in dyne-seconds per square centimeter. Since 1 lb = 444 800 dynes and 1 ft = 30.48 cm, $1 \text{ lb} \cdot \text{s/ft}^2 = 444\,800 \text{ dyne} \cdot \text{s}/(30.48 \text{ cm})^2 = 478.8 \text{ poises}$

$$\mu = \frac{0.01008}{478.8} = 2.11 \times 10^{-5} \text{ lb} \cdot \text{s/ft}^2 \qquad \nu = \frac{\mu}{\rho} = \frac{\mu}{\gamma/g} = \frac{\mu g}{\gamma} = \frac{(2.11 \times 10^{-5})(32.2)}{(0.998)(62.4)} = 1.09 \times 10^{-5} \text{ ft}^2/\text{s}$$

**1.66** Convert 15.14 poises to kinematic viscosity in square feet per second if the liquid has a specific gravity of 0.964.

▌ $$1 \text{ lb} \cdot \text{s/ft}^2 = 478.8 \text{ poises} \qquad \text{(from Prob. 1.65)}$$

$$\mu = 15.14/478.8 = 0.03162 \text{ lb} \cdot \text{s/ft}^2 \qquad \nu = \mu g/\gamma = (0.03162)(32.2)/[(0.964)(62.4)] = 0.0169 \text{ ft}^2/\text{s}$$

**1.67** The fluid flowing in Fig. 1-3 has an absolute viscosity ($\mu$) of 0.0010 lb · s/ft$^2$ and specific gravity of 0.913. Calculate the velocity gradient and intensity of shear stress at the boundary and at points 1 in, 2 in, and 3 in from the boundary, assuming (*a*) a straight-line velocity distribution and (*b*) a parabolic velocity distribution. The parabola in the sketch has its vertex at *A* and origin at *B*.

▌(*a*) For the straight-line assumption, the relation between velocity $v$ and distance $y$ is $v = 15y$, $dv = 15dy$. The velocity gradient $= dv/dy = 15$. Since $\mu = \tau/(dv/dy)$, $\tau = \mu\,(dv/dy)$. For $y = 0$ (i.e., at the boundary), $v = 0$ and $dv/dy = 15\,\text{s}^{-1}$; $\tau = (0.0010)(15) = 0.015\,\text{lb/ft}^2$. For $y = 1$ in, 2 in, and 3 in, $dv/dy$ and $\tau$ are also $15\,\text{s}^{-1}$ and $0.015\,\text{lb/ft}^2$, respectively. (*b*) For the parabolic assumption, the parabola passes through the points $v = 0$ when $y = 0$ and $v = 45$ when $y = 3$. The equation of this parabola is $v = 45 - 5(3 - y)^2$, $dv/dy = 10(3 - y)$, $\tau = 0.0010\,(dv/dy)$. For $y = 0$ in, $v = 0$ in/s, $dv/dy = 30\,\text{s}^{-1}$, and $\tau = 0.030\,\text{lb/ft}^2$. For $y = 1$ in, $v = 25$ in/s, $dv/dy = 20\,\text{s}^{-1}$, and $\tau = 0.020\,\text{lb/ft}^2$. For $y = 2$ in, $v = 40$ in/s, $dv/dy = 10\,\text{s}^{-1}$, and $\tau = 0.010\,\text{lb/ft}^2$. For $y = 3$ in, $v = 45$ in/s, $dv/dy = 0\,\text{s}^{-1}$, and $\tau = 0\,\text{lb/ft}^2$.

**Fig. 1-3**

**1.68** A cylinder of 0.40-ft radius rotates concentrically inside a fixed cylinder of 0.42-ft radius. Both cylinders are 1.00 ft long. Determine the viscosity of the liquid that fills the space between the cylinders if a torque of 0.650 lb · ft is required to maintain an angular velocity of 60 rpm.

▌The torque is transmitted through the field layers to the outer cylinder. Since the gap between the cylinders is small, the calculations may be made without integration. The tangential velocity $v_t$ of the inner cylinder $= r\omega$, where $r = 0.40$ ft and $\omega = 2\pi$ rad/s. Hence, $v_t = (0.40)(2\pi) = 2.51$ ft/s. For the small space between cylinders, the velocity gradient may be assumed to be a straight line and the mean radius can be used. Then, $dv/dy = (2.51 - 0)/(0.42 - 0.40) = 125.5\,\text{s}^{-1}$. Since applied torque equals resisting torque, applied torque $= (\tau)(\text{area})(\text{arm})$, $0.650 = \tau[(1.00)(2\pi)(0.40 + 0.42)/2][(0.40 + 0.42)/2]$, $\tau = 0.615\,\text{lb/ft}^2 = \mu\,(dv/dy)$, $0.615 = (\mu)(125.5)$, $\mu = 0.00490\,\text{lb} \cdot \text{s/ft}^2$.

**1.69** Water is moving through a pipe. The velocity profile at some section is shown in Fig. 1-4 and is given mathematically as $v = (\beta/4\mu)(d^2/4 - r^2)$, where $v =$ velocity of water at any position $r$, $\beta =$ a constant, $\mu =$ viscosity of water, $d =$ pipe diameter, and $r =$ radial distance from centerline. What is the shear stress at the wall of the pipe due to the water? What is the shear stress at a position $r = d/4$? If the given profile persists a distance $L$ along the pipe, what drag is induced on the pipe by the water in the direction of flow over this distance?

▌
$$v = (\beta/4\mu)(d^2/4 - r^2) \qquad dv/dr = (\beta/4\mu)(-2r) = -2\beta r/4\mu$$

$$\tau = \mu\,(dv/dr) = \mu(-2\beta r/4\mu) = -2\beta r/4$$

At the wall, $r = d/2$. Hence,

$$\tau_{\text{wall}} = \frac{-2\beta(d/2)}{4} = -\frac{\beta d}{4} \qquad \tau_{r=d/4} = \frac{-2\beta(d/4)}{4} = -\frac{\beta d}{8}$$

$$\text{Drag} = (\tau_{\text{wall}})(\text{area}) = (\tau_{\text{wall}})(\pi d L) = (\beta d/4)(\pi d L) = \beta d^2 \pi L/4$$

**Fig. 1-4**

**1.70** A large plate moves with speed $v_0$ over a stationary plate on a layer of oil (see Fig. 1-5). If the velocity profile is that of a parabola, with the oil at the plates having the same velocity as the plates, what is the shear stress on the moving plate from the oil? If a linear profile is assumed, what is the shear stress on the upper plate?

▌ For a parabolic profile, $v^2 = ay$. When $y = d$, $v = v_0$. Hence, $v_0^2 = ad$, $a = v_0^2/d$. Therefore,

$$v^2 = (v_0^2/d)(y) = (v_0^2)(y/d) \qquad v = v_0\sqrt{y/d} \qquad dv/dy = [(v_0)(1/\sqrt{d})(\tfrac{1}{2})(y^{-1/2})]$$

$$\tau = \mu\,(dv/dy) = \mu[(v_0)(1/\sqrt{d})(\tfrac{1}{2})(y^{-1/2})]$$

For $y = d$, $\tau = \mu[(v_0)(1/\sqrt{d})(\tfrac{1}{2})(d^{-1/2})] = \mu v_0/(2d)$. For a linear profile, $dv/dy = v_0/d$, $\tau = \mu(v_0/d)$.

Fig. 1-5

**1.71** A square block weighing 1.1 kN and 250 mm on an edge slides down an incline on a film of oil 6.0 $\mu$m thick (see Fig. 1-6). Assuming a linear velocity profile in the oil, what is the terminal speed of the block? The viscosity of the oil is 7 mPa · s.

▌ $\tau = \mu\,(dv/dy) = (7 \times 10^{-3})[v_T/(6.0 \times 10^{-6})] = 1167v_T \qquad F_f = \tau A = (1167v_T)(0.250)^2 = 72.9v_T$

At the terminal condition, equilibrium occurs. Hence, $1100 \sin 20° = 72.9v_T$, $v_T = 5.16$ m/s.

Fig. 1-6(a)

Fig. 1-6(b)

**1.72** A piston of weight 21 lb slides in a lubricated pipe, as shown in Fig. 1-7. The clearance between piston and pipe is 0.001 in. If the piston decelerates at 2.1 ft/s$^2$ when the speed is 21 ft/s, what is the viscosity of the oil?

▌ $$\tau = \mu\,(dv/dy) = \mu[v/(0.001/12)] = 12\,000\mu v$$

$$F_f = \tau A = 12\,000\mu v[(\pi)(\tfrac{6}{12})(\tfrac{5}{12})] = 7854\mu v$$

$$\Sigma F = ma \qquad 21 - (7854)(\mu)(21) = (21/32.2)(-2.1) \qquad \mu = 1.36 \times 10^{-4}\ \text{lb} \cdot \text{s/ft}^2$$

6 in

5 in

Fig. 1-7

**1.73** A piston is moving through a cylinder at a speed of 19 ft/s, as shown in Fig. 1-8. The film of oil separating the piston from the cylinder has a viscosity of 0.020 lb · s/ft$^2$. What is the force required to maintain this motion?

▌ Assume a cylindrically symmetric, linear velocity profile for the flow of oil in the film. To find the frictional resistance, compute the shear stress at the piston surface.

$$\tau = \mu\frac{dv}{dr} = 0.020\left[\frac{19}{(5.000 - 4.990)/2}\right](12) = 912\ \text{lb/ft}^2 \qquad F_f = \tau A = 912\left[\pi\left(\frac{4.990}{12}\right)\left(\frac{3}{12}\right)\right] = 298\ \text{lb}$$

**Fig. 1-8(a)**

**Fig. 1-8(b)**

**1.74** To damp oscillations, the pointer of a galvanometer is fixed to a circular disk which turns in a container of oil (see Fig. 1-9). What is the damping torque for $\omega = 0.3$ rad/s if the oil has a viscosity of $8 \times 10^{-3}$ Pa · s? Neglect edge effects.

**▌** Assume at any point that the velocity profile of the oil is linear $dv/dn = r\omega/(0.5/1000) = (r)(0.3)/(0.5/1000) = 600r$; $\tau = \mu \, (dv/dn) = \mu(600r) = (8 \times 10^{-3})(600r) = 4.80r$. The force $dF_f$ on $dA$ on the upper face of the disc is then $dF_f = \tau \, dA = (4.80r)(r \, d\theta \, dr) = 4.80r^2 \, d\theta \, dr$. The torque $dT$ for $dA$ on the upper face is then $dT = r \, dF_f = r(4.80r^2 \, d\theta \, dr) = 4.80r^3 \, d\theta \, dr$. The total resisting torque on both faces is

$$T = 2\left[\int_0^{0.075/2} \int_0^{2\pi} 4.80r^3 \, d\theta \, dr\right] = (9.60)(2\pi)\left[\frac{r^4}{4}\right]_0^{0.075/2} = 2.98 \times 10^{-5} \, \text{N} \cdot \text{m}$$

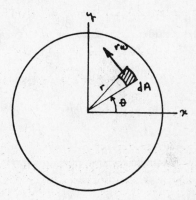

**Fig. 1-9(a)**

**Fig. 1-9(b)**

**1.75** For angular velocity 0.3 rad/s of the mechanism of Prob. 1.74, express the damping torque (in N · m) as a function of displacement $x$ (in mm) of the disk from its center position (Fig. 1-10).

**▌** Assume at any point that the velocity profile of the oil is linear; $\tau = \mu \, (dv/dn)$. For the upper face, $dv/dn = r\omega/[(0.5 - x)/1000] = (r)(0.3)/[(0.5 - x)/1000]$; $\tau = (8 \times 10^{-3})\{(r)(0.3)/[(0.5 - x)/1000]\} = 2.40r/(0.5 - x)$. The force $dF_f$ on $dA$ on the upper face of the disc is then $dF_f = \tau \, dA = [2.40r/(0.5 - x)](r \, d\theta \, dr) = [2.40r^2/(0.5 - x)] \, (d\theta \, dr)$. The torque $dT$ for $dA$ on the upper face is then $dT = r \, dF_f = r[2.40r^2/(0.5 - x)] \, (d\theta \, dr) = [2.40r^3/(0.5 - x)] \, (d\theta \, dr)$. For the lower face, $dv/dn = r\omega/[(0.5 + x)/1000] = r(0.3)/[(0.5 + x)/1000]$; $\tau = (8 \times 10^{-3})\{r(0.3)/[(0.5 + x)/1000]\} = 2.40r/(0.5 + x)$. The force $dF_f$ on $dA$ on the lower face of the disc is then $dF_f = \tau \, dA = [2.40r/(0.5 + x)](r \, d\theta \, dr) = [2.40r^2/(0.5 + x)] \, (d\theta \, dr)$. The torque $dT$ for $dA$ on the lower face is then $dT = r \, dF_f = r[2.40r^2/(0.5 + x)] \, (d\theta \, dr) = [2.40r^3/(0.5 + x)] \, (d\theta \, dr)$. The total resisting torque on both faces is

$$T = \int_0^{0.075/2} \int_0^{2\pi} \frac{2.40r^3}{0.5 - x} \, d\theta \, dr + \int_0^{0.075/2} \int_0^{2\pi} \frac{2.40r^3}{0.5 + x} \, d\theta \, dr$$

$$= \left(\frac{1}{0.5 - x} + \frac{1}{0.5 + x}\right)(2.40)(2\pi)\left[\frac{r^4}{4}\right]_0^{0.075/2} = \left(\frac{0.5 + x + 0.5 - x}{0.25 - x^2}\right)(7.46 \times 10^{-6})$$

$$= \frac{7.46 \times 10^{-6}}{0.25 - x^2}$$

**Fig. 1-10**

**1.76** A conical body turns in a container, as shown in Fig. 1-11, at constant speed 11 rad/s. A uniform 0.01-in film of oil with viscosity $3.125 \times 10^{-7}$ lb · s/in² separates the cone from the container. What torque is required to maintain this motion, if the cone has a 2-in radius at its base and is 4 in tall?

▮ Consider the conical surface first ($r/2 = z/4$, $r = z/2$). The stress on this element is $\tau = \mu \, (dv/dx) = \mu(r\omega/0.01) = (3.125 \times 10^{-7})[(z/2)(11)/0.01] = 1.719 \times 10^{-4}z$. The area of the strip shown is $dA = 2\pi r \, ds = (2\pi z/2)[dz/(4/\sqrt{20})] = 3.512z \, dz$. The torque on the strip is $dT = \tau \, (dA)(r) = (1.719 \times 10^{-4}z)(3.512z \, dz)(z/2) = 3.019 \times 10^{-4}z^3 \, dz$.

$$T_1 = \int_0^4 3.019 \times 10^{-4}z^3 \, dz = 3.019 \times 10^{-4}\left[\frac{z^4}{4}\right]_0^4 = 0.01932 \text{ in} \cdot \text{lb}$$

Next consider the base: $dF_f = \tau \, dA$, $\tau = \mu(r\omega/0.01) = (3.125 \times 10^{-7})[(r)(11)/0.01] = 3.438 \times 10^{-4}r$, $dF_f = (3.438 \times 10^{-4}r)(r \, d\theta \, dr) = 3.438 \times 10^{-4}r^2 \, d\theta \, dr$, $dT_2 = (3.438 \times 10^{-4}r^2 \, d\theta \, dr)(r) = 3.438 \times 10^{-4}r^3 \, d\theta \, dr$.

$$T_2 = \int_0^2 \int_0^{2\pi} 3.438 \times 10^{-4}r^3 \, d\theta dr = (3.438 \times 10^{-4})(2\pi)\left[\frac{r^4}{4}\right]_0^2 = 0.00864 \text{ in} \cdot \text{lb}$$

$$T_{\text{tot}} = 0.01932 + 0.00864 = 0.0280 \text{ in} \cdot \text{lb}$$

**Fig. 1-11(a)**

**Fig. 1-11(b)**

**1.77** In Fig. 1-12, if the fluid is SAE 30 oil at 20 °C and $D = 7$ mm, what shear stress is required to move the upper plate at 3.5 m/s? Compute the Reynolds number based on $D$.

▮
$$\tau = \mu \, (dv/dh) = (0.440)[3.5/(\tfrac{7}{1000})] = 220 \text{ Pa}$$
$$N_R = \rho Dv/\mu = (888)(\tfrac{7}{1000})(3.5)/0.440 = 49.4$$

**Fig. 1-12**

**1.78** Benzene at 20 °C has a viscosity of 0.000651 Pa · s. What shear stress is required to deform this fluid at a strain rate of 4900 s$^{-1}$?

$$\tau = \mu \, (dv/dx) = (0.000651)(4900) = 3.19 \, \text{Pa}$$

**1.79** SAE 30 oil at 20 °C is sheared between two parallel plates 0.005 in apart with the lower plate fixed and the upper plate moving at 13 ft/s. Compute the shear stress in the oil.

$$\tau = \mu \, (dv/dh) = (9.20 \times 10^{-3})[13/(0.005/12)] = 287 \, \text{lb/ft}^2$$

**1.80** An 18-kg slab slides down a 15° inclined plane on a 3-mm-thick film of SAE 10 oil at 20 °C; the contact area is 0.3 m². Find the terminal velocity of the slab.

See Fig. 1-13.

$$\Sigma F_x = 0 \qquad W \sin \theta - \tau A_{\text{bottom}} = 0$$

$$\tau = \mu \, (dv/dy) = (8.14 \times 10^{-2})(v_T/0.003) = 27.1 v_T$$

$$[(18)(9.81)](\sin 15°) - (27.1 v_T)(0.3) = 0 \qquad v_T = 5.62 \, \text{m/s}$$

**Fig. 1-13**

**1.81** A shaft 70.0 mm in diameter is being pushed at a speed of 400 mm/s through a bearing sleeve 70.2 mm in diameter and 250 mm long. The clearance, assumed uniform, is filled with oil at 20 °C with $v = 0.005$ m²/s and s.g. = 0.9. Find the force exerted by the oil on the shaft.

$$F = \tau A \qquad \tau = \mu \, (dv/dr) \qquad \mu = \rho v = [(0.9)(998)](0.005) = 4.49 \, \text{kg/(m · s)}$$

$$dr = (0.0702 - 0.0700)/2 = 0.0001 \, \text{m} \qquad \tau = (4.49)(0.4/0.0001) = 17\,960 \, \text{N/m}^2$$

$$A = (\pi)(7.00/100)(25/100) = 0.05498 \, \text{m}^2 \qquad F = (17\,960)(0.05498) = 987 \, \text{N}$$

**1.82** If the shaft in Prob. 1.81 is fixed axially and rotated inside the sleeve at 2000 rpm, determine the resisting torque exerted by the oil and the power required to rotate the shaft.

$$T = \tau A r \qquad \tau = \mu \, (dv/dr)$$

$$v = r\omega = [(7.00/2)/100][(2000)(2\pi/60)] = 7.330 \, \text{m/s} \qquad dr = 0.0001 \, \text{m}$$

$$\tau = (4.49)(7.330/0.0001) = 329.1 \times 10^3 \, \text{N/m}^2 \qquad A = (\pi)(7.00/100)(\tfrac{25}{100}) = 0.05498 \, \text{m}^2$$

$$T = (329.1 \times 10^3)(0.05498)[(7.00/2)/100] = 633 \, \text{N · m}$$

$$P = \omega T = [(2000)(2\pi/60)](633) = 132.6 \times 10^3 \, \text{W} \quad \text{or} \quad 132.6 \, \text{kW}$$

**1.83** A steel (7850-kg/m³) shaft 40.0 mm in diameter and 350 mm long falls of its own weight inside a vertical open tube 40.2 mm in diameter. The clearance, assumed uniform, is a film of SAE 30 oil at 20 °C. What speed will the cylinder ultimately reach?

$$W_{shaft} = \tau A = [(7850)(9.81)][(0.350)(\pi)(0.0400)^2/4] = 33.87 \text{ N}$$

$$dr = (0.0402 - 0.0400)/2 = 0.0001 \text{ m}$$

$$\tau = \mu (dy/dr) = (0.440)(v_T/0.0001) = 4400 v_T$$

$$A = (\pi)(4.00/100)(\tfrac{35}{100}) = 0.04398 \text{ m}^2 \qquad 33.87 = (4400 v_T)(0.04398) \qquad v_T = 0.1750 \text{ m/s}$$

**1.84** Air at 20 °C forms a boundary layer near a solid wall, in which the velocity profile is sinusoidal (see Fig. 1-14). The boundary-layer thickness is 7 mm and the peak velocity is 9 m/s. Compute the shear stress in the boundary layer at $y$ equal to (a) 0, (b) 3.5 mm, and (c) 7 mm.

$$\tau = \mu (dv/dy) \qquad v = v_{max} \sin [\pi y/(2\delta)]$$

$$dv/dy = [\pi v_{max}/(2\delta)] \cos [\pi y/(2\delta)] = \{(\pi)(9)/[(2)(0.007)]\} \cos \{\pi y/[(2)(0.007)]\} = 2020 \cos (224.4y)$$

**Note:** "224.4y" in the above equation is in radians.

$$\tau = (1.81 \times 10^{-5})[2020 \cos (224.4y)] = 0.03656 \cos (224.4y)$$

(a) At $y = 0$, $\tau = 0.03656 \cos [(224.4)(0)] = 0.0366$ Pa. (b) At $y = 0.0035$ m, $\tau = 0.03656 \cos [(224.4)(0.0035)] = 0.0259$ Pa. (c) At $y = 0.007$ m, $\tau = 0.03656 \cos [(224.4)(0.007)] = 0$.

**Fig. 1-14**

**1.85** A disk of radius $r_0$ rotates at angular velocity $\omega$ inside an oil bath of viscosity $\mu$, as shown in Fig. 1-15. Assuming a linear velocity profile and neglecting shear on the outer disk edges, derive an expression for the viscous torque on the disk.

$$\tau = \mu (dv/dy) = \mu(r\omega/h) \qquad \text{(on both sides)}$$

$$dT = (2)(r\tau \, dA) = (2)\{(r)[\mu(r\omega/h)](2\pi r \, dr)\} = (4\mu\omega\pi/h)(r^3 \, dr)$$

$$T = \int_0^{r_0} \frac{4\mu\omega\pi}{h} (r^3 \, dr) = \frac{4\mu\omega\pi}{h} \left[\frac{r^4}{4}\right]_0^{r_0} = \frac{\pi\mu\omega r_0^4}{h}$$

**Fig. 1-15**

**1.86** A 35-cm-by-55-cm block slides on oil ($\mu = 0.81$ Pa · s) over a large plane surface. What force is required to drag the block at 3 m/s, if the separating oil film is 0.6 mm thick?

$$\tau = \mu (dv/dx) = (0.81)[3/(0.6/1000)] = 4050 \text{ N/m}^2 \qquad F = \tau A = (4050)[(\tfrac{35}{100})(\tfrac{55}{100})] = 780 \text{ N}$$

**1.87** The 1.5-in (0.125-ft) gap between two large plane surfaces is filled with SAE 30 oil at 80 °F ($\mu = 0.0063$ lb · s/ft$^2$). What force is required to drag a very thin plate of 5-ft$^2$ area between the surfaces at a speed of 0.5 ft/s if this plate is equally spaced between the two surfaces?

$$\tau = \mu (dv/dx) = (0.0063)[0.5/(0.125/2)] = 0.0504 \text{ lb/ft}^2 \qquad F = \tau A = (0.0504)(5) = 0.252 \text{ lb}$$

Since there are two sides, $F_{required} = (2)(0.252)$, or 0.504 lb.

**1.88** Rework Prob. 1.87 if the plate is at a distance of 0.50 in (0.0417 ft) from one surface.

$$\tau = \mu\,(dv/dx) \qquad \tau_1 = (0.0063)(0.5/0.0417) = 0.0755\ \text{lb/ft}^2$$

$$F = \tau A \qquad F_1 = (0.0755)(5) = 0.3775\ \text{lb} \qquad \tau_2 = (0.0063)[0.5/(0.125 - 0.0417)] = 0.0378\ \text{lb/ft}^2$$

$$F_2 = (0.0378)(5) = 0.1890\ \text{lb} \qquad F_{\text{required}} = F_1 + F_2 = 0.3775 + 0.1890 = 0.566\ \text{lb}$$

**1.89** A 10.000-in-diameter plunger slides in a 10.006-in-diameter cylinder, the annular space being filled with oil having a kinematic viscosity of 0.004 ft²/s and specific gravity of 0.85. If the plunger moves at 0.6 ft/s, find the frictional resistance when 9 ft is engaged in the cylinder.

$$\tau = \mu\,(dv/dx) \qquad \rho = \gamma/g = [(0.85)(62.4)]/32.2 = 1.647\ \text{slugs/ft}^3$$

$$\mu = \rho v = (1.647)(0.004) = 0.006588\ \text{lb}\cdot\text{s/ft}^2 \qquad dx = [(10.006 - 10.000)/2]/12 = 0.000250\ \text{ft}$$

$$\tau = (0.006588)(0.6/0.000250) = 15.81\ \text{lb/ft}^2 \qquad F_f = \tau A = (15.81)[(9)(\pi)(\tfrac{10}{12})] = 373\ \text{lb}$$

**1.90** A 6.00-in shaft rides in a 6.01-in sleeve 8 in long, the clearance space (assumed to be uniform) being filled with lubricating oil at 100 °F ($\mu = 0.0018\ \text{lb}\cdot\text{s/ft}^2$). Calculate the rate at which heat is generated when the shaft turns at 90 rpm.

$$dv = \omega(\text{circumference}) = \tfrac{90}{60}[\pi(6.00/12)] = 2.356\ \text{ft/s}$$

$$dx = [(6.01 - 6.00)/2]/12 = 0.0004167\ \text{ft}$$

$$\tau = \mu\,(dv/dx) = (0.0018)(2.356/0.0004167) = 10.18\ \text{lb/ft}^2$$

$$F_f = \tau A = 10.18[\pi(8.00/12)(\tfrac{6}{12})] = 10.66\ \text{lb}$$

$$\text{Rate of energy loss} = F_f v = (10.66)(2.356) = 25.11\ \text{ft}\cdot\text{lb/s}$$

$$\text{Rate of heat generation} = (25.11)(3600)/778 = 116\ \text{Btu/h}$$

**1.91** A 10.00-cm shaft rides in an 10.03-cm sleeve 12 cm long, the clearance space (assumed to be uniform) being filled with lubricating oil at 40 °C ($\mu = 0.11\ \text{Pa}\cdot\text{s}$). Calculate the rate at which heat is generated when the shaft turns at 100 rpm.

$$dv = \omega(\text{circumference}) = \tfrac{100}{60}[\pi(0.10)] = 0.5236\ \text{m/s} \qquad dx = (0.1003 - 0.1000)/2 = 0.00015\ \text{m}$$

$$\tau = \mu\,(dv/dx) = (0.11)(0.5236)/0.00015 = 384.0\ \text{N/m}^2$$

$$F_f = \tau A = 384.0[\pi(0.12)(0.10)] = 14.48\ \text{N}$$

$$\text{Rate of energy loss} = F_f v = (14.48)(0.5236) = 7.582\ \text{N}\cdot\text{m/s} = 7.582\ \text{W}$$

**1.92** In using a rotating-cylinder viscometer, a bottom correction must be applied to account for the drag on the flat bottom of the cylinder. Calculate the theoretical amount of this torque correction, neglecting centrifugal effects, for a cylinder of diameter $d$, rotated at a constant angular velocity $\omega$, in a liquid of viscosity $\mu$, with a clearance $\Delta h$ between the bottom of the inner cylinder and the floor of the outer one.

Let $r$ = variable radius. $T = \int r\tau\,dA$, $\tau = \mu\,(dv/dx) = \mu(r\omega/\Delta h)$, $dA = 2\pi r\,dr$.

$$T = \int_0^{d/2} r\left[\mu\left(\frac{r\omega}{\Delta h}\right)\right](2\pi r\,dr) = \frac{2\pi\mu\omega}{\Delta h}\int_0^{d/2} r^3\,dr = \frac{2\pi\mu\omega}{\Delta h}\left[\frac{r^4}{4}\right]_0^{d/2} = \frac{\pi\mu\omega d^4}{32\,\Delta h}$$

**1.93** Assuming a boundary-layer velocity distribution as shown in Fig. 1-16, which is a parabola having its vertex 3 in from the wall, calculate the shear stresses for $y = 0$, 1 in, 2 in, and 3 in. Use $\mu = 0.00835\ \text{lb}\cdot\text{s/ft}^2$.

$\tau = \mu\,(dv/dy)$. At $y = 0$, $v = 0$ and at $y = 3$ in, $v = 6$ ft/s, or 72 in/s. The equation of the parabola is $v = 72 - (8)(3 - y)^2$ ($y$ in inches gives $v$ in inches per second); $dv/dy = (16)(3 - y)$; $\tau = (0.00835)[(16)(3 - y)] = 0.4008 - 0.1336y$. At $y = 0$, $\tau = 0.4008 - (0.1336)(0) = 0.401\ \text{lb/ft}^2$. At $y = 1$ in, $\tau = 0.4008 - (0.1336)(1) = 0.267\ \text{lb/ft}^2$. At $y = 2$ in, $\tau = 0.4008 - (0.1336)(2) = 0.134\ \text{lb/ft}^2$. At $y = 3$ in, $\tau = 0.4008 - (0.1336)(3) = 0$.

**Fig. 1-16**

**1.94** In Fig. 1-17a, oil of viscosity $\mu$ fills the small gap of thickness $Y$. Determine an expression for the torque $T$ required to rotate the truncated cone at constant speed $\omega$. Neglect fluid stress exerted on the circular bottom.

❚ See Fig. 1-17b. $\tau = \mu (dv/dy)$, $v = r\omega = (y \tan \alpha)(\omega)$, $dv/dy = (y \tan \alpha)(\omega)/Y$.

$$\tau = \mu \left[ \frac{(y \tan \alpha)(\omega)}{Y} \right] = \frac{\mu y \omega \tan \alpha}{Y}$$

$$dA = 2\pi r\, ds = 2\pi (y \tan \alpha)(dy/\cos \alpha) = 2\pi y (\tan \alpha / \cos \alpha)(dy)$$

$$dF = \tau\, dA = \left( \frac{\mu y \omega \tan \alpha}{Y} \right) \left[ 2\pi y \left( \frac{\tan \alpha}{\cos \alpha} \right)(dy) \right] = \left( \frac{2\pi \mu \omega \tan^2 \alpha}{Y \cos \alpha} \right) y^2\, dy$$

$$dT = r\, dF = (y \tan \alpha) \left( \frac{2\pi \mu \omega \tan^2 \alpha}{Y \cos \alpha} \right) y^2\, dy = \left( \frac{2\pi \mu \omega \tan^3 \alpha}{Y \cos \alpha} \right) y^3\, dy$$

$$T = \int_a^{a+b} \left( \frac{2\pi \mu \omega \tan^3 \alpha}{Y \cos \alpha} \right) y^3\, dy = \left( \frac{2\pi \mu \omega \tan^3 \alpha}{Y \cos \alpha} \right) \left[ \frac{y^4}{4} \right]_a^{a+b} = \left( \frac{2\pi \mu \omega \tan^3 \alpha}{Y \cos \alpha} \right) \left[ \frac{(a+b)^4}{4} - \frac{a^4}{4} \right]$$

$$= \left( \frac{\pi \mu \omega \tan^3 \alpha}{2Y \cos \alpha} \right)[(a+b)^4 - a^4]$$

**Fig. 1-17(a)**

$r = y \tan d$

**Fig. 1-17(b)**

$dy = ds \cos \alpha$

$ds = \dfrac{dy}{\cos \alpha}$

**1.95** A Newtonian fluid fills the gap between a shaft and a concentric sleeve. When a force of 788 N is applied to the sleeve parallel to the shaft, the sleeve attains a speed of 2 m/s. If a 1400-N force is applied, what speed will the sleeve attain? The temperature of the sleeve remains constant.

❚ $\tau = F/A = \mu (dv/dx)$; $F/dv = \mu A/dx = $ constant. Therefore, $F_1/dv_1 = F_2/dv_2$, $\frac{788}{2} = 1400/dv_2$, $dv_2 = 3.55$ m/s.

**1.96** A plate separated by 0.5 mm from a fixed plate moves at 0.50 m/s under a force per unit area of 4.0 N/m². Determine the viscosity of the fluid between the plates.

❚ $\tau = \mu (dv/dx)$    $4.0 = \mu[0.50/(0.0005)]$    $\mu = 0.00400$ N $\cdot$ s/m² $= 4.00$ mPa $\cdot$ s

**1.97** Determine the viscosity of fluid between shaft and sleeve in Fig. 1-18.

❚ $\tau = F/A = \mu (dv/dx)$    $25/[(\pi)(\frac{4}{12})(\frac{9}{12})] = \mu[0.5/(0.004/12)]$    $\mu = 0.0212$ lb $\cdot$ s/ft²

**Fig. 1-18**

**1.98** A 1-in-diameter steel cylinder 10 in long falls at 0.6 ft/s inside a tube of slightly larger diameter. A castor-oil film of constant thickness is between the cylinder and the tube. Determine the clearance between the cylinder and the tube, if the temperature is 100 °F, s.g. = 7.85 for steel, and $\mu = 6 \times 10^{-3}$ lb · s/ft$^2$ for castor oil.

**❚** $$\tau = F/A = \mu \, (dv/dx)$$

$$F = W = \gamma V = [(7.85)(62.4)][(\tfrac{10}{12})(\pi)(\tfrac{1}{12})^2/4] = 2.226 \text{ lb} \quad 2.226/[(\tfrac{10}{12})(\pi)(\tfrac{1}{12})] = (6 \times 10^{-3})(0.6/dx)$$

$$dx = 0.0003528 \text{ ft} \quad \text{or} \quad 0.00423 \text{ in}$$

**1.99** A piston of diameter 70.00 mm moves inside a cylinder of diameter 70.10 mm. Determine the percent decrease in force necessary to move the piston when the lubricant warms from 0 to 120 °C. Values of $\mu$ for the lubricant are 0.01820 Pa · s at 0 °C and 0.00206 Pa · s at 120 °C.

**❚** $\tau = F/A = \mu \, (dv/dx)$; $F/\mu = A \, (dv/dx) = $ constant. Therefore, $\Delta F/F_{0\,°C} = \Delta\mu/\mu_{0\,°C} = (0.01820 - 0.00206)/0.01820 = 0.887$, or 88.7%.

**1.100** A body weighing 100 lb with a flat surface area of 3 ft$^2$ slides down a lubricated inclined plane making a 35° angle with the horizontal. For viscosity of 0.002089 lb · s/ft$^2$ and a body speed of 3.5 ft/s, determine the lubricant film thickness.

**❚** $$F = \text{weight of body along inclined plane} = 100 \sin 35° = 57.4 \text{ lb}$$

$$\tau = F/A = \mu \, (dv/dx) \qquad 57.4/3 = (0.002089)(3.5/dx) \qquad dx = 0.0003821 \text{ ft} \quad \text{or} \quad 0.00459 \text{ in}$$

**1.101** A small drop of water at 80 °F is in contact with the air and has a diameter of 0.0200 in. If the pressure within the droplet is 0.082 psi greater than the atmosphere, what is the value of the surface tension?

**❚** $$p(\pi d^2/4) = (\pi d)(\sigma) \qquad \sigma = pd/4 = [(0.082)(144)](0.0200/12)/4 = 0.00492 \text{ lb/ft}$$

**1.102** Estimate the height to which water at 70 °F will rise in a capillary tube of diameter 0.120 in.

**❚** $h = 4\sigma \cos\theta/(\gamma d)$. From Table A-1, $\sigma = 0.00500$ lb/ft and $\gamma = 62.3$ lb/ft$^3$ at 70 °F. Assume $\theta = 0°$ for a clean tube. $h = (4)(0.00500)(\cos 0°)/[(62.3)(0.120/12)] = 0.0321$ ft, or 0.385 in.

**1.103** The shape of a hanging drop of liquid is expressible by the following formulation developed from photographic studies of the drop: $\sigma = (\gamma - \gamma_0)(d_e)^2/H$, where $\sigma =$ surface tension, i.e., force per unit length, $\gamma =$ specific weight of liquid drop, $\gamma_0 =$ specific weight of vapor around it, $d_e =$ diameter of drop at its equator, and $H = $ a function determined by experiment. For this equation to be dimensionally homogeneous, what dimensions must $H$ possess?

**❚** Dimensionally, $(F/L) = (F/L^3)(L^2)/\{H\}$, $\{H\} = (1)$. Therefore, $H$ is dimensionless.

**1.104** Two clean, parallel glass plates, separated by a distance $d = 1.5$ mm, are dipped in a bath of water. How far does the water rise due to capillary action, if $\sigma = 0.0730$ N/m?

**❚** Because the plates are clean, the angle of contact between water and glass is taken as zero. Consider the free-body diagram of a unit width of the raised water (Fig. 1-19). Summing forces in the vertical direction gives $(2)[(\sigma)(0.0015)] - (0.0015)^2(h)(\gamma) = 0$, $(2)[(0.0730)(0.0015)] - (0.0015)^2(h)(9790) = 0$, $h = 0.00994$ m, or 9.94 mm.

**Fig. 1-19(c)**

**1.105** A glass tube is inserted in mercury (Fig. 1-20); the common temperature is 20 °C. What is the upward force on the glass as a result of surface effects?

▌ $F = (\sigma)(\pi d_o)(\cos 50°) + (\alpha)(\pi d_i)(\cos 50°) = (0.514)[(\pi)(0.035)](\cos 50°) + (0.514)[(\pi)(0.025)](\cos 50°) = 0.0623$ N

$d_o = 35$ mm
$d_i = 25$ mm
50°
50°

**Fig. 1-20**

**1.106** In Fig. 1-21a estimate the depression $h$ for mercury in the glass capillary tube. Angle $\theta$ is 40°.

▌ Consider the meniscus of the mercury as a free body (see Fig. 1-21b) of negligible weight. Summing forces in the vertical direction gives $-(\sigma)(\pi d)(\cos \theta) + (p)(\pi d^2/4) = 0$, $-(0.514)[(\pi)(0.002)](\cos 40°) + [(13.6)(9790)(h)][(\pi)(0.002)^2/4] = 0$, $h = 0.00591$ m, or 5.91 mm. Actual $h$ must be larger because the weight of the meniscus was neglected.

$d = 2$ mm
Free surface
$h$
Mercury
$\theta$

**Fig. 1-21(a)**

$\sigma$ $p = \gamma h$ $\sigma$
$\theta$

**Fig. 1-21(b)**

**1.107** A narrow trough (Fig. 1-22) is filled with water at 20 °C to the maximum extent. If the gage measures a gage pressure of 2.8458 kPa, what is the radius of curvature of the water surface (away from the ends)?

▌ $\qquad p = \sigma/r = p_{\text{gage}} - \gamma d = 2845.8 - (9790)(0.290) = 6.70$ Pa gage

$\qquad 6.70 = 0.0728/r \qquad r = 0.01087$ m   or   10.87 mm

290 mm

**Fig. 1-22**

**1.108** Water at 10 °C is poured into a region between concentric cylinders until water appears above the top of the open end (see Fig. 1-23). If the pressure measured by a gage 42 cm below the open end is 4147.38 Pa gage, what is the curvature of the water at the top?

▌ $\qquad p = \sigma/r = p_{\text{gage}} - \gamma d = 4147.38 - (9810)(0.42) = 27.18$ Pa gage

$\qquad 27.18 = 0.0742/r \qquad r = 0.00273$ m   or   2.73 mm

Top view        **Fig. 1-23**

**1.109** The rate of twist $\alpha$ of a shaft of any shape may be found by using *Prandtl's soap-film analogy*. A soap film is attached to a sharp edge having the shape of the outside boundary of the shaft cross section (a rectangle here, as shown in Fig. 1-24). Air pressure is increased under the film so that it forms an elevated curved surface above the boundary. Then

$$\alpha = \frac{M_x \Delta p}{4\sigma G V} \quad \text{(radians per unit length)}$$

where $\Delta p$ = gage air pressure under the soap film, $M_x$ = torque transmitted by actual shaft, $G$ = shear modulus of actual shaft, and $V$ = volume of air under the soap film and above the cross section formed by the sharp edge. For the case at hand, $\Delta p = 0.4$ lb/ft² gage and $V = 0.5$ in³. The angle $\theta$ along the long edge of the cross section is measured optically to be 30°. For a torque of 600 lb · ft on a shaft having $G = 10 \times 10^6$ lb/in², what angle of twist does this analogy predict?

**∎**
$$\alpha = \frac{M_x \Delta p}{4\sigma G V}$$

To get $\sigma$, consider a unit length of the long side of the shaft cross section away from the ends (see Fig. 1-24c). For equilibrium of the film in the vertical direction (remembering there are two surfaces on each side)
$(-4)[(\sigma)(L)(\cos \theta)] + pA = 0$, $(-4)[(\sigma)(\frac{1}{12})(\cos 30°)] + (0.4)[(0.5)(1)/144] = 0$, $\sigma = 0.00481$ lb/ft;

$$\alpha = \frac{(600)(0.4)}{(4)(0.00481)[(10 \times 10^6)(144)](0.5/1728)} = 0.0299 \text{ rad/ft}$$

**Fig. 1-24(a)**        **Fig. 1-24(b)**        **Fig. 1-24(c)**

**1.110** In using Prandtl's soap-film analogy (see Prob. 1.109), we wish to check the mechanism for measuring the pressure $\Delta p$ under the soap film. Accordingly, we use a circular cross section (Fig. 1-25) for which we have an accurate theory for determining the rate of twist $\alpha$. The surface tension for the soap film is 0.1460 N/m and volume $V$ under the film is measured to be 0.001120 m³. Compute $\Delta p$ from consideration of the soap film and from solid mechanics using the equation given in Prob. 1.109 and the well-known formula from strength of materials

$$\alpha = \frac{M_x}{GJ}$$

where $J$, the polar moment of inertia, is $\pi r^4/2$. Compare the results.

▮ From consideration of the film (see Fig. 1-25), $-2\sigma\pi d \cos 45° + (\Delta p)(\pi d^2)/4 = 0$,
$-(2)(0.1460)(\pi)(\frac{200}{1000})(\cos 45°) + (\Delta p)[(\pi)(\frac{200}{1000})^2/4] = 0$, $\Delta p = 4.13$ Pa gage. From strength of materials, equate $\alpha$'s for the equations given in this problem and in Prob. 1.109.

$$\frac{M_x \Delta p}{4\sigma GV} = \frac{M_x}{GJ} \qquad J = \frac{\pi[(\frac{200}{2})/1000]^4}{2} = 0.0001571 \text{ m}^4 \qquad \frac{\Delta p}{(4)(0.1460)(0.001120)} = \frac{1}{0.0001571} \qquad \Delta p = 4.16 \text{ Pa gage}$$

The pressure measurement is quite close to what is expected from theory.

| Fig. 1-25(a) | Fig. 1-25(b) | Fig. 1-25(c) |

**1.111** Find the capillary rise in the tube shown in Fig. 1-26 for a water–air–glass interface ($\theta = 0°$) if the tube radius is 1 mm and the temperature is 20 °C.

▮
$$h = \frac{2\sigma \cos \theta}{\rho g r} = \frac{(2)(0.0728)(\cos 0°)}{(1000)(9.81)(\frac{1}{1000})} = 0.0148 \text{ m} \quad \text{or} \quad 14.8 \text{ mm}$$

Fig. 1-26

**1.112** Find the capillary rise in the tube shown in Fig. 1-26 for a mercury–air–glass interface with $\theta = 130°$ if the tube radius is 1 mm and the temperature is 20 °C.

▮
$$h = \frac{2\sigma \cos \theta}{\rho g r} = \frac{(2)(0.514)(\cos 130°)}{(13\,570)(9.81)(\frac{1}{1000})} = -0.0050 \text{ m} \quad \text{or} \quad -5.0 \text{ mm}$$

**1.113** If a bubble is equivalent to an air–water interface with $\sigma = 0.005$ lb/ft, what is the pressure difference between the inside and outside of a bubble of diameter 0.003 in?

▮
$$p = 2\sigma/r = (2)(0.005)/[(0.003/2)/12] = 80.0 \text{ lb/ft}^2$$

**1.114** A small circular jet of mercury 200 μm in diameter issues from an opening. What is the pressure difference between the inside and outside of the jet at 20 °C?

▮ See Fig. 1-27. Equating the force due to surface tension ($2\sigma L$) and the force due to pressure ($pDL$),
$2\sigma L = pDL$, $p = 2\sigma/D = (2)(0.514)/(200 \times 10^{-6}) = 5140$ Pa.

Fig. 1-27

**1.115** The surface tensions of mercury and water at 60 °C are 0.47 N/m and 0.0662 N/m, respectively. What capillary-height changes will occur in these two fluids when they are in contact with air in a glass tube of radius 0.30 mm? Use $\theta = 130°$ for mercury, and 0° for water; $\gamma = 132.3$ kN/m$^3$ for mercury, and 9.650 kN/m$^3$ for water.

**▮**
$$h = \frac{2\sigma \cos \theta}{\rho g r}$$

For mercury:

$$h = \frac{(2)(0.47)(\cos 130°)}{(132\,300)(0.30/1000)} = -0.0152 \text{ m} \quad \text{or} \quad -15.2 \text{ mm}$$

For water:

$$h = \frac{(2)(0.0662)(\cos 0°)}{(9650)(0.30/1000)} = 0.0457 \text{ m} \quad \text{or} \quad 45.7 \text{ mm}$$

**1.116** At 30 °C what diameter glass tube is necessary to keep the capillary-height change of water less than 2 mm?

**▮**
$$h = \frac{2\sigma \cos \theta}{\rho g r} \qquad \frac{2}{1000} = \frac{(2)(0.0712)(\cos 0°)}{(996)(9.81)(r)}$$
$$r = 0.00729 \text{ m} \quad \text{or} \quad 7.29 \text{ mm} \qquad d = (2)(7.29) = 14.6 \text{ mm (or greater)}$$

**1.117** A 1-in-diameter soap bubble has an internal pressure 0.0045 lb/in$^2$ greater than that of the outside atmosphere. Compute the surface tension of the soap–air interface. Note that a soap bubble has two interfaces with air, an inner and outer surface of nearly the same radius.

**▮** $\qquad p = 4\sigma/r \qquad (0.0045)(144) = (4)(\sigma)/[(\tfrac{1}{2})/12] \qquad \sigma = 0.00675 \text{ lb/ft}$

**1.118** What force is required to lift a thin wire ring 6 cm in diameter from a water surface at 20 °C?

**▮** Neglecting the weight of the wire, $F = \sigma L$. Since there is resistance on the inside and outside of the ring, $F = (2)(\sigma)(\pi d) = (2)(0.0728)[(\pi)(0.06)] = 0.0274$ N.

**1.119** The glass tube in Fig. 1-28 is used to measure pressure $p_1$ in the water tank. The tube diameter is 1 mm and the water is at 30 °C. After correcting for surface tension, what is the true water height in the tube?

**▮**
$$h = \frac{2\sigma \cos \theta}{\rho g r} = \frac{(2)(0.0712)(\cos 0°)}{(996)(9.81)[(\tfrac{1}{2})/1000]} = 0.029 \text{ m} \quad \text{or} \quad 2.9 \text{ cm}$$

True water height in the tube $= 17 - 2.9 = 14.1$ cm.

Fig. 1-28

**1.120** An atomizer forms water droplets 45 $\mu$m in diameter. Find the excess pressure within these droplets for water at 30 °C?

$$p = 2\sigma/r = (2)(0.0712)/[(45 \times 10^{-6})/2] = 6329 \text{ Pa}$$

**1.121** Rework Prob. 1.120 for droplets of 0.0018 in diameter and at 68 °F.

$$p = 2\sigma/r = (2)(0.005)/[(0.0018/2)/12] = 133 \text{ lb/ft}^2 \quad \text{or} \quad 0.93 \text{ lb/in}^2$$

**1.122** What is the pressure difference between the inside and outside of a cylindrical water jet when the diameter is 2.2 mm and the temperature is 10 °C? (See Fig. 1-27.)

$$p = \sigma/r = 0.0742/0.0011 = 67.5 \text{ Pa}$$

**1.123** Find the angle the surface tension film leaves the glass for a vertical tube immersed in water if the diameter is 0.25 in and the capillary rise is 0.08 in. Use $\sigma = 0.005$ lb/ft.

$$h = \frac{2\sigma \cos \theta}{\rho g r} \qquad \frac{0.08}{12} = \frac{(2)(0.005)(\cos \theta)}{(1.94)(32.2)[(0.25/2)/12]} \qquad \cos \theta = 0.433806 \qquad \theta = 64.3°$$

**1.124** Develop a formula for capillary rise between two concentric glass tubes of radii $r_o$ and $r_i$ and contact angle $\theta$.

See Fig. 1-29. Equating the force due to pressure and the force due to surface tension,

$$(h)(\gamma)(\pi r_o^2 - \pi r_i^2) = \sigma(2\pi r_i + 2\pi r_o)(\cos \theta)$$

$$h = \frac{(2)(\sigma)(r_i + r_o)(\cos \theta)}{\gamma(r_o^2 - r_i^2)} = \frac{2\sigma \cos \theta}{\gamma(r_o - r_i)}$$

**Fig. 1-29**

**1.125** Distilled water at 10 °C stands in a glass tube of 9.0-mm diameter at a height of 24.0 mm. What is the true static height?

$$h = \frac{2\sigma \cos \theta}{\rho g r} = \frac{(2)(0.0742)(\cos 0°)}{(1000)(9.81)[(9.0/2)/1000]} = 0.0034 \text{ m} \quad \text{or} \quad 3.4 \text{ mm}$$

True static height = 24.0 − 3.4 = 20.6 mm.

**1.126** What capillary depression of mercury ($\theta = 140°$) may be expected in a 0.08-in-diameter tube at 68 °F?

$$h = \frac{2\sigma \cos \theta}{\rho g r} = \frac{(2)(0.0352)(\cos 140°)}{(26.34)(32.2)[(0.08/2)/12]} = -0.01908 \text{ ft} \quad \text{or} \quad -0.23 \text{ in}$$

**1.127** At the top of Mount Olympus water boils at 85 °C. Approximately how high is the mountain?

From Table A-2, water boiling at 85 °C corresponds to a vapor pressure of 58.8 kPa. From Table A-8, this corresponds to a standard atmosphere elevation of approximately 4200 m.

**1.128** At approximately what temperature will water boil at an elevation of 12 500 ft?

❚ From Table A-7, the pressure of the standard atmosphere at 12 500-ft evevation is 9.205 psia, or 1326 lb/ft² abs. From Table A-1, the saturation pressure of water is 1326 lb/ft² abs at about 189 °F. Hence, the water will boil at 193 °F; this explains why it takes longer to cook at high altitudes.

**1.129** At approximately what temperature will water boil in Denver (elevation 5280 ft)?

❚ From Table A-7, the pressure of the standard atmosphere at 5280-ft elevation is 12.12 psia, or 1745 lb/ft² abs. From Table A-1, the saturation pressure of water is 1745 lb/ft² abs at about 202 °F. Hence, the water will boil at 198 °F.

**1.130** Water at 105 °F is placed in a beaker within an airtight container. Air is gradually pumped out of the container. What reduction below standard atmospheric pressure of 14.7 psia must be achieved before the water boils?

❚ From Table A-1, $p_v = 162$ lb/ft² abs, or 1.12 psia at 105 °F. Hence, pressure must be reduced by $14.7 - 1.12$, or 13.58 psi.

**1.131** At what pressure will 50 °C water boil?

❚ From Table A-2, $p_v = 12.3$ kPa at 50 °C. Hence, water will boil at 12.3 kPa.

**1.132** At what pressure will cavitation occur at the inlet of a pump that is drawing water at 25 °C?

❚ Cavitation occurs when the internal pressure drops to the vapor pressure. From Table A-2, the vapor pressure of water at 25 °C is 3.29 kPa.

**1.133** For low-speed (laminar) flow through a circular pipe, as shown in Fig. 1-30, the velocity distribution takes the form $v = (B/\mu)(r_0^2 - r^2)$, where $\mu$ is the fluid viscosity. What are the units of the constant $B$?

❚ Dimensionally, $(L/T) = [\{B\}/(M/LT)](L^2)$, $\{B\} = ML^{-2}T^{-2}$. In SI units, $B$ could be kg/(m² · s²), or Pa/m.

**Fig. 1-30**

**1.134** The mean free path $L$ of a gas is defined as the mean distance traveled by molecules between collisions. According to kinetic theory, the mean free path of an ideal gas is given by $L = 1.26(\mu/\rho)(RT)^{-1/2}$, where $R$ is the gas constant and $T$ is the absolute temperature. What are the units of the constant 1.26?

❚ Dimensionally, $L = \{1.26\}[(M/LT)/(M/L^3)][(L^2/T^2D)(D)]^{-1/2}$, $L = \{1.26\}(L)$, $\{1.26\} = 1$. Therefore, the constant 1.26 is dimensionless.

**1.135** The Stokes–Oseen formula for the drag force $F$ on a sphere of diameter $d$ in a fluid stream of low velocity $v$ is $F = 3\pi\mu dv + (9\pi/16)(\rho v^2 d^2)$. Is this formula dimensionally consistent?

❚ Dimensionally, $(F) = (1)(M/LT)(L)(L/T) + (1)(M/L^3)(L/T)^2(L)^2 = (ML/T^2) + (ML/T^2) = (F) + (F)$. Therefore, the formula is dimensionally consistent.

**1.136** The speed of propagation $C$ of waves traveling at the interface between two fluids is given by $C \doteq (\pi\sigma/\rho_a\lambda)^{1/2}$, where $\lambda$ is the wavelength and $\rho_a$ is the average density of the two fluids. If the formula is dimensionally consistent, what are the units of $\sigma$? What might it represent?

❚ Dimensionally, $(L/T) = [(1)\{\sigma\}/(M/L^3)(L)]^{1/2} = [\{\sigma\}(L^2/M)]^{1/2}$, $\{\sigma\} = M/T^2 = F/L$. In SI units, $\sigma$ could be N/m. (In this formula, $\sigma$ is actually the surface tension.)

**1.137** Is the following equation dimensionally homogeneous? $a = 2d/t^2 - 2v_0/t$, where $a$ = acceleration, $d$ = distance, $v_0$ = velocity, and $t$ = time.

▌ $L/T^2 = (L)/(T^2) - (L/T)/(T) = (L/T^2) - (L/T^2)$. Therefore, the equation is homogeneous.

**1.138** A popular formula in the hydraulics literature is the Hazen-Williams formula for volume flow rate $Q$ in a pipe of diameter $D$ and pressure gradient $dp/dx$: $Q = 61.9D^{2.63}(dp/dx)^{0.54}$. What are the dimensions of the constant 61.9?

▌ $$\frac{L^3}{T} = \{61.9\}(L)^{2.63}\left(\frac{M}{L^2T^2}\right)^{0.54} \qquad \{61.9\} = L^{1.45}T^{0.08}M^{-0.54}$$

# CHAPTER 2
# Fluid Statics

**2.1** For the dam shown in Fig. 2-1, find the horizontal pressure acting at the face of the dam at 20-ft depth.

$$p = \gamma h = (62.4)(20) = 1248 \, \text{lb/ft}^2$$

**Fig. 2-1.** Dam.

**2.2** For the vessel containing glycerin under pressure as shown in Fig. 2-2, find the pressure at the bottom of the tank.

$$p = 50 + \gamma h = 50 + (12.34)(2.0) = 74.68 \, \text{kN/m}^2 \quad \text{or} \quad 74.68 \, \text{kPa}$$

**Fig. 2-2**

**2.3** If the pressure in a tank is 50 psi, find the equivalent pressure head of (*a*) water, (*b*) mercury, and (*c*) heavy fuel oil with a specific gravity of 0.92.

$$h = p/\gamma$$

(*a*) $\qquad\qquad h = [(50)(144)]/62.4 = 115.38 \, \text{ft}$

(*b*) $\qquad\qquad h = [(50)(144)]/847.3 = 8.50 \, \text{ft}$

(*c*) $\qquad\qquad h = [(50)(144)]/[(0.92)(62.4)] = 125.42 \, \text{ft}$

**2.4**   A weather report indicates the barometric pressure is 29.75 in of mercury. What is the atmospheric pressure in pounds per square inch?

▮   $p = \gamma h = [(13.6)(62.4)][(29.75/12)]/144 = 14.61 \text{ lb/in}^2$   or   14.61 psi

**2.5**   Find the atmospheric pressure in kilopascals if a mercury barometer reads 742 mm.

▮   $p = \gamma h = (133.1)(\frac{742}{1000}) = 98.8 \text{ kN/m}^2$   or   98.8 kPa

**2.6**   A pressure gage 7.0 m above the bottom of a tank containing a liquid reads 64.94 kPa; another gage at height 4.0 m reads 87.53 kPa. Compute the specific weight and mass density of the fluid.

▮   $\gamma = \Delta p / \Delta h = (87.53 - 64.94)/(7.0 - 4.0) = 7.53 \text{ kN/m}^3$   or   7530 N/m³

$$\rho = \gamma/g = 7530/9.81 = 786 \text{ kg/m}^3$$

**2.7**   A pressure gage 19.0 ft above the bottom of a tank containing a liquid reads 13.19 psi; another gage at height 14.0 ft reads 15.12 psi. Compute the specific weight, mass density, and specific gravity of the liquid.

▮   $\Delta p = \gamma(\Delta h)$      $(15.12 - 13.19)(144) = (\gamma)(19.0 - 14.0)$      $\gamma = 55.6 \text{ lb/ft}^3$

$$\rho = \gamma/g = 55.6/32.2 = 1.73 \text{ slug/ft}^3 \qquad \text{s.g.} = 55.6/62.4 = 0.891$$

**2.8**   An open tank contains 5.7 m of water covered with 2.8 m of kerosene ($\gamma = 8.0 \text{ kN/m}^3$). Find the pressure at the interface and at the bottom of the tank.

▮   $$p_{\text{int}} = \gamma h = (8.0)(2.8) = 22.4 \text{ kPa}$$
$$p_{\text{bot}} = 22.4 + (9.79)(5.7) = 78.2 \text{ kPa}$$

**2.9**   An open tank contains 9.4 ft of water beneath 1.8 ft of oil (s.g. = 0.85). Find the pressure at the interface and at the bottom of the tank.

▮   $$p_{\text{int}} = \gamma h = [(0.85)(62.4)](1.8)/144 = 0.663 \text{ psi}$$
$$p_{\text{bot}} = 0.663 + (62.4)(9.4)/144 = 4.74 \text{ psi}$$

**2.10**   If air had a constant specific weight of 0.076 lb/ft³ and were incompressible, what would be the height of the atmosphere if sea-level pressure were 14.92 psia?

▮   $h = p/\gamma = (14.92)(144)/0.076 = 28\,270 \text{ ft}$

**2.11**   If the weight density of mud is given by $\gamma = 65.0 + 0.2h$, where $\gamma$ is in lb/ft³ and depth $h$ is in ft, determine the pressure, in psi, at a depth of 17 ft.

▮   $dp = \gamma \, dh = (65.0 + 0.2h) \, dh$. Integrating both sides: $p = 65.0h + 0.1h^2$. For $h = 17$ ft: $p = (65.0)(17)/144 + (0.1)(17)^2/144 = 7.87 \text{ psi}$.

**2.12**   If the absolute pressure in a gas is 40.0 psia and the atmospheric pressure is 846 mbar abs, find the gage pressure in (**a**) lb/in²; (**b**) kPa; (**c**) bar.

▮   (**a**)          $p_{\text{atm}} = (846)(0.0145) = 12.3 \text{ lb/in}^2$      $p_{\text{gage}} = 40.0 - 12.3 = 27.7 \text{ lb/in}^2$

(**b**) $p_{\text{abs}} = (40.0)(6.894) = 276 \text{ kPa}$      $p_{\text{atm}} = (846)(0.100) = 85 \text{ kPa}$      $p_{\text{gage}} = 276 - 85 = 191 \text{ kPa}$

(**c**)          $p_{\text{abs}} = 40.0/14.5 = 2.759 \text{ bar}$      $p_{\text{gage}} = 2.759 - 0.846 = 1.913 \text{ bar}$

**2.13**   If the atmospheric pressure is 0.900 bar abs and a gage attached to a tank reads 390 mmHg vacuum, what is the absolute pressure within the tank?

▮   $$p = \gamma h \qquad p_{\text{atm}} = 0.900 \times 100 = 90.0 \text{ kPa}$$

$$p_{\text{gage}} = [(13.6)(9.79)](\tfrac{390}{1000}) = 51.9 \text{ kPa vacuum} \quad \text{or} \quad -51.9 \text{ kPa}$$

$$p_{\text{abs}} = 90.0 + (-51.9) = 38.1 \text{ kPa}$$

**2.14** If atmospheric pressure is 13.99 psia and a gage attached to a tank reads 7.4 inHg vacuum, find the absolute pressure within the tank.

$$p = \gamma h \qquad p_{gage} = [(13.6)(62.4)][(7.4/12)/144] = 3.63 \text{ psi vacuum} \quad \text{or} \quad -3.63 \text{ psi}$$

$$p_{abs} = 13.99 + (-3.63) = 10.36 \text{ psia}$$

**2.15** The closed tank in Fig. 2-3 is at 20 °C. If the pressure at point $A$ is 98 kPa abs, what is the absolute pressure at point $B$? What percent error results from neglecting the specific weight of the air?

$p_A + \gamma_{air}h_{AC} - \gamma_{H_2O}h_{DC} - \gamma_{air}h_{DB} = p_B$, $98 + (0.0118)(5) - (9.790)(5-3) - (0.0118)(3) = p_B = 78.444 \text{ kPa}$. Neglecting air, $p_B = 98 - (9.790)(5-3) = 78.420 \text{ kPa}$; error $= (78.444 - 78.420)/78.444 = 0.00031$, or $0.031\%$.

**Fig. 2-3**

**2.16** The system in Fig. 2-4 is at 70 °F. If the pressure at point $A$ is 2900 lb/ft², determine the pressures at points $B$, $C$, and $D$.

$$p_B = 2900 - (62.4)(4-3) = 2838 \text{ lb/ft}^2 \qquad p_D = 2900 + (62.4)(6) = 3274 \text{ lb/ft}^2$$

$$p_C = 2900 + (62.4)(6-2) - (0.075)(5+3) = 3149 \text{ lb/ft}^2$$

**Fig. 2-4**

**2.17** The system in Fig. 2-5 is at 20 °C. If atmospheric pressure is 101.03 kPa and the absolute pressure at the bottom of the tank is 231.3 kPa, what is the specific gravity of olive oil?

$$101.03 + (0.89)(9.79)(1.5) + (9.79)(2.5) + (\text{s.g.})(9.79)(2.9) + (13.6)(9.79)(0.4) = 231.3 \qquad \text{s.g.} = 1.39$$

**Fig. 2-5**

**2.18** Find the pressures at *A*, *B*, *C*, and *D* in Fig. 2-6.

▌ $p_A = (62.4)(4+2) = 374$ lb/ft$^2$, $p_B = -(62.4)(2) = -125$ lb/ft$^2$. Neglecting air, $p_C = p_B = -125$ lb/ft$^2$; $p_D = -125 - (62.4)(4+2+2) = -624$ lb/ft$^2$.

Fig. 2-6

**2.19** The tube shown in Fig. 2-7 is filled with oil. Determine the pressure heads at *A* and *B* in meters of water.

▌ $(h_{H_2O})(\gamma_{H_2O}) = (h_{oil})(\gamma_{oil}) = (h_{oil})[(s.g._{oil})(\gamma_{H_2O})]$; therefore, $h_{H_2O} = (h_{oil})(s.g._{oil})$. Thus, $h_A = -(2.2+0.6)(0.85) = -2.38$ m H$_2$O and $h_B = (-0.6)(0.85) = -0.51$ m H$_2$O.

Fig. 2-7

**2.20** Calculate the pressure, in kPa, at *A*, *B*, *C*, and *D* in Fig. 2-8.

▌ $p_A = -(0.4+0.4)(9.790) = -7.832$ kPa; $p_B = (0.5)(9.790) = 4.895$ kPa. Neglecting air, $p_C = p_B = 4.895$ kPa; $p_D = 4.895 + (0.9)(9.790)(1+0.5+0.4) = 21.636$ kPa.

Fig. 2-8

**2.21** Convert 9 psi to (*a*) inches of mercury, (*b*) feet of water, (*c*) feet of ichor (s.g. = 2.94).

▌ (*a*) $h = p/\gamma = [(9)(144)]/[(13.6)(62.4)] = 1.527$ ft, or 18.33 inHg
  (*b*) $h = [(9)(144)]/62.4 = 20.77$ ft of water
  (*c*) $h = [(9)(144)]/[(2.94)(62.4)] = 7.06$ ft ichor

**2.22** Express an absolute pressure of 5 atm in meters of water gage when the barometer reads 760 mmHg.

▌         $p_{abs} = (5)(101.3)/9.79 = 51.74$ m of water     $p_{atm} = (0.760)(13.6) = 10.34$ m of water

$$p_{gage} = 51.74 - 10.34 = 41.40 \text{ m of water}$$

**2.23** Figure 2-9 shows one pressurized tank inside another. If the sum of the readings of Bourdan gages $A$ and $B$ is 34.1 psi, and an aneroid barometer reads 29.90 inHg, what is the absolute pressure at $A$, in inHg?

$$h = p/\gamma \qquad h_A + h_B = 34.1/[(13.6)(62.4)/(12)^3] = 69.44 \text{ inHg}$$

$$(h_A)_{\text{abs}} = 29.90 + 69.44 = 99.34 \text{ inHg}$$

**Fig. 2-9**

**2.24** Determine the heights of columns of water, kerosene (ker), and nectar (s.g. = 2.94) equivalent to 277 mmHg.

$$(h_{\text{Hg}})(\gamma_{\text{Hg}}) = (H_{\text{H}_2\text{O}})(\gamma_{\text{H}_2\text{O}}) = (h_{\text{ker}})(\gamma_{\text{ker}}) = (h_{\text{nectar}})(\gamma_{\text{nectar}})$$

$$0.277[(13.6)(9.79)] = (h_{\text{H}_2\text{O}})(9.79) \qquad h_{\text{H}_2\text{O}} = 3.77 \text{ m}$$

$$0.277[(13.6)(9.79)] = (h_{\text{ker}})[(0.82)(9.79)] \qquad h_{\text{ker}} = 4.59 \text{ m}$$

$$0.277[(13.6)(9.79)] = (h_{\text{nectar}})[(2.94)(9.79)] \qquad h_{\text{nectar}} = 1.28 \text{ m}$$

**2.25** In Fig. 2-10, if $h = 25.5$ in, determine the pressure at $A$. The liquid has a specific gravity of 1.85.

$$p = \gamma h = [(1.85)(62.4)][25.5/12] = 245.3 \text{ lb/ft}^2 \quad \text{or} \quad 1.70 \text{ psi}$$

**Fig. 2-10**

**2.26** For the pressure vessel containing glycerin, with piezometer attached, as shown in Fig. 2-11, what is the pressure at point $A$?

$$p = \gamma h = [(1.26)(62.4)](40.8/12) = 267 \text{ lb/ft}^2$$

Open to atmosphere

40.8 in

Glycerin —— $A$

**Fig. 2-11**

**2.27** For the open tank, with piezometers attached on the side, containing two different immiscible liquids, as shown in Fig. 2-12, find the (a) elevation of the liquid surface in piezometer A, (b) elevation of the liquid surface in piezometer B, and (c) total pressure at the bottom of the tank.

▌ (a) Liquid A will simply rise in piezometer A to the same elevation as liquid A in the tank (i.e., to elevation 2 m). (b) Liquid B will rise in piezometer B to elevation 0.3 m (as a result of the pressure exerted by liquid B) plus an additional amount as a result of the overlying pressure of liquid A. The overlying pressure can be determined by $p = \gamma h = [(0.72)(9.79)](2 - 0.3) = 11.98 \text{ kN/m}^2$. The height liquid B will rise in piezometer B as a result of the overlying pressure of liquid A can be determined by $h = p/\gamma = 11.98/[(2.36)(9.79)] = 0.519 \text{ m}$. Hence, liquid B will rise in piezometer B to an elevation of 0.3 m + 0.519 m, or 0.819 m.
(c) $p_{\text{bottom}} = [(0.72)(9.79)](2 - 0.3) + [(2.36)(9.79)](0.3) = 18.9 \text{ kPa}$.

El 2 m

Liquid A
(s.g. = 0.72)

El 0.3 m

Liquid B
(s.g. = 2.36)

El .0 m

**Fig. 2-12**

**2.28** The air–oil–water system shown in Fig. 2-13 is at 70 °F. If gage A reads 16.1 lb/in² abs and gage B reads 2.00 lb/in² less than gage C, compute (a) the specific weight of the oil and (b) the reading of gage C.

▌ (a) $(16.1)(144) + (0.0750)(3) + (\gamma_{\text{oil}})(2) = p_B$, $p_B + (\gamma_{\text{oil}})(2) + (62.4)(3) = p_C$. Since $p_C - p_B = 2.00$, $(\gamma_{\text{oil}})(2) + (62.4)(3) = (2.00)(144)$, $\gamma_{\text{oil}} = 50.4 \text{ lb/ft}^3$. (b) $(16.1)(144) + (0.0750)(3) + (50.4)(2) = p_B$, $p_B = 2419 \text{ lb/ft}^2$; $p_C = 2419 + (2.00)(144) = 2707 \text{ lb/ft}^2$, or 18.80 lb/in².

16.1 lb /in² abs

Air    3ft

Oil    2ft
       —
       2ft

Water  3ft

**Fig. 2-13**

**2.29** For a gage reading at A of −2.50 psi, determine the (a) elevations of the liquids in the open piezometer columns E, F, and G and (b) deflection of the mercury in the U-tube gage in Fig. 2-14. Neglect the weight of the air.

▌ (a) The liquid between the air and the water would rise to elevation 49.00 ft in piezometer column E as a result of its weight. The actual liquid level in the piezometer will be lower, however, because of the vacuum in the air above the liquid. The amount the liquid level will be lowered (h in Fig. 2-14) can be determined by

$(-2.50)(144) + [(0.700)(62.4)](h) = 0$, $h = 8.24$ ft. Elevation at $L = 49.00 - 8.24 = 40.76$ ft; $(-2.50)(144) +$ $[(0.700)(62.4)][49.00 - 38.00] = p_M$, $p_M = 120.5$ lb/ft². Hence, pressure head at $M = 120.5/62.4 = 1.93$ ft of water. Elevation at $N = 38.00 + 1.93 = 39.93$ ft; $120.5 + (62.4)(38.00 - 26.00) = p_O$, $p_O = 869.3$ lb/ft². Hence, pressure head at $O = 869.3/[(1.600)(62.4)] = 8.71$ ft (of the liquid with s.g. = 1.600). Elevation at $Q = 26.00 + 8.71 = 34.71$ ft. **(b)** $869.3 + (62.4)(26.00 - 14.00) - [(13.6)(62.4)](h_1) = 0$, $h_1 = 1.91$ ft.

**Fig. 2-14**

**2.30** A vessel containing oil under pressure is shown in Fig. 2-15. Find the elevation of the oil surface in the attached piezometer.

❙   Elevation of oil surface in piezometer $= 2 + 35/[(0.83)(9.79)] = 6.31$ m

**Fig. 2-15**

**2.31** The reading of an automobile fuel gage is proportional to the gage pressure at the bottom of the tank (Fig. 2-16). If the tank is 32 cm deep and is contaminated with 3 cm of water, how many centimeters of air remains at the top when the gage indicates "full"? Use $\gamma_{gasoline} = 6670$ N/m³ and $\gamma_{air} = 11.8$ N/m³.

❙   When full of gasoline, $p_{gage} = (6670)(0.32) = 2134$ Pa. With water added, $2134 = (9790)(0.03) + (6670)[(0.32 - 0.03) - h] + (11.8)(h)$, $h = 0.0141$ m, or 1.41 cm.

**Fig. 2-16**

**2.32** The hydraulic jack shown in Fig. 2-17 is filled with oil at 55 lb/ft³. Neglecting the weight of the two pistons, what force $F$ on the handle is required to support the 2200-lb weight?

▮ The pressure against the large and the small piston is the same. $p = W/A_{large} = 2200/[\pi(\frac{3}{12})^2/4] = 44\,818$ lb/ft². Let $P$ be the force from the small piston onto the handle. $P = pA_{small} = (44\,818)[\pi(\frac{1}{12})^2/4] = 244$ lb. For the handle, $\Sigma M_A = 0 = (16+1)(F) - (1)(244)$, $F = 14.4$ lb.

**Fig. 2-17**

**2.33** Figure 2-18 shows a setup with a vessel containing a plunger and a cylinder. What force $F$ is required to balance the weight of the cylinder if the weight of the plunger is negligible?

▮ $\qquad 10\,000/500 - [(0.78)(62.4)](15)/144 = F/5 \qquad F = 74.6$ lb

$F = ?$

Plunger area = 5 in²

Cylinder
Weight = 10,000 lb
Cross-sectional
area = 500 in²

15 ft

Oil (s.g. = 0.78)

**Fig. 2-18**

**2.34** For the vertical pipe with manometer attached, as shown in Fig. 2-19, find the pressure in the oil at point $A$.

▮ $\qquad p_A + [(0.91)(62.4)](7.22) - [(13.6)(62.4)](1.00) = 0 \qquad p_A = 438.7$ lb/ft² or 3.05 lb/in²

Oil
(s.g. = 0.91)

*A*

7.22 ft

*C*
1.00 ft
*B* — — *B*

Mercury
(s.g. = 13.6)   **Fig. 2-19**

**2.35** A monometer is attached to a tank containing three different fluids, as shown in Fig. 2-20. What will be the difference in elevation of the mercury column in the manometer (i.e., $y$ in Fig. 2-20)?

$$30 + [(0.82)(9.79)](5 - 2) + (9.79)(2 - 0) + (9.79)(1.00) - [(13.6)(9.79)]y = 0 \qquad y = 0.627 \text{ m}$$

Elev. 6 m
Elev. 5 m        Air pressure = 30 kPa   ▽

Oil
(s.g. = 0.82)

Elev. 2 m

Water
Elev. 0 m

1.00 m   $y$

Mercury
(s.g. = 13.6)

**Fig. 2-20**

**2.36** Oil of specific gravity 0.750 flows through the nozzle shown in Fig. 2-21 and deflects the mercury in the U-tube gage. Determine the value of $h$ if the pressure at $A$ is 20.0 psi.

$$20.0 + [(0.750)(62.4)](2.75 + h)/144 - [(13.6)(62.4)](h)/144 = 0 \qquad h = 3.75 \text{ ft}$$

Fig. 2-21

**2.37** Determine the reading $h$ in Fig. 2-22 for $p_A = 39$ kPa vacuum if the liquid is kerosene (s.g. = 0.83).

$$-39 + [(0.83)(9.79)]h = 0 \qquad h = 4.800 \text{ m}$$

Fig. 2-22

**2.38** In Fig. 2-22, the liquid is water. If $h = 9$ in and the barometer reading is 29.8 inHg, find $p_A$ in feet of water absolute.

$$p_A + \tfrac{9}{12} = (13.6)(29.8/12) \qquad p_A = 33.0 \text{ ft of water absolute}$$

**2.39** In Fig. 2-23, s.g.$_1$ = 0.84, s.g.$_2$ = 1.0, $h_2 = 96$ mm, and $h_1 = 159$ mm. Find $p_A$ in mmHg gage. If the barometer reading is 729 mmHg, what is $p_A$ in mmH$_2$O absolute?

$$p_A + (0.84)(96) - (1.0)(159) = 0$$

$$p_A = 78.4 \text{ mmH}_2\text{O gage} = 78.4/13.6 = 5.76 \text{ mmHg gage}$$
$$= 78.4 + (13.6)(729) = 9993 \text{ mmH}_2\text{O absolute}$$

Fig. 2-23

**2.40** At 20 °C, gage $A$ in Fig. 2-24 reads 290 kPa abs. What is the height $h$ of water? What does gage $B$ read?

$$290 - [(13.6)(9.79)](\tfrac{70}{100}) - 9.79h = 175 \qquad h = 2.227 \text{ m}$$
$$p_B - (9.79)(\tfrac{70}{100} + 2.227) = 175 \qquad p_B = 204 \text{ kPa}$$

**Fig. 2-24**

**2.41** The U-tube shown in Fig. 2-25$a$ is 10 mm in diameter and contains mercury. If 12.0 mL of water is poured into the right-hand leg, what are the ultimate heights in the two legs?

▌ After the water is poured, the orientation of the liquids will be as shown in Fig. 2-25$b$; $h = (12.0 \times 10^3 \text{ mm}^3)/\pi(5 \text{ mm})^2 = 152.8$ mm, $(13.6)(240 - L) = 13.6L + 152.8$, $L = 114.4$ mm. Left leg height above bottom of U-tube $= 240 - 114.4 = 125.6$ mm; right leg height above bottom of U-tube $= 114.4 + 152.8 = 267.2$ mm.

**Fig. 2-25($a$)**

**Fig. 2-25($b$)**

**2.42** Assuming sea water to have a constant specific weight of 10.05 kN/m³, what is the absolute pressure at a depth of 10 km?

▌
$$p = 1 + (10.05)(10\,000)/101.3 = 993 \text{ atm}$$

**2.43** In Fig. 2-26, fluid 2 is carbon tetrachloride and fluid 1 is benzene. If $p_{\text{atm}}$ is 101.5 kPa, determine the absolute pressure at point $A$.

▌
$$101.5 + (15.57)(0.35) - (8.62)(0.12) = p_A \qquad p_A = 105.9 \text{ kPa}$$

**Fig. 2-26**

**2.44** In Fig. 2-27$a$, the manometer reads 4 in when atmospheric pressure is 14.7 psia. If the absolute pressure at $A$ is doubled, what is the new manometer reading?

▌ $p_A + (62.4)(3.5) - [(13.6)(62.4)](\frac{4}{12}) = (14.7)(144)$, $p_A = 2181$ lb/ft². If $p_A$ is doubled to 4362 lb/ft², the mercury level will fall $x$ inches on the left side of the manometer and will rise by that amount on the right side of the manometer (see Fig. 2-27$b$). Hence, $4362 + (62.4)(3.5 + x/12) - [(13.6)(62.4)][(4 + 2x)/12] = (14.7)(144)$, $x = 16.0$ in. New manometer reading $= 4 + (2)(16.0) = 36.0$ in.

Fig. 2-27(a)    Fig. 2-27(b)

**2.45** In Fig. 2-28a, A contains water, and the manometer fluid has density 2900 kg/m³. When the left meniscus is at zero on the scale, $p_A = 100$ mm of water. Find the reading of the right meniscus for $p_A = 10$ kPa with no adjustment of the U-tube or scale.

▮ First, determine the reading of the right meniscus for $p_A = 100$ mm of water (see Fig. 2-28b): $100 + 500 - 2.90h = 0$, $h = 206.9$ mm. When $p_A = 10$ kPa, the mercury level will fall some amount, $d$, on the left side of the manometer and will rise by that amount on the right side of the manometer (see Fig. 2-28b). Hence, $10/9.79 + (500 + d)/1000 - [(206.9 + 2d)/1000](2.90) = 0$, $d = 192.0$ mm. Scale reading for $p_A = 10$ kPa is 206.9 + 192.0, or 398.9 mm.

Fig. 2-28(a)    Fig. 2-28(b)

**2.46** A manometer is attached to a conduit, as shown in Fig. 2-29. Calculate the pressure at point A.

▮    $$p_A + (62.4)[(5 + 15)/12] - [(13.6)(62.4)](\tfrac{15}{12}) = 0 \qquad p_A = 957 \text{ lb/ft}^2$$

Fig. 2-29

**2.47** A manometer is attached to a pipe containing oil, as shown in Fig. 2-30. Calculate the pressure at point A.

▮    $$p_A + [(0.85)(9.79)](0.2) - (9.79)(1.5) = 0 \qquad p_A = 13.02 \text{ kN/m}^2$$

Fig. 2-30

**2.48** A monometer is attached to a pipe to measure pressure, as shown in Fig. 2-31. Calculate the pressure at point $A$.

$$p_A + (62.4)(\tfrac{18}{12}) - [(13.6)(62.4)](\tfrac{6}{12}) = 0 \qquad p_A = 331 \text{ lb/ft}^2$$

Fig. 2-31

**2.49** A glass U-tube open to the atmosphere at both ends is shown in Fig. 2-32. if the U-tube contains oil and water as shown, determine the specific gravity of the oil.

$$[(\text{s.g.}_{\text{oil}})(9.79)](0.35) - (9.79)(0.30) = 0 \qquad \text{s.g.}_{\text{oil}} = 0.86$$

Fig. 2-32

**2.50** A differential manometer is shown in Fig. 2-33. Calculate the pressure difference between points $A$ and $B$.

$$p_A + [(0.92)(62.4)][(x+12)/12] - [(13.6)(62.4)](\tfrac{12}{12}) - [(0.92)(62.4)][(x+24)/12] = p_B$$

$$p_A - p_B = 906 \text{ lb/ft}^2$$

Fig. 2-33

**2.51** A differential manometer is attached to a pipe, as shown in Fig. 2-34. Calculate the pressure difference between points $A$ and $B$.

$$p_A + [(0.91)(62.4)](y/12) - [(13.6)(62.4)](\tfrac{4}{12}) - [(0.91)(62.4)][(y-4)/12] = p_B$$

$$p_A - p_B = 264 \text{ lb/ft}^2$$

Fig. 2-34

**2.52** A differential manometer is attached to a pipe, as shown in Fig. 2-35. Calculate the pressure difference between points $A$ and $B$.

$$p_A - [(0.91)(62.4)](y/12) - [(13.6)(62.4)](\tfrac{4}{12}) + [(0.91)(62.4)][(y+4)/12] = p_B$$

$$p_A - p_B = 264 \text{ lb/ft}^2$$

Fig. 2-35

**2.53** For the configuration shown in Fig. 2-36, calculate the weight of the piston if the gage pressure reading is 70.0 kPa.

▮ Let $W$ = weight of the piston. $W/[(\pi)(1)^2/4] - [(0.86)(9.79)](1) = 70.0$, $W = 61.6$ kN.

Fig. 2-36

**2.54** A manometer is attached to a horizontal oil pipe, as shown in Fig. 2-37. If the pressure at point $A$ is 10 psi, find the distance between the two mercury surfaces in the manometer (i.e., determine the distance $y$ in Fig. 2-37).

▮ $(10)(144) + [(0.90)(62.4)](3 + y) - [(13.6)(62.4)]y = 0$    $y = 2.03$ ft  or  24.4 in

Fig. 2-37

**2.55** A vertical pipe with attached gage and manometer is shown in Fig. 2-38. What will be the gage reading in pounds per square inch if there is no flow in the pipe?

▮ Gage reading $+ [(0.85)(62.4)](2 + 8)/144 - [(13.6)(62.4)](\frac{18}{12})/144 = 0$    Gage reading $= 5.16$ psi

**2.56** A monometer is attached to a vertical pipe, as shown in Fig. 2-39. Calculate the pressure difference between points $A$ and $B$.

▮ $p_A - (62.4)(5 + 1) - [(13.6)(62.4)](2) + (62.4)(2 + 1) = p_B$

$p_A - p_B = 1884$ lb/ft$^2$  or  13.1 lb/in$^2$

**2.57** A manometer is attached to a water tank, as shown in Fig. 2-40. Find the height of the free water surface above the bottom of the tank.

▮ $(9.79)(H - 0.15) - [(13.6)(9.79)](0.20) = 0$    $H = 2.87$ m

Gage

2 ft

8 ft

Oil
(s.g. = 0.85)

18 in

Mercury
(s.g. = 13.6)

**Fig. 2-38**

Mercury (s.g. = 13.6)

2 ft

1 ft

B ●

5 ft

A ●

Water

**Fig. 2-39**

Water

H = ?

0.15 m

0.15 m

0.20 m

Mercury
(s.g. = 13.6)   **Fig. 2-40**

**2.58**   A differential manometer is attached to two tanks, as shown in Fig. 2-41. Calculate the pressure difference between chambers $A$ and $B$.

$$p_A + [(0.89)(9.79)](1.1) + [(13.6)(9.79)](0.3) - [(1.59)(9.79)](0.8) = p_B$$

$$p_A - p_B = -37.1 \text{ kN/m}^2 \qquad (\text{i.e., } p_B > p_A)$$

Chamber $A$

SAE 30 oil

1.1 m

Chamber $B$

Carbon tetrachloride

0.8 m

0.3 m

Mercury

**Fig. 2-41**

**2.59**   Calculate the pressure difference between $A$ and $B$ for the setup shown in Fig. 2-42.

$$p_A + (62.4)(66.6/12) - [(13.6)(62.4)](40.3/12) + (62.4)(22.2/12)$$
$$- [(13.6)(62.4)](30.0/12) - (62.4)(10.0/12) = p_B$$

$$p_A - p_B = 4562 \text{ lb/ft}^2 \quad \text{or} \quad 31.7 \text{ lb/in}^2$$

**2.60**   Calculate the pressure difference between $A$ and $B$ for the setup shown in Fig. 2-43.

$$p_A - (9.79)x - [(0.8)(9.79)](0.70) + (9.79)(x - 0.80) = p_B \qquad p_A - p_B = 13.3 \text{ kN/m}^2$$

**2.61**   Calculate the pressure difference between $A$ and $B$ for the setup shown in Fig. 2-44.

$$p_A + (62.4)(x + 4) - [(13.6)(62.4)](4) + (62.4)(7 - x) = p_B$$

$$p_A - p_B = 2708 \text{ lb/ft}^2 \quad \text{or} \quad 18.8 \text{ lb/in}^2$$

Water

66.6 in

30.0 in

22.2 in

40.3 in

10.0 in

*A*

*B*

Mercury
(s.g. = 13.6)

**Fig. 2-42**

Oil (s.g. = 0.8)

$x$ m + 0.70 m − 1.50 m = ($x$ − 0.80) m

0.70 m

*B*

Water

1.50 m

$x$ m

*A*

Water

**Fig. 2-43**

Fig. 2-44

**2.62** Vessels $A$ and $B$ in Fig. 2-45 contain water under pressures of 40.0 psi and 20.0 psi, respectively. What is the deflection of the mercury in the differential gage?

$\blacksquare$  $(40.0)(144) + (62.4)(x + h) - [(13.6)(62.4)]h + 62.4y = (20.0)(144)$. Since $x + y = 16.00 - 10.00$, or $6.00$ ft, $h = 4.14$ ft.

Fig. 2-45

**2.63** For a gage pressure at $A$ in Fig. 2-46 of $-1.58$ psi, find the specific gravity of gage liquid $B$.

$\blacksquare$  $(-1.58)(144) + [(1.60)(62.4)](10.50 - 9.00) - (0.0750)(11.25 - 9.00) + [(\text{s.g.}_{\text{liq. }B})(62.4)](11.25 - 10.00) = 0$

$$\text{s.g.}_{\text{liq. }B} = 1.00$$

Air → | 11.25 ft

10.50 ft → A

| 10.00 ft

← Liquid B

9.00

← s.g. = 1.60

**Fig. 2-46**

**2.64** In Fig. 2-47, liquid $A$ weighs $53.5 \text{ lb/ft}^3$ and liquid $B$ weighs $78.8 \text{ lb/ft}^3$. Manometer liquid $M$ is mercury. If the pressure at $B$ is 30 psi, find the pressure at $A$.

▮ $$p_A - (53.5)(6.5 + 1.3) + [(13.6)(62.4)](1.3) + (78.8)(6.5 + 10.0) = (30)(144)$$

$$p_A = 2334 \text{ lb/ft}^2 \quad \text{or} \quad 16.2 \text{ lb/in}^2$$

1.3 ft

6.5 ft

M

10.0 ft

d

B **Fig. 2-47**

**2.65** What would be the manometer reading in Fig. 2-47 if $p_B - p_A$ is 165 kPa?

▮ Converting to $\text{lb/ft}^2$, $p_B - p_A = 3446 \text{ lb/ft}^2$. The mercury level will rise some amount, $x$, on the left side of the manometer and will fall by that amount on the right side of the manometer (see Fig. 2.48). Hence, taking weight densities from Prob. 2.64, $p_A - (53.5)(6.5 + 1.3 + x) + [(13.6)(62.4)](1.3 + 2x) + (78.8)(6.5 + 10.0 - x) = p_B$, $1644x + 1986 = p_B - p_A = 3446$, $x = 0.89 \text{ ft}$; manometer reading $= 1.3 + (2)(0.89) = 3.08 \text{ ft}$.

1.3 ft
6.5 ft
A
10.0 ft
B **Fig. 2-48**

**2.66** In Fig. 2-49, water is contained in $A$ and rises in the tube to a level 85 in above $A$; glycerin is contained in $B$. The inverted U-tube is filled with air at 23 psi and 70 °F. Atmospheric pressure is 14.6 psia. Determine the difference in pressure (psi) between $A$ and $B$ if $y$ is 16 in. What is the absolute pressure in $B$ in inches of mercury and in feet of glycerin?

$$p_A - (62.4)(\tfrac{85}{12}) = (23)(144) \qquad p_A = 3754.0 \text{ lb/ft}^2$$

$$p_B - [(1.26)(62.4)][(85-16)/12] = (23)(144) \qquad p_B = 3764.1 \text{ lb/ft}^2$$

$$p_A - p_B = 3754.0 - 3764.1 = -10.1 \text{ lb/ft}^2 \quad \text{or} \quad -0.070 \text{ lb/in}^2$$

$$(p_{\text{abs}})_B = (3764.1/144 + 14.6)/[(13.6)(62.4)/(12)^3] = 83.0 \text{ inHg}$$

$$(p_{\text{abs}})_B = (3764.1/144 + 14.6)/[(1.26)(62.4)/(12)^3] = 895.4 \text{ in} \quad \text{or} \quad 74.6 \text{ ft of glycerin}$$

Fig. 2-49

**2.67** Gas confined in a rigid container exerts a gage pressure of 150 kPa when its temperature is 7 °C. What pressure would the gas exert at 67 °C? Barometric pressure remains constant at 719 mmHg.

$$p_{\text{atm}} = [(13.6)(9.79)](0.719) = 95.7 \text{ kPa} \qquad p_{\text{abs}} = 95.7 + 150 = 245.7 \text{ kPa}$$

$$p_1 V_1/T_1 = p_2 V_2/T_2 \qquad (245.7)(V)/(273+7) = (p_2)(V)/(273+67) \qquad [V \text{ (volume) is constant}]$$

$$p_2 = 298.4 \text{ kPa (absolute)} = 298.4 - 95.7 = 202.7 \text{ kPa (gage)}$$

**2.68** In Fig. 2-50, atmospheric pressure is 14.6 psia, the gage reading at $A$ is 6.1 psi, and the vapor pressure of the alcohol is 1.7 psia. Compute $x$ and $y$.

I   Working in terms of absolute pressure heads, $[(6.1 + 14.6)(144)]/[(0.90)(62.4)] - x = (1.7)(144)/[(0.90)(62.4)]$, $x = 48.72 \text{ ft}$; $[(6.1 + 14.6)(144)]/[(0.90)(62.4)] + (y + 4.2) - (4.2)(13.6/0.90) = 0$, $y = 6.19 \text{ ft}$.

Fig. 2-50

**2.69** In Fig. 2-50, assume the following: atmospheric pressure = 858 mbar abs, vapor pressure of the alcohol = 160 mbar abs, $x = 2.90$ m, $y = 2.10$ m. Compute the reading on the pressure gage ($p_A$) and on the manometer ($z$).

▮ Working in terms of absolute pressure heads, $[(p_A)_{\text{gage}} + 858](0.100)/[(0.90)(9.79)] - 2.90 = (160)(0.100)/[(0.90)(9.79)]$, $(p_A)_{\text{gage}} = -442$ mbar; $[(-442 + 858)(0.100)]/[(0.90)(9.79)] + (2.10 + z) - (z)(13.6/0.90) = 0$, $z = 0.483$ m.

**2.70** A pipeline contains an incompressible gas ($\gamma = 0.05$ lb/ft³) at rest; at point $A$ the pressure is 4.69 in of water. What is the pressure, in inches of water, at point $B$, 492 ft higher than $A$?

▮ The change in pressure in the atmosphere must be considered; assume, however, that $\gamma_{\text{air}} = 0.076$ lb/ft³ is constant.

$$(p_A/\gamma)_{\text{abs}} = (p_A/\gamma)_{\text{atm}} + 4.69/12 \text{ ft of water} \tag{1}$$

$$(p_B/\gamma)_{\text{abs}} = (p_B/\gamma)_{\text{atm}} + x/12 \text{ ft of water} \tag{2}$$

Subtracting Eq. (2) from Eq. (1),

$$(p_A/\gamma)_{\text{abs}} - (p_B/\gamma)_{\text{abs}} = (p_A/\gamma)_{\text{atm}} - (p_B/\gamma)_{\text{atm}} + 4.69/12 - x/12 \tag{3}$$

$$(p_A/\gamma)_{\text{atm}} - (p_B/\gamma)_{\text{atm}} = 492 \text{ ft of air} = (492)(0.076/62.4) = 0.599 \text{ ft of water}$$

$$(p_A/\gamma)_{\text{abs}} - (p_B/\gamma)_{\text{abs}} = 492 \text{ ft of gas} = (492)(0.05/62.4) = 0.394 \text{ ft of water}$$

Substituting these relationships into Eq. (3), $0.394 = 0.599 + 4.69/12 - x/12$, $x = 7.15$ in of water.

**2.71** Determine the pressure difference between points $A$ and $B$ in Fig. 2-51.

▮ $p_A + [(0.88)(9.79)](0.21) - [(13.6)(9.79)](0.09) - [(0.82)(9.79)](0.41 - 0.09)$
$$+ (9.79)(0.41 - 0.15) - (0.0118)(0.10) = p_B$$

$$p_A - p_B = 10.2 \text{ kPa}$$

**Fig. 2-51**

**2.72** In Fig. 2-52, if $p_B - p_A = 97.4$ kPa, calculate $H$.

▮
$$p_A - (9.79)(H/100) - [(0.827)(9.79)](\tfrac{17}{100}) + [(13.6)(9.79)][(34 + H + 17)/100] = p_B$$

$$1.234H + 66.53 = p_B - p_A = 97.4 \qquad H = 25.0 \text{ cm}$$

**Fig. 2-52**

**2.73** For Fig. 2-53, if fluid 1 is water and fluid 2 is mercury, and $z_A = 0$ and $z_1 = -11$ cm, what is level $z_2$ at which $p_A = p_{atm}$?

$$0 + (9.79)[0 - (-11)]/100 - [(13.6)(9.79)][z_2 - (-11)]/100 = 0 \qquad z_2 = -10.19 \text{ cm}$$

**Fig. 2-53**

**2.74** The inclined manometer in Fig. 2-54a contains Meriam red manometer oil (s.g. = 0.827). Assume the reservoir is very large. What should the angle $\theta$ be if each inch along the scale is to represent a change of 0.8 lb/ft$^2$ in gage pressure $p_A$?

From Fig. 2-54b, $\Delta p = \gamma \Delta z$, or

$$0.8 \text{ lb/ft}^2 = [(0.827)(62.4 \text{ lb/ft}^3)](\tfrac{1}{12} \text{ ft})(\sin \theta)$$

from which $\theta = 10.72°$.

**Fig. 2-54(a)**

**Fig. 2-54(b)**

**2.75** The system in Fig. 2-55 is at 20 °C. Compute the absolute pressure at point $A$.

$$p_A + [(0.85)(62.4)](\tfrac{7}{12}) - [(13.6)(62.4)](\tfrac{9}{12}) + (62.4)(\tfrac{6}{12}) = (14.7)(144) \qquad p_A = 2691 \text{ lb/ft}^2 \text{ abs}$$

**Fig. 2-55**

**2.76** Very small pressure differences $p_A - p_B$ can be measured accurately by the two-fluid differential manometer shown in Fig. 2-56. Density $\rho_2$ is only slightly larger than the upper fluid $\rho_1$. Derive an expression for the proportionality between $h$ and $p_A - p_B$ if the reservoirs are very large.

$p_A + \rho_1 g h_1 - \rho_2 g h - \rho_1 g (h_1 - h) = p_B$, $p_A - p_B = (\rho_2 - \rho_1)gh$. If $(\rho_2 - \rho_1)$ is small, $h$ will be large (sensitive).

**Fig. 2-56**

**2.77** Water flows downward in a pipe at 35°, as shown in Fig. 2-57. The pressure drop $p_1 - p_2$ is partly due to gravity and partly due to friction. The mercury manometer reads a 5-in height difference. What is the total pressure drop $p_1 - p_2$? What is the pressure drop due to friction only between 1 and 2? Does the manometer reading correspond only to friction drop?

$$p_1 + (62.4)(6 \sin 35° + x/12 + \tfrac{5}{12}) - [(13.6)(62.4)](\tfrac{5}{12}) - (62.4)(x/12) = p_2$$

$$p_1 - p_2 = 112.9 \text{ lb/ft}^2 \qquad \text{(total pressure drop)}$$

Pressure drop due to friction only $= [(13.6)(62.4) - 62.4](\tfrac{5}{12}) = 327.6 \text{ lb/ft}^2$

Manometer reads only the friction loss.

**Fig. 2-57**

**2.78** Determine the gage pressure at point $A$ in Fig. 2-58.

$p_A - (9.79)(0.50) + (0.0118)(0.33) + [(13.6)(9.79)](0.17) - [(0.83)(9.79)](0.44) = 0 \qquad p_A = -14.17 \text{ kPa}$

**Fig. 2-58**

**2.79** In Fig. 2-59, calculate level $h$ of the oil in the right-hand tube. Both tubes are open to the atmosphere.

$0 + (9.79)(0.110 + 0.240) - [(0.83)(9.79)](0.240 + h) = 0 \qquad h = 0.1817 \text{ m} = 181.7 \text{ mm}$

**Fig. 2-59**

**2.80** In Fig. 2-60a the inclined manometer measures the excess pressure at A over that at B. The reservoir diameter is 2.5 in and that of the inclined tube is $\frac{1}{4}$ in. For $\theta = 32°$ and gage fluid with s.g. = 0.832, calibrate the scale in psi per ft.

❚          $p_A = \gamma(\Delta h + \Delta y) + p_B$    (see Fig. 2-60b)    $p_A - p_B = \gamma(\Delta h + \Delta y)$

From Fig. 2-60b, $(A_A)(\Delta y) = (A_B)(R)$ or $\Delta y = A_B R/A_A$, $\Delta h = R \sin \theta$, $p_A - p_B = \gamma(R \sin \theta + A_B R/A_A) = \gamma R(\sin \theta + A_B/A_A)$, $A_B/A_A = [\pi(\frac{1}{4})^2/4]/[\pi(2.5)^2/4] = \frac{1}{100}$; $p_A - p_B = [(0.832)(62.4)](R)(\sin 32° + \frac{1}{100})/144 = 0.1947R$. The scale factor is thus 0.1947 psi/ft.

**Fig. 2-60(a)**

**Fig. 2-60(b)**

**2.81** Determine the weight W that can be equilibrated by the force acting on the piston of Fig. 2-61.

❚          $p_1 = p_2 = F_1/A_1 = F_2/A_2$    $1.25/[\pi(35)^2/4] = W/[\pi(250)^2/4]$    $W = 63.8$ kN

**Fig. 2-61**

**2.82** Neglecting the container's weight in Fig. 2-62, find the force tending to lift the circular top CD.

❚          $p_{CD} - [(0.8)(62.4)](4) = 0$    $p_{CD} = 199.7$ lb/ft²    $F = pA = (199.7)[\pi(2.5)^2/4] = 980$ lb

**Fig. 2-62**

**2.83** Find the force of oil on the top surface $CD$ of Fig. 2-62 if the liquid level in the open pipe is reduced by 1.3 m.

$$p_{CD} - [(0.8)(62.4)][4 - (1.3)(3.281)] = 0 \qquad p_{CD} = -13.24 \text{ lb/ft}^2 \quad \text{(i.e., a downward pressure by } CD\text{)}$$

$$F = pA = (-13.24)[\pi(2.5)^2/4] = -65.0 \text{ lb}$$

**2.84** A drum 2.25 ft in diameter filled with water has a vertical pipe of 0.70-in diameter attached to the top. How many pounds of water must be poured into the pipe to exert a force of 1500 lb on the top of the drum?

$$p = F/A = 1500/[\pi(2.25)^2/4] = 377.3 \text{ lb/ft}^2 \qquad h = p/\gamma = 377.3/62.4 = 6.05 \text{ ft}$$

$$W_{\text{H}_2\text{O}} = (6.05)[\pi(0.70/12)^2/4](62.4) = 1.01 \text{ lb}$$

**2.85** In Fig. 2-63, the liquid at $A$ and $B$ is water and the manometer liquid is oil with s.g. = 0.80, $h_1 = 300$ mm, $h_2 = 200$ mm, and $h_3 = 600$ mm. (**a**) Determine $p_A - p_B$. (**b**) If $p_B = 50$ kPa and the barometer reading is 730 mmHg, find the absolute pressure at $A$ in meters of water.

(**a**) $$p_A - (9.79)(\tfrac{300}{1000}) - [(0.80)(9.79)](\tfrac{200}{1000}) + (9.79)(\tfrac{600}{1000}) = p_B \qquad p_A - p_B = -1.37 \text{ kPa}$$

(**b**) $$p_A - (9.79)(\tfrac{300}{1000}) - [(0.80)(9.79)](\tfrac{200}{1000}) + (9.79)(\tfrac{600}{1000}) = 50$$

$$p_A = 48.63 \text{ kPa} \quad \text{(gage)} = 48.63/9.79 + \tfrac{730}{1000}(13.6) = 14.90 \text{ m water} \quad \text{(absolute)}$$

**Fig. 2-63**

**2.86** In Fig. 2-63, s.g.$_1$ = 1.0, s.g.$_2$ = 0.96, s.g.$_3$ = 1.0, $h_1 = h_2 = 269$ mm, and $h_3 = 1.2$ m. Compute $p_A - p_B$ in millimeters of water.

$$p_A - (1.0)(269) - (0.96)(269) + (1.0)(1200) = p_B \qquad p_A - p_B = -673 \text{ mm of water}$$

**2.87** In Fig. 2-63, s.g.$_1$ = 1.0, s.g.$_2$ = 0.94, s.g.$_3$ = 1.0, $h_1 = 300$ mm, $h_3 = 1.1$ m, and $p_A - p_B = -360$ mm of water. Find the gage difference ($h_2$).

$$p_A - (1.0)(300) - (0.94)(h_2) + (1.0)(1100) = p_B \qquad p_A - p_B = -360 = -800 + (0.94)(h_2) \qquad h_2 = 468 \text{ mm}$$

**2.88** What is the pressure difference, in pounds per square inch, of a 1000-ft water column?

$$p = \gamma h = (62.4)(1000)/144 = 433 \text{ psi}$$

**2.89** Find the pressure at a point 9.5 m below the free surface in a fluid whose density varies with depth $h$ (in m) according to

$$\rho = (450 \text{ kg/m}^3) + (11 \text{ kg/m}^4)h$$

$dp = \gamma \, dh = \rho g \, dh = (g)(450 + 11h) \, dh$. Integrating both sides: $p = (g)(450h + 11h^2/2)$. For $h = 9.5$ m: $p = (9.81)[(450)(9.5) + (11)(9.5)^2/2] = 46.807$ kPa.

**2.90** If atmospheric pressure is 29.72 inHg, what will be the height of water in a water barometer if the temperature of the water is (**a**) 50 °F, (**b**) 100 °F, and (**c**) 150 °F?

$$p = \gamma h = [(13.6)(62.4)](29.72/12) = 2102 \text{ lb/ft}^2 \quad \text{or} \quad 14.60 \text{ lb/in}^2$$

(**a**) At 50 °F, $\gamma = 62.4$ lb/ft$^3$ and $p_{\text{vapor}} = 25.7/144$, or 0.178 lb/in$^2$, $h_{\text{H}_2\text{O}} = (14.60 - 0.178)(144)/62.4 = 33.28$ ft.
(**b**) At 100 °F, $\gamma = 62.0$ lb/ft$^3$ and $p_{\text{vapor}} = \tfrac{135}{144}$, or 0.938 lb/in$^2$, $h_{\text{H}_2\text{O}} = (14.60 - 0.938)(144)/62.0 = 31.73$ ft.
(**c**) At 150 °F, $\gamma = 61.2$ lb/ft$^3$ and $p_{\text{vapor}} = \tfrac{545}{144}$, or 3.78 lb/in$^2$, $h_{\text{H}_2\text{O}} = (14.60 - 3.78)(144)/61.2 = 25.46$ ft.

**2.91** A bicycle tire is inflated at sea level (where atmospheric pressure is 14.6 psia and the temperature is 69 °F) to 65.0 psi. Assuming the tire does not expand, what is the gage pressure within the tire on the top of Everest (altitude 30 000 ft), where atmospheric pressure is 4.3 psia and the temperature is −38 °F?

I Let subscript 1 indicate sea level and subscript 2 indicate altitude 30 000 ft.

$$(p_1)_{abs} = 14.6 + 65.0 = 79.6 \text{ psia} \qquad p_1 V_1 / T_1 = p_2 V_2 / T_2$$

$$(79.6)(V)/(460 + 69) = (p_2)(V)/[460 + (-38)] \qquad (V \text{ is constant})$$

$$(p_2)_{abs} = 63.5 \text{ psia} \qquad (p_2)_{gage} = 63.5 - 4.3 = 59.2 \text{ psi}$$

**2.92** Find the difference in pressure between tanks $A$ and $B$ in Fig. 2-64 if $d_1 = 330$ mm, $d_2 = 160$ mm, $d_3 = 480$ mm, and $d_4 = 230$ mm.

I
$$p_A + (9.79)(0.330) - [(13.6)(9.79)](0.480 + 0.230 \sin 45°) = p_B \qquad p_A - p_B = 82.33 \text{ kPa}$$

**Fig. 2-64**

**2.93** A cylindrical tank contains water at a height of 55 mm, as shown in Fig. 2-65. Inside is a smaller open cylindrical tank containing cleaning fluid (s.g. = 0.8) at height $h$. If $p_B = 13.40$ kPa gage and $p_C = 13.42$ kPa gage, what are gage pressure $p_A$ and height $h$ of cleaning fluid? Assume that the cleaning fluid is prevented from moving to the top of the tank.

I
$$p_A + (9.79)(0.055) = 13.42 \qquad p_A = 12.88 \text{ kPa}$$

$$12.88 + (9.79)(0.055 - h) + [(0.8)(9.79)]h = 13.40 \qquad h = 0.00942 \text{ m} = 9.42 \text{ mm}$$

**Fig. 2-65**

**2.94** An open tube is attached to a tank, as shown in Fig. 2-66. If the water rises to a height of 800 mm in the tube, what are the pressures $p_A$ and $p_B$ of the air above the water? Neglect capillary effects in the tube.

I
$$p_A - (9.79)[(800 - 300 - 100)/1000] = 0 \qquad p_A = 3.92 \text{ kPa}$$

$$p_B - (9.79)[(800 - 300)/1000] = 0 \qquad p_B = 4.90 \text{ kPa}$$

Fig. 2-66

**2.95** For the setup shown in Fig. 2-67, what is the pressure $p_A$ if the specific gravity of the oil is 0.82?

I    $p_A + [(0.82)(9.79)](3) + (9.79)(4-3) - [(13.6)(9.79)](0.320) = 0$    $p_A = 8.73$ kPa

Fig. 2-67

**2.96** For the setup shown in Fig. 2-68, calculate the absolute pressure at $a$. Assume standard atmospheric pressure, 101.3 kPa.

I    $101.3 + (9.79)(0.600 - 0.200) - [(13.6)(9.79)](0.140) + [(0.83)(9.79)](0.140 + 0.090) = p_A$

$$p_A = 88.44 \text{ kPa}$$

Fig. 2-68

**2.97** A force of 460 N is exerted on lever $AB$, as shown in Fig. 2-69. End $B$ is connected to a piston which fits into a cylinder having a diameter of 60 mm. What force $F_D$ acts on the larger piston, if the volume between $C$ and $D$ is filled with water?

I    Let $F_C$ = force exerted on smaller piston at $C$: $F_C = (460)(\frac{220}{120}) = 843$ N. $F_C/A_C = F_D/A_D$, $(843)/[\pi(\frac{60}{1000})^2/4] = F_D/[\pi(\frac{260}{1000})^2/4]$, $F_D = 15\,830$ N, or 15.83 kN.

Fig. 2-69

# CHAPTER 3
# Forces on Submerged Plane Areas

**3.1** If a triangle of height $d$ and base $b$ is vertical and submerged in liquid with its vertex at the liquid surface (see Fig. 3-1), derive an expression for the depth to its center of pressure.

$$h_{cp} = h_{cg} + \frac{I_{cg}}{h_{cg}A} = \frac{2d}{3} + \frac{bd^3/36}{(2d/3)(bd/2)} = \frac{3d}{4}$$

Fig. 3-1

**3.2** If a triangle of height $d$ and base $b$ is vertical and submerged in liquid with its vertex a distance $a$ below the liquid surface (see Fig. 3-2), derive an expression for the depth to its center of pressure.

$$h_{cp} = h_{cg} + \frac{I_{cg}}{h_{cg}A} = \left(a + \frac{2d}{3}\right) + \frac{bd^3/36}{(a + 2d/3)(bd/2)} = \left(a + \frac{2d}{3}\right) + \frac{d^2}{18(a + 2d/3)}$$

$$= \frac{18(a^2 + 4ad/3 + 4d^2/9) + d^2}{18(a + 2d/3)} = \frac{6a^2 + 8ad + 3d^2}{6(a + 2d/3)}$$

Fig. 3-2

**3.3** If a triangle of height $d$ and base $b$ is vertical and submerged in liquid with its base at the liquid surface (see Fig. 3-3), derive an expression for the depth to its center of pressure.

$$h_{cp} = h_{cg} + \frac{I_{cg}}{h_{cg}A} = \frac{d}{3} + \frac{bd^3/36}{(d/3)(bd/2)} = \frac{d}{3} + \frac{d}{6} = \frac{d}{2}$$

Fig. 3-3

**3.4** A circular area of diameter $d$ is vertical and submerged in a liquid. Its upper edge is coincident with the liquid surface (see Fig. 3-4). Derive an expression for the depth to its center of pressure.

▮

$$h_{cp} = h_{cg} + \frac{I_{cg}}{h_{cg}A} = \frac{d}{2} + \frac{\pi d^4/64}{(d/2)(\pi d^2/4)} = \frac{d}{2} + \frac{d}{8} = \frac{5d}{8}$$

Fig. 3-4

**3.5** A vertical semicircular area of diameter $d$ and radius $r$ is submerged and has its diameter in a liquid surface (see Fig. 3-5). Derive an expression for the depth to its center of pressure.

▮

$$h_{cp} = h_{cg} + \frac{I_{cg}}{h_{cg}A} \qquad h_{cg} = \frac{4r}{3\pi} \qquad I_x = \frac{1}{2}\left(\frac{\pi d^4}{64}\right) = \frac{1}{2}\left[\frac{\pi(2r)^4}{64}\right] = \frac{\pi r^4}{8}$$

$$I_{cg} = \frac{\pi r^4}{8} - \left(\frac{\pi r^2}{2}\right)\left(\frac{4r}{3\pi}\right)^2 = \left(\frac{\pi}{8} - \frac{8}{9\pi}\right)(r^4) \qquad h_{cp} = \frac{4r}{3\pi} + \frac{[\pi/8 - 8/(9\pi)](r^4)}{[4r/(3\pi)][(\pi r^2/2)]} = 0.589r$$

Fig. 3-5

**3.6** A dam 20 m long retains 7 m of water, as shown in Fig. 3-6. Find the total resultant force acting on the dam and the location of the center of pressure.

▮ $F = \gamma h A = (9.79)[(0 + 7)/2][(20)(7/\sin 60°)] = 5339$ kN. The center of pressure is located at two-thirds the total water depth of 7 m, or 4.667 m below the water surface (i.e., $h_{cp} = 4.667$ m in Fig. 3-6).

Fig. 3-6

**3.7** A vertical, rectangular gate with water on one side is shown in Fig. 3-7. Determine the total resultant force acting on the gate and the location of the center of pressure.

$$F = \gamma h_{cg} A = (9.79)(3 + 1.2/2)[(2)(1.2)] = 84.59 \text{ kN}$$

$$h_{cp} = h_{cg} + \frac{I_{cg}}{h_{cg}A} = \left(3 + \frac{1.2}{2}\right) + \frac{(2)(1.2)^3/12}{(3 + 1.2/2)[(2)(1.2)]} = 3.633 \text{ m}$$

**Fig. 3-7**

**3.8** Solve Prob. 3.7 by the integration method.

$$F = \int \gamma h \, dA = \int_0^{1.2} (9.79)(3 + y)(2 \, dy) = (19.58)\left[3y + \frac{y^2}{2}\right]_0^{1.2} = 84.59 \text{ kN}$$

$$h_{cp} = \frac{\int \gamma h^2 \, dA}{F} = \frac{\int_0^{1.2} (9.79)(3 + y)^2(2 \, dy)}{84.59} = \frac{(19.58)[9y + 3y^2 + y^3/3]_0^{1.2}}{84.59} = 3.633 \text{ m}$$

**3.9** A vertical, triangular gate with water on one side is shown in Fig. 3-8. Determine the total resultant force acting on the gate and the location of the center of pressure.

$$F = \gamma h_{cg} A = (62.4)(6 + 3/3)[(2)(3)/2] = 1310 \text{ lb}$$

$$h_{cp} = h_{cg} + \frac{I_{cg}}{h_{cg}A} = \left(6 + \frac{3}{3}\right) + \frac{(2)(3)^3/36}{(6 + 3/3)[(2)(3)/2]} = 7.07 \text{ ft}$$

**Fig. 3-8**

**3.10** Solve Prob. 3.9 by the integration method.

$$F = \gamma h_{cg} A = [(0.82)(9.79)][4 + (1 + 1.2/2)(\sin 40°)][(0.8)(1.2)] = 38.75 \text{ kN}$$

$$F = \int_0^3 (62.4)(6 + y)[(2 - 2y/3) \, dy] = \int_0^3 (62.4)(12 - 2y - 2y^2/3) \, dy = (62.4)[12y - y^2 - 2y^3/9]_0^3 = 1310 \text{ lb}$$

$$h_{cp} = \frac{\int \gamma h^2 \, dA}{F} = \frac{\int_0^3 (62.4)(6 + y)^2(2 - 2y/3) \, dy}{1310} = \frac{\int_0^3 (62.4)(72 - 6y^2 - 2y^3/3) \, dy}{1310}$$

$$= \frac{(62.4)[72y - 2y^3 - y^4/6]_0^3}{1310} = 7.07 \text{ ft}$$

**3.11** An inclined, rectangular gate with water on one side is shown in Fig. 3-9. Determine the total resultant force acting on the gate and the location of the center of pressure.

$$F = \gamma h_{cg} A = (62.4)[8 + \tfrac{1}{2}(4 \cos 60°)][(4)(5)] = 11\,230 \text{ lb}$$

$$z_{cp} = z_{cg} + \frac{I_{cg}}{z_{cg} A} = \left(\frac{8}{\cos 60°} + \frac{4}{2}\right) + \frac{(5)(4)^3/12}{(8/\cos 60° + \tfrac{4}{2})[(4)(5)]} = 18.07 \text{ ft}$$

Fig. 3-9

**3.12** Solve Prob. 3.11 by the integration method.

$$F = \int \gamma h \, dA = \int_0^4 (62.4)(8 + y \cos 60°)(5 \, dy) = (312)\left[8y + \frac{y^2}{4}\right]_0^4 = 11\,230 \text{ lb}$$

$$h_{cp} = \frac{\int \gamma h^2 \, dA}{F} = \frac{\int_0^4 (62.4)(8 + y \cos 60°)^2(5 \, dy)}{11\,230} = \frac{\int_0^4 (312)(64 + 8y + y^2/4) \, dy}{11\,230}$$

$$= \frac{(312)[64y + 4y^2 + y^3/12]_0^4}{11\,230} = 9.04 \text{ ft}$$

**Note:** $h_{cp}$ is the vertical distance from the water surface to the center of pressure. The distance from the water surface to the center of pressure as measured along the inclination of the gate ($z_{cp}$) would be 9.04/cos 60°, or 18.08 ft.

**3.13** An inclined, circular gate with water on one side is shown in Fig. 3-10. Determine the total resultant force acting on the gate and the location of the center of pressure.

$$F = \gamma h_{cg} A = (9.79)[1.5 + \tfrac{1}{2}(1.0 \sin 60°)][\pi(1.0)^2/4] = 14.86 \text{ kN}$$

$$z_{cp} = z_{cg} + \frac{I_{cg}}{z_{cg} A} = \left[\frac{1.5}{\sin 60°} + \frac{1}{2}(1.0)\right] + \frac{\pi(1.0)^4/64}{[1.5/\sin 60° + \tfrac{1}{2}(1.0)][\pi(1.0)^2/4]} = 2.260 \text{ m}$$

**Fig. 3-10**

**3.14** A vertical, triangular gate with water on one side is shown in Fig. 3-11. Determine the total resultant force acting on the gate and the location of the center of pressure.

$$F = \gamma h_{cg} A = (9.79)[3 + \tfrac{2}{3}(1)][(1.2)(1)/2] = 21.54 \text{ kN}$$

$$h_{cp} = h_{cg} + \frac{I_{cg}}{h_{cg}A} = [3 + (\tfrac{2}{3})(1)] + \frac{(1.2)(1)^3/36}{[3 + \tfrac{2}{3}(1)][(1.2)(1)/2]} = 3.68 \text{ m}$$

**Fig. 3-11**

**3.15** Solve Prob. 3.14 by the integration method.

$F = \int \gamma h \, dA$. From Fig. 3-11, $y/x = 1/1.2$. Therefore, $x = 1.2y$.

$$F = \int_0^1 (9.79)(3+y)(1.2y\,dy) = \int_0^1 (11.75)(3y+y^2)\,dy = (11.75)\left[\frac{3y^2}{2} + \frac{y^3}{3}\right]_0^1 = 21.54 \text{ kN}$$

$$h_{cp} = \frac{\int \gamma h^2 \, dA}{F} = \frac{\int_0^1 (9.79)(3+y)^2(1.2y\,dy)}{21.54} = \frac{\int_0^1 (11.75)(9y+6y^2+y^3)\,dy}{21.54}$$

$$= \frac{(11.75)[9y^2/2 + 2y^3 + y^4/4]_0^3}{21.54} = 3.68 \text{ m}$$

**3.16** A tank containing water is shown in Fig. 3-12. Calculate the total resultant force acting on side *ABCD* of the container and the location of the center of pressure.

▮

$$F = \gamma h A = (62.4)[(0 + 6)/2][(20)(6)] = 22\,500 \text{ lb}$$

$$h_{cp} = (\tfrac{2}{3})(6) = 4.00 \text{ ft} \qquad \text{(vertically below the water surface)}$$

**Fig. 3-12**

**3.17** The gate in Fig. 3-13 is 4 ft wide, is hinged at point *B*, and rests against a smooth wall at *A*. Compute (*a*) the force on the gate due to seawater pressure, (*b*) the (horizontal) force *P* exerted by the wall at point *A*, and (*c*) the reaction at hinge *B*.

▮ (*a*)

$$F = \gamma h_{cg} A = (64)(17 - \tfrac{7.2}{2})[(4)(12)] = 30\,106 \text{ lb}$$

(*b*)

$$y_{cp} = \frac{-I_{xx} \sin \theta}{h_{cg} A} = \frac{-[(4)(12)^3/12](\tfrac{7.2}{12})}{(17 - \tfrac{7.2}{2})[(4)(12)]} = -0.537 \text{ ft}$$

$$\sum M_B = 0 \qquad (P)(7.2) - (30\,106)(12 - 6 - 0.537) = 0 \qquad P = 22\,843 \text{ lb}$$

(*c*)

$$\sum F_x = 0 \qquad B_x + (30\,106)(\tfrac{7.2}{12}) - 22\,843 = 0 \qquad B_x = 4779 \text{ lb}$$

$$\sum F_y = 0 \qquad B_y - (30\,106)(\tfrac{9.6}{12}) = 0 \qquad B_y = 24\,085 \text{ lb}$$

**Fig. 3-13(*a*)**        **Fig. 3-13(*b*)**

**3.18** Repeat Prob. 3.17, but instead let the hinge be at point *A* and let point *B* rest against a smooth bottom.

▮ (*a*) From Prob. 3.17, $F = 30\,106$ lb. (*b*) From Prob. 3.17, $y_{cp} = -0.537$ ft; $\sum M_A = 0$; $(B_y)(9.6) - (30\,106)(6 + 0.537) = 0$, $B_y = 20\,500$ lb.

(*c*)

$$\sum F_x = 0 \qquad (30\,106)(\tfrac{7.2}{12}) - A_x = 0 \qquad A_x = 18\,064 \text{ lb}$$

$$\sum F_y = 0 \qquad A_y - (30\,106)(\tfrac{9.6}{12}) + 20\,500 = 0 \qquad A_y = 3585 \text{ lb}$$

**3.19** A tank of dye has a right-triangular panel near the bottom as shown in Fig. 3-14a. Calculate the resultant force on the panel and locate its center of pressure.

$$F = \gamma h_{cg}A = \rho g h_{cg}A = (820)(9.81)(6+8)[\tfrac{1}{2}(8+16)(8+4)] = 16.22 \text{ MN}$$

$$I_{xx} = \frac{bh^3}{36} = \frac{(4+8)(8+16)^3}{36} = 4608 \text{ m}^4 \qquad y_{cp} = \frac{-I_{xx}\sin\theta}{h_{cg}A} = \frac{-(4608)(\sin 30°)}{(6+8)[\tfrac{1}{2}(8+16)(8+4)]} = -1.143 \text{ m}$$

$$I_{xy} = b(b-2s)(h)^2/72 = (4+8)[(4+8)-(2)(4+8)](8+16)^2/72 = -1152 \text{ m}^4$$

$$x_{cp} = \frac{-I_{xy}\sin\theta}{h_{cg}A} = \frac{-(-1152)(\sin 30°)}{(6+8)[\tfrac{1}{2}(8+16)(8+4)]} = +0.286 \text{ m}$$

(The resultant force acts at 1.143 m down and 0.286 m to the right of the centroid.)

**Fig. 3-14(a)**        **Fig. 3-14(b)**

**3.20** Gate AB in Fig. 3-15 is 1.0 m long and 0.9 m wide. Calculate force F on the gate and the position X of its center of pressure.

$$F = \gamma h_{cg}A = [(0.81)(9.79)][3 + (1 + 1.0/2)(\sin 50°)][(0.9)(1.0)] = 29.61 \text{ kN}$$

$$y_{cp} = \frac{-I_{XX}\sin\theta}{h_{cg}A} = \frac{-[(0.9)(1.0)^3/12](\sin 50°)}{[3 + (1 + 1.0/2)(\sin 50°)][(0.9)(1.0)]}$$

$$= -0.015 \text{ m from the centroid}$$

$$X = 1.0/2 + 0.015 = 0.515 \text{ m from point } A$$

**Fig. 3-15**

**3.21** A fishpond gate 6 ft wide and 9 ft high is hinged at the top and held closed by water pressure as shown in Fig. 3-16. What horizontal force applied at the bottom of the gate is required to open it?

$$F = \gamma h_{cg} A = (62.4)(8 + 4.5)[(6)(9)] = 42\,120 \text{ lb}$$

$$h_{cp} = h_{cg} + \frac{I_{cg}}{h_{cg}A} = (8 + 4.5) + \frac{(6)(9)^3/12}{(8 + 4.5)[(6)(9)]} = 13.04 \text{ ft}$$

$$\sum M_A = 0 \qquad (P)(9) - (42\,120)(13.04 - 8) = 0 \qquad P = 23\,587 \text{ lb}$$

**Fig. 3-16**

**3.22** A vat holding paint (s.g. = 0.80) is 8 m long and 4 m deep and has a trapezoidal cross section 3 m wide at the bottom and 5 m wide at the top (see Fig. 3-17). Compute (a) the weight of the paint, (b) the force on the bottom of the vat, and (c) the force on the trapezoidal end panel.

**(a)** $$W = \gamma V = [(0.80)(9.79)][(8)(4)(5 + 3)/2] = 1002 \text{ kN}$$

**(b)** $$F = \gamma h_{cg} A \qquad F_{bottom} = [(0.80)(9.79)](4)[(3)(8)] = 752 \text{ kN}$$

**(c)** $$F_{end} = F_{square} + 2F_{triangle} = [(0.80)(9.79)][(0 + 4)/2][(4)(3)] + (2)[(0.80)(9.79)](\tfrac{4}{3})[(4)(1)/2] = 230 \text{ kN}$$

**Fig. 3-17**

**3.23** Gate $AB$ in Fig. 3-18 is 5 ft wide, hinged at point $A$, and restrained by a stop at point $B$. Compute the force on the stop and the components of the reaction at $A$ if water depth $h$ is 9 ft.

$$F = \gamma h_{cg} A = (62.4)(9 - \tfrac{3}{2})[(3)(5)] = 7020 \text{ lb}$$

$$y_{cp} = \frac{-I_{xx} \sin \theta}{h_{cg}A} = \frac{-[(5)(3)^3/12](\sin 90°)}{(9 - 3/2)[(3)(5)]} = -0.100 \text{ ft}$$

$$\sum M_A = 0 \qquad (B_x)(3) - (7020)(1.5 + 0.100) = 0 \qquad B_x = 3744 \text{ lb}$$

$$\sum F_x = 0 \qquad 7020 - 3744 - A_x = 0 \qquad A_x = 3276 \text{ lb}$$

If gate weight is neglected, $A_y = 0$.

Fig. 3-18(a)          Fig. 3-18(b)

**3.24** In Fig. 3-18, stop $B$ will break if the force on it reaches 9000 lb. Find the critical water depth.

$$F = \gamma h_{cg} A = (62.4)(h_{cg})[(3)(5)] = 936 h_{cg} \qquad y_{cp} = \frac{-I_{xx}\sin\theta}{h_{cg}A} = \frac{-[(5)(3)^3/12](\sin 90°)}{(h_{cg})[(3)(5)]} = -\frac{0.750}{h_{cg}}$$

$$\sum M_A = 0 \qquad (9000)(3) - (936 h_{cg})(1.5 + 0.750/h_{cg}) = 0$$

$$h_{cg} = 18.73 \text{ ft} \qquad h_{crit} = 18.73 + 1.5 = 20.23 \text{ ft}$$

**3.25** In Fig. 3-18, hinge $A$ will break if its horizontal reaction becomes equal to 8000 lb. Find the critical water depth.

From Prob. 3.24, $F = 936 h_{cg}$ and $y_{cp} = -0.750/h_{cg}$; $\sum M_B = 0$; $(936 h_{cg})(1.5 - 0.750/h_{cg}) - (8000)(3) = 0$, $h_{cg} = 17.59$ ft; $h_{crit} = 17.59 + 1.5 = 19.09$ ft.

**3.26** Calculate the resultant force on triangular window $ABC$ in Fig. 3-19 and locate its center of pressure.

$$F = \gamma h_{cg} A = (10.08)[0.25 + (\tfrac{2}{3})(0.60)][(0.40)(0.60)/2] = 0.786 \text{ kN}$$

$$I_{xx} = bh^3/36 = (0.40)(0.60)^3/36 = 0.00240 \text{ m}^4$$

$$y_{cp} = \frac{-I_{xx}\sin\theta}{h_{cg}A} = \frac{-(0.00240)(\sin 90°)}{[0.25 + (\tfrac{2}{3})(0.60)][(0.40)(0.60)/2]} = -31 \text{ mm} \qquad \text{(i.e., below the centroid)}$$

$$I_{xy} = b(b - 2s)(h)^2/72 = 0.40[0.40 - (2)(0.40)](0.60)^2/72 = -0.000800 \text{ m}^4$$

$$x_{cp} = \frac{-I_{xy}\sin\theta}{h_{cg}A} = \frac{-(-0.000800)(\sin 90°)}{[0.25 + (\tfrac{2}{3})(0.60)][(0.40)(0.60)/2]} = +10 \text{ mm} \qquad \text{(i.e., right of the centroid)}$$

Fig. 3-19

**3.27** Freshly poured concrete approximates a fluid with s.g. = 2.40. Figure 3-20 shows a slab poured between wooden forms which are connected by four corner bolts $A$, $B$, $C$, and $D$. Neglecting end effects, compute the forces in the four bolts.

$$F = \gamma h_{cg} A = [(2.40)(62.4)](\tfrac{12}{2})[(9)(12)] = 97\,044\ \text{lb}$$

$$y_{cp} = \frac{-I_{xx}\sin\theta}{h_{cg}A} = \frac{-[(9)(12)^3/12](\sin 90°)}{[(\tfrac{12}{2})][(9)(12)]} = -2.00\ \text{ft}$$

$$\sum M_A = 0 \qquad (2)(F_C)(12) - (97\,044)(6 + 2.00) = 0 \qquad F_C = F_D = 32\,348\ \text{lb}$$

$$\sum M_C = 0 \qquad (97\,044)(6 - 2.00) - (2)(F_A)(12) = 0 \qquad F_A = F_B = 16\,174\ \text{lb}$$

Fig. 3-20

**3.28** Find the net hydrostatic force per unit width on rectangular panel $AB$ in Fig. 3-21 and determine its line of action.

$$F_{H_2O} = (9.79)(2 + 1 + \tfrac{2}{2})[(2)(1)] = 78.32\ \text{kN} \qquad F_{glyc} = (12.36)(1 + \tfrac{2}{2})[(2)(1)] = 49.44\ \text{kN}$$

$$F_{net} = F_{H_2O} - F_{glyc} = 78.32 - 49.44 = 28.88\ \text{kN}$$

$$y_{cp} = \frac{-I_{xx}\sin\theta}{h_{cg}A}$$

$$(y_{cp})_{H_2O} = \frac{-[(1)(2)^3/12](\sin 90°)}{(2 + 1 + \tfrac{2}{2})[(2)(1)]} = -0.0833\ \text{m}$$

$$(y_{cp})_{glyc} = \frac{-[(1)(2)^3/12](\sin 90°)}{[(1 + \tfrac{2}{2})][(2)(1)]} = -0.1667\ \text{m}$$

$$\sum M_B = 0 \qquad (78.32)(1 - 0.0833) - (49.44)(1 - 0.1667) = 28.88 D$$

$$D = 1.059\ \text{m} \qquad \text{(above point } B \text{, as shown in Fig. 3-21}c\text{)}$$

Fig. 3-21(a)          Fig. 3-21(b)          Fig. 3-21(c)

**3.29** A cylindrical, wooden-stave barrel is 3.5 ft in diameter and 5 ft high, as shown in Fig. 3-22. It is held together by steel hoops at the top and bottom, each with a cross section of 0.40 in². If the barrel is filled with orange juice (s.g. = 1.04), compute the tension stress in each hoop.

$$F = \gamma h_{cg} A = [(1.04)(62.4)](\tfrac{5}{2})[(3.5)(5)] = 2839 \text{ lb}$$

$$y_{cp} = \frac{-I_{xx} \sin \theta}{h_{cg} A} = \frac{-[(3.5)(5)^3/12](\sin 90°)}{\tfrac{5}{2}[(3.5)(5)]} = -0.833 \text{ ft}$$

$$\sum M_B = 0 \qquad 2839(\tfrac{5}{2} - 0.833) - 2(F_{upper})(5) = 0 \qquad F_{upper} = 473 \text{ lb}$$

$$\sum M_A = 0 \qquad 2(F_{lower})(5) - 2839(\tfrac{5}{2} + 0.833) = 0 \qquad F_{lower} = 946 \text{ lb}$$

$$\sigma_{upper} = 473/0.40 = 1182 \text{ psi} \qquad \sigma_{lower} = 946/0.40 = 2365 \text{ psi}$$

Fig. 3-22(a)          Fig. 3-22(b)

**3.30** Gate AB in Fig. 3-23a is 16 ft long and 8 ft wide. Neglecting the weight of the gate, compute the water level h for which the gate will start to fall.

$$F = \gamma h_{cg} A = (62.4)(h/2)[(8)(h/\sin 60°)] = 288.2h^2$$

$$y_{cp} = \frac{-I_{xx} \sin \theta}{h_{cg} A} = \frac{-[8(h/\sin 60°)^3/12](\sin 60°)}{(h/2)[8(h/\sin 60°)]} = -0.1925h$$

$$\sum M_B = 0 \qquad (11\,000)(16) - (288.2h^2)[(h/\sin 60°)/2 - 0.1925h] = 0 \qquad h = 11.7 \text{ ft}$$

Fig. 3-23(a)          Fig. 3-23(b)

**3.31** Repeat Prob. 3.30, including the weight of the 2-in-thick steel (s.g. = 7.85) gate. (See Fig. 3-24.)

▌ $W_{gate} = [(7.85)(62.4)][(16)(8)(\frac{2}{12})] = 10\,450$ lb.  From Prob. 3.30, $F = 288.2h^2$; $\sum M_B = 0$, $(11\,000)(16) - (288.2h^2)[(h/\sin 60°)/2 - 0.1925h] - 10\,450(\frac{16}{2}\cos 60°) = 0$, $h = 10.7$ ft.

**Fig. 3-24**

**3.32** A horizontal duct coming from a large dam is 2.5 m in diameter; it is closed by a circular door whose center or centroid is 45 m below the dam's water level. Compute the force on the door and locate its center of pressure.

▌ $\qquad F = \gamma h_{cg}A = (9.79)(45)[\pi(2.5)^2/4] = 2163$ kN $\qquad I_{xx} = \pi r^4/4 = \pi(\frac{2.5}{2})^4/4 = 1.917$ m$^4$

$$y_{cp} = \frac{-I_{xx}\sin\theta}{h_{cg}A} = \frac{-(1.917)(\sin 90°)}{(45)[\pi(2.5)^2/4]} = -0.0087 \text{ m}$$

Line of action of $F$ is 8.7 mm below the centroid of the door.

**3.33** Gate $AB$ in Fig. 3-25 is semicircular, hinged at $B$. What horizontal force $P$ is required at $A$ for equilibrium?

▌ $\qquad 4r/(3\pi) = (4)(4)/(3\pi) = 1.698$ m $\qquad F = \gamma h_{cg}A = (9.79)(6 + 4 - 1.698)[\pi(4)^2/2] = 2043$ kN

$$y_{cp} = \frac{-I_{xx}\sin\theta}{h_{cg}A} = \frac{-[(0.10976)(4)^4](\sin 90°)}{(6 + 4 - 1.698)[\pi(4)^2/2]} = -0.1347 \text{ m}$$

$$\sum M_B = 0 \qquad (2043)(1.698 - 0.1347) - 4P = 0 \qquad P = 798 \text{ kN}$$

$$\boxed{\begin{array}{l} I_{xx} = 0.10976r^4 \\ I_{xy} = 0 \end{array}}$$

**Fig. 3-25(a)**       **Fig. 3-25(b)**       **Fig. 3-25(c)**

**3.34** Dam $ABC$ in Fig. 3-26 is 38 m wide and made of concrete weighing 22 kN/m$^3$. Find the hydrostatic force on surface $AB$ and its moment about $C$. Could this force tip the dam over?

❚ $F = \gamma h_{cg} A = (9.79)(\frac{64}{2})[(38)(80)] = 952\,371$ kN. $F$ acts at $(\frac{2}{3})(80)$, or 53.33 m from $A$ along surface $AB$ (see Fig. 3-26b). For the given triangular shape, the altitude from $C$ to $AB$ intersects $AB$ 51.2 m from $A$ (see Fig. 3-26b). Hence, $M_C = (952\,371)(53.33 - 51.2) = 2\,028\,550$ kN. Since the moment of $F$ about point $C$ is counterclockwise, there is no danger of tipping.

Fig. 3-26(a)

Fig. 3-26(b)

**3.35** Isosceles triangular gate $AB$ in Fig. 3-27 is hinged at $A$. Compute the horizontal force $P$ required at point $B$ for equilibrium, neglecting the weight of the gate.

❚   $AB = 3/\sin 60° = 3.464$ m    $F = \gamma h_{cg} A = [(0.82)(9.79)](2 + 1.00)[(1.2)(3.464)/2] = 50.05$ kN

$$y_{cp} = \frac{-I_{xx} \sin \theta}{h_{cg} A} = \frac{-[(1.2)(3.464)^3/36](\sin 60°)}{(2 + 1.00)[(1.2)(3.464)/2]} = -0.1924 \text{ m}$$

$$\sum M_A = 0 \qquad 3P - (50.05)(3.464/3 + 0.1924) = 0 \qquad P = 22.47 \text{ kN}$$

Fig. 3-27(a)

Fig. 3-27(b)

**3.36** The tank in Fig. 3-28 is 40 cm wide. Compute the hydrostatic forces on horizontal panels $BC$ and $AD$. Neglect atmospheric pressure.

❚            $p = \gamma h$    $p_{BC} = [(0.84)(9.79)](0.35 + 0.40) + (9.79)(0.25) = 8.615$ kPa

$$F = pA \qquad F_{BC} = (8.615)[(1.20)(0.40)] = 4.135 \text{ kN}$$

$$p_{AD} = [(0.84)(9.79)](0.40) = 3.289 \text{ kPa} \qquad F_{AD} = (3.289)[(0.55)(0.40)] = 0.724 \text{ kN}$$

**Fig. 3-28**

**3.37** Water in a tank is pressurized to 85 cmHg (Fig. 3-29). Determine the hydrostatic force per meter width on panel *AB*.

▎ On panel *AB*, $p_{cg} = [(13.6)(9.79)](0.85) + (9.79)(4 + \frac{3}{2}) = 167.0$ kPa, $F_{AB} = (167.0)[(3)(1)] = 501$ kN.

**Fig. 3-29**

**3.38** Calculate the force and center of pressure on one side of the vertical triangular panel *ABC* in Fig. 3-30.

▎
$$F = \gamma h_{cg} A = (62.4)(1 + 6)[(9)(6)/2] = 11\ 794\ \text{lb} \qquad I_{xx} = (6)(9)^3/36 = 121.5\ \text{ft}^4$$
$$y_{cp} = \frac{-I_{xx}\sin\theta}{h_{cg}A} = \frac{-(121.5)(\sin 90°)}{(1 + 6)[(9)(6)/2]} = -0.64\ \text{ft}$$
$$I_{xy} = \frac{6[6 - (2)(6)](9)^2}{72} = -40.5\ \text{ft}^4 \qquad x_{cp} = \frac{-I_{xy}\sin\theta}{h_{cg}A} = \frac{-(-40.5)(\sin 90°)}{(1 + 6)[(9)(6)/2]} = 0.21\ \text{ft}$$

Thus, the center of pressure is $6 + 0.64$, or 6.64 ft below point *A* and $2 + 0.21$, or 2.21 ft to the right of point *B*.

**3.39** In Fig. 3-31, gate *AB* is 4 m wide and is connected by a rod and pulley to a massive sphere (s.g. = 2.40). What is the smallest radius that will keep the gate closed?

▎
$$F = \gamma h_{cg} A = (9.79)(9 + \tfrac{3}{2})[(4)(3)] = 1234\ \text{kN}$$
$$y_{cp} = \frac{-I_{xx}\sin\theta}{h_{cg}A} = \frac{-[(4)(3)^3/12](\sin 90°)}{(9 + \tfrac{3}{2})[(4)(3)]} = -0.071\ \text{m}$$
$$\sum M_B = 0 \qquad (W_{sphere})(7 + 9 + 3) - (1234)(3 - 1.5 - 0.071) = 0 \qquad W_{sphere} = 92.8\ \text{kN}$$
$$W_{sphere} = \gamma(4\pi r^3/3) \qquad 92.8 = [(2.40)(9.79)](4\pi r^3/3) \qquad r = 0.98\ \text{m}$$

**Fig. 3-30(a)**

$$I_{xx} = \frac{bh^3}{36}$$

$$I_{xy} = \frac{b(b-2a)h^2}{72}$$

**Fig. 3-30(b)**

**Fig. 3-31**

**3.40**  The triangular trough in Fig. 3-32 is hinged at $A$ and held together by cable $BC$ at the top. If cable spacing is 1 m into the paper, what is the cable tension?

$$F = \gamma h_{cg}A = (9.79)(\tfrac{5}{2})[(8.717)(1)] = 213.3 \text{ kN}$$

$$y_{cp} = \frac{-I_{xx}\sin\theta}{h_{cg}A} = \frac{-[(1)(8.717)^3/12](\sin 35°)}{\tfrac{5}{2}[(8.717)(1)]} = -1.453 \text{ m}$$

$$\sum M_A = 0 \qquad (T)(2+5) - (213.3)(4.359 - 1.453) = 0 \qquad T = 88.5 \text{ kN}$$

**Fig. 3-32(a)**

**Fig. 3-32(b)**

**3.41** In Fig. 3-33, gate $AB$ is 4 ft wide and opens to let fresh water out when the ocean tide is falling. The hinge at $A$ is 3 ft above the fresh-water surface. At what ocean depth $h$ will the gate open? Neglect the gate's weight.

▮ $$F = \gamma h_{cg} A \qquad F_1 = (62.4)(\tfrac{12}{2})[(12)(4)] = 17\,971 \text{ lb} \qquad F_2 = [(1.025)(62.4)](h/2)[(4)(h)] = 127.9h^2$$

$$\sum M_A = 0 \qquad (127.9h^2)(12 + 3 - h/3) - (17\,971)(3 + 8) = 0 \qquad h = 11.8 \text{ ft}$$

Fig. 3-33(a)

Fig. 3-33(b)

**3.42** Show that depth $h$ in Prob. 3.41 is independent of gate width $b$ (perpendicular to the paper).

▮ Areas, and hence pressure forces, are directly proportional to $b$. Thus $b$ will cancel out of the equation $\sum M_A = 0$ that determines $h$.

**3.43** Compute the force on one side of parabolic panel $ABC$ in Fig. 3-34 and the vertical distance down to the center of pressure.

▮ $$F = \gamma h_{cg} A = (9.79)(1 + 6)[(\tfrac{2}{3})(10)(6)] = 2741 \text{ kN}$$

$$I_{xx} = I_{x'} - A(\Delta h)^2 = \tfrac{2}{7}(bh^3) - [\tfrac{2}{3}(bh)][\tfrac{3}{5}(h)]^2 = (\tfrac{2}{7})(6)(10)^3 - [(\tfrac{2}{3})(6)(10)][(\tfrac{3}{5})(10)]^2 = 274.3 \text{ m}^4$$

$$y_{cp} = \frac{-I_{xx} \sin \theta}{h_{cg} A} = \frac{-(274.3)(\sin 90°)}{(1 + 6)[(\tfrac{2}{3})(10)(6)]} = -0.980 \text{ m}$$

Hence, the center of pressure is $6 + 0.980$, or 6.980 m below point $A$.

Fig. 3-34(a)

Fig. 3-34(b)

**3.44** Circular gate $ABC$ in Fig. 3-35 is 4 m in diameter and is hinged at $B$. Compute the force $P$ just sufficient to keep the gate from opening when $h$ is 8 m.

$$F = \gamma h_{cg}A = (9.79)(8)[\pi(4)^2/4] = 984.2 \text{ kN} \qquad I_{xx} = \pi d^4/64 = \pi(4)^4/64 = 12.57 \text{ m}^4$$

$$y_{cp} = \frac{-I_{xx}\sin\theta}{h_{cg}A} = \frac{-(12.57)(\sin 90°)}{(8)[(\pi)(2)^2]} = -0.125 \text{ m}$$

$$\sum M_B = 0 \qquad (P)(2) - (984.2)(0.125) = 0 \qquad P = 61.5 \text{ kN}$$

Fig. 3-35(a)  Fig. 3-35(b)

**3.45** For the conditions given in Prob. 3.44, derive an analytical expression for $P$ as a function of $h$.

$$F = \gamma h_{cg}A = \gamma h_{cg}[\pi(r)^2] \qquad I_{xx} = \pi(r)^4/4$$

$$y_{cp} = \frac{-I_{xx}\sin\theta}{h_{cg}A} = \frac{-[(\pi)(r)^4/4](\sin 90°)}{h[(\pi)(r)^2]} = \frac{-r^2}{4h}$$

$$\sum M_B = 0 \qquad Pr - [\gamma h_{cg}(\pi)(r)^2][(r)^2/(4r)] = 0 \qquad P = \gamma\pi r^3/4$$

(Note that force $P$ is independent of depth $h$.)

**3.46** Gate $ABC$ in Fig. 3-36 is 2 m square and hinged at $B$. How large must $h$ be for the gate to open?

The gate will open when resultant force $F$ acts above point $B$—i.e., when $|y_{cp}| < 0.2$ m. (Note in Fig. 3-36$b$ that $y_{cp}$ is the distance between $F$ and the centroid of gate $ABC$.)

$$y_{cp} = \frac{-I_{xx}\sin\theta}{h_{cg}A} = \frac{-[(2)(2)^3/12](\sin 90°)}{(h+1.0)[(2)(2)]} = \frac{-1.333}{4h+4}$$

For $|y_{cp}| < 0.2$, $1.333/(4h+4) < 0.2$, $h > 0.666$ m. (Note that this result is independent of fluid weight.)

Fig. 3-36(a)  Fig. 3-36(b)

**3.47** Gate *AB* in Fig. 3-37 is 6 ft wide and weighs 2000 lb when submerged. It is hinged at *B* and rests against a smooth wall at *A*. Determine the water level *h* which will just cause the gate to open.

$$F = \gamma h_{cg}A \qquad F_1 = 62.4(h + \tfrac{8}{2})[(10)(6)] = 3744h + 14\,976 \qquad F_2 = 62.4(5 + \tfrac{8}{2})[(10)(6)] = 33\,696 \text{ lb}$$

$$y_{cp} = \frac{-I_{xx}\sin\theta}{h_{cg}A} \qquad (y_{cp})_1 = \frac{-[(6)(10)^3/12](\tfrac{8}{10})}{(h + \tfrac{8}{2})[(10)(6)]} = \frac{-6.67}{h+4}$$

$$(y_{cp})_2 = \frac{-[(6)(10)^3/12](\tfrac{8}{10})}{(5 + \tfrac{8}{2})[(10)(6)]} = -0.741 \text{ ft}$$

$$\sum M_B = 0 \qquad (3744h + 14\,976)[5 - 6.67/(h+4)] - (33\,696)(5 - 0.741) - (2000)(\tfrac{6}{2}) = 0 \qquad h = 5.32 \text{ ft}$$

Fig. 3-37(a)

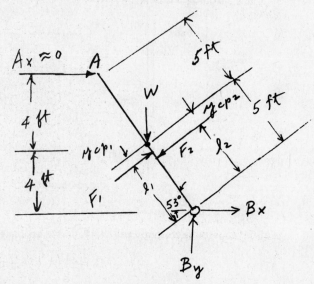

Fig. 3-37(b)

**3.48** The tank in Fig. 3-38 contains oil and water as shown. Find the resultant force on side *ABC*, which is 4 ft wide.

$$F = \gamma h_{cg}A \qquad F_{AB} = [(0.80)(62.4)](\tfrac{10}{2})[(10)(4)] = 9980 \text{ lb}$$

$F_{AB}$ acts at a point $(\tfrac{2}{3})(10)$, or 6.67 ft below point *A*. Water is acting on area *BC*, and any superimposed liquid can be converted to an equivalent depth of water. Employ an imaginary water surface (IWS) for this calculation, locating IWS by changing 10 ft of oil to $(0.80)(10)$, or 8 ft of water. Thus, $F_{BC} = (62.4)(8 + \tfrac{6}{2})[(6)(4)] = 16\,470$ lb.

$$y_{cp} = \frac{-I_{xx}\sin\theta}{h_{cg}A} = \frac{-[(4)(6)^3/12](\sin 90°)}{(8 + \tfrac{6}{2})[(6)(4)]} = -0.27 \text{ ft} \qquad \text{(i.e., below the centroid of } BC\text{)}$$

$F_{BC}$ acts at a point $(2 + 8 + \tfrac{6}{2} + 0.27)$, or 13.27 ft below *A*. $\sum M_A = 0$; $(9980 + 16\,470)(h_{cp}) - (9980)(6.67) - (16\,470)(13.27) = 0$, $h_{cp} = 10.78$ ft from *A*. Thus, the total resultant force on side *ABC* is $9980 + 16\,470$, or 26 450 lb acting 10.78 ft below *A*.

Fig. 3-38

**3.49** Gate $AB$ in Fig. 3-39 is 4 ft wide and hinged at $A$. Gage $G$ reads $-2.17$ psi, while oil (s.g. $= 0.75$) is in the right tank. What horizontal force must be applied at $B$ for equilibrium of gate $AB$?

$$F = \gamma h_{cg}A \qquad F_{oil} = [(0.75)(62.4)](\tfrac{6}{2})[(6)(4)] = 3370 \text{ lb}$$

$F_{oil}$ acts $(\tfrac{2}{3})(6)$, or 4.0 ft from $A$. For the left side, the negative pressure due to the air can be converted to its equivalent head in feet of water. $h = p/\gamma = (-2.17)(144)/62.4 = -5.01$ ft. This negative pressure head is equivalent to having 5.01 ft less water above $A$. Hence, $F_{H_2O} = (62.4)(6.99 + \tfrac{6}{2})[(6)(4)] = 14\,960$ lb.

$$y_{cp} = \frac{-I_{xx}\sin\theta}{h_{cg}A} = \frac{-[(4)(6)^3/12](\sin 90°)}{(6.99 + \tfrac{6}{2})[(6)(4)]} = -0.30 \text{ ft}$$

$F_{H_2O}$ acts at $(0.30 + \tfrac{6}{2})$, or 3.30 ft below $A$. $\sum M_A = 0$; $(3370)(4.0) + 6F - (14\,960)(3.30) = 0$, $F = 5980$ lb (acting leftward).

Fig. 3-39

**3.50** A vertical circular disk 1.1 m in diameter has its highest point 0.4 m below the surface of a pond. Find the magnitude of the hydrostatic force on one side and the depth to the center of pressure.

$$F = \gamma h_{cg}A = (9.79)(0.4 + 1.1/2)[(\pi)(1.1)^2/4] = 8.84 \text{ kN}$$

$$h_{cp} = h_{cg} + \frac{I_{cg}}{h_{cg}A} = \left(0.4 + \frac{1.1}{2}\right) + \frac{(\pi)(1.1)^4/64}{(0.4 + 1.1/2)[(\pi)(1.1)^2/4]} = 1.03 \text{ m}$$

**3.51** The vertical plate shown in Fig. 3-40 is submerged in vinegar (s.g. $= 0.80$). Find the magnitude of the hydrostatic force on one side and the depth to the center of pressure.

$$F = \gamma h_{cg}A \qquad F_1 = [(0.80)(9.79)](2 + \tfrac{7}{2})[(3)(7)] = 905 \text{ kN} \qquad h_{cp} = h_{cg} + \frac{I_{cg}}{h_{cg}A}$$

$$(h_{cp})_1 = 2 + \tfrac{7}{2} + \frac{(3)(7)^3/12}{(2 + \tfrac{7}{2})[(3)(7)]} = 6.24 \text{ m} \qquad F_2 = [(0.80)(9.79)][2 + 3 + 4/2][(2)(4)] = 439 \text{ kN}$$

$$(h_{cp})_2 = [2 + 3 + 4/2] + \frac{(2)(4)^3/12}{(2 + 3 + 4/2)[(2)(4)]} = 7.19 \text{ m}$$

$$F = 905 + 439 = 1344 \text{ kN} \qquad 1344h_{cp} = (905)(6.24) + (439)(7.19) \qquad h_{cp} = 6.55 \text{ m}$$

Fig. 3-40

**3.52** The irrigation head gate shown in Fig. 3-41$a$ is a plate which slides over the opening to a culvert. The coefficient of friction between the gate and its sliding ways is 0.5. Find the force required to slide open this 1000-lb gate if it is set (**a**) vertically and (**b**) on a 2:1 slope ($n = 2$ in Fig. 3-41$a$), as is common.

▌ **(a)** $F = \gamma h_{cg}A = (62.4)[14 + (\frac{60}{12})/2][(\frac{60}{12})(\frac{60}{12})] = 25\,740$ lb. Let $T =$ force parallel to gate required to open it. $\sum F_y = 0$; $T - 1000 - (0.5)(25\,740) = 0$, $T = 13\,870$ lb. **(b)** See Fig. 3-41*b*. $F = (62.4)[14 + \frac{60}{12}(1/\sqrt{5})/2][(\frac{60}{12})(\frac{60}{12})] = 23\,584$ lb. Let $N =$ total force normal to gate; $N = 23\,584 + (1000)(2/\sqrt{5}) = 24\,478$ lb. $\sum F_y = 0$; $T - (1000)(1/\sqrt{5}) - (0.5)(24\,478) = 0$, $T = 12\,686$ lb.

**Fig. 3-41(a)**          **Fig. 3-41(b)**

**3.53** A 65-in-square floodgate, weighing 2200 lb, is hinged 44.5 in above the center, as shown in Fig. 3-42, and the face is inclined 5° to the vertical. Find the depth to which water will rise behind the gate before it will open.

▌      Closing moment of gate about hinge $= (2200)[(\frac{44.5}{12})(\sin 5°)] = 711$ lb · ft

$$F = \gamma h_{cg}A = (62.4)(h/2)[(\frac{65}{12})(h)/\cos 5°] = 169.6h^2$$

$$\sum M_{hinge} = 0 \qquad (169.6h^2)[(65 + 12)/12 - (h/\cos 5°)/3] - 711 = 0 \qquad h = 0.826 \text{ ft}$$

**Fig. 3-42**

**3.54** Gate *MN* in Fig. 3-43 rotates about an axis through *N*. If the width of the gate is 5 ft, what torque applied to the shaft through *N* is required to hold the gate closed?

▌   $F = \gamma h_{cg}A$     $F_1 = 62.4[6 + (3 + 4)/2][(3 + 4)(5)] = 20\,748$ lb     $F_2 = (62.4)(\frac{4}{2})[(5)(4)] = 2496$ lb

$$y_{cp} = \frac{-I_{xx}\sin\theta}{h_{cg}A} \qquad (y_{cp})_1 = \frac{-[(5)(3+4)^3/12](\sin 90°)}{[6 + (3+4)/2][(3+4)(5)]} = 0.430 \text{ ft}$$

$F_2$ acts at $(\frac{1}{3})(4)$, or 1.333 ft from *N*. $\sum M_N = 0$; $(20\,748)[(3 + 4)/2 - 0.430] - (2496)(1.333) - \text{torque}_N = 0$, $\text{torque}_N = 60\,369$ lb · ft.

**Fig. 3-43**

**3.55** Find the minimum depth of $z$ for which the gate in Fig. 3-44 will open, if the gate is (a) square and (b) isosceles triangular, with base = height.

**▌** *(a)*
$$F = \gamma h_{cg} A \qquad F_{H_2O} = (62.4)(z - \tfrac{3}{2})[(3)(3)] = (561.6)(z - 1.5)$$

$$y_{cp} = \frac{-I_{xx} \sin\theta}{h_{cg} A} \qquad (y_{cp})_{H_2O} = \frac{-[(3)(3)^3/12](\sin 90°)}{(z - \tfrac{3}{2})[(3)(3)]} = \frac{-0.750}{z - 1.5}$$

Moment due to water = $[(561.6)(z - 1.5)][\tfrac{3}{2} + 0.750/(z - 1.5)] = (561.6)(1.5z - 1.500)$

$F_{gas} = pA = [(5)(144)][(3)(3)] = 6480$ lb. $F_{gas}$ acts at $\tfrac{3}{2}$, or 1.5 ft below hinge. Moment due to gas = $(6480)(1.5) =$ 9720 lb · ft. Equating moments gives $(561.6)(1.5z - 1.500) = 9720$, $z = 12.54$ ft.

*(b)*
$$F_{H_2O} = (62.4)[z - (\tfrac{2}{3})(3)][(3)(3)/2] = (280.8)(z - 2.000)$$

$$(y_{cp})_{H_2O} = \frac{-[(3)(3)^3/36](\sin 90°)}{[z - (\tfrac{2}{3})(3)][(3)(3)/2]} = \frac{0.500}{z - 2.000}$$

Moment due to water = $[(280.8)(z - 2.000)][\tfrac{3}{3} + 0.500/(z - 2.000)] = 280.8z - 421.2$

$F_{gas} = [(5)(144)][(3)(3)/2] = 3240$ lb. $F_{gas}$ acts at $\tfrac{3}{3}$, or 1.000 ft below hinge. Moment due to gas = $(3240)(1.000) = 3240$ lb · ft. Equating moments gives $(280.8z - 421.2) = 3240$, $z = 13.04$ ft.

End view    **Fig. 3-44**

**3.56** The triangular gate $CDE$ in Fig. 3-45 is hinged along $CD$ and is opened by a normal force $P$ applied at $E$. It holds a liquid of specific gravity 0.82 above it and is open to the atmosphere on its lower side. Neglecting the weight of the gate, find (a) the magnitude of force exerted on the gate, by direct integration; (b) the location of the center of pressure; and (c) the force $P$ needed to open the gate.

**▌** *(a)* $F = \int \gamma h \, dA = \int \gamma(y \sin\theta)(x \, dy)$. When $y = 8$, $x = 0$, and when $y = 8 + \tfrac{12}{2}$, or 14, $x = 6$, with $x$ varying linearly with $y$. Hence, $x = y - 8$. When $y = 14$, $x = 6$, and when $y = 8 + 12$, or 20, $x = 0$, with $x$ varying linearly with $y$. Hence, $x = 20 - y$.

$$F = \int_8^{14} [(0.82)(62.4)](y \sin 30°)[(y - 8) \, dy] + \int_{14}^{20} [(0.82)(62.4)](y \sin 30°)[(20 - y) \, dy]$$

$$= [(0.82)(62.4)](\sin 30°)\left\{\left[\frac{y^3}{3} - 4y^2\right]_8^{14} + \left[10y^2 - \frac{y^3}{3}\right]_{14}^{20}\right\} = 12\,894 \text{ lb}$$

*(b)*
$$x_{cp} = \frac{-I_{xy} \sin\theta}{h_{cg} A}$$

Since $I_{xy} = 0$, $x_{cp} = 0$,

$$y_{cp} = \frac{-I_{xx} \sin\theta}{h_{cg} A} = \frac{-[(2)(6)(\tfrac{12}{2})^3/12](\sin 30°)}{[(8 + \tfrac{12}{2})(\sin 30°)][(12)(6)/2]} = -0.43 \text{ ft}$$

(i.e., the pressure center is 0.43 ft below the centroid, measured in the plane of the area).

*(c)*
$$\sum M_{CD} = 0 \qquad 6P = (12\,894)(\tfrac{6}{3}) \qquad P = 4298 \text{ lb}$$

Fig. 3-45

**3.57** Determine the force acting on one side of vertical surface $OACO$ in Fig. 3-46 and the location of the center of pressure, if $\gamma = 8.4$ kN/m³. The curved edge is an arc of the parabola $y = x^2/8$.

▮ $$F = \int \gamma y \, dA = \int_0^1 (8.4)(y)(2x\,dy) = \int_0^1 (8.4)(y)(2\sqrt{8y})\,dy = \int_0^1 47.52y^{3/2}\,dy = [19.01y^{5/2}]_0^1 = 19.01 \text{ kN}$$

$$y_{cp} = \frac{\int \gamma y^2\,dA}{F} = \frac{\int_0^1 (8.4)(y)^2(2x\,dy)}{19.01} = \frac{\int_0^1 (8.4)(y)^2(2\sqrt{8y})\,dy}{19.01} = \frac{\int_0^1 47.52y^{5/2}\,dy}{19.01} = \frac{[13.58y^{7/2}]_0^1}{19.01} = 0.714 \text{ m}$$

Fig. 3-46

**3.58** Find the force exerted by water on one side of the vertical annular disk shown in Fig. 3-47. Also locate the center of pressure.

▮ $$F = \gamma h_{cg}A = (9.79)(3)[(\pi)(1)^2 - (\pi)(\tfrac{600}{1000})^2] = 59.05 \text{ kN}$$

$$I_{cg} = (\pi)(1)^4/4 - (\pi)(\tfrac{600}{1000})^4/4 = 0.6836 \text{ m}^4$$

$$h_{cp} = h_{cg} + \frac{I_{cg}}{h_{cg}A} = 3 + \frac{0.6836}{3[(\pi)(1)^2 - (\pi)(\tfrac{600}{1000})^2]} = 3.113 \text{ m}$$

Fig. 3-47

**3.59**    Determine $y$ in Fig. 3-48 so that the flashboards will tumble only when the water reaches their top.

▌ The flashboards will tumble when $y$ is at the center of pressure. Hence, $y = \frac{4}{3}$, or 1.333 m.

Fig. 3-48

**3.60**    Determine the pivot location $y$ of the square gate in Fig. 3-49 so that it will rotate open when the liquid surface is as shown.

▌ The gate will open when the pivot location is at the center of pressure.

$$h_{cp} = h_{cg} + \frac{I_{cg}}{h_{cg}A} = \left(3 - \frac{2}{2}\right) + \frac{(1)(2)^3/12}{(3 - \frac{2}{2})[(2)(1)]} = 2.167 \text{ m} \qquad y = 3 - 2.167 = 0.833 \text{ m}$$

Fig. 3-49

**3.61**    The gate in Fig. 3-50a (shown in raised position) weighs 350 lb for each foot normal to the paper. Its center of gravity is 1.5 ft from the left face and 2.0 ft above the lower face. For what water level below the hinge at $O$ does the gate just begin to swing up (rotate counterclockwise)?

▌ Refer to Fig. 3-50b and consider 1 ft of length. $F = \gamma h A = (62.4)[(h_O/2)][(h_O)(1)] = 31.2h_O^2$; $\sum M_O = 0$; $(2)(350) - (5 - h_O/3)(31.2h_O^2) = 0$, $h_O = 2.30$ ft.

Fig. 3-50(a)

$h_o/3$

Fig. 3-50(b)

**3.62** For the gate described in Prob. 3.61 and Fig. 3-50a, find $h$ for the gate just to come up to the vertical position shown in Fig. 3-50a.

▌ See Fig. 3-51. $F_1 = \gamma h A = (62.4)(h)[(5)(1)] = 312h$, $F_2 = (62.4)(h/2)[(h)(1)] = 31.2h^2$; $\sum M_O = 0$; $(1.5)(350) + (h/3)(31.2h^2) - (2.5)(312h) = 0$, $h = 0.68$ ft.

**Fig. 3-51**

**3.63** For the gate described in Prob. 3.61 and Fig. 3-50a, find $h$ and the force against the stop when this force is a maximum for the gate.

▌ See Fig. 3-51. $F_1 = \gamma h A = (62.4)(h)[(5)(1)] = 312h$, $F_2 = (62.4)(h/2)[(h)(1)] = 31.2h^2$; $\sum M_O = 0$; $(1.5)(350) + (h/3)(31.2h^2) - (2.5)(312h) + (5)(F_{stop}) = 0$, $F_{stop} = 156h - 2.08h^3 - 105$.

$$\frac{dF_{stop}}{dh} = 156 - 6.24h^2 = 0 \qquad h = 5.00 \text{ ft}$$

$$F_{stop} = (156)(5.00) - (2.08)(5.00)^3 - 105 = 415 \text{ lb}$$

**3.64** Compute the air pressure required to keep the gate of Fig. 3-52 closed. The gate is a circular plate of diameter 0.8 m and weight 2.0 kN.

▌
$$F = \gamma h A \qquad F_{liq} = [(2)(9.79)][1.7 + (\tfrac{1}{2})(0.8)(\sin 45°)][\pi(0.8)^2/4] = 19.52 \text{ kN}$$

$$z_{cp} = z_{cg} + \frac{I_{cg}}{z_{cg}A} = \left[\frac{1.7}{\cos 45°} + \left(\frac{1}{2}\right)(0.8)\right] + \frac{\pi[(\tfrac{1}{2})(0.8)]^4/4}{[1.7/\cos 45° + (\tfrac{1}{2})(0.8)][\pi(0.8)^2/4]} = 2.818 \text{ m}$$

$$\sum M_{hinge} = 0 \qquad (19.52)(2.818 - 1.7/\cos 45°) + 2.0[(\tfrac{1}{2})(0.8)(\cos 45°)] - [\pi(0.8)^2/4](p_{air})[(\tfrac{1}{2})(0.8)] = 0$$

$$p_{air} = 42.99 \text{ kPa}$$

**Fig. 3-52**

**4.1** In Fig. 4-1, calculate the width of concrete dam that is necessary to prevent the dam from sliding. The specific weight of the concrete is 150 lb/ft³, and the coefficient of friction between the base of the dam and the foundation is 0.42. Use 1.5 as the factor of safety (F.S.) against sliding. Will it also be safe against overturning?

▎ Working with a 1-ft "slice" (i.e., dimension perpendicular to the paper) of the dam, $W_{dam} = (20)(w)(1)(150) = 3000w$, $F = \gamma hA$, $F_H = (62.4)[(0 + 15)/2][(15)(1)] = 7020$ lb.

$$\text{F.S.}_{\text{sliding}} = \frac{\text{sliding resistance}}{\text{sliding force}} \qquad 1.5 = \frac{(0.42)(3000w)}{7020} \qquad w = 8.36 \text{ ft}$$

$$\text{F.S.}_{\text{overturning}} = \frac{\text{total righting moment}}{\text{overturning moment}} = \frac{[(3000)(8.36)](8.36/2)}{(7020)(\frac{15}{3})} = 2.99$$

Therefore, it should be safe against overturning.

**Fig. 4-1**

**4.2** Figure 4-2 is the cross section of an earthwork (s.g. = 2.5) dam. Assuming that hydrostatic uplift varies linearly from one-half the hydrostatic head at the upstream edge of the dam to zero at the downstream edge, find the maximum and minimum pressure intensity in the base of the dam.

▎ $$F = \gamma hA \qquad F_H = (62.4)[(0 + 97)/2][(97)(1)] = 293\ 561 \text{ lb}$$

For equilibrium, $R_x = 293\ 561$ lb.

$$W_1 = [(2.5)(62.4)][(1)(10)(90 + 30)] = 187\ 200 \text{ lb} \qquad W_2 = [(2.5)(62.4)][(1)(60)(90)/2] = 421\ 200 \text{ lb}$$

$$F_U = [(62.4)(48.5 + 0)/2][(60 + 10)(1)] = 105\ 924 \text{ lb} \qquad R_y = 187\ 200 + 421\ 200 - 105\ 924 = 502\ 476 \text{ lb}$$

$$\sum M_0 = 0 \qquad (293\ 561)(32.33) + (187\ 200)(5) + (421\ 200)(30) - (105\ 924)[(60 + 10)/3] - 502\ 476x = 0$$

$$x = 40.98 \text{ ft} \qquad \text{Eccentricity} = 40.98 - (60 + 10)/2 = 5.98 \text{ ft}$$

Since the eccentricity is less than one-sixth the base of the dam, the resultant acts within the middle third of the base.

$$p = \frac{F}{A} \pm \frac{M_y x}{I_y} \pm \frac{M_x y}{I_x} = \frac{502\ 476}{(60 + 10)(1)} \pm \frac{[(502\ 476)(5.98)](60 + 10)/2}{(1)(60 + 10)^3/12} \pm 0 = 7178 \pm 3679$$

$$p_{\max} = 7178 + 3679 = 10\ 857 \text{ lb/ft}^2 \qquad p_{\min} = 7178 - 3679 = 3499 \text{ lb/ft}^2$$

**Fig. 4-2**

**4.3** For linear stress variation over the base of the dam of Fig. 4-3a, find where the resultant crosses the base and compute the maximum and minimum pressure intensity at the base. Neglect hydrostatic uplift.

▌ Figure 4-3b shows the forces acting on the dam. $F_1 = \gamma[(19+6)/2][(19+6)(1)] = 312\gamma$, $F_2 = \gamma[(6)(3)(1)] = 18\gamma$, $F_3 = \gamma[(1)(19)(3)/2] = 28.5\gamma$, $F_4 = [(2.5)(\gamma)][(4)(19+6)(1)] = 250\gamma$, $F_5 = [(2.5)(\gamma)][(1)(19)(3)/2] = 71.25\gamma$, $F_6 = [(2.5)(\gamma)][(1)(19)(11)/2] = 261\gamma$; $R_y = 18\gamma + 28.5\gamma + 250\gamma + 71.25\gamma + 261\gamma = 628.75\gamma$. $\sum M_A = 0$; $(628.75\gamma)(x) - (312\gamma)[(19+6)/3] - (18\gamma)(1.5) - (28.5\gamma)(1) - (250\gamma)(3+2) - (71.25\gamma)(3-1) - (261\gamma)(4+3+\frac{11}{3})$ : $x = 10.87$ m. Eccentricity $= 10.87 - (11+4+3)/2 = 1.87$ ft. Since the eccentricity is less than one-sixth the base of the dam, the resultant acts within the middle third of the base.

$$p = \frac{F}{A} \pm \frac{M_y x}{I_y} \pm \frac{M_x y}{I_x} = \frac{(628.75)(9.79)}{(11+4+3)(1)} \pm \frac{[(628.75)(9.79)(1.87)][(11+4+3)/2]}{(1)(11+4+3)^3/12} \pm 0 = 342 \pm 213 \text{ kPa}$$

$$p_{max} = 342 + 213 = 555 \text{ kPa} \qquad p_{min} = 342 - 213 = 129 \text{ kPa}$$

**Fig. 4-3(a)**

**Fig. 4-3(b)**

**4.4**  For the conditions given in Prob. 4.3 with the addition that hydrostatic uplift varies linearly from 19 m at $A$ to zero at the toe of the dam, would the resultant still act within the middle third of the base?

$F_U = \gamma[(19+0)/2][(4+3+11)(1)] = 171\gamma \qquad R_y = 18\gamma + 28.5\gamma + 250\gamma + 71.25\gamma + 261\gamma - 171\gamma = 457.75\gamma$

$\sum M_A = 0 \qquad (457.75\gamma)(x) - (312\gamma)[(19+6)/3] - (18\gamma)(1.5) - (28.5\gamma)(1) - (250\gamma)(3+2)$

$$- (71.25\gamma)(3-1) - (261\gamma)(4+3+\tfrac{11}{3}) + (171\gamma)[(4+3+11)/3] = 0$$

$x = 12.68 \text{ m} \qquad \text{Eccentricity} = 12.68 - (11+4+3)/2 = 3.68 \text{ ft}$

Since the eccentricity is greater than one-sixth the base of the dam, the resultant acts outside the middle third of the base.

**4.5**  A concrete dam retaining water is shown in Fig. 4-4a. If the specific weight of the concrete is 150 lb/ft$^3$, find the factor of safety against sliding, the factor of safety against overturning, and the pressure intensity on the base. Assume the foundation soil is impermeable and that the coefficient of friction between dam and foundation soil is 0.45.

The forces acting on the dam are shown in Fig. 4.4b. $F = \gamma hA$, $F_x = (62.4)[(0+42)/2][(42)(1)] = 55\,040$ lb. From Fig. 4-4b, $CD/42 = \tfrac{10}{50}$, $CD = 8.40$ ft; $F_y = (62.4)[(8.40)(42)/2](1) = 11\,010$ lb.

| component | weight of component (kips) | moment arm from toe, $B$ (ft) | righting moment about toe, $B$ (kip · ft) |
|---|---|---|---|
| 1 | $(\tfrac{1}{2})(10 \times 50)(0.15)(1) = 37.50$ | $20 + \tfrac{10}{3} = 23.33$ | 875 |
| 2 | $(10 \times 50)(0.15)(1) = 75.00$ | $10 + \tfrac{10}{2} = 15.00$ | 1125 |
| 3 | $(\tfrac{1}{2})(10 \times 50)(0.15)(1) = 37.50$ | $(\tfrac{2}{3})(10) = 6.67$ | 250 |
| $F_y$ | 11.01 | $30 - (\tfrac{1}{3})(8.40) = 27.20$ | 299 |
| | $\sum V = 161.01$ **kips** | | $\sum M_r = 2549$ **kip · ft** |

$$M_{\text{overturning}} = (55.04)(\tfrac{42}{3}) = 771 \text{ kip-ft} \qquad \text{F.S.}_{\text{sliding}} = \frac{\text{sliding resistance}}{\text{sliding force}} = \frac{(0.45)(161.01)}{55.04} = 1.32$$

$$\text{F.S.}_{\text{overturning}} = \frac{\text{total righting moment}}{\text{overturning moment}} = \frac{2549}{771} = 3.31$$

$R_x = F_x = 55.04$ kips and $R_y = \sum V = 161.01$ kips; hence, $R = \sqrt{55.04^2 + 161.01^2} = 170.16$ kips.

$$x = \frac{\sum M_B}{R_y} = \frac{\sum M_r - M_0}{\sum V} = \frac{2549 - 771}{161.01} = 11.04 \text{ ft} \qquad \text{Eccentricity} = \tfrac{30}{2} - 11.04 = 3.96 \text{ ft}$$

Since the eccentricity is less than one-sixth the base of the dam, the resultant acts within the middle third of the base.

$$p = \frac{F}{A} \pm \frac{M_y x}{I_y} \pm \frac{M_x y}{I_x} = \frac{161.01}{(30)(1)} \pm \frac{[(161.01)(3.96)](15)}{(1)(30)^3/12} \pm 0 = 5.37 \pm 4.25$$

$$p_B = 5.37 + 4.25 = 9.62 \text{ kips/ft}^2 \qquad p_A = 5.37 - 4.25 = 1.12 \text{ kips/ft}^2$$

The complete pressure distribution on the base of the dam is given in Fig. 4-4c.

Fig. 4-4(a)

Fig. 4-4(b)

Fig. 4-4(c)

**4.6** A concrete dam retaining water is shown in Fig. 4-5a. If the specific weight of the concrete is 23.5 kN/m³, find the factor of safety against sliding, the factor of safety against overturning, and the pressure intensity on the base. Assume there is a hydrostatic uplift that varies uniformly from full hydrostatic head at the heel of the dam to zero at the toe and that the coefficient of friction between dam and foundation soil is 0.45.

▌ The forces acting on the dam are shown in Fig. 4-5b. $F = \gamma hA$, $F_x = (9.79)[(0 + 14)/2][(14)(1)] = 959.4$ kN, $F_y = (9.79)[(3)(14 - 3)(1)] = 323.1$ kN. Hydrostatic uplift varies from $(14)(9.79)$, or $137.1$ kN/m² at the heel to zero at the toe, as shown in Fig. 4-5b. $F_U = (137.1/2)(15)(1) = 1028$ kN. It acts at $(\frac{1}{3})(15)$, or 5.0 m from point $A$, as shown in Fig. 4-5b.

| component | weight of component (kN) | moment arm from toe, $B$ (m) | righting moment about toe, $B$ (kN · m) |
|---|---|---|---|
| 1 | $(\frac{1}{2})(15-3-4)(12)(23.5)(1) = 1128$ | $(\frac{2}{3})(15-3-4) = 5.333$ | 6 016 |
| 2 | $(4)(12+3)(23.5)(1) = 1410$ | $(15-3-\frac{4}{2}) = 10.000$ | 14 100 |
| 3 | $(15)(3)(23.5)(1) = 1058$ | $\frac{15}{2} = 7.500$ | 7 935 |
| $F_y$ | $= 323$ | $(15-\frac{3}{2}) = 13.500$ | 4 360 |
| | $\sum V = 3919 \text{ kN}$ | | $\sum M_r = 32\,411 \text{ kN} \cdot \text{m}$ |

$$M_{\text{overturning}} = (959.4)(\tfrac{14}{3}) + (1028)(10) = 14\,760 \text{ kN}$$

$$\text{F.S.}_{\text{sliding}} = \frac{\text{sliding resistance}}{\text{sliding force}} = \frac{(0.45)(3919-1028)}{959.4} = 1.36$$

$$\text{F.S.}_{\text{overturning}} = \frac{\text{total righting moment}}{\text{overturning moment}} = \frac{32\,411}{14\,760} = 2.20$$

$R_x = F_x = 959.4$ kN and $R_y = \sum V - F_U = 3919 - 1028 = 2891$ kN; hence, $R = \sqrt{959.4^2 + 2891^2} = 3046$ kN.

$$x = \frac{\sum M_B}{R_y} = \frac{\sum M_r - M_0}{\sum V} = \frac{32\,411 - 14\,760}{2891} = 6.105 \text{ m} \qquad \text{Eccentricity} = \tfrac{15}{2} - 6.105 = 1.395 \text{ m}$$

Since the eccentricity is less than one-sixth the base of the dam, the resultant acts within the middle third of the base.

$$p = \frac{F}{A} \pm \frac{M_y x}{I_y} \pm \frac{M_x y}{I_x} = \frac{2891}{(15)(1)} \pm \frac{[(2891)(1.395)](\frac{15}{2})}{(1)(15)^3/12} \pm 0 = 192.7 \pm 107.5$$

$$p_B = 192.7 + 107.5 = 300.2 \text{ kN/m}^2 \qquad p_A = 192.7 - 107.5 = 85.2 \text{ kN/m}^2$$

The complete pressure distribution on the base of the dam is given in Fig. 4-5c.

Fig. 4-5(a)

Fig. 4-5(b)

Fig. 4-5(c)

**4.7** A concrete dam retaining water is shown in Fig. 4-6a. If the specific weight of the concrete is 23.5 kN/m³, find the factor of safety against sliding, the factor of safety against overturning, and the maximum and minimum pressure intensity on the base. Assume there is no hydrostatic uplift and that the coefficient of friction between dam and foundation soil is 0.48.

▌ The forces acting on the dam are shown in Fig. 4-6b. $F = \gamma h A$, $F_H = (9.79)[(0 + 6)/2][(6)(1)] = 176.2$ kN.

| component | weight of component (kN) | moment arm from toe, $A$ (m) | righting moment about toe, $A$ (kN · m) |
|---|---|---|---|
| 1 | $(\frac{1}{2})(2)(7)(23.5) = 164.5$ | $(\frac{2}{3})(2) = 1.333$ | 219 |
| 2 | $(2)(7)(23.5) = 329.0$ | $2 + \frac{2}{2} = 3.000$ | 987 |
| | $\sum V = 493.5$ kN | | $\sum M_r = 1206$ kN · m |

$$M_{overturning} = (176.2)(\tfrac{6}{3}) = 352.4 \text{ kN}$$

$$\text{F.S.}_{sliding} = \frac{\text{slinding resistance}}{\text{sliding force}} = \frac{(0.48)(493.5)}{176.2} = 1.34$$

$$\text{F.S.}_{overturning} = \frac{\text{total righting moment}}{\text{overturning moment}} = \frac{1206}{352.4} = 3.42$$

$$R_x = F_H = 176.2 \text{ kN and } R_y = \Sigma V = 493.5 \text{ kN; hence, } R = \sqrt{176.2^2 + 493.5^2} = 524 \text{ kN.}$$

$$x = \frac{\Sigma M_A}{R_y} = \frac{\Sigma M_r - M_0}{\Sigma V} = \frac{1206 - 352.4}{493.5} = 1.730 \text{ m} \qquad \text{Eccentricity} = \tfrac{4}{2} - 1.730 = 0.270 \text{ m}$$

Since the eccentricity is less than one-sixth the base of the dam, the resultant acts within the middle third of the base.

$$p = \frac{F}{A} \pm \frac{M_y x}{I_y} \pm \frac{M_x y}{I_x} = \frac{493.5}{(4)(1)} \pm \frac{[(493.5)(0.270)](\tfrac{4}{2})}{(1)(4)^3/12} \pm 0 = 123.4 \pm 50.0$$

$$p_{max} = 123.4 + 50.0 = 173.4 \text{ kN/m}^2 \qquad p_{min} = 123.4 - 50.0 = 73.4 \text{ kN/m}^2$$

Fig. 4-6(a)

Fig. 4-6(b)

**4.8** For the dam shown in Fig. 4-7, what is the minimum width $b$ for the base of a dam 100 ft high if hydrostatic uplift is assumed to vary uniformly from full hydrostatic head at the heel to zero at the toe, and also assuming an ice thrust $P_1$ of 12 480 lb per linear foot of dam at the top? For this study, make the resultant of the reacting forces cut the base at the downstream edge of the middle third of the base (i.e., at $O$ in Fig. 4-7) and take the weight of the masonry as $2.50\gamma$.

$$F = \gamma h A \qquad F_H = (62.4)[(100 + 0)/2][(100)(1)] = 312\,000 \text{ lb} \qquad F_U = [(100)(62.4)/2][(1)(b)] = 3120b$$

$$W_1 = [(2.50)(62.4)][(20)(100)(1)] = 312\,000 \text{ lb} \qquad W_2 = [(2.50)(62.4)][(b - 20)(100)(1)/2] = 7800b - 156\,000$$

$$\sum M_O = 0$$

$$(312\,000)(\tfrac{100}{3}) + (3120b)(b/3) - (312\,000)[(\tfrac{2}{3})(b) - \tfrac{20}{2}] - (7800b - 156\,000)[(\tfrac{2}{3})(b - 20) - b/3] + (12\,480)(100) = 0$$

$$3b^2 + 100b - 24\,400 = 0 \qquad b = 75.0 \text{ ft}$$

Fig. 4-7

# CHAPTER 5

# Forces on Submerged Curved Areas

**5.1** The submerged, curved surface $AB$ in Fig. 5-1$a$ is one-quarter of a circle of radius 4 ft. The tank's length (distance perpendicular to the plane of the figure) is 6 ft. Find the horizontal and vertical components of the total resultant force acting on the curved surface and their locations.

▌ The horizontal component of the total resultant force acting on the curved surface is equal to the total resultant force, $F_H$, acting on the vertical projection of curved surface $AB$ (i.e., $BF$ in Fig. 5-1$b$). This projection is a rectangle 6 ft long and 4 ft high. For the portion of $F_H$ resulting from horizontal pressure of $BHEF$ in Fig. 5-1$b$, $p_1 = (8)(62.4) = 499$ lb/ft$^2$, $A = (6)(4) = 24$ ft$^2$, $F_1 = (499)(24) = 11\,980$ lb. For the portion of $F_H$ resulting from horizontal pressure of $HGE$ in Fig. 5-1$b$, $p_2 = (62.4)[(0+4)/2] = 125$ lb/ft$^2$, $F_2 = (125)(24) = 3000$ lb; $F_H = F_1 + F_2 = 11\,980 + 3000 = 14\,980$ lb. The vertical component of the total resultant force acting on the curved surface is equal to the weight of the volume of water vertically above curved surface $AB$. This volume consists of a rectangular area ($AFCD$ in Fig. 5-1$c$) 4 ft by 8 ft and a quarter-circular area ($ABF$ in Fig. 5-1$c$) of radius 4 ft, both areas being 6 ft long. This volume ($V$) is $V = [(4)(8) + (\pi)(4)^2/4](6) = 267.4$ ft$^3$, $F_V = $ weight of water in $V = (267.4)(62.4) = 16\,690$ lb. The location of the horizontal component ($F_H$) is along a (horizontal) line through the center of pressure for the vertical projection (i.e., the center of gravity of $EFBG$ in Fig. 5-1$b$). This can be determined by equating the sum of the moments of $F_1$ and $F_2$ about point $C$ to the moment of $F_H$ about the same point. $(11\,980)(8 + \frac{4}{2}) + (3000)[8 + (\frac{2}{3})(4)] = 14\,980h_{cp}$, $h_{cp} = 10.13$ ft. (This is the depth from the water surface to the location of the horizontal component. Stated another way, the horizontal component acts at a distance of $12 - 10.13$, or 1.87 ft above point $B$ in Fig. 5-1$b$.) The location of the vertical component ($F_V$) is

**Fig. 5-1(a)**

**Fig. 5-1(b)**

**Fig. 5-1(c)**

along a (vertical) line through the center of gravity of the liquid volume vertically above surface $AB$ (i.e., the center of gravity of $ABCD$ in Fig. 5-1$c$). This can be determined by referring to Fig. 5-1$c$ and equating the sum of the moments of the rectangular area ($AFCD$ in Fig. 5-1$c$) and of the quarter-circular area ($ABF$ in Fig. 5-1$c$) about a vertical line through point $B$ to the moment of the total area about the same line. $(x)[(8)(4) + (\pi)(4)^2/4] = [(8)(4)](\frac{4}{2}) + [(\pi)(4)^2/4][(4)(4)/(3\pi)]$, $x = 1.91$ ft. (This is the distance from point $B$ to the line of action of the vertical component.)

**5.2** Solve Prob. 5.1 for the same given conditions except that water is on the other side of curved surface $AB$, as shown in Fig. 5-2.

▌ If necessary, refer to the solution of Prob. 5.1 for a more detailed explanation of the general procedure for solving this type of problem. $p = p_{avg} = (\gamma)[(h_1 + h_2)/2] = (62.4)[(8 + 12)/2] = 624$ lb/ft$^2$, $A = (6)(4) = 24$ ft$^2$, $F_H = pA = (624)(24) = 14\,980$ lb. The vertical component ($F_V$) is equal to the weight of the imaginary volume of water vertically above surface $AB$. Hence, $F_V = [(4)(8) + (\pi)(4)^2/4](6)(62.4) = 16\,690$ lb. The location of the horizontal component is 10.13 ft below the water surface (same as in Prob. 5.1 except that $F_H$ acts toward the left). The location of the vertical component is 1.91 ft from point $B$ (same as in Prob. 5.1 except that $F_V$ acts upward).

Fig. 5-2

**5.3** The submerged sector gate $AB$ shown in Fig. 5-3$a$ is one-sixth of a circle of radius 6 m. The length of the gate is 10 m. Determine the amount and location of the horizontal and vertical components of the total resultant force acting on the gate.

▌ If necessary, refer to the solution of Prob. 5.1 for a more detailed explanation of the general procedure for solving this type of problem. Refer to Fig. 5-3$b$. $F_H = \gamma \bar{h} A = (9.79)[(0 + 5.196)/2][(10)(5.196)] = 1322$ kN, Area$_{ABC}$ = area$_{ACBD}$ + area$_{BDO}$ − area$_{ABO}$ = $(5.196)(3) + (3.000)(5.196)/2 − (\pi)(6)^2/6 = 4.532$ m$^2$, $F_V = $ (area$_{ABC}$)(length of gate)$(\gamma) = (4.532)(10)(9.79) = 444$ kN. The location of the horizontal component ($F_H$) is along a (horizontal) line 5.196/3, or 1.732 m above the bottom of the gate ($A$). The location of the vertical component ($F_V$) is along a (vertical) line through the center of gravity of section $ABC$. Taking area moments about $AC$, $4.532x = [(5.196)(3)](\frac{3}{2}) + [(\frac{1}{2})(3.000)(5.196)](3 + 3.000/3) − [(\pi)(6)^2/6]\{6 − [\cos{(60°/2)}](2)(6)/\pi\}$, $x = 0.842$ m.

(a)

Fig. 5-3($a$)

**Fig. 5-3(b)**

**5.4** The curved surface $AB$ shown in Fig. 5-4a is a quarter of a circle of radius 5 ft. Determine, for an 8-ft length perpendicular to the paper, the amount and location of the horizontal and vertical components of the total resultant force acting on surface $AB$.

▌ If necessary, refer to the solution of Prob. 5.1 for a more detailed explanation of the general procedure for solving this type of problem. Refer to Fig. 5-4b. $F_H = \gamma \bar{h} A = (62.4)[(0+5)/2][(5)(8)] = 6240$ lb, area$_{ABD}$ = area$_{ACBD}$ − area$_{ABC}$ = (5)(5) − $(\pi)(5)^2/4 = 5.365$ ft$^2$, $F_V$ = (area$_{ABD}$)(length)($\gamma$) = (5.365)(8)(62.4) = 2678 lb. $F_H$ is located at $\frac{5}{3}$, or 1.67 ft above $C$. $F_V$ is located at $x$ from line $AD$. $5.365x = [(5)(5)](\frac{5}{2}) − [(\pi)(5)^2/4][5 − (4)(5)/(3\pi)]$, $x = 1.12$ ft.

**Fig. 5-4(a)**

**5.5** Determine the value and location of the horizontal and vertical components of the force due to water acting on curved surface $AB$ in Fig. 5-5, per foot of its length.

▌ If necessary, refer to the solution of Prob. 5.1 for a more detailed explanation of the general procedure for solving this type of problem. $F_H = \gamma \bar{h} A = (62.4)[(0+6)/2][(6)(1)] = 1123$ lb, $F_V$ = (area)(length)($\gamma$) = $[(\pi)(6)^2/4](1)(62.4) = 1764$ lb. $F_H$ is located at $(\frac{2}{3})(6)$, or 4.00 ft below $C$. $F_V$ is located at the center of gravity of area $ABC$, or distance $x$ from line $CB$. $x = 4r/(3\pi) = (4)(6)/(3\pi) = 2.55$ ft.

**5.6** The 6-ft-diameter cylinder in Fig. 5-6 weighs 5000 lb and is 5 ft long. Determine the reactions at $A$ and $B$, neglecting friction.

Fig. 5-4(b)

Fig. 5-5

▮ The reaction at $A$ is due to the horizontal component of the liquid force acting on the cylinder ($F_H$).
$F_H = \gamma \bar{h} A = [(0.800)(62.4)][(0+6)/2][(3+3)(5)] = 4493$ lb. $F_H$ acts to the right; hence, the reaction at $A$ is
4493 lb to the left. The reaction at $B$ is the algebraic sum of the weight of the cylinder and the net vertical
component of the force due to the liquid. $(F_V)_{up} = (area_{ECOBDE})(length)(\gamma)$, $(F_V)_{down} = (area_{ECDE})(length)(\gamma)$,
$(F_V)_{net} = (F_V)_{up} - (F_V)_{down} = (area_{COBDC})(length)(\gamma) = [(\pi)(3)^2/2](5)[(0.800)(62.4)] = 3529$ lb (upward). The
reaction at $B$ is $5000 - 3529$, or 1471 lb upward.

Fig. 5-6

**5.7**  Referring to Fig. 5-7, determine the horizontal and vertical forces due to the water acting on the cylinder per foot of its length.

**∎**
$$(F_H)_{CDA} = 62.4\{[4 + (4 + 4.24 + 0.88)]/2\}[(2.12 + 3)(1)] = 2096 \text{ lb}$$

$$(F_H)_{AB} = (62.4)\{[(4 + 4.24) + (4 + 4.24 + 0.88)]/2\}[(0.88)(1)] = 477 \text{ lb}$$

$$(F_H)_{net} = (F_H)_{CDA} - (F_H)_{AB} = 2096 - 477 = 1619 \text{ lb (right)}$$

$$(F_V)_{net} = (F_V)_{DAB} - (F_V)_{DC} = \text{weight of volume}_{DABFED} - \text{weight of volume}_{DCGED} = \text{weight of volume}_{DABFGCD}$$

$$= \text{weight of (rectangle}_{GFJC} + \text{triangle}_{CJB} + \text{semicircle}_{CDAB})$$

$$= 62.4[(4)(4.24) + (4.24)(4.24)/2 + (\pi)(3)^2/2](1) = 2501 \text{ lb (upward)}$$

**Fig. 5-7**

**5.8**  In Fig. 5-8, an 8-ft-diameter cylinder plugs a rectangular hole in a tank that is 3 ft long. With what force is the cylinder pressed against the bottom of the tank due to the 9-ft depth of water?

**∎**
$$(F_V)_{net} = (F_V)_{CDE} - (F_V)_{CA} - (F_V)_{BE} = 62.4[(4 + 4)(7) - (\pi)(4)^2/2](3)$$

$$- 62.4[(7)(0.54) + (\tfrac{30}{360})(\pi)(4)^2 - (2)(3.46)/2](3)$$

$$- 62.4[(7)(0.54) + (\tfrac{30}{360})(\pi)(4)^2 - (2)(3.46)/2](3) = 4090 \text{ lb (downward)}$$

**Fig. 5-8**

**5.9**  In Fig. 5-9, the 8-ft-diameter cylinder weighs 500 lb and rests on the bottom of a tank that is 3 ft long. Water and oil are poured into the left- and right-hand portions of the tank to depths of 2 ft and 4 ft, respectively. Find the magnitudes of the horizontal and vertical components of the force that will keep the cylinder touching the tank at $B$.

**∎**
$$(F_H)_{net} = (F_H)_{AB} - (F_H)_{CB} = [(0.750)(62.4)][(0 + 4)/2][(4)(3)] - (62.4)[(0 + 2)/2][(2)(3)] = 749 \text{ lb (left)}$$

$$(F_V)_{net} = (F_V)_{AB} + (F_V)_{CB} = [(0.750)(62.4)][(\pi)(4)^2/4](3) + (62.4)[(\tfrac{60}{360})(\pi)(4)^2 - (2)(\sqrt{12})/2](3)$$

$$= 2684 \text{ lb (upward)}$$

The components to hold the cylinder in place are 749 lb to the right and 2684 − 500, or 2184 lb down.

Fig. 5-9

**5.10** The half-conical buttress $ABE$ shown in Fig. 5-10 is used to support a half-cylindrical tower $ABCD$. Calculate the horizontal and vertical components of the force due to water acting on the buttress.

$$F_H = \gamma h_{cg} A = (62.4)(3 + \tfrac{6}{3})[(6)(2+2)/2] = 3744 \text{ lb (right)}$$

$$F_V = \text{weight of (imaginary) volume of water above curved surface}$$

$$= (62.4)[(\tfrac{1}{2})(6)(\pi)(2)^2/3 + (\tfrac{1}{2})(\pi)(2)^2(3)] = 1960 \text{ lb (up)}$$

Fig. 5-10

**5.11** A dam has a parabolic shape $z = z_0(x/x_0)^2$, as shown in Fig. 5-11a. The fluid is water and atmospheric pressure may be neglected. If $x_0 = 10$ ft and $z_0 = 24$ ft, compute forces $F_H$ and $F_V$ on the dam and the position c.p. where they act. The width of the dam is 50 ft.

$F_H = \gamma \bar{h} A = 62.4[(24+0)/2][(24)(50)] = 898\,600$ lb. The location of $F_H$ is along a (horizontal) line $\tfrac{24}{3}$, or 8.00 ft above the bottom of the dam. $F_V = (\text{area}_{0AB})(\text{width of dam})(\gamma)$. (See Fig. 5-11b.) $\text{Area}_{0AB} = 2x_0 Z_0/3 = (2)(10)(24)/3 = 160$ ft$^2$, $F_V = (160)(50)(62.4) = 499\,200$ lb. The location of $F_V$ is along a (vertical) line through the center of gravity of area$_{0AB}$. From Fig. 5-11b, $x = 3x_0/8 = (3)(10)/8 = 3.75$ ft, $z = 3z_0/5 = (3)(24)/5 = 14.4$ ft, $F_{\text{resultant}} = \sqrt{499\,200^2 + 898\,600^2} = 1\,028\,000$ lb. As seen in Fig. 5-11c, $F_{\text{resultant}}$ acts down and to the right at an angle of arctan $(499\,200/898\,600)$, or 29.1°. $F_{\text{resultant}}$ passes through the point $(x, z) = (3.75$ ft, 8 ft). If we move down alone the 29.1° line until we strike the dam, we find an equivalent center of pressure on the dam at $x = 5.43$ ft and $z = 7.07$ ft. This definition of c.p. is rather artificial, but this is an unavoidable complication of dealing with a curved surface.

Fig. 5-11(a)

Fig. 5-11(b)

Fig. 5-11(c)

**5.12** The canal shown in cross section in Fig. 5-12a runs 40 m into the paper. Determine the horizontal and vertical components of the hydrostatic force against the quarter-circle wall and the point c.p. where the resultant strikes the wall.

▌ $F_H = \gamma \bar{h} A = 9.79[(18+0)/2][(18)(40)] = 63\,439$ kN. The location of $F_H$ is along a (horizontal) line $\frac{18}{3}$, or 6.00 m above the bottom of the wall. $F_V = 9.79[(40)(\pi)(18)^2/4] = 99\,650$ kN. The location of $F_V$ is along a (vertical) line through the center of gravity of area$_{OAB}$. $x = 4r/(3\pi) = (4)(18)/(3\pi) = 7.64$ m, $F_{resultant} = \sqrt{63\,439^2 + 99\,650^2} = 118\,130$ kN. As seen in Fig. 5-12b, $F_{resultant}$ acts down and to the right at an angle of arctan $(99\,650/63\,439)$, or 57.5°. $F_{resultant}$ passes through the point $(x, z) = (7.64$ m, $6.00$ m). If we move down along the 57.5° line until we strike the wall, we find an equivalent center of pressure at $x = 8.33$ m and $z = 2.82$ m.

Fig. 5-12(a)

Fig. 5-12(b)

**5.13** Gate $AB$ in Fig. 5-13a is a quarter circle 8 ft wide into the paper. Find the force $F$ just sufficient to prevent rotation about hinge $B$. Neglect the weight of the gate.

▌ $F_H = \gamma \bar{h} A = 62.4[(7+0)/2][(7)(8)] = 12\,230$ lb (left). The location of $F_H$ is along a (horizontal) line $\frac{7}{3}$, or 2.333 ft above point $B$. (See Fig. 5-13b.) $F_V = F_1 - F_2 = 62.4[(8)(7)(7)] - 62.4[(8)(\pi)(7)^2/4] = 24\,461 - 19\,211 = 5250$ lb (up). The location of $F_V$ can be determined by taking moments about point $B$ in Fig. 5-13b. $5250x = (24\,461)(\frac{7}{2}) - (19\,211)[7 - (4)(7)/(3\pi)]$, $x = 1.564$ ft. The forces acting on the gate are shown in Fig. 5-13c. $\sum M_B = 0$; $7F - (2.333)(12\,230) - (1.564)(5250) = 0$, $F = 5249$ lb (down).

Fig. 5-13(a)

Fig. 5-13(b)

Fig. 5-13(c)

**5.14** Repeat Prob. 5.13 if the gate is steel weighing 3000 lb.

❚ The weight of the gate acts at the center of gravity of the gate shown in Fig. 5-14. $2r/\pi = (2)(7)/\pi = 4.456$ ft; $\sum M_B = 0$. From Prob. 5.14, $7F - (2.333)(12\,230) - (1.564)(5250) + (3000)(7 - 4.456) = 0$, $F = 4159$ lb.

Fig. 5-14

**5.15** Compute the horizontal and vertical components of the hydrostatic force on the quarter-circle face of the tank shown in Fig. 5-15a.

❚
$$F_H = \gamma h_{cg} A = 9.79[4 + \tfrac{1}{2}][(1)(7)] = 308 \text{ kN}$$

$$F_V = F_1 - F_2 = (9.79)[(7)(1)(5)] - (9.79)[(7)(\pi)(1)^2/4] = 289 \text{ kN} \quad \text{(See Fig. 5-15b.)}$$

Fig. 5-15(a)

Fig. 5-15(b)

**5.16** Compute the horizontal and vertical components of the hydrostatic force on the hemispherical boulder shown in Fig. 5-16a.

▮ From symmetry, $F_H = 0$, $F_V = F_1 - F_2$ (see Fig. 5-16b). $F_V = 62.4[(\pi)(3)^2(12)] - (62.4)[(\frac{1}{2})(\frac{4}{3})(\pi)(3)^3] = 17\,643$ lb.

Fig. 5-16(a)

Fig. 5-16(b)

**5.17** The bottled cider (s.g. = 0.96) in Fig. 5-17 is under pressure, as shown by the manometer reading. Compute the net force on the 2-in-radius concavity in the bottom of the bottle.

▮ From symmetry, $F_H = 0$, $p_{AA} + [(0.96)(62.4)](\frac{3}{12}) - [(13.6)(62.4)](\frac{5}{12}) = p_{atm} = 0$, $p_{AA} = 339$ lb/ft$^2$ (gage);
$F_V = p_{AA}A_{bottom} +$ weight of liquid below $AA = 339[(\pi)(\frac{4}{12})^2/4] + [(0.96)(62.4)][(\frac{7}{12})(\pi)(\frac{4}{12})^2/4] - [(0.96)(62.4)][(\frac{1}{2})(\frac{4}{3})(\pi)(\frac{2}{12})^3] = 32.1$ lb.

**Fig. 5-17**

**5.18** Half-cylinder $ABC$ in Fig. 5-18$a$ is 9 ft wide into the paper. Calculate the net moment of the pressure forces on the body about point $C$.

▎ From symmetry, the horizontal forces balance and produce no net moment about point $C$. (See Fig. 5-18$b$.) $F_V = F_1 - F_2 = F_{\text{buoyancy of body }ABC} = [(0.85)(62.4)][(9)(\pi)(\frac{9}{2})^2/2] = 15\,184$ lb, $x = 4r/(3\pi) = (4)(\frac{9}{2})/(3\pi) = 1.910$ ft, $M_C = (15\,184)(1.910) = 29\,001$ lb · ft (clockwise).

**Fig. 5-18($a$)**

**Fig. 5-18($b$)**

**5.19** Compute the hydrostatic force and its line of action on semicylindrical indentation $ABC$ in Fig. 5-19$a$ per meter of width into the paper.

▎
$$F_H = \gamma h_{cg}A = [(0.88)(9.79)][(2 + 2 + \tfrac{2.5}{2})][(2.5)(1)] = 113.1 \text{ kN}$$

$$y_{cp} = \frac{-I_{xx}\sin\theta}{h_{cg}A} = \frac{-[(1)(2.5)^3/12](\sin 90°)}{(2 + 2 + \tfrac{2.5}{2})[(2.5)(1)]} = -0.099 \text{ m}$$

As demonstrated in Prob. 5.18, $F_V = F_{\text{buoyancy of body }ABC}$ and it acts at $4r/(3\pi)$ from point $C$. $F_V = [(0.88)(9.79)][(1)(\pi)(\frac{2.5}{2})^2/2] = 21.14$ kN, $x = 4r/(3\pi) = (4)(\frac{2.5}{2})/(3\pi) = 0.531$ m. The forces acting on the indentation are shown in Fig. 5-19$b$. $F_{\text{resultant}} = \sqrt{21.14^2 + 113.1^2} = 115.1$ kN. As shown in Fig. 5-19$b$, $F_{\text{resultant}}$ passes through point $O$ and acts up and to the right at an angle of arctan $(21.14/113.1)$, or $10.59°$.

**Fig. 5-19(a)**

**Fig. 5-19(b)**

**5.20**   Find the force on the conical plug in Fig. 5-20. Neglect the weight of the plug.

▮       $F_v = pA_{\text{hole}} + \text{weight of water above cone} = [(4.5)(144)][(\pi)(1)^2/4] + (62.4)[(4)(\pi)(1)^2/4]$

$$- (62.4)[(\tfrac{1}{3})(1.207)(\pi)(1)^2/4] = 685 \text{ lb}$$

**Fig. 5-20**

**5.21**   The hemispherical dome in Fig. 5-21 is filled with water and is attached to the floor by two diametrically opposed bolts. What force in either bolt is required to hold the dome down, if the dome weighs 25 kN?

▮       $F_v = \text{weight of (imaginary) water above the container}$

$$= 9.79[(5 + 1.5)(\pi)(1.5)^2] - 9.79[(5)(\pi)(0.04)^2/4] - 9.79[(\tfrac{1}{2})(\tfrac{4}{3})(\pi)(1.5)^3] = 380.5 \text{ kN (up)}$$

net upward force on dome $= 380.5 - 25 = 355.5 \text{ kN}$

force per bolt $= 355.5/2 = 177.7 \text{ kN}$

Fig. 5-21

**5.22** A 3-m-diameter water tank consists of two half-cylinders, each weighing 3.5 kN/m, bolted together as shown in Fig. 5-22a. If support of the end caps is neglected, determine the force induced in each bolt.

▌ See Fig. 5-22b. Assuming the bottom half is properly supported, only the top half affects the bolt force. $p_1 = (9.79)(1.5 + 1) = 24.48$ kN/m²; $\sum F_y = p_1 A_1 - 2F_{bolt} - W_{H_2O} - W_{tank\ half} = 0$, $24.48[(3)(\frac{25}{100})] - 2F_{bolt} - 9.79[(\frac{25}{100})(\pi)(1.5)^2/2] - 3.5/4 = 0$, $F_{bolt} = 4.42$ kN.

Fig. 5-22(a)                        Fig. 5-22(b)

**5.23** The cylinder in Fig. 5-23a extends 5 ft into the paper. Compute the horizontal and vertical components of the pressure force on the cylinder.

▌ See Fig. 5-23b. Note that the net horizontal force is based on the projected vertical area with depth $AB$. $F_H = \gamma h_{cg}A = 62.4[(4 + 2.828)/2][(4 + 2.828)(5)] = 7273$ lb; $F_V$ = equivalent weight of fluid in regions 1, 2, 3, and 4 = $(62.4)(5)[(\pi)(4)^2/2 + (2.828)(4) + (2.828)(2.828)/2 + (\pi)(4)^2/8] = 14\,579$ lb.

Fig. 5-23(a)                        Fig. 5-23(b)

**5.24** A 3-ft-diameter log (s.g. = 0.82) divides two shallow ponds as shown in Fig. 5-24a. Compute the net vertical and horizontal reactions at point $C$, if the log is 12 ft long.

▮ $F = \gamma hA$. Figure 5-24b shows the forces acting on the log.

$$(F_H)_1 = 62.4[(0+3)/2][(1.5+1.5)(12)] = 3370 \text{ lb} \qquad (F_H)_2 = 62.4[(0+1.5)/2][(1.5)(12)] = 842 \text{ lb}$$

$$(F_V)_1 = 62.4[(12)(\pi)(1.5)^2/2] = 2646 \text{ lb} \qquad (F_V)_2 = 62.4[(12)(\pi)(1.5)^2/4] = 1323 \text{ lb}$$

$$\sum F_x = 0 \qquad 3370 - 842 - C_x = 0 \qquad C_x = 2528 \text{ lb (left)}$$

$$\sum F_y = 0 \qquad 2646 + 1323 - [(0.82)(62.4)][(12)(\pi)(1.5)^2] + C_y = 0 \qquad C_y = 371 \text{ lb (up)}$$

Fig. 5-24(a)           Fig. 5-24(b)

**5.25** The 1-m-diameter cylinder in Fig. 5-25a is 8 m long into the paper and rests in static equilibrium against a frictionless wall at point $B$. Compute the specific gravity of the cylinder.

▮ See Fig. 5-25b. The wall reaction at $B$ is purely horizontal. Then the log weight must exactly balance the vertical hydrostatic force, which equals the equivalent weight of water in the shaded area. $W_{\log} = F_V = (9.79)(8)[(\frac{3}{4})(\pi)(0.5)^2 + (0.5)(0.5)] = 65.71 \text{ kN}$, $\gamma_{\log} = 65.71/[(8)(\pi)(0.5)^2] = 10.46 \text{ kN/m}^3$, s.g. $= 10.46/9.79 = 1.07$.

Fig. 5-25(a)      r = 0.5 m      Fig. 5-25(b)

**5.26** The tank in Fig. 5-26a is 3 m wide into the paper. Neglecting atmospheric pressure, compute the hydrostatic horizontal, vertical, and resultant force on quarter-circle panel $BC$.

▮ $F_H = \gamma h_{cg} A = (9.79)(4 + \frac{5}{2})[(5)(3)] = 954.5 \text{ kN}$, $F_V =$ weight of water above panel $BC = (9.79)[(3)(5)(4)] + (9.79)[(3)(\pi)(5)^2/4] = 1164 \text{ kN}$, $F_{\text{resultant}} = \sqrt{954.5^2 + 1164^2} = 1505 \text{ kN}$. As seen in Fig. 5-26b, $F_{\text{resultant}}$ passes through point $O$ and acts down and to the right at an angle of arctan (1164/954.5), or 50.6°.

Fig. 5-26(a)           Fig. 5-26(b)

**5.27** Gate $AB$ in Fig. 5-27a is a quarter circle 7 ft wide, hinged at $B$ and resting against a smooth wall at $A$. Compute the reaction forces at $A$ and $B$.

▌ $\quad F_H = \gamma h_{cg} A = (64)(11 - \frac{6}{2})[(7)(6)] = 21\,504 \text{ lb} \qquad y_{cp} = \frac{-I_{xx} \sin \theta}{h_{cg} A} = \frac{-[(7)(6)^3/12](\sin 90°)}{(11 - \frac{6}{2})[(7)(6)]} = -0.375 \text{ ft}$

Thus, $F_H$ acts at $\frac{6}{2} - 0.375$, or 2.625 ft above point $B$. $F_V$ = weight of seawater above gate $AB$ = $(64)(7)[(11)(6)] - (64)(7)[(\pi)(6)^2/4] = 29\,568 - 12\,667 = 16\,901 \text{ lb}$. The location of $F_V$ can be determined by taking moments about point $A$ in Fig. 5-27b. $(29\,568)(\frac{6}{2}) - (12\,667)[(4)(6)/(3\pi)] = 16\,901x$, $x = 3.340$ ft. The forces acting on the gate are shown in Fig. 5-27c.

$$\sum M_B = 0 \qquad (21\,504)(2.625) + (16\,901)(6 - 3.340) - 6A_x = 0 \qquad A_x = 16\,901 \text{ lb}$$

$$\sum F_x = 0 \qquad 21\,504 - B_x - 16\,901 = 0 \qquad B_x = 4603 \text{ lb}$$

$$\sum F_y = 0 \qquad B_y - 16\,901 = 0 \qquad B_y = 16\,901 \text{ lb}$$

**Fig. 5-27(a)**

**Fig. 5-27(b)**

**Fig. 5-27(c)**

**5.28** Curved wall $ABC$ in Fig. 5-28$a$ is a quarter circle 9 ft wide into the paper. Compute the horizontal and vertical hydrostatic forces on the wall and the line of action of the resultant force.

▐ See Fig. 5-28$b$. $F_H = \gamma h_{cg}A = (62.4)(3.536)[(7.072)(9)] = 14\,044$ lb, $F_v =$ weight of (imaginary) water in crosshatched area in Fig. 5-28$b$ = $(62.4)(9)[(\pi)(5)^2/4 - (2)(5\sin 45°)(5\cos 45°)/2] = 4007$ lb; $F_{resultant} = \sqrt{4007^2 + 14\,044^2} = 14\,604$ lb. $F_{resultant}$ passes through point $O$ and acts at an angle of arctan $\frac{4007}{14\,044}$, or 15.9°, as shown in Fig. 5-28$c$.

**Fig. 5-28(b)**

**Fig. 5-28(c)**

**5.29** Pressurized water fills the tank in Fig. 5-29$a$. Compute the net hydrostatic force on conical surface $ABC$.

▐ From symmetry, $F_H = 0$. The gage pressure of 100 kPa corresponds to a fictitious water level at 100/9.79, or 10.215 m above the gage or 10.215 − 7, or 3.215 m above $AC$ (see Fig. 5-29$b$). $F_v =$ weight of fictitious water above cone $ABC$ = $9.79[(3.215)(\pi)(3)^2/4 + (\frac{1}{3})(6)(\pi)(3)^2/4] = 361$ kN (up).

**Fig. 5-29(a)**

**Fig. 5-29(b)**

**5.30** Gate $AB$ in Fig. 5-30$a$ is 7 m wide into the paper. Compute the force $F$ required to prevent rotation about the hinge at $B$. Neglect atmospheric pressure.

▌ $F_H = \gamma \bar{h} A = 9.79[(9.6 + 0)/2][(9.6)(7)] = 3158$ kN. $F_H$ acts at $\frac{9.6}{3}$, or 3.200 m above $B$ (see Fig. 5-30$b$). $F_V$ = weight of water above the gate = $9.79[(\frac{2}{3})(6)(9.6)(7)] = 2632$ kN. $F_V$ acts at $\frac{18}{8}$, or 2.250 m right of $B$ (see Fig. 5-30$b$). $\sum M_B = 0$; $(3.200)(3158) + (2.250)(2632) - 9.6F = 0$, $F = 1670$ kN.

**Fig. 5-30($a$)**　　　　　　　　　**Fig. 5-30($b$)**

**5.31** The cylindrical tank in Fig. 5-31 has a hemispherical end cap $ABC$. Compute the total horizontal and vertical forces exerted on $ABC$ by the oil and water.

▌
$$F = \gamma h_{cg} A \qquad (F_H)_1 = [(0.9)(9.79)][(3 + \tfrac{2}{2})][(\pi)(2)^2/2] = 221 \text{ kN} \quad \text{(left)}$$

$$(F_H)_2 = \{[(0.9)(9.79)](3 + 2) + (9.79)(\tfrac{2}{2})\}[(\pi)(2)^2]/2 = 338 \text{ kN} \quad \text{(left)}$$

$$(F_H)_{\text{total}} = 221 + 338 = 559 \text{ kN} \quad \text{(left)}$$

$$F_V = \text{weight of fluid within hemisphere} = [(0.9)(9.79)][(\tfrac{1}{4})(\tfrac{4}{3})(\pi)(2)^3] + (9.79)[(\tfrac{1}{4})(\tfrac{4}{3})(\pi)(2)^3] = 156 \text{ kN} \quad \text{(down)}$$

**Fig. 5-31**

**5.32** A cylindrical barrier holds water, as shown in Fig. 5-32. The contact between cylinder and wall is smooth. Consider a 1-m length of cylinder and determine its weight and the force exerted against the wall.

▌
$$(F_V)_{BCD} = (9.79)(1)[(\pi)(2)^2/2 + (2)(2) + (2)(2)] = 139.8 \text{ kN} \quad \text{(up)}$$

$$(F_V)_{AB} = (9.79)(1)[(2)(2) - (\pi)(2)^2/4] = 8.4 \text{ kN} \quad \text{(down)}$$

$$\sum F_y = 0 \qquad 139.8 - W_{\text{cylinder}} - 8.4 = 0 \qquad W_{\text{cylinder}} = 131.4 \text{ kN}$$

$$F_H = \gamma h_{cg} A \qquad (F_H)_{ABC} = (9.79)(2)[(2 + 2)(1)] = 78.3 \text{ kN} \quad \text{(right)}$$

$$(F_H)_{DC} = (9.79)(2 + \tfrac{2}{2})[(2)(1)] = 58.7 \text{ kN} \quad \text{(left)} \qquad (F_H)_{\text{against wall}} = 78.3 - 58.7 = 19.6 \text{ kN} \quad \text{(right)}$$

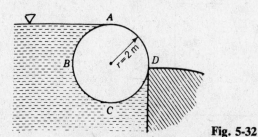

**Fig. 5-32**

**5.33** The revolving gate in Fig. 5-33 is a quarter-cylinder with pivot through $O$. What force $F$ is required to open it? (Treat the gate as weightless.)

**I** At each point of $\overarc{ABC}$ the line of action of the pressure force passes through $O$; hence the pressure has no moment about $O$. It follows that any $F$, no matter how small, suffices to produce a net opening moment.

**Fig. 5-33**

**5.34** Find the vertical component of force on the parabolic gate of Fig. 5-34a and its line of action.

**I**
$$F_V = \text{weight of imaginary liquid above gate} = \gamma L \int (H - y)\, dx \qquad \text{(see Fig. 5-34b)}$$

$$= (9.00)(3) \int_0^{0.8} (2 - \sqrt{5x})\, dx = (9.00)(3)\left[2x - \frac{\sqrt{5}x^{3/2}}{\frac{3}{2}}\right]_0^{0.8} = 14.40 \text{ kN}$$

$$x_{cp} = \frac{\gamma L \int (H - y)x\, dx}{F_V} \qquad \text{(see Fig. 5-34b)}$$

$$= \frac{(9.00)(3) \int_0^{0.8} (2 - \sqrt{5x})x\, dx}{14.40} = \frac{(9.00)(3) \int_0^{0.8} (2x - \sqrt{5}x^{3/2})\, dx}{14.40}$$

$$= (9.00)(3)[x^2 - (\sqrt{5}x^{5/2})/\tfrac{5}{2}]_0^{0.8}/14.40 = 0.240 \text{ m}$$

**Fig. 5-34(a)**

**Fig. 5-34(b)**

**5.35** Determine the moment $M$ needed to hold the gate of Fig. 5-34a shut. Neglect its weight.

**I** $F_H = \gamma \bar{h} A = 9.00[(0 + 2)/2][(2)(3)] = 54.0$ kN (left). $F_H$ acts at $\frac{2}{3}$, or 0.667 m above point 0. $F_V = 14.40$ kN (up) and $x_{cp} = 0.240$ m (from Prob. 5.34 and Fig. 5-34b). $\sum M_0 = 0$; $M - (14.40)(0.240) - (54.0)(0.667) = 0$, $M = 39.5$ kN · m.

**5.36** Find the force on the body (part of a parabolic cylinder) of Fig. 5-35. The length normal to the paper is $L = 4.5$ m, and $\gamma$ is 9.20 kN/m³.

$$F_H = \gamma \bar{h} A = (9.20)(\tfrac{1}{2})[(1)(4.5)] = 20.70 \text{ kN}$$

$$F_V = \text{weight of liquid above } OA = \int \gamma L y \, dx = \int_0^{\sqrt{8}} (9.20)(4.5)\left(\frac{x^2}{8}\right) dx = (9.20)(4.5)\left[\frac{x^3}{24}\right]_0^{\sqrt{8}} = 39.03 \text{ kN}$$

$$F_{\text{resultant}} = \sqrt{39.03^2 + 20.70^2} = 44.18 \text{ kN}$$

Fig. 5-35

**5.37** The curved plate in Fig. 5-36 is an octant of a sphere. Find the resultant force, including its line of action, acting on the outer surface, if the radius of the sphere is 600 mm and its center is 2 m below the water surface.

See Fig. 5-36. $F_H = \gamma \bar{h} A = \gamma[H - 4r/(3\pi)](\pi r^2/4)$, $F_x = F_z = F_H = 9.79[2 - (4)(0.6)/(3\pi)][(\pi)(0.6)^2/4] = 4.831$ kN (both $F_x$ and $F_z$ act toward 0); $F_y = F_V = $ weight of water above curved surface $= \gamma[(H)(\pi)(r)^2/4 - (\tfrac{4}{3})(\pi)(r)^3/8] = 9.79[(2)(\pi)(0.6)^2/4 - (\tfrac{4}{3})(\pi)(0.6)^3/8] = 4.429$ kN. $F_{\text{resultant}}$ acts on a line through 0 making a 45° angle with the $x$ and $z$ axes because of symmetry; $F_{\text{resultant}} = \sqrt{4.429^2 + 4.831^2 + 4.831^2} = 8.142$ kN. It acts at an angle $\theta = \arccos(4.429/8.142) = 57.0°$.

Fig. 5-36

**5.38** Find the horizontal and vertical components of the force per unit width exerted by fluids on the horizontal cylinder in Fig. 5-37a if the fluid to the left of the cylinder is (a) a gas confined in a closed tank at a pressure of 35.0 kN/m² and (b) water with a free surface at an elevation coincident with the uppermost part of the cylinder. Assume in both instances that atmospheric pressure occurs to the right of the cylinder.

(a) The "net vertical projection" (see Fig. 5-37a) of the portion of the cylinder surface under consideration is $4 - (2 - 2\cos 30°)$, or 3.732 m. $F_H = pA = 35.0[(1)(3.732)] = 130.6$ kN (right). Note that the vertical force of the gas on surface $ab$ is equal and opposite to that on surface $bc$. Hence, the "net horizontal projection" with regard to the gas is $ae$ (see Fig. 5-38b), which is $2\sin 30°$, or 1.000 m. $F_V = 35.0[(1)(1.000)] = 35.0$ kN (up).

(b) $$F_H = \gamma \bar{h} A = (9.79)(3.732/2)[(1)(3.732)] = 68.2 \text{ kN} \quad \text{(right)}$$

$$F_V = \text{weight of crosshatched volume of water} \quad \text{(Fig. 5-37b)}$$

$$= (9.79)(1)[(\tfrac{210}{360})(\pi)(4)^2/4 + (\tfrac{1}{2})(1.000)(3.732 - \tfrac{4}{2}) + (1)(\tfrac{4}{2})] = 99.8 \text{ kN} \quad \text{(up)}$$

**Fig. 5-37(a)**

**Fig. 5-37(b)**

**5.39** A vertical-thrust bearing consists of an 8-in-radius bronze hemisphere mating into a steel hemispherical shell. At what pressure must grease be supplied to the bearing so that an unbroken film is present when the vertical thrust on the bearing is 600 000 lb?

❚ $\qquad$ Projected area $= \pi r^2 = (\pi)(8)^2 = 201.1 \text{ in}^2 \qquad p = F/A = 600\,000/201.1 = 2984 \text{ lb/in}^2$

**5.40** Find horizontal and vertical forces per foot of width on the Tainter gate shown in Fig. 5-38.

❚ $F_H = \gamma \bar{h} A = (62.4)[(0 + 25)/2][(25)(1)] = 19\,500$ lb. $F_H$ acts at $(\frac{2}{3})(25)$, or 16.67 ft below the water surface. $F_V = $ weight of imaginary water in $ACBA = (62.4)(1)[(\pi)(25)^2/5 - (2)(25\cos 36°)(25\sin 36°)/2] = 5959$ lb. $F_V$ acts through the centroid of segment $ABCA$.

**Fig. 5-38**

**5.41** The tank indicated in cross section in Fig. 5-39 is 6 m long normal to the paper. Curved panel $MN$ is one-quarter of an ellipse with semiaxes $b$ and $d$. If $b = 5$ m, $d = 7$ m, and $a = 1.0$ m, calculate the horizontal and vertical components of force and the resultant force on the panel.

❚ $$F_H = \gamma h_{cg} A = 9.79(1.0 + \tfrac{7}{2})[(6)(7)] = 1850 \text{ kN}$$

$$h_{cp} = h_{cg} + \frac{I_{cg}}{h_{cg}A} = (1.0 + \tfrac{7}{2}) + \frac{(6)(7)^3/12}{(1.0 + \tfrac{7}{2})[(6)(7)]} = 5.407 \text{ m below water surface}$$

$$F_V = \text{weight of water above surface } MN = (9.79)(6)[(\pi)(5)(7)/4 + (1.0)(5)] = 1908 \text{ kN}$$

$$x_{cp} = 4b/(3\pi) = (4)(5)/(3\pi) = 2.122 \text{ m to the right of } N \qquad F_{\text{resultant}} = \sqrt{1908^2 + 1850^2} = 2658 \text{ kN}$$

$F_{\text{resultant}}$ acts through the intersection of $F_H$ and $F_V$ at an angle of arctan (1908/1850), or 45.9°.

$$\text{Area}_{\text{one quarter of elipse}} = \frac{\pi\, b\, d}{4}$$

$$\text{Area}_{\text{parabola}} = \frac{2\, b\, d}{3}$$

**Fig. 5-39**

**5.42** Solve Prob. 5.41 if $a = 1.0$ ft, $b = 5$ ft, $d = 7$ ft, the tank is 6 ft long, and $MN$ represents a parabola with vertex at $N$.

❚
$$F_H = \gamma h_{cg} A = (62.4)(1.0 + \tfrac{7}{2})[(6)(7)] = 11\ 794\ \text{lb}$$

$$h_{cp} = h_{cg} + \frac{I_{cg}}{h_{cg}A} = (1.0 + \tfrac{7}{2}) + \frac{(6)(7)^3/12}{(1.0 + \tfrac{7}{2})[(6)(7)]} = 5.407\ \text{ft below water surface}$$

$$F_V = \text{weight of water above surface } MN = (62.4)(6)[(\tfrac{2}{3})(7)(5) + (1.0)(5)] = 10\ 608\ \text{lb}$$

$$x_{cp} = (\tfrac{3}{8})(b) = (\tfrac{3}{8})(5) = 1.88\ \text{ft to the right of } N \qquad F_{\text{resultant}} = \sqrt{10\ 608^2 + 11\ 794^2} = 15\ 863\ \text{lb}$$

$F_{\text{resultant}}$ acts through the intersection of $F_H$ and $F_V$ at an angle of arctan $(10\ 608/11\ 794)$, or $42.0°$.

**5.43** In the cross section shown in Fig. 5-40, $BC$ is a quarter-circle. If the tank contains water to a depth of 6 ft, determine the magnitude and location of the horizontal and vertical components on wall $ABC$ per 1 ft width.

❚
$$F_H = \gamma \bar{h} A = (62.4)[(0 + 6)/2][(1)(6)] = 1123\ \text{lb} \qquad h_{cp} = (\tfrac{2}{3})(6) = 4.00\ \text{ft}$$

$$F_V = \text{weight of water above surface } BC = (62.4)(1)[(6)(5)] - (62.4)(1)[(\pi)(5)^2/4] = 1872 - 1225 = 647\ \text{lb}$$

The location of $F_V$ can be determined by taking moments about point $B$. $(1872)(\tfrac{5}{2}) - (1225)[(4)(5)/(3\pi)] = 647x_{cp}$, $x_{cp} = 3.22$ ft.

Fig. 5-40

**5.44** Rework Prob. 5.43 where the tank is closed and contains gas at a pressure of 10 psi.

❚
$$F_H = pA_V = [(10)(144)][(1)(6)] = 8640\ \text{lb} \qquad h_{cp} = \tfrac{6}{2} = 3.00\ \text{ft}$$

$$F_V = pA_H = [(10)(144)][(1)(5)] = 7200\ \text{lb} \qquad x_{cp} = \tfrac{5}{2} = 2.50\ \text{ft}$$

**5.45** A spherical steel tank of 22 m diameter contains gas under a pressure of 300 kPa. The tank consists of two half-spheres joined together with a weld. What will be the tensile force across the weld? If the steel is 25.0 mm thick, what is the tensile stress in the steel?

❚
$$F = pA = 300[(\pi)(22)^2/4] = 114\ 040\ \text{kN} \qquad \sigma = \frac{\text{force/length}}{\text{thickness}} = \frac{114\ 040/(22\pi)}{25.0/1000} = 66\ 000\ \text{kPa}$$

**5.46** Determine the force required to hold the cone shown in Fig. 5-41a in position.

❚ Figure 5-41b shows the vertical projection above the opening. $p_{\text{air}} = 0.6 - [(0.83)(62.4)](5.1)/144 = -1.23$ psi, $F_{\text{air}} = [(1.23)(144)][(\pi)(0.804)^2] = 360$ lb, $F_{\text{cylinder}} = (62.4)(0.83)[(\pi)(0.804)^2(5.1 + 3)] = 852$ lb, $F_{\text{cone}} = (62.4)(0.83)[(3)(\pi)(0.804)^2/3] = 105$ lb; $\sum F_y = 0$, $360 - 852 + 105 + F = 0$, $F = 387$ lb.

Fig. 5-41(a)

Fig. 5-41(b)

**5.47** The cross section of the gate in Fig. 5-42 is given by $10x = 3y^2$; its dimension normal to the plane of the paper is 7 m. The gate is pivoted about $O$. Find the horizontal and vertical forces and the clockwise moment acting on the gate if the water depth is 1.8 m.

$$F_H = \gamma \bar{h} A = 9.79[(0 + 1.8)/2][(7)(1.8)] = 111.0 \text{ kN}$$

$$F_V = \text{weight of water above the gate} = \int_0^{1.8} (9.79)(7)(x\,dy) = (9.79)(7)\int_0^{1.8} 0.3y^2\,dy = (9.79)(7)\left[\frac{0.3y^3}{3}\right]_0^{1.8} = 40.0 \text{ kN}$$

$$M_O = (111.0)(\tfrac{1.8}{3}) + \int_0^{1.8} (9.79)(7)\left(\frac{x}{2}\right)(x\,dy) = 66.6 + (9.79)(7)\int_0^{1.8} \frac{(0.3y^2)^2}{2}\,dy$$

$$= 66.6 + (9.79)(7)\left[\frac{0.09y^5}{10}\right]_0^{1.8} = 78.3 \text{ kN} \cdot \text{m}$$

Water

**Fig. 5-42**

**5.48** Find the wall thickness of steel pipe needed to resist the static pressure in a 36-in-diameter steel pipe carrying water under a head of 750 ft of water. Use an allowable working stress for steel pipe of 16 000 psi.

$$p = \gamma h = (62.4)(750) = 46\,800 \text{ lb/ft}^2 \quad \text{or} \quad 325 \text{ lb/in}^2$$

$$T = pd/2 = (325)(36)/2 = 5850 \text{ lb/in of pipe length} \qquad t = 5850/16\,000 = 0.366 \text{ in}$$

**5.49** A vertical cylindrical tank is 6 ft in diameter and 10 ft high. Its sides are held in position by means of two steel hoops, one at the top and one at the bottom. The tank is filled with water up to 9 ft high. Determine the tensile stress in each hoop.

See Fig. 5-43. $F = \gamma \bar{h} A = 62.4[(0 + 9)/2][(9)(6)] = 15\,163 \text{ lb}$, $T = F/2 = 15\,163/2 = 7582 \text{ lb}$; stress in top hoop $= (7582)(\tfrac{3}{10}) = 2275 \text{ lb}$, stress in bottom hoop $= (7582)[(10 - 3)/10] = 5307 \text{ lb}$.

**Fig. 5-43(a)**   **Fig. 5-43(b)**

**5.50** A 48-in-diameter steel pipe, $\frac{1}{4}$ in thick, carries oil of s.g. = 0.822 under a head of 400 ft of oil. Compute the (*a*) stress in the steel and (*b*) thickness of steel required to carry a pressure of 250 psi with an allowable stress of 18 000 psi.

▌ $p = \gamma h = [(0.822)(62.4)](400) = 20\,517\ \text{lb/ft}^2$ or $142.5\ \text{lb/in}^2$ $\qquad \sigma = \dfrac{pr}{t}$

(*a*) $\qquad\qquad\qquad\qquad \sigma = \dfrac{(142.5)(48/2)}{\frac{1}{4}} = 13\,680\ \text{psi}$

(*b*) $\qquad\qquad\qquad\qquad 18\,000 = \dfrac{(250)(48/2)}{t} \qquad t = 0.333\ \text{in}$

**5.51** A wooden storage vat, 20 ft in outside diameter, is filled with 24 ft of brine, s.g. = 1.06. The wood staves are bound by flat steel bands, 2 in wide by $\frac{1}{4}$ in thick, whose allowable stress is 16 000 psi. What is the spacing of the bands near the bottom of the vat, neglecting any initial stress? Refer to Fig. 5-44.

▌ Force $P$ represents the sum of all the horizontal components of small forces $dP$ acting on length $y$ of the vat, and forces $T$ represent the total tension carried in a band loaded by the same length $y$.

$$\sum F_x = 0 \qquad 2T - P = 0 \qquad T = A_{\text{steel}}\sigma_{\text{steel}} = [(2)(\tfrac{1}{4})](16\,000) = 8000\ \text{lb}$$

$$p = \gamma hA = [(1.06)(62.4)](24)(20y) = 31\,749y \qquad (2)(8000) - 31\,749y = 0 \qquad y = 0.504\ \text{ft} \quad \text{or} \quad 6.05\ \text{in}$$

**Fig. 5-44**

**5.52** A 4.0-in-ID steel pipe has a $\frac{1}{4}$-in wall thickness. For an allowable tensile stress of 10 000 psi, what is the maximum pressure?

▌ $\qquad\qquad \sigma = \dfrac{pr}{t} \qquad 10\,000 = \dfrac{(p)(4.0/2)}{\frac{1}{4}} \qquad p = 1250\ \text{lb/in}^2$

**5.53** A thin-walled hollow sphere 3.5 m in diameter holds gas at 1700 kPa. For an allowable stress of 50 000 kPa, determine the minimum wall thickness.

▌ Considering half a sphere of diameter $d$ (3.5 m) and thickness $t$, $(\pi dt)(\sigma) = (p)(\pi d^2/4)$, $[(\pi)(3.5)(t)](50\,000) = 1700[(\pi)(3.5)^2/4]$, $t = 0.02975$ m, or 29.75 mm.

**5.54** A cylindrical container 8 ft high and 3 ft in diameter is reinforced with two hoops a foot from each end. When it is filled with water, what is the tension in each hoop due to the water?

▌ See Fig. 5-45. $F = \gamma \bar{h}A = 62.4[(0 + 8)/2][(8)(3)] = 5990$ lb. $F$ acts at $(\frac{2}{3})(8)$, or 5.333 ft from the top of the container.

$$\sum F_x = 0$$

$$2T_1 + 2T_2 - 5990 = 0 \qquad\qquad\qquad (1)$$

$$\sum M_A = 0 \qquad (2T_2)(1.667) - (2T_1)(4.333) = 0$$

$$T_2 = 2.60T_1 \qquad\qquad\qquad (2)$$

Solve simultaneous equations (1) and (2). $2T_1 + (2)(2.60T_1) - 5990 = 0$, $T_1 = 832$ lb, $T_2 = (2.60)(832) = 2163$ lb.

**Fig. 5-45**

5.55  A 30-mm-diameter steel ($\gamma = 77.0$ kN/m$^3$) ball covers a 15-mm-diameter hole in the bottom of a chamber in which gas pressure is 27 000 kPa. What force is required to push the ball up into the chamber?

$$F = pA + \text{weight of ball} = 27\,000[(\pi)(\tfrac{15}{1000})^2/4] + [(\tfrac{4}{3})(\pi)(\tfrac{30}{1000})^3](77.0) = 4.780 \text{ kN}$$

# CHAPTER 6
# Buoyancy and Flotation

**6.1** A stone weighs 105 lb in air. When submerged in water, it weighs 67.0 lb. Find the volume and specific gravity of the stone.

   Buoyant force $(F_b)$ = weight of water displaced by stone $(W)$ = $105 - 67.0 = 38.0$ lb

$$W = \gamma V = 62.4V \qquad 38.0 = 62.4V \qquad V = 0.609 \text{ ft}^3$$

$$\text{s.g.} = \frac{\text{weight of stone in air}}{\text{weight of equal volume of water}} = \frac{105}{(0.609)(62.4)} = 2.76$$

**6.2** A piece of irregularly shaped metal weighs 300.0 N in air. When the metal is completely submerged in water, it weighs 232.5 N. Find the volume of the metal.

$$F_b = W \qquad 300.0 - 232.5 = [(9.79)(1000)](V) \qquad V = 0.00689 \text{ m}^3$$

**6.3** A cube of timber 1.25 ft on each side floats in water as shown in Fig. 6-1. The specific gravity of the timber is 0.60. Find the submerged depth of the cube.

$$F_b = W \qquad 62.4[(1.25)(1.25)(D)] = [(0.60)(62.4)][(1.25)(1.25)(1.25)] \qquad D = 0.750 \text{ ft}$$

**Fig. 6-1**

**6.4** Determine the magnitude and direction of the force necessary to hold a concrete cube, 0.300 m on each side, in equilibrium and completely submerged (*a*) in mercury (Hg) and (*b*) in water. Use s.g.$_\text{concrete}$ = 2.40.

   (*a*) Since s.g.$_\text{Hg}$ = 13.6 and s.g.$_\text{concrete}$ = 2.40, it is evident that the concrete will float in mercury. Therefore, a force $F$ acting downward will be required to hold the concrete in equilibrium and completely submerged in mercury. The forces acting on the concrete are shown in Fig. 6-2*a*, where $F$ is the force required to hold the concrete cube in equilibrium and completely submerged, $W$ is the weight of the concrete cube in air, and $F_b$ is the buoyant force. $\sum F_y = 0$, $F + W - F_b = 0$, $F + [(2.40)(9.79)][(0.300)(0.300)(0.300)] - [(13.6)(9.79)][(0.300)(0.300)(0.300)] = 0$, $F = 2.96$ kN (downward). (*b*) Since s.g.$_\text{concrete}$ = 2.40, it will sink in water. Therefore, a force $F$ acting upward will be required to hold the concrete in equilibrium and completely submerged in water. The forces acting on the concrete in this case are shown in Fig. 6-2*b*. $\sum F_y = 0$, $W - F - F_b = 0$, $[(2.40)(9.79)][(0.300)(0.300)(0.300)] - F - 9.79[(0.300)(0.300)(0.300)] = 0$, $F = 0.370$ kN (upward).

**6.5** A concrete cube 10.0 in on each side is to be held in equilibrium under water by attaching a lightweight foam buoy to it, as shown in Fig. 6-3. (In theory, the attached foam buoy and concrete cube, when placed under water, will neither rise nor sink.) If the specific weight of concrete and foam are 150 lb/ft$^3$ and 5.0 lb/ft$^3$, respectively, what minimum volume of foam is required?

108

Fig. 6-2(a)          Fig. 6-2(b)

**❚** The forces acting in this problem are shown in Fig. 6-3, where $W_f$ and $W_c$ are the respective weights of the foam and the concrete, and $F_{bf}$ and $F_{bc}$ are the respective buoyant forces on the foam and the concrete. $\Sigma F_y = 0$, $W_f - F_{bf} + W_c - F_{bc} = 0$, $5.0V_{\text{foam}} - 62.4V_{\text{foam}} + 150[(\frac{10}{12})(\frac{10}{12})(\frac{10}{12})] - 62.4[(\frac{10}{12})(\frac{10}{12})(\frac{10}{12})] = 0$, $V_{\text{foam}} = 0.883 \text{ ft}^3$.

Fig. 6-3

**6.6** A barge is loaded with 150 tons of coal. The weight of the empty barge in air is 35 tons. If the barge is 18 ft wide, 52 ft long, and 9 ft high, what is its draft (i.e., its depth below the water surface)?

**❚** $F_b = W$     $62.4[(18)(52)(D)] = (150 + 35)(2000)$     $D = 6.33$ ft

**6.7** Determine the submerged depth of a cube of steel 0.30 m on each side floating in mercury. The specific gravities of steel and mercury are 7.8 and 13.6, respectively.

**❚** $F_b = W$     $[(13.6)(9.79)][(0.3)(0.3)(D)] = [(7.8)(9.79)][(0.3)(0.3)(0.3)]$     $D = 0.172$ m

**6.8** A cube of wood (s.g. = 0.60) has 9-in sides. Compute the magnitude and direction of the force $F$ required to hold the wood completely submerged in water.

**❚** Since s.g.$_{\text{wood}} = 0.60$, it is evident that the wood will float in water. Therefore, a force $F$ acting downward will be required to hold the wood in equilibrium and completely submerged. The forces acting on the wood are essentially the same as those shown acting on the concrete cube in Fig. 6-2a: $\Sigma F_y = 0$, $F + W - F_b = 0$, $F + [(0.60)(62.4)][(\frac{9}{12})(\frac{9}{12})(\frac{9}{12})] - 62.4[(\frac{9}{12})(\frac{9}{12})(\frac{9}{12})] = 0$, $F = 10.5$ lb (downward).

**6.9** A hollow cube 1.0 m on each side weighs 2.4 kN. The cube is tied to a solid concrete block weighing 10.0 kN. Will these two objects tied together float or sink in water? The specific gravity of the concrete is 2.40.

**∎** Let $W$ = weight of hollow cube plus solid concrete block, $(F_b)_1$ = buoyant force on hollow cube, and $(F_b)_2$ = buoyant force on solid concrete block. $W = 2.4 + 10.0 = 12.4$ kN, $(F_b)_1 = 9.79[(1)(1)(1)] = 9.79$ kN, $V_{block} = 10/[(2.40)(9.79)] = 0.4256$ m³, $(F_b)_2 = (9.79)(0.4256) = 4.17$ kN, $(F_b)_1 + (F_b)_2 = 9.79 + 4.17 = 13.96$ kN. Since $[W = 12.4] < [(F_b)_1 + (F_b)_2 = 13.96$ kN$]$, the two objects tied together will float in water.

**6.10** A concrete cube 0.5 m on each side is to be held in equilibrium under water by attaching a light foam buoy to it. What minimum volume of foam is required? The specific weights of concrete and foam are 23.58 kN/m³ and 0.79 kN/m³, respectively.

**∎** Let $W_f$ = weight of foam in air, $(F_b)_f$ = buoyant force on foam, $W_c$ = weight of concrete in air, and $(F_b)_c$ = buoyant force on concrete. $\sum F_y = 0$, $W_f - (F_b)_f + W_c - (F_b)_c = 0$, $0.79V_{foam} - 9.79V_{foam} + 23.58[(0.5)(0.5)(0.5)] - 9.79[(0.5)(0.5)(0.5)] = 0$, $V_{foam} = 0.192$ m³.

**6.11** A prismatic object 8 in thick by 8 in wide by 16 in long is weighed in water at a depth of 20 in and found to weigh 11.0 lb. What is its weight in air and its specific gravity?

**∎** The forces acting on the object are shown in Fig. 6-4. $\sum F_y = 0$, $T + F_b - W = 0$, $F_b$ = weight of displaced water $= 62.4[(8)(8)(16)/1728] = 37.0$ lb, $11.0 + 37.0 - W = 0$, $W = 48.0$ lb, s.g. $= 48.0/37.0 = 1.30$.

**Fig. 6-4**

**6.12** A hydrometer weighs 0.00485 lb and has a stem at the upper end which is cylindrical and 0.1100 in in diameter. How much deeper will it float in oil of s.g. 0.780 than in alcohol of s.g. 0.821?

**∎** $W_{hydrometer} = W_{displaced\ liquid}$. For position 1 in Fig. 6-5 in the alcohol, $0.00485 = [(0.821)(62.4)](V_1)$, $V_1 = 0.0000947$ ft³ (in alcohol). For position 2 in Fig. 6-5 in the oil, $0.00485 = [(0.780)(62.4)][0.0000947 + (h)(\pi)(0.1100/12)^2/4]$, $h = 0.0750$ ft, or 0.900 in.

s.g. = 0.821    s.g. = 0.780    **Fig. 6-5**

**6.13** A piece of wood of s.g. 0.651 is 3 in square and 5 ft long. How many pounds of lead weighing 700 lb/ft³ must be fastened at one end of the stick so that it will float upright with 1 ft out of water?

**∎**
$$W_{wood\ and\ lead} = W_{displaced\ water}$$
$$[(0.651)(62.4)][(5)(\tfrac{3}{12})(\tfrac{3}{12})] + 700V = 62.4[(5-1)(\tfrac{3}{12})(\tfrac{3}{12}) + V]$$
$$V = 0.00456\ \text{ft}^3 \qquad W_{lead} = (0.00456)(700) = 3.19\ \text{lb}$$

**6.14** What fraction of the volume of a solid piece of metal of s.g. 7.25 floats above the surface of a container of mercury?

**∎** Let $V$ = volume of the metal and $V'$ = volume of mercury displaced. $F_b = W$, $[(13.6)(62.4)](V') = [(7.25)(62.4)](V)$, $V'/V = 0.533$. Fraction of volume above mercury $= 1 - 0.533 = 0.467$.

**6.15** A rectangular open box 25 ft by 10 ft in plan and 12 ft deep weighs 40 tons and is launched in fresh water. (*a*) How deep will it sink? (*b*) If the water is 12 ft deep, what weight of stone placed in the box will cause it to rest on the bottom?

**I** $$F_b = W$$

(*a*) $$62.4[(25)(10)(D)] = (40)(2000) \qquad D = 5.13 \text{ ft}$$

(*b*) $$62.4[(25)(10)(12)] = (40 + W_{stone})(2000) \qquad W_{stone} = 53.6 \text{ tons}$$

**6.16** A block of wood floats in water with 2.0 in projecting above the water surface. When placed in glycerin of s.g. 1.35, the block projects 3.0 in above the liquid surface. Determine the specific gravity of the wood.

**I** Let $A$ = area of block and $h$ = height of block. $W_{block} = [(s.g.)(62.4)](Ah/12)$, $W_{displaced\ water} = 62.4[(A)(h-2)/12]$, $W_{displaced\ glycerin} = [(1.35)(62.4)][(A)(h-3)/12]$. Since the weight of each displaced liquid equals the weight of the block, $W_{displaced\ water} = W_{displaced\ glycerin}$: $62.4[(A)(h-2)/12] = [(1.35)(62.4)][(A)(h-3)/12]$, $h = 5.86$ in. Also, $W_{block} = W_{displaced\ water}$: $[(s.g.)(62.4)][(A)(5.86/12)] = 62.4[(A)(5.86-2)/12]$, s.g. $= 0.659$.

**6.17** To what depth will an 8-ft-diameter log 15 ft long and of s.g. 0.425 sink in fresh water?

**I** The log is sketched in Fig. 6-6 with center $O$ of the log above the water surface because the specific gravity is less than 0.5. (Had the specific gravity been equal to 0.5, the log would be half submerged.) $F_b = W$, $F_b$ = weight of displaced liquid = $62.4\{(15)[(2\theta/360)(\pi 4^2) - (2)(\frac{1}{2})(4\sin\theta)(4\cos\theta)]\} = 261.4\theta - (14\,976)(\sin\theta)(\cos\theta)$, $W = [(0.425)(62.4)][(15)(\pi 4^2)] = 19\,996$.

$$261.4\theta - (14\,976)(\sin\theta)(\cos\theta) = 19\,996$$

This equation can be solved by successive trials.

Try $\theta = 85°$: $\qquad (261.4)(85) - (14\,976)(\sin 85°)(\cos 85°) = 20\,919 \qquad (\neq 19\,996)$

Try $\theta = 83°$: $\qquad (261.4)(83) - (14\,976)(\sin 83°)(\cos 83°) = 19\,885 \qquad (\neq 19\,996)$

Try $\theta = 83.2°$: $\qquad (261.4)(83.2) - (14\,976)(\sin 83.2°)(\cos 83.2°) = 19\,988 \qquad (\neq 19\,996)$

Try $\theta = 83.22°$: $\qquad (261.4)(83.22) - (14\,976)(\sin 83.22°)(\cos 83.22°) = 19\,998 \qquad$ (close enough)

$$\text{Depth of flotation} = DC = OC - OD = 4.00 - 4.00\cos 83.22° = 3.53 \text{ ft}$$

**Fig. 6-6**

**6.18** (*a*) Neglecting the thickness of the tank walls in Fig. 6-7*a*, if the tank floats in the position shown what is its weight? (*b*) If the tank is held so that the top is 10 ft below the water surface, as shown in Fig. 6-7*b*, what is the force on the inside top of the tank? Use an atmospheric pressure equivalent to a 34.0-ft head of water.

**I** (*a*) $$W_{tank} = W_{displaced\ liquid} = 62.4[(1)(\pi 4^2/4)] = 784 \text{ lb}$$

(*b*) The space occupied by the air will be less at the new depth shown in Fig. 6-7*b*. Assuming that the temperature of the air is constant, then for positions *a* and *b*, $p_A V_A = p_D V_D$, $[62.4(34.0 + 1)][(4)(\pi 4^2/4)] = [(62.4)(34.0 + 10 + y)][(y)(\pi 4^2/4)]$, $y^2 + 44.0y - 140 = 0$, $y = 2.98$ ft. The pressure at $D$ is $10 + 2.98$, or $12.98$ ft of water (gage), which is essentially the same as the pressure on the inside top of the cylinder. Hence, the force on the inside top of the cylinder is given by $F = \gamma h A = (62.4)(12.98)(\pi 4^2/4) = 10\,180$ lb.

**6.19** A ship, with vertical sides near the water line, weighs 4000 tons and draws 22 ft in salt water ($\gamma = 64.0$ lb/ft³)(see Fig. 6-8). Discharge of 200 tons of water ballast decreases the draft to 21 ft. What would be the draft $d$ of the ship in fresh water?

Fig. 6-7(a)                    Fig. 6-7(b)

▍ Because the shape of the underwater section of the ship is not known, it is best to solve this problem on the basis of volumes displaced. A 1-ft decrease in draft was caused by a reduction in weight of 200 tons, or $\gamma V_d = 64.0[(1)(A)] = (200)(2000)$ where $V_d$ represents the volume between drafts 22 ft and 21 ft, and $[(1)(A)]$ represents the water-line area times 1 ft, or the same volume $V_d$. From the equation above, $V_d = (200)(2000)/64.0 = 6250$ ft$^3$ (this is per foot depth), $F_b =$ weight of displaced liquid $= \gamma V_d$, $V_d = F_b/\gamma$. In Fig. 6-8, the vertically crosshatched volume is the difference in displaced fresh water and salt water. This difference in volume can be expressed as $W/\gamma_{fresh\ H_2O} - W/\gamma_{salt\ H_2O}$, or $(4000 - 200)(2000)/62.4 - (4000 - 200)(2000)/64.0$. Since $V_d = 6250$ ft$^3$/ft depth, the vertically crosshatched volume can also be expressed as $6250y$. Hence, $6250y = (4000 - 200)(2000)/62.4 - (4000 - 200)(2000)/64.0$, $y = 0.49$ ft; $d = 21 + 0.49 = 21.49$ ft.

Fig. 6-8

**6.20**   A barrel containing water weighs 283.5 lb. What will be the reading on the scales if a 2 in by 2 in piece of wood is held vertically in the water to a depth of 2.0 ft?

▍ For every acting force, there must be an equal and opposite reacting force. The buoyant force exerted by the water upward against the bottom of the piece of wood is opposed by the 2 in by 2 in area of wood acting downward on the water with equal magnitude. This force will measure the increase in scale reading. $F_b = 62.4[(2)(\frac{2}{12})(\frac{2}{12})] = 3.5$ lb, new scale reading $= 283.5 + 3.5 = 287.0$ lb.

**6.21**   Find the weight of the floating can in Fig. 6-9.

▍          $F_b = W$      $9.79[(\frac{7}{100})(\pi)(\frac{8}{100})^2/4] = W$      $W = 0.00344$ kN   or   3.44 N

Fig. 6-9

**6.22** The weight of a certain crown in air was found to be 14.0 N and its weight in water, 12.7 N. Was it gold (s.g. = 19.3)?

❚ $F_b = 14.0 - 12.7 = 1.3$ N   $V_{\text{displaced H}_2\text{O}} = V_{\text{crown}} = 1.3/[(9.79)(1000)] = 0.0001328$ m$^3$

$\gamma_{\text{crown}} = 14.0/0.0001328 = 105\,422$ N/m$^3$   or   105.4 kN/m$^3$   s.g.$_{\text{crown}} = 105.4/9.79 = 10.77$

Thus the crown was not pure gold.

**6.23** Repeat Prob. 6.22 assuming the crown is an alloy of gold (s.g. = 19.3) and silver (s.g. = 10.5). For the same measured weights, compute the fraction of silver in the crown.

❚ From Prob. 6.22, s.g.$_{\text{crown}} = 10.77$. Let $\alpha$ = fraction of silver in crown. $(\alpha)(10.5) + (1 - \alpha)(19.3) = 10.77$, $10.5\alpha + 19.3 - 19.3\alpha = 10.77$, $\alpha = 0.969$.

**6.24** A plastic sphere is immersed in sea water ($\gamma = 64.0$ lb/ft$^3$) and moored at the bottom. The sphere radius is 15 in. The mooring line has a tension of 160 lb. What is the specific weight of the sphere?

❚ The mooring line tension ($T$) and sphere weight ($W$) act downward on the sphere, while the buoyant force ($F_b$) acts upward. $\sum F_y = 0$; $F_b - T - W = 0$, $64.0[(\frac{4}{3})(\pi)(\frac{15}{12})^3] - 160 - (\gamma_{\text{sphere}})[(\frac{4}{3})(\pi)(\frac{15}{12})^3] = 0$, $\gamma_{\text{sphere}} = 44.4$ lb/ft$^3$.

**6.25** If the total weight of the hydrometer in Fig. 6-10 is 0.035 lb and the stem diameter is 0.35 in, compute the elevation $h$ for a fluid of specific gravity 1.4.

❚ Let $\Delta V$ = submerged volume between s.g. = 1 and s.g. = 1.4, $V_0$ = submerged total volume when s.g. = 1.0, $\gamma$ = specific weight of pure water, and $W$ = weight of hydrometer. $W = \gamma V_0 = (\text{s.g.})(\gamma)(V_0 - \Delta V) = (\text{s.g.})(\gamma)(V_0) - (\text{s.g.})(\gamma)(\Delta V)$. Since $(\gamma)(V_0) = W$ and $\Delta V = hA = h(\pi d^2/4)$, $W = (\text{s.g.})(W) - (\text{s.g.})(\gamma)[(h)(\pi d^2/4)]$, $0.035 = (1.4)(0.035) - (1.4)(62.4)[(h)(\pi)(0.35/12)^2/4]$, $h = 0.240$ ft, or 2.88 in.

**Fig. 6-10**

**6.26** For the hydrometer of Fig. 6-10, derive a formula for float position $h$ as a function of s.g., $W$, $d$, and the specific weight $\gamma$ of pure water. Are the scale markings linear or nonlinear as a function of s.g.?

❚ From Prob. 6.25, $W = (\text{s.g.})(W) - (\text{s.g.})(\gamma)[(h)(\pi d^2/4)]$.

$$h = \frac{(\text{s.g.})(W) - W}{(\text{s.g.})(\gamma)(\pi d^2/4)} = \frac{(W)(\text{s.g.} - 1)}{(\text{s.g.})(\gamma)(\pi d^2/4)}$$

When plotted in Fig. 6-11 (in arbitrary units), it is slightly nonlinear.

**6.27** A hydrometer weighs of 0.17 N and has a stem diameter of 11 mm. What is the distance between scale markings for s.g. = 1.0 and s.g. = 1.1? Between 1.1 and 1.2?

❚ Let $h_1$ = distance between markings for s.g. = 1.0 and s.g. = 1.1 and $h_2$ = distance between scale markings for s.g. = 1.1 and s.g. = 1.2. From Prob. 6.26,

$$h = \frac{(W)(\text{s.g.} - 1)}{(\text{s.g.})(\gamma)(\pi d^2/4)} \qquad h_1 = \frac{(0.17)(1.1 - 1)}{1.1[(9.79)(1000)][(\pi)(0.011)^2/4]} = 0.0166 \text{ m} \quad \text{or} \quad 16.6 \text{ mm}$$

$$h_1 + h_2 = \frac{0.17(1.2 - 1)}{1.2[(9.79)(1000)][(\pi)(0.011)^2/4]} = 0.0305 \text{ m} \quad \text{or} \quad 30.5 \text{ mm}$$

$$h_2 = 30.5 - 16.6 = 13.9 \text{ mm}$$

**Fig. 6-11**

**6.28** A square pole (s.g. = 0.68), 80 mm by 80 mm by 6 m long, is suspended by a wire so that 4 m is submerged in water and 2 m is above the surface. What is the tension in the wire?

▮ Tension $(T)$ and buoyant force $(F_b)$ act upward on the pole, while pole weight $(W)$ acts downward. $\sum F_y = 0$; $T + F_b - W = 0$, $T + 9.79[(0.080)(0.080)(4)] - [(0.68)(9.79)][(0.080)(0.080)(6)] = 0$, $T = 0.00501$ kN, or 5.01 N.

**6.29** The spar in Fig. 6-12 is wood (s.g. = 0.62), 2 in by 2 in by 10 ft, and floats in sea water (s.g. = 1.025). How many pounds of steel (s.g. = 7.85) should be attached to the bottom to make a buoy that floats with exactly $h = 1.5$ ft of the spar exposed?

▮
$$V_{spar} = (\tfrac{2}{12})(\tfrac{2}{12})(10) = 0.2778 \text{ ft}^3 \qquad V_{submerged} = (\tfrac{2}{12})(\tfrac{2}{12})(8.5) = 0.2361 \text{ ft}^3$$

$$V_{steel} = W_{steel}/[(7.85)(62.4)] = 0.002041 W_{steel} \qquad F_b = W_{wood} + W_{steel}$$

$$[(1.025)(62.4)](0.2361 + 0.002041 W_{steel}) = [(0.62)(62.4)](0.2778) + W_{steel} \qquad W_{steel} = 5.01 \text{ lb}$$

$W'_{steel}$   **Fig. 6-12**

**6.30** A right circular cone is 50 mm in radius and 170 mm high and weighs 1.5 N in air. How much force is required to push this cone vertex-downward into ethanol so that its base is exactly at the surface? How much additional force will push the base 6.5 mm below the surface?

▮ Downward force $(F)$ and cone weight $(W)$ act downward on the cone, while buoyant force $(F_b)$ acts upward. $\sum F_y = 0$; $F_b - F - W = 0$, $[(0.79)(9.79)(1000)][(\pi)(0.050)^2(0.170)/3] - F - 1.5 = 0$, $F = 1.94$ N. Once the cone is fully submerged, $F$ is constant at 1.94 N.

**6.31** A 2-in by 2-in by 10-ft spar has 7 lb of steel weight attached (Fig. 6-12); the buoy has lodged against a rock 7 ft deep, as depicted in Fig. 6-13. Compute the angle $\theta$ at which the buoy will lean, assuming the rock exerts no moment on the buoy.

▮ From Prob. 6.29, $V_{spar} = 0.2778$ ft³. $W_{wood} = [(0.62)(62.4)](0.2778) = 10.75$ lb and $F_b = 62.4[(\tfrac{2}{12})(\tfrac{2}{12})(L)] = 1.733L$. $W_{wood}$ acts downward at a distance of $5 \sin \theta$ to the right of $A$, and $F_b$ acts upward at a distance of $(L/2)(\sin \theta)$ to the right of $A$; while the steel force passes through point $A$. Hence, $\sum M_A = 0$, $10.75(5 \sin \theta) - (1.733L)[(L/2)(\sin \theta)] = 0$, $L = 7.876$ ft; $\cos \theta = 7/L = 7/7.876 = 0.88878$, $\theta = 27.3°$.

**6.32** The submerged brick in Fig. 6-14 is balanced by a 2.54-kg mass on the beam scale. What is the specific weight of the brick, if it displaces 2.197 liters of water?

Fig. 6-13

■ $F = mg = (2.54)(9.81) = 24.92$ N. Upward force ($F$) and buoyant force ($F_b$) act upward on the brick, while its weight ($W$) acts downward. $\sum F_y = 0$; $F_b + F - W = 0$, $[(9.79)(1000)](2.197 \times 10^{-3}) + 24.92 - W = 0$, $W = 46.43$ N; $\gamma = 46.43/(2.197 \times 10^{-3}) = 21\,133$ N/m³, or 21.13 kN/m³.

Fig. 6-14

**6.33** The balloon in Fig. 6-15 is filled with helium pressurized to 111 kPa. Compute the tension in the mooring line.

■ $\gamma = p/RT$; $\gamma_{air} = [(101)(1000)]/[(29.3)(273 + 20)] = 11.76$ N/m³, $\gamma_{He} = [(111)(1000)]/[(212.0)(273 + 20)] = 1.787$ N/m³. Weight of helium ($W$) and tension in mooring line ($T$) act downward on the balloon, while buoyant force ($F_b$) acts upward. $\sum F_y = 0$; $F_b - W - T = 0$, $11.76[(\frac{4}{3})(\pi)(\frac{9}{2})^3] - 1.787[(\frac{4}{3})(\pi)(\frac{9}{2})^3] - T = 0$, $T = 3807$ N.

Fig. 6-15

**6.34** A 1.1-ft-diameter hollow sphere is made of steel (s.g. = 7.85) with 0.015-ft wall thickness. How deep will the sphere sink in water (i.e., find $h$ in Fig. 6-16)? How much weight must be added inside to make the sphere neutrally buoyant?

■
$$F_b = W = \text{weight of displaced water} = \gamma[(\pi/3)(h)^2(3r - h)]$$
$$= 62.4\{(\pi/3)(h^2)[(3)(\tfrac{1.1}{2}) - h)]\} = 107.8h^2 - 65.35h^3$$
$$W = (\gamma_{steel})(V_{steel}) \qquad \gamma_{steel} = (7.85)(62.4) = 489.8 \text{ lb/ft}^3$$
$$V_{steel} = (\tfrac{4}{3})(\pi)(\tfrac{1.1}{2})^3 - (\tfrac{4}{3})(\pi)\{[1.1 - (2)(0.01500)]/2\}^3 = 0.05548 \text{ ft}^3$$
$$W = (489.8)(0.05548) = 27.17 \text{ lb} \qquad 107.8h^2 - 65.35h^3 = 27.17$$

Two roots of this equation are complex. The other, obtained by trial and error, is $h = 0.643$ ft. For neutral buoyancy, the total weight of the sphere plus added weight must equal the weight of water displaced by the entire sphere. Hence, $27.17 + W_{added} = 62.4[(\frac{4}{3})(\pi)(\frac{1.1}{2})^3]$, $W_{added} = 16.32$ lb.

Fig. 6-16

**6.35** When a 6-lb weight is placed on the end of a floating 5-in by 5-in by 10-ft wooden beam, the beam tilts at 1.6° with the weight at the surface, as shown in Fig. 6-17. What is the specific weight of the wood?

$$\tan 1.6° = h/10 \qquad h = 0.2793 \text{ ft} \qquad V_{wood} = (\tfrac{5}{12})(\tfrac{5}{12})(10) = 1.736 \text{ ft}^3$$

$$F_b = W = 62.4[1.736 - (\tfrac{1}{2})(0.2793)(\tfrac{5}{12})(10)] = 72.02 \text{ lb}$$

$$W = (\gamma_{wood})(1.736) + 6 \qquad 72.02 = (\gamma_{wood})(1.736) + 6 \qquad \gamma_{wood} = 38.0 \text{ lb/ft}^3$$

Fig. 6-17

**6.36** A wooden beam (s.g. = 0.64) is 140 mm by 140 mm by 5 m and is hinged at $A$, as shown in Fig. 6-18. At what angle $\theta$ will the beam float in water?

■ The forces acting on the beam are shown in Fig. 6-18. $W_{beam} = [(0.64)(9.79)][(0.140)(0.140)(5)] = 0.6140$ kN and $F_b = 9.79[(0.140)(0.140)(L)] = 0.1919L$. $\sum M_A = 0$; $(0.1919L)[(5 - L/2)(\cos \theta)] - (0.6140)[(\tfrac{5}{2})(\cos \theta)] = 0$, $-0.0960L^2 + 0.9595L - 1.535 = 0$, $L = 2.000$ m; $\sin \theta = 1/(5 - 2.000) = 0.33333$, $\theta = 19.5°$.

Fig. 6-18

**6.37** A barge weighs 45 tons empty and is 18 ft wide, 45 ft long, and 9 ft high. What will be its draft when loaded with 125 tons of gravel and floating in sea water (s.g. = 1.025)?

$$F_b = W \qquad [(1.025)(62.4)][(18)(45)(h)] = (45 + 125)(2000) \qquad h = 6.56 \text{ ft}$$

**6.38** A block of steel (s.g. = 7.85) will "float" at a mercury–water interface as in Fig. 6-19. What will be the ratio of distances $a$ and $b$ for this condition?

■ Let $w$ = width of block and $L$ = length of block. $F_b = W$, $(\gamma_{H_2O})(aLw) + (13.6)(\gamma_{H_2O})(bLw) = (7.85)(\gamma_{H_2O})(a + b)(Lw)$, $a + 13.6b = 7.85a + 7.85b$, $a/b = 0.839$.

**Fig. 6-19**

**6.39** A balloon weighing 3.2 lb is 5.5 ft in diameter. Upon release it is filled with hydrogen at 17 psia and 65° F. At what altitude in the standard atmosphere will this balloon be neutrally buoyant?

$$F_b = W \qquad \gamma = p/RT \qquad \gamma_H = (17)(144)/[(765.5)(460 + 65)] = 0.006091 \text{ lb/ft}^3$$

$$(\gamma_{air})[(\tfrac{4}{3})(\pi)(\tfrac{5.5}{2})^3] = 3.2 + (0.006091)[(\tfrac{4}{3})(\pi)(\tfrac{5.5}{2})^3] \qquad \gamma_{air} = 0.04282 \text{ lb/ft}^3$$

From Table A-7, altitude = approximately 18 600 ft.

**6.40** A rectangular barge 18 ft wide by 46 ft long by 9 ft deep floats empty with a draft of 4 ft in a canal lock 28 ft wide by 56 ft long and water depth 7 ft when the empty barge is present. If 170 000 lb of steel is loaded onto the barge, what are the new draft of the barge ($h$) and water depth in the lock ($H$)?

The weight of the barge ($W_b$) is equal to the buoyant force when the draft is 4 ft. $W_b = 62.4[(18)(46)(4)] = 206\ 669$ lb; $F_b = W$, $62.4[(18)(46)(h)] = 206\ 669 + 170\ 000$, $h = 7.290$ ft. Volume of water in lock $= (7)(28)(56) - (4)(18)(46) = 7664$ ft³. After steel is added, $(H)(28)(56) - (7.290)(18)(46) = 7664$, $H = 8.74$ ft.

**6.41** A 4-in-diameter solid cylinder of height 3.75 in weighing 0.85 lb is immersed in liquid ($\gamma = 52.0$ lb/ft³) contained in a tall, upright metal cylinder having a diameter of 5 in. Before immersion the liquid was 3.0 in deep. At what level will the solid cylinder float? See Fig. 6-20.

Let $x$ = distance solid cylinder falls below original liquid surface, $y$ = distance liquid rises above original liquid surface, and $x + y$ = depth of submergence. $V_A = V_B$, $x[(\pi)(4)^2/4] = y[(\pi)(5)^2/4] - y[(\pi)(4)^2/4]$, $x = 0.5625y$. $F_b = W$, $52.0[(\pi)(\tfrac{4}{12})^2/4][(x + y)/12] = 0.85$, $x + y = 2.248$, $0.5625y + y = 2.248$, $y = 1.44$ in, $x = (0.5625)(1.44) = 0.81$ in. The bottom of the solid cylinder will be $3.0 - 0.81$, or 2.19 in above the bottom of the hollow cylinder.

**Fig. 6-20**

**6.42** An iceberg in the ocean floats with one-seventh of its volume above the surface. What is its specific gravity relative to ocean water? What portion of its volume would be above the surface if ice were floating in pure water? $\gamma_{\text{ocean H}_2\text{O}} = 64.0$ lb/ft³.

$$(V_{iceberg})_{submerged} = (1 - \tfrac{1}{7})V_{iceberg} = 0.857 V_{iceberg}$$

$$F_b = W \qquad (\gamma_{\text{ocean H}_2\text{O}})(V_{iceberg})_{submerged} = (\gamma_{iceberg})(V_{iceberg})$$

$$\text{s.g.}_{iceberg} = \gamma_{iceberg}/\gamma_{\text{ocean H}_2\text{O}} = (V_{iceberg})_{submerged}/V_{iceberg}$$

$$= 00.857 V_{iceberg}/V_{iceberg} = 0.857 \qquad \text{(relative to ocean water)}$$

$$\gamma_{iceberg} = (0.857)(64.0) = 54.85 \text{ lb/ft}^3 \qquad \text{s.g.}_{iceberg} = 54.85/62.4 = 0.879 \qquad \text{(relative to pure water)}$$

Therefore, $1 - 0.879 = 0.121$, or 12.1 percent of its volume would be above the water surface in pure water.

**6.43** A hydrometer consists of an 11-mm-diameter cylinder of length 220 mm attached to a 26-mm-diameter weighted sphere. The cylinder has a mass of 1.5 g, and the mass of the sphere is 13.0 g. At what level will this device float in liquids having specific gravities of 0.8, 1.0, and 1.2? Is the scale spacing on the cylindrical stem uniform? Why or why not?

▐ Let $y$ = submerged length of cylinder in millimeters. $V_{sphere} = (\frac{4}{3})(\pi)(\frac{26}{2})^3 = 9203$ mm$^3$, $V_{submerged\ cylinder}$ = $(y)(\pi)(\frac{11}{2})^2 = 95.03y$, $F_b = W = (s.g.)[(9.79)(1000)/1000^3](9203 + 95.03y) = (0.09010)(s.g.) + (0.0009303)(s.g.)(y)$, $W = mg = (1.5 + 13.0)(9.81) = 142.2$ g $\cdot$ m/s$^2$, or 0.1422 N, $(0.09010)(s.g.) + (0.0009303)(s.g.)(y) = 0.1422$, $y = [0.1422 - (0.09010)(s.g.)]/[(0.0009303)(s.g.)]$.

For s.g. = 0.8          $y = [0.1422 - (0.09010)(0.8)]/[(0.0009303)(0.8)] = 94.2$ mm

For s.g. = 1.0          $y = [0.1422 - (0.09010)(1.0)]/[(0.0009303)(1.0)] = 56.0$ mm

For s.g. = 1.2          $y = [0.1422 - (0.09010)(1.2)]/[(0.0009303)(1.2)] = 30.5$ mm

Scale spacing is not uniform because buoyant force is not directly proportional to submergence.

**6.44** A typewriter weighs 6 lb in water and 8 lb in oil of specific gravity 0.86. Find its specific weight.

▐ $F_b = W$, $62.4V = W - 6$, $[(0.86)(62.4)](V) = W - 8$. Subtracting the second equation from the first gives $62.4V - [(0.86)(62.4)](V) = -6 - (-8)$, $V = 0.229$ ft$^3$; $(62.4)(0.229) = W - 6$, $W = 20.3$ lb. $\gamma = 20.3/0.229 = 88.6$ lb/ft$^3$.

**6.45** A balloon weighs 270 lb and has a volume of 14 900 ft$^3$. It is filled with helium, which weighs 0.0112 lb/ft$^3$ at the temperature and pressure of the air, which weighs 0.0807 lb/ft$^3$. What load will the balloon support?

$$F_b = W \qquad (0.0807)(14\,900) = 270 + (0.0112)(14\,900) + load \qquad Load = 766\ lb$$

**6.46** A small cylindrical drum 32 cm in diameter and 52 cm high, weighing 27.0 N, contains perfume (s.g. = 0.83) to a depth of 22 cm. (**a**) When placed in water, what will be the depth $y$ to the bottom of the drum? (**b**) How much perfume can the drum hold and still float?

▐ (**a**)        $F_b = W$     $9.79[(y)(\pi)(\frac{32}{100})^2/4] = 27.0/1000 + [(0.83)(9.79)][(0.22)(\pi)(0.32)^2/4]$

$$y = 0.217\ m \quad or \quad 21.7\ cm$$

(**b**) $9.79[(0.52)(\pi)(0.32)^2/4] = 27.0/1000 + [(0.83)(9.79)][(h)(\pi)(0.32)^2/4]$      $h = 0.585$ m   or   58.5 cm

Since $h = 58.5$ cm is greater than the height of the drum (52 cm), the drum will float when full. Therefore, $V_{max} = (0.52)(\pi)(0.32)^2/4 = 0.0418$ m$^3$, or 41.8 L.

**6.47** A block ($\gamma = 124$ lb/ft$^3$) 1 ft square and 9 in deep floats on a stratified liquid composed of a 7-in layer of water above a layer of mercury. (**a**) Determine the position of the bottom of the block.
(**b**) If a downward vertical force of 260 lb is applied to the center of mass of this block, what is the new position of the bottom of the block?

▐ (**a**) $F_b = W$. Let $x$ = depth into mercury below water–mercury interface. $[(13.6)(62.4)][(1)(1)(x)] + 62.4[(1)(1)(\frac{7}{12})] = (124)[(1)(1)(\frac{9}{12})]$, $x = 0.0667$ ft, or 0.800 in. (**b**) In this case the top of the block will be below the water surface. Hence, $[(13.6)(62.4)][(1)(1)(x)] + 62.4[(1)(1)(\frac{9}{12} - x)] = 124[(1)(1)(\frac{9}{12})] + 260$, $x = 0.389$ ft, or 4.67 in.

**6.48** Two spheres, each 1.3 m in diameter, weigh 5 kN and 13 kN, respectively. They are connected with a short rope and placed in water. What is the tension ($T$) in the rope and what portion of the lighter sphere protrudes from the water?

▐ For the lower (heavier) sphere, the buoyant force and $T$ act upward and its weight acts downward. Hence, $\sum F_y = 0$, $F_b = 9.79[(\frac{4}{3})(\pi)(1.3/2)^3] = 11.26$ kN, $11.26 + T - 13 = 0$, $T = 1.74$ kN. For the upper (lighter) sphere, the buoyant force acts upward and its weight and $T$ act downward. Hence, $F_b - 5 - 1.74 = 0$, $F_b = 6.74$ kN. Portion above water = $(11.26 - 6.74)/11.26 = 0.401$, or 40.1 percent of volume.

**6.49** A board weighing 2.2 lb/ft and of cross-sectional area 8 in$^2$ dips into oil as shown in Fig. 6-21. If the hinge is frictionless, find $\theta$.

▐ The forces acting on the board are shown in Fig. 6-21. $W = (2.2)(11) = 24.2$ lb; $F_b = (53)[(\frac{8}{144})(x)] = 2.944x$, $\sum M_{hinge} = 0$, $(24.2)[(\frac{11}{2})(\sin \theta)] - (2.944x)[(11 - x/2)(\sin \theta)] = 0$, $1.472x^2 - 32.38x + 133.1 = 0$; $x_1 = 16.53$ ft and

$x_2 = 5.47$ ft. Using $x = 5.47$ ft (since $x = 16.53$ ft is impossible for this situation), $\cos \theta = 5/(11 - 5.47) = 0.90416$, $\theta = 25.3°$.

**Fig. 6-21**

**6.50** A cube 2.2 ft on an edge has its lower half of s.g. = 1.6 and upper half of s.g. = 0.7. It rests in a two-layer fluid, with lower s.g. = 1.4 and upper s.g. = 0.8. Determine the height $h$ of the top of the cube above the interface (see Fig. 6-22).

$$F_b = W$$

$$[(1.4)(62.4)][(2.2)(2.2)(2.2 - h)] + [(0.8)(62.4)][(2.2)(2.2)(h)]$$
$$= [(1.6)(62.4)][(2.2)(2.2)(\tfrac{2.2}{2})] + [(0.7)(62.4)][(2.2)(2.2)(\tfrac{2.2}{2})]$$

$$h = 0.917 \text{ ft}$$

**Fig. 6-22**

**6.51** Determine the volume and density of an object that weighs 4 N in water and 5 N in an alcohol of s.g. 0.80.

$$F_b = W \qquad [(9.79)(1000)](V_{object}) = W_{object} - 4 \qquad [(0.80)(9.79)(1000)](V_{object}) = W_{object} - 5$$

Subtracting the second equation from the first gives $9790V_{object} - 7832V_{object} = 1$, $V_{object} = 0.0005107 \text{ m}^3$.
$[(9.79)(1000)](0.0005107) = W_{object} - 4$, $W_{object} = 9.000 \text{ N}$; $\gamma = 9.000/0.0005107 = 17\,623 \text{ N/m}^3$.

$$\rho = \frac{\gamma}{g} = \frac{17\,623}{9.81} = 1796 \text{ kg/m}^3 \qquad V_s = \frac{1}{\rho} = \frac{1}{1796} = 0.000557 \text{ m}^3/\text{kg}$$

**6.52** With how many pounds of concrete ($\gamma = 25 \text{ kN/m}^3$) must a beam of volume of 0.2 m³ and s.g. = 0.67 be coated to insure that it sinks in water?

$$F_b = W \qquad (9.79)(0.2) + 9.79V_{concrete} = [(0.67)(9.79)](0.2) + 25V_{concrete} \qquad V_{concrete} = 0.04248 \text{ m}^3$$

$$W_{concrete} = (0.04248)(25) = 1.062 \text{ kN} \quad \text{or} \quad 1062 \text{ N} \quad \text{or} \quad 1062/4.448 = 239 \text{ lb}$$

**6.53** The gate of Fig. 6-23 weighs 160 lb/ft normal to the page. It is in equilibrium as shown. Neglecting the weight of the arm and brace supporting the counterweight, find $W$ (weight in air). The weight is made of concrete, s.g. = 2.50.

$$F_H = \gamma h A = (62.4)(\tfrac{6}{2})[(6)(1)] = 1123 \text{ lb} \qquad \sum M_{hinge} = 0 \qquad (1123)(\tfrac{6}{3}) - (W)(5 \sin 30°) = 0$$

$$W = 898 \text{ lb}$$

This is the submerged weight.

$$F_b = W \qquad 62.4V_{concrete} = [(2.50)(62.4)](V_{concrete}) - 898 \qquad V_{concrete} = 9.594 \text{ ft}^3$$

$$W_{concrete} = [(2.50)(62.4)](9.594) = 1497 \text{ lb}$$

**Fig. 6-23**

**6.54** A wooden pole (s.g. = 0.55), 550 mm in diameter, has a concrete cylinder (s.g. = 2.50), 550 mm long and of the same diameter, attached to one end. Determine the minimum length of pole for the system to float vertically in static equilibrium.

▌ The system will float at minimum length of wooden cylinder as shown in Fig. 6-24. $F_b = W$, $(\gamma)(A)(L + 0.550) = [(0.55)(\gamma)](L)(A) + [(2.50)(\gamma)](A)(0.550)$, $L = 1.833$ m.

**Fig. 6-24**

**6.55** A hydrometer weighs 0.040 N and has a stem 7 mm in diameter. Compute the distance between specific gravity markings 1.0 and 1.1.

▌ From Prob. 6.26,

$$h = \frac{(W)(\text{s.g.} - 1)}{(\text{s.g.})(\gamma)(\pi d^2/4)} = \frac{(0.040)(1.1 - 1)}{(1.1)[(9.79)(1000)][(\pi)(0.007)^2/4]} = 0.0097 \text{ m} \quad \text{or} \quad 9.7 \text{ mm}$$

**6.56** What is the weight of the loaded barge in Fig. 6-25? The barge is 7 m in width.

▌ $\qquad F_b = W \qquad 9.79\{(7)[(14)(2.4) + (2)(2.4)(2.4)/2]\} = W \qquad W = 2359 \text{ kN}$

**Fig. 6-25**

**6.57** In Fig. 6-26, a wedge of wood having specific gravity 0.66 supports a 160-lb mermaid (not shown). The wedge is 3 ft in width. What is depth $d$?

▌ The 160-lb force and the weight of the wood ($W$) act downward on the wedge, while the buoyant force ($F_b$) acts upward. $\sum F_y = 0$, $F_b - 160 - W = 0$, $62.4[(2)(3)(d)(d \tan 30°)/2] - 160 - [(0.66)(62.4)]\{(2)(3)(\frac{3}{2})[(\frac{3}{2})/\tan 30°]/2\} = 0$, $d = 2.44$ ft.

**6.58** The tank in Fig. 6-27 is filled brimfull with water. If a cube 700 mm on an edge and weighing 530 N is lowered slowly into the water until it floats, how much water flows over the edge of the tank? Neglect sloshing, etc.

▌ $F_b = W$. Let $h$ = the depth to which the cube will sink in the water. $[(9.79)(1000)][(0.700)(0.700)(h)] = 530$, $h = 0.120$ m, $V_{\text{displaced}} = [(0.700)(0.700)(0.120)] = 0.0588$ m³. This is the amount of water that will overflow.

Fig. 6-26

Fig. 6-27

**6.59** A stone cube 280 mm on a side and weighing 425 N is lowered into a tank containing a layer of water over a layer of mercury. Determine the position of the block when it has reached equilibrium.

▌ $\gamma_{cube} = 425/(0.280)^3 = 19.360$ kN/m³. Since the cube is heavier than water but lighter than mercury, it will sink beneath the water surface and come to rest at the water–mercury interface, as shown in Fig. 6-28. $F_b = W$, $9.79[(0.280)(0.280)(0.280 - x)] + [(13.6)(9.79)][(0.280)(0.280)(x)] = 0.425$, $x = 0.0217$ m, or 21.7 mm. Thus, the bottom of the cube will come to rest 21.7 mm below the water–mercury interface.

Fig. 6-28

**6.60** An iceberg ($\gamma = 9$ kN/m³) floats in ocean water ($\gamma = 10$ kN/m³) with 3000 m³ of the iceberg protruding above the free surface. What is the volume of the iceberg below the free surface?

▌ $$F_b = W \qquad 10\,000 V_{below} = 9000(V_{below} + 3000) \qquad V_{below} = 27\,000 \text{ m}^3$$

**6.61** A rectangular tank of internal width 7 m, partitioned as shown in Fig. 6-29, contains oil and water. (*a*) If the oil's specific gravity is 0.84, find its depth *h*. (*b*) If a 900-N block of wood is floated in the oil, what is the rise in free surface of the water in contact with air?

▌ (*a*) $p_{atm} + [(0.84)(9.79)](h) + (9.79)(3) - (9.79)(4) = p_{atm}$, $h = 1.190$ m. (*b*) Let $h'$ = the new value of $h$ with the 900-N block in flotation. Since the volume of oil does not change, $(1.190)(0.5)(7) = (h')(0.5)(7) - 900/[(0.84)(9.79)(1000)]$, $h' = 1.221$ m. If the oil–water interface drops by a distance $\delta$, the free surface of water with air will rise by $\delta/2$. $p_{atm} + [(0.84)(9.79)](1.221) + 9.79(3 - \delta) - 9.79(4 + \delta/2) = p_{atm}$, $\delta = 0.01709$ m, or 17.09 mm. The free surface of the water will rise by 17.09/2, or 8.54 mm.

Fig. 6-29

**6.62** A balloon is filled with 3000 m³ of hydrogen having specific weight 1.1 N/m³. What lift does the balloon exert (*a*) at the earth's surface, if the balloon weighs 1400 N and the temperature is 15 °C; (*b*) at an elevation of 10 km, assuming that the volume has increased 6 percent?

▮ From Table A-8, $\gamma_{air} = 12.01$ N/m³ at elevation 0 and 4.04 N/m³ at elevation 10 000 m. $\sum F_y = 0$, $F_b - W_{balloon} - W_H = 0$.

(*a*) $(12.01)(3000) - 1400 - (1.1)(3000) - \text{lift} = 0$    Lift = 31 330 N  or  31.33 kN

(*b*) $4.04[(1.06)(3000)] - 1400 - (1.1)(3000) - \text{lift} = 0$    Lift = 8147 N  or  8.15 kN

**6.63** A wooden rod weighing 4 lb is hinged at one end (Fig. 6-30). The rod is 9 ft long and uniform in cross section, and the support is 4 ft below the free surface of a freshwater pond. At what angle $\alpha$ will it come to rest when allowed to drop from a vertical position? The cross section of the stick is 1.4 in² in area.

▮ The forces acting on the beam are shown in Fig. 6-30.

$$F_b = 62.4[(9 - e)(1.4/144)] = 5.460 - 0.6067e \qquad \sum M_A = 0$$

$$4(4.5\cos\alpha) - (5.460 - 0.6067e)[(9-e)/2](\cos\alpha) = 0 \qquad -0.303e^2 + 5.46e - 6.57 = 0 \qquad e = 1.297\text{ ft}$$

$$\sin\alpha = 4/(9 - e) = 4/(9 - 1.297) = 0.51928 \qquad \alpha = 31.3°$$

A

**Fig. 6-30**

**6.64** A block of wood having a volume of 0.034 m³ and weighing 300 N is suspended in water as shown in Fig. 6-31. A wooden rod of length 3.4 m and cross section of 2000 mm² is attached to the weight and also to the wall. If the rod weighs 16 N, what will angle $\theta$ be for equilibrium?

▮ $(F_b)_{block} = [(9.79)(1000)](0.034) = 333$ N    $(F_b)_{rod} = [(9.79)(1000)][(AC)(2000/10^6)] = 19.58AC$ N

$$\sum M_B = 0$$

$$333(3.4\cos\theta) + (19.58AC)[(AC/2) + (\tfrac{310}{1000})/\sin\theta](\cos\theta) - 300(3.4\cos\theta) - (16)(3.4/2)(\cos\theta) = 0$$

$$AC = 3.4 - (\tfrac{310}{1000})/\sin\theta$$

$$333(3.4\cos\theta) + 19.58[3.4 - (\tfrac{310}{1000})/\sin\theta]$$
$$\times \{[3.4 - (\tfrac{310}{1000})/\sin\theta]/2 + (\tfrac{310}{1000})/\sin\theta\}(\cos\theta) - 300(3.4\cos\theta) - (16)(3.4/2)(\cos\theta) = 0$$

$$4.341 = [3.4 - (\tfrac{310}{1000})/\sin\theta][1.700 + (\tfrac{310}{1000})/(2\sin\theta)] \qquad 4.341 = 5.780 - 0.048/\sin^2\theta$$

$$\sin^2\theta = 0.033357 \qquad \sin\theta = 0.18264 \qquad \theta = 10.5°$$

**Fig. 6-31**

**6.65** A barge with a flat bottom and square ends has a draft of 6.0 ft when fully loaded and floating in an upright position, as shown in Fig. 6-32a. The center of gravity ($CG$) of the barge when fully loaded is on the axis of symmetry and 1.0 ft above the water surface. Is the barge stable? If it is stable, what is the righting moment when the angle of heel is 12°?

▌ $\overline{MB} = I/V_d = [(42)(25)^3/12]/[(25)(42)(6)] = 8.68$ ft. Therefore, the metacenter (mc) is located 8.68 ft above the center of buoyancy ($CB$), as shown in Fig. 6-32b. Hence, it (the metacenter) is located $8.68 - 3 - 1$, or 4.68 ft above the barge's center of gravity and the barge is stable. The end view of the barge when the angle of heel is 12° is shown in Fig. 6-32c. Righting moment = $(F_b)(x)$, $F_b = 62.4[(25)(42)(6)] = 393\,120$ lb, $x = (\sin 12°)(\text{distance from mc to } CG) = (\sin 12°)(4.68) = 0.973$ ft, righting moment = $(393\,120)(0.973) = 382\,500$ lb · ft.

(a) Top view

(b) End view

Fig. 6-32(a)

Fig. 6-32(b)

**Fig. 6-32(c)**

Water

**6.66** Would the wooden cylinder (s.g. = 0.61) in Fig. 6-33a be stable if placed vertically in oil as shown in the figure?

❙ The first step is to determine the submerged depth of the cylinder when placed in the oil. $F_b = W$,
$[(0.85)(9.79)][(D)(\pi)(0.666)^2/4] = [(0.61)(9.79)][(1.300)(\pi)(0.666)^2/4]$, $D = 0.9333$ m. The center of buoyancy
is located at a distance of 0.933/2, or 0.466 m from the bottom of the cylinder (see Fig. 6-33b).
$\overline{MB} = I/V_d = [(\pi)(0.666)^4/64]/[(0.933)(\pi)(0.666)^2/4] = 0.030$ m. The metacenter is located 0.030 m above the
center of buoyancy, as shown in Fig. 6-33b. This places the metacenter 1.300/2 − 0.466 − 0.030, or 0.154 m
below the center of gravity. Therefore, the cylinder is not stable.

**Fig. 6-33(a)**

**Fig. 6-33(b)**

**6.67** Figure 6-34a shows the cross section of a boat. The hull of the boat is solid. Show if the boat is stable or not. If the boat is stable, compute the righting moment when the angle of heel is 10°?

▌ $\overline{MB} = I/V_d = [(20)(10)^3/12]/[(10)(5)(20)] = 1.67$ ft. Therefore, the metacenter is located $1.67 - 0.5$, or 1.17 ft above the center of gravity, as shown in Fig. 6-34b, and the barge is stable. The end view of the barge when the angle of heel is 10° is shown in Fig. 6-34c. Righting moment $= (F_b)(x)$, $F_b = 62.4[(10)(5)(20)] = 62\,400$ lb, $x = (\sin 10°)(1.17) = 0.203$ ft, righting moment $= (62\,400)(0.203) = 12\,670$ lb · ft.

**Fig. 6-34(a)**

**Fig. 6-34(b)**

**Fig. 6-34(c)**

**6.68** A solid wood cylinder has a diameter of 2.0 ft and a height of 4.0 ft. The specific gravity of the wood is 0.60. If the cylinder is placed vertically in oil (s.g. = 0.85), would it be stable?

▮ $F_b = W$, $[(0.85)(62.4)][(D)(\pi)(2)^2/4] = [(0.60)(62.4)][(4)(\pi)(2)^2/4]$, $D = 2.82$ ft. The center of buoyancy is located at a distance of 2.82/2, or 1.41 ft from the bottom of the cylinder (see Fig. 6-35). $\overline{MB} = I/V_d = [(\pi)(2)^4/64]/[(2.82)(\pi)(2)^2/4] = 0.09$ ft. The metacenter is located $2 - 1.41 - 0.09$, or 0.50 ft below the center of gravity, as shown in Fig. 6-35. Therefore, the cylinder is not stable.

**Fig. 6-35**

**6.69** A wood cone floats in water in the position shown in Fig. 6-36a. The specific gravity of the wood is 0.60. Would it be stable?

▌ The center of gravity is located $\frac{10}{4}$, or 2.50 in from the base of the cone or 7.50 in from the tip, as shown in Fig. 6-36b. $W_{\text{cone}} = [(0.60)(62.4)][(10)(\pi)(7)^2/12]/1728 = 2.779$ lb. Let $x =$ submerged depth.

$$D_x = 0.700x \qquad V_d = (x)(\pi)(D_x)^2/12 = (x)(\pi)(0.700x)^2/12 = 0.1283x^3$$

$$F_b = W \qquad 62.4(0.1283x^3) = 2.779 \qquad x = 0.703 \text{ ft} \quad \text{or} \quad 8.44 \text{ in}$$

$$D_x = (0.700)(8.44) = 5.91 \text{ in} \qquad V_d = (0.1283)(0.703)^3 = 0.0446 \text{ ft}^3 \quad \text{or} \quad 77.1 \text{ in}^3$$

$$\overline{MB} = I/V_d = [(\pi)(5.91)^4/64]/77.1 = 0.78 \text{ in}$$

The metacenter is located 0.78 in above the center of buoyancy. Hence, the metacenter is located $7.50 - 6.33 - 0.78$, or 0.39 in below the cone's center of gravity, and the cone is not stable.

**6.70** A block of wood 6 ft by 8 ft floats on oil of specific gravity 0.751. A clockwise couple holds the block in the position shown in Fig. 6-37. Determine the (a) buoyant force acting on the block and its position, (b) magnitude of the couple acting on the block, and (c) location of the metacenter for the tilted position.

▌ (a) $F_b = W = [(0.751)(62.4)][(10)(4+4)(4.618)/2] = 8656$ lb. $F_b$ acts upward through the center of gravity $O'$ of the displaced oil. The center of gravity lies 5.333 ft from $A$ and 1.540 ft from $D$, as shown in Fig. 6-37. $AC = AR + RC = AR + LO' = (5.333)(\cos 30°) + (1.540)(\sin 30°) = 5.388$ ft. Hence, the buoyant force of 8650 lb acts upward through the center of gravity of the displaced oil, which is 5.388 ft to the right of $A$.

(b) One method of obtaining the magnitude of the righting couple (which must equal the magnitude of the external couple for equilibrium) is to find the eccentricity $e$. This dimension is the distance between the two parallel, equal forces $W$ and $F_b$, which form the righting couple. $e = FC = AC - AF$, $AF = AR + RF = (5.333)(\cos 30°) + GR \sin 30° = 4.619 + (0.691)(\sin 30°) = 4.964$ ft, $e = 5.388 - 4.964 = 0.424$ ft; couple $= (8656)(0.424) = 3670$ ft · lb.

(c) Metacentric distance $MG = MR - GR = RC/\sin 30° - GR = 0.770/\sin 30° - 0.691 = 0.85$ ft

Water

$d = 7$ in

$h = 10$ in

**Fig. 6-36(a)**

7 in

5.91 in

c.g.

0.39 in

mc

0.78 in

c.b.

6.33 in.

7.5 in

8.44 in

10 in

**Fig. 6-36(b)**

All lengths in feet.

**Fig. 6-37**

**6.71** A rectangular scow 7 ft by 18 ft by 32 ft long; its load of garbage has its center of gravity 2 ft above the waterline, as shown in Fig. 6-38. Is the scow stable for this configuration?

**❚** $\overline{MB} = I/V_d = [(32)(18)^3/12]/[(5)(18)(32)] = 5.40$ ft. The metacenter is located 5.40 ft above the center of buoyancy, which is 2.5 ft above the bottom of the scow. Hence, the metacenter is located at $5.40 - 4.5$, or 0.90 ft above the center of gravity, and the scow is stable.

**Fig. 6-38**

**6.72** In Fig. 6-39, a scow 20 ft wide and 60 ft long has a gross weight of 225 tons. Its center of gravity is 1.0 ft above the water surface. Find the metacentric height and restoring couple when $\Delta y = 1.0$ ft.

**❚** $F_b = W$, $62.4[(60)(20)(h)] = (225)(2000)$, $h = 6.01$ ft. To locate $CB'$, the center of buoyancy in the tipped position, take moments about $AB$ and $BC$. $(6.01)(20)(x) = (6.01 - 1.0)(20)(10) + [(1.0 + 1.0)(20)/2](\frac{20}{3})$, $x = 9.45$ ft; $(6.01)(20)(y) = (6.01 - 1.0)(20)[(6.01 - 1.0)/2] + [(1.0 + 1.0)(20)/2][(6.01 - 1.0) + (1.0 + 1.0)/3]$, $y = 3.03$ ft. By similar triangles $AEO$ and $\overline{CB'}PM$,

$$\frac{\Delta y}{b/2} = \frac{\overline{CB'}P}{MP} \qquad \frac{1.0}{20/2} = \frac{10 - 9.45}{MP} \qquad MP = 5.50 \text{ ft}$$

$CG$ is $6.01 + 1.0$, or 7.01 ft from the bottom. Hence, $\overline{CGP} = 7.01 - 3.03 = 3.98$ ft, $M\overline{CG} = MP - \overline{CGP} = 5.50 - 3.98 = 1.52$ ft. The scow is stable, since $M\overline{CG}$ is positive. Righting moment $= W(M\overline{CG}) \sin \theta = [(225)(2000)](1.52)[1/\sqrt{(\frac{20}{2})^2 + 1.0^2}] = 68\,060$ ft · lb.

**Fig. 6-39**

**6.73** What are the proportions of radius to height $(r_0/h)$ of a right-circular cylinder of specific gravity s.g. so that it will float in water with end faces horizontal in stable equilibrium?

▮ See Fig. 6-40.

$$h_G = h/2 \qquad h_B = h_1/2 \qquad F_b = W \qquad \gamma(h_1 \pi r_0^2) = [(s.g.)(\gamma)](h \pi r_0^2) \qquad h_1 = (s.g.)(h)$$

$$h_B = (s.g.)(h)/2 \qquad MG = MB - GB$$

$$\overline{MB} = I/V_d = (\pi r_0^4/4)/(h_1 \pi r_0^2) = r_0^2/(4h_1) = r_0^2/[(4)(s.g.)(h)]$$

$$GB = h_G - h_B = h/2 - (s.g.)(h)/2 = (h)(1 - s.g.)/2 \qquad MG = r_0^2[(4)(s.g.)(h)] - (h)(1 - s.g.)/2$$

For stable equilibrium, $MG \geq 0$, in which case $r_0^2/[(4)(s.g.)(h)] \geq (h)(1 - s.g.)/2$, $r_0/h \geq \overline{\sqrt{(2)(s.g.)(1 - s.g.)}}$.

**Fig. 6-40**

**6.74**   The plane gate in Fig. 6-41a weighs 2.1 kN/m normal to the paper, and its center of gravity is 2 m from the hinge at $O$. (**a**) Find $h$ as a function of $\theta$ for equilibrium of the gate. (**b**) Is the gate in stable equilibrium for any values of $\theta$?

▮ Consider a unit width of gate, as shown in Fig. 6-41b.

(**a**)   $$F = \gamma \bar{h} A \qquad F_x = [(9.79)(1000)](h/2)(h/\sin\theta) = 4895h^2/\sin\theta \qquad \sum M_O = 0$$

$$(4895h^2/\sin\theta)[(h/\sin\theta)/3] - 2100[(\tfrac{4}{2})(\cos\theta)] = 0 \qquad h^3 = 2.574\sin^2\theta\cos\theta \qquad h = 1.370(\sin^2\theta\cos\theta)^{1/3}$$

(**b**) From part (**a**) $\sum M_O = (1632h^3)/\sin^2\theta - 4200\cos\theta$, $dM/d\theta = -3264h^3\sin^{-3}\theta\cos\theta + 4200\sin\theta$. Substituting $h = 1.370(\sin^2\theta\cos\theta)^{1/3}$ [from part (**a**)], $dM/d\theta = -(3264)(1.370)^3(\cos^2\theta/\sin\theta) + 4200\sin\theta = -(8393)(\cos^2\theta/\sin\theta) + 4200\sin\theta$. For stability, $dM/d\theta < 0$, in which case $4200\sin\theta < 8393(\cos^2\theta/\sin\theta)$, $0.500\sin\theta < \cos^2\theta/\sin\theta$, $\tan^2\theta < (1/0.500 = 2.00)$. This occurs for $\theta \leq 54.7°$ (upper limit). For the lower limit (when water spills over the top of the gate), $h = 4\sin\theta$, $\sum M_O = (1632h^3)/\sin^2\theta - 4200\cos\theta$. Substituting $h = 4\sin\theta$, $\sum M_O = 1632(4\sin\theta)^3/\sin^2\theta - 4200\cos\theta = 104\,448\sin\theta - 4200\cos\theta$. In this case, $\sum M_O = 0$, $\tan\theta = 4200/104\,448 = 0.040211$, $\theta = 2.3°$. Thus for stable equilibrium, $\theta$ must be between 2.3° and 54.7°.

**Fig. 6-41(a)**          **Fig. 6-41(b)**

**6.75**   The barge shown in Fig. 6-42 has the form of a parallelopiped having dimensions 10 m by 26.7 m by 3 m. The barge weighs 4450 kN when loaded and has a center of gravity 4 m from the bottom. Find the metacentric height for a rotation about its longest centerline, and determine whether or not the barge is stable.

▮ First, find the center of buoyancy of the barge. $F_b = W$, $9.79[(10)(26.7)(D)] = 4450$, $D = 1.702$ m. Hence, the center of buoyancy (CB) is at a distance 1.702/2, or 0.851 m above the bottom of the barge. $\overline{MB} = I/V_d = [(26.7)(10)^3/12]/[(10)(26.7)(1.702)] = 4.896$. The distance from CB to CG is $4 - 0.851$, or 3.149 m. Therefore, the metacenter is located $4.896 - 3.149$, or 1.747 m above the CG, and the barge is stable.

Fig. 6-42

**6.76** A plastic cube of dimension $L$ and specific gravity 0.82 floats in water. Is the cube stable?

▮ The cube's center of gravity is at $0.5L$ above its bottom. If the cube has a s.g. $= 0.82$, it will float at a submerged depth of $0.82L$, and its center of buoyancy will be at $0.41L$ above its bottom. $\overline{MB} = I/V_d = [(L)(L)^3/12]/[(L)(L)(0.82L)] = 0.1016L$. Therefore, the metacenter is located $0.1016L$ above the center of buoyancy and $0.1016L + 0.41L - 0.5L$, or $0.0116L$ above the center of gravity, and the cube is stable (although just barely).

**6.77** For the cube specified in Prob. 6.76, determine the range of values of specific gravity between 0 and 1.0 for which the cube is stable.

▮ The cube's center of gravity is at $0.5L$ above its bottom. For any specific gravity s.g., the cube will float at a submerged depth of $(\text{s.g.})(L)$, and its center of buoyancy will be at $(\text{s.g.})(L)/2$ above its bottom. $\overline{MB} = I/V_d = [(L)(L)^3/12]/\{(L)(L)[(\text{s.g.})(L)]\} = 0.08333L/(\text{s.g.})$. Therefore, if the cube is stable, the metacenter must be located $0.08333L/(\text{s.g.})$ above the center of buoyancy and $0.08333L/(\text{s.g.}) + (\text{s.g.})(L)/2 - 0.5L$ above the center of gravity. For this to occur, $0.08333L/(\text{s.g.}) + (\text{s.g.})(L)/2 - 0.5L > 0$, $(\text{s.g.})^2/2 - (0.5)(\text{s.g.}) + 0.08333 > 0$. This condition is true (i.e., the cube is stable) for s.g. $> 0.789$ and s.g. $< 0.211$.

# CHAPTER 7
# Kinematics of Fluid Motion

**7.1** A nozzle with base diameter 75-mm and a 35-mm-diameter tip discharges 12 L/s of fluid. Derive an expression for fluid velocity along the nozzle's axis. Measure distance $x$ along the axis from the plane of the larger diameter.

▮ Let $L$ = length of nozzle and $D$ = diameter of nozzle at any point. $D = \frac{75}{1000} - (\frac{75}{1000} - \frac{35}{1000})(x/L) = 0.075 - 0.040x/L$, $v = Q/A = 0.012/[\pi(0.075 - 0.040x/L)^2/4] = 1.528/(0.70 - 0.40x/L)^2$. *Note*: $x$ and $L$ in millimeters gives $v$ in m/s.

**7.2** What angle $\alpha$ of jet is required to reach the roof of the building in Fig. 7-1 with minimum jet velocity $v_0$ at the nozzle? What is the value of $v_0$?

▮ $d^2y/dt^2 = -g$, $dy/dt = -gt + c_1$. At $t = 0$, $dy/dt = v_0 \sin \alpha$. Therefore, $c_1 = v_0 \sin \alpha$, and $dy/dt = -gt + v_0 \sin \alpha$, $y = -gt^2/2 + tv_0 \sin \alpha + c_2$. At $t = 0$, $y = 0$. Therefore $c_2 = 0$, and $y = -gt^2/2 + tv_0 \sin \alpha$, $L = tv_0 \cos \alpha$, $t = L/(v_0 \cos \alpha)$.

$$H = -g[L/(v_0 \cos \alpha)]^2/2 + [L/(v_0 \cos \alpha)](v_0 \sin \alpha) \tag{1}$$

Let $F = gL^2/(2v_0^2)$. Then, from Eq. (1), $F = (\cos \alpha)(L \sin \alpha - H \cos \alpha) = L \cos \alpha \sin \alpha - H \cos^2 \alpha$. Find maximum $F$ for minimum $v_0$.

$$dF/d\alpha = L(\cos^2 \alpha - \sin^2 \alpha) + 2H \sin \alpha \cos \alpha = 0 \qquad 2H/L = -(\cos^2 \alpha - \sin^2 \alpha)/(\sin \alpha \cos \alpha) = -2 \cot 2\alpha$$

$$(2)(28)/24 = -2 \cot 2\alpha \qquad \alpha = 69.7°$$

Substituting into Eq. (1), $28 = -(9.807)[24/(v_0 \cos 69.7°)]^2/2 + [24/(v_0 \cos 69.7°)](v_0 \sin 69.7°)$, $v_0 = 25.2$ m/s.

**Fig. 7-1**

**7.3** Given the velocity field, $\mathbf{V}(x, y, z, t) = (5xy^2 + t)\mathbf{i} + (2z + 8)\mathbf{j} + 18\mathbf{k}$ m/s, with $x$, $y$, $z$ in meters and $t$ in seconds. Calculate $\mathbf{V}(9, -2, 1, 4)$. What is the magnitude of this velocity?

▮
$$\mathbf{V} = [(5)(9)(-2)^2 + 4]\mathbf{i} + [(2)(1) + 8]\mathbf{j} + 18\mathbf{k} = 184\mathbf{i} + 10\mathbf{j} + 18\mathbf{k} \quad \text{m/s}$$
$$|\mathbf{V}| = \sqrt{184^2 + 10^2 + 18^2} = 185 \text{ m/s}$$

**Note:** Boldface letters are used herein and hereafter to denote vectors.

**7.4** The velocity components in a flow of fluid are specified as $v_x = 4xt + y^2z + 14$ m/s, $v_y = 2xy^2 + t^2 + y$ m/s, and $v_z = 3 + 2ty$ m/s, where $x$, $y$, and $z$ are given in meters and $t$ in seconds. What is the velocity vector at $(2, 4, 3)$ m at time $t = 4$ s? What is the magnitude of this vector at this point and time?

▮
$$\mathbf{V} = [(4)(2)(4) + (4)^2(3) + 14]\mathbf{i} + [(2)(2)(4)^2 + 4^2 + 4]\mathbf{j} + [3 + (2)(4)(4)]\mathbf{k} = 94\mathbf{i} + 84\mathbf{j} + 35\mathbf{k} \text{ m/s}$$
$$|\mathbf{V}| = \sqrt{94^2 + 84^2 + 35^2} = 131 \text{ m/s}$$

**7.5** Given the velocity field $\mathbf{V} = (5x)\mathbf{i} + (15y + 11)\mathbf{j} + (19t^2)\mathbf{k}$ m/s, determine the path of a particle which is at $(4, 6, 2)$ m at time $t = 3$ s.

$$v_x = dx/dt = 5x \tag{1}$$

$$v_y = dy/dt = 15y + 11 \tag{2}$$

$$v_z = dz/dt = 19t^2 \tag{3}$$

From (1), $dx/x = 5\,dt$, $\ln x = 5t + c_1$. At $t = 3$, $x = 4$. Hence, $\ln 4 = (5)(3) + c_1$, $c_1 = -13.6$.

$$\ln x = 5t - 13.6 \tag{4}$$

From (2), $dy/(15y + 11) = dt$, $\ln(15y + 11) = 15t + c_2$. At $t = 3$, $y = 6$. Hence, $\ln[(15)(6) + 11] = (15)(3) + c_2$, $c_2 = -40.4$.

$$\ln(15y + 11) = 15t - 40.4 \tag{5}$$

From (3), $dz = 19t^2\,dt$, $z = 19t^3/3 + c_3$. At $t = 3$, $z = 2$. Hence, $2 = (19)(3)^3/3 + c_3$, $c_3 = -169$.

$$z = 19t^3/3 - 169 \tag{6}$$

Add Eqs. (4) and (5) to get

$$\ln x + \ln(15y + 11) = 20t - 54.0 \tag{7}$$

Solve for $t$ in Eq. (6): $t = [(z + 169)(\frac{3}{19})]^{1/3}$. Substitute this value of $t$ into Eq. (7): $\ln x + \ln(15y + 11) = 20[(z + 169)(\frac{3}{19})]^{1/3} - 54.0$, $\ln[(x)(15y + 11)] = 10.81(z + 169)^{1/3} - 54.0$.

**7.6** An incompressible ideal fluid flows at 0.5 cfs through a circular pipe into a conically converging nozzle, as shown in Fig. 7-2. Determine the average velocity of flow at sections $A$ and $B$.

❚ As a first step, an approximate flow net is sketched to provide a general picture of the flow. Since this is an axially symmetric flow, the net is not a true two-dimensional flow net. At section $A$, the streamlines are parallel; hence, the area at right angles to the velocity vectors is a circle. Thus, $v_A = Q/A_A = 0.5/[(\pi)(\frac{8}{12})^2/4] = 1.43$ ft/s. At section $B$, however, the area at right angles to the streamlines is not clearly defined; it is a curved, dish-shaped section. As a rough approximation, it might be assumed to be the portion of the surface of a sphere of radius 2.0 in that is intersected by a circle of diameter 2.82 in. $v_B = Q/A_B = Q/(2\pi rh) = 0.5/[(2)(\pi)(\frac{2}{12})(0.59/12)] = 9.71$ ft/s.

**Fig. 7-2**

**7.7** Water flows at 6 gal/min through a small circular hole in the bottom of a large tank. Assuming the water in the tank approaches the hole radially, find the velocity in the tank at 2, 4, and 8 in from the hole.

❚ The area through which flow occurs is a hemispherical surface, with $A = 2\pi r^2$. $Q = 6/[(7.48)(60)] = 0.01337$ ft³/s, $v = Q/A$. At 2 in from the hole, $v = 0.01337/[(2)(\pi)(\frac{2}{12})^2] = 0.0766$ ft/s. At 4 in from the hole, $v = 0.01337/[(2)(\pi)(\frac{4}{12})^2] = 0.0192$ ft/s. At 8 in from the hole, $v = 0.01337/[(2)(\pi)(\frac{8}{12})^2] = 0.00479$ ft/s.

**7.8** Given the eulerian velocity-vector field $\mathbf{V}(x, y, z, t) = 3t\mathbf{i} + xz\mathbf{j} + ty^2\mathbf{k}$, find the acceleration of a particle.

$$\frac{dV}{dt} = \frac{\partial V}{\partial t} + \left( u\frac{\partial V}{\partial x} + v\frac{\partial V}{\partial y} + w\frac{\partial V}{\partial z} \right) \qquad u = 3t \qquad v = xz \qquad w = ty^2$$

$$\frac{\partial V}{\partial t} = \mathbf{i}\frac{\partial u}{\partial t} + \mathbf{j}\frac{\partial v}{\partial t} + \mathbf{k}\frac{\partial w}{\partial t} = 3\mathbf{i} + y^2\mathbf{k} \qquad \frac{\partial V}{\partial x} = z\mathbf{j} \qquad \frac{\partial V}{\partial y} = 2ty\mathbf{k} \qquad \frac{\partial V}{\partial z} = x\mathbf{j}$$

$$\frac{dV}{dt} = (3\mathbf{i} + y^2\mathbf{k}) + (3t)(z\mathbf{j}) + (xz)(2ty\mathbf{k}) + (ty^2)(x\mathbf{j}) = 3\mathbf{i} + (3tz + txy^2)\mathbf{j} + (y^2 + 2xyzt)\mathbf{k}$$

If $V$ is valid everywhere as given, this acceleration applies to all positions and times within the flow field.

**7.9** Flow through a converging nozzle can be approximated by a one-dimensional velocity distribution $u = u(x)$. For the nozzle shown in Fig. 7-3, assume the velocity varies linearly from $u = v_0$ at the entrance to $u = 3v_0$ at the exit: $u(x) = v_0(1 + 2x/L)$; $\partial u/\partial x = 2v_0/L$. (**a**) Compute the acceleration $du/dt$ as a general function of $x$, and (**b**) evaluate $du/dt$ at the entrance and exit if $v_0 = 10$ ft/s and $L = 1$ ft.

(**a**) $$\frac{du}{dt} = \frac{\partial u}{\partial t} + u\frac{\partial u}{\partial x} + v\frac{\partial u}{\partial y} + w\frac{\partial u}{\partial z} \qquad \frac{\partial u}{\partial t} = v\frac{\partial u}{\partial y} = w\frac{\partial u}{\partial z} = 0 \qquad u = v_0\left(1 + \frac{2x}{L}\right) \qquad \frac{\partial u}{\partial x} = \frac{2v_0}{L}$$

$$\frac{du}{dt} = 0 + \left[ v_0\left(1 + \frac{2x}{L}\right)\right]\left(\frac{2v_0}{L}\right) + 0 + 0 = \left(\frac{2v_0^2}{L}\right)\left(1 + \frac{2x}{L}\right)$$

(**b**) At the entrance, where $x = 0$, $du/dt = [(2)(10)^2(1)][1 + (2)(0)/(1)] = 200$ ft/s$^2$. At the exit, where $x = 1$ ft, $du/dt = [(2)(10)^2(1)][1 + (2)(1)/(1)] = 600$ ft/s$^2$.

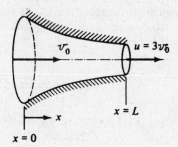

**Fig. 7-3**

**7.10** A two-dimensional velocity field is given by $u = 2y^2$, $v = 3x$, $w = 0$. At $(x, y, z) = (1, 2, 0)$, compute the (**a**) velocity, (**b**) local acceleration, and (**c**) convective acceleration.

(**a**) $$\mathbf{V} = \mathbf{i}[(2)(2)^2] + \mathbf{j}[(3)(1)] = 4\mathbf{i} + 3\mathbf{j}$$

(**b**) $$\frac{\partial V}{\partial t} = 0$$

(**c**) $$\frac{du}{dt} = \frac{\partial u}{\partial t} + u\frac{\partial u}{\partial x} + v\frac{\partial u}{\partial y} + w\frac{\partial u}{\partial z} = 0 + (2y^2)(0) + (3x)(4y) + (0)(0) = 12xy$$

$$\frac{dv}{dt} = \frac{\partial v}{\partial t} + u\frac{\partial v}{\partial x} + v\frac{\partial v}{\partial y} + w\frac{\partial v}{\partial z} = 0 + (2y^2)(3) + (3x)(0) + (0)(0) = 6y^2$$

$$\mathbf{a} = (12)(1)(2)\mathbf{i} + (6)(2)^2\mathbf{j} = 24\mathbf{i} + 24\mathbf{j}$$

**7.11** For the velocity field described in Prob. 7.10, at $(1, 2, 0)$ compute the (**a**) acceleration component parallel to the velocity vector and (**b**) component normal to the velocity vector.

From Prob. 7.10, $\mathbf{V} = 4\mathbf{i} + 3\mathbf{j}$ and $\mathbf{a} = 24\mathbf{i} + 24\mathbf{j}$ at $(1, 2, 0)$.
(**a**) Tangential acceleration:

$$\mathbf{n}_v = \mathbf{V}/|\mathbf{V}| = \tfrac{4}{5}\mathbf{i} + \tfrac{3}{5}\mathbf{j} \qquad a_t = \mathbf{a} \cdot \mathbf{n}_v = (24\mathbf{i} + 24\mathbf{j}) \cdot (\tfrac{4}{5}\mathbf{i} + \tfrac{3}{5}\mathbf{j}) = 19.2 + 14.4 = 33.6 \text{ units parallel to } \mathbf{V}$$

(b) From Fig. 7-4, the angle $\theta$ between $\mathbf{V}$ and $\mathbf{a}$ is determined by $\cos\theta = a_t/|\mathbf{a}| = 33.6/(24^2 + 24^2)^{1/2} = 0.98995$, $\theta = 8.13°$, $a_n = |\mathbf{a}|\sin\theta = (24^2 + 24^2)^{1/2}(\sin 8.13°) = 4.80$ units normal to $\mathbf{V}$.

**Fig. 7-4**

**7.12** Is the flow with velocity field $\mathbf{V} = 2tx\mathbf{i} - t^2y\mathbf{j} + 3xz\mathbf{k}$ steady or unsteady? Is it two- or three-dimensional? At the point $(x, y, z) = (2, -2, 0)$, compute the (a) total acceleration vector and (b) unit vector normal to the acceleration.

Flow is unsteady because time $t$ appears explicitly. Flow is three-dimensional because $u, v, w \neq 0$.

$$\frac{du}{dt} = \frac{\partial u}{\partial t} + u\frac{\partial u}{\partial x} + v\frac{\partial u}{\partial y} + w\frac{\partial u}{\partial z} = 2x + (2tx)(2t) + (-t^2y)(0) + (3xz)(0) = 2x + 4t^2x$$

$$\frac{dv}{dt} = \frac{\partial v}{\partial t} + u\frac{\partial v}{\partial x} + v\frac{\partial v}{\partial y} + w\frac{\partial v}{\partial z} = -2ty + (2tx)(0) + (-t^2y)(-t^2) + (3xz)(0) = -2ty + t^4y$$

$$\frac{dw}{dt} = \frac{\partial w}{\partial t} + u\frac{\partial w}{\partial x} + v\frac{\partial w}{\partial y} + w\frac{\partial w}{\partial z} = 0 + (2tx)(3z) + (-t^2y)(0) + (3xz)(3x) = 6txz + 9x^2z$$

(a)
$$\mathbf{a} = \mathbf{i}\frac{du}{dt} + \mathbf{j}\frac{dv}{dt} + \mathbf{k}\frac{dw}{dt}$$

At point $(2, -2, 0)$, $du/dt = (2)(2) + (4)(t^2)(2) = 4 + 8t^2$, $dv/dt = -(2)(t)(-2) + (t^4)(-2) = 4t - 2t^4$, $dw/dt = (6)(t)(2)(0) + (9)(2)^2(0) = 0$. Hence, $\mathbf{a} = (4 + 8t^2)\mathbf{i} + (4t - 2t^4)\mathbf{j}$. (b) The unit vector normal to $\mathbf{a}$ must satisfy $\mathbf{a}\cdot\mathbf{n} = 0 = n_x(4 + 8t^2) + n_y(4t - 2t^4) + n_z(0)$ plus $n_x^2 + n_y^2 + n_z^2 = 1$. A special case solution is $n = \pm\mathbf{k}$.

**7.13** For steady flow through a conical nozzle, the axial velocity is approximately $u = U_0(1 - x/L)^{-2}$, where $U_0$ is the entrance velocity and $L$ is the distance to the geometrical vertex of the cone. Compute (a) a general expression for the axial acceleration $du/dt$ and (b) its values at the entrance and at $x = 2$ m, if $U_0 = 4$ m/s and $L = 3$ m.

(a) $$\frac{du}{dt} = \frac{\partial u}{\partial t} + u\frac{\partial u}{\partial x} + v\frac{\partial u}{\partial y} + w\frac{\partial u}{\partial z} = 0 + \left[U_0\left(1 - \frac{x}{L}\right)^{-2}\right]\left[\left(1 - \frac{x}{L}\right)^{-3}(-2U_0)\left(-\frac{1}{L}\right)\right] + 0 + 0 = \left(1 - \frac{x}{L}\right)^{-5}\left(\frac{2U_0^2}{L}\right)$$

(b) At entrance $(x = 0)$:
$$\frac{du}{dt} = (1 - \tfrac{0}{3})^{-5}[(2)(4)^2/3] = 10.7 \text{ m/s}^2$$

At $x = 2$ m:
$$\frac{du}{dt} = (1 - \tfrac{2}{3})^{-5}[(2)(4)^2/3] = 2592 \text{ m/s}^2$$

**7.14** A two-dimensional velocity field is given by $\mathbf{V} = (x^2 - 2y^2 + 2x)\mathbf{i} - (3xy + y)\mathbf{j}$. At $x = 2$ and $y = 2$, compute the (a) accelerations $a_x$ and $a_y$, (b) velocity component in the direction $\theta = 32°$, and (c) directions of maximum acceleration and maximum velocity.

$$\frac{du}{dt} = a_x = \frac{\partial u}{\partial t} + u\frac{\partial u}{\partial x} + v\frac{\partial u}{\partial y} + w\frac{\partial u}{\partial z} = 0 + (x^2 - 2y^2 + 2x)(2x + 2) + (-3xy - y)(-4y) + 0$$

$$\frac{dv}{dt} = a_y = \frac{\partial v}{\partial t} + u\frac{\partial v}{\partial x} + v\frac{\partial v}{\partial y} + w\frac{\partial v}{\partial z} = 0 + (x^2 - 2y^2 + 2x)(-3y) + (-3xy - y)(-3x - 1) + 0$$

(a) $\quad a_x = [2^2 - (2)(2)^2 + (2)(2)][(2)(2) + 2] + [(-3)(2)(2) - 2][(-4)(2)] = 112$

$\quad a_y = [2^2 - (2)(2)^2 + (2)(2)][(-3)(2)] + [(-3)(2)(2) - 2][(-3)(2) - 1] = 98$

(b) $\quad v_{30°} = \mathbf{V} \cdot \mathbf{n}_{30°} \qquad \mathbf{V} = [2^2 - (2)(2)^2 + (2)(2)]\mathbf{i} - [(3)(2)(2) + 2]\mathbf{j} = 0\mathbf{i} - 14\mathbf{j}$

$\quad n_{32°} = 0.848\mathbf{i} + 0.530\mathbf{j} \qquad v_{32°} = (0\mathbf{i} - 14\mathbf{j})(0.848\mathbf{i} + 0.530\mathbf{j}) = -7.42$

(c) Direction of $\mathbf{a}$: $\alpha = \arctan(98/112) = \arctan 0.87500 = 41.2°$. Direction of $\mathbf{V}$ (direction of $-\mathbf{j}$): $\beta = -90°$.

**7.15** The velocity field in the neighborhood of a stagnation point is given by $u = U_0 x/L$, $v = -U_0 y/L$, $w = 0$. (a) Show that the acceleration vector is purely radial. (b) If $L = 3$ ft, what is the magnitude of $U_0$ if the total acceleration at $(x, y) = (L, L)$ is 29 ft/s²?

$$\frac{du}{dt} = a_x = \frac{\partial u}{\partial t} + u\frac{\partial u}{\partial x} + v\frac{\partial u}{\partial y} + w\frac{\partial u}{\partial z} = 0 + \left(\frac{U_0 x}{L}\right)\left(\frac{U_0}{L}\right) + 0 + 0 = \frac{U_0^2 x}{L^2}$$

$$\frac{dv}{dt} = a_y = \frac{\partial v}{\partial t} + u\frac{\partial v}{\partial x} + v\frac{\partial v}{\partial y} + w\frac{\partial v}{\partial z} = 0 + 0 + \left(\frac{-U_0 y}{L}\right)\left(\frac{-U_0}{L}\right) + 0 = \frac{U_0^2 y}{L^2}$$

(a) $\mathbf{a} = a_x\mathbf{i} + a_y\mathbf{j} = (U_0^2/L^2)(x\mathbf{i} + y\mathbf{j}) = (U_0^2/L^2)(\mathbf{r})$. (Hence, purely radial.)
(b) $|\mathbf{a}| = a(L, L) = (U_0^2/L^2)|L\mathbf{i} + L\mathbf{j}| = U_0^2\sqrt{2}/L$. If $L = 3$ ft and $|\mathbf{a}| = 29$ ft/s², $29 = U_0^2\sqrt{2}/3$, $U_0 = 7.84$ ft/s.

**7.16** A particle moves around the circular path $x^2 + y^2 = 9$ m² at a uniform speed of 4 m/s. Express the $u$ and $v$ components as functions of time, assuming $\theta = 0$ at $t = 0$. See Fig. 7-5.

$\quad u = u_r \cos\theta - u_\theta \sin\theta = -4\sin\theta$ m/s $\qquad v = v_r \sin\theta + v_\theta \cos\theta = +4\cos\theta$ m/s

But $v_\theta = r\dot{\theta}$, $4 = 3\dot{\theta}$, $\theta = \frac{4}{3}t$; hence $u = -4\sin\frac{4}{3}t$ and $v = +4\cos\frac{4}{3}t$.

**Fig. 7-5**

**7.17** A perfect fluid flows from the bottom of a large tank through a small hole at the rate $Q = 0.9$ L/s. If the fluid flows radially toward the hole with the same volume flow across every section, compute the convective acceleration at points 100 mm and 200 mm from the hole.

$\quad$ Consider the radial velocity $(v_r)$: $v_r = -Q/A_r = -Q/(2\pi r^2)$ ($A_r$ is the area of a hemisphere).

$$a_r = v_r\frac{\partial v_r}{\partial r} = \left(\frac{-Q}{2\pi r^2}\right)\left(\frac{Q}{\pi r^3}\right) = -\frac{Q^2}{2\pi^2 r^5}$$

At $r = 0.100$ m, $a_r = -(0.9 \times 10^{-3})^2/[(2)(\pi)^2(0.100)^5] = -0.0041$ m/s², or $-4.1$ mm/s². At $r = 0.200$ m, $a_r = -0.0041/32 = -0.000128$ m/s², or $-0.128$ mm/s².

**7.18** Given the velocity field $\mathbf{V}(x, y, z, t) = 10x^2\mathbf{i} - 20yx\mathbf{j} + 100t\mathbf{k}$, determine the velocity and acceleration of a particle at position $x = 1$ m, $y = 2$ m, $z = 5$ m, and $t = 0.1$ s.

**I**
$$\mathbf{V} = (10)(1)^2\mathbf{i} - (20)(2)(1)\mathbf{j} + (100)(0.1)\mathbf{k} = 10\mathbf{i} - 40\mathbf{j} + 10\mathbf{k} \text{ m/s}$$

$$\mathbf{a}(x, y, z, t) = \frac{\partial \mathbf{V}}{\partial t} + \left( v_x \frac{\partial \mathbf{V}}{\partial x} + v_y \frac{\partial \mathbf{V}}{\partial y} + v_z \frac{\partial \mathbf{V}}{\partial z} \right)$$

$$\mathbf{a} = 100\mathbf{k} + [(10x^2)(20x\mathbf{i} - 20y\mathbf{j}) + (-20yx)(-20x\mathbf{j}) + (100t)(0)] = 200x^3\mathbf{i} + (-200x^2y + 400yx^2)\mathbf{j} + 100\mathbf{k}$$

At position $x = 1$ m, $y = 2$ m, $z = 5$ m, and $t = 0.1$ s, $\mathbf{a} = (200)(1)^3\mathbf{i} + [(-200)(1)^2(2) + (400)(2)(1)^2]\mathbf{j} + 100\mathbf{k} = 200\mathbf{i} + 400\mathbf{j} + 100\mathbf{k}$ m/s$^2$.

**7.19** If the flow in Fig. 7-2 is steady at 0.50 cfs, find the acceleration in the flow at sections $A$ and $B$.

**I** Since the flow at section $A$ is uniform and also steady, $\mathbf{a}_A = 0$.

$$\mathbf{a} = \frac{\partial \mathbf{V}}{\partial t} + u\frac{\partial \mathbf{V}}{\partial x} + v\frac{\partial \mathbf{V}}{\partial y} + w\frac{\partial \mathbf{V}}{\partial z} \qquad \mathbf{a}_B = 0 + u\frac{\partial \mathbf{V}}{\partial x} + v\frac{\partial \mathbf{V}}{\partial y} + 0 = u\frac{\partial \mathbf{V}}{\partial x} + v\frac{\partial \mathbf{V}}{\partial y}$$

For point $B$ on the axis of the pipe at section $B$, $v = 0$; hence,

$$\mathbf{a}_B = u\frac{\partial \mathbf{V}}{\partial x}$$

The effective area through which the flow is occurring in the converging section of the nozzle may be expressed approximately as $A = 2\pi hr$, where $h = r(1 - \cos 45°) = 0.293r$ and $r$ is the distance from point $C$. Thus $A = (2\pi)(0.293r^2) = 1.84r^2$, and the velocity in the converging nozzle (assuming the streamlines flow radially toward $C$) may be expressed approximately as $v = Q/A = 0.50/(1.84r^2)$. At section $B$, $r = 2$ in $= 0.167$ ft; hence, $v = 0.50/[(1.84)(0.167)^2] = 9.744$ fps.

$$\frac{\partial \mathbf{V}}{\partial x} = -\frac{\partial \mathbf{V}}{\partial r} = -\left[ \frac{(-2)(0.50)}{1.84r^3} \right] = \frac{(2)(0.50)}{(1.84)(0.167^3)} = 116.7 \text{ fps/ft} \qquad a_B = u\frac{\partial \mathbf{V}}{\partial x} = (9.744)(116.7) = 1137 \text{ ft/s}^2$$

**7.20** A two-dimensional flow field is given by $u = 2y$, $v = x$. Sketch the flow field. Derive a general expression for the velocity and acceleration ($x$ and $y$ are in units of length $L$; $u$ and $v$ are in units of $L/T$). Find the acceleration in the flow field at point $A$ ($x = 3.5$, $y = 1.2$).

**I** The flow field is sketched in Fig. 7-6$a$. Velocity components $u$ and $v$ are plotted to scale, and streamlines are sketched tangentially to the resultant velocity vectors. This gives a general picture of the flow field.

$$V = (u^2 + v^2)^{1/2} = (4y^2 + x^2)^{1/2} \qquad a_x = u\frac{\partial u}{\partial x} + v\frac{\partial u}{\partial y} = 2y(0) + x(2) = 2x \qquad a_y = u\frac{\partial v}{\partial x} + v\frac{\partial v}{\partial y} = 2y(1) + x(0) = 2y$$

$$a = (a_x^2 + a_y^2)^{1/2} = (4x^2 + 4y^2)^{1/2} \qquad (a_A)_x = 2x = 7.0L/T^2 \qquad (a_A)_y = 2y = 2.4L/T^2$$

$$a_A = [(a_A)_x^2 + (a_A)_y^2]^{1/2} = [(7.0)^2 + (2.4)^2]^{1/2} = 7.4L/T^2$$

To get a rough check on the acceleration imagine a velocity vector at point $A$. This vector would have a magnitude approximately midway between that of the adjoining vectors, or $V_A \approx 4L/T$. The radius of curvature of the sketched streamline at $A$ is roughly $3L$. Thus $(a_A)_n \approx 4^2/3 \approx 5.3L/T^2$. The tangential acceleration of the particle at $A$ may be approximated by noting that the velocity along the streamline increases from about $3.2L/T$, where it crosses the $x$ axis, to about $8L/T$ at $B$. The distance along the streamline between these two points is roughly $4L$. Hence a very approximate value of the tangential acceleration at $A$ is

$$(a_A)_t = V\frac{\partial V}{\partial s} \approx 4\left( \frac{8 - 3.2}{4} \right) \approx 4.8L/T^2$$

Vector diagrams of these roughly computed normal and tangential acceleration components are plotted (Fig. 7-6$b$) for comparison with the true acceleration as given by the analytic expressions (Fig. 7-6$c$).

**7.21** The velocity along a streamline coincident with the $x$ axis is $u = 9 + x^{1/3}$. What is the convective acceleration at $x = 3.2$? Specify units in terms of $L$ and $T$. Assuming the fluid is incompressible, is the flow converging or diverging?

**I**
$$a_x = \frac{\partial u}{\partial t} + u\frac{\partial u}{\partial x} + v\frac{\partial u}{\partial y} + w\frac{\partial u}{\partial z}$$

Fig. 7-6(a)

Fig. 7-6(b)

Fig. 7-6(c)

Since

$$\frac{\partial u}{\partial t} = v\frac{\partial u}{\partial y} = w\frac{\partial u}{\partial z} = 0 \qquad a_x = u\frac{\partial u}{\partial x} = (9 + x^{1/3})\left(\frac{x^{-2/3}}{3}\right) = 3x^{-2/3} + \tfrac{1}{3}x^{-1/3}$$

At $x = 3.2$, $a_x = (3)(3.2)^{-2/3} + (\tfrac{1}{3})(3.2)^{-1/3} = 1.61 L/T^2$. For incompressible flow, the flow is converging.

**7.22** A large hemispherical vat has a small taphole centered on its lowest point. Ideal liquid drains through the hole according to $Q = 11 - 0.5t$, where $Q$ is in cubic feet per second and $t$ is in seconds. Find the total acceleration at a point 3 ft from the center of the hole at $t = 16$ s. Assume that liquid approaches the center of the hole radially.

▮ $v = Q/A$. The area through which flow occurs is a hemispherical surface, so $v = (11 - 0.5t)/(2\pi r^2)$.

$$a_{\text{total}} = v\frac{\partial v}{\partial r} + \frac{\partial v}{\partial t} = \left[\frac{(11 - 0.5t)}{2\pi r^2}\right]\left[\frac{-(11 - 0.5t)}{\pi r^3}\right] - \frac{0.5}{2\pi r^2}$$

At $r = 3$ ft and $t = 16$ s, $a_{\text{total}} = \{[11 - (0.5)(16)]/[(2)(\pi)(3)^2]\}\{-[11 - (0.5)(16)]/[(\pi)(3)^3]\} - 0.5/[(2)(\pi)(3)^2] = -0.0107$ ft/s².

**7.23** Under what conditions does the velocity field $\mathbf{V} = (a_1x + b_1y + c_1z)\mathbf{i} + (a_2x + b_2y + c_2z)\mathbf{j} + (a_3x + b_3y + c_3z)\mathbf{k}$, where $a_1$, $a_2$, etc. = constant, represent an incompressible flow which conserves mass?

▮ $$\frac{\partial u}{\partial x} + \frac{\partial v}{\partial y} + \frac{\partial w}{\partial z} = 0 \qquad \frac{\partial}{\partial x}(a_1x + b_1y + c_1z) + \frac{\partial}{\partial y}(a_2x + b_2y + c_2z) + \frac{\partial}{\partial z}(a_3x + b_3y + c_3z) = 0$$

or $a_1 + b_2 + c_3 = 0$. At least two of the constants $a_1$, $b_2$, and $c_3$ must have opposite signs. Continuity imposes no restrictions whatever on the constants $b_1$, $c_1$, $a_2$, $c_2$, $a_3$, and $b_3$, which do not contribute to a mass increase or decrease of a differential element.

**7.24**  An incompressible velocity field is given by $u = a(x^2 - y^2)$, $v$ unknown, $w = b$, where $a$ and $b$ are constants. What must the form of the velocity component $v$ be?

∎
$$\frac{\partial u}{\partial x} + \frac{\partial v}{\partial y} + \frac{\partial w}{\partial z} = 0 \qquad \frac{\partial}{\partial x}(ax^2 - ay^2) + \frac{\partial v}{\partial y} + \frac{\partial b}{\partial z} = 0 \qquad \frac{\partial v}{\partial y} = -2ax$$

This is integrated partially with respect to $y$: $v(x, y, z, t) = -2axy + f(x, z, t)$. This is the only possible form for $v$ that satisfies the incompressible continuity equation. The function of integration $f$ is entirely arbitrary since it vanishes when $v$ is differentiated with respect to $y$.

**7.25**  An incompressible flow field has $u = xz^3$ and $w = xe^{-y}$ (dimensional factors omitted). What form does continuity imply for the velocity component $v$?

∎
$$\frac{\partial u}{\partial x} + \frac{\partial v}{\partial y} + \frac{\partial w}{\partial z} = 0 \qquad z^3 + \frac{\partial v}{\partial y} + 0 = 0 \qquad \frac{\partial v}{\partial y} = -z^3 \qquad v = -z^3 y + f(x, z)$$

**7.26**  A two-dimensional incompressible velocity field has $u = K(1 - e^{-ay})$, for $x \le L$ and $0 \le L \le \infty$. What is the most general form of $v(x, y)$ for which continuity is satisfied and $v = v_0$ at $y = 0$? What are the proper dimensions for the constants $K$ and $a$?

∎  Dimensions of constants: $\{K\} = \{L/T\}$, $\{a\} = \{1/L\}$.

$$\frac{\partial u}{\partial x} + \frac{\partial v}{\partial y} + \frac{\partial w}{\partial z} = 0 \qquad 0 + \frac{\partial v}{\partial y} + 0 = 0 \qquad \frac{\partial v}{\partial y} = 0 \qquad v = f(x) \quad \text{only}$$

If $v = v_0$ at $y = 0$ for all $x$, then $v = v_0$ everywhere.

**7.27**  Which of the following velocity fields satisfies conservation of mass for incompressible plane flow?
(**a**) $u = -x$, $v = y$  (**b**) $u = 3y$, $v = 3x$  (**c**) $u = 4x$, $v = -4y$
(**d**) $u = 3xt$, $v = 3yt$  (**e**) $u = xy + y^2t$, $v = xy + x^4t$  (**f**) $u = 4x^2y^3$, $v = -2xy^4$

Ignore dimensional inconsistencies.

∎  In order to satisfy continuity,
$$\frac{\partial u}{\partial x} + \frac{\partial v}{\partial t} = 0 \qquad \text{or} \qquad \frac{\partial u}{\partial x} = -\frac{\partial v}{\partial y}$$

(**a**) $\qquad \dfrac{\partial u}{\partial x} = -1 \quad$ and $\quad \dfrac{\partial v}{\partial y} = 1 \qquad$ therefore, it does satisfy continuity.

(**b**) $\qquad \dfrac{\partial u}{\partial x} = 0 \quad$ and $\quad \dfrac{\partial v}{\partial y} = 0 \qquad$ therefore, it does satisfy continuity.

(**c**) $\qquad \dfrac{\partial u}{\partial x} = 4 \quad$ and $\quad \dfrac{\partial v}{\partial y} = -4 \qquad$ therefore, it does satisfy continutiy.

(**d**) $\qquad \dfrac{\partial u}{\partial x} = 3t \quad$ and $\quad \dfrac{\partial v}{\partial y} = 3t \qquad$ therefore, it does not satisfy continuity.

(**e**) $\qquad \dfrac{\partial u}{\partial x} = y \quad$ and $\quad \dfrac{\partial v}{\partial y} = x \qquad$ therefore, it does not satisfy continuity.

(**f**) $\qquad \dfrac{\partial u}{\partial x} = 8xy^3 \quad$ and $\quad \dfrac{\partial v}{\partial y} = -8xy^3 \qquad$ therefore, it does satisfy continuity.

**7.28**  If the radial velocity for incompressible flow is given by $v_r = b \cos \theta / r^2$, $b = $ constant, what is the most general form of $v_\theta(r, \theta)$ that satisfies continuity?

∎
$$\frac{1}{r}\frac{\partial}{\partial r}(rv_r) + \frac{1}{r}\frac{\partial}{\partial \theta}(v_\theta) + \frac{\partial v_z}{\partial z} = 0 \qquad \frac{1}{r}\frac{\partial}{\partial r}\left[(r)\left(\frac{b \cos \theta}{r^2}\right)\right] + \frac{1}{r}\frac{\partial v_\theta}{\partial \theta} + 0 = 0$$

$$\frac{\partial v_\theta}{\partial \theta} = \frac{b \cos \theta}{r^2} \qquad v_\theta = \frac{b \sin \theta}{r^2} + f(r)$$

**7.29** A two-dimensional velocity field is given by

$$u = -\frac{Ky}{x^2 + y^2} \qquad v = \frac{Kx}{x^2 + y^2}$$

where $K$ is constant. Does this field satisfy incompressible continuity? Transform these velocities into polar components $v_r$ and $v_\theta$. What might the flow represent?

■ $$\frac{\partial u}{\partial x} + \frac{\partial v}{\partial y} + \frac{\partial w}{\partial z} = 0 \qquad \frac{2xKy}{(x^2 + y^2)^2} - \frac{2yKx}{(x^2 + y^2)^2} + 0 = 0$$

Therefore, continuity is satisfied. $x^2 + y^2 = r^2$, $\cos\theta = x/r$, $\sin\theta = y/r$.

$$v_r = u\cos\theta + v\sin\theta = -\left(\frac{Ky}{x^2 + y^2}\right)\left(\frac{x}{r}\right) + \left(\frac{Kx}{x^2 + y^2}\right)\left(\frac{y}{r}\right) = -\frac{Kyx}{r^3} + \frac{Kxy}{r^3} = 0$$

$$v = -u\sin\theta + v\cos\theta = -\left[-\left(\frac{Ky}{x^2 + y^2}\right)\right]\left(\frac{y}{r}\right) + \left(\frac{Kx}{x^2 + y^2}\right)\left(\frac{x}{r}\right) = \frac{Ky^2}{r^3} + \frac{Kx^2}{r^3} = (y^2 + x^2)\left(\frac{K}{r^3}\right) = \frac{K}{r}$$

Hence, in polar coordinates $v_r(r, \theta)$ and $v_\theta(r, \theta)$, $v_r = 0$, $v_\theta = K/r$. (This represents a potential vortex.)

**7.30** For incompressible polar coordinate flow, what is the most general form of a purely circulatory motion, $v_\theta = v(r, \theta, t)$ and $v_r = 0$, which satisfies continuity?

■ $$\frac{1}{r}\frac{\partial}{\partial r}(rv_r) + \frac{1}{r}\frac{\partial}{\partial\theta}(v_\theta) + \frac{\partial v_z}{\partial z} = 0 \qquad \frac{1}{r}\frac{\partial}{\partial r}[(r)(0)] + \frac{1}{r}\frac{\partial}{\partial\theta}(v_\theta) + 0 = 0 \qquad \frac{1}{r}\frac{\partial}{\partial\theta}(v_\theta) = 0 \qquad v_\theta = f(r)$$

**7.31** What is the most general form of a purely radial polar coordinate incompressible flow pattern, $v_r = v_r(r, \theta, t)$ and $v_\theta = 0$ that satisfies continuity?

■ $$\frac{1}{r}\frac{\partial}{\partial r}(rv_r) + \frac{1}{r}\frac{\partial}{\partial\theta}(v_\theta) + \frac{\partial v_z}{\partial z} = 0 \qquad \frac{1}{r}\frac{\partial}{\partial r}(rv_r) + \left(\frac{1}{r}\right)(0) + 0 = 0 \qquad \frac{1}{r}\frac{\partial}{\partial r}(rv_r) = 0 \qquad v_r = \left(\frac{1}{r}\right)f(\theta)$$

**7.32** An incompressible steady flow pattern is given by $u = x^4 + 3z^4$ and $w = y^4 - 3yz$. What is the most general form of the third component, $v(x, y, z)$, that satisfies continuity?

■ $$\frac{\partial u}{\partial x} + \frac{\partial v}{\partial y} + \frac{\partial w}{\partial z} = 0 \qquad 4x^3 + \frac{\partial v}{\partial y} - 3y = 0 \qquad \frac{\partial v}{\partial y} = 3y - 4x^3 \qquad v = \tfrac{3}{2}y^2 - 4x^3 y + f(x, z)$$

**7.33** A certain two-dimensional shear flow near a wall, as in Fig. 7-7, has the velocity component

$$u = U\left(\frac{3y}{ax} - \frac{y^2}{a^2 x^2}\right)$$

where $a$ and $U$ are constants. Derive from continuity the velocity component $v(x, y)$ assuming that $v = 0$ at the wall, $y = 0$.

■ $$\frac{\partial u}{\partial x} + \frac{\partial v}{\partial y} + \frac{\partial w}{\partial z} = 0 \qquad \frac{\partial u}{\partial x} + \frac{\partial v}{\partial y} + 0 = 0 \qquad \frac{\partial v}{\partial y} = -\frac{\partial u}{\partial x} = -U\left(\frac{-3y}{ax^2} + \frac{2y^2}{a^2 x^3}\right) \qquad v = U\left(\frac{3y^2}{2ax^2} - \frac{2y^3}{3a^2 x^3}\right) + f(x)$$

Enforce no-slip condition: $v(x, 0) = U(0 - 0) + f(x) = 0$, $f(x) = 0$.

**Fig. 7-7**

**7.34** Consider the flat-plate boundary-layer flow in Fig. 7-8. From the no-slip condition $v = 0$ all along the wall $y = 0$, and $u = U = $ constant outside the layer. If the layer thickness $\delta$ increases with $x$ as shown, prove with incompressible two-dimensional continuity that (a) the component $v(x, y)$ is everywhere positive within the layer; (b) $v$ increases parabolically with $y$ very near the wall; and (c) $v$ reaches a positive maximum at $y = \delta$.

▮ (a) If $\delta$ increases with $x$, the streamlines in the shear layer must everywhere move upward to satisfy continuity. Therefore, $\partial u/\partial x < 0$ everywhere inside the shear layer. Since continuity requires

$$\frac{\partial u}{\partial x} + \frac{\partial v}{\partial y} = 0$$

everywhere, it follows that $\partial v/\partial y > 0$ everywhere in the shear layer.
(b) Near the wall, $u = y f(x)$, $\partial v/\partial y = -\partial u/\partial x = -f'(x)$; therefore,

$$v = -\frac{y^2}{2}f'(x) \qquad \text{(parabolic)}$$

(c) At $y \ge \delta(x)$, $\partial u/\partial x \approx 0$; therefore $\partial v/\partial y = 0$, and $v = \text{maximum}$.

Layer thickness $\delta(x)$

Fig. 7-8

**7.35** The axial velocity field for fully developed laminar flow in a pipe is $v_z = u_{max}(1 - r^2/R^2)$ and there is no swirl, $v_\theta = 0$. Determine the radial velocity field $v_r(r, z)$ from the incompressible relation if $u_{max}$ is constant and $v_r = 0$ at $r = R$. ($r$ denotes radial distance from the pipe's center; $R$ denotes the pipe's radius.)

▮ $\dfrac{1}{r}\dfrac{\partial}{\partial r}(rv_r) + \dfrac{1}{r}\dfrac{\partial}{\partial \theta}(v_\theta) + \dfrac{\partial v_z}{\partial z} = 0 \qquad \dfrac{1}{r}\dfrac{\partial}{\partial r}(rv_r) + 0 + 0 = 0 \qquad \dfrac{\partial}{\partial r}(rv_r) = 0 \qquad v_r = \dfrac{f(\theta, z)}{r}$

if $v_r(R) = 0$ for all $\theta$, $z$, $v_r = 0$.

**7.36** An incompressible flow field has the cylindrical components $v_\theta = Cr$, $v_z = K(R^2 - r^2)$, $v_r = 0$, where $C$ and $K$ are constants and $r \le R$, $z \le L$. Does this flow satisfy continuity? What might it represent physically?

▮ $\dfrac{1}{r}\dfrac{\partial}{\partial r}(rv_r) + \dfrac{1}{r}\dfrac{\partial}{\partial \theta}(v_\theta) + \dfrac{\partial v_z}{\partial z} = 0 \qquad \dfrac{1}{r}\dfrac{\partial}{\partial r}[(r)(0)] + \dfrac{1}{r}\dfrac{\partial}{\partial \theta}(Cr) + \dfrac{\partial}{\partial z}[K(R^2 - r^2)] = 0$

$$0 + 0 + 0 = 0 \qquad \text{(satisfies continuity)}$$

This flow represents pressure-driven, laminar, steady flow in a rotating tube (fully developed).

**7.37** An incompressible flow in polar coordinates is given by $v_r = K \cos \theta(1 - b/r^2)$, $v_\theta = -K \sin \theta(1 + b/r^2)$. Does this field satisfy continuity? For consistency, what should the dimensions of the constants $K$ and $b$ be?

▮ $\dfrac{1}{r}\dfrac{\partial}{\partial r}(rv_r) + \dfrac{1}{r}\dfrac{\partial}{\partial \theta}(v_\theta) + \dfrac{\partial v_z}{\partial z} = 0 \qquad \dfrac{1}{r}\dfrac{\partial}{\partial r}\left[rK \cos \theta\left(1 - \dfrac{b}{r^2}\right)\right] + \dfrac{1}{r}\dfrac{\partial}{\partial \theta}\left[-K \sin \theta\left(1 + \dfrac{b}{r^2}\right)\right] + 0 = 0$

$$\frac{1}{r}K \cos \theta\left(1 + \frac{b}{r^2}\right) - \frac{1}{r}K \cos \theta\left(1 + \frac{b}{r^2}\right) = 0 \qquad 0 = 0 \qquad \text{(satisfies continuity)}$$

Dimensions of constants: $\{K\} = \{L/T\}$, $\{b\} = \{L^2\}$.

**7.38** The $x$ component of velocity is $u = x^3 + z^4 + 6$, and the $y$ component is $v = y^3 + z^4$. Find the simplest $z$ component of velocity that satisfies continuity.

▮ $\dfrac{\partial u}{\partial x} + \dfrac{\partial v}{\partial y} + \dfrac{\partial w}{\partial z} = 0 \qquad 3x^2 + 3y^2 + \dfrac{\partial w}{\partial z} = 0 \qquad \dfrac{\partial w}{\partial z} = -3(x^2 + y^2) \qquad w = -3z(x^2 + y^2)$

**7.39** Is the continuity equation for steady, incompressible flow satisfied if the following velocity components are involved?

$$u = 2x^2 - xy + z^2 \qquad v = x^2 - 4xy + y^2 \qquad w = -2xy - yz + y^2$$

▮ $\dfrac{\partial u}{\partial x} + \dfrac{\partial v}{\partial y} + \dfrac{\partial w}{\partial z} = 0 \qquad (4x - y) + (-4x + 2y) + (-y) = 0 \qquad$ (satisfies continuity)

**7.40** For steady, incompressible flow, are the following values of $u$ and $v$ possible?
(a) $u = 4xy + y^2$, $v = 6xy + 3x$  (b) $u = 2x^2 + y^2$, $v = -4xy$

▮ $$\frac{\partial u}{\partial x} + \frac{\partial v}{\partial y} + \frac{\partial w}{\partial z} = 0$$

(a) $4y + 6x + 0 \neq 0$ (Flow is not possible.)
(b) $4x - 4x + 0 = 0$ (Flow is possible.)

**7.41** Determine whether the velocity field $\mathbf{V} = 3t\mathbf{i} + xz\mathbf{j} + ty^2\mathbf{k}$ is incompressible, irrotational, both, or neither.

▮ The divergence of this velocity field is

$$\nabla \cdot V = \frac{\partial}{\partial x}(3t) + \frac{\partial}{\partial y}(xz) + \frac{\partial}{\partial z}(ty^2) = 0$$

Therefore, this velocity field is incompressible. The curl of this velocity field is

$$\nabla \times \mathbf{V} = \begin{vmatrix} \mathbf{i} & \mathbf{j} & \mathbf{k} \\ \dfrac{\partial}{\partial x} & \dfrac{\partial}{\partial y} & \dfrac{\partial}{\partial z} \\ 3t & xz & ty^2 \end{vmatrix} = (2ty - x)\mathbf{i} + z\mathbf{k}$$

This is not zero; hence, the flow field is rotational, not irrotational.

**7.42** If a velocity potential exists for the velocity field $u = a(x^2 - y^2)$, $v = -2axy$, $w = 0$, find it and plot it.

▮ Since $w = 0$, the curl of $\mathbf{V}$ has only one ($z$) component, and we must show that it is zero.

$$(\nabla \times \mathbf{V})_z = 2\omega_z = \frac{\partial v}{\partial x} - \frac{\partial u}{\partial y} = \frac{\partial}{\partial x}(-2axy) - \frac{\partial}{\partial y}(ax^2 - ay^2) = -2ay + 2ay = 0 \qquad \text{checks}$$

The flow is indeed irrotational. A potential exists. To find $\phi(x, y)$, set

$$u = \frac{\partial \phi}{\partial x} = ax^2 - ay^2 \tag{1}$$

$$v = \frac{\partial \phi}{\partial y} = -2axy \tag{2}$$

Integrate (1)

$$\phi = \frac{ax^3}{3} - axy^2 + f(y) \tag{3}$$

Differentiate (3) and compare with (2)

$$\frac{\partial \phi}{\partial y} = -2axy + f'(y) = -2axy \tag{4}$$

Therefore $f' = 0$, or $f = $ constant. The velocity potential is $\phi = ax^3/3 - axy^2 + C$. Letting $C = 0$, we can plot the $\phi$ lines as shown in Fig. 7-9.

**7.43** Given the velocity field $\mathbf{V} = 13x^2y\mathbf{i} + 18(yz + x)\mathbf{j} + 15\mathbf{k}$, find the angular velocity vector of a fluid particle at $(2, 3, 4)$ m.

▮ $\omega_x = \dfrac{1}{2}\left(\dfrac{\partial V_z}{\partial y} - \dfrac{\partial V_y}{\partial z}\right) = \frac{1}{2}(0 - 18y) = -9y \qquad \omega_y = \dfrac{1}{2}\left(\dfrac{\partial V_x}{\partial z} - \dfrac{\partial V_z}{\partial x}\right) = \frac{1}{2}(0 - 0) = 0$

$\omega_z = \dfrac{1}{2}\left(\dfrac{\partial V_y}{\partial x} - \dfrac{\partial V_x}{\partial y}\right) = \frac{1}{2}(18 - 13x^2) = 9 - 6.5x^2 \qquad \omega = -9y\mathbf{i} + 0\mathbf{j} + (9 - 6.5x^2)\mathbf{k}$

At point $(2, 3, 4)$ m, $\omega = (-9)(3)\mathbf{i} + [9 - (6.5)(2)^2]\mathbf{k} = -27\mathbf{i} - 17\mathbf{k}$ rad/s.

**Fig. 7-9**

**7.44** Given the velocity field $\mathbf{V} = 6x^2y\mathbf{i} - (4x - 4z)\mathbf{j} + 12z^2\mathbf{k}$ m/s, compute the angular velocity field $\boldsymbol{\omega}(x, y, z)$.

$$\omega_x = \frac{1}{2}\left(\frac{\partial V_z}{\partial y} - \frac{\partial V_y}{\partial z}\right) = \tfrac{1}{2}(0 - 4) = -2 \qquad \omega_y = \frac{1}{2}\left(\frac{\partial V_x}{\partial z} - \frac{\partial V_z}{\partial x}\right) = \tfrac{1}{2}(0 - 0) = 0$$

$$\omega_z = \frac{1}{2}\left(\frac{\partial V_y}{\partial x} - \frac{\partial V_x}{\partial y}\right) = \tfrac{1}{2}(-4 - 6x^2) = -(2 + 3x^2) \qquad \boldsymbol{\omega} = -2\mathbf{i} + 0\mathbf{j} - (2 + 3x^2)\mathbf{k} \text{ rad/s}$$

**7.45** Show that any velocity field $\mathbf{V}$ expressible as the gradient of a scalar $\phi$ must be an irrotational field.

Show curl (grad $\phi$) = $\mathbf{0}$

$$\text{curl}\left(\frac{\partial \phi}{\partial x}\mathbf{i} + \frac{\partial \phi}{\partial y}\mathbf{j} + \frac{\partial \phi}{\partial z}\mathbf{k}\right) = 0$$

$$\left[\frac{\partial}{\partial y}\left(\frac{\partial \phi}{\partial z}\right) - \frac{\partial}{\partial z}\left(\frac{\partial \phi}{\partial y}\right)\right]\mathbf{i} + \left[\frac{\partial}{\partial z}\left(\frac{\partial \phi}{\partial x}\right) - \frac{\partial}{\partial x}\left(\frac{\partial \phi}{\partial z}\right)\right]\mathbf{j} + \left[\frac{\partial}{\partial x}\left(\frac{\partial \phi}{\partial y}\right) - \frac{\partial}{\partial y}\left(\frac{\partial \phi}{\partial x}\right)\right]\mathbf{k} = 0$$

Since $\partial^2\phi/\partial y\,\partial z = \partial^2\phi/\partial z\,\partial y$, etc., we see that we have proven our point provided the partial derivatives of $\phi$ are continuous.

**7.46** Is the following flow field irrotational or not? $\mathbf{V} = 12x^3y\mathbf{i} + 3x^4\mathbf{j} + 10\mathbf{k}$ ft/s.

$$\frac{\partial V_x}{\partial y} - \frac{\partial V_y}{\partial x} = 12x^3 - 12x^3 = 0 \qquad \frac{\partial V_y}{\partial z} - \frac{\partial V_z}{\partial y} = 0 \qquad \frac{\partial V_z}{\partial x} - \frac{\partial V_x}{\partial z} = 0$$

Therefore, the flow is irrotational.

**7.47** For the velocity vector $\mathbf{V} = 3t\mathbf{i} + xz\mathbf{j} + ty^2\mathbf{k}$ evaluate the volume flow and the average velocity through the square surface whose vertices are at $(0, 1, 0)$, $(0, 1, 2)$, $(2, 1, 2)$, and $(2, 1, 0)$. See Fig. 7-10.

The surface $S$ is shown in Fig. 7-10 and is such that $\mathbf{n} = \mathbf{j}$ and $dA = dx\,dz$ everywhere. The velocity field is $\mathbf{V} = 3t\mathbf{i} + xz\mathbf{j} + ty^2\mathbf{k}$. The normal component to $S$ is $\mathbf{V} \cdot \mathbf{n} = \mathbf{V} \cdot \mathbf{j} = v$, the $y$ component, which equals $xz$. The limits on the integral for $Q$ are 0 to 2 for both $dx$ and $dz$. The volume flow is thus

$$Q = \int_S V_n\,dA = \int_0^2 \int_0^2 xz\,dx\,dz = 4.0 \text{ units}$$

The area of the surface is $(2)(2) = 4$ units. Then the average velocity is $V_{av} = Q/A = 4.0/4.0 = 1.0$ unit.

**7.48** At low velocities, the flow through a long circular tube has a paraboloid velocity distribution $u = u_{max}(1 - r^2/R^2)$, where $R$ is the tube radius and $u_{max}$ is the maximum velocity, which occurs at the tube centerline. (*a*) Find a general expression for volume flow and average velocity through the tube; (*b*) compute the volume flow if $R = 3$ cm and $u_{max} = 8$ m/s; and (*c*) compute the mass flow if $\rho = 1000$ kg/m³.

**Fig. 7-10**

**(a)** The area $S$ is the cross section of the tube, and $\mathbf{n} = \mathbf{i}$. The normal component $\mathbf{V} \cdot \mathbf{n} = \mathbf{V} \cdot \mathbf{i} = u$. Since $u$ varies only with $r$, the element $dA$ can be taken to be the annular strip $dA = 2\pi r \, dr$. The volume flow becomes

$$Q = \int_S u \, dA = \int_0^R u_{max}\left(1 - \frac{r^2}{R^2}\right)2\pi r \, dr$$

Carrying out the integration over $r$, we obtain $Q = \frac{1}{2}u_{max}\pi R^2$. The average velocity is $u_{av} = Q/A = \frac{1}{2}u_{max}\pi R^2/\pi R^2 = \frac{1}{2}u_{max}$. The average velocity is half the maximum, which is an accepted result for low-speed, or *laminar*, flow through a long tube. **(b)** For the given numerical values $Q = \frac{1}{2}(8)\pi(0.03)^2 = 0.0113 \text{ m}^3/\text{s}$. **(c)** For the given density, assumed constant, $\dot{m} = \rho Q = (1000)(0.0113) = 11.3 \text{ kg/s}$.

**7.49** For low-speed (laminar) flow through a circular pipe, as shown in Fig. 7-11, the velocity distribution takes the form $u = (B/\mu)(r_0^2 - r^2)$, where $\mu$ is the fluid viscosity. Determine **(a)** the maximum velocity in terms of $B$, $\mu$, and $r_0$ and **(b)** the mass flow rate in terms of $B$, $\mu$, and $r_0$.

**(a)** $u_{max}$ occurs when $du/dr = 0$. $du/dr = -2Br/\mu = 0$, $r = 0$, $u_{max} = Br_0^2/\mu$.

**(b)**
$$\dot{m} = \int \rho v_n \, dA = \int_0^{r_0} \rho \frac{B}{\mu}(r_0^2 - r^2)(2\pi r \, dr) = 2\pi\rho\frac{B}{\mu}\left[\frac{r_0^2 r^2}{2} - \frac{r^4}{4}\right]_0^{r_0}$$

$$= \frac{\rho}{2}\left(\frac{Br_0^2}{\mu}\right)\pi r_0^2 = \left(\frac{\rho}{2}\right)u_{max}(\pi r_0^2)$$

**Fig. 7-11**

**7.50** If the fluid in Fig. 7-11 is water at 20 °C and 1 atm, what is the centerline velocity $U_0$ if the tube radius is 20 mm and the mass flow through the tube is 1.3 kg/s?

From Prob. 7.49, $\dot{m} = (\rho/2)u_{max}(\pi r_0^2)$, $1.3 = (\frac{998}{2})(u_{max})[(\pi)(0.020)^2]$, $u_{max} = U_0 = 2.07 \text{ m/s}$. (Actually, this is unrealistic. At this $\mu$, $N_R > 2000$, so the flow is probably turbulent.)

**7.51** A velocity field in arbitrary units is given by $\mathbf{V} = 3x^2\mathbf{i} - xy\mathbf{j} - 6xz\mathbf{k}$. Find the volume flow $Q$ passing through the square with corners $(x, y, z) = (1, 0, 0)$, $(1, 1, 0)$, $(1, 1, 1)$, and $(1, 0, 1)$. See Fig. 7-12.

$Q = \iint (u)_{x=1} \, dy \, dz$. Since $\mathbf{n} = \mathbf{i}$, $\mathbf{V} \cdot \mathbf{n} = u = 3x^2$. $Q = \int_0^1 \int_0^1 (3) \, dy \, dz = 3$ units.

**Fig. 7-12**

**7.52** For an incompressible fluid, express the volume flow $Q$ across the upper surface in Fig. 7-13 in terms of the inlet velocity $U_0$ and the height $\delta$ of the fluid region shown in the figure?

$$Q = \int_0^\delta U_0 b \, dy - \int_0^\delta U_0 \sin\left(\frac{\pi y}{\delta}\right) b \, dy = U_0 b [y]_0^\delta - U_0 b\left(\frac{\delta}{\pi}\right)\left[-\cos\left(\frac{\pi y}{\delta}\right)\right]_0^\delta$$

$$= U_0 b \delta - U_0 b\left(\frac{\delta}{\pi}\right)[1+1] = U_0 b \delta - 2 U_0 b\left(\frac{\delta}{\pi}\right)$$

$$= U_0 b \delta (1 - 2/\pi) = 0.363 U_0 b \delta$$

**Plate (width $b$ into paper)**     **Fig. 7-13**

**7.53** The velocity profile in water flow down a spillway is given approximately by $u = (U_0)(y/h)^{1/7}$ where $y = 0$ denotes the bottom and the depth is $h$ (see Fig. 7-14). If $U_0 = 1.4$ m/s, $h = 3$ m, and the width is 17 m, how long will it take $10^5$ m$^3$ of water to pass this section of the spillway?

$$Q = \int u \, dA = \int_0^h (U_0)\left(\frac{y}{h}\right)^{1/7} (b \, dy) = U_0 b h^{-1/7}\left[\frac{y^{8/7}}{\frac{8}{7}}\right]_0^h = \frac{7}{8} U_0 b h = (\tfrac{7}{8})(1.4)(17)(3) = 62.5 \text{ m}^3/\text{s}$$

$$t = V/Q = 10^5/62.5 = 1603 \text{ s} \quad \text{or} \quad 26.7 \text{ min}$$

**Fig. 7-14**

**7.54** Find the total derivative, $d\rho/dt$, at $x = 1$ and $y = 3$ of the density field $\rho = 3x^3 - 4y^2$ corresponding to the velocity field $\mathbf{V} = (x^2 - y^2 + x)\mathbf{i} - (3xy + y)\mathbf{j}$.

▌ $$\frac{d\rho}{dt} = \frac{\partial\rho}{\partial t} + u\frac{\partial\rho}{\partial x} + v\frac{\partial\rho}{\partial y} + w\frac{\partial\rho}{\partial z} = 0 + (x^2 - y^2 + x)(9x^2) - (3xy + y)(-8y) + 0$$

At $x = 1$ and $y = 3$, $d\rho/dt = (1^2 - 3^2 + 1)[(9)(1)^2] - [(3)(1)(3) + 3][(-8)(3)] = 279$ units.

**7.55** A frictionless, incompressible ($\rho = \rho_0$) steady flow field is given by $\mathbf{V} = 3xy\mathbf{i} - 2y^2\mathbf{j}$ in arbitrary units. Neglecting gravity, calculate the pressure gradient and evaluate this gradient at $(3, 1, 0)$.

▌ $$\rho\frac{d\mathbf{V}}{dt} = \rho_0\left(u\frac{\partial\mathbf{V}}{\partial x} + v\frac{\partial\mathbf{V}}{\partial y} + w\frac{\partial\mathbf{V}}{\partial z}\right) = \rho_0[(3xy)(3y\mathbf{i}) + (-2y^2)(3x\mathbf{i} - 4y\mathbf{j}) + 0] = -\nabla p \qquad \nabla p = \rho_0(6xy^2\mathbf{i} + 4y^3\mathbf{j})$$

At $(3, 1, 0)$, $\nabla p = \rho_0[(6)(3)(1)^2\mathbf{i} + (4)(1)^3\mathbf{j}] = \rho_0(18\mathbf{i} + 4\mathbf{j})$.

**7.56** A temperature field $T = 5xy^2$ is associated with a velocity field given by $u = 2y^2$, $v = 3x$, $w = 0$. Compute the rate of change $dT/dt$ at the point $(x, y) = (3, 4)$.

▌ $$\frac{dT}{dt} = u\frac{\partial T}{\partial x} + v\frac{\partial T}{\partial y} + w\frac{\partial T}{\partial z} = (2y^2)(5y^2) + (3x)(10xy) + 0 = 10y^4 + 30x^2y$$

At $(3, 4)$,

$$\frac{dT}{dt} = (10)(4)^4 + (30)(3)^2(4) = 3640 \text{ units}$$

**7.57** Take the velocity field $u = a(x^2 - y^2)$, $v = -2axy$, $w = 0$ and determine under what conditions it is a solution to the Navier–Stokes momentum equation. Assuming that these conditions are met, determine the resulting pressure distribution when $z$ is "up" ($g_x = 0$, $g_y = 0$, $g_z = -g$).

▌ $$\rho g_x - \frac{\partial p}{\partial x} + \mu\left(\frac{\partial^2 u}{\partial x^2} + \frac{\partial^2 u}{\partial y^2} + \frac{\partial^2 u}{\partial z^2}\right) = \rho\frac{du}{dt} \qquad \rho g_y - \frac{\partial p}{\partial y} + \mu\left(\frac{\partial^2 v}{\partial x^2} + \frac{\partial^2 v}{\partial y^2} + \frac{\partial^2 v}{\partial z^2}\right) = \rho\frac{dv}{dt}$$

$$\rho g_z - \frac{\partial p}{\partial z} + \mu\left(\frac{\partial^2 w}{\partial x^2} + \frac{\partial^2 w}{\partial y^2} + \frac{\partial^2 w}{\partial z^2}\right) = \rho\frac{dw}{dt}$$

Make a direct substitution of $u$, $v$, $w$.

$$\rho(0) - \frac{\partial p}{\partial x} + \mu(2a - 2a) = 2a^2\rho(x^3 + xy^2) \tag{1}$$

$$\rho(0) - \frac{\partial p}{\partial y} + \mu(0) = 2a^2\rho(x^2y + y^3) \tag{2}$$

$$\rho(-g) - \frac{\partial p}{\partial z} + \mu(0) = 0 \tag{3}$$

The viscous terms vanish identically (although $\mu$ is *not* zero). Equation (3) can be integrated partially to obtain

$$p = -\rho gz + f_1(x, y) \tag{4}$$

i.e., the pressure is hydrostatic in the $z$ direction, which follows anyway from the fact that the flow is two-dimensional ($w = 0$). Now the question is: Do Eqs. (1) and (2) show that the given velocity field *is* a solution? One way to find out is to form the mixed derivative $\partial^2 p/(\partial x\,\partial y)$ from (1) and (2) separately and then compare them.

Differentiate Eq. (1) with respect to $y$

$$\frac{\partial^2 p}{\partial x\,\partial y} = -4a^2\rho xy \tag{5}$$

Now differentiate Eq. (2) with respect to $x$

$$-\frac{\partial^2 p}{\partial x\,\partial y} = \frac{\partial}{\partial x}[2a^2\rho(x^2y + y^3)] = -4a^2\rho xy \tag{6}$$

Since these are identical, the given velocity field is an *exact* solution to the Navier–Stokes equation.

To find the pressure distribution, substitute Eq. (4) into Eqs. (1) and (2), which will enable us to find $f_1(x, y)$

$$\frac{\partial f_1}{\partial x} = -2a^2\rho(x^3 + xy^2) \tag{7}$$

$$\frac{\partial f_1}{\partial y} = -2a^2\rho(x^2y + y^3) \tag{8}$$

Integrate Eq. (7) partially with respect to $x$

$$f_1 = -\tfrac{1}{2}a^2\rho(x^4 + 2x^2y^2) + f_2(y) \tag{9}$$

Differentiate this with respect to $y$ and compare with Eq. (8)

$$\frac{\partial f_1}{\partial y} = -2a^2\rho x^2 y + f_2'(y) \tag{10}$$

Comparing (8) and (10), we see they are equivalent if

$$f_2'(y) = -2a^2\rho y^3$$

or

$$f_2(y) = \tfrac{1}{2}a^2\rho y^4 + C \tag{11}$$

where $C$ is a constant. Combine Eqs. (4), (9), and (11) to give the complete expression for pressure distribution

$$p(x, y, z) = -\rho gz - \tfrac{1}{2}a^2\rho(x^4 + y^4 + 2x^2y^2) + C \tag{12}$$

This is the desired solution. Do you recognize it? Not unless you go back to the beginning and square the velocity components:

$$u^2 + v^2 + w^2 = V^2 = a^2(x^4 + y^4 + 2x^2y^2) \tag{13}$$

Comparing with Eq. (12), we can rewrite the pressure distribution as

$$p + \tfrac{1}{2}\rho V^2 + \rho gz = C \tag{14}$$

**7.58** The sprinkler shown in Fig. 7-15 on p. 148 discharges water upward and outward from the horizontal plane so that it makes an angle of $\theta°$ with the $t$ axis when the sprinkler arm is at rest. It has a constant cross-sectional flow area of $A_0$ and discharges $q$ cfs starting with $\omega = 0$ and $t = 0$. The resisting torque due to bearings and seals is the constant $T_0$, and the moment of inertia of the rotating empty sprinkler head is $I_s$. Determine the equation for $\omega$ as a function of time.

▌ The control volume is the cylindrical area enclosing the rotating sprinkler head. The inflow is along the axis, so that it has no moment of momentum; hence, the torque $-T_0$ due to friction is equal to the time rate of change of moment of momentum of sprinkler head and fluid within the sprinkler head plus the net efflux of moment of momentum from the control volume. Let $V_r = q/2A_0$.

$$-T_0 = 2\frac{d}{dt}\int_0^{r_0} A_0\rho\omega r^2\, dr + I_s\frac{d\omega}{dt} - \left[\frac{2\rho q r_0}{2}(V_r\cos\theta - \omega r_0)\right]$$

The total derivative can be used. Simplifying gives

$$\frac{d\omega}{dt}(I_s + \tfrac{2}{3}\rho A_0 r_0^3) = \rho q r_0(V_r\cos\theta - \omega r_0) - T_0$$

For rotation to start, $\rho q r_0 V_r\cos\theta$ must be greater than $T_0$. The equation is easily integrated to find $\omega$ as a function of $t$. The final value of $\omega$ is obtained by setting $d\omega/dt = 0$ in the equation.

**7.59** A turbine discharging $10\text{ m}^3/\text{s}$ is to be so designed that a torque of $10\,000\text{ N}\cdot\text{m}$ is to be exerted on an impeller turning at 200 rpm that takes all the moment of momentum out of the fluid. At the outer periphery of the impeller, $r = 1$ m. What must the tangential component of velocity be at this location?

▌ $\qquad T = \rho Q[(rv_t)_2 - (rv_t)_1] \qquad 10\,000 = (1000)(10)[(1)(v_t)_{in} - 0] \qquad (v_t)_{in} = 1.00\text{ m/s}$

**7.60** The sprinkler of Fig. 7-16 discharges 0.01 cfs through each nozzle. Neglecting friction, find its speed of rotation. The area of each nozzle opening is $0.001\text{ ft}^2$.

▌ The fluid entering the sprinkler has no moment of momentum, and no torque is exerted on the system externally; hence the moment of momentum of fluid leaving must be zero. Let $\omega$ be the speed of rotation; then

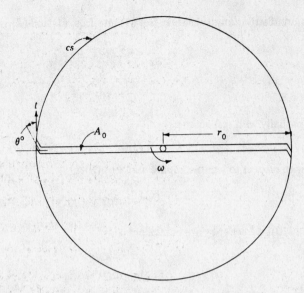

**Fig. 7-15**

the moment of momentum leaving is $\rho Q_1 r_1 v_{t1} + \rho Q_2 r_2 v_{t2}$ in which $v_{t1}$ and $v_{t2}$ are absolute velocities. Then $v_{t1} = v_{r1} - \omega r_1 = Q_1/0.001 - \omega r_1 = 10 - \omega$ and $v_{t2} = v_{r2} - \omega r_2 = 10 - \frac{2}{3}\omega$. For the moment of momentum to be zero, $\rho Q(r_1 v_{t1} + r_2 v_{t2}) = 0$ or $(1)(10 - \omega) + (\frac{2}{3})(10 - \frac{2}{3}\omega) = 0$ and $\omega = 11.54$ rad/s, or 110.2 rpm.

**Fig. 7-16**

**7.61** The velocity profile for laminar flow between two plates, as in Fig. 7-17, is

$$u = \frac{4u_{max}y(h - y)}{h^2} \qquad v = w = 0$$

If the wall temperature is $T_w$ at both walls, use the incompressible-flow energy equation to solve for the temperature distribution $T(y)$ between the walls for steady flow.

$$\rho c_v \frac{dT}{dt} = K\frac{d^2T}{dy^2} + \mu\left(\frac{du}{dy}\right)^2 \qquad 0 = K\frac{d^2T}{dy^2} + \mu\left(\frac{du}{dy}\right)^2$$

$$\frac{d^2T}{dy^2} = -\frac{\mu}{K}\left(\frac{du}{dy}\right)^2 = -\frac{\mu}{K}\left[\frac{4u_{max}}{h^2}(h - 2y)\right]^2 = \left(\frac{-16\mu u_{max}^2}{Kh^4}\right)(h^2 - 4hy + 4y^2)$$

$$\frac{dT}{dy} = \left(\frac{-16\mu u_{max}^2}{Kh^4}\right)[h^2y - 2hy^2 + (\tfrac{4}{3})(y)^3 + C_1]$$

Since $dT/dy = 0$ at $y = h/2$, $C_1 = -h^3/6$.

$$\frac{dT}{dy} = \left(\frac{-16\mu u_{max}^2}{Kh^4}\right)\left[h^2y - 2hy^2 + (\tfrac{4}{3})(y)^3 - \frac{h^3}{6}\right] \qquad T = \left(\frac{-16\mu u_{max}^2}{Kh^4}\right)\left[\frac{h^2y^2}{2} - (\tfrac{2}{3})(hy^3) + \frac{y^4}{3} - \frac{yh^3}{6} + C_2\right]$$

If $T = T_w$ at $y = 0$ and $y = h$, then $C_2 = T_w$.

$$T = T_w + \left(\frac{8\mu u_{max}^2}{K}\right)\left[\left(\frac{1}{3}\right)\left(\frac{y}{h}\right) - \frac{y^2}{h^2} + \left(\frac{4}{3}\right)\left(\frac{y^3}{h^3}\right) - \left(\frac{2}{3}\right)\left(\frac{y^4}{h^4}\right)\right]$$

**Fig. 7-17**

**7.62**  Consider a viscous, steady flow through a pipe (Fig. 7-18$a$). The velocity profile forms a paraboloid about the pipe centerline, given as

$$V = -C(r^2 - D^2/4) \qquad \text{m/s} \tag{1}$$

where $C$ is a constant. **(a)** What is the flow of mass through the left end of the control surface shown dashed? **(b)** What is the flow of kinetic energy through the left end of the control surface? Assume that the velocity profile does not change along the pipe.

❚  In Fig. 7-18$b$, we have shown a cross section of the pipe. For an infinitesimal strip, we can say noting that $\mathbf{V}$ and $d\mathbf{A}$ are collinear but of opposite sense: $\rho\mathbf{V} \cdot d\mathbf{A} = \rho[C(r^2 - D^2/4)]2\pi r\, dr$. For the whole cross section, we have

$$\iint \rho\mathbf{V} \cdot d\mathbf{A} = \rho \int_0^{D/2} C\left(r^2 - \frac{D^2}{4}\right)2\pi r\, dr = 2\pi\rho C\left[\frac{r^4}{4} - \frac{D^2}{4}\frac{r^2}{2}\right]_0^{D/2} = -\frac{\rho C\pi D^4}{32} \qquad \text{kg/s} \tag{2}$$

We now turn to the flow of kinetic energy through the left end of the control surface. The kinetic energy for an element of fluid is $\frac{1}{2}dm V^2$. This corresponds to an infinitesimal amount of an extensive property $N$. To get $\eta$, the corresponding intensive property, we divide by $dm$ to get

$$\eta = \tfrac{1}{2}V^2 \tag{3}$$

We accordingly wish to compute $\iint \eta\rho\mathbf{V} \cdot d\mathbf{A} = \iint (\tfrac{1}{2}V^2)\{\rho\mathbf{V} \cdot d\mathbf{A}\}$. Employing Eq. (1) for $V$, and noting again that $\mathbf{V}$ and $d\mathbf{A}$ are collinear but of opposite sense, we get

$$\iint \eta\rho\mathbf{V} \cdot d\mathbf{A} = \int_0^{D/2} \frac{1}{2}C^2\left(r^2 - \frac{D^2}{4}\right)^2 \left\{\rho\left[C\left(r^2 - \frac{D^2}{4}\right)2\pi r\, dr\right]\right\}$$

$$= \rho C^3\pi \int_0^{D/2} \left(r^2 - \frac{D^2}{4}\right)^3 r\, dr = \frac{\rho C^3\pi D^8}{2048}\,\text{N}\cdot\text{m/s} \tag{4}$$

**Fig. 7-18($a$)**　　　　　　　　　　**Fig. 7-18($b$)**

**7.63**  In Prob. 7.62, assume a *one-dimensional* model with the same mass flow. Compute the kinetic energy flow through a section of the pipe for this model. That is, compute kinetic energy flow with an average constant velocity. What is the ratio of the actual kinetic energy to the kinetic energy flow for the one-dimensional model flow?

❚  We first compute the constant velocity at a section for the one-dimensional model. Hence, using Eq. (2) of Prob. 7.62,

$$-(V_{av})\left(\frac{\rho\pi D^2}{4}\right) = -\frac{\rho C D^4\pi}{32}$$

$$V_{av} = \frac{CD^2}{8} \qquad \text{m/s} \tag{1}$$

The kinetic energy flow for the one-dimensional model is then

$$\iint \frac{V^2}{2}(\rho\mathbf{V} \cdot d\mathbf{A}) = -\frac{\rho}{2}\left(\frac{CD^2}{8}\right)^3\left(\frac{\pi D^2}{4}\right) = -\frac{\rho C^3 D^8\pi}{4096} \qquad \text{N}\cdot\text{m/s} \tag{2}$$

We now define the *kinetic-energy correction* factor $\alpha$ as the ratio of the actual flow of kinetic energy through a cross section to the flow of kinetic energy for one-dimensional model for the same mass flow. That is

$$\alpha = \frac{\text{KE flow for section}}{\text{KE flow for 1-}D\text{ model}} \qquad (3)$$

For the case at hand, we have from Eq. (2) of this problem and Eq. (4) of Prob. 7.62

$$\alpha = \frac{-\rho C^3 \pi D^8/2048}{-\rho C^3 \pi D^8/4096} = 2 \qquad (4)$$

The factor $\alpha$ exceeds unity, so there is an underestimation of kinetic energy flow for a one-dimensional model.

**7.64** The velocity field in a diffuser is $u = U_0 e^{-2x/L}$, and the density field is $\rho = \rho_0 e^{-x/L}$. Find the rate of change of density at $x = L$.

■ $$\frac{d\rho}{dt} = \frac{\partial\rho}{\partial t} + u\frac{\partial\rho}{\partial x} + v\frac{\partial\rho}{\partial y} + w\frac{\partial\rho}{\partial z} = 0 + (U_0 e^{-2x/L})\left(\frac{-\rho_0 e^{-x/L}}{L}\right) + 0 + 0 = -\frac{\rho_0 U_0}{L}e^{-3x/L}$$

At $x = L$,

$$\frac{d\rho}{dt} = -\frac{\rho_0 U_0}{L}e^{-3L/L} = -\frac{0.0498\rho_0 U_0}{L}$$

**7.65** Gas is flowing in a long 4-in-diameter pipe from $A$ to $B$. At section $A$ the flow is 0.30 lb/s, while at the same instant at section $B$ the flow is 0.33 lb/s. The distance between $A$ and $B$ is 700 ft. Find the mean value of the time rate of change of the specific weight of the gas between sections $A$ and $B$ at that instant.

■ $$\gamma_1 A_1 v_1 - \gamma_2 A_2 v_2 = \left(\frac{\partial\gamma}{\partial t}\right)_{avg}\text{(volume of section)}.\text{ Since }G = \gamma A v,$$

$$0.30 - 0.33 = \left(\frac{\partial\gamma}{\partial t}\right)_{avg}[(700)(\pi)(\tfrac{4}{12})^2/4] \qquad \left(\frac{\partial\gamma}{\partial t}\right)_{avg} = -0.000491 \text{ lb/ft}^3/\text{s}$$

**7.66** An incompressible flow field is given by $\mathbf{V} = x^2\mathbf{i} - z^2\mathbf{j} - 3xz\mathbf{k}$ with $V$ in meters per second and $(x, y, z)$ in meters. If the fluid viscosity is 0.04 Pa · s, evaluate the entire viscous stress tensor at the point $(x, y, z) = (3, 2, 1)$.

■ $$\tau_{ij} = \begin{vmatrix} \tau_{xx} & \tau_{yx} & \tau_{zx} \\ \tau_{xy} & \tau_{yy} & \tau_{zy} \\ \tau_{xz} & \tau_{yz} & \tau_{zz} \end{vmatrix} \quad \tau_{xx} = 2\mu\frac{\partial u}{\partial x} = 4\mu x \quad \tau_{yy} = 2\mu\frac{\partial v}{\partial y} = 0 \quad \tau_{zz} = 2\mu\frac{\partial w}{\partial z} = -6\mu x$$

$$\tau_{xy} = \tau_{yx} = (\mu)\left(\frac{\partial u}{\partial y} + \frac{\partial v}{\partial x}\right) = 0 \quad \tau_{yz} = \tau_{zy} = (\mu)\left(\frac{\partial v}{\partial z} + \frac{\partial w}{\partial y}\right) = -2\mu z \quad \tau_{xz} = \tau_{zx} = (\mu)\left(\frac{\partial u}{\partial z} + \frac{\partial w}{\partial x}\right) = -3\mu z$$

At $(x, y, z) = (3, 2, 1)$ for $\mu = 0.04$ kg/(m · s):

$$\tau_{ij} = \begin{vmatrix} 0.48 & 0 & -0.12 \\ 0 & 0 & -0.08 \\ -0.12 & -0.08 & -0.72 \end{vmatrix} \text{ Pa}$$

**7.67** Given the velocity distribution

$$u = Kx \qquad v = -Ky \qquad w = 0 \qquad (1)$$

where $k$ is constant, compute and plot the streamlines of flow, including directions, and give some possible interpretations of the pattern.

■ Since time does not appear explicitly in Eqs. (1), the motion is steady, so that streamlines, path lines, and streaklines will coincide. Since $w = 0$ everywhere, the motion is two-dimensional, in the $xy$ plane. The streamlines can be computed by substituting the expressions for $u$ and $v$ into

$$\frac{dx}{u} = \frac{dy}{v} = \frac{dz}{w} = \frac{dr}{V} \quad \frac{dx}{Kx} = -\frac{dy}{Ky} \quad \text{or} \quad \int\frac{dx}{x} = -\int\frac{dy}{y}$$

Integrating, we obtain $\ln x = -\ln y + \ln C$, or

$$xy = C \qquad (2)$$

This is the general expression for the streamlines, which are hyperbolas. The complete pattern is plotted in Fig. 7-19 by assigning various values to the constant $C$. The arrowheads can be determined only by returning to Eqs. ($1$) to ascertain the velocity component directions, assuming $K$ is positive. For example, in the upper right quadrant ($x > 0$, $y > 0$), $u$ is positive and $v$ is negative; hence the flow moves down and to the right, establishing the arrowheads as shown.

Note that the streamline pattern is entirely independent of the constant $K$. It could represent the impingement of two opposing streams, or the upper half could simulate the flow of a single downward stream against a flat wall. Taken in isolation, the upper right quadrant is similar to the flow in a 90° corner.

Finally note the peculiarity that the two streamlines ($C = 0$) have opposite directions and intersect each other. This is possible only at a point where $u = v = w = 0$, which occurs at the origin in this case. Such a point of zero velocity is called a *stagnation point*.

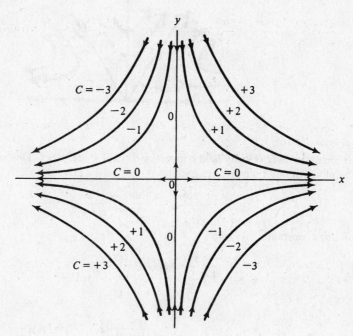

**Fig. 7-19**

**7.68** A velocity field is given by $u = V \cos \theta$, $v = V \sin \theta$, and $w = 0$, where $V$ and $\theta$ are constants. Find an expression for the streamlines of this flow.

$$\frac{dx}{u} = \frac{dy}{v} = \frac{dz}{w} = \frac{dr}{V} \qquad \frac{dx}{V \cos \theta} = \frac{dy}{V \sin \theta} = \frac{dz}{0}$$

(*Note*: $dz/0$ indicates that the streamlines do not vary with $z$.)

$$\frac{dy}{dx} = \frac{V \sin \theta}{V \cos \theta} = \tan \theta \qquad y = x \tan \theta + C$$

Hence, the streamlines are straight and inclined at angle $\theta$, as illustrated in Fig. 7-20.

**Fig. 7-20**

**7.69** A two-dimensional steady velocity field is given by $u = 3x^2 - 2y^2$, $v = -6xy$. Derive the streamline pattern and sketch a few streamlines in the upper half-plane.

$$\frac{dx}{u} = \frac{dy}{v} = \frac{dz}{w} = \frac{dr}{V} \qquad \frac{dx}{3x^2 - 2y^2} = \frac{dy}{-6xy}$$

$$-6xy\,dx = (3x^2 - 2y^2)\,dy \qquad df = 6xy\,dx + (3x^2 - 2y^2)\,dy \qquad f(x, y) = 3x^2y - 2y^3/3 = \text{const.}$$

Hence, the streamlines represent inviscid flow in three corners, as illustrated in Fig. 7-21.

Fig. 7-21

**7.70** A two-dimensional unsteady velocity field is given by $u = x(1 + 3t)$, $v = y$. Determine the one-parameter $(t)$ family of streamlines through the point $(x_0, y_0)$.

$$\frac{dx}{u} = \frac{dy}{v} = \frac{dz}{w} = \frac{dr}{V} \qquad \frac{dx}{x(1 + 3t)} = \frac{dy}{y}$$

Integrate, holding $t$ constant.

$$\frac{\ln x}{1 + 3t} = \ln y + C \qquad y = Cx^{1/(1+3t)}$$

If $y = y_0$ at $x = x_0$, $y_0 = Cx_0^{1/(1+3t)}$.

$$C = \frac{y_0}{x_0^{1/(1+3t)}} \qquad y = \left[\frac{y_0}{x_0^{1/(1+3t)}}\right][x^{1/(1+3t)}] = (y_0)\left(\frac{x}{x_0}\right)^{1/(1+3t)}$$

Some streamlines of the family are sketched in Fig. 7-22.

Fig. 7-22

**7.71** Investigate the stream function in polar coordinates

$$\psi = U \sin \theta (r - R^2/r) \qquad (1)$$

where $U$ and $R$ are constants, a velocity and a length, respectively. Plot the streamlines. What does the flow represent? Is it a realistic solution to the basic equations?

**▮** The streamlines are lines of constant $\psi$, which has units of square meters per second. Note that $\psi/UR$ is dimensionless. Rewrite Eq. (1) in dimensionless form

$$\psi/UR = \sin \theta (\eta - 1/\eta) \qquad \eta = r/R \qquad (2)$$

Of particular interest is the special line $\psi = 0$. From Eq. (1) or (2) this occurs when (a) $\theta = 0°$ or $180°$ and (b) $r = R$. Case (a) is the $x$ axis and case (b) is a circle of radius $R$, both of which are plotted in Fig. 7-23.

For any other nonzero value of $\psi$ it is easiest to pick a value of $r$ and solve for $\theta$:

$$\sin \theta = \frac{\psi/UR}{r/R - R/r} \qquad (3)$$

In general, there will be two solutions for $\theta$ because of the symmetry about the $y$ axis. For example take $\psi/UR = +1.0$:

| Guess $r/R$ | 3.0 | 2.5 | 2.0 | 1.8 | 1.7 | 1.618 |
|---|---|---|---|---|---|---|
| Compute $\theta$ | 22° | 28° | 42° | 54° | 64° | 90° |
|  | 158° | 152° | 138° | 126° | 116° |  |

This line is plotted in Fig. 7-23 and passes over the circle $r = R$. You have to watch it, though, because there is a second curve for $\psi/UR = +1.0$ for small $r < R$ below the $x$ axis:

| Guess $r/R$ | 0.618 | 0.6 | 0.5 | 0.4 | 0.3 | 0.2 | 0.1 |
|---|---|---|---|---|---|---|---|
| Compute $\theta$ | −90° | −70° | −42° | −28° | −19° | −12° | −6° |
|  |  | −110° | −138° | −152° | −161° | −168° | −174° |

This second curve plots as a closed curve inside the circle $r = R$. There is a singularity of infinite velocity and indeterminate flow direction at the origin. Figure 7-23 shows the full pattern.

The given stream function, Eq. (1), is an exact and classic solution to the momentum equation for frictionless flow. Outside the circle $r = R$ it represents two-dimensional inviscid flow of a uniform stream past a circular cylinder. Inside the circle it represents a rather unrealistic trapped circulating motion of what is called a *line doublet*.

Fig. 7-23

**7.72** In two-dimensional, incompressible steady flow around an airfoil, the streamlines are drawn so that they are 10 mm apart at a great distance from the airfoil, where the velocity is 40 m/s. What is the velocity near the airfoil, where the streamlines are 7.5 mm apart?

**▮**
$$Q = Av = (\tfrac{10}{1000})(40) = 0.40 \frac{\text{m}^3/\text{s}}{\text{m}} \qquad 0.40 = (7.5/1000)(v) \qquad v = 53.3 \text{ m/s}$$

**7.73** A three-dimensional velocity distribution is given by $u = -x$, $v = 2y$, $w = 6 - z$. Find the equation of the streamline through $(1, 2, 3)$.

$$\frac{dx}{u} = \frac{dy}{v} = \frac{dz}{w} \qquad \frac{dx}{-x} = \frac{dy}{2y} \qquad -\ln x = \ln \sqrt{y} + \ln C_1 \qquad x\sqrt{y} = -C_1$$

At $x = 1$, $y = 2$, $1\sqrt{2} = -C_1$, $C_1 = -1.414$; $x\sqrt{y} = 1.414$.

$$\frac{dx}{-x} = \frac{dz}{5-z} \qquad -\ln x = -\ln(6-z) + \ln C_2 \qquad \frac{6-z}{x} = C_2$$

At $x = 1$, $z = 3$, $(6-3)/1 = C_2$, $C_2 = 3$; $(6-z)/x = 3$. Therefore, $x\sqrt{y} = 1.414$ and $(6-z)/x = 3$ is the equation of the streamline.

**7.74** A two-dimensional flow can be described by $u = -y/b^2$, $v = x/a^2$. Verify that this is the flow of an incompressible fluid and that the ellipse $x^2/a^2 + y^2/b^2 = 1$ is a streamline.

$$\frac{\partial u}{\partial x} + \frac{\partial v}{\partial y} = 0 + 0 = 0 \qquad \text{(Therefore, continuity is satisfied.)}$$

$$\frac{dx}{u} = \frac{dy}{v} = \frac{dz}{w} \qquad \frac{dx}{-y/b^2} = \frac{dy}{x/a^2} \qquad \frac{x\,dx}{a^2} = \frac{-y\,dy}{b^2}$$

$$x^2/a^2 + y^2/b^2 = \text{constant} \qquad \text{(Therefore, ellipse } x^2/a^2 + y^2/b^2 = 1 \text{ is a streamline.)}$$

**7.75** A velocity potential in two-dimensional flow is $\phi = xy + x^2 - y^2$. Find the stream function for this flow.

$$\frac{\partial \phi}{\partial x} = \frac{\partial \psi}{\partial y} \qquad \frac{\partial \phi}{\partial x} = y + 2x \qquad \frac{\partial \psi}{\partial y} = y + 2x \qquad \psi = \tfrac{1}{2}y^2 + 2xy + f(x)$$

$$\frac{\partial \phi}{\partial y} = -\frac{\partial \psi}{\partial x} \qquad \frac{\partial \phi}{\partial y} = x - 2y \qquad -\frac{\partial \psi}{\partial x} = x - 2y \qquad -2y - f'(x) = x - 2y$$

Therefore, $f(x) = -\tfrac{1}{2}x^2(+C)$ and $\psi = 2xy + \tfrac{1}{2}y^2 - \tfrac{1}{2}x^2(+C)$.

**7.76** For the steady two-dimensional flow shown in Fig. 7-24, the scalar components of the velocity field are $V_x = -x$, $V_y = y$, $V_z = 0$. Find the equations of the streamlines and the components of acceleration.

$(dy/dx)_{\text{stream}} = V_y/V_x = -y/x$, $dy/y = -dx/x$, $\ln y = -\ln x + \ln C$. Hence, $xy = C$. Note that the streamlines form a family of rectangular hyperbolas. The wetted boundaries are part of the family, as is to be expected.

$$a_x = \frac{\partial V_x}{\partial t} + \left( V_x \frac{\partial V_x}{\partial x} + V_y \frac{\partial V_x}{\partial y} + V_z \frac{\partial V_x}{\partial z} \right) = 0 + (-x)(-1) + (y)(0) + 0 = x$$

$$a_y = \frac{\partial V_y}{\partial t} + \left( V_x \frac{\partial V_y}{\partial x} + V_y \frac{\partial V_y}{\partial y} + V_z \frac{\partial V_y}{\partial z} \right) = 0 + (-x)(0) + (y)(1) + 0 = y$$

$$a_z = 0 \qquad \mathbf{a} = x\mathbf{i} + y\mathbf{j}$$

$xy = K$

**Fig. 7-24**

**7.77** A two-dimensional dipole source at the origin produces steady incompressible flow with stream function

$$\psi = \frac{y}{x^2 + y^2}$$

Find the direction of motion of a fluid particle at the point $x = 6$, $y = 9$.

**❚** Along the streamline $\psi = $ const. through the given point,

$$0 = d\psi = \frac{(x^2 + y^2)\,dy - y\,(2x\,dx + 2y\,dy)}{(x^2 + y^2)^2}$$

from which

$$\frac{dy}{dx} = \frac{2xy}{x^2 - y^2} = \frac{2(6)(9)}{6^2 - 9^2} = -2.4$$

Therefore the particle is moving at an angle of arctan $(-2.4) = -67.4°$ with the positive $x$-axis.

**7.78** Sketch the streamlines for Prob. 7.77.

**❚** The streamline $\psi = 1/2\lambda$ has the equation

$$\frac{y}{x^2 + y^2} = \frac{1}{2\lambda} \qquad \text{or} \qquad x^2 + (y - \lambda)^2 = \lambda^2$$

It is thus a circle of radius $|\lambda|$ centered at $(0, \lambda)$; see Fig. 7-25.

**Fig. 7-25**

**7.79** In Prob. 7.76 find the normal acceleration of a fluid particle as it moves through the position $x = 3$, $y = 5$.

**❚** At $(3, 5)$, $\mathbf{V} = -3\mathbf{i} + 5\mathbf{j}$ and $\mathbf{a} = 3\mathbf{i} + 5\mathbf{j}$. Therefore (see Fig. 7-26),

$$\mathbf{a}_V = \frac{\mathbf{a} \cdot \mathbf{V}}{|\mathbf{V}|^2}\mathbf{V} = \frac{(3)(-3) + (5)(5)}{(-3)^2 + (5)^2}(-3\mathbf{i} + 5\mathbf{j}) = -\tfrac{24}{17}\mathbf{i} + \tfrac{40}{17}\mathbf{j}$$

$$\mathbf{a}_N = \mathbf{a} - \mathbf{a}_V = \tfrac{15}{17}(5\mathbf{i} + 3\mathbf{j})$$

**Fig. 7-26**

**7.80** Repeat Prob. 7.70 to find the equation of the path line that passes through $(x_0, y_0)$ at time $t = 0$. Sketch it.

▮ $$u = dx/dt = x(1 + 3t) \qquad dx/x = (1 + 3t)\,dt \qquad \ln x = t + 1.5t^2 + C_1$$
$$x = \exp\,(t + 1.5t^2 + C_1) = [\exp\,(t + 1.5t^2)](C_2)$$

At $x = x_0$ and $t = 0$, $x_0 = e^0 C_2$, $C_2 = x_0$; $x = x_0 \exp\,(t + 1.5t^2)$, $v = dy/dt = y$, $dy/y = dt$, $\ln y = t + C_3$, $y = \exp\,(t + C_3) = e^t C_4$.

At $y = y_0$ and $t = 0$, $y_0 = e^0 C_4$, $C_4 = y_0$; $y = y_0 e^t$, $t = \ln\,(y/y_0)$, $x = x_0 \exp\,(t + 1.5t^2) = x_0 \exp\,[\ln\,(y/y_0) + 1.5 \ln^2\,(y/y_0)]$. This pathline is sketched in Fig. 7-27.

Fig. 7-27

# CHAPTER 8
# Fundamentals of Fluid Flow

**8.1** Water flows through a 3-in-diameter pipe at a velocity of 10 ft/s. Find the (*a*) volume flow rate in cfs and gpm, (*b*) weight flow rate, and (*c*) mass flow rate.

■ (*a*) $$Q = Av = [(\pi)(\tfrac{3}{12})^2/4](10) = 0.4909 \text{ cfs} = 0.4909/0.002228 = 220 \text{ gpm}$$

(*b*) $$W = \gamma Av = 62.4[(\pi)(\tfrac{3}{12})^2/4](10) = 30.6 \text{ lb/s}$$

(*c*) $$M = \rho Av = 1.94[(\pi)(\tfrac{3}{12})^2/4](10) = 0.952 \text{ slug/s}$$

**8.2** Benzene flows through a 100-mm-diameter pipe at a mean velocity of 3.00 m/s. Find the (*a*) volume flow rate in $m^3$/s and L/min, (*b*) weight flow rate, and (*c*) mass flow rate.

■ (*a*) $$Q = Av = [(\pi)(\tfrac{100}{1000})^2/4](3.00) = 0.0236 \text{ m}^3\text{/s} = 0.0236/0.00001667 = 1416 \text{ L/min}$$

(*b*) $$W = \gamma Av = 8.62[(\pi)(\tfrac{100}{1000})^2/4](3.00) = 0.203 \text{ kN/s}$$

(*c*) $$M = \rho Av = 879[(\pi)(\tfrac{100}{1000})^2/4](3.00) = 20.7 \text{ kg/s}$$

**8.3** The flow rate of air moving through a square 0.50-m by 0.50-m duct is 160 $m^3$/min. What is the mean velocity of the air?

■ $$v = Q/A = 160/[(0.50)(0.50)] = 640 \text{ m/min} \quad \text{or} \quad 10.7 \text{ m/s}$$

**8.4** Assume the conduit shown in Fig. 8-1 has (inside) diameters of 12 in and 18 in at sections 1 and 2, respectively. If water is flowing in the conduit at a velocity of 16.6 ft/s at section 2, find the (*a*) velocity at section 1, (*b*) volume flow rate at section 1, (*c*) volume flow rate at section 2, (*d*) weight flow rate, and (*e*) mass flow rate.

■ (*a*) $$A_1v_1 = A_2v_2 \quad [(\pi)(\tfrac{12}{12})^2/4](v_1) = [(\pi)(\tfrac{18}{12})^2/4](16.6) \quad v_1 = 37.3 \text{ ft/s}$$

(*b*) $$Q_1 = A_1v_1 = [(\pi)(\tfrac{12}{12})^2/4](37.3) = 29.3 \text{ ft}^3\text{/s}$$

(*c*) $Q_2 = A_2v_2 = [(\pi)(\tfrac{18}{12})^2/4](16.6) = 29.3 \text{ ft}^3\text{/s}$. (Since the flow in incompressible, the flow rate is the same at sections 1 and 2.)

(*d*) $$W = \gamma A_1v_1 = 62.4[(\pi)(\tfrac{12}{12})^2/4](37.3) = 1828 \text{ lb/s}$$

(*e*) $$M = \rho A_1v_1 = 1.94[(\pi)(\tfrac{12}{12})^2/4](37.3) = 56.8 \text{ slugs/s}$$

Fluid flow →

**Fig. 8-1**

**8.5** A gas flows through a square conduit. At one point along the conduit, the conduit sides are 0.100 m, the velocity is 7.55 m/s, and the gas's mass density is (for its particular pressure and temperature) 1.09 $kg/m^3$. At a second point, the conduit sides are 0.250 m, and the velocity is 2.02 m/s. Find the mass flow rate of the gas and its mass density at the second point.

■ $$M = \rho_1 A_1 v_1 = 1.09[(0.100)(0.100)](7.55) = 0.0823 \text{ kg/s} \qquad \rho_1 A_1 v_1 = \rho_2 A_2 v_2$$

$$1.09[(0.100)(0.100)](7.55) = (\rho_2)[(0.250)(0.250)](2.02) \qquad \rho_2 = 0.652 \text{ kg/m}^3$$

**8.6** Water enters the mixing device shown in Fig. 8-2 at 150 L/s through pipe A, while oil with specific gravity 0.8 is forced in at 30 L/s through pipe B. If the liquids are incompressible and form a homogeneous mixture of oil globules in water, find the average velocity and density of the mixture leaving through the 30-cm-diameter pipe C.

■ $$M = \rho Av = \rho Q \qquad \sum(\text{mass flow in unit time})_{\text{in}} = \sum(\text{mass flow in unit time})_{\text{out}}$$

$$(1000)(0.15) + [(0.8)(1000)](0.03) = (\rho)[(\pi)(0.30)^2/4](v) \qquad \rho v = 2462 \text{ kg/m}^2 \cdot \text{s}$$

We can assume no chemical reaction between oil and water and its mixture is incompressible; it is clear that volume is conserved. Hence, $Q = 0.15 + 0.03 = 0.18 \text{ m}^3/\text{s}$; $Q = Av$, $0.18 = [(\pi)(0.30)^2/4](v)$, $v_C = 2.55 \text{ m/s}$; $\rho_C = 2462/2.55 = 965 \text{ kg/m}^3$.

**Fig. 8-2**

**8.7** Water flows into a cylindrical tank (Fig. 8-3) through pipe 1 at the rate of 25 ft/s and leaves through pipes 2 and 3 at 10 ft/s and 12 ft/s, respectively. At 4 is an open air vent. Inside pipe diameters are: $D_1 = 3 \text{ in}$, $D_2 = 2 \text{ in}$, $D_3 = 2.5 \text{ in}$, $D_4 = 2 \text{ in}$. Calculate (a) $dh/dt$; (b) the average velocity of airflow through vent 4, assuming that the flow is incompressible.

▎ (a) With the entire volume of the tank as control volume,

$$M = \rho Av = \rho Q \qquad \sum (\text{mass flow in unit time})_{in} = \sum (\text{mass flow in unit time})_{out}$$

$$(\rho)[(\pi)(\tfrac{3}{12})^2/4](25) = (\rho)[(\pi)(\tfrac{2}{12})^2/4](10) + (\rho)[(\pi)(2.5/12)^2/4](12) + (\rho)[(\pi)(2)^2/4](dh/dt)$$

$$dh/dt = 0.1910 \text{ ft/s}$$

(b) Consider only air in the control volume. It must be conserved. Hence, $(\rho_{air})[(\pi)(\tfrac{2}{12})^2/4](v) = (\rho_{air})[(\pi)(2)^2/4](0.1484)$, $v = 21.4 \text{ ft/s}$.

**Fig. 8-3**

**8.8** The piston of a hypodermic apparatus (Fig. 8-4) is being withdrawn at 0.30 in/s; air leaks in around the piston at the rate $0.0012 \text{ in}^3/\text{s}$. What is the average speed of blood flow in the needle?

▎ Choose as a control volume the region between the piston and the tip of the needle.

$$M = \rho Av = \rho Q \qquad \sum (\text{mass flow in unit time})_{in} = \sum (\text{mass flow in unit time})_{out}$$

$$(\rho_{blood})[(\pi)(0.02/12)^2/4](v) + (\rho_{blood})(0.0012/1728) = (\rho_{blood})[(\pi)(0.2/12)^2/4](0.30/12) \qquad v = 2.18 \text{ ft/s}$$

**Fig. 8-4**

**8.9** Air at 30 °C and 110 kPa flows at 16 N/s through a rectangular duct that measures 160 mm by 320 mm. Compute the average velocity and volume flux.

$$W = \gamma Av \qquad \gamma = p/RT = (110)(1000)/[(29.3)(30 + 273)] = 12.39 \text{ N/m}^3$$

$$16 = 12.39[(0.160)(0.320)](v) \qquad v = 25.2 \text{ m/s} \qquad Q = Av = [(0.160)(0.320)](25.2) = 1.29 \text{ m}^3/\text{s}$$

**8.10** Oil (s.g. = 0.86) flows through a 30-in-diameter pipeline at 8000 gpm. Compute the (a) volume flux, (b) average velocity, and (c) mass flux.

**(a)** $$Q = 8000/[(7.48)(60)] = 17.8 \text{ ft}^3/\text{s}$$

**(b)** $$Q = Av \qquad 17.8 = [(\pi)(\tfrac{30}{12})^2/4](v) \qquad v = 3.63 \text{ ft/s}$$

**(c)** $$M = \rho Av = [(0.86)(1.94)][(\pi)(\tfrac{30}{12})^2/4](3.63) = 29.7 \text{ slugs/s}$$

**8.11** In the rectilinear chamber of Fig. 8-5, section 1 has a diameter of 4 in and the flow in is 2 cfs. Section 2 has a diameter of 3 in and the flow out is 36 fps average velocity. Compute the average velocity and volume flux at section 3 if $D_3 = 1$ in. Is the flow at 3 in or out?

$$Q_1 = Q_2 + Q_3 \qquad \text{(assuming } Q_3 \text{ is out)}$$

$$2 = [(\pi)(\tfrac{3}{12})^2/4](36) + Q_3 \qquad Q_3 = 0.233 \text{ cfs} \qquad \text{(out)} \qquad v = Q/A = 0.233/[(\pi)(\tfrac{1}{12})^2/4] = 42.7 \text{ fps}$$

Fig. 8-5

**8.12** The water tank in Fig. 8-6 is being filled through section 1 at $v_1 = 5$ m/s and through section 3 at $Q_3 = 0.012$ m³/s. If water level $h$ is constant, determine exit velocity $v_2$.

$$Q_1 + Q_3 = Q_2 \qquad [(\pi)(0.040)^2/4](5) + 0.012 = Q_2$$

$$Q_2 = 0.01828 \text{ m}^3/\text{s} \qquad v_2 = Q_2/A_2 = 0.01828/[(\pi)(0.060)^2/4] = 6.47 \text{ m/s}$$

Fig. 8-6

**8.13** If the water level varies in Prob. 8.12 and $v_2 = 8$ m/s, find rate of change $dh/dt$. Assume $d = 1.0$ m.

$$Q_1 + Q_3 = Q_2 + \frac{d}{dt}\frac{h\pi d^2}{4} \qquad [(\pi)(\tfrac{4}{100})^2/4](5) + 0.012 = [(\pi)(\tfrac{6}{100})^2/4](8) + \frac{dh}{dt}[(\pi)(1.0)^2/4]$$

$$dh/dt = -5.52 \text{ mm/s} \qquad \text{(i.e., falling)}$$

**8.14** For the general case of the flow depicted in Fig. 8-6, derive an expression for $dh/dt$ in terms of tank size and volume flows $Q_1$, $Q_2$, and $Q_3$ at the three ports.

$$Q_1 + Q_3 = Q_2 + \frac{d}{dt}\frac{h\pi d^2}{4} \qquad \frac{dh}{dt} = \frac{4(Q_1 - Q_2 + Q_3)}{\pi d^2}$$

**8.15** Water at 20 °C flows steadily through the nozzle in Fig. 8-7 at 60 kg/s. The diameters are $D_1 = 220$ mm and $D_2 = 80$ mm. Compute the average velocities at sections 1 and 2.

$$Q = M/\rho = \tfrac{60}{998} = 0.0601 \text{ m}^3/\text{s}$$

$$v_1 = Q/A_1 = 0.0601/[(\pi)(0.220)^2/4] = 1.58 \text{ m/s} \qquad v_2 = Q/A_2 = 0.0601/[(\pi)(0.080)^2/4] = 12.0 \text{ m/s}$$

**Fig. 8-7**

**8.16** The inseminator in Fig. 8-8 contains fluid of s.g. = 1.04. If the plunger is pushed in steadily at 1.0 in/s, what is exit velocity $V_2$? Assume no leakage past the plunger.

$$\gamma_1 A_1 V_1 = \gamma_2 A_2 V_2$$

$$[(1.04)(62.4)][(\pi)(0.80/12)^2/4](1.0/12) = [(1.04)(62.4)][(\pi)(0.04/12)^2/4](V_2)$$

$$V_2 = 33.3 \text{ ft/s}$$

(Note that the answer is independent of the fluid's specific gravity.)

**Fig. 8-8**

**8.17** Repeat Prob. 8.16 assuming there is leakage back past the plunger equal to 1/4 of the volume flux out of the needle. Compute $V_2$ and the average leakage velocity relative to the needle walls if the plunger diameter is 0.796 in.

$$Q = Q_1 = A_1 V_1 = [(\pi)(0.750/12)^2/4](1.0/12) = 0.000256 \text{ ft}^3/\text{s}$$

$$Q_2 = (\tfrac{3}{4})(0.000256) = 0.000192 \text{ ft}^3/\text{s}$$

$$V_2 = Q_2/A_2 = 0.000192/[(\pi)(0.040/12)^2/4] = 22.0 \text{ ft/s}$$

$$Q_{\text{leak}} = (\tfrac{1}{4})(0.000256) = 0.000064 \text{ ft}^3/\text{s}$$

$$V_{\text{leak}} = Q_{\text{leak}}/A_{\text{leak}} = 0.000064/[(\pi)(0.80/12)^2/4 - (\pi)(0.796/12)^2/4] = 1.84 \text{ ft/s}$$

**8.18** A 100-mm-diameter plunger (1) is being pushed at 60 mm/s into a tank filled with a fluid of s.g. = 0.68. If the fluid is incompressible, how many pounds per second is being forced out at section 2, $D_2 = 20$ mm?

$$A_1 v_1 = A_2 v_2 \qquad [(\pi)(0.100)^2/4](0.060) = [(\pi)(0.020)^2/4](v_2) \qquad v_2 = 1.500 \text{ m/s}$$

$$W = \gamma A v = [(0.68)(9.79)][(\pi)(0.020)^2/4](1.500) = 0.003137 \text{ kN/s} \quad \text{or} \quad 3.137 \text{ N/s}$$

$$= 3.137/4.448 = 0.705 \text{ lb/s}$$

**8.19** A gasoline pump fills a 80-L tank in 1 min 15 s. If the pump exit diameter is 4 cm, what is the average pump-flow exit velocity?

$$Q = V/t = (\tfrac{80}{1000})/(60 + 15) = 0.001067 \text{ m}^3/\text{s} \qquad v = Q/A = 0.001067/[(\pi)(0.04)^2/4] = 0.85 \text{ m/s}$$

**8.20** The tank in Fig. 8-9 is admitting water at 100 N/s and ejecting gasoline (s.g. = 0.69) at 52 N/s. If all three fluids are incompressible, how much air is passing through the vent? In which direction?

❚ $$Q_1 = Q_2 + Q_3 \quad \text{(assuming airflow is out)}$$

$$Q_1 = W_1/\gamma_{H_2O} = 100/[(9.79)(1000)] = 0.01021 \text{ m}^3/\text{s}$$

$$Q_2 = W_2/\gamma_{gas} = 52/[(0.69)(9.79)(1000)] = 0.007698 \text{ m}^3/\text{s}$$

$$0.01021 = 0.007698 + Q_3 \qquad Q_3 = 0.002512 \text{ m}^3/\text{s} \quad \text{(out)}$$

$$\gamma_{air} = p/RT = (1)(101.3)/[(29.3)(20 + 273)] = 0.01180 \text{ kN/m}^3$$

$$W_3 = (\gamma_{air})(Q_3) = (0.01180)(0.002512) = 0.00002964 \text{ kN/s} \quad \text{or} \quad 0.0296 \text{ N/s}$$

Fig. 8-9

**8.21** Air at 72 °F and 16 psia enters a chamber at section 1 at velocity 210 fps and leaves section 2 at 1208 °F and 202 psia. What is the exit velocity if $D_1 = 8$ in and $D_2 = 3$ in? Assume the flow is steady.

❚ $$\rho_1 A_1 v_1 = \rho_2 A_2 v_2 \qquad \rho = p/RT \qquad \rho_1 = (16)(144)/[(1716)(460 + 72)] = 0.002524 \text{ slug/ft}^3$$

$$\rho_2 = (202)(144)/[(1716)(460 + 1208)] = 0.01016 \text{ slug/ft}^3$$

$$0.002524[(\pi)(\tfrac{8}{12})^2/4](210) = 0.01016[(\pi)(\tfrac{3}{12})^2/4](v_2) \qquad v_2 = 371 \text{ fps}$$

**8.22** Kerosene (s.g. = 0.88) enters the cylindrical arrangement of Fig. 8-10 at section 1, at 0.08 N/s. The 80-mm-diameter plates are 2 mm apart. Assuming steady flow, compute the inlet average velocity $v_1$, outlet average velocity $v_2$ assuming radial flow, and outlet volume flux.

❚ $$W_1 = \gamma_{oil} A_1 v_1 \qquad 0.08 = [(0.88)(9.79)(1000)][(\pi)(0.004)^2/4](v_1) \qquad v_1 = 0.739 \text{ m/s}$$

$$Q_1 = A_1 v_1 = [(\pi)(0.004)^2/4](0.739) = 0.00000929 \text{ m}^3/\text{s} \qquad Q_2 = Q_1 = 0.00000929 \text{ m}^3/\text{s} \quad \text{or} \quad 9.29 \text{ mL/s}$$

$$v_2 = Q_2/A_2 = 0.00000929/[(\pi)(0.080)(0.003)] = 0.0123 \text{ m/s} \quad \text{or} \quad 12.3 \text{ mm/s}$$

Fig. 8-10

**8.23** In Fig. 8-11, pipes 1 and 2 are of diameter 3 cm; $D_3 = 4$ cm. Alcohol (s.g. = 0.80) enters section 1 at 6 m/s while water enters section 2 at 10 m/s. Assuming ideal mixing of incompressible fluids, compute the exit velocity and density of the mixture at section 3. The temperature is 20 °C.

❚ $$Q_1 + Q_2 = Q_3 \qquad [(\pi)(0.03)^2/4](6) + [(\pi)(0.03)^2/4](10) = Q_3 \qquad Q_3 = 0.01131 \text{ m}^3/\text{s}$$

$$v_3 = Q_3/A_3 = 0.01131/[(\pi)(0.04)^2/4] = 9.00 \text{ m/s} \qquad M_1 + M_2 = M_3$$

$$\rho_{alcohol} A_1 v_1 + \rho_{H_2O} A_2 v_2 = \rho_{mixture} A_3 v_3$$

$$[(0.80)(998)][(\pi)(0.03)^2/4](6) + 998[(\pi)(0.03)^2/4](10) = (\rho_{mixture})[(\pi)(0.04)^2/4](9.00) \qquad \rho_{mixture} = 923 \text{ kg/m}^3$$

Fig. 8-11

**8.24** In the wind tunnel of Fig. 8-12, the test-section wall is porous; fluid is sucked out to provide a thin viscous boundary layer. The wall contains 800 holes of 7-mm diameter per square meter of area. The suction velocity out each hole is $V_s = 10$ m/s, and the test section entrance velocity is $V_1 = 46$ m/s. Assuming incompressible flow of air at 20 °C and 1 atm, compute (**a**) $V_0$, (**b**) the total wall suction volume flow, (**c**) $V_2$, and (**d**) $V_f$.

**▮** (**a**) $\qquad A_0 V_0 = A_1 V_1 \qquad [(\pi)(2.6)^2/4](V_0) = [(\pi)(0.9)^2/4](45) \qquad V_0 = 5.4$ m/s

(**b**) $\qquad Q_{suction} = N_{holes} Q_{hole} \qquad N_{holes} = 800[(\pi)(0.9)(4)] = 9048$

$$Q_{hole} = A_{hole} V_{hole} = [(\pi)(\tfrac{7}{1000})^2/4](10) = 0.0003848 \text{ m}^3/\text{s}$$

$$Q_{suction} = (9048)(0.0003848) = 3.48 \text{ m}^3/\text{s}$$

(**c**) $\qquad Q_1 = Q_2 + Q_{suction} \qquad [(\pi)(0.9)^2/4](46) = Q_2 + 3.48 \qquad Q_2 = 25.78 \text{ m}^3/\text{s}$

$$v_2 = Q_2/A_2 = 25.78/[(\pi)(0.9)^2/4] = 40.5 \text{ m/s}$$

(**d**) $\qquad A_f V_f = A_2 V_2 \qquad [(\pi)(2.4)^2/4](V_f) = [(\pi)(0.9)^2/4](40.5) \qquad V_f = 5.70 \text{ m/s}$

Fig. 8-12

**8.25** A rocket motor is operating steadily as shown in Fig. 8-13. The exhaust products may be considered an ideal gas of molecular weight 26. Calculate $v_2$.

**▮** $\qquad M_2 = M_1 + M_3 = 0.7 + 0.1 = 0.8 \text{ slug/s} = \rho_2 A_2 v_2 \qquad R = 49\,709/26 = 1912 \text{ lb} \cdot \text{ft}/(\text{slug} \cdot \text{°R})$

$$\rho_2 = p/RT = (16)(144)/[(1912)(1105 + 460)] = 0.000770 \text{ slug/ft}^3$$

$$0.8 = 0.000770[(\pi)(6.0/12)^2/4](v_2) \qquad v_2 = 5291 \text{ ft/s}$$

Fig. 8-13

**8.26** For the solid-propellant rocket in Fig. 8-14, compute the rate of mass loss of the propellant, assuming the exit gas has a molecular weight of 30.

$$M_{in} = M_{out} + \frac{d}{dt}(m_{propellant}) \qquad 0 = M_{out} + \frac{d}{dt}(m_{propellant})$$

$$\frac{d}{dt}(m_{propellant}) = -M_{out} = -\rho_e A_e V_e$$

$$R = \frac{8312}{30} = 277 \text{ N} \cdot \text{m}/(\text{kg} \cdot \text{K}) \qquad \rho_e = p_e/RT_e = (105)(1000)/[(277)(800)] = 0.4738 \text{ kg/m}^3$$

$$\frac{d}{dt}(m_{propellant}) = -(0.4738)\left[\frac{(\pi)(0.200)^2}{4}\right](1100) = -16.4 \text{ kg/s}$$

Fig. 8-14

**8.27** The water-jet pump in Fig. 8-15 injects water at $U_1 = 80$ ft/s through a 4-in pipe which is surrounded by a secondary flow of water at $U_2 = 8$ ft/s. The two flows become fully mixed downstream, where $U_3$ is approximately constant. If the flow is steady and incompressible, compute $U_3$.

$$Q_1 + Q_2 = Q_3 \qquad [(\pi)(\tfrac{4}{12})^2/4](80) + \{(\pi)[(\tfrac{12}{12})^2 - (\tfrac{4}{12})^2]/4\}(8) = [(\pi)(\tfrac{12}{12})^2/4](U_3) \qquad U_3 = 16.0 \text{ ft/s}$$

Fig. 8-15

**8.28** The flow in the inlet between parallel plates in Fig. 8-16 is uniform at $U_0 = 50$ mm/s, while downstream the flow develops into the parabolic laminar profile $u = az(z_0 - z)$, where $a$ is a constant. If $z_0 = 20$ mm, compute $v_{max}$.

Let $b$ = width of plates (into paper).

$$Q_{in} = Q_{out} \qquad z_0 b U_0 = \int u \, dA$$

$$z_0 b U_0 = \int_0^{z_0} az(z_0 - z)b \, dz = ab\left[\frac{z_0 z^2}{2} - \frac{z^3}{3}\right]_0^{z_0} = ab\left(\frac{z_0 z_0^2}{2} - \frac{z_0^3}{3}\right) = \frac{abz_0^3}{6}$$

$$a = 6U_0/z_0^2 \qquad u = az(z_0 - z) = (6U_0/z_0^2)(z)(z_0 - z)$$

$u_{max}$ occurs at $z = z_0/2 = 0.020/2 = 0.010$ m: $u_{max} = [(6)(0.050)/(0.020)^2](0.01)(0.020 - 0.010) = 0.0750$ m/s or 75.0 mm/s.

Fig. 8-16

**8.29** Assuming the container in Fig. 8-17a is large and losses are negligible, derive an expression for the distance $X$ where the free jet leaving horizontally will strike the floor, as a function of $h$ and $H$. Sketch the three trajectories for $h/H = 0.25, 0.50$, and $0.75$.

$$v_0 = \sqrt{2g(H-h)} \qquad h = gt^2/2 \qquad t = \sqrt{2h/g} \qquad X = v_0 t = \sqrt{2g(H-h)}\sqrt{2h/g} = 2\sqrt{h(H-h)}$$

For $h/H = 0.25$, or $h = 0.25H$, $X = 2\sqrt{(0.25H)(H-0.25H)} = 0.866H$. For $h/H = 0.50$, or $h = 0.50H$, $X = 2\sqrt{(0.50H)(H-0.50H)} = H$. For $h/H = 0.75$, or $h = 0.75H$, $X = 2\sqrt{(0.75H)(H-0.75)} = 0.866H$. These three trajectories are sketched in Fig. 8-17b.

**Fig. 8-17(a)**          **Fig. 8-17(b)**

**8.30** In Fig. 8-18 what should the water level $h$ be for the free jet just to clear the wall?

$$v_0 = \sqrt{2gh} \qquad \text{Fall distance} = gt^2/2 = 0.40 \qquad t = 0.8944/\sqrt{g}$$

$$\text{Horizontal distance} = v_0 t = (\sqrt{2gh})(0.8944/\sqrt{g}) = 0.50 \qquad h = 0.156 \text{ m} = 15.6 \text{ cm}$$

**Fig. 8-18**

**8.31** When 500 gpm flows through a 12-in pipe which later reduces to a 6-in pipe, calculate the average velocities in the two pipes.

$$Q = 500/[(7.48)(60)] = 1.114 \text{ ft}^3/\text{s}$$

$$v_{12} = Q/A_{12} = 1.114/[(\pi)(\tfrac{12}{12})^2/4] = 1.42 \text{ ft/s} \qquad v_6 = Q/A_6 = 1.114/[(\pi)(\tfrac{6}{12})^2/4] = 5.67 \text{ ft/s}$$

**8.32** If the velocity in a 12-in pipe is 1.65 ft/s, what is the velocity in a 3-in-diameter jet issuing from a nozzle attached to the pipe?

$$A_1 v_1 = A_2 v_2 \qquad [(\pi)(\tfrac{12}{12})^2/4](1.65) = [(\pi)(\tfrac{3}{12})^2/4](v_2) \qquad v_2 = 26.4 \text{ ft/s}$$

**8.33** Air flows in a 6-in pipe at a pressure of 30.0 psig and a temperature of 100 °F. If barometric pressure is 14.7 psia and velocity is 10.5 ft/s, how many pounds of air per second are flowing?

$$\gamma = p/RT = (30.0 + 14.7)(144)/[(53.3)(100 + 460)] = 0.2157 \text{ lb/ft}^3$$

$$W = \gamma A v = 0.2157[(\pi)(\tfrac{6}{12})^2/4](10.5) = 0.445 \text{ lb/s}$$

**8.34** Carbon dioxide passes point $A$ in a 3-in pipe at a velocity of 15.0 ft/s. The pressure at $A$ is 30 psig and the temperature is 70 °F. At point $B$ downstream, the pressure is 20 psig and the temperature is 90 °F. For a barometric pressure reading of 14.7 psia, calculate the velocity at $B$ and compare the flows at $A$ and $B$.

$$\gamma = p/RT \qquad \gamma_A = (30 + 14.7)(144)/[(35.1)(70 + 460)] = 0.3460 \text{ lb/ft}^3$$

$$\gamma_B = (20 + 14.7)(144)/[(35.1)(90 + 460)] = 0.2588 \text{ lb/ft}^3 \qquad \gamma_A A_A v_A = \gamma_B A_B v_B$$

Since $A_A = A_B$, $(0.3460)(15.0) = (0.2588)(v_B)$, $v_B = 20.1$ ft/s. The number of pounds per second flowing is constant, but the flow in cubic feet per second will differ because the specific weight is not constant. $Q_A = A_A v_A = [(\pi)(\frac{3}{12})^2/4](15.0) = 0.736$ ft³/s; $Q_B = A_B v_B = [(\pi)(\frac{3}{12})^2/4](20.1) = 0.987$ ft³/s.

**8.35** What minimum diameter of pipe is necessary to carry 0.500 lb/s of air with a maximum velocity of 18.5 ft/s? The air is at 80 °F and under an absolute pressure of 34.0 psi.

$\blacksquare$ $\qquad W = \gamma A v \qquad \gamma_{air} = p/RT = (34.0)(144)/[(53.3)(80 + 460)] = 0.170 \text{ lb/ft}^3$

$\qquad 0.500 = (0.170)[(\pi)(d)^2/4](18.5) \qquad d = 0.450 \text{ ft} \quad \text{or} \quad 5.40 \text{ in}$

**8.36** In the laminar flow of a fluid in a circular pipe, the velocity profile is exactly a true parabola. The rate of discharge is then represented by the volume of a paraboloid. Prove that for this case the ratio of the mean velocity to the maximum velocity is 0.5.

$\blacksquare$ See Fig. 8-19. For a paraboloid, $u = u_{max}[1 - (r/r_0)^2]$.

$$Q = \int u \, dA = \int_0^{r_0} u_{max}\left[1 - \left(\frac{r}{r_0}\right)^2\right](2\pi r \, dr) = 2\pi u_{max}\left[\frac{r^2}{2} - \frac{r^4}{4r_0^2}\right]_0^{r_0} = 2\pi u_{max}\left[\frac{r_0^2}{2} - \frac{r_0^4}{4r_0^2}\right] = u_{max}\left(\frac{\pi r_0^2}{2}\right)$$

$$V_{mean} = Q/A = u_{max}(\pi r_0^2/2)/(\pi r_0^2) = u_{max}/2. \text{ Thus } V_{mean}/u_{max} = 0.5.$$

**Fig. 8-19**

**8.37** A gas ($\gamma = 0.05$ lb/ft³) flows at the rate of 1.6 lb/s past section $A$ through a long rectangular duct of uniform cross section 2 ft by 2 ft. At section $B$, the gas weighs 0.060 lb/ft³. Find the average velocities of flow at sections $A$ and $B$.

$\blacksquare$ $\quad W = \gamma A v \quad 1.6 = (0.05)[(2)(2)](v_A) \quad v_A = 8.00 \text{ ft/s} \quad 1.6 = (0.060)[(2)(2)](v_B) \quad v_B = 6.67 \text{ ft/s}$

**8.38** The velocity of a liquid (s.g. = 1.26) in a 6-in pipeline is 1.6 ft/s. Calculate the flow in: (a) ft³/s, (b) slug/s.

$\blacksquare$ (a) $\qquad\qquad Q = Av = [(\pi)(\frac{6}{12})^2/4](1.6) = 0.314 \text{ ft}^3/\text{s}$

(b) $\qquad\qquad M = \rho A v = [(1.26)(1.94)][(\pi)(\frac{6}{12})^2/4](1.6) = 0.768 \text{ slug/s}$

**8.39** Oxygen flows in a 3-in by 3-in duct at a pressure of 42 psi and a temperature of 105 °F. If atmospheric pressure is 13.4 psia and the velocity of flow is 18 fps, calculate the weight-flow rate.

$\blacksquare$ $\qquad\qquad \gamma = p/RT = (42 + 13.4)(144)/[(48.2)(460 + 105)] = 0.2929 \text{ lb/ft}^3$

$\qquad\qquad W = \gamma A v = (0.2929)[(\frac{3}{12})(\frac{3}{12})](18) = 0.330 \text{ lb/s}$

**8.40** Air at 42 °C and at 3 bar absolute pressure flows in a 200-mm-diameter conduit at a mean velocity of 12 m/s. Find the mass flow rate.

$\blacksquare$ $\quad \rho = p/RT = 3 \times 10^5/[(287)(273 + 42)] = 3.318 \text{ kg/m}^3 \quad M = \rho A v = 3.318[(\pi)(0.200)^2/4](12) = 1.25 \text{ kg/s}$

**8.41** A 120-mm-diameter pipe enlarges to a 180-mm-diameter pipe. At section 1 of the smaller pipe, the density of a gas in steady flow is 200 kg/m³ and the velocity is 20 m/s; at section 2 of the larger pipe the velocity is 14 m/s. Find the density of the gas at section 2.

$\blacksquare$ $\quad \rho_1 A_1 v_1 = \rho_2 A_2 v_2 \quad 200[(\pi)(0.120)^2/4](20) = (\rho_2)[(\pi)(0.180)^2/4](14) \quad \rho_2 = 127 \text{ kg/m}^3$

**8.42** The Peconic River is spanned by Noyack Bridge and Smith's Bridge. At noon on July 4, the measured flows under the two bridges were $Q_N = 50 \text{ m}^3/\text{s}$ and $Q_S = 40 \text{ m}^3/\text{s}$. Neglecting losses, compute the instantaneous rate of water storage between the two bridges.

■ $$Q_N - Q_S = dS/dt \qquad 50 - 40 = dS/dt \qquad dS/dt = 10 \text{ m}^3/\text{s}$$

**8.43** A worker in a children's playground is cleaning a slide with a hose. She observes that a horizontal stream directed into the low end climbs to a point 12 ft above the nozzle. What is the nozzle velocity of the stream?

■ $$v = \sqrt{2gh} = \sqrt{(2)(32.2)(12)} = 27.8 \text{ ft/s}$$

**8.44** At section 1 of a pipe system carrying water the velocity is 3.0 fps and the diameter is 2.0 ft. At section 2 the diameter is 3.0 ft. Find the discharge and velocity at section 2.

■ $$Q_1 = Q_2 = Av = [(\pi)(2.0)^2/4](3.0) = 9.42 \text{ cfs} \qquad v_2 = Q_2/A_2 = 9.42/[(\pi)(3.0)^2/4] = 1.33 \text{ fps}$$

**8.45** In two-dimensional flow around a circular cylinder (Fig. 8-20), the discharge between streamlines is 34.56 in³/s per foot of depth. At a great distance the streamlines are 0.25 in. apart, and at a point near the cylinder they are 0.12 in. apart. Calculate the magnitudes of the velocity at these two points.

■ $v = (Q/d)/w$. At great distance, $v = (34.56/12)/0.25 = 11.52$ in/s. Near the cylinder, $v = (34.56/12)/0.12 = 24.0$ in/s.

**Fig. 8-20**

**8.46** A pipeline carries oil (s.g. $= 0.86$) at $v = 2 \text{ m/s}$ through a 20-cm-ID pipe. At another section the diameter is 8 cm. Find the velocity at this section and the mass flow rate.

■ $$Q = A_1 v_1 = [(\pi)(0.20)^2/4](2) = 0.06283 \text{ m}^3/\text{s} \qquad v_2 = Q/A_2 = 0.06283/[(\pi)(0.08)^2/4] = 12.5 \text{ m/s}$$
$$M = \rho Av = [(0.86)(1000)][(\pi)(0.20)^2/4](2) = 54.0 \text{ kg/s}$$

**8.47** Hydrogen is flowing in a 3.0-in-diameter pipe at the steady rate of 0.03 lbm/s. Calculate the average velocity over a section where the pressure is 30 psia and the temperature is 80 °F.

■ $$M = \rho Av \qquad \rho = p/RT = (30)(144)/[(765.5)(460 + 80)] = 0.01045 \text{ lbm/ft}^3$$
$$0.03 = (0.01045)[(\pi)(3.0/12)^2/4](v) \qquad v = 58.5 \text{ ft/s}$$

**8.48** If a jet is inclined upward 30° from the horizontal, what must be its velocity to reach over a 10-ft wall at a horizontal distance of 60 ft, neglecting friction?

■ $(v_x)_0 = v_0 \cos 30° = 0.8660 v_0$, $(v_z)_0 = v_0 \sin 30° = 0.5000 v_0$. From Newton's laws, $x = (0.8660 v_0)(t) = 60$, $z = 0.5000 v_0 t - 32.2t^2/2 = 10$. From the first equation, $t = 69.28/v_0$. Substituting this into the second equation, $(0.5000)(v_0)(69.28/v_0) - (32.2)(69.28/v_0)^2/2 = 10$, $v_0 = 56.0$ fps.

**8.49** Water flows at 10 m³/s in a 150-cm-diameter pipe; the head loss in a 1000-m length of this pipe is 20 m. Find the rate of energy loss due to pipe friction.

■ $$\text{Rate of energy loss} = \gamma QH = (9.79)(10)(20) = 1958 \text{ kW}$$

**8.50** Oil with specific gravity 0.750 is flowing through a 6-in pipe under a pressure of 15.0 psi. If the total energy relative to a datum plane 8.00 ft below the center of the pipe is 58.6 ft · lb/lb, determine the flow rate of the oil.

■ $$H = z + v^2/2g + p/\gamma \qquad 58.6 = 8.00 + v^2/[(2)(32.2)] + (15)(144)/[(0.750)(62.4)]$$
$$v = 16.92 \text{ ft/s} \qquad Q = Av = [(\pi)(\tfrac{6}{12})^2/4](16.92) = 3.32 \text{ ft}^3/\text{s}$$

**8.51** In Fig. 8-21, water flows from $A$, where the diameter is 12 in, to $B$, where the diameter is 24 in, at the rate of 13.2 cfs. The pressure head at $A$ is 22.1 ft. Considering no loss of energy from $A$ to $B$, find the pressure head at $B$.

$$p_A/\gamma + v_A^2/2g + z_A = p_B/\gamma + v_B^2/2g + z_B \qquad v_A = Q/A_A = 13.2/[(\pi)(\tfrac{12}{12})^2/4] = 16.81 \text{ ft/s}$$

$$v_B = Q/A_B = 13.2/[(\pi)(\tfrac{24}{12})^2/4] = 4.202 \text{ ft/s}$$

$$22.1 + 16.81^2/[(2)(32.2)] + 0 = p_B/\gamma + 4.202^2/[(2)(32.2)] + (25.0 - 10.0) \qquad p_B/\gamma = 11.2 \text{ ft of water}$$

Fig. 8-21

**8.52** A pipe carrying oil with specific gravity 0.877 changes in size from 6 in at section $E$ to 18 in at section $R$. Section $E$ is 12 ft lower than $R$, and the pressures are 13.2 psi and 8.75 psi, respectively. If the discharge is 5.17 cfs, determine the lost head and the direction of flow.

$H = z + v^2/2g + p/\gamma$. Use the lower section ($E$) as the datum plane. $v_E = Q/A_E = 5.17/[(\pi)(\tfrac{6}{12})^2/4] = 26.33$ ft/s, $v_R = Q/A_R = 5.17/[(\pi)(\tfrac{18}{12})^2/4] = 2.926$ ft/s; $H_E = 0 + 26.33^2/[(2)(32.2)] + (13.2)(144)/[(0.877)(62.4)] = 45.50$ ft, $H_R = 12 + 2.926^2/[(2)(32.2)] + (8.75)(144)/[(0.877)(62.4)] = 35.16$ ft. Since the energy at $E$ exceeds that at $R$, flow occurs from $E$ to $R$. The lost head is $45.50 - 35.16$, or 10.34 ft, $E$ to $R$.

**8.53** A horizontal air duct is reduced in cross-sectional area from 0.75 ft² to 0.20 ft². Assuming no losses, what pressure change will occur when 1.50 lb/s of air flows? Use $\gamma = 0.200$ lb/ft³ for the pressure and temperature conditions involved.

$$Q = 1.50/0.200 = 7.500 \text{ ft}^3/\text{s} \qquad p_A/\gamma + v_A^2/2g + z_A = p_B/\gamma + v_B^2/2g + z_B$$

$$p_A/\gamma + (7.500/0.75)^2/[(2)(32.2)] + 0 = p_B/\gamma + (7.500/0.20)^2/[(2)(32.2)] + 0$$

$$p_A/\gamma - p_B/\gamma = 20.28 \text{ ft of air} \qquad p_A - p_B = (20.28)(0.200)/144 = 0.0282 \text{ psi}$$

**8.54** A turbine is rated at 600 hp when the flow of water through it is 21.5 cfs. Assuming an efficiency of 87 percent, what head is acting on the turbine?

$$\text{Rated horsepower} = (\text{extracted horsepower})(\text{efficiency}) = (\gamma Q H/550)(\text{efficiency})$$

$$600 = [(62.4)(21.5)(H)/550](0.87) \qquad H = 283 \text{ ft}$$

**8.55** A standpipe 20 ft in diameter and 40 ft high is filled with water. Calculate the potential energy of the water if the elevation datum is taken 10 ft below the base of the standpipe.

$$\text{PE} = Wz = [(62.4)(40)(\pi)(20)^2/4](10 + 40/2) = 2.35 \times 10^7 \text{ ft} \cdot \text{lb}$$

**8.56** How much work could be obtained from the water in Prob. 8.55 if run through a 50-percent-efficient turbine that discharged into a reservoir 30 ft below the base of the standpipe?

$$\text{Work} = \eta \text{PE} = \eta Wz = (0.50)[(62.4)(40)(\pi)(20)^2/4](30 + \tfrac{40}{2}) = 1.96 \times 10^7 \text{ ft} \cdot \text{lb}$$

**8.57** Determine the kinetic-energy flux of 0.01 m³/s of oil (s.g. = 0.80) discharging through a 40-mm-diameter nozzle.

$$v = Q/A = 0.01/[(\pi)(0.040)^2/4] = 7.96 \text{ m/s}$$

$$\dot{\text{KE}} = \dot{m}v^2/2 = \rho Q v^2/2 = [(0.80)(1000)](0.01)(7.96)^2/2 = 253 \text{ W}$$

**8.58** Neglecting air resistance, determine the height a vertical jet of water will rise if projected with velocity 58 ft/s.

$$\text{PE} = \text{KE} \qquad Wz = mv^2/2 = (W/32.2)(58)^2/2 \qquad z = 52.2 \text{ ft}$$

**8.59** If the water jet of Prob. 8.58 is directed upward 45° with the horizontal and air resistance is neglected, how high will it rise?

▎ At 45°, $v_H = v_v = (58)(0.7071) = 41.01$ ft/s; $Wz = mv^2/2 = (W/32.2)(41.01)^2/2$, $z = 26.1$ ft.

**8.60** Show that the work a liquid can do by virtue of its pressure is $\int p \, dV$, in which $V$ is the volume of liquid displaced.

▎ Work $= \int F \, ds$. Since $F = pA$, work $= \int pA \, ds$. Since $A \, ds = dV$, work $= \int p \, dV$.

**8.61** A fluid is flowing in a 6-in-diameter pipe at a pressure of 4.00 lb/in² with a velocity of 8.00 ft/s. As shown in Fig. 8-22, the elevation of the center of the pipe above a given datum is 10.0 ft. Find the total energy head above the given datum if the fluid is (a) water, (b) oil with a specific gravity of 0.82, and (c) gas with a specific weight of 0.042 lb/ft³.

▎
$$H = z + v^2/2g + p/\gamma$$

(a) $$H = 10.0 + 8.00^2/[(2)(32.2)] + (4.00)(144)/62.4 = 20.22 \text{ ft}$$

(b) $$H = 10.0 + 8.00^2/[(2)(32.2)] + (4.00)(144)/[(0.82)(62.4)] = 22.25 \text{ ft}$$

(c) $$H = 10.0 + 8.00^2/[(2)(32.2)] + (4.00)(144)/(0.042) = 13\,725 \text{ ft}$$

6-in diameter

10 ft

Datum

**Fig. 8-22**

**8.62** A 100-mm-diameter suction pipe leading to a pump, as shown in Fig. 8-23, carries a discharge of 0.0300 m³/s of oil (s.g. = 0.85). If the pressure at point $A$ in the suction pipe is a vacuum of 180 mmHg, find the total energy head at point $A$ with respect to a datum at the pump.

▎ $v = Q/A = 0.0300/[(\pi)(\frac{100}{1000})^2/4] = 3.820$ m/s   $p = \gamma h = [(13.6)(9.79)](-\frac{180}{1000}) = -23.97$ kN/m²

$H = z + v^2/2g + p/\gamma = -1.200 + 3.820^2/[(2)(9.807)] + (-23.97)/[(0.85)(9.79)] = -3.337$ m

**8.63** Figure 8-24 shows a pump drawing water from a reservoir and discharging it into the air at point $B$. The pressure at point $A$ in the suction pipe is a vacuum of 10 in mercury, and the discharge is 3.00 ft³/s. Determine the total head at point $A$ and at point $B$ with respect to a datum at the base of the reservoir.

▎ $H = z + v^2/2g + p/\gamma$   $v_A = Q/A_A = 3.00/[(\pi)(\frac{10}{12})^2/4] = 5.50$ ft/s

$H_A = 25 + 5.50^2/[(2)(32.2)] + [(13.6)(62.4)](-\frac{10}{12})/62.4 = 14.14$ ft

$v_B = Q/A_B = 3.00/[(\pi)(\frac{8}{12})^2/4] = 8.59$ ft/s   $H_B = (25 + 15 + 40) + 8.59^2/[(2)(32.2)] + 0 = 81.15$ ft

**8.64** If the total available head of a stream flowing at a rate of 300 ft³/s is 25.0 ft, what is the theoretical horsepower available?

▎ $$P = Q\gamma H = (300)(62.4)(25.0) = 468\,000 \text{ ft} \cdot \text{lb/s} = 468\,000/550 = 851 \text{ hp}$$

**8.65** A 150-mm-diameter jet of water is discharging from a nozzle into the air at a velocity of 36.0 m/s. Find the power in the jet with respect to a datum at the jet.

▎
$$Q = Av = [(\pi)(\tfrac{150}{1000})^2/4](36.0) = 0.6262 \text{ m}^3/\text{s}$$

$$H = z + v^2/2g + p/\gamma = 0 + 36.0^2/[(2)(9.807)] + 0 = 66.08 \text{ m}$$

$$P = Q\gamma H = (0.6362)(9.79)(66.08) = 412 \text{ kN} \cdot \text{m/s} \quad \text{or} \quad 412 \text{ kW}$$

**Fig. 8-23**

**Fig. 8-24**

**8.66** Oil (s.g. $= 0.84$) is flowing in a pipe under the conditions shown in Fig. 8-25. If the total head loss ($h_L$) from point 1 to point 2 is 3.0 ft, find the pressure at point 2.

$$p_1/\gamma + v_1^2/2g + z_1 = p_2/\gamma + v_2^2/2g + z_2 + h_L$$

$$v_1 = Q/A_1 = 2.08/[(\pi)(\tfrac{6}{12})^2/4] = 10.59 \text{ ft/s} \qquad v_2 = Q/A_2 = 2.08/[(\pi)(\tfrac{9}{12})^2/4] = 4.71 \text{ ft/s}$$

$$(65)(144)/[(0.84)(62.4)] + 10.59^2/[(2)(32.2)] + 10.70 = p_2/\gamma + 4.71^2/[(2)(32.2)] + 4.00 + 3.00$$

$$p_2/\gamma = 183.67 \text{ ft} \qquad p_2 = [(0.84)(62.4)](183.67) = 9627 \text{ lb/ft}^2 \quad \text{or} \quad 66.9 \text{ lb/in}^2$$

Fig. 8-25

8.67    An 8-in-diameter horizontal pipe is attached to a reservoir, as shown in Fig. 8-26. If the total head loss between the water surface in the reservoir and the water jet at the end of the pipe is 6.0 ft, what are the velocity and flow rate of the water being discharged from the pipe?

▌    $p_1/\gamma + v_1^2/2g + z_1 = p_2/\gamma + v_2^2/2g + z_2 + h_L$        $0 + 0 + 15 = 0 + v_2^2/[(2)(32.2)] + 0 + 6.0$

   $v_2 = 24.1$ ft/s        $Q = A_2 v_2 = [(\pi)(\frac{8}{12})^2/4](24.1) = 8.41$ ft³/s

Fig. 8-26

8.68    A 50-mm-diameter siphon is drawing oil (s.g. = 0.82) from an oil reservoir, as shown in Fig. 8-27. If the head loss from point 1 to point 2 is 1.50 m and from point 2 to point 3 is 2.40 m, find the discharge of oil from the siphon and the oil pressure at point 2.

▌    $p_1/\gamma + v_1^2/2g + z_1 = p_3/\gamma + v_3^2/2g + z_3 + h_L$        $0 + 0 + 5.00 = 0 + v_3^2/[(2)(9.807)] + 0 + 3.90$

   $v_3 = 4.645$ m/s        $Q = A_3 v_3 = [(\pi)(\frac{50}{1000})^2/4](4.645) = 0.00912$ m³/s

   $p_1/\gamma + v_1^2/2g + z_1 = p_2/\gamma + v_2^2/2g + z_2 + h_L$        $0 + 0 + 5.00 = p_2/\gamma + 4.645^2/[(2)(9.807)] + 7.00 + 1.50$

   $p_2/\gamma = -4.60$ m        $p_2 = [(0.82)(9.79)](-4.60) = -36.9$ kN/m²   or   $-36.9$ kPa

8.69    Figure 8-28 shows a siphon discharging oil (s.g. = 0.84) from a reservoir into open air. If the velocity of flow in the pipe is $v$, the head loss from point 1 to point 2 is $2.0v^2/2g$, and the head loss from point 2 to point 3 is

**Fig. 8-27**

3.0$v^2/2g$, determine the volume flow rate in the siphon pipe and the absolute pressure at point 2. Assume an atmospheric pressure of 14.70 psia.

$\blacksquare$ $p_1/\gamma + v_1^2/2g + z_1 = p_3/\gamma + v_3^2/2g + z_3 + h_L$ $\qquad$ $0 + 0 + 10 = 0 + v_3^2/[(2)(32.2)] + 0 + 5\{v_3^2/[(2)(32.2)]\}$

$v_3 = 10.36$ ft/s $\qquad$ $Q = A_3 v_3 = [(\pi)(\frac{3}{12})^2/4](10.36) = 0.509$ ft³/s $\qquad$ $p_1/\gamma + v_1^2/2g + z_1 = p_2/\gamma + v_2^2/2g + z_2 + h_L$

$0 + 0 + 10 = p_2/\gamma + 10.36^2/[(2)(32.2)] + 15 + 2\{10.36^2/[(2)(32.2)]\}$ $\qquad$ $p_2/\gamma = -10.0$ ft of oil

$p_2 = [(0.84)(62.4)](-10.0) = -524$ lb/ft² $\quad$ or $\quad$ $-3.64$ lb/in² $\qquad$ $p_2 = 14.70 - 3.64 = 11.06$ lb/in² abs

**Fig. 8-28**

**8.70** Once it has been started by sufficient suction, the siphon in Fig. 8-29 will run continuously as long as reservoir fluid is available. Using Bernoulli's equation with no losses, show (a) that the exit velocity $v_2$ depends only upon gravity and the distance $H$ and (b) that the lowest (vacuum) pressure occurs at point 3 and depends on the distance $L + H$.

$\blacksquare$ $p_1/\gamma + v_1^2/2g + z_1 = p_2/\gamma + v_2^2/2g + z_2 + h_L$, $0 + 0 + z_1 = 0 + v_2^2/2g + z_2 + 0$, $v_2 = \sqrt{2g(z_1 - z_2)} = \sqrt{2gH}$. For any point $B$ in the tube, $p_B/\gamma + v_B^2/2g + z_B = p_2/\gamma + v_2^2/2g + z_2 + h_L$. Since $v_B = v_2$ and $p_2 = p_{atm}$, $p_B = p_{atm} - \gamma(z_B - z_2)$. The lowest pressure occurs at the highest $z_B$, or $p_{min} = p_3 = p_{atm} - \gamma(L + H)$.

**Fig. 8-29**

**8.71** The siphon of Fig. 8-30 is filled with water and discharging at 150 L/s. Find the losses from point 1 to point 3 in terms of velocity head $v^2/2g$. Find the pressure at point 2 if two-thirds of the losses occur between points 1 and 2.

$$p_1/\gamma + v_1^2/2g + z_1 = p_3/\gamma + v_3^2/2g + z_3 + h_L \qquad 0 + 0 + 1.5 = 0 + v_3^2/2g + 0 + (K)(v_3^2/2g)$$

$$v_3 = Q/A_3 = (\tfrac{150}{1000})/[(\pi)(\tfrac{200}{1000})^2/4] = 4.775 \text{ m/s} \qquad 1.5 = 4.775^2/[(2)(9.807)] + K\{4.775^2/[(2)(9.807)]\}$$

$$K = 0.2904 \qquad h_L = (0.2904)\{4.775^2/[(2)(9.807)]\} = 0.338 \text{ m}$$

$$p_1/\gamma + v_1^2/2g + z_1 = p_2/\gamma + v_2^2/2g + z_2 + h_L \qquad 0 + 0 + 0 = p_2/\gamma + 4.775^2/[(2)(9.807)] + (\tfrac{2}{3})(0.338)$$

$$p_2/\gamma = -3.388 \text{ m of water} \qquad p_2 = (-3.388)(9.79) = -33.2 \text{ kN/m}^2$$

Fig. 8-30

**8.72** For the water shooting out of the pipe and nozzle under the conditions shown in Fig. 8-31, find the height above the nozzle to which the water jet will "shoot" (i.e., distance $h$ in Fig. 8-31). Assume negligible head loss.

$$p_A/\gamma + v_A^2/2g + z_A = p_{top}/\gamma + v_{top}^2/2g + z_{top} + h_L$$

$$55.0/9.79 + v_A^2/2g + 0 = 0 + 0 + (1.00 + h) + 0 \qquad h = 4.518 + v_A^2/2g$$

$$p_A/\gamma + v_A^2/2g + z_A = p_{nozzle}/\gamma + v_{nozzle}^2/2g + z_{nozzle} + h_L \qquad 55.0/9.79 + v_A^2/2g + 0 = 0 + v_{nozzle}^2/2g + 1.100 + 0$$

$$A_A v_A = A_{nozzle} v_{nozzle} \qquad [(\pi)(\tfrac{200}{1000})^2/4]v_A = [(\pi)(\tfrac{100}{1000})^2/4]v_{nozzle} \qquad v_{nozzle} = 4.00 v_A$$

$$55.0/9.79 + v_A^2/[(2)(9.807)] + 0 = 0 + (4.00 v_A)^2/[(2)(9.807)] + 1.100 + 0$$

$$v_A = 2.431 \text{ m/s} \qquad h = 4.518 + 2.431^2/[(2)(9.807)] = 4.82 \text{ m}$$

**8.73** Water flows from section 1 to section 2 in the pipe shown in Fig. 8-32. Determine the velocity of flow and the fluid pressure at section 2. Assume the total head loss from section 1 to section 2 is 3.00 m.

$$Q = A_1 v_1 = A_2 v_2 \qquad [(\pi)(\tfrac{100}{1000})^2/4](2.0) = [(\pi)(\tfrac{50}{1000})^2/4](v_2) \qquad v_2 = 8.00 \text{ m/s}$$

$$p_1/\gamma + v_1^2/2g + z_1 = p_2/\gamma + v_2^2/2g + z_2 + h_L$$

$$300/9.79 + 2.0^2/[(2)(9.807)] + 2 = p_2/9.79 + 8.00^2/[(2)(9.807)] + 0 + 3.00 \qquad p_2 = 260 \text{ kPa}$$

**8.74** A nozzle is attached to a pipe as shown in Fig. 8-33. The inside diameter of the pipe is 100 mm, while the water jet exiting from the nozzle has a diameter of 50 mm. If the pressure at section 1 is 500 kPa, determine the water jet's velocity. Assume head loss in the jet is negligible.

$$Q = A_1 v_1 = A_2 v_2 \qquad [(\pi)(\tfrac{100}{1000})^2/4](v_1) = [(\pi)(\tfrac{50}{1000})^2/4](v_2) \qquad v_1 = 0.250 v_2$$

$$p_1/\gamma + v_1^2/2g + z_1 = p_2/\gamma + v_2^2/2g + z_2 + h_L$$

$$500/9.79 + (0.250 v_2)^2/[(2)(9.807)] + 0 = 0 + v_2^2/[(2)(9.807)] + 0 + 0 \qquad v_2 = 32.7 \text{ m/s}$$

Nozzle
D = 100 mm

1.100 m

Point A
D = 200 mm
$p_A$ = 55.0 kPa

Water

**Fig. 8-31**

100-mm
diameter

Water

$v_1$ = 2 m/s
$p_1$ = 300 kPa

2 m

50-mm diameter

$v_2$ = ?
$p_2$ = ?
$h_{L(1-2)}$ = 3 m

**Fig. 8-32**

① $p_1$ = 500 kPa

② $v_2$ = ?

100-mm diameter

Water

50-mm
diameter

**Fig. 8-33**

**8.75** Oil flows from a tank through 500 ft of 6-in-diameter pipe and then discharges into the air, as shown in Fig. 8-34. If the head loss from point 1 to point 2 is 1.95 ft of oil, determine the pressure needed at point 1 to cause 0.60 ft³ of oil to flow.

$$v_2 = Q/A = 0.60/[(\pi)(\tfrac{6}{12})^2/4] = 3.06 \text{ ft/s} \qquad p_1/\gamma + v_1^2/2g + z_1 = p_2/\gamma + v_2^2/2g + z_2 + h_L$$

$$p_1/\gamma + 0 + 80 = 0 + 3.06^2/[(2)(32.2)] + 100 + 1.95 \qquad p_1/\gamma = 22.10 \text{ ft of oil}$$

$$p_1 = [(0.84)(62.4)](22.10)/144 = 8.04 \text{ lb/in}^2$$

**Fig. 8-34**

**8.76** Water is to be delivered from a reservoir through a pipe to a lower level and discharged into the air, as shown in Fig. 8-35. If head loss in the entire system is 11.58 m, determine the vertical distance between the point of water discharge and the water surface in the reservoir.

$$v_2 = Q/A_2 = 0.00631/[(\pi)(\tfrac{50}{1000})^2/4] = 3.214 \text{ m/s} \qquad p_1/\gamma + v_1^2/2g + z_1 = p_2/\gamma + v_2^2/2g + z_2 + h_L$$

$$0 + 0 + z_1 = 0 + 3.214^2/[(2)(9.807)] + 0 + 11.58 \qquad z_1 = 12.11 \text{ m}$$

**Fig. 8-35**

**8.77** Determine the velocity and pressure at section 2 and section 3 if water flows steadily through the pipe system shown in Fig. 8-36. Assume a head loss of 6.0 ft from section 1 to section 2 and of 15.0 ft from section 2 to section 3.

$$A_1 v_1 = A_2 v_2 \qquad [(\pi)(\tfrac{16}{12})^2/4](5.0) = [(\pi)(\tfrac{8}{12})^2/4](v_2) \qquad v_2 = 20.0 \text{ ft/s}$$

$$p_1/\gamma + v_1^2/2g + z_1 = p_2/\gamma + v_2^2/2g + z_2 + h_L$$

$$(25)(144)/62.4 + 5.0^2/[(2)(32.2)] + 20 = (p_2)(144)/62.4 + 20.0^2/[(2)(32.2)] + 15 + 6.0$$

$$p_2 = 22.0 \text{ lb/in}^2 \qquad A_1 v_1 = A_3 v_3$$

$$[(\pi)(\tfrac{16}{12})^2/4](5.0) = [(\pi)(\tfrac{12}{12})^2/4](v_3) \qquad v_3 = 8.99 \text{ ft/s}$$

$$p_1/\gamma + v_1^2/2g + z_1 = p_3/\gamma + v_3^2/2g + z_3 + h_L$$

$$(25)(144)/62.4 + 5.0^2/[(2)(32.2)] + 20 = (p_3)(144)/62.4 + 8.89^2/[(2)(32.2)] + 10 + (15.0 + 6.0)$$

$$p_3 = 19.9 \text{ lb/in}^2$$

$D_1 = 16$ in
$p_1 = 25$ psi
$v_1 = 5$ ft/s

$D_2 = 8$ in
$p_2 = ?$
$v_2 = ?$

$D_3 = 12$ in
$p_3 = ?$
$v_3 = ?$

20 ft

15 ft

10 ft

Datum     **Fig. 8-36**

**8.78**   Compute the ideal flow rate through the pipe system shown in Fig. 8-37.

$$p_1/\gamma + v_1^2/2g + z_1 = p_2/\gamma + v_2^2/2g + z_2 + h_L \qquad p_1/\gamma + v_1^2/[(2)(9.807)] + 0.6 \sin 30° = p_2/\gamma + 0 + 0 + 0$$

$$v_1^2/[(2)(9.807)] = p_2/\gamma - p_1/\gamma - 0.300$$

From the manometer reading, $p_1 - 9.79(1.2 \sin 60°) = p_2$, $p_1 - p_2 = 10.17$ kN/m²; $v_1^2/[(2)(9.807)] = 10.17/9.79 - 0.300$, $v_1 = 3.807$ m/s; $Q = A_1 v_1 = [(\pi)(\frac{200}{1000})^2/4](3.807) = 0.120$ m³/s.

1.2 m

200 mm

−100 mm

0.6 m

30°     **Fig. 8-37**

**8.79**   A large tank with a well-rounded, small opening as an outlet is shown in Fig. 8-38. What is the velocity of a jet issuing from the tank?

$$p_1/\gamma + v_1^2/2g + z_1 = p_2/\gamma + v_2^2/2g + z_2 + h_L \qquad 0 + 0 + h = 0 + v_2^2/2g + 0 + 0 \qquad v_2 = \sqrt{2gh}$$

1

h

2

Datum

**Fig. 8-38**

**8.80**   Neglecting friction, find the velocity and volumetric discharge at the exit 2 in Fig. 8-39.

$$p_1/\gamma + v_1^2/2g + z_1 = p_2/\gamma + v_2^2/2g + z_2 + h_L \qquad 0 + 0 + (3.0 + 0.6 + 1.5) = 0 + v_2^2/[(2)(9.807)] + 0 + 0$$

$$v_2 = 10.0 \text{ m/s} \qquad Q = Av = [(\pi)(\tfrac{150}{1000})^2/4](10.0) = 0.177 \text{ m}^3/\text{s}$$

**Fig. 8-39**

**8.81** One end of a U-tube is oriented directly into the flow (Fig. 8-40) so that the velocity of the stream is zero at this point. The pressure at a point in the flow that has been stopped in this way is called the *stagnation pressure*. The other end of the U-tube measures the undisturbed pressure at that section in the flow. Neglecting friction, determine the volume flow of water in the pipe.

▎ $p_1/\gamma + v_1^2/2g + z_1 = p_2/\gamma + v_2^2/2g + z_2 + h_L$, $p_1/\gamma + 0 + 0 = p_2/\gamma + v_2^2/2g + 0 + 0$, $v_2^2/2g = p_1/\gamma - p_2/\gamma$. From the manometer reading, $p_1 + (62.4)(\frac{2.5}{12}) - [(13.6)(62.4)](\frac{2.5}{12}) = p_2$, $p_1 - p_2 = 163.8$ lb/ft²; $v_2^2/[(2)(32.2)] = 163.8/62.4$, $v_2 = 13.00$ ft/s; $Q = Av = [(\pi)(\frac{8}{12})^2/4](13.00) = 4.54$ ft³/s.

**Fig. 8-40**

**8.82** A cylindrical tank contains air, oil, and water, as shown in Fig. 8-41; the air is under gage pressure $p = 4$ lb/in². Find the exit velocity at 2, neglecting any friction and the kinetic energy of the fluid above elevation $A$. The jet of water leaving has a diameter of 1 ft.

▎ $p_1/\gamma + v_1^2/2g + z_1 = p_2/\gamma + v_2^2/2g + z_2 + h_L$    $p_1 = (4)(144) + [(0.8)(62.4)](2) = 675.8$ lb/ft²

$675.8/62.4 + 0 + 8 = 0 + v_2^2/[(2)(32.2)] + 0 + 0$    $v_2 = 34.8$ ft/s

**Fig. 8-41**

**8.83** A large tank contains compressed air, gasoline at specific gravity 0.68, light oil at specific gravity 0.80, and water, as shown in Fig. 8-42. The pressure $p$ of the air is 120 kPa gage. If we neglect friction, what is the mass flow of oil from a 20-mm-diameter jet?

▎ $p_1/\gamma + v_1^2/2g + z_1 = p_2/\gamma + v_2^2/2g + z_2 + h_L$    $p_1 = 120 + [(0.68)(9.79)](2) = 133.3$ kN/m²

$133.3/[(0.80)(9.79)] + 0 + 0 = 0 + v_2^2/[(2)(9.807)] + 4 + 0$    $v_2 = 15.98$ m/s

$M = \rho Av = [(0.80)(1000)][(\pi)(\frac{20}{1000})^2/4](15.98) = 4.02$ kg/s

Oil specific gravity = 0.80

20 mm

Air

Gasoline specific gravity = 0.68

2 m

4 m

2 m

Water

**Fig. 8-42**

**8.84** A flow nozzle is a device inserted into a pipe as shown in Fig. 8-43. If $A_2$ is the exit area of the flow nozzle, show that for incompressible flow we get for $Q$,

$$Q = C_d \left[ \frac{A_2}{\sqrt{1 - (A_2/A_1)^2}} \sqrt{2g \left( \frac{p_1 - p_2}{\gamma} \right)} \right]$$

where $C_d$ is the *coefficient of discharge*, which takes into account frictional effects and is determined experimentally.

∎ $\dfrac{p_A}{\gamma} + \dfrac{v_A^2}{2g} + z_A = \dfrac{p_B}{\gamma} + \dfrac{v_B^2}{2g} + z_B + h_L$   $\dfrac{p_A}{\gamma} + \dfrac{v_1^2}{2g} + 0 = \dfrac{p_B}{\gamma} + \dfrac{v_2^2}{2g} + 0 + 0$   $v_1^2 = v_2^2 + 2g \left( \dfrac{p_B - p_A}{\gamma} \right)$

But $p_B - p_A = p_2 - p_1$ and $v_2 = (v_1)(A_1/A_2)$; hence,

$$v_1^2 = \left[ v_1 \left( \frac{A_1}{A_2} \right) \right]^2 + 2g \left( \frac{p_2 - p_1}{\gamma} \right) \qquad v_1 = \sqrt{\frac{1}{1 - (A_1/A_2)^2}} \sqrt{2g \left( \frac{p_2 - p_1}{\gamma} \right)}$$

$$Q = Av = C_d A_1 \sqrt{\frac{1}{1 - (A_1/A_2)^2}} \sqrt{2g \left( \frac{p_2 - p_1}{\gamma} \right)} = C_d \sqrt{\frac{A_1^2}{1 - (A_1/A_2)^2}} \sqrt{2g \left( \frac{p_2 - p_1}{\gamma} \right)}$$

Multiplying by $A_2^2/A_1^2$ in the numerator and denominator of the radical gives

$$Q = C_d \left[ \frac{A_2}{\sqrt{1 - (A_2/A_1)^2}} \sqrt{2g \left( \frac{p_1 - p_2}{\gamma} \right)} \right]$$

Flow nozzle

A

$A_2$ B

h

Hg

**Fig. 8-43**

**8.85** In Prob. 8.84, express $Q$ in terms of $h$, the height of the mercury column (Fig. 8-43), and the diameters of the pipe and flow nozzle.

∎ From Prob. 8.84,

$$Q = C_d \left[ \frac{A_2}{\sqrt{1 - (A_2/A_1)^2}} \sqrt{2g \left( \frac{p_1 - p_2}{\gamma} \right)} \right]$$

From the manometer, $p_1 - p_2 = (\gamma_{Hg} - \gamma_{H_2O})(h)$.

$$Q = C_d\left[\frac{\pi d_2^2/4}{\sqrt{1 - (d_2^2/d_1^2)^2}} \sqrt{\frac{2g(\gamma_{Hg} - \gamma_{H_2O})(h)}{\gamma_{H_2O}}}\right]$$

**8.86** A hump of height $\delta$ is placed on the channel bed in a rectangular channel of uniform width over its entire width (see Fig. 8-44). The free surface has a dip $d$ as shown. If we neglect friction, we can consider that we have one-dimensional flow. Compute the flow $q$ for the channel per unit width. This system is called a *venturi flume*.

**▮**
$$p_1/\gamma + v_1^2/2g + z_1 = p_2/\gamma + v_2^2/2g + z_2 + h_L \qquad 0 + v_1^2/2g + h = 0 + v_2^2/2g + (h - d) + 0$$

$$v_1^2 = v_2^2 - 2gd \qquad A_1 v_1 = A_2 v_2 \qquad [(1)(h)](v_1) = [(1)(h - d - \delta)](v_2)$$

$$v_2 = \left(\frac{h}{h - d - \delta}\right)(v_1) \qquad v_1^2 = (v_1)^2\left(\frac{h}{h - d - \delta}\right)^2 - 2gd \qquad v_1^2\left[1 - \left(\frac{h}{h - d - \delta}\right)^2\right] = -2gd$$

$$v_1 = \sqrt{\frac{-2gd}{1 - [h/(h - d - \delta)]^2}} \qquad q = hv_1 = \sqrt{\frac{-2gd}{1/h^2 - [1/(h - d - \delta)]^2}}$$

Fig. 8-44

**8.87** In the fountain of Fig. 8-45, water flows steadily up the vertical pipe, enters the annular region between the circular plates, and emerges as a free sheet. Find the volume flow of water through the pipe, if the pressure at $A$ is 70 kPa gage and friction is negligible.

**▮**
$$p_A/\gamma + v_A^2/2g + z_A = p_E/\gamma + v_E^2/2g + z_E + h_L$$

$$70/9.79 + v_A^2/[(2)(9.807)] + 0 = 0 + v_E^2/[(2)(9.807)] + 1.5 + 0$$

$$A_A v_A = A_E v_E \qquad [(\pi)(0.200)^2/4](v_A) = [(0.013)(\pi)(0.3 + 0.3)](v_E) \qquad v_A = 0.780 v_E$$

$$70/9.79 + (0.780 v_E)^2/[(2)(9.807)] + 0 = 0 + v_E^2/[(2)(9.807)] + 2.0 + 0 \qquad v_E = 16.06 \text{ m/s}$$

$$Q = A_E v_E = [(0.015)(\pi)(0.5 + 0.5)](16.06) = 0.757 \text{ m}^3/\text{s}$$

Fig. 8-45

**8.88** If the velocity at point $A$ in Fig. 8-46 is 18 m/s, what is the pressure at point $B$ if we neglect friction?

$$p_B/\gamma + v_B^2/2g + z_B = p_A/\gamma + v_A^2/2g + z_A + h_L$$

$$p_B/9.79 + v_B^2/[(2)(9.807)] + 0 = 0 + 18^2/[(2)(9.807)] + (0.5 + 15) + 0$$

$$p_B = -0.4991v_B^2 + 313.5 \qquad p_C/\gamma + v_C^2/2g + z_C = p_A/\gamma + v_A^2/2g + z_A + h_L$$

$$0 + v_C^2/[(2)(9.807)] + 0 = 0 + 18^2/[(2)(9.807)] + 15 + 0 \qquad v_C = 24.86 \text{ m/s} \qquad A_B v_B = A_C v_C$$

$$[(\pi)(\tfrac{200}{1000})^2/4](v_B) = [(\pi)(\tfrac{75}{1000})^2/4](24.86) \qquad v_B = 3.496 \text{ m/s}$$

$$p_B = (-0.4991)(3.496)^2 + 313.5 = 319.6 \text{ kN/m}^2$$

Fig. 8-46

**8.89** A rocket-powered sled is used in the training of astronauts (Fig. 8-47). For braking, scoops are lowered to deflect water from a stationary tank of water. To what height $h$ does a sled traveling at 100 km/h deflect water?

▌ In an inertial frame fixed to the sled, apply Bernoulli's equation between the scoop (point 1) and the highest point in the trajectory (point 2); $p_1/\gamma + v_1^2/2g + z_1 = p_2/\gamma + v_2^2/2g + z_2 + h_L$, $0 + v_1^2/[(2)(9.807)] + 0 = 0 + (v_1 \cos 20°)^2/[(2)(9.807)] + (h - 0.150) + 0$. From the data, $v_1 = (100)(1000)/3600 = 27.78$ m/s, $27.78^2/[(2)(9.807)] = (27.78 \cos 20°)^2/[(2)(9.807)] + (h - 0.150)$, $h = 4.80$ m.

(a)  Rail

(b)  Fig. 8-47

**8.90** A *venturi meter* is a device which is inserted into a pipe line to measure incompressible flow rates. As shown in Fig. 8-48, it consists of a convergent section which reduces the diameter to between one-half and one-fourth the pipe diameter. This is followed by a divergent section. The pressure difference between the position just before the venturi and at the throat of the venturi is measured by a differential manometer as shown. Show that

$$Q = C_d\left[\frac{A_2}{\sqrt{1 - (A_2/A_1)^2}} \sqrt{2g\left(\frac{p_1 - p_2}{\gamma}\right)}\right]$$

where $C_d$ is the *coefficient of discharge*, which takes into account frictional effects and is determined experimentally.

$$\frac{p_1}{\gamma} + \frac{v_1^2}{2g} + z_1 = \frac{p_2}{\gamma} + \frac{v_2^2}{2g} + z_2 + h_L \qquad \frac{p_1}{\gamma} + \frac{v_1^2}{2g} + 0 = \frac{p_2}{\gamma} + \frac{v_2^2}{2g} + 0 + 0 \qquad v_1^2 - v_2^2 = 2g\left(\frac{p_2 - p_1}{\gamma}\right)$$

$$A_1 v_1 = A_2 v_2 \qquad v_1 = (v_2)\left(\frac{A_2}{A_1}\right) \qquad \left[(v_2)\left(\frac{A_2}{A_1}\right)\right]^2 - v_2^2 = 2g\left(\frac{p_2 - p_1}{\gamma}\right) \qquad \left[\left(\frac{A_2}{A_1}\right)^2 - 1\right](v_2^2) = 2g\left(\frac{p_2 - p_1}{\gamma}\right)$$

$$v_2 = \sqrt{\frac{1}{1 - (A_2/A_1)^2}} \sqrt{2g\left(\frac{p_1 - p_2}{\gamma}\right)}$$

$$Q = Av = C_d A_2 \sqrt{\frac{1}{1 - (A_2/A_1)^2}} \sqrt{2g\left(\frac{p_1 - p_2}{\gamma}\right)} = C_d\left[\frac{A_2}{\sqrt{1 - (A_2/A_1)^2}} \sqrt{2g\left(\frac{p_1 - p_2}{\gamma}\right)}\right]$$

![Venturi meter diagram]

Venturi meter

**Fig. 8-48**

**8.91** A necked-down, or venturi, section of a pipe flow develops a low pressure which can be used to aspirate fluid upward from a reservoir, as shown in Fig. 8-49. Using Bernoulli's equation with no losses, derive an expression for the exit velocity $v_2$ that is just sufficient to cause the reservoir fluid to rise in the tube up to section 1.

$$p_1/\gamma + v_1^2/2g + z_1 = p_2/\gamma + v_2^2/2g + z_2 + h_L \qquad p_1/\gamma + v_1^2/2g + 0 = p_{\text{atm}}/\gamma + v_2^2/2g + 0 + 0$$

$$A_1 v_1 = A_2 v_2 \qquad (\pi d_1^2/4)(v_1) = (\pi d_2^2/4)(v_2) \qquad v_1 = (v_2)(d_2/d_1)^2$$

$$p_1/\gamma + [(v_2)(d_2/d_1)^2]^2/2g = p_{\text{atm}}/\gamma + v_2^2/2g \qquad p_{\text{atm}} - p_1 = (\gamma/2g)(v_2^2)[(d_2/d_1)^4 - 1]$$

For fluid to rise in the tube, $p_{\text{atm}} - p_1 \geq \gamma h$; hence, $(\gamma/2g)(v_2^2)[(d_2/d_1)^4 - 1] \geq \gamma h$,

$$v_2 \geq \sqrt{\frac{2gh}{(d_2/d_1)^4 - 1}}$$

**Fig. 8-49**

**8.92** Neglecting losses, find the discharge through the venturi meter of Fig. 8-50.

$p_1/\gamma + v_1^2/2g + z_1 = p_2/\gamma + v_2^2/2g + z_2 + h_L$. From the manometer, $p_1/\gamma - (k + 0.200) + (z_1 - z_2 + k) = p_2/\gamma$; $A_1 v_1 = A_2 v_2$, $[(\pi)(0.300)^2/4](v_1) = [(\pi)(0.150)^2/4](v_2)$, $v_1 = 0.250v_2$; $p_1/\gamma + (0.250v_2)^2/[(2)(9.807)] + z_1 = [p_1/\gamma - (k + 0.250) + (z_1 - z_2 + k)] + v_2^2/[(2)(9.807)] + z_2 + 0$, $v_2 = 2.287 \text{ m/s}$; $Q = A_2 v_2 = [(\pi)(0.150)^2/4](2.287) = 0.0404 \text{ m}^3/\text{s}$.

**Fig. 8-50**

**8.93** With losses of $0.2v_1^2/2g$ between sections 1 and 2 of Fig. 8-50, calculate the flow in gallons per minute.

▐ From Prob. 8.92, $v_1 = 0.250v_2 = (0.250)(2.287) = 0.5718$ m/s; $p_1/\gamma + (0.250v_2)^2/[(2)(9.807)] + z_1 = [p_1/\gamma - (k + 0.250) + (z_1 - z_2 + k)] + v_2^2/[(2)(9.807)] + z_2 + 0$. For Prob. 8.93, add a term $0.2v_1^2/2g$ to the previous equation, giving $p_1/\gamma + (0.250v_2)^2/[(2)(9.807)] + z_1 = [p_1/\gamma - (k + 0.250) + (z_1 - z_2 + k)] + v_2^2/[(2)(9.807)] + z_2 + (0.2)\{0.5718^2/[(2)(9.807)]\}$, $v_2 = 2.272$ m/s; $Q = A_2v_2 = [(\pi)(0.150)^2/4](2.272) = 0.0401$ m³/s $= [0.0401/(0.3048)^3](7.48)(60) = 636$ gpm.

**8.94** The device shown in Fig. 8-51 is used to determine the velocity of liquid at point 1. It is a tube with its lower end directed upstream and its other leg vertical and open to the atmosphere. The impact of liquid against opening 2 forces liquid to rise in the vertical leg to the height $z$ above the free surface. Determine the velocity at 1.

▐ $\quad p_1/\gamma + v_1^2/2g + z_1 = p_2/\gamma + v_2^2/2g + z_2 + h_L \quad\quad k + v_1^2/2g + 0 = 0 + (k + \Delta z) + 0 + 0 \quad\quad v_1 = \sqrt{2g\,\Delta z}$

**Fig. 8-51**

**8.95** In Fig. 8-52 the losses in the exit pipe equal $Kv^2/2g$, where $K = 5.0$. The tank reservoir is large. Compute the flow rate in cubic feet per minute.

▐
$$p_1/\gamma + v_1^2/2g + z_1 = p_2/\gamma + v_2^2/2g + z_2 + h_L$$
$$(15)(144)/[(0.86)(62.4)] + 0 + 6 = (14.7)(144)/[(0.86)(62.4)] + v_2^2/[(2)(32.2)] + 0 + (5.0)\{v_2^2/[(2)(32.2)]\}$$
$$v_2 = 8.54 \text{ ft/s} \quad\quad Q = Av = [(\pi)(\tfrac{1}{12})^2/4](8.54) = 0.0466 \text{ ft}^3/\text{s} \quad \text{or} \quad 2.80 \text{ ft}^3/\text{min}$$

**Fig. 8-52**

**8.96** The manometer fluid in Fig. 8-53 is mercury. Neglecting losses, calculate the flow rate in the tube if the flowing fluid is (a) water, (b) air. Use 60 °F as the fluid temperature.

▐ $p_1/\gamma + v_1^2/2g + z_1 = p_2/\gamma + v_2^2/2g + z_2 + h_L$, $p_1/\gamma + v_1^2/2g + 0 = p_2/\gamma + 0 + 0 + 0$, $p_2 - p_1 = (\gamma)(v_1^2/2g)$. From the manometer, $p_1 + (62.4)(y/12) + [(13.6)(62.4)](\tfrac{1}{12}) - (62.4)(\tfrac{1}{12}) - (62.4)(y/12) = p_2$, $p_2 - p_1 = 65.52$ lb/ft²; $(\gamma)(v_1^2/2g) = 65.52$.

(a) $\quad (62.4)\{v_1^2/[(2)(32.2)]\} = 65.52 \quad\quad v_1 = 8.223 \text{ ft/s} \quad\quad Q = A_1v_1 = [(\pi)(\tfrac{4}{12})^2/4](8.223) = 0.718 \text{ ft}^3/\text{s}$

(b) $\quad (0.0763)\{v_1^2/[(2)(32.2)]\} = 65.52 \quad\quad v_1 = 235.2 \text{ ft/s} \quad\quad Q = A_1v_1 = [(\pi)(\tfrac{4}{12})^2/4](235.2) = 20.5 \text{ ft}^3/\text{s}$

**Fig. 8-53**

**8.97** In Fig. 8-54 the fluid is air ($\gamma = 12\,\text{N/m}^3$), and the manometer fluid has s.g. = 0.827. Assuming no losses, compute the flow rate in L/s.

▌ $p_1/\gamma + v_1^2/2g + z_1 = p_2/\gamma + v_2^2/2g + z_2 + h_L$, $p_1/\gamma + 0 + 0 = p_2/\gamma + v_2^2/2g + 0 + 0$, $p_1 - p_2 = (\gamma)(v_2^2/2g)$. From the manometer, $p_1 + (12 \times 10^{-3})(y + 0.080) - [(0.827)(9.79)](0.080) - (12 \times 10^{-3})y = p_2$, $p_1 - p_2 = 0.6467\,\text{kN/m}^2$; $(\gamma)(v_2^2/2g) = 0.6467$, $(12 \times 10^{-3})\{v_2^2/[(2)(9.807)]\} = 0.6467$, $v_2 = 32.51\,\text{m/s}$; $Q = A_2 v_2 = [(\pi)(0.050)^2/4](32.51) = 0.0638\,\text{m}^3/\text{s} = 63.8\,\text{L/s}$.

**Fig. 8-54**

**8.98** The flow from two reservoirs mixes together and flows through a common pipe. The elevations and pipe diameters are indicated in Fig. 8-55. Both reservoirs contain the same liquid and are open to the atmosphere. The common pipe empties to the atmosphere. Neglecting any frictional effects, find the flow rate through the common pipe.

▌
$$p_1/\gamma + v_1^2/2g + z_1 = p_2/\gamma + v_2^2/2g + z_2 + h_L \qquad p_{\text{atm}}/\gamma + 0 + h_1 = p_2/\gamma + v_2^2/2g + 0 + 0$$

$$P_{\text{atm}} + \gamma h_1 = p_2 + \gamma v_2^2/2g \qquad (1)$$

$$p_3/\gamma + v_3^2/2g + z_3 = p_4/\gamma + v_4^2/2g + z_4 + h_L \qquad p_{\text{atm}}/\gamma + 0 + h_2 = p_4/\gamma + v_4^2/2g + 0 + 0$$

$$p_{\text{atm}} + \gamma h_2 = p_4 + \gamma v_4^2/2g \qquad (2)$$

$$p_5/\gamma + v_5^2/2g + z_5 = p_6/\gamma + v_6^2/2g + z_6 + h_L \qquad p_5/\gamma + v_5^2/2g + 0 = p_{\text{atm}}/\gamma + v_6^2/2g + (-h_3) + 0$$

Since $v_5 = v_6$,

$$p_5 = p_{\text{atm}} - \gamma h_3 \qquad (3)$$

Assume $p_5 = p_2 = p_4$. Substituting this common value of pressure back into Eqs. (1) and (2) and solving for the velocity in each branch, we get $v_2 = \sqrt{2g(h_1 + h_3)}$, $v_4 = \sqrt{2g(h_2 + h_3)}$ $Q = Av = (\pi d_1^2/4)\sqrt{2g(h_1 + h_3)} + (\pi d_2^2/4)\sqrt{2g(h_2 + h_3)} = (\pi/4)[d_1^2\sqrt{2g(h_1 + h_3)} + d_2^2\sqrt{2g(h_2 + h_3)}]$.

**8.99** A steady jet of water comes from a hydrant and hits the ground some distance away, as shown in Fig. 8-56. If the water outlet is 1 m above the ground and the hydrant water pressure is 862 kPa, what distance from the hydrant does the jet hit the ground? Atmospheric pressure is 101 kPa.

▌ The magnitude of $v_x$ can be obtained by noting that at the hydrant outlet the flow is entirely in the $x$ direction, $v_x = V_2$. Applying the Bernoulli equation between the interior of the hydrant and the outlet gives $p_1 + \rho g y_1 + \frac{1}{2}\rho V_1^2 = p_2 + \rho g y_2 + \frac{1}{2}\rho V_2^2$. The pressure in the hydrant $p_1$ is given, and the outlet is open to the atmosphere, $p_2 = p_{\text{atm}}$. The elevation of points 1 and 2 is the same, $y_1 = y_2$. We assume that the outlet area is small enough compared with the hydrant cross-sectional area for the hydrant to be essentially a reservoir, $V_1^2 \ll V_2^2$. Neglecting $V_1$, we get $v_x = \sqrt{(2/\rho)(p_1 - p_{\text{atm}})}$. Since $v_x$ is constant, it can be brought outside the integral for $l$, giving $l = v_x T$.

To find the time $T$ required for a fluid particle to hit the ground, we apply the Bernoulli equation between point 1 and some arbitrary point on the jet having elevation $y$ and velocity $V$: $p_1 + \rho g h = p_{\text{atm}} + \rho g y + \frac{1}{2}\rho V^2$. Now $V^2 = v_x^2 + v_y^2$. When we use the previously determined value of $v_x$ and note that $v_y = dy/dt$, the Bernoulli equation becomes $(dy/dt)^2 = 2g(h - y)$. We take the square root (the negative root is the appropriate one since $dy/dt$ must be negative): $dy/dt = -\sqrt{2g(h - y)}$. Then we separate variables and integrate between the limits of $y = h$, $t = 0$ and $y = 0$, $t = T$:

$$\int_h^0 \frac{dy}{\sqrt{h - y}} = -\sqrt{2g} \int_0^T dt$$

Integrating and solving for $T$ gives $T = \sqrt{2h/g}$. The $y$ component of the fluid motion is that of a body freely falling under the influence of gravity.

Finally, we substitute numerical values to get $v_x = \sqrt{\frac{2}{1000}[(862 - 101)(1000)]} = 39.0\,(\text{m} \cdot \text{N/kg})^{1/2}$, or 39.0 m/s; $T = \sqrt{(2)(1.0)/9.807} = 0.452\,\text{s}$; $L = vt = (39.0)(0.452) = 17.6\,\text{m}$.

**Fig. 8-55**

**Fig. 8-56**

**8.100** Water flows between the two reservoirs in Fig. 8-57 at the rate of 16 L/s. What is the head loss in the pipe? If atmospheric pressure is 100 kPa and the vapor pressure is 8 kPa, for what constriction diameter $d$ will cavitation occur? Assume no additional losses due to changes in the constriction.

▮ $p_1/\gamma + v_1^2/2g + z_1 = p_2/\gamma + v_2^2/2g + z_2 + h_L$, $0 + 0 + 20 = 0 + 0 + 5 + h_L$, $h_L = 15$ m; $v_{\text{throat}} = Q/A_{\text{throat}} = 0.016/(\pi d^2/4)$. Assume a central constriction, with $\frac{15}{2}$, or 7.5-m head loss on each side. Apply Bernoulli's equation between point 1 and the constriction, with $p_1 = p_{\text{atm}} = 100$ kPa and $p_v = 8$ kPa at the constriction.

$100/9.79 + 0 + 20 = 8/9.79 + v_{max}^2/[(2)(9.807)] + 0 + 7.5$, $v_{max} = 20.7$ m/s; $20.7 = 0.016/(\pi d^2/4)$, $d = 0.0314$ m, or 2.98 cm.

**Fig. 8-57**

**8.101** The horizontal wye fitting in Fig. 8-58 splits $Q_1$ into two equal flow rates. At section 1, $Q_1 = 4$ ft$^3$/s and $p_1 = 20$ psig. Neglecting losses, compute pressures $p_2$ and $p_3$.

▌ $p_1/\gamma + v_1^2/2g + z_1 = p_2/\gamma + v_2^2/2g + z_2 + h_L$ $\quad v_1 = Q_1/A_1 = 4/[(\pi)(\frac{6}{12})^2/4] = 20.37$ ft/s

$$v_2 = Q_2/A_2 = \tfrac{4}{2}/[(\pi)(\tfrac{4}{12})^2/4] = 22.92 \text{ ft/s}$$

$$(20)(144)/62.4 + 20.37^2/[(2)(32.2)] + 0 = (p_2)(144)/62.4 + 22.92^2/[(2)(32.2)] + 0 + 0$$

$p_2 = 19.3$ psig $\quad p_1/\gamma + v_1^2/2g + z_1 = p_3/\gamma + v_3^2/2g + z_3 + h_L$ $\quad v_3 = Q_3/A_3 = \tfrac{4}{2}/[(\pi)(\tfrac{3}{12})^2/4] = 40.74$ ft/s

$$(20)(144)/62.4 + 20.37^2/[(2)(32.2)] + 0 = (p_3)(144)/62.4 + 40.74^2/[(2)(32.2)] + 0 + 0 \qquad p_3 = 11.6 \text{ psig}$$

**Fig. 8-58**

**8.102** A cylindrical tank of diameter $d_0$ contains liquid to an initial height $h_0$. At time $t = 0$ a small stopper of diameter $d$ is removed from the bottom. Using Bernoulli's equation with no losses, derive a differential equation for the free-surface height $h$ during draining and an expression for the time $t_0$ to drain the entire tank.

▌ Letting point 1 be the liquid surface and point 2 at the exit, $p_1/\gamma + v_1^2/2g + z_1 = p_2/\gamma + v_2^2/2g + z_2 + h_L$,
$0 + v_1^2/2g + h = 0 + v_2^2/2g + 0 + 0$, $v_1^2 = v_2^2 - 2gh$; $A_1 v_1 = A_2 v_2$, $(\pi d_0^2/4)(v_1) = (\pi d^2/4)(v_2)$, $v_2 = (d_0/d)^2(v_1)$;
$v_1^2 = [(d_0/d)^2(v_1)]^2 - 2gh$, $v_1 = \sqrt{2gh/(d_0^4/d^4 - 1)}$. Or $v_1 = \sqrt{Kh}$ where $K = 2g/[(d_0/d)^4 - 1]$. But also,
$v_1 = -dh/dt$, $dh/dt = -\sqrt{Kh}$

$$\int_{h_0}^h \frac{dh}{\sqrt{h}} = \int_0^t -\sqrt{K}\,dt \qquad (2)[h^{1/2}]_{h_0}^h = -\sqrt{K}\,[t]_0^t \qquad (2)(h^{1/2} - h_0^{1/2}) = -\sqrt{K}\,t \qquad h = \left(h_0^{1/2} - \frac{\sqrt{K}\,t}{2}\right)^2$$

Or, $h = \{h_0^{1/2} - [g/(2)(d_0^4/d^4 - 1)]^{1/2}t\}^2$.

**8.103** In the water flow over the spillway in Fig. 8-59, the velocity is uniform at sections 1 and 2 and the pressure approximately hydrostatic. Neglecting losses, compute $v_1$ and $v_2$. Assume unit width.

▌ $\qquad p_1/\gamma + v_1^2/2g + z_1 = p_2/\gamma + v_2^2/2g + z_2 + h_L \qquad 0 + v_1^2/2g + 6 = 0 + v_2^2/2g + 1 + 0$

$$A_1 v_1 = A_2 v_2 \qquad [(6)(1)](v_1) = [(1)(1)](v_2) \qquad v_2 = 6v_1$$

$$v_1^2/[(2)(9.807)] + 6 = (6v_1)^2/[(2)(9.807)] + 1 \qquad v_1 = 1.67 \text{ m/s} \qquad v_2 = (6)(1.67) = 10.02 \text{ m/s}$$

Fig. 8-59

**8.104** For the water channel flow down the sloping ramp of Fig. 8-60, $h_1 = 1$ m, $H = 3$ m, and $v_1 = 4$ m/s. The flow is uniform at 1 and 2. Neglecting losses, find the downstream depth $h_2$ and show that three solutions are possible, of which only two are realistic. Neglect friction.

❚ $p_1/\gamma + v_1^2/2g + z_1 = p_2/\gamma + v_2^2/2g + z_2 + h_L$ $\quad 0 + 4^2/[(2)(9.807)] + (3+1) = 0 + v_2^2/[(2)(9.807)] + h_2 + 0$

$A_1 v_1 = A_2 v_2$ $\quad [(1)(1)](4) = [(h_2)(1)](v_2)$ $\quad v_2 = 4/h_2$

$4^2/[(2)(9.807)] + 4 = (4/h_2)^2/[(2)(9.807)] + h_2$ $\quad h_2^3 - 4.816 h_2^2 + 0.8157 = 0$

There are three mathematical solutions to this equation:

$$h_2 = 4.78 \text{ m} \quad \text{(subcritical)}$$
$$h_2 = 0.432 \text{ m} \quad \text{(supercritical)}$$
$$h_2 = -0.396 \text{ m} \quad \text{(impossible)}$$

Fig. 8-60

**8.105** For water flow up the sloping channel in Fig. 8-61, $h_1 = 0.5$ ft, $v_1 = 15$ ft/s, and $H = 2$ ft. Neglect losses and assume uniform flow at 1 and 2. Find the downstream depth $h_2$ and show that three solutions are possible, of which only two are realistic.

❚ $p_1/\gamma + v_1^2/2g + z_1 = p_2/\gamma + v_2^2/2g + z_2 + h_L$ $\quad 0 + 15^2/[(2)(32.2)] + 0.5 = 0 + v_2^2/[(2)(32.2)] + (2 + h_2) + 0$

$A_1 v_1 = A_2 v_2$ $\quad [(0.5)(1)](15) = [(h_2)(1)](v_2)$ $\quad v_2 = 7.5/h_2$

$15^2/[(2)(32.2)] + 0.5 = (7.5/h_2)^2/[(2)(32.2)] + (2 + h_2)$ $\quad h_2^3 - 1.994 h_2^2 + 0.8734 = 0$

There are three mathematical solutions to this equation:

$$h_2 = 1.69 \text{ ft} \quad \text{(subcritical)}$$
$$h_2 = 0.887 \text{ ft} \quad \text{(supercritical)}$$
$$h_2 = -0.582 \text{ ft} \quad \text{(impossible)}$$

Fig. 8-61

**8.106** A constant 12-lb force is applied to the piston in Fig. 8-62, and flow losses are negligible. Compute the water-jet exit velocity $v_2$.

▌ $p_1/\gamma + v_1^2/2g + z_1 = p_{atm}/\gamma + v_2^2/2g + z_2 + h_L$. Considering the force acting on the piston, $12 = (p_1 - p_{atm})[(\pi)(\frac{8}{12})^2/4]$, $p_1 - p_{atm} = 34.38$ lb/ft$^2$; $A_1 v_1 = A_2 v_2$, $[(\pi)(\frac{8}{12})^2/4](v_1) = [(\pi)(\frac{4}{12})^2/4](v_2)$, $v_1 = 0.250v_2$; $(p_1 - p_{atm})/\gamma + (0.250v_2)^2/[(2)(32.2)] + 0 = v_2^2/[(2)(32.2)] + 0 + 0$, $34.38/62.4 + (0.250v_2)^2/[(2)(32.2)] + 0 = v_2^2/[(2)(32.2)] + 0 + 0$, $v_2 = 6.15$ ft/s.

Fig. 8-62

Fig. 8-63

**8.107** The horizontal lawn sprinkler of Fig. 8-63 is fed water through the center at 1.2 L/s. If collar friction is negligible, what is the steady rotation rate for (**a**) $\theta = 0°$ and (**b**) $\theta = 30°$?

▌
$$v_0 = \frac{Q}{A} = \frac{(1.2 \times 10^{-3})/3}{(\pi)(0.007)^2/4} = 10.39 \text{ m/s}$$

Choose an inertial (nonrotating) frame with origin at the center of the sprinkler; let $(\rho, \phi)$ be polar coordinates relative to this frame.

An emergent water jet has velocity components

$$v_\rho = v_0 \sin \theta \qquad v_\phi = v_0 \cos \theta - r\omega$$

For zero reactive torque—the criterion for the steady state—$v_\phi = 0$, or $\omega = (v_0 \cos \theta)/r$.

(**a**) $\qquad \omega = (10.39)(\cos 0°)/0.20 = 51.95$ rad/s or 496 rpm

(**b**) $\qquad \omega = (10.39)(\cos 30°)/0.20 = 44.99$ rad/s or 430 rpm

**8.108** Water flows at 6 ft/s through a pipe 500 ft long with diameter 1 in. The inlet pressure $p_1 = 200$ psig, and the exit section is 100 ft higher than the inlet. What is the exit pressure $p_2$ if the friction head loss is 350 ft?

▌ $\qquad\qquad p_1/\gamma + v_1^2/2g + z_1 = p_2/\gamma + v_2^2/2g + z_2 + h_L$

$(200)(144)/62.4 + v_1^2/2g + 0 = (p_2)(144)/62.4 + v_2^2/2g + 100 + 350 \qquad v_1^2/2g = v_2^2/2g \qquad p_2 = 5.00$ psig

**8.109** A 30-in-diameter pipeline carries oil (s.g. = 0.86) at 600 000 barrels per day. The friction head loss is 10 ft per 1000 ft of pipe. Compute the pressure drop per mile.

▌ $\qquad\qquad p_1/\gamma + v_1^2/2g + z_1 = p_2/\gamma + v_2^2/2g + z_2 + h_L$

$p_1/[(0.86)(62.4)] + v_1^2/2g + 0 = p_2/[(0.86)(62.4)] + v_2^2/2g + 0 + (10/1000)(5280)$

$v_1^2/2g = v_2^2/2g \qquad p_1 - p_2 = 2833$ lb/ft$^2$ or 19.7 lb/in$^2$

**8.110** The long pipe in Fig. 8-64 is filled with water. When valve $A$ is closed, $p_2 - p_1 = 12$ psi. When the valve is open and water flows at 10 ft$^3$/s, $p_1 - p_2 = 25$ psi. What is the friction head loss between 1 and 2 for the flowing condition?

▌ $\qquad\qquad p_1/\gamma + v_1^2/2g + z_1 = p_2/\gamma + v_2^2/2g + z_2 + h_L \qquad v_1^2/2g = v_2^2/2g$

Valve closed:
$$(p_2 - p_1)(144)/62.4 = z_1 - z_2 \qquad z_1 - z_2 = (12)(144)/62.4 = 27.69 \text{ ft}$$

Valve open:

$$(p_1 - p_2)(144)/62.4 + (z_1 - z_2) = h_L \qquad (25)(144)/62.4 + 27.69 = h_L \qquad h_L = 85.4 \text{ ft}$$

Constant-
diameter
pipe

**Fig. 8-64**

**8.111**  Find the manometer reading in the lossless system of Fig. 8-65.

$$p_1/\gamma + v_1^2/2g + z_1 = p_2/\gamma + v_2^2/2g + z_2 + h_L$$

$$A_1 v_1 = A_2 v_2 \qquad [(\pi)(\tfrac{3}{12})^2/4](2) = [(\pi)(\tfrac{1}{12})^2/4](v_2) \qquad v_2 = 18.0 \text{ ft/s}$$

$$p_1/\gamma + 2^2/[(2)(32.2)] + 0 = 0 + 18.0^2/[(2)(32.2)] + 8 + 0 \qquad p_1/\gamma = 12.97 \text{ ft}$$

For the manometer, $12.97 + 2.5 - 13.6h = 0$, $h = 1.14$ ft.

1 in → ② 

$v_1 = 2\text{ft/s}$

8 ft

3 in
① 

→ Water

2.5 ft

$h$

Mercury  **Fig. 8-65**

**8.112**  In Fig. 8-66 on p. 188 the pipe exit losses are $(1.5)v^2/2g$, where $v$ is the exit velocity. What is the exit weight flux of water?

$$p_1/\gamma + v_1^2/2g + z_1 = p_2/\gamma + v_2^2/2g + z_2 + h_L \qquad p_1 = (20)(144) + [(0.68)(62.4)](4) = 3050 \text{ lb/ft}^2$$

$$3050/62.4 + 0 + 5 = 0 + v_2^2/[(2)(32.2)] + 0 + (1.5)\{v_2^2/[(2)(32.2)]\} \qquad v_2 = 37.25 \text{ ft/s}$$

$$W = \gamma A v = (62.4)[(\pi)(\tfrac{2}{12})^2/4](37.25) = 50.7 \text{ lb/s}$$

**8.113**  In Fig. 8-67 the fluid is water, and the pressure gage reads $p_1 = 180$ kPa gage. If the mass flux is 15 kg/s, what is the head loss between 1 and 2?

$$p_1/\gamma + v_1^2/2g + z_1 = p_2/\gamma + v_2^2/2g + z_2 + h_L$$

$$M = \rho A v \qquad 15 = 1000[(\pi)(0.08)^2/4](v_1) \qquad v_1 = 2.984 \text{ m/s} \qquad 15 = 1000[(\pi)(0.05)^2/4](v_2) \qquad v_2 = 7.639 \text{ m/s}$$

$$180/9.79 + 2.984^2/[(2)(9.807)] + 0 = 0 + 7.639^2/[(2)(9.807)] + 12 + h_L \qquad h_L = 3.86 \text{ m}$$

Fig. 8-66

Fig. 8-67

**8.114** Oil at specific gravity 0.761 flows from tank $A$ to tank $E$, as shown in Fig. 8-68. Lost head items may be assumed to be as follows: $A$ to $B = 0.60v_{12}^2/2g$; $B$ to $C = 9.0v_{12}^2/2g$; $C$ to $D = 0.40v_6^2/2g$; $D$ to $E = 9.0v_6^2/2g$. Find the flow rate and the pressure at $C$.

$$p_A/\gamma + v_A^2/2g + z_A = p_E/\gamma + v_E^2/2g + z_E + h_L$$

$$0 + 0 + 40.0 = 0 + 0 + 0 + (0.60v_{12}^2 + 9.0v_{12}^2 + 0.40v_6^2 + 9.0v_6^2)/[(2)(32.2)]$$

$$9.60v_{12}^2 + 9.40v_6^2 = 2576 \qquad A_1v_1 = A_2v_2 \qquad [(\pi)(\tfrac{12}{12})^2/4](v_{12}) = [(\pi)(\tfrac{6}{12})^2/4](v_6) \qquad v_6 = 4.00v_{12}$$

$$9.60v_{12}^2 + (9.40)(4.00v_{12})^2 = 2576 \qquad v_{12} = 4.012 \text{ ft/s}$$

$$Q = Av = [(\pi)(\tfrac{12}{12})^2/4](4.012) = 3.15 \text{ ft}^3/\text{s}$$

$$p_A/\gamma + v_A^2/2g + z_A = p_C/\gamma + v_C^2/2g + z_C + h_L$$

$$0 + 0 + 40.0 = (p_C)(144)/[(0.761)(62.4)] + 4.012^2/[(2)(32.2)]$$

$$+ (40.0 + 2) + [(0.60)(4.012)^2 + (9.0)(4.012)^2]/[(2)(32.2)]$$

$$p_C = -1.53 \text{ lb/in}^2$$

Fig. 8-68

**8.115** (a) What is the pressure on the nose of a torpedo moving in salt water at 100 ft/s at a depth of 30.0 ft? (b) If the pressure at point $C$ on the side of the torpedo at the same elevation as the nose is 10.0 psig, what is the relative velocity at that point?

$\blacksquare$ (a) In this case, greater clarity in the application of the Bernoulli equation may be attained by considering the relative motion of a stream of water past the stationary torpedo. The velocity of the nose of the torpedo will then be zero. Assume no lost head in the streamtube from a point $A$ in the undisturbed water just ahead of the torpedo to a point $B$ on the nose of the torpedo: $p_A/\gamma + v_A^2/2g + z_A = p_B/\gamma + v_B^2/2g + z_B + h_L$, $30.0 + 100^2/[(2)(32.2)] + 0 = p_B/\gamma + 0 + 0 + 0$, $p_B/\gamma = 185.3$ ft, $p_B = (64.2)(185.3)/144 = 82.6$ psi. This pressure is called the stagnation pressure and may be expressed as $p_s = p_0 + \rho v_0^2/2$.

**(b)**
$$p_A/\gamma + v_A^2/2g + z_A = p_C/\gamma + v_C^2/2g + z_C + h_L$$

$$30.0 + 100^2/[(2)(32.2)] + 0 = (10.0)(144)/64.2 + v_C^2/[(2)(32.2)] + 0 + 0 \qquad v_C = 102.4 \text{ ft/s}$$

**8.116** A sphere is placed in an air stream which is at atmospheric pressure and is moving at 100.0 ft/s. Using the density of air constant at 0.00238 slug/ft$^3$, calculate the stagnation pressure and the pressure on the surface of the sphere at a point $B$, 75° from the stagnation point, if the velocity there is 220.0 ft/s.

**❚** From Prob. 8.115, $p_s = p_0 + \rho v_0^2/2 = (14.7)(144) + (0.00238)(100.0)^2/2 = 2129$ lb/ft$^2$, or 14.8 lb/in$^2$. $p_S/\gamma + v_S^2/2g + z_S = p_B/\gamma + v_B^2/2g + z_B + h_L$, $2129/[(0.00238)(32.2)] + 0 + 0 = p_B/\gamma + 220.0^2/[(2)(32.2)] + 0 + 0$, $p_B/\gamma = 27\,029$ ft of air, $p_B = [(0.00238)(32.2)](27\,029) = 2071$ lb/ft$^2$, or 14.4 lb/in$^2$.

**8.117** A large closed tank is filled with ammonia (NH$_3$) under a pressure of 5.30 psig and at 65 °F. The ammonia discharges into the atmosphere through a small opening in the side of the tank. Neglecting friction losses, calculate the velocity of the ammonia leaving the tank assuming constant density. The gas constant for ammonia is 89.5 ft/°R.

**❚** Apply Bernoulli's equation between the tank (1) and the atmosphere (2). $p_1/\gamma + v_1^2/2g + z_1 = p_2/\gamma + v_2^2/2g + z_2 + h_L$, $\gamma = p/RT$, $\gamma_{\text{NH}_3} = (5.30 + 14.7)(144)/[(89.5)(460 + 65)] = 0.06129$ lb/ft$^3$, $(5.30)(144)/0.06129 + 0 + 0 = 0 + v_2^2/[(2)(32.2)] + 0 + 0$, $v_2 = 896$ ft/s.

**8.118** Water at 90 °F is to be lifted from a sump at a velocity of 6.50 ft/s through the suction pipe of a pump. Calculate the theoretical maximum height of the pump setting under the following conditions: $p_{\text{atm}} = 14.25$ psia, $p_v = 0.70$ psia, and $h_L$ in the suction pipe = 3 velocity heads.

**❚** The minimum pressure at the entrance to the pump cannot be less than the vapor pressure of the liquid. Apply Bernoulli's equation between the water surface outside the suction pipe and the entrance to the pump: $p_1/\gamma + v_1^2/2g + z_1 = p_2/\gamma + v_2^2/2g + z_2 + h_L$, $(14.25)(144)/62.1 + 0 + 0 = (0.70)(144)/62.1 + 6.50^2/[(2)(32.2)] + z_2 + 3\{6.50^2/[(2)(32.2)]\}$, $z_2 = 28.8$ ft. (Under these conditions, serious damage due to cavitation will probably occur.)

**8.119** For the Venturi meter shown in Fig. 8-69, the deflection of mercury in the differential gage is 14.3 in. Determine the flow of water through the meter if no energy is lost between $A$ and $B$.

**❚**
$$p_A/\gamma + v_A^2/2g + z_A = p_B/\gamma + v_B^2/2g + z_B + h_L$$

$$p_A/\gamma + v_A^2/[(2)(32.2)] + 0 = p_B/\gamma + v_B^2/[(2)(32.2)] + 30.0/12 + 0 \qquad p_A/\gamma - p_B/\gamma = 0.01553(v_B^2 - v_A^2) + 2.500$$

$$A_A v_A = A_B v_B \qquad [(\pi)(\tfrac{12}{12})^2/4](v_A) = [(\pi)(\tfrac{6}{12})^2/4](v_B) \qquad v_A = 0.250 v_B$$

From the manometer, $p_A/\gamma + z + 14.3/12 - (13.6)(14.3/12) - z - 30.0/12 = p_B/\gamma$, $p_A/\gamma - p_B/\gamma = 17.52$ ft, $17.52 = 0.01553[v_B^2 - (0.250 v_B)^2] + 2.500$, $v_B = 32.12$ ft/s; $Q = Av = [(\pi)(\tfrac{6}{12})^2/4](32.12) = 6.31$ ft$^3$/s.

**Fig. 8-69**

**8.120** For the meter in Fig. 8-69, consider air at 80 °F with the pressure at $A = 37.5$ psig. Consider a deflection of the gage of 14.3 in of water. Assuming that the specific weight of the air does not change between $A$ and $B$ and that the energy loss is negligible, determine the amount of air flowing in pounds per second.

$$p_A/\gamma + v_A^2/2g + z_A = p_B/\gamma + v_B^2/2g + z_B + h_L$$

$$p_A/\gamma + v_A^2/[(2)(32.2)] + 0 = p_B/\gamma + v_B^2/[(2)(32.2)] + 30.0/12 + 0 \qquad p_A/\gamma - p_B/\gamma = 0.01553(v_B^2 - v_A^2) + 2.500$$

$$A_A v_A = A_B v_B \qquad [(\pi)(\tfrac{12}{12})^2/4](v_A) = [(\pi)(\tfrac{6}{12})^2/4](v_B) \qquad v_A = 0.250 v_B$$

$$\gamma = p/RT \qquad \gamma_{\text{air}} = (37.5 + 14.7)(144)/[(53.3)(460 + 80)] = 0.2612 \text{ lb/ft}^3$$

From the manometer, $p_A/\gamma + z + 14.3/12 - (62.4/0.2612)(14.3/12) - z - 30.0/12 = p_B/\gamma$, $p_A/\gamma - p_B/\gamma = 286.0$ ft of air; $286.0 = 0.01553[v_B^2 - (0.250 v_B)^2] + 2.500$, $v_B = 139.5$ ft/s; $W = \gamma A v = 0.2612[(\pi)(\tfrac{6}{12})^2/4](139.5) = 7.15$ lb/s.

**8.121** Given a frictionless flow of water at 125.6 ft³/s in a long, horizontal, conical pipe, of diameter 2 ft at one end and 6 ft at the other. The pressure head at the smaller end is 18 ft of water. Find the velocities at the two ends and the pressure head at the larger end.

$$v_1 = Q/A_1 = 125.6/[(\pi)(2)^2/4] = 39.98 \text{ ft/s} \qquad v_2 = Q/A_2 = 125.6/[(\pi)(6)^2/4] = 4.44 \text{ ft/s}$$

$$p_1/\gamma + v_1^2/2g + z_1 = p_2/\gamma + v_2^2/2g + z_2 + h_L$$

$$18 + 39.98^2/[(2)(32.2)] + 0 = p_2/\gamma + 4.44^2/[(2)(32.2)] + 0 + 0 \qquad p_2/\gamma = 42.5 \text{ ft of water}$$

**8.122** Water flows through a long, horizontal, conical diffuser at the rate of 4.0 m³/s. The diameter of the diffuser varies from 1.0 m to 2.0 m; the pressure at the smaller end is 8.0 kPa. Find the pressure at the downstream end of the diffuser, assuming frictionless flow and no separation from the walls.

$$p_1/\gamma + v_1^2/2g + z_1 = p_2/\gamma + v_2^2/2g + z_2 + h_L$$

$$v_1 = Q/A_1 = 4.0/[(\pi)(1.0)^2/4] = 5.093 \text{ m/s} \qquad v_2 = Q/A_2 = 4.0/[(\pi)(2.0)^2/4] = 1.273 \text{ m/s}$$

$$8.0/9.79 + 5.093^2/[(2)(9.807)] + 0 = p_2/9.79 + 1.273^2/[(2)(9.807)] + 0 + 0 \qquad p_2 = 20.13 \text{ kPa}$$

**8.123** A vertical pipe 3 ft in diameter and 30 ft long has a pressure head at the upper end of 22 ft of water. When water flows through it with mean velocity 15 fps, the friction loss is 6 ft. Find the pressure head at the lower end of the pipe when the flow is (a) downward and (b) upward.

**(a)**
$$p_1/\gamma + v_1^2/2g + z_1 = p_2/\gamma + v_2^2/2g + z_2 + h_L$$

$$22 + 15^2/[(2)(32.2)] + 30 = p_2/\gamma + 15^2/(2)(32.2)] + 0 + 6 \qquad p_2/\gamma = 46.0 \text{ ft}$$

**(b)**
$$p_2/\gamma + v_2^2/2g + z_2 = p_1/\gamma + v_1^2/2g + z_1 + h_L$$

$$p_2/\gamma + 15^2/[(2)(32.2)] + 0 = 22 + 15^2/[(2)(32.2)] + 30 + 6 \qquad p_2/\gamma = 58.0 \text{ ft}$$

**8.124** A vertical conical pipe has diameter 1.5 ft at the top and 3.0 ft at the bottom, and is 60 ft long. The friction loss is 10 ft for flow in either direction when the velocity at the top is 30 fps and the pressure head there is 6.5 ft of water. Find the pressure head at the bottom when the flow is (a) downward and (b) upward.

$$Q = A_1 v_1 = [(\pi)(1.5)^2/4](30) = 53.01 \text{ ft}^3/s \qquad v_2 = Q/A_2 = 53.01/[(\pi)(3.0)^2/4] = 7.50 \text{ ft/s}$$

**(a)**
$$p_1/\gamma + v_1^2/2g + z_1 = p_2/\gamma + v_2^2/2g + z_2 + h_L$$

$$6.5 + 30^2/[(2)(32.2)] + 60 = p_2/\gamma + 7.50^2/[(2)(32.2)] + 0 + 10 \qquad p_2/\gamma = 69.6 \text{ ft}$$

**(b)**
$$p_2/\gamma + v_2^2/2g + z_2 = p_1/\gamma + v_1^2/2g + z_1 + h_L$$

$$p_2/\gamma + 7.5^2/[(2)(32.2)] + 0 = 6.5 + 30^2/[(2)(32.2)] + 60 + 10 \qquad p_2/\gamma = 89.6 \text{ ft}$$

**8.125** The inclined pipe in Fig. 8-70 is of uniform diameter. The pressure at $A$ is 20 psi and at $B$, 30 psi. In which direction is the flow, and what is the friction loss of the fluid, if the liquid has specific weight (a) 30 lb/ft³ and (b) 100 lb/ft³?

$p_A/\gamma + v_A^2/2g + z_A = p_B/\gamma + v_B^2/2g + z_B + h_L$. Assume flow is from $A$ to $B$.

(a) $(20)(144)/30 + v_A^2/2g + 25 = (30)(144)/30 + v_B^2/2g + 0 + h_L$, $v_A^2/2g = v_B^2/2g$, $h_L = -23.0$ ft. Since $h_L$ is negative, flow is actually from $B$ to $A$.

(b) $(20)(144)/100 + v_A^2/2g + 25 = (30)(144)/100 + v_B^2/2g + 0 + h_L$, $v_A^2/2g = v_B^2/2g$, $h_L = 10.6$ ft. Since $h_L$ is positive, flow is from $A$ to $B$, as assumed.

Fig. 8-70

**8.126** In Fig. 8-70, if the difference in elevation between $A$ and $B$ is 10 m and the pressures at $A$ and $B$ are 150 kPa and 250 kPa, respectively, find the direction of flow and the head loss. The liquid has specific gravity 0.85.

**❚** $p_A/\gamma + v_A^2/2g + z_A = p_B/\gamma + v_B^2/2g + z_B + h_L$. Assume flow is from $A$ to $B$. $150/[(0.85)(9.79)] + v_A^2/2g + 10 = 250/[(0.85)(9.79)] + v_B^2/2g + 0 + h_L$, $v_A^2/2g = v_B^2/2g$, $h_L = -2.02$ m. Since $h_L$ is negative, flow is actually from $B$ to $A$.

**8.127** An irrigation line carries water from a lake down into an arid valley floor 810 ft below the surface of the lake. The water is discharged through a nozzle with a jet velocity of 220 fps; the diameter of the jet is 4 in. Find the power of the jet and the power lost in friction.

**❚**
$$p_A/\gamma + v_A^2/2g + z_A = p_B/\gamma + v_B^2/2g + z_B + h_L \qquad 0 + 0 + 810 = 0 + 220^2/[(2)(32.2)] + 0 + h_L$$

$$h_L = 58.45 \text{ ft} \qquad Q = Av = [(\pi)(\tfrac{4}{12})^2/4](220) = 19.2 \text{ ft}^3/\text{s}$$

$$P_{jet} = Q\gamma v_2^2/2g = (19.2)(62.4)\{220^2/[(2)(32.2)]\} = 900\,000 \text{ ft} \cdot \text{lb/s} = 900\,000/550 = 1636 \text{ hp}$$

$$P_{lost} = Q\gamma h_L = (19.2)(62.4)(58.45) = 70\,000 \text{ ft} \cdot \text{lb/s} = 70\,000/550 = 127 \text{ hp}$$

**8.128** Water is flowing in a channel, as shown in Fig. 8-71. Neglecting all losses, determine the two possible depths of flow $y_1$ and $y_2$.

**❚**
$$p_A/\gamma + v_A^2/2g + z_A = p_B/\gamma + v_B^2/2g + z_B + h_L$$

$$Q = A_A v_A = [(4)(10)](16.1) = 644 \text{ ft}^3/\text{s} \qquad v_B = Q/A_B = 644/(10y) = 64.4/y$$

$$0 + 16.1^2/[(2)(32.2)] + (8 + 4) = 0 + (64.4/y)^2/[(2)(32.2)] + y + 0 \qquad 64.4/y^2 + y - 16.02 = 0$$

$$y^3 - 16.02y^2 + 64.4 = 0 \qquad y_1 = 2.16 \text{ ft} \qquad y_2 = 15.8 \text{ ft}$$

Fig. 8-71

**8.129** Neglecting all losses, in Fig. 8-71 the channel narrows in the drop to 6 ft wide at section $B$. For uniform flow across section $B$, determine the two possible depths of flow.

**❚**
$$p_A/\gamma + v_A^2/2g + z_A = p_B/\gamma + v_B^2/2g + z_B + h_L$$

$$Q = A_A v_A = [(4)(10)](16.1) = 644 \text{ ft}^3/\text{s} \qquad v_B = Q/A_B = 644/(6y) = 107.3/y$$

$$0 + 16.1^2/[(2)(32.2)] + (8 + 4) = 0 + (107.3/y)^2/[(2)(32.2)] + y + 0 \qquad 178.8/y^2 + y - 16.02 = 0$$

$$y^3 - 16.02y^2 + 178.8 = 0 \qquad y_1 = 3.83 \text{ ft} \qquad y_2 = 15.3 \text{ ft}$$

**8.130** If the losses from section $A$ to section $B$ of Fig. 8-71 are 1.9 ft, determine the two possible depths at section $B$.

**❚** 
$$p_A/\gamma + v_A^2/2g + z_A = p_B/\gamma + v_B^2/2g + z_B + h_L$$

$$Q = A_A v_A = [(4)(10)](16.1) = 644 \text{ ft}^3/\text{s} \qquad v_B = Q/A_B = 644/(10y) = 64.4/y$$

$$0 + 16.1^2/[(2)(32.2)] + (8+4) = 0 + (64.4/y)^2/[(2)(32.2)] + y + 1.9 \qquad 64.4/y^2 + y - 14.12 = 0$$

$$y^3 - 14.12y^2 + 64.4 = 0 \qquad y_1 = 2.34 \text{ ft} \qquad y_2 = 13.8 \text{ ft}$$

**8.131** High-velocity water flows up an inclined plane, as shown in Fig. 8-72. Neglecting all losses, calculate the two possible depths of flow at section $B$.

**❚** 
$$p_A/\gamma + v_A^2/2g + z_A = p_B/\gamma + v_B^2/2g + z_B + h_L$$

$$Q = A_A v_A = [(\tfrac{500}{1000})(2)](9.806) = 9.806 \text{ m}^3/\text{s} \qquad v_B = Q/A_B = 9.806/(2y) = 4.903/y$$

$$0 + 9.806^2/[(2)(9.807)] + \tfrac{500}{1000} = 0 + (4.903/y)^2/[(2)(9.807)] + (2.5 + y) + 0 \qquad 1.226/y^2 + y - 2.903 = 0$$

$$y^3 - 2.903y^2 + 1.226 = 0 \qquad y_1 = 0.775 \text{ m} \qquad y_2 = 2.74 \text{ m}$$

9.806 m/s    500 mm    2.5 m    Channel 2 m wide    $A$    $B$    **Fig. 8-72**

**8.132** In Fig. 8-72, the channel changes in width from 2 m at section $A$ to 3 m at section $B$. For losses of 0.3 m between sections $A$ and $B$, find the two possible depths at section $B$.

**❚** 
$$p_A/\gamma + v_A^2/2g + z_A = p_B/\gamma + v_B^2/2g + z_B + h_L \qquad Q = A_A v_A = [(\tfrac{500}{1000})(2)](9.806) = 9.806 \text{ m}^3/\text{s}$$

$$v_B = Q/A_B = 9.806/(3y) = 3.269/y$$

$$0 + 9.806^2/[(2)(9.807)] + \tfrac{500}{1000} = 0 + (3.269/y)^2/[(2)(9.807)] + (2.5 + y) + 0.3$$

$$0.5448/y^2 + y - 2.603 = 0 \qquad y^3 - 2.603y^2 + 0.5448 = 0 \qquad y_1 = 0.510 \text{ m} \qquad y_2 = 2.52 \text{ m}$$

**8.133** For losses of $0.05H$ through the nozzle of Fig. 8-73, what is the gage difference $R$ in terms of $H$?

**❚** 
$$1.2H + 1.2y + 1.2R - 3.0R - 1.2y = (0.95)(1.2)(H) \qquad R = 0.0333H$$

$H$    3 in diam    $s.g = 1.2$    $y$    $R$    $s.g = 3.0$    **Fig. 8-73**

**8.134** Neglecting losses, calculate $H$ in terms of $R$ for Fig. 8-73.

**❚** $1.2H + 1.2y + 1.2R - 3.0R - 1.2y = 1.2H$. Therefore, $R = 0$ for all $H$.

**8.135** At point $A$ in a pipeline carrying water, the diameter is 1 m, the pressure 100 kPa, and the velocity 1 m/s. At point $B$, 2 m higher than $A$, the diameter is 0.5 m and the pressure is 20 kPa. Determine the head loss and the direction of flow.

▮ Assume the direction of flow is from $A$ to $B$. $p_A/\gamma + v_A^2/2g + z_A = p_B/\gamma + v_B^2/2g + z_B + h_L$, $Q = A_1v_1 = [(\pi)(1)^2/4](1) = 0.7854 \text{ m}^3/\text{s}$, $v_2 = Q/A_2 = 0.7854/[(\pi)(0.5)^2/4] = 4.00 \text{ m/s}$, $100/9.79 + 1^2/[(2)(9.807)] + 0 = 20/9.79 + 4.00^2/[(2)(9.807)] + 2 + h_L$, $h_L = 5.40 \text{ m}$. Since $h_L$ is positive, flow is from $A$ to $B$ as assumed.

**8.136** Water is flowing in an open channel at a depth of 2 m and velocity of 3 m/s, as shown in Fig. 8-74. It then flows down a contracting chute into another channel where the depth is 1 m and the velocity is 10 m/s. Assuming frictionless flow, determine the difference in elevation of the channel floors.

▮ The velocities are assumed to be uniform over the cross sections, and the pressures hydrostatic.
$p_A/\gamma + v_A^2/2g + z_A = p_B/\gamma + v_B^2/2g + z_B + h_L$, $0 + 3^2/[(2)(9.807)] + (y + 2) = 0 + 10^2/[(2)(9.807)] + 1 + 0$, $y = 3.64 \text{ m}$.

**Fig. 8-74**

**8.137** For losses of 0.1 m, find the velocity at $A$ in Fig. 8-75. The barometer reading is 750 mmHg.

▮
$$p_B/\gamma + v_B^2/2g + z_B = p_A/\gamma + v_A^2/2g + z_A + h_L$$

$75/9.79 + 0 + 3 = [(13.6)(9.79)](0.750)/9.79 + v_A^2/[(2)(9.807)] + 0 + 0.1 \qquad v_A = 2.66 \text{ m/s}$

**Fig. 8-75**

**8.138** For flow of 375 gpm in Fig. 8-76, determine $H$ for losses of $5v^2/2g$.

▮
$$\frac{p_B}{\gamma} + \frac{v_B^2}{2g} + z_B = \frac{p_A}{\gamma} + \frac{v_A^2}{2g} + z_A + h_L \qquad v = \frac{Q}{A} = \frac{(375/7.48)/60}{(\pi)(\frac{6}{12})^2/4} = 4.255 \text{ ft/s}$$

$0 + 0 + H = 0 + 4.255^2/[(2)(32.2)] + 0 + 5\{4.255^2/[(2)(32.2)]\} \qquad H = 1.69 \text{ ft}$

**Fig. 8-76**

**8.139** For 1500-gpm flow and $H = 30$ ft in Fig 8-76, calculate the losses through the system in velocity heads, $Kv^2/2g$.

▮
$$\frac{p_B}{\gamma} + \frac{v_B^2}{2g} + z_B = \frac{p_A}{\gamma} + \frac{v_A^2}{2g} + z_A + h_L \qquad v = \frac{Q}{A} = \frac{(1500/7.48)/60}{(\pi)(\frac{6}{12})^2/4} = 17.02 \text{ ft/s}$$

$0 + 0 + 30 = 0 + 17.02^2/[(2)(32.2)] + 0 + K\{17.02^2/[(2)(32.2)]\} \qquad K = 5.67 \qquad \text{(i.e., 5.67 velocity heads)}$

**8.140** The losses in Fig. 8-76 for $H = 20$ ft are $8(v^2/2g)$. What is the discharge?

$\blacksquare$   $p_B/\gamma + v_B^2/2g + z_B = p_A/\gamma + v_A^2/2g + z_A + h_L$     $0 + 0 + 20 = 0 + v_A^2/[(2)(32.2)] + 0 + 8\{v_A^2/[(2)(32.2)]\}$

$$v_A = 11.96 \text{ ft/s} \qquad Q = Av_A = [(\pi)(\tfrac{6}{12})^2/4](11.96) = 2.35 \text{ ft}^3/\text{s}$$

**8.141** In Fig. 8-77, the losses up to section $A$ are $5v_1^2/2g$ and the nozzle losses are $0.05v_2^2/2g$. Determine the discharge and pressure at $A$, if $H = 8$ m.

$\blacksquare$
$$p_B/\gamma + v_B^2/2g + z_B = p_C/\gamma + v_C^2/2g + z_C + h_L$$

$$0 + 0 + 8 = 0 + v_2^2/[(2)(9.807)] + 0 + 5\{v_1^2/[(2)(9.807)]\} + 0.05\{v_2^2/[(2)(9.807)]\}$$

$$0.05353v_2^2 + 0.2549v_1^2 - 8 = 0 \qquad A_1v_1 = A_2v_2 \qquad [(\pi)(\tfrac{150}{1000})^2/4](v_1) = [(\pi)(\tfrac{50}{1000})^2/4](v_2) \qquad v_1 = 0.1111v_2$$

$$0.05353v_2^2 + (0.2549)(0.1111v_2)^2 - 8 = 0 \qquad v_2 = 11.88 \text{ m/s} \qquad v_1 = (0.1111)(11.88) = 1.320 \text{ m/s}$$

$$Q = A_2v_2 = [(\pi)(\tfrac{50}{1000})^2/4](11.88) = 0.0233 \text{ m}^3/\text{s} \qquad p_B/\gamma + v_B^2/2g + z_B = p_A/\gamma + v_A^2/2g + z_A + h_L$$

$$0 + 0 + 8 = p_A/9.79 + 1.320^2/[(2)(9.807)] + 0 + 5\{1.320^2/[(2)(9.807)]\} \qquad p_A = 73.1 \text{ kN/m}^2$$

**8.142** For pressure at $A$ of 25 kPa in Fig. 8-77 with the losses given in Prob. 8.141, determine the discharge and the head $H$.

$\blacksquare$
$$p_B/\gamma + v_B^2/2g + z_B = p_A/\gamma + v_A^2/2g + z_A + h_L$$

$$0 + 0 + H = 25/9.79 + v_1^2/[(2)(9.807)] + 0 + 5\{v_1^2/[(2)(9.807)]\}$$

$$H = 0.3059v_1^2 + 2.554 \qquad\qquad (1)$$

$$p_B/\gamma + v_B^2/2g + z_B = p_C/\gamma + v_C^2/2g + z_C + h_L$$

$$0 + 0 + H = 0 + v_2^2/[(2)(9.807)] + 0 + 5\{v_1^2/[(2)(9.807)]\} + 0.05\{v_2^2/[(2)(9.807)]\}$$

$$0.05353v_2^2 + 0.2549v_1^2 - H = 0 \qquad A_1v_1 = A_2v_2 \qquad [(\pi)(\tfrac{150}{1000})^2/4](v_1) = [(\pi)(\tfrac{50}{1000})^2/4](v_2) \qquad v_2 = 9.000v_1$$

$$(0.05353)(9.000v_1)^2 + 0.2549v_1^2 = H \qquad\qquad (2)$$

Solving Eqs. (1) and (2) simultaneously, $(0.05353)(9.000v_1)^2 + 0.2549v_1^2 = 0.3059v_1^2 + 2.554$, $v_1 = 0.7720$ m/s; $Q = A_1v_1 = [(\pi)(\tfrac{150}{1000})^2/4](0.7720) = 0.0136$ m$^3$/s, $H = (0.3059)(0.7720)^2 + 2.554 = 2.736$ m.

**8.143** The system shown in Fig. 8-78 involves 6-in.-i.d. pipe. The exit nozzle diameter is 3 in. What is the velocity $v_e$ of flow leaving the nozzle? Neglect losses.

$\blacksquare$
$$p_1/\gamma + v_1^2/2g + z_1 = p_A/\gamma + v_A^2/2g + z_A + h_L$$

$$(14.7)(144)/62.4 + 0 + 40 = 4000/62.4 + v_A^2/[(2)(32.2)] + 0 + 0$$

$$v_A = 25.14 \text{ ft/s} \qquad A_Av_A = A_Dv_D \qquad [(\pi)(\tfrac{6}{12})^2](25.14) = [(\pi)(\tfrac{3}{12})^2](v_e) \qquad v_e = 100.56 \text{ ft/s}$$

**Fig. 8-78**

**8.144** In Fig. 8-79, $H = 6$ m and $h = 5.75$ m. Calculate the head loss.

$v_2 = \sqrt{2gh} = \sqrt{(2)(9.807)(5.75)} = 10.62$ m/s     $p_1/\gamma + v_1^2/2g + z_1 = p_2/\gamma + v_2^2/2g + z_2 + h_L$

$0 + 0 + 6 = 0 + 10.62^2/[(2)(9.807)] + 0 + h_L$     $h_L = 0.250$ m

Fig. 8-79

**8.145** In Fig. 8-80, $0.1$ m$^3$/s of water flows from section 1 to section 2 with losses of $0.4(v_1 - v_2)^2/2g$. If $p_1 = 100$ kPa, find $p_2$.

$$p_1/\gamma + v_1^2/2g + z_1 = p_2/\gamma + v_2^2/2g + z_2 + h_L$$

$v_1 = Q/A_1 = (0.1)/[(\pi)(0.300)^2/4] = 1.415$ m/s     $v_2 = Q/A_2 = (0.1)/[(\pi)(0.450)^2/4] = 0.629$ m/s

$100/9.79 + 1.415^2/[(2)(9.807)] + 0 = p_2/9.79 + 0.629^2/[(2)(9.807)] + 0 + \dfrac{0.4(1.415 - 0.629)^2}{(2)(9.807)}$     $p_2 = 100.8$ kPa

Fig. 8-80

**8.146** Neglecting losses, determine the discharge in Fig. 8-81.

$p_1/\gamma + v_1^2/2g + z_1 = p_2/\gamma + v_2^2/2g + z_2 + h_L$     $[(0.86)(62.4)](3)/62.4 + 0 + 4 = 0 + v_2^2/[(2)(32.2)] + 0 + 0$

$v_2 = 20.59$ ft/s     $Q = Av = [(\pi)(\tfrac{2}{12})^2/4](20.59) = 0.45$ ft$^3$/s

Fig. 8-81

**8.147** A pipeline leads from one reservoir to another which has its water surface 10 m lower. For a discharge of $1.0$ m$^3$/s, determine the losses in meters and in kilowatts.

$p_1/\gamma + v_1^2/2g + z_1 = p_2/\gamma + v_2^2/2g + z_2 + h_L$     $0 + 0 + 10 = 0 + 0 + 0 + h_L$     $h_L = 10$ m

Losses $= Q\gamma h_L = (1.0)(9.79)(10) = 97.9$ kW

**8.148** In the siphon of Fig. 8-82, $h_1 = 1$ m, $h_2 = 3$ m, $d_1 = 3$ m, and $d_2 = 5$ m, and the losses to section 2 are $2.6v_2^2/2g$, with 10 percent of the losses occurring before section 1. Find the discharge and the pressure at section 1.

**∎**
$$p_B/\gamma + v_B^2/2g + z_B = p_2/\gamma + v_2^2/2g + z_2 + h_L$$
$$0 + 0 + (1 + 3) = 0 + v_2^2/[(2)(9.807)] + 0 + 2.6v_2^2/[(2)(9.807)] \qquad v_2 = 4.668 \text{ m/s}$$
$$Q = A_2 v_2 = [(\pi)(5)^2/4](4.668) = 91.7 \text{ m}^3/\text{s} \qquad v_1 = Q/A_1 = 91.7/[(\pi)(3)^2/4] = 12.97 \text{ m/s}$$
$$p_B/\gamma + v_B^2/2g + z_B = p_1/\gamma + v_1^2/2g + z_1 + h_L$$
$$0 + 0 + (1 + 3) = p_1/9.79 + 12.97^2/[(2)(9.807)] + 3 + 0.10\{(2.6)(4.668)^2/[(2)(9.807)]\}$$
$$p_1 = -77.0 \text{ kPa}$$

**Fig. 8-82**

**8.149** Find the pressure at $A$ of Prob. 8.148 if it is a stagnation point (velocity zero).

**∎** $\quad p_B/\gamma + v_B^2/2g + z_B = p_A/\gamma + v_A^2/2g + z_A + h_L \qquad 0 + 0 + 0 = p_A/9.79 + 0 + 4 + 0 \qquad p_A = -39.2 \text{ kPa}$

**8.150** In the friction-free siphon shown in Fig. 8-83, what are the pressures of the water in the tube at $B$ and at $A$?

**∎**
$$p_D/\gamma + v_D^2/2g + z_D = p_C/\gamma + v_C^2/2g + z_C + h_L$$
$$0 + 0 + 3.0 = 0 + v_C^2/[(2)(9.807)] + 0 + 0 \qquad v_C = 7.67 \text{ m/s} = v_B = v_A$$
$$p_D/\gamma + v_D^2/2g + z_D = p_B/\gamma + v_B^2/2g + z_B + h_L$$
$$0 + 0 + 3.0 = p_B/9.79 + 7.67^2/[(2)(9.807)] + (3.0 + 1.5) + 0 \qquad p_B = -44.0 \text{ kPa}$$
$$p_D/\gamma + v_D^2/2g + z_D = p_A/\gamma + v_A^2/2g + z_A + h_L$$
$$0 + 0 + 3.0 = p_A/9.79 + 7.67^2/[(2)(9.807)] + 3.0 + 0 \qquad p_A = -29.4 \text{ kPa}$$

**Fig. 8-83**

**8.151** If the vapor pressure of water is 0.1799 m of water, how high ($h$) above the free surface can point $B$ be in Prob. 8.150 before the siphon action breaks down? Assume atmospheric pressure is 101 kPa.

**∎** $p_D/\gamma + v_D^2/2g + z_D = p_B/\gamma + v_B^2/2g + z_B + h_L$. Using absolute pressures and considering that $v_B = 0$ at maximum $h$ when the siphon action breaks down, $101/9.79 + 0 + 3.0 = 0.1799 + 0 + (3.0 + h) + 0$, $h = 10.14$ m.

# CHAPTER 9

# Flow in Closed Conduits

**9.1** Water at 10 °C flows in a 150-mm-diameter pipe at a velocity of 5.5 m/s. Is this flow laminar or turbulent?

▋ $N_R = dv/v = (0.150)(5.5)/(1.30 \times 10^{-6}) = 634\,615$. Since $634\,615 > 4000$, the flow is turbulent.

**9.2** SAE10 oil at 68 °F flows in a 9-in-diameter pipe. Find the maximum velocity for which the flow will be laminar.

▋ $N_R = \rho dv/\mu$. For laminar flow, assume $N_R \le 2000$. $2000 = (1.68)(\frac{9}{12})(v)/(1.70 \times 10^{-3})$, $v = 2.70$ ft/s.

**9.3** The accepted transition Reynolds number for flow past a smooth sphere is 250 000. At what velocity will this occur for airflow at 20 °C past a 10-cm-diameter sphere?

▋ $\qquad N_R = dv/v \qquad 250\,000 = (0.10)(v)/(1.51 \times 10^{-5}) \qquad v = 37.8$ m/s

**9.4** Repeat Prob. 9.3 if the fluid is (a) water at 20° and (b) hydrogen at 20 °C ($v = 1.08 \times 10^{-4}$ m²/s).

▋ (a)  $\qquad N_R = dv/v \qquad 250\,000 = (0.10)(v)/(1.02 \times 10^{-6}) \qquad v = 2.55$ m/s
(b)  $\qquad N_R = dv/v \qquad 250\,000 = (0.10)(v)/(1.08 \times 10^{-4}) \qquad v = 270$ m/s

**9.5** A $\frac{1}{2}$-in-diameter water pipe is 60 ft long and delivers water at 5 gpm at 20 °C. What fraction of this pipe is taken up by the entrance region?

▋ $Q = (5)(0.002228) = 0.01114$ ft³/s $\qquad V = Q/A = 0.01114/[(\pi)(0.5/12)^2/4] = 8.170$ ft/s $\qquad N_R = dv/v$

From Table A-2, $v = 1.02 \times 10^{-6}$ m²/s at 20 °C, which equals $1.10 \times 10^{-5}$ ft²/s; hence, $N_R = (0.5/12)(8.170)/(1.10 \times 10^{-5}) = 30\,947$. Since $30\,947 > 4000$, the flow is turbulent and for entrance length, $L_e/d = 4.4 N_R^{1/6} = (4.4)(30\,947)^{1/6} = 25$. The actual pipe has $L/d = 60/[(\frac{1}{2})/12] = 1440$; hence,

$$\frac{L_e/d}{L/d} = \frac{L_e}{L} = \tfrac{25}{1440} = 0.017 \quad \text{or} \quad 1.7 \text{ percent}$$

**9.6** An oil with $\rho = 900$ kg/m³ and $v = 0.0002$ m²/s flows upward through an inclined pipe as shown in Fig. 9-1. Assuming steady laminar flow, (a) verify that the flow is up and find the (b) head loss between section 1 and section 2, (c) flow rate, (d) velocity, and (e) Reynolds number. Is the flow really laminar?

▋ (a)  $\qquad \text{HGL} = z + p/\rho g \qquad \text{HGL}_1 = 0 + 350\,000/[(900)(9.807)] = 39.65$ m

$$\text{HGL}_2 = (10)(\sin 40°) + 250\,000/[(900)(9.807)] = 34.75 \text{ m}$$

Since $\text{HGL}_1 > \text{HGL}_2$, the flow is upward.

(b)  $\qquad h_f = \text{HGL}_1 - \text{HGL}_2 = 39.65 - 34.75 = 4.90$ m
(c)  $\qquad \mu = \rho v = (900)(0.0002) = 0.180$ kg/(m · s)

$$Q = \frac{\pi \rho g d^4 h_f}{128 \mu L} = \frac{(\pi)(900)(9.807)(\frac{6}{100})^4(4.90)}{(128)(0.180)(10)} = 0.00764 \text{ m}^3/\text{s}$$

(d)  $\qquad v = Q/A = 0.00764/[(\pi)(\frac{6}{100})^2/4] = 2.70$ m/s
(e)  $\qquad N_R = dv/v = (\frac{6}{100})(2.70)/0.0002 = 810$

This value of $N_R$ is well within the laminar range; hence, the flow is most likely laminar.

**Fig. 9-1**

**9.7** For flow of SAE10 oil through a 100-mm-diameter pipe, for what flow rate in cubic meters per hour would we expect transition to turbulence at **(a)** 20 °C [$\mu = 0.104$ Pa · s], and **(b)** 100 °C [$\mu = 0.0056$ Pa · s]?

  ❚  Assume transition to turbulence occurs at $N_R = 2300$. $N_R = \rho d v / \mu$.

**(a)**  $2300 = (869)(0.100)(v)/0.104$    $v = 2.753$ m/s

$$Q = Av = [(\pi)(0.100)^2/4](2.753) = 0.0216 \text{ m}^3/\text{s} \quad \text{or} \quad 77.76 \text{ m}^3/\text{h}$$

**(b)**  $2300 = (869)(0.100)(v)/0.0056$    $v = 0.1482$ m/s

$$Q = Av = [(\pi)(0.100)^2/4](0.1482) = 0.00116 \text{ m}^3/\text{s} \quad \text{or} \quad 4.18 \text{ m}^3/\text{h}$$

**9.8** A fluid at 20 °C flows at 0.8 L/s through an 100-mm-diameter pipe. Determine whether the flow is laminar or turbulent if the fluid is **(a)** hydrogen ($v = 1.08 \times 10^{-4}$ m²/s), **(b)** air, **(c)** gasoline ($v = 4.06 \times 10^{-7}$ m²/s), **(d)** water, **(e)** mercury ($v = 1.15 \times 10^{-7}$ m²/s), or **(f)** glycerin.

  ❚        $N_R = dv/v$    $v = Q/A = (0.8 \times 10^{-3})/[(\pi)(0.100)^2/4] = 0.1019$ m/s

**(a)**            $N_R = (0.100)(0.1019)/(1.08 \times 10^{-4}) = 94$    (laminar)

**(b)**            $N_R = (0.100)(0.1019)/(1.51 \times 10^{-5}) = 675$    (laminar)

**(c)**            $N_R = (0.100)(0.1019)/(4.06 \times 10^{-7}) = 25\,099$    (turbulent)

**(d)**            $N_R = (0.100)(0.1019)/(1.02 \times 10^{-6}) = 9990$    (turbulent)

**(e)**            $N_R = (0.100)(0.1019)/(1.15 \times 10^{-7}) = 88\,609$    (turbulent)

**(f)**            $N_R = (0.100)(0.1019)/(1.18 \times 10^{-3}) = 9$    (laminar)

**9.9** Oil (s.g. = 0.9, $v = 0.0003$ m²/s) enters a 50-mm-diameter tube. Estimate the entrance length if the flow rate is 1 L/s.

  ❚  $N_R = dv/v$

$$v = Q/A = 0.001/[(\pi)(0.050)^2/4] = 0.5093 \text{ m/s} \quad N_R = (0.050)(0.5093)/0.0003 = 85 \quad \text{(laminar)}$$

$$L_e/d = 0.06 N_R \quad L_e = (0.050)(0.06)(85) = 0.255 \text{ m}$$

**9.10** What is the Reynolds number for a flow of oil (s.g. = 0.8, $\mu = 0.00200$ lb · s/ft²) in a 6-in-diameter pipe at a flow rate of 10 ft³/s? Is the flow laminar or turbulent?

  ❚        $v = Q/A = 10/[(\pi)(\frac{6}{12})^2/4] = 50.9$ ft/s

$$N_R = \rho dv/\mu = [(0.8)(1.94)](\tfrac{6}{12})(50.9)/0.00200 = 19\,749 \quad \text{(turbulent)}$$

**9.11** Gasoline at a temperature of 20 °C flows at the rate of 2 L/s through a pipe of inside diameter 60 mm. Find the Reynolds number.

  ❚        $v = Q/A = (2 \times 10^{-3})/[(\pi)(0.060)^2/4] = 0.707$ m/s

$$N_R = \rho dv/\mu = (719)(0.060)(0.707)/(2.92 \times 10^{-4}) = 104\,452$$

**9.12** The Reynolds number for fluid in a pipe of 10 in diameter is 2000. What will be the Reynolds number in a 6-in-diameter pipe forming an extension of the 10-in pipe? Take the flow as incompressible.

  ❚  $N_R = dv/v$. Since $v$ is constant, $[N_R/(dv)]_1 = [N_R/(dv)]_2$, $A_1 v_1 = A_2 v_2$, $[(\pi)(\frac{10}{12})^2/4](v_1) = [(\pi)(\frac{6}{12})^2/4](v_2)$, $v_1 = 0.360 v_2$, $2000/[(\frac{10}{12})(0.360 v_2)] = (N_R)_2/[(\frac{6}{12})(v_2)]$, $(N_R)_2 = 3333$.

**9.13** Water is flowing through capillary tubes $A$ and $B$ into tube $C$, as shown in Fig. 9-2. If $Q_A = 3$ mL/s in tube $A$, what is the largest $Q_B$ allowable in tube $B$ for laminar flow in tube $C$? The water is at a temperature of 40 °C. With the calculated $Q_B$, what kind of flow exists in tubes $A$ and $B$?

  ❚  For laminar flow, assume $N_R \le 2300$. $N_R = dv/v$. In tube $C$, $2300 = (0.006)(v_C)/(6.56 \times 10^{-7})$, $v_C = 0.2515$ m/s; $Q_C = A_C v_C = [(\pi)(0.006)^2/4](0.2515) = 7.11 \times 10^{-6}$ m³/s, or 7.11 mL/s, $Q_B = 7.11 - 3 = 4.11$ mL/s. In tube $A$, $v_A = Q_A/A_A = (3 \times 10^{-6})/[(\pi)(0.005)^2/4] = 0.1528$ m/s,

$$N_R = (0.005)(0.1528)/(6.56 \times 10^{-7}) = 1165 \quad \text{(laminar)}$$

In tube $B$, $v_B = Q_B/A_B = (4.11 \times 10^{-6})/[(\pi)(0.004)^2/4] = 0.3271$ m/s,

$$N_R = (0.004)(0.3271)/(6.56 \times 10^{-7}) = 1995 \qquad \text{(turbulent)}$$

**Fig. 9-2**

**9.14** Incompressible steady flow of water occurs in a tube of constant cross section, as shown in Fig. 9-3. What is the head loss between sections $A$ and $B$?

$$p_A/\gamma + v_A^2/2g + z_A = p_B/\gamma + v_B^2/2g + z_B + h_L$$
$$(90)(144)/62.4 + v_A^2/2g + 0 = (30)(144)/62.4 + v_B^2/2g + 100 + h_L$$
$$v_A^2/2g = v_B^2/2g \qquad h_L = 38.5 \text{ ft}$$

**Fig. 9-3**

**9.15** Water flows through a pipe at 5 L/s, as shown in Fig. 9-4. If gage pressures of 12.5 kPa, 11.5 kPa, and 10.3 kPa are measured for $p_1$, $p_2$ and $p_3$, respectively, what are the head losses between 1 and 2 and 1 and 3?

$$p_1/\gamma + v_1^2/2g + z_1 = p_2/\gamma + v_2^2/2g + z_2 + (h_L)_{1-2} \qquad 12.5/9.79 + v_1^2/2g + 10 = 11.5/9.79 + v_2^2/2g + 10 + (h_L)_{1-2}$$

$$v_1^2/2g = v_2^2/2g \qquad (h_L)_{1-2} = 0.1021 \text{ m} \qquad p_1/\gamma + v_1^2/2g + z_1 = p_3/\gamma + v_3^2/2g + z_3 + (h_L)_{1-3}$$

$$v_1 = Q/A_1 = (5 \times 10^{-3})/[(\pi)(0.050)^2/4] = 2.546 \text{ m/s} \qquad v_3 = Q/A_3 = (5 \times 10^{-3})/[(\pi)(0.030)^2/4] = 7.074 \text{ m/s}$$

$$12.5/9.79 + 2.546^2/[(2(9.807)] + 10 = 10.3/9.79 + 7.074^2/[(2)(9.807)] + 0 + (h_L)_{1-3} \qquad (h_L)_{1-3} = 8.00 \text{ m}$$

**Fig. 9-4**

**9.16** A large oil reservoir has a pipe of 3 in diameter and 7000-ft length connected to it, as shown in Fig. 9-5. Assuming laminar flow through the pipe, compute the amount of oil issuing out of the pipe as a free jet.

Compute the velocity and Reynolds number to see if the flow is laminar. $\nu_{oil} = 1 \times 10^{-4}$ ft²/s. Neglect entrance losses to the pipe.

▌ $p_2/\gamma + v_2^2/2g + z_2 = p_3/\gamma + v_3^2/2g + z_3 + h_L$    $p_2/62.4 + v_2^2/2g + 0 = 0 + v_3^2/2g + 0 + h_f$    $v_2^2/2g = v_3^2/2g$

$$p_2/62.4 = h_f \tag{1}$$

$$Q = \frac{\pi \rho g d^4 h_f}{128 \mu L}$$

$$h_f = \frac{128 Q \mu L}{\pi \rho g d^4} = \frac{128[(\pi d^2/4)(v_2)]\mu L}{\pi \rho g d^4} = \frac{32 v_2 \nu L}{g d^2} = \frac{(32)(v_2)(1 \times 10^{-4})(7000)}{(32.2)(3/12)^2} = 11.13 v_2 \tag{2}$$

$p_1/\gamma + v_1^2/2g + z_1 = p_2/\gamma + v_2^2/2g + z_2 + h_L$    $0 + 0 + 10 = p_2/62.4 + v_2^2/[(2)(32.2)] + 0 + 0$

$$p_2/62.4 = 10 - v_2^2/[(2)(32.2)] \tag{3}$$

Equating $h_f$ from Eqs. (1) and (2),

$$p_2/62.4 = 11.13 v_2 \tag{4}$$

Equating $p_2/62.4$ from Eqs. (3) and (4), $11.13 v_2 = 10 - v_2^2/[(2)(32.2)]$, $v_2^2 + 716.8 v_2 - 644 = 0$, $v_2 = 0.8973$ ft/s; $Q = Av = [(\pi)(\frac{3}{12})^2/4](0.8973) = 0.0440$ ft³/s; $N_R = dv/\nu = (\frac{3}{12})(0.8973)/(1 \times 10^{-4}) = 2243$ (barely laminar).

Fig. 9-5

**9.17**    If 140 L/s of water flows through the system shown in Fig. 9-6, calculate the total head loss between 2 and 3.

▌ $$p_2/\gamma + v_2^2/2g + z_2 = p_3/\gamma + v_3^2/2g + z_3 + h_L$$

$v_2 = Q/A_2 = (140 \times 10^{-3})/[(\pi)(0.300)^2/4] = 1.981$ m/s    $v_3 = Q/A_3 = (140 \times 10^{-3})/[(\pi)(0.150)^2/4] = 7.922$ m/s

$p_2/9.79 + 1.981^2/[(2)(9.807)] + 0 = 0 + 7.922^2/[(2)(9.807)] + 15 + h_L$    $h_L = p_2/9.79 - 18.00$

$p_1/\gamma + v_1^2/2g + z_1 = p_2/\gamma + v_2^2/2g + z_2 + h_L$    $0 + 0 + 30 = p_2/9.79 + 1.981^2/[(2)(9.807)] + 0 + 0$

$p_2/9.79 = 29.80$ m    $h_L = 29.80 - 18.00 = 11.80$ m

Fig. 9-6

**9.18**    Determine the maximum velocity for laminar flow for (a) medium fuel oil at 60 °F ($\nu = 4.75 \times 10^{-5}$ ft²/s) flowing through a 6-in pipe and (b) water at 60 °F flowing in the 6-in pipe.

▌ For laminar flow, assume $N_R \le 2000$. $N_R = dv/\nu$.

(a)                    $2000 = (\frac{6}{12})(v)/(4.75 \times 10^{-5})$    $v = 0.190$ ft/s

(b)                    $2000 = (\frac{6}{12})(v)(1.21 \times 10^{-5})$    $v = 0.0484$ ft/s

**9.19** Determine the type of flow occurring in a 12-in pipe when (*a*) water at 60 °F flows at a velocity of 3.50 ft/s and (*b*) heavy fuel oil at 60 °F ($v = 221 \times 10^{-5}$ ft²/s) flows at the same velocity.

▌ $N_R = dv/v$

(*a*) $\qquad\qquad\qquad\qquad N_R = (\frac{12}{12})(3.50)/(1.21 \times 10^{-5}) = 289\,256 \qquad$ (turbulent)

(*b*) $\qquad\qquad\qquad\qquad N_R = (\frac{12}{12})(3.50)/(221 \times 10^{-5}) = 1584 \qquad$ (laminar)

**9.20** For laminar flow conditions, what size pipe will deliver 90 gpm of medium fuel oil at 40 °F ($v = 6.55 \times 10^{-5}$ ft²/s)?

▌ $Q = (90)(0.002228) = 0.2005$ ft³/s. For laminar flow, assume $N_R \le 2000$. $N_R = dv/v$, $v = Q/A = 0.2005/(\pi d^2/4) = 0.2553d^{-2}$, $2000 = (d)(0.2553d^{-2})/(6.55 \times 10^{-5})$, $d = 1.95$ ft.

**9.21** What is the Reynolds number of flow of 0.4 m³/s of oil (s.g. = 0.86, $\mu = 0.025$ Pa · s) through a 450-mm-diameter pipe?

▌ $\qquad\qquad\qquad\qquad v = Q/A = 0.4/[(\pi)(0.450)^2/4] = 2.515$ m/s

$\qquad\qquad N_R = \rho\,dv/\mu = [(0.86)(1000)](0.450)(2.515)/0.025 = 38\,932$

**9.22** An oil with s.g. = 0.85 and $v = 1.8 \times 10^{-5}$ m²/s flows in a 10-cm-diameter pipe at 0.50 L/s. Is the flow laminar or turbulent?

▌ $\qquad\qquad\qquad v = Q/A = (0.50)(1000)/[(\pi)(\frac{10}{100})^2/4] = 0.06366$ m/s

$\qquad\qquad N_R = dv/v = (\frac{10}{100})(0.06366)/(1.8 \times 10^{-5}) = 354 \qquad$ (laminar)

**9.23** Fluid with kinematic viscosity 0.00015 ft²/s flows through a pipe of diameter 9 in. What is the maximum velocity for laminar flow?

▌ For laminar flow, assume $N_R \le 2000$. $N_R = dv/v$, $2000 = (\frac{9}{12})(v)/0.00015$, $v = 0.400$ ft/s.

**9.24** An oil with $v = 0.005$ ft²/s flows through a 6-in-diameter pipe at 10 ft/sec. Is the flow laminar or turbulent?

▌ $\qquad\qquad\qquad\qquad N_R = dv/v = (\frac{6}{12})(10)/0.005 = 1000 \qquad$ (laminar)

**9.25** Hydrogen at atmospheric pressure and 50 °F has a kinematic viscosity of 0.0011 ft²/s. Determine the maximum mass flow rate for laminar flow in a 3-in-diameter pipe. $\gamma = 0.00540$ lb/ft³.

▌ For laminar flow, assume $N_R \le 2000$. $N_R = dv/v$, $2000 = (\frac{3}{12})(v)/0.0011$, $v = 8.80$ ft/s; $W = \gamma Av = (0.00540)[(\pi)(\frac{2}{12})^2/4](8.80) = 0.001037$ lb/s.

**9.26** Air at 1500 kPa abs and 100 °C flows in a 20-mm-diameter tube. What is the maximum laminar flow rate?

▌ For laminar flow, assume $N_R \le 2000$. $N_R = \rho dv/\mu$, $\rho = p/RT = (1.5 \times 10^6)/[(287)(273 + 100)] = 14.01$ kg/m³, $2000 = (14.01)(0.020)(v)/(2.17 \times 10^{-5})$, $v = 0.1549$ m/s; $Q = Av = [(\pi)(0.020)^2/4](0.1549) = 0.0000487$ m³/s, or 0.0487 L/s.

**9.27** What is the hydraulic radius of a rectangular air duct 8 in by 14 in?

▌ $\qquad\qquad\qquad R_h = A/p_w = [(8)(14)]/(8 + 8 + 14 + 14) = 2.55$ in $\quad$ or $\quad$ 2.55 ft

**9.28** What is the percentage difference between the hydraulic radii of 30-cm-diameter circular and 30-cm square ducts?

▌ $\qquad\qquad\qquad\qquad\qquad\qquad R_h = A/p_w$

$(R_h)_{\text{circular}} = [(\pi)(30)^2/4]/[(\pi)(30)] = 7.50$ cm $\qquad (R_h)_{\text{square}} = (30)(30)/(30 + 30 + 30 + 30) = 7.50$ cm

Since they are equal, the percentage difference is zero. Note that the hydraulic radius of a circular section is one-fourth its diameter.

**9.29** Two pipes, one circular and one square, have the same cross-sectional area. Which has the larger hydraulic radius, and by what percentage?

▮ Let $d$ = diameter of the circular pipe and $a$ = the side of the square one. Since they have the same cross-sectional area, $\pi d^2/4 = a^2$, $a = \sqrt{\pi}d/2$; $(R_h)_{\text{circular}} = d/4 = 0.2500d$, $(R_h)_{\text{square}} = A/p_w = a^2/4a = a/4$. Since $a = \sqrt{\pi}d/2$, $(R_h)_{\text{square}} = (\sqrt{\pi}d/2)/4 = 0.2216d$, hence, the circular pipe has the larger hydraulic radius by $(0.2500 - 0.2216)/0.2216 = 0.128$, or 12.8 percent.

**9.30** Steam of weight density 0.26 lb/ft³ flows at 100 fps through a circular pipe. What is the shearing stress at the wall, if the friction factor is 0.015?

▮ $$\tau_0 = (f/4)(\gamma)(v^2/2g) = (0.015/4)(0.26)\{100^2/[(2)(32.2)]\} = 0.151 \text{ lb/ft}^2$$

**9.31** Glycerin at 68 °F flows 120 ft through a 6-in-diameter new wrought iron pipe at a velocity of 10.0 ft/s. Determine the head loss due to friction.

▮ $$h_f = (f)(L/d)(v^2/2g) \qquad N_R = \rho dv/\mu = (2.44)(\tfrac{6}{12})(10.0)/(3.11 \times 10^{-2}) = 392$$

Since $N_R < 2000$, the flow is laminar and $f = 64/N_R = \frac{64}{392} = 0.1633$, $h_f = 0.1633[120/(\tfrac{6}{12})]\{10.0^2/[(2)(32.2)]\} = 60.9$ ft.

**9.32** SAE10 oil flows through a cast iron pipe at a velocity of 1.0 m/s. The pipe is 45.0 m long and has a diameter of 150 mm. Find the head loss due to friction.

▮ $$h_f = (f)(L/d)(v^2/2g) \qquad N_R = \rho dv/\mu = (869)(\tfrac{150}{1000})(1.0)/0.0814 = 1601$$

Since $N_R < 2000$, the flow is laminar and $f = 64/N_R = \frac{64}{1601} = 0.0400$, $h_f = 0.0400[45.0/(\tfrac{150}{1000})]\{1.0^2/[(2)(9.807)]\} = 0.612$ m.

**9.33** A 60-mm-diameter pipe (Fig. 9-7) contains glycerin at 20 °C flowing at 8.5 m³/h. Verify that the flow is laminar. For the pressure measurements shown, is the flow ascending or descending? What is the head loss for these pressures?

▮ $$v = Q/A = (8.5/3600)/[(\pi)(0.060)^2/4] = 0.835 \text{ m/s}$$

$$N_R = \rho dv/\mu = (1258)(0.060)(0.835)/1.49 = 63 \qquad (\text{laminar})$$

$$\text{HGL} = z + p/\rho g \qquad \text{HGL}_A = 0 + (2.0)(101\,400)/[(1258)(9.807)] = 16.44 \text{ m}$$

$$\text{HGL}_B = 12 + (3.8)(101\,400)/[(1258)(9.807)] = 43.23 \text{ m}$$

Hence, the flow is from $B$ to $A$ (i.e., descending). Head loss = $43.23 - 16.44 = 26.79$ m.

**Fig. 9-7**

**9.34** For the data of Prob. 9.33, compute the theoretical head loss if the pipe length is 30 m between $A$ and $B$. Compare with the head loss corresponding to the measured pressures.

▮ $$Q = \frac{\pi \rho g d^4 h_f}{128\mu L} \qquad \frac{8.5}{3600} = \frac{(\pi)(1258)(9.807)(0.060)^4(h_f)}{(128)(1.49)(30)} \qquad h_f = 26.89 \text{ m}$$

which is only 10 cm greater than the value found in Prob. 9.33.

**9.35** Two horizontal infinite plates keep a distance $h$ apart as the upper plate moves at speed $V$, as in Fig. 9-8. There is a fluid of constant viscosity and constant pressure between the plates. If $V = 5$ m/s and $h = 20$ mm, compute the shear stress at the plates, given that the fluid is SAE 30 oil at 20 °C.

$\blacksquare$ $N_R = \rho h V / \mu = (888)(0.020)(5)/0.440 = 202$. Since the flow is laminar,

$$\tau = \mu \frac{\partial u}{\partial y}$$

where $u = (V/h)(y)$, $\tau = (\mu)(V/h) = 0.440[5/(2.0/100)] = 110$ Pa.

**Fig. 9-8**

**9.36** Find the head loss per unit length when a fluid of s.g. 0.86 and kinematic viscosity 0.008 ft²/s flows in a 3-in-diameter pipe at a rate of 5 gpm.

$\blacksquare$ $Q = (5)(0.002228) = 0.01114 \text{ ft}^3/\text{s}$   $h_f = (f)(L/d)(v^2/2g)$   $v = Q/A = 0.01114/[(\pi)(\frac{3}{12})^2/4] = 0.2269 \text{ ft/s}$

$$N_R = dv/\nu = (\tfrac{3}{12})(0.2269)/0.008 = 7.09$$

Since $N_R < 2000$, the flow is laminar and $f = 64/N_R = 64/7.09 = 9.03$, $h_f = 9.03[1/(\frac{3}{12})]\{0.2269^2/[(2)(32.2)]\} = 0.0289$ ft per foot of length.

**9.37** Tests made on a certain 12-in-diameter pipe showed that, when $V = 10$ fps, $f = 0.015$. The fluid used was water at 60 °F. Find the unit shear at the wall and at radii of 0, 0.2, 0.3, 0.5, and 0.75 times the pipe radius.

$\blacksquare$          $\tau_0 = (f/4)(\gamma)(V^2/2g) = (0.015/4)(62.4)\{10^2/[(2)(32.2)]\} = 0.3634 \text{ lb/ft}^2$

The stress distribution is linear; hence,

| $r/r_o$ | $\tau$, lb/ft² |
|---------|----------------|
| 0       | 0              |
| 0.2     | 0.0727         |
| 0.3     | 0.1090         |
| 0.5     | 0.1817         |
| 0.75    | 0.2726         |

**9.38** If oil with a kinematic viscosity of 0.005 ft²/s weighs 54 lb/ft³, what will be the flow rate and head loss in a 3600-ft length of 4-in-diameter pipe when the Reynolds number is 800?

$\blacksquare$    $N_R = dv/\nu$    $800 = (\frac{4}{12})(v)/0.005$    $v = 12.00 \text{ ft/s}$    $Q = Av = [(\pi)(\frac{4}{12})^2/4](12.00) = 1.047 \text{ ft}^3/\text{s}$

$f = 64/N_R = 64/800 = 0.0800$    $h_f = (f)(L/d)(v^2/2g) = 0.0800[3600/(\frac{4}{12})]\{12.00^2/[(2)(32.2)]\} = 1932 \text{ ft}$

**9.39** How much power is lost per kilometer of length when a viscous fluid ($\mu = 0.20 \text{ Pa} \cdot \text{s}$) flows in a 200-mm-diameter pipeline at 1.00 L/s? The fluid has a density of 840 kg/m³.

$\blacksquare$          $v = Q/A = (1.00 \times 10^{-3})/[(\pi)(0.200)^2/4] = 0.03183 \text{ m/s}$

$$N_R = \rho dv/\mu = (840)(0.200)(0.03183)/0.20 = 26.74 \qquad h_f/L = (f)(1/d)(v^2/2g)$$

Since $N_R < 2000$, the flow is laminar and $f = 64/N_R = 64/26.74 = 2.393$, $h_f/L = 2.393[1/(0.200)]\{0.03183^2/[(2)(9.807)]\} = 0.0006180 \text{ m}$, $P/L = Q\gamma h_f/L = Q\rho g h_f/L = (1.00 \times 10^{-3})(840)(9.807)(0.0006180) = 0.00509 \text{ W/m} = 5.09 \text{ W/km}$.

**9.40** Calculate the discharge of the system in Fig. 9-9, neglecting all losses except through the pipe.

$\blacksquare$ Assume laminar flow and use the conversion 1.0 centipoise = 0.0002089 lb · s/ft².

$$v = \frac{\gamma \, \Delta h \, d^2}{32\mu L} = \frac{(55)(18)[(\frac{1}{4})/12]^2}{(32)(0.0002089)(16)} = 4.017 \text{ ft/s}$$

$$N_R = \rho dv/\mu = (\gamma/g)(d)(v)/\mu = (55/32.2)[(\tfrac{1}{4})/12](4.017)/0.0002089 = 684 \qquad \text{(laminar)}$$

$$Q = Av = \{(\pi)[(\tfrac{1}{4})/12]^2/4\}(4.017) = 0.00137 \text{ ft}^3/\text{s}$$

**Fig. 9-9**

**9.41** In Fig. 9-10, $H = 25$ m, $L = 40$ m, $\theta = 30°$, $d = 8$ mm, $\gamma = 10$ kN/m$^3$, and $\mu = 0.08$ Pa $\cdot$ s. Find the head loss per unit length of pipe and the discharge in liters per minute.

▌ Assuming laminar flow,

$$v = \frac{\gamma \, \Delta h \, d^2}{32 \mu L} = \frac{(10^4)(25)(8 \times 10^{-3})^2}{(32)(0.08)(40)} = 0.1563 \text{ m/s}$$

$$N_R = \rho d v / \mu = (\gamma/g)(d)(v)/\mu = [10^4/9.807](8 \times 10^{-3})(0.1563)/0.08 = 16 \qquad \text{(laminar)}$$

$$Q = Av = [(\pi)(8 \times 10^{-3})^2/4](0.1563) = 7.857 \times 10^{-6} \text{ m}^3/\text{s} = 0.471 \text{ L/min}$$

$$\Delta h / L = \tfrac{25}{40} = 0.625 \text{ m/m}$$

**Fig. 9-10**

**9.42** For the data of Prob. 9.41, find $H$ if the velocity is 0.1 m/s.

▌ By proportion, $H = (0.1/0.1563)(25) = 16.0$ m.

**9.43** Water flows at 0.20 m$^3$/s through a 300-mm-diameter, 120-m-long pipe, under a pressure difference of 280 mmHg. Find the friction factor.

▌ $h_f = (f)(L/d)(v^2/2g)$. From the pressure gradient, $h_f = (13.6/1)(0.280) = 3.808$ m; $v = Q/A = 0.20/[(\pi)(0.300)^2/4] = 2.829$ m/s, $3.808 = (f)[120/(0.300)]\{2.829^2/[(2)(9.807)]\}$, $f = 0.0233$.

**9.44** Use the Blasius equation for determination of friction factor to find the horsepower per mile required to pump 3.0 ft$^3$/s of liquid ($v = 3.3 \times 10^{-4}$ ft$^2$/s, $\gamma = 60$ lb/ft$^3$) through an 18-in pipeline.

▌
$$h_f = (f)(L/d)(v^2/2g) \qquad f = 0.316/N_R^{1/4} \qquad v = Q/A = 3.0/[(\pi)(\tfrac{18}{12})^2/4] = 1.698 \text{ ft/s}$$

$$N_R = dv/v = (\tfrac{18}{12})(1.698)/(3.3 \times 10^{-4}) = 7718 \qquad f = 0.316/7718^{1/4} = 0.03371$$

$$h_f = 0.03371[5280/(\tfrac{18}{12})]\{1.698^2/[(2)(32.2)]\} = 5.312 \text{ ft}$$

$$P = Q\gamma h_f = (3.0)(60)(5.312) = 956.2 \text{ ft} \cdot \text{lb/s per mile} \qquad 956.2/550 = 1.74 \text{ hp per mile}$$

**9.45** Determine the head loss per kilometer required to maintain a velocity of 3 m/s in a 20-mm-diameter pipe, if $v = 4 \times 10^{-5}$ m$^2$/s.

▌
$$h_f = (f)(L/d)(v^2/2g) \qquad N_R = dv/v = (0.020)(3)/(4 \times 10^{-5}) = 1500 \qquad \text{(laminar)}$$

$$f = 64/N_R = 64/1500 = 0.04267 \qquad h_f = 0.04267[1000/(0.020)]\{3^2/[(2)(9.807)]\} = 979.0 \text{ m per km}$$

**9.46** Fluid flows through a 10-mm-diameter tube at a Reynolds number of 1800. The head loss is 30 m in a 120-m length of tubing. Calculate the discharge in liters per minute.

$\blacksquare$ $h_f = (f)(L/d)(v^2/2g)$. Since $N_R < 2000$, flow is laminar and $f = 64/N_R = 64/1800 = 0.03556$,
$30 = 0.03556[120/(0.010)]\{v^2/[(2)(9.807)]\}$, $v = 1.174$ m/s; $Q = Av = [(\pi)(0.010)^2/4](1.174) = 92.21 \times 10^{-6}$ m$^3$/s $= 5.53$ L/min.

**9.47** Oil of absolute viscosity 0.00210 lb · s/ft$^2$ and specific gravity 0.850 flows through 10 000 ft of 12-in-diameter cast iron pipe at the rate of 1.57 cfs. What is the lost head in the pipe?

$\blacksquare$
$$h_f = (f)(L/d)(v^2/2g) \qquad v = Q/A = 1.57/[(\pi)(\tfrac{12}{12})^2/4] = 1.999 \text{ ft/s}$$
$$N_R = \rho dv/\mu = [(0.850)(1.94)](\tfrac{12}{12})(1.999)/0.00210 = 1570 \qquad \text{(laminar)}$$
$$f = 64/N_R = \tfrac{64}{1570} = 0.04076 \qquad h_f = 0.04076[10\,000/(\tfrac{12}{12})]\{1.999^2/[(2)(32.2)]\} = 25.3 \text{ ft}$$

**9.48** When first installed between two reservoirs, a 4-in-diameter metal pipe of length 6000 ft conveyed 0.20 cfs of water. (*a*) If after 15 years a chemical deposit had reduced the effective diameter of the pipe to 3.0 in, what then would be the flow rate? Assume $f$ remains constant. Assume no change in reservoir levels. (*b*) What would be the flow rate if in addition to the diamater change, $f$ had doubled in value?

$\blacksquare$ (*a*) $(f_1)(L_1/d_1)(v_1^2/2g) = (f_2)(L_2/d_2)(v_2^2/2g)$. Since $f$, $L$, and $g$ are constant and $v = Q/A = Q/(\pi d^2/4)$,
$Q_1^2/d_1^5 = Q_2^2/d_2^5$, $0.20^2/4^5 = Q_2^2/3.0^5$, $Q_2 = 0.0974$ cfs.
(*b*) $(f_1)(Q_1^2/d_1^5) = (f_2)(Q_2^2/d_2^5)$. Since $f_2 = 2f_1$, $Q_1^2/d_1^5 = (2)(Q_2^2/d_2^5)$, $0.20^2/4^5 = (2)(Q_2^2/3.0^5)$, $Q_2 = 0.0689$ cfs.

**9.49** A liquid with $\gamma = 58$ lb/ft$^3$ flows by gravity through a 1-ft tank and a 1-ft capillary tube at a rate of 0.15 ft$^3$/h, as shown in Fig. 9-11. Sections 1 and 2 are at atmospheric pressure. Neglecting entrance effects, compute the viscosity of the liquid in slugs per foot-second.

$\blacksquare$ $p_1/\gamma + v_1^2/2g + z_1 = p_2/\gamma + v_2^2/2g + z_2 + h_L \qquad v_2 = Q/A_2 = (0.15/3600)/[(\pi)(0.004)^2/4] = 3.316$ ft/s
$$0 + 0 + (1+1) = 0 + 3.316^2/[(2)(32.2)] + 0 + h_f \qquad h_f = 1.829 \text{ ft}$$

Assuming laminar flow,

$$h_f = \frac{32\mu L v}{\gamma d^2} \qquad 1.829 = \frac{(32)(\mu)(1)(3.316)}{(58)(0.004)^2} \qquad \mu = 1.600 \times 10^{-5} \text{ slug}/(\text{ft} \cdot \text{s})$$
$$N_R = \rho dv/\mu = (\gamma/g)(d)(v)/\mu = (58/32.2)(0.004)(3.316)/(1.600 \times 10^{-5}) = 1493 \qquad \text{(laminar)}$$

(1)

1 ft

1 ft $\quad \rightarrow | \leftarrow d = 0.004$ ft

(2)

$Q = 0.15$ ft$^3$/h $\qquad$ **Fig. 9-11**

**9.50** In Prob. 9.49, suppose the flow rate is unknown but the liquid viscosity is $2.1 \times 10^{-5}$ slug/(ft · s). What will be the flow rate in cubic feet per hour? Is the flow still laminar?

$\blacksquare$
$$p_1/\gamma + v_1^2/2g + z_1 = p_2/\gamma + v_2^2/2g + z_2 + h_L$$
$$0 + 0 + (1+1) = 0 + v^2/[(2)(32.2)] + 0 + h_f \qquad h_f = 2 - 0.01553v^2$$

Assuming laminar flow,

$$h_f = 32\mu Lv/\gamma d^2 \qquad 2 - 0.01553v^2 = (32)(2.1 \times 10^{-5})(1)(v)/[(58)(0.004)^2]$$

$$v^2 + 46.63v - 128.8 = 0 \qquad v = 2.616 \text{ ft/s}$$

$$Q = Av = [(\pi)(0.004)^2/4](2.616) = 0.00003287 \text{ ft}^3/\text{s} = 0.118 \text{ ft}^3/\text{h}$$

$$N_R = \rho dv/\mu = (\gamma/g)(d)(v)/\mu = (58/32.2)(0.004)(2.616)/(2.1 \times 10^{-5}) = 897 \qquad \text{(laminar)}$$

**9.51** In the syringe of Fig. 9-12 the drug has $\rho = 900 \text{ kg/m}^3$ and $\mu = 0.002 \text{ Pa} \cdot \text{s}$. What steady force $F$ is required to produce a flow of 0.4 mL/s through the needle? Neglect head loss in the larger cylinder.

❚

$$p_A/\rho g + v_A^2/2g + z_A = p_B/\rho g + v_B^2/2g + z_B + h_L$$

$$v_B = Q/A_B = 0.4 \times 10^{-6}/[(\pi)(0.25 \times 10^{-3})^2/4] = 8.149 \text{ m/s}$$

$$N_R = \rho dv/\mu = (900)(0.25 \times 10^{-3})(8.149)/0.002 = 917 \qquad \text{(laminar)}$$

$$h_f = \frac{32\mu Lv}{\rho g d^2} = \frac{(32)(0.002)(0.020)(8.149)}{(900)(9.807)(0.25 \times 10^{-3})^2} = 18.91 \text{ m}$$

$$p_A/[(900)(9.807)] + 0 + 0 = p_B/[(900)(9.807)] + (8.149)^2/[(2)(9.807)] + 0 + 18.91$$

$$p_A - p_B = 196\,788 \text{ N/m}^2$$

$$F = (p_A - p_B)(A_{\text{piston}}) = 196\,788[(\pi)(0.010)^2/4] = 15.5 \text{ N}$$

**Fig. 9-12**

**9.52** Paint issues from the tank in Fig. 9-13 at $Q = 45 \text{ ft}^3/\text{h}$. Find the kinematic viscosity. Is the flow laminar?

❚ $$p_A/\rho g + v_A^2/2g + z_A = p_B/\rho g + v_B^2/2g + z_B + h_L \qquad v_B = Q/A_B = (\tfrac{45}{3600})/[(\pi)(0.5/12)^2/4] = 9.167 \text{ ft/s}$$

$$0 + 0 + 9 = 0 + 9.167^2/[(2)(32.2)] + 0 + h_f \qquad h_f = 7.695 \text{ ft}$$

Assuming laminar flow,

$$h_f = \frac{128\nu LQ}{\pi g d^4} \qquad 7.695 = \frac{(128)(\nu)(6)(\tfrac{45}{3600})}{(\pi)(32.2)(0.5/12)^4} \qquad \nu = 0.0002444 \text{ ft}^2/\text{s}$$

$$N_R = dv/\nu = (0.5/12)(9.167)/0.0002444 = 1563 \qquad \text{(laminar)}$$

**Fig. 9-13**

**9.53** In Prob. 9.52, what will the flow rate be if the paint properties are $\rho = 1.78 \text{ slugs/ft}^3$ and $\mu = 0.00217 \text{ lb} \cdot \text{s/ft}^2$?

❚ $$p_A/\rho g + v_A^2/2g + z_A = p_B/\rho g + v_B^2/2g + z_B + h_L \qquad 0 + 0 + 9 = 0 + v^2/[(2)(32.2)] + 0 + h_f$$

Assuming laminar flow,

$$h_f = \frac{32\mu L v}{\rho g d^2} = \frac{(32)(0.00217)(6)(v)}{(1.78)(32.2)(0.5/12)^2} = 4.187v$$

$$0 + 0 + 9 = 0 + v^2/[(2)(32.2)] + 0 + 4.187v \qquad v^2 + 269.6v - 579.6 = 0 \qquad v = 2.133 \text{ ft/s}$$

$$Q = Av = [(\pi)(0.5/12)^2/4](2.133) = 0.002908 \text{ ft}^3/\text{s} \quad \text{or} \quad 10.47 \text{ ft}^3/\text{h}$$

$$N_R = \rho dv/\mu = (1.78)(0.5/12)(2.133)/0.00217 = 73 \qquad \text{(laminar)}$$

**9.54** The smaller tank in Fig. 9-14 is 50 m in diameter. If the fluid is ethanol at 20 °C, find the flow rate.

⬛ $$p_A/\gamma + v_A^2/2g + z_A = p_B/\gamma + v_B^2/2g + z_B + h_L \qquad 0 + 0 + (0.4 + 0.6) = 0 + 0 + 0 + h_f$$

$$h_f = 1.000 \text{ m} = \frac{128\mu LQ}{\pi \rho g d^4} \qquad 1.000 = \frac{(128)(1.20 \times 10^{-3})(0.8 + 0.4)(Q)}{(\pi)(788)(9.807)(0.002)^4}$$

$$Q = 2.107 \times 10^{-6} \text{ m}^3/\text{s} \quad \text{or} \quad 7.59 \text{ L/h}$$

**Fig. 9-14**

**9.55** For the system in Fig. 9-14, if the fluid has density of 920 kg/m³ and the flow rate is unknown, for what value of viscosity will the capillary Reynolds number exactly equal the critical value 2300?

⬛ $$h_f = 1.000 \text{ m} = \frac{32\mu L v}{\rho g d^2} \qquad \text{(from Prob. 9.54)}$$

$$1.000 = \frac{(32)(\mu)(0.8 + 0.4)(v)}{(920)(9.807)(0.002)^2} \qquad v = \frac{0.0009398}{\mu}$$

$$N_R = \rho dv/\mu \qquad 2300 = (920)(0.002)(0.0009398/\mu)/\mu \qquad \mu = 0.000867 \text{ Pa} \cdot \text{s}$$

**9.56** For the pressure measurements shown in Fig. 9-15, determine (a) whether the flow is up or down, and (b) the flow rate. Use $\rho = 917$ kg/m³ and $\mu = 0.290$ Pa · s.

⬛ HGL $= z + p/\rho g$
(a)
$$\text{HGL}_B = 15 + (200)(1000)/[(917)(9.807)] = 37.24 \text{ m}$$

$$\text{HGL}_A = 0 + (600)(1000)/[(917)(9.807)] = 66.72 \text{ m}$$

Since HGL$_A$ > HGL$_B$, the flow is from A to B (i.e., up).
(b) Assume flow is laminar.

$$h_f = \frac{128\mu LQ}{\pi \rho g d^4} = 66.72 - 37.24 = 29.48 \text{ m} \qquad L = \sqrt{15^2 + 20^2} = 25.00 \text{ m} \qquad 29.48 = \frac{(128)(0.290)(25.00)(Q)}{(\pi)(917)(9.807)(0.030)^4}$$

$$Q = 0.000727 \text{ m}^3/\text{s} \quad \text{or} \quad 2.617 \text{ m}^3/\text{h} \qquad v = Q/A = 0.000727/[(\pi)(0.030)^2/4] = 1.028 \text{ m/s}$$

$$N_R = \rho dv/\mu = (917)(0.030)(1.028)/0.290 = 98 \qquad \text{(laminar)}$$

**Fig. 9-15**

**9.57** Repeat Prob. 9.56 if the pressures are the same but there is a pump between $A$ and $B$ which adds a 10-m head rise in the flow direction. Is the flow still laminar?

❚ $h_f = \text{HGL}_A - \text{HGL}_B + h_{\text{pump}}$. Using values of $\text{HGL}_A$ and $\text{HGL}_B$ from Prob. 9.56,

$$h_f = 66.72 - 37.24 + 10 = 39.48 \text{ m} = \frac{128\mu LQ}{\pi\rho g d^4} \qquad L = 25.00 \text{ m} \qquad \text{(from Prob. 9.56)}$$

$$39.48 = \frac{(128)(0.290)(25.00)(Q)}{(\pi)(917)(9.807)(0.030)^4} \qquad Q = 0.000974 \text{ m}^3/\text{s} \quad \text{or} \quad 3.51 \text{ m}^3/\text{h}$$

$$v = Q/A = 0.000974/[(\pi)(0.030)^2/4] = 1.378 \text{ m/s}$$

$$N_R = \rho dv/\mu = (917)(0.030)(1.378)/0.290 = 131 \qquad \text{(laminar)}$$

**9.58** Water at 40 °C flows from tank $A$ to tank $B$ as shown in Fig. 9-16. Find the volumetric flow, neglecting entrance losses to the capillary tube as well as exit losses.

❚ $p_A/\gamma + v_A^2/2g + z_A = p_B/\gamma + v_B^2/2g + z_B + h_L \qquad 0 + 0 + (0.22 + 0.1) = 0 + 0 + 0 + h_f \qquad h_f = 0.32$
Assume laminar flow.

$$h_f = \frac{128\mu LQ}{\pi\rho g d^4} \qquad 0.32 = \frac{(128)(6.51 \times 10^{-4})(0.22 + 0.08)(Q)}{(\pi)(992)(9.807)(0.001)^4} \qquad Q = 3.912 \times 10^{-7} \text{ m}^3/\text{s} = 1.41 \text{ L/h}$$

$$v = Q/A = 3.912 \times 10^{-7}/[(\pi)(0.001)^2/4] = 0.4981 \text{ m/s}$$

$$N_R = \rho dv/\mu = (992)(0.001)(0.4981)/(6.51 \times 10^{-4}) = 759 \qquad \text{(laminar)}$$

**Fig. 9-16**

**9.59** In Prob. 9.58, what should the internal diameter of the tube be to permit a flow of 2.16 L/h?

❚ Assuming laminar flow, $Q \propto d^4$ or $d \propto Q^{1/4}$; hence

$$d = \left(\frac{2.16}{1.41}\right)^{1/4} (1 \text{ mm}) = 1.534 \text{ mm}$$

Computation of $N_R$ shows the flow to be indeed laminar.

**9.60** A hypodermic needle has an inside diameter of 0.3 mm and is 60 mm in length, as shown in Fig. 9-17. If the piston moves to the right at a speed of 18 mm/s and there is no leakage, what force $F$ is needed on the piston? The medicine in the hypodermic has a viscosity $\mu$ of $0.980 \times 10^{-3}$ Pa · s and its density $\rho$ is 800 kg/m³. Consider flows in both needle and cylinder. Neglect exit losses from the needle as well as losses at the juncture of the needle and cylinder.

■ For cylinder:

$$Q = Av = [(\pi)(0.005)^2/4](0.018) = 3.534 \times 10^{-7} \text{ m}^3/\text{s}$$

$$N_R = \rho dv/\mu = (800)(0.005)(0.018)/(0.980 \times 10^{-3}) = 73 \quad \text{(laminar)}$$

$$p = \frac{128\mu LQ}{\pi d^4} \qquad p_1 = \frac{(128)(0.980 \times 10^{-3})(0.050)(3.534 \times 10^{-7})}{(\pi)(0.005)^4} = 1.129 \text{ Pa}$$

For needle:

$$v = Q/A = 3.534 \times 10^{-7}/[(\pi)(0.3/1000)^2/4] = 5.000 \text{ m/s}$$

$$N_R = (800)(0.3/1000)(5.000)/(0.980 \times 10^{-3}) = 1224 \quad \text{(laminar)}$$

$$p_2 = \frac{(128)(0.980 \times 10^{-3})(0.060)(3.534 \times 10^{-7})}{(\pi)(0.3/1000)^4} = 104\,525 \text{ Pa}$$

$$F = (\Delta p)(A_{\text{cylinder}}) = (104\,525 - 1.129)[(\pi)(0.005)^2/4] = 2.05 \text{ N}$$

Fig. 9-17

**9.61** In Prob. 9.60, suppose that medicine is drawn from a bottle at atmospheric pressure. What is the largest flow of fluid if the fluid has a vapor pressure of 4.8 kPa abs? Neglect losses in the cylinder.

■ $$\Delta p = 101\,400 - 4800 = 96\,600 \text{ Pa} = \frac{128\mu LQ}{\pi d^4} \qquad 96\,600 = \frac{(128)(0.980 \times 10^{-3})(0.060)(Q_{\text{max}})}{(\pi)(3 \times 10^{-4})^4}$$

$$Q_{\text{max}} = 3.27 \times 10^{-7} \text{ m}^3/\text{s} = 0.327 \text{ mL/s}$$

**9.62** In Prob. 9.60, it took a force of 2.05 N to move the piston to the right at a speed of 18 mm/s. What should the inside diameter be for the cylinder if the force needed is only 1.2 N for the same piston speed? Neglect losses in cylinder.

■ $$p_1 = \frac{F}{A} = \frac{1.2}{(\pi)(0.005)^2/4} = 61\,116 \text{ Pa} \qquad p = \frac{128\mu LQ}{\pi d^4} \qquad 61\,116 = \frac{(128)(0.980 \times 10^{-3})(0.060)(3.534 \times 10^{-7})}{\pi d^4}$$

$$d = 0.343 \text{ mm} \qquad N_R = \rho dv/\mu = (800)(0.343/1000)(0.018)/(0.980 \times 10^{-3}) = 5 \quad \text{(laminar)}$$

**9.63** Water at 70 °F flows through a new cast iron pipe at a velocity of 9.7 ft/s. The pipe is 1200 ft long and has a diameter of 6 in. Find the head loss due to friction.

■ $$h_f = (f)(L/d)(v^2/2g) \qquad N_R = dv/v = (\tfrac{6}{12})(9.7)/(1.05 \times 10^{-5}) = 461\,905$$

From Table A-9, $\epsilon = 0.00085$ ft for new cast iron pipe; $\epsilon/d = 0.00085/(\tfrac{6}{12}) = 0.0017$. From Fig. A-5, $f = 0.0230$; $h_f = 0.0230[1200/(\tfrac{6}{12})]\{9.7^2/[(2)(32.2)]\} = 80.6$ ft.

**9.64** A 96-in-diameter new cast iron pipe carries water at 60 °F. The head loss due to friction is 1.5 ft per 1000 ft of pipe. What is the discharge capacity of the pipe?

■ $$h_f = (f)(L/d)(v^2/2g) \qquad 1.5 = (f)[1000/(\tfrac{96}{12})]\{v^2/[(2)(32.2)]\} \qquad fv^2 = 0.7728$$

Assume $f = 0.0150$; $(0.0150)(v^2) = 0.7728$, $v = 7.178$ ft/s; $N_R = dv/v = (\tfrac{96}{12})(7.178)/(1.21 \times 10^{-5}) = 4.75 \times 10^6$.

From Table A-9, $\epsilon = 0.00085$ ft for new cast iron pipe, $\epsilon/d = 0.00085/(\frac{96}{12}) = 0.000106$. From Fig. A-5, $f = 0.0124$. Evidently, the assumed value of $f$ of 0.0150 was not the correct one. Try a value of $f$ of 0.0124. $(0.0124)(v^2) = 0.7728$, $v = 7.894$ ft/s; $N_R = (\frac{96}{12})(7.894)/(1.21 \times 10^{-5}) = 5.22 \times 10^6$. From Fig. A-5, $f = 0.0124$. Hence, 0.0124 must be the correct value of $f$, and $v = 7.894$ ft/s. $Q = Av = [(\pi)(\frac{96}{12})^2/4](7.894) = 397$ ft³/s.

**9.65** Water at 70 °F is being drained from an open tank through a 24-in-diameter, 130-ft-long new cast iron pipe, as shown in Fig. 9-18. Find the flow rate at which water is being discharged from the pipe. Neglect minor losses.

$$p_1/\gamma + v_1^2/2g + z_1 = p_2/\gamma + v_2^2/2g + z_2 + h_L$$

$$h_L = h_f = (f)(L/d)(v^2/2g) = (f)[130/(\tfrac{24}{12})]\{v_2^2/[(2)(32.2)]\} = 1.009 f v_2^2$$

$$0 + 0 + 150.5 = 0 + v_2^2/[(2)(32.2)] + 98.4 + 1.009 f v_2^2$$

Assume $f = 0.0240$.

$$150.5 = v_2^2/[(2)(32.2)] + 98.4 + (1.009)(0.0240)(v_2^2) \qquad v_2 = 36.21 \text{ ft/s}$$

$$N_R = dv/\nu = (\tfrac{24}{12})(36.21)/(1.05 \times 10^{-5}) = 6.90 \times 10^6$$

From Table A-9, $\epsilon = 0.00085$, $\epsilon/d = 0.00085/(\frac{24}{12}) = 0.000425$. From Fig. A-5, $f = 0.0162$. Evidently, the assumed value of $f$ of 0.0240 was not the correct one. Try a value of $f$ of 0.0162.

$$150.5 = v_2^2/[(2)(32.2)] + 98.4 + (1.009)(0.0162)(v_2^2) \qquad v_2 = 40.43 \text{ ft/s}$$

$$N_R = (\tfrac{24}{12})(40.43)/(1.05 \times 10^{-5}) = 7.70 \times 10^6$$

From Fig. A-5, $f = 0.0162$. Hence, 0.0162 must be the correct value of $f$, and $v = 40.43$ ft/s. $Q = Av = [(\pi)(\frac{24}{12})^2/4](40.43) = 127$ ft³/s.

Elev. 150.5 ft

Water

Elev. 98.4 ft

**Fig. 9-18**

**9.66** Gasoline is being discharged from a pipe, as shown in Fig. 9-19. The pipe roughness ($\epsilon$) is 0.500 mm, and the pressure at point 1 is 2500 kPa. Find the pipe diameter needed to discharge gasoline at a rate of 0.10 m³/s. Neglect any minor losses.

$$p_1/\gamma + v_1^2/2g + z_1 = p_2/\gamma + v_2^2/2g + z_2 + h_L$$

$$h_L = h_f = (f)(L/d)(v^2/2g) = (f)(965.5/d)\{v_2^2/[(2)(9.807)]\} = 49.23 f v_2^2/d$$

$$2.500/7.05 + v_1^2/2g + 82.65 = 0 + v_2^2/2g + 66.66 + 49.23 f v_2^2/d \qquad v_1^2/2g = v_2^2/2g$$

$$f v_2^2/d = 0.3320 \qquad v_2 = Q/A_2 = 0.10/(\pi d^2/4) = 0.1273/d^2 \qquad (f)(0.1273/d^2)^2/d = 0.3320 \qquad d = (0.04881 f)^{1/5}$$

Assume $f = 0.0200$. $d = [(0.04881)(0.0200)]^{1/5} = 0.2500$ m, $v_2 = 0.1273/0.2500^2 = 2.037$ m/s; $N_R = \rho dv/\mu = (719)(0.2500)(2.037)/(2.92 \times 10^{-4}) = 1.25 \times 10^6$. From Table A-9, $\epsilon = 0.00050$ m. $\epsilon/d = 0.00050/0.2500 = 0.0020$. From Fig. A-5, $f = 0.0235$. Evidently, the assumed value of $f$ of 0.0200 was not the correct one. Try a value of $f$ of 0.0235.

$$d = [(0.04881)(0.0235)]^{1/5} = 0.2582 \text{ m} \qquad v = 0.1273/0.2582^2 = 1.909 \text{ m/s}$$

$$N_R = (719)(0.2582)(1.909)/(2.92 \times 10^{-4}) = 1.21 \times 10^6$$

$$\epsilon/d = 0.00050/0.2582 = 0.00194 \qquad f = 0.0235$$

Hence, 0.0235 must be the correct value of $f$, and $d = 0.2582$ m.

Elev. 82.65 m

Elev. 66.66 m

965.5 m

**Fig. 9-19**

**9.67**  Water at 20 °C flows through a new cast iron pipe at a velocity of 4.2 m/s. The pipe is 400 m long and has a diameter of 150 mm. Determine the head loss due to friction.

$$h_f = (f)(L/d)(v^2/2g) \qquad N_R = dv/v = (\tfrac{150}{1000})(4.2)/(1.02 \times 10^{-6}) = 6.18 \times 10^5$$

From Table A-9, $\epsilon = 0.00026$ m. $\epsilon/d = 0.00026/0.150 = 0.00173$. From Fig. A-5, $f = 0.0226$. $h_f = 0.0226[400/(\tfrac{150}{1000})]\{4.2^2/[(2)(9.807)]\} = 54.20$ m.

**9.68**  SAE10 oil at 68 °F is to be pumped at a flow rate of 2.0 ft³/s through a level 6-in-diameter new wrought iron pipe. Determine the pressure loss in pounds per square inch per mile of pipe and compute the horsepower lost to friction.

$$h_f = (f)(L/d)(v^2/2g) \qquad v = Q/A = 2.0/[(\pi)(\tfrac{6}{12})^2/4] = 10.19 \text{ ft/s}$$

$$N_R = \rho dv/\mu = (1.68)(\tfrac{6}{12})(10.19)/(1.70 \times 10^{-3}) = 5035 \qquad \text{(turbulent)}$$

From Table A-9, $\epsilon = 0.00015$ ft. $\epsilon/d = 0.00015/(\tfrac{6}{12}) = 0.00030$. From Fig. A-5, $f = 0.038$. $h_f = 0.038[5280/(\tfrac{6}{12})]\{10.19^2/[(2)(32.2)]\} = 647$ ft of oil; $p = \gamma h = (54.2)(647)/144 = 244$ psi/mile.

**9.69**  Water at 20 °C flows in a 100-mm-diameter new cast iron pipe with a velocity of 5.0 m/s. Determine the pressure drop in kilopascals per 100 m of pipe and the power lost to friction.

$$h_f = (f)(L/d)(v^2/2g) \qquad N_R = dv/v = (\tfrac{100}{1000})(5.0)/(1.02 \times 10^{-6}) = 4.90 \times 10^5$$

From Table A-9, $\epsilon = 0.00026$ m. $\epsilon/d = 0.00026/(\tfrac{100}{1000}) = 0.0026$. From Fig. A-5, $f = 0.0252$.

$$h_f = 0.0252[100/(\tfrac{100}{1000})]\{5.0^2/[(2)(9.807)]\} = 32.12 \text{ m} \qquad p = (9.79)(32.12) = 314 \text{ kN/m}^2 \text{ per 100 m of pipe}$$

$$Q = Av = [(\pi)(\tfrac{100}{1000})^2/4](5.0) = 0.03927 \text{ m}^3/\text{s}$$

$$\text{Power lost} = Q\gamma h_f = (0.03927)(9.79)(32.12) = 12.35 \text{ kW per 100 m of pipe}$$

**9.70**  Determine the discharge capacity of a 150-mm-diameter new wrought iron pipe to carry water at 20 °C if the pressure loss due to friction may not exceed 35 kPa per 100 m of level pipe.

$$N_R = dv/v = (\tfrac{150}{1000})(v)/(1.02 \times 10^{-6}) = 1.47 \times 10^5 v$$

**Trial No. 1**
Assume $v = 3.0$ m/s: $N_R = (1.47 \times 10^5)(3.0) = 4.41 \times 10^5$, $\epsilon/d = 0.000046/(\tfrac{150}{1000}) = 0.000307$. From Fig. A-5, $f = 0.0164$, $h_f = (f)(L/d)(v^2/2g) = p/\gamma = 35/9.79 = 3.575$ m, $3.575 = 0.0164[100/(\tfrac{150}{1000})]\{v^2/[(2)(9.807)]\}$, $v = 2.53$ m/s.

**Trial No. 2**
Assume $v = 2.53$ m/s: $N_R = (1.47 \times 10^5)(2.53) = 3.72 \times 10^5$, $f = 0.0166$, $3.575 = 0.0166[100/(\tfrac{150}{1000})]\{v^2/[(2)(9.807)]\}$, $v = 2.52$ m/s; $Q = Av = [(\pi)(\tfrac{150}{1000})^2/4](2.52) = 0.0445$ m³/s.

**9.71**  SAE30 oil at 68 °F is to be pumped at a flow rate of 3.0 ft³/s through a level new cast iron pipe. Allowable pipe friction loss is 10 psi per 1000 ft of pipe. What size commercial pipe should be used?

$$p = \gamma h \qquad (10)(144) = 55.4 h_f \qquad h_f = 26.0 \text{ ft of oil per 1000 ft of pipe}$$

**Trial No. 1**
Assume $v = 5.0$ ft/s:

$$Q = A/v \qquad 3.0 = (\pi d^2/4)(5.0) \qquad d = 0.874 \text{ ft}$$

$$N_R = \rho dv/\mu = (1.72)(0.874)(5.0)/(9.2 \times 10^{-3}) = 817 \qquad \text{(laminar)}$$

$$h_f = (f)(L/d)(v^2/2g) \qquad f = 64/N_R = \tfrac{64}{817} = 0.0783$$

$$26.0 = (0.0783)(1000/0.874)\{v^2/[(2)(32.2)]\} \qquad v = 4.32 \text{ ft/s}$$

**Trial No. 2**
Assume $v = 4.32$ ft/s:

$$3.0 = (\pi d^2/4)(4.32) \qquad d = 0.940 \text{ ft} \qquad N_R = (1.72)(0.940)(4.32)/(9.2 \times 10^{-3}) = 759 \qquad \text{(laminar)}$$

$$f = \tfrac{64}{759} = 0.0843 \qquad 26.0 = (0.0843)(1000/0.940)\{v^2/[(2)(32.2)]\} \qquad v = 4.32 \text{ ft/s}$$

Hence, a pipe diameter of 0.940 ft, or 11.28 in, would be required. A 12-in-diameter commercial pipe should be used, which would result in a pipe friction loss somewhat less than the allowable 10 psi per 1000 ft of pipe.

**9.72** SAE10 oil at 20 °C is to flow through a 300-m level concrete pipe. What size pipe will carry 0.0142 m³/s with a pressure drop due to friction of 23.94 kPa?

**▌** $$p = \gamma h \qquad 23.94 = 8.52 h_f \qquad h_f = 2.81 \text{ m}$$

**Trial No. 1**
Assume $v = 1.5$ m/s:

$$Q = A/v \qquad 0.0142 = (\pi d^2/4)(1.5) \qquad d = 0.110 \text{ m}$$

$$N_R = \rho dv/\mu = (869)(0.110)(1.5)/(8.14 \times 10^{-2}) = 1761 \qquad \text{(laminar)}$$

$$h_f = (f)(L/d)(v^2/2g) \qquad f = 64/N_R = \tfrac{64}{1761} = 0.0363$$

$$2.81 = (0.0363)(300/0.110)\{v^2/[(2)(9.807)]\} \qquad v = 0.746 \text{ m/s}$$

**Trial No. 2**
Assume $v = 0.746$ m/s:

$$0.0142 = (\pi d^2/4)(0.746) \qquad d = 0.156 \text{ m} \qquad N_R = (869)(0.156)(0.746)/(8.14/10^{-2}) = 1242 \qquad \text{(laminar)}$$

$$f = \tfrac{64}{1242} = 0.0515 \qquad 2.81 = (0.0515)(300/0.156)\{v^2/[(2)(9.807)]\} \qquad v = 0.746 \text{ m/s}$$

Hence, a pipe diameter of 0.156 m, or 156 mm, would be required.

**9.73** Compute the friction factor for flow having a Reynolds number of $5 \times 10^3$ and relative roughness ($\epsilon/d$) of 0.015 (transition zone). Use the Colebrook formula, the Swamee–Jain formula, and the Moody diagram.

**▌ Colebrook formula:**

$$1/\sqrt{f} = 1.14 - 2.0 \log{[\epsilon/d + 9.35/(N_F\sqrt{f})]} = 1.14 - 2.0 \log{[0.015 + 9.35/(5 \times 10^3 \sqrt{f})]}$$

$$f = 0.0515 \qquad \text{(by trial and error)}$$

**Swamee–Jain formula:**

$$f = 0.25/[\log{(\epsilon/3.7d)} + (5.47/N_R^{0.9})]^2 = 0.25/\{\log{(0.015/3.7)} + [5.47/(5 \times 10^3)^{0.9}]\}^2 = 0.0438$$

**Moody diagram** (Fig. A-5): $$f = 0.0512$$

**9.74** Repeat Prob. 9.73 for flow having a Reynolds number of $4 \times 10^6$ and relative roughness ($\epsilon/d$) of 0.0001 (rough-pipe zone).

**▌ Colebrook formula:**

$$f = 1/[1.14 - 2.0 \log{(\epsilon/d)}]^2 = 1/[1.14 - 2.0 \log{(0.0001)}]^2 = 0.0120$$

**Swamee–Jain formula:**

$$f = 0.25/[\log{(\epsilon/3.7d)} + (5.74/N_R^{0.9})]^2 = 0.25/\{\log{(0.0001/3.7)} + [5.74/(4 \times 10^6)^{0.9}]\}^2 = 0.0120$$

**Moody diagram** (Fig. A-5):

$$f = 0.0125$$

**9.75** We have oil of kinematic viscosity $8 \times 10^{-5}$ ft²/s going through an 80-ft horizontal pipe. If the initial pressure is 5.0 psig and the final pressure is 3.5 psig, compute the mass flow if the pipe has a diameter of 3 in. At a point 10 ft from the end of the pipe a vertical tube is attached to be flush with the inside radius of the pipe. How high will the oil rise in the tube? $\rho = 50$ lbm/ft³. Pipe is commercial steel ($\epsilon = 0.000145$ ft).

$$p_1/\gamma + v_1^2/2g + z_1 = p_2/\gamma + v_2^2/2g + z_2 + h_L$$

$$(5.0)(144)/(50/32.2) + v_1^2/2g + 0 = (3.5)(144)/(50/32.2) + v_2^2/2g + 0 + h_f$$

$$v_1^2/2g = v_2^2/2g \qquad h_f = 139.1 \text{ ft} = (f)(L/d)(v^2/2) \qquad 139.1 = (f)[80/(\tfrac{3}{12})](v^2/2) \qquad fv^2 = 0.8694$$

Try $f = 0.020$: $0.020v^2 = 0.8694$, $v = 6.593$ ft/s; $N_R = dv/v = (\tfrac{3}{12})(6.593)/(8 \times 10^{-5}) = 2.06 \times 10^4$; $\epsilon/d = 0.000145/(\tfrac{3}{12}) = 0.000580$. From Fig. A-5, $f = 0.0265$. Try $f = 0.0265$: $0.0265v^2 = 0.8694$, $v = 5.728$ ft/s; $N_R = (\tfrac{3}{12})(5.728)/(8 \times 10^{-5}) = 1.79 \times 10^4$; $f = 0.0267$, $0.0267v^2 = 0.8694$, $v = 5.706$ ft/s; $M = \rho A v = 50[(\pi)(\tfrac{3}{12})^2/4](5.706) = 14.0$ lbm/s. To find the pressure at the point 10 ft from the end of the pipe (call it point $A$), apply the Bernoulli equation between point 1 and point $A$:

$$(5.0)(144)/(50/32.2) + v_1^2/2g + 0 = p_A/(50/32.2) + v_2^2/2g + 0 + h_f$$

$$h_f = 463.7 - 0.6440p_A = 0.0267[70/(\tfrac{3}{12})](5.706^2/2) = 121.7 \text{ ft}$$

$$121.7 = 463.7 - 0.6440p_A \qquad p_A = 531.1 \text{ lbm/ft}^2 \qquad h = p/\rho = 531.1/50 = 10.62 \text{ ft}$$

**9.76** How much water is flowing through the pipe shown in Fig. 9-20? Take $v = 0.114 \times 10^{-5}$ m²/s and $\epsilon = 0.0000442$ m.

$$p_1/\gamma + v_1^2/2g + z_1 = p_2/\gamma + v_2^2/2g + z_2 + h_L \qquad 1.6 + v_1^2/2g + 0 = 0.3 + v_2^2/2g + 0 + h_f$$

$$v_1^2/2g = v_2^2/2g \qquad h_f = 1.3 \text{ m} = (f)(L/d)(v^2/2g)$$

Try $f = 0.015$:

$$1.3 = 0.015[10/(0.150)]\{v^2/[(2)(9.807)]\} \qquad v = 5.050 \text{ m/s}$$

$$N_R = dv/v = (0.150)(5.050)/(0.114 \times 10^{-5}) = 6.64 \times 10^5 \qquad \epsilon/d = 0.0000442/(0.150) = 0.000295$$

From Fig. A-5, $f = 0.016$.

$$1.3 = 0.016[10/(0.150)]\{v^2/[(2)(9.807)]\} \qquad v = 4.889 \text{ m/s}$$

$$M = \rho A v = 1000[(\pi)(0.150)^2/4](4.889) = 86.4 \text{ kg/s}$$

1.6 m

$d = 150$ mm   0.3 m

1   2

10 m   **Fig. 9-20**

**9.77** Whiskey (s.g. $= 0.6$, $v = 5.0 \times 10^{-7}$ m²/s) is drawn from a tank through a hose of inside diameter 25 mm (see Fig. 9-21). The relative roughness for the hose is 0.0004. Calculate the volumetric flow and the minimum pressure in the hose. The total length of hose is 9 m and the length of hose to point $A$ is 3.25 m. Neglect minor losses at head entrance.

$$p_1/\gamma + v_1^2/2g + z_1 = p_2/\gamma + v_2^2/2g + z_2 + h_L \qquad 0 + 0 + (5 - 1.5) = 0 + v_2^2/[(2)(9.807)] + 0 + h_f$$

$$h_f = 3.5 - 0.05098v_2^2 = (f)(L/d)(v^2/2g)$$

Try $f = 0.016$:

$$h_f = 0.016[9/(0.025)]\{v_2^2/[(2)(9.807)]\} = 0.2937v_2^2 \qquad 0.2937v_2^2 = 3.5 - 0.05098v_2^2 \qquad v = 3.187\,\text{m/s}$$

$$N_R = dv/v = (0.025)(3.187)/(5.0 \times 10^{-7}) = 1.59 \times 10^5$$

From Fig. A-5 with $\epsilon/d = 0.0004$, $f = 0.019$. Try $f = 0.019$:

$$h_f = 0.019[9/(0.025)]\{v_2^2/[(2)(9.807)]\} = 0.3487v_2^2 \qquad 0.3487v_2^2 = 3.5 - 0.05098v_2^2 \qquad v = 2.959\,\text{m/s}$$

$$N_R = dv/v = (0.025)(2.959)/(5.0 \times 10^{-7}) = 1.48 \times 10^5 \qquad f = 0.019$$

$$Q = Av = [(\pi)(0.025)^2/4](2.959) = 1.45\,\text{L/s} \qquad p_1/\gamma + v_1^2/2g + z_1 = p_A/\gamma + v_A^2/2g + z_A + h_L$$

$$0 + 0 + (5 - 1.5) = p_A/[(0.6)(9.79)] + 2.959^2/[(2)(9.807)] + 5 + h_f \qquad p_A = -11.43 - 5.874 h_f$$

$$h_f = 0.019[3.25/(0.025)]\{2.959^2/[(2)(9.807)]\} = 1.103\,\text{m} \qquad p_A = -11.43 - (5.874)(1.103) = -17.91\,\text{kPa}$$

**Fig. 9-21**

**9.78** In using the Darcy–Weisbach equation for flow in a pressure conduit, what percentage error is introduced in $Q$ when $f$ is misjudged by 20 percent?

$$\blacksquare \qquad h_f = (f)(L/d)(v^2/2g) = K_1 f Q^2 \qquad \text{(where } K_1 \text{ is a constant)}$$

Assume $h_f$ is constant. $Q = K_2/\sqrt{f} = K_2 f^{-1/2}$, $dQ = -\frac{1}{2}(K_2)(f^{-3/2})(df)$

$$\frac{dQ}{Q} = \frac{-\frac{1}{2}(K_2)(f^{-3/2})(df)}{K_2 f^{-1/2}} = -\frac{1}{2}\frac{df}{f} = -(\tfrac{1}{2})(0.20) = -0.10 \quad \text{or} \quad -10 \text{ percent}$$

**9.79** For the system in Fig. 9-13, find the flow rate if the liquid is water at 68 °F.

$\blacksquare$ Assume smooth-wall turbulent flow. $p_A/\gamma + v_A^2/2g + z_A = p_B/\gamma + v_B^2/2g + z_B + h_L$, $0 + 0 + 10 = 0 +$
$v_B^2/[(2)(32.2)] + 0 + h_f$, $h_f = 10 - 0.01553 v_B^2 = (f)(L/d)(v^2/2g)$. Try $f = 0.02$:

$$h_f = 0.02[6/(0.5/12)]\{v_B^2/[(2)(32.2)]\} = 0.04472 v_B^2 \qquad 10 - 0.01553 v_B^2 = 0.04472 v_B^2 \qquad v_B = 12.88\,\text{ft/s}$$

$$N_R = \rho dv/\mu = (1.93)(0.5/12)(12.88)/(2.04 \times 10^{-5}) = 5.08 \times 10^4 \qquad \text{(turbulent)}$$

From Fig. A-5, $f = 0.0208$. Try $f = 0.0208$:

$$h_f = 0.0208[6/(0.5/12)]\{v_B^2/[(2)(32.2)]\} = 0.04651 v_B^2 \qquad 10 - 0.01553 v_B^2 = 0.04651 v_B^2 \qquad v_B = 12.70\,\text{ft/s}$$

$$N_R = \rho dv/\mu = (1.93)(0.5/12)(12.70)/(2.04 \times 10^{-5}) = 5.01 \times 10^4 \qquad \text{(turbulent)}$$

$$f = 0.0208 \qquad Q = Av = [(\pi)(0.5/12)^2/4](12.70) = 0.0173\,\text{ft}^3/\text{s}$$

**9.80** If 1 mile of 3-in-diameter wrought iron pipe carries water at 68 °F and $v = 23$ ft/s, compute the head loss and the pressure drop.

$$\blacksquare \qquad h_f = (f)(L/d)(v^2/2g) \qquad N_R = \rho dv/\mu = (1.93)(\tfrac{3}{12})(7/0.3048)/(2.04 \times 10^{-5}) = 5.43 \times 10^5$$

$$\epsilon/d = 0.00015/(\tfrac{3}{12}) = 0.000600$$

From Fig. A-5, $f = 0.0182$.

$$h_f = 0.0182[5280/(\tfrac{3}{12})]\{(23)^2/[(2)(32.2)]\} = 3157 \text{ ft} \qquad p = \gamma h_f = (62.4)(3157)/144 = 1368 \text{ lb/in}^2$$

**9.81** Mercury at 20 °C flows through 3 m of 6-mm-diameter glass tubing with average velocity 2.0 m/s. Compute the head loss and the pressure drop.

$\blacksquare \qquad h_f = (f)(L/d)(v^2/2g) \qquad N_R = \rho dv/\mu = (13\,570)(0.006)(2.0)/(1.56 \times 10^{-3}) = 1.04 \times 10^5$

From Fig. A-5, $f = 0.0180$ (assuming glass to be "smooth").

$$h_f = 0.0180[3/(0.006)]\{2.0^2/[(2)(9.807)]\} = 1.835 \text{ m} \qquad p = \gamma h_f = [(13.6)(9.79)](1.835) = 244 \text{ kPa}$$

**9.82** Gasoline at 20 °C is pumped at 0.2 m³/s through 16 km of 180-mm-diameter cast iron pipe. Compute the power required if the pumps are 75 percent efficient.

$\blacksquare \qquad h_f = (f)(L/d)(v^2/2g) \qquad v = Q/A = 0.2/[(\pi)(0.180)^2/4] = 7.860 \text{ m/s}$

$N_R = \rho dv/\mu = (719)(0.180)(7.860)/(2.92 \times 10^{-4}) = 3.48 \times 10^6 \qquad \epsilon/d = 0.00026/(0.180) = 0.00144$

From Fig. A-5, $f = 0.0216$.

$$h_f = 0.0216[(16)(1000)/(0.180)]\{7.860^2/[(2)(9.807)]\} = 6048 \text{ m}$$

$$P = \rho g Q h_f/\eta = (719)(9.807)(0.2)(6048)/0.75 = 11.37 \times 10^6 \text{ W} \quad \text{or} \quad 11.37 \text{ MW}$$

**9.83** Vinegar (s.g. = 0.86, $v = 0.00003$ ft²/s) flows at 1 ft³/s through a 6-in asphalted cast iron pipe. The pipe is 2000 ft long and slopes upward at 10° in the flow direction. Compute the head loss and the pressure change.

$\blacksquare \qquad h_f = (f)(L/d)(v^2/2g) \qquad v = Q/A = 1/[(\pi)(\tfrac{6}{12})^2/4] = 5.093 \text{ ft/s}$

$N_R = dv/v = (\tfrac{6}{12})(5.093)/0.00003 = 8.49 \times 10^4 \qquad \epsilon/d = 0.0004/(\tfrac{6}{12}) = 0.000800$

From Fig. A-5, $f = 0.0219$.

$$h_f = 0.0219[2000/(\tfrac{6}{12})]\{5.093^2/[(2)(32.2)]\} = 35.28 \text{ ft} \qquad h_L = 35.28 + 2000 \sin 10° = 382.6 \text{ ft}$$

$$p = \gamma h_f = [(0.86)(62.4)](382.6)/144 = 142.6 \text{ lb/in}^2$$

**9.84** The pipe flow in Fig. 9-22 is driven by pressurized air in the tank. What gage pressure $p_1$ is needed to provide a flow rate of 50 m³/h of water? Assume a "smooth" pipe.

$\blacksquare \qquad p_1/\gamma + v_1^2/2g + z_1 = p_2/\gamma + v_2^2/2g + z_2 + h_L \qquad v_2 = Q/A_2 = (50/3600)/[(\pi)(0.050)^2/4] = 7.074 \text{ m/s}$

$h_L = h_f = (f)(L/d)(v^2/2g) \qquad N_R = \rho dv/\mu = (998)(0.050)(7.074)/(1.02 \times 10^{-3}) = 3.46 \times 10^5$

From Fig. A-5, $f = 0.0140$.

$$h_L = 0.0140[(40 + 80 + 20)/0.050]\{7.074^2/[(2)(9.807)]\} = 100.0 \text{ m}$$

$$p_1/9.79 + 0 + 10 = 0 + 7.074^2/[(2)(9.807)] + 80 + 100.0 \qquad p_1 = 1689 \text{ kPa gage}$$

Smooth pipe:
$d = 50\,\text{mm}$

20 m

2 $Q$

Open jet

80 m

$p_1$ 1

10 m

40 m

**Fig. 9-22**

**9.85** In Fig. 9-22 suppose the fluid is methanol at 20 °C and $p_1 = 900$ kPa gage. What flow rate $Q$ results?

❚ $p_1/\gamma + v_1^2/2g + z_1 = p_2/\gamma + v_2^2/2g + z_2 + h_L$     $900/7.73 + 0 + 10 = 0 + v_2^2/[(2)(9.807)] + 80 + h_L$

$$h_L = 46.43 - 0.05098v_2^2 = h_f = (f)(L/d)(v^2/2g)$$

Try $f = 0.02$:

$$h_L = 0.02[(40 + 80 + 20)/0.050]\{v_2^2/[(2)(9.807)]\} = 2.855v_2^2 \qquad 2.855v_2^2 = 46.43 - 0.05098v_2^2 \qquad v_2 = 4.000 \text{ m/s}$$

$$N_R = \rho dv/\mu = (788)(0.050)(4.000)/(5.98 \times 10^{-4}) = 2.64 \times 10^5$$

From Fig. A-5, $f = 0.0150$. Try $f = 0.0150$:

$$h_L = 0.0150[(40 + 80 + 20)/0.050]\{v_2^2/[(2)(9.807)]\} = 2.141v_2^2 \qquad 2.141v_2^2 = 46.43 - 0.05098v_2^2 \qquad v_2 = 4.602 \text{ m/s}$$

$$N_R = (788)(0.050)(4.602)/(5.98 \times 10^{-4}) = 3.03 \times 10^5$$

Try $f = 0.0145$:

$$h_L = 0.0145[(40 + 80 + 20)/0.050]\{v_2^2/[(2)(9.807)]\} = 2.070v_2^2 \qquad 2.070v_2^2 = 46.43 - 0.05098v_2^2 \qquad v_2 = 4.679 \text{ m/s}$$

$$N_R = (788)(0.050)(4.679)/(5.98 \times 10^{-4}) = 3.08 \times 10^5 \qquad f = 0.0145$$

At this Reynolds number, $Q = Av = [(\pi)(0.050)^2/4](4.679) = 0.00919 \text{ m}^3/\text{s}$ or $33.1 \text{ m}^3/\text{h}$.

**9.86** In Fig. 9-22 suppose the fluid is carbon tetrachloride at 20 °C and $p_1 = 1300$ kPa. Calculate the pipe diameter needed for a volumetric flow of 5.555 L/s.

❚ $p_1/\gamma + v_1^2/2g + z_1 = p_2/\gamma + v_2^2/2g + z_2 + h_L$     $p_1/\gamma + 0 + z_1 = 0 + v_2^2/2g + z_2 + (f)(L/d)(v_2^2/2g)$

$$v_2^2 = \frac{(2g)(p_1/\gamma + z_1 - z_2)}{1 + fL/d} = \frac{(2)(9.807)(1300/15.57 + 10 - 80)}{1 + (f)(40 + 80 + 20)/d} = \frac{264.7}{1 + 140f/d}$$

$$= (Q/A_2)^2 = [(5.555 \times 10^{-3})/(\pi d^2/4)]^2 = 0.00005004/d^4$$

$$v_2 = 0.007074/d^2$$

$$\frac{0.00005004}{d^4} = \frac{264.7}{1 + 140f/d} \qquad d = \frac{(1 + 140f/d)^{1/4}}{47.96}$$

Try $d = 50$ mm, or 0.050 m: $N_R = \rho dv/\mu = (1588)(0.050)(0.007074/0.050^2)/(9.67 \times 10^{-4}) = 2.32 \times 10^5$. From Fig. A-5, $f = 0.0151$. $d = [1 + (140)(0.0151/0.050)]^{1/4}/47.96 = 0.0535$ m. Try $d = 0.0535$ m:

$$N_R = (1588)(0.0535)(0.007074/0.0535^2)/(9.67 \times 10^{-4}) = 2.17 \times 10^5 \qquad f = 0.0155$$

$$d = [1 + (140)(0.0155/0.0535)]^{1/4}/47.96 = 0.0529 \text{ m}$$

Try $d = 0.0529$ m:

$$N_R = (1588)(0.0529)(0.007074/0.0529^2)/(9.67 \times 10^{-4}) = 2.20 \times 10^5 \qquad f = 0.0155$$

$$d = [1 + (140)(0.0155/0.0529)]^{1/4}/47.96 = 0.0531 \text{ m}$$

Try $d = 0.0531$ m:

$$N_R = (1588)(0.0531)(0.007074/0.0531^2)/(9.67 \times 10^{-4}) = 2.19 \times 10^5 \qquad f = 0.0155$$

$$d = [1 + (140)(0.0155/0.0531)]^{1/4}/47.96 = 0.05301 \text{ m}$$

Hence, use $d = 0.053$ m, or 53 mm.

**9.87** The reservoirs in Fig. 9-23 contain water at 20 °C. If the pipe is smooth, with $L = 7$ km and $d = 50$ mm, what will the flow rate be for $\Delta z = 98$ m?

❚ $p_1/\gamma + v_1^2/2g + z_1 = p_2/\gamma + v_2^2/2g + z_2 + h_L$     $0 + 0 + 100 = 0 + 0 + 0 + h_f$     $h_f = 98 \text{ m} = (f)(L/d)(v^2/2g)$

$$98 = (f)[7000/0.050]\{v^2/[(2)(9.807)]\} \qquad v = 0.1172/\sqrt{f}$$

Try $f = 0.02$: $v = 0.1172/\sqrt{0.02} = 0.8287$ m/s, $N_R = \rho dv/\mu = (998)(0.050)(0.8287)/(1.02 \times 10^{-3}) = 4.05 \times 10^4$. From Fig. A-5, $f = 0.022$. Try $f = 0.022$:

$$v = 0.1172/\sqrt{0.022} = 0.7902 \text{ m/s} \qquad N_R = (998)(0.050)(0.7902)/(1.02 \times 10^{-3}) = 3.87 \times 10^4$$

$$f = 0.022 \qquad \text{(O.K.)}$$

$$Q = Av = [(\pi)(0.050)^2/4](0.7902) = 0.00155 \text{ m}^3/\text{s} \quad \text{or} \quad 5.58 \text{ m}^3/\text{h}$$

**Fig. 9-23**

**9.88** Repeat Prob. 9.87 to find $Q$ if $L = 2500$ ft, $d = 3$ in, and $\Delta z = 82$ ft.

$\blacksquare$ $p_1/\gamma + v_1^2/2g + z_1 = p_2/\gamma + v_2^2/2g + z_2 + h_L$    $0 + 0 + 82 = 0 + 0 + 0 + h_f$    $h_f = 82 \text{ ft} = (f)(L/d)(v^2/2g)$

$$82 = (f)[2500/(\tfrac{3}{12})]\{v^2/[(2)(32.2)]\}    v = 0.7267/\sqrt{f}$$

Try $f = 0.02$:

$$v = 0.7267/\sqrt{0.02} = 5.139 \text{ ft/s}    N_R = \rho dv/\mu = (1.93)(\tfrac{3}{12})(5.139)/(2.04 \times 10^{-5}) = 1.22 \times 10^5$$

From Fig. A-5, $f = 0.0175$. Try $f = 0.0175$:

$$v = 0.7267/\sqrt{0.0175} = 5.493 \text{ ft/s}    N_R = (1.93)(\tfrac{3}{12})(5.493)/(2.04 \times 10^{-5}) = 1.30 \times 10^5    f = 0.0170$$

Try $f = 0.0170$:

$$v = 0.7267/\sqrt{0.0170} = 5.574 \text{ ft/s}    N_R = (1.93)(\tfrac{3}{12})(5.574)/(2.04 \times 10^{-5}) = 1.32 \times 10^5$$

$$f = 0.0170    \text{(O.K.)}    Q = Av = [(\pi)(\tfrac{3}{12})^2/4](5.574) = 0.2736 \text{ ft}^3/\text{s}    \text{or}    985 \text{ ft}^3/\text{h}$$

**9.89** Repeat Prob. 9.88 if the pipe has a roughness of 0.2 mm.

$\blacksquare$ From Prob. 9.88, $v = 0.7267/\sqrt{f}$. Try $f = 0.02$:

$$N_R = 1.22 \times 10^5    \epsilon/d = 0.2/[(3)(25.4)] = 0.00262$$

From Fig. A-5, $f = 0.0265$. Try $f = 0.0265$:

$$v = 0.7267/\sqrt{0.0265} = 4.464 \text{ ft/s}    N_R = (1.93)(\tfrac{3}{12})(4.464)/(2.04 \times 10^{-5}) = 1.06 \times 10^5$$

$$f = 0.0265    \text{(O.K.)}    Q = Av = [(\pi)(\tfrac{3}{12})^2/4](4.464) = 0.2191 \text{ ft}^3/\text{s}    \text{or}    789 \text{ ft}^3/\text{h}$$

This is $(985 - 789)/985 = 0.199$, or 19.9 percent less than when the pipe is smooth.

**9.90** Water at 20 °C flows through a 598-m pipe 150 mm in diameter at 60 L/s. Determine the pipe roughness if the head loss is 49 m.

$\blacksquare$ $h_L = (f)(L/d)(v^2/2g)$    $v = Q/A = 0.06/[(\pi)(0.150)^2/4] = 3.395 \text{ m/s}$

$$49 = (f)[598/(0.150)]\{3.395^2/[(2)(9.807)]\}    f = 0.0209$$

$$N_R = \rho dv/\mu = (998)(0.150)(3.395)/(1.02 \times 10^{-3}) = 4.98 \times 10^5$$

From Fig. A-5 with $f = 0.0209$ and $N_R = 4.98 \times 10^5$, $\epsilon/d = 0.0012$; $\epsilon = (150)(0.0012) = 0.180$ mm.

**9.91** A 4-in-diameter commercial steel pipe is to be sloped so that 198 gpm of water at 20 °C passes through it in gravity flow. Find the declination $\theta$ of the pipe.

$\blacksquare$ $Q = (198)(0.002228) = 0.4411 \text{ ft}^3/\text{s}$    $v = Q/A = 0.4411/[(\pi)(\tfrac{4}{12})^2/4] = 5.055 \text{ ft/s}$

$$\epsilon/d = 0.00015/(\tfrac{4}{12}) = 0.00045    N_R = dv/\nu = (\tfrac{4}{12})(5.055)/(1.05 \times 10^{-5}) = 1.60 \times 10^5$$

From Fig. A-5, $f = 0.0190$. $\sin \theta = \Delta z/L = h_f/L = (f)(1/d)(v^2/2g)$, $\sin \theta = 0.0190[1/(\frac{4}{12})]\{5.055^2/[(2)(32.2)]\}$, $\theta = 1.30°$.

**9.92** In Prob. 9.91 find the volume flow corresponding to $\theta = 3°$.

**▌** From Prob. 9.91, $(f)(1/d)(v^2/2g) = \sin \theta$, $(f)[1/(\frac{4}{12})]\{v^2/[(2)(32.2)]\} = \sin 3°$, $v = 1.060/\sqrt{f}$. Try $f = 0.02$:

$$v = 1.060/\sqrt{0.02} = 7.495 \text{ ft/s} \qquad N_R = dv/\nu = (\frac{4}{12})(7.495)/(1.05 \times 10^{-5}) = 2.38 \times 10^5$$

$$\epsilon/d = 0.00045 \qquad \text{(from Prob. 9.91)}$$

From Fig. A-5, $f = 0.0182$. Try $f = 0.0182$:

$$v = 1.060/\sqrt{0.0182} = 7.857 \text{ ft/s} \qquad N_R = (\frac{4}{12})(7.857)/(1.05/10^{-5}) = 2.49 \times 10^5$$

$$f = 0.0182 \quad \text{(O.K.)} \qquad Q = Av = [(\pi)(\frac{4}{12})^2/4](7.857) = 0.6857 \text{ ft}^3/\text{s} = 0.6857/0.002228 = 308 \text{ gpm}$$

**9.93** A tank containing $1 \text{ m}^3$ of water at $20 °C$ has an outlet tube at the bottom, as shown in Fig. 9-24. Find the instantaneous volume flux $Q$, if the roughness of the tank is $\epsilon = 1.5 \, \mu\text{m}$.

**▌**
$$p_1/\gamma + v_1^2/2g + z_1 = p_2/\gamma + v_2^2/2g + z_2 + h_L$$

$$h_L = h_f = (f)(L/d)(v^2/2g) = (f)[0.82/0.04]\{v_2^2/[(2)(9.807)]\} = 1.045 f v_2^2$$

$$0 + 0 + (1 + 0.82) = 0 + v_2^2/[(2)(9.807)] + 0 + 1.045 f v_2^2 \qquad v_2 = 1.349/\sqrt{0.05098 + 1.045 f}$$

Try $f = 0.02$:

$$v_2 = 1.349/\sqrt{0.05098 + (1.045)(0.02)} = 5.032 \text{ m/s}$$

$$N_R = dv/\nu = (0.04)(5.032)/(1.02 \times 10^{-6}) = 1.97 \times 10^5 \qquad \epsilon/d = 0.0000015/0.04 = 0.0000375$$

From Fig. A-5, $f = 0.016$. Try $f = 0.016$:

$$v_2 = 1.349/\sqrt{0.05098 + (1.045)(0.016)} = 5.185 \text{ m/s} \qquad N_R = (0.04)(5.185)/(1.02 \times 10^{-6}) = 2.03 \times 10^5$$

$$f = 0.016 \quad \text{(O.K.)} \qquad Q = Av = [(\pi)(0.04)^2/4](5.185) = 0.006516 \text{ m}^3/\text{s} = 6.516 \text{ L/s}$$

$L = 82$ cm
$d = 4$ cm

**Fig. 9-24**

**9.94** Repeat Prob. 9.93 for a fluid with $\rho = 917 \text{ kg/m}^3$ and $\mu = 0.29 \text{ Pa} \cdot \text{s}$.

**▌** $p_1/\gamma + v_1^2/2g + z_1 = p_2/\gamma + v_2^2/2g + z_2 + h_L$. Assume laminar flow.

$$h_L = h_f = \frac{32\mu L v}{\rho g d^2} = \frac{(32)(0.29)(0.82)(v)}{(917)(9.807)(0.04)^2} = 0.5288 v_2$$

$$0 + 0 + (1 + 0.82) = 0 + v_2^2/[(2)(9.807)] + 0 + 0.5288 v_2 \qquad v_2^2 + 10.37 v_2 - 35.70 = 0 \qquad v = 2.726 \text{ m/s}$$

$$N_R = \rho dv/\mu = (917)(0.04)(2.726)/0.29 = 345 \qquad \text{(laminar)}$$

$$Q = Av = [(\pi)(0.04)^2/4](2.726) = 0.003426 \text{ m}^3/\text{s} = 3.426 \text{ L/s}$$

**9.95** What depth of water behind a dam (Fig. 9-25) will yield a flow rate of 0.02 ft³/s through the 0.5-in commercial steel exit pipe?

$$p_1/\gamma + v_1^2/2g + z_1 = p_2/\gamma + v_2^2/2g + z_2 + h_L \qquad v_2 = Q/A = 0.02/[(\pi)(0.5/12)^2/4] = 14.67 \text{ ft/s}$$

$$h_L = h_f = (f)(L/d)(v^2/2g) \qquad N_R = dv/\nu = (0.5/12)(14.67)/(1.05 \times 10^{-5}) = 5.82 \times 10^4$$

$$\epsilon/d = 0.00015/(0.5/12) = 0.00360$$

From Fig. A-5, $f = 0.0295$: $h_L = 0.0295[100/(0.5/12)]\{14.67^2/[(2)(32.2)]\} = 236.6$ ft, $0 + 0 + H = 0 + 14.67^2/[(2)(32.2)] + 0 + 236.6$, $H = 239.9$ ft.

Water
20 °C

$L$ = 100 ft
$d = \frac{1}{2}$ in

**Fig. 9-25**

**9.96** In Fig. 9-25, imagine the fluid to be benzene at 20 °C and $H = 100$ ft. What pipe diameter is required for the flow rate to be 0.02 ft³/s?

$$p_1/\gamma + v_1^2/2g + z_1 = p_2/\gamma + v_2^2/2g + z_2 + h_L$$

$$h_L = h_f = (f)(L/d)(v^2/2g) = (f)(100/d)\{v_2^2/[(2)(32.2)]\} = 1.553fv_2^2/d$$

$$0 + 0 + 100 = 0 + v_2^2/[(2)(32.2)] + 0 + 1.553fv_2^2/d \qquad v_2 = Q/A_2 = 0.02/(\pi d^2/4) = 0.02546/d^2$$

$$100 = (0.02546/d^2)^2/[(2)(32.2)] + (1.553)(f)(0.02546/d^2)^2/d \qquad d = [(1.007 \times 10^{-7}) + (1.007 \times 10^{-5}f/d)]^{1/4}$$

Try $d = 0.05$ ft: $N_R = \rho dv/\mu = (1.70)(0.05)(0.02546/0.05^2)/(1.36 \times 10^{-5}) = 6.36 \times 10^4$, $\epsilon/d = 0.00015/0.05 = 0.00300$. From Fig. A-5, $f = 0.028$. $d = [(1.007 \times 10^{-7}) + (1.007 \times 10^{-5})(0.028)/0.05]^{1/4} = 0.0489$ ft. Try $d = 0.00489$ ft:

$$N_R = (1.70)(0.0489)(0.02546/0.0489^2)/(1.36 \times 10^{-5}) = 6.51 \times 10^4 \qquad f = 0.028 \qquad \text{(O.K.)}$$

$$d = 0.0489 \text{ ft} \quad \text{or} \quad 0.59 \text{ in}$$

**9.97** Water at 70 °F flows through 102 ft of 0.5-in-diameter piping whose roughness is 0.01 in. Find the volumetric flow if the pressure drop is 60 psi.

$$p_1/\gamma + v_1^2/2g + z_1 = p_2/\gamma + v_2^2/2g + z_2 + h_L$$

$$h_L = h_f = (f)(L/d)(v^2/2g) = (f)[102/(0.5/12)]\{v^2/[(2)(32.2)]\} = 38.01fv^2$$

$$(60)(144)/62.3 + v_1^2/2g + 0 = 0 + v_2^2/2g + 0 + 38.01fv^2 \qquad v_1^2/2g = v_2^2/2g \qquad v_2 = 1.910/\sqrt{f}$$

Try $f = 0.05$:

$$v_2 = 1.910/\sqrt{0.05} = 8.542 \text{ ft/s} \qquad N_R = dv/\nu = (0.5/12)(8.542)/(1.05 \times 10^{-5}) = 3.39 \times 10^4$$

$$\epsilon/d = 0.01/0.5 = 0.0200$$

$$f = 0.050 \qquad \text{(O.K.)} \qquad Q = Av = [(\pi)(0.5/12)^2/4](8.542) = 0.0116 \text{ ft}^3/\text{s}$$

**9.98** Ethanol at 20 °C flows at 3.2 m/s through a 10-cm-diameter pipeline ($\epsilon = 1.5\ \mu$m). Compute (a) the head loss per kilometer of tube and (b) the wall shear stress.

**(a)** $$h_f = (f)(L/d)(v^2/2g) \qquad N_R = \rho dv/\mu = (788)(0.10)(3.2)/(1.20 \times 10^{-3}) = 2.10 \times 10^5$$

$$\epsilon/d = 0.0000015/0.10 = 0.0000150$$

From Fig. A-5, $f = 0.0158$. $h_f = 0.0158[1000/0.10]\{(3.2)^2/[(2)(9.807)]\} = 82.5$ m per 1000 m.

**(b)** $$u^* = v\sqrt{f/8} = (3.2)\sqrt{0.0158/8} = 0.1422 \text{ m/s} \qquad \tau_{\text{wall}} = (\rho)(u^*)^2 = (788)(0.1422)^2 = 15.9 \text{ Pa}$$

**9.99** In Fig. 9-26, the 50-m duct is 60 mm in diameter. Compute the flow rate if the fluid has $\rho = 917$ kg/m$^3$ and $\mu = 0.29$ Pa · s.

**I** Assume laminar flow from 1 to 2.

$$p_1/\rho g + v_1^2/2g + z_1 = p_2/\rho g + v_2^2/2g + z_2 + h_L$$

$$h_L = h_f = \frac{128\mu LQ}{\pi\rho g d^4} = \frac{(128)(0.29)(50)(Q)}{(\pi)(917)(9.807)(0.060)^4} = 5069Q$$

$$0 + 0 + 12 = (205)(1000)/[(917)(9.807)] + 0 + 0 + 5069Q \qquad Q = -0.002130 \text{ m}^3/\text{s} \quad \text{or} \quad -7.67 \text{ m}^3/\text{h}$$

Since $Q$ is negative, flow is from 2 to 1.

$$v = Q/A = 0.002130/[(\pi)(0.060)^2/4] = 0.7533 \text{ m/s}$$
$$N_R = \rho d v/\mu = (917)(0.060)(0.7533)/0.29 = 143 \qquad \text{(laminar)}$$

$L = 50$ m     **Fig. 9-26**

**9.100** Compute the diameter of duct in Fig. 9-26 required to maintain a flow rate of 25 m$^3$/h.

**I**

$$\frac{p_2}{\rho g} + \frac{v_2^2}{2g} + z_2 = \frac{p_1}{\rho g} + \frac{v_1^2}{2g} + z_1 + h_L \qquad h_L = h_f = \frac{128\mu LQ}{\pi\rho g d^4} = \frac{(128)(0.29)(50)(25/3600)}{(\pi)(917)(9.807)(d)^4} = \frac{4.5621 \times 10^{-4}}{d^4}$$

$$(205)(1000)/[(917)(9.807)] + 0 + 0 = 0 + 0 + 12 + 4.5621 \times 10^{-4}/d^4 \qquad d = 0.0806 \text{ m}$$

$$v = Q/A = (25/3600)/[(\pi)(0.0806)^2/4] = 1.361 \text{ m/s}$$

$$N_R = \rho d v/\mu = (917)(0.0806)(1.361)/0.29 = 347 \qquad \text{(laminar)}$$

**9.101** A 40-ft-long conduit, having an annular cross section with $a = 1$ in and $b = 0.5$ in, connects two reservoirs which differ in surface height by 21 ft. Compute the flow rate if the fluid is water at 20 °C and the conduit is made of commercial steel.

**I** Hydraulic diameter = $D_h = (2)(a - b) = (2)(1 - 0.5) = 1.00$ in    $p_1/\gamma + v_1^2/2g + z_1 = p_2/\gamma + v_2^2/2g + z_2 + h_L$

$$0 + 0 + 21 = 0 + 0 + 0 + h_L \qquad h_L = 21 \text{ ft} = h_f = (f)(L/d)(v^2/2g)$$

$$21 = (f)[40/(1.00/12)]\{v^2/[(2)(32.2)]\} \qquad v = 1.679/\sqrt{f}$$

Try $f = 0.02$:

$$v = 1.679/\sqrt{0.02} = 11.87 \text{ ft/s} \qquad N_R = dv/\nu = (1.00/12)(11.87)/(1.05 \times 10^{-5}) = 9.42 \times 10^4$$

$$\epsilon/d = 0.00015/(1.00/12) = 0.00180$$

From Fig. A-5, $f = 0.0246$. Try $f = 0.0246$:

$$v = 1.679/\sqrt{0.0246} = 10.70 \text{ ft/s} \qquad N_R = (1.00/12)(10.70)/(1.05 \times 10^{-5}) = 8.49 \times 10^4 \qquad f = 0.0250$$

Try $f = 0.0250$:

$$v = 1.679/\sqrt{0.0250} = 10.62 \text{ ft/s} \qquad N_R = (1.00/12)(10.62)/(1.05 \times 10^{-5}) = 8.43 \times 10^4$$

$$f = 0.0250 \qquad \text{(O.K.)} \qquad Q = Av = \{(\pi)[(1.00/12)^2 - (0.5/12)^2]\}(10.62) = 0.174 \text{ ft}^3/\text{s}$$

**9.102** An annular capillary causes a very large pressure drop and is useful in an accurate measurement of viscosity. If a smooth annulus 1 m long with $a = 50$ mm and $b = 49$ mm carries an oil flow at 0.0012 m³/s, what is the oil viscosity if the pressure drop is 260 kPa?

**I**
$$Q = (\pi/8\mu)(\Delta p/L)[a^4 - b^4 - (a^2 - b^2)^2/\ln(a/b)]$$

$$0.0012 = (\pi/8\mu)(260\,000/1)[0.050^4 - 0.049^4 - (0.050^2 - 0.049^2)^2/\ln(0.050/0.049)] \qquad \mu = 0.0738 \text{ Pa} \cdot \text{s}$$

**9.103** A metal ventilation duct ($\epsilon = 0.00015$ ft) carries air at 20 °C and 1 atm. The duct section is an equilateral triangle 1 ft on a side, and its length is 100 ft. If a blower can deliver 1 hp to the air, what flow rate can occur?

**I** From Fig. 9-27,

$$A = \tfrac{1}{2}(1)(\tfrac{1}{2}\sqrt{3}) = 0.4330 \text{ ft}^2$$

$$P = Q\gamma h_f \qquad Q = Av = 0.4330v \qquad h_f = (f)(L/d)(v^2/2g) \qquad d = D_h = 4A/p_w$$

$$D_h = (4)(0.4330)/[(12 + 12 + 12)/12] = 0.5773 \text{ ft} \qquad h_f = (f)(100/0.5773)\{v^2/[(2)(32.2)]\} = 2.69fv^2$$

$$(1)(550) = (0.4330v)(0.0750)(2.69fv^2) \qquad v = 18.47/f^{1/3}$$

Try $f = 0.02$:

$$v = 18.47/0.02^{1/3} = 68.04 \text{ ft/s} \qquad N_R = dv/v = (0.5773)(68.04)/(1.64 \times 10^{-4}) = 2.40 \times 10^5$$

$$\epsilon/D_h = 0.00015/0.5773 = 0.000260$$

From Fig. A-5, $f = 0.0176$. Try $f = 0.0176$:

$$v = 18.47/0.0176^{1/3} = 71.00 \text{ ft/s} \qquad N_R = (0.5773)(71.00)/(1.64 \times 10^{-4}) = 2.60 \times 10^5$$

$$f = 0.0176 \qquad \text{(O.K.)} \qquad Q = (0.4330)(71.00) = 30.7 \text{ ft}^3/\text{s}$$

**Fig. 9-27**

**9.104** A fluid with $\rho = 917$ kg/m³ and $\mu = 0.29$ Pa · s flows between two smooth parallel plates 5 cm apart ($h = 5$ cm) at an average velocity of 3 m/s. Compute the pressure drop, centerline velocity, and head loss for each 100 m of length.

**I**
$$d = D_h = 2h = (2)(0.05) = 0.1000 \text{ m} \qquad N_R = \rho dv/\mu = (917)(0.1000)(3)/0.29 = 949 \qquad \text{(laminar)}$$

$$v = (h^2/12\mu)(\Delta p/L) \qquad 3 = \{(0.05)^2/[(12)(0.29)]\}(\Delta p/100) \qquad \Delta p = 417\,600 \text{ Pa}$$

$$h_f = \Delta p/\rho g = 417\,600/[(917)(9.807)] = 46.44 \text{ m} \qquad u_{\mathcal{C}} = 1.5v = (1.5)(3) = 4.50 \text{ m/s}$$

**9.105** Air at STP is forced at 25 m/s through a 30-cm-square steel duct 148 m long. Compute the head loss, pressure drop, and power required if the blower efficiency is 60 percent. Use $\epsilon = 0.000046$ m.

**I**
$$d = D_h = 4A/p_w = (4)[(0.30)(0.30)]/[(4)(0.30)] = 0.3000 \text{ m}$$

$$h_f = (f)(L/d)(v^2/2g) = (f)(148/0.3000)\{25^2/[(2)(9.807)]\} = 15\,720f$$

$$N_R = \rho dv/\mu = (1.20)(0.3000)(25)/(1.81 \times 10^{-5}) = 4.97 \times 10^5 \qquad \epsilon/d = 0.000046/0.3000 = 0.000153$$

From Fig. A-5, $f = 0.0150$.

$$h_f = (15\,720)(0.0150) = 235.8 \text{ m} \qquad \Delta p = \rho g h_f = (1.20)(9.807)(235.8) = 2775 \text{ Pa}$$

$$Q = Av = [(0.30)(0.30)](25) = 2.25 \text{ m}^3/\text{s}$$

$$P = \rho g Q h_f/\eta = (1.20)(9.807)(2.25)(235.8)/0.60 = 10\,406 \text{ W} \qquad \text{or} \qquad 10.4 \text{ kW}$$

**9.106** A 50-m-long wind tunnel has a wooden ($\epsilon = 0.0001$ m) rectangular section 40 cm by 1 m. The average flow velocity is 45 m/s for air at STP. Compute the pressure drop, assuming fully developed conditions, and the power required if the fan has 70 percent efficiency.

$$d = D_h = 4A/p_w = (4)[(\tfrac{40}{100})(1)]/(\tfrac{40}{100} + \tfrac{40}{100} + 1 + 1) = 0.5714 \text{ m}$$

$$h_f = (f)(L/d)(v^2/2g) = (f)(50/0.5714)\{45^2/[(2)(9.807)]\} = 9034f$$

$$N_R = \rho dv/\mu = (1.20)(0.5714)(45)/(1.81 \times 10^{-5}) = 1.70 \times 10^6 \qquad \epsilon/d = 0.0001/0.5714 = 0.000175$$

From Fig. A-5, $f = 0.0140$.

$$h_f = (9034)(0.0140) = 126.5 \text{ m} \qquad \Delta p = \rho g h_f = (1.20)(9.807)(126.5) = 1489 \text{ Pa}$$

$$Q = Av = [(\tfrac{40}{100})(1)](45) = 18.00 \text{ m}^3/\text{s}$$

$$P = \rho g Q h_f/\eta = (1.20)(9.807)(18.00)(126.5)/0.70 = 38\,281 \text{ W} \quad \text{or} \quad 38.3 \text{ kW}$$

**9.107** A smooth rectangular duct 82 m long, of aspect ratio 6 : 1, is designed to transport 0.6 m³/s of hydrogen at STP. If the pressure drop is 75 Pa, calculate the dimensions of the cross section. For hydrogen, $\rho = 0.0838$ kg/m³ and $\mu = 9.05 \ \mu\text{Pa} \cdot \text{s}$.

Let $h$ = height of duct; then duct width = $6h$.

$$d = D_h = 4A/p_w = (4)[(6h)(h)]/(6h + 6h + h + h) = 1.714h \qquad v = Q/A = 0.6/[(6h)(h)] = 0.1000/h^2$$

$$h_f = (f)(L/d)(v^2/2g) = (f)(82/1.714h)\{(0.1000/h^2)^2/[(2)(9.807)]\} = 0.02439f/h^5$$

$$\Delta p = \rho g h_f \qquad 75 = (0.0838)(9.807)(0.02439f/h^5) \qquad h = 0.1929f^{1/5}$$

Try $f = 0.02$: $h = (0.1929)(0.02)^{1/5} = 0.08821$ m, $N_R = \rho dv/\mu =$
$0.0838[(1.714)(0.08821)](0.1000/0.08821^2)/(9.05 \times 10^{-6}) = 1.80 \times 10^4$. From Fig. A-5, $f = 0.0265$. Try $f = 0.0265$: $h = (0.1929)(0.0265)^{1/5} = 0.09332$ m, $N_R = 0.0838[(1.714)(0.09332)](0.1000/0.09332^2)/(9.05 \times 10^{-6}) = 1.70 \times 10^4$, $f = 0.0267$. Try $f = 0.0267$: $h = (0.1929)(0.0267)^{1/5} = 0.09346$ m, $N_R = 0.0838[(1.714)(0.09346)](0.1000/0.09346^2)/(9.05 \times 10^{-6}) = 1.70 \times 10^4$.

$$f = 0.0267 \qquad \text{(O.K.)}$$

Hence, $h = 9.35$ cm and width = $6h = 56.1$ cm.

**9.108** Kerosene at 10 °C flows steadily at 15 L/min through a 150-m-long horizontal length of 5.5-cm-diameter cast iron pipe. Compare the pressure drop of the kerosene flow with that of the same flow rate of benzene at 10 °C through the same pipe. At 10 °C, $\rho = 820$ kg/m³ and $\mu = 0.0025$ N · s/m² for kerosene, and $\rho = 899$ kg/m³ and $\mu = 0.0008$ N · s/m² for benzene.

For kerosene:

$$v = Q/A = [(\tfrac{15}{1000})/60]/[(\pi)(5.5/100)^2/4] = 0.1052 \text{ m/s}$$

$$N_R = \rho dv/\mu = (820)(5.5/100)(0.1052)/0.0025 = 1898 \qquad \text{(laminar)}$$

$$\Delta p = 8\mu v L/r_0^2 = (8)(0.0025)(0.1052)(150)/[(5.5/100)/2]^2 = 417 \text{ Pa}$$

For benzene:

$$N_R = (899)(5.5/100)(0.1052)/0.0008 = 6502 \qquad \text{(turbulent)}$$

$$\Delta p = (f)(L/d)(\rho)(v^2/2) \qquad \epsilon/d = 0.00026/(5.5/100) = 0.00473$$

From Fig. A-5, $f = 0.040$: $\Delta p = (0.040)[150/(5.5/100)](899)(0.1052^2/2) = 543$ Pa. The pressure drop is greater for the benzene than the kerosene, although the benzene has a lower viscosity. If both flows had been laminar or both turbulent, the lower-viscosity fluid, benzene, would have had the lower pressure drop. However, the viscosity of the kerosene is high enough to give laminar flow, while the lower viscosity of the benzene causes a high enough Reynolds number for turbulent flow.

**9.109** Water at 60 °F flows through a 250-ft length of horizontal 2-in-diameter drawn tubing. If the pressure drop across the tubing is 10 psi, what is the flow rate?

Assume turbulent flow. $h_f = (f)(L/d)(v^2/2g)$, $h_f = (10)(144)/62.4 = 23.08$ ft, $23.08 = (f)[250/(\tfrac{2}{12})]\{v^2/[(2)(32.2)]\}$, $v = 0.9954/\sqrt{f}$. Try $v = 5$ ft/s: $N_R = \rho dv/\mu = (1.94)(\tfrac{2}{12})(5)/(2.35 \times 10^{-5}) = 6.88 \times 10^4$ (turbulent). $\epsilon/d = 0.000005/(\tfrac{2}{12}) = 0.000030$. From Fig. A-5, $f = 0.0195$. $v = 0.9954/\sqrt{0.0195} = 7.13$ ft/s.

Try $v = 7.13$ ft/s: $N_R = (1.94)(\frac{2}{12})(7.13)/(2.35 \times 10^{-5}) = 9.81 \times 10^4$, $f = 0.0180$, $v = 0.9954/\sqrt{0.0180} = 7.42$ ft/s.
Try $v = 7.42$ ft/s: $N_R = (1.94)(\frac{2}{12})(7.42)/(2.35 \times 10^{-5}) = 1.02 \times 10^5$, $f = 0.0180$ (O.K.); $Q = Av =$
$[(\pi)(\frac{2}{12})^2/4](7.42) = 0.162$ ft$^2$/s.

**9.110** Air at 200 °F and 15 psig is to be passed through a 150-ft length of new galvanized iron pipe at a rate of 15 ft$^3$/s. If the maximum allowable pressure drop is 5 psi, estimate the minimum pipe diameter.

$\blacksquare$      $h_f = (f)(L/d)(v^2/2g) = p/\gamma$      $\gamma = p/RT = (15 + 14.7)(144)/[(53.3)(200 + 460)] = 0.1216$ lb/ft$^3$

$$h_f = (5)(144)/0.1216 = 5921 \text{ ft} \qquad v = Q/A = 15/(\pi d^2/4) = 19.10/d^2$$

$$5921 = (f)(150/d)\{(19.10/d^2)^2/[(2)(32.2)]\} \qquad d = 0.6782 f^{1/5}$$

Try $d = 0.5$ ft:

$$N_R = \rho dv/\mu = (\gamma/g)(dv)/\mu = (0.1216/32.2)(0.5)(19.10/0.5^2)/(4.49 \times 10^{-7}) = 3.21 \times 10^5$$

$$\epsilon/d = 0.0005/0.5 = 0.0010$$

From Fig. A-5, $f = 0.0205$; $d = (0.6782)(0.0205)^{1/5} = 0.3117$ ft.

Try $d = 0.3117$ ft:

$$N_R = (0.1216/32.2)(0.3117)(19.10/0.3117^2)/(4.49 \times 10^{-7}) = 5.15 \times 10^5 \qquad \epsilon/d = 0.0005/0.3117 = 0.0016$$

$$f = 0.0225 \qquad d = (0.6782)(0.0225)^{1/5} = 0.3175 \text{ ft}$$

Try $d = 0.3175$ ft:

$$N_R = (0.1216/32.2)(0.3175)(19.10/0.3175^2)/(4.49 \times 10^{-7}) = 5.06 \times 10^5 \qquad \epsilon/d = 0.0005/0.3175 = 0.0016$$

$$f = 0.0225 \qquad \text{(O.K.)}$$

Hence, $d = 0.3175$ ft, or 3.81 in.

**9.111** Compute the loss of head and pressure drop in 200 ft of horizontal 6-in-diameter asphalted cast iron pipe carrying water with a mean velocity of 6 ft/s.

$\blacksquare$   $h_f = (f)(L/d)(v^2/2g)$      $N_R = dv/v = (\frac{6}{12})(6)/(1.05 \times 10^{-5}) = 2.86 \times 10^5$      $\epsilon/d = 0.0004/(\frac{6}{12}) = 0.00080$

From Fig. A-5, $f = 0.020$:

$$h_f = (0.020)[200/(\tfrac{6}{12})]\{6^2/[(2)(32.2)]\} = 4.47 \text{ ft} \qquad \Delta p = \gamma h_f = (62.4)(4.47) = 279 \text{ lb/ft}^2$$

**9.112** Oil with $\rho = 900$ kg/m$^3$ and $v = 0.00001$ m$^2$/s flows at 0.2 m$^3$/s through 500 m of 200-mm-diameter cast iron pipe. Determine (a) the head loss and (b) the pressure drop if the pipe slopes down at 10° in the flow direction.

$\blacksquare$ **(a)**      $h_f = (f)(L/d)(v^2/2g)$      $v = Q/A = 0.2/[(\pi)(\tfrac{200}{1000})^2/4] = 6.366$ m/s

$$N_R = dv/v = (\tfrac{200}{1000})(6.366)/0.00001 = 1.27 \times 10^5 \qquad \epsilon/d = 0.00026/(\tfrac{200}{1000}) = 0.00130$$

From Fig. A-5, $f = 0.0225$, $h_f = 0.0225[500/(\tfrac{200}{1000})]\{6.366^2/[(2)(9.807)]\} = 116.2$ m.

     **(b)**      $h_f = \Delta p/(\rho g) + L \sin 10°$      $116.2 = \Delta p/[(900)(9.807)] + (500)(\sin 10°)$

$$\Delta p = 259\,300 \text{ Pa} \quad \text{or} \quad 259.3 \text{ kPa}$$

**9.113** Oil with $\rho = 950$ kg/m$^3$ and $v = 0.00002$ m$^2$/s flows through a 30-cm-diameter pipe 100 m long with a head loss of 8 m. The roughness ratio $\epsilon/d$ is 0.0002. Find the average velocity and flow rate.

$\blacksquare$      $h_f = (f)(L/d)(v^2/2g)$      $8 = (f)[100/(\tfrac{30}{100})]\{v^2/[(2)(9.807)]\}$      $v = 0.6861/\sqrt{f}$

Try $v = 5$ m/s: $N_R = dv/v = (\tfrac{30}{100})(5)/0.00002 = 7.50 \times 10^4$. From Fig. A-5, $f = 0.020$, $v = 0.6861/\sqrt{0.020} = 4.851$ m/s. Try $v = 4.851$ m/s: $N_R = (\tfrac{30}{100})(4.851)/0.00002 = 7.28 \times 10^4$, $f = 0.020$ (O.K.). $Q = Av = [(\pi)(\tfrac{30}{100})^2/4](4.851) = 0.343$ m$^3$/s.

**9.114** Fluid flows at an average velocity of 6 ft/s between horizontal parallel plates a distance of 2.4 in apart ($h = 2.4$ in). Estimate the head loss and pressure drop for each 100 ft of length for $\rho = 1.9$ slugs/ft$^3$ and $v = 0.00002$ ft$^2$/s. Assume smooth walls.

$\blacksquare$      $h_f = (f)(L/d)(v^2/2g)$      $d = D_h = 2h$      $D_h = (2)(2.4/12) = 0.400$ ft

$$N_R = dv/v = (0.400)(6)/0.00002 = 1.20 \times 10^5 \qquad \text{(turbulent)}$$

From Fig. A-5, $f = 0.0173$.

$$h_f = (0.0173)(100/0.400)\{6^2/[(2)(32.2)]\} = 2.42 \text{ ft} \qquad \Delta p = \rho g h_f = (1.9)(32.2)(2.42) = 148 \text{ lb/ft}^2$$

**9.115** Repeat Prob. 9.114 if $\nu = 0.002 \text{ ft}^2/\text{s}$.

▮   $$N_R = dv/\nu = (0.400)(6)/0.002 = 1200 \qquad \text{(laminar)} \qquad f = 96/N_R = \tfrac{96}{1200} = 0.0800$$

$$h_f = (f)(L/d)(v^2/2g) = (0.0800)(100/0.400)\{6^2/[(2)(32.2)]\} = 11.18 \text{ ft}$$

$$\Delta p = \rho g h_f = (1.9)(32.2)(11.18) = 684 \text{ lb/ft}^2$$

**9.116** Estimate the reservoir level $h$ needed to maintain a flow of 0.01 m³/s through the commercial steel annulus 30 m long shown in Fig. 9-28. Neglect entrance effects and take $\rho = 1000 \text{ kg/m}^3$ and $\nu = 1.02 \times 10^{-6} \text{ m}^2/\text{s}$ for water.

▮   $$p_1/\gamma + v_1^2/2g + z_1 = p_2/\gamma + v_2^2/2g + z_2 + h_f \qquad v = Q/A = 0.01/\{(\pi)[(\tfrac{5}{100})^2 - (\tfrac{3}{100})^2]\} = 1.989 \text{ m/s}$$

$$h_f = (f)(L/d)(v^2/2g) \qquad d = D_h = (2)(a - b) \qquad D_h = (2)(\tfrac{5}{100} - \tfrac{3}{100}) = 0.0400 \text{ m}$$

$$h_f = (f)(30/0.0400)\{1.989^2/[(2)(9.807)]\} = 151.3f \qquad 0 + 0 + h = 0 + 1.989^2/[(2)(9.807)] + 0 + 151.3f$$

$$h = 0.2017 + 151.3f \qquad N_R = dv/\nu = (0.0400)(1.989)/(1.02 \times 10^{-6}) = 7.80 \times 10^4$$

$$\epsilon/d = 0.000046/0.0400 = 0.00115$$

From Fig. A-5, $f = 0.0232$, $h = 0.2017 + (151.3)(0.0232) = 3.71 \text{ m}$.

**Fig. 9-28**

**9.117** Air with $\rho = 0.00237 \text{ slug/ft}^3$ and $\nu = 0.000157 \text{ ft}^2/\text{s}$ is forced through a horizontal square 9-in by 9-in duct 100 ft long at 25 ft³/s. Find the presssure drop if $\epsilon = 0.0003 \text{ ft}$.

▮   $$h_f = (f)(L/d)(v^2/2g) \qquad N_R = dv/\nu \qquad D_h = 4A/p_w = 4[(\tfrac{9}{12})(\tfrac{9}{12})]/[(4)(\tfrac{9}{12})] = 0.7500 \text{ ft}$$

From Table A-10, for $b/a = \tfrac{9}{9}$, or 1.0, the effective diameter is $d = D_{\text{eff}} = (64/56.91)(D_h) = (64/56.91)(0.7500) = 0.8434 \text{ ft}$, $v = Q/A = 25/[(\tfrac{9}{12})(\tfrac{9}{12})] = 44.44 \text{ ft/s}$, $N_R = (0.8434)(44.44)/0.000157 = 2.39 \times 10^5$, $\epsilon/d = 0.0003/0.8434 = 0.000354$. From Fig. A-5, $f = 0.0177$; $h_f = (0.0177)(100/0.7500)\{44.44^2/[(2)(32.2)]\} = 72.37 \text{ ft}$, $\Delta p = \rho g h_f = (0.00237)(32.2)(72.37) = 5.52 \text{ lb/ft}^2$.

**9.118** Find the head loss in 1 ft of 6-in-diameter pipe ($\epsilon = 0.042 \text{ in}$) when turpentine (s.g. $= 0.86$) having a viscosity of $0.0008 \text{ lb} \cdot \text{s/ft}^2$ flows at a rate of 16 cfs. Also find the shear stress at the wall of the pipe.

▮   $$h_f = (f)(L/d)(v^2/2g) \qquad v = Q/A = 16/[(\pi)(\tfrac{6}{12})^2/4] = 81.49 \text{ ft/s}$$

$$N_R = \rho dv/\mu = [(0.86)(1.94)](\tfrac{6}{12})(81.49)/0.0008 = 8.50 \times 10^4 \qquad \epsilon/d = 0.042/6 = 0.00700$$

From Fig. A-5, $f = 0.0324$.

$$h_f = 0.0324[1/(\tfrac{6}{12})]\{81.49^2/[(2)(32.2)]\} = 6.68 \text{ ft}$$

$$\tau_0 = f\rho v^2/8 = 0.0324[(0.86)(1.94)](81.49)^2/8 = 44.9 \text{ lb/ft}^2$$

**9.119** When water at 150 °F flows in a 0.75-in-diameter copper tube at 1.0 gpm, find the head loss per 1000 ft. Also find the centerline velocity.

▮ $\quad Q = (1.0)(0.002228) = 0.002228 \text{ ft}^3/\text{s} \qquad v = Q/A = 0.002228/[(\pi)(0.75/12)^2/4] = 0.7262 \text{ ft/s}$

$$h_f = (f)(L/d)(v^2/2g) \qquad N_R = dv/\nu = (0.75/12)(0.7262)/(4.68 \times 10^{-6}) = 9.70 \times 10^3$$

$$\epsilon/d = 0.000005/(0.75/12) = 0.0000800$$

From Fig. A-5, $f = 0.030$.

$$h_f = (0.030)[1000/(0.75/12)]\{0.7262^2/[(2)(32.2)]\} = 3.93 \text{ ft}$$

$$u_{\max} = (v)(1 + 1.33f^{1/2}) = 0.7262[1 + (1.33)(0.030)^{1/2}] = 0.893 \text{ ft/s}$$

$$\delta_1 = 32.8\nu/(vf^{1/2}) = (32.8)(4.68 \times 10^{-6})/[(0.7262)(0.030)^{1/2}] = 0.00122 \text{ ft}$$

**9.120** Refer to Prob. 9.119. How thick is the viscous boundary layer for a flow of 20 gpm?

▮ The Reynolds number is now $(20/1.0)(9.70 \times 10^3) = 1.94 \times 10^5$. From Fig. A-5, $f = 0.016$. Then,

$$\delta_1 = 32.8\nu/(vf^{1/2}) = (32.8)(4.68 \times 10^{-6})/[(20)(0.7262)(0.016)^{1/2}] = 8.36 \times 10^{-5} \text{ ft}$$

**9.121** Find the maximum velocity in a 100-mm-diameter pipe ($\epsilon = 0.00085$ m) when oil (s.g. $= 0.84$), of viscosity 0.0052 Pa · s flows at a rate of 40 L/s.

▮ $$v = Q/A = (40 \times 10^{-3})/[(\pi)(0.100)^2/4] = 5.093 \text{ m/s}$$

$$N_R = \rho dv/\mu = [(0.84)(1000)](0.100)(5.093)/0.0052 = 8.23 \times 10^4 \qquad \epsilon/d = 0.00085/(0.100) = 0.0085$$

From Fig. A-5, $f = 0.0365$. Hence,

$$u_{\max} = (v)(1 + 1.33f^{1/2}) = 5.093[1 + (1.33)(0.0365)^{1/2}] = 6.387 \text{ m/s}$$

**9.122** The head loss in 240 ft of 6-in-diameter pipe is known to be 30 ft when oil (s.g. $= 0.90$) flows at 2.0 ft$^3$/s. Determine the shear stress at the wall of the pipe.

▮ $$v = Q/A = 2.0/[(\pi)(\tfrac{6}{12})^2/4] = 10.19 \text{ ft/s}$$

$$N_R = \rho dv/\mu = [(0.90)(1.94)](\tfrac{6}{12})(10.19)/0.0008 = 1.11 \times 10^4 \qquad \text{(turbulent)}$$

$$h_f = (f)(L/d)(v^2/2g) \qquad 30 = (f)[240/(\tfrac{6}{12})]\{10.19^2/(2)(32.2)]\} \qquad f = 0.0388$$

$$\tau_0 = f\rho v^2/8 = 0.0388[(0.90)(1.94)](10.19)^2/8 = 0.879 \text{ lb/ft}^2$$

**9.123** If, in a 1-m-diameter pipe, velocities are 5.03 m/s on the centerline and 4.85 m/s at $r = 100$ mm, what is the flow rate?

▮ $$u = u_{\max} - (5.75)(\tau_0/\rho)^{1/2} \log [r_0/(r_0 - r)]$$

$$4.85 = 5.03 - (5.75)(\tau_0/\rho)^{1/2} \log [500/(500 - 100)] \qquad (\tau_0/\rho)^{1/2} = 0.3230$$

$$v = u_{\max} - (\tfrac{3}{2})(2.5\sqrt{\tau_0/\rho}) = 5.03 - (\tfrac{3}{2})(2.5)(0.3230) = 3.819 \text{ m/s} \qquad Q = Av = [(\pi)(1)^2/4](3.819) = 3.00 \text{ m}^3/\text{s}$$

**9.124** The velocities in a 30-in-diameter pipe are measured as 15.0 and 14.5 ft/s at $r = 0$ and $r = 4$ in, respectively. Approximate the flow rate.

$$u = u_{max} - (5.75)(\tau_0/\rho)^{1/2} \log [r_0/(r_0 - r)] \qquad 14.5 = 15.0 - (5.75)(\tau_0/\rho)^{1/2} \log [15/(15 - 4)]$$

$$(\tau_0/\rho)^{1/2} = 0.6456 \qquad \tau_0/\rho = fv^2/8 \qquad (\tau_0/\rho)^{1/2} = (fv^2/8)^{1/2} = 0.6456 \qquad f = 3.3344/v^2$$

$$u_{max} = (v)(1 + 1.33f^{1/2}) \qquad 15.0 = (v)(1 + 1.33f^{1/2}) \qquad 15.0 = (v)[1 + (1.33)(3.3344/v^2)^{1/2}]$$

$$15.0 = v + 2.4286 \qquad v = 12.57 \text{ ft/s} \qquad Q = Av = [(\pi)(\tfrac{30}{12})^2/4](12.57) = 61.7 \text{ ft}^3/\text{s}$$

**9.125** With turbulent flow in a circular pipe, prove that the mean velocity occurs at a distance of approximately $0.78r_0$ from the centerline of the pipe.

$$u = (1 + 1.33f^{1/2})(v) - (2.04)(f^{1/2})(v) \log [r_0/(r_0 - r)] \qquad 0 = 1.33f^{1/2}v - (2.04f^{1/2}v)\{\log [r_0/(r_0 - r)]\}$$

$$\log [r_0/(r_0 - r)] = 0.65196 \qquad r_0/(r_0 - r) = \text{antilog } 0.65196 = 4.487 \qquad r = 0.776r_0$$

**9.126** The flow rate in a 12-in-diameter pipe is 8 cfs. The flow is known to be turbulent, and the centerline velocity is 12.0 fps. Determine the velocity profile, and determine the head loss per foot of pipe.

$$v = Q/A = 8/[(\pi)(\tfrac{12}{12})^2/4] = 10.19 \text{ ft/s} \qquad u_{max} = (v)(1 + 1.33f^{1/2}) \qquad 12.0 = (10.19)(1 + 1.33f^{1/2})$$

$$f = 0.01784 \qquad h_f = (f)(L/d)(v^2/2g) = 0.01784[1/(\tfrac{12}{12})]\{10.19^2/[(2)(32.2)]\} = 0.0288 \text{ ft per foot}$$

$$\tau_0 = (f/4)(\rho)(v^2/2) = (0.01784/4)(\rho)(10.19^2/2) \qquad (\tau_0/\rho)^{1/2} = 0.4812$$

$$u = u_{max} - 5.75(\tau_0/\rho)^{1/2} \log [r_0/(r_0 - r)]$$

For $r = 0$, $u = 12.0 - (5.75)(0.4812) \log [6/(6 - 0)] = 12.0$ ft/s. For $r = 2$, $u = 12.0 - (5.75)(0.4812) \log [6/(6 - 2)] = 11.5$ ft/s. For $r = 4$, $u = 12.0 - (5.75)(0.4812) \log [6/(6 - 4)] = 10.7$ ft/s. For $r = 5$, $u = 12.0 - (5.75)(0.4812) \log [6/(6 - 5)] = 9.85$ ft/s. For $r = 5.5$, $u = 12.0 - (5.75)(0.4812) \log [6/(6 - 5.5)] = 9.01$ ft/s. For $r = 5.9$, $u = 12.0 - (5.75)(0.4812) \log [6/(6 - 5.9)] = 7.08$ ft/s. For $r = 5.99$, $u = 12.0 - (5.75)(0.4812) \log [6/(6 - 5.99)] = 4.31$ ft/s.

**9.127** Tung oil (s.g. = 0.82) flows at a temperature of 80 °F ($v = 2.21 \times 10^{-5} \text{ ft}^2/\text{s}$) in a 2-in-diameter smooth brass pipeline at 10 gpm. Find the head loss per mile.

$$Q = (10)(0.002228) = 0.02228 \text{ ft}^3/\text{s} \qquad v = Q/A = 0.02228/[(\pi)(\tfrac{2}{12})^2/4] = 1.021 \text{ ft/s}$$

$$h_f = (f)(L/d)(v^2/2g) \qquad N_R = dv/v = (\tfrac{2}{12})(1.021)/(2.21 \times 10^{-5}) = 7.70 \times 10^3$$

From Fig. A-5 for smooth pipe, $f = 0.0333$. $h_f = 0.0333[5280/(\tfrac{2}{12})]\{1.021^2/[(2)(32.2)]\} = 17.1$ ft per mile.

**9.128** Water at 40 °C flows in a 20-cm-diameter pipe with $v = 5.1$ m/s. Head loss measurements give $f = 0.022$. Determine the value of $\epsilon$ and find the shear stress at the wall of the pipe and at $r = 3$ cm.

$N_R = dv/v = (0.20)(5.1)/(6.56 \times 10^{-7})$. From Fig. A-5, $\epsilon/d = 0.0015$, $\epsilon = (20)(0.0015) = 0.0300$ cm; $\tau_0 = f\rho v^2/8 = (0.022)(992)(5.1)^2/8 = 71.0$ Pa; $\tau = (\tau_0)(r/r_0) = (71.0)(0.03) = 21.3$ Pa.

**9.129** Water at 15 °C flows through a 200-mm-diameter pipe with $\epsilon = 0.01$ mm. (a) If the mean velocity is 3.6 m/s, what is the nominal thickness $\delta_1$ of the viscous boundary layer? (b) What will be the boundary layer thickness if the velocity is increased to 6.0 m/s?

(a)
$$\delta_1 = 32.8v/(vf^{1/2}) \qquad N_R = dv/v = (0.200)(3.6)/(1.16 \times 10^{-6}) = 6.21 \times 10^5$$

$$\epsilon/d = (0.01/10)/20 = 0.0000500$$

From Fig. A-5, $f = 0.0133$. $\delta_1 = (32.8)(1.16 \times 10^{-6})/[(3.6)(0.0133)^{1/2}] = 9.16 \times 10^{-5}$ m, or 91.6 $\mu$m.

(b)
$$N_R = (0.200)(6.0)/(1.16 \times 10^{-6}) = 1.03 \times 10^6 \qquad f = 0.0126$$

$$\delta_1 = (32.8)(1.16 \times 10^{-6})/[(6.0)(0.0126)^{1/2}] = 5.65 \times 10^{-5} \text{ m} \quad \text{or} \quad 56.5 \ \mu\text{m}$$

**9.130** When water at 50 °F flows at 3.2 cfs in a 2-ft pipeline, the head loss is 0.0004 ft per foot. What will be the head loss when glycerin at 68 °F flows through the same pipe at the same rate?

$\blacksquare$ $\qquad h_f = (f)(L/d)(v^2/2g) \qquad v = Q/A = 3.2/[(\pi)(\tfrac{24}{12})^2/4] = 1.0186 \text{ ft/s}$

**For water:**

$$0.0004 = (f)[1/(2)]\{1.0186^2/[(2)(32.2)]\} \qquad f = 0.04966$$

$$N_R = \rho dv/\mu = (1.94)(2)(1.0186)/(2.72 \times 10^{-5}) = 1.45 \times 10^5 \qquad \text{(turbulent)}$$

From Fig. A-5, $\epsilon/d = 0.021$.

**For glycerin:**

$$N_R = (2.44)(2)(1.0186)/(3.11 \times 10^{-2}) = 160 \qquad \text{(laminar)}$$

$$f = 64/N_R = 64/160 = 0.4000 \qquad h_f = 0.4000[1/(2)]\{1.0186^2/[(2)(32.2)]\} = 0.00322 \text{ ft per ft}$$

**9.131** Air flows at 50 lb/min in a 4-in-diameter welded steel pipe ($\epsilon = 0.0018$ in) at 100 psia and 60 °F. Determine the head loss and pressure drop in 200 ft of this pipe. Assume the air to be of constant density.

$\blacksquare$ $\quad h_f = (f)(L/d)(v^2/2g) \qquad N_R = \rho dv/\mu \qquad \rho = p/RT = (100)(144)/[(1716)(460 + 60)] = 0.01614 \text{ slug/ft}^3$

$$W = \gamma Av = \rho g Av \qquad \tfrac{50}{60} = (0.01614)(32.2)[(\pi)(\tfrac{4}{12})^2/4](v) \qquad v = 18.37 \text{ ft/s}$$

$$N_R = (0.01614)(\tfrac{4}{12})(18.37)/(3.74 \times 10^{-7}) = 2.64 \times 10^5 \qquad \epsilon/d = 0.0018/4 = 0.00045$$

From Fig. A-5, $f = 0.018$. $h_f = 0.018[200/(\tfrac{4}{12})]\{18.37^2/[(2)(32.2)]\} = 56.6$ ft of air, $\Delta p = \rho g h_f = (0.01614)(32.2)(56.6) = 29.4 \text{ lb/ft}^2$, or $0.204 \text{ lb/in}^2$.

**9.132** Air flows at an average velocity of 0.5 m/s through a long, 3.2-m-diameter, circular tunnel. Find the head-loss gradient at a point where the air temperature and pressure are 16 °C and 109 kPa abs, respectively. Assume $\epsilon = 2$ mm. Find also the shear stress at the wall and the thickness of the viscous sublayer.

$\blacksquare$ $\qquad h_f/L = (f/d)(v^2/2g) \qquad \rho = p/RT = (109)(1000)/[(287)(273 + 16)] = 1.314 \text{ kg/m}^3$

$$N_R = \rho dv/\mu = (1.314)(3.2)(0.5)/(1.79 \times 10^{-5}) = 1.17 \times 10^5 \qquad \epsilon/d = (0.002)/3.2 = 0.000625$$

From Fig. A-5, $f = 0.021$.

$$h_f/L = (0.021/3.2)\{0.5^2/[(2)(9.807)]\} = 8.36 \times 10^{-5} \text{ m/m} \quad \text{or} \quad 83.6 \text{ } \mu\text{m/m}$$

$$\tau_0 = f\rho v^2/8 = (0.021)(1.314)(0.5)^2/8 = 0.862 \text{ mPa}$$

$$\delta_1 = 32.8\nu/(vf^{1/2}) = (32.8)(\mu/\rho)/(vf^{1/2}) = (32.8)(1.79 \times 10^{-5}/1.314)/[(0.5)(0.021)^{1/2}] = 0.00620 \text{ m} \quad \text{or} \quad 6.20 \text{ mm}$$

**9.133** Repeat Prob. 9.132 for the average velocity 5.0 m/s.

$\blacksquare$ From Fig. A-5, for $N_R = (5.0/0.5)(1.17 \times 10^5) = 1.17 \times 10^6$ and $\epsilon/d = 0.000625$, $f = 0.018$.

$$h_f/L = (f/d)(v^2/2g) = (0.018/3.2)\{5.0^2/[(2)(9.807)]\} = 7.17 \times 10^{-3} \text{ m/m}$$

$$\tau_0 = f\rho v^2/8 = (0.018)(1.314)(5.0)^2/8 = 73.9 \text{ mPa}$$

$$\delta_1 = 32.8\nu/(vf^{1/2}) = (32.8)(\mu/\rho)/(vf^{1/2})$$

$$= (32.8)(1.79 \times 10^{-5}/1.314)/[(5.0)(0.018)^{1/2}] = 6.66 \times 10^{-4} \text{ m} \quad \text{or} \quad 0.660 \text{ mm}$$

**9.134** Air at 20 °C and atmospheric pressure flows with a velocity of 6 m/s through a 50-mm-diameter pipe. Find the head loss per meter of pipe if $\epsilon = 0.0025$ mm.

$\blacksquare$ $\qquad h_f/L = (f)(1/d)(v^2/2g) \qquad N_R = dv/\nu = (0.050)(6)/(1.51 \times 10^{-5}) = 1.99 \times 10^4$

$$\epsilon/d = 0.0025/50 = 0.000050$$

From Fig. A-5, $f = 0.026$. $h_f/L = 0.026[1/0.050]\{6^2/[(2)(9.807)]\} = 0.954 \text{ m/m}$.

**9.135** What is the head loss per foot of pipe when oil (s.g. = 0.88) having a viscosity of $1.9 \times 10^{-4}$ lb · s/ft² flows in a 2-in-diameter welded steel pipe at 0.15 cfs?

**I** $$h_f/L = (f/d)(v^2/2g) \qquad v = Q/A = 0.15/[(\pi)(\tfrac{2}{12})^2/4] = 6.875 \text{ ft/s}$$

$$N_R = \rho dv/\mu = [(0.88)(1.94)](\tfrac{2}{12})(6.875)/(1.9 \times 10^{-4}) = 1.03 \times 10^4 \qquad \epsilon/d = 0.00015/(\tfrac{2}{12}) = 0.00090$$

From Fig. A-5, $f = 0.033$. $h_f/L = [0.033/(\tfrac{2}{12})]\{6.875^2/[(2)(32.2)]\} = 0.145$ ft/ft.

**9.136** Water at 50 °F flows in a 36-in-diameter concrete pipe ($\epsilon = 0.02$ in). For a flow rate of 202 cfs, determine $N_R$ and $\tau_0$.

**I** $$v = Q/A = 202/[(\pi)(\tfrac{36}{12})^2/4] = 28.58 \text{ ft/s} \qquad N_R = dv/v = (\tfrac{36}{12})(28.58)/(1.40 \times 10^{-5}) = 6.12 \times 10^6$$

$$\tau_0 = f\rho v^2/8 \qquad \epsilon/d = 0.02/36 = 0.000556$$

From Fig. A-5, $f = 0.0175$.

$$\tau_0 = (0.0175)(1.94)(28.58)^2/8 = 3.47 \text{ lb/ft}^2$$

**9.137** What is the flow regime in Prob. 9.136?

**I** $$\delta_1 = 32.8v/(vf^{1/2}) = (32.8)(1.40 \times 10^{-5})/[(28.58)(0.0175)^{1/2}] = 0.000121 \text{ ft}$$

Since $[\delta_1 = 0.000121] < [0.3\epsilon = (0.3)(0.02/12) = 0.000500]$, regime is "rough."

**9.138** Find the flow rate if water at 60 °F experiences a head loss of $\frac{1}{408}$ ft/ft in 6-in cast iron pipe.

**I** $$h_f/L = (f)(1/d)(v^2/2g) \qquad \tfrac{1}{408} = (f)[1/(\tfrac{6}{12})]\{v^2/[(2)(32.2)]\} \qquad v = 0.2809/\sqrt{f}$$

Try $v = 2$ ft/s: $N_R = dv/v = (\tfrac{6}{12})(2)/(1.21 \times 10^{-5}) = 8.26 \times 10^4$, $\epsilon/d = 0.00085/(\tfrac{6}{12}) = 0.00170$. From Fig. A-5, $f = 0.0245$. $v = 0.2809/\sqrt{0.0245} = 1.79$ ft/s. Try $v = 1.79$ ft/s: $N_R = (\tfrac{6}{12})(1.79)/(1.21 \times 10^{-5}) = 7.40 \times 10^4$, $f = 0.025$, $v = 0.2809/\sqrt{0.025} = 1.78$ ft/s. Try $v = 1.78$ ft/s: $N_R = (\tfrac{6}{12})(1.78)/(1.21 \times 10^{-5}) = 7.40 \times 10^4$, $f = 0.025$ (O.K.); $Q = Av = [(\pi)(\tfrac{6}{12})^2/4](1.78) = 0.350$ ft³/s.

**9.139** Kerosene with kinematic viscosity $5.1 \times 10^{-7}$ m²/s flows in a 30-cm-diameter smooth pipe. Find the flow rate when the head loss is 0.4 m per 100 m.

**I** $$h_f = (f)(L/d)(v^2/2g) \qquad 0.4 = (f)[100/(\tfrac{30}{100})]\{v^2/[(2)(9.807)]\} \qquad v = 0.1534/\sqrt{f}$$

Try $v = 1$ m/s: $N_R = dv/v = (\tfrac{30}{100})(1)/(5.1 \times 10^{-7}) = 5.88 \times 10^5$. From Fig. A-5, $f = 0.0128$. $v = 0.1534/\sqrt{0.0128} = 1.36$ m/s. Try $v = 1.36$ m/s: $N_R = (\tfrac{30}{100})(1.36)/(5.1 \times 10^{-7}) = 8.00 \times 10^5$, $f = 0.0122$, $v = 0.1534/\sqrt{0.0122} = 1.39$ m/s. Try $v = 1.39$ m/s: $N_R = (\tfrac{30}{100})(1.39)/(5.1 \times 10^{-7}) = 8.18 \times 10^5$, $f = 0.0122$ (O.K.); $Q = Av = [(\pi)(\tfrac{30}{100})^2/4](1.39) = 0.0983$ m³/s.

**9.140** A pipe with $\epsilon = 0.00015$ ft is required to carry fluid of kinematic viscosity 0.00021 ft²/s at the rate of 8.0 cfs. If the head loss is to be 0.004 ft/ft, calculate the pipe diameter.

**I** $$h_f/L = (f/d)(v^2/2g) \qquad v = Q/A = 8.0/(\pi d^2/4) = 10.19/d^2 \qquad 0.004 = (f/d)\{(10.19/d^2)^2/[(2)(32.2)]\}$$

$$d = 3.320f^{1/5} \qquad N_R = dv/v = (d)(10.19/d^2)/0.00021 = 48\,524/d$$

Try $d = 1$ ft: $N_R = 48\,524/1 = 4.85 \times 10^4$, $\epsilon/d = 0.00015/1 = 0.00015$. From Fig. A-5, $f = 0.0215$. $d = (3.320)(0.0215)^{1/5} = 1.54$ ft. Try $d = 1.54$ ft: $N_R = 48\,524/1.54 = 3.15 \times 10^4$, $\epsilon/d = 0.00015/1.54 = 0.0000974$, $f = 0.0235$, $d = (3.320)(0.0235)^{1/5} = 1.57$ ft. Try $d = 1.57$ ft: $N_R = 50\,950/1.57 = 3.25 \times 10^4$, $\epsilon/d = 0.00015/1.57 = 0.0000955$, $f = 0.0235$ (O.K.). Hence, $d = 1.57$ ft, or 18.8 in.

**9.141** (a) Find the shear stress on 40-in-diameter asphalted iron pipe if the fluid is water at 72 °F and the average velocity is 10 fps. (b) What will be the shear stress if the average velocity is reduced to 5 fps?

$$\tau_0 = f\rho v^2/8 \qquad N_R = dv/\nu$$

(a) $$N_R = (\tfrac{40}{12})(10)/(1.02 \times 10^{-5}) = 3.27 \times 10^6 \qquad \epsilon/d = 0.0004/(\tfrac{40}{12}) = 0.00012$$

From Fig. A-5, $f = 0.0127$.

$$\tau_0 = (0.0127)(1.93)(10)^2/8 = 0.306\ \text{lb/ft}^2$$

(b) $$N_R = (\tfrac{40}{12})(5)/(1.02 \times 10^{-5}) = 1.63 \times 10^6 \qquad f = 0.0132$$

$$\tau_0 = (0.0132)(1.93)(5)^2/8 = 0.0796\ \text{lb/ft}^2$$

**9.142** A straight steel pipeline ($\epsilon = 0.00015$ ft) slopes downward at a small angle $\theta$, where $\sin \theta \approx \theta = 0.01523$. For gravity flow of oil ($\nu = 0.0006$ ft$^2$/s) at 10 ft$^3$/s, what pipe size is needed?

$$\sin \theta = h_f/L = (f/d)(v^2/2g) \qquad v = Q/A = 10/(\pi d^2/4) = 12.73/d^2$$

$$0.01523 = (f/d)\{(12.73/d^2)^2/[(2)(32.2)]\}$$

$$d = 2.777f^{1/5} \qquad N_R = dv/\nu = (d)(12.73/d^2)/0.0006 = 21\,217/d$$

Try $d = 1$ ft: $N_R = 21\,217/1 = 2.12 \times 10^4$, $\epsilon/d = 0.00015/1 = 0.00015$. From Fig. A-5, $f = 0.0265$. $d = (2.777)(0.0265)^{1/5} = 1.34$ ft. Try $d = 1.34$ ft: $N_R = 21\,217/1.34 = 1.58 \times 10^4$, $\epsilon/d = 0.00015/1.34 = 0.00011$, $f = 0.0284$, $d = (2.777)(0.0284)^{1/5} = 1.36$ ft. Try $d = 1.36$ ft: $N_R = 21\,217/1.36 = 1.56 \times 10^4$, $\epsilon/d = 0.00015/1.36 = 0.00011$, $f = 0.0284$ (O.K.). Hence, $d = 1.36$ ft, or 16.3 in.

**9.143** Water at 140 °F flows in a 0.824-in-diameter iron pipe ($\epsilon = 0.00015$ ft) of length 400 ft between points $A$ and $B$. At point $A$ the elevation of the pipe is 104.0 ft and the pressure is 8.50 psi. At point $B$ the elevation of the pipe is 100.3 ft and the pressure is 9.00 psi. Compute the flow rate.

$$p_A/\gamma + v_A^2/2g + z_A = p_B/\gamma + v_B^2/2g + z_B + h_L$$

$$(8.50)(144)/61.4 + v_A^2/2g + 104.0 = (9.00)(144)/61.4 + v_B^2/2g + 100.3 + h_L \qquad v_A^2/2g = v_B^2/2g$$

$$h_L = 2.52\ \text{ft} \qquad h_f = h_L = (f)(L/d)(v^2/2g) \qquad 2.52 = (f)[400/(0.824/12)]\{v^2/[(2)(32.2)]\} \qquad v = 0.1669/\sqrt{f}$$

Try $v = 1$ ft/s: $N_R = dv/\nu = (0.824/12)(1)/(5.03 \times 10^{-6}) = 1.37 \times 10^4$, $\epsilon/d = 0.00015/(0.824/12) = 0.0022$. From Fig. A-5, $f = 0.0325$. $v = 0.1669/\sqrt{0.0325} = 0.926$ ft/s. Try $v = 0.926$ ft/s: $N_R = (0.824/12)(0.926)/(5.03 \times 10^{-6}) = 1.26 \times 10^4$, $f = 0.0330$, $v = 0.1669/\sqrt{0.0330} = 0.919$ ft/s. Try $v = 0.919$ ft/s: $N_R = (0.824/12)(0.919)/(5.03 \times 10^{-6}) = 1.25 \times 10^4$, $f = 0.0330$ (O.K.); $Q = Av = [(\pi)(0.824/12)^2/4](0.917) = 0.00339$ ft$^3$/s.

**9.144** Air at 50 psia and 160 °F flows in a 1-ft by 1.5-ft by 1000-ft duct at the rate of 1 lb/min. Find the head loss if $\epsilon = 0.0005$ in.

$$h_f = (f)(L/d)(v^2/2g) \qquad d = 4R_h \qquad R_h = A/p_w = (1)(1.5)/(1+1+1.5+1.5) = 0.300\ \text{ft}$$

$$d = (4)(0.300) = 1.20\ \text{ft} \qquad W = \gamma Av \qquad \gamma = p/RT = (50)(144)/[(53.3)(460+160)] = 0.2179\ \text{lb/ft}^3$$

$$1/60 = 0.2179[(1)(1.5)](v) \qquad v = 0.05099\ \text{ft/s} \qquad h_f = (f)(1000/1.20)\{0.05099^2/[(2)(32.2)]\} = 0.03364f$$

$$N_R = dv/\nu = (1.20)(0.05099)/(2.06 \times 10^{-4}) = 297 \qquad \text{(laminar)}$$

$$f = 64/N_R = 64/297 = 0.2155 \qquad h_f = (0.03364)(0.2155) = 0.00725\ \text{ft}$$

**9.145** Find the approximate rate at which 60 °F water will flow in a conduit shaped in the form of an equilateral triangle if the head loss is 2 ft per 100 ft. The cross-sectional area of the duct is 120 in$^2$, and $\epsilon = 0.0018$ in.

First, find the length of each side ($s$) of the cross section (see Fig. 9-29):

$$A = bh/2 \qquad 120 = (x+x)(\sqrt{3}x)/2 \qquad x = 8.324\ \text{in} \qquad s = 2x = (2)(8.324) = 16.65\ \text{in}$$

$$h_f = (f)(L/d)(v^2/2g) \qquad R_h = A/p_w = 120/[(3)(16.65)] = 2.402\ \text{in} \quad \text{or} \quad 0.2002\ \text{ft}$$

$$d = 4R_h = (4)(0.2002) = 0.8008\ \text{ft} \qquad 2 = (f)(100/0.8008)\{v^2/[(2)(32.2)]\} \qquad v = 1.016/\sqrt{f}$$

Try $v = 10$ ft/s: $N_R = dv/\nu = (0.8008)(10)/(1.21 \times 10^{-5}) = 6.62 \times 10^5$, $\epsilon/d = (0.0018/12)/0.8008 = 0.000187$. From Fig. A-5, $f = 0.0150$. $v = 1.016/\sqrt{0.0150} = 8.30$ ft/s. Try $v = 8.30$ ft/s: $N_R = (0.8008)(8.30)/(1.21 \times 10^{-5}) = 5.49 \times 10^5$, $f = 0.0155$, $v = 1.016/\sqrt{0.0155} = 8.16$ ft/s. Try $v = 8.16$ ft/s: $N_R = (0.8008)(8.16)/(1.21 \times 10^{-5}) = 5.40 \times 10^5$, $f = 0.0155$ (O.K.); $Q = Av = (\frac{120}{144})(8.16) = 6.80$ ft$^3$/s.

Fig. 9-29

**9.146** When fluid of weight density 50 lb/ft$^3$ flows in a 6-in-diameter pipe, the frictional stress between the fluid and the pipe wall is 0.5 lb/ft$^2$. Calculate the head loss per mile of pipe. If the flow rate is 2.5 cfs, how much power is lost per mile?

$$R_h = d/4 = (\tfrac{6}{12})/4 = 0.1250 \text{ ft} \qquad h_f = (\tau_0)(L/R_h\gamma) = 0.5\{5280/[(0.1250)(50)]\} = 422.4 \text{ ft per mile}$$

$$P = Q\gamma h_f = (2.5)(50)(422.4) = 52\,800 \text{ ft} \cdot \text{lb/s per mile} = 52\,800/550 = 96.0 \text{ hp per mile}$$

**9.147** Prove that for a constant rate of discharge and a constant value of $f$ the friction head loss in a pipe varies inversely as the fifth power of the diameter.

$$v = Q/A = 4Q/\pi d^2 \qquad h_f = (f)(L/d)(v^2/2g) = (f)(L/d)[(4Q/\pi d^2)^2/2g] = (f)(L/d^5)(8Q^2/\pi^2 g)$$

Thus for constant $f$ and constant $Q$, $h_f \propto 1/d^5$.

**9.148** Two long pipes are used to convey water between two reservoirs whose water surfaces are at different elevations. One pipe has a diameter twice that of the other. If both pipes have the same value of $f$ and if minor losses are neglected, what is the ratio of the flow rates through the two pipes?

$$h_f = (f)(L/d)(v^2/2g) = \Delta \text{ elevation} \qquad h_f \propto Q^2/d^5 \qquad \text{(from Prob. 9.147)}$$

$$(h_f)_1 = (h_f)_2$$

Therefore, $Q_1^2/d_1^5 = Q_2^2/d_2^5$, $Q_2/Q_1 = (d_2/d_1)^{5/2} = 2^{5/2} = 5.66$. Thus the flow in the larger pipe will be 5.66 times that in the smaller one.

**9.149** Points $C$ and $D$, at the same elevation, are 500 ft apart in an 8-in pipe and are connected to a differential gage by means of small tubing. When the flow of water is 6.31 cfs, the deflection of mercury in the gage is 6.43 ft. Determine the friction factor $f$.

$p_C/\gamma + v_C^2/2g + z_C = p_D/\gamma + v_D^2/2g + z_D + h_f$. Since $v_C^2/2g = v_D^2/2g$ and $z_C = z_D$, $p_C/\gamma - p_D/\gamma = h_f = (6.43)(13.6 - 1) = 81.02$ ft, $v = Q/A = 6.31/[(\pi)(\frac{8}{12})^2/4] = 18.08$ ft/s, $h_f = (f)(L/d)(v^2/2g) = (f)[500/(\frac{8}{12})]\{18.08^2/[(2)(32.2)]\} = 3807f$, $81.02 = 3807f$, $f = 0.0213$.

**9.150** Oil flows from tank $A$ through 500 ft of 6-in new asphalt-dipped cast iron pipe to point $B$, as shown in Fig. 9-30. What pressure in pounds per square inch will be needed at $A$ to cause 0.450 cfs of oil to flow? (s.g. = 0.840; $\nu = 2.27 \times 10^{-5}$ ft$^2$/s; $\epsilon = 0.0004$ ft.)

$$p_A/\gamma + v_A^2/2g + z_A = p_B/\gamma + v_B^2/2g + z_B + h_L \qquad v_B = Q/A_B = 0.450/[(\pi)(\tfrac{6}{12})^2/4] = 2.292 \text{ ft/s}$$

$$h_L = h_f = (f)(L/d)(v^2/2g) = (f)[500/(\tfrac{6}{12})]\{2.292^2/[(2)(32.2)]\} = 81.57f$$

$$p_A/[(0.840)(62.4)] + 0 + 80.0 = 0 + 2.292^2/[(2)(32.2)] + 100.0 + 81.57f \qquad p_A = 1053 + 4276f$$

$$N_R = dv/\nu = (\tfrac{6}{12})(2.292)/(2.27 \times 10^{-5}) = 5.05 \times 10^4 \qquad \epsilon/d = 0.0004/(\tfrac{6}{12}) = 0.000800$$

From Fig. A-5, $f = 0.0235$. $p_A = 1053 + (4276)(0.0235) = 1153$ lb/ft$^2$, or 8.01 lb/in$^2$.

El. 100.0 ft
B

Air  El. 80.0 ft  500 ft
A  6 in

**Fig. 9-30**

**9.151** An old 12-in by 18-in rectangular duct carries air at 15.2 psia and 68 °F through 1500 ft with an average velocity of 9.75 ft/s. Determine the loss of head and the pressure drop, assuming the duct to be horizontal and the size of the surface imperfections is 0.0018 ft.

■ $\quad R_h = A/p_w = (\frac{12}{12})(\frac{18}{12})/(\frac{12}{12} + \frac{12}{12} + \frac{18}{12} + \frac{18}{12}) = 0.300 \text{ ft} \qquad d = 4R_h = (4)(0.300) = 1.20 \text{ ft}$

$$h_f = (f)(L/d)(v^2/2g) = (f)(1500/1.20)\{9.75^2/[(2)(32.2)]\} = 1845f$$

$N_R = dv/v = (1.20)(9.75)/[(14.7/15.2)(1.64 \times 10^{-4})] = 7.38 \times 10^4 \quad \text{(turbulent)} \qquad \epsilon/d = 0.0018/1.20 = 0.00150$

From Fig. A-5, $f = 0.024$.

$$h_f = (1845)(0.024) = 44.28 \text{ ft of air} \qquad \Delta p = \gamma h_f = [(15.2/14.7)(0.0750)](44.28)/144 = 0.0238 \text{ lb/in}^2$$

**9.152** What size of new cast iron pipe, 8000 ft long, will deliver 37.5 cfs of water at 70 °F with a drop in the hydraulic grade line of 215 ft?

■ $\quad p_A/\gamma + v_A^2/2g + z_A = p_B/\gamma + v_B^2/2g + z_B + h_L \qquad [(p_A/\gamma + z_A) - (p_B/\gamma + z_B)] = h_L$

$\qquad [(p_A/\gamma + z_A) - (p_B/\gamma + z_B)] = \text{HGL} = 215 \text{ ft} \qquad v = Q/A = 37.5/(\pi d^2/4) = 47.75/d^2$

$\qquad h_L = h_f = (f)(L/d)(v^2/2g) = (f)(8000/d)\{(47.75/d^2)^2/[(2)(32.2)]\} = 283\,238f/d^5$

$\quad 215 = 283\,238f/d^5 \qquad d = 4.207f^{1/5} \qquad N_R = dv/v = (d)(47.75/d^2)/(1.05 \times 10^{-5}) = 4.55 \times 10^6/d$

Try $d = 2$ ft: $N_R = (4.55 \times 10^6)/2 = 2.28 \times 10^6$, $\epsilon/d = 0.00085/2 = 0.000425$. From Fig. A-5, $f = 0.0164$. $d = (4.207)(0.0164)^{1/5} = 1.85$ ft. Try $d = 1.85$ ft: $N_R = (4.55 \times 10^6)/1.85 = 2.46 \times 10^6$, $\epsilon/d = 0.00085/1.85 = 0.000459$, $f = 0.0164$ (O.K.). Hence, $d = 1.85$ ft, or 22.2 in.

**9.153** What rate of flow of air at 68 °F will be carried by a new horizontal 2-in-diameter steel pipe at an absolute pressure of 3 atm and with a drop of 0.150 psi in 100 ft of pipe? Use $\epsilon = 0.00025$ ft.

■ At 68 °F and standard atmospheric pressure, $\gamma = 0.0752$ lb/ft$^3$ and $v = 1.60 \times 10^{-4}$ ft$^2$/s. At a pressure of 3 atm, $\gamma = (0.0752)(3) = 0.2256$ lb/ft$^3$ and $v = (1.60 \times 10^{-4})/3 = 5.333 \times 10^{-5}$ ft$^2$/s. $h_f = (f)(L/d)(v^2/2g)$, $(0.150)(144)/0.2256 = (f)[100/(\frac{2}{12})]\{v^2/[(2)(32.2)]\}$, $v = 3.206/\sqrt{f}$. Try $v = 10$ ft/s: $N_R = dv/v = (\frac{2}{12})(10)/(5.333 \times 10^{-5}) = 3.13 \times 10^4$, $\epsilon/d = 0.00025/(\frac{2}{12}) = 0.00150$. From Fig. A-5, $f = 0.027$. $v = 3.206/\sqrt{0.027} = 19.51$ ft/s. Try $v = 19.51$ ft/s: $N_R = (\frac{2}{12})(19.51)/(5.333 \times 10^{-5}) = 6.10 \times 10^4$, $f = 0.0248$, $v = 3.206/\sqrt{0.0248} = 20.36$ ft/s. Try $v = 20.36$ ft/s: $N_R = (\frac{2}{12})(20.36)/(5.333 \times 10^{-5}) = 6.36 \times 10^4$, $f = 0.0248$ (O.K.); $Q = Av = [(\pi)(\frac{2}{12})^2/4](20.36) = 0.444$ ft$^3$/s.

**9.154** Determine the nature of the distribution of shear stress at a cross section in a horizontal, circular pipe under steady flow conditions.

■ For the free body in Fig. 9-31a, since the flow is steady, each particle moves to the right without acceleration. Hence, the summation of the forces in the x direction must equal zero. $(p_1)(\pi r^2) - (p_2)(\pi r^2) - (\tau)(2\pi r L) = 0$ or

$$\tau = (p_1 - p_2)(r)/(2L) \tag{1}$$

When $r = 0$, the shear stress $\tau$ is zero; and when $r = r_0$, the stress $\tau_0$ at the wall is a maximum. The variation is linear and is indicated in Fig. 9-31b. Equation (1) holds for laminar and turbulent flows as no limitations concerning flow were imposed in the derivation. Since $(p_1 - p_2)/\gamma$ represents the drop in the energy line, or the lost head $h_L$, multiplying Eq. (1) by $\gamma/\gamma$ yields $\tau = (\gamma r/2L)[(p_1 - p_2)/\gamma]$ or

$$\tau = (\gamma h_L/2L)(r) \tag{2}$$

**9.155** Develop the expression for shear stress at a pipe wall.

■ $h_L = (f)(L/d)(v^2/2g)$. From Prob. 9.154, $h_L = 2\tau_0 L/\gamma r_0 = 4\tau_0 L/\gamma d$, $4\tau_0 L/\gamma d = (f)(L/d)(v^2/2g)$, $\tau_0 = f\gamma v^2/8g = f\rho v^2/8$.

**Fig. 9-31**

**9.156** For steady, laminar flow (a) what is the relationship between the velocity at a point in the cross section and the velocity at the center of the pipe, and (b) what is the equation for velocity distribution?

**I** (a) $$\tau = -(\mu)(dv/dr) = (p_1 - p_2)(r)/(2L) \qquad \text{(from Prob. 9.154)}$$

$$-(\mu)(dv/dr) = (p_1 - p_2)(r)/(2L)$$

Since $(p_1 - p_2)/L$ is not a function of $r$,

$$-\int_{v_{\mathbb{C}}}^{v} dv = \frac{p_1 - p_2}{2\mu L} \int_0^r r\, dr \qquad -[v]_{v_{\mathbb{C}}}^{v} = \frac{p_1 - p_2}{4\mu L}[r^2]_0^r \qquad -(v - v_{\mathbb{C}}) = \frac{(p_1 - p_2)r^2}{4\mu L}$$

$$v = v_{\mathbb{C}} - \frac{(p_1 - p_2)r^2}{4\mu L} \tag{1}$$

But the lost head in $L$ feet is $h_L = (p_1 - p_2)/\gamma$; hence,

$$v = v_{\mathbb{C}} - \frac{\gamma h_L r^2}{4\mu L} \tag{2}$$

(b) Since the velocity at the boundary is zero, when $r = r_o$, $v = 0$ in Eq. (1), we have

$$v_c = \frac{(p_1 - p_2)r_o^2}{4\mu L} \qquad \text{(at centerline)} \tag{3}$$

Thus, in general,

$$v = \left(\frac{p_1 - p_2}{4\mu L}\right)(r_o^2 - r^2) \tag{4}$$

**9.157** Develop the expression for the loss of head in a pipe for steady, laminar flow of an incompressible fluid. Refer to Fig. 9-31d.

**I** $$v_{\text{av}} = \frac{Q}{A} = \frac{\int v\, dA}{\int dA} = \frac{\int_0^{r_o} (v)(2\pi r\, dr)}{\pi r_o^2} = \frac{(2\pi)(p_1 - p_2)}{(\pi r_o^2)(4\mu L)} \int_0^{r_o} (r_o^2 - r^2)(r\, dr)$$

$$v_{\text{av}} = \frac{(p_1 - p_2)(r_o^2)}{8\mu L} \tag{1}$$

Thus for laminar flow, the average velocity is half the maximum velocity $v_c$ in Eq. (3) of Prob. 9.156. Rearranging Eq. (1), we obtain

$$h_L = \frac{p_1 - p_2}{\gamma} = \frac{8\mu L v_{\text{av}}}{\gamma r_o^2} = \frac{32\mu L v_{\text{av}}}{\gamma d^2} \tag{2}$$

These expressions apply for laminar flow of all fluids in all pipes and conduits.

**9.158** Determine (a) the shear stress at the walls of a 12-in-diameter pipe when water flowing causes a measured head loss of 15 ft in 300 ft of pipe length, (b) the shear stress 2 in from the centerline of the pipe, (c) the shear velocity, (d) the average velocity for an $f$ value of 0.50, and (e) the ratio $v/v_*$.

**(a)** $\tau_o = \gamma h_L r_o/2L = (62.4)(15)[(\frac{12}{12})/2]/[(2)(300)] = 0.780 \text{ lb/ft}^2$, or $0.00542 \text{ lb/in}^2$.

**(b)** Since $\tau$ varies linearly from centerline to wall, $\tau = (0.00542)(\frac{2}{6}) = 0.00181 \text{ lb/in}^2$.

**(c)** $v_* = \sqrt{\tau_o/\rho} = \sqrt{0.780/1.94} = 0.634 \text{ ft/s}$.

**(d)** $\tau_o = f\rho v^2/8$, $0.780 = (0.050)(1.94)(v^2)/8$, $v = 8.02 \text{ ft/s}$.

**(e)** $\tau_o = (\mu)(v/y)$, $v = \mu/\rho$, $\tau_0 = \rho v(v/y)$, $\tau_o/\rho = (v)(v/y) = v_*^2 = (v)(v/y)$, $v/v_*^2 = y/v$, $v/v_* = v_* y/v$.

**9.159** If in Prob. 9.158 the water is flowing through a 3-ft by 4-ft rectangular conduit of the same length with the same lost head, what is the shear stress between the water and the pipe wall?

$$R_h = A/p_w = (3)(4)/(3 + 3 + 4 + 4) = 0.8571 \text{ ft}$$

$$\tau = (\gamma h_L/L)(R_h) = [(62.4)(15)/300](0.8571) = 2.67 \text{ lb/ft}^2 \quad \text{or} \quad 0.0186 \text{ lb/in}^2$$

**9.160** Medium lubricating oil (s.g. = 0.860) is pumped through 1000 ft of horizontal 2-in pipe at the rate of 0.0436 cfs. If the drop in pressure is 30.0 psi, what is the absolute viscosity of the oil?

Assuming laminar flow,

$$\frac{p_1 - p_2}{\gamma} = \frac{32\mu L v_{av}}{\gamma d^2} \quad \text{(from Prob. 9.157)}$$

$$v_{av} = Q/A = 0.0436/[(\pi)(\tfrac{2}{12})^2/4] = 1.998 \text{ ft/s}$$

$$(30.0)(144)/[(0.860)(62.4)] = (32)(\mu)(1000)(1.998)/\{[(0.860)(62.4)](\tfrac{2}{12})^2\} \quad \mu = 0.00188 \text{ lb} \cdot \text{s/ft}^2$$

$$N_R = \rho dv/\mu = [(0.860)(1.94)](\tfrac{2}{12})(1.998)/0.00188 = 296 \quad \text{(laminar)}$$

**9.161** A horizontal wrought iron pipe, 6-in inside diameter and somewhat corroded, is transporting 4.50 lb of air per second from $A$ to $B$. At $A$ the pressure is 70 psia and at $B$ the pressure must be 65 psia. Flow is isothermal at 68 °F. What is the length of pipe from $A$ to $B$? Use $\epsilon = 0.0013$ ft.

$$\frac{p_1 - p_2}{\gamma_1} = \frac{2[2 \ln (v_2/v_1) + (f)(L/d)](v_1^2/2g)}{1 + p_2/p_1}$$

$$\gamma_1 = (0.0752)(70/14.7) = 0.3581 \text{ lb/ft}^3 \qquad \gamma_2 = (0.0752)(65/14.7) = 0.3325 \text{ lb/ft}^3$$

$$W = \gamma A v \qquad 4.50 = (0.3581)[(\pi)(\tfrac{6}{12})^2/4](v_1) \qquad v_2 = 64.00 \text{ ft/s}$$

$$4.50 = 0.3325[(\pi)(\tfrac{6}{12})^2/4](v_2) \qquad v_1 = 68.93 \text{ ft/s}$$

$$N_R = dv/v = (\tfrac{6}{12})(64.00)/[(14.7/70.0)(1.60 \times 10^{-4})] = 9.52 \times 10^5 \qquad \epsilon/d = 0.0013/(\tfrac{6}{12}) = 0.0026$$

From Fig. A-5, $f = 0.025$.

$$\frac{(70 - 65)(144)}{0.3581} = \frac{2\{2 \ln (68.93/64.00) + 0.025[L/(\tfrac{6}{12})]\}\{64.00^2/[(2)(32.2)]\}}{1 + \tfrac{65}{70}} \qquad L = 607 \text{ ft}$$

**9.162** Heavy fuel oil flows from $A$ to $B$ through 3000 ft of horizontal 6-in steel pipe. The pressure at $A$ is 155 psi and at $B$ is 5.0 psi. The kinematic viscosity is 0.00444 ft²/s and the specific gravity is 0.918. What is the flow rate?

Assuming laminar flow, from Eq. (2) of Prob. 9.157,

$$\frac{p_1 - p_2}{\gamma} = \frac{32\mu L v_{av}}{\gamma d^2} = \frac{(32)(v\rho)(L v_{av})}{\gamma d^2} \qquad \frac{(155 - 5.0)(144)}{(0.918)(62.4)} = \frac{32\{(0.00444)[(0.918)(1.94)]\}(3000)(v_{av})}{[(0.918)(62.4)](\tfrac{6}{12})^2}$$

$$v_{av} = 7.11 \text{ ft/s} \qquad N_R = dv/v = (\tfrac{6}{12})(7.11)/0.00444 = 808 \quad \text{(laminar)}$$

$$Q = Av = [(\pi)(\tfrac{6}{12})^2/4](7.11) = 1.40 \text{ ft}^3/\text{s}$$

**9.163** What size pipe should be installed to carry 0.785 cfs of heavy fuel oil ($v = 0.00221$ ft²/s, s.g. = 0.912) at 60 °F if the available lost head in the 1000-ft length of horizontal pipe is 22.0 ft?

Assuming laminar flow,

$$h_f = \frac{32\mu L v}{\gamma d^2} \qquad \mu = \rho v = [(0.912)(1.94)](0.00221) = 0.003910 \text{ lb} \cdot \text{s/ft}^2$$

$$v = Q/A = 0.785/(\pi d^2/4) = 0.9995/d^2 \qquad 22.0 = (32)(0.003910)(1000)(0.9995/d^2)/\{[(0.912)(62.4)](d^2)\}$$

$$d = 0.562 \text{ ft} \quad \text{or} \quad 6.75 \text{ in} \qquad N_R = dv/v = (0.562)(0.9995/0.562^2)/0.00221 = 805 \quad \text{(laminar)}$$

**9.164** Determine the head loss in 1000 ft of new, uncoated 12-in-ID cast iron pipe when water at 60 °F flows at 5.00 ft/s. Use $\epsilon/d = 0.0008$.

$$h_f = (f)(L/d)(v^2/2g) \qquad N_R = dv/v = (\tfrac{12}{12})(5.00)/(1.21 \times 10^{-5}) = 4.13 \times 10^5$$

From Fig. A-5, $f = 0.0194$. $h_f = 0.0194[1000/(\tfrac{12}{12})]\{5.00^2/[(2)(32.2)]\} = 7.53$ ft.

**9.165** Rework Prob. 9.164 if the liquid is medium fuel oil at 60 °F ($v = 4.75 \times 10^{-5}$ ft²/s) flowing at the same velocity.

$$h_f = (f)(L/d)(v^2 2g) \qquad N_R = dv/v = (\tfrac{12}{12})(5.00)/(4.75 \times 10^{-5}) = 1.05 \times 10^5$$

From Fig. A-5, $f = 0.0213$. $h_f = 0.0213[1000/(\tfrac{12}{12})]\{5.00^2/[(2)(32.2)]\} = 8.27$ ft.

**9.166** Points $A$ and $B$ are 4000 ft apart along a new 6-in-ID steel pipe. Point $B$ is 50.5 ft higher than $A$ and the pressures at $A$ and $B$ are 123 psi and 48.6 psi, respectively. How much medium fuel oil at 70 °F will flow from $A$ to $B$? Use s.g. = 0.854, $v = 4.12 \times 10^{-5}$ ft²/s, $\epsilon = 0.0002$ ft.

$$p_A/\gamma + v_A^2/2g + z_A = p_B/\gamma + v_B^2/2g + z_B + h_L$$

$$(123)(144)/[(0.854)(62.4)] + v_A^2/2g + 0 = (48.6)(144)/[(0.854)(62.4)] + v_B^2/2g + 50.5 + h_L$$

$$v_A^2/2g = v_B^2/2g \qquad h_L = 150.5 \text{ ft} = h_f = (f)(L/d)(v^2/2g)$$

$$150.5 = (f)[4000/(\tfrac{6}{12})]\{v^2/[(2)(32.2)]\} \qquad v = 1.101/\sqrt{f}$$

Try $v = 10$ ft/s: $N_R = dv/v = (\tfrac{6}{12})(10)/(4.12 \times 10^{-5}) = 1.21 \times 10^5$, $\epsilon/d = 0.0002/(\tfrac{6}{12}) = 0.000400$. From Fig. A-5, $f = 0.0195$. $v = 1.101/\sqrt{0.0195} = 7.884$ ft/s. Try $v = 7.884$ ft/s: $N_R = (\tfrac{6}{12})(7.884)/(4.12 \times 10^{-5}) = 9.57 \times 10^4$, $f = 0.0195$ (O.K.); $Q = Av = [(\pi)(\tfrac{6}{12})^2/4](7.884) = 1.55$ ft³/s.

**9.167** How much water (60 °F) would flow under the conditions of Prob. 9.166.

$$p_A/\gamma + v_A^2/2g + z_A = p_B/\gamma + v_B^2/2g + z_B + h_L$$

$$(123)(144)/62.4 + v_A^2/2g + 0 = (48.6)(144)/62.4 + v_B^2/2g + 50.5 + h_L$$

$$v_A^2/2g = v_B^2/2g \qquad h_L = 121.2 \text{ ft} = h_f = (f)(L/d)(v^2/2g)$$

$$121.2 = (f)[4000/(\tfrac{6}{12})]\{v^2/[(2)(32.2)]\} \qquad v = 0.9878/\sqrt{f}$$

Try $v = 10$ ft/s: $N_R = dv/v = (\tfrac{6}{12})(10)/(1.21 \times 10^{-5}) = 4.13 \times 10^5$, $\epsilon/d = 0.0002/(\tfrac{6}{12}) = 0.000400$. From Fig. A-5, $f = 0.0172$. $v = 0.9878/\sqrt{0.0172} = 7.532$ ft/s. Try $v = 7.532$ ft/s: $N_R = (\tfrac{6}{12})(7.532)/(1.21 \times 10^{-5}) = 3.11 \times 10^5$, $f = 0.0176$, $v = 0.9878/\sqrt{0.0176} = 7.446$ ft/s. Try $v = 7.446$ ft/s: $N_R = (\tfrac{6}{12})(7.446)/(1.21 \times 10^{-5}) = 3.08 \times 10^5$, $f = 0.0176$ (O.K.); $Q = Av = [(\pi)(\tfrac{6}{12})^2/4](7.446) = 1.46$ ft³/s.

**9.168** To transport 300 cfs of air, $p = 16$ psia, $T = 68$ °F, with a head loss of 3 in of water per 1000 ft, what size galvanized pipe is needed? ($\epsilon = 0.0005$ ft.)

$$d = 0.66[(\epsilon^{1.25})(LQ^2/gh_f)^{4.75} + (v)(Q)^{9.4}(L/gh_f)^{5.2}]^{0.04}$$

$$\gamma = p/RT = (16)(144)/[(53.3)(460 + 68)] = 0.08187 \text{ lb/ft}^3 \qquad h_f = (\tfrac{3}{12})(62.4/0.08187) = 190.5 \text{ ft of air}$$

$$d = 0.66\left\{(0.0005^{1.25})\left[\frac{(1000)(300)^2}{(32.2)(190.5)}\right]^{4.75} + (1.64 \times 10^{-4})(300)^{9.4}\left[\frac{1000}{(32.2)(190.5)}\right]^{5.2}\right\}^{0.04} = 2.84 \text{ ft}$$

**9.169** Two tanks of a solvent ($\mu = 0.05$ Pa · s, $\gamma = 8$ kN/m³) are connected by 300 m of commercial steel pipe. What size must the pipe be to convey 50 L/s, if one tank is 4 m higher than the other?

$$d = 0.66[(\epsilon^{1.25})(LQ^2/gh_f)^{4.75} + (v)(Q)^{9.4}(L/gh_f)^{5.2}]^{0.04}$$

$$v = \mu/\rho = \mu g/\gamma = (0.05)(9.807)/[8 \times 10^3] = 6.129 \times 10^{-5} \text{ m}^2/\text{s}$$

$$d = 0.66\left\{(0.000046^{1.25})\left[\frac{(300)(\tfrac{50}{1000})^2}{(9.807)(4)}\right]^{4.75} + (6.129 \times 10^{-5})(\tfrac{50}{1000})^{9.4}\left[\frac{300}{(9.807)(4)}\right]^{5.2}\right\}^{0.04} = 0.222 \text{ m}$$

**9.170** Calculate the diameter of a wooden conduit ($\epsilon = 0.006$ ft) that is to carry 300 ft³/s of water at 60 °F a distance of 1000 ft with a head loss of 1.1 ft.

$$d = 0.66[(\epsilon^{1.25})(LQ^2/gh_f)^{4.75} + (v)(Q)^{9.4}(L/gh_f)^{5.2}]^{0.04}$$

$$d = 0.66\left\{(0.0006^{1.25})\left[\frac{(1000)(300)^2}{(32.2)(1.1)}\right]^{4.75} + (1.21 \times 10^{-5})(300)^{9.4}\left[\frac{1000}{(32.2)(1.1)}\right]^{5.2}\right\}^{0.04} = 7.59 \text{ ft}$$

**9.171** An old pipe 2 m in diameter has a roughness of $\epsilon = 30$ mm. A 12-mm-thick lining would reduce the roughness to $\epsilon = 1$ mm. How much would pumping costs be reduced per kilometer of pipe for water at 20 °C with discharge of 6 m³/s? The pumps are 75 percent efficient, and the cost of energy is $1 per 72 MJ.

$$v_1 = Q/A_1 = 6/[(\pi)(2)^2/4] = 1.910 \text{ m/s} \qquad N_R = dv/\nu$$

$$(N_R)_1 = (2)(1.910)/(1.02 \times 10^{-6}) = 3.75 \times 10^6 \qquad \epsilon_1/d_1 = (0.030)/2 = 0.015$$

From Fig. A-5, $f_1 = 0.044$.

$$d_2 = [2 - (2)(0.012)] = 1.976 \text{ m} \qquad v_2 = Q/A_2 = 6/[(\pi)(1.976)^2/4] = 1.957 \text{ m/s}$$

$$(N_R)_2 = (1.976)(1.957)/(1.02 \times 10^{-6}) = 3.79 \times 10^6 \qquad \epsilon_2/d_2 = (0.001)/1.976 = 0.000506$$

$$f_2 = 0.017 \qquad h_f = (f)(L/d)(v^2/2g)$$

$$(h_f)_1 = 0.044[1000/2]\{1.910^2/[(2)(9.807)]\} = 4.902 \text{ m}$$

$$(h_f)_2 = 0.017[1000/1.976]\{1.957^2/[(2)(9.807)]\} = 1.680 \text{ m}$$

$$\text{Saving in head} = 4.092 - 1.680 = 2.412 \text{ m} \qquad P = Q\gamma h_f/\eta = (6)(9.79 \times 10^3)(2.412)/0.75 = 0.1889 \text{ MJ/s}$$

$$\text{Savings per year} = (0.1889)[(365)(24)(3600)]/72 = \$82\,738$$

**9.172** What size of new cast iron pipe is needed to transport 0.5 m³/s of water at 25 °C for 1000 m with head loss of 2 m?

$$d = 0.66[(\epsilon^{1.25})(LQ^2/gh_f)^{4.75} + (\nu)(Q)^{9.4}(L/gh_f)^{5.2}]^{0.04}$$

$$d = 0.66\left\{(0.00026^{1.25})\left[\frac{(1000)(0.5)^2}{(9.807)(2)}\right]^{4.75} + (9.02 \times 10^{-7})(0.5)^{9.4}\left[\frac{1000}{(9.807)(2)}\right]^{5.2}\right\}^{0.04} = 0.713 \text{ m}$$

**9.173** Pure water at 70 °F runs at 2.778 lb/s through a smooth tube between two reservoirs 30 ft apart and having a difference in elevation of 4.1 ft. What size tubing is needed?

$$Q = 2.778/62.3 = 0.04459 \text{ ft}^3/\text{s} \qquad h_f = (f)(L/d)(v^2/2g)$$

$$v = Q/A = 0.04459/(\pi d^2/4) = 0.05677/d^2 \qquad 4.1 = (f)(30/d)\{(0.05677/d^2)^2/[(2)(32.2)]\} \qquad d = 0.2055f^{1/5}$$

$$N_R = dv/\nu = (d)(0.05677/d^2)/(1.05 \times 10^{-5}) = 5.41 \times 10^3/d$$

Try $f = 0.020$: $d = (0.2055)(0.020)^{1/5} = 0.09398$ ft, $N_R = 5.41 \times 10^3/0.09398 = 5.76 \times 10^4$. From Fig. A-5, $f = 0.0205$. Try $f = 0.0205$: $d = (0.2055)(0.0205)^{1/5} = 0.09444$ ft, $N_R = 5.41 \times 10^3/0.09444 = 5.73 \times 10^4$, $f = 0.0205$ (O.K.). Hence, $d = 0.09444$ ft, or 1.13 in.

**9.174** In Fig. 9-10, $H = 20$ m, $L = 150$ m, $d = 50$ mm, s.g. $= 0.85$, $\mu = 0.400$ N · s/m², and $\epsilon = 1$ mm. Find the newtons per second flowing. Neglect minor losses.

$$h_f = (f)(L/d)(v^2/2g) \qquad 20 = (f)[150/(\tfrac{50}{1000})]\{v^2/[(2)(9.807)]\} \qquad v = 0.03616/\sqrt{f}$$

$$N_R = \rho dv/\mu = [(0.85)(1000)](\tfrac{50}{1000})(v)/0.004 = 10\,625v$$

Try $f = 0.050$: $v = 0.3616/\sqrt{0.050} = 1.617$ m/s, $N_R = (10\,625)(1.617) = 1.72 \times 10^4$, $\epsilon/d = (\tfrac{1}{1000})/(\tfrac{50}{1000}) = 0.0200$. From Fig. A-5, $f = 0.051$.

$$v = 0.3616/\sqrt{0.051} = 1.601 \text{ m/s} \qquad N_R = (10\,625)(1.601) = 1.70 \times 10^4 \qquad f = 0.051 \quad \text{(O.K.)}$$

$$W = \gamma Av = [(0.85)(9.79)][(\pi)(\tfrac{50}{1000})^2/4](1.601) = 0.0262 \text{ kN/s} \quad \text{or} \quad 26.2 \text{ N/s}$$

**9.175** Determine the head loss for flow of 140 L/s of oil, $\nu = 0.00001$ m²/s, through 400 m of 200-mm-diameter cast iron pipe.

$$h_f = (f)(L/d)(v^2/2g) \qquad v = Q/A = (\tfrac{140}{1000})/[(\pi)(\tfrac{200}{1000})^2/4] = 4.456 \text{ m/s}$$

$$N_R = dv/\nu = (\tfrac{200}{1000})(4.456)/0.00001 = 8.91 \times 10^4 \qquad \epsilon/d = 0.00026/(\tfrac{200}{1000}) = 0.00130$$

From Fig. A-5, $f = 0.023$. $h_f = 0.023[400/(\tfrac{200}{1000})]\{4.456^2/[(2)(9.807)]\} = 46.6$ m.

**9.176** Water at 15 °C flows through a 300-mm-diameter riveted steel pipe, $\epsilon = 3$ mm, with a head loss of 6 m in 300 m. Determine the flow.

▮ $h_f = (f)(L/d)(v^2/2g)$. Try $f = 0.040$: $6 = 0.040[300/(\frac{300}{1000})]\{v^2/[(2)(9.807)]\}$, $v = 1.715$ m/s; $N_R = dv/v = (\frac{300}{1000})(1.715)/(1.16 \times 10^{-6}) = 4.44 \times 10^5$; $\epsilon/d = \frac{3}{300} = 0.0100$. From Fig. A-5, $f = 0.038$. Try $f = 0.038$: $6 = (0.038)[300/(\frac{300}{1000})]\{v^2/[(2)(9.807)]\}$, $v = 1.760$ m/s; $N_R = (\frac{300}{1000})(1.760)/(1.16 \times 10^{-6}) = 4.55 \times 10^5$; $f = 0.038$ (O.K.); $Q = Av = [(\pi)(\frac{300}{1000})^2/4](1.760) = 0.124$ m$^3$/s.

**9.177** Determine the size of clean wrought iron pipe required to convey 4000 gpm of oil, $v = 0.0001$ ft$^2$/s, 10 000 ft with a head loss of 75 ft.

▮ $$Q = (4000)(0.002228) = 8.912 \text{ ft}^3/\text{s} \qquad h_f = (f)(L/d)(v^2/2g)$$
$$v = Q/A = 8.912/(\pi d^2/4) = 11.35/d^2 \qquad 75 = (f)(10\,000/d)\{(11.35/d^2)^2/[(2)(32.2)]\} \qquad d = 3.056f^{1/5}$$

Try $f = 0.020$: $d = (3.056)(0.020)^{1/5} = 1.398$ ft, $N_R = dv/v = (1.398)(11.35/1.398^2)/0.0001 = 8.12 \times 10^4$, $\epsilon/d = 0.00015/1.398 = 0.000107$. From Fig. A-5, $f = 0.019$. Try $f = 0.019$: $d = (3.056)(0.019)^{1/5} = 1.383$ ft, $N_R = (1.383)(11.35/1.383^2)/0.0001 = 8.21 \times 10^4$, $f = 0.019$ (O.K.). Hence, $d = 1.383$ ft, or 16.6 in.

**9.178** In Prob. 9.177, for $d = 16.6$ in, if the specific gravity is 0.85, $p_1 = 40$ psi, $z_1 = 200$ ft, and $z_2 = 50$ ft, determine the pressure at point 2.

▮ $$p_1/\gamma + v_1^2/2g + z_1 = p_2/\gamma + v_2^2/2g + z_2 + h_L$$
$$(40)(144)/[(0.85)(62.4)] + v_1^2/2g + 200 = (p_2)(144)/[(0.85)(62.4)] + v_2^2/2g + 50 + 75$$
$$v_1^2/2g = v_2^2/2g \qquad p_2 = 67.6 \text{ lb/in}^2$$

**9.179** What size galvanized iron pipe is needed to be "hydraulically smooth" at $N_R = 3.5 \times 10^5$? (A pipe is said to be hydraulically smooth when it has the same losses as a smoother pipe under the same conditions.)

▮ From Fig. A-5, $\epsilon/d = 0.00001$ is equivalent to smooth pipe. $0.0005/d = 0.00001$, $d = 50.0$ ft.

**9.180** Above what Reynolds number is the flow through a 3-m-diameter riveted steel pipe, $\epsilon = 30$ mm, independent of the viscosity of the fluid?

▮ $\epsilon/d = 0.030/3 = 0.010$. From Fig. A-5, $N_R = 1.0 \times 10^5$ at complete turbulence.

**9.181** Determine the absolute roughness of a 1-ft-diameter pipe that has a friction factor of 0.032 when $N_R = 1.0 \times 10^6$.

▮ From Fig. A-5, $\epsilon/d = 0.006$. $\epsilon/1 = 0.006$, $\epsilon = 0.006$ ft.

**9.182** Galvanized iron pipe of diameter $d$ has the same friction factor for $N_R = 100\,000$ as 400-mm-diameter cast iron pipe. Evaluate $d$.

▮ **For cast iron:** $\epsilon/d = 0.00026/0.400 = 0.00065$.

**For galvanized iron:** $\epsilon/d = 0.00015/d = 0.00065$, $d = 0.231$ m, or 231 mm.

**9.183** Calculate the friction factor for atmospheric air at 80 °F, flowing at $v = 40$ ft/s in a 2-ft-diameter galvanized pipe.

▮ $$f = 1.325/[\ln(\epsilon/3.7d + 5.74/N_R^{0.9})]^2 \qquad N_R = dv/v = (3)(40)/(1.69 \times 10^{-4}) = 710\,059$$
$$f = 1.325/\{\ln[0.0005/(3.7)(2) + 5.74/710\,059^{0.9}]\}^2 = 0.0156$$

**9.184** Atmospheric air at 92 °F is conducted through a 4-ft-diameter, 1000-ft-long wrought iron pipe. Find the head loss corresponding to a flow of 266.7 cfs.

▮ $$h_f = (f)(L/d)(v^2/2g) \qquad \rho = p/RT = (14.7)(144)/[(1716)(460 + 92)] = 0.002234 \text{ slug/ft}^3$$
$$v = Q/A = 266.7/[(\pi)(4)^2/4] = 21.22 \text{ ft/s}$$
$$N_R = \rho dv/\mu = (0.002234)(4)(21.22)/(3.90 \times 10^{-7}) = 4.86 \times 10^5 \qquad \epsilon/d = 0.00015/4 = 0.0000375$$

From Fig. A-5, $f = 0.013$. $h_f = (0.013)(1000/4)\{21.22^2/[(2)(32.2)]\} = 22.72$ ft.

**9.185** A smooth toroidal wind tunnel is 60 m around at the centerline and has cross-sectional diameter 2 m; the fluid is air at 1 atm and 20 °C. Determine the horsepower rating of a fan that will produce a 500-km/h airstream.

■ $$h_f = (f)(L/d)(v^2/2g) \qquad v = 5 \times 10^5/3600 = 138.9 \text{ m/s}$$
$$N_R = dv/\nu = (2)(138.9)/(1.46 \times 10^{-5}) = 1.90 \times 10^7$$

From Fig. A-5, $f = 0.0072$ (extrapolated).

$$h_f = (0.0072)(60/2)\{138.9^2/[(2)(9.807)]\} = 212.5 \text{ m} \qquad Q = Av = [(\pi)(2^2/4)](138.9) = 436.4 \text{ m}^3/\text{s}$$
$$P = Q\gamma h_f = (436.4)(11.8)(212.5) = 1.09 \times 10^6 \text{ W} \quad \text{or} \quad 1460 \text{ hp}$$

**9.186** Assume that 2.0 cfs of oil ($\mu = 0.0003342$ slug/ft · s, $\rho = 1.677$ slug/ft³) is pumped through a 12-in pipeline of cast iron. If each pump produces 100 psi, how far apart can they be placed?

■ $$h_f = (f)(L/d)(v^2/2g) = p/\rho g = (100 \times 144)/(1.677)(32.2) = 266.7 \text{ft}$$
$$v = Q/A = 2.0/[(\pi)(\tfrac{12}{12})^2/4] = 2.546 \text{ ft/s}$$
$$N_R = \rho dv/\mu = (1.667)(\tfrac{12}{12})(2.546)/0.0003342 = 1.27 \times 10^4 \qquad \epsilon/d = 0.00085/(\tfrac{12}{12}) = 0.00085$$

From Fig. A-5, $f = 0.031$. $266.7 = 0.031[L/(\tfrac{12}{12})]\{2.546^2/[(2)(32.2)]\}$, $L = 85\,473$ ft, or 16.2 miles.

**9.187** A 60-mm-diameter smooth pipe 160 m long conveys 36 m³/h of water at 25 °C from a sidewalk hydrant to the top of a building 25 m tall. What pressure can be maintained at the top of the building, if the hydrant pressure is 1.6 MPa?

■ $$p_1/\gamma + v_1^2/2g + z_1 = p_2/\gamma + v_2^2/2g + z_2 + h_L \qquad h_L = h_f = (f)(L/d)(v^2/2g)$$
$$v = Q/A = (36/3600)/[(\pi)(0.060)^2/4] = 3.537 \text{ m/s} \qquad N_R = dv/\nu = (0.060)(3.537)/(9.10 \times 10^{-7}) = 2.33 \times 10^5$$

From Fig. A-5, $f = 0.016$.

$$h_L = 0.016[160/0.060]\{3.537^2/[(2)(9.807)]\} = 27.21 \text{ m}$$

$$(1.6)(1000)/9.79 + v_1^2/2g + 0 = (p_2)(1000)/9.79 + v_2^2/2g + 25 + 27.21 \qquad v_1^2/2g = v_2^2/2g \qquad p_2 = 1.09 \text{ MPa}$$

**9.188** Calculate the discharge for the pipe of Fig. 9-32; the fluid is water at 150 °F.

■ $$h_L = h_f = (f)(L/d)(v^2/2g) \qquad 258 = (f)[238/(\tfrac{2}{12})]\{v_2^2/[(2)(32.2)]\} \qquad v = 3.411/\sqrt{f}$$

Try $f = 0.019$: $v = 3.411/\sqrt{0.019} = 24.74$ ft/s, $\epsilon/d = 0.00015/(\tfrac{2}{12}) = 0.000900$, $N_R = dv/\nu = (\tfrac{2}{12})(24.74)/(4.68 \times 10^{-6}) = 8.81 \times 10^5$. From Fig. A-5, $f = 0.019$ (O.K.); $Q = Av = [(\pi)(\tfrac{2}{12})^2/4](24.74) = 0.540$ ft³/s.

2-in-diam
wrought iron

258 ft
238 ft

**Fig. 9-32**

**9.189** In Fig. 9-32 how much power would be required to pump 160 gpm of water at 60 °F from a reservoir at the bottom of the pipe to the reservoir shown?

■ $$Q = (160)(0.002228) = 0.3565 \text{ ft}^3/\text{s} \qquad p_1/\gamma + v_1^2/2g + z_1 = p_2/\gamma + v_2^2/2g + z_2 + h_L$$
$$v_1 = Q/A_1 = 0.3565/[(\pi)(\tfrac{2}{12})^2/4] = 16.34 \text{ ft/s} \qquad h_L = h_f = (f)(L/d)(v^2/2g)$$
$$N_R = dv/\nu = (\tfrac{2}{12})(16.34)/(1.21 \times 10^{-5}) = 2.25 \times 10^5 \qquad \epsilon/d = 0.00015/(\tfrac{2}{12}) = 0.000900$$

From Fig. A-5, $f = 0.0205$.

$$h_L = 0.0205[238/(\tfrac{2}{12})]\{16.34^2/[(2)(32.2)]\} = 121.4 \text{ ft} \qquad p_1/\gamma + 16.34^2/[(2)(32.2)] + 0 = 0 + 0 + 260 + 121.4$$

$$p_1/\gamma = 377.3 \text{ ft} \qquad P = Q\gamma(\Delta p/\gamma) = (0.3565)(62.4)(377.3) = 8393 \text{ ft} \cdot \text{lb/s} = 15.3 \text{ hp}$$

**9.190** A 12-mm-diameter commercial steel pipe 16 m long is used to drain an oil tank. Determine the discharge when the oil level in the tank is 2 m above the exit end of the pipe. ($\mu = 0.10 \text{ Pa} \cdot \text{s}$, $\gamma = 8 \text{ kN/m}^3$.)

▌ Assuming laminar flow,

$$v = \frac{h_L \gamma d^2}{32\mu L} = \frac{2[(8)(1000)](0.012)^2}{(32)[(0.1)(0.10)](16)} = 0.4500 \text{ m/s}$$

$$N_R = \rho dv/\mu = (\gamma/g)(dv)/\mu = [(8)(1000)/9.807](0.012)(0.4500)/[(0.1)(0.10)] = 440 \qquad \text{(laminar)}$$

$$Q = Av = [(\pi)(0.012)^2/4](0.4500) = 0.0000509 \text{ m}^3/\text{s} \quad \text{or} \quad 0.0509 \text{ L/s}$$

**9.191** Two liquid reservoirs are connected by 198 ft of 2-in-diameter smooth tubing. What is the flow rate when the difference in elevation is 50 ft? ($\nu = 0.001 \text{ ft}^2/\text{s}$).

▌ $\qquad p_1/\gamma + v_1^2/2g + z_1 = p_2/\gamma + v_2^2/2g + z_2 + h_L \qquad h_L = h_f = (f)(L/d)(v^2/2g)$

Assuming laminar flow,

$$f = 64/N_R \qquad N_R = dv/\nu = (\tfrac{2}{12})(v)/0.001 = 166.7v \qquad h_L = (64/166.7v)[198/(\tfrac{2}{12})]\{v^2/[(2)(32.2)]\} = 7.082v$$

$$0 + 0 + 50 = 0 + 0 + 0 + 7.082v \qquad v = 7.060 \text{ ft/s} \qquad N_R = (\tfrac{2}{12})(7.060)/0.001 = 1177 \qquad \text{(laminar)}$$

$$Q = Av = [(\pi)(\tfrac{2}{12})^2/4](7.060) = 0.154 \text{ ft}^3/\text{s}$$

**9.192** Atmospheric air at 16 °C flows for 200 m through a 1.25-m-diameter duct ($\epsilon = 1 \text{ mm}$). Calculate the flow volume, if the head loss measures 80 mmH$_2$O.

▌ $h_f = (f)(L/d)(v^2/2g) \qquad (\gamma h_f)_{\text{air}} = (\gamma h_f)_{\text{H}_2\text{O}} \qquad \gamma_{\text{air}} = p/RT = 101.4/[(29.3)(273 + 16)] = 0.01197 \text{ kN/m}^3$

$$(0.01197)(h_f)_{\text{air}} = (9.79)(0.080) \qquad (h_f)_{\text{air}} = 65.43 \text{ m} \qquad 65.43 = (f)(200/1.25)\{v^2/[(2)(9.807)]\} \qquad v = 2.832/\sqrt{f}$$

Try $f = 0.020$: $v = 2.832/\sqrt{0.020} = 20.03 \text{ m/s}$, $N_R = dv/\nu = (1.25)(20.03)/(1.46 \times 10^{-5}) = 1.71 \times 10^6$, $\epsilon/d = 0.001/1.25 = 0.000800$. From Fig. A-5, $f = 0.0205$. Try $f = 0.0205$: $v = 2.832/\sqrt{0.0205} = 19.78 \text{ m/s}$, $N_R = (1.25)(19.78)/(1.46 \times 10^{-5}) = 1.69 \times 10^6$, $f = 0.0205$ (O.K.); $Q = Av = [(\pi)(1.25^2/4)](19.78) = 24.3 \text{ m}^3/\text{s}$.

**9.193** Water at 20 °C is to be pumped through 2 km of 200-mm-diameter wrought iron pipe at the rate of 60 L/s. Compute the head loss and power required.

▌ $\qquad h_f = (f)(L/d)(v^2/2g) \qquad v = Q/A = (60 \times 10^{-3})/[(\pi)(0.200)^2/4] = 1.910 \text{ m/s}$

$$N_R = dv/\nu = (0.200)(1.910)/(1.02 \times 10^{-6}) = 3.75 \times 10^5 \qquad \epsilon/d = 0.000046/(0.200) = 0.000230$$

From Fig. A-5, $f = 0.016$.

$$h_f = 0.016[2000/0.200]\{1.910^2/[(2)(9.807)]\} = 29.76 \text{ m} \qquad P = Q\gamma h_f = (60 \times 10^{-3})(9.79)(29.76) = 17.48 \text{ kW}$$

**9.194** An industrial ventilation system contains 4000 ft of 12-in-diameter galvanized pipe. Neglecting minor losses, what head must a blower produce to furnish 3 ton/h of air at $p = 14$ psia, $T = 90$ °F?

▌ $h_f = (f)(L/d)(v^2/2g) \qquad N_R = \rho dv/\mu \qquad \rho = p/RT = (14)(144)/[(1716)(460 + 90)] = 0.002136 \text{ slug/ft}^3$

$$M = \rho Av \qquad 6000/3600 = [(0.002136)(32.2)][(\pi)(\tfrac{12}{12})^2/4](v) \qquad v = 30.85 \text{ ft/s}$$

$$N_R = (0.002136)(\tfrac{12}{12})(30.85)/(3.90 \times 10^{-7}) = 1.69 \times 10^5 \qquad \epsilon/d = 0.0005/(\tfrac{12}{12}) = 0.000500$$

From Fig. A-5, $f = 0.019$.

$$h_f = 0.019[4000/(\tfrac{12}{12})]\{30.85^2/[(2)(32.2)]\} = 1123.2 \text{ ft of air} \quad \text{or} \quad 1.24 \text{ ft of water}$$

**9.195** A 2.0-m-diameter pipe of length 1560 m for which $\epsilon = 1.5$ mm conveys water at 12 °C between two reservoirs at a rate of 8.0 m³/s. What must be the difference in water-surface elevations between the two reservoirs? Neglect minor losses.

$$p_1/\gamma + v_1^2/2g + z_1 = p_2/\gamma + v_2^2/2g + z_2 + h_L \qquad 0 + 0 + z_1 = 0 + 0 + z_2 + h_f$$

$$z_1 - z_2 = h_f = (f)(L/d)(v^2 2g)$$

$$v = Q/A = 8.0/[(\pi)(2.0)^2/4] = 2.546 \text{ m/s} \qquad N_R = dv/v = (2.0)(2.546)/(1.24 \times 10^{-6}) = 4.11 \times 10^6$$

$$\epsilon/d = (1.5/1000)/2.0 = 0.000750$$

From Fig. A-5, $f = 0.018$. $h_f = (0.018)(1560/2.0)\{2.546^2/[(2)(9.807)]\} = 4.64$ m. Hence, the difference in water-surface elevations between the two reservoirs is 4.64 m.

**9.196** Water flows from reservoir 1 to reservoir 2 through a 4-in-diameter, 500-ft-length pipe, as shown in Fig. 9-33. Assume an initial friction factor ($f$) of 0.037 and a roughness ($\epsilon$) of 0.003 ft for the pipe. Find the flow rate.

$$p_1/\gamma + v_1^2/2g + z_1 = p_2/\gamma + v_2^2/2g + z_2 + h_L \qquad 0 + 0 + 700.6 = 0 + 0 + 655.5 + h_L \qquad h_L = 45.2 \text{ ft} = h_f + h_m$$

(I) **Friction loss:**    $h_f = (f)(L/d)(v^2/2g) = 0.037[500/(\frac{4}{12})](v^2/2g) = 55.50v^2/2g$.

(II) **Minor losses:**
  (a) **Due to entrance:** From Fig. A-7, take $K_1 = 0.45$.
  (b) **Due to globe valve:** From Table A-11, $K_{\text{open}} = 5.7$. From Table A-12, take $K_2/K_{\text{open}} = 1.75$. Hence, $K_2 = (5.7)(1.75) = 9.98$.
  (c) **Due to bend:** $R/D = \frac{12}{4} = 3.0$, $\epsilon/D = 0.003/(\frac{4}{12}) = 0.00900$. From Fig. A-12, $K_3 = 0.45$.
  (d) **Due to elbow:** From Table A-11, $K_4 = 0.23$.
  (e) **Due to exit:** From Fig. A-7, $K_5 = 1.0$.
Thus,

$$h_f + h_m = \{v^2/[(2)(32.2)]\}(55.50 + 0.45 + 9.98 + 0.45 + 0.23 + 1.0) = 1.050v^2$$

$$1.050v^2 = 45.2 \qquad v = 6.561 \text{ ft/s}$$

$$N_R = Dv/v = (\tfrac{4}{12})(6.561)/(1.90 \times 10^{-5}) = 1.15 \times 10^5 \qquad \epsilon/D = 0.003/(\tfrac{4}{12}) = 0.00900$$

From Fig. A-5, $f = 0.037$. (Assumed value of $f$ O.K.) $Q = Av = [(\pi)(\tfrac{4}{12})^2/4](6.561) = 0.573$ ft³/s.

Elev. 700.6 ft

12-in-bend radius

Elev. 655.5 ft

Sharp-edged entrance

Globe valve,
25 % closed

Sharp-edged exit

90° long-radius elbow

Fig. 9-33

**9.197** Determine the head loss in the system shown in Fig. 9-34 and the discharge in the pipe.

$$p_1/\gamma + v_1^2/2g + z_1 = p_2/\gamma + v_2^2/2g + z_2 + h_L \qquad 0 + 0 + 100 = 0 + v_2^2/[(2)(32.2)] + 0 + h_L$$

$$h_L = h_f + h_m \qquad h_f = (f)(L/D)(v^2/2g) = f[500/(\tfrac{6}{12})]\{v_2^2/[(2)(32.2)]\} = 15.53fv_2^2 \qquad h_m = (K)(v^2/2g) \qquad K = 0.5$$

$$h_m = 0.5\{v_2^2/[(2)(32.2)]\} = 0.007764v_2^2 \qquad h_L = 15.53fv_2^2 + 0.007764v_2^2$$

$$100 = 0.01553v_2^2 + 15.53fv_2^2 + 0.007764v_2^2 \qquad v_2 = \sqrt{100/(0.02329 + 15.53f)}$$

Try $v_2 = 10$ ft/s: $N_R = Dv/v = (\tfrac{6}{12})(10)/(1.21 \times 10^{-5}) = 4.13 \times 10^5$, $\epsilon/D = 0.00015/(\tfrac{6}{12}) = 0.00030$. From Fig. A-5, $f = 0.0165$. $v_2 = \sqrt{100/[(0.02329 + (15.53)(0.0165)]} = 18.91$ ft/s. Try $v_2 = 18.91$ ft/s: $N_R = (\tfrac{6}{12})(18.91)/(1.21 \times 10^{-5}) = 7.81 \times 10^5$, $f = 0.0157$, $v_2 = \sqrt{100/[0.02329 + (15.53)(0.0157)]} = 19.35$ ft/s. Try $v_2 = 19.35$ ft/s: $N_R = (\tfrac{6}{12})(19.35)/(1.21 \times 10^{-5}) = 8.00 \times 10^5$, $f = 0.0157$ (O.K.); $h_L = (15.53)(0.0157)(19.35)^2 + (0.007764)(19.35)^2 = 94.2$ ft of water, $Q = Av = [(\pi)(\tfrac{6}{12})^2/4](19.35) = 3.80$ ft³/s.

Water at 60°F

100 ft

6-in-diameter new wrought iron pipe

500 ft

**Fig. 9-34**

**9.198** Figure 9-35 shows two reservoirs containing water at 60 °F. The water in the upper reservoir is to be drained to the other reservoir at a lower level as shown. The total length of commercial steel pipe is 100 ft, and the diameter of the pipe is 4 in. What will be the flow rate of water through the pipe when the water surface elevations are as shown in Fig. 9-35?

$$p_1/\gamma + v_1^2/2g + z_1 = p_2/\gamma + v_2^2/2g + z_2 + h_L \qquad 0 + 0 + 335 = 0 + 0 + 300 + h_L$$

$$h_L = h_f + h_m \qquad h_f = (f)(L/D)(v^2/2g) = f[100/(\tfrac{4}{12})]\{v_2^2/[(2)(32.2)]\} = 4.658fv_2^2 \qquad h_m = (K)(v^2/2g)$$

Due to entrance, take $K_1 = 0.45$. Due to elbow, $K_2 = 0.64$. Due to gate valve, $K_{\text{open}} = 0.11$.
Take $K_3/K_{\text{open}} = 17$. Hence, $K_3 = (0.11)(17) = 1.87$. Due to exit, $K_4 = 1.0$.

$$h_m = (0.45 + 0.64 + 1.87 + 1.0)\{v^2/[(2)(32.2)]\} = 0.06149v^2 \qquad h_L = 4.658fv^2 + 0.06149v^2$$

$$335 = 300 + 4.658fv^2 + 0.06149v^2 \qquad v = \sqrt{35/(4.658f + 0.06149)}$$

Try $f = 0.019$: $v = \sqrt{35/[(4.658)(0.019) + 0.06149]} = 15.28$ ft/s, $N_R = Dv/v = (\tfrac{4}{12})(15.28)/(1.21 \times 10^{-5}) = 4.21 \times 10^5$. From Fig. A-5, $f = 0.0175$. Try $f = 0.0175$: $v = \sqrt{35/[(4.658)(0.0175) + 0.06149]} = 15.64$ ft/s, $N_R = (\tfrac{4}{12})(15.64)/(1.21 \times 10^{-5}) = 4.31 \times 10^5$, $f = 0.0175$ (O.K.); $Q = Av = [(\pi)(\tfrac{4}{12})^2/4](15.64) = 1.36$ ft³/s.

**9.199** A 15-in-diameter new cast iron pipe connecting two reservoirs as shown in Fig. 9-36 carries water at 60 °F. The pipe is 120 ft long, and the discharge is 20 ft/s. Determine the difference in elevation between water surfaces in the two reservoirs.

$$p_1/\gamma + v_1^2/2g + z_1 = p_2/\gamma + v_2^2/2g + z_2 + h_L \qquad 0 + 0 + z_1 = 0 + 0 + z_2 + h_L$$

Since $z_1 - z_2 = H$ and $h_L = h_f + h_m$, $H = h_f + h_m$, $h_f = (f)(L/D)(v^2/2g)$.

$$v = Q/A = 20/[(\pi)(\tfrac{15}{12})^2/4] = 16.30 \text{ ft/s}$$

$$N_R = Dv/v = (\tfrac{15}{12})(16.30)/(1.21 \times 10^{-5}) = 1.68 \times 10^6 \qquad \epsilon/D = 0.00085/(\tfrac{15}{12}) = 0.00068$$

From Fig. A-5, $f = 0.018$. $h_f = 0.018[120/(\tfrac{15}{12})]\{16.30^2/[(2)(32.2)]\} = 7.13$ ft, $h_m = (K)(v^2/2g)$. For entrance, take $K_1 = 0.45$. For exit, $K_2 = 1.0$. $h_m = (0.45 + 1.0)\{16.30^2/[(2)(32.2)]\} = 5.98$ ft, $H = 7.13 + 5.98 = 13.11$ ft.

Fig. 9-35

Fig. 9-36

**9.200** Repeat Prob. 9.87 by including losses due to a sharp-edged entrance, the exit, and a fully open flanged globe valve.

▮ For entrance, $K_1 = 0.5$. For exit, $K_2 = 1.0$. For globe valve, $K_3 = 8.5$. From Prob. 9.87, $h_L = h_f + h_f = [(f)(L/d) + K_1 + K_2 + K_3](v^2/2g)$, $98 = \{f[7000/0.050] + 0.5 + 1.0 + 8.5\}\{v^2/[(2)(9.807)]\}$, $1922 = (140\,000f + 10.0)(v^2)$, $v = [1922/(140\,000f + 10.0)]^{1/2}$. Try $f = 0.02$: $v = \{1922/[(140\,000)(0.02) + 10.0]\}^{1/2} = 0.8270$ m/s, $N_R = \rho dv/\mu = (998)(0.050)(0.8270)/(1.02 \times 10^{-3}) = 4.05 \times 10^4$. From Fig. A-5, $f = 0.022$. Try $f = 0.022$: $v = \{1922/[(140\,000)(0.022) + 10.0]\}^{1/2} = 0.7887$ m/s, $N_R = (998)(0.050)(0.7887)/(1.02 \times 10^{-3}) = 3.86 \times 10^4$, $f = 0.022$ (O.K.); $Q = Av = [(\pi)(0.050)^2/4](0.7887) = 0.001548$ m³/s, or 5.57 m³/h.

**9.201** Repeat Prob. 9.95 by including losses due to a sharp entrance and a fully open screwed swing-check valve.

▮ For sharp entrance, $K_1 = 0.5$. For swing-check valve, $K_2 = 5.1$. $h_m = (K_1 + K_2)(v^2/2g)$. From Prob. 9.95, $v = 14.67$ ft/s. $h_m = (0.5 + 5.1)\{14.67^2/[(2)(32.2)]\} = 18.71$ ft. From Prob. 9.95, $0 + 0 + h = 0 + 14.67^2/[(2)(32.2)] + 0 + (236.6 + 18.71)$, $h = 258.7$ ft.

**9.202** Two lakes (Fig. 9-37) communicate via two 20-ft-long wrought iron pipes joined abruptly. The entrance and exit are sharp-edged. Including minor losses, compute the flow rate if lake 1 stands 59 ft higher than lake 2, and the average temperature of the system is 20 °C.

▮
$$p_1/\gamma + v_1^2/2g + z_1 = p_2/\gamma + v_2^2/2g + z_2 + h_L \qquad h_L = h_f + h_m$$

$$h_f = (f)(L/D)(v^2/2g) = (f_a)[20/(\tfrac{1}{12})]\{v_a^2/[(2)(32.2)]\} + f_b[20/(\tfrac{2}{12})]\{v_b^2/[(2)(32.2)]\}$$

$$A_a v_a = A_b v_b \qquad [(\pi)(\tfrac{1}{12})^2/4](v_a) = [(\pi)(\tfrac{2}{12})^2/4](v_b) \qquad v_b = v_a/4$$

$$h_f = 3.727 f_a v_a^2 + (1.863)(f_b)(v_a/4)^2 = 3.727 f_a v_a^2 + 0.01164 f_b v_a^2$$

For sharp entrance, $K_1 = 0.5$. For sharp exit, $K_2 = 1.0$. For sudden expansion, $K_2 = (1 - 0.5)^2 = 0.5625$.
$h_m = (0.5 + 0.5625)\{v_a^2/[(2)(32.2)]\} + (1.0)\{v_b^2/[(2)(32.2)]\} = 0.01650v_a^2 + (0.01553)(v_a/4)^2 = 0.01747v_a^2$, $0 + 0 + 59 = 0 + 0 + 0 + [(3.727f_a v_a^2 + 0.1164f_b v_a^2) + 0.01747v_a^2]$. Try $f_a = f_b = 0.020$:

$$59 = (3.727)(0.020)(v_a^2) + (0.1164)(0.020)(v_a^2) + 0.01747v_a^2 \qquad v_a = 25.01 \text{ ft/s} \qquad v_b = 25.01/4 = 6.253 \text{ ft/s}$$

$$N_R = Dv/v \qquad (N_R)_a = (\tfrac{1}{12})(25.01)/(1.11 \times 10^{-5}) = 1.88 \times 10^5 \qquad (\epsilon/D)_a = 0.00015/(\tfrac{1}{12}) = 0.00180$$

From Fig. A-5, $f_a = 0.024$. $(N_R)_b = (\tfrac{2}{12})(6.253)/(1.11 \times 10^{-5}) = 9.39 \times 10^4$, $(\epsilon/D)_b = 0.00015/(\tfrac{2}{12}) = 0.000900$, $f_b = 0.022$. Try $f_a = 0.024$ and $f_b = 0.022$:

$$59 = (3.727)(0.024)(v_a^2) + (0.1164)(0.022)(v_a^2) + 0.01747v_a^2 \qquad v_a = 23.21 \text{ ft/s} \qquad v_b = 23.21/4 = 5.803 \text{ ft/s}$$

$$(N_R)_a = (\tfrac{1}{12})(23.41)/(1.11 \times 10^{-5}) = 1.76 \times 10^5 \qquad f_a = 0.024$$

$$(N_R)_b = (\tfrac{2}{12})(5.852)/(1.11 \times 10^{-5}) = 8.79 \times 10^4 \qquad f_b = 0.022$$

Therefore, $f_a = 0.024$ and $f_b = 0.022$ is O.K. $Q = Av = [(\pi)(\tfrac{1}{12})^2/4](23.21) = 0.127 \text{ ft}^3/\text{s}$.

$D = 2$ in
$L = 20$ ft

(1)

(2)

$D = 1$ in
$L = 20$ ft

**Fig. 9-37**

**9.203** Two reservoirs containing water at 20 °C are connected by 800 m of 180-mm cast iron pipe, including a sharp entrance, a submerged exit, a gate valve 75 percent open, two 1-m-radius bends, and six regular 90° elbows. If the flow rate is 9 m³/min, find the difference in reservoir elevations.

$$p_1/\gamma + v_1^2/2g + z_1 = p_2/\gamma + v_2^2/2g + z_2 + h_L \qquad h_L = h_f + h_m \qquad h_f = (f)(L/D)(v^2/2g)$$

$$v = Q/A = (9/60)/[(\pi)(0.180)^2/4] = 5.895 \text{ m/s} \qquad N_R = Dv/v = (0.180)(5.895)/(1.02 \times 10^{-6}) = 1.04 \times 10^6$$

$$\epsilon/D = 0.00026/0.180 = 0.00144$$

From Fig. A-5, $f = 0.0217$. $h_f = 0.0217[800/0.180]\{5.895^2/[(2)(9.807)]\} = 170.9 \text{ m}$. For sharp entrance, $K_1 = 0.5$. For exit, $K_2 = 1.0$. For gate valve 75 percent open, $K_3 = 0.3$. For bends, $K_4 = (2)(0.15) = 0.30$. For elbows, $K_5 = (6)(0.27) = 1.62$. $h_m = (0.5 + 1.0 + 0.3 + 0.30 + 1.62)\{5.895^2/[(2)(9.807)]\} = 6.6 \text{ m}$, $h_L = 170.9 + 6.6 = 177.5 \text{ m}$, $0 + 0 + z_1 = 0 + 0 + z_2 + 177.5$, $z_1 - z_2 = 177.5 \text{ m}$.

**9.204** The system in Fig. 9-38 consists of 1000 m of 50-mm cast iron pipe, two 45° and four 90° flanged long-radius elbows, a fully open flanged globe valve, and a sharp exit into a reservoir. What gage pressure is required at point 1 to deliver 5 L/s of water at 20 °C into the reservoir, whose free surface lies 100 m above point 1?

$$p_1/\gamma + v_1^2/2g + z_1 = p_2/\gamma + v_2^2/2g + z_2 + h_L \qquad h_L = h_f + h_m \qquad h_f = (f)(L/D)(v^2/2g)$$

$$v = Q/A = 0.005/[(\pi)(0.050)^2/4] = 2.546 \text{ m/s} \qquad N_R = Dv/v = (0.050)(2.546)/(1.02 \times 10^{-6}) = 1.25 \times 10^5$$

$$\epsilon/D = 0.00026/0.050 = 0.00520$$

From Fig. A-5, $f = 0.0315$. $h_f = 0.0315[1000/0.050]\{2.546^2/[(2)(9.807)]\} = 208.2 \text{ m}$. For 45° elbows, $K_1 = (2)(0.20) = 0.40$. For 90° elbows, $K_2 = (4)(0.30) = 1.20$. For the open valve, $K_3 = 8.5$. For exit, $K_4 = 1.0$. $h_m = (0.40 + 1.20 + 8.5 + 1.0)\{2.546^2/[(2)(9.807)]\} = 3.7 \text{ m}$, $h_L = 208.2 + 3.7 = 211.9 \text{ m}$, $p_1/9.79 + 2.546^2/[(2)(9.807)] + 0 = 0 + 0 + 100 + 211.9$, $p_1 = 3056 \text{ kPa gage}$.

45°

Sharp
exit

Open
globe

1

45°

**Fig. 9-38**

**9.205** A 1-in-diameter smooth water pipe slopes upward at 30° (Fig. 9-39); the flanged globe valve is fully open. Calculate the flow rate.

▮ $$p_1/\gamma + v_1^2/2g + z_1 = p_2/\gamma + v_2^2/2g + z_2 + h_L \qquad v_1^2/2g = v_2^2/2g \qquad h_L = h_f + h_m$$

Therefore, $p_1/\gamma + z_1 = p_2/\gamma + z_2 + h_f + h_m$,

$$(p_1 - p_2)/\gamma = z_2 - z_1 + h_f + h_m \tag{1}$$

From manometer,

$$p_1 - p_2 = (\gamma_{Hg} - \gamma_{H_2O})(\tfrac{7}{12}) + (\gamma_{H_2O})(z_2 - z_1) \tag{2}$$

Combining Eqs. (1) and (2), $[(\gamma_{Hg} - \gamma_{H_2O})/\gamma_{H_2O}](\tfrac{7}{12}) = h_f + h_m$, $h_f = (f)(L/D)(v^2/2g) = (f)[(9/\cos 30°)/(\tfrac{1}{12})]\{v^2/[(2)(32.2)]\} = 1.936fv^2$, $h_m = (K)(v^2/2g)$. For globe valve, $K = 13$. $h_m = 13\{v^2/[(2)(32.2)]\} = 0.2019v^2$, $\{[(13.6)(62.4) - 62.4]/62.4\}(\tfrac{7}{12}) = 1.936fv^2 + 0.2019v^2$, $v = \sqrt{7.350/(1.936f + 0.2019)}$. Try $f = 0.02$: $v = \sqrt{7.350/[(1.936)(0.02) + 0.2019]} = 5.527$ ft/s, $N_R = Dv/\nu = (\tfrac{1}{12})(5.527)/(1.05 \times 10^{-5}) = 4.39 \times 10^4$. From Fig. A-5, $f = 0.0217$. Try $f = 0.0217$: $v = \sqrt{7.350/[(1.936)(0.0217) + 0.2019]} = 5.489$ ft/s, $N_R = (\tfrac{1}{12})(5.489)/(1.05 \times 10^{-5}) = 4.36 \times 10^4$, $f = 0.0217$ (O.K.); $Q = Av = [(\pi)(\tfrac{1}{12})^2/4](5.489) = 0.0299$ ft³/s.

**Fig. 9-39**

**9.206** A pipe system carries water from a reservoir and discharges it as a free jet, as shown in Fig. 9-40. How much flow is to be expected through a 200-mm steel commercial pipe with the fittings shown?

▮ $$p_1/\gamma + v_1^2/2g + z_1 = p_2/\gamma + v_2^2/2g + z_2 + h_L \qquad h_L = h_f + h_m$$

$$h_f = (f)(L/D)(v^2/2g) = f[(60 + 20 + 60)/(\tfrac{200}{1000})]\{v_2^2/[(2)(9.807)]\} = 35.69fv_2^2$$

$$h_m = (K)(v^2/2g) = [0.05 + (2)(0.40)]\{v_2^2/[(2)(9.807)]\} = 0.04334v_2^2$$

$$0 + 0 + 30 = 0 + v_2^2/[(2)(9.807)] + 20 + (35.69fv_2^2 + 0.04334v_2^2) \qquad v_2 = \sqrt{10/(35.69f + 0.09432)}$$

Try $f = 0.014$: $v_2 = \sqrt{10/[(35.69)(0.014) + 0.09432]} = 4.103$ m/s, $N_R = Dv/\nu = (\tfrac{200}{1000})(4.103)/(0.0113 \times 10^{-4}) = 7.26 \times 10^5$, $\epsilon/D = 0.000046/(\tfrac{200}{1000}) = 0.000230$. From Fig. A-5, $f = 0.0152$. Try $f = 0.0152$: $v_2 = \sqrt{10/[(35.69)(0.0152) + 0.09432]} = 3.963$ m/s, $N_R = (\tfrac{200}{1000})(3.963)/(0.0113 \times 10^{-4}) = 7.01 \times 10^5$, $f = 0.0152$ (O.K.); $Q = Av = [(\pi)(\tfrac{200}{1000})^2/4](3.963) = 0.125$ m³/s.

$\nu = .0113 \times 10^{-4}$ m²/s
$\rho = 999$ kg/m³  **Fig. 9-40**

**9.207** A pipe system having a given centerline geometry as shown in Fig. 9-41 is to be chosen to transport a maximum of 1 ft³/s of oil from tank $A$ to tank $B$. What is a pipe size that will do the job?

▮ $$p_c/\gamma + v_c^2/2g + z_c = p_p/\gamma + v_p^2/2g + z_p + h_L \qquad h_L = h_f + h_m$$

$$h_f = (f)(L/D)(v^2/2g) = f[(100 + 130 + 300)/D]\{v_p^2/[(2)(32.2)]\} = 8.230fv_p^2/D$$

$$h_m = (K)(v^2/2g) = (0.05 + 0.5 + 0.5 + 1)\{v_p^2/[(2)(32.2)]\} = 0.03183v_p^2$$

$$(100)(144)/50 + 0 + 130 = (50)(144)/50 + v_p^2/[(2)(32.2)] + 0 + 8.230fv_p^2/D + 0.03183v_p^2$$

$$0.04736v_p^2 + 8.230fv_p^2/D = 274 \qquad v_p^2 = 274/(0.04736 + 8.230f/D) \qquad v_p = Q/A_p = 1/(\pi D^2/4) = 1.273/D^2$$

$$(1.273/D^2)^2 = 274/(0.04736 + 8.230f/D)$$

Try $f = 0.015$: $(1.273/D^2)^2 = 274/[0.04736 + (8.230)(0.015)/D]$. By trial and error, $D = 0.240$ ft.

$$v = 1.273/0.240^2 = 22.1 \text{ ft/s} \qquad N_R = \rho Dv/\mu = (\gamma/g)(Dv)/\mu = (50/32.2)(0.240)(22.1)/(50 \times 10^{-5}) = 1.65 \times 10^4$$

$$\epsilon/D = 0.00015/0.240 = 0.000625$$

From Fig. A-5, $f = 0.0285$. Try $f = 0.0285$: $(1.273/D^2)^2 = 274/[0.04736 + (8.230)(0.0285)/D]$. By trial and error, $D = 0.272$ ft.

$$v = 1.273/0.272^2 = 17.2 \text{ ft/s} \qquad N_R = (50/32.2)(0.272)(17.2)/(50 \times 10^{-5}) = 1.45 \times 10^4$$

$$f = 0.0285 \qquad \text{(O.K.)}$$

Hence, $D = 0.272$ ft, or 3.26 in.

**Fig. 9-41**

**9.208** What gage pressure $p_1$ is required to cause 5 ft³/s of water to flow through the system shown in Fig. 9-42? Assume that the reservoir is large, that minor losses are negligible, and that $\nu = 2.11 \times 10^{-5}$ ft²/s.

$$p_1/\gamma + v_1^2/2g + z_1 = p_2/\gamma + v_2^2/2g + z_2 + h_L \qquad v_2 = Q/A_2 = 5/[(\pi)(\tfrac{6}{12})^2/4] = 25.46 \text{ ft/s}$$

$$h_L = h_f = (f)(L/D)(v^2/2g) \qquad N_R = dv/\nu = (\tfrac{6}{12})(25.46)/(2.11 \times 10^{-5}) = 6.03 \times 10^5$$

$$\epsilon/d = 0.00015/(\tfrac{6}{12}) = 0.00030$$

From Fig. A-5, $f = 0.016$.

$$h_L = 0.016[(400 + 160 + 250)/(\tfrac{6}{12})]\{25.46^2/[(2)(32.2)]\} = 260.9 \text{ ft}$$

$$(p_1)(144)/62.4 + 0 + 120 = 0 + 25.46^2/[(2)(32.2)] + 160 + 260.9 \qquad p_1 = 135 \text{ lb/in}^2 \text{ gage}$$

**Fig. 9-42**

**9.209** In Prob. 9.208, let 6 in be the *nominal* diameter of the pipe. For the entrance fitting, $r/d = 0.06$. Calculate the pressure $p_1$. The elbows are screwed elbows and there is now an open globe valve in the pipe system. Include minor losses.

Use $d = 6.065$ in

$$p_1/\gamma + v_1^2/2g + z_1 = p_2/\gamma + v_2^2/2g + z_2 + h_L \qquad v_2 = Q/A_2 = 5/[(\pi)(6.065/12)^2/4] = 24.92 \text{ ft/s}$$

$$h_L = h_f + h_m \qquad h_f = (f)(L/d)(v^2/2g) \qquad N_R = dv/\nu = (6.065/12)(24.92)/(2.11 \times 10^{-5}) = 5.97 \times 10^5$$

$$\epsilon/d = 0.00015/(6.065/12) = 0.000297$$

From Fig. A-5, $f = 0.016$. $h_f = 0.016[(400 + 160 + 250)/(6.065/12)]\{24.92^2/[(2)(32.2)]\} = 247.3$ ft, $h_m = (K)(v^2/2g)$. For entrance, $K_1 = 0.15$. For elbows, $K_2 = (2)(0.45) = 0.90$. For globe valve, $K_3 = 5.1$.

$$h_m = (0.15 + 0.90 + 5.1)\{24.92^2/[(2)(32.2)]\} = 59.3 \text{ ft} \qquad h_L = 247.3 + 59.3 = 306.6 \text{ ft}$$

$$(p_1)(144)/62.4 + 0 + 120 = 0 + 24.92^2/[(2)(32.2)] + 160 + 306.6 \qquad p_1 = 154 \text{ lb/in}^2 \text{ gage}$$

**9.210** The industrial scrubber $B$ of Fig. 9-43 consumes water ($v = 0.113 \times 10^{-5} \text{ m}^2/\text{s}$) at the rate of $0.1 \text{ m}^3/\text{s}$. If the pipe is 150-mm commercial pipe, compute the necessary tank pressure $p_1$.

$\blacksquare$ $\qquad p_A/\gamma + v_A^2/2g + z_A = p_B/\gamma + v_B^2/2g + z_B + h_L \qquad v_B = Q/A_B = 0.1/[(\pi)(0.150)^2/4] = 5.659 \text{ m/s}$

$$h_L = h_f + h_m \qquad h_f = (f)(L/d)(v^2/2g) \qquad N_R = dv/v = (0.150)(5.659)/(0.113 \times 10^{-5}) = 7.51 \times 10^5$$

$$\epsilon/d = 0.000046/0.150 = 0.000307$$

From Fig. A-5, $f = 0.016$.

$$h_f = 0.016[(300 + 150 + 295)/0.150]\{5.659^2/[(2)(9.807)]\} = 129.7 \text{ m}$$

$$h_m = (K)(v^2/2g) = (0.4 + 0.9 + 0.9 + 1)\{5.659^2/[(2)(9.807)]\} = 5.2 \text{ m} \qquad h_L = 129.7 + 5.2 = 134.9 \text{ m}$$

$$p_1/9.79 + 0 + 25 = 40/9.79 + 5.659^2/[(2)(9.807)] + 150 + 134.9 \qquad p_1 = 2600 \text{ kPa gage}$$

Fig. 9-43

**9.211** What pressure $p_1$ is required in Fig. 9-44 to cause $1 \text{ ft}^3/\text{s}$ of water to flow, given $p_2 = 5$ psig? Take $\mu = 2.11 \times 10^{-5} \text{ lb} \cdot \text{s/ft}^2$ for water.

$\blacksquare$ $\qquad p_A/\gamma + v_A^2/2g + z_A = p_B/\gamma + v_B^2/2g + z_B + h_L \qquad p_A/\gamma = (p_1)(144)/62.4 + (10)(0.8) = 2.308p_1 + 8.00$

$$v_B = Q/A_B = 1/[(\pi)(\tfrac{6}{12})^2/4] = 5.093 \text{ ft/s} \qquad h_L = h_f + h_m \qquad h_f = (f)(L/d)(v^2/2g)$$

$$N_R = \rho dv/\mu = (1.94)(\tfrac{6}{12})(5.093)/(2.11 \times 10^{-5}) = 2.34 \times 10^5 \qquad \epsilon/d = 0.00015/(\tfrac{6}{12}) = 0.00030$$

From Fig. A-5, $f = 0.0175$.

$$h_f = 0.0175[2800/(\tfrac{6}{12})]\{5.093^2/[(2)(32.2)]\} = 39.47 \text{ ft} \qquad h_m = (K)(v^2/2g) = (0.5 + 1)\{5.093^2/[(2)(32.2)]\} = 0.60 \text{ ft}$$

$$h_L = 39.47 + 0.60 = 40.07 \text{ ft}$$

$$(2.308p_1 + 8.00) + 0 + 25 = (5)(144)/62.4 + 5.093^2/[(2)(32.2)] + 0 + 40.07 \qquad p_1 = 8.2 \text{ psig}$$

Fig. 9-44

**9.212** Find the flow through the system of Fig. 9-45. Use $\epsilon = 0.00015$ ft.

█ $p_A/\gamma + v_A^2/2g + z_A = p_B/\gamma + v_B^2/2g + z_B + h_L$ $\quad 0 + 0 + (100 + 100) = 0 + 0 + 100 + h_L$ $\quad h_L = 100$ ft $= h_f + h_m$

$$h_f = (f)(L/D)(v^2/2g) = f[(500 + 100 + 102)/(\tfrac{6}{12})]\{v^2/[(2)(32.2)]\} = 21.80fv^2$$

$$h_m = (K)(v^2/2g) = (0.4 + 0.8 + 0.8 + 1)\{v^2/[(2)(32.2)]\} = 0.04658v^2 \quad h_L = 21.80fv^2 + 0.04658v^2$$

$$100 = 21.80fv^2 + 0.04658v^2 \quad v = \sqrt{100/(21.80f + 0.04658)}$$

Try $f = 0.015$: $v = \sqrt{100/[(21.80)(0.015) + 0.04658]} = 17.49$ ft/s, $N_R = Dv/v = (\tfrac{6}{12})(17.49)/(1.21 \times 10^{-5}) =$
$7.23 \times 10^5$, $\epsilon/D = 0.00015/(\tfrac{6}{12}) = 0.00030$. From Fig. A-5, $f = 0.016$. Try $f = 0.016$: $v =$
$\sqrt{100/[(21.74)(0.016) + 0.04658]} = 15.92$ ft/s, $N_R = (\tfrac{6}{12})(15.92)/(1.21 \times 10^{-5}) = 6.58 \times 10^5$, $f = 0.016$ (O.K.);
$Q = Av = [(\pi)(\tfrac{6}{12})^2/4](15.92) = 3.13$ ft³/s.

**Fig. 9-45**

**9.213** A flow of 170 L/s is to go from tank $A$ to tank $B$ in Fig. 9-46. If $v = 0.113 \times 10^{-5}$ m²/s, what should the diameter be for the horizontal section of pipe?

█ $p_A/\gamma + v_A^2/2g + z_A = p_B/\gamma + v_B^2/2g + z_B + h_L$ $\quad 0 + 0 + (35 + 35) = 0 + 0 + 16 + h_L$ $\quad h_L = 54$ m $= h_f + h_m$

$$v_1 = Q/A_1 = (\tfrac{170}{1000})/(\pi D^2/4) = 0.2165/D^2 \quad v_2 = Q/A_2 = (\tfrac{170}{1000})/[(\pi)(\tfrac{150}{1000})^2/4] = 9.620 \text{ m/s}$$

$$h_f = (f)(L/D)(v^2/2g) = (f_1)(65/D)\{(0.2165/D^2)^2/[(2)(9.807)]\}$$
$$+ f_2[35/(\tfrac{150}{1000})]\{9.620^2/[(2)(9.807)]\} = 0.1553f_1/D^5 + 1101f_2$$

$$h_m = (K)(v^2/2g) = (0.4 + 0.9)\{(0.2165/D^2)^2/[(2)(9.807)]\}$$
$$+ 1.0\{9.620^2/[(2)(9.807)]\} = 0.003107/D^4 + 4.718$$

$$h_L = 0.1553f_1/D^5 + 1101f_2 + 0.003107/D^4 + 4.718$$

$$54 = 0.1553f_1/D^5 + 1101f_2 + 0.003107/D^4 + 4.718$$

Try $f_1 = f_2 = 0.015$: $54 = (0.1553)(0.015)/D^5 + (1101)(0.015) + 0.003107/D^4 + 4.718$.
By trial and error, $D = 0.154$ m.

$$v_1 = 0.2165/0.154^2 = 9.129 \text{ m/s} \quad N_R = Dv/v$$

$$(N_R)_1 = (0.154)(9.129)/(0.113 \times 10^{-5}) = 1.24 \times 10^6 \quad (\epsilon/D)_1 = 0.000046/0.154 = 0.00030$$

From Fig. A-5, $f_1 = 0.0155$. $(N_R)_2 = (\tfrac{150}{1000})(9.620)/(0.113 \times 10^{-5}) = 1.28 \times 10^6$, $(\epsilon/D)_2 = 0.000046/(\tfrac{150}{1000}) =$
$0.00031$, $f_2 = 0.0155$. Try $f_1 = f_2 = 0.0155$: $54 = (0.1553)(0.0155)/D^5 + (1101)(0.0155) + 0.003107/D^4 + 4.718$. By
trial and error, $D = 0.155$ m.

$$v_1 = 0.2165/0.155^2 = 9.011 \text{ m/s} \quad (N_R)_1 = (0.155)(9.011)/(0.113 \times 10^{-5}) = 1.24 \times 10^6$$

$$(\epsilon/D)_1 = 0.000046/0.155 = 0.00030 \quad f_1 = 0.0155$$

$$(N_R)_2 = (\tfrac{150}{1000})(9.620)/(0.113 \times 10^{-5}) = 1.28 \times 10^6 \quad f_2 = 0.0155 \quad \text{(O.K.)}$$

Therefore, $D = 0.155$ m, or 155 mm.

$$A$$

$$35 \text{ m}$$

$$K = 0.4 \quad 65 \text{ m} \quad D = ?$$

$$K = 0.9$$

$$D = 150 \text{ mm} \quad 35 \text{ m}$$

$$B$$

$$16 \text{ m}$$

$$K = 1$$

$$q = 170 \text{ L/s}$$
$$\nu = 0.113 \times 10^{-5} \text{ m}^2/\text{s}$$
commercial steel pipe

**Fig. 9-46**

**9.214** In Prob. 9.208, if $p_1 = 200 \text{ lb/in}^2$ gage, what should the inside pipe diameter be to enable transport of 12 ft³/s of water? Again neglect minor losses.

**❚**
$$p_1/\gamma + v_1^2/2g + z_1 = p_2/\gamma + v_2^2/2g + z_2 + h_L \qquad v_2 = Q/A_2 = 12/(\pi d^2/4) = 15.28/d^2$$

$$h_L = h_f = (f)(L/d)(v^2/2g)$$

$$h_f = (f)[(400 + 160 + 250)/d]\{(15.28/d^2)^2/[(2)(32.2)]\} = 2937f/d^5$$

$$(200)(144)/62.4 + 0 + 120 = 0 + (15.28/d^2)^2/[(2)(32.2)] + 160 + 2937f/d^5 \qquad 421.5 = 3.625/d^4 + 2937f/d^5$$

Try $f = 0.015$: $421.5 = 3.625/d^4 + (2937)(0.015)/d^5$. By trial and error, $d = 0.643$ ft. $v = 15.28/0.643^2 = 36.96$ ft/s, $N_R = dv/\nu = (0.643)(36.96)/(2.11 \times 10^{-5}) = 1.12 \times 10^6$, $\epsilon/d = 0.00015/0.652 = 0.00023$. From Fig. A-5, $f = 0.015$ (O.K.). Therefore, $d = 0.643$ ft, or 7.72 in.

**9.215** Under the pressures established in Prob. 9.210, what size pipe is needed to double the flow?

**❚**
$$p_A/\gamma + v_A^2/2g + z_A = p_B/\gamma + v_B^2/2g + z_B + h_L \qquad v_B = Q/A_B = 0.2/(\pi d^2/4) = 0.2546/d^2$$

$$h_f = (f)(L/d)(v^2/2g) = f[(300 + 150 + 295)/d]\{(0.2546/d^2)^2/[(2)(9.807)]\} = 2.462f/d^5$$

$$h_m = (K)(v^2/2g) = (0.4 + 0.9 + 0.9 + 1)\{(0.2546/d^2)^2/[(2)(9.807)]\} = 0.01058/d^4$$

$$h_L = h_f + h_m = 2.462f/d^5 + 0.01058/d^4$$

$$2600/9.79 + 0 + 25 = 40/9.79 + (0.2546/d^2)^2/[(2)(9.807)] + 150 + 2.462f/d^5 + 0.01058/d^4$$

$$136.5 = 2.462f/d^5 + 0.01388/d^4$$

Try $f = 0.015$: $136.5 = (2.462)(0.015)/d^5 + 0.01388/d^4$. By trial and error, $d = 0.196$ m. $v = 0.2546/0.196^2 = 6.627$ m/s, $N_R = dv/\nu = (0.196)(6.627)/(0.113 \times 10^{-5}) = 1.15 \times 10^6$, $\epsilon/d = 0.000046/0.196 = 0.000235$. From Fig. A-5, $f = 0.015$ (O.K.). Therefore, $d = 0.196$ m, or 196 mm.

**9.216** A 12-in-diameter pipe with a friction factor of 0.02 conducts fluid between two tanks at 10 fps. The ends of the pipe are flush with the tank walls. Find the ratio of the minor losses to the pipe friction loss if the length of the pipe is (**a**) 10 ft, (**b**) 250 ft, and (**c**) 1500 ft.

**❚**
$$h_f = (f)(L/d)(v^2/2g) \qquad h_m = (K)(v^2/2g)$$

(**a**)
$$h_f = 0.02[(10/\tfrac{12}{12})](v^2/2g) = 0.20v^2/2g \qquad h_m = (0.5 + 1.0)(v^2/2g) = 1.5v^2/2g$$

$$h_m/h_f = (1.5v^2/2g)/(0.20v^2/2g) = 7.5 \qquad (7.5:1)$$

(**b**)
$$h_f = 0.02[250/(\tfrac{12}{12})](v^2/2g) = 5.0v^2/2g \qquad h_m = (0.5 + 1.0)(v^2/2g) = 1.5v^2/2g$$

$$h_m/h_f = (1.5v^2/2g)/(5.0v^2/2g) = 0.3 \qquad (0.3:1)$$

(**c**)
$$h_f = 0.02[1500/(\tfrac{12}{12})](v^2/2g) = 30v^2/2g \qquad h_m = (0.5 + 1.0)(v^2/2g) = 1.5v^2/2g$$

$$h_m/h_f = (1.5v^2/2g)/(30v^2/2g) = 0.05 \qquad (0.05:1)$$

**9.217** A smooth pipe 30 cm in diameter and 100 m long has a flush entrance and a submerged discharge. The velocity is 3 m/s. If the fluid is water at 15 °C, what is the total loss of head?

∎ $h_L = h_f + h_m$    $h_f = (f)(L/d)(v^2/2g)$    $N_R = dv/v = (0.30)(3)/(1.16 \times 10^{-6}) = 7.76 \times 10^5$

From Fig. A-5, $f = 0.0122$. $h_f = 0.0122[100/0.30]\{3^2/[(2)(9.807)]\} = 1.87$ m, $h_m = (K)(v^2/2g) =$
$(0.5 + 1.0)\{3^2/[(2)(9.807)]\} = 0.69$ m, $h_L = 1.87 + 0.69 = 2.56$ m.

**9.218** The water of Prob. 9.217 is replaced by oil with kinematic viscosity $9.3 \times 10^{-5}$ m²/s and specific gravity 0.925. Determine the pressure loss.

∎ $h_L = h_f + h_m$    $h_f = (f)(L/d)(v^2/2g)$    $N_R = dv/v = (0.30)(3)/(9.3 \times 10^{-5}) = 9.68 \times 10^3$

From Fig. A-5, $f = 0.031$.

$h_f = 0.031[100/0.30]\{3^2/[(2)(9.807)]\} = 4.74$ m    $h_m = (K)(v^2/2g) = (0.5 + 1.0)\{3^2/[(2)(9.807)]\} = 0.69$ m

$h_L = 4.74 + 0.69 = 5.43$ m of oil    $\Delta p = \gamma h_L = [(0.925)(9.79)](5.43) = 49.2$ kPa

**9.219** A smooth pipe consists of 100 ft of 8-in pipe followed by 200 ft of 24-in pipe, with an abrupt change of cross section at the junction. It has a flush entrance and a submerged exit. If it carries water at 60 °F in the smaller leg with a velocity of 18 fps, what is the total head loss?

∎ $h_L = h_f + h_m$    $h_f = (f)(L/d)(v^2/2g)$    $N_R = dv/v$    $(N_R)_{d=8} = (\frac{8}{12})(18)/(1.21 \times 10^5) = 9.92 \times 10^5$

From Fig. A-5, $(f)_{d=8} = 0.0117$.

$A_1v_1 = A_2v_2$    $A_1 = [(\pi)(\frac{8}{12})^2/4] = 0.3491$ ft²    $A_2 = [(\pi)(\frac{24}{12})^2/4] = 3.142$ ft²

$(0.3491)(18) = (3.142)(v_2)$    $v_2 = 2.000$ ft/s    $(N_R)_{d=24} = (\frac{24}{12})(2.000)/(1.21 \times 10^{-5}) = 3.31 \times 10^5$

$(f)_{d=24} = 0.0142$    $h_f = 0.0117[100/(\frac{8}{12})]\{18^2/[(2)(32.2)]\} + 0.0142[200/(\frac{24}{12})]\{2.000^2/[(2)(32.2)]\} = 8.92$ ft

$$h_m = (K)(v^2/2g)$$

For entrance, $K_1 = 0.5$. For abrupt change, $K_2 = (1 - A_1/A_2)^2 = (1 - 0.3491/3.142)^2 = 0.7901$. For exit, $K_3 = 1.0$.
$h_m = (0.5 + 0.7901)\{18^2/[(2)(32.2)]\} + 1.0\{2.000^2/[(2)(32.2)]\} = 6.55$ ft, $h_L = 8.92 + 6.55 = 15.47$ ft.

**9.220** A 6-in-diameter pipe ($f = 0.032$) of length 110 ft connects two reservoirs whose water-surface elevations differ by 10 ft. The pipe entrance is flush, and the discharge is submerged. (a) Compute the flow rate. (b) If the last 10 ft of pipe were replaced with a conical diffuser with a cone angle of 10°, compute the flow rate.

∎ $$p_1/\gamma + v_1^2/2g + z_1 = p_2/\gamma + v_2^2/2g + z_2 + h_L$$

(a) $$h_f = (f)(L/d)(v^2/2g) = 0.032[110/(\frac{6}{12})]\{v^2/[(2)(32.2)]\} = 0.1093v^2$$

$$h_m = (K)(v^2/2g) = (0.5 + 1.0)\{v^2/[(2)(32.2)]\} = 0.02329v^2$$

$$h_L = h_f + h_m = 0.1093v^2 + 0.02329v^2 = 0.1326v^2$$

$$0 + 0 + 10 = 0 + 0 + 0 + 0.1326v^2    v = 8.684 \text{ ft/s}$$

$$Q = Av = [(\pi)(\frac{6}{12})^2/4](8.684) = 1.71 \text{ ft}^3/\text{s}$$

(b) $h_f = 0.032[(110 - 10)/(\frac{6}{12})]\{v^2/[(2)(32.2)]\} = 0.09938v^2$    $h_m = (0.5 + 0.40)\{v^2/[(2)(32.2)]\} = 0.01398v^2$

$$h_L = 0.09938v^2 + 0.01398v^2 = 0.1134v^2    0 + 0 + 10 = 0 + 0 + 0 + 0.1134v^2$$

$$v = 9.391 \text{ ft/s}    Q = Av = [(\pi)(\frac{6}{12})^2/4](9.391) = 1.84 \text{ ft}^3/\text{s}$$

**9.221** Given two pipes in series with a diameter ratio of 1:2 and flow velocity of 15 fps in the smaller pipe, find the loss of head due to abrupt (a) contraction and (b) enlargement.

∎ $$h_m = (K)(v^2/2g)$$

(a) $d/D = 0.5$    $K = 0.33$    $h_m = 0.33\{15^2/[(2)(32.2)]\} = 1.15$ ft

(b) $K = 0.55$    $h_m = 0.55\{15^2/[(2)(32.2)]\} = 1.92$ ft

**9.222** In a 50-ft length of 4-in-diameter wrought iron pipe there are one open globe valve ($K = 10$), one 45° regular elbow ($K = 0.75$), and one pipe bend with a radius of curvature of 40 in ($K = 0.10$). The bend is 90°, and its length is not included in the 50 ft. No entrance or discharge losses are involved. If the fluid is water at 72 °F and the velocity is 6 fps, what is the total head loss?

$$h_L = h_f + h_m \qquad h_f = (f)(L/d)(v^2/2g)$$

$$N_R = dv/\nu = (\tfrac{4}{12})(6)/(1.02 \times 10^{-5}) = 1.96 \times 10^5 \qquad \epsilon/d = 0.00015/(\tfrac{4}{12}) = 0.00045$$

From Fig. A-5, $f = 0.0185$.

$$L = 50 + \tfrac{1}{4}[(2)(\pi)(\tfrac{40}{12})] = 55.2 \text{ ft} \qquad h_f = 0.0185[55.2/(\tfrac{4}{12})]\{6^2/[(2)(32.2)]\} = 1.71 \text{ ft}$$

$$h_m = (K)(v^2/2g) = (10 + 0.75 + 0.10)\{6^2/[(2)(32.2)]\} = 6.07 \text{ ft} \qquad h_L = 1.71 + 6.07 = 7.78 \text{ ft}$$

**9.223** Compute the friction head per 1000 ft of pipe for laminar flow at Reynolds number 50 000 (the empirical upper limit). Consider two situations: one where the fluid is water at 60 °F, the other where the fluid is SAE10 oil at 150 °F ($\nu = 0.00016 \text{ ft}^2/\text{s}$). Pipe diameter is 2.0 in.

$$h_f = (f)(L/d)(v^2/2g) \qquad f = 64/N_R = 64/50\,000 = 0.00128 \qquad N_R = dv/\nu$$

**For water:**

$$50\,000 = (2.0/12)(v)/(1.21 \times 10^{-5}) \qquad v = 3.630 \text{ ft/s}$$

$$h_f = 0.00128[1000/(2.0/12)]\{3.630^2/[(2)(32.2)]\} = 1.57 \text{ ft}$$

**For oil:**

$$50\,000 = (2.0/12)(v)/0.00016 \qquad v = 48.00 \text{ ft/s}$$

$$h_f = 0.00128[1000/(2.0/12)]\{48.00^2/[(2)(32.2)]\} = 275 \text{ ft}$$

**9.224** Repeat Prob. 9.223 if the flow is turbulent in a smooth pipe.

$h_f = (f)(L/d)(v^2/2g)$. From Fig. A-5, $f = 0.0207$.

**For water:** From Prob. 9.223, $v = 3.630 \text{ ft/s}$. $h_f = 0.0207[1000/(2.0/12)]\{3.630^2/[(2)(32.2)]\} = 25.4 \text{ ft}$.

**For oil:** From Prob. 9.223, $v = 48.00 \text{ ft/s}$. $h_f = 0.0207[1000/(2.0/12)]\{48.00^2/[(2)(32.2)]\} = 4443 \text{ ft}$.

**9.225** Repeat Prob. 9.223 if the flow is turbulent in a rough pipe with $\epsilon/d = 0.05$.

$h_f = (f)(L/d)(v^2/2g)$. From Fig. A-5, $f = 0.072$.

**For water:** From Prob. 9.223, $v = 3.630 \text{ ft/s}$. $h_f = 0.072[1000/(2.0/12)]\{3.630^2/[(2)(32.2)]\} = 88.4 \text{ ft}$.

**For oil:** From Prob. 9.223, $v = 48.00 \text{ ft/s}$. $h_f = 0.072[1000/(2.0/12)]\{48.00^2/[(2)(32.2)]\} = 15\,455 \text{ ft}$.

**9.226** Water at 60 °F flows through 10 000 ft of 12-in-diameter pipe between two reservoirs whose water-surface elevation difference is 200 ft. Find the flow rate if $\epsilon = 0.0018$ in.

$$p_1/\gamma + v_1^2/2g + z_1 = p_2/\gamma + v_2^2/2g + z_2 + h_L \qquad 0 + 0 + 200 = 0 + 0 + 0 + h_L$$

$$h_L = 200 \text{ ft} = h_f = (f)(L/d)(v^2/2g) = f[10\,000/(\tfrac{12}{12})]\{v^2/[(2)(32.2)]\} = 155.3fv^2$$

$$200 = 155.3fv^2 \qquad v = 1.135/\sqrt{f}$$

Try $f = 0.03$: $v = 1.135/\sqrt{0.03} = 6.553 \text{ ft/s}$, $N_R = dv/\nu = (\tfrac{12}{12})(6.553)/(1.21 \times 10^{-5}) = 5.42 \times 10^5$, $\epsilon/d = 0.0018/12 = 0.00015$. From Fig. A-5, $f = 0.015$. Try $f = 0.015$: $v = 1.135/\sqrt{0.015} = 9.267 \text{ ft/s}$, $N_R = (\tfrac{12}{12})(9.267)/(1.21 \times 10^{-5}) = 7.66 \times 10^5$, $f = 0.0145$. Try $f = 0.0145$: $v = 1.135/\sqrt{0.0145} = 9.426 \text{ ft/s}$, $N_R = (\tfrac{12}{12})(9.426)/(1.21 \times 10^{-5}) = 7.79 \times 10^5$, $f = 0.0145$ (O.K.); $Q = Av = [(\pi)(\tfrac{12}{12})^2/4](9.426) = 7.40 \text{ ft}^3/\text{s}$.

**9.227** Repeat Prob. 9.226 if $\epsilon$ is twenty times larger than in Prob. 9.226.

$v = 1.135/\sqrt{f}$ (from Prob. 9.226). Try $f = 0.03$: $v = 1.135/\sqrt{0.03} = 6.553 \text{ ft/s}$, $N_R = dv/\nu = (\tfrac{12}{12})(6.553)/(1.21 \times 10^{-5}) = 5.42 \times 10^5$, $\epsilon/d = (20)(0.0018)/12 = 0.00300$. From Fig. A-5, $f = 0.027$. Try $f = 0.027$: $v = 1.135/\sqrt{0.027} = 6.907 \text{ ft/s}$, $N_R = (\tfrac{12}{12})(6.907)/(1.21 \times 10^{-5}) = 5.71 \times 10^5$, $f = 0.027$ (O.K.); $Q = Av = [(\pi)(\tfrac{12}{12})^2/4](6.907) = 5.42 \text{ ft}^3/\text{s}$.

**9.228** How large a wrought iron pipe is required to convey oil (s.g. $= 0.9$, $\mu = 0.0008$ lb $\cdot$ s/ft$^2$) from one tank to another at a rate of 1.0 cfs if the pipe is 3000 ft long and the difference in elevation of the free liquid surfaces is 40 ft?

▌
$$p_1/\gamma + v_1^2/2g + z_1 = p_2/\gamma + v_2^2/2g + z_2 + h_L \qquad 0 + 0 + 40 = 0 + 0 + 0 + h_L$$

$$h_L = 40 \text{ ft} = h_f = (f)(L/d)(v^2/2g) \qquad v = Q/A = 1.0/(\pi d^2/4) = 1.273/d^2$$

$$h_L = (f)(3000/d)\{(1.273/d^2)^2/[(2)(32.2)]\} = 75.49f/d^5 \qquad 40 = 75.49f/d^5 \qquad d = 1.135f^{1/5}$$

Try $f = 0.03$: $d = (1.135)(0.03)^{1/5} = 0.5629$ ft, $N_R = \rho dv/\mu = [(0.9)(1.94)](0.5629)(1.273/0.5629^2)/0.0008 = 4936$, $\epsilon/d = 0.00015/0.5629 = 0.00027$. From Fig. A-5, $f = 0.0375$. Try $f = 0.0375$: $d = (1.135)(0.0375)^{1/5} = 0.5886$ ft, $N_R = [(0.9)(1.94)](0.5886)(1.273/0.5886^2)/0.0008 = 4720$, $f = 0.038$. Try $f = 0.038$: $d = (1.135)(0.038)^{1/5} = 0.5901$ ft, $N_R = [(0.9)(1.94)](0.5901)(1.273/0.5901^2)/0.0008 = 4708$, $f = 0.038$   (O.K.). Therefore, $d = 0.5901$ ft, or 7.08 in.

**9.229** If the diameter of a pipe is doubled, what effect does this have on the flow rate for a given head loss if the flow is laminar?

▌
$$h_L = (32)(v)(L/gd^2)(v) = (\text{constant})\{v/d^2\} \qquad v = kd^2 \qquad Q = Av = k'd^4$$

Thus, doubling the diameter will increase the flow rate by a factor of $2^4$, or 16.

**9.230** If the diameter of a pipe is doubled, what effect does this have on the flow rate for a given head loss if the flow is turbulent?

▌ If $f = $ constant (complete turbulence): $h_L = k_1v^2/d$, $v = k_2d^{1/2}$, $Q = Av = k_3d^{5/2}$. Thus, doubling the diameter will increase the flow rate by a factor of $2^{5/2}$, or 5.66.
   For smooth pipe with $N_R < 100\,000$: $f = 0.316/N_R^{0.25}$, $h_L = (f)(L/d)(v^2/2g) = (k_1)(v^{7/4}/d^{5/4})$, $v = k_2d^{5/7}$, $Q = Av = k_3d^{19/7}$. Thus, doubling the diameter will increase the flow rate by a factor of $2^{19/7}$, or 6.56.

**9.231** A 150-mm-diameter pipeline 100 m long discharges a 50-mm-diameter jet of water into the atmosphere at a point 60 m below the water surface at intake. The entrance to the pipe is a projecting one, with $K = 0.9$, and the nozzle loss coefficient is 0.05. Find the flow rate and the pressure head at the base of the nozzle, assuming $f = 0.03$.

▌
$$p_1/\gamma + v_1^2/2g + z_1 = p_2/\gamma + v_2^2/2g + z_2 + h_L \qquad\qquad (1)$$

$$h_f = (f)(L/d)(v^2/2g) = 0.03[100/0.150]\{v^2/[(2)(9.807)]\} = 1.020v^2$$

$$h_m = (K)(v^2/2g) = (0.9)\{v^2/[(2)(9.907)]\} + (0.05)\{v_{\text{jet}}^2/[(2)(9.907)]\} = 0.04589v^2 + 0.002549v_{\text{jet}}^2$$

$$h_L = h_f + h_m = 1.020v^2 + 0.04589v^2 + 0.002549v_{\text{jet}}^2 = 1.066v^2 + 0.002549v_{\text{jet}}^2$$

$$0 + 0 + 60 = 0 + v_{\text{jet}}^2/[(2)(9.807)] + 0 + 1.066v^2 + 0.002549v_{\text{jet}}^2$$

Since velocity varies as the square of the diameter, $v_{\text{jet}} = (\frac{15}{5})^2(v) = 9v$, $60 = (9v)^2/[(2)(9.807)] + 1.066v^2 + (0.002549)(9v)^2$, $v = 3.333$ m/s; $Q = Av = [(\pi)(0.150)^2/4](3.333) = 0.0589$ m$^3$/s. Applying Eq. (1) to the nozzle, $p_1/9.79 + 3.333^2/[(2)(9.807)] + 0 = 0 + [(9)(3.333)]^2/[(2)(9.807)] + 0 + 0.05\{[(9)(3.333)]^2/[(2)(9.807)]\}$, $p_1 = 46.9$ kPa.

**9.232** A 2.0-m-diameter, 1600-m-long concrete pipe ($\epsilon = 1.5$ mm) carries water at 12 °C between two reservoirs at 8.0 m$^3$/s. Find the difference in water-surface elevation between the two reservoirs, considering minor losses at entrance and exit.

▌
$$p_1/\gamma + v_1^2/2g + z_1 = p_2/\gamma + v_2^2/2g + z_2 + h_L \qquad h_L = h_f + h_m \qquad h_f = (f)(L/d)(v^2/2g)$$

$$v = Q/A = 8.0/[(\pi)(2.0)^2/4] = 2.546 \text{ m/s} \qquad N_R = dv/\nu = (2.0)(2.546)/(1.24 \times 10^{-6}) = 4.11 \times 10^6$$

$$\epsilon/d = 0.0015/2.0 = 0.00075$$

From Fig. A-5, $f = 0.0185$.

$$h_f = (0.0185)(1600/2.0)\{2.546^2/[(2)(9.807)]\} = 4.89 \text{ m}$$

$$h_m = (K)(v^2/2g) = (0.5 + 1.0)\{2.546^2/[(2)(9.807)]\} = 0.50 \text{ m}$$

$$h_L = 4.89 + 0.50 = 5.39 \text{ m} \qquad 0 + 0 + z_1 = 0 + 0 + z_2 + 5.39 \text{ m} \qquad z_1 - z_2 = 5.39 \text{ m}$$

**9.233** A pipe of mean diameter 5 ft and length 6000 ft delivers water to a facility 1300 ft below the water surface at intake. Assume $f = 0.025$. When the pipe delivers 300 cfs, what is the horsepower delivered?

$$v_2 = Q/A_2 = 300/[(\pi)(\tfrac{60}{12})^2/4] = 15.28 \text{ ft/s}$$

$$h_f = (f)(L/d)(v^2/2g) = 0.025[6000/(\tfrac{60}{12})]\{15.28^2/[(2)(32.2)]\} = 108.76 \text{ ft}$$

$$h_m = (K)(v_2/2g) = 0.50\{15.28^2/[(2)(32.2)]\} = 1.81 \text{ ft}$$

$$h_L = h_f + h_m = 108.76 + 1.81 = 110.57 \text{ ft}$$

$$P = Q\gamma(\Delta z - h_L) = (300)(62.4)(1300 - 110.57) = 2.227 \times 10^7 \text{ ft} \cdot \text{lb/s} = (2.227 \times 10^7)/550 = 40\,491 \text{ hp}$$

**9.234** Find the kilowatt loss in 500 m of 50-cm-diameter pipe for which $\epsilon = 0.05$ mm when dye at 45 °C (s.g. = 0.86, $v = 4.4 \times 10^{-6}$ ft$^2$/s) flows at 0.22 m$^3$/s. Neglect minor losses.

$$h_f = (f)(L/d)(v^2/2g) \qquad v = Q/A = 0.22/[(\pi)(\tfrac{50}{100})^2/4] = 1.120 \text{ m/s}$$

$$N_R = dv/v = (0.50)(1.120)/(4.4 \times 10^{-6}) = 1.27 \times 10^5 \qquad \epsilon/d = (5 \times 10^{-5})/(5 \times 10^{-1}) = 0.0001$$

From Fig. A-5, $f = 0.018$. $h_f = 0.018[500/0.50]\{1.120^2/[(2)(9.807)]\} = 1.151$ m, $P = Q\gamma h_f = 0.22[(0.86)(9.79)](1.151) = 2.13$ kW.

**9.235** Linseed oil, of kinematic viscosity 0.0005 ft$^2$/s and weight density 59.8 lb/ft$^3$, is pumped through a 3-in pipe ($\epsilon = 0.001$ in). (a) At what maximum velocity would the flow still be laminar? (b) What would then be the loss in energy head per 1000 ft of pipe?

**(a)** Assume laminar flow exists for $N_R \le 2000$. $N_R = dv/v$, $2000 = (\tfrac{3}{12})(v)/0.0005$, $v = 4.00$ ft/s.

**(b)** $f = 64/N_R = \tfrac{64}{2000} = 0.032$   $h_f = (f)(L/d)(v^2/2g) = 0.032[1000/(\tfrac{3}{12})]\{4.00^2/[(2)(32.2)]\} = 31.80$ ft

$$p = \gamma h_f = (59.8)(31.80)/144 = 13.2 \text{ psi per 1000 ft}$$

**9.236** Repeat Prob. 9.235 if the velocity is three times the maximum velocity for laminar flow.

$$v = (3)(4.00) = 12.00 \text{ ft/s} \qquad h_f = (f)(L/d)(v^2/2g)$$

$$N_R = dv/v = (\tfrac{3}{12})(12.00)/0.0005 = 6000 \qquad \epsilon/d = 0.001/2 = 0.00050$$

From Fig. A-5, $f = 0.036$.

$$h_f = 0.036[1000/(\tfrac{3}{12})]\{12.00^2/[(2)(32.2)]\} = 322.0 \text{ ft}$$

$$p = \gamma h_f = (59.8)(322.0)/144 = 134 \text{ psi per 1000 ft}$$

**9.237** Water flows upward at 3 m/s through a vertical 150-mm-diameter pipe standing in a body of water with its lower end 1.0 m below the surface. Considering all losses and with $f = 0.022$, find the pressure at a point 3 m above the surface of the water.

$p_1/\gamma + v_1^2/2g + z_1 = p_2/\gamma + v_2^2/2g + z_2 + h_L$. Let point 1 be at the water surface and point 2 be 3 m above the water surface. $h_L = h_f + h_m$, $h_f = (f)(L/d)(v^2/2g) = 0.022[(3 + 1.0)/0.150]\{3^2/[(2)(9.807)]\} = 0.269$ m, $h_m = (K)(v^2/2g)$. For entrance loss, assume $K = 0.8$. $h_m = 0.8\{3^2/[(2)(9.807)]\} = 0.367$ m, $h_L = 0.269 + 0.367 = 0.636$ m, $0 + 0 + 0 = p_2/9.79 + 3^2/[(2)(9.807)] + 3 + 0.636$, $p_2 = -40.1$ kPa.

**9.238** Work Prob. 9.237 if the flow is downward.

$$p_2/\gamma + v_2^2/2g + z_2 = p_1/\gamma + v_1^2/2g + z_1 + h_L \qquad h_L = h_f + h_m$$

$$h_f = 0.269 \text{ m} \quad \text{(from Prob. 9.237)} \qquad h_m = (K)(v^2/2g)$$

For exit loss, assume $K = 1.0$.

$$h_m = 1.0\{3^2/[(2)(9.807)]\} = 0.459 \text{ m} \qquad h_L = 0.269 + 0.459 = 0.728 \text{ m}$$

$$p_2/9.79 + 3^2/[(2)(9.807)] + 3 = 0 + 0 + 0 + 0.728 \qquad p_2 = -26.7 \text{ kPa}$$

**9.239** A horizontal pipe 20 cm in diameter and for which $f = 0.030$ projects into a pond 1 m below the surface. Considering all losses, find the pressure at a point 4 m from the end of the pipe if the flow is at 3 m/s from the pipe into the pond.

▌ $p_1/\gamma + v_1^2/2g + z_1 = p_2/\gamma + v_2^2/2g + z_2 + h_L$. Let point 1 be 4 m from the end of the pipe and point 2 be at the water surface. $h_L = h_f + h_m$, $h_f = (f)(L/d)(v^2/2g) = 0.030[4/0.20]\{3^2/[(2)(9.807)]\} = 0.275$ m, $h_m = (K)(v^2/2g)$. For exit loss, $K = 1.0$. $h_m = 1.0\{3^2/[(2)(9.807)]\} = 0.459$ m, $h_L = 0.275 + 0.459 = 0.734$ m, $p_1/9.79 + 3^2/[(2)(9.807)] + 0 = 0 + 0 + 1 + 0.734$, $p_1 = 12.5$ kPa.

**9.240** Repeat Prob. 9.239 if the flow is from the pond into the pipe.

▌ $$p_2/\gamma + v_2^2/2g + z_2 = p_1/\gamma + v_1^2/2g + z_1 + h_L \qquad h_L = h_f + h_m$$

$$h_f = 0.275 \text{ m} \qquad \text{(from Prob. 9.239)} \qquad h_m = (K)(v^2/2g)$$

For entrance, $K = 0.8$.

$$h_m = 0.8\{3^2/[(2)(9.807)]\} = 0.367 \text{ m} \qquad h_L = 0.275 + 0.367 = 0.642 \text{ m}$$

$$0 + 0 + 1 = p_1/9.79 + 3^2/[(2)(9.807)] + 0 + 0.642 \qquad p_1 = -0.99 \text{ kPa}$$

**9.241** A pipe runs from one reservoir to another, both ends of the pipe being under water. The intake is nonprojecting; the length of pipe is 480 ft; its diameter is 10.25 in; and the difference in the two water levels is 106 ft. If $f = 0.02$, what will be the pressure at a point 300 ft from the intake and 120 ft below the surface of the water in the upper reservoir?

▌ $$p_1/\gamma + v_1^2/2g + z_1 = p_2/\gamma + v_2^2/2g + z_2 + h_L \qquad (1)$$

Let points 1 and 2 be at the water surface in the upper and lower reservoirs, respectively.

$h_f = (f)(L/d)(v^2/2g) = 0.02[(480)/(10.25/12)]\{v^2/[(2)(32.2)]\} = 0.1745v^2$

$$h_m = (K)(v^2/2g) = (0.5 + 1.0)\{v^2/[(2)(32.2)]\} = 0.02329v^2$$

$h_L = h_f + h_m = 0.1745v^2 + 0.02329v^2 = 0.1978v^2 \qquad 0 + 0 + 106 = 0 + 0 + 0 + 0.1978v^2 \qquad v = 23.15 \text{ ft/s}$

Now apply Eq. (1) between the upper reservoir (point 1) and the point 300 ft from the intake (point 2).

$h_f = 0.02[(300)/(10.25/12)]\{23.15^2/[(2)(32.2)]\} = 58.46 \text{ ft} \qquad h_m = 0.5\{23.15^2/[(2)(32.2)]\} = 4.16 \text{ ft}$

$h_L = 58.46 + 4.16 = 62.62 \text{ ft} \qquad 0 + 0 + 120 = (p_2)(144)/62.4 + 23.15^2/[(2)(32.2)] + 0 + 62.62 \qquad p_2 = 21.3 \text{ lb/in}^2$

**9.242** A 9.5-in-diameter pipeline runs from one reservoir to another, both ends being under water, and the intake end is nonprojecting. The difference in water levels between the two reservoirs is 110 ft, and the length of pipe is 1000 ft. What is the discharge if $f = 0.06$?

▌ $$p_1/\gamma + v_1^2/2g + z_1 = p_2/\gamma + v_2^2/2g + z_2 + h_L$$

$$h_f = (f)(L/d)(v^2/2g) = 0.06[(1000)/(9.5/12)]\{v^2/[(2)(32.2)]\} = 1.1769v^2$$

$$h_m = (K)(v^2/2g) = (0.5 + 1.0)\{v^2/[(2)(32.2)]\} = 0.02329v^2$$

$$h_L = h_f + h_m = 1.1769v^2 + 0.02329v^2 = 1.200v^2 \qquad 0 + 0 + 110 = 0 + 0 + 0 + 1.200v^2$$

$$v = 9.57 \text{ ft/s} \qquad Q = Av = [(\pi)(9.5/12)^2/4](9.57) = 4.71 \text{ ft}^3/\text{s}$$

**9.243** A jet of water is discharged through a nozzle at a point 200 ft below the water level at intake. The jet is 4 in in diameter, and the loss coefficient of the nozzle is 0.15. If the pipeline is 12 in in diameter, 600 ft long, with a nonprojecting entrance, what is the pressure at the base of the nozzle? Assume $f = 0.0125$.

$$p_1/\gamma + v_1^2/2g + z_1 = p_2/\gamma + v_2^2/2g + z_2 + h_L \qquad (1)$$

$$h_f = (f)(L/d)(v^2/2g) = 0.0125[600/(\tfrac{12}{12})]\{v^2/[(2)(32.2)]\} = 0.1165v^2$$

$$h_m = (K)(v^2/2g) = 0.5\{v^2/[(2)(32.2)]\} + (0.15)\{v_{\text{jet}}^2/[(2)(32.2)]\} = 0.007764v^2 + 0.002329v_{\text{jet}}^2$$

$$h_L = h_f + h_m = 0.1165v^2 + 0.007764v^2 + 0.002329v_{\text{jet}}^2 = 0.1243v^2 + 0.002329v_{\text{jet}}^2$$

$$0 + 0 + 200 = 0 + v_{\text{jet}}^2/[(2)(32.2)] + 0 + 0.1243v^2 + 0.002329v_{\text{jet}}^2$$

Since velocity varies with the square of diameter, $v_{\text{jet}} = (\tfrac{12}{4})^2(v) = 9v$, $200 = (9v)^2/[(2)(32.2)] + 0.1243v^2 + (0.002329)(9v)^2$, $v = 11.28$ ft/s. Applying Eq. (1) to the nozzle, $(p_1)(144)/62.4 + 11.28^2/[(2)(32.2)] + 0 = 0 + [(9)(11.28)]^2/[(2)(32.2)] + 0 + (0.15)\{[(9)(11.28)]^2/[(2)(32.2)]\}$, $p_1 = 78.9$ lb/in$^2$.

**9.244** Compute the losses due to flow of 25 m$^3$/min of air, $p = 1$ atm, $T = 20\,°$C, through a sudden expansion from 300-mm pipe to 900-mm pipe. How much head would be saved by using a 10° conical diffuser?

**For sudden expansion:** $h_m = [1 - (D_1/D_2)^2]^2(v_1^2/2g)$, $v_1 = Q/A_1 = (\tfrac{25}{60})/[(\pi)(\tfrac{300}{1000})^2/4] = 5.895$ m/s, $h_m = [1 - (\tfrac{300}{900})^2]^2\{5.895^2/[(2)(9.807)]\} = 1.400$ m.

**For conical diffuser:** $h_m = 0.152[(v_1 - v_2)^2/2g]$, $v_2 = Q/A_2 = (\tfrac{25}{60})/[(\pi)(\tfrac{900}{1000})^2/4] = 0.655$ m/s, $h_m = 0.152\{(5.895 - 0.655)^2/[(2)(9.807)]\} = 0.213$ m.

$$\text{Saving in head} = 1.400 - 0.213 = 1.187 \text{ m} \quad \text{or} \quad 1.187\,\text{N} \cdot \text{m/N} \quad \text{or} \quad 1.187\,\text{J/N}$$

**9.245** Calculate the value of $H$ in Fig. 9-47 for 125 L/s of water at 15 °C through commercial steel pipe.

$$p_1/\gamma + v_1^2/2g + z_1 = p_2/\gamma + v_2^2/2g + z_2 + h_L \qquad h_L = h_f + h_m \qquad h_f = (f)(L/d)(v^2/2g)$$

$$v = Q/A = (125 \times 10^{-3})/[(\pi)(0.30)^2/4] = 1.768 \text{ m/s}$$

$$N_R = dv/\nu = (0.30)(1.768)/(1.16 \times 10^{-6}) = 4.57 \times 10^5 \qquad \epsilon/d = 0.000046/0.30 = 0.00015$$

From Fig. A-5, $f = 0.015$.

$$h_f = 0.015[50/0.30]\{1.768^2/[(2)(9.807)]\} = 0.398 \text{ m}$$

$$h_m = (K)(v^2/2g) = (0.5 + 1.0)\{1.768^2/[(2)(9.807)]\} = 0.239 \text{ m}$$

$$h_L = 0.398 + 0.239 = 0.637 \text{ m} \qquad 0 + 0 + z_1 = 0 + 0 + z_2 + 0.637 \qquad z_1 - z_2 = H = 0.637 \text{ m}$$

**Fig. 9-47**

50m,30-cm-diam.

**9.246** In Fig. 9-47, for $H = 3$ m and a fluid with s.g. $= 0.8$ and $\mu = 0.007$ Pa · s, calculate the discharge through smooth pipe.

$$p_1/\gamma + v_1^2/2g + z_1 = p_2/\gamma + v_2^2/2g + z_2 + h_L$$

$$h_f = (f)(L/d)(v^2/2g) = (f)[50/0.30]\{v^2/[(2)(9.807)]\} = 8.497fv^2$$

$$h_m = (K)(v^2/2g) = (0.5 + 1.0)\{v^2/[(2)(9.807)]\} = 0.07648v^2 \qquad h_L = h_f + h_m = 8.497fv^2 + 0.07648v^2$$

$$0 + 0 + 3 = 0 + 0 + 0 + 8.497fv^2 + 0.07648v^2 \qquad v = \sqrt{3/(8.497f + 0.07648)}$$

Try $f = 0.02$: $v = \sqrt{3/[(8.497)(0.02) + 0.07648]} = 3.489$ m/s, $N_R = \rho dv/\mu = [(0.8)(1000)](0.30)(3.489)/0.007 = 1.20 \times 10^5$. From Fig. A-5, $f = 0.017$. Try $f = 0.017$: $v = \sqrt{3/[(8.497)(0.017) + 0.07648]} = 3.685$ m/s, $N_R = [(0.8)(1000)](0.30)(3.685)/0.007 = 1.26 \times 10^5$, $f = 0.017$ (O.K.); $Q = Av = [(\pi)(0.30)^2/4](3.685) = 0.260$ m$^3$/s.

**9.247** Evaluate $K$ for a valve that, placed in the line in Prob. 9.246, would reduce the discharge by 50%.

❚ From Prob. 9.246, $H = 3 = [(f)(50)/0.30 + 0.5 + 1.0 + K_{valve}]\{(3.685/2)^2/[(2)(9.807)]\}$, $K_{valve} = 15.83 - 166.7f$, $N_R = \rho dv/\mu = [(0.8)(1000)](0.30)(3.685/2)/0.007 = 6.32 \times 10^4$. From Fig. A-5, $f = 0.0197$. $K_{valve} = 15.83 - (166.7)(0.0197) = 12.5$.

**9.248** A water line connecting two reservoirs at 70 °F has 4860 ft of 24-in-diameter steel pipe, three standard elbows ($K = 0.9$), a globe valve ($K = 10$), a re-entrant pipe entrance ($K = 1.0$), and a submerged pipe exit ($K = 1.0$). What is the difference in reservoir elevations for 20 cfs?

❚ $$p_1/\gamma + v_1^2/2g + z_1 = p_2/\gamma + v_2^2/2g + z_2 + h_L \qquad h_L = h_f + h_m \qquad h_f = (f)(L/d)(v^2/2g)$$

$$v = Q/A = 20/[(\pi)(\tfrac{24}{12})^2/4] = 6.366 \text{ ft/s} \qquad N_R = dv/v = (\tfrac{24}{12})(6.366)/(1.05 \times 10^{-5}) = 1.21 \times 10^6$$

$$\epsilon/d = 0.00015/(\tfrac{24}{12}) = 0.000075$$

From Fig. A-5, $f = 0.013$.

$$h_f = 0.013[4860/(\tfrac{24}{12})]\{6.366^2/[(2)(32.2)]\} = 19.88 \text{ ft}$$

$$h_m = (K)(v^2/2g) = [(3)(0.9) + 10 + 1.0 + 1.0]\{6.366^2/[(2)(32.2)]\} = 9.25 \text{ ft}$$

$$h_L = 19.88 + 9.25 = 29.13 \text{ ft} \qquad 0 + 0 + z_1 = 0 + 0 + z_2 + 29.13 \qquad z_1 - z_2 = 29.13 \text{ ft}$$

**9.249** For the conditions given in Prob. 9.248, determine the discharge if the difference in elevations is 40 ft.

❚ From Prob. 9.248, $40 = h_f + h_m = \{v^2/[(2)(32.2)]\}\{(f)[4860/(\tfrac{24}{12})] + [(3)(0.9) + 10 + 1.0 + 1.0]\}$, $v = \sqrt{2576/(2430f + 14.7)}$. Try $f = 0.013$: $v = \sqrt{2576/[(2430)(0.013) + 14.7]} = 7.46$ ft/s, $N_R = dv/v = (\tfrac{24}{12})(7.46)/(1.05 \times 10^{-5}) = 1.42 \times 10^6$, $\epsilon/d = 0.000075$ (from Prob. 9.248). From Fig. A-5, $f = 0.0125$. Try $f = 0.0125$: $v = \sqrt{2576/[(2430)(0.0125) + 14.7]} = 7.560$ ft/s, $N_R = (\tfrac{24}{12})(7.560)/(1.05 \times 10^{-5}) = 14.4 \times 10^6$, $f = 0.0125$ (O.K.); $Q = Av = [(\pi)(\tfrac{24}{12})^2/4](7.560) = 23.8$ ft$^3$/s.

**9.250** What size commercial steel pipe is needed to convey 200 L/s of water at 20 °C a distance of 5000 m with a head drop of 4 m? The line connects two reservoirs, has a re-entrant ($K = 1.0$), a submerged outlet ($K = 1.0$), four standard elbows ($K = 0.9$), and a globe valve ($K = 10$).

❚ $$p_1/\gamma + v_1^2/2g + z_1 = p_2/\gamma + v_2^2/2g + z_2 + h_L \qquad 0 + 0 + 4 = 0 + 0 + 0 + h_L \qquad h_L = h_f + h_m$$

$$v = Q/A = (\tfrac{200}{1000})/(\pi d^2/4) = 0.2546/d^2$$

$$h_f = (f)(L/d)(v^2/2g) = (f)(5000/d)\{(0.2546/d^2)^2/[(2)(9.807)]\} = 16.52f/d^5$$

$$h_m = (K)(v^2/2g) = [1.0 + 1.0 + (4)(0.9) + 10]\{(0.2546/d^2)^2/[(2)(9.807)]\} = 0.05156/d^4$$

$$4 = 16.52f/d^5 + 0.05156/d^4$$

Try $f = 0.02$: $4 = (16.52)(0.02)/d^5 + 0.05156/d^4$. By trial and error, $d = 0.619$ m.

$$v = 0.2546/0.619^2 = 0.6645 \text{ m/s} \qquad N_R = dv/v = (0.619)(0.6645)/(1.02 \times 10^{-6}) = 4.03 \times 10^5$$

$$\epsilon/d = 0.000046/0.619 = 0.000074$$

From Fig. A-5, $f = 0.0145$. Try $f = 0.0145$: $4 = (16.52)(0.0145)/d^5 + 0.05156/d^4$, $d = 0.588$ m, $v = 0.2546/0.588^2 = 0.7364$ m/s, $N_R = (0.588)(0.7364)/(1.02 \times 10^{-6}) = 4.25 \times 10^5$, $f = 0.0145$. Therefore, $d = 0.588$ m, or 588 mm.

**9.251** Find the equivalent lengths of 167-mm-diameter pipe, $f = 0.024$, for (a) a re-entrant pipe entrance ($K = 1.0$), (b) a sudden expansion from 167 mm to 334 mm diameter, and (c) a globe valve and a standard tee?

❚ $$L_e = KD/f$$

(a) $$L_e = (1.0)(0.167)/0.024 = 6.94 \text{ m}$$

(b) $$K = [1 - (D_1/D_2)^2]^2 = [1 - (\tfrac{1}{2})^2]^2 = 0.5625 \qquad L_e = (0.5625)(0.167)/0.024 = 3.91 \text{ m}$$

(c) $$K = 10 + 1.8 = 11.8 \qquad L_e = (11.8)(0.167)/0.024 = 81.9 \text{ m}$$

**9.252** Find $H$ in Fig. 9-48 for 200 gpm of oil ($\mu = 1$ cP, $\gamma = 60$ lb/ft³) for the angle valve wide open ($K = 5.0$).

$p_1/\gamma + v_1^2/2g + z_1 = p_2/\gamma + v_2^2/2g + z_2 + h_L$     $h_L = h_f + h_m$     $h_f = (f)(L/d)(v^2/2g)$

$Q = (200)(0.002228) = 0.4456$ ft³/s     $v = Q/A = 0.4456/[(\pi)(\frac{3}{12})^2/4] = 9.078$ ft/s

By Prob. 9.40, $\mu = 0.0002089$ lb · s/ft²; $N_R = (\gamma/g)(dv)/\mu = (60/32.2)(\frac{3}{12})(9.078)/0.0002089 = 2.02 \times 10^4$, $\epsilon/d = 0.00015/(\frac{3}{12}) = 0.000600$. From Fig. A-5, $f = 0.0275$.

$$h_f = 0.0275[212/(\tfrac{3}{12})]\{9.078^2/[(2)(32.2)]\} = 29.84 \text{ ft}$$

$$h_m = (K)(v^2/2g) = [0.5 + 5.0 + 1.0]\{9.078^2/[(2)(32.2)]\} = 8.32 \text{ ft}$$

$$h_L = 29.84 + 8.32 = 38.16 \text{ ft} \qquad 0 + 0 + z_1 = 0 + 0 + z_2 + 38.16 \qquad z_1 - z_2 = H = 38.16 \text{ ft}$$

**Fig. 9-48**

**9.253** Find $K$ for the angle valve in Prob. 9.252 for a flow of 10 L/s at the same $H$.

$$v = Q/A = [(\tfrac{10}{1000})/0.3048^3]/[(\pi)(\tfrac{3}{12})^2/4] = 7.194 \text{ ft/s}$$

$$N_R = \rho dv/\mu = (\gamma/g)(dv)/\mu = (60/32.2)(\tfrac{3}{12})(7.194)/0.0002089 = 1.60 \times 10^4$$

$$\epsilon/d = 0.000600 \qquad \text{(from Prob. 9.252)}$$

From Fig. A-5, $f = 0.0285$. From Prob. 9.252, $H = [(f)(L/d) + K_1 + K_2 + K_3](v^2/2g)$, $38.16 = \{(0.0285)[212/(\frac{3}{12})] + 0.5 + K_2 + 1.0\}\{7.194^2/[(2)(32.2)]\}$, $K_2 = 21.8$.

**9.254** Calculate the discharge through the system of Fig. 9-48 for water at 25 °C and $H = 8$ m.

$$p_1/\gamma + v_1^2/2g + z_1 = p_2/\gamma + v_2^2/2g + z_2 + h_L$$

In British Engineering units, we have

$$h_f = (f)(L/d)(v^2/2g) = (f)[212/(\tfrac{3}{12})]\{v^2/[(2)(32.2)]\} = 13.17fv^2$$

$$h_m = (K)(v^2/2g) = (0.5 + 5.0 + 1.0)\{v^2/[(2)(32.2)]\} = 0.1009v^2 \qquad h_L = h_f + h_m = 13.17fv^2 + 0.1009v^2$$

$$0 + 0 + 8/0.3048 = 0 + 0 + 0 + 13.17fv^2 + 0.1009v^2 \qquad v = \sqrt{26.25/(13.17f + 0.1009)}$$

Try $f = 0.02$: $v = \sqrt{26.25/[(13.17)(0.02) + 0.1009]} = 8.489$ ft/s, $N_R = dv/\nu = (\frac{3}{12})(8.489)/(9.56 \times 10^{-6}) = 2.22 \times 10^5$, $\epsilon/d = 0.00015/(\frac{3}{12}) = 0.000600$. From Fig. A-5, $f = 0.019$. Try $f = 0.019$: $v = \sqrt{26.25/[(13.17)(0.019) + 0.1009]} = 8.646$ ft/s, $N_R = (\frac{3}{12})(8.677)/(9.56 \times 10^{-6}) = 2.27 \times 10^5$, $f = 0.019$ (O.K.); $Q = Av = [(\pi)(\frac{3}{12})^2/4](8.646) = 0.4244$ ft³/s $= (0.4244)(0.3048)^3 = 0.0120$ m³/s, or 12.0 L/s.

**9.255** Find the discharge through the pipeline in Fig. 9-49 for $H = 10$ m, as shown. Use minor loss coefficients for the entrance, elbows, and globe valve of 0.5, 0.9 (each), and 10, respectively.

$$p_1/\gamma + v_1^2/2g + z_1 = p_2/\gamma + v_2^2/2g + z_2 + h_L$$

$$h_f = (f)(L/d)(v^2/2g) = f[(30 + 12 + 60)/(\tfrac{150}{1000})]\{v_2^2/[(2)(9.807)]\} = 34.67fv_2^2$$

$$h_m = (K)(v^2/2g) = [0.5 + (2)(0.9) + 10]\{v_2^2/[(2)(9.807)]\} = 0.6271v_2^2$$

$$h_L = h_f + h_m = 34.67fv_2^2 + 0.6271v_2^2 \qquad 0 + 0 + 10 = 0 + v_2^2/[(2)(9.807)] + 0 + 34.67fv_2^2 + 0.6271v_2^2$$

$$v = \sqrt{10/(34.67f + 0.6781)}$$

Try $f = 0.02$: $v = \sqrt{10/[(34.67)(0.02) + 0.6781]} = 2.700$ m/s, $N_R = dv/\nu = (\frac{150}{1000})(2.700)/(1.02 \times 10^{-6}) = 3.97 \times 10^5$, $\epsilon/d = 0.00026/(\frac{150}{1000}) = 0.00173$. From Fig. A-5, $f = 0.023$. Try $f = 0.023$: $v = \sqrt{10/[(34.67)(0.023) + 0.6781]} = 2.603$ m/s, $N_R = (\frac{150}{1000})(2.603)/(1.02 \times 10^{-6}) = 3.83 \times 10^5$, $f = 0.023$ (O.K.); $Q = Av = [(\pi)(\frac{150}{1000})^2/4](2.603) = 0.0460$ m³/s, or 46.0 L/s.

**Fig. 9-49**

**9.256** Rework Prob. 9.255 to find $H$ if $Q = 60$ L/s.

$$p_1/\gamma + v_1^2/2g + z_1 = p_2/\gamma + v_2^2/2g + z_2 + h_L \qquad h_L = h_f + h_m \qquad h_f = (f)(L/d)(v^2/2g)$$

$$v = Q/A = (\tfrac{60}{1000})/[(\pi)(\tfrac{150}{1000})^2/4] = 3.395 \text{ m/s} \qquad N_R = dv/\nu = (\tfrac{150}{1000})(3.395)/(1.02 \times 10^{-6}) = 4.99 \times 10^5$$

$$\epsilon/d = 0.00173 \qquad \text{(from Prob. 9.255)}$$

From Fig. A-5, $f = 0.0225$.

$$h_f = 0.0225[(30 + 12 + 60)/(\tfrac{150}{1000})]\{3.395^2/[(2)(9.807)]\} = 8.99 \text{ m}$$

$$h_m = (K)(v^2/2g) = [0.5 + (2)(0.9) + 10]\{3.395^2/[(2)(9.807)]\} = 7.23 \text{ m}$$

$$h_L = 8.99 + 7.23 = 16.22 \text{ m} \qquad 0 + 0 + z_1 = 0 + 3.395^2/[(2)(9.807)] + z_2 + 16.22 \qquad z_1 - z_2 = H = 16.81 \text{ m}$$

**9.257** Assume that water at 10 °C is to be conveyed at 300 L/s through 500 m of commercial steel pipe with a total head drop of 6 m. Minor losses are $12v^2/2g$. Determine the required diameter.

$$N_R = R_5/D \tag{1}$$

$$f = R_7/[\ln (R_3/D + R_2/N_R^{0.9})]^2 \tag{2}$$

$$x = R_6 + R_4 D/f \tag{3}$$

$$D = (R_0)(x^{4.75} + R_1 x^{5.2})^{0.04} \tag{4}$$

where $R_0 = (0.66)(\epsilon^{1.25}Q^{9.5})^{0.04}$, $R_1 = \nu/\epsilon^{1.25}Q^{0.1}$, $R_2 = 5.74$, $R_3 = \epsilon/3.7$, $R_4 = K/gh_f$, $R_5 = 4Q/\pi\nu$, $R_6 = L/gh_f$, $R_7 = 1.325$, $\epsilon = 0.000046$ m, $Q = 300$ L/s, or $0.300$ m³/s, $\nu = 1.30 \times 10^{-6}$ m²/s.

$$K = 12 \qquad h_f = 6 \text{ mm} \qquad R_0 = (0.66)(0.000046^{1.25}0.300^{9.5})^{0.04} = 0.25351$$

$$R_1 = 1.30 \times 10^{-6}/(0.000046^{1.25}0.300^{0.1}) = 0.38707 \qquad R_3 = 0.000046/3.7 = 1.2432 \times 10^{-5}$$

$$R_4 = 12/[(9.807)(6)] = 0.20394 \qquad R_5 = (4)(0.300)/[(\pi)(1.30 \times 10^{-6})] = 2.9382 \times 10^5$$

$$R_6 = 500/[(9.807)(6)] = 8.4973$$

Assume $D = 1$ m. Substituting into Eqs. (1), (2), (3), and (4),

$$N_R = 2.9382 \times 10^5/1 = 2.9382 \times 10^5 \qquad f = 1.325/\{\ln [(1.2432 \times 10^{-5})/1 + 5.74/(2.9382 \times 10^5)^{0.9}]\}^2 = 0.014938$$

$$x = 8.4973 + (0.20394)(1)/0.014938 = 22.150 \qquad D = (0.25351)[22.150^{4.75} + (0.38707)(22.150^{5.2})]^{0.04} = 0.47418 \text{ m}$$

Try $D = 0.47418$ m: $N_R = 2.9382 \times 10^5/0.47418 = 6.1964 \times 10^5$, $f = 1.325/\{\ln [(1.2432 \times 10^{-5})/0.47418 + 5.74/(6.1964 \times 10^5)^{0.9}]\}^2 = 0.014086$, $x = 8.4973 + (0.20394)(0.47418)/0.014086 = 15.363$, $D = (0.25351)[15.363^{4.75} + (0.38707)(15.363^{5.2})]^{0.04} = 0.44063$ m. Try $D = 0.44063$ m: $N_R = 2.9382 \times 10^5/0.44063 = 6.6682 \times 10^5$, $f = 1.325/\{\ln [(1.2432 \times 10^{-5})/0.44063 + 5.74/(6.6682 \times 10^5)^{0.9}]\}^2 = 0.014075$, $x = 8.4973 + (0.20394)(0.44063)/0.014075 = 14.882$, $D = (0.25351)[14.882^{4.75} + (0.38707)(14.882^{5.2})]^{0.04} = 0.43783$ m. Try $D = 0.43783$ m: $N_R = 2.9382 \times 10^5/0.43783 = 6.7108 \times 10^5$, $f = 1.325/\{\ln [(1.2432 \times 10^{-5})/0.43783 + 5.74/(6.7108 \times 10^5)^{0.9}]\}^2 = 0.014074$, $x = 8.4973 + (0.20394)(0.43783)/0.014074 = 14.842$, $D = (0.25351)[14.842^{4.75} + (0.38707)(14.842^{5.2})]^{0.04} = 0.43759$ m. Therefore, $D = 0.438$ m, or 438 mm.

**9.258** One equation for determining the friction factor is $1/\sqrt{f} = 0.869 \ln (N_R\sqrt{f}) - 0.8$. Compare the smooth pipe curve on the Moody diagram (Fig. A-5) with results from the equation above for values of Reynolds number of $10^5$, $10^6$, and $10^7$.

For $N_R = 10^5$, from Fig. A-5, $f = 0.0178$. From the equation, $1/\sqrt{f} = 0.869\{\ln [(10^5)(\sqrt{f})]\} - 0.8$. By trial and error, $f = 0.0183$. For $N_R = 10^6$, from Fig. A-5, $f = 0.0116$. From the equation, $1/\sqrt{f} = 0.869\{\ln [(10^6)(\sqrt{f})]\} - 0.8$. By trial and error, $f = 0.0116$. For $N_R = 10^7$, from Fig. A-5, $f = 0.0082$. From the equation, $1/\sqrt{f} = 0.869\{\ln [(10^7)(\sqrt{f})]\} - 0.8$. By trial and error, $f = 0.00810$.

**9.259** An equation for determining the friction factor developed by Colebrook is $1/\sqrt{f} = -0.869 \ln[(\epsilon/D)/3.7 +$ $2.523/(N_R\sqrt{f})]$. Check the location of line $\epsilon/D = 0.0002$ on the Moody diagram (Fig. A-5) with the equation above for a Reynolds number of $10^5$.

▮ From Fig. A-5, $f = 0.0190$. From the equation, $1/\sqrt{f} = -0.869 \ln\{0.0002/3.7 + 2.523/[(10^5)(\sqrt{f})]\}$. By trial and error, $f = 0.0192$.

**9.260** Find the head loss in a pipeline consisting of 200 ft of 4-in steel pipe, a 90° bend on 24-in radius ($K = 0.15$), 4-in gate valve (fully open) ($K = 0.20$), 100 ft of 4-in steel pipe, expansion to 6 in with a 20° taper ($K = 0.4$), 300 ft of 6-in steel pipe, abrupt contraction to 3-in diameter ($K = 0.35$), and 50 ft of 3-in steel pipe. The discharge rate is 1.5 cfs.

▮ $h_f = (f)(L/d)(v^2/2g)$, $h_m = (K)(v^2/2g)$. For 200 ft of 4-in pipe: $v_1 = Q/A = 1.5/[(\pi)(\frac{4}{12})^2/4] = 17.19$ ft/s, $N_R = dv_1/v = (\frac{4}{12})(17.19)/(1.05 \times 10^{-5}) = 5.46 \times 10^5$, $\epsilon/d = 0.00015/(\frac{4}{12}) = 0.00045$. From Fig. A-5, $f = 0.0175$. $h_f = 0.0175[200/(\frac{4}{12})]\{17.19^2/[(2)(32.2)]\} = 48.2$ ft.

**For bend:** $h_m = 0.15\{17.19^2/[(2)(32.2)]\} = 0.7$ ft.

**For gate valve:** $h_m = 0.20\{17.19^2/[(2)(32.2)]\} = 0.9$ ft.

**For 100 ft of 4-in pipe:** $h_f = 0.0175[100/(\frac{4}{12})]\{17.19^2/[(2)(32.2)]\} = 24.1$ ft.

**For expansion:** $v_2 = 1.5/[(\pi)(\frac{6}{12})^2/4] = 7.639$ ft/s, $h_m = 0.4\{17.19 - 7.639)^2/[(2)(32.2)]\} = 0.6$ ft.

**For 300 ft of 6-in pipe:** $N_R = (\frac{6}{12})(7.639)/(1.05 \times 10^{-5}) = 3.64 \times 10^5$, $\epsilon/d = 0.00015/(\frac{6}{12}) = 0.00030$, $f = 0.0170$, $h_f = 0.0170[300/(\frac{6}{12})]\{7.639^2/[(2)(32.2)]\} = 9.2$ ft.

**For abrupt contraction:** $v_3 = 1.5/[(\pi)(\frac{3}{12})^2/4] = 30.56$ ft/s, $h_m = 0.35\{30.56^2/[(2)(32.2)]\} = 5.1$ ft.

**For 50 ft of 3-in pipe:** $N_R = (\frac{3}{12})(30.56)/(1.05 \times 10^{-5}) = 7.28 \times 10^5$, $\epsilon/d = 0.00015/(\frac{3}{12}) = 0.00060$, $f = 0.0180$, $h_f = 0.0180[50/(\frac{3}{12})]\{30.56^2/[(2)(32.2)]\} = 52.2$ ft, $h_L = 48.2 + 0.7 + 0.9 + 24.1 + 0.6 + 9.2 + 5.1 + 52.2 = 141.0$ ft.

**9.261** Using the Darcy–Weisbach formula, find the head loss in 1000 ft of 6-ft-diameter smooth concrete pipe carrying 80 cfs of water at 50 °F.

▮
$$h_f = (f)(L/D)(v^2/2g) \qquad v = Q/A = 80/[(\pi)(6)^2/4] = 2.829 \text{ ft/s}$$
$$N_R = Dv/v = (6)(2.829)/(1.40 \times 10^{-5}) = 1.21 \times 10^6 \qquad \epsilon/D = 0.001/6 = 0.000167$$

From Fig. A-5, $f = 0.014$.
$$h_f = (0.014)(\tfrac{1000}{6})\{2.829^2/[(2)(32.2)]\} = 0.290 \text{ ft}$$

**9.262** Solve Prob. 9.261 using the Manning formula.

▮
$$v = (1.486/n)(R)^{2/3}(s)^{1/2} = 2.829 \text{ ft/s} \qquad \text{(from Prob. 9.261)}$$
$$n = 0.013 \quad \text{(from Table A-13)} \qquad R = D/4 = \tfrac{6}{4} = 1.500 \text{ ft}$$
$$2.829 = (1.486/0.013)(1.500)^{2/3}(s)^{1/2} \qquad s = 0.0003567 \qquad h_f = (0.0003567)(1000) = 0.357 \text{ ft}$$

**9.263** Solve Prob. 9.261 using the Hazen–Williams formula.

▮
$$v = 1.318CR^{0.63}s^{0.54} = 2.829 \text{ ft/s} \qquad \text{(from Prob. 9.261)}$$
$$C = 120 \quad \text{(from Table A-14)} \qquad R = 1.500 \text{ ft} \quad \text{(from Prob. 9.262)}$$
$$2.829 = (1.318)(120)(1.500)^{0.63}(s)^{0.54} \qquad s = 0.0003618 \qquad h_f = (1000)(0.0003618) = 0.362 \text{ ft}$$

**9.264** Using the Darcy–Weisbach formula, find the head loss in 100 ft of 3-ft-diameter welded steel pipe carrying 15 cfs of water at 60 °F.

▮
$$h_f = (f)(L/D)(v^2/2g) \qquad v = Q/A = 15/[(\pi)(3)^2/4] = 2.122 \text{ ft/s}$$
$$N_R = Dv/v = (3)(2.122)/(1.21 \times 10^{-5}) = 5.26 \times 10^5 \qquad \epsilon/D = 0.00015/3 = 0.000050$$

From Fig. A-5, $f = 0.0135$. $h_f = (0.0135)(\tfrac{100}{3})\{2.122^2/[(2)(32.2)]\} = 0.031$ ft.

**9.265** Solve Prob. 9.264 using the Manning formula with $n = 0.012$.

$$v = (1.486/n)(R)^{2/3}(s)^{1/2} = 2.122 \text{ ft/s} \quad \text{(from Prob. 9.264)}$$

$$R = D/4 = \tfrac{3}{4} = 0.7500 \text{ ft} \quad 2.122 = (1.486/0.012)(0.7500)^{2/3}(s)^{1/2} \quad s = 0.0004309$$

$$h_f = (0.0004309)(100) = 0.043 \text{ ft}$$

**9.266** Solve Prob. 9.264 using the Hazen–Williams formula with $C = 120$.

$$v = 1.318CR^{0.63}s^{0.54} = 2.122 \text{ ft/s} \quad \text{(from Prob. 9.264)}$$

$$R = 0.7500 \text{ ft} \quad \text{(from Prob. 9.265)}$$

$$2.122 = (1.318)(120)(0.7500)^{0.63}(s)^{0.54} \quad s = 0.0004769 \quad h_f = (100)(0.0004769) = 0.048 \text{ ft}$$

**9.267** A 36-in-diameter concrete pipe is 4000 ft long and has a head loss of 12.7 ft. Find the discharge capacity of water for this pipe by the Hazen–Williams formula

$$v = 1.318CR^{0.63}s^{0.54} = (1.318)(120)[(\tfrac{36}{12})/4]^{0.63}(12.7/4000)^{0.54} = 5.906 \text{ ft/s}$$

$$Q = Av = [(\pi)(\tfrac{36}{12})^2/4](5.906) = 41.7 \text{ ft}^3/\text{s}$$

**9.268** Solve Prob. 9.267 using the Manning formula.

$$v = (1.486/n)(R)^{2/3}(s)^{1/2} = (1.486/0.013)[(\tfrac{36}{12})/4]^{2/3}(12.7/4000)^{1/2} = 5.317 \text{ ft/s}$$

$$Q = Av = [(\pi)(\tfrac{36}{12})^2/4](5.317) = 37.6 \text{ ft}^3/\text{s}$$

**9.269** A 1-m-diameter new cast iron pipe is 845 m long and has a head loss of 1.11 m. Find the discharge capacity of water for this pipe by the Hazen–Williams formula.

$$v = 0.8492CR^{0.63}s^{0.54} = (0.8492)(130)(\tfrac{1}{4})^{0.63}(1.11/845)^{0.54} = 1.281 \text{ m/s}$$

$$Q = Av = [(\pi)(1)^2/4](1.281) = 1.01 \text{ m}^3/\text{s}$$

**9.270** Solve Prob. 9.269 using the Manning formula.

$$v = (1.0/n)(R)^{2/3}(s)^{1/2} = (1.0/0.013)(\tfrac{1}{4})^{2/3}(1.11/845)^{1/2} = 1.106 \text{ m/s}$$

$$Q = Av = [(\pi)(1)^2/4](1.106) = 0.869 \text{ m}^3/\text{s}$$

**9.271** A riveted steel pipe must transport 2.4 ft³/s of water a distance of 190 ft with a head loss of 2.7 ft. Find the necessary pipe diameter using the Hazen–Williams formula.

$$v = 1.318CR^{0.63}s^{0.54} = Q/A = 2.4/(\pi D^2/4) = 3.056/D^2$$

$$3.056/D^2 = (1.318)(110)(D/4)^{0.63}(2.7/190)^{0.54} \quad D = 0.7695 \text{ ft} \quad \text{or} \quad 9.23 \text{ in}$$

**9.272** A square concrete conduit must transport 4.0 m³/s of water a distance of 45 m with a head loss of 1.80 m. Find the necessary conduit size using the Hazen–Williams formula.

$v = 0.8492CR^{0.63}s^{0.54}$. Let $a =$ length of conduit side. $v = Q/A = 4.0/a^2$, $4.0/a^2 = (0.8492)(120)(a^2/4a)^{0.63}(1.80/45)^{0.54}$, $a = 0.788$ m.

**9.273** Solve Prob. 9.267 using the Hazen–Williams pipe diagram.

Let $h_1 =$ unit head loss. $h_1 = 12.7/4000 = 0.003175$. From Fig. A-13 with $h_1 = 0.003175$ and $D = 3$ ft, or 36 in, $Q = 41.5$ ft³/s.

**9.274** Solve Prob. 9.269 using the Hazen–Williams pipe diagram.

$h_1 = 1.11/845 = 0.001314$. From Fig. A-14 with $h_1 = 0.001314$ and $D = 1$ m, or 1000 mm, $Q = 0.91$ m³/s. This value of $Q$ is for $C = 120$ (since the pipe diagram is for $C = 120$). Inasmuch as $C = 130$ for new cast iron (Table A-14), this value of $Q$ must be adjusted. Since $Q$ varies directly with $C$, $0.91/(Q)_{C=130} = \tfrac{120}{130}$, $(Q)_{C=130} = 0.99$ m³/s.

**9.275** Water is flowing in a 500-mm-diameter new cast iron pipe at a velocity of 2.0 m/s. Using the Hazen–Williams pipe diagram, find the pipe friction loss per 100 m of pipe.

$C = 130$ for new cast iron pipe. In order to use the pipe diagram (Fig. A-14) for which $C = 120$, the given velocity must be adjusted for a value of $C$ of 120. Since velocity varies directly with $C$, $2.0/(v)_{C=120} = \frac{130}{120}$, $(v)_{c=120} = 1.85$ m/s. From Fig. A-14 with $v = 1.85$ m/s and $D = 500$ mm, $h_1 = 0.0067$ m/m, or 0.67 m per 100 m. $p = \gamma h = (9.79)(0.67) = 6.6$ kPa per 100 m.

**9.276** A new cast iron pipe must carry 30 cfs of water at a head loss of 19.0 ft per mile of pipe length. Find the required pipe diameter using the Hazen–Williams pipe diagram.

▌ $h_1 = 19.0/5280 = 0.003598$. $C = 130$ for new cast iron pipe. In order to use the pipe diagram (Fig. A-13) for which $C = 120$, the given discharge must be adjusted for a value of $C$ of 120. Since discharge varies directly with $C$, $30/(Q)_{C=120} = \frac{130}{120}$, $(Q)_{C=120} = 27.7$ cfs. With $Q = 27.7$ cfs and $h_1 = 0.003598$, a pipe diameter of 30 in is determined from Fig. A-13.

**9.277** Solve Prob. 9.270 using the Manning pipe diagram with $n = 0.012$.

▌ $h_1 = 1.11/845 = 0.001314$. With $D = 1$ m, or 1000 mm, and $h_1 = 0.001314$, a discharge of 0.88 m³/s is determined from Fig. A-16. Inasmuch as Fig. A-16 is for $n = 0.013$ and $n$ is 0.012 in this problem, this discharge must be adjusted. Since discharge varies inversely with $n$, $0.88/(Q)_{n=0.012} = 0.012/0.013$, $(Q)_{n=0.012} = 0.95$ m³/s.

**9.278** A concrete pipe must carry 80 cfs of water at a head loss of 1.5 ft per 100 ft of pipe length. Find the required pipe diameter using the Manning pipe diagram.

▌ $h_1 = 1.5/100 = 0.015$. With $Q = 80$ cfs and $h_1 = 0.015$, a pipe diameter of about 36 in is determined from Fig. A-15.

**9.279** Solve Prob. 9.278 for the same given conditions except that the pipe has an $n$ value of 0.015.

▌ $$h_1 = 0.015 \quad \text{(from Prob. 9.278)}$$

The Manning pipe diagram (Fig. A-15) is for a value of $n$ of 0.013. The given discharge (80 cfs) is for a value of $n$ of 0.015. In order to use Fig. A-15, the given discharge must be adjusted for a value of $n$ of 0.013. Since discharge varies inversely with $n$, $80/(Q)_{n=0.013} = 0.013/0.015$, $(Q)_{n=0.013} = 92.3$ cfs. With $Q = 92.3$ cfs and $h_1 = 0.015$, a pipe diameter of about 38 in is determined from Fig. A-15.

**9.280** A capillary tube of inside diameter 6 mm connects tank $A$ and open container $B$, as shown in Fig. 9-50. The liquid in $A$, $B$, and capillary $CD$ is water having a specific weight of 9780 N/m³ and a viscosity of 0.0008 kg/(m·s). The pressure $p_A = 34.5$ kPa gage. Which direction will the water flow? What is the flow rate?

▌ Assume laminar flow from $B$ to $A$.

$$\frac{p_B}{\gamma} + \frac{v_B^2}{2g} + z_B = \frac{p_A}{\gamma} + \frac{v_A^2}{2g} + z_A + h_L \qquad h_f = \frac{128\mu LQ}{\pi\gamma D^4} = \frac{(128)(0.0008)(4.3)(Q)}{(\pi)(9780)(\frac{6}{1000})^4} = 11\,058Q$$

$$0 + 0 + (1.4 + 4.3 \sin 45°) = 34.5/9.79 + 0 + 1 + 11\,058Q$$

$$Q = -7.55 \times 10^{-6}\ \text{m}^3/\text{s} \quad \text{or} \quad -7.55 \times 10^{-3}\ \text{L/s}$$

Since $Q$ is negative, the flow must be from $A$ to $B$. Check for laminar flow: $v = Q/A = (7.55 \times 10^{-6})/[(\pi)(\frac{6}{1000})^2/4] = 0.2670$ m/s, $N_R = \rho Dv/\mu = (9780/9.807)(\frac{6}{1000})(0.2670)/0.0008 = 1997$ (laminar).

Fig. 9-50

**9.281** An *equivalent length* of pipe is one whose head loss for the same value of flow is equal to that of some other system of different geometry for which it is equivalent. Consider a steel pipe of inside diameter 10.02 in having in it an open globe valve ($K = 4.8$) and four screwed 90° elbows ($K = 0.42$ each). The length of the pipe is 100 ft, and 5 ft³/s of water at 60 °F flows through the pipe. What is the equivalent length of pipe with inside diameter 13.00 in?

**❚ For the given pipe:** $h_L = h_f + h_m$, $h_f = (f)(L/d)(v^2/2g)$, $v = Q/A = 5/[(\pi)(10.02/12)^2/4] = 9.131$ ft/s, $N_R = dV/v = (10.02/12)(9.131)/(1.21 \times 10^{-5}) = 6.30 \times 10^5$, $\epsilon/d = 0.00015/(10.02/12) = 0.00018$. From Fig. A-5, $f = 0.015$. $h_f = 0.015[100/(10.02/12)]\{9.131^2/[(2)(32.2)]\} = 2.33$ ft, $h_m = (K)(v^2/2g) = [4.8 + (4)(0.42)]\{9.131^2/[(2)(32.2)]\} = 8.39$ ft, $h_L = 2.33 + 8.39 = 10.72$ ft.

**For the equivalent pipe:** $v = 5/[(\pi)(13.00/12)^2/4] = 5.424$ ft/s, $N_R = (13.00/12)(5.424)/(1.21 \times 10^{-5}) = 4.86 \times 10^5$, $\epsilon/d = 0.00015/(13.00/12) = 0.000138$, $f = 0.015$, $10.72 = 0.015[L_e/(13.00/12)]\{5.424^2/[(2)(32.2)]\}$, $L_e = 1695$ ft.

**9.282** A 100-ft-long duct transports air at 50 °F (assumed constant) at the rate of 8000 ft³/min. The duct is made of galvanized iron and has a cross section of 2 ft by 1 ft. Find the outlet pressure, if the inlet pressure is 16.7 psia.

**❚** Assume incompressible, isothermal, fully turbulent flow.

$$p_1/\gamma + v_1^2/2g + z_1 = p_2/\gamma + v_2^2/2g + z_2 + h_L \qquad h_L = h_f = (f)(L/d)(v^2/2g) \qquad N_R = \rho dv/\mu$$

$$\rho = p/RT = (16.7)(144)/[(1716)(460 + 50)] = 0.002748 \text{ slug/ft}^3 \qquad \gamma = \rho g = (0.002748)(32.2) = 0.08849 \text{ lb/ft}^3$$

$$d = D_h = 4A/p_w = 4[(2)(1)]/[(2)(2 + 1)] = 1.333 \text{ ft} \qquad v = Q/A = (\tfrac{8000}{60})/[(2)(1)] = 66.67 \text{ ft/s}$$

$$N_R = (0.002748)(1.333)(66.67)/(3.68 \times 10^{-7}) = 6.64 \times 10^5 \quad \text{(O.K.)} \qquad \epsilon/d = 0.0005/1.333 = 0.000375$$

From Fig. A-5, $f = 0.0165$.

$$h_L = (0.0165)(100/1.333)\{66.67^2/[(2)(32.2)]\} = 85.43 \text{ ft}$$

$$p_1/0.08849 + v_1^2/2g + 0 = p_2/0.08849 + v_2^2/2g + 85.43$$

$$v_1^2/2g = v_2^2/2g \qquad p_1 - p_2 = 7.56 \text{ lb/ft}^2 \qquad p_2 = 16.7 - 7.56 = 9.1 \text{ psia}$$

**9.283** In a heating system, there is a run of insulated duct of 50 m carrying air at a temperature of 35 °C at a pressure at the inlet of 100 kPa. The duct has a rectangular cross section of 650 mm by 320 mm. If there is a pressure drop from inlet to outlet of 5 mm of mercury, what is the volumetric flow? For such a small pressure drop in the duct, treat the flow as incompressible. Take $R = 287$ J/(kg · K). The duct is galvanized iron. Consider that the flow is entirely turbulent.

**❚** $$p_1/\gamma + v_1^2/2g + z_1 = p_2/\gamma + v_2^2/2g + z_2 + h_L \qquad h_L = h_f = (f)(L/d)(v^2/2g) \qquad N_R = \rho dv/\mu$$

$$\rho = p/RT = (100)(1000)/[(287)(273 + 35)] = 1.131 \text{ kg/m}^3 \qquad \gamma = \rho g = (1.131)(9.807) = 11.09 \text{ N/m}^3$$

$$d = D_h = 4A/p_w = 4[(\tfrac{650}{1000})(\tfrac{320}{1000})]/[(2)(\tfrac{650}{1000} + \tfrac{320}{1000})] = 0.4289 \text{ m} \qquad h_L = (f)(50/0.4289)\{v^2/[(2)(9.807)]\} = 5.944fv^2$$

$$p_1 - p_2 = \tfrac{5}{1000}[(13.6)(9.79)] = 0.6657 \text{ kPa} \quad \text{or} \quad 665.7 \text{ Pa} \qquad p_1/\gamma - p_2/\gamma + v_1^2/2g + z_1 = v_2^2/2g + z_2 + h_L$$

$$v_1^2/2g = v_2^2/2g \qquad 665.7/11.09 + 0 = 0 + 5.944fv^2 \qquad v = 3.178/\sqrt{f}$$

Try $f = 0.016$: $v = 3.178/\sqrt{0.016} = 25.12$ m/s, $N_R = (1.131)(0.4289)(25.12)/(1.88 \times 10^{-5}) = 6.48 \times 10^5$, $\epsilon/d = 0.00015/0.4289 = 0.000350$. From Fig. A-5, $f = 0.0165$. Try $f = 0.0165$: $v = 3.178/\sqrt{0.0165} = 24.74$ m/s, $N_R = (1.131)(0.4289)(24.74)/(1.88 \times 10^{-5}) = 6.38 \times 10^5$, $f = 0.0165$ (O.K.); $Q = Av = [(\tfrac{650}{1000})(\tfrac{320}{1000})](24.74) = 5.15$ m³/s.

**9.284** Water at 60 °F flows through a smooth pipe of 12 in diameter at 9 ft³/s. Estimate the shear stress at the wall and the thickness of the viscous sublayer.

**❚** $$v = Q/A = 9/[(\pi)(\tfrac{12}{12})^2/4] = 11.46 \text{ ft/s}$$

$$\tau_0 = 0.03325\rho v^2(v/Rv)^{1/4} = (0.03325)(1.94)(11.46)^2\{(1.21 \times 10^{-5})/[(\tfrac{6}{12})(11.46)]\}^{1/4} = 0.323 \text{ lb/ft}^2$$

$$\delta_1 = 5v/v_* = 5v\sqrt{\rho/\tau_0} = (5)(1.21 \times 10^{-5})\sqrt{1.94/0.323} = 0.000148 \text{ ft} \quad \text{or} \quad 0.00178 \text{ in}$$

**9.285** Solve Prob. 9.284 if the shear stress is given by $\tau_0 = (f/4)(\rho v^2/2)$.

$$N_R = dv/\nu = (\tfrac{12}{12})(11.46)/(1.21 \times 10^{-5}) = 9.47 \times 10^5$$

From Fig. A-5, $f = 0.0118$.

$$\tau_0 = (0.0118/4)[(1.94)(11.46)^2/2] = 0.376 \text{ lb/ft}^2$$

$$\delta_1 = 5\nu/v_* = 5\nu\sqrt{\rho/\tau_0} = (5)(1.21 \times 10^{-5})\sqrt{1.94/0.376} = 0.0001374 \text{ ft} \quad \text{or} \quad 0.00165 \text{ in}$$

**9.286** What head $H$ is needed in Fig. 9-51 to produce a discharge of 0.3 m³/s? Use minor loss coefficients of 0.5 for the entrance and 0.12 for the diffusers.

$$p_A/\gamma + v_A^2/2g + z_A = p_B/\gamma + v_B^2/2g + z_B + h_L \qquad h_L = h_f + h_m \qquad h_f = (f)(L/d)(v^2/2g) \qquad N_R = dv/(\mu/\rho)$$

$$v_1 = Q/A_1 = 0.3/[(\pi)(0.210)^2/4] = 9.549 \text{ m/s} \qquad (N_R)_1 = (0.210)(9.549)/[(0.04)(0.1)/1000] = 5.01 \times 10^5$$

From Fig. A-5, $f_1 = 0.013$.

$$v_2 = Q/A_2 = 0.3/[(\pi)(0.305)^2/4] = 4.244 \text{ m/s} \qquad (N_R)_2 = (0.305)(4.244)/[(0.04)(0.1)/1000] = 3.24 \times 10^5$$

$$f_2 = 0.014 \qquad h_f = 0.013[30/(0.210)]\{9.549^2/[(2)(9.907)]\} + 0.014[60/(0.305)]\{4.244^2/[(2)(9.907)]\} = 11.64 \text{ m}$$

**For entrance:** $h_m = (K)(v^2/2g)$, $(h_m)_1 = 0.5\{9.549^2/[(2)(9.907)]\} = 2.32 \text{ m}$.

**For first diffuser:** $h_m = (K)(v_1 - v_2)^2/2g$, $(h_m)_2 = (0.12)(9.549 - 4.244)^2/[(2)(9.807)] = 0.17 \text{ m}$.

**For second diffuser:** $v_3 = Q/A_3 = 0.3/[(\pi)(0.450)^2/4] = 1.886 \text{ m/s}$, $(h_m)_3 = (0.12)(4.244 - 1.886)^2/[(2)(9.807)] = 0.03 \text{ m}$.

**For exit:** $(h_m)_4 = 1.0\{1.886^2/[(2)(9.807)]\} = 0.18 \text{ m}$,
$H = h_L = 11.64 + 2.32 + 0.17 + 0.03 + 0.18 = 14.34 \text{ m}$.

**Fig. 9-51**

**9.287** Calculate the discharge through the siphon of Fig. 9-52 ($H = 4$ ft) without the conical diffuser. Use minor loss coefficients of 1.0 for both the entrance and exit, and 0.9 for each elbow.

$$p_A/\gamma + v_A^2/2g + z_A = p_B/\gamma + v_B^2/2g + z_B + h_L$$

$$h_f = (f)(L/d)(v^2/2g) = f[(10 + 18 + 12)/(\tfrac{8}{12})]\{v^2/[(2)(32.2)]\} = 0.9317fv^2$$

$$h_m = (K)(v^2/2g) = [(1.0 + (2)(0.9) + 1.0]\{v^2/[(2)(32.2)]\} = 0.05901v^2 \qquad h_L = h_f + h_m = 0.9317fv^2 + 0.05901v^2$$

$$0 + 0 + 4 = 0 + 0 + 0 + 0.9317fv^2 + 0.05901v^2 \qquad v = \sqrt{4/(0.9317f + 0.05901)}$$

Try $f = 0.016$: $v = \sqrt{4/[(0.9317)(0.016) + 0.05901]} = 7.356$ ft/s, $N_R = dv/\nu = (\tfrac{8}{12})(7.356)/(1.21 \times 10^{-5}) = 4.05 \times 10^5$, $\epsilon/d = 0.00015/(\tfrac{8}{12}) = 0.000225$. From Fig. A-5, $f = 0.016$ (O.K.). $Q = Av = [(\pi)(\tfrac{8}{12})^2/4](7.356) = 2.57$ ft³/s.

**Fig. 9-52**

**9.288** Calculate the discharge in the siphon of Fig. 9-52 for $H = 8$ ft and with diffuser included. Use minor loss coefficients of 1.0 for both the entrance and exit, 0.9 for each elbow, and 0.15 for the diffuser.

$$p_A/\gamma + v_A^2/2g + z_A = p_B/\gamma + v_B^2/2g + z_B + h_L \tag{1}$$

$$h_L = h_f + h_m \qquad h_f = (f)(L/d)(v^2/2g) = f[(10 + 18 + 12)/(\tfrac{8}{12})]\{v_1^2/[(2)(32.2)]\} = 0.9317fv_1^2$$

$h_m = (K)(v^2/2g)$ for entrance, exit, and elbows. $h_m = (K)(v_1 - v_2)^2/2g$ for diffuser.

$$A_1v_1 = A_2v_2 \qquad [(\pi)(\tfrac{8}{12})^2/4](v_1) = \{(\pi)[(1.5)(\tfrac{8}{12})]^2/4\}(v_2) \qquad v_2 = 0.4444v_1$$

$$h_m = [1.0 + (2)(0.9)]\{v_1^2/[(2)(32.2)]\} + 0.15(v_1 - 0.4444v_1)^2/[(2)(32.2)]$$

$$+ 1.0\{(0.4444v_1)^2/[(2)(32.2)]\} = 0.04726v_1^2$$

$$h_L = 0.9317fv_1^2 + 0.04726v_1^2 \qquad 0 + 0 + 8 = 0 + 0 + 0 + 0.9317fv^2 + 0.04726v^2$$

$$v = \sqrt{8/(0.9317f + 0.04726)}$$

Try $f = 0.015$: $v = \sqrt{8/[(0.9317)(0.015) + 0.04726]} = 11.43$ ft/s, $N_R = dv/v = (\tfrac{8}{12})(11.43)/(1.21 \times 10^{-5}) = 6.30 \times 10^5$, $\epsilon/d = 0.00015/(\tfrac{8}{12}) = 0.000225$. From Fig. A-5, $f = 0.0155$. Try $f = 0.0155$: $v = \sqrt{8/[(0.9317)(0.0155) + 0.04726]} = 11.39$ ft/s, $N_R = (\tfrac{8}{12})(11.39)/(1.21 \times 10^{-5}) = 6.28 \times 10^5$, $f = 0.0155$ (O.K.); $Q = Av = [(\pi)(\tfrac{8}{12})^2/4](11.39) = 3.98$ ft³/s.

**9.289** Find the discharge through the siphon of Fig. 9-53, as well as the minimum pressure in the system. Use minor loss coefficients of 1.0 for the entrance, 0.1 for the nozzle, and 2.2 for the bend.

Let $v$ = velocity of water in the pipe and $v_2$ = velocity through the nozzle. Since velocity varies with the square of diameter, $v_2 = 4v$.

$$p_1/\gamma + v_1^2/2g + z_1 = p_2/\gamma + v_2^2/2g + z_2 + h_L \tag{1}$$

$$h_f = (f)(L/d)(v^2/2g) = f[(6 + 1.8 + 3.6)/0.100]\{v^2/[(2)(9.807)]\} = 5.812fv^2$$

$$h_m = (K)(v^2/2g) = (1.0 + 2.2)\{v^2/[(2)(9.807)]\} + (0.1)\{v_2^2/[(2)(9.807)]\}$$

$$= (1.0 + 2.2)\{v^2/[(2)(9.807)]\} + (0.1)\{(4v)^2/[(2)(9.807)]\} = 0.2447v^2$$

$$h_L = h_f + h_m = 5.812fv^2 + 0.2447v^2 \qquad 0 + 0 + (6 - 1.8) = 0 + (4v)^2/[(2)(9.807)]$$

$$+ (6 - 1.8 - 3.6) + 5.812fv^2 + 0.2447v^2 \qquad v = \sqrt{3.6/(5.812f + 1.060)}$$

Try $f = 0.03$: $v = \sqrt{3.6/[(5.812)(0.03) + 1.060]} = 1.708$ m/s, $N_R = \rho dv/\mu = [(0.81)(1000)](0.100)(1.708)/0.01 = 1.38 \times 10^4$. From Fig. A-5, $f = 0.028$. Try $f = 0.028$: $v = \sqrt{3.6/[(5.812)(0.028) + 1.060]} = 1.716$ m/s, $N_R = [(0.81)(1000)](0.100)(1.716)/0.01 = 1.38 \times 10^4$, $f = 0.028$ (O.K.); $Q = Av = [(\pi)(0.100)^2/4](1.716) = 0.0135$ m³/s, or 13.5 L/s.

Minimum pressure would occur at point 3. Apply Eq. (1) between points 1 and 3: $h_f = 0.028[6/0.100]\{1.716^2/[(2)(9.807)]\} = 0.2522$ m, $h_m = (1.0 + 2.2)\{1.716^2/[(2)(9.807)]\} = 0.4804$ m, $h_L = 0.2522 + 0.4804 = 0.7326$ m, $0 + 0 + 0 = p_3/[(0.81)(9.79)] + 1.716^2/[(2)(9.807)] + 1.8 + 0.7326$, $p_3 = -21.3$ kPa.

Fig. 9-53

**9.290** With reference to Fig. 9-54, what is the maximum height of point $A$ for no cavitation? Barometer reading is 29.5 in of mercury. Use minor loss coefficients of 1.0 for both entrance and exit, and 4.5 for the globe valve.

▌ First calculate the flow velocity $v$.

$$p_1/\gamma + v_1^2/2g + z_1 = p_2/\gamma + v_2^2/2g + z_2 + h_L \tag{1}$$

$$h_f = (f)(L/d)(v^2/2g) = f[(100 + 70 + 105)/(\tfrac{8}{12})]\{v^2/[(2)(32.2)]\} = 6.405fv^2$$

$$h_m = (K)(v^2/2g) = (1.0 + 4.5 + 1.0)\{v^2/[(2)(32.2)]\} = 0.1009v^2$$

$$h_L = h_f + h_m = 6.405fv^2 + 0.1009v^2 \qquad 0 + 0 + 12 = 0 + 0 + 0 + 6.405fv^2 + 0.1009v^2$$

$$v = \sqrt{12/(6.405f + 0.1009)}$$

Try $f = 0.013$: $v = \sqrt{12/[(6.405)(0.013) + 0.1009]} = 8.072$ ft/s, $N_R = dv/\nu = (\tfrac{8}{12})(8.072)/(1.21 \times 10^{-5}) = 4.45 \times 10^5$. From Fig. A-5, $f = 0.0135$. Try $f = 0.0135$: $v = \sqrt{12/[(6.405)(0.0135) + 0.1009]} = 8.003$ ft/s, $N_R = (\tfrac{8}{12})(8.003)/(1.21 \times 10^{-5}) = 4.41 \times 10^5$, $f = 0.0135$ (O.K.). Now, to find the maximum height of point $A$, apply Eq. (1) between points 1 and $A$: $h_f = 0.0135[100/(\tfrac{8}{12})]\{8.003^2/[(2)(32.2)]\} = 2.014$ ft, $h_m = 0$, $h_L = 2.014 + 0 = 2.014$ ft. From Table A-1, $p_v = 36.5$ lb/ft². $p_{\text{atm}} = [(13.6)(62.4)](29.5/12) = 2086.2$ lb/ft². At cavitation, $p = (2086.2 - 36.5)/62.4 = 32.85$ ft (vacuum), or $-32.85$ ft, $0 + 0 + 0 = -32.85 + 0.995 + z_A + 2.014$, $z_A = 29.8$ ft. ($z_A$ is the height of point $A$ above the water surface in the left reservoir.)

**Fig. 9-54**

**9.291** What diameter smooth pipe is required to convey 8 L/s of kerosene ($\nu = 1.93 \times 10^{-6}$ m²/s) at 32 °C, 150 m with a head of 5 m? There are a valve and other minor losses with total $K$ of 7.6.

▌
$$p_1/\gamma + v_1^2/2g + z_1 = p_2/\gamma + v_2^2/2g + z_2 + h_L \qquad v = Q/A = (\tfrac{8}{1000})/(\pi D^2/4) = 0.01019/D^2$$

$$h_f = (f)(L/d)(v^2/2g) = (f)(150/D)\{(0.01019/D^2)^2/[(2)(9.807)]\} = 0.0007941f/D^5$$

$$h_m = (K)(v^2/2g) = 7.6\{(0.01019/D^2)^2/[(2)(9.807)]\} = 0.00004023/D^4$$

$$h_L = h_f + h_m = 0.0007941f/D^5 + 0.00004023/D^4 \qquad p_1/\gamma + z_1 - p_2/\gamma - z_2 = 5$$

$$v_1^2/2g = v_2^2/2g \qquad 5 = 0.0007941f/D^5 + 0.00004023/D^4$$

Try $f = 0.02$: $5 = (0.0007941)(0.02)/D^5 + 0.00004023/D^4$. By trial and error, $D = 0.0826$ m. $N_R = Dv/\nu = (0.0826)(0.01019/0.0826^2)/(1.93 \times 10^{-6}) = 6.39 \times 10^4$. From Fig. A-5, $f = 0.0195$. Try $f = 0.0195$: $5 = (0.0007941)(0.0195)/D^5 + 0.00004023/D^4$, $D = 0.0822$ m, $N_R = (0.0822)(0.01019/0.0822^2)/(1.93 \times 10^{-6}) = 6.42 \times 10^4$, $f = 0.0195$. Therefore, $D = 0.0822$ m, or 82.2 mm.

**9.292** Find the value of the Hazen–Williams coefficient for the water flow in Prob. 9.130.

▌
$$v = 1.318CR^{0.63}s^{0.54} = Q/A = 3.2/[(\pi)(\tfrac{24}{12})^2/4] = 1.0186 \text{ ft/s}$$

$$1.0186 = (1.318)(C)[(\tfrac{24}{12})/4]^{0.63}(0.0003/1)^{0.54} \qquad C = 95.5$$

**9.293** Find the value of the Hazen–Williams coefficient for the case where water flows at 0.20 m³/s in a 1-m-diameter pipeline with a head loss of 0.0012 m/m.

▌
$$v = 0.8492CR^{0.63}s^{0.54} = Q/A = 0.20/[(\pi)(1)^2/4] = 0.2546 \text{ m/s}$$

$$0.2546 = (0.8492)(C)[1/4]^{0.63}(0.0012)^{0.54} \qquad C = 27.1$$

**9.294** Manning's $n$ equals 0.0132 when water at 50 °F flows at a Reynolds number of $1.10 \times 10^7$ through a tunnel that is 16 ft in diameter. Determine the average value of $\epsilon$.

▌
$$v = \left(\frac{1.486}{n}\right)(R^{2/3})(s^{1/2}) \qquad s = \left(\frac{nv}{1.486R^{2/3}}\right)^2 \qquad h_L = (f)\left(\frac{L}{d}\right)\left(\frac{v^2}{2g}\right) \qquad \frac{h_L}{L} = \left(\frac{f}{d}\right)\left(\frac{v^2}{2g}\right)$$

Equating $s$ and $h_L/L$ gives

$$\left(\frac{nv}{1.486R^{2/3}}\right)^2 = \left(\frac{f}{d}\right)\left(\frac{v^2}{2g}\right) \qquad \left[\frac{0.0132v}{(1.486)(\frac{16}{4})^{2/3}}\right]^2 = \left(\frac{f}{16}\right)\left\{\frac{v^2}{(2)(32.2)}\right\} \qquad f = 0.0128$$

From Fig. A-5 with $f = 0.0128$ and $N_R = 1.10 \times 10^7$, $\epsilon/d = 0.00015$; $\epsilon = (16)(0.00015) = 0.00240$ ft, or 0.0288 in.

**9.295** Prepare a computer program that will determine for a single pipe with constant diameter either the flow rate or the required conduit diameter for closed conduit, incompressible flow. The program must be usable for problems in both the USCS and the SI.

```
C    THIS PROGRAM DETERMINES THE FLOW RATE OR THE REQUIRED CONDUIT
C    DIAMETER FOR CLOSED CONDUIT, INCOMPRESSIBLE FLOW.  IT CAN BE USED
C    FOR PROBLEMS IN BOTH THE ENGLISH SYSTEM OF UNITS AND THE INTER-
C    NATIONAL SYSTEM OF UNITS.  THE APPLICATION OF THE PROGRAM IS
C    LIMITED TO CASES INVOLVING A SINGLE PIPE WITH CONSTANT DIAMETER.
C
C    THIS PROGRAM IS BASED ON THE SOLUTION OF THE BERNOULLI EQUATION;
C    ACCORDINGLY, CERTAIN DATA MUST BE ENTERED FOR EACH OF TWO POINTS
C    IN A PIPE SYSTEM.  HOWEVER, BOTH THE VELOCITY AT POINT 1 AND THE
C    VELOCITY AT POINT 2 ARE NOT CONSIDERED AS KNOWN VALUES INITIALLY.
C    IF EITHER VALUE OF VELOCITY IS ESSENTIALLY ZERO (SUCH AS AT A
C    RESERVOIR OR TANK SURFACE), ENTER 0 (ZERO) FOR THAT VELOCITY (OR
C    FOR BOTH VELOCITIES IF BOTH ARE ESSENTIALLY ZERO).  OTHERWISE,
C    ENTER 1.0 FOR ALL OTHER VELOCITIES.
C
C    INPUT DATA MUST BE SET UP AS FOLLOWS.
C
C    CARD 1    COLUMN 1        ENTER 0 (ZERO) OR BLANK IF ENGLISH SYSTEM
C                              OF UNITS IS TO BE USED.  ENTER 1 (ONE)
C                              IF INTERNATIONAL SYSTEM OF UNITS IS TO
C                              BE USED.
C              COLUMNS 2-79    ENTER TITLE, DATE, AND OTHER INFORMATION,
C                              IF DESIRED.
C    CARD 2    COLUMNS 1-10    ENTER NUMBER INCLUDING DECIMAL GIVING
C                              GAUGE PRESSURE AT POINT 1   (IN POUNDS
C                              PER SQUARE INCH OR KILOPASCALS).
C              COLUMNS 11-20   ENTER NUMBER INCLUDING DECIMAL GIVING
C                              GAUGE PRESSURE AT POINT 2   (IN POUNDS
C                              PER SQUARE INCH OR KILOPASCALS).
C              COLUMNS 21-30   ENTER 0 (ZERO) OR BLANK IF VELOCITY AT
C                              POINT 1 IS ESSENTIALLY ZERO; OTHERWISE,
C                              ENTER 1.0.
C              COLUMNS 31-40   ENTER 0 (ZERO) OR BLANK IF VELOCITY AT
C                              POINT 2 IS ESSENTIALLY ZERO; OTHERWISE,
C                              ENTER 1.0.
C              COLUMNS 41-50   ENTER NUMBER INCLUDING DECIMAL GIVING
C                              ELEVATION AT POINT 1 (IN FEET OR METERS).
C              COLUMNS 51-60   ENTER NUMBER INCLUDING DECIMAL GIVING
C                              ELEVATION AT POINT 2 (IN FEET OR METERS).
C              COLUMNS 61-70   ENTER NUMBER INCLUDING DECIMAL GIVING
C                              ACTUAL ENERGY ADDED (IN HORSEPOWER OR
C                              KILOWATTS).
C              COLUMNS 71-80   ENTER NUMBER INCLUDING DECIMAL GIVING
C                              ACTUAL ENERGY REMOVED (IN HORSEPOWER OR
C                              KILOWATTS).
C    CARD 3    COLUMNS 1-10    ENTER NUMBER INCLUDING DECIMAL GIVING
C                              TOTAL "MINOR LOSSES" (IN FEET OR METERS).
C              COLUMNS 11-20   ENTER NUMBER INCLUDING DECIMAL GIVING
C                              DIAMETER OF CONDUIT (IN INCHES OR
C                              MILLIMETERS).
C              COLUMNS 21-30   ENTER NUMBER INCLUDING DECIMAL GIVING
C                              LENGTH OF CONDUIT (IN FEET OR METERS).
C              COLUMNS 31-40   ENTER NUMBER INCLUDING DECIMAL GIVING
C                              KINEMATIC VISCOSITY OF FLUID (IN SQUARE
C                              FEET PER SECOND OR SQUARE METERS PER
C                              SECOND).
C              COLUMNS 41-50   ENTER NUMBER INCLUDING DECIMAL GIVING
C                              ROUGHNESS OF CONDUIT MATERIAL (IN FEET
C                              OR METERS).  ENTER A VALUE OF 0 (ZERO)
C                              FOR "SMOOTH" CONDUITS.
```

```
C                    COLUMNS 51-60    ENTER NUMBER INCLUDING DECIMAL GIVING
C                                     SPECIFIC (OR UNIT) WEIGHT OF FLUID (IN
C                                     POUNDS PER CUBIC FOOT OR KILO-
C                                     NEWTONS PER CUBIC METER).
C                    COLUMNS 61-70    ENTER NUMBER INCLUDING DECIMAL GIVING
C                                     FLOW RATE OF FLUID (IN CUBIC FEET PER
C                                     SECOND OR CUBIC METERS PER SECOND).
C         CARD 4     COLUMNS 1-24     ENTER TYPE OF FLUID
C                    COLUMNS 25-48    ENTER TYPE OF CONDUIT.
C
C  *************************************************************************
C  *                                                                      *
C  *    NOTE WELL....EITHER THE FLOW RATE (COLUMNS 61-70) OR THE          *
C  *    CONDUIT DIAMETER (COLUMNS 11-20), WHICHEVER ONE IS TO BE DETER-   *
C  *    MINED BY THIS PROGRAM, SHOULD BE LEFT BLANK ON CARD 3.            *
C  *                                                                      *
C  *************************************************************************
C
C      MULTIPLE DATA SETS FOR SOLVING ANY NUMBER OF PROBLEMS MAY BE
C      INCLUDED FOR PROCESSING.
C
       REAL L
       DIMENSION TITLE(13),FLUID(4),PIPE(4)
       INTEGER UNITS
       COMMON F,ED,RN
       PI=3.14159265
    1  READ(5,100,END=2)UNITS,TITLE
       FACTOR=12.0
       IF(UNITS.EQ.1)FACTOR=1000.0
  100  FORMAT(I1,13A6)
       WRITE(6,101)TITLE
  101  FORMAT('1',13A6,////)
       READ(5,102)P1,P2,V1,V2,Z1,Z2,HA,HR,HM,D,L,VIS,E,SW,Q,FLUID,PIPE
  102  FORMAT(8F10.0/7F10.0/8A6)
       P1SW=P1/SW*FACTOR**2
       IF(UNITS.EQ.1)P1SW=P1/SW
       G=32.2
       IF(UNITS.EQ.1)G=9.807
       P2SW=P2/SW*FACTOR**2
       IF(UNITS.EQ.1)P2SW=P2/SW
       FF=0.02
       IF(Q.GT.0.0001)GO TO 117
       IF(V1.GT.0.0001)V1=1.0/2.0/G
       IF(V2.GT.0.0001)V2=1.0/2.0/G
  105  HF=FF*L/D*FACTOR/2.0/G
       HAT=550.*HA/SW
       IF(UNITS.EQ.1)HAT=HA/SW
       HRT=550.*HR/SW
       IF(UNITS.EQ.1)HRT=HR/SW
       Q=0.001
       VTRY=(Q/(PI*(D/FACTOR)**2/4.0))**2
       TRY1=P1SW+VTRY*V1+Z1+HAT/Q-HRT/Q-(P2SW+VTRY*V2+Z2+HF*VTRY)
  116  Q=Q+0.001
       VTRY=(Q/(PI*(D/FACTOR)**2/4.0))**2
       TRY2=P1SW+VTRY*V1+Z1+HAT/Q-HRT/Q-(P2SW+VTRY*V2+Z2+HF*VTRY)
       IF(TRY1*TRY2)114,114,115
  115  TRY1=TRY2
       GO TO 116
  114  Q=Q-0.0005
       V=Q/(PI*(D/FACTOR)**2/4.0)
       RN=D/FACTOR*V/VIS
       ED=E/D*FACTOR
       CALL ROUGH
       DIFF=ABS(F-FF)
       IF(DIFF.LT.0.0001)GO TO 104
       FF=F
       GO TO 105
  104  IF(V1.GT.0.0001)V1=V
       IF(V2.GT.0.0001)V2=V
       IF(UNITS.EQ.0)WRITE(6,106)P1,P2,Z1,Z2,HA,HR,HM,D,L,FLUID,PIPE,Q,
      *V1,V2
  106  FORMAT(1X,'GIVEN DATA FOR A CIRCULAR CLOSED CONDUIT CARRYING INCOM
      *PRESSIBLE FLOW',//5X,'PRESSURE AT POINT 1 =',F7.1,' PSI',//5X,'PRE
```

```
      *SSURE AT POINT 2 =',F7.1,' PSI',//5X,'ELEVATION AT POINT 1 =',
      *F7.1,' FT',//5X,'ELEVATION AT POINT 2 =',F7.1,' FT',//5X,'ACTUAL E
      *NERGY ADDED BETWEEN POINTS 1 AND 2 =',F5.1,' HP',//5X,'ACTUAL ENER
      *GY REMOVED BETWEEN POINTS 1 AND 2 =',F5.1,' HP',//5X,'MINOR HEAD L
      *OSSES BETWEEN POINTS 1 AND 2 =',F5.1,' FT',//5X,'DIAMETER OF CONDU
      *IT =',F6.2,' IN',//5X,'LENGTH OF CONDUIT =',F8.1,' FT',//5X,'FLUID
      * FLOWING IS ',4A6,//5X,'CONDUIT MATERIAL IS ',4A6,////1X,'THE FLOW
      * RATE WILL BE',F7.1,' CU FT/S',//5X,'VELOCITY AT POINT 1 =',F6.2,
      *' FT/S',//5X,'VELOCITY AT POINT 2 =',F6.2,' FT/S')
        IF(UNITS.EQ.1)WRITE(6,107)P1,P2,Z1,Z2,HA,HR,HM,D,L,FLUID,PIPE,Q,
      *V1,V2
  107 FORMAT(1X,'GIVEN DATA FOR A CIRCULAR CLOSED CONDUIT CARRYING INCOM
      *PRESSIBLE FLOW',//5X,'PRESSURE AT POINT 1 =',F7.1,' KPA',//5X,'PRE
      *SSURE AT POINT 2 =',F7.1,' KPA',//5X,'ELEVATION AT POINT 1 =',
      *F7.1,' M ',//5X,'ELEVATION AT POINT 2 =',F7.1,' M ',//5X,'ACTUAL E
      *NERGY ADDED BETWEEN POINTS 1 AND 2 =',F5.1,' KW',//5X,'ACTUAL ENER
      *GY REMOVED BETWEEN POINTS 1 AND 2 =',F5.1,' KW',//5X,'MINOR HEAD L
      *OSSES BETWEEN POINTS 1 AND 2 =',F5.1,' M ',//5X,'DIAMETER OF CONDU
      *IT =',F7.1,' MM',//5X,'LENGTH OF CONDUIT =',F8.1,' M ',//5X,'FLUID
      * FLOWING IS ',4A6,//5X,'CONDUIT MATERIAL IS ',4A6,////1X,'THE FLOW
      * RATE WILL BE',F7.3,' CU M/S',//5X,'VELOCITY AT POINT 1 =',F6.2,
      *' M/S ',//5X,'VELOCITY AT POINT 2 =',F6.2,' M/S ')
        GO TO 1
  117 V1=V1*(Q/PI*4.0)
        V2=V2*(Q/PI*4.0)
  103 HF=FF*L*(Q/PI*4.0)**2/2.0/G
        HAT=550.*HA/SW/Q
        IF(UNITS.EQ.1)HAT=HA/SW/Q
        HRT=550.*HR/SW/Q
        IF(UNITS.EQ.1)HRT=HR/SW/Q
        D=0.001
        TRY1=HF/D**5+(V2**2/2.0/G)/D**4-(V1**2/2.0/G)/D**4-P1SW-Z1-HAT
      *+HRT*P2SW+Z2+HM
  110 D=D+0.001
        TRY2=HF/D**5+(V2**2/2.0/G)/D**4-(V1**2/2.0/G)/D**4-P1SW-Z1-HAT
      *+HRT+P2SW+Z2+HM
        IF(TRY1*TRY2)108,108,109
  109 TRY1=TRY2
        GO TO 110
  108 D=D-0.0005
        RN=D*Q/(PI*D**2/4.0)/VIS
        ED=E/D
        CALL ROUGH
        DIFF=ABS(F-FF)
        IF(DIFF.LT.0.0001)GO TO 111
        FF=F
        GO TO 103
  111 V1=V1/D**2
        V2=V2/D**2
        D=D*FACTOR
        IF(UNITS.EQ.0)WRITE(6,112)P1,P2,Z1,Z2,HA,HR,HM,Q,L,FLUID,PIPE,D,
      *V1,V2
  112 FORMAT(1X,'GIVEN DATA FOR A CIRCULAR CLOSED CONDUIT CARRYING INCOM
      *PRESSIBLE FLOW',//5X,'PRESSURE AT POINT 1 =',F7.1,' PSI',//5X,'PRE
      *SSURE AT POINT 2 =',F7.1,' PSI',//5X,'ELEVATION AT POINT 1 =',
      *F7.1,' FT',//5X,'ELEVATION AT POINT 2 =',F7.1,' FT',//5X,'ACTUAL E
      *NERGY ADDED BETWEEN POINTS 1 AND 2 =',F5.1,' HP',//5X,'ACTUAL ENER
      *GY REMOVED BETWEEN POINTS 1 AND 2 =',F5.1,' HP',//5X,'MINOR LOSSES
      * BETWEEN POINTS 1 AND 2 =',F5.1,' FT',//5X,'FLOW RATE =',F7.1,
      *' CU FT/S',//5X,'LENGTH OF CONDUIT =',F8.1,' FT',//5X,'FLUID FLOWI
      *NG IS ',4A6,//5X,'CONDUIT MATERIAL IS ',4A6,///1X,'THE CONDUIT DIA
      *METER REQUIRED WILL BE',F6.2,' IN',//5X,'VELOCITY AT POINT 1 =',
      *F6.2,' FT/S',//5X,'VELOCITY AT POINT 2 =',F6.2,' FT/S')
        IF(UNITS.EQ.1)WRITE(6,113)P1,P2,Z1,Z2,HA,HR,HM,Q,L,FLUID,PIPE,D,
      *V1,V2
  113 FORMAT(1X,'GIVEN DATA FOR A CIRCULAR CLOSED CONDUIT CARRYING INCOM
      *PRESSIBLE FLOW',//5X,'PRESSURE AT POINT 1 =',F7.1,' KPA',//5X,'PRE
      *SSURE AT POINT 2 =',F7.1,' KPA',//5X,'ELEVATION AT POINT 1 =',
      *F7.1,' M ',//5X,'ELEVATION AT POINT 2 =',F7.1,' M ',//5X,'ACTUAL E
      *NERGY ADDED BETWEEN POINTS 1 AND 2 =',F5.1,' KW',//5X,'ACTUAL ENER
      *GY REMOVED BETWEEN POINTS 1 AND 2 =',F5.1,' KW',//5X,'MINOR LOSSES
      * BETWEEN POINTS 1 AND 2 =',F5.1,' M ',//5X,'FLOW RATE =',F7.3,
      *' CU M/S ',//5X,'LENGTH OF CONDUIT =',F8.1,' M ',//5X,'FLUID FLOWI
```

```
               *NG IS ',4A6,//5X,'CONDUIT MATERIAL IS ',4A6,///1X,'THE CONDUIT DIA
               *METER REQUIRED WILL BE',F7.1,' MM',//5X,'VELOCITY AT POINT 1 =',
               *F6.2,' M/S ',//5X,'VELOCITY AT POINT 2 =',F6.2,' M/S ')
               GO TO 1
      2        STOP
               END
               SUBROUTINE ROUGH
               COMMON F,ED,RN
               IF(RN.LE.2000.0)F=64.0/RN
               IF(RN.LE.2000.0)RETURN
               IF(RN.LT.4000.0)WRITE(6,103)
    103 FORMAT(1X,'A REYNOLDS NUMBER IS IN THE CRITICAL ZONE, FOR WHICH TH
               *E FRICTION FACTOR IS UNCERTAIN.  HENCE, PROGRAM EXECUTION WAS TERM
               *INATED.')
               IF(RN.LT.4000.0)STOP
               F=0.006
               TRY1=1.0/SQRT(F)+2.0*ALOG10(ED/3.7+2.51/RN/SQRT(F))
    102 F=F+0.00001
               TRY2=1.0/SQRT(F)+2.0*ALOG10(ED/3.7+2.51/RN/SQRT(F))
               IF(TRY1*TRY2)100,100,101
    101 TRY1=TRY2
               GO TO 102
    100 F=F-0.000005
               RETURN
               END
```

**9.296**    For the data given in Prob. 9.65, find the flow rate at which water is being discharged from the pipe, utilizing the computer program of Prob. 9.295.

### INPUT

```
1 2 3 4 5 6 7 8 9 10 11 12 13 14 15 16 17 18 19 20 21 22 23 24 25 26 27 28 29 30 31 32 33 34 35 36 37 38 39 40 41 42 43 44 45 46 47 48 49 50 51 52 53 54 55 56 57 58 59 60 61 62 63 64 65 66 67 68 69 70 71 72 73 74 75 76 77 78 79 80
0SAMPLE ANALYSIS ØF INCØMPRESSIBLE FLØW
                                    1.0        150.5      98.4
             24.0       130.0      0.0000105 0.00085    62.4
WATER                    NEW CAST IRØN
```

### OUTPUT

```
SAMPLE ANALYSIS OF INCOMPRESSIBLE FLOW


GIVEN DATA FOR A CIRCULAR CLOSED CONDUIT CARRYING INCOMPRESSIBLE FLOW

     PRESSURE AT POINT 1 =    0.0 PSI

     PRESSURE AT POINT 2 =    0.0 PSI

     ELEVATION AT POINT 1 =  150.5 FT

     ELEVATION AT POINT 2 =   98.4 FT

     ACTUAL ENERGY ADDED BETWEEN POINTS 1 AND 2 =  0.0 HP

     ACTUAL ENERGY REMOVED BETWEEN POINTS 1 AND 2 =  0.0 HP

     MINOR HEAD LOSSES BETWEEN POINTS 1 AND 2 =  0.0 FT

     DIAMETER OF CONDUIT = 24.00 IN

     LENGTH OF CONDUIT =   130.0 FT

     FLUID FLOWING IS WATER

     CONDUIT MATERIAL IS NEW CAST IRON


  THE FLOW RATE WILL BE  127.0 CU FT/S

     VELOCITY AT POINT 1 =  0.00 FT/S

     VELOCITY AT POINT 2 = 40.44 FT/S
```

**9.297**    For the data given in Prob. 9.66, find the pipe diameter, utilizing the computer program of Prob. 9.295.

### INPUT

```
1 2  3  4  5  6  7  8  9 10 11 12 13 14 15 16 17 18 19 20 21 22 23 24 25 26 27 28 29 30 31 32 33 34 35 36 37 38 39 40 41 42 43 44 45 46 47 48 49 50 51 52 53 54 55 56 57 58 59 60 61 62 63 64 65 66 67 68 69 70 71 72 73 74 75 76 77 78 79 80
1SAMPLE ANALYSIS ØF INCØMPRESSIBLE FLØW
2.5                 1.0       1.0       82.65     66.66
                    965.5     .000000406.00050    7.05       .10
GASØLINE                      (RØUGHNESS = 0.500 MM)
```

### OUTPUT

```
SAMPLE ANALYSIS OF INCOMPRESSIBLE FLOW


GIVEN DATA FOR A CIRCULAR CLOSED CONDUIT CARRYING INCOMPRESSIBLE FLOW

    PRESSURE AT POINT 1 =    2.5 KPA

    PRESSURE AT POINT 2 =    0.0 KPA

    ELEVATION AT POINT 1 =   82.7 M

    ELEVATION AT POINT 2 =   66.7 M

    ACTUAL ENERGY ADDED BETWEEN POINTS 1 AND 2 =  0.0 KW

    ACTUAL ENERGY REMOVED BETWEEN POINTS 1 AND 2 =  0.0 KW

    MINOR LOSSES BETWEEN POINTS 1 AND 2 =  0.0 M

    FLOW RATE =  0.100 CU M/S

    LENGTH OF CONDUIT =   965.5 M

    FLUID FLOWING IS GASOLINE

    CONDUIT MATERIAL IS (ROUGHNESS = 0.500 MM)



THE CONDUIT DIAMETER REQUIRED WILL BE   257.5 MM

    VELOCITY AT POINT 1 =  1.92 M/S

    VELOCITY AT POINT 2 =  1.92 M/S
```

**10.1** For a 12-in-diameter concrete pipe 12 000 ft long, find the diameter of a 1000-ft-long equivalent pipe.

▌ Assume a flow rate of 3.0 cfs. (The result should be the same regardless of the flow rate assumed.) From Fig. A-13, with $D = 12$ in and $Q = 3.0$ ft$^3$/s, $h_1 = 0.0052$ ft/ft. Therefore, $h_f = (0.0052)(12\,000) = 62.40$ ft. For a 1000-ft-long equivalent pipe with the same head loss, $h_1 = 62.40/1000 = 0.06240$ ft/ft. From Fig. A-13, with $h_1 = 0.06240$ ft/ft and $Q = 3.0$ ft$^3$/s, $D = 7.3$ in.

**10.2** A 480-ft-long, 18-in-diameter concrete pipe and a 590-ft-long, 12-in-diameter concrete pipe are connected in series. Find the length of an equivalent pipe of 10 in diameter.

▌ Assume a flow rate of 5 cfs through the two given pipes. For the 18-in-diameter pipe, from Fig. A-13, $h_1 = 0.00180$ ft/ft. For the 12-in-diameter pipe, $h_1 = 0.0137$ ft/ft. The total head loss for both pipes is $h_f = (0.00180)(480) + (0.0137)(590) = 8.947$ ft. For a 10-in-diameter pipe with $Q = 5.0$ cfs, from Fig. A-13, $h_1 = 0.032$ ft/ft. Since the equivalent pipe must have the same head loss as that of the system it replaces (i.e., 8.947 ft), the required length of a 10-in-diameter equivalent pipe can now be determined by $0.032L = 8.947$, $L = 280$ ft. Note that the required length of pipe can be determined in a single computation as follows: $L = (0.00180)(480)/0.032 + (0.0137)(590)/0.032 = 280$ ft.

**10.3** A 225-m-long, 300-mm-diameter concrete pipe and a 400-m-long, 500-mm-diameter concrete pipe are connected in series. Find the diameter of a 625-m-long equivalent pipe.

▌ Assume a flow rate of 0.1 m$^3$/s. For the 300-mm-diameter pipe, from Fig. A-14, $h_1 = 0.0074$ m/m. For the 500-mm-diameter pipe, $h_1 = 0.00064$ m/m. The total head loss for both pipes is $h_f = (0.0074)(225) + (0.00064)(400) = 1.921$ m. For a 625-m-long equivalent pipe with this head loss, $h_1 = 1.921/625 = 0.00307$ m/m. From Fig. A-14, $D = 360$ mm.

**10.4** Water flows at a rate of 0.020 m$^3$/s from reservoir $A$ to reservoir $B$ through three concrete pipes connected in series, as shown in Fig. 10-1. Find the difference in water-surface elevations in the reservoirs. Neglect minor losses.

▌ $$p_A/\gamma + v_A^2/2g + z_A = p_B/\gamma + v_B^2/2g + z_B + h_L \qquad 0 + 0 + z_A = 0 + 0 + z_B + h_L \qquad h_L = h_f = z_A - z_B$$

With $Q = 0.020$ m$^3$/s and $D = 160$ mm, from Fig. A-14, $h_1 = 0.0082$ m/m. With $Q = 0.020$ m$^3$/s and $D = 200$ mm, $h_1 = 0.0028$ m/m. With $Q = 0.020$ m$^3$/s and $D = 180$ mm, $h_1 = 0.0046$ m/m. $H = h_f = (0.0082)(1000) + (0.0028)(1600) + (0.0046)(850) = 16.59$ m.

**Fig. 10-1**

**10.5** Compute the flow rate of water through the three concrete pipes connected in series as shown in Fig. 10-2 by the equivalent length method.

▐ $h_f = z_A - z_B = 20$ ft/ft. Assume $Q = 1$ cfs. From Fig. A-13, $(h_1)_{12\text{-in}} = 0.00067$ ft/ft; $(h_1)_{10\text{-in}} = 0.0016$ ft/ft; $(h_1)_{8\text{-in}} = 0.00475$ ft/ft.

Consider a 10-in-diameter equivalent pipe: $L_e = 120 + (0.00475)(150)/0.0016 + (0.00067)(100)/0.0016 = 607$ ft. With $D = 10$ in and $h_1 = \frac{20}{607} = 0.0329$ ft/ft, from Fig. A-13, $Q = 5.1$ cfs.

Fig. 10-2

**10.6** Solve Prob. 10.5 by the equivalent diameter method.

▐ Using values obtained from Prob. 10.5, $h_f = (0.00067)(100) + (0.00475)(150) + (0.0016)(120) = 0.972$ ft. With $Q = 1$ cfs and $h_1 = 0.972/(100 + 150 + 120) = 0.00263$ ft/ft, $D = 8.9$ in. With $D = 8.9$ in and $h_1 = \frac{20}{370} = 0.0541$ ft/ft, $Q = 5.1$ cfs.

**10.7** Two concrete pipes are connected in series. The flow rate of water through the pipes is 0.14 m³/s with a total friction loss of 14.10 m for both pipes. Each pipe has a length of 300 m. If one pipe has a diameter of 300 mm, what is the diameter of the other one? Neglect minor losses.

▐ For first pipe: With $Q = 0.14$ m³/s and $D = 300$ mm, $h_1 = 0.014$ m/m. $h_f = (0.014)(300) = 4.20$ m. For second pipe: $h_f = 14.10 - 4.20 = 9.90$ m, $h_1 = 9.90/300 = 0.033$ m/m. With $Q = 0.14$ m³/s and $h_1 = 0.033$ m/m, $D = 250$ mm.

**10.8** Three concrete pipes are connected in series, as shown in Fig. 10-3. Determine the length of an 8-in-diameter equivalent pipe.

▐ Assume $Q = 1$ cfs. $(h_1)_{8\text{-in}} = 0.00475$ ft/ft; $(h_1)_{6\text{-in}} = 0.0195$ ft/ft; $(h_1)_{10\text{-in}} = 0.0016$ ft/ft; $L_e = 1200 + (0.0195)(1000)/0.00475 + (0.0016)(2000)/0.00475 = 5979$ ft.

Fig. 10-3

**10.9** For the three pipes in Fig. 10-3, determine the diameter of a 4200-ft-long equivalent pipe.

▐ Using values obtained from Prob. 10.8, for a 4200-ft-long equivalent pipe, $h_f = (0.00475)(1200) + (0.0195)(1000) + (0.0016)(2000) = 28.4$ ft, $h_1 = 28.4/4200 = 0.00676$ ft/ft. With $Q = 1$ cfs and $h_1 = 0.00676$ ft/ft, $D = 7.5$ in.

**10.10** For three pipes in series, the total pressure drop is $p_A - p_B = 150\,000$ Pa, and the elevation drop is $z_A - z_B = 5$ m. The pipe data are

| pipe | $L$, m | $d$, cm | $\epsilon$, mm | $\epsilon/d$ |
|------|--------|---------|----------------|--------------|
| 1 | 100 | 8 | 0.24 | 0.003 |
| 2 | 150 | 6 | 0.12 | 0.002 |
| 3 | 80 | 4 | 0.20 | 0.005 |

The fluid is water, $\rho = 1000$ kg/m$^3$, and $\nu = 1.02 \times 10^{-6}$ m$^2$/s. Calculate the flow rate in cubic meters per hour through the system. Neglect minor losses.

▌ The total head loss in the system is $h_L = (p_A - p_B)/\rho g + (z_A - z_B) = 150\,000/[(1000)(9.807)] + 5 = 20.30$ m $= h_f = (f)(L/d)(v^2/2g)$. From the continuity relation, $v_2 = (d_1/d_2)^2(v_1) = (\frac{8}{6})^2(v_1) = 1.778v_1$, $v_3 = (d_1/d_3)^2(v_1) = (\frac{8}{4})^2(v_1) = 4.000v_1$, and

$$(N_R)_2 = \left(\frac{v_2 d_2}{v_1 d_1}\right)(N_R)_1 = \left[\frac{(1.778v_1)(6)}{(v_1)(8)}\right](N_R)_1 = (1.333)(N_R)_1$$

$$(N_R)_3 = \left(\frac{v_3 d_3}{v_1 d_1}\right)(N_R)_1 = \left[\frac{(4.000v_1)(4)}{(v_1)(8)}\right](N_R)_1 = (2.000)(N_R)_1$$

$$20.30 = f_1[100/(\tfrac{8}{100})]\{v_1^2/[(2)(9.807)]\} + f_2[150/(\tfrac{6}{100})]\{(1.778v_1)^2/[(2)(9.807)]\} + f_3[80/(\tfrac{4}{100})]\{(4.000v_1)^2/[(2)(9.807)]\}$$

$$20.30 = (63.73f_1 + 402.9f_2 + 1631f_3)(v_1)^2$$

From Fig. A-5 from the fully rough regime, estimate $f_1 = 0.0262$, $f_2 = 0.0234$, and $f_3 = 0.0304$.

$$20.35 = [(63.73)(0.0262) + (402.9)(0.0234) + (1631)(0.0304)](v_1)^2 \qquad v_1 = 0.5791 \text{ m/s} \qquad N_R = dv/\nu$$

$$(N_R)_1 = (\tfrac{8}{100})(0.5791)/(1.02 \times 10^{-6}) = 4.54 \times 10^4$$

$$(N_R)_2 = (1.333)(4.54 \times 10^{-4}) = 6.05 \times 10^4 \qquad (N_R)_3 = (2.000)(4.54 \times 10^{-4}) = 9.08 \times 10^4$$

From Fig. A-5, $f_1 = 0.0288$, $f_2 = 0.0260$, and $f_3 = 0.0314$. $20.35 = [(63.73)(0.0288) + (402.9)(0.0260) + (1631)(0.0314)](v_1)^2$, $v_1 = 0.5660$ m/s; $Q = Av = [(\pi)(\tfrac{8}{100})^2/4](0.5660) = 0.002845$ m$^3$/s, or 10.2 m$^3$/h. An additional iteration (not shown) gives essentially the same result.

**10.11** For a head loss of 5.0 ft/1000 ft, and using $C = 120$ for all pipes, how many 8-in pipes are equivalent to a 16-in pipe? To a 24-in pipe?

▌ From Fig. A-13, for $h_1 = 5.0/1000$, or 0.005, and $d = 8$ in, $Q = 1.0$ cfs. For $d = 16$ in, $Q = 6.6$ cfs. For $d = 24$ in, $Q = 17$ cfs. Thus it would take 6.6/1.0, or 6.6 eight-in pipes to be hydraulically equivalent to a 16-in pipe of the same relative roughness. Likewise, 17/1.0, or 17 eight-in pipes are equivalent to a 24-in pipe for a head loss of 5.0 ft/1000 ft, or for any other head loss condition.

**10.12** A series piping system consists of 6000 ft of 20-in new cast iron pipe, 4000 ft of 16-in, and 2000 ft of 12-in pipe. Convert the system to an equivalent length of 16-in pipe.

▌ Assume a value of $Q$ of 3.8 cfs. For new cast iron, $C = 130$. In order to use Fig. A-13, change $Q_{130}$ to $Q_{120}$. $Q_{120} = (\frac{120}{130})(3.8) = 3.5$ cfs. From Fig. A-13, $(h_1)_1 = 0.00064$ ft/ft, $(h_1)_2 = 0.00187$ ft/ft, and $(h_1)_3 = 0.0070$ ft/ft. $h_L = (0.00064)(6000) + (0.00187)(4000) + (0.0070)(2000) = 25.32$ ft. The equivalent 16-in pipe must carry 3.8 cfs with a head loss of 25.32 ft ($C = 130$). Hence, $25.32/L_e = 0.00187$, $L_e = 13\,540$ ft.

**10.13** Convert the system of Prob. 10.12 to an equivalent size pipe 12 000 ft long.

▌ The 12 000 ft of pipe, $C = 130$, must carry 3.8 cfs with a head loss of 25.32 ft. Hence, $h_1 = 25.32/12\,000 = 0.00211$. From Fig. A-13, using $Q = 3.5$ cfs, $d = 15.5$ in.

**10.14** Suppose in Fig. 10-4 pipes 1, 2, and 3 are 300 m of 30-cm-diameter, 150 m of 20-cm-diameter, and 250 m of 25-cm-diameter, respectively, of new cast iron and are conveying water at 15 °C. If $h = 10$ m, find the rate of flow from $A$ to $B$ by the equivalent velocity method. Neglect minor losses.

▌ $(\epsilon/d)_1 = 0.00026/(\tfrac{30}{100}) = 0.000867 \qquad (\epsilon/d)_2 = 0.00026/(\tfrac{20}{100}) = 0.00130 \qquad (\epsilon/d)_3 = 0.00026/(\tfrac{25}{100}) = 0.00104$

From Fig. A-5, assume $f_1 = 0.019$, $f_2 = 0.021$, and $f_3 = 0.020$. $h_f = (f)(L/d)(v^2/2g) = 10 =$
$0.019[300/(\frac{30}{100})]\{v_1^2/[(2)(9.807)]\} + 0.021[150/(\frac{20}{100})]\{v_2^2/[(2)(9.807)]\} + 0.020[250/(\frac{25}{100})]\{v_3^2/[(2)(9.807)]\}$,
$10 = 0.9687v_1^2 + 0.8030v_2^2 + 1.020v_3^2$. From the continuity relation, $v_2 = (d_1/d_2)^2(v_1) = (\frac{30}{20})^2(v_1) = 2.250v_1$,
$v_3 = (d_1/d_3)^2(v_1) = (\frac{30}{25})^2(v_1) = 1.440v_1$, and

$$(N_R)_2 = \left(\frac{v_2 d_2}{v_1 d_1}\right)(N_R)_1 = \left[\frac{(2.250v_1)(20)}{(v_1)(30)}\right](N_R)_1 = (1.500)(N_R)_1$$

$$(N_R)_3 = \left(\frac{v_3 d_3}{v_1 d_1}\right)(N_R)_1 = \left[\frac{(1.440v_1)(25)}{(v_1)(30)}\right](N_R)_1 = (1.200)(N_R)_1$$

$$10 = 0.9687v_1^2 + (0.8030)(2.250v_1)^2 + (1.020)(1.440v_1)^2 \qquad v_1 = 1.183 \text{ m/s} \qquad N_R = dv/v$$

$$(N_R)_1 = (\tfrac{30}{100})(1.183)/(1.16 \times 10^{-6}) = 3.06 \times 10^5 \qquad (N_R)_2 = (1.500)(3.06 \times 10^5) = 4.59 \times 10^5$$

$$(N_R)_3 = (1.200)(3.06 \times 10^5) = 3.67 \times 10^5$$

From Fig. A-5, $f_1 = 0.021$, $f_2 = 0.021$, and $f_3 = 0.020$.

$$h_f = 10 = 0.021[300/(\tfrac{30}{100})]\{v_1^2/[(2)(9.807)]\} + 0.021[150/(\tfrac{20}{100})]\{v_2^2/[(2)(9.807)]\} + 0.020[250/(\tfrac{25}{100})]\{v_3^2/[(2)(9.807)]\}$$

$$10 = 1.071v_1^2 + 0.8030v_2^2 + 1.020v_3^2 = 1.071v_1^2 + (0.8030)(2.250v_1)^2 + (1.020)(1.440v_1)^2$$

$$v_1 = 1.174 \text{ m/s} \qquad Q = Av = [(\pi)(\tfrac{30}{100})^2/4](1.174) = 0.0830 \text{ m}^3/\text{s}$$

**Fig. 10-4**

**10.15** Solve Prob. 10.14 by the equivalent length method.

▌ Using data from Prob. 10.14 and choosing a 30-cm pipe as the standard,

$$(L_e)_2 = (150)(0.021/0.020)(\tfrac{30}{20})^5 = 1196 \text{ m of 30-cm pipe}$$

$$(L_e)_3 = (250)(0.020/0.020)(\tfrac{30}{25})^5 = 622 \text{ m of 30-cm pipe}$$

$$(L_e)_{\text{total}} = 300 + 1196 + 622 = 2118 \text{ m of 30-cm pipe} \qquad h = 10 = 0.020[2118/(\tfrac{30}{100})]\{v^2/[(2)(9.807)]\}$$

$$v = 1.179 \text{ m/s} \qquad Q = [(\pi)(\tfrac{30}{100})^2/4](1.179) = 0.0833 \text{ m}^3/\text{s}$$

**10.16** In Fig. 10-4 pipes 1, 2, and 3 are 500 ft of 3.068-in, 200 ft of 2.067-in, and 400 ft of 2.469-in wrought iron pipe. Given a head loss of 19.5 ft from $A$ to $B$, find the flow of water at 60 °F.

▌ $\qquad (\epsilon/d)_1 = 0.00015/(3.068/12) = 0.000587 \qquad (\epsilon/d)_2 = 0.00015/(2.067/12) = 0.000871$

$$(\epsilon/d)_3 = 0.00015/(2.469/12) = 0.000729$$

From Fig. A-5, assume $f_1 = 0.0172$, $f_2 = 0.0190$, and $f_3 = 0.0181$.

$$h_f = (f)(L/d)(v^2/2g) = 19.5 = 0.0172[500/(3.068/12)]\{v_1^2/[(2)(32.2)]\} + 0.0190[200/(2.067/12)]\{v_2^2/[(2)(32.2)]\}$$
$$+ 0.0181[400/(2.469/12)]\{v_3^2/[(2)(32.2)]\}$$

$$19.5 = 0.5223v_1^2 + 0.3426v_2^2 + 0.5464v_3^2$$

From the continuity relation, $v_2 = (d_1/d_2)^2(v_1) = (3.068/2.067)^2(v_1) = 2.203v_1$, $v_3 = (d_1/d_3)^2(v_1) = (3.068/2.469)^2(v_1) = 1.544v_1$, and

$$(N_R)_2 = \left(\frac{v_2 d_2}{v_1 d_1}\right)(N_R)_1 = \left[\frac{(2.203v_1)(2.067)}{(v_1)(3.068)}\right](N_R)_1 = (1.484)(N_R)_1$$

$$(N_R)_3 = \left(\frac{v_3 d_3}{v_1 d_1}\right)(N_R)_1 = \left[\frac{(1.544v_1)(2.469)}{(v_1)(3.068)}\right](N_R)_1 = (1.243)(N_R)_1$$

$$20 = 0.5223v_1^2 + (0.3426)(2.203v_1)^2 + (0.5464)(1.544v_1)^2 \qquad v_1 = 2.365 \text{ ft/s}$$

$$N_R = dv/v \qquad (N_R)_1 = (3.068/12)(2.365)/(1.21 \times 10^{-5}) = 5.00 \times 10^4$$

$$(N_R)_2 = (1.484)(5.00 \times 10^4) = 7.42 \times 10^4 \qquad (N_R)_3 = (1.243)(5.00 \times 10^4) = 6.22 \times 10^4$$

From Fig. A-5, $f_1 = 0.0228$, $f_2 = 0.0225$, and $f_3 = 0.0225$.

$$h_f = 19.5 = 0.0228[500/(3.068/12)]\{v_1^2/[(2)(32.2)]\} + 0.0225[200/(2.067/12)]\{v_2^2/[(2)(32.2)]\}$$
$$+ 0.0225[400/(2.469/12)]\{v_3^2/[(2)(32.2)]\}$$
$$19.5 = 0.6924v_1^2 + 0.4057v_2^2 + 0.6792v_3^2 = 0.6924v_1^2 + (0.4057)(2.203v_1)^2 + (0.6792)(1.544v_1)^2 \qquad v_1 = 2.134 \text{ ft/s}$$
$$Q = Av = [(\pi)(3.068/12)^2/4](2.134) = 0.110 \text{ ft}^3/\text{s}$$

An additional iteration (not shown) produces an insignificant difference.

**10.17** Suppose that 100 ft of 1-in ($\epsilon = 0.000005$ ft) pipe and 500 ft of 6-in ($\epsilon = 0.0008$ ft) pipe are connected in series. For a total head loss of 24 ft, find the flow of water at 60 °F.

**I**  $\qquad (\epsilon/d)_1 = 0.000005/(\frac{1}{12}) = 0.0000600 \qquad\qquad (\epsilon/d)_2 = 0.0008/(\frac{6}{12}) = 0.00160$

From Fig. A-5, assume $f_1 = 0.0110$ and $f_2 = 0.0215$.

$$h_f = (f)(L/d)(v^2/2g) = 25 = 0.0110[100/(\frac{1}{12})]\{v_1^2/[(2)(32.2)]\} + 0.0215[500/(\frac{6}{12})]\{v_2^2/[(2)(32.2)]\}$$
$$25 = 0.2050v_1^2 + 0.3339v_2^2$$

From the continuity relation, $v_2 = (d_1/d_2)^2(v_1) = (\frac{1}{6})^2(v_1) = 0.02778v_1$ and

$$(N_R)_2 = \left(\frac{v_2 d_2}{v_1 d_1}\right)(N_R)_1 = \left[\frac{(0.02778v_1)(6)}{(v_1)(1)}\right](N_R)_1 = (0.1667)(N_R)_1$$
$$24 = 0.2050v_1^2 + (0.3339)(0.02778v_1)^2 \qquad v_1 = 11.04 \text{ ft/s} \qquad N_R = dv/\nu$$
$$(N_R)_1 = (\tfrac{1}{12})(11.04)/(1.21 \times 10^{-5}) = 7.60 \times 10^4 \qquad (N_R)_2 = (0.1667)(7.60 \times 10^4) = 1.27 \times 10^4$$

From Fig. A-5, $f_1 = 0.0195$ and $f_2 = 0.032$.

$$h_f = 24 = 0.0195[100/(\tfrac{1}{12})]\{v_1^2/[(2)(32.2)]\} + 0.032[500/(\tfrac{6}{12})]\{v_2^2/[(2)(32.2)]\}$$
$$24 = 0.3634v_1^2 + 0.4969v_2^2 = 0.3634v_1^2 + (0.4969)(0.02778v_1)^2 \qquad v_1 = 8.122 \text{ ft/s}$$
$$(N_R)_1 = (\tfrac{1}{12})(8.122)/(1.21 \times 10^{-5}) = 5.59 \times 10^4 \qquad (N_R)_2 = (0.1667)(5.59 \times 10^4) = 9.32 \times 10^3$$

From Fig. A-5, $f_1 = 0.0203$ and $f_2 = 0.033$.

$$h_f = 24 = 0.0203[100/(\tfrac{1}{12})]\{v_1^2/[(2)(32.2)]\} + 0.033[500/(\tfrac{6}{12})]\{v_2^2/[(2)(32.2)]\}$$
$$24 = 0.3783v_1^2 + 0.5124v_2^2 = 0.3783v_1^2 + (0.5124)(0.02778v_1)^2 \qquad v_1 = 7.961 \text{ ft/s}$$
$$Q = Av = [(\pi)(\tfrac{1}{12})^2/4](7.961) = 0.0434 \text{ ft}^3/\text{s}$$

An additional iteration (not shown) produces an insignificant difference.

**10.18** Repeat Prob. 10.16 for the case where the fluid has s.g. = 0.9 and $\mu = 0.0008$ lb · s/ft².

**I**  First apply $h_f = (f)(L/d)(v^2/2g)$, using values from Prob. 10.16 and assuming $f = 0.03$ for all pipes.

$$19.5 = 0.03[500/(3.068/12)]\{v_1^2/[(2)(32.2)]\} + 0.03[200/(2.067/12)]\{v_2^2/[(2)(32.2)]\}$$
$$+ 0.03[400/(2.469/12)]\{v_3^2/[(2)(32.2)]\}$$
$$= 0.9110v_1^2 + 0.5409v_2^2 + 0.9056v_3^2$$
$$= 0.9110v_1^2 + (0.5409)(2.203v_1)^2 + (0.9056)(1.544v_1)^2$$
$$v_1 = 1.850 \text{ ft/s} \qquad N_R = \rho dv/\mu \qquad (N_R)_1 = [(0.9)(1.94)](3.068/12)(1.850)/0.0008 = 1032$$
$$(N_R)_2 = (1.484)(1046) = 1532 \qquad (N_R)_3 = (1.243)(1046) = 1283$$

Therefore, the flow is laminar and $h_L = (32)(\mu/\gamma)(L/d^2)(v)$.

$$19.5 = 32\{0.0008/[(0.9)(62.4)]\}[500/(3.068/12)^2](v_1)$$
$$+ 32\{0.0008/[(0.9)(62.4)]\}[200/(2.067/12)^2](2.203v_1)$$
$$+ 32\{0.0008/[(0.9)(62.4)]\}[400/(2.469/12)^2](1.544v_1)$$
$$v_1 = 1.074 \text{ ft/s} \qquad Q = Av = [(\pi)(3.068/12)^2/4](1.074) = 0.0551 \text{ ft}^3/\text{s}$$

**10.19** One end of a 150-m-long, 300-mm-diameter pipe is submerged in a reservoir; the other end abuts on a 90-m-long, 200-mm-diameter pipe at a point 30 m below the reservoir surface. Water discharges freely ($K = 1.0$) from the free end of the shorter pipe, which is 15 m below the junction. [This implies a 2° bend at the junction.]

Determine the pressure heads just above and just below the junction, if $f = 0.04$, $K = 0.8$ for the entrance, and $K = 0.24$ for the contraction-bend at the junction.

▮
$$p_1/\gamma + v_1^2/2g + z_1 = p_2/\gamma + v_2^2/2g + z_2 + h_L \qquad (1)$$

$$h_f = (f)(L/d)(v^2/2g) = 0.04[150/0.300]\{v_1^2/[(2)(9.807)]\} + 0.04[90/0.200]\{v_2^2/[(2)(9.807)]\} = 1.020v_1^2 + 0.9177v_2^2$$

$$h_m = (K)(v^2/2g) = 0.8\{v_1^2/[(2)(9.807)]\} + 0.24\{v_2^2/[(2)(9.807)]\} = 0.04079v_1^2 + 0.01224v_2^2$$

$$h_L = h_f + h_m = (1.020v_1^2 + 0.9177v_2^2) + (0.04079v_1^2 + 0.01224v_2^2) = 1.061v_1^2 + 0.9299v_2^2$$

$$A_1v_1 = A_2v_2 \qquad [(\pi)(0.300)^2/4](v_1) = [(\pi)(0.200)^2/4](v_2) \qquad v_1 = 0.4444v_2$$

$$h_L = (1.061)(0.4444v_2)^2 + 0.9299v_2^2 = 1.139v_2^2 \qquad 0 + 0 + 45 = 0 + v_2^2/[(2)(9.807)] + 0 + 1.139v_2^2$$

$$v_2 = 6.149 \text{ m/s} \qquad v_1 = (0.4444)(6.149) = 2.733 \text{ m/s} \qquad Q = A_2v_2 = [(\pi)(0.200)^2/4](6.149) = 0.193 \text{ m}^3/\text{s}$$

Now apply Eq. (1) between the water level at intake and a point just above the junction. $0 + 0 + 30 = p_2/\gamma + 2.733^2/[(2)(9.807)] + 0 + (1.020v_1^2 + 0.04079v_1^2)$, $30 = p_2/\gamma + 0.3808 + (1.020 + 0.04079)(2.733^2)$, $p_2/\gamma = 21.70$ m. Finally apply Eq. (1) between the water level at intake and a point just below the junction. $0 + 0 + 30 = p_2/\gamma + 6.149^2/[(2)(9.807)] + 0 + (1.020v_1^2 + 0.04079v_1^2 + 0.01224v_2^2)$, $30 = p_2/\gamma + 1.928 + [(1.020 + 0.04079)(2.733^2) + (0.01224)(6.149^2)]$, $p_2/\gamma = 19.69$ m.

**10.20** Repeat Prob. 10.19 neglecting minor losses.

▮
$$p_1/\gamma + v_1^2/2g + z_1 = p_2/\gamma + v_2^2/2g + z_2 + h_L \qquad (1)$$

$$h_L = h_f + h_m \qquad h_m = 0$$

Using data from Prob. 10.19, $h_f = 1.020v_1^2 + 0.9177v_2^2$, $h_L = (1.020)(0.4444v_2)^2 + 0.9177v_2^2 = 1.119v_2^2$, $0 + 0 + 45 = 0 + v_2^2/[(2)(9.807)] + 0 + 1.119v_2^2$, $v_2 = 6.202$ m/s, $v_1 = (0.4444)(6.202) = 2.756$ m/s; $Q = A_2v_2 = [(\pi)(0.200)^2/4](6.202) = 0.195 \text{ m}^3/\text{s}$. For the pressure head just above the junction, $0 + 0 + 30 = p_2/\gamma + 2.756^2/[(2)(9.807)] + 0 + 1.020v_1^2$, $30 = p_2/\gamma + 0.3873 + (1.020)(2.756^2)$, $p_2/\gamma = 21.87$ m. For the pressure head just below the junction, $0 + 0 + 30 = p_2/\gamma + 6.202^2/[(2)(9.807)] + 0 + 1.020v_1^2$, $30 = p_2/\gamma + 1.961 + (1.020)(2.756^2)$, $p_2/\gamma = 20.29$ m.

**10.21** Three new cast iron pipes, having diameters of 30 in, 24 in, and 18 in, respectively, each 500 ft long, are connected in series. The 30-in pipe leads from a reservoir (flush entrance), and the 18-in pipe discharges into the air at a point 11.5 ft below the water surface in the reservoir. Assuming all changes in section to be abrupt, find the rate of discharge of water at 60 °F.

▮
$$p_1/\gamma + v_1^2/2g + z_1 = p_2/\gamma + v_2^2/2g + z_2 + h_L \qquad h_L = h_f + h_m \qquad h_f = (f)(L/d)(v^2/2g)$$

Assume $f = 0.016$ for each pipe. $v_2 =$ velocity for 18-in pipe, velocity for 24-in pipe $= (\frac{18}{24})^2(v_2) = 0.5625v_2$, velocity for 30-in pipe $= (\frac{18}{30})^2(v_2) = 0.3600v_2$.

$$h_f = 0.016[500/(\tfrac{30}{12})]\{(0.3600v_2)^2/[(2)(32.2)]\} + 0.016[500/(\tfrac{24}{12})]\{(0.5625v_2)^2/[(2)(32.2)]\}$$
$$+ 0.016[500/(\tfrac{18}{12})]\{v_2^2/[(2)(32.2)]\} = 0.1089v_2^2$$

$$h_m = (K)(v^2/2g)$$

For entrance, take $K = 0.5$ (Fig. A-7). For sudden contractions, with $d/D = \frac{24}{30}$, or 0.80, $K = 0.15$ and with $d/D = \frac{18}{24}$, or 0.75, $K = 0.18$ (Fig. A-9). For exit, $K = 1.0$ (Fig. A-7).

$$h_m = (0.5)(0.3600v_2)^2 + (0.15)(0.5625v_2)^2 + 0.18v_2^2 = 0.2923v_2^2 \qquad h_L = 0.1089v_2^2 + 0.2923v_2^2 = 0.4012v_2^2$$

$$0 + 0 + 11.5 = 0 + v_2^2/[(2)(32.2)] + 0 + 0.4012v_2^2 \qquad v_2 = 5.253 \text{ ft/s} \qquad N_R = dv/\nu$$

$$(N_R)_{30\text{-in}} = \tfrac{30}{12}[(0.3600)(5.253)]/(1.21 \times 10^{-5}) = 3.91 \times 10^5 \quad (N_R)_{24\text{-in}} = \tfrac{24}{12}[(0.5625)(5.253)]/(1.21 \times 10^{-5}) = 4.88 \times 10^5$$

$$(N_R)_{18\text{-in}} = (\tfrac{18}{12})(5.253)/(1.21 \times 10^{-5}) = 6.51 \times 10^5 \qquad (\epsilon/d)_{30\text{-in}} = 0.00085/(\tfrac{30}{12}) = 0.000340$$

$$(\epsilon/d)_{24\text{-in}} = 0.00085/(\tfrac{24}{12}) = 0.000425 \qquad (\epsilon/d)_{18\text{-in}} = 0.00085/(\tfrac{18}{12}) = 0.000567$$

From Fig. A-5, $f_{30\text{-in}} = 0.0168$, $f_{24\text{-in}} = 0.0172$, and $f_{18\text{-in}} = 0.0176$.

$$h_f = 0.0168[500/(\tfrac{30}{12})]\{(0.3600v_2)^2/[(2)(32.2)]\} + 0.0172[500/(\tfrac{24}{12})]\{(0.5625v_2)^2/[(2)(32.2)]\}$$
$$+ 0.0176[500/(\tfrac{18}{12})]\{v_2^2/[(2)(32.2)]\} = 0.1190v_2^2$$

$$h_L = 0.1190v_2^2 + 0.2923v_2^2 = 0.4113v_2^2 \qquad 0 + 0 + 11.5 = 0 + v_2^2/[(2)(32.2)] + 0 + 0.4113v_2^2 \qquad v_2 = 5.191 \text{ ft/s}$$

An additional iteration (not shown) gives no significant change in $v_2$. $Q = Av = [(\pi)(\tfrac{18}{12})^2/4](5.191) = 9.17 \text{ ft}^3/\text{s}$.

**10.22** In Fig. 10-5, $K_e = 0.5$, $L_1 = 300$ m, $D_1 = 600$ mm, $\epsilon_1 = 2$ mm, $L_2 = 240$ m, $D_2 = 1$ m, $\epsilon_2 = 0.3$ mm, $\nu = 3 \times 10^{-6}$ m²/s, and $H = 6$ m. Determine the discharge through the system.

∎ $$p_1/\gamma + v_1^2/2g + z_1 = p_2/\gamma + v_2^2/2g + z_2 + h_L \qquad h_L = h_f + h_m \qquad h_f = (f)(L/d)(v^2/2g)$$

Assume $f_1 = 0.026$ and $f_2 = 0.015$. $v_2 = [(\frac{600}{1000})/1]^2(v_1) = 0.3600v_1$, $h_f = 0.026[300/(\frac{600}{1000})]\{v_1^2/[(2)(9.807)]\} + (0.015)(240/1)\{(0.3600v_1)^2/[(2)(9.807)]\} = 0.6866v_1^2$.

**For entrance:** $h_m = (K)(v^2/2g) = 0.5\{v_1^2/[(2)(9.807)]\} = 0.02549v_1^2$.

**For exit:** $h_m = 1.0\{v_2^2/[(2)(9.807)]\} = 0.05098v_2^2 = (0.05098)(0.3600v_1)^2 = 0.006607v_1^2$.

**For sudden contraction:**

$$h_m = (v_1 - v_2)^2/2g = (v_1 - 0.3600v_1)^2/[(2)(9.807)]$$

$$= 0.02088v_1^2 (h_m)_{\text{total}} = 0.02549v_1^2 + 0.006607v_1^2 + 0.02088v_1^2 = 0.05298v_1^2$$

$$h_L = 0.6866v_1^2 + 0.05298v_1^2 = 0.7396v_1^2 \qquad 0 + 0 + 6 = 0 + 0 + 0 + 0.7396v_1^2 \qquad v_1 = 2.848 \text{ m/s}$$

$$v_2 = (0.3600)(2.848) = 1.025 \text{ m/s}$$

$$N_R = dv/\nu \qquad (N_R)_1 = (\tfrac{600}{1000})(2.848)/(3 \times 10^{-6}) = 5.70 \times 10^5 \qquad (N_R)_2 = (1)(1.025)/(3 \times 10^{-6}) = 3.42 \times 10^5$$

$$(\epsilon/d)_1 = \tfrac{2}{600} = 0.00333 \qquad (\epsilon/d)_2 = (0.3/1000)/1 = 0.000300$$

From Fig. A-5, $f_1 = 0.0265$ and $f_2 = 0.0168$.

$$h_f = 0.0265[300/(\tfrac{600}{1000})]\{v_1^2/[(2)(9.807)]\} + (0.0168)(240/1)\{(0.3600v_1)^2/[(2)(9.807)]\} = 0.7022v_1^2$$

$$h_L = 0.7022v_1^2 + 0.05298v_1^2 = 0.7552v_1^2 \qquad 0 + 0 + 6 = 0 + 0 + 0 + 0.7552v_1^2 \qquad v_1 = 2.819 \text{ m/s}$$

An additional iteration (not shown) gives no significant change in $v_1$. $Q = Av = [(\pi)(\tfrac{600}{1000})^2/4](2.819) = 0.797$ m³/s.

**Fig. 10-5**

**10.23** Solve Prob. 10.22 by means of equivalent pipes.

∎ Expressing the minor losses in terms of equivalent lengths gives for pipe 1: $K_1 = 0.5 + [1 - (\frac{600}{1000})^2]^2 = 0.9096$, $(L_e)_1 = K_1 D_1/f_1 = (0.9096)(\frac{600}{1000})/0.026 = 20.99$ m; and for pipe 2: $K_2 = 1.0$, $(L_e)_2 = K_2 D_2/f_2 = (1.0)(1)/0.015 = 66.67$ m. The values of $f_1$ and $f_2$ are selected for the fully turbulent range as an approximation. The problem is now reduced to $300 + 20.99$, or 320.99 m of 600-mm pipe and $240 + 66.67$, or 306.67 m of 1-m pipe. Expressing the 1-m pipe in terms of an equivalent length of 600-m pipe, $L_e = (f_2/f_1)(L_2)(D_1/D_2)^5 = (0.015/0.026)(306.67)[(\frac{600}{1000})/1]^5 = 13.76$ m. By adding to the 600-mm pipe, the problem is reduced to finding the discharge through $320.99 + 13.76$, or 334.75 m of 600-mm pipe, $\epsilon_1 = 2$ mm, $H = 6$ m: $h_f = (f)(L/D)(v^2/2g)$, $6 = (f)[334.75/(\frac{600}{1000})]\{v^2/[(2)(9.807)]\}$, $v = 0.4593/\sqrt{f}$. Try $f = 0.026$: $v = 0.4593/\sqrt{0.026} = 2.848$ m/s, $N_R = Dv/\nu = (\frac{600}{1000})(2.848)/(3 \times 10^{-6}) = 5.70 \times 10^5$. From Fig. A-5, with $N_R = 5.70 \times 10^5$ and $\epsilon/D = 0.00333$, $f = 0.0265$. Try $f = 0.0265$: $v = 0.4593/\sqrt{0.0265} = 2.821$ m/s, $N_R = (\frac{600}{1000})(2.821)/(3 \times 10^{-6}) = 5.64 \times 10^5$, $f = 0.0265$ (O.K.); $Q = Av = [(\pi)(\frac{600}{1000})^2/4](2.821) = 0.798$ m³/s.

**10.24** Two reservoirs are connected by three clean cast iron pipes in series: $L_1 = 300$ m, $D_1 = 200$ mm; $L_2 = 400$ m, $D_2 = 300$ mm; $L_3 = 1200$ m, $D_3 = 450$ mm. If the flow is 360 m³/h of water at 20 °C, determine the difference in elevation of the reservoirs.

∎ $$v = Q/A \qquad v_1 = (360/3600)/[(\pi)(0.200)^2/4] = 3.183 \text{ m/s}$$

$$v_2 = (360/3600)/[(\pi)(0.300)^2/4] = 1.415 \text{ m/s} \qquad v_3 = (360/3600)/[(\pi)(0.450)^2/4] = 0.6288 \text{ m/s}$$

$$N_R = Dv/\nu \qquad (N_R)_1 = (0.200)(3.183)/(1.02 \times 10^{-6}) = 6.24 \times 10^5$$

$$(N_R)_2 = (0.300)(1.415)/(1.02 \times 10^{-6}) = 4.16 \times 10^5 \qquad (N_R)_3 = (0.450)(0.6288)/(1.02 \times 10^{-6}) = 2.77 \times 10^5$$

$$(\epsilon/D)_1 = 0.00026/0.200 = 0.00130 \qquad (\epsilon/D)_2 = 0.0026/0.300 = 0.000867$$

$$(\epsilon/D)_3 = 0.00026/0.450 = 0.000578$$

From Fig. A-5, $f_1 = 0.0215$, $f_2 = 0.020$, and $f_3 = 0.0185$. $H = h_f = (f)(L/D)(v^2/2g) = 0.0215[300/0.200]\{3.183^2/[(2)(9.807)]\} + 0.020[400/0.300]\{1.415^2/[(2)(9.807)]\} + 0.0185[1200/0.450]\{0.6288^2/[(2)(9.807)]\} = 20.37$ m.

**10.25** Solve Prob. 10.24 by the method of equivalent lengths.

▮ Express pipes 2 and 3 in terms of pipe 1:
$$L_e = (f_2/f_1)(L_2)(D_1/D_2)^5$$
$$(L_e)_2 = (0.020/0.0215)(400)(\tfrac{200}{300})^5 = 49.00 \text{ m} \qquad (L_e)_3 = (0.0185/0.0215)(1200)(\tfrac{200}{450})^5 = 17.91 \text{ m}$$
$$(L_e)_{\text{total}} = 300 + 49.00 + 17.91 = 366.9 \text{ m}$$
$$H = h_f = (f)(L/D)(v^2/2g) = 0.0215[366.9/0.200]\{3.183^2/[(2)(9.807)]\} = 20.37 \text{ m}$$

**10.26** Air at atmospheric pressure and 60 °F is carried through two horizontal pipes ($\epsilon = 0.06$ in) in series. The upstream pipe is 360 ft of 24 in diameter, and the downstream pipe is 120 ft of 36 in diameter. Estimate the equivalent length of 18-in ($\epsilon = 0.003$ in) pipe. Neglect minor losses.

▮ $$\epsilon/D_1 = 0.06/24 = 0.0025 \qquad \epsilon/D_2 = 0.06/36 = 0.00167$$

From Fig. A-5, assuming high Reynolds numbers, $f_1 = 0.025$ and $f_2 = 0.022$.
$$L_e = (0.025/f_3)(360)(\tfrac{18}{24})^5 + (0.022/f_3)(120)(\tfrac{18}{36})^5 = 2.218/f_3$$
$$\epsilon_3/D_3 = 0.003/18 = 0.000167 \qquad f_3 = 0.013 \qquad L_e = 2.218/0.013 = 171 \text{ ft}$$

**10.27** What pressure drop is required for flow of 6000 ft³/min in Prob. 10.26? Include losses due to sudden expansion.

▮ $$h_L = h_f + h_m \qquad h_f = (f)(L/D)(v^2/2g)$$

From Prob. 10.26, $\epsilon/D_1 = 0.0025$, $\epsilon/D_2 = 0.00167$.
$$v_1 = Q/A_1 = (\tfrac{6000}{60})/[(\pi)(\tfrac{24}{12})^2/4] = 31.83 \text{ ft/s} \qquad v_2 = Q/A_2 = (\tfrac{6000}{60})/[(\pi)(\tfrac{36}{12})^2/4] = 14.15 \text{ ft/s}$$
$$N_R = Dv/\nu \qquad (N_R)_1 = (\tfrac{24}{12})(31.83)/(1.58 \times 10^{-4}) = 4.03 \times 10^5 \qquad (N_R)_2 = (\tfrac{36}{12})(14.15)/(1.58 \times 10^{-4}) = 2.69 \times 10^5$$

From Fig. A-5, $f_1 = 0.025$ and $f_2 = 0.022$.
$$h_f = 0.025[360/(\tfrac{24}{12})]\{31.83^2/[(2)(32.2)]\} + 0.022[120/(\tfrac{36}{12})]\{14.15^2/[(2)(32.2)]\} = 73.53 \text{ ft}$$
$$h_m = (v_1 - v_2)^2/2g = (31.83 - 14.15)^2/[(2)(32.2)] = 4.85 \text{ m} \qquad h_L = 73.53 + 4.85 = 78.38 \text{ ft}$$
$$\Delta p = \gamma h \qquad \gamma = p/RT$$
$$\gamma = (14.7)(144)/[(53.3)(460 + 60)] = 0.07637 \text{ lb/ft}^3 \qquad \Delta p = (0.07637)(78.38) = 5.99 \text{ lb/ft}^2$$

**10.28** Two pipes, $D_1 = 3$ in, $L_1 = 300$ ft, and $D_2 = 2$ in, $L_2 = 180$ ft, are joined in series with a total pressure drop of 3 lb/in². What is the flow rate of syrup ($\rho = 1.78$ slugs/ft³, $\mu = 0.00606$ lb · s/ft²) at 20 °C?

▮ $$h_f = \Delta p/\gamma = \Delta p/\rho g = (3)(144)/[(1.78)(32.2)] = 7.537 \text{ ft}$$

Assume laminar flow
$$h_f = \frac{128\mu LQ}{\pi\rho g D^4} \qquad 7.537 = \frac{(128)(0.00606)(300)(Q)}{(\pi)(1.78)(32.2)(\tfrac{3}{12})^4} + \frac{(128)(0.00606)(180)(Q)}{(\pi)(1.78)(32.2)(\tfrac{2}{12})^4} \qquad Q = 0.00564 \text{ ft}^3/\text{s}$$

Computation of $N_R$ (not shown) indicates the flow is indeed laminar.

**10.29** Convert the piping system shown in Fig. 10-6 to an equivalent length of 6-in pipe.

▮ $$p_A/\gamma + V_A^2/2g + z_A = p_M/\gamma + V_M^2/2g + z_M + h_L \qquad h_L = h_f + h_m$$
$$h_f = (f)(L/d)(V^2/2g) = 0.025[150/(\tfrac{12}{12})]\{V_{12}^2/[(2)(32.2)]\} + 0.020[100/(\tfrac{6}{12})]\{V_6^2/[(2)(32.2)]\}$$
$$= 0.05823V_{12}^2 + 0.06211V_6^2$$
$$h_m = (K)(V^2/2g) = [8.0 + (2)(0.5) + 0.7 + 1.0]\{V_{12}^2/[(2)(32.2)]\}$$
$$+ [0.7 + 6.0 + (2)(0.5) + 3.0 + 1.0]\{V_6^2/[(2)(32.2)]\}$$
$$= 0.1661V_{12}^2 + 0.1817V_6^2$$
$$h_L = 0.05823V_{12}^2 + 0.06211V_6^2 + 0.1661V_{12}^2 + 0.1817V_6^2 = 0.2243V_{12}^2 + 0.2438V_6^2$$
$$0 + 0 + h = 0 + 0 + 0 + 0.2243V_{12}^2 + 0.2438V_6^2 \qquad V_{12} = (\tfrac{6}{12})^2(V_6) = 0.2500V_6$$
$$h = (0.2243)(0.2500V_6)^2 + 0.2438V_6^2 = 0.2578V_6^2$$

For a 6-in equivalent pipe, $h = 0.020[L_e/(\tfrac{6}{12})]\{V_6^2/[(2)(32.2)]\} = 0.0006211L_e V_6^2$, $0.2578V_6^2 = 0.0006211L_e V_6^2$, $L_e = 415$ ft.

| | |
|---|---|
| Strainer $B$ | = 8.0 |
| 12" Bends $C$, $F$ (each) | = 0.5 |
| 12" Tee $D$ | = 0.7 |
| 12" Valve $E$ | = 1.0 |
| 12" × 6" Cross $G$ ($\times V_6^2/2g$) | = 0.7 |
| 6" Meter $H$ | = 6.0 |
| 6" Bends $J$, $K$ (each) | = 0.5 |
| 6" Valve $L$ | = 3.0 |

**Fig. 10-6**

**10.30** For the compound piping system described in Prob. 10.12, what flow will be produced for a total head loss of 70.0 ft?

**∎** From Prob. 10.12, 13 540 ft of 16-in pipe is equivalent to the compound system. For a head loss of 70.0 ft, $h_1 = 70.0/13\,540 = 0.00517$. From Fig. A-13, $Q = 6.3$ cfs. This is for $C = 120$. For new cast iron pipe ($C = 130$), $Q = (\frac{130}{120})(6.3) = 6.8$ cfs.

**10.31** For the system shown in Fig. 10-7, when the flow from reservoir $A$ to main $D$ is 3.25 mgd, the pressure at $D$ is 20.0 psi. The flow to $D$ must be increased to 4.25 mgd with the pressure at 40.0 psi. What size pipe, 5000 ft long, should be laid from $B$ to $C$ (shown dashed) parallel to the existing 12-in pipe to accomplish this?

**∎** The elevation of reservoir $A$ can be determined as follows: $Q = (3.25 \times 10^6)/[(7.48)(86\,400)] = 5.03$ cfs. From Fig. A-13, $(h_1)_{16} = 0.0038$ ft/ft, $(h_1)_{12} = 0.0150$ ft/ft. These values of $h_1$ are for $C = 120$; they must be adjusted for $C = 100$:

$$(h_1)_{16} = (\tfrac{120}{100})(0.0038) = 0.0046 \text{ ft/ft} \qquad (h_1)_{12} = (\tfrac{120}{100})(0.0150) = 0.018 \text{ ft/ft}$$

$$(h_f)_{16} = (0.0046)(8000) = 36.8 \text{ ft} \qquad (h_f)_{12} = (0.018)(5000) = 90.0 \text{ ft} \qquad (h_f)_{\text{total}} = 36.8 + 90.0 = 126.8 \text{ ft}$$

The hydraulic grade line drops 126.8 ft to an elevation of 46.2 ft above $D$ (the equivalent of 20.0 psi). Thus reservoir $A$ is $126.8 + 46.2$, or 173.0 ft above point $D$. For a pressure of 40.0 psi, the elevation of the hydraulic grade line at $D$ will be 92.4 ft above $D$, or the available head for the flow of 4.25 mgd will be $173.0 - 92.4$, or 80.6 ft. In the 16-in pipe with $Q = 4.25$ mgd, or 6.58 cfs, $(h_1)_{16} = 0.0062$ ft/ft for $C = 120$.
**For $C = 100$:** $(h_1)_{16} = (\tfrac{120}{100})(0.0062) = 0.0074$ ft/ft, $(h_f)_{16} = (0.0074)(8000) = 59.2$ ft, $(h_f)_{B \text{ to } C} = 80.6 - 59.2 = 21.4$ ft.
**For the existing 12-in pipe:** $(h_1)_{12} = 21.4/5000 = 0.0043$ ft/ft, $Q_{12} = 2.7$ cfs for $C = 120$.
For $C = 100$, $Q_{12} = (\tfrac{100}{120})(2.7) = 2.3$ cfs. The flow in the new pipe must be $6.58 - 2.3$, or 4.28 cfs with an available head (drop in the hydraulic grade line) of 21.4 ft from $B$ to $C$. $h_1 = 21.4/5000 = 0.0043$ ft/ft. From Fig. A-13, $D = 14$ in, approximately.

**Fig. 10-7**

# CHAPTER 11
# Parallel Pipeline Systems

**11.1** Figure 11-1 shows a looping pipe system. Pressure heads at points $A$ and $E$ are 70.0 m and 46.0 m, respectively. Compute the flow rate of water through each branch of the loop. Assume $C = 120$ for all pipes.

▮ $(h_f)_{A-E} = 70.0 - 46.0 = 24.0$ m. From Fig. A-14, for pipe $ABE$, with $h_1 = 24.0/3000$, or 0.0080 m/m, and $D = 300$ mm, $Q_{ABE} = 0.105$ m³/s. For pipe $ACE$, with $h_1 = 24.0/1300$, or 0.0185 m/m, and $D = 200$ mm, $Q_{ACE} = 0.056$ m³/s. For pipe $ADE$, with $h_1 = 24.0/2600$, or 0.0092 m/m, and $D = 250$ mm, $Q_{ADE} = 0.070$ m³/s.

3000 m, 300-mm diameter

1300 m, 200-mm diameter

2600 m, 250-mm diameter

**Fig. 11-1**

**11.2** A looping concrete pipe system is shown in Fig. 11-2. The total flow rate of water is 18.0 cfs. Determine the division of flow and the loss of head from point $B$ to point $E$.

▮ Assume a head loss of 20 ft from point $B$ to point $E$. With $(h_1)_{BCE} = \frac{20}{5000}$, or 0.0040 ft/ft and $D = 15$ in, $Q_{BCE} = 4.75$ cfs. With $(h_1)_{BDE} = \frac{20}{4000}$, or 0.0050 ft/ft and $D = 18$ in, $Q_{BDE} = 8.60$ cfs. Fraction of flow through pipe $BCE = 4.75/(4.75 + 8.60) = 0.356$. Fraction of flow through pipe $BDE = 8.60/(4.75 + 8.60) = 0.644$. $Q_{BCE} = (18.0)(0.356) = 6.4$ cfs, $Q_{BDE} = (18.0)(0.644) = 11.6$ cfs. With $Q_{BCE} = 6.4$ cfs and $D = 15$ in, $(h_1)_{B-E} = 0.0070$ ft/ft. $(h_f)_{B-E} = (0.0070)(5000) = 35.0$ ft.

5000 ft, 15-in diameter

3000 ft, 30-in diameter
$Q = 18$ ft³/s

4000 ft, 24-in diameter
$Q = 18$ ft³/s

4000 ft, 18-in diameter

**Fig. 11-2**

**11.3** The discharge of water in the concrete pipe looping system shown in Fig. 11-3 if 15.0 ft³/s. Compute the head loss from point $A$ to point $G$.

▮ With $Q_{A-B} = 15.0$ cfs and $D = 30$ in, $(h_1)_{A-B} = 0.00116$ ft/ft. With $Q_{FG} = 15.0$ cfs and $D = 24$ in, $(h_1)_{FG} = 0.00345$ ft/ft. $(h_f)_{AB} = (0.00116)(2500) = 2.90$ ft, $(h_f)_{FG} = (0.00345)(3000) = 10.35$ ft. Assume $(h_f)_{BF} = 30$ ft. With $D = 18$ in and $(h_1)_{BCF} = \frac{30}{1500} = 0.020$ ft/ft, $Q_{BCF} = 18.1$ cfs. With $D = 12$ in and $(h_1)_{BDF} = \frac{30}{1000} = 0.030$ ft/ft, $Q_{BDF} = 7.7$ cfs. With $D = 15$ in and $(h_1)_{BEF} = \frac{30}{2000} = 0.015$ ft/ft, $Q_{BEF} = 9.6$ cfs. Fraction of flow through pipe $BCF = 18.1/(18.1 + 7.7 + 9.6) = 0.511$, fraction of flow through pipe $BDF = 7.7/(18.1 + 7.7 + 9.6) = 0.218$, fraction of flow through pipe $BEF = 9.6/(18.1 + 7.7 + 9.6) = 0.271$; $Q_{BCF} = (15.0)(0.511) = 7.67$ cfs. With $Q_{BCF} = 7.67$ and $D = 18$ in, $(h_1)_{BCF} = 0.0040$ ft/ft. $(h_f)_{B-F} = (0.0040)(1500) = 6.00$ ft; $(h_f)_{A-G} = (h_f)_{AB} + (h_f)_{B-F} + (h_f)_{FG} = 2.90 + 6.00 + 10.35 = 19.25$ ft.

**Fig. 11-3**

**11.4** If the flow rate of water through the pipe system shown in Fig. 11-4 is 0.050 m³/s under total head loss of 9.0 m, determine the diameter of pipe $C$. Assume a $C$ coefficient of 120 for all pipes.

▎ With $Q_A = 0.050$ m³/s and $D_A = 200$ mm, $(h_1)_A = 0.015$ m/m and $(h_f)_A = (0.015)(300) = 4.50$ m. With $Q_D = 0.050$ m³/s and $D_D = 250$ mm, $(h_1)_D = 0.0049$ m/m and $(h_f)_D = (0.0049)(510) = 2.50$ m. $(h_f)_B = (h_f)_C = 9.0 - 4.50 - 2.50 = 2.00$ m. With $(h_1)_B = 2.00/266 = 0.0075$ m/m, $Q_B = 0.019$ m³/s. $Q_C = 0.050 - 0.019 = 0.031$ m³/s. With $Q_C = 0.031$ m³/s and $(h_1)_C = 2.0/190 = 0.0105$ m/m, $D_C = 180$ mm.

**Fig. 11-4**

**11.5** For the looping pipe system shown in Fig. 11-5, the head loss between points $A$ and $F$ is 160.0 ft of water. Determine the flow rate of water in the system. Assume $C = 120$ for all pipes.

▎ Assume a head loss of 12 ft from point $B$ to point $E$. With $(h_1)_{BCE} = \frac{12}{4000} = 0.0030$ ft/ft and $D = 15$ in, $Q_{BCE} = 4.0$ cfs. With $(h_1)_{BDE} = \frac{12}{3000} = 0.0040$ ft/ft and $D = 12$ in, $Q_{BDE} = 2.6$ cfs. Fraction of flow through pipe $BCE = 4.0/(4.0 + 2.6) = 0.606$; fraction of flow through pipe $BDE = 2.6/(4.0 + 2.6) = 0.394$; $Q_{AB} = Q_{EF} = 4.0 + 2.6 = 6.6$ cfs. With $Q_{AB} = 6.6$ cfs and $D = 24$ in, $(h_1)_{AB} = 0.00075$ ft/ft; $(h_f)_{AB} = (0.00075)(8000) = 6.0$ ft. With $Q_{EF} = 6.6$ cfs and $D = 18$ in, $(h_1)_{EF} = 0.0030$ ft/ft; $(h_f)_{EF} = (0.0030)(6000) = 18.0$ ft. Fraction of total head loss through pipe $AB = 6.0/(6.0 + 12.0 + 18.0) = 0.167$, fraction of total head loss from $B$ to $E = 12.0/(6.0 + 12.0 + 18.0) = 0.333$, fraction of total head loss through pipe $EF = 18.0/(6.0 + 12.0 + 18.0) = 0.500$. Actual head losses are, therefore, $(h_f)_{AB} = (160)(0.167) = 26.72$ ft, $(h_f)_{B-E} = (160)(0.333) = 53.28$ ft, $(h_f)_{EF} = (160)(0.500) = 80.00$ ft. With $(h_1)_{AB} = 26.72/8000 = 0.00334$ ft/ft and $D = 24$ in, $Q_{AB} = 14.5$ cfs. With $(h_1)_{EF} = 80.00/6000 = 0.0133$ ft/ft and $D = 18$ in, $Q_{EF} = 14.5$ cfs. $Q_{BCE} = (14.5)(0.606) = 8.8$ cfs, $Q_{BDE} = (14.5)(0.394) = 5.7$ cfs.

**Fig. 11-5**

**11.6** A flow of 570 L/s is proceeding through the pipe network shown in Fig. 11-6. For a pressure of 690 kPa at node $A$, what pressure may be expected at node $B$? Neglect minor losses. Take $\rho = 1000$ kg/m³.

▎ $p_A/\gamma + v_A^2/2g + z_A = p_B/\gamma + v_B^2/2g + z_B + h_L$. Assume that a flow $Q_1'$ of 170 L/s is proceeding through branch 1.

$$h_f = (f)(L/D)(v^2/2g) \qquad v_1 = Q_1' = (\tfrac{170}{1000})/[(\pi)(\tfrac{300}{1000})^2/4] = 2.405 \text{ m/s} \qquad N_R = Dv/v$$

$$(N_R)_1 = (\tfrac{300}{1000})(2.405)/(0.0113 \times 10^{-4}) = 6.38 \times 10^5 \qquad \epsilon/D_1 = 0.00026/(\tfrac{300}{1000}) = 0.00087$$

From Fig. A-5, $f = 0.0198$. $h_L = (h_f)_1 = 0.0198[600/(\tfrac{300}{1000})]\{2.405^2/[(2)(9.807)]\} = 11.68$ m. For branch 1,

$$p_A/\gamma + (v_1^2)_A/2g + 6 = p_B/\gamma + (v_1^2)_B/2g + 15 + 11.68 \qquad (v_1^2)_A/2g = (v_1^2)_B/2g \qquad [(p_A - p_B)/\gamma]_1 = 20.68 \text{ m}$$

Using this pressure head, which must be the same for each loop, for branch 2,

$$20.68 + (v_2^2)_A/2g + 6 = (v_2^2)_B/2g + 15 + (h_f)_2 \qquad (v_2^2)_A/2g = (v_2^2)_B/2g \qquad (h_f)_2 = 11.68 \text{ m}$$

$$11.68 = f_2[460/(\tfrac{470}{1000})]\{v_2^2/[(2)(9.807)]\} = 49.90 f_2 v_2^2 \qquad (\epsilon/D)_2 = 0.00026/(\tfrac{470}{1000}) = 0.00055$$

Estimate $f_2 = 0.018$.

$$11.68 = (49.90)(0.018)(v_2^2) \qquad v_2 = 3.606 \text{ m/s}$$

$$Q_2' = A_2 v_2 = [(\pi)(\tfrac{470}{1000})^2/4](3.606) = 0.6256 \text{ m}^3/\text{s} \quad \text{or} \quad 625.6 \text{ L/s}$$

Now the desired actual flows $Q_1$ and $Q_2$ may be computed so as to maintain the ratio $Q_1'/Q_2'$ and to satisfy continuity so that $Q_1 + Q_2 = 570 \text{ L/s}$.

$$Q_1 = [170/(625.6 + 170)](570) = 121.8 \text{ L/s} \qquad Q_2 = [625.6/(625.6 + 170)](570) = 448.2 \text{ L/s}$$

$$v_1 = (121.8/1000)/[(\pi)(\tfrac{300}{1000})^2/4] = 1.723 \text{ m/s} \qquad v_2 = (448.2/1000)/[(\pi)(\tfrac{470}{1000})^2/4] = 2.583 \text{ m/s}$$

$$(N_R)_1 = (\tfrac{300}{1000})(1.723)/(0.0113 \times 10^{-4}) = 4.57 \times 10^5 \qquad (N_R)_2 = (\tfrac{470}{1000})(2.583)/(0.0113 \times 10^{-4}) = 1.07 \times 10^6$$

$$f_1 = 0.0198 \qquad f_2 = 0.0180 \qquad (h_f)_1 = 0.0198[600/(\tfrac{300}{1000})]\{1.723^2/[(2)(9.807)]\} = 5.99 \text{ m}$$

$$p_A/\gamma + (v_1^2)_A/2g + 6 = p_B/\gamma + (v_1^2)_B/2g + 15 + 5.99 \qquad [(p_A - p_B)/\gamma]_1 = 14.99 \text{ m}$$

$$(h_f)_2 = 0.0180[460/(\tfrac{470}{1000})]\{2.583^2/[(2)(9.807)]\} = 5.99 \text{ m}$$

$$p_A/\gamma + (v_2^2)_A/2g + 6 = p_B/\gamma + (v_2^2)_B/2g + 15 + 5.99$$

$$[(p_A - p_B)/\gamma]_2 = 14.99 \text{ m} \qquad [(690 - p_B)/9.79] = 14.99 \qquad p_B = 543 \text{ kPa}$$

Branch 1, $L = 600$ m, $D = 300$ mm

Cast iron pipe throughout

$\nu = 0.0113 \times 10^{-4}$ m²/s

$A$     $y_A = 6$ m     $y_B = 15$ m     $B$

Branch 2, $L = 460$ m, $D = 470$ mm

**Fig. 11-6**

**11.7** The system shown in Fig. 11-7 receives at $A$ 66 ft³/min of water at a pressure of 95 psig. Neglecting minor losses, find the pressure at $B$. The pipe is commercial steel and $\mu = 2.11 \times 10^{-5}$ lb · s/ft².

$$p_A/\gamma + v_A^2/2g + z_A = p_C/\gamma + v_C^2/2g + z_C + h_L.$$

Assume that a flow $Q_1'$ of 30 ft³/min = 0.5 ft³/s is proceeding through branch 1.

$$h_f = (f)(L/D)(v^2/2g) \qquad v_1 = Q_1'/A_1 = 0.5/[(\pi)(\tfrac{6}{12})^2/4] = 2.546 \text{ ft/s} \qquad N_R = \rho D v/\mu$$

$$(N_R)_1 = (1.94)(\tfrac{6}{12})(2.546)/(2.11 \times 10^{-5}) = 1.17 \times 10^5 \qquad \epsilon/D_1 = 0.00015/(\tfrac{6}{12}) = 0.00030$$

From Fig. A-5, $f = 0.0195$. $h_L = (h_f)_1 = 0.0195[(35 + 1200 + 35)/(\tfrac{6}{12})]\{2.546^2/[(2)(32.2)]\} = 4.985 \text{ ft}$.

For branch 1, $p_A/\gamma + (v_1^2)_A/2g + 0 = p_B/\gamma + (v_1^2)_B/2g + 0 + 4.985$, $(v_1^2)_A/2g = (v_1^2)_B/2g$, $[(p_A - p_B)/\gamma]_1 = 4.985$ ft. Using this pressure head, which must be the same for each loop, for branch 2,

$$4.985 + (v_2^2)_A/2g + 0 = (v_2^2)_B/2g + 0 + (h_f)_2 \qquad (v_2^2)_A/2g = (v_2^2)_B/2g \qquad (h_f)_2 = 4.985 \text{ ft}$$

$$4.985 = f_2[(35 + 1200 + 35)/(\tfrac{8}{12})]\{v_2^2/[(2)(32.2)]\} = 29.58 f_2 v_2^2 \qquad (\epsilon/D)_2 = 0.00015/(\tfrac{8}{12}) = 0.000225$$

Estimate $f_2 = 0.018$.

$$4.985 = (29.58)(0.018)(v_2^2) \qquad v_2 = 3.060 \text{ ft/s} \qquad Q_2' = A_2 v_2 = [(\pi)(\tfrac{8}{12})^2/4](3.060) = 1.068 \text{ ft}^3/\text{s}$$

Now the desired actual flows $Q_1$ and $Q_2$ may be computed so as to maintain the ratio $Q_1'/Q_2'$ and to satisfy continuity so that $Q_1 + Q_2 = 66 \text{ ft}^3/\text{min} = 1.1 \text{ ft}^3/\text{s}$.

$$Q_1 = [0.5/(0.5 + 1.068)](1.1) = 0.3508 \text{ ft}^3/\text{s} \qquad Q_2 = [1.068/(0.5 + 1.068)](1.1) = 0.7492 \text{ ft}^3/\text{s}$$

$$v_1 = (0.3508)/[(\pi)(\tfrac{6}{12})^2/4] = 1.787 \text{ ft/s} \qquad v_2 = (0.7492)/[(\pi)(\tfrac{8}{12})^2/4] = 2.146 \text{ ft/s}$$

$$(N_R)_1 = (1.94)(\tfrac{6}{12})(1.787)/(2.11 \times 10^{-5}) = 8.22 \times 10^4 \qquad (N_R)_2 = (1.94)(\tfrac{8}{12})(2.146)/(2.11 \times 10^{-5}) = 1.32 \times 10^5$$

$$f_1 = 0.0205 \qquad f_2 = 0.0190 \qquad (h_f)_1 = 0.0205[(35 + 1200 + 35)/(\tfrac{6}{12})]\{1.787^2/[(2)(32.2)]\} = 2.582 \text{ ft}$$

$$p_A/\gamma + (v_1^2)_A/2g + 0 = p_C/\gamma + (v_1^2)_C/2g + 0 + 2.582 \qquad [(p_A - p_B)/\gamma]_1 = 2.582 \text{ ft}$$

$$(h_f)_2 = 0.0190[(35 + 1200 + 35)/(\tfrac{8}{12})]\{2.146^2/[(2)(32.2)]\} = 2.588 \text{ ft}$$

$$p_A/\gamma + (v_2^2)_A/2g + 0 = p_C/\gamma + (v_2^2)_C/2g + 0 + 2.588 \qquad [(p_A - p_C)/\gamma]_2 = 2.588 \text{ ft}$$

$$[(95 - p_C)(144)/62.4] = 2.588 \text{ ft} \qquad p_C = 93.9 \text{ lb/in}^2$$

For pipe $CB$,

$$v = Q/A = 1.1/[(\pi)(\tfrac{8}{12})^2/4] = 3.151 \text{ ft/s} \qquad N_R = (1.94)(\tfrac{8}{12})(3.151)/(2.11 \times 10^{-5}) = 1.93 \times 10^5 \qquad f = 0.0178$$

$$h_f = 0.0178[3200/(\tfrac{8}{12})]\{3.151^2/[(2)(32.2)]\} = 13.17 \text{ ft} \qquad p_B = 93.9 - (13.17)(62.4)/144 = 88.2 \text{ psig}$$

**Fig. 11-7**

**11.8** Repeat Prob. 11.7 using *nominal* pipe sizes (6.065 in and 7.981 in diameters) and supposing an open globe valve in the 8-in pipe just before $B$. The fittings are all screwed fittings. Use $K = 0.45$ for a 6-in standard tee, 0.30 for elbows, and 4.8 for the globe valve.

❚ $p_A/\gamma + v_A^2/2g + z_A = p_C/\gamma + v_C^2/2g + z_C + h_L$. Assume that a flow $Q_1'$ of 0.5 ft³/s is proceeding through branch 1.

$$h_L = h_f + h_m \qquad h_f = (f)(L/D)(v^2/2g) \qquad v_1 = Q_1'/A_1 = 0.5/[(\pi)(6.065/12)^2/4] = 2.492 \text{ ft/s}$$

$$N_R = \rho D v/\mu \qquad (N_R)_1 = (1.94)(6.065/12)(2.492)/(2.11 \times 10^{-5}) = 1.16 \times 10^5 \qquad \epsilon/D_1 = 0.00015/(\tfrac{6.065}{12}) = 0.00030$$

From Fig. A-5, $f = 0.0198$.

$$(h_f)_1 = 0.0198[(35 + 1200 + 35)/(6.065/12)]\{2.492^2/[(2)(32.2)]\} = 4.798 \text{ ft}$$

$$h_m = (K)(v^2/2g) \qquad (h_m)_1 = [(2)(0.30) + (2)(0.45)]\{2.492^2/[(2)(32.2)]\} = 0.145 \text{ ft}$$

$$(h_L)_1 = 4.798 + 0.145 = 4.943 \text{ ft}$$

For branch 1,

$$p_A/\gamma + (v_1^2)_A/2g + 0 = p_B/\gamma + (v_1^2)_B/2g + 0 + 4.943 \qquad (v_1^2)_A/2g = (v_1^2)_B/2g \qquad [(p_A - p_B)/\gamma]_1 = 4.943 \text{ ft}$$

Using this pressure head, which must be the same for each loop, for branch 2,

$$4.943 + (v_2^2)_A/2g + 0 = (v_2^2)_B/2g + 0 + (h_L)_2 \qquad (v_2^2)_A/2g = (v_2^2)_B/2g \qquad (h_L)_2 = 4.943 \text{ ft}$$

$$4.943 = f_2[(35 + 1200 + 35)/(7.981/12)]\{v_2^2/[(2)(32.2)]\} + [(2)(0.30) + (2)(0.45)]\{v_2^2/[(2)(32.2)]\}$$

$$= 29.65 f_2 v_2^2 + 0.02329 v_2^2$$

$$(\epsilon/D)_2 = 0.00015/(7.981/12) = 0.000226$$

Estimate $f_2 = 0.018$.

$$4.943 = (29.65)(0.018)(v_2^2) + 0.02329 v_2^2 \qquad v_2 = 2.979 \text{ ft/s} \qquad Q_2' = A_2 v_2 = [(\pi)(7.981/12)^2/4](2.979) = 1.035 \text{ ft}^3/\text{s}$$

Now the desired actual flows $Q_1$ and $Q_2$ may be computed so as to maintain the ratio $Q_1'/Q_2'$ and to satisfy continuity so that $Q_1 + Q_2 = 1.1$ ft³/s.

$$Q_1 = [0.5/(0.5 + 1.035)](1.1) = 0.3583 \text{ ft}^3/\text{s} \qquad Q_2 = [1.035/(0.5 + 1.035)](1.1) = 0.7417 \text{ ft}^3/\text{s}$$

$$v_1 = (0.3583)/[(\pi)(6.065/12)^2/4] = 1.786 \text{ ft/s} \qquad v_2 = (0.7417)/[(\pi)(7.981/12)^2/4] = 2.135 \text{ ft/s}$$

$$(N_R)_1 = (1.94)(6.065/12)(1.786)/(2.11 \times 10^{-5}) = 8.30 \times 10^4$$

$$(N_R)_2 = (1.94)(7.981/12)(2.135)/(2.11 \times 10^{-5}) = 1.31 \times 10^5$$

$$f_1 = 0.0205 \qquad f_2 = 0.019$$

$$(h_f)_1 = 0.0205[(35 + 1200 + 35)/(6.065/12)]\{1.786^2/[(2)(32.2)]\} = 2.551 \text{ ft}$$

$$(h_m)_1 = [(2)(0.30) + (2)(0.45)]\{1.786^2/[(2)(32.2)]\} = 0.074 \text{ ft} \qquad (h_L)_1 = 2.551 + 0.074 = 2.625 \text{ ft}$$

$$p_A/\gamma + (v_1^2)_A/2g + 0 = p_C/\gamma + (v_1^2)_C/2g + 0 + 2.625 \qquad [(p_A - p_B)/\gamma]_1 = 2.625 \text{ ft}$$

$$(h_f)_2 = 0.019[(35 + 1200 + 35)/(7.981/12)]\{2.135^2/[(2)(32.2)]\} = 2.568 \text{ ft}$$

$$(h_m)_2 = [(2)(0.30) + (2)(0.45)]\{2.135^2/[(2)(32.2)]\} = 0.106 \text{ ft}$$

$$(h_L)_1 = 2.568 + 0.106 = 2.674 \text{ ft} \qquad p_A/\gamma + (v_2^2)_A/2g + 0 = p_C/\gamma + (v_2^2)_C/2g + 0 + 2.674$$

$$[(p_A - p_C)/\gamma]_2 = 2.674 \text{ ft} \qquad [(95 - p_C)(144)/62.4] = 2.674 \text{ ft} \qquad p_C = 93.8 \text{ lb/in}^2$$

For pipe $CB$,

$$v = Q/A = 1.1/[(\pi)(7.981/12)^2/4] = 3.166 \text{ ft/s} \qquad N_R = (1.94)(7.981/12)(3.166)/(2.11 \times 10^{-5}) = 1.94 \times 10^5$$

$$f = 0.0178 \qquad h_f = 0.0178[3200/(7.981/12)]\{3.166^2/[(2)(32.2)]\} = 13.33 \text{ ft}$$

$$h_m = 4.8\{3.166^2/[(2)(32.2)]\} = 0.75 \text{ ft} \qquad h_L = 13.33 + 0.75 = 14.08 \text{ ft}$$

$$p_B = 93.8 - (14.08)(62.4)/144 = 87.7 \text{ lb/in}^2$$

**11.9** A two-branch, commercial steel pipe system (Fig. 11-8) delivers 0.42 m³/s of water at 5 °C. The pressure at $B$ is 22 kPa gage. What is the pressure at $A$? Neglect minor losses.

▌ $p_A/\gamma + v_A^2/2g + z_A = p_B/\gamma + v_B^2/2g + z_B + h_L$. Assume that a flow $Q_1'$ of 0.20 m³/s is proceeding through branch 1.

$$h_f = (f)(L/D)(v^2/2g) \qquad (v_1)_{250} = Q_1'/A_1 = 0.20/[(\pi)(0.250)^2/4] = 4.074 \text{ m/s}$$

$$(v_1)_{200} = Q_1'/A_1 = 0.20/[(\pi)(0.200)^2/4] = 6.366 \text{ m/s} \qquad N_R = Dv/\nu$$

$$[(N_R)_1]_{250} = (0.250)(4.074)/(1.52 \times 10^{-6}) = 6.70 \times 10^5 \qquad (\epsilon/D_1)_{250} = 0.000046/0.250 = 0.000184$$

From Fig. A-5, $(f_1)_{250} = 0.015$.

$$[(N_R)_1]_{200} = (0.200)(6.366)/(1.52 \times 10^{-6}) = 8.38 \times 10^5 \qquad (\epsilon/D_1)_{200} = 0.000046/0.200 = 0.00023 \qquad (f_1)_{200} = 0.015$$

$$h_L = 0.015[(60 + 110 + 40)/0.250]\{4.074^2/[(2)(9.807)]\} + 0.015[(110 + 40)/0.200]\{6.366^2/[(2)(9.807)]\} = 33.91 \text{ m}$$

For branch 1,

$$p_A/\gamma + (v_1^2)_A/2g + 0 = p_B/\gamma + (v_1^2)_B/2g + 0 + 33.91 \qquad (v_1^2)_A/2g = (v_1^2)_B/2g \qquad [(p_A - p_B)/\gamma]_1 = 33.91 \text{ m}$$

Using this pressure head, which must be the same for each loop, for branch 2,

$$33.91 + (v_2^2)_A/2g + 0 = (v_2^2)_B/2g + 0 + (h_f)_2 \qquad (v_2^2)_A/2g = (v_2^2)_B/2g \qquad (h_f)_2 = 33.91 \text{ m}$$

$$33.91 = f_2[(20 + 220)/0.200]\{v_2^2/[(2)(9.807)]\} = 61.18 f v_2^2 \qquad (\epsilon/D)_2 = 0.000046/0.200 = 0.00023$$

Estimate $f_2 = 0.015$.

$$33.91 = (61.18)(0.015)(v_2^2) \qquad v_2 = 6.079 \text{ m/s} \qquad Q_2' = A_2 v_2 = [(\pi)(0.200)^2/4](6.079) = 0.1910 \text{ m}^3/\text{s}$$

Now the desired actual flows $Q_1$ and $Q_2$ may be computed so as to maintain the ratio $Q_1'/Q_2'$ and to satisfy continuity so that $Q_1 + Q_2 = 0.42 \text{ m}^3/\text{s}$.

$$Q_1 = [0.20/(0.20 + 0.1910)](0.42) = 0.2148 \text{ m}^3/\text{s} \qquad Q_2 = [0.1910/(0.20 + 0.1910)](0.42) = 0.2052 \text{ m}^3/\text{s}$$

$$(v_1)_{250} = 0.2148/[(\pi)(0.250)^2/4] = 4.376 \text{ m/s} \qquad (v_1)_{200} = (0.2148)/[(\pi)(0.200)^2/4] = 6.837 \text{ m/s}$$

$$v_2 = (205.2/1000)/[(\pi)(0.200)^2/4] = 6.532 \text{ m/s}$$

Since the assumed value of $Q_1$ is close to 0.2148 m³/s, the values of $f$ of 0.015 should be close enough.

$$(h_f)_1 = 0.015[(60 + 110 + 40)/0.250]\{4.376^2/[(2)(9.807)]\}$$

$$+ 0.015[(110 + 40)/0.200]\{6.837^2/[(2)(9.807)]\} = 39.11 \text{ m}$$

$$p_A/\gamma + (v_1^2)_A/2g + 0 = p_B/\gamma + (v_1^2)_B/2g + 0 + 39.11 \qquad [(p_A - p_B)/\gamma]_1 = 39.11 \text{ m}$$

$$(h_f)_2 = 0.015[(20 + 220)/0.200]\{6.532^2/[(2)(9.807)]\} = 39.16 \text{ m}$$

$$p_A/\gamma + (v_2^2)_A/2g + 0 = p_B/\gamma + (v_2^2)_B/2g + 0 + 39.16 \qquad [(p_A - p_B)/\gamma]_2 = 39.16 \text{ m}$$

These values of $[(p_A - p_B)/\gamma]$ are close enough, so take an average value of $(39.11 + 39.16)/2$, or 39.14 m:
$(p_A - 22)/9.79 = 39.14$, $p_A = 405 \text{ kPa}$.

110 m

Branch I

$D_i$ = 250 mm

60 m          40 m

110 m

$A$        $D_i$ = 200 mm        40 m

$D_i$ = 200 mm          $B$

20 m

Branch II

220 m

**Fig. 11-8**

**11.10** The vertical, commercial steel pipe system shown in Fig. 11-9 ($D$ = 300 mm throughout) carries 0.82 m³/s of water. Find the pressure drop between $A$ and $B$ if $B$ is 100 m higher than $A$. Neglect minor losses. The water is at 5 °C.

$$p_A/\gamma + v_A^2/2g + z_A = p_B/\gamma + v_B^2/2g + z_B + h_L$$

**For branch I:** Estimate $Q_I' = 0.25$ m³/s, $h_f = (f)(L/D)(v^2/2g)$. Let $f = 0.014$: $v_I' = Q_I'/A_I = 0.25/[(\pi)(0.300)^2/4] = 3.537$ m/s, $h_f = 0.014[(250 + 550 + 250)/0.300]\{3.537^2/[(2)(9.807)]\} = 31.25$ m, $p_A/\gamma + v_A^2/2g + 0 = p_B/\gamma + v_B^2/2g + 100 + 31.25$, $v_A^2/2g = v_B^2/2g$, $(p_A - p_B)/\gamma = 131.25$ m.

**For branch II:** Let $f = 0.015$: $h_f = 0.015[550/0.300]\{(v_{II}')^2/[(2)(9.807)]\} = (1.402)(v_{II}')^2$, $p_A/\gamma + v_A^2/2g + 0 = p_B/\gamma + v_B^2/2g + 100 + (1.402)(v_{II}')^2$, $(p_A - p_B)/\gamma = 100 + (1.402)(v_{II}')^2$. Use $(p_A - p_B)/\gamma = 131.25$ m. $131.25 = 100 + (1.402)(v_{II}')^2$, $v_{II}' = 4.721$ m/s.

**For branch III:** Let $f = 0.013$: $h_f = 0.013[(350 + 550 + 350)/0.300]\{(v_{III}')^2/[(2)(9.807)]\} = (2.762)(v_{III}')^2$, $p_A/\gamma + v_A^2/2g + 0 = p_B/\gamma + v_B^2/2g + 100 + (2.762)(v_{III}')^2$, $(p_A - p_B)/\gamma = 100 + (2.762)(v_{III}')^2$. Use $(p_A - p_B)/\gamma = 131.25$ m. $131.25 = 100 + (2.762)(v_{III}')^2$, $v_{III}' = 3.364$ m/s. The new values of $Q$ are $Q_I = 250$ L/s; $Q_{II} = [(\pi)(0.300)^2/4](4.721) = 0.3337$ m³/s, or 333.7 L/s; $Q_{III} = [(\pi)(0.300)^2/4](3.364) = 0.2378$ m³/s, or 237.8 L/s. Find new values of $Q$: $Q_I = [250/(250 + 333.7 + 237.8)](820) = 249.5$ L/s, $Q_{II} = [333.7/(250 + 333.7 + 237.8)](820) = 333.1$ L/s, $Q_{III} = [237.8/(250 + 383.4 + 237.8)](820) = 237.4$ L/s.

**For branch I:** $v_I = (249.5/1000)/[(\pi)(0.300)^2/4] = 3.530$ m/s, $N_R = Dv/\nu = (0.300)(3.530)/(1.52 \times 10^{-6}) = 6.97 \times 10^5$, $\epsilon/D = 0.000046/0.300 = 0.00015$. From Fig. A-5, $f = 0.0146$.

**For branch II:** $v_{II} = (333.1/1000)/[(\pi)(0.300)^2/4] = 4.712$ m/s, $N_R = (0.300)(4.712)/(1.52 \times 10^{-6}) = 9.30 \times 10^5$, $f = 0.0141$.

**For branch III:** $v_{III} = (237.4/1000)/[(\pi)(0.300)^2/4] = 3.359$ m/s, $N_R = (0.300)(3.359)/(1.52 \times 10^{-6}) = 6.63 \times 10^5$, $f = 0.0146$. Compute $(p_A - p_B)/\gamma$ for each branch.

**For branch I:** $h_f = 0.0146[(250 + 550 + 250)/0.300]\{3.530^2/[(2)(9.807)]\} = 32.46$ m, $p_A/\gamma + v_A^2/2g + 0 = p_B/\gamma + v_B^2/2g + 100 + 32.46$, $(p_A - p_B)/\gamma = 132.46$ m.

**For branch II:** $h_f = 0.0141[550/0.300]\{4.712^2/[(2)(9.807)]\} = 29.26$ m, $p_A/\gamma + v_A^2/2g + 0 = p_B/\gamma + v_B^2/2g + 100 + 29.26$, $(p_A - p_B)/\gamma = 129.26$ m.

**For branch III:** $h_f = 0.0146[(350 + 550 + 350)/0.300]\{3.359^2/[(2)(9.807)]\} = 34.99$ m, $p_A/\gamma + v_A^2/2g + 0 = p_B/\gamma + v_B^2/2g + 100 + 34.99$, $(p_A - p_B)/\gamma = 134.99$ m.

We could go through another cycle. Instead, we stop here and take an average value of $(p_A - p_B)/\gamma$: $(132.46 + 129.26 + 134.99)/3$, or 132.24 m. $(p_A - p_B) = (132.24)(9.79) = 1295$ kPa gage.

**11.11** In Fig. 11-10, $L_1 = 3000$ ft, $D_1 = 1$ ft, $\epsilon_1 = 0.001$ ft; $L_2 = 2000$ ft, $D_2 = 8$ in, $\epsilon_2 = 0.0001$ ft; $L_3 = 4000$ ft, $D_3 = 16$ in, $\epsilon_3 = 0.0008$ ft; $\rho = 2.00$ slugs/ft³, $\nu = 0.00003$ ft²/s, $p_A = 80$ psi, $z_A = 100$ ft, $z_B = 80$ ft. For a total flow of 12 cfs, determine the flow through each pipe and the pressure at $B$.

$h_f = (f)(L/D)(v^2/2g)$. Assume $Q_1' = 3$ cfs.

$$v_1' = Q_1'/A_1 = 3/[(\pi)(1)^2/4] = 3.820 \text{ ft/s} \qquad N_R = Dv/\nu$$

$$(N_R)_1' = (1)(3.820)/0.00003 = 1.27 \times 10^5 \qquad \epsilon/D_1 = 0.001/1 = 0.0010$$

From Fig. A-5, $f_1' = 0.022$. $(h_f)_1' = (0.022)(\frac{3000}{1})\{3.820^2/[(2)(32.2)]\} = 14.95$ ft. For pipe 2, assume $f_2' = 0.020$.

$$14.95 = 0.020[2000/(\tfrac{8}{12})]\{v_2'^2/[(2)(32.2)]\} \qquad v_2' = 4.006 \text{ ft/s} \qquad (N_R)_2' = (\tfrac{8}{12})(4.006)/0.00003 = 8.90 \times 10^4$$

$$\epsilon_2/D_2 = 0.0001/(\tfrac{8}{12}) = 0.00015 \qquad f_2' = 0.019 \qquad 14.95 = 0.019[2000/(\tfrac{8}{12})]\{v_2'^2/[(2)(32.2)]\}$$

$$v_2' = 4.110 \text{ ft/s} \qquad Q_2' = [(\pi)(\tfrac{8}{12})^2/4](4.110) = 1.43 \text{ ft}^3/\text{s}$$

**Fig. 11-9**

For pipe 3, assume $f_3' = 0.020$.

$$14.95 = 0.020[4000/(\tfrac{16}{12})]\{v_3'^2/[(2)(32.2)]\} \qquad v_3' = 4.006 \text{ ft/s} \qquad (N_R)_2' = (\tfrac{16}{12})(4.006)/0.00003 = 1.78 \times 10^5$$

$$\epsilon_3/D_3 = 0.0008/(\tfrac{16}{12}) = 0.00060 \qquad f_3' = 0.020 \qquad Q_3' = [(\pi)(\tfrac{16}{12})^2/4](4.006) = 5.59 \text{ ft}^3/\text{s}$$

$$\Sigma Q = 3 + 1.43 + 5.59 = 10.02 \text{ ft}^3/\text{s} \qquad Q_1 = (3/10.02)(12) = 3.59 \text{ ft}^3/\text{s}$$

$$Q_2 = (1.43/10.02)(12) = 1.71 \text{ ft}^3/\text{s} \qquad Q_3 = (5.59/10.02)(12) = 6.69 \text{ ft}^3/\text{s}$$

Check the values of $h_1$, $h_2$, and $h_3$:

$$v_1 = 3.59/[(\pi)(1)^2/4] = 4.571 \text{ ft/s} \qquad (N_R)_1 = (1)(4.571)/0.00003 = 1.52 \times 10^5$$

$$f_1 = 0.021 \qquad (h_f)_1 = (0.021)(\tfrac{3000}{1})\{4.571^2/[(2)(32.2)]\} = 20.44 \text{ ft} \qquad v_2 = 1.71/[(\pi)(\tfrac{8}{12})^2/4] = 4.899 \text{ ft/s}$$

$$(N_R)_2 = (\tfrac{8}{12})(4.899)/0.00003 = 1.09 \times 10^5 \qquad f_2 = 0.018 \qquad (h_f)_2 = 0.018[2000/(\tfrac{8}{12})]\{4.899^2/[(2)(32.2)]\} = 20.12 \text{ ft}$$

$$v_3 = 6.69/[(\pi)(\tfrac{16}{12})^2/4] = 4.791 \text{ ft/s} \qquad (N_R)_3 = (\tfrac{16}{12})(4.791)/0.00003 = 2.13 \times 10^5$$

$$f_3 = 0.019 \qquad (h_f)_3 = 0.019[4000/(\tfrac{16}{12})]\{4.791^2/[(2)(32.2)]\} = 20.32 \text{ ft}$$

Since these values of $h_f$ are close, accept the flow values determined above and take an average value of $(20.44 + 20.12 + 20.32)/3$, or 20.29 ft for calculating $p_B$: $p_A/\gamma + v_A^2/2g + z_A = p_B/\gamma + v_B^2/2g + z_B + h_L$, $(80)(144)/62.4 + v_A^2/2g + 100 = (p_B)(144)/62.4 + v_B^2/2g + 80 + 20.29$, $v_A^2/2g = v_B^2/2g$, $p_B = 79.9 \text{ lb/in}^2$.

**Fig. 11-10**

**11.12** Prepare a computer program to solve for the head drop and flow distribution through $N$ pipes in parallel. Check it out with the data in Prob. 11.11.

```
10 REM B:                          SOLUTION OF PARALLEL PIPE PROBLEMS
20 DEFINT I,N:    DEF FNQ(D1,D2,D3,DH)=D1*DH*LOG(D2+D3/DH)
30 READ N,QQ,RHO,KVIS,PA,ZA,ZB,G,II
40 DATA 3,12.,2.,3E-5,80.,100.,80.,32.2,15
50 LPRINT:      LPRINT"N,QQ,RHO,KVIS=";N;QQ;RHO;KVIS
60 LPRINT:      LPRINT"PA,ZA,ZB,G,II=";PA;ZA;ZB;G;II:      LPRINT
70 FOR I=1 TO N:     READ L(I),D(I),EP(I):     NEXT I
80 DATA 3000.,1.,.001,2000.,.6667,.0001,4000.,1.3333,.0008
90 FOR I=1 TO N:     LPRINT"I,L,D,EPS=";I;L(I);D(I);EP(I):     NEXT I:     LPRINT
100 FOR I=1 TO N:     C=SQR(G*D(I)/L(I)):     E1(I)=-.9650001*D(I)^2*C
110 E2(I)=EP(I)/(3.7*D(I)):     E3(I)=1.784*KVIS/(D(I)*C):     NEXT I
120 X=.02*L(1)*QQ^2/(.7854*D(1)^5*2!*G) ' MAX POSS. HEAD WITH FLOW THRU PIPE 1
130 HMA=SQR(X):     HMI=0!
140 FOR II=1 TO II:     S=0!:     HF=.5*(HMA+HMI)     ' START OF BISECTION METHOD
150 FOR I=1 TO N:     S=S+FNQ(E1(I),E2(I),E3(I),HF):     NEXT I
160 IF S-QQ>0! THEN HMA=HF ELSE HMI=HF
170 PRINT"S,QQ,HF=";S;QQ;HF
```

```
180 NEXT I1:    HFR=.5$(HMA+HMI):    HFF=HFR^2    ' FINAL VALUE OF HEAD DROP
190 FOR I=1 TO N:    LPRINT"I,HFF,Q=";I;HFF;FNQ(E1(I),E2(I),E3(I),HFR):  NEXT I
200 LPRINT:  LPRINT"PB=";:   LPRINT USING" ###.### ";PA+(ZA-ZB-HFF)$RHO$G/144!

N,QQ,RHO,KVIS= 3  12  2  .00003

PA,ZA,ZB,G,II= 80  100  80  32.2  15

I,L,D,EPS= 1  3000  1     .001
I,L,D,EPS= 2  2000  .6667  .0001
I,L,D,EPS= 3  4000  1.3333  .0008

I,HFF,Q= 1  20.68776  3.576553
I,HFF,Q= 2  20.68776  1.710832
I,HFF,Q= 3  20.68776  6.713136

PB=  79.692
```

**11.13** Two pipes are connected in parallel between two reservoirs: $L_1 = 2600$ m, $D_1 = 1.3$ m, $C = 90$; $L_2 = 2400$ m, $D_2 = 0.9$ m, $C = 100$. For a difference in elevation of 3.8 m, determine the total flow of water at 20 °C.

**∎**
$$v = 0.8492CR^{0.63}s^{0.54}$$
$$v_1 = (0.8492)(90)(1.3/4)^{0.63}(3.8/2600)^{0.54} = 1.109 \text{ m/s}$$
$$v_2 = (0.8492)(100)(\tfrac{0.9}{4})^{0.63}(3.8/2400)^{0.54} = 1.020 \text{ m/s}$$
$$Q = Av = [(\pi)(1.3)^2/4](1.109) + [(\pi)(0.9)^2/4](1.020) = 2.12 \text{ m}^3/\text{s}$$

**11.14** For 5.4 m³/s of flow in the system of Prob. 11.13, determine the difference in elevation of reservoir surfaces.

**∎**
$$Q = Av = 0.8492ACR^{0.63}s^{0.54}$$
$$5.4 = 0.8492[(\pi)(1.3)^2/4](90)(1.3/4)^{0.63}s^{0.54} + 0.8492[(\pi)(0.9)^2/4](100)(\tfrac{0.9}{4})^{0.63}s^{0.54}$$
$$s = 0.008455 \text{ m/m} \qquad \Delta \text{ elevation} = (0.008455)(2400) = 20.3 \text{ m}$$

**11.15** Three smooth tubes are connected in parallel: $L_1 = 45$ ft, $D_1 = \tfrac{1}{2}$ in; $L_2 = 55$ ft, $D_2 = 1$ in; $L_3 = 60$ ft, $D_3 = \tfrac{3}{4}$ in. Given $Q = 33$ gpm, $\gamma = 55$ lb/ft³, and $\mu = 0.13$ Pa · s, what is the drop in hydraulic grade line between junctions?

**∎**
$$h_f = \frac{128\mu LQ}{\pi\gamma D^4}$$

Therefore, $L_1Q_1/D_1^4 = L_2Q_2/D_2^4 = L_3Q_3/D_3^4$.

$$Q = Q_1 + Q_2 + Q_3 = (33)(0.002228) = 0.07352 \text{ ft}^3/\text{s} \qquad 45Q_1/[(\tfrac{1}{2})/12]^4 = 55Q_2/(\tfrac{1}{12})^4 \qquad Q_1 = 0.07639Q_2$$
$$55Q_2/(\tfrac{1}{12})^4 = 60Q_3/[(\tfrac{3}{4})/12]^4 \qquad Q_3 = 0.2900Q_2 \qquad 0.07639Q_2 + Q_2 + 0.2900Q_2 = 0.07352 \qquad Q_2 = 0.05381 \text{ ft}^3/\text{s}$$
$$\mu = \frac{(0.13)(0.3048)^2}{4.448} = 0.002716 \text{ lb} \cdot \text{s/ft}^2 \qquad h_f = \frac{(128)(0.002716)(55)(0.05381)}{(\pi)(55)(\tfrac{1}{12})^4} = 123.4 \text{ ft}$$

**11.16** If $H = 11$ m, $\mu = 8 \times 10^{-3}$ Pa · s, and s.g. = 0.9 in Fig. 11-16, find the discharges through pipes 1, 2, and 3.

**∎**
$$h_f = (f)(L/D)(v^2/2g) \qquad (\epsilon/D)_1 = 0.006/0.050 = 0.120 \qquad f_1 = 0.114$$
$$(\epsilon/D)_2 = 0.009/0.120 = 0.075 \qquad f_2 = 0.088 \qquad (\epsilon/D)_3 = 0.012/0.100 = 0.120 \qquad f_3 = 0.114$$
$$(h_f)_1 = (h_f)_2 \qquad v_1 = Q_1/A_1 = Q_1/[(\pi)(0.050)^2/4] = 509.3Q_1 \qquad v_2 = Q_2/A_2 = Q_2/[(\pi)(0.120)^2/4] = 88.42Q_2$$
$$0.114[70/0.050]\{(509.3Q_1)^2/[(2)(9.807)]\} = 0.088[80/0.120]\{(88.42Q_2)^2/[(2)(9.807)]\}$$
$$Q_2 = 9.500Q_1 \qquad Q_3 = Q_1 + Q_2 = Q_1 + 9.500Q_1 = 10.500Q_1$$
$$v_3 = Q_3/A_3 = 10.500Q_1/[(\pi)(0.100)^2/4] = 1337Q_1 \qquad H = (h_f)_1 + (h_f)_3$$
$$11 = 0.114[70/0.050]\{(509.3Q_1)^2/[(2)(9.807)]\} + 0.114[110/0.100]\{1337Q_1)^2/[(2)(9.807)]\}$$
$$Q_1 = 0.000901 \text{ m}^3/\text{s} = 0.901 \text{ L/s} \qquad Q_2 = (9.500)(0.901) = 8.56 \text{ L/s} \qquad Q_3 = (10.500)(0.901) = 9.46 \text{ L/s}$$
$$N_R = \frac{4Q\rho}{\pi D\mu} = \frac{(4)(0.00946)(0.9)(1000)}{(\pi)(0.100)(8 \times 10^{-3})} = 13\,550$$

Hence, both $f$'s are O.K.

**Fig. 11-11**

**11.17** Find $H$ in Fig. 11-11 for $Q_3 = 20$ L/s, $\mu = 5 \times 10^{-3}$ Pa·s, s.g. $= 0.9$.

❚ $Q_1$ and $Q_2$ will be in the same proportion as in Prob. 11.16. Hence, $Q_1 = (0.901/9.46)(20) = 1.90$ L/s, $Q_2 = (9.500)(1.90) = 18.0$ L/s.

$$H = (h_f)_1 + (h_f)_3 \qquad h_f = (f)(L/D)(v^2/2g)$$

$$v_1 = Q_1/A_1 = 0.00190/[(\pi)(0.050)^2/4] = 0.9677 \text{ m/s} \qquad v_3 = Q_3/A_3 = 0.020/[(\pi)(0.100)^2/4] = 2.546 \text{ m/s}$$

Using values of $f$ from Prob. 11.16,

$$H = 0.114[70/0.050]\{0.9677^2/[(2)(9.907)]\} + 0.114[110/0.100]\{2.546^2/[(2)(9.907)]\} = 49.06 \text{ m}$$

**11.18** What length of 300-mm-diameter, clean cast-iron pipe could replace the system of Fig. 11-12?

❚ $(\epsilon/D)_1 = 0.00026/0.200 = 0.0013 \qquad (\epsilon/D)_2 = 0.00026/0.300 = 0.00087 \qquad (\epsilon/D)_3 = 0.00026/0.500 = 0.00052$

$$(\epsilon/D)_4 = 0.00026/0.300 = 0.00087 \qquad (\epsilon/D)_5 = 0.00026/0.300 = 0.00087$$

Try $f_1 = 0.023$, $f_2 = 0.021$, $f_3 = 0.018$, $f_4 = 0.021$, and $f_5 = 0.021$: Assume a head loss in pipes 1 and 2 of 3 m.

$$h_f = (f)(L/D)(v^2/2g) \qquad 3 = 0.023[310/0.200]\{v_1^2/[(2)(9.807)]\} \qquad v_1 = 1.285 \text{ m/s}$$

$$Q_1 = A_1 v_1 = [(\pi)(0.200)^2/4](1.285) = 0.04037 \text{ m}^3/\text{s} \qquad 3 = 0.021[310/0.300]\{v_2^2/[(2)(9.807)]\} \qquad v_2 = 1.647 \text{ m/s}$$

$$Q_2 = A_2 v_2 = [(\pi)(0.300)^2/4](1.647) = 0.1164 \text{ m}^3/\text{s}$$

To find equivalent for 1 and 2, $v_e = (0.04037 + 0.1164)/[(\pi)(0.300)^2/4] = 2.218$ m/s, $3 = 0.021[L_e/0.300]\{2.218^2/[(2)(9.807)]\}$, $L_e = 170.9$ m (for pipes 1 and 2).
For pipe 3:

$$0.018\left(\frac{290}{0.500}\right)\frac{\{Q/[(\pi)(0.500)^2/4]\}^2}{(2)(9.807)} = 0.021\left(\frac{L_e}{0.300}\right)\frac{\{Q/[(\pi)(0.300)^2/4]\}^2}{(2)(9.807)}$$

$$L_e = 53.7 \text{ m} \qquad \text{(for pipe 3)}$$

Assume a head loss in pipes 4 and 5 of 3 m.

$$3 = 0.021[620/0.300]\{v_4^2/[(2)(9.807)]\} \qquad v_4 = 1.164 \text{ m/s}$$

$$Q_4 = [(\pi)(0.300)^2/4](1.164) = 0.08228 \text{ m}^3/\text{s} \qquad 3 = 0.021[790/0.300]\{v_5^2/[(2)(9.807)]\}$$

$$v_5 = 1.032 \text{ m/s} \qquad Q_5 = [(\pi)(0.300)^2/4](1.032) = 0.07295 \text{ m}^3/\text{s}$$

To find equivalent for 4 and 5,

$$v_e = (0.08228 + 0.07295)/[(\pi)(0.300)^2/4] = 2.196 \text{ m/s} \qquad 3 = 0.021[L_e/0.300]\{2.196^2/[(2)(9.807)]\}$$

$$L_e = 174.3 \text{ m} \qquad \text{(for pipes 4 and 5)} \qquad (L_e)_{\text{total}} = 170.9 + 53.7 + 174.3 = 398.9 \text{ m}$$

**Fig. 11-12**

**11.19** For a velocity of 1.1 m/s in pipe 1 of Fig. 11-12, calculate the flow through the system and the head $H$ required. The fluid is water at 15 °C.

❚ Using values from Prob. 11.18, $h_f = (f)(L/D)(v^2/2g)$, $(h_f)_1 = 0.023[310/0.200]\{1.1^2/[(2)(9.807)]\} = 2.199$ m, $(h_f)_2 = (h_f)_1 = 0.021[310/0.300]\{v_2^2/[(2)(9.807)]\} = 2.199$ m, $v_2 = 1.410$ m/s; $Q = Av = [(\pi)(0.200)^2/4](1.1) + [(\pi)(0.300)^2/4](1.410) = 0.134$ m$^3$/s. (This is the flow through the system.) $v_3 = Q_3/A_3 = 0.134/[(\pi)(0.500)^2/4] = 0.6825$ m/s, $(h_f)_3 = 0.018[290/0.500]\{0.6825^2/[(2)(9.807)]\} = 0.363$ m. The flow of 0.134 m$^3$/s will split in pipes 4 and 5 in the same proportions as determined in Prob. 11.18. Hence, $Q_4 = [0.08228/(0.08228 + 0.07295)](0.134) = 0.0710$ m$^3$/s, $v_4 = 0.0710/[(\pi)(0.300)^2/4] = 1.004$ m/s, $(h_f)_4 = 0.021[620/0.300]\{1.004^2/[(2)(9.807)]\} = 2.230$ m, $H = 2.199 + 0.363 + 2.230 = 4.792$ m.

**11.20** Three pipes are in parallel with a total head loss of 20.3 m. The pipe data are

| pipe | $L$, m | $d$, cm | $\epsilon$, mm | $\epsilon/d$ |
|------|--------|---------|----------------|--------------|
| 1 | 100 | 8 | 0.24 | 0.003 |
| 2 | 150 | 6 | 0.12 | 0.002 |
| 3 | 80 | 4 | 0.20 | 0.005 |

The fluid is water, $\rho = 1000$ kg/m$^3$, and $v = 1.02 \times 10^{-6}$ m$^2$/s. Calculate the total flow rate in cubic meters per hour, neglecting minor losses.

❚ $h_f = (f)(L/D)(v^2/2g)$, $(h_f)_1 = (h_f)_2 = (h_f)_3$. Guess fully rough flow in pipe 1: $f_1 = 0.0262$.

$$(h_f)_1 = 20.3 = 0.0262[100/(\tfrac{8}{100})]\{v_1^2/[(2)(9.807)]\} \qquad v_1 = 3.487 \text{ m/s}$$
$$N_R = Dv/v \qquad (N_R)_1 = (\tfrac{8}{100})(3.487)/(1.02 \times 10^{-6}) = 2.73 \times 10^5$$

From Fig. A-5, $f_1 = 0.0267$.

$$20.3 = 0.0267[100/(\tfrac{8}{100})]\{v_1^2/[(2)(9.807)]\} \qquad v_1 = 3.454 \text{ m/s}$$
$$Q_1 = A_1 v_1 = [(\pi)(\tfrac{8}{100})^2/4](3.454) = 0.01736 \text{ m}^3/\text{s}$$

Guess $f_2 = 0.0234$:

$$(h_f)_2 = 20.3 = 0.0234[150/(\tfrac{6}{100})]\{v_2^2/[(2)(9.807)]\} \qquad v_2 = 2.609 \text{ m/s}$$
$$(N_R)_2 = (\tfrac{6}{100})(2.609)/(1.02 \times 10^{-6}) = 1.53 \times 10^5 \qquad f_2 = 0.0246 \qquad 20.3 = 0.0246[150/(\tfrac{6}{100})]\{v_2^2/[(2)(9.807)]\}$$
$$v_2 = 2.544 \text{ m/s} \qquad Q_2 = [(\pi)(\tfrac{6}{100})^2/4](2.544) = 0.007193 \text{ m}^3/\text{s}$$

Guess $f_3 = 0.0304$:

$$(h_f)_3 = 20.3 = 0.0304[80/(\tfrac{4}{100})]\{v_3^2/[(2)(9.807)]\} \qquad v_3 = 2.559 \text{ m/s}$$
$$(N_R)_3 = (\tfrac{4}{100})(2.559)/(1.02 \times 10^{-6}) = 1.00 \times 10^5$$
$$f_3 = 0.0313 \qquad 20.3 = 0.0313[80/(\tfrac{4}{100})]\{v_3^2/[(2)(9.807)]\} \qquad v_3 = 2.522 \text{ m/s}$$
$$Q_3 = [(\pi)(\tfrac{4}{100})^2/4](2.522) = 0.003169 \text{ m}^3/\text{s}$$

This is satisfactory convergence. Hence, $Q_{\text{total}} = 0.01736 + 0.007193 + 0.003169 = 0.02772$ m$^3$/s, or 99.8 m$^3$/h.

**11.21** For the parallel pipe system of Fig. 11-13, each pipe is cast iron, and the pressure drop $p_1 - p_2$ equals 2.5 psi. Compute the total flow rate between 1 and 2 if the fluid properties are $\gamma = 57.32$ lb/ft$^3$, $\mu = 0.00606$ lb · s/ft$^2$.

❚ $h_f = (p_1 - p_2)/\gamma = (2.5)(144)/57.32 = 6.281$ ft. Assume laminar flow.

$$h_f = \frac{128\mu L Q}{\pi\gamma D^4}$$

**For the 3-in-diameter pipe:**

$$6.281 = \frac{(128)(0.00606)(225)(Q_3)}{(\pi)(57.32)(\tfrac{3}{12})^4} \qquad Q_3 = 0.02531 \text{ ft}^3/\text{s} \qquad N_R = \frac{\gamma Dv}{\mu g}$$

$v_3 = Q_3/A_3 = 0.02531/[(\pi)(\tfrac{3}{12})^2/4] = 0.5156$ ft/s $\qquad (N_R)_3 = (57.32)(\tfrac{3}{12})(0.5156)/(0.00606)(32.2) = 38$ (laminar)

**For the 2-in-diameter pipe:**

$$6.281 = \frac{(128)(0.00606)(185)(Q_2)}{(\pi)(57.32)(\frac{2}{12})^4} \qquad Q_2 = 0.006081 \text{ ft}^3/\text{s} \qquad v_2 = 0.006081/[(\pi)(\tfrac{2}{12})^2/4] = 0.2787 \text{ ft/s}$$

$$(N_R)_3 = (57.32)(\tfrac{2}{12})(0.2787)/(0.00606)(32.2) = 14 \qquad \text{(laminar)} \qquad Q_{\text{total}} = 0.02531 + 0.006081 = 0.0314 \text{ ft}^3/\text{s}$$

**Fig. 11-13**

**11.22** The system of Fig. 11-14 consists of 60-mm-diameter, wrought iron pipes containing water at 20 °C. If the total flow rate from 1 to 2 is 13 L/s, compute the total pressure drop $p_1 - p_2$. Neglect minor losses.

**❙ For pipe c:**

$$v_c = Q_C/A_C = 0.013/[(\pi)(0.060)^2/4] = 4.598 \text{ m/s} \qquad N_R = dv/\nu$$

$$(N_R)_c = (0.060)(4.598)/(1.02 \times 10^{-6}) = 2.70 \times 10^5 \qquad \epsilon/d = 0.000046/0.060 = 0.00077$$

From Fig. A-5, $f_c = 0.020$.

$$h_f = (f)(L/d)(v^2/2g) \qquad (h_f)_c = 0.020[175/0.060]\{4.598^2/[(2)(9.807)]\} = 62.88 \text{ m}$$

$$(\Delta p)_c = (\gamma)(h_f)_c = (9.79)(62.88) = 616 \text{ kPa}$$

**For pipes a and b:** $(h_f)_a = (h_f)_b$, $v_a = Q_a/A_a = Q_a/[(\pi)(\tfrac{6}{100})^2/4] = 353.7Q_a$, $v_b = Q_b/A_b = Q_b/[(\pi)(\tfrac{6}{100})^2/4] = 353.7Q_b$.

$$f_a[240/0.060]\{(353.7Q_a)^2/[(2)(9.807)]\} = f_b[120/0.060]\{(353.7Q_b)^2/[(2)(9.807)]\} \qquad (1)$$

If $f_a = f_b$, $Q_a = 0.7071Q_b$, $Q_a + Q_b = 1.7071Q_b = 0.013 \text{ m}^3/\text{s}$, $Q_b = 0.007615 \text{ m}^3/\text{s}$, $Q_a = (0.7071)(0.007615) = 0.005385 \text{ m}^3/\text{s}$; $(N_R)_a = \frac{6}{100}[(353.7)(0.005385)]/(1.02 \times 10^{-6}) = 1.12 \times 10^5$, $f_a = 0.0219$; $(N_R)_b = \frac{6}{100}[(353.7)(0.007615)]/(1.02 \times 10^{-6}) = 1.58 \times 10^5$, $f_b = 0.0208$. Substituting these values of $f$ into Eq. (1), $0.0219[240/(\tfrac{6}{100})]\{353.7Q_a)^2/[(2)(9.807)]\} = 0.0208[120/(\tfrac{6}{100})]\{(353.7Q_b)^2/[(2)(9.807)]\}$, $Q_a = 0.6891Q_b$, $Q_a + Q_b = 0.6891Q_b + Q_b = 0.013$, $Q_b = 0.007696 \text{ m}^3/\text{s}$, $Q_a = (0.6891)(0.007696) = 0.005303 \text{ m}^3/\text{s}$.

$$(N_R)_a = 0.060[(353.7)(0.005303)]/(1.02 \times 10^{-6}) = 1.10 \times 10^5 \qquad f_a = 0.0219 \qquad \text{(O.K.)}$$

$$(N_R)_b = 0.060[(353.7)(0.007696)]/(1.02 \times 10^{-6}) = 1.60 \times 10^5 \qquad f_b = 0.0208 \qquad \text{(O.K.)}$$

Hence, $Q_a = 0.005303 \text{ m}^3/\text{s}$ and $Q_b = 0.007696 \text{ m}^3/\text{s}$.

$$(h_f)_a = (h_f)_b = 0.0219[240/0.060]\{[(353.7)(0.005303)]^2/[(2)(9.807)]\} = 15.71 \text{ m}$$

$$(\Delta p)_{a,b} = (9.79)(15.71) = 154 \text{ kPa} \qquad (\Delta p)_{\text{total}} = 616 + 154 = 770 \text{ kPa}$$

**Fig. 11-14**

**11.23** Reconsider Prob. 11.22 with the flow rate unknown but with the total pressure drop given as 850 kPa. Find the resulting flow rate $Q$.

**❙** Using data from Prob. 11.22, guess $Q_a = 0.006 \text{ m}^3/\text{s}$: $Q_b = 0.006/0.6891 = 0.008707 \text{ m}^3/\text{s}$, $Q_c = 0.006 + 0.008707 = 0.014707 \text{ m}^3/\text{s}$, $v_c = Q_c/A_c = 0.014707/[(\pi)(0.060)^2/4] = 5.202 \text{ m/s}$, $N_R = dv/\nu$, $(N_R)_c = (0.060)(5.202)/(1.02 \times 10^{-6}) = 3.06 \times 10^5$, $\epsilon/d = 0.000046/0.060 = 0.00077$. From Fig. A-5, $f_c = 0.0195$: $h_f = (f)(L/d)(v^2/2g)$, $(h_f)_c = 0.0195[175/0.060]\{5.202^2/[(2)(9.807)]\} = 78.47 \text{ m}$.

**For pipes $a$ and $b$:**

$$(h_f)_a = (h_f)_b \qquad v_a = Q_a/A_a = 0.006/[(\pi)(0.060)^2/4] = 2.122\ \text{m/s}$$

$$v_b = Q_b/A_b = 0.008707/[(\pi)(0.060)^2/4] = 3.079\ \text{m/s}$$

$$(N_R)_a = (0.060)(2.122)/(1.02 \times 10^{-6}) = 1.25 \times 10^5 \qquad f_a = 0.0205$$

$$(N_R)_b = (0.060)(3.079)/(1.02 \times 10^{-6}) = 1.81 \times 10^5 \qquad f_b = 0.020$$

$$(h_f)_a = 0.0205[240/0.060]\{2.122^2/[(2)(9.807)]\} = 18.83\ \text{m}$$

$$(h_f)_b = 0.020[120/0.060]\{3.079^2/[(2)(9.807)]\} = 19.33\ \text{m}$$

Since these are fairly close, forego another iteration and take an average value of $(h_f)_{a,b}$:
$(h_f)_{a,b} = (18.83 + 19.33)/2 = 19.08\ \text{m}$, $(h_f)_{\text{total}} = 78.47 + 19.08 = 97.55\ \text{m}$, $(\Delta p)_{\text{total}} = (\gamma)(h_f)_{\text{total}} = (9.79)(97.55) = 955\ \text{kPa}$. Since this is slightly larger than the required 850 kPa, reduce $Q_a$ a small amount; e.g., try $Q_a = 0.00565\ \text{m}^3/\text{s}$:

$$Q_b = 0.00565/0.6891 = 0.008199\ \text{m}^3/\text{s} \qquad Q_c = 0.00565 + 0.008199 = 0.013849\ \text{m}^3/\text{s}$$

$$v_c = Q_c/A_c = 0.013849/[(\pi)(0.060)^2/4] = 4.898\ \text{m/s} \qquad (N_R)_c = (0.060)(4.898)/(1.02 \times 10^{-6}) = 2.88 \times 10^5$$

$$f_c = 0.0195 \qquad (h_f)_c = 0.0195[175/0.060]\{4.898^2/[(2)(9.807)]\} = 69.57\ \text{m}$$

**For pipes $a$ and $b$:**

$$v_a = Q_a/A_a = 0.00565/[(\pi)(0.060)^2/4] = 1.998\ \text{m/s} \qquad v_b = Q_b/A_b = 0.008199/[(\pi)(0.060)^2/4] = 2.900\ \text{m/s}$$

$$(N_R)_a = (0.060)(1.998)/(1.02 \times 10^{-6}) = 1.18 \times 10^5 \qquad f_a = 0.0205$$

$$(N_R)_b = (0.060)(2.900)/(1.02 \times 10^{-6}) = 1.71 \times 10^5 \qquad f_b = 0.020$$

$$(h_f)_a = 0.0205[240/0.060]\{1.998^2/[(2)(9.807)]\} = 16.69\ \text{m}$$

$$(h_f)_b = 0.020[120/0.060]\{2.900^2/[(2)(9.807)]\} = 17.15\ \text{m}$$

Since these are fairly close, forego another iteration and take an average value of $(h_f)_{a,b}$:
$(h_f)_{a,b} = (16.69 + 17.15)/2 = 16.92\ \text{m}$, $(h_f)_{\text{total}} = 69.57 + 16.92 = 86.49\ \text{m}$, $(\Delta p)_{\text{total}} = (\gamma)(h_f)_{\text{total}} = (9.79)(86.49) = 847\ \text{kPa}$ (close enough). Hence, $Q = Q_c = 0.0138\ \text{m}^3/\text{s} = 13.8\ \text{L/s}$.

**11.24** For the piping system of Fig. 11-15, all pipes are concrete with roughness of 0.03 in. If the flow rate is 23 cfs of water at 20 °C, compute the pressure drop $p_1 - p_2$ and the three volumetric flows.

**▌ For pipe $a$:**

$$v_a = Q_a/A_a = 23/[(\pi)(\tfrac{12}{12})^2/4] = 29.28\ \text{ft/s} \qquad N_R = Dv/\nu$$

$$(N_R)_a = (\tfrac{12}{12})(29.28)/(1.08 \times 10^{-5}) = 2.71 \times 10^6 \qquad \epsilon/D = 0.03/12 = 0.0025$$

From Fig. A-5, $f_a = 0.0249$. $h_f = (f)(L/D)(v^2/2g)$, $(h_f)_a = 0.0249[1200/(\tfrac{12}{12})]\{29.28^2/[(2)(32.2)]\} = 397.8\ \text{ft}$.

**For pipes $b$, $c$, and $d$:**

$$(h_f)_b = (h_f)_c = (h_f)_d \qquad v_b = Q_b/A_b = Q_b/[(\pi)(\tfrac{8}{12})^2/4] = 2.865Q_b$$

$$v_c = Q_c/A_c = Q_c/[(\pi)(\tfrac{12}{12})^2/4] = 1.273Q_c \qquad v_d = Q_d/A_d = Q_d/[(\pi)(\tfrac{15}{12})^2/4] = 0.8149Q_d$$

$$f_b\left(\frac{1400}{\tfrac{8}{12}}\right)\left[\frac{(2.865Q_b)^2}{(2)(32.2)}\right] = f_c\left(\frac{900}{\tfrac{12}{12}}\right)\left[\frac{(1.273Q_c)^2}{(2)(32.2)}\right] = f_d\left(\frac{1300}{\tfrac{15}{12}}\right)\left[\frac{(0.8149Q_d)^2}{(2)(32.2)}\right] \tag{1}$$

If $f_b = f_c = f_d$,

$$Q_b = 0.2209Q_c \qquad Q_d = 1.453Q_c \qquad Q_b + Q_c + Q_d = 0.2209Q_c + Q_c + 1.453Q_c = 23 \qquad Q_c = 8.602\ \text{ft}^3/\text{s}$$

$$Q_b = (0.2209)(8.602) = 1.900\ \text{ft}^3/\text{s} \qquad Q_d = (1.453)(8.602) = 12.50\ \text{ft}^3/\text{s}$$

$$(N_R)_b = \tfrac{8}{12}[(2.865)(1.900)]/(1.08 \times 10^{-5}) = 3.36 \times 10^5 \qquad (\epsilon/D)_b = 0.03/8 = 0.0038 \qquad f_b = 0.028$$

$$(N_R)_c = \tfrac{12}{12}[(1.273)(8.602)]/(1.08 \times 10^{-5}) = 1.01 \times 10^6 \qquad (\epsilon/D)_c = 0.03/12 = 0.0025 \qquad f_c = 0.025$$

$$(N_R)_d = \tfrac{15}{12}[(0.8149)(12.50)]/(1.08 \times 10^{-5}) = 1.18 \times 10^6 \qquad (\epsilon/D)_d = 0.03/15 = 0.0020 \qquad f_d = 0.0235$$

Substituting these values of $f$ into Eq. (1),

$$0.028[1400/(\tfrac{8}{12})]\{(2.865Q_b)^2/[(2)(32.2)]\} = 0.025[900/(\tfrac{12}{12})]\{(1.273Q_c)^2/[(2)(32.2)]\}$$

$$= 0.0235[1300/(\tfrac{15}{12})]\{(0.8149Q_d)^2/[(2)(32.2)]\}$$

$$Q_b = 0.2749Q_c \qquad Q_d = 1.499Q_c \qquad Q_b + Q_c + Q_d = 0.2749Q_c + Q_c + 1.499Q_c = 23 \qquad Q_c = 8.292\ \text{ft}^3/\text{s}$$

$$Q_b = (0.2749)(8.292) = 2.279\ \text{ft}^3/\text{s} \qquad Q_d = (1.499)(8.292) = 12.43\ \text{ft}^3/\text{s}$$

$$(N_R)_b = \tfrac{8}{12}[(2.865)(2.279)]/(1.08 \times 10^{-5}) = 4.03 \times 10^5 \qquad f_b = 0.028 \qquad \text{(O.K.)}$$

$$(N_R)_c = \tfrac{12}{12}[(1.273)(8.292)]/(1.08 \times 10^{-5}) = 9.77 \times 10^5 \qquad f_c = 0.025 \qquad \text{(O.K.)}$$

$$(N_R)_d = \tfrac{15}{12}[(0.8149)(12.43)]/(1.08 \times 10^{-5}) = 1.17 \times 10^6 \qquad f_d = 0.0235 \qquad \text{(O.K.)}$$

Hence, $Q_b = 2.28 \text{ ft}^3/\text{s}$, $Q_c = 8.29 \text{ ft}^3/\text{s}$, and $Q_d = 12.43 \text{ ft}^3/\text{s}$.

$$(h_f)_b = (h_f)_c = (h_f)_d \qquad (h_f)_b = 0.028[1400/(\tfrac{8}{12})]\{[(2.865)(2.28)]^2/[(2)(32.2)]\} = 39.0 \text{ ft}$$

$$(h_f)_{\text{total}} = 397.8 + 39.0 = 436.8 \text{ ft} \qquad (\Delta p)_{\text{total}} = (\gamma)(h_f)_{\text{total}} = (62.4)(436.8) = 27\,256 \text{ lb/ft}^2 \quad \text{or} \quad 189 \quad \text{lb/in}^2$$

Fig. 11-15

**11.25** For the system of Fig. 11-15, with vegetable oil ($\nu = 4.62 \times 10^{-6} \text{ ft}^2/\text{s}$, $\gamma = 42.5 \text{ lb/ft}^3$) as the fluid, compute the flow rate in all pipes if the pressure drop $p_1 - p_2$ is 44 psi. Neglect minor losses.

▊ Using data from Prob. 11.24, note that pipe $a$ takes $397.8/436.8 = 0.9107$, or 91.07 percent of the total pressure drop. Hence, take $(\Delta p)_a = (44)(144)(0.9107) = 5770 \text{ lb/ft}^2$, $(h_f)_a = (\Delta p)_a/\gamma = 5770/42.5 = 135.8 \text{ ft}$, $h_f = (f)(L/d)(v^2/2g)$, $135.8 = f_a[1200/(\tfrac{12}{12})]\{v_a^2/[(2)(32.2)]\}$, $v_a = \sqrt{7.288/f_a}$. Try $f_a = 0.025$: $v_a = \sqrt{7.288/0.025} = 17.07 \text{ ft/s}$, $N_R = dv/\nu$, $(N_R)_a = (\tfrac{12}{12})(17.07)/(4.62 \times 10^{-6}) = 3.69 \times 10^6$, $(\epsilon/d)_a = 0.03/12 = 0.0025$. From Fig. A-5, $f_a = 0.0245$. Try $f_a = 0.0245$: $v_a = \sqrt{7.288/0.0245} = 17.25 \text{ ft/s}$, $(N_R)_a = (\tfrac{12}{12})(17.25)/(4.62 \times 10^{-6}) = 3.73 \times 10^6$, $f_a = 0.0245$ (O.K.); $Q_a = A_a v_a = [(\pi)(\tfrac{12}{12})^2/4](17.25) = 13.55 \text{ ft}^3/\text{s}$. This flow of 13.55 ft³/s for $Q_a$ will divide among the three parallel pipes in the same proportions as found in Prob. 11.24. Hence,

$$Q_b = [2.28/(2.28 + 8.29 + 12.43)](13.55) = 1.343 \text{ ft}^3/\text{s} \qquad Q_c = [8.29/(2.28 + 8.29 + 12.43)](13.55) = 4.884 \text{ ft}^3/\text{s}$$

$$Q_d = [12.43/(2.28 + 8.29 + 12.43)](13.55) = 7.323 \text{ ft}^3/\text{s} \qquad v_b = 1.343/[(\pi)(\tfrac{8}{12})^2/4] = 3.847 \text{ ft/s}$$

$$(\epsilon/d)_b = 0.03/8 = 0.0038 \qquad (N_R)_b = (\tfrac{8}{12})(3.847)/(4.62 \times 10^{-6}) = 5.55 \times 10^5 \qquad f_b = 0.028$$

$$(h_f)_{b,c,d} = 0.028[1400/(\tfrac{8}{12})]\{3.847^2/[(2)(32.2)]\} = 13.51 \text{ ft} \qquad (\Delta p)_{b,c,d} = (42.5)(13.51) = 574 \text{ lb/ft}^2$$

$$(\Delta p)_{\text{total}} = 5770 + 574 = 6344 \text{ lb/ft}^2 \quad \text{or} \quad 44.1 \text{ lb/in}^2 \quad \text{(O.K.)}$$

Hence, $Q_a = 13.55 \text{ cfs}$, $Q_b = 1.34 \text{ cfs}$, $Q_c = 4.88 \text{ cfs}$, and $Q_d = 7.32 \text{ cfs}$.

**11.26** In Fig. 11-16 all pipes are cast iron and $p_1 - p_2 = 49$ psi. Compute the total flow rate of water at 20 °C.

▊
$$h_f = (49)(144)/62.4 = 113.1 \text{ ft} = (f)(L/d)(v^2/2g)$$

**For pipe $c$:** $(h_f)_c = 113.1 = f_c[2000/(\tfrac{3}{12})]\{v_c^2/[(2)(32.2)]\}$, $v_c = \sqrt{0.9105/f_c}$. Try $f_c = 0.03$: $v_c = \sqrt{0.9105/0.03} = 5.509 \text{ ft/s}$, $N_R = dv/\nu$, $(N_R)_c = (\tfrac{3}{12})(5.509)/(1.08 \times 10^{-5}) = 1.28 \times 10^5$, $(\epsilon/d)_c = 0.00085/(\tfrac{3}{12}) = 0.0034$. From Fig. A-5, $f_c = 0.0283$. Try $f_c = 0.0283$: $v_c = \sqrt{0.9105/0.0283} = 5.672 \text{ ft/s}$, $(N_R)_c = (\tfrac{3}{12})(5.672)/(1.08 \times 10^{-5}) = 1.31 \times 10^5$, $f_c = 0.0283$ (O.K.); $Q_c = A_c v_c = [(\pi)(\tfrac{3}{12})^2/4](5.672) = 0.2784 \text{ ft}^3/\text{s}$.

**For pipes $a$ and $b$:** $(h_f)_{a,b} = 113.1 = f_a[1200/(\tfrac{2}{12})]\{v_a^2/[(2)(32.2)]\} + f_b[1300/(\tfrac{4}{12})]\{v_b^2/[(2)(32.2)]\}$. From continuity, $v_b = v_a/4$. $113.1 = f_a[1200/(\tfrac{2}{12})]\{v_a^2/[(2)(32.2)]\} + f_b[1300/(\tfrac{4}{12})]\{(v_a/4)^2/[(2)(32.2)]\}$, $7284 = (v_a^2)(7200f_a + 243.8f_b)$. Try $f_a = f_b = 0.03$:

$$7284 = (v_a^2)[(7200)(0.03) + (243.8)(0.03)] \qquad v_a = 5.711 \text{ ft/s} \qquad (N_R)_a = (\tfrac{2}{12})(5.711)/(1.08 \times 10^{-5}) = 8.81 \times 10^4$$

$$(\epsilon/d)_a = 0.00085/(\tfrac{2}{12}) = 0.0051 \qquad f_a = 0.031 \qquad v_b = 5.711/4 = 1.428 \text{ ft/s}$$

$$(N_R)_b = (\tfrac{4}{12})(1.428)/(1.08 \times 10^{-5}) = 4.41 \times 10^4$$

$$(\epsilon/d)_b = 0.00085/(\tfrac{4}{12}) = 0.0026 \qquad f_b = 0.028$$

Try $f_a = 0.031$ and $f_b = 0.028$:

$$7284 = (v_a^2)[(7200)(0.031) + (243.8)(0.028)] \qquad v_1 = 5.627 \text{ ft/s}$$

$$(N_R)_a = (\tfrac{2}{12})(5.627)/(1.08 \times 10^{-5}) = 8.68 \times 10^4 \qquad f_a = 0.031 \qquad \text{(O.K.)}$$

$$v_b = 5.627/4 = 1.407 \text{ ft/s} \qquad (N_R)_b = (\tfrac{4}{12})(1.407)/(1.08 \times 10^{-5}) = 4.34 \times 10^4 \qquad f_b = 0.028 \qquad \text{(O.K.)}$$

$$Q_a = [(\pi)(\tfrac{2}{12})^2/4](5.627) = 0.1228 \text{ ft}^3/\text{s} \qquad Q_{\text{total}} = 0.2784 + 0.1228 = 0.4012 \text{ ft}^3/\text{s}$$

**Fig. 11-16**

**11.27** Refer to Prob. 11.26. Given $Q_{\text{total}} = 0.7$ ft³/s, compute the pressure drop $p_1 - p_2$ and the division of flow between the two paths.

▌ The flow will divide between the two paths in the same proportions as found in Prob. 11.26. Hence,

$$Q_c = [0.2784/(0.2784 + 0.1228)](0.7) = 0.4857 \text{ ft}^3/\text{s}$$
$$Q_a = Q_b = [0.1228/(0.2784 + 0.1228)](0.7) = 0.2143 \text{ ft}^3/\text{s}$$
$$h_f = (f)(L/d)(v^2/2g) \qquad v_c = Q_c/A_c = 0.4857/[(\pi)(\tfrac{3}{12})^2/4] = 9.895 \text{ ft/s} \qquad N_R = dv/v$$
$$(N_R)_c = (\tfrac{3}{12})(9.895)/(1.08 \times 10^{-5}) = 2.29 \times 10^5 \qquad (\epsilon/d)_c = 0.00085/(\tfrac{3}{12}) = 0.0034$$

From Fig. A-5, $f_c = 0.028$. $(h_f)_c = 0.028[2000/(\tfrac{3}{12})]\{9.895^2/[(2)(32.2)]\} = 340.6$ ft, $p_1 - p_2 = (\gamma)(h_f)_c = (62.4)(340.6) = 21\,253$ lb/ft², or 147.6 lb/in².

**11.28** Two cisterns are connected by cast iron pipes as in Fig. 11-17. Neglecting minor losses, compute the flow rate in each pipe for water at 20 °C.

▌ Assume $(h_f)_c = 20$ ft.

$$\alpha = gd^3h_f/Lv^2 \qquad N_R = -\sqrt{8\alpha}\,\log\,[(\epsilon/d)/3.7 + 2.51/\sqrt{2\alpha}]$$
$$\alpha_c = (32.2)(\tfrac{4}{12})^3(20)/[(1300)(1.08 \times 10^{-5})^2] = 1.573 \times 10^8 \qquad (\epsilon/d)_c = 0.00085/(\tfrac{4}{12}) = 0.00255$$
$$(N_R)_c = -\sqrt{(8)(1.573 \times 10^8)}\,\log\,[0.00255/3.7 + 2.51/\sqrt{(2)(1.573 \times 10^8)}] = 1.093 \times 10^5 \qquad N_R = 4Q/\pi vd$$
$$1.093 \times 10^5 = 4Q_c/[(\pi)(1.08 \times 10^{-5})(\tfrac{4}{12})] \qquad Q_c = 0.3090 \text{ ft}^3/\text{s} \qquad (h_f)_a = (h_f)_b = 59 - 20 = 39 \text{ ft}$$
$$\alpha_a = (32.2)(\tfrac{3}{12})^3(39)/[(900)(1.08 \times 10^{-5})^2] = 1.869 \times 10^8 \qquad (\epsilon/d)_a = 0.00085/(\tfrac{3}{12}) = 0.00340$$
$$(N_R)_a = -\sqrt{(8)(1.869 \times 10^8)}\,\log\,[0.00340/3.7 + 2.51/\sqrt{(2)(1.869 \times 10^8)}] = 1.152 \times 10^5$$
$$1.152 \times 10^5 = 4Q_a/[(\pi)(1.08 \times 10^{-5})(\tfrac{3}{12})] \qquad Q_a = 0.2443 \text{ ft}^3/\text{s}$$
$$\alpha_b = (32.2)(\tfrac{2}{12})^3(39)/[(600)(1.08 \times 10^{-5})^2] = 8.308 \times 10^7 \qquad (\epsilon/d)_b = 0.00085/(\tfrac{2}{12}) = 0.00510$$
$$(N_R)_b = -\sqrt{(8)(8.308 \times 10^7)}\,\log\,[0.00510/3.7 + 2.51/\sqrt{(2)(8.308 \times 10^7)}] = 7.227 \times 10^4$$
$$7.227 \times 10^4 = 4Q_b/[(\pi)(1.08 \times 10^{-5})(\tfrac{2}{12})] \qquad Q_b = 0.1022 \text{ ft}^3/\text{s} \qquad Q_a + Q_b = 0.2443 + 0.1022 = 0.3465 \text{ ft}^3/\text{s}$$

Since this value of $Q_a + Q_b = 0.3465$ ft/s is not equal to $Q_c = 0.3090$ ft³/s, the assumed value of $(h_f)_c = 20$ ft is incorrect. Try $(h_f)_c = 23.3$ ft:

$$\alpha_c = (32.2)(\tfrac{4}{12})^3(23.3)/[(1300)(1.08 \times 10^{-5})^2] = 1.833 \times 10^8$$
$$(N_R)_c = -\sqrt{(8)(1.833 \times 10^8)}\,\log\,[0.00255/3.7 + 2.51/\sqrt{(2)(1.833 \times 10^8)}] = 1.182 \times 10^5$$
$$1.182 \times 10^5 = 4Q_c/[(\pi)(1.08 \times 10^{-5})(\tfrac{4}{12})] \qquad Q_c = 0.3341 \text{ ft}^3/\text{s}$$
$$(h_f)_a = (h_f)_b = 59 - 23.3 = 35.7 \text{ ft} \qquad \alpha_a = (32.2)(\tfrac{3}{12})^3(35.7)/[(900)(1.08 \times 10^{-5})^2] = 1.711 \times 10^8$$
$$(N_R)_a = -\sqrt{(8)(1.711 \times 10^8)}\,\log\,[0.00340/3.7 + 2.51/\sqrt{(2)(1.711 \times 10^8)}] = 1.101 \times 10^5$$
$$1.101 \times 10^5 = 4Q_a/[(\pi)(1.08 \times 10^{-5})(\tfrac{3}{12})] \qquad Q_a = 0.2336 \text{ ft}^3/\text{s}$$
$$\alpha_b = (32.2)(\tfrac{2}{12})^3(35.7)/[(600)(1.08 \times 10^{-5})^2] = 7.604 \times 10^7$$
$$(N_R)_b = -\sqrt{(8)(7.604 \times 10^7)}\,\log\,[0.00510/3.7 + 2.51/\sqrt{(2)(7.604 \times 10^7)}] = 6.908 \times 10^4$$
$$6.908 \times 10^4 = 4Q_b/[(\pi)(1.08 \times 10^{-5})(\tfrac{2}{12})] \qquad Q_b = 0.0977 \text{ ft}^3/\text{s} \qquad Q_a + Q_b = 0.2336 + 0.0977 = 0.3313 \text{ ft}^3/\text{s}$$

This value of $Q_a + Q_b = 0.3313$ ft³/s is practically the same as $Q_c = 0.3341$ ft³/s. Hence, take $Q_a = 0.235$ ft³/s, $Q_b = 0.098$ ft³/s, $Q_c = 0.333$ ft³/s.

**Fig. 11-17**

2 in, 600 ft     4 in, 1300 ft

**11.29** For the pipe system in Fig. 11-18, the pressure head at $A$ is 120.0 ft of water and the pressure head at $E$ is 72.0 ft of water. Assuming the pipes are in a horizontal plane, what are the flows in each branch of the loop? Neglect minor losses.

$\blacksquare$
$$h_f = 120.0 - 72.0 = 48.0 \text{ ft} \qquad Q = 1.318ACR^{0.63}s^{0.54}$$
$$Q_B = 1.318[(\pi)(\tfrac{12}{12})^2/4](100)[(\tfrac{12}{12})/4]^{0.63}(48.0/12\,000)^{0.54} = 2.19 \text{ ft}^3/\text{s}$$
$$Q_C = 1.318[(\pi)(\tfrac{8}{12})^2/4](100)[(\tfrac{8}{12})/4]^{0.63}(48.0/4000)^{0.54} = 1.37 \text{ ft}^3/\text{s}$$
$$Q_B = 1.318[(\pi)(\tfrac{10}{12})^2/4](100)[(\tfrac{10}{12})/4]^{0.63}(48.0/8000)^{0.54} = 1.69 \text{ ft}^3/\text{s}$$

**Fig. 11-18**

**11.30** In Fig. 11-18, if the total flow is 6.50 mgd, how much head loss occurs between $A$ and $E$ and how does $Q$ divide in the loop?

$\blacksquare$ The total flow will divide in the loop in the same proportions as in Prob. 11.29. Hence, $Q_B =$ $[2.19/(2.19 + 1.37 + 1.69)](6.50) = 2.71$ mgd, or 4.20 cfs; $Q_C = [1.37/(2.19 + 1.37 + 1.69)](6.50) = 1.70$ mgd, or 2.62 cfs; $Q_D = [1.69/(2.19 + 1.37 + 1.69)](6.50) = 2.09$ mgd, or 3.24 cfs; $Q = 1.318ACR^{0.63}s^{0.54}$. For pipe $B$, $4.20 = 1.318[(\pi)(\tfrac{12}{12})^2/4](100)[(\tfrac{12}{12})/4]^{0.63}s^{0.54}$, $s = 0.01334$; $(h_f)_B = (0.01334)(12\,000) = 160$ ft. This is the head loss between $A$ and $E$. It should, of course, be the same when computed by pipes $C$ and $D$. To confirm this, for pipe $C$, $2.62 = 1.318[(\pi)(\tfrac{8}{12})^2/4](100)[(\tfrac{8}{12})/4]^{0.63}s^{0.54}$, $s = 0.04010$; $(h_f)_C = (0.04010)(4000) = 160$ ft. For pipe $D$, $3.24 = 1.318[(\pi)(\tfrac{10}{12})^2/4](100)[(\tfrac{10}{12})/4]^{0.63}s^{0.54}$, $s = 0.02005$; $(h_f)_B = (0.02005)(8000) = 160$ ft.

**11.31** For the system shown in Fig. 11-19, what flow will occur when the drop in the hydraulic grade line from $A$ to $B$ is 200 ft?

$\blacksquare$ Assume $(h_f)_{WZ} = 30$ ft. Using Fig. A-13: For pipe 2, with $d = 12$ in and $h_1 = \frac{30}{5000} = 0.0060$, $Q_2 = 3.2$ ft$^3$/s. For pipe 3, with $d = 16$ in and $h_1 = \frac{30}{3000} = 0.0100$, $Q_3 = 9.0$ ft$^3$/s. For pipe 1, with $d = 24$ in and $Q = 3.2 + 9.0 = 12.2$ ft$^3$/s, $(h_1)_1 = 0.0024$ ft/ft; $(h_f)_{AW} = (0.0024)(10\,000) = 24.0$ ft. For pipe 4, with $d = 20$ in and $Q = 12.2$ ft$^3$/s, $(h_1)_4 = 0.0060$ ft/ft; $(h_f)_{ZB} = (0.0060)(8000) = 48.0$ ft, $(h_f)_{A-B} = 24.0 + 30 + 48.0 = 102.0$ ft. This value of $(h_f)_{A-B} = 24.0 + 30 + 48.0 = 102.0$ ft is not equal to the given value of 200 ft, but the actual head losses will be in the same proportions as those above. Hence, $(h_f)_{AW} = (24.0/102.0)(200) = 47.06$ ft. For pipe 1, with $d = 24$ in and $h_1 = 47.06/10\,000 = 0.0047$, $Q = 18$ cfs.

**Fig. 11-19**

**11.32** In Prob. 11.31, what length of 20-in pipe with $C = 120$ is equivalent to section $A–B$?

$\blacksquare$ From Prob. 11.31, it is known that a flow of 18 cfs is produced when the drop in the hydraulic grade line from $A$ to $B$ is 200 ft. For 18 cfs in a 20-in pipe, from Fig. A-13, $h_1 = 0.012$ ft/ft, $0.012L_e = 200$, $L_e = 16\,700$ ft.

**11.33**  In Fig. 11-20, when pump $YA$ delivers 5.00 cfs, find the pressure heads at $A$ and $B$. Draw the hydraulic grade lines.

**▌** Reduce loop $BC$ to an equivalent pipe, 16 in in diameter, $C = 100$. By so doing, a single-size pipe of the same relative roughness is readily handled for all conditions of flow. Assume a drop in the grade line of 22 ft ftom $B$ to $C$. $Q = 1.318AR^{0.63}s^{0.54}$.

**For the 10-in pipe:** $Q_{10} = 1.318[(\pi)(\frac{10}{12})^2/4](90)[(\frac{10}{12})/4]^{0.63}(22/10\,000)^{0.54} = 0.8843$ cfs.

**For the 8-in pipe:** $Q_8 = 1.318[(\pi)(\frac{8}{12})^2/4](100)[(\frac{8}{12})/4]^{0.63}(22/11\,000)^{0.54} = 0.5190$ cfs. $Q_{\text{total}} = 0.8843 + 0.5190 = 1.403$ cfs. For a 16-in-diameter equivalent pipe with $C = 100$, $1.403 = 1.318[(\pi)(\frac{16}{12})^2/4](100)[(\frac{16}{12})/4]^{0.63}s^{0.54}$, $s = 0.0004313$ ft/ft; $0.0004313L_e = 22$, $L_e = 51\,000$ ft. For a 16-in-diameter pipe from $A$ to $C$ with a length of $16\,000 + 51\,000 = 67\,000$ ft and carrying 5.00 cfs, $5.00 = 1.318[(\pi)(\frac{16}{12})^2/4](100)[(\frac{16}{12})/4]^{0.63}s^{0.54}$, $s = 0.004537$ ft/ft; $h_f = (0.004537)(67\,000) = 304.0$ ft. Thus the elevation of the hydraulic grade line at $A$ is $217.0 + 304.0 = 521.0$ ft, as shown in the figure. The drop from $A$ to $B$ is $(0.004537)(16\,000) = 72.6$ ft and the elevation of the hydraulic grade line at $B$ becomes $521.0 - 72.6 = 448.4$ ft. Pressure head at $A = 521.0 - 50.0 = 471.0$ ft, pressure head at $B = 448.4 - 50.0 = 398.4$ ft.

**Fig. 11-20**

**11.34**  In Fig. 11-21, which system has the greater capacity, $ABCD$ or $EFGH$? Use $C = 120$ for all pipes.

**▌** Assume $Q = 3$ cfs in $ABCD$. Using Fig. A-13,

$$(h_1)_{AB} = 0.0014 \text{ ft/ft} \qquad (h_f)_{AB} = (0.0014)(9000) = 12.6 \text{ ft}$$
$$(h_1)_{BC} = 0.0053 \text{ ft/ft} \qquad (h_f)_{BC} = (0.0053)(6000) = 31.8 \text{ ft}$$
$$(h_1)_{CD} = 0.013 \text{ ft/ft} \quad (h_f)_{AB} = (0.013)(3000) = 39.0 \text{ ft} \qquad (h_f)_{\text{total}} = 12.6 + 31.8 + 39.0 = 83.4 \text{ ft} \quad \text{(for } ABCD\text{)}$$

For $EFGH$, assume $(h_f)_{FG} = 24$ ft.

$$(h_1)_{FIG} = \tfrac{24}{5000} = 0.00480 \text{ ft/ft} \qquad (h_1)_{FJG} = \tfrac{24}{7000} = 0.00343 \text{ ft/ft} \qquad Q_{FIG} = 0.97 \text{ cfs} \qquad Q_{FJG} = 1.6 \text{ cfs}$$

Hence, pipe $FIG$ carries $0.97/(0.97 + 1.6) = 0.3774$, or 37.74 percent of the flow and pipe $FJG$ carries $1.6/(0.97 + 1.6) = 0.6226$, or 62.26 percent. For $Q = 3.0$ cfs in pipe $EF$,

$$(h_1)_{EF} = 0.00074 \text{ ft/ft} \qquad (h_f)_{EF} = (0.00074)(11\,000) = 8.1 \text{ ft} \qquad Q_{FIG} = (0.3774)(3.0) = 1.13 \text{ cfs}$$
$$(h_1)_{FIG} = 0.0060 \text{ ft/ft} \qquad (h_f)_{FIG} = (0.0060)(5000) = 30.0 \qquad \text{ft}$$

For $Q = 3.0$ cfs in pipe $GH$,

$$(h_1)_{GH} = 0.013 \text{ ft/ft} \qquad (h_f)_{GH} = (0.013)(2500) = 32.5 \text{ ft} \qquad (h_f)_{\text{total}} = 8.1 + 30.0 + 32.5 = 70.6 \text{ ft} \quad \text{(for } EFGH\text{)}$$

Since $EFGH$ carries the assumed flow of 3.0 cfs with a lesser head loss than $ABCD$, it ($EFGH$) has the greater capacity.

**Fig. 11-21**

**11.35** Three pipes *A*, *B*, and *C* are interconnected as shown in Fig. 11-22. The pipe characteristics are as follows:

| pipe | $D$, in | $L$, ft | $f$ |
|------|---------|---------|-------|
| $A$ | 6 | 2000 | 0.020 |
| $B$ | 4 | 1600 | 0.032 |
| $C$ | 8 | 4000 | 0.024 |

Find the rate at which water will flow in each pipe. Find also the pressure at point *P*. Neglect minor losses.

▌

$$p_1/\gamma + v_1^2/2g + z_1 = p_2/\gamma + v_2^2/2g + z_2 + h_L \tag{1}$$

$h_L = h_f = (f)(L/d)(v^2/2g) = 0.020[2000/(\tfrac{6}{12})]\{v_A^2/[(2)(32.2)]\}$

$$+ 0.024[4000/(\tfrac{8}{12})]\{v_C^2/[(2)(32.2)]\} = 1.242v_A^2 + 2.236v_C^2$$

$$0 + 0 + 200 = 0 + v_C^2/[(2)(32.2)] + 50 + 1.242v_A^2 + 2.236v_C^2$$

$$150 = 1.242v_A^2 + 2.252v_C^2 \tag{2}$$

$$Q_A + Q_B = Q_C \qquad [(\pi)(\tfrac{6}{12})^2/4](v_A) + [(\pi)(\tfrac{4}{12})^2/4](v_B) = [(\pi)(\tfrac{8}{12})^2/4](v_C)$$

$$0.1963v_A + 0.08727v_B = 0.3491v_C \tag{3}$$

$(h_f)_A = (h_f)_B \qquad 0.020[2000/(\tfrac{6}{12})]\{v_A^2/[(2)(32.2)]\} = 0.032[1600/(\tfrac{4}{12})]\{v_B^2/[(2)(32.2)]\} \qquad v_B = 0.7217v_A$

Substituting into Eq. (3), $0.1963v_A + (0.08727)(0.7217v_A) = 0.3491v_C$, $v_A = 1.346v_C$. Substituting into Eq. (2), $150 = (1.242)(1.346v_C)^2 + 2.252v_C^2$, $v_C = 5.772$ ft/s; $Q_C = A_Cv_C = [(\pi)(\tfrac{8}{12})^2/4](5.772) = 2.01$ ft³/s, $v_A = (1.346)(5.772) = 7.769$ ft/s, $Q_A = [(\pi)(\tfrac{6}{12})^2/4](7.769) = 1.53$ ft³/s. Substituting into Eq. (3), $(0.1963)(7.769) + 0.08727v_B = (0.3491)(5.772)$, $v_B = 5.614$ ft/s; $Q_B = [(\pi)(\tfrac{4}{12})^2/4](5.614) = 0.490$ ft³/s. To find $p_P$, apply Eq. (1) between points *P* and 2. $(p_P)(144)/62.4 + v_P^2/2g + 120 = 0 + v_2^2/2g + 50 + 0.024[4000/(\tfrac{8}{12})]\{5.772^2/[(2)(32.2)]\}$, $v_P^2/2g = v_2^2/2g$, $p_P = 1.95$ lb/in².

**Fig. 11-22**

**11.36** Figure 11-23 shows an element of a sugar refinery. Smooth brass pipes 1, 2, and 3 are 550 ft of 2-in diameter, 350 ft of 3-in diameter, and 600 ft of 4-in diameter, respectively. If the net flow of molasses ($v = 6.5 \times 10^{-5}$ ft²/s) is 0.7 ft³/s, find the head loss from *A* to *B* and the flow in each pipe.

▌

$$h_f = (f)(L/d)(v^2/2g) \qquad (h_f)_1 = (h_f)_2 = (h_f)_3$$

$$f_1[550/(\tfrac{2}{12})]\{v_1^2/[(2)(32.2)]\} = f_2[350/(\tfrac{3}{12})]\{v_2^2/[(2)(32.2)]\} = f_3[600/(\tfrac{4}{12})]\{v_3^2/[(2)(32.2)]\} \tag{1}$$

If $f_1 = f_2 = f_3$,

$$51.24v_1^2 = 21.74v_2^2 = 27.95v_3^2 \qquad v_2 = 1.535v_1 \qquad v_3 = 1.354v_1 \qquad Q_{total} = Q_1 + Q_2 + Q_3 = A_1v_1 + A_2v_2 + A_3v_3$$

$$0.7 = [(\pi)(\tfrac{2}{12})^2/4](v_1) + [(\pi)(\tfrac{3}{12})^2/4](1.535v_1) + [(\pi)(\tfrac{4}{12})^2/4](1.354v_1)$$

$$v_1 = 3.251 \text{ ft/s} \qquad v_2 = (1.535)(3.251) = 4.990 \text{ ft/s} \qquad v_3 = (1.354)(3.251) = 4.402 \text{ ft/s}$$

$N_R = dv/v$, $(N_R)_1 = (\tfrac{2}{12})(3.251)/(6.5 \times 10^{-5}) = 8336$, $(N_R)_2 = (\tfrac{3}{12})(4.990)/(6.5 \times 10^{-5}) = 19\,192$, $(N_R)_3 = (\tfrac{4}{12})(4.402)/(6.5 \times 10^{-5}) = 22\,547$. For Reynolds numbers in this range the Blasius formula can be used to determine friction factors: $f = 0.316/N_R^{0.25}$.

$$f_1 = 0.316/8336^{0.25} = 0.0331 \qquad f_2 = 0.316/19\,192^{0.25} = 0.0268 \qquad f_3 = 0.316/22\,547^{0.25} = 0.0258$$

Substituting these values of *f* into Eq. (1),

$$0.0331[550/(\tfrac{2}{12})]\{v_1^2/[(2)(32.2)]\} = 0.0268[350/(\tfrac{3}{12})]\{v_2^2/[(2)(32.2)]\} = 0.0258[600/(\tfrac{4}{12})]\{v_3^2/[(2)(32.2)]\}$$

$$1.696v_1^2 = 0.5826v_2^2 = 0.7211v_3^2 \qquad v_2 = 1.706v_1 \qquad v_3 = 1.534v_1$$

$$0.7 = [(\pi)(\tfrac{2}{12})^2/4](v_1) + [(\pi)(\tfrac{3}{12})^2/4](1.706v_2) + [(\pi)(\tfrac{4}{12})^2/4](1.534v_1)$$

$$v_1 = 2.918 \text{ ft/s} \qquad v_2 = (1.706)(2.918) = 4.978 \text{ ft/s} \qquad v_3 = (1.534)(2.918) = 4.476 \text{ ft/s}$$

$$(N_R)_1 = (\tfrac{2}{12})(2.918)/(6.5 \times 10^{-5}) = 7482 \qquad (N_R)_2 = (\tfrac{3}{12})(4.978)/(6.5 \times 10^{-5}) = 19\,146$$

$$(N_R)_3 = (\tfrac{4}{12})(4.476)/(6.5 \times 10^{-5}) = 22\,954$$

$$f_1 = 0.316/7482^{0.25} = 0.0340 \qquad f_2 = 0.316/19\,146^{0.25} = 0.0269 \qquad f_3 = 0.316/22\,954^{0.25} = 0.0257$$

$$0.0340[550/(\tfrac{2}{12})]\{v_1^2/[(2)(32.2)]\} = 0.0269[350/(\tfrac{3}{12})]\{v_2^2/[(2)(32.2)]\}$$

$$= 0.0257[600/(\tfrac{4}{12})]\{v_3^2/[(2)(32.2)]\}$$

$$1.742v_1^2 = 0.5848v_2^2 = 0.7183v_3^2 \qquad v_2 = 1.726v_1 \qquad v_3 = 1.557v_1$$

$$0.7 = [(\pi)(\tfrac{2}{12})^2/4](v_1) + [(\pi)(\tfrac{3}{12})^2/4](1.726v_1) + [(\pi)(\tfrac{4}{12})^2/4](1.557v_1)$$

$$v_1 = 2.888 \text{ ft/s} \qquad v_2 = (1.726)(2.888) = 4.985 \text{ ft/s} \qquad v_3 = (1.557)(2.888) = 4.497 \text{ ft/s}$$

$$(N_R)_1 = (\tfrac{2}{12})(2.888)/(6.5 \times 10^{-5}) = 7405 \qquad (N_R)_2 = (\tfrac{3}{12})(4.985)/(6.5 \times 10^{-5}) = 19\,173$$

$$(N_R)_3 = (\tfrac{4}{12})(4.497)/(6.5 \times 10^{-5}) = 23\,062$$

$$f_1 = 0.316/7405^{0.25} = 0.0341 \quad \text{(O.K.)} \qquad f_2 = 0.316/19\,173^{0.25} = 0.0269 \quad \text{(O.K.)}$$

$$f_3 = 0.316/23\,062^{0.25} = 0.0256 \quad \text{(O.K.)}$$

$$Q_1 = [(\pi)(\tfrac{2}{12})^2/4](2.888) = 0.0630 \text{ ft}^3/\text{s} \qquad Q_2 = [(\pi)(\tfrac{3}{12})^2/4](4.985) = 0.245 \text{ ft}^3/\text{s}$$

$$Q_3 = [(\pi)(\tfrac{4}{12})^2/4](4.497) = 0.392 \text{ ft}^3/\text{s}$$

$$(h_f)_{AB} = (h_f)_1 = 0.0341[550/(\tfrac{2}{12})]\{2.888^2/[(2)(32.2)]\} = 14.6 \text{ ft}$$

**Fig. 11-23**

**11.37** Repeat Prob. 11.36 for a flow rate of 0.07 ft³/s.

▌ Assume this flow will be laminar. $h_f = (32v)(L/gD^2)(v)$. Since $(h_f)_1 = (h_f)_2 = (h_f)_3$,

$$L_1v_1/D_1^2 = L_2v_2/D_2^2 = L_3v_3/D_3^2 \qquad 550v_1/(\tfrac{2}{12})^2 = 350v_2/(\tfrac{3}{12})^2 = 600v_3/(\tfrac{4}{12})^2 \qquad v_2 = 3.536v_1 \qquad v_3 = 3.667v_1$$

$$Q_{\text{total}} = Q_1 + Q_2 + Q_3 = A_1v_1 + A_2v_2 + A_3v_3$$

$$0.07 = [(\pi)(\tfrac{2}{12})^2/4](v_1) + [(\pi)(\tfrac{3}{12})^2/4](3.536v_1) + [(\pi)(\tfrac{4}{12})^2/4](3.667v_1)$$

$$v_1 = 0.1358 \text{ ft/s} \qquad v_2 = (3.536)(0.1358) = 0.4802 \text{ ft/s} \qquad v_3 = (3.667)(0.1358) = 0.4980 \text{ ft/s} \qquad N_R = dv/v$$

From Fig. A-2, $v = 6.5 \times 10^{-5}$ ft²/s;

$$(N_R)_1 = (\tfrac{2}{12})(0.1358)/(6.5 \times 10^{-5}) = 348 \quad \text{(laminar)}$$

$$(N_R)_2 = (\tfrac{3}{12})(0.4802)/(6.5 \times 10^{-5}) = 1847 \quad \text{(laminar)}$$

$$(N_R)_3 = (\tfrac{4}{12})(0.4980)/(6.5 \times 10^{-5}) = 2554 \quad \text{(laminar)}$$

$$Q_1 = [(\pi)(\tfrac{2}{12})^2/4](0.1358) = 0.00296 \text{ ft}^3/\text{s} \qquad Q_2 = [(\pi)(\tfrac{3}{12})^2/4](0.4802) = 0.0236 \text{ ft}^3/\text{s}$$

$$Q_3 = [(\pi)(\tfrac{4}{12})^2/4](0.4980) = 0.0435 \text{ ft}^3/\text{s}$$

$$(h_f)_{AB} = (h_f)_1 = (32)(6.5 \times 10^{-5})\{550/[(32.2)(\tfrac{2}{12})^2]\}(0.1358) = 0.174 \text{ ft}$$

**11.38** Rework Prob. 11.36 if the parameters are changed to: 90 m of 30-mm (pipe 1), 120 m of 40-mm (pipe 2), 90 m of 50-mm (pipe 3); $v = 4.8 \times 10^{-6}$ m²/s; 6 L/s net flow.

▮ Assume Reynolds numbers between 3000 and 100 000, in which case the Blasius equation can be used to determine $f$: $h_f = (f)(L/d)(v^2/2g)$, $f = 0.316/N_R^{0.25}$, $N_R = dv/\nu$. Therefore,

$$h_f = \left[\frac{0.316}{(dv/\nu)^{0.25}}\right]\left(\frac{L}{d}\right)\left(\frac{v^2}{2g}\right) = \frac{0.316\nu^{0.25}Lv^{1.75}}{2gd^{1.25}} \qquad h_f \propto \frac{Lv^{1.75}}{d^{1.25}} \qquad (h_f)_1 = (h_f)_2 = (h_f)_3$$

$$(90)(v_1^{1.75})/(0.030)^{1.25} = (120)(v_2^{1.75})/(0.040)^{1.25} = (90)(v_3^{1.75})/(0.050)^{1.25} \qquad v_1 = 0.9597v_2 \qquad v_3 = 1.382v_2$$

$$Q_{\text{total}} = Q_1 + Q_2 + Q_3 = A_1v_1 + A_2v_2 + A_3v_3$$

$$6 \times 10^{-3} = [(\pi)(0.030)^2/4](0.9597v_2) + [(\pi)(0.040)^2/4](v_2) + [(\pi)(0.050)^2/4](1.382v_2)$$

$$v_2 = 1.291 \text{ m/s} \qquad v_1 = (0.9597)(1.291) = 1.239 \text{ m/s} \qquad v_3 = (1.382)(1.291) = 1.784 \text{ m/s}$$

$$(N_R)_1 = (0.030)(1.239)/(4.8 \times 10^{-6}) = 7.74 \times 10^3 \qquad (N_R)_2 = (0.040)(1.291)/(4.8 \times 10^{-6}) = 1.08 \times 10^4$$

$$(N_R)_3 = (0.050)(1.784)/(4.8 \times 10^{-6}) = 1.86 \times 10^4$$

Hence, use of the Blasius equation is O.K.

$$Q_1 = [(\pi)(0.030)^2/4](1.239) = 0.00088 \text{ m}^3/\text{s} \quad \text{or} \quad 0.88 \text{ L/s}$$

$$Q_2 = [(\pi)(0.040)^2/4](1.291) = 0.00162 \text{ m}^3/\text{s} \quad \text{or} \quad 1.62 \text{ L/s}$$

$$Q_3 = [(\pi)(0.050)^2/4](1.784) = 0.00350 \text{ m}^3/\text{s} \quad \text{or} \quad 3.50 \text{ L/s}$$

$$(h_f)_{AB} = (h_f)_1 = \frac{(0.316)(4.8 \times 10^{-6})^{0.25}(90)(1.239)^{1.75}}{(2)(9.807)(0.030)^{1.25}} = 7.91 \text{ m}$$

**11.39** Repeat Prob. 11.38 for a net flow of 0.5 L/s; other parameters as in Prob. 11.38.

▮ Assume laminar flow, in which case the following equation can be used to determine head loss: $h_f = (32\nu)(L/gd^2)(v)$, $h_f \propto Lv/d^2$.

$$(h_f)_1 = (h_f)_2 = (h_f)_3 \qquad 90v_1/(0.030)^2 = 120v_2/(0.040)^2 = 90v_3/(0.050)^2 \qquad v_1 = 0.7500v_2 \qquad v_3 = 2.083v_2$$

$$Q_{\text{total}} = Q_1 + Q_2 + Q_3 = A_1v_1 + A_2v_2 + A_3v_3$$

$$0.5 \times 10^{-3} = [(\pi)(0.030)^2/4](0.7500v_2) + [(\pi)(0.040)^2/4](v_2) + [(\pi)(0.050)^2/4](2.083v_2)$$

$$v_2 = 0.08508 \text{ m/s} \qquad v_1 = (0.7500)(0.08508) = 0.06381 \text{ m/s} \qquad v_3 = (2.083)(0.08508) = 0.1772 \text{ m/s}$$

$$(N_R)_1 = (0.030)(0.06381)/(4.8 \times 10^{-6}) = 399 \qquad \text{(laminar)}$$

$$(N_R)_2 = (0.040)(0.08508)/(4.8 \times 10^{-6}) = 709 \qquad \text{(laminar)}$$

$$(N_R)_3 = (0.050)(0.1772)/(4.8 \times 10^{-6}) = 1846 \qquad \text{(laminar)}$$

$$Q_1 = [(\pi)(0.030)^2/4](0.06381) = 0.0000451 \text{ m}^3/\text{s} \quad \text{or} \quad 0.0451 \text{ L/s}$$

$$Q_2 = [(\pi)(0.040)^2/4](0.08508) = 0.0001069 \text{ m}^3/\text{s} \quad \text{or} \quad 0.1069 \text{ L/s}$$

$$Q_3 = [(\pi)(0.050)^2/4](0.1772) = 0.0003479 \text{ m}^3/\text{s} \quad \text{or} \quad 0.3479 \text{ L/s}$$

$$(h_f)_{AB} = (h_f)_1 = (32)(4.8 \times 10^{-6})\{90/[(9.807)(0.030)^2]\}(0.06381) = 0.0999 \text{ m} \quad \text{or} \quad 1 \text{ cm}$$

**11.40** Figure 11-24 shows a network of cast iron pipes. Given a flow of 24 cfs, find the head loss from $A$ to $D$. Neglect minor losses.

▮ $Q_n = A_n\sqrt{2gh_L/(f_n)(L_n/d_n)}$ or $Q_n = C_n\sqrt{h_L}$, where $C_n = A_n\sqrt{2gd_n/f_nL_n}$. Assume $f_1 = f_2 = f_3 = 0.019$.

$$C_2 = [(\pi)(\tfrac{14}{12})^2/4]\sqrt{(2)(32.2)(\tfrac{14}{12})/[(0.019)(3300)]} = 1.170$$

$$C_3 = [(\pi)(\tfrac{12}{12})^2/4]\sqrt{(2)(32.2)(\tfrac{12}{12})/[(0.019)(2600)]} = 0.8967$$

$$C_4 = [(\pi)(\tfrac{16}{12})^2/4]\sqrt{(2)(32.2)(\tfrac{16}{12})/[(0.019)(3200)]} = 1.659$$

$$Q_2 = [1.170/(1.170 + 0.8967 + 1.659)](24) = 7.537 \text{ ft}^3/\text{s}$$

$$Q_3 = [0.8967/(1.170 + 0.8967 + 1.659)](24) = 5.776 \text{ ft}^3/\text{s}$$

$$Q_4 = [1.659/(1.170 + 0.8967 + 1.659)](24) = 10.687 \text{ ft}^3/\text{s}$$

$$v_2 = Q_2/A_2 = 7.537/[(\pi)(\tfrac{14}{12})^2/4] = 7.050 \text{ ft/s} \qquad v_3 = 5.776/[(\pi)(\tfrac{12}{12})^2/4] = 7.354 \text{ ft/s}$$

$$v_4 = 10.687/[(\pi)(\tfrac{16}{12})^2/4] = 7.654 \text{ ft/s} \qquad N_R = dv/\nu$$

$$(N_R)_2 = (\tfrac{14}{12})(7.050)/(1.05 \times 10^{-5}) = 7.83 \times 10^5$$

$$(N_R)_3 = (\tfrac{12}{12})(7.354)/(1.05 \times 10^{-5}) = 7.00 \times 10^5$$

$$(N_R)_4 = (\tfrac{16}{12})(7.654)/(1.05 \times 10^{-5}) = 9.72 \times 10^5$$

$$(\epsilon/d)_2 = 0.00085/(\tfrac{14}{12}) = 0.00073 \qquad (\epsilon/d)_3 = 0.00085/(\tfrac{12}{12}) = 0.00085 \qquad (\epsilon/d)_4 = 0.00085/(\tfrac{16}{12}) = 0.00064$$

From Fig. A-5, $f_2 = 0.0185$, $f_3 = 0.0195$, and $f_4 = 0.0185$.

$$C_2 = [(\pi)(\tfrac{14}{12})^2/4]\sqrt{(2)(32.2)(\tfrac{14}{12})/[(0.0185)(3300)]} = 1.186$$

$$C_3 = [(\pi)(\tfrac{12}{12})^2/4]\sqrt{(2)(32.2)(\tfrac{12}{12})/[(0.0195)(2600)]} = 0.8852$$

$$C_4 = [(\pi)(\tfrac{16}{12})^2/4]\sqrt{(2)(32.2)(\tfrac{16}{12})/[(0.0185)(3200)]} = 1.585$$

$$Q_2 = [1.186/(1.186 + 0.8852 + 1.585)](24) = 7.785 \text{ ft}^3/\text{s}$$

$$Q_3 = [0.8852/(1.186 + 0.8852 + 1.585)](24) = 5.811 \text{ ft}^3/\text{s}$$

$$Q_4 = [1.585/(1.186 + 0.8852 + 1.585)](24) = 10.404 \text{ ft}^3/\text{s}$$

$$v_2 = Q_2/A_2 = 7.785/[(\pi)(\tfrac{14}{12})^2/4] = 7.282 \text{ ft/s} \qquad v_3 = 5.811/[(\pi)(\tfrac{12}{12})^2/4] = 7.399 \text{ ft/s}$$

$$v_4 = 10.404/[(\pi)(\tfrac{16}{12})^2/4] = 7.451 \text{ ft/s}$$

$$(N_R)_2 = (\tfrac{14}{12})(7.282)/(1.05 \times 10^{-5}) = 8.09 \times 10^5$$

$$(N_R)_3 = (\tfrac{12}{12})(7.399)/(1.05 \times 10^{-5}) = 7.05 \times 10^5$$

$$(N_R)_4 = (\tfrac{16}{12})(7.451)/(1.05 \times 10^{-5}) = 9.46 \times 10^5$$

$$f_2 = 0.0185 \qquad f_3 = 0.0195 \quad \text{and} \quad f_4 = 0.0185 \quad \text{(O.K.)}$$

$$Q_{\text{total}} = \sqrt{h_L}(C_2 + C_3 + C_4) \qquad 24 = \sqrt{(h_L)_{BC}}(1.186 + 0.8852 + 1.585)$$

$$(h_L)_{BC} = 43.09 \text{ ft} \qquad h_L = h_f = (f)(L/d)(v^2/2g)$$

$$v_1 = 24/[(\pi)(\tfrac{24}{12})^2/4] = 7.639 \text{ ft/s} \qquad (N_R)_1 = (\tfrac{24}{12})(7.639)/(1.05 \times 10^{-5}) = 1.46 \times 10^6$$

$$(\epsilon/d)_1 = 0.00085/(\tfrac{24}{12}) = 0.00042 \qquad f_1 = 0.0168$$

$$(h_L)_1 = 0.0168[3300/(\tfrac{24}{12})]\{7.639^2/[(2)(32.2)]\} = 25.12 \text{ ft} \qquad v_5 = 24/[(\pi)(\tfrac{30}{12})^2/4] = 4.889 \text{ ft/s}$$

$$(N_R)_5 = (\tfrac{30}{12})(4.889)/(1.05 \times 10^{-5}) = 1.16 \times 10^6 \qquad (\epsilon/d)_5 = 0.00085/(\tfrac{30}{12}) = 0.00034 \qquad f_5 = 0.016$$

$$(h_L)_5 = 0.016[5200/(\tfrac{30}{12})]\{4.889^2/[(2)(32.2)]\} = 12.35 \text{ ft} \qquad (h_L)_{AD} = 25.12 + 43.09 + 12.35 = 80.56 \text{ ft}$$

Fig. 11-24

**11.41** Find the diameter of the series equivalent of the three parallel pipes in Prob. 11.40.

$\blacksquare$ $h_f = (f)(L/d)(v^2/2g)$. Assume $f = 0.016$:

$$v = Q/A = 24/(\pi d^2/4) = 30.56/d^2 \qquad 43.09 = (0.016)(2600/d)\{(30.56/d^2)^2/[(2)(32.2)]\} \qquad d = 1.695 \text{ ft}$$

$$N_R = dv/v = (1.695)(30.56/1.695^2)/(1.05 \times 10^{-5}) = 1.72 \times 10^6 \qquad \epsilon/d = 0.00085/1.707 = 0.00050$$

From Fig. A-5, $f = 0.017$. Try $f = 0.017$:

$$43.09 = (0.017)(2600/d)\{(30.56/d^2)^2/[(2)(32.2)]\} \qquad d = 1.716 \text{ ft}$$

$$N_R = (1.716)(30.56/1.716^2)/(1.05 \times 10^{-5}) = 1.70 \times 10^6 \qquad f = 0.017 \quad \text{(O.K.)}$$

Therefore, $d = 1.716 \text{ ft} = 20.6 \text{ in}$.

**11.42** For the network of Fig. 11-24, determine the five pipe flows, given the head loss from $A$ to $D$ is 91 ft and all pipes have $f = 0.017$.

$\blacksquare$ $$h_f = (f)(L/d)(v^2/2g) = (fL/d)(Q^2/2gA^2) = (fL/2gd)[Q/(\pi d^2/4)]^2 = (8f/\pi^2 g)(LQ^2/d^5)$$

Hence,

$$h_f \propto LQ^2/d^5 \qquad (h_f)_2 = (h_f)_3 = (h_f)_4 \qquad 3300Q_2^2/(\tfrac{14}{12})^5 = 2600Q_3^2/(\tfrac{12}{12})^5 = 3200Q_4^2/(\tfrac{16}{12})^5$$

$$Q_2 = 1.305Q_3 \qquad Q_4 = 1.850Q_3 \qquad Q_1 = Q_5 = Q_2 + Q_3 + Q_4 = 1.305Q_3 + Q_3 + 1.850Q_3 = 4.155Q_3$$

$$(h_f)_{AD} = (h_f)_1 + (h_f)_3 + (h_f)_5$$

$$91 = \{(8)(0.017)/[(\pi)^2(32.2)]\}[(3300)(4.155Q_3)^2/(\tfrac{24}{12})^5 + (2600)(Q_3)^2/(\tfrac{12}{12})^5 + (5200)(4.155Q_3)^2/(\tfrac{30}{12})^5]$$

$$Q_3 = 6.33 \text{ ft}^3/\text{s} \qquad Q_2 = (1.305)(6.33) = 8.26 \text{ ft}^3/\text{s}$$

$$Q_4 = (1.850)(6.33) = 11.71 \text{ ft}^3/\text{s} \qquad Q_1 = Q_5 = (4.155)(6.33) = 26.30 \text{ ft}^3/\text{s}$$

**11.43** In Prob. 11.42, what percentage increase in the capacity of the system would be achieved by adding another 12-in pipe 2600 ft long between $B$ and $C$?

▮ The additional pipe from $B$ to $C$ is identical to pipe 3. As in Prob. 11.42,

$$h_f = (8f/\pi^2 g)(LQ^2/d^5) \qquad Q_2 = 1.305Q_3 \qquad Q_4 = 1.850Q_3$$

$$Q_1 = Q_5 = Q_2 + 2Q_3 + Q_4 = 1.305Q_3 + 2Q_3 + 1.850Q_3 = 5.155Q_3 \qquad (h_f)_{AD} = (h_f)_1 + (h_f)_3 + (h_f)_5$$

$$91 = \{(8)(0.017)/[(\pi)^2(32.2)]\}[(3300)(5.155Q_3)^2/(\tfrac{24}{12})^5 + (2600)(Q_3)^2/(\tfrac{12}{12})^5 + (5200)(5.155Q_3)^2/(\tfrac{30}{12})^5]$$

$$Q_3 = 5.61 \text{ ft}^3/\text{s} \qquad Q_1 = Q_5 = (5.155)(5.61) = 28.92 \text{ ft}^3/\text{s}$$

Increase in capacity $= (28.92 - 26.30)/26.30 = 0.100$, or 10.0 percent.

**11.44** Compute the flows in all pipes of the system shown in Fig. 11-25. Pipe $AB$ is 800 ft long, 6 in. in diameter, $f = 0.03$; pipe $u$ is 500 ft long, 4 in. in diameter, $f = 0.02$; pipe $l$ is 700 ft long, 2 in. in diameter, $f = 0.04$; pipe $CD$ is 300 ft long, 4 in. in diameter, $f = 0.02$. The tank surface is 84 ft above the free outlet $D$.

▮ $h_f = (f)(L/d)(v^2/2g)$. For $B$ to $C$, $(h_f)_u = (h_f)_l$.

$$0.02[500/(\tfrac{4}{12})]\{v_u^2/[(2)(32.2)]\} = 0.04[700/(\tfrac{2}{12})]\{v_l^2/[(2)(32.2)]\} \qquad v_u = 2.366v_l$$

$$Q = Q_u + Q_l = A_u v_u + A_l v_l = [(\pi)(\tfrac{4}{12})^2/4](v_u) + [(\pi)(\tfrac{2}{12})^2/4](v_l) = [(\pi)(\tfrac{4}{12})^2/4](2.366v_l) + [(\pi)(\tfrac{2}{12})^2/4](v_l)$$

$$= 0.2283v_l \qquad Q = Q_u + A_l v_l = Q_u + [(\pi)(\tfrac{2}{12})^2/4](Q/0.2283)$$

$$Q_u = 0.9044Q \qquad Q_l = (1 - 0.9044)(Q) = 0.0956Q \qquad v_u = Q_u/A_u = 0.9044Q/[(\pi)(\tfrac{4}{12})^2/4] = 10.36Q$$

$$v_{AB} = Q/A_{AB} = Q/[(\pi)(\tfrac{6}{12})^2/4] = 5.093Q \qquad v_u = (10.36)(v_{AB}/5.093) = 2.034v_{AB} \qquad h_f = (h_f)_{AB} + (h_f)_u + (h_f)_{CD}$$

$$84 = 0.03[800/(\tfrac{6}{12})]\{v_{AB}^2/[(2)(32.2)]\} + 0.02[500/(\tfrac{4}{12})]\{v_u^2/[(2)(32.2)]\} + 0.02[300/(\tfrac{4}{12})]\{v_{CD}^2/[(2)(32.2)]\}$$

$$[(\pi)(\tfrac{6}{12})^2/4](v_{AB}) = [(\pi)(\tfrac{4}{12})^2/4](v_{CD}) \qquad v_{CD} = 2.250v_{AB}$$

$$84 = 0.03[800/(\tfrac{6}{12})]\{v_{AB}^2/[(2)(32.2)]\} + 0.02[500/(\tfrac{4}{12})]\{(2.034v_{AB})^2/[(2)(32.2)]\}$$

$$+ 0.02[300/(\tfrac{4}{12})]\{(2.250v_{AB})^2/[(2)(32.2)]\}$$

$$v_{AB} = 4.533 \text{ ft/s} \qquad Q = Q_{AB} = [(\pi)(\tfrac{6}{12})^2/4](4.533) = 0.890 \text{ ft}^3/\text{s} \qquad Q_{CD} = Q_{AB} = 0.890 \text{ ft}^3/\text{s}$$

$$Q_u = (0.9044)(0.890) = 0.805 \text{ ft}^3/\text{s} \qquad Q_l = (0.0956)(0.890) = 0.085 \text{ ft}^3/\text{s}$$

**Fig. 11-25**

**11.45** Rework Prob. 11.44 if pipe $AB$ is 400 m long, 200 mm in diameter, and $f = 0.03$; pipe $u$ is 300 m long, 100 mm in diameter, and $f = 0.02$; pipe $l$ is 200 m long, 150 mm in diameter, and $f = 0.025$; pipe $CD$ is 700 m long, 300 mm in diameter, and $f = 0.018$; the tank surface is 94 ft above $D$.

▮ $(h_f)_u = (h_f)_l$. $\quad 0.02[300/(0.100)]\{v_u^2/[(2)(9.807)]\} = 0.025[200/(0.150)]\{v_l^2/[(2)(9.807)]\} \qquad v_u = 0.7454v_l$

$$Q = Q_u + Q_l = A_u v_u + A_l v_l = [(\pi)(0.100)^2/4](v_u) + [(\pi)(0.150)^2/4](v_l)$$

$$= [(\pi)(0.100)^2/4](0.7454v_l) + [(\pi)(0.150)^2/4](v_l)$$

$$Q = 0.02353v_l \qquad Q = Q_u + A_l v_l = Q_u + [(\pi)(0.150)^2/4](Q/0.02353)$$

$$Q_u = 0.2490Q \qquad Q_l = (1 - 0.2490)(Q) = 0.7510Q \qquad v_u = Q_u/A_u = 0.2490Q/[(\pi)(0.100)^2/4] = 31.70Q$$

$$v_{AB} = Q/A_{AB} = Q/[(\pi)(0.200)^2/4] = 31.83Q \qquad v_u = (31.70)(v_{AB}/31.83) = 0.9959v_{AB}$$

$$h_f = (h_f)_{AB} + (h_f)_u + (h_f)_{CD}$$

$$94 = 0.03[400/0.200]\{v_{AB}^2/[(2)(9.807)]\} + 0.02[300/0.100]\{v_u^2/[(2)(9.807)]\}$$
$$+ 0.018[700/0.300]\{v_{CD}^2/[(2)(9.807)]\}$$

$$[(\pi)(0.200)^2/4](v_{AB}) = [(\pi)(0.300)^2/4](v_{CD}) \qquad v_{CD} = 0.4444v_{AB}$$

$$94 = 0.03[400/0.200]\{v_{AB}^2/[(2)(9.807)]\} + 0.02[300/0.100]\{(0.9959v_{AB})^2/[(2)(9.807)]\}$$
$$+ 0.018[700/0.300]\{(0.4444v_{AB})^2/[(2)(9.807)]\}$$

$$v_{AB} = 3.798 \text{ m/s} \qquad Q = Q_{AB} = [(\pi)(0.200)^2/4](3.798) = 0.1193 \text{ m}^3/\text{s} \qquad Q_{CD} = Q_{AB} = 0.01193 \text{ m}^3/\text{s}$$
$$Q_u = (0.2490)(0.1193) = 0.0297 \text{ m}^3/\text{s} \qquad Q_l = (0.7510)(0.1193) = 0.0896 \text{ m}^3/\text{s}$$

Here, as in Prob. 11.44, the question remains open as to whether the computed flows are physically realizable. It all depends on the elevations of junctions $B$ and $C$.

**11.46** Pump $P$ in Fig. 11-26 develops 27 ft of head, producing a velocity of flow in pipe $C$ of 5 fps. Neglecting minor losses, find the flow rates in pipes $A$ and $B$. The parameters are: pipe $A$—4200 ft long, 2 ft diameter, $f = 0.03$; pipe $B$—4200 ft long, 1 ft in diameter, $f = 0.03$; pipe $C$—4200 ft long, 2 ft in diameter, $f = 0.02$.

$\blacksquare$ $\quad h_f = (f)(L/d)(v^2/2g) \qquad (h_f)_A = (0.03)(\frac{4200}{2})\{v_A^2/[(2)(32.2)]\} = 0.9783v_A^2 \qquad (h_L)_C = (h_f)_C - h_{\text{pump}}$

$$(h_f)_C = (0.02)(\tfrac{4200}{2})\{5^2/[(2)(32.2)]\} = 16.30 \text{ ft} \qquad (h_L)_C = 16.30 - 27 = -10.70 \text{ ft}$$

Hence the energy is greater at the right end of $C$ and flow will be to the left in $A$.

$$0.9783v_A^2 = 10.70 \qquad v_A = 3.307 \text{ ft/s} \qquad Q_A = A_A v_A = [(\pi)(2)^2/4](3.307) = 10.39 \text{ ft}^3/\text{s} \qquad \text{(to the left)}$$
$$Q_C = [(\pi)(2)^2/4](5) = 15.71 \text{ ft}^3/\text{s} \qquad \text{(to the right)}$$
$$Q_B = Q_A + Q_C = -10.39 + 15.71 = -5.32 \text{ ft}^3/\text{s} \qquad \text{(to the right)}$$

Fig. 11-26

**11.47** In Prob. 11.46, find the elevation (El) of pipe $B$ at discharge.

$\blacksquare$
$$\text{El}_B = \text{El}_{\text{reservoir surface}} + h_{\text{pump}} - \sum h_f - v_B^2/2g$$

$$v_B = Q_B/A_B = 5.32/[(\pi)(1)^2/4] = 6.774 \text{ ft/s}$$

$$(h_f)_B = (0.03)(\tfrac{4200}{1})\{6.774^2/[(2)(32.2)]\} = 89.78 \text{ ft} \qquad \sum h_f = 16.30 + 89.78 = 106.08 \text{ ft}$$

$$\text{El}_B = 21.5 + 27 - 106.08 - 6.774^2/[(2)(32.2)] = -58.3 \text{ ft}$$

**11.48** Repeat Prob. 11.46 given that the velocity in pipe $C$ is 3 fps and all other data remain the same.

$\blacksquare$ $\quad h_f = (f)(L/d)(v^2/2g) \qquad (h_f)_A = (0.03)(\frac{4200}{2})\{v_A^2/[(2)(32.2)]\} = 0.9783v_A^2 \qquad (h_L)_C = (h_f)_C - h_{\text{pump}}$

$$(h_f)_C = (0.02)(\tfrac{4200}{2})\{3^2/[(2)(32.2)]\} = 5.87 \text{ ft} \qquad (h_L)_C = 5.87 - 27 = -21.13 \text{ ft}$$

Hence the energy is greater at the right end of $C$ and flow will be to the left in $A$.

$$0.9783v_A^2 = 21.13 \qquad v_A = 4.647 \text{ ft/s} \qquad Q_A = A_A v_A = [(\pi)(2)^2/4](4.647) = 14.60 \text{ ft}^3/\text{s} \qquad \text{(to the left)}$$
$$Q_C = [(\pi)(2)^2/4](3) = 9.42 \text{ ft}^3/\text{s} \qquad \text{(to the right)}$$
$$Q_B = Q_A + Q_C = -14.60 + 9.42 = -5.18 \text{ ft}^3/\text{s} \qquad \text{(to the left)}$$

**11.49** In Prob. 11.48, find the elevation of pipe $B$ at discharge.

$\blacksquare$
$$\text{El}_B = \text{El}_{\text{reservoir surface}} + h_{\text{pump}} - \sum h_f + v_B^2/2g \qquad v_B = Q_B/A_B = 5.18/[(\pi)(1)^2/4] = 6.595 \text{ ft/s}$$

$$(h_f)_B = (0.03)(\tfrac{4200}{1})\{6.595^2/[(2)(32.2)]\} = 85.10 \text{ ft} \qquad \sum h_f = 5.87 - 85.10 = -79.23 \text{ ft}$$

$$\text{El}_B = 21.5 + 27 - (-79.23) + 6.595^2/[(2)(32.2)] = 127.8 \text{ ft}$$

**11.50** Refer to Fig. 11-26. Assume the water surface in the reservoir is at elevation 94 m. Pipes $A$, $B$, and $C$ are all 840 m long, and they all have diameter of 0.7 m, with $f = 0.022$. Neglecting minor losses, find the flow rate in all pipes, supposing that the pump develops 9 m of head when the velocity in pipe $C$ is 3.6 m/s.

**❚** $h_f = (f)(L/d)(v^2/2g)$ $\quad (h_f)_A = 0.022[840/(0.7)]\{v_A^2/[(2)(9.807)]\} = 1.346v_A^2$ $\quad (h_L)_C = (h_f)_C - h_{\text{pump}}$

$$(h_f)_C = 0.022[840/(0.7)]\{3.6^2/[(2)(9.807)]\} = 17.44 \text{ m} \qquad (h_L)_C = 17.44 - 9 = 8.44 \text{ m}$$

$$1.346v_A^2 = 8.44 \qquad v_A = 2.50 \text{ m/s}$$

$$Q_A = A_A v_A = [(\pi)(0.7)^2/4](2.50) = 0.962 \text{ m}^2/\text{s} \qquad \text{(to the right)}$$

$$Q_C = [(\pi)(0.7)^2/4](3.6) = 1.385 \text{ m}^3/\text{s} \qquad \text{(to the right)}$$

$$Q_B = Q_A + Q_C = 0.962 + 1.385 = 2.347 \text{ m}^3/\text{s} \qquad \text{(to the right)}$$

**11.51** In Prob. 11.50, find the elevation of pipe $B$ at discharge.

**❚** $\quad \text{El}_B = \text{El}_{\text{reservoir surface}} + h_{\text{pump}} - \sum h_f - v_B^2/2g \qquad v_B = Q_B/A_B = 2.347/[(\pi)(0.7)^2/4] = 6.099 \text{ m/s}$

$$(h_f)_B = 0.022[840/(0.7)]\{6.099^2/[(2)(9.807)]\} = 50.07 \text{ m} \qquad \sum h_f = 17.44 + 50.07 = 67.51 \text{ m}$$

$$\text{El}_B = 94 + 9 - 67.51 - 6.099^2/[(2)(9.807)] = 33.6 \text{ m}$$

**11.52** Using $n = 0.013$ and neglecting minor losses, express the head loss through the pipe system of Fig. 11-27 in the form of $h_L = KQ^x$.

**❚** $\quad h_L = \dfrac{n^2 v^2 L}{(1.486)^2 (R^{2/3})^2} = \dfrac{n^2 Q^2 L}{(1.486)^2 R^{4/3} A^2} \qquad (h_L)_{AB} = \dfrac{(0.013)^2 (Q)^2 (500)}{(1.486)^2 [(\frac{24}{12})/4]^{4/3} [(\pi)(\frac{24}{12})^2/4]^2} = 0.009770 Q^2$

$$(h_L)_{CD} = \frac{(0.013)^2 (Q)^2 (1000)}{(1.486)^2 [(\frac{18}{12})/4]^{4/3} [(\pi)(\frac{18}{12})^2/4]^2} = 0.09063 Q^2 \qquad (h_L)_{BEC} = (h_L)_{BFC}$$

$$\frac{Q_1^2 L_1}{R_1^{4/3} A_1^2} = \frac{Q_2^2 L_2}{R_2^{4/3} A_2^2} \qquad \frac{(Q_1^2)(4000)}{[(\frac{18}{12})/4]^{4/3}[(\pi)(\frac{18}{12})^2/4]^2} = \frac{(Q_2^2)(5000)}{[(\frac{12}{12})/4]^{4/3}[(\pi)(\frac{12}{12})^2/4]^2}$$

$$Q_1 = 3.269 Q_2 \qquad Q = Q_1 + Q_2 = 3.269 Q_2 + Q_2 \qquad Q_2 = 0.2342 Q$$

$$(h_L)_{BFC} = \frac{(0.013)^2 (Q_2)^2 (5000)}{(1.486)^2 [(\frac{12}{12})/4]^{4/3} [(\pi)(\frac{12}{12})^2/4]^2} = 3.939 Q_2^2$$

$$(h_L)_{BFC} = (3.939)(0.2342 Q)^2 = 0.05485 Q^2 \qquad (h_L)_{AD} = 0.009770 Q^2 + 0.09063 Q^2 + 0.05485 Q^2 = 0.155 Q^2$$

**Fig. 11-27**

**11.53** A pipe system, connecting two reservoirs whose difference in surface elevation is 13 m, consists of 320 m of 60-cm concrete pipe (pipe $A$), branching into 640 m of 30 cm (pipe $B$) and 640 m of 45 cm (pipe $C$) in parallel, which join again to a single 60-cm line 1300 m long (pipe $D$). $f = 0.032$ for all pipes, calculate all flows.

**❚** $\qquad\qquad (h_f)_B = (h_f)_C \qquad h_f = (f)(L/D)(v^2/2g)$

$$(f)[640/0.30]\{v_B^2/[(2)(9.807)]\} = f[640/0.45]\{v_C^2/[(2)(9.807)]\} \qquad v_C = 1.225 v_B \qquad Q_A = Q_B + Q_C$$

$$[(\pi)(0.60)^2/4](v_A) = [(\pi)(0.30)^2/4(v_B)] + [(\pi)(0.45)^2/4](v_C) \qquad 0.2827 v_A = 0.07069 v_B + 0.1590 v_C$$

$$0.2827 v_A = 0.07069 v_B + (0.1590)(1.225 v_B) \qquad v_B = 1.065 v_A$$

Convert the parallel pipes to a single equivalent 60-cm-diameter pipe.

$$f[640/0.30]\{v_B^2/[(2)(9.807)]\} = f[L_e/0.60]\{v_A^2/[(2)(9.807)]\}$$

$$108.8 V_B^2 = 0.08497 L_e v_A^2 \qquad (108.8)(1.065 v_A)^2 = 0.08497 L_e v_A^2 \qquad L_e = 1452 \text{ m} \qquad \Delta \text{El} = \sum h_f$$

$$13 = 0.032[(320 + 1452 + 1300)/0.60]\{v_A^2/[(2)(9.807)]\} \qquad v_A = 1.248 \text{ m/s}$$

$$Q_A = A_A v_A = [(\pi)(0.60)^2/4](1.248) = 0.353 \text{ m}^3/\text{s} \qquad Q_D = Q_A = 0.353 \text{ m}^3/\text{s} \qquad v_B = (1.065)(1.248) = 1.329 \text{ m/s}$$

$$Q_B = [(\pi)(0.30)^2/4](1.329) = 0.094 \text{ m}^3/\text{s} \qquad Q_C = Q_A - Q_B = 0.353 - 0.094 = 0.259 \text{ m}^3/\text{s}$$

**11.54** Refer to Fig. 11-27. Suppose $p_A/\gamma = 6.5$ ft, $p_D/\gamma = 20$ ft, and $z_A = z_D$. A pump in the 4000-ft pipe (flow from left to right) develops 30 ft of head. Find the flow rate in each pipe. Assume $n = 0.013$ for all pipes.

$$Q_{AB} = Q_{BEC} - Q_{BFC} = Q_{CD} \tag{1}$$

$$6.5 - (h_f)_{AB} = p_B/\gamma \tag{2}$$

$$p_B/\gamma + h_{\text{pump}} - (h_f)_{BEC} = p_C/\gamma \tag{3}$$

$$p_C/\gamma - (h_f)_{BFC} = p_B/\gamma \tag{4}$$

$$p_C/\gamma - (h_f)_{CD} = 20 \tag{5}$$

$$h_f = n^2 v^2 L/(1.486 R^{2/3})^2 \qquad v_{AB} = Q_{AB}/A_{AB} = Q_{AB}/[(\pi)(\tfrac{24}{12})^2/4] = 0.3183 Q_{AB}$$

$$(h_f)_{AB} = (0.013)^2(0.3183 Q_{AB})^2(500)/\{(1.486)[(\tfrac{24}{12})/4]^{2/3}\}^2 = 0.009769 Q_{AB}^2 \tag{6}$$

$$v_{BEC} = Q_{BEC}/[(\pi)(\tfrac{18}{12})^2/4] = 0.5659 Q_{BEC}$$

$$(h_f)_{BEC} = (0.013)^2(0.5659 Q_{BEC})^2(4000)/\{(1.486)[(\tfrac{18}{12})/4]^{2/3}\}^2 = 0.3625 Q_{BEC}^2 \tag{7}$$

$$v_{BFC} = Q_{BFC}/[(\pi)(\tfrac{12}{12})^2/4] = 1.273 Q_{BFC}$$

$$(h_f)_{BFC} = (0.013)^2(1.273 Q_{BFC})^2(5000)/\{(1.486)[(\tfrac{12}{12})/4]^{2/3}\}^2 = 3.938 Q_{BFC}^2 \tag{8}$$

$$v_{CD} = Q_{CD}/[(\pi)(\tfrac{18}{12})^2/4] = 0.5659 Q_{CD}$$

$$(h_f)_{CD} = (0.013)^2(0.5659 Q_{CD})^2(1000)/\{(1.486)[(\tfrac{18}{12})/4]^{2/3}\}^2 = 0.09063 Q_{CD}^2 \tag{9}$$

Substitute Eqs. (6), (7), (8), and (9) into Eqs. (2), (3), (4), and (5), respectively, and then solve simultaneously between Eqs. (2), (3), (4), and (5), introducing also Eq. (1). Two equations result, such as

$$Q_{BEC}^2 - 2Q_{BEC}^2 Q_{BFC} - 37.40 Q_{BFC}^2 = -132 \qquad Q_{BEC}^2 + 10.7 Q_{BFC}^2 = 82$$

By trial and error, $Q_{BEC} = 6.25$ cfs and $Q_{BFC} = 2.00$ cfs. Hence, $Q_{AB} = 6.25 - 2.00 = 4.25$ cfs.

# CHAPTER 12
# Branching Pipeline Systems

**12.1** In Fig. 12-1, find the flows for the following data: $L_1 = 200$ m, $D_1 = 300$ mm, $\epsilon_1/D_1 = 0.0002$, $z_1 = 700$ m, $p_1 = 7$ atm; $L_2 = 300$ m, $D_2 = 350$ mm, $\epsilon_2/D_2 = 0.00015$, $z_2 = 400$ m, $p_2 = 2$ atm; $L_3 = 400$ m, $D_3 = 400$ mm, $\epsilon_3/D_3 = 0.0001$, $z_3 = 100$ m, $p_3 = 3$ atm. The fluid is water with $\nu = 0.113 \times 10^{-5}$ m²/s.

▐ We first calculate $H_1$, $H_2$, and $H_3$. Thus, $H_1 = z_1 + (7)(p_{atm})/\gamma = 700 + (7)(101\,325)/9806 = 772.3$ m, $H_2 = z_2 + (2)(p_{atm})/\gamma = 400 + (2)(101\,325)/9806 = 420.7$ m, $H_3 = z_3 + (3)(p_{atm})/\gamma = 100 + (3)(101\,325)/9806 = 131.0$ m. Now estimate $H_J = 380$ m. We will hence use the continuity equation for test purposes. We can then say that $(772.3 - 380) = (1/g)f_1(200/0.300)(V_1^2/2)$, $(420.7 - 380) = (1/g)f_2(300/0.350)(V_2^2/2)$, $(380 - 131.0) = (1/g)f_3(400/0.400)(V_3^2/2)$. Estimating $f_1 = 0.014$, $f_2 = 0.013$, and $f_3 = 0.012$, we can compute velocities and the flows $q_i$. We get

$$V_1 = 28.72 \text{ m/s} \qquad q_1 = (\pi/4)(0.300^2)(28.72) = 2.03 \text{ m}^3/\text{s}$$

$$V_2 = 8.465 \text{ m/s} \qquad q_2 = (\pi/4)(0.350^2)(8.465) = 0.814 \text{ m}^3/\text{s}$$

$$V_3 = 20.18 \text{ m/s} \qquad q_3 = (\pi/4)(0.400^2)(20.18) = 2.536 \text{ m}^3/\text{s}$$

We see that $(q_1 + q_2) > q_3$. As a second estimate, we increase $H_J$ and use more accurate friction factors. Using the preceding velocities, we now find a second set of friction factors: $(N_R)_1 = (28.72)(0.300)/(0.0113 \times 10^{-4}) = 7.625 \times 10^6$, $f_1 = 0.014$; $(N_R)_2 = (8.465)(0.350)/(0.0113 \times 10^{-4}) = 2.622 \times 10^6$, $f_2 = 0.0134$; $(N_R)_3 = (20.18)(0.400)/(0.0113 \times 10^{-4}) = 7.143 \times 10^6$, $f_3 = 0.012$. Suppose we next choose $H_J$ to be 400 m. Thus, we still use the continuity equation. We get the following results: $V_1 = 27.98$ m/s, $q_1 = 1.977$ m³/s; $V_2 = 5.95$ m/s, $q_2 = 0.5721$ m³/s; $V_3 = 20.97$ m/s, $q_3 = 2.635$ m³/s.

Note that $(q_1 + q_2) < q_3$. We now *interpolate* to get the final result. That is, $H_J = 380$ m gave us a value $(q_1 + q_2) - q_3 = 0.308$, while $H_J = 400$ m gave us a value $(q_1 + q_2) - q_3 = -0.0859$. Hence we choose a final $H_J$ to be $H_J = 380 + [0.308/(0.308 + 0.0859)](400 - 380) = 396$ m. For this we get $q_1 = 1.988$ m³/s = 1988 L/s, $q_2 = 0.6249$ m³/s = 624.9 L/s, $q_3 = 2.616$ m³/s = 2616 L/s. We come very close to satisfying the continuity equation so the above are the desired flows.

Datum  **Fig. 12-1**

**12.2** The three-cistern system of Fig. 12-2 contains water at 5 °C, compute the three volumetric flows, given: $L_1 = 2000$ m, $L_2 = 2300$ m, $L_3 = 2500$ m; $D_1 = 1$ m, $D_2 = 0.60$ m, $D_3 = 1.20$ m; $\epsilon_1/D_1 = 0.00015$, $\epsilon_2/D_2 = 0.001$, $\epsilon_3/D_3 = 0.002$.

▐ We assume initially a junction head $h_J = z_J + (p_J/\gamma) = 80$ m. Using $h_f = f(L/d)(v^2/2g)$ for each branch we get:

$$(120 - 80) = f_1(2000/1)(v_1^2/2g) \qquad (100 - 80) = f_2(2300/0.60)(v_2^2/2g) \qquad (80 - 28) = f_3(2500/1.20)(v_3^2/2g)$$

Guessing $f_1 = 0.013$, $f_2 = 0.020$, $f_3 = 0.023$, we obtain

$$v_1 = 5.494 \text{ m/s} \qquad Q_1 = (\pi/4)(1^2)(5.494) = 4.315 \text{ m}^3/\text{s}$$
$$v_2 = 2.262 \text{ m/s} \qquad Q_2 = (\pi/4)(0.60)^2(2.262) = 0.6396 \text{ m}^3/\text{s}$$
$$v_3 = 4.6136 \text{ m/s} \qquad Q_3 = (\pi/4)(1.2)^2(4.6136) = 5.218 \text{ m}^3/\text{s}$$

Checking with the continuity equation,

$$\left. \begin{array}{l} Q_1 + Q_2 = 4.955 \\ Q_3 = 5.2179 \end{array} \right\} \qquad \Delta = (Q_1 + Q_2) - Q_3 - 0.2629$$

we decrease $h_J$ to 76 m. Then

$$(120 - 76) = 0.013(\tfrac{2000}{1})(v_1^2/2g) \qquad v_1 = 5.7613 \text{ m/s} \qquad Q_1 = 4.5249 \text{ m}^3/\text{s}$$
$$(100 - 76) = 0.020(2300/0.6)(v_2^2/2g) \qquad v_2 = 2.4779 \text{ m/s} \qquad Q_2 = 0.7006 \text{ m}^3/\text{s}$$
$$(76 - 28) = 0.023(2500/1.20)(v_3^2/2g) \qquad v_3 = 4.4326 \text{ m/s} \qquad Q_3 = 5.0132 \text{ m}^3/\text{s}$$

$$\left. \begin{array}{l} Q_1 + Q_2 = 5.226 \\ Q_3 = 5.013 \end{array} \right\} \qquad \Delta = 0.213 \text{ m/s}$$

Interpolation now yields:

$$h_J = 80 - [0.2629/(0.213 + 0.2629)](4) = 77.7903 \text{ m}$$
$$v_1 = 5.643 \text{ m/s} \qquad Q_1 = 4.432 \text{ m}^3/\text{s}$$
$$v_2 = 2.3837 \text{ m/s} \qquad Q_2 = 0.674 \text{ m}^3/\text{s}$$
$$v_3 = 4.5145 \text{ m/s} \qquad Q_3 = 5.106 \text{ m}^3/\text{s}$$

Checking:

$$Q_1 + Q_2 = 4.432 + 0.674 = 5.106 \qquad q_3 = 5.106 \qquad \text{O.K.}$$

Therefore, $Q_1 = 4.39 \text{ m}^3/\text{s}$, $Q_2 = 0.661 \text{ m}^3/\text{s}$, $Q_3 = 5.048 \text{ m}^3/\text{s}$.

**Fig. 12-2**

**12.3** The system of Fig. 12-3(a) holds water at 60 °F. Data are: $p = 50$ psig; $z_1 = 700$ ft, $L_1 = 2000$ ft, $D_1 = 3$ ft, $(\epsilon/D)_1 = 0.001$; $z_2 = 650$ ft, $L_2 = 2500$ ft $D_2 = 3.5$ ft, $(\epsilon/D)_2 = 0.002$; $z_3 = 100$ ft, $L_3 = 2200$ ft, $D_3 = 4$ ft, $(\epsilon/D)_3 = 0.002$. Determine the three pipe flows.

▌ The heads for the free surfaces are $h_1 = 700$ ft, $h_2 = 650$ ft, $h_3 = 100 + (50)(144)/62.4 = 215.4$ ft. Initially we take $h_J$ to be 450 ft. Applying $h_f = f(L/d)(v^2/2g)$,

$$(700 - 450) = f_1(2000/3)(v_1^2/2g) \qquad (650 - 450) = f_2(2500/3.5)(v_2^2/2g)$$
$$(450 - 215.4) = f_3(2200/4)(v_3^2/2g)$$

Guessing $f_1 = 0.02$, $f_2 = 0.023$, $f_3 = 0.023$, we solve for the velocities and flows:

$$v_1 = 34.75 \text{ ft/s} \qquad Q_1 = 245 \text{ cfs} \qquad v_2 = 28.0 \text{ ft/s} \qquad Q_2 = 269.4 \text{ cfs}$$
$$v_3 = 34.56 \text{ ft/s} \qquad Q_3 = 434.3 \text{ cfs}$$

Checking continuity: $Q_1 + Q_2 = 245 + 269.4 = 514$ cfs, $Q_3 = 434.3$ cfs; $\Delta = 79.7$ cfs. We increase $h_J$ to 475 ft.

$$(700 - 475) = (0.02)(2000/3)(v_1^2/2g) \qquad (650 - 475) = (0.023)(2500/3.5)v_2^2/2g$$

$$(475 - 215.4) = (0.023)(2200/4)(v_3^2/2g)$$

from which

$$v_1 = 32.96 \text{ ft/s} \qquad Q_1 = 233 \text{ cfs} \qquad v_2 = 26.19 \text{ ft/s} \qquad Q_2 = 252 \text{ cfs} \qquad v_3 = 36.35 \text{ ft/s} \qquad Q_3 = 456 \text{ cfs}$$

Check continuity again: $Q_1 + Q_2 = 485$, $Q_3 = 456$ ft$^3$/s; $\Delta = 29.0$ cfs. Extrapolating $\Delta$ to zero [Fig. 12-3(b)], we get $h_J = 487$ ft:

$$(700 - 487) = (0.02)(2000/3)(v_1^2/2g) \qquad (650 - 487) = (0.023)(2500/3.5)(v_2^2/2g)$$

$$(487 - 215.4) = (0.023)(2200/4)(v_3^2/2g)$$

$$v_1 = 32.07 \text{ ft/s} \qquad Q_1 = 226.7 \text{ cfs} \qquad v_2 = 25.27 \text{ ft/s} \qquad Q_2 = 243.2 \text{ cfs}$$

$$v_3 = 37.185 \text{ ft/s} \qquad Q_3 = 467.3 \text{ cfs}$$

Check continuity: $Q_1 + Q_2 = 469.9$, $Q_3 = 467.3$ ft$^3$/s; close enough. Therefore, $Q_1 = 227$ cfs, $Q_2 = 243$ cfs, $Q_3 = 467$ cfs.

**Fig. 12-3**

**12.4**  In Fig. 12-4, valve $F$ is partly closed, creating a head loss of 3.60 ft when the flow through the valve is 0.646 mgd (1.00 cfs). What is the length of 10-in pipe to reservoir $A$?

$$Q = 1.318 A C R^{0.63} s^{0.54} \qquad 1.00 = 1.318[(\pi)(\tfrac{12}{12})^2/4](80)[(\tfrac{12}{12})/4]^{0.63}(s_{DB})^{0.54}$$

$$s_{DB} = 0.001414 \text{ ft/ft}$$

$$(h_L)_{DB} = (0.001414)(1000) + 3.60 = 5.01 \text{ ft}$$

If $\text{El}_E = 0$, the grade line elevation at $B = 20 - 5.01 = 14.99$ ft.

$$s_{BE} = (14.99 - 0)/5000 = 0.002998 \text{ ft/ft}$$

$$Q_{BE} = 1.318[(\pi)(\tfrac{12}{12})^2/4](120)[(\tfrac{12}{12})/4]^{0.63}(0.002998)^{0.54} = 2.25 \text{ ft}^3/\text{s}$$

$$Q_{AB} = Q_{BE} - Q_{DB} = 2.25 - 1.00 = 1.25 \text{ ft}^3/\text{s}$$

$$1.25 = 1.318[(\pi)(\tfrac{10}{12})^2/4](100)[(\tfrac{10}{12})/4]^{0.63}(s_{AB})^{0.54}$$

$$s_{AB} = 0.003436 \text{ ft/ft} = (h_f)_{AB}/L_{AB} \qquad 0.003436 = (20 - 15 - 2)/L_{AB} \qquad L_{AB} = 873 \text{ ft}$$

**Fig. 12-4**

**12.5**  For the constant elevations of the water surfaces shown in Fig. 12-5a, what flows will occur?

▌ Because the elevation of the hydraulic grade line at $C$ cannot be computed (all flows unknown), the problem will be solved by successive trials. A convenient assumption is to choose the elevation of the hydraulic grade line at $C$ at 190.0 ft. By so assuming, flow to or from reservoir $B$ will be zero, reducing the number of calculations thereby. With this assumption,

$$Q = 1.318ACR^{0.63}s^{0.54} \qquad s_{AC} = (212.0 - 190.0)/8000 = 0.002750 \text{ ft/ft}$$

$$Q_{AB} = 1.318[(\pi)(\tfrac{24}{12})^2/4](100)[(\tfrac{24}{12})/4]^{0.63}(0.002750)^{0.54} = 11.1 \text{ ft}^3/\text{s}$$

$$s_{CD} = (190.0 - 100.0)/4000 = 0.02250 \text{ ft/ft}$$

$$Q_{CD} = 1.318[(\pi)(\tfrac{12}{12})^2/4](100)[(\tfrac{12}{12})/4]^{0.63}(0.02250)^{0.54} = 5.57 \text{ ft}^3/\text{s} \qquad \text{(away from } C\text{)}$$

Examination of these values of flow indicates that the grade line at $C$ must be higher, thereby reducing the flow from $A$ and increasing the flow to $D$ as well as adding flow to $B$. In an endeavor to "straddle" the correct elevation at $C$, assume a value of 200.0 ft. Thus, for elevation at $C = 200.0$ ft,

$$s_{AC} = (212.0 - 200.0)/8000 = 0.001500 \text{ ft/ft}$$

$$Q_{AB} = 1.318[(\pi)(\tfrac{24}{12})^2/4](100)[(\tfrac{24}{12})/4]^{0.63}(0.001500)^{0.54} = 7.99 \text{ ft}^3/\text{s}$$

$$s_{CD} = (200.0 - 100.0)/4000 = 0.02500 \text{ ft/ft}$$

$$Q_{CD} = 1.318[(\pi)(\tfrac{12}{12})^2/4](100)[(\tfrac{12}{12})/4]^{0.63}(0.02500)^{0.54} = 5.90 \text{ ft}^3/\text{s} \qquad \text{(away from } C\text{)}$$

$$s_{CB} = (200.0 - 190.0)/4000 = 0.002500 \text{ ft/ft}$$

$$Q_{CB} = 1.318[(\pi)(\tfrac{16}{12})^2/4](120)[(\tfrac{16}{12})/4]^{0.63}(0.002500)^{0.54} = 4.35 \text{ ft}^3/\text{s} \qquad \text{(away from } C\text{)}$$

The flow away from $C$ is $5.90 + 4.35$, or $10.25$ ft$^3$/s compared with the flow to $C$ of $7.90$ ft$^3$/s. Using Fig. 12-5b to obtain a guide regarding a reasonable third assumption, connect plotted points $R$ and $S$. The line so drawn intersects the $(Q_{to} - Q_{away})$ zero abscissa at approximately $Q_{to} = 8.8$ ft$^3$/s. Inasmuch as the values plotted do not vary linearly, use a flow to $C$ slightly larger, say $9.2$ ft$^3$/s. For $Q = 9.2$ ft$^3$/s to $C$ (i.e., $Q_{AB} = 9.2$ ft$^3$/s), $9.2 = 1.318[(\pi)(\tfrac{24}{12})^2/4](100)[(\tfrac{24}{12})/4]^{0.63}s^{0.54}$, $s_{AC} = 0.001948$ ft/ft; $(h_f)_{AC} = (0.001948)(8000) = 15.6$ ft. Then, the hydraulic grade line at $C$ will be $212.0 - 15.6 = 196.4$ ft.

$$s_{CD} = (196.4 - 100.0)/4000 = 0.02410 \text{ ft/ft}$$

$$Q_{CD} = 1.318[(\pi)(\tfrac{12}{12})^2/4](100)[(\tfrac{12}{12})/4]^{0.63}0.02410^{0.54} = 5.78 \text{ ft}^3/\text{s} \qquad \text{(away from } C\text{)}$$

$$s_{CB} = (196.4 - 190.0)/4000 = 0.001600 \text{ ft/ft}$$

$$Q_{CB} = 1.318[(\pi)(\tfrac{16}{12})^2/4](120)[(\tfrac{16}{12})/4]^{0.63}0.001600^{0.54} = 3.42 \text{ ft}^3/\text{s} \qquad \text{(away from } C\text{)}$$

The flow away from $C$ is $5.78 + 3.42$, or $9.20$ ft$^3$/s compared with the flow to $C$ of $9.2$ ft$^3$/s; hence these values of $Q_{AC}$, $Q_{CD}$, and $Q_{CB}$ must be O.K.

**12.6**  Three pipes connect three reservoirs as shown in Fig. 12-6 at these surface elevations: $z_1 = 20$ m, $z_2 = 100$ m, and $z_3 = 40$ m. The pipe data are

| pipe | L, m | d, m | ε, mm | ε/d |
|------|------|------|-------|-----|
| 1 | 100 | 8 | 0.24 | 0.003 |
| 2 | 150 | 6 | 0.12 | 0.002 |
| 3 | 80 | 4 | 0.20 | 0.005 |

**Fig. 12-5**

The fluid is water, $\rho = 1000 \text{ kg/m}^3$, and $\nu = 1.02 \times 10^{-6} \text{ m}^2/\text{s}$. Find the resulting flow in each pipe, neglecting minor losses.

▌ As a first guess, take $h_J$ equal to the middle reservoir height, $z_3 = h_J = 40$ m. This saves one calculation ($Q_3 = 0$) and enables us to get the lay of the land:

| reservoir | $h_J$, m | $z_i - h_J$, m | $f_i$ | $V_i$, m/s | $Q_i$, m³/h | $L_i/d_i$ |
|-----------|----------|----------------|-------|------------|-------------|-----------|
| 1 | 40 | −20 | 0.0267 | −3.43 | −62.1 | 1250 |
| 2 | 40 | 60 | 0.0241 | 4.42 | 45.0 | 2500 |
| 3 | 40 | 0 | | 0 | 0 | 2000 |
| | | | | | $\Sigma Q = -17.1$ | |

Since the sum of the flow rates toward the junction is negative, we guessed $h_J$ too high. Reduce $h_J$ to 30 m and repeat:

| reservoir | $h_J$, m | $z_i - h_J$, m | $f_i$ | $V_i$, m/s | $Q_i$, m³/h |
|-----------|----------|----------------|-------|------------|-------------|
| 1 | 30 | −10 | 0.0269 | −2.42 | −43.7 |
| 2 | 30 | 70 | 0.0241 | 4.78 | 48.6 |
| 3 | 30 | 10 | 0.0317 | 1.76 | 8.0 |
| | | | | | $\Sigma Q = 12.9$ |

This is positive $\Sigma Q$, and so we can linearly interpolate to get an accurate guess: $h_J \approx 34.3$ m. Make one final list:

| reservoir | $h_J$, m | $z_i - h_J$, m | $f_i$ | $V_i$, m/s | $Q_i$, m³/h |
|-----------|----------|----------------|-------|------------|-------------|
| 1 | 34.3 | −14.3 | 0.0268 | −2.90 | −52.4 |
| 2 | 34.3 | 65.7 | 0.0241 | 4.63 | 47.1 |
| 3 | 34.3 | 5.7 | 0.0321 | 1.32 | 6.0 |
| | | | | | $\Sigma Q = 0.7$ |

This is close enough; hence we calculate that the flow rate is 52.4 m³/h toward reservoir 3, balanced by 47.1 m³/h away from reservoir 1 and 6.0 m³/h away from reservoir 3.

One further iteration with this problem would give $h_J = 34.53$ m, resulting in $Q_1 = -52.8$, $Q_2 = 47.0$, and $Q_3 = 5.8$ m³/h, so that $\Sigma Q = 0$ to three-place accuracy.

**Fig. 12-6**

**12.7** For the three-reservoir system of Fig. 12-7, $z_1 = 29$ m, $L_1 = 80$ m, $z_2 = 129$ m, $L_2 = 150$ m, $z_3 = 69$ m, and $L_3 = 110$ m. All pipes are 250-mm-diameter concrete with roughness height 0.5 mm. Compute the flow rates for water at 20 °C.

$\blacksquare$ $h_f = (f)(L/d)(v^2/2g)$. Guess $h_a = 69$ m: $(h_f)_1 = z_1 - h_a = 29 - 69 = -40$ m, $\epsilon/d = 0.5/250 = 0.00200$ (for all pipes), $N_R = dv/\nu$. Try $v_1 = 10$ m/s: $(N_R)_1 = (0.250)(10)/(1.02 \times 10^{-6}) = 2.45 \times 10^6$. From Fig. A-5, $f_1 = 0.0235$.

$$40 = 0.0235[80/0.250]\{v_1^2/[(2)(9.807)]\} \qquad v_1 = 10.21 \text{ m/s}$$
$$(N_R)_1 = (0.250)(10.21)/(1.02 \times 10^{-6}) = 2.50 \times 10^6 \qquad f_1 = 0.0235 \qquad \text{(O.K.)}$$
$$Q_1 = A_1 v_1 = [(\pi)(0.250)^2/4](10.21) = 0.5012 \text{ m}^3/\text{s} \qquad \text{(away from } a\text{)}$$
$$(h_f)_2 = z_2 - h_a = 129 - 69 = 60 \text{ m}$$

Try $v_2 = 10$ m/s:

$$(N_R)_2 = (0.250)(10)/(1.02 \times 10^{-6}) = 2.45 \times 10^6 \qquad f_2 = 0.0235$$
$$60 = 0.0235[150/0.250]\{v_2^2/[(2)(9.807)]\} \qquad v_2 = 9.136 \text{ m/s}$$
$$(N_R)_2 = (0.250)(9.136)/(1.02 \times 10^{-6}) = 2.24 \times 10^6 \qquad f_1 = 0.0235 \qquad \text{(O.K.)}$$
$$Q_2 = [(\pi)(0.250)^2/4](9.136) = 0.4485 \text{ m}^3/\text{s} \qquad \text{(toward } a\text{)}$$
$$(h_f)_3 = z_3 - h_a = 69 - 69 = 0 \text{ m}$$

Hence, $Q_3 = 0$; $Q_{\text{to } a} - Q_{\text{from } a} = 0.4485 - 0.5012 = -0.0527$ m³/s. Hence $h_a$ must be a little lower. Try $h_a = 68.5$ m: $(h_f)_1 = z_1 - h_a = 29 - 68.5 = -39.5$ m. Try $v_1 = 10$ m/s:

$$(N_R)_1 = (0.250)(10)/(1.02 \times 10^{-6}) = 2.45 \times 10^6 \qquad f_1 = 0.0235$$
$$39.5 = (0.0235)[80/0.250]\{v_1^2/[(2)(9.807)]\} \qquad v_1 = 10.15 \text{ m/s}$$
$$(N_R)_1 = (0.250)(10.15)/(1.02 \times 10^{-6}) = 2.49 \times 10^6 \qquad f_1 = 0.0235 \qquad \text{(O.K.)}$$
$$Q_1 = [(\pi)(0.250)^2/4](10.15) = 0.4982 \text{ m}^3/\text{s} \qquad \text{(away from } a\text{)}$$
$$(h_f)_2 = z_2 - h_a = 129 - 68.5 = 60.5 \text{ m}$$

Try $v_2 = 10$ m/s:

$$(N_R)_2 = (0.250)(10)/(1.02 \times 10^{-6}) = 2.45 \times 10^6 \qquad f_2 = 0.0235$$
$$60.5 = 0.0235[150/0.250]\{v_2^2/[(2)(9.807)]\} \qquad v_2 = 9.174 \text{ m/s}$$
$$(N_R)_2 = (0.250)(9.174)/(1.02 \times 10^{-6}) = 2.25 \times 10^6 \qquad f_2 = 0.0235 \qquad \text{(O.K.)}$$
$$Q_2 = [(\pi)(0.250)^2/4](9.174) = 0.4503 \text{ m}^3/\text{s} \qquad \text{(toward } a\text{)}$$
$$(h_f)_3 = z_3 - h_a = 69 - 68.5 = 0.5 \text{ m}$$

Try $v_3 = 1$ m/s:

$$(N_R)_3 = (0.250)(1)/(1.02 \times 10^{-6}) = 2.45 \times 10^5 \qquad f_3 = 0.0244$$
$$0.5 = 0.0244[110/0.250]\{v_3^2/[(2)(9.807)]\} \qquad v_3 = 0.9558 \text{ m/s}$$
$$(N_R)_3 = (0.250)(0.9558)/(1.02 \times 10^{-6}) = 2.34 \times 10^5 \qquad f_3 = 0.0244 \qquad \text{(O.K.)}$$
$$Q_3 = [(\pi)(0.250)^2/4](0.9558) = 0.0469 \text{ m}^3/\text{s} \qquad \text{(toward } a\text{)}$$
$$Q_{\text{to } a} - Q_{\text{from } a} = (0.4503 + 0.0469) - 0.4982 = -0.001 \text{ m}^3/\text{s}$$

Hence, $Q_1 = 0.498$ m³/s from $a$, and $Q_2 = 0.450$ m³/s and $Q_3 = 0.047$ m³/s toward $a$.

**Fig. 12-7**

**12.8** The three pipes in Fig. 12-8 are cast iron: $d_1 = 7$ in, $L_1 = 2000$ ft; $d_2 = 5$ in, $L_2 = 1000$ ft; $d_3 = 8$ in, $L_3 = 1500$ ft. Compute the flow rates for water at 68 °F.

▌ $h_f = (f)(L/d)(v^2/2g)$. Guess $h_J = 48$ ft: $(h_f)_1 = z_1 - h_J = 18 - 48 = -30$ m, $(\epsilon/d)_1 = 0.00085/(\frac{7}{12}) = 0.00146$, $N_R = dv/\nu$. Try $v_1 = 5$ ft/s: $(N_R)_1 = (\frac{7}{12})(5)/(1.02 \times 10^{-5}) = 2.86 \times 10^5$. From Fig. A-5, $f_1 = 0.0225$.

$$30 = 0.0225[2000/(\tfrac{7}{12})]\{v_1^2/[(2)(32.2)]\} \qquad v_1 = 5.004 \text{ ft/s} \qquad \text{(value of } f_1 \text{ O.K.)}$$

$$Q_1 = A_1 v_1 = [(\pi)(\tfrac{7}{12})^2/4](5.004) = 1.337 \text{ ft}^3/\text{s} \qquad \text{(away from } J)$$

$$(h_f)_2 = z_2 - h_J = 98 - 48 = 50 \text{ ft}$$

Try $v_2 = 7$ ft/s:

$$(N_R)_2 = (\tfrac{5}{12})(7)/(1.02 \times 10^{-5}) = 2.86 \times 10^5$$

$$(\epsilon/d)_2 = 0.00085/(\tfrac{5}{12}) = 0.00204 \qquad f_2 = 0.024$$

$$50 = 0.024[1000/(\tfrac{5}{12})]\{v_2^2/[(2)(32.2)]\} \qquad v_2 = 7.477 \text{ ft/s}$$

$$(N_R)_2 = (\tfrac{5}{12})(7.477)/(1.02 \times 10^{-5}) = 3.05 \times 10^5 \qquad f_2 = 0.024 \qquad \text{(O.K.)}$$

$$Q_2 = [(\pi)(\tfrac{5}{12})^2/4](7.477) = 1.020 \text{ ft}^3/\text{s} \qquad \text{(toward } J) \qquad (h_f)_3 = z_3 - h_J = 48 - 48 = 0 \text{ ft}$$

Hence, $Q_3 = 0$; $Q_{\text{to } J} - Q_{\text{from } J} = 1.020 - 1.337 = -0.317$ ft³/s. Hence $h_J$ must be a little lower. Try $h_J = 47.35$ ft: $(h_f)_1 = z_1 - h_J = 18 - 47.35 = -29.35$ m. Try $v_1 = 5$ ft/s:

$$(N_R)_1 = (\tfrac{7}{12})(5)/(1.02 \times 10^{-5}) = 2.86 \times 10^5 \qquad f_1 = 0.0225$$

$$29.35 = 0.0225[2000/(\tfrac{7}{12})]\{v_1^2/[(2)(32.2)]\} \qquad v_1 = 4.950 \text{ ft/s} \qquad \text{(value of } f_1 \text{ O.K.)}$$

$$Q_1 = [(\pi)(\tfrac{7}{12})^2/4](4.950) = 1.323 \text{ ft}^3/\text{s} \qquad \text{(away from } J)$$

$$(h_f)_2 = z_2 - h_J = 98 - 47.35 = 50.65 \text{ ft}$$

Try $v_2 = 7.5$ ft/s:

$$(N_R)_2 = (\tfrac{5}{12})(7.5)/(1.02 \times 10^{-5}) = 3.06 \times 10^5 \qquad f_2 = 0.024$$

$$50.65 = 0.024[1000/(\tfrac{5}{12})]\{v_2^2/[(2)(32.2)]\} \qquad v_2 = 7.525 \text{ ft/s} \qquad \text{(value of } f_2 \text{ O.K.)}$$

$$Q_2 = [(\pi)(\tfrac{5}{12})^2/4](7.525) = 1.026 \text{ ft}^3/\text{s} \qquad \text{(toward } J)$$

$$(h_f)_3 = z_3 - h_J = 48 - 47.35 = 0.65 \text{ ft}$$

Try $v_3 = 1$ ft/s:

$$(N_R)_3 = (\tfrac{8}{12})(1)/(1.02 \times 10^{-5}) = 6.54 \times 10^4$$

$$(\epsilon/d)_3 = 0.00085/(\tfrac{8}{12}) = 0.00128 \qquad f_3 = 0.024$$

$$0.65 = 0.024[1500/(\tfrac{8}{12})]\{v_3^2/[(2)(32.2)]\} \qquad v_3 = 0.8804 \text{ ft/s}$$

$$(N_R)_3 = (\tfrac{8}{12})(0.8804)/(1.02 \times 10^{-5}) = 5.75 \times 10^4 \qquad f_3 = 0.024 \qquad \text{(O.K.)}$$

$$Q_3 = [(\pi)(\tfrac{8}{12})^2/4](0.8804) = 0.307 \text{ ft}^3/\text{s} \qquad \text{(toward } J)$$

$$Q_{\text{to } J} - Q_{\text{from } J} = (1.026 + 0.307) - 1.323 = 0.010 \text{ ft}^3/\text{s}$$

Hence, $Q_1 = 1.323$ ft³/s from $J$, and $Q_2 = 1.026$ ft³/s and $Q_3 = 0.307$ ft³/s toward $J$.

**Fig. 12-8**

**12.9** In Fig. 12-9 find the discharges for water at 20 °C with the following pipe data and reservoir elevations: $L_1 = 3000$ m, $D_1 = 1$ m, $\epsilon_1/D_1 = 0.0002$; $L_2 = 600$ m, $D_2 = 0.45$ m, $\epsilon_2/D_2 = 0.002$; $L_3 = 1000$ m, $D_3 = 0.6$ m, $\epsilon_3/D_3 = 0.001$; $z_1 = 30$ m, $z_2 = 18$ m, $z_3 = 9$ m.

▌ Assume $z_J + p_J/\gamma = 23$ m. Then $7 = f_1(\frac{3000}{1})(V_1^2/2g)$, $f_1 = 0.014$, $V_1 = 1.75$ m/s, $Q_1 = 1.380$ m³/s; $5 = f_2(600/0.45)(V_2^2/2g)$, $f_2 = 0.024$, $V_2 = 1.75$ m/s, $Q_2 = 0.278$ m³/s; $14 = f_3(1000/0.60)(V_3^2/2g)$, $f_3 = 0.020$, $V_3 =$

2.87 m/s, $Q_3 = 0.811$ m³/s; so that the inflow is greater than the outflow by $1.380 - 0.278 - 0.811 = 0.291$ m³/s. Assume $z_J + p_J/\gamma = 24.6$ m. Then $5.4 = f_1(\frac{3000}{1})(V_1^2/2g)$, $f_1 = 0.015$, $V_1 = 1.534$ m/s, $Q_1 = 1.205$ m³/s; $6.6 = f_2(600/0.45)(V_2^2/2g)$, $f_2 = 0.024$, $V_2 = 2.011$ m/s, $Q_2 = 0.320$ m³/s; $15.6 = f_3(1000/0.60)(V_3^2/2g)$, $f_3 = 0.020$, $V_3 = 3.029$ m/s, $Q_3 = 0.856$ m³/s. The inflow is still greater by 0.029 m³/s. By extrapolating linearly, $z_J + p_J/\gamma = 24.8$ m, $Q_1 = 1.183$, $Q_2 = 0.325$, $Q_3 = 0.862$ m³/s.

**Fig. 12-9**

**12.10** Prepare a computer program to balance the flows in a system of $N$ reservoirs connected by a common junction. Use the data of Prob. 12.9 to check out the program.

```
10 REM B:   N reservoirs with a common junction
20 DEFINT I,N: DEF FNQ(D1,D2,D3,DH)=D1*DH*LOG(D2+D3/DH)
30 READ N,G,KVIS,II:     DATA 3,9.806,1.007E-6,15
40 LPRINT:     LPRINT"N,G,KVIS,II=";N;G;KVIS;II
50 ZMA=-1000!:     ZMI=1000!
60 FOR I=1 TO N:     READ L(I),D(I),EP(I),Z(I)
70 LPRINT"L,D,EP,Z=";L(I);D(I);EP(I);Z(I)
80 NEXT I
90 DATA 3000.,1.,.0002,30.,600.,.45,.0009,18.,1000.,.6,.0006,9.
100 FOR I=1 TO N:     C=SQR(G*D(I)/L(I)):     E1(I)=-.9650001*D(I)^2*C
110 E2(I)=EP(I)/(3.7*D(I)):     E3(I)=1.784*KVIS/(D(I)*C)     ' FOR EQ.(5.8.15)
120 IF Z(I)>ZMA THEN ZMA=Z(I)                ' ELEV. OF HIGHEST RESERVOIR
130 IF Z(I)<ZMI THEN ZMI=Z(I)                ' ELEV. OF LOWEST RESERVOIR
140 NEXT I
150 FOR I1=1 TO 15:     HJUN=.5*(ZMA+ZMI):     S=0!     ' START BISECTION METHOD
160 FOR I=1 TO N    :HF=Z(I)-HJUN:     HFS=SQR(ABS(HF))
170 Q(I)=FNQ(E1(I),E2(I),E3(I),HFS)*SGN(HF)
180 S=S+Q(I):     PRINT"HF,Q,S=";HF;Q(I);S
190 NEXT I
200 IF S>0! THEN ZMI=HJUN ELSE ZMA=HJUN
210 NEXT I1
220 LPRINT"ELEVATION OF JUNCTION IS ";HJUN
230 LPRINT"PIPE DISCHARGES ARE (POSITIVE INTO THE JUNCTION)"
240 FOR I=1 TO N:     LPRINT"Q(";I;")=";:     LPRINT USING" ###.### ";Q(I);
250 NEXT I:     LPRINT

N,G,KVIS,II= 3  9.806001  1.007E-06  15
L,D,EP,Z= 3000  1  .0002  30
L,D,EP,Z= 600  .45  .0009  18
L,D,EP,Z= 1000  .6  .0006  9
ELEVATION OF JUNCTION IS  24.8801
PIPE DISCHARGES ARE (POSITIVE INTO THE JUNCTION)
Q( 1 )=  1.198 Q( 2 )=  -0.329 Q( 3 )=  -0.869
```

**12.11** Calculate the three volumetric flows in Fig. 12-10.

$$(\epsilon/D)_{AJ} = 1/200 = 0.0050 \qquad (\epsilon/D)_{BJ} = 1/200 = 0.0050 \qquad (\epsilon/D)_{CJ} = 3/300 = 0.0100$$

$$f_{AJ} = 0.032 \qquad f_{BJ} = 0.032 \qquad f_{CJ} = 0.038 \qquad h_f = (f)(L/D)(v^2/2g) = (f)(L/D^5)(8Q^2/g\pi^2)$$

$$28 - h_J = 0.032[1000/(0.200)^5]\{8Q_{AJ}^2/[(9.807)(\pi)^2]\} = 8265Q_{AJ}^2$$

$$25 - h_J = 0.032[300/(0.200)^5]\{8Q_{BJ}^2/[(9.807)(\pi)^2]\} = 2480Q_{BJ}^2$$

$$h_J - 15 = 0.038[600/(0.300)^5]\{8Q_{CJ}^2/[(9.807)(\pi)^2]\} = 775.5Q_{CJ}^2$$

Try $h_J = 19$ m:

$$28 - 19 = 8265Q_{AJ}^2 \qquad Q_{AJ} = 0.03300 \text{ m}^3/\text{s} \qquad 25 - 19 = 2480Q_{BJ}^2 \qquad Q_{BJ} = 0.04919 \text{ m}^3/\text{s}$$

$$19 - 15 = 775.5Q_{CJ}^2 \qquad Q_{CJ} = 0.07182 \text{ m}^3/\text{s}$$

$$Q_{\text{to }J} - Q_{\text{from }J} = 0.03300 + 0.04919 - 0.07182 = 0.01037 \text{ m}^3/\text{s}$$

Try $h_J = 20$ m:

$$28 - 20 = 8265Q_{AJ}^2 \qquad Q_{AJ} = 0.03111 \text{ m}^3/\text{s} \qquad 25 - 20 = 2480Q_{BJ}^2 \qquad Q_{BJ} = 0.04490 \text{ m}^3/\text{s}$$
$$20 - 15 = 775.5Q_{CJ}^2 \qquad Q_{CJ} = 0.08030 \text{ m}^3/\text{s}$$
$$Q_{\text{to }J} - Q_{\text{from }J} = 0.03111 + 0.04490 - 0.08030 = -0.00429 \text{ m}^3/\text{s}$$

Try $h_J = 19.7$ m:

$$28 - 19.7 = 8265Q_{AJ}^2 \qquad Q_{AJ} = 0.03169 \text{ m}^3/\text{s} \qquad 25 - 19.7 = 2480Q_{BJ}^2 \qquad Q_{BJ} = 0.04623 \text{ m}^3/\text{s}$$
$$19.7 - 15 = 775.5Q_{CJ}^2 \qquad Q_{CJ} = 0.07785 \text{ m}^3/\text{s}$$
$$Q_{\text{to }J} - Q_{\text{from }J} = 0.03169 + 0.04623 - 0.07785 = 0.00007 \text{ m}^3/\text{s}$$

Hence, $Q_{CJ} = 77.8$ L/s from $J$, and $Q_{AJ} = 31.7$ L/s and $Q_{BJ} = 46.2$ L/s toward $J$.

**Fig. 12-10**

**12.12** Calculate the five volumetric flows in Fig. 12-11.

$$h_f = (f)(L/D)(v^2/2g) = (f)(L/D^5)(8Q^2/g\pi^2)$$
$$h_J - 78 = 0.025[600/(0.600)^5]\{8Q_A^2/[(9.807)(\pi)^2]\} = 15.945Q_A^2$$
$$h_J - 88 = 0.030[600/(0.450)^5]\{8Q_B^2/[(9.807)(\pi)^2]\} = 80.624Q_B^2$$
$$h_j - h_J = 0.030[900/(0.450)^5]\{8Q_{JJ}^2/[(9.807)(\pi)^2]\} = 120.94Q_{JJ}^2$$
$$98 - h_j = 0.030[300/(0.450)^5]\{8Q_C^2/[(9.807)(\pi)^2]\} = 40.312Q_C^2$$
$$88 - h_j = 0.030[300/(0.450)^5]\{8Q_B^2/[(9.807)(\pi)^2]\} = 40.312Q_D^2$$

Try $h_J = 82$ m:

$$82 - 78 = 15.945Q_A^2 \qquad Q_A = 0.5009 \text{ m}^3/\text{s} \qquad 82 - 88 = 80.624Q_B^2 \qquad Q_B = -0.2728 \text{ m}^3/\text{s}$$
$$h_j - 82 = (120.94)(0.5009 - 0.2728)^2 \qquad h_j = 88.29 \text{ m}$$
$$98 - 88.29 = 40.312Q_C^2 \qquad Q_C = 0.4908 \text{ m}^3/\text{s}$$
$$88 - 88.29 = 40.312Q_D^2 \qquad Q_D = -0.0848 \text{ m}^3/\text{s}$$

If the above values are correct, $Q_{Jj}$ $(0.5009 - 0.2728 = 0.2281 \text{ m}^3/\text{s})$ must equal $Q_C + Q_D$ $(0.4908 - 0.0848 = 0.4060 \text{ m}^3/\text{s})$. Since they are not equal, try $h_J = 83$ m and note that $Q_{Jj} < Q_C + Q_D$:

$$83 - 78 = 15.945Q_A^2 \qquad Q_A = 0.5600 \text{ m}^3/\text{s} \qquad 83 - 88 = 80.624Q_B^2 \qquad Q_B = -0.2490 \text{ m}^3/\text{s}$$
$$h_j - 83 = (120.94)(0.5600 - 0.2490)^2 \qquad h_j = 94.70 \text{ m}$$
$$98 - 94.70 = 40.312Q_C^2 \qquad Q_C = 0.2861 \text{ m}^3/\text{s}$$
$$88 - 94.70 = 40.312Q_D^2 \qquad Q_D = -0.4077 \text{ m}^3/\text{s}$$

If these values are correct, $Q_{Jj}$ $(0.5600 - 0.2490 = 0.3110 \text{ m}^3/\text{s})$ must equal $Q_C + Q_D$ $(0.2861 - 0.4077 = -0.1216 \text{ m}^3/\text{s})$. Since this time, $Q_{Jj} > Q_C + Q_D$, the correct value of $h_J$ must be between 82 m and 83 m. Try $h_J = 82.24$ m:

$$82.24 - 78 = 15.945Q_A^2 \qquad Q_A = 0.5157 \text{ m}^3/\text{s} \qquad 82.24 - 88 = 80.624Q_B^2 \qquad Q_B = -0.2673 \text{ m}^3/\text{s}$$
$$h_j - 82.24 = (120.94)(0.5157 - 0.2673)^2 \qquad h_j = 89.70 \text{ m}$$
$$98 - 89.70 = 40.312Q_C^2 \qquad Q_C = 0.4538 \text{ m}^3/\text{s}$$
$$88 - 89.70 = 40.312Q_D^2 \qquad Q_D = -0.2054 \text{ m}^3/\text{s}$$

If these values are correct, $Q_{Jj}$ $(0.5157 - 0.2673 = 0.2484 \text{ m}^3/\text{s})$ must equal $Q_C + Q_D$

$(0.4538 - 0.2054 = 0.2484 \text{ m}^3/\text{s})$. Since they are in fact equal, the correct values must be $Q_A = 516 \text{ L/s}$, $Q_B = 267 \text{ L/s}$, $Q_{Jj} = 248 \text{ L/s}$, $Q_C = 454 \text{ L/s}$, and $Q_D = 205 \text{ L/s}$.

**Fig. 12-11**

**12.13** In Fig. 12-12, pipe 1 is 36-in smooth concrete, 5000 ft long; pipe 2 is 24-in cast iron, 1500 ft long; and pipe 3 is 18-in cast iron, 4000 ft long. The elevations of water surfaces in reservoirs $A$ and $B$ are 300 ft and 250 ft, respectively, and discharge $Q_1$ is 50 ft$^3$/s. Find the elevation of the surface of reservoir $C$.

$$h_f = (f)(L/d)(v^2/2g) \qquad (\epsilon/d)_1 = 0.001/(\tfrac{36}{12}) = 0.000333 \qquad N_R = dv/v$$
$$v_1 = Q_1/A_1 = 50/[(\pi)(\tfrac{36}{12})^2/4] = 7.074 \text{ ft/s}$$
$$(N_R)_1 = (\tfrac{36}{12})(7.074)/(1.05 \times 10^{-5}) = 2.02 \times 10^6$$

From Fig. A-5, $f_1 = 0.0157$. $(h_f)_1 = 0.0157[5000/(\tfrac{36}{12})]\{7.074^2/[(2)(32.2)]\} = 20.33 \text{ ft}$, $h_J = 300 - 20.33 = 279.67 \text{ ft}$, $(h_f)_2 = 279.67 - 250 = 29.67 \text{ ft}$. Assume $f_2 = 0.0162$:

$$29.67 = 0.0162[1500/(\tfrac{24}{12})]\{v_2^2/[(2)(32.2)]\} \qquad v_2 = 12.54 \text{ ft/s}$$
$$(N_R)_2 = (\tfrac{24}{12})(12.54)/(1.05 \times 10^{-5}) = 2.39 \times 10^6$$
$$(\epsilon/d)_2 = 0.00085/(\tfrac{24}{12}) = 0.000425 \qquad f_2 = 0.0162 \qquad \text{(O.K.)}$$
$$Q_2 = A_2 v_2 = [(\pi)(\tfrac{24}{12})^2/4](12.54) = 39.40 \text{ ft}^3/\text{s} \qquad Q_3 = 50 - 39.40 = 10.6 \text{ ft}^3/\text{s}$$
$$v_3 = 10.6/[(\pi)(\tfrac{18}{12})^2/4] = 5.998 \text{ ft/s} \qquad (N_R)_3 = (\tfrac{18}{12})(5.998)/(1.05 \times 10^{-5}) = 8.57 \times 10^5$$
$$(\epsilon/d)_3 = 0.00085/(\tfrac{18}{12}) = 0.000567 \qquad f_3 = 0.0177$$
$$(h_f)_3 = 0.0177[4000/(\tfrac{18}{12})]\{5.998^2/[(2)(32.2)]\} = 26.37 \text{ ft}$$
$$z_C = 279.67 - 26.37 = 253.30 \text{ ft}$$

**Fig. 12-12**

**12.14** In Prob. 12.13, suppose that the surface elevations of reservoirs $A$ and $C$ are 300 ft and 225 ft, respectively, and discharge $Q_2$ is 20 cfs into $B$; other data are unchanged. Find the surface elevation of reservoir $B$.

$$h_f = (f)(L/d)(v^2/2g) \qquad v_2 = Q_2/A_2 = 20/[(\pi)(\tfrac{24}{12})^2/4] = 6.366 \text{ ft/s} \qquad N_R = dv/v$$
$$(N_R)_2 = (\tfrac{24}{12})(6.366)/(1.05 \times 10^{-5}) = 1.21 \times 10^6$$
$$(\epsilon/d)_2 = 0.000425 \qquad \text{(from Prob. 12.13)}$$

From Fig. A-5, $f_2 = 0.0168$. $(h_f)_2 = 0.0168[1500/(\frac{24}{12})]\{6.366^2/[(2)(32.2)]\} = 7.93$ ft.

$$Q_1 - Q_3 = 20 \tag{1}$$

$$(h_f)_1 + (h_f)_3 = 300 - 225 = 75 \text{ ft} \tag{2}$$

We need to find an $h_f$ in line 1 or 3 such that both Eqs. (1) and (2) are satisfied. Try $(h_f)_3 = 64$ ft and $f_3 = 0.0175$:

$$64 = 0.0175[4000/(\tfrac{18}{12})]\{v_3^2/[(2)(32.2)]\} \qquad v_3 = 9.398 \text{ ft/s}$$

$$(N_R)_3 = \tfrac{18}{12}(9.398)/(1.05 \times 10^{-5}) = 1.34 \times 10^6$$

$$(\epsilon/d)_3 = 0.000567 \qquad \text{(from Prob. 12.13)}$$

$$f_3 = 0.0175 \qquad \text{(O.K.)} \qquad Q_3 = A_3 v_3 = [(\pi)(\tfrac{18}{12})^2/4](9.398) = 16.61 \text{ ft}^3/\text{s}$$

$$Q_1 - 16.61 = 20 \qquad Q_1 = 36.61 \text{ ft}^3/\text{s}$$

$$v_1 = 36.61/[(\pi)(\tfrac{36}{12})^2/4] = 5.179 \text{ ft/s} \qquad (N_R)_1 = (\tfrac{36}{12})(5.179)/(1.05 \times 10^{-5}) = 1.48 \times 10^6$$

$$(\epsilon/d)_1 = 0.000333 \qquad \text{(from Prob. 12.13)} \qquad f_1 = 0.0158$$

$$(h_f)_1 = 0.0158[5000/(\tfrac{36}{12})]\{5.179^2/[(2)(32.2)]\} = 10.97 \text{ ft}$$

$$z_{BB} = z_A - (h_f)_1 - (h_f)_2 = 300 - 10.97 - 7.93 = 281.1 \text{ ft}$$

**12.15** Given, in Prob. 12.14, that discharge $Q_2$ is 20 cfs *out of* reservoir B, find the elevation of the surface of B.

▌ As in Prob. 12.14, $(h_f)_2 = 7.93$ ft, but Eq. (1) becomes

$$Q_3 - Q_1 = 20 \tag{1}$$

Equation (2) is the same as in Prob. 12.14; that is,

$$(h_f)_1 + (h_f)_3 = 300 - 225 = 75 \text{ ft} \tag{2}$$

We need to find an $h_f$ in line 1 or 3 such that both Eqs. (1) and (2) are satisfied. Try $(h_f)_3 = 77$ ft and $f_3 = 0.0175$:

$$77 = 0.0175[4000/(\tfrac{18}{12})]\{v_3^2/[(2)(32.2)]\} \qquad v_3 = 10.31 \text{ ft/s}$$

$$(N_R)_3 = (\tfrac{18}{12})(10.31)/(1.05 \times 10^{-5}) = 1.47 \times 10^6$$

$$(\epsilon/d)_3 = 0.000567 \qquad \text{(from Prob. 12.13)}$$

$$f_3 = 0.0175 \qquad \text{(O.K.)}$$

$$Q_3 = A_3 v_3 = [(\pi)(\tfrac{18}{12})^2/4](10.31) = 18.22 \text{ ft}^3/\text{s} \qquad 18.22 - Q_1 = 20$$

$$Q_1 = -1.78 \text{ ft}^3/\text{s} \qquad \text{(i.e., flow is into reservoir } A)$$

$$v_1 = 1.78/[(\pi)(\tfrac{36}{12})^2/4] = 0.2518 \text{ ft/s}$$

$$(N_R)_1 = (\tfrac{36}{12})(0.2518)/(1.05 \times 10^{-5}) = 7.19 \times 10^4$$

$$(\epsilon/d)_1 = 0.000333 \qquad \text{(from Prob. 12.13)} \qquad f_1 = 0.0208$$

$$(h_f)_1 = 0.0208[5000/(\tfrac{36}{12})]\{0.2518^2/[(2)(32.2)]\} = 0.03 \text{ ft}$$

$$z_B = z_A + (h_f)_1 + (h_f)_2 = 300 + 0.03 + 7.93 = 308 \text{ ft}$$

**12.16** In Fig. 12-12, assume that 1, 2, and 3 represent 900 m of 60-cm, 300 m of 45-cm, and 1200 m of 40-cm, respectively, of new welded steel pipe. The surface elevations of A, B, and C are 32 m, 20 m, and 2 m, respectively. Estimate the water flows in all pipes.

▌ $$(\epsilon/d)_1 = 0.000046/(0.60) = 0.0000767 \qquad (\epsilon/d)_2 = 0.000046/(0.45) = 0.0001022$$

$$(\epsilon/d)_3 = 0.000046/(0.40) = 0.0001150$$

Assuming complete turbulence (high Reynolds numbers), friction factors for these values of $(\epsilon/d)$ will be $f_1 = 0.0115$, $f_2 = 0.0122$, and $f_3 = 0.0126$.

$$h_f = (f)(L/d)(v^2/2g) = (f)(L/d^5)(8Q^2/g\pi^2)$$

$$(h_f)_1 = 0.0115[900/(0.60)^5]\{8Q_1^2/[(9.807)(\pi)^2]\} = 11.00Q_1^2$$

$$(h_f)_2 = 0.0122[300/(0.45)^5]\{8Q_2^2/[(9.807)(\pi)^2]\} = 16.39Q_2^2$$

$$(h_f)_3 = 0.0126[1200/(0.40)^5]\{8Q_3^2/[(9.807)(\pi)^2]\} = 122.0Q_3^2$$

$$32 - 11.00Q_1^2 = h_J \tag{1}$$

$$h_J - 16.39 Q_2^2 = 20 \quad \text{or} \quad 20 - 16.39 Q_2^2 = h_J \tag{2}$$

$$h_J - 122.0 Q_3^2 = 2 \tag{3}$$

$$Q_1 = Q_2 + Q_3 \quad \text{or} \quad Q_1 + Q_2 = Q_3 \tag{4}$$

Assume no flow in pipe 2, in which case $h_J = 20$ m. $32 - 11.00 Q_1^2 = 20$, $Q_1^2 = 1.091$; $20 - 122.0 Q_3^2 = 2$, $Q_3^2 = 0.1475$. Since $Q_1 > Q_3$, flow must be into reservoir $B$, and $h_J > 20$ m. By trial and error, $h_J = 23.5$ m. $32 - 11.00 Q_1^2 = 23.5$, $Q_1 = 0.879$ m³/s; $23.5 - 16.39 Q_2^2 = 20$, $Q_2 = 0.462$ m³/s; $23.5 - 122.0 Q_3^2 = 2$, $Q_3 = 0.420$ m³/s. Check Eq. (4): $0.879 = 0.462 + 0.420$, $0.879 = 0.882$ (close enough). Check the Reynolds numbers for normal temperature:

$$N_R = dv/\nu \qquad v_1 = 0.879/[(\pi)(0.60)^2/4] = 3.109 \text{ m/s}$$

$$(N_R)_1 = (0.60)(3.109)/(1.02 \times 10^{-6}) = 1.83 \times 10^6$$

$$v_2 = 0.462/[(\pi)(0.45)^2/4] = 2.905 \text{ m/s}$$

$$(N_R)_2 = (0.45)(2.905)/(1.02 \times 10^{-6}) = 1.28 \times 10^6$$

$$v_3 = 0.420/[(\pi)(0.40)^2/4] = 3.342 \text{ m/s}$$

$$(N_R)_1 = (0.40)(3.342)/(1.02 \times 10^{-6}) = 1.31 \times 10^6$$

From Fig. A-5, $f = 0.0135$ approximately for all pipes; hence, the above flows should be multiplied by $(0.012/0.0135)$, giving $Q_1 = (0.012/0.0135)(0.879) = 0.781$ m³/s, $Q_2 = (0.012/0.0135)(0.462) = 0.411$ m³/s, $Q_3 = (0.012/0.0135)(0.420) = 0.373$ m³/s. Further refinement of these approximations is not justified.

**12.17** Suppose that, in Fig. 12-12, pipe 1 is 1500 ft of 12-in new cast iron, pipe 2 is 800 ft of 6-in wrought iron, and pipe 3 is 1200 ft of 8-in wrought iron. The water surface of reservoir $B$ is 20 ft below that of $A$, while junction $J$ is 40 ft below the surface of $A$. When the pressure head at $J$ is 30 ft, find the flow in each pipe.

**∎** $\qquad (h_f)_1 = 40 - 30 = 10 \text{ ft} \qquad (h_f)_2 = (40 - 20) - 30 = -10 \text{ ft} \qquad h_f = (f)(L/d)(v^2/2g)$

Assume $f_1 = 0.019$:

$$10 = 0.019[1500/(\tfrac{12}{12})]\{v_1^2/[(2)(32.2)]\} \qquad v_1 = 4.754 \text{ ft/s} \qquad N_R = dv/\nu$$

$$(N_R)_1 = (\tfrac{12}{12})(4.754)/(1.05 \times 10^{-5}) = 4.53 \times 10^5 \qquad (\epsilon/d)_1 = 0.00085/(\tfrac{12}{12}) = 0.00085$$

From Fig. A-5, $f_1 = 0.0198$. Try $f_1 = 0.0198$:

$$10 = 0.0198[1500/(\tfrac{12}{12})]\{v_1^2/[(2)(32.2)]\} \qquad v_1 = 4.657 \text{ ft/s}$$

$$(N_R)_1 = (\tfrac{12}{12})(4.657)/(1.05 \times 10^{-5}) = 4.44 \times 10^5 \qquad f_1 = 0.0198 \quad \text{(O.K.)}$$

$$Q_1 = A_1 v_1 = [(\pi)(\tfrac{12}{12})^2/4](4.657) = 3.658 \text{ ft}^3/\text{s}$$

Assume $f_2 = 0.015$:

$$10 = 0.015[800/(\tfrac{6}{12})]\{v_2^2/[(2)(32.2)]\} \qquad v_2 = 5.180 \text{ ft/s}$$

$$(N_R)_2 = (\tfrac{6}{12})(5.180)/(1.05 \times 10^{-5}) = 2.47 \times 10^5 \qquad (\epsilon/d)_2 = 0.00015/(\tfrac{6}{12}) = 0.00030 \qquad f = 0.0175$$

Try $f_2 = 0.0175$:

$$10 = 0.0175[800/(\tfrac{6}{12})]\{v_2^2/[(2)(32.2)]\} \qquad v_2 = 4.796 \text{ ft/s}$$

$$(N_R)_2 = (\tfrac{6}{12})(4.796)/(1.05 \times 10^{-5}) = 2.28 \times 10^5 \qquad f_2 = 0.0175 \quad \text{(O.K.)}$$

$$Q_2 = [(\pi)(\tfrac{6}{12})^2/4](4.796) = 0.942 \text{ ft}^3/\text{s} \qquad Q_3 = Q_1 - Q_2 = 3.658 - 0.942 = 2.716 \text{ ft}^3/\text{s}$$

**12.18** In the reservoir system of Fig. 12-13, $z_A = 210$ ft, $z_B = 240$ ft, $z_C = 130$ ft, $z_D = 150$ ft, $BD = 3000$ ft of 4-in cast iron pipe, $AD = 2000$ ft of 1-in steel pipe, and $DC = 500$ ft of 6-in cast iron pipe. Using $f = 0.025$ and neglecting minor losses, determine the flow in each pipe.

**∎** $h_f = (f)(L/d)(v^2/2g)$. Let $p_D/\gamma = -15$ ft:

$$(h_f)_{AD} = 210 - 150 - (-15) = 0.025[2000/(\tfrac{1}{12})]\{v_{AD}^2/[(2)(32.2)]\} \qquad v_{AD} = 2.837 \text{ ft/s}$$

$$(h_f)_{BD} = 240 - 150 - (-15) = 0.025[3000/(\tfrac{4}{12})]\{v_{BD}^2/[(2)(32.2)]\} \qquad v_{BD} = 5.482 \text{ ft/s}$$

$$(h_f)_{CD} = 130 - 150 - (-15) = 0.025[500/(\tfrac{6}{12})]\{v_{CD}^2/[(2)(32.2)]\} \qquad v_{CD} = 3.589 \text{ ft/s} \quad \text{(away from } D\text{)}$$

If these velocities are correct, $Q_{AD} + Q_{BD}$ must equal $Q_{CD}$: $Q_{AD} + Q_{BD} = [(\pi)(\tfrac{1}{12})^2/4](2.837) + [(\pi)(\tfrac{4}{12})^2/4](5.482) = 0.4939$ ft³/s, $Q_{CD} = [(\pi)(\tfrac{6}{12})^2/4](3.589) = 0.7047$ ft³/s. Since they are unequal, try

$p_D/\gamma = -17$ ft:

$$(h_f)_{AD} = 210 - 150 - (-17) = 0.025[2000/(\tfrac{1}{12})]\{v_{AD}^2/[(2)(32.2)]\} \qquad v_{AD} = 2.875 \text{ ft/s}$$

$$(h_f)_{BD} = 240 - 150 - (-17) = 0.025[3000/(\tfrac{4}{12})]\{v_{BD}^2/[(2)(32.2)]\} \qquad v_{BD} = 5.534 \text{ ft/s}$$

$$(h_f)_{CD} = 130 - 150 - (-17) = 0.025[500/(\tfrac{6}{12})]\{v_{CD}^2/[(2)(32.2)]\} \qquad v_{CD} = 2.789 \text{ ft/s} \qquad (\text{away from } D)$$

$$Q_{AD} + Q_{BD} = [(\pi)(\tfrac{1}{12})^2/4](2.875) + [(\pi)(\tfrac{4}{12})^2/4](5.534) = 0.4986 \text{ ft}^3/\text{s}$$

$$Q_{CD} = [(\pi)(\tfrac{6}{12})^2/4](2.789) = 0.5476 \text{ ft}^3/\text{s}$$

$Q_{AD} + Q_{BD}$ is still not equal to $Q_{CD}$; try $p_D/\gamma = -17.4$ ft:

$$(h_f)_{AD} = 210 - 150 - (-17.5) = 0.025[2000/(\tfrac{1}{12})]\{v_{AD}^2/[(2)(32.2)]\} \qquad v_{AD} = 2.884 \text{ ft/s}$$

$$(h_f)_{BD} = 240 - 150 - (-17.5) = 0.025[3000/(\tfrac{4}{12})]\{v_{BD}^2/[(2)(32.2)]\} \qquad v_{BD} = 5.547 \text{ ft/s}$$

$$(h_f)_{CD} = 130 - 150 - (-17.5) = 0.025[500/(\tfrac{6}{12})]\{v_{CD}^2/[(2)(32.2)]\} \qquad v_{CD} = 2.538 \text{ ft/s} \qquad (\text{away from } D)$$

$$Q_{AD} + Q_{BD} = [(\pi)(\tfrac{1}{12})^2/4](2.884) + [(\pi)(\tfrac{4}{12})^2/4](5.547) = 0.4998 \text{ ft}^3/\text{s}$$

$$Q_{CD} = [(\pi)(\tfrac{6}{12})^2/4](2.538) = 0.4983 \text{ ft}^3/\text{s}$$

Since $Q_{AD} + Q_{BD}$ is close enough to $Q_{CD}$, take

$$Q_{AD} = [(\pi)(\tfrac{1}{12})^2/4](2.884) = 0.015 \text{ ft}^3/\text{s} \qquad (\text{toward } D)$$

$$Q_{BD} = [(\pi)(\tfrac{4}{12})^2/4](5.547) = 0.483 \text{ ft}^3/\text{s} \qquad (\text{toward } D)$$

$$Q_{CD} = 0.498 \text{ ft}^3/\text{s} \qquad (\text{away from } D)$$

Fig. 12-13

**12.19** Three reservoirs $A$, $B$, and $C$ whose water-surface elevations are $z_A = 60$ ft, $z_B = 50$ ft, and $z_C = 32$ ft, are interconnected by a pipe system with a common junction $D$, $z_D = 35$ ft. The pipes are as follows: from $A$ to junction, $L = 800$ ft, $d = 3$ in; from $B$ to junction, $L = 500$ ft, $d = 10$ in; from $C$ to junction, $L = 1000$ ft, $d = 4$ in. Assume $f = 0.02$ for all pipes and neglect minor losses and velocity heads. Determine whether the flow is into or out of reservoir $B$.

▌ Assume no flow between $B$ and $D$.

$$h_f = (f)(L/d)(v^2/2g) \qquad (h_f)_A = 0.02[800/(\tfrac{3}{12})]\{v_A^2/[(2)(32.2)]\} = 0.9938v_A^2$$

$$(h_f)_C = 0.02[1000/(\tfrac{4}{12})]\{v_C^2/[(2)(32.2)]\} = 0.9317v_C^2$$

$$A_A v_A = A_C v_C \qquad [(\pi)(\tfrac{3}{12})^2/4](v_A) = [(\pi)(\tfrac{4}{12})^2/4](v_C)$$

$$v_A = 1.778v_C \qquad (h_f)_A/(h_f)_C = 0.9938v_A^2/0.9317v_C^2 = (0.9938)(1.778v_C)^2/0.9317v_C^2 = 3.372$$

Hence, $(h_f)_A = [3.372/(3.372 + 1)](60 - 32) = 21.60$ ft, and $z_D = 60 - 21.60 = 38.40$ ft. Since our no-flow assumption makes $D$ too high (by 3.40 ft) relative to $B$, the flow must be out of $B$ into $D$.

**13.1** Develop the expression utilized to study flows in a pipe network (Hardy Cross method).

▋ The method of attack, developed by Professor Hardy Cross, consists of assuming flows throughout the network, and then balancing the calculated head losses (LH). In the simple looping pipe system shown in Fig. 13-1, for the correct flow in each branch of the loop,

$$\text{LH}_{ABC} = \text{LH}_{ADC} \quad \text{or} \quad \text{LH}_{ABC} - \text{LH}_{ADC} = 0 \tag{1}$$

In order to utilize this relationship, the flow formula to be used must be written in the form $\text{LH} = kQ^n$. For the Hazen–Williams formula, this expression is $\text{LH} = kQ^{1.85}$.

But, since we are assuming flows $Q_o$, the correct flow $Q$ in any pipe of a network can be expressed as $Q = Q_o + \Delta$, where $\Delta$ is the correction to be applied to $Q_o$. Then, using the binomial theorem, $kQ^{1.85} = k(Q_o + \Delta)^{1.85} = k(Q_o^{1.85} + 1.85Q_o^{1.85-1}\Delta + \cdots)$. Terms beyond the second can be neglected since $\Delta$ is small compared with $Q_o$.

For the loop above, substituting in expression ($1$) we obtain $k(Q_o^{1.85} + 1.85Q_o^{0.85}\Delta) - k(Q_{o'}^{1.85} + 1.85Q_{o'}^{0.85}\Delta) = 0$, $k(Q_o^{1.85} - Q_{o'}^{1.85}) + 1.85k(Q_o^{0.85} - Q_{o'}^{0.85})\Delta = 0$. Solving for $\Delta$,

$$\Delta = -k(Q_o^{1.85} - Q_{o'}^{1.85})/[1.85k(Q_o^{0.85} - Q_{o'}^{0.85})] \tag{2}$$

In general, we may write for a more complicated loop,

$$\Delta = -\sum kQ_o^{1.85}/(1.85 \sum kQ_o^{0.85}) \tag{3}$$

But $kQ_o^{1.85} = \text{LH}$ and $kQ_o^{0.85} = (\text{LH})/Q_o$. Therefore

$$\Delta = -\sum (\text{LH})/[1.85 \sum (\text{LH}/Q_o)] \quad \text{for each loop of a network} \tag{4}$$

In utilizing expression ($4$), care must be exercised regarding the sign of the numerator. Expression ($1$) indicates that clockwise flows may be considered as producing clockwise losses, and counterclockwise flows, counterclockwise losses. This means that the minus sign is assigned to all counterclockwise conditions in a loop, namely flow $Q$ and lost head LH. Hence to avoid mistakes, this sign notation must be observed in carrying out a solution. On the other hand, the denominator of ($4$) is always positive.

Fig. 13-1

**13.2** In Fig. 13-2, for $Q = 11.7$ mgd total, how much flow occurs in each branch of the loop, using the Hardy Cross method?

▋ Values of $Q_{12}$ and $Q_{16}$ are assumed to be 4.0 mgd and 7.7 mgd, respectively. The tabulation below is prepared, (note the $-7.70$ mgd), the values of $S$ calculated by Fig. A-17, then $\text{LH} = S \times L$, and $\text{LH}/Q_o$ can be calculated. Note that the large $\sum \text{LH}$ indicates that the $Q$'s are not well-balanced. (The values were assumed deliberately to produce this large $\sum \text{LH}$, to illustrate the procedure.)

| $D$ | $L$ | assumed $Q_o$ mgd | $S$, ft/1000 ft | LH, ft | LH/$Q_o$ | $\Delta$ | $Q_1$ |
|------|------|------|------|------|------|------|------|
| 12 in | 5000 ft | 4.00 | 19.5 | 97.5 | 24.4 | −0.85 | 3.15 |
| 16 in | 3000 ft | −7.70 | −16.3 | −48.9 | 6.4 | −0.85 | −8.55 |
| | | $|\Sigma| = 11.70$ | | $\Sigma = +48.6$ | 30.8 | | 11.70 |

$$\Delta = -\sum \text{LH}/[1.85 \sum (\text{LH}/Q)] = -(+48.6)/[1.85(30.8)] = -0.85 \text{ mgd}$$

Then the $Q_1$ values become $(4.00 - 0.85) \doteq 3.15$ mgd and $(-7.70 - 0.85) = -8.55$ mgd. Repeating the calculation produces

| $S$ | LH | LH/$Q_1$ | $\Delta$ | $Q_2$ |
|---|---|---|---|---|
| 12.5 | 62.5 | 19.80 | −0.06 | 3.09 |
| −19.8 | −59.4 | 6.95 | −0.06 | −8.61 |
| | $\Sigma = +3.1$   26.75 | | | 11.70 |

No further calculation is necessary, since the slide rule or chart cannot be read to the accuracy of 0.06 mgd. Ideally, $\Sigma$ LH should equal zero, but this goal is seldom attained.

**Fig. 13-2**

**13.3** For the pipe network giving pipe diameters and lengths and external flows entering and leaving the network as shown in Fig. 13-3a, find the flow rate in each pipe in the network.

❚ The first iteration in carrying out the Hardy Cross analysis is given in the table below. A detailed explanation corresponding to steps in carrying out the procedure is given after the table.

**Iteration No. 1**

**Loop I**

| (1) pipe no. | (2) length, ft | (3) diameter, in | (4) flow rate ($Q$), ft³/s | (5) unit head loss, ft/ft | (6) head loss ($h_f$), ft | (7) $h_f/Q$ s/ft² |
|---|---|---|---|---|---|---|
| given | given | given | estimated or assumed | from pipe diagram | (2) × (5) | (6)/(4) |
| 1 | 2000 | 18 | 7.00 | 0.0034 | 6.800 | 0.971 |
| 2 | 900 | 24 | 3.50 | 0.000235 | 0.212 | 0.061 |
| 3 | 2800 | 12 | −2.19 | −0.0029 | −8.120 | 3.708 |
| 4 | 1100 | 24 | −7.00 | −0.00085 | −0.935 | 0.134 |
| | | | | | −2.043 | 4.874 |

**Loop II**

| (1) pipe no. | (2) length, ft | (3) diameter, in | (4) flow rate ($Q$), ft³/s | (5) unit head loss, ft/ft | (6) head loss ($h_f$), ft | (7) $h_f/Q$, s/ft² |
|---|---|---|---|---|---|---|
| given | given | given | estimated or assumed | from pipe diagram | (2) × (5) | (6)/(4) |
| 5 | 2200 | 12 | 3.50 | 0.0070 | 15.400 | 4.400 |
| 6 | 750 | 18 | −3.14 | −0.00078 | −0.585 | 0.186 |
| 7 | 2600 | 24 | −5.69 | −0.00057 | −1.482 | 0.260 |
| 2 | 900 | 24 | −3.50 | −0.00023 | −0.207 | 0.059 |
| | | | | | 13.126 | 4.905 |

**Step 1:** Columns (1), (2), and (3) in the table above record given data. Flow rates recorded in column (4) are the initial flow rate estimates. These are listed on the sketch of Fig. 13-3b, but they are not "given" data. The flow rate of 14.00 cfs entering joint A was estimated to separate such that 7.00 cfs goes through pipe 1 and an equal amount goes through pipe 4. At joint F, since 7.00 cfs enters the joint from pipe 4 and 4.81 cfs leaves the joint externally, the flow rate in pipe 3 has to be 2.19 cfs (that is, 7.00 cfs − 4.81 cfs) in the direction away from joint F in order to satisfy the principle that the total flow entering a joint must be equal to the total flow leaving that joint. Remaining flow rates were estimated in a similar manner, making sure that for each joint the total flow entering equaled the total flow leaving. It should be noted that clockwise flows in each loop (such as in pipes 1 and 5) are indicated as positive, while counterclockwise flows (such as in pipes 3 and 6) are indicated as negative. It should particularly be noted that the flow in pipe 2 is clockwise with respect to loop I but counterclockwise with respect to loop II; hence, it is indicated as positive when listed in loop I and negative when listed in loop II.

**Step 2:** Unit head losses in column (5) are determined from Fig. A-13, based on diameters [column (3)] and flow rates [column (4)]. For example, pipe 1 has a diameter of 18 in and flow rate of 7.00 cfs; hence, the unit head loss is determined from Fig. A-13 to be 0.0034 ft/ft. Head losses in column (6) are determined by multiplying pipe lengths [column (2)] by unit head losses [column (5)]. For example, pipe 1 has a length of 2000 ft and unit head loss of 0.0034 ft/ft; hence, the head loss is (2000 ft)(0.0034 ft/ft), or 6.800 ft. It should be noted that head losses are positive if their corresponding flow rates are positive and negative if their corresponding flow rates are negative. The $h_f/Q$ fractions in column (7) are determined by dividing head losses [column (6)] by flow rates [column (4)]. For example, pipe 1 has a head loss of 6.800 ft and a flow rate of 7.00 cfs; hence, $h_f/Q$ is 6.800/7.00, or 0.971 s/ft$^2$.

**Step 3:** Algebraic sums of head losses are observed from the table to be −2.043 ft in loop I and 13.126 ft in loop II. Since these are not zero, the original estimated flows are not correct.

**Step 4:** A flow rate correction ($\Delta Q$) can be computed for each loop: $\Delta Q = -\sum h_f/[n \sum (h_f/Q)]$. (Since the Hazen–Williams formula is being used, $n = 1.85$.) $(\Delta Q)_{\text{loop I}} = -(-2.043)/[(1.85)(4.874)] = 0.23$ cfs, $(\Delta Q)_{\text{loop II}} = -(13.126)/[(1.85)(4.905)] = -1.45$ cfs.

**Step 5:** Adjusted flow rates for each pipe are determined by adding flow rate corrections to the previous rate for each pipe. These are as follows:

### Loop I

| pipe | old $Q$, cfs | $\Delta Q$, cfs | new $Q$, cfs |
|------|--------------|-----------------|--------------|
| 1 | 7.00 | +0.23 | 7.23 |
| 2 | 3.50 | +0.23 + 1.45 | 5.18 |
| 3 | −2.19 | +0.23 | −1.96 |
| 4 | −7.00 | +0.23 | −6.77 |

### Loop II

| pipe | old $Q$, cfs | $\Delta Q$, cfs | new $Q$, cfs |
|------|--------------|-----------------|--------------|
| 5 | 3.50 | −1.45 | 2.05 |
| 6 | −3.14 | −1.45 | −4.59 |
| 7 | −5.69 | −1.45 | −7.14 |
| 2 | −3.50 | −1.45 − 0.23 | −5.18 |

These adjusted flow rates are shown in Fig. 13-3c. It should be noted that the flow rate in pipe 2 was adjusted using flow rate corrections for both loops, since this pipe is common to both. It should be particularly noted that the sign of the flow rate correction for loop II was reversed when it was applied in loop I (and vice versa).

**Step 6:** Return to step 2 and repeat the entire procedure using adjusted flow rates. This is given in the table below.

**Iteration No. 2**

### Loop I

| (1)<br>pipe<br>no. | (2)<br>length,<br>ft | (3)<br>diameter,<br>in | (4)<br>flow rate ($Q$),<br>ft³/s | (5)<br>unit head loss,<br>ft/ft | (6)<br>head loss ($h_f$),<br>ft | (7)<br>$h_f/Q$,<br>s/ft² |
|---|---|---|---|---|---|---|
| given | given | given | estimate or<br>assumed | from pipe<br>diagram | (2) × (5) | (6)/(4) |
| 1 | 2000 | 18 | 7.23 | 0.0036 | 7.200 | 0.996 |
| 2 | 900 | 24 | 5.18 | 0.00047 | 0.423 | 0.082 |
| 3 | 2800 | 12 | −1.96 | −0.0024 | −6.720 | 3.429 |
| 4 | 1100 | 24 | −6.77 | −0.00080 | −0.880 | 0.130 |
| | | | | | 0.023 | 4.637 |

### Loop II

| (1)<br>pipe<br>no. | (2)<br>length,<br>ft | (3)<br>diameter,<br>in | (4)<br>flow rate ($Q$),<br>ft³/s | (5)<br>unit head loss,<br>ft/ft | (6)<br>head loss ($h_f$),<br>ft | (7)<br>$h_f/Q$<br>s/ft |
|---|---|---|---|---|---|---|
| given | given | given | estimate or<br>assumed | from pipe<br>diagram | (2) × (5) | (6)/(4) |
| 5 | 2200 | 12 | 2.05 | 0.0026 | 5.720 | 2.790 |
| 6 | 750 | 18 | −4.59 | −0.00156 | −1.170 | 0.255 |
| 7 | 2600 | 24 | −7.14 | −0.00088 | −2.288 | 0.320 |
| 2 | 900 | 24 | −5.18 | −0.00047 | −0.423 | 0.082 |
| | | | | | 1.839 | 3.447 |

Since the algebraic sums of head losses are not both zero, new flow rate corrections must be computed. $(\Delta Q)_{\text{loop I}} = -(0.023)/[(1.85)(4.637)] = 0.00$ cfs, $(\Delta Q)_{\text{loop II}} = -(1.839)/[(1.85)(3.447)] = -0.29$ cfs. Revised flow rates for each pipe are determined using these flow rate corrections. These are as follows:

### Loop I

| pipe | old $Q$, cfs | $\Delta Q$, cfs | new $Q$, cfs |
|---|---|---|---|
| 1 | 7.23 | 0.00 | 7.23 |
| 2 | 5.18 | 0.00 + 0.29 | 5.47 |
| 3 | −1.96 | 0.00 | −1.96 |
| 4 | −6.77 | 0.00 | −6.77 |

### Loop II

| pipe | old $Q$, cfs | $\Delta Q$, cfs | new $Q$, cfs |
|---|---|---|---|
| 5 | 2.05 | −0.29 | 1.76 |
| 6 | −4.59 | −0.29 | −4.88 |
| 7 | −7.14 | −0.29 | −7.43 |
| 2 | −5.18 | −0.29 | −5.47 |

These adjusted flow rates are shown in Fig. 13-3d. Return to step 2 and repeat the entire procedure using these revised flow rates. This is given in the table below.

**Iteration No. 3**

Loop I

| (1)<br>pipe<br>no. | (2)<br>length,<br>ft | (3)<br>diameter,<br>in | (4)<br>flow rate $(Q)$,<br>ft$^3$/s | (5)<br>unit head loss,<br>ft/ft | (6)<br>head loss $(h_f)$,<br>ft | (7)<br>$h_f/Q$,<br>s/ft$^2$ |
|---|---|---|---|---|---|---|
| given | given | given | estimate or<br>assumed | from pipe<br>diagram | (2) × (5) | (6)/(4) |
| 1 | 2000 | 18 | 7.23 | 0.0036 | 7.200 | 0.996 |
| 2 | 900 | 24 | 5.47 | 0.00053 | 0.477 | 0.087 |
| 3 | 2800 | 12 | −1.96 | −0.0024 | −6.720 | 3.429 |
| 4 | 1100 | 24 | −6.77 | −0.00080 | −0.880 | 0.130 |
| | | | | | 0.777 | 4.642 |

Loop II

| (1)<br>pipe<br>no. | (2)<br>length,<br>ft | (3)<br>diameter,<br>in | (4)<br>flow rate $(Q)$,<br>ft$^3$/s | (5)<br>unit head loss,<br>ft/ft | (6)<br>head loss $(h_f)$,<br>ft | (7)<br>$h_f/Q$,<br>s/ft$^2$ |
|---|---|---|---|---|---|---|
| given | given | given | estimate or<br>assumed | from pipe<br>diagram | (2) × (5) | (6)/(4) |
| 5 | 2200 | 12 | 1.76 | 0.0019 | 4.180 | 2.375 |
| 6 | 750 | 18 | −4.88 | −0.00175 | −1.312 | 0.269 |
| 7 | 2600 | 24 | −7.43 | −0.00094 | −2.444 | 0.329 |
| 2 | 900 | 24 | −5.47 | −0.00053 | −0.477 | 0.087 |
| | | | | | −0.053 | 3.060 |

Since the algebraic sums of head losses are not both zero, new flow rate corrections must be computed. $(\Delta Q)_{\text{loop I}} = -(0.077)/[(1.85)(4.642)] = -0.01$ cfs, $(\Delta Q)_{\text{loop II}} = -(-0.053)/[(1.85)(3.060)] = 0.01$ cfs. Revised flow rates for each pipe are determined using these flow rate corrections. These are as follows:

Loop I

| pipe | old $Q$, cfs | $\Delta Q$, cfs | new $Q$, cfs |
|---|---|---|---|
| 1 | 7.23 | −0.01 | 7.22 |
| 2 | 5.47 | −0.01 − 0.01 | 5.45 |
| 3 | −1.96 | −0.01 | −1.97 |
| 4 | −6.77 | −0.01 | −6.78 |

Loop II

| pipe | old $Q$, cfs | $\Delta Q$, cfs | new $Q$, cfs |
|---|---|---|---|
| 5 | 1.76 | +0.01 | 1.77 |
| 6 | −4.88 | +0.01 | −4.87 |
| 7 | −7.43 | +0.01 | −7.42 |
| 2 | −5.47 | +0.01 + 0.01 | −5.45 |

These adjusted flow rates are shown in Fig. 13-3e. It would be appropriate to return to step 2 and repeat the entire procedure using these revised flow rates. However, an additional iteration (not shown) indicates that the next flow rate corrections would be no greater than 0.01 cfs, and further computation would appear to be wasted effort. Hence, the "new $Q$" values just before this paragraph are taken to be the correct flow rates for these pipes.

$Q_{in} = 14.00$ ft³/s

2000 ft
18-in diameter

2200 ft
12-in diameter

$Q_{out} = 6.64$ ft³/s

1100 ft
24-in diameter

900 ft
24-in diameter

750 ft
18-in diameter

$Q_{out} = 4.81$ ft³/s

2800 ft
12-in diameter

2600 ft
24-in diameter

$Q_{out} = 2.55$ ft³/s

(a) Pipe network

14.00 ft³/s

7.00 ft³/s

3.50 ft³/s

6.64 ft³/s

7.00 ft³/s

3.50 ft³/s

3.14 ft³/s

4.81 ft³/s

2.19 ft³/s

5.69 ft³/s

2.55 ft³/s

(b) Initial flow rate estimates

14.00 ft³/s

7.23 ft³/s

2.05 ft³/s

6.64 ft³/s

6.77 ft³/s

5.18 ft³/s

4.59 ft³/s

1.96 ft³/s

7.14 ft³/s

4.81 ft³/s

2.55 ft³/s

(c) Adjusted flow rates after Iteration No. 1

14.00 ft³/s

7.23 ft³/s

1.76 ft³/s

6.64 ft³/s

6.77 ft³/s

5.47 ft³/s

4.88 ft³/s

1.96 ft³/s

7.43 ft³/s

4.81 ft³/s

2.55 ft³/s

(d) Adjusted flow rates after Iteration No. 2

14.00 ft³/s

7.22 ft³/s

1.77 ft³/s

6.64 ft³/s

6.78 ft³/s

5.45 ft³/s

4.87 ft³/s

1.97 ft³/s

7.42 ft³/s

4.81 ft³/s

2.55 ft³/s

(e) Adjusted flow rates after Iteration No. 3

**Fig. 13-3**

**13.4** For the pipe network shown in Fig. 13-4, find the rate of flow of water in each pipe. Assume $C = 120$ for all pipes.

**⌐**

| pipe | $D$, in | $L$, ft | $Q_{assumed}$, cfs | $h_1$, ft/ft | $h_f$, ft | $h_f/Q$ | $\Delta Q$ | $Q_{new}$ |
|------|---------|---------|---------------------|--------------|-----------|---------|-----------|-----------|
| $AB$ | 18 | 1500 | 8.0 | 0.0044 | 6.60 | 0.825 | −0.11 | 7.89 |
| $BC$ | 12 | 500 | 3.0 | 0.0053 | 2.65 | 0.883 | −0.11 | 2.89 |
| $CF$ | 15 | 1500 | −4.0 | −0.00295 | −4.42 | 1.105 | 2.23 | −1.77 |
| $FA$ | 15 | 500 | −7.0 | −0.0083 | −4.15 | 0.593 | −0.11 | −7.11 |
| | | | | | 0.68 | 3.406 | | |
| $FC$ | 15 | 1500 | 4.0 | 0.00295 | 4.42 | 1.105 | −2.23 | 1.77 |
| $CD$ | 10 | 500 | 7.0 | 0.059 | 29.50 | 4.214 | −2.34 | 4.66 |
| $DE$ | 12 | 1500 | 2.0 | 0.0025 | 3.75 | 1.875 | −2.34 | −0.34 |
| $EF$ | 12 | 500 | −3.0 | −0.0053 | −2.65 | 0.883 | −2.34 | −5.34 |
| | | | | | 35.02 | 8.077 | | |

$$\Delta Q_{\text{I}} = -\sum h_f / [n \sum (h_f/Q)] = -0.68/[(1.85)(3.406)] = -0.11 \text{ cfs} \qquad \Delta Q_{\text{II}} = -35.02/[(1.85)(8.077)] = -2.34 \text{ cfs}$$

| pipe | $h_1$ | $h_f$ | $h_f/Q$ | $\Delta Q$ | $Q_{new}$ |
|------|-------|-------|---------|-----------|-----------|
| $AB$ | 0.0042 | 6.30 | 0.798 | −0.66 | 7.23 |
| $BC$ | 0.0049 | 2.45 | 0.848 | −0.66 | 2.23 |
| $CF$ | 0.00067 | −1.00 | 0.565 | 0.08 | −1.69 |
| $FA$ | 0.0086 | −4.30 | 0.605 | −0.66 | −7.77 |
| | | 3.45 | 2.816 | | |
| $FC$ | 0.00067 | 1.00 | 0.565 | −0.08 | 1.69 |
| $CD$ | 0.028 | 14.00 | 3.000 | −0.74 | 3.92 |
| $DE$ | 0.000095 | −0.14 | 0.412 | −0.74 | −1.08 |
| $EF$ | 0.015 | −7.50 | 1.404 | −0.74 | −6.08 |
| | | 7.36 | 5.381 | | |

$$\Delta Q_{\text{I}} = -3.45/[(1.85)(2.816)] = -0.66 \text{ cfs} \qquad \Delta Q_{\text{II}} = -7.36/[(1.85)(5.381)] = -0.74 \text{ cfs}$$

| pipe | $h_1$ | $h_f$ | $h_f/Q$ | $\Delta Q$ | $Q_{new}$ |
|------|-------|-------|---------|-----------|-----------|
| $AB$ | 0.0036 | 5.40 | 0.747 | −0.20 | 7.03 |
| $BC$ | 0.00295 | 1.48 | 0.664 | −0.20 | 2.03 |
| $CF$ | −0.00060 | −0.90 | 0.533 | −0.20 | −1.89 |
| $FA$ | −0.0100 | −5.00 | 0.644 | −0.20 | −7.97 |
| | | 0.98 | 2.588 | | |
| $FC$ | 0.00060 | 0.90 | 0.533 | 0.20 | 1.89 |
| $CD$ | 0.0200 | 10.00 | 2.551 | 0.00 | 3.92 |
| $DE$ | −0.00080 | −1.20 | 1.111 | 0.00 | −1.08 |
| $EF$ | −0.0195 | −9.75 | 1.604 | 0.00 | −6.08 |
| | | −0.05 | 5.799 | | |

$$\Delta Q_{\text{I}} = -0.98/[(1.85)(2.588)] = -0.20 \text{ cfs} \qquad \Delta Q_{\text{II}} = -(-0.05)/[(1.85)(5.799)] = 0.00 \text{ cfs}$$

| pipe | $h_1$ | $h_f$ | $h_f/Q$ | $\Delta Q$ | $Q_{new}$ |
|------|-------|-------|---------|-----------|-----------|
| AB | 0.0034 | 5.10 | 0.725 | 0.02 | 7.05 |
| BC | 0.0026 | 1.30 | 0.640 | 0.02 | 2.05 |
| CF | −0.00075 | −1.11 | 0.587 | 0.03 | −1.86 |
| FA | −0.0108 | −5.40 | 0.678 | 0.02 | −7.95 |
| | | −0.11 | 2.630 | | |
| FC | 0.00074 | 1.11 | 0.587 | −0.03 | 1.86 |
| CD | 0.0200 | 10.00 | 2.551 | −0.01 | 3.91 |
| DE | −0.00080 | −1.20 | 1.111 | −0.01 | −1.09 |
| EF | −0.0195 | −9.75 | 1.604 | −0.01 | −6.09 |
| | | 0.16 | 5.853 | | |

$$\Delta Q_I = -(-0.11)/[(1.85)(2.630)] = 0.02 \text{ cfs} \qquad \Delta Q_{II} = -0.16/[(1.85)(5.853)] = -0.01 \text{ cfs}$$

15 ft³/s

Fig. 13-4

**13.5** The pipe network shown in Fig. 13-5 represents a spray rinse system. Find the flow rate of water in each pipe. Assume $C = 120$ for all pipes.

| pipe | D, mm | L, m | $Q_{assumed}$, m³/s | $h_1$, m/m | $h_f$, m | $h_f/Q$ | $\Delta Q$ | $Q_{new}$ |
|------|-------|------|---------------------|------------|----------|---------|-----------|-----------|
| AB | 300 | 600 | 0.200 | 0.027 | 16.20 | 81.0 | 0.011 | 0.211 |
| BG | 250 | 400 | 0.100 | 0.0175 | 7.00 | 70.0 | −0.003 | 0.097 |
| GH | 300 | 600 | −0.100 | −0.0074 | −4.44 | 44.4 | 0.011 | −0.089 |
| HA | 250 | 400 | −0.200 | −0.064 | −25.60 | 128.0 | 0.011 | −0.189 |
| | | | | | −6.84 | 323.4 | | |
| BC | 300 | 600 | 0.100 | 0.0074 | 4.44 | 44.4 | 0.014 | 0.114 |
| CF | 250 | 400 | 0.050 | 0.0049 | 1.96 | 39.2 | 0.014 | 0.064 |
| FG | 300 | 600 | −0.100 | −0.0074 | −4.44 | 44.4 | 0.014 | −0.086 |
| GB | 250 | 400 | −0.100 | −0.0175 | −7.00 | 70.0 | 0.003 | −0.097 |
| | | | | | −5.04 | 198.0 | | |

| pipe | $D$, mm | $L$, m | $Q_{assumed}$, m³/s | $h_1$, m/m | $h_f$, m | $h_f/Q$ | $\Delta Q$ | $Q_{new}$ |
|------|---------|--------|---------------------|------------|----------|---------|------------|-----------|
| CD | 300 | 600 | 0.050 | 0.0020 | 1.20 | 24.0 | 0.000 | 0.050 |
| DE | 250 | 400 | 0.050 | 0.0049 | 1.96 | 39.2 | 0.000 | 0.050 |
| EF | 300 | 600 | −0.050 | −0.0020 | −1.20 | 24.0 | 0.000 | −0.050 |
| FC | 250 | 400 | −0.050 | −0.0049 | −1.96 | 39.2 | −0.014 | −0.064 |
|    |     |     |       |        | 0.00 | 126.4 |       |       |

$$\Delta Q_I = -\sum h_f/[n\sum(h_f/Q)] = -(-6.84)/[(1.85)(323.4)] = 0.011$$

$$\Delta Q_{II} = -(-5.04)/[(1.85)(198.0)] = 0.014 \qquad \Delta Q_{III} = -0.00/[(1.85)(126.4)] = 0.00$$

| pipe | $h_1$ | $h_f$ | $h_f/Q$ | $\Delta Q$ | $Q_{new}$ |
|------|-------|-------|---------|------------|-----------|
| AB | 0.0295 | 17.70 | 83.9 | 0.004 | 0.215 |
| BG | 0.017 | 6.80 | 70.1 | 0.001 | 0.098 |
| GH | −0.0059 | −3.54 | 39.8 | 0.004 | −0.085 |
| HA | −0.058 | −23.20 | 122.8 | 0.004 | −0.185 |
|    |        | −2.24 | 316.6 |       |       |
| BC | 0.0095 | 5.70 | 50.0 | 0.003 | 0.117 |
| CF | 0.0079 | 3.16 | 49.4 | −0.002 | 0.062 |
| FG | −0.0056 | −3.36 | 39.1 | 0.003 | −0.083 |
| GB | −0.017 | −6.80 | 70.1 | −0.001 | −0.098 |
|    |        | −1.30 | 208.6 |       |       |
| CD | 0.0020 | 1.20 | 24.0 | 0.005 | 0.055 |
| DE | 0.0049 | 1.96 | 39.2 | 0.005 | 0.055 |
| EF | −0.0020 | −1.20 | 24.0 | 0.005 | −0.045 |
| FC | −0.0079 | −3.16 | 49.4 | 0.002 | −0.062 |
|    |        | −1.20 | 136.6 |       |       |

$$\Delta Q_I = -(-2.24)/[(1.85)(316.6)] = 0.004 \qquad \Delta Q_{II} = -(-1.30)/[(1.85)(208.6)] = 0.003$$

$$\Delta Q_{III} = -(-1.20)/[(1.85)(136.6)] = 0.005$$

| pipe | $h_1$ | $h_f$ | $h_f/Q$ | $\Delta Q$ | $Q_{new}$ |
|------|-------|-------|---------|------------|-----------|
| AB | 0.031 | 18.60 | 86.5 | 0.000 | 0.215 |
| BG | 0.0172 | 6.88 | 70.2 | −0.003 | 0.095 |
| GH | 0.0055 | −3.30 | 38.8 | 0.000 | −0.085 |
| HA | 0.056 | −22.40 | 121.1 | 0.000 | −0.185 |
|    |        | −0.22 | 316.6 |       |       |
| BC | 0.010 | 6.00 | 51.3 | 0.003 | 0.120 |
| CF | 0.0075 | 3.00 | 48.4 | 0.002 | 0.064 |
| FG | 0.0052 | −3.12 | 37.6 | 0.003 | −0.080 |
| GB | 0.0172 | −6.88 | 70.2 | 0.003 | −0.095 |
|    |        | −1.00 | 207.5 |       |       |
| CD | 0.0024 | 1.44 | 26.2 | 0.001 | 0.056 |
| DE | 0.0059 | 2.36 | 42.9 | 0.001 | 0.056 |
| EF | 0.0017 | −1.02 | 22.7 | 0.001 | −0.044 |
| FC | 0.0075 | −3.00 | 48.4 | −0.002 | −0.064 |
|    |        | −0.22 | 140.2 |       |       |

$$\Delta Q_I = -(-0.22)/[(1.85)(316.6)] = 0.000 \qquad \Delta Q_{II} = -(-1.00)/[(1.85)(207.5)] = 0.003$$

$$\Delta Q_{III} = -(-0.22)/[(1.85)(140.2)] = 0.001$$

**Fig. 13-5**

**13.6** The pipe network shown in Fig. 13-6 represents a small industrial park. Find the flow rate of water in each pipe. Assume $C = 120$ for all pipes.

| pipe | $D$, mm | $L$, m | $Q_{assumed}$, m³/s | $h_1$, m/m | $h_f$, m | $h_f/Q$ | $\Delta Q$ | $Q_{new}$ |
|------|---------|--------|---------------------|------------|----------|---------|------------|-----------|
| AB | 500 | 1000 | 0.250 | 0.0034 | 3.40 | 13.6 | 0.028 | 0.278 |
| BE | 400 | 1200 | 0.120 | 0.0026 | 3.12 | 26.0 | −0.005 | 0.115 |
| EF | 300 | 1000 | −0.130 | −0.012 | −12.00 | 92.3 | 0.063 | −0.067 |
| FA | 600 | 1200 | −0.250 | −0.0014 | −1.68 | 6.7 | 0.028 | −0.222 |
|    |     |      |       |        | −7.16 | 138.6 |        |        |
| BC | 500 | 1000 | 0.130 | 0.00102 | 1.02 | 7.8 | 0.033 | 0.163 |
| CD | 400 | 1200 | 0.030 | 0.00020 | 0.24 | 8.0 | 0.033 | 0.063 |
| DE | 400 | 1000 | −0.100 | −0.0018 | −1.8 | 18.0 | 0.050 | −0.050 |
| EB | 400 | 1200 | −0.120 | −0.0026 | −3.12 | 26.0 | 0.005 | −0.115 |
|    |     |      |       |        | −3.66 | 59.8 |        |        |
| ED | 400 | 1000 | 0.100 | 0.0018 | 1.80 | 18.0 | −0.050 | 0.050 |
| DI | 300 | 1200 | 0.080 | 0.0048 | 5.76 | 72.0 | −0.017 | 0.063 |
| IH | 300 | 1000 | −0.020 | −0.00037 | −0.37 | 18.5 | −0.017 | −0.037 |
| HE | 300 | 1200 | −0.050 | −0.0020 | −2.40 | 48.0 | 0.018 | −0.032 |
|    |     |      |       |        | 4.79 | 156.5 |        |        |
| FE | 300 | 1000 | 0.130 | 0.012 | 12.00 | 92.3 | −0.063 | 0.067 |
| EH | 300 | 1200 | 0.050 | 0.0020 | 2.40 | 48.0 | −0.018 | 0.032 |
| HG | 400 | 1000 | −0.020 | −0.000095 | −0.10 | 5.0 | −0.035 | −0.055 |
| GF | 400 | 1200 | −0.120 | −0.0026 | −3.12 | 26.0 | −0.035 | −0.155 |
|    |     |      |       |        | 11.18 | 171.3 |        |        |

$$\Delta Q_{\mathrm{I}} = -\sum h_f/[n \sum (h_f/Q)] = -(-7.16)/[(1.85)(138.6)] = 0.028 \qquad \Delta Q_{\mathrm{II}} = -(-3.66)/[(1.85)(59.8)] = 0.033$$

$$\Delta Q_{\mathrm{III}} = -4.79/[(1.85)(156.5)] = -0.017 \qquad \Delta Q_{\mathrm{IV}} = -11.18/[(1.85)(171.3)] = -0.035$$

| pipe | $h_1$ | $h_f$ | $h_f/Q$ | $\Delta Q$ | $Q_{new}$ |
|------|-------|-------|---------|------------|-----------|
| AB | 0.0043 | 4.30 | 15.5 | −0.013 | 0.265 |
| BE | 0.0023 | 2.76 | 24.0 | −0.020 | 0.095 |
| EF | −0.0034 | −3.40 | 50.7 | −0.018 | −0.085 |
| FA | −0.0011 | −1.32 | 5.9 | −0.013 | −0.235 |
|    |        | 2.34 | 96.1 |        |        |

| pipe | $h_1$ | $h_f$ | $h_f/Q$ | $\Delta Q$ | $Q_{new}$ |
|------|-------|-------|---------|------------|-----------|
| BC | 0.0016 | 1.60 | 9.8 | 0.007 | 0.170 |
| CD | 0.00078 | 0.94 | 14.9 | 0.007 | 0.070 |
| DE | −0.00050 | −0.50 | 10.0 | 0.015 | −0.035 |
| EB | −0.0023 | −2.76 | 24.0 | 0.020 | −0.095 |
|    |         | −0.72 | 58.7 |       |       |
| ED | 0.00050 | 0.50 | 10.0 | −0.015 | 0.035 |
| DI | 0.0031 | 3.72 | 59.0 | −0.008 | 0.055 |
| IH | −0.00115 | −1.15 | 31.1 | −0.008 | −0.045 |
| HE | −0.00087 | −1.04 | 32.5 | −0.013 | −0.045 |
|    |         | 2.03 | 132.6 |       |       |
| FE | 0.0034 | 3.40 | 50.7 | 0.018 | 0.085 |
| EH | 0.00087 | 1.04 | 32.5 | 0.013 | 0.045 |
| HG | −0.00060 | −0.60 | 10.9 | 0.005 | −0.050 |
| GF | −0.0041 | −4.92 | 31.7 | 0.005 | −0.150 |
|    |         | −1.08 | 125.8 |       |       |

$$\Delta Q_{\mathrm{I}} = -2.34/[(1.85)(96.1)] = -0.013 \qquad \Delta Q_{\mathrm{II}} = -(-0.72)/[(1.85)(58.7)] = 0.007$$
$$\Delta Q_{\mathrm{III}} = -2.03/[(1.85)(132.6)] = -0.008 \qquad \Delta Q_{\mathrm{IV}} = -(-1.08)/[(1.85)(125.8)] = 0.005$$

| pipe | $h_1$ | $h_f$ | $h_f/Q$ | $\Delta Q$ | $Q_{new}$ |
|------|-------|-------|---------|------------|-----------|
| AB | 0.0039 | 3.90 | 14.7 | 0.006 | 0.271 |
| BE | 0.00165 | 1.98 | 20.8 | 0.012 | 0.107 |
| EF | −0.0055 | −5.50 | 64.7 | 0.014 | −0.071 |
| FA | −0.00125 | −1.50 | 6.4 | 0.006 | −0.229 |
|    |         | −1.12 | 106.6 |       |       |
| BC | 0.0017 | 1.70 | 10.0 | −0.006 | 0.164 |
| CD | 0.00094 | 1.13 | 16.1 | −0.006 | 0.064 |
| DE | −0.000265 | −0.26 | 7.4 | −0.008 | −0.043 |
| EB | −0.00165 | −1.98 | 20.8 | −0.012 | −0.107 |
|    |         | 0.59 | 54.3 |       |       |
| ED | 0.000265 | 0.26 | 7.4 | 0.008 | 0.043 |
| DI | 0.0024 | 2.88 | 52.4 | 0.002 | 0.057 |
| IH | −0.0017 | −1.70 | 37.8 | 0.002 | −0.043 |
| HE | −0.0017 | −2.04 | 45.3 | 0.010 | −0.035 |
|    |         | −0.60 | 142.9 |       |       |
| FE | 0.0055 | 5.50 | 64.7 | −0.014 | 0.071 |
| EH | 0.0017 | 2.04 | 45.3 | −0.010 | 0.035 |
| HG | −0.00051 | −0.51 | 10.2 | −0.008 | −0.058 |
| GF | −0.0039 | −4.68 | 31.2 | −0.008 | −0.158 |
|    |         | 2.35 | 151.4 |       |       |

$$\Delta Q_{\mathrm{I}} = -(-1.12)/[(1.85)(106.6)] = 0.006 \qquad \Delta Q_{\mathrm{II}} = -0.59/[(1.85)(54.3)] = -0.006$$
$$\Delta Q_{\mathrm{III}} = -(-0.60)/[(1.85)(142.9)] = 0.002 \qquad \Delta Q_{\mathrm{IV}} = -2.35/[(1.85)(151.4)] = -0.008$$

| pipe | $h_1$ | $h_f$ | $h_f/Q$ | $\Delta Q$ | $Q_{new}$ |
|------|-------|-------|---------|------------|-----------|
| $AB$ | 0.0040 | 4.00 | 14.8 | −0.006 | 0.265 |
| $BE$ | 0.0021 | 2.52 | 23.6 | −0.009 | 0.098 |
| $EF$ | −0.0039 | −3.90 | 54.9 | −0.008 | −0.079 |
| $FA$ | −0.0012 | −1.44 | 6.3 | −0.006 | −0.235 |
| | | 1.18 | 99.6 | | |
| $BC$ | 0.0016 | 1.60 | 9.8 | 0.003 | 0.167 |
| $CD$ | 0.00080 | 0.96 | 15.0 | 0.003 | 0.067 |
| $DE$ | −0.00038 | −0.38 | 8.8 | 0.006 | −0.037 |
| $EB$ | −0.0021 | −2.52 | 23.6 | 0.009 | −0.098 |
| | | −0.34 | 57.2 | | |
| $ED$ | 0.00038 | 0.38 | 8.8 | −0.006 | 0.037 |
| $DI$ | 0.0026 | 3.12 | 54.7 | −0.003 | 0.054 |
| $IH$ | −0.0015 | −1.50 | 34.9 | −0.003 | −0.046 |
| $HE$ | −0.00105 | −1.26 | 36.0 | −0.005 | −0.040 |
| | | 0.74 | 134.4 | | |
| $FE$ | 0.0039 | 3.90 | 54.9 | 0.008 | 0.079 |
| $EH$ | 0.00105 | 1.26 | 36.0 | 0.005 | 0.040 |
| $HG$ | −0.00067 | −0.67 | 11.6 | 0.002 | −0.056 |
| $GF$ | −0.0041 | −4.92 | 31.1 | 0.002 | −0.156 |
| | | −0.43 | 133.6 | | |

$$\Delta Q_{\text{I}} = -1.18/[(1.85)(99.6)] = -0.006 \qquad \Delta Q_{\text{II}} = -(-0.34)/[(1.85)(57.2)] = 0.003$$

$$\Delta Q_{\text{III}} = -0.74/[(1.85)(134.4)] = -0.003 \qquad \Delta Q_{\text{IV}} = -(-0.43)/[(1.85)(133.6)] = 0.002$$

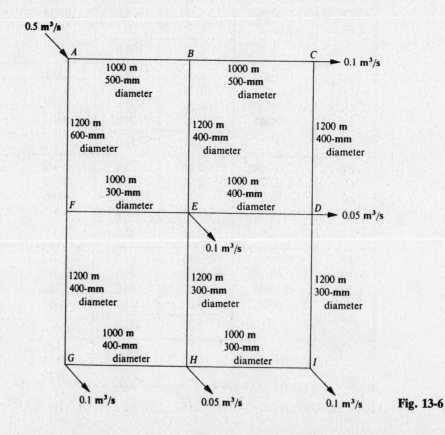

Fig. 13-6

**13.7** Compute the flow rate of water in each pipe of the network shown in Fig. 13-7. Assume $C = 120$ for all pipes.

| pipe | D, in. | L, ft | $Q_{assumed}$, cfs | $h_1$, ft/ft | $h_f$, ft | $h_f/Q$ | $\Delta Q$ | $Q_{new}$ |
|------|--------|-------|-------------------|-------------|----------|---------|-----------|----------|
| AB | 24 | 2000 | 5.0 | 0.00045 | 0.90 | 0.180 | 0.29 | 5.29 |
| BG | 15 | 3000 | 1.0 | 0.00022 | 0.66 | 0.660 | 0.42 | 1.42 |
| GH | 18 | 2000 | −2.5 | −0.00051 | −1.02 | 0.408 | 0.34 | −2.16 |
| HA | 24 | 3000 | −5.0 | −0.00045 | −1.35 | 0.270 | 0.29 | −4.71 |
|    |    |    |    |    | −0.81 | 1.518 |    |    |
| BC | 24 | 2000 | 4.0 | 0.00030 | 0.60 | 0.150 | −0.13 | 3.87 |
| CF | 15 | 3000 | 1.0 | 0.00022 | 0.66 | 0.660 | 0.53 | 1.53 |
| FG | 18 | 2000 | −1.0 | −0.000095 | −0.19 | 0.190 | −0.48 | −1.48 |
| GB | 15 | 3000 | −1.0 | −0.00022 | −0.66 | 0.660 | −0.42 | −1.42 |
|    |    |    |    |    | 0.41 | 1.660 |    |    |
| CD | 15 | 2000 | 3.0 | 0.0017 | 3.40 | 1.133 | −0.66 | 2.34 |
| DE | 12 | 3000 | 1.0 | 0.00067 | 2.01 | 2.010 | −0.66 | 0.34 |
| EF | 18 | 2000 | −0.3 | −0.000010 | −0.02 | 0.067 | −0.33 | −0.63 |
| FC | 15 | 3000 | −1.0 | −0.00022 | −0.66 | 0.660 | −0.53 | −1.53 |
|    |    |    |    |    | 4.73 | 3.870 |    |    |
| FE | 18 | 2000 | 0.3 | 0.000010 | 0.02 | 0.067 | 0.33 | 0.63 |
| EL | 12 | 3000 | 1.3 | 0.0011 | 3.30 | 2.538 | −0.33 | 0.97 |
| LK | 15 | 2000 | −1.7 | −0.00059 | −1.18 | 0.694 | −0.33 | −2.03 |
| KF | 15 | 3000 | −0.2 | −0.000011 | −0.03 | 0.150 | −0.68 | −0.88 |
|    |    |    |    |    | 2.11 | 3.449 |    |    |
| GF | 18 | 2000 | 1.0 | 0.000095 | 0.19 | 0.190 | 0.48 | 1.48 |
| FK | 15 | 3000 | 0.2 | 0.000011 | 0.03 | 0.150 | 0.68 | 0.88 |
| KJ | 18 | 2000 | −1.5 | −0.00019 | −0.38 | 0.253 | 0.35 | −1.15 |
| JG | 15 | 3000 | −1.0 | −0.00022 | −0.66 | 0.660 | 0.40 | −0.60 |
|    |    |    |    |    | −0.82 | 1.253 |    |    |
| HG | 18 | 2000 | 2.5 | 0.00051 | 1.02 | 0.408 | −0.34 | 2.16 |
| GJ | 15 | 3000 | 1.0 | 0.00022 | 0.66 | 0.660 | −0.40 | 0.60 |
| JI | 24 | 2000 | −0.5 | −0.0000063 | −0.01 | 0.020 | −0.05 | −0.55 |
| IH | 18 | 3000 | −2.5 | −0.00050 | −1.50 | 0.600 | −0.05 | −2.55 |
|    |    |    |    |    | 0.17 | 1.688 |    |    |

$$\Delta Q_{I} = -(-0.81)/[(1.85)(1.518)] = 0.29 \qquad \Delta Q_{II} = -0.41/[(1.85)(1.660)] = -0.13$$

$$\Delta Q_{III} = -4.73/[(1.85)(3.870)] = -0.66 \qquad \Delta Q_{IV} = -2.11/[(1.85)(3.449)] = -0.33$$

$$\Delta Q_{V} = -(-0.82)/[(1.85)(1.253)] = 0.35 \qquad \Delta Q_{VI} = -0.17/[(1.85)(1.688)] = -0.05$$

| pipe | $h_1$ | $h_f$ | $h_f/Q$ | $\Delta Q$ | $Q_{new}$ |
|------|-------|-------|---------|-----------|----------|
| AB | 0.00050 | 1.00 | 0.189 | −0.10 | 5.19 |
| BG | 0.00043 | 1.29 | 0.908 | 0.00 | 1.42 |
| GH | −0.00039 | −0.78 | 0.361 | −0.31 | −2.47 |
| HA | −0.00040 | −1.20 | 0.255 | −0.10 | −4.81 |
|    |    | 0.31 | 1.713 |    |    |

| pipe | $h_1$ | $h_f$ | $h_f/Q$ | $\Delta Q$ | $Q_{new}$ |
|------|-------|-------|---------|-----------|-----------|
| BC | 0.00029 | 0.58 | 0.150 | −0.10 | 3.77 |
| CF | 0.00050 | 1.50 | 0.980 | 0.07 | 1.60 |
| FG | −0.00019 | −0.38 | 0.257 | 0.04 | −1.44 |
| GB | −0.00043 | −1.29 | 0.908 | 0.00 | −1.42 |
|    |          | 0.41 | 2.295 |      |      |
| CD | 0.0011 | 2.20 | 0.940 | −0.17 | 2.17 |
| DE | 0.000090 | 0.27 | 0.794 | −0.17 | 0.17 |
| EF | −0.000039 | −0.08 | 0.127 | −0.20 | −0.83 |
| FC | −0.00050 | −1.50 | 0.980 | −0.07 | −1.60 |
|    |          | 0.89 | 2.841 |      |      |
| FE | 0.000039 | 0.08 | 0.127 | 0.20 | 0.83 |
| EL | 0.00064 | 1.92 | 1.979 | 0.03 | 1.00 |
| LK | −0.00082 | −1.64 | 0.808 | 0.03 | −2.00 |
| KF | −0.00018 | −0.54 | 0.614 | 0.17 | −0.71 |
|    |          | −0.18 | 3.528 |      |      |
| GF | 0.00019 | 0.38 | 0.257 | −0.04 | 1.44 |
| FK | 0.00018 | 0.54 | 0.614 | −0.17 | 0.71 |
| KJ | −0.00012 | −0.24 | 0.209 | −0.14 | −1.29 |
| JG | −0.000090 | −0.27 | 0.450 | −0.35 | −0.95 |
|    |          | 0.41 | 1.530 |      |      |
| HG | 0.00039 | 0.78 | 0.361 | 0.31 | 2.47 |
| GJ | 0.000090 | 0.27 | 0.450 | 0.35 | 0.95 |
| JI | −0.0000075 | −0.02 | 0.036 | 0.21 | −0.34 |
| IH | −0.00053 | −1.59 | 0.624 | 0.21 | −2.34 |
|    |          | −0.56 | 1.471 |      |      |

$$\Delta Q_{\mathrm{I}} = -0.31/[(1.85)(1.713)] = -0.10 \qquad \Delta Q_{\mathrm{II}} = -0.41/[(1.85)(2.295)] = -0.10$$
$$\Delta Q_{\mathrm{III}} = -0.89/[(1.85)(2.841)] = -0.17 \qquad \Delta Q_{\mathrm{IV}} = -(-0.18)/[(1.85)(3.528)] = 0.03$$
$$\Delta Q_{\mathrm{V}} = -0.41/[(1.85)(1.530)] = -0.14 \qquad \Delta Q_{\mathrm{VI}} = -(-0.56)/[(1.85)(1.471)] = 0.21$$

| pipe | $h_1$ | $h_f$ | $h_f/Q$ | $\Delta Q$ | $Q_{new}$ |
|------|-------|-------|---------|-----------|-----------|
| AB | 0.00048 | 0.96 | 0.185 | 0.00 | 5.19 |
| BG | 0.00043 | 1.29 | 0.908 | 0.11 | 1.53 |
| GH | −0.00050 | −1.00 | 0.405 | 0.08 | −2.39 |
| HA | −0.00042 | −1.26 | 0.262 | 0.00 | −4.81 |
|    |          | −0.01 | 1.760 |      |      |
| BC | 0.00027 | 0.54 | 0.143 | −0.11 | 3.66 |
| CF | 0.00053 | 1.59 | 0.994 | −0.05 | 1.55 |
| FG | −0.00018 | −0.36 | 0.250 | −0.17 | −1.61 |
| GB | −0.00043 | −1.29 | 0.908 | −0.11 | −1.53 |
|    |          | 0.48 | 2.295 |      |      |
| CD | 0.00097 | 1.94 | 0.894 | −0.06 | 2.11 |
| DE | 0.000025 | 0.08 | 0.471 | −0.06 | 0.11 |
| EF | −0.000066 | −0.13 | 0.157 | −0.04 | −0.87 |
| FC | −0.00053 | −1.59 | 0.994 | 0.05 | −1.55 |
|    |          | 0.30 | 2.156 |      |      |

| pipe | $h_1$ | $h_f$ | $h_f/Q$ | $\Delta Q$ | $Q_{new}$ |
|------|-------|-------|---------|------------|-----------|
| FE | 0.000066 | 0.13 | 0.157 | 0.04 | 0.87 |
| EL | 0.00067 | 2.01 | 2.010 | −0.02 | 0.98 |
| LK | −0.00081 | −1.62 | 0.810 | −0.02 | −2.02 |
| KF | −0.00012 | −0.36 | 0.507 | −0.08 | −0.79 |
| | | 0.16 | 3.484 | | |
| GF | 0.00018 | 0.36 | 0.250 | 0.17 | 1.61 |
| FK | 0.00012 | 0.36 | 0.507 | 0.08 | 0.79 |
| KJ | −0.00015 | −0.30 | 0.233 | 0.06 | −1.23 |
| JG | −0.00020 | −0.60 | 0.632 | 0.14 | −0.81 |
| | | −0.18 | 1.622 | | |
| HG | 0.00050 | 1.00 | 0.405 | −0.08 | 2.39 |
| GJ | 0.00020 | 0.60 | 0.632 | −0.14 | 0.81 |
| JI | −0.0000031 | −0.01 | 0.029 | −0.08 | −0.42 |
| IH | −0.00045 | −1.35 | 0.577 | −0.08 | −2.42 |
| | | 0.24 | 1.643 | | |

$$\Delta Q_{I} = -(-0.01)/[(1.85)(1.760)] = 0.00 \qquad \Delta Q_{II} = -0.48/[(1.85)(2.295)] = -0.11$$

$$\Delta Q_{III} = -0.30/[(1.85)(2.516)] = -0.06 \qquad \Delta Q_{IV} = -0.16/[(1.85)(3.484)] = -0.02$$

$$\Delta Q_{V} = -(-0.18)/[(1.85)(1.622)] = 0.06 \qquad \Delta Q_{VI} = -0.24/[(1.85)(1.643)] = -0.08$$

Fig. 13-7

**13.8** Water flows through the piping system shown in Fig. 13-8 with certain measured flows indicated on the sketch. Determine the flows throughout the network. Use $C = 120$ throughout.

| line | $D$, in | $L$, ft | assumed $Q_1$, mgd | $S$, ft/1000 ft | LH, ft | $\dfrac{LH}{Q_1}$ | $\Delta$ | $Q_2$ |
|------|---------|---------|--------------------|-----------------|--------|-------------------|----------|-------|
| AB | 20 | 3000 | 4.0 | 1.62 | 4.86 | 1.22 | +0.31 | 4.31 |
| BE | 16 | 4000 | 1.0 | 0.37 | 1.48 | 1.48 | +0.31 − (0.13) = +0.18 | 1.18 |
| EF | 16 | 3000 | −2.0 | −1.33 | −3.99 | 2.00 | +0.31 − (0.50) = −0.19 | −2.19 |
| FA | 24 | 4000 | −6.0 | −1.41 | −5.64 | 0.94 | +0.31 | −5.69 |
|    |    |    |    |    | $\Sigma = -3.29$ | 5.64 | | |
| BC | 20 | 3000 | 3.0 | 0.95 | 2.85 | 0.95 | +0.13 | 3.13 |
| CD | 16 | 4000 | 2.0 | 1.33 | 5.32 | 2.66 | +0.13 | 2.13 |
| DE | 12 | 3000 | −1.5 | −3.15 | −9.45 | 6.30 | +0.13 − (−0.12) = +0.25 | −1.25 |
| EB | 16 | 4000 | −1.0 | −0.37 | −1.48 | 1.48 | +0.13 − (0.31) = −0.18 | −1.18 |
|    |    |    |    |    | $\Sigma = -2.76$ | 11.39 | | |
| FE | 16 | 3000 | 2.0 | 1.33 | 3.99 | 2.00 | +0.50 − (+0.31) = +0.19 | 2.19 |
| EH | 12 | 4000 | 1.0 | 1.48 | 5.92 | 5.92 | +0.50 − (−0.12) = +0.62 | 1.62 |
| HG | 16 | 3000 | −2.0 | −1.33 | −3.99 | 2.00 | +0.50 | −1.50 |
| GF | 16 | 4000 | −4.0 | −4.85 | −19.40 | 4.85 | +0.50 | −3.50 |
|    |    |    |    |    | $\Sigma = -13.48$ | 14.77 | | |
| ED | 12 | 3000 | 1.5 | 3.15 | 9.45 | 6.30 | −0.12 − (0.13) = −0.25 | 1.25 |
| DI | 12 | 4000 | 1.0 | 1.48 | 5.92 | 5.92 | −0.12 | 0.88 |
| IH | 12 | 3000 | −1.0 | −1.48 | −4.44 | 4.44 | −0.12 | −1.12 |
| HE | 12 | 4000 | −1.0 | −1.48 | −5.92 | 5.92 | −0.12 − (0.50) = −0.62 | −1.62 |
|    |    |    |    |    | $\Sigma = +5.01$ | 22.58 | | |

$$\Delta_\text{I} = -(-3.29)/[1.85(5.64)] = +0.31 \qquad \Delta_\text{II} = -(-2.76)/[1.85(11.39)] = +0.13$$

$$\Delta_\text{III} = -(-13.48)/[1.85(14.77)] = +0.50 \qquad \Delta_\text{IV} = -(+5.01)/[1.85(22.58)] = -0.12$$

For line $EF$ in loop I, its net $\Delta$ term is $(\Delta_\text{I} - \Delta_\text{III})$ or $[+0.31 - (+0.50)] = -0.19$. It will be observed that the $\Delta$ for loop I is combined with that of the $\Delta$ for loop III since the line $EF$ occurs in each loop. In similar fashion, for line $FE$ in loop III, the net $\Delta$ term is $(\Delta_\text{III} - \Delta_\text{I})$ or $[+0.50 - (+0.31)] = +0.19$. Note that the net $\Delta$'s have the same magnitude but *opposite signs*. This can readily be understood since flow in $EF$ is counterclockwise for loop I, whereas flow in $FE$ in loop III is clockwise.

Fig. 13-8

In determining the values of $Q_2$ for the second calculation, $Q_{AB} = (4.00 + 0.31) = 4.31$ mgd whereas $Q_{EF} = (-2.00 - 0.19) = -2.19$ mgd and $Q_{FA} = (-6.00 + 0.31) = -5.69$ mgd.

| line | $Q_2$ | S | LH | LH/Q | Δ |
|------|-------|---|-----|------|---|
| AB | 4.31 | 1.86 | 5.58 | 1.29 | +0.20 |
| BE | 1.18 | 0.51 | 2.04 | 1.72 | +0.20 + negl = +0.20 |
| EF | -2.19 | -1.57 | -4.71 | 2.15 | +0.20 - (-0.06) = +0.26 |
| FA | -5.69 | -1.28 | -5.12 | 0.90 | +0.20 |
|  |  |  | $\Sigma = -2.21$ | 6.06 |  |
| BC | 3.13 | 1.02 | 3.06 | 0.98 | negl |
| CD | 2.13 | 1.48 | 5.92 | 2.79 | negl |
| DE | -1.25 | -2.28 | -6.84 | 5.50 | negl - 0.19 = -0.19 |
| EB | -1.18 | -0.51 | -2.04 | 1.72 | negl - 0.20 = -0.20 |
|  |  |  | $\Sigma = +0.10$ | 10.99 |  |
| FE | 2.19 | 1.57 | 4.71 | 2.15 | -0.06 - 0.20 = -0.26 |
| EH | 1.62 | 3.65 | 14.60 | 9.02 | -0.06 - 0.19 = -0.25 |
| HG | -1.50 | -0.79 | -2.37 | 1.58 | -0.06 |
| GF | -3.50 | -3.75 | -15.00 | 4.28 | -0.06 |
|  |  |  | $\Sigma = +1.94$ | 17.03 |  |
| ED | 1.25 | 2.28 | 6.84 | 5.42 | +0.19 + negl = +0.19 |
| DI | 0.88 | 1.18 | 4.72 | 5.38 | +0.19 |
| IH | -1.12 | -1.83 | -5.49 | 4.90 | +0.19 |
| HE | -1.62 | -3.65 | -14.60 | 9.02 | +0.19 - (-0.06) = +0.25 |
|  |  |  | $\Sigma = -8.53$ | 24.72 |  |

| line | $Q_3$ | S | LH | LH/Q | Δ | $Q_4$ |
|------|-------|---|-----|------|---|-------|
| AB | 4.51 | 2.02 | 6.06 | 1.34 | -0.02 | 4.49 |
| BE | 1.39 | 0.68 | 2.72 | 1.95 | -0.02 - 0.12 = -0.14 | 1.25 |
| EF | -1.93 | -1.25 | -3.75 | 1.94 | -0.02 - 0.12 = -0.14 | -2.07 |
| FA | -5.49 | -1.20 | -4.80 | 0.88 | -0.02 | -5.51 |
|  |  |  | $\Sigma = +0.23$ | 6.11 |  |  |
| BC | 3.12 | 1.02 | 3.06 | 0.98 | +0.12 | 3.24 |
| CD | 2.12 | 1.49 | 5.96 | 2.81 | +0.12 | 2.24 |
| DE | -1.45 | -2.97 | -8.91 | 6.15 | +0.12 + 0.02 = +0.14 | -1.31 |
| EB | -1.39 | -0.68 | -2.72 | 1.95 | +0.12 + 0.02 = +0.14 | -1.25 |
|  |  |  | $\Sigma = -2.61$ | 11.89 |  |  |
| FE | 1.93 | 1.25 | 3.75 | 1.94 | +0.12 + 0.02 = +0.14 | 2.07 |
| EH | 1.37 | 2.68 | 10.72 | 7.83 | +0.12 + 0.02 = +0.14 | 1.51 |
| HG | -1.56 | -0.84 | -2.52 | 1.62 | +0.12 | -1.44 |
| GF | -3.56 | -3.90 | -15.60 | 4.38 | +0.12 | -3.44 |
|  |  |  | $\Sigma = -3.65$ | 15.77 |  |  |
| ED | 1.45 | 2.97 | 8.91 | 6.15 | -0.02 - 0.12 = -0.14 | 1.31 |
| DI | 1.07 | 1.68 | 6.72 | 6.28 | -0.02 | 1.05 |
| IH | -0.93 | -1.31 | -3.93 | 4.23 | -0.02 | -0.95 |
| HE | -1.37 | -2.68 | -10.72 | 7.83 | -0.02 - 0.12 = -0.14 | -1.51 |
|  |  |  | $\Sigma = +0.98$ | 24.49 |  |  |

**13.9** For the piping system of Prob. 13.8, if the elevation at point $A$ is 200.0 ft and the pressure head is 150.0 ft and the elevation at $I$ is 100.0 ft, find the pressure head at $I$.

❚ The elevation of the hydraulic grade line at $A$ is $(200.0 + 150.0) = 350.0$. The lost head to $I$ can be calculated by any route from $A$ to $I$, adding the losses in the usual manner, i.e., in the direction of flow. Using $ABEHI$ we obtain $LH_{A \text{ to } I} = (6.06 + 2.72 + 10.72 + 3.93) = 23.43$ ft. As a check, using $ABEDI$, $LH = (6.06 + 2.72 + 8.91 + 6.72) = 24.41$ ft. Using a value of 24 ft, the elevation of the hydraulic grade line at $I = (350.0 - 24.0) = 326.0$ ft. Hence the pressure head at $I = (326.0 - 100.0) = 226.0$ ft.

**13.10** If the flows into and out of a two-loop pipe system are as shown in Fig. 13-9, determine the flow in each pipe. The $K$ values for each pipe were calculated from the pipe and minor loss characteristics and from an assumed value of $f$.

❚ As a first step, assume flow in each pipe such that continuity is satisfied at all junctions. Calculate $\Delta Q$ for each loop, make corrections to the assumed $Q$'s and repeat several times until the $\Delta Q$'s are quite small. As a final step the values of $f$ for each pipe should be checked against the Moody diagram and modified, if necessary.

Fig. 13-9

| | left loop | | right loop | |
|---|---|---|---|---|
| | $\sum KQ_0^n$ | $\sum \|KnQ_0^{n-1}\|$ | $\sum KQ_0^n$ | $\sum \|KnQ_0^{n-1}\|$ |
| | $1 \times 60^2 = 3\,600$ | $1 \times 2 \times 60 = 120$ | $4 \times 50^2 = 10\,000$ | $4 \times 2 \times 50 = 400$ |
| | $4 \times 10^2 = \quad 400$ | $4 \times 2 \times 10 = \quad 80$ | $2 \times 25^2 = \ 1\,250$ | $2 \times 2 \times 25 = 100$ |
| | $\overline{\phantom{xx}4\,000}$ ↄ | | $\overline{\phantom{x}11\,250}$ ↄ | |
| | $3 \times 40^2 = 4\,800$ ↄ | $3 \times 2 \times 40 = 240$ | $4 \times 10^2 = \quad 400$ | $4 \times 2 \times 10 = \quad 80$ |
| | $\overline{\phantom{xxxx}800}$ ↄ | $\overline{\phantom{xxx}440}$ | $5 \times 25^2 = \ 3\,125$ | $5 \times 2 \times 25 = 250$ |
| | | | $\overline{\phantom{x}3\,525}$ ↄ | $\overline{\phantom{xx}830}$ |
| | | | $\overline{\phantom{x}7\,725}$ ↄ | |
| | $\Delta Q_1 = \dfrac{800}{440} \approx 2$ ↄ | | $\Delta Q_1 = \dfrac{7\,725}{830} \approx 9$ ↄ | |
| | $1 \times 62^2 = 3\,844$ | $1 \times 2 \times 62 = 124$ | $4 \times 41^2 = \ 6\,724$ | $4 \times 2 \times 41 = 328$ |
| | $4 \times 21^2 = 1\,764$ | $4 \times 2 \times 21 = 168$ | $2 \times 16^2 = \quad 512$ | $2 \times 2 \times 16 = \quad 64$ |
| | $\overline{\phantom{xx}5\,608}$ ↄ | | $\overline{\phantom{x}7\,236}$ ↄ | |
| | $3 \times 38^2 = 4\,332$ ↄ | $3 \times 2 \times 38 = 228$ | $4 \times 21^2 = \ 1\,764$ | $4 \times 2 \times 21 = 168$ |
| | $\overline{\phantom{xx}1\,276}$ ↄ | $\overline{\phantom{xxx}520}$ | $5 \times 34^2 = \ 5\,780$ | $5 \times 2 \times 34 = 340$ |
| | | | $\overline{\phantom{x}7\,544}$ ↄ | $\overline{\phantom{xx}900}$ |
| | | | $\overline{\phantom{xxx}308}$ ↄ | |
| | $\Delta Q_2 = \dfrac{1\,276}{520} \approx 2$ ↄ | | $\Delta Q_2 = \dfrac{308}{900} \approx 0$ | |

**13.11** Determine the flows in the cast iron piping of Fig. 13-10. Take $n = 2.0$ and use the values of $f$ for complete turbulence, as given in Fig. A-5. Use four stages of approximation.

▌ With $V = Q/(\pi D^2/4)$: $h_L = f(L/D)(16Q^2/\pi^2 D^4)/2g$, that is, $h_L = KQ^n$ with $K = 16fL/(2\pi^2 gD^5) = 0.025fL/D^5$, $n = 2$. For complete turbulence, $f = $ const, $K = $ const (for a given pipe). Table A-9 for cast iron gives $\epsilon = 0.00085$ ft. Thus for pipe $ab$, $\epsilon/D = 0.00085$. Figure A-5 for complete turbulence gives $f = 0.0189$; $K = 0.476$. Similarly $K = 1.606$ for $ac$ and $be$, $K = 2.01$ for $cd$ and $de$, $K = 1.722$ for $cf$, $dg$, and $eh$, $K = 0.620$ for $fg$ and $gh$.

**First Approximation and Correction**

| loop A | | | |
|---|---|---|---|
| pipe | $K$ | $KQ_0^n$ | $\lvert KnQ_0^{n-1}\rvert$ |
| ab | 0.476 | 0.476 | 0.952 |
| be | 1.606 | 1.606 | 3.21 |
| ed | 2.01 | 0.502 | 2.01 |
| | | 2.584 | |
| ac | 1.606 | 1.606 | 3.21 |
| cd | 2.01 | 0.502 | 2.01 |
| | | 2.108 | 11.39 |

$$\Delta Q = (2.584 - 2.108)/11.39 = 0.0418 \, \circlearrowright$$

| loop B | | | | loop C | | | |
|---|---|---|---|---|---|---|---|
| pipe | $K$ | $KQ_0^n$ | $\lvert KnQ_0^{n-1}\rvert$ | pipe | $K$ | $KQ_0^n$ | $\lvert KnQ_0^{n-1}\rvert$ |
| cd | 2.01 | 0.502 $\circlearrowleft$ | 2.01 | eh | 7.22 | 1.804 $\circlearrowleft$ | 7.22 |
| cf | 7.22 | 1.804 $\circlearrowright$ | 7.22 | ed | 2.01 | 0.502 $\circlearrowright$ | 2.01 |
| | | | 9.22 | | | | 9.22 |

$$\Delta Q = (1.804 - 0.502)/9.22 = 0.1412 \, \circlearrowleft \qquad \Delta Q = (1.804 - 0.502)/9.22 = 0.1412 \, \circlearrowright$$

(a)

Initial flow assumptions:

(b)

Resulting final flows:

(c)

**Fig. 13-10**

| pipe | Q(2d approximation), cfs | Q(4th and final approx), cfs |
|------|--------------------------|------------------------------|
| ab | $1.00 - 0.04 = 0.96$ | 0.95 |
| be | $1.00 - 0.04 = 0.96$ | 0.95 |
| ed | $0.50 - 0.04 + 0.14 = 0.60$ | 0.59 |
| ac | $1.00 + 0.04 = 1.04$ | 1.06 |
| cd | $0.50 + 0.04 + 0.14 = 0.68$ | 0.64 |
| dg | $0 + 0.14 + 0.14 = 0.28$ | 0.22 |
| gf | $0 + 0.14 = 0.14$ | 0.09 |
| cf | $0.50 - 0.14 = 0.36$ | 0.42 |
| eh | $0 + 0.14 = 0.14$ | 0.37 |
| gh | $0.50 - 0.14 = 0.36$ | 0.14 |

**13.12** If in Prob. 13.11 the pressure head at $a$ is 200 ft, find the pressure head at $d$. Neglect velocity heads.

▌ From the final approximations in Prob. 13.11,

In pipe $ac$                $h_L = KQ^2 = 1.606 \times 1.06^2 = 1.805$ ft
In pipe $cd$,             $h_L = 2.01 \times 0.64^2 \qquad = 0.823$ ft

$$\text{2.628 ft}$$

In pipes $ac + cd$, $h_L = p_d/\gamma = p_a/\gamma - (h_L)_{ad} = 200 - 2.63 = 197.37$ ft.

**13.13** In the network of Fig. 13-11, the 12-in and 16-in pipes are cast iron, while the 18-in and 24-in sizes are concrete ($\epsilon = 0.003$ ft). Assume complete turbulence with $n = 2$. Terminate the solution after five trials.

▌        $h_L = KQ^n = (f)(L/D)(V^2/2g) = (f)(L/D)(Q^2/[2g(\pi/4)^2 D^4]) = (fL/39.7D^5)Q^2$

Values of $f$ for fully turbulent flow are obtained from Fig. A-5 ($n = 2$):

| pipe | material | length, ft | diam., ft | $\epsilon/D$ | $f$ | $D^5$ | $K = \dfrac{fL}{39.7D^5}$ |
|------|----------|-----------|-----------|--------------|-----|-------|---------------------------|
| ab | Avg. conc. | 998 | 1.50 | 0.00200 | 0.0233 | 7.59 | 0.0772 |
| bc | Avg. conc. | 499 | 1.50 | 0.00200 | 0.0233 | 7.59 | 0.0386 |
| cd | Avg. conc. | 499 | 1.50 | 0.00200 | 0.0233 | 7.59 | 0.0386 |
| ef | New C.I. | 499 | 1.00 | 0.00085 | 0.0188 | 1.00 | 0.237 |
| fg | New C.I. | 499 | 1.00 | 0.00085 | 0.0188 | 1.00 | 0.237 |
| hi | Avg. conc. | 998 | 2.00 | 0.00150 | 0.0218 | 32.00 | 0.01715 |
| ij | Avg. conc. | 998 | 2.00 | 0.00150 | 0.0218 | 32.00 | 0.01715 |
| ah | Avg. conc. | 1596 | 2.00 | 0.00150 | 0.0218 | 32.00 | 0.0274 |
| be | New C.I. | 798 | 1.33 | 0.00062 | 0.0175 | 4.21 | 0.0836 |
| ei | Avg. conc. | 798 | 1.50 | 0.00200 | 0.0233 | 7.59 | 0.0618 |
| cf | New C.I. | 798 | 1.33 | 0.00062 | 0.0175 | 4.21 | 0.0836 |
| dg | New C.I. | 798 | 1.33 | 0.00062 | 0.0175 | 4.21 | 0.0836 |
| gi | Avg. conc. | 798 | 1.50 | 0.00200 | 0.0233 | 7.59 | 0.0618 |

| | loop A | | | | loop B | | |
|------|-------|-------|-----------|------|-------|-------|-----------|
| pipe | $Q_0$ | $KQ_0^2$ | $\|h_L/Q_0\|$ | pipe | $Q_0$ | $KQ_0^2$ | $\|h_L/Q_0\|$ |
| ha | 3 | 0.247 | 0.0823 | bc | 2 | 0.1545 | 0.0772 |
| ab | 3 | 0.695 | 0.232 | cf | 2 | 0.335 ⌐ | 0.1673 |
| | | | | | | 0.489 ↵ | |
| be | 1 | 0.0836 ⌐ | 0.0836 | | | | |
| | | 1.026 ↵ | | | | | |
| hi | 7 | 0.840 | 0.1200 | be | 1 | 0.0836 | 0.0836 |
| ie | 4 | 0.989 ↵ | 0.247 | ef | 2 | 0.947 ↵ | 0.473 |
| | | 1.829 ⌐ | 0.765 | | | 1.030 ↵ | 0.801 |

| loop A | loop B |
|--------|--------|
| $\Delta Q = \dfrac{1.026 - 1.829}{2 \times 0.765} = -0.525\,\wr$ | $\Delta Q = \dfrac{0.489 - 1.030}{2 \times 0.801} = -0.338\,\wr$ |

<table>
<tr><td colspan="4" align="center">loop C</td><td colspan="4" align="center">loop D</td></tr>
<tr><td>pipe</td><td>$Q_0$</td><td>$KQ_0^2$</td><td>$|h_L/Q_0|$</td><td>pipe</td><td>$Q_0$</td><td>$KQ_0^2$</td><td>$|h_L/Q_0|$</td></tr>
<tr><td>cd</td><td>0</td><td>0.000</td><td>0.000</td><td>ie</td><td>4</td><td>0.989</td><td>0.247</td></tr>
<tr><td>gf</td><td>1</td><td>0.237</td><td>0.237</td><td rowspan="2">ef</td><td rowspan="2">2</td><td>0.947</td><td>0.473</td></tr>
<tr><td></td><td></td><td>0.237</td><td></td><td>1.935</td><td></td></tr>
<tr><td>cf</td><td>2</td><td>0.335</td><td>0.1673</td><td>ij</td><td>3</td><td>0.1543</td><td>0.0514</td></tr>
<tr><td>gd</td><td>2</td><td>0.335</td><td>0.1673</td><td>jg</td><td>3</td><td>0.556</td><td>0.1854</td></tr>
<tr><td></td><td></td><td>0.669</td><td>0.571</td><td></td><td></td><td></td><td></td></tr>
<tr><td></td><td></td><td></td><td></td><td>gf</td><td>1</td><td>0.237</td><td>0.237</td></tr>
<tr><td></td><td></td><td></td><td></td><td></td><td></td><td>0.947</td><td>1.194</td></tr>
</table>

| | |
|---|---|
| $\Delta Q = \dfrac{0.237 - 0.669}{2 \times 0.573} = -0.379\,\wr$ | $\Delta Q = \dfrac{1.935 - 0.947}{2 \times 1.194} = +0.414\,\curvearrowright$ |

After five trials, the flows (cfs) within approximately 1 percent are $ab = 3.54$; $bc = 2.48$; $cd = 0.40$; $ef = 1.47$; $gf = 1.45$; $hi = 6.46$; $ij = 3.05$; $ha = 3.54$; $be = 1.06$; $ie = 3.41$; $cf = 2.08$; $gd = 1.60$; $jg = 3.05$.

**Fig. 13-11**

**13.14** Calculate the pressure drop from $h$ to $f$ in Prob. 13.13.

▌ Head loss from $h$ to $f$ (any path):

$$hi = 0.716 \text{ ft}$$
$$ie = 0.719 \text{ ft}$$
$$ef = 0.511 \text{ ft}$$
$$\overline{\phantom{ef = }1.946 \text{ ft}}$$

Hence $\Delta p = (1.946)(62.4)/144 = 0.8$ psi

**13.15** Fit an equation of the form $h_f = KQ^n$ to flow of 60-°F water through 1000 ft of 10-in cast iron pipe. The equation should hold over the velocity range 2 to 8 ft/s.

▌ $\qquad h_f = (f)(L/D)(v^2/2g) \qquad \epsilon/D = 0.00085/(\tfrac{10}{12}) = 0.00128 \qquad N_R = Dv/\nu$

For $v = 2$ fps: $N_R = (\tfrac{10}{12})(2)/(1.21 \times 10^{-5}) = 1.38 \times 10^5$. From Fig. A-5, $f = 0.0230$: $h_f = 0.0230[1000/(\tfrac{10}{12})]\{2^2/[(2)(32.2)]\} = 1.714$ ft, $Q = Av = [(\pi)(\tfrac{10}{12})^2/4](2) = 1.091$ ft³/s. For $v = 8$ fps: $N_R = (\tfrac{10}{12})(8)/(1.21 \times 10^{-5}) = 5.51 \times 10^5$, $f = 0.0213$, $h_f = 0.0213[1000/(\tfrac{10}{12})]\{8^2/[(2)(32.2)]\} = 25.40$ ft, $Q = Av = [(\pi)(\tfrac{10}{12})^2/4](8) = 4.363$ ft³/s. Given $h_f = KQ^n$, at 2 fps, $1.714 = (K)(1.091)^n$. At 8 fps, $25.4 = (K)(4.363)^n$:

$$\log 1.714 = \log K + n \log 1.091 \qquad\qquad (1)$$

$$\log 25.4 = \log K + n \log 4.363 \qquad\qquad (2)$$

Subtracting Eq. (*1*) from Eq. (*2*) gives $1.171 = 0.6022n$, $n = 1.945$. Substituting into Eq. (*1*), $\log 1.714 = \log K + (1.945)(\log 1.091)$, $K = 1.447$; $h_f = 1.447Q^{1.945}$.

**13.16** The distribution of flow through the network of Fig. 13-12 is desired for the inflows and outflows as given. For simplicity $n$ has been given the value 2.0.

▌ The assumed distribution is shown in Fig. 13-12*a*. At the upper left the term $\sum rQ_0 |Q_0|^{n-1}$ is computed for the lower circuit number 1. Next to the diagram on the left is the computation of $\sum nr |Q_0|^{n-1}$ for the same circuit. The same format is used for the second circuit in the upper right of the figure. The corrected flow after the first step for the top horizontal pipe is determined as $15 + 11.06 = 26.06$ and for the diagonal as $35 + (-21.17) + (-11.06) = 2.77$. Figure 13-12*b* shows the flows after one correction and Fig. 13-12*c* the values after four corrections.

$$70^2 \times 6 = 29,400 \qquad 2 \times 70 \times 6 = 840$$
$$35^2 \times 3 = 3,675 \qquad 2 \times 35 \times 3 = 210$$
$$-30^2 \times 5 = \underline{-4,500} \qquad 2 \times 30 \times 5 = \underline{300}$$
$$28,575 \qquad\qquad 1,350$$

$$\Delta Q_1 = -\frac{28,575}{1,350} = -21.17$$

$$15^2 \times 1 = 225 \qquad 2 \times 15 \times 1 = 30$$
$$-35^2 \times 2 = -2,450 \qquad 2 \times 35 \times 2 = 140$$
$$-13.83^2 \times 3 = \underline{-574} \qquad 2 \times 13.83 \times 3 = \underline{83}$$
$$-2,799 \qquad\qquad 253$$

$$\Delta Q_2 = \frac{2,799}{253} = 11.06$$

(*a*)

$$48.83^2 \times 6 = 14,308 \qquad 2 \times 48.83 \times 6 = 586$$
$$2.77^2 \times 3 = 23 \qquad 2 \times 2.77 \times 3 = 17$$
$$-51.17^2 \times 5 = \underline{-13,090} \qquad 2 \times 51.17 \times 5 = \underline{511}$$
$$1,241 \qquad\qquad 1,114$$

$$\Delta Q_1 = -\frac{1,241}{1,114} = -1.114$$

$$26.06^2 \times 1 = 679 \qquad 2 \times 26.06 \times 1 = 52$$
$$-23.94^2 \times 2 = -1,146 \qquad 2 \times 23.94 \times 2 = 96$$
$$-1.656^2 \times 3 = \underline{-8} \qquad 2 \times 1.656 \times 3 = \underline{10}$$
$$-475 \qquad\qquad 158$$

$$\Delta Q_2 = \frac{475}{158} = 3.006$$

(*b*)

$$\Delta Q_1 = 0.0079$$
$$\Delta Q_1 = 0.0013$$

$$\Delta Q_2 = 0.169$$
$$\Delta Q_2 = 0.0003$$

(*c*)

**Fig. 13-12**

**13.17** Calculate the flow through each of the pipes of the network shown in Fig. 13-13 ($n = 2$).

▌

| | | | |
|---|---|---|---|
| $1 \times 60^2 = 3600$ | $2 \times 1 \times 60 = 120$ | $2 \times 30^2 = 1800$ | $2 \times 2 \times 30 = 120$ |
| $3 \times 5^2 = 75$ | $2 \times 3 \times 5 = 30$ | $3 \times 5^2 = -75$ | $2 \times 3 \times 5 = 30$ |
| $2 \times 40^2 = -3200$ | $2 \times 2 \times 40 = 160$ | $1 \times 45^2 = -2025$ | $2 \times 1 \times 45 = 90$ |
| $475$ | $310$ | $-300$ | $240$ |
| $\Delta Q = -\dfrac{475}{310} \approx -1.5$ | | $\Delta Q = \dfrac{300}{240} \approx 1$ | Fig. 13-13*b* |

| | | | |
|---|---|---|---|
| $1 \times 58.5^2 = 3422.25$ | $2 \times 1 \times 58.5 = 117$ | $2 \times 31^2 = 1922$ | $2 \times 2 \times 31 = 124$ |
| $3 \times 2.5^2 = 18.75$ | $2 \times 3 \times 2.5 = 15$ | $3 \times 2.5^2 = -19$ | $2 \times 3 \times 2.5 = 15$ |
| $2 \times 41.5^2 = -3444.5$ | $2 \times 2 \times 41.5 = 166$ | $1 \times 44^2 = -1936$ | $2 \times 1 \times 44 = 88$ |
| $-3.5$ | $298$ | $-33$ | $227$ |
| $\Delta Q = \dfrac{3.5}{298} = 0.012$ | | $\Delta Q = \dfrac{33}{227} = 0.145$ | Fig. 13-13*c* |

**Fig. 13-13**

**13.18** Determine the flow through each line of Fig. 13-14 ($n = 2$).

| $2 \times 19^2 =$ | 722 | $2 \times 2 \times 19 =$ | 76 | $4 \times 37^2 =$ | 5476 | $2 \times 4 \times 37 = 296$ | |
|---|---|---|---|---|---|---|---|
| $1 \times 32^2 =$ | 1024 | $2 \times 1 \times 32 =$ | 64 | $1 \times 32^2 =$ | $-1024$ | $2 \times 1 \times 32 = 64$ | |
| $2 \times 31^2 =$ | $-1922$ | $2 \times 2 \times 31 =$ | 124 | $3 \times 38^2 =$ | $-4332$ | $2 \times 3 \times 38 = 228$ | Fig. 13-14*b* |
| | $-176$ | | 264 | | 120 | 588 | |

$$\Delta Q = \frac{176}{264} = 0.67 \qquad\qquad \Delta Q = \frac{-120}{588} = -0.20$$

| $2 \times 19.67^2 =$ | 773.82 | $2 \times 2 \times 19.67 =$ | 78.68 | $4 \times 36.8^2 =$ | 5417.0 | $2 \times 4 \times 36.8 = 294.4$ |
|---|---|---|---|---|---|---|
| $1 \times 32.87^2 =$ | 1080.44 | $2 \times 1 \times 32.87 =$ | 65.74 | $1 \times 32.87^2 =$ | $-1080.4$ | $2 \times 1 \times 32.87 = 65.74$ |
| $2 \times 30.33^2 =$ | $-1839.82$ | $2 \times 2 \times 30.33 =$ | 121.32 | $3 \times 38.2^2 =$ | $-4377.7$ | $2 \times 3 \times 38.2 = 229.2$ |
| | 14.4 | | 265.74 | | $-41.12$ | 589.3 |

$$\Delta Q = -\frac{14.44}{265.74} = -0.05 \qquad\qquad \Delta Q = -\frac{41.12}{589.3} = 0.07$$

**Fig. 13-14**

**13.19** Determine the flow rates in Fig. 13-15. The fluid is water at 20 °C and all five pipes have $f = 0.0201$.

$$h_f = (f)\left(\frac{L}{D}\right)\left(\frac{v^2}{2g}\right) = \frac{8fLQ^2}{\pi^2 g D^5} = KQ^2 \qquad K_1 = \frac{8f_1 L_1}{\pi^2 g D_1^5} = \frac{(8)(0.0201)(3000)}{(\pi)^2(32.2)(\frac{6}{12})^5} = 48.57$$

$$K_2 = \frac{(8)(0.0201)(4000)}{(\pi)^2(32.2)(\frac{8}{12})^5} = 15.37 \qquad K_3 = \frac{(8)(0.0201)(5000)}{(\pi)^2(32.2)(\frac{9}{12})^5} = 10.66$$

$$K_4 = \frac{(8)(0.0201)(4000)}{(\pi)^2(32.2)(\frac{8}{12})^5} = 15.37 \qquad K_5 = \frac{(8)(0.0201)(3000)}{(\pi)^2(32.2)(\frac{3}{12})^5} = 1554$$

Two head-loss loops:

Loop *ABC*: $\qquad\qquad 15.37Q_2^2 + 10.66Q_3^2 - 48.57Q_1^2 = 0 \qquad\qquad (1)$

Loop *BCD*: $\qquad\qquad 10.66Q_3^2 + 15.37Q_4^2 - 1554Q_5^2 = 0 \qquad\qquad (2)$

Three junctions:

Junction $A$: $\qquad\qquad\qquad\qquad\qquad Q_1 + Q_2 = 2.0$ $\qquad\qquad\qquad\qquad$ (3)

Junction $B$: $\qquad\qquad\qquad\qquad\qquad Q_2 = Q_3 + Q_5$ $\qquad\qquad\qquad\qquad$ (4)

Junction $C$: $\qquad\qquad\qquad\qquad\qquad Q_1 + Q_3 = Q_4$ $\qquad\qquad\qquad\qquad$ (5)

Solving these simultaneous equations by trial and error gives $Q_1 = 0.81$ cfs, $Q_2 = 1.19$ cfs, $Q_3 = 0.99$ cfs, $Q_4 = 1.80$ cfs, $Q_5 = 0.20$ cfs.

**Fig. 13-15**

**13.20**  In Prob. 13.19 the pressure at $A$ is given as 200 psig. Find the pressures at $B$, $C$, and $D$.

$\qquad p_B = p_A - \rho g K_2 Q_2^2 = (200)(144) - (1.94)(32.2)(15.37)(1.19)^2 = 27\,440$ lb/ft$^2$ or 190.6 psig

$\qquad p_C = p_A - \rho g K_1 Q_1^2 = (200)(144) - (1.94)(32.2)(48.57)(0.81)^2 = 26\,809$ lb/ft$^2$ or 186.2 psig

$\qquad p_D = p_C - \rho g K_4 Q_4^2 = (186.2)(144) - (1.94)(32.2)(15.37)(1.80)^2 = 23\,702$ lb/ft$^2$ or 164.6 psig

**13.21**  Repeat Prob. 13.19 for the boundary flows indicated in Fig. 13-16.

$$h_f = (f)(L/D)(v^2/2g) = 8fLQ^2/\pi^2 g D^5 = KQ^2$$

From Prob. 13.19, $K_1 = 48.57$, $K_2 = 15.37$, $K_3 = 10.66$, $K_4 = 15.37$, and $K_5 = 1554$.
 Two head-loss loops:

Loop $ABC$: $\qquad\qquad\qquad 15.37Q_2^2 + 10.66Q_3^2 - 48.57Q_1^2 = 0$ $\qquad\qquad\qquad$ (1)

Loop $BCD$: $\qquad\qquad\qquad 10.66Q_3^2 + 15.37Q_4^2 - 1554Q_5^2 = 0$ $\qquad\qquad\qquad$ (2)

Three junctions:

Junction $A$: $\qquad\qquad\qquad\qquad\qquad Q_1 + Q_2 = 2.0$ $\qquad\qquad\qquad\qquad$ (3)

Junction $B$: $\qquad\qquad\qquad\qquad\qquad Q_2 = Q_3 + Q_5 + 1.0$ $\qquad\qquad\qquad\qquad$ (4)

Junction $C$: $\qquad\qquad\qquad\qquad\qquad Q_1 + Q_3 = Q_4 + 0.5$ $\qquad\qquad\qquad\qquad$ (5)

Solving these simultaneous equations by trial and error gives $Q_1 = 0.725$ cfs, $Q_2 = 1.275$ cfs, $Q_3 = 0.226$ cfs, $Q_4 = 0.451$ cfs, $Q_5 = 0.049$ cfs.

**Fig. 13-16**

**13.22** Repeat Prob. 13.20 for the flows found in Prob. 13.21.

$$p_B = p_A - \rho g K_2 Q_2^2 = (200)(144) - (1.94)(32.2)(15.37)(1.275)^2 = 27\,239\ \text{lb/ft}^2 \quad \text{or} \quad 189.2\ \text{lb/in}^2$$
$$p_C = p_A - \rho g K_1 Q_1^2 = (200)(144) - (1.94)(32.2)(48.57)(0.725)^2 = 27\,205\ \text{lb/ft}^2 \quad \text{or} \quad 188.9\ \text{lb/in}^2$$
$$p_D = p_C - \rho g K_4 Q_4^2 = (188.9)(144) - (1.94)(32.2)(15.37)(0.451)^2 = 27\,006\ \text{lb/ft}^2 \quad \text{or} \quad 187.5\ \text{lb/in}^2$$

**13.23** Compute the junction flows in Fig. 13-17, taking flow *into* the junction as positive. Each pipe is 30 m of 60-mm-diameter cast iron, with $f = 0.0294$.

$$h_f = \Delta p/\gamma = 8fLQ^2/\pi^2 g D^5$$

$(800 - 500)/9.79 = (8)(0.0294)(30)(Q_1)^2/(\pi)^2(9.807)(0.060)^5 \qquad Q_1 = +0.0181\ \text{m}^3/\text{s}$
$(500 - 200)/9.79 = (8)(0.0294)(30)(Q_4)^2/(\pi)^2(9.807)(0.060)^5 \qquad Q_4 = -0.0181\ \text{m}^3/\text{s}$
$(500 - 400)/9.79 = (8)(0.0294)(30)(Q_2)^2/(\pi)^2(9.807)(0.060)^5 \qquad Q_2 = -0.0104\ \text{m}^3/\text{s}$
$(600 - 500)/9.79 = (8)(0.0294)(30)(Q_3)^2/(\pi)^2(9.807)(0.060)^5 \qquad Q_3 = +0.0104\ \text{m}^3/\text{s}$

**Fig. 13-17**

**13.24** Determine the flow in each pipe of the network shown in Fig. 13-18, using $f = 0.02$ throughout.

Taking $x = 2$, $h_L = f(L/D)(V^2/2g) = f(L/D)(1/2g)(4Q/\pi D^2)^2 = (8fL/\pi^2 g D^5)(Q^2) = KQ^2$. Hence $K = 0.81fL/gD^5$, and the $K$ value for each pipe is

| Diameter, in | 3 | 4 | 5 | 6 | 7 | 8 |
|---|---|---|---|---|---|---|
| $K$ | 1030 | 368 | 160 | 80.4 | 22.4 | 11.5 |

The assumed flows are indicated on the figure in parentheses. For loop *AEDB*,

$$\Delta_1 = -[(1030 \times 0.5^2) + (11.5 \times 0.1^2) - (22.4 \times 0.2^2) - (368 \times 0.7^2)]/$$
$$2[(1030 \times 0.5) + (11.5 \times 0.1) + (22.4 \times 0.2) + (368 \times 0.7)] = -0.05\ \text{cfs}$$

and for loop *BDC*,

$$\Delta_2 = -[(22.4 \times 0.2^2) + (80.4 \times 0.3^2) - (160 \times 0.5^2)]/2[(22.4 \times 0.2) + (80.4 \times 0.3) + (160 \times 0.5)] = +0.15\ \text{cfs}$$

The corrected flows appear on the figure below the first assumed flows. Recomputing $\Delta$ for each loop yields $\Delta_1 = +0.001$ cfs and $\Delta_2 = -0.001$ cfs.

**Fig. 13-18**

**13.25** Replace the quadrilateral network of Fig. 13-19 by a single 500-ft-long pipe $AC$. Assume $f$ is the same for all pipes.

❚
$$(h_f)_{ABC} = (h_f)_{ADC} \qquad h_f = (f)(L/D)(v^2/2g)$$
$$f(500/\tfrac{8}{12})(V_8^2/2g) + f(\tfrac{500}{2})(V_{24}^2/2g) = f(\tfrac{800}{1})(V_{12}^2/2g) + f(\tfrac{1000}{3})(V_{36}^2/2g)$$
$$750V_8^2 + 250V_{24}^2 = 800V_{12}^2 + 333V_{36}^2 \qquad V_8 = 9V_{24} \qquad V_{12} = 9V_{36}$$
$$(750 \times 81 + 250)V_{24}^2 = (800 \times 81 + 333)V_{36}^2 \qquad V_{36} = 0.96V_{24}$$
$$D^2V = 2^2V_{24} + 3^2V_{36} = 4V_{24} + 9(0.96V_{24}) \qquad D^2V = 12.63V_{24}$$
$$(500/\tfrac{8}{12})V_8^2 + \tfrac{500}{2}V_{24}^2 = (1000/D)V^2$$
$$750V_8^2 + 250V_{24}^2 = 61\,000V_{24}^2 = (1000/D)V^2 \tag{1}$$

But
$$V = 12.63V_{24}/D^2 \tag{2}$$

Substituting (2) into (1) yields $D^5 = 1.308$. Hence $D = 1.055$ ft.

**Fig. 13-19**

**13.26** The pipe network in Fig. 13-20 consists of pipes as follows: $AB$, 5000 ft, 12 in; $BC$, 3000 ft, 6 in; $CD$, 8000 ft, 24 in; $DE$, 7000 ft, 8 in; $EA$, 4000 ft, 10 in; $BD$, 7000 ft, 12 in. A flow of 4000 gpm enters the system at $D$, while outflow at the junctions is as follows: $A$, 500 gpm; $B$, 300 gpm; $C$, 1000 gpm; $E$, 2200 gpm. Find the flow in each pipe and the pressure at each junction if the head at $D$ is 400 ft. Take $f = 0.023$.

❚
$$h_f = 8fLQ^2/\pi^2gD^5 = (8)(0.023)(LQ^2)/(\pi)^2(32.2)D^5 = 0.000579LQ^2/D^5$$

| $D^5$ | pipe | $D$, in | $k$ |
|---|---|---|---|
| 1.0 | $AB$ | 12 | 2.89 |
| 0.031 | $BC$ | 6 | 56.0 |
| 32 | $CD$ | 24 | 0.143 |
| 0.132 | $DE$ | 8 | 30.6 |
| 1.0 | $BD$ | 12 | 4.05 |
| 0.402 | $AE$ | 10 | 5.75 |

Loop 1:
$$\Delta_1 = [-(2.9)(3.62)^2 - (4.05)(3.40)^2 + (30.6)(2.43)^2 - (5.75)(2.5)^2]/$$
$$2[(2.9)(3.62) + (4.05)(3.40) + (30.6)(2.43) + (5.75)(2.5)] = -0.27$$

Loop 2:
$$\Delta_2 = -[-(56)(0.89)^2 - (0.143)(3.13)^2 + (4.05)(3.4)^2]/2[(56)(0.89) + (0.143)(3.13) + (4.05)(3.4)] = -0.01$$
$$\Delta_1 = [-(2.9)(3.89)^2 - (4.05)(3.66)^2 + (30.6)(2.16)^2 - (5.75)(2.77)^2]/$$
$$2[(2.9)(3.89) + (4.05)(3.66) + (30.6)(2.16) + (5.75)(2.77)] = 0$$
$$\Delta_2 = -(-45.3 - 1.4 + 54.1)/2(50.5 + 0.45 + 14.8) = -0.056$$
$$\Delta_1 = -(-44 - 52.5 + 142.5 - 44)/2(11.3 + 14.6 + 66.1 + 15.9) = -0.01$$
$$\Delta_2 = -[-(56)(0.96)^2 - (0.143)(3.20)^2 + (4.05)(3.6)^2]/2(53.6 + 0.46 + 14.6) = +0.01$$
$$\Delta_1 = -44.1 - 53.1 + 140.2 - 44.2 = -1.2 \qquad \Delta_2 = -50.5 - 1.46 + 53.2 = +1.2$$

Flow in 1000's gpm
f = 0.023

1000 gpm = 2.24 cfs

First trial

Second trial

Third trial

**Fig. 13-20**

**13.27** The pipe network in Fig. 13-21 consists of the following pipes: $AB$, 3000 ft, 8 in; $BC$, 3000 ft, 12 in; $CD$, 10 000 ft, 36 in; $DE$, 8000 ft, 24 in; $EF$, 5000 ft, 6 in; $FA$, 4000 ft, 8 in; $BE$, 10 000 ft, 6 in; $BF$, 8000 ft, 12 in. Inflow at $D$ = 6 cfs. Outflows are $A$, 2 cfs; $B$, 1 cfs; $E$, 1.5 cfs; $C$, 0.5 cfs; $F$, 1 cfs. Assume Manning $n = 0.015$, and find the flow in each pipe and the pressure at each junction if the pressure at $D$ is 120 psi.

▮ $n = 0.015$ $\qquad S = n^2V^2/2.21R^{4/3} = n^2Q^2/[2.21(D/4)^{4/3}(\pi^2D^4/16)]$

$$= [(0.015)^2Q^2(16)(6.4)]/(2.21\pi^2D^{16/3}) = 0.00106(Q^2/D^{16/3})$$

$$h_L = kQ^2 = (0.00106L/D^{16/3})(Q^2) \qquad k = 0.00106L/D^{16/3}$$

|  | $D$ | $D^{16/3}$ | $0.00106L$ | $k$ |
|---|---|---|---|---|
| $AB$ | 0.67 | 0.116 | 3.18 | 27.4 |
| $BC$ | 1.00 | 1.0 | 3.18 | 3.2 |
| $CD$ | 3.0 | 355 | 10.60 | 0.03 |
| $DE$ | 2.0 | 40 | 8.48 | 0.21 |
| $EF$ | 0.5 | 0.0245 | 5.30 | 216 |
| $FA$ | 0.67 | 0.116 | 4.24 | 36.5 |
| $BF$ | 1.0 | 1.0 | 8.48 | 8.48 |
| $BE$ | 0.5 | 0.0245 | 10.60 | 432 |

Loop 1:

$$\Delta_1 = -\frac{-(27.4)(1.3)^2 + (8.48)(1.6)^2 + (36.5)(0.7)^2}{2[(27.4)(1.3) + (8.48)(1.6) + (36.5)(0.7)]} = +0.044$$

Loop 2:

$$\Delta_2 = -\frac{(216)(0.1)^2 - (8.48)(1.6)^2}{2[0 + (216)(0.1) + (8.48)(1.6)]} = +0.277$$

Loop 3:

$$\Delta_3 = -\frac{-(3.2)(3.9)^2 - (0.03)(4.4)^2 + (0.21)(1.6)^2}{2[(3.2)(3.9) + (0.03)(4.4) + (0.21)(1.6)]} = +1.88$$

$$\Delta_1 = -\frac{-43.5 + 15.8 + 20}{2[(27.4)(1.26) + (8.48)(1.36) + (36.5)(0.74)]} = +0.053$$

$$\Delta_2 = -\frac{-1110 + 31 - 15.6}{2[432(1.60) + 216(0.38) + (8.48)(1.36)]} = +0.695$$

$$\Delta_3 = -\frac{-13.1 - 0.19 + 2.56 + 1110}{2[(3.2)(2.02) + (0.03)(2.52) + (0.21)(3.48) + (432)(1.6)]} = -0.72$$

$$\Delta_1 = -\frac{-40 + 4.2 + 22.8}{2[(27.4)(1.21) + (8.48)(0.71) + (36.5)(0.79)]} = +0.096$$

$$\Delta_2 = -\frac{-14.1 + 251 - 4.2}{2[(432)(0.18) + (216)(1.08) + (8.48)(0.71)]} = -0.37$$

$$\Delta_3 = -\frac{-24.2 - 0.3 + 1.6 + 14.1}{2[(3.2)(2.74) + (0.03)(3.24) + (0.21)(2.76) + (432)(0.18)]} = +0.05$$

$$\Delta_1 = -\frac{-33.8 + 11.8 + 28.9}{2[(27.4)(1.11) + (8.48)(1.18) + (36.5)(0.89)]} = -0.05$$

$$\Delta_2 = -\frac{-155 + 109 - 11.8}{2[(432)(0.60) + (216)(0.71) + (8.48)(1.18)]} = +0.07$$

$$\Delta_3 = -\frac{-23 - 0.3 + 1.6 + 155}{2[(3.2)(2.69) + (0.03)(3.19) + (0.21)(2.81) + (432)(0.6)]} = -0.25$$

$$\Delta_1 = -\frac{-37 + 9.5 + 25.7}{2[(27.4)(1.16) + (8.5)(1.06) + (36.5)(0.84)]} = +0.01$$

$$\Delta_2 = -\frac{-33.8 + 132 - 9.6}{2[(432)(0.28) + (216)(0.78) + (8.5)(1.06)]} = -0.15$$

$$\Delta_3 = -\frac{-27.5 - 0.4 + 1.4 + 33.8}{2[(3.2)(2.94) + (0.03)(3.44) + (0.21)(2.56) + (432)(0.28)]} = -0.03$$

And after several more trials, final results are as shown on the last sketch of Fig. 13-21.

| pressures (psi) | |
|---|---|
| A | 87.5 |
| B | 104.5 |
| C | 119.8 |
| D | 120.0 |
| E | 119.5 |
| F | 98.0 |

First trial

Second trial

Third trial

**Fig. 13-21**

Fourth trial

Fifth trial

Fig. 13-21 (continued)

**13.28** Prepare a computer program written in Fortran to determine the flow in each pipe in a pipe network by the Hardy Cross method. The program must be usable in both the English system of units and the International system of units.

```
C    THIS PROGRAM DETERMINES THE FLOW RATE IN EACH PIPE IN A PIPE NET-
C    WORK BY THE HARDY CROSS METHOD.  IT CAN BE USED FOR PROBLEMS IN
C    BOTH THE ENGLISH SYSTEM OF UNITS AND THE INTERNATIONAL SYSTEM OF
C    UNITS.  EACH LOOP MUST BE NUMBERED AND EACH INDIVIDUAL PIPE MUST
C    ALSO BE NUMBERED.
C
C    INPUT DATA MUST BE SET UP AS FOLLOWS.
C
C    CARD 1    COLUMN 1         ENTER 0 (ZERO) OR BLANK IF ENGLISH SYSTEM
C                               OF UNITS IS TO BE USED.  ENTER 1 (ONE) IF
C                               INTERNATIONAL SYSTEM OF UNITS IS TO BE
C                               USED.
C              COLUMN 2         ENTER 0 (ZERO) OR BLANK IF ONLY THE
C                               RESULTS AFTER THE FINAL ITERATION ARE TO
C                               BE PRINTED.  ENTER 1 (ONE) IF THE RESULTS
C                               AFTER EACH ITERATION ARE TO BE PRINTED.
C              COLUMNS 3-5      ENTER INTEGER NUMBER (RIGHT ADJUSTED)
C                               TELLING HOW MANY LOOPS ARE IN THE PIPE
C                               NETWORK.
C              COLUMNS 6-80     ENTER TITLE, DATE, AND OTHER INFORMATION,
C                               IF DESIRED.
C    CARD 2    COLUMNS 1-2      ENTER INTEGER NUMBER (RIGHT ADJUSTED)
C                               TELLING HOW MANY PIPES ARE IN THE FIRST
C                               LOOP.
C              COLUMNS 3-4      ENTER INTEGER NUMBER (RIGHT ADJUSTED)
C                               TELLING HOW MANY PIPES ARE IN THE SECOND
C                               LOOP.
C              COLUMNS 5-6      ENTER INTEGER NUMBER (RIGHT ADJUSTED)
C                               TELLING HOW MANY PIPES ARE IN THE THIRD
C                               LOOP.
C              (CONTINUE PATTERN FOR ALL LOOPS.)
C    CARD 3    COLUMNS 1-3      ENTER INTEGER NUMBER (RIGHT ADJUSTED) TO
C                               IDENTIFY AND TO REPRESENT A PARTICULAR
C                               PIPE IN THE FIRST LOOP.
```

```
C                     COLUMNS 4-6      IF THIS PIPE IS IN COMMON WITH ANY OTHER
C                                      LOOP, ENTER INTEGER NUMBER (RIGHT AD-
C                                      JUSTED) IDENTIFYING THAT LOOP.  OTHER-
C                                      WISE, ENTER 0 (ZERO) OR BLANK.
C                     COLUMNS 7-16     ENTER NUMBER INCLUDING DECIMAL GIVING
C                                      DIAMETER OF THIS PIPE (IN INCHES OR
C                                      MILLIMETERS).
C                     COLUMNS 17-26    ENTER NUMBER INCLUDING DECIMAL GIVING
C                                      LENGTH OF THIS PIPE (IN FEET OR METERS).
C                     COLUMNS 27-36    ENTER NUMBER INCLUDING DECIMAL GIVING
C                                      INITIAL ESTIMATE OF FLOW RATE FOR THIS
C                                      PIPE (IN CUBIC FEET PER SECOND OR CUBIC
C                                      METERS PER SECOND).  (ENTER CLOCKWISE
C                                      FLOW RATES WITH RESPECT TO THIS LOOP AS
C                                      POSITIVE AND COUNTERCLOCKWISE FLOW RATES
C                                      AS NEGATIVE.)
C                     COLUMNS 37-46    ENTER NUMBER INCLUDING DECIMAL GIVING
C                                      HAZEN-WILLIAMS ROUGHNESS COEFFICIENT FOR
C                                      THIS PIPE.
C                     (ENTER ADDITIONAL CARDS LIKE CARD 3 FOR EACH REMAINING
C                     PIPE IN THE FIRST LOOP, THEN FOR EACH PIPE IN THE SECOND
C                     LOOP, ETC. FOR ALL LOOPS.)
C
C     MULTIPLE DATA SETS FOR SOLVING ANY NUMBER OF PROBLEMS MAY BE
C     INCLUDED FOR PROCESSING.
C
      DIMENSION TITLE(13),DIAM(100,10),Q(100,10),ROUGH(100,10),
     *      HLOSS(100,10),QDEL(100)
      REAL LENGTH(100,10)
      INTEGER UNITS,LOOPS,PIPES(100),PPLOOP,PIPENO(100,10),OLOOP(100,10)
1     READ(5,100,END=2)UNITS,IWRITE,LOOPS,TITLE
100   FORMAT(2I1,I3,12A6,A3)
      WRITE(6,105)TITLE
105   FORMAT('1',12A6,A3,////)
      COEFF=1.318
      FACTOR=12.0
      ERROR=.01
      IF(UNITS.EQ.1)COEFF=.8492
      IF(UNITS.EQ.1)FACTOR=1000.0
      IF(UNITS.EQ.1)ERROR=.001
      READ(5,101)(PIPES(J),J=1,LOOPS)
101   FORMAT(40I2)
      DO 200 J=1,LOOPS
      PPLOOP=PIPES(J)
      DO 200 K=1,PPLOOP
200   READ(5,102)PIPENO(J,K),OLOOP(J,K),DIAM(J,K),LENGTH(J,K),Q(J,K),
     *      ROUGH(J,K)
102   FORMAT(2I3,4F10.0)
      NIT=1
205   IF(IWRITE.EQ.1)WRITE(6,106)NIT
106   FORMAT(//38X,'ITERATION NO.',I3,/,38X,'========= === ==',//)
      IF(IWRITE.EQ.1.AND.UNITS.EQ.0)WRITE(6,107)
107   FORMAT(' LOOP NO.  PIPE NO.  DIAMETER (IN)  LENGTH (FT)  ROUGHNESS
     * FLOW RATE (CFS)  HEAD LOSS (FT)',/' --------  --------  --------
     *-----  -----------  ---------  --------------  --------------'/)
      IF(IWRITE.EQ.1.AND.UNITS.EQ.1)WRITE(6,108)
108   FORMAT(' LOOP NO.  PIPE NO.  DIAMETER (MM)  LENGTH ( M)  ROUGHNESS
     * FLOW RATE (CMS)  HEAD LOSS ( M)',/' --------  --------  --------
     *-----  -----------  ---------  --------------  --------------'/)

      DO 201 J=1,LOOPS
      IF(J.EQ.1)NIT=NIT+1
      SHLOSS=0.0
      SHQ=0.0
      PPLOOP=PIPES(J)
      DO 202 K=1,PPLOOP
      HLOSS(J,K)=(ABS(Q(J,K))*4.0**1.63/3.14159265/(DIAM(J,K)/FACTOR)
     *      **2.63/COEFF/ROUGH(J,K))**(1.0/.54)*LENGTH(J,K)
      IF(Q(J,K).LT.0.0)HLOSS(J,K)=-HLOSS(J,K)
      HQ=HLOSS(J,K)/Q(J,K)
      SHLOSS=SHLOSS+HLOSS(J,K)
      SHQ=SHQ+HQ
      IF(IWRITE.EQ.1)WRITE(6,109)J,PIPENO(J,K),DIAM(J,K),LENGTH(J,K),
     *      ROUGH(J,K),Q(J,K),HLOSS(J,K)
```

```
109 FORMAT(3X,I3,7X,I3,7X,F7.1,6X,F8.0,7X,F5.0,7X,F8.3,7X,F9.3)
202 CONTINUE
    QDEL(J)=-SHLOSS/1.85/SHQ
    IF(IWRITE.EQ.1.AND.UNITS.EQ.0)WRITE(6,110)SHLOSS,J,QDEL(J)
110 FORMAT(78X,'-----------',/,78X,F9.3,/1X,'FLOW RATE ADJUSTMENT FOR
   * LOOP',I3,' = ',F7.3,' CFS',/)
    IF(IWRITE.EQ.1.AND.UNITS.EQ.1)WRITE(6,111)SHLOSS,J,QDEL(J)
111 FORMAT(78X,'-----------',/,78X,F9.3,/1X,'FLOW RATE ADJUSTMENT FOR
   * LOOP',I3,' = ',F7.3,' CMS',/)
201 CONTINUE
    DO 203 J=1,LOOPS
    PPLOOP=PIPES(J)
    DO 203 K=1,PPLOOP
    Q(J,K)=Q(J,K)+QDEL(J)
    L=OLOOP(J,K)
    IF(OLOOP(J,K).NE.0)Q(J,K)=Q(J,K)-QDEL(L)
203 CONTINUE
    DO 204 J=1,LOOPS
    IF(ABS(QDEL(J)).GT.ERROR)GO TO 205
204 CONTINUE
    IF(IWRITE.EQ.1)GO TO 1
    IF(UNITS.EQ.0)WRITE(6,107)
    IF(UNITS.EQ.1)WRITE(6,108)
    DO 206 J=1,LOOPS
    PPLOOP=PIPES(J)
    DO 206 K=1,PPLOOP
    IF(K.NE.PPLOOP)WRITE(6,109)J,PIPENO(J,K),DIAM(J,K),LENGTH(J,K),
   *     ROUGH(J,K),Q(J,K),HLOSS(J,K)
    IF(K.EQ.PPLOOP)WRITE(6,112)J,PIPENO(J,K),DIAM(J,K),LENGTH(J,K),
   *     ROUGH(J,K),Q(J,K),HLOSS(J,K)
112 FORMAT(3X,I3,7X,I3,7X,F7.1,6X,F8.0,7X,F5.0,7X,F8.3,7X,F9.3,/)
206 CONTINUE
    NIT=NIT-1
    WRITE(6,104)NIT
104 FORMAT(//,1X,I3,' ITERATIONS WERE REQUIRED.')
    GO TO 1
2   STOP
    END
```

**13.29** Use the computer program written for Prob. 13.28 to solve for the flow rate in each pipe of the network of Prob. 13.3.

**Input**

```
01  2SAMPLE PIPE NETWØRK ANALYSIS
 4 4
 1           18.    2000.     7.00     120.
 2  2        24.     900.     3.50     120.
 3           12.    2800.    -2.19     120.
 4           24.    1100.    -7.00     120.
 5           12.    2200.     3.50     120.
 6           18.     750.    -3.14     120.
 7           24.    2600.    -5.69     120.
 2  1        24.     900.    -3.50     120.
```

**Output**

SAMPLE PIPE NETWORK ANALYSIS

ITERATION NO.  1
========= === ==

| LOOP NO. | PIPE NO. | DIAMETER (IN) | LENGTH (FT) | ROUGHNESS | FLOW RATE (CFS) | HEAD LOSS (FT) |
|---|---|---|---|---|---|---|
| 1 | 1 | 18.0 | 2000. | 120. | 7.000 | 6.803 |
| 1 | 2 | 24.0 | 900. | 120. | 3.500 | 0.209 |
| 1 | 3 | 12.0 | 2800. | 120. | -2.190 | -7.978 |
| 1 | 4 | 24.0 | 1100. | 120. | -7.000 | -0.922 |

-1.888

FLOW RATE ADJUSTMENT FOR LOOP  1 =   0.212 CFS

| | | | | | | |
|---|---|---|---|---|---|---|
| 2 | 5 | 12.0 | 2200. | 120. | 3.500 | 14.936 |
| 2 | 6 | 18.0 | 750. | 120. | -3.140 | -0.578 |
| 2 | 7 | 24.0 | 2600. | 120. | -5.690 | -1.484 |
| 2 | 2 | 24.0 | 900. | 120. | -3.500 | -0.209 |
| | | | | | | -------------- |
| | | | | | | 12.665 |

FLOW RATE ADJUSTMENT FOR LOOP  2 = -1.435 CFS


### ITERATION NO.  2
========= === ==

| LOOP NO. | PIPE NO. | DIAMETER (IN) | LENGTH (FT) | ROUGHNESS | FLOW RATE (CFS) | HEAD LOSS (FT) |
|---|---|---|---|---|---|---|
| 1 | 1 | 18.0 | 2000. | 120. | 7.212 | 7.190 |
| 1 | 2 | 24.0 | 900. | 120. | 5.147 | 0.427 |
| 1 | 3 | 12.0 | 2800. | 120. | -1.978 | -6.605 |
| 1 | 4 | 24.0 | 1100. | 120. | -6.788 | -0.870 |
| | | | | | | -------------- |
| | | | | | | 0.141 |

FLOW RATE ADJUSTMENT FOR LOOP  1 = -0.017 CFS

| | | | | | | |
|---|---|---|---|---|---|---|
| 2 | 5 | 12.0 | 2200. | 120. | 2.065 | 5.624 |
| 2 | 6 | 18.0 | 750. | 120. | -4.575 | -1.160 |
| 2 | 7 | 24.0 | 2600. | 120. | -7.125 | -2.251 |
| 2 | 2 | 24.0 | 900. | 120. | -5.147 | -0.427 |
| | | | | | | -------------- |
| | | | | | | 1.786 |

FLOW RATE ADJUSTMENT FOR LOOP  2 = -0.286 CFS


### ITERATION NO.  3
========= === ==

| LOOP NO. | PIPE NO. | DIAMETER (IN) | LENGTH (FT) | ROUGHNESS | FLOW RATE (CFS) | HEAD LOSS (FT) |
|---|---|---|---|---|---|---|
| 1 | 1 | 18.0 | 2000. | 120. | 7.196 | 7.159 |
| 1 | 2 | 24.0 | 900. | 120. | 5.416 | 0.469 |
| 1 | 3 | 12.0 | 2800. | 120. | -1.994 | -6.709 |
| 1 | 4 | 24.0 | 1100. | 120. | -6.804 | -0.874 |
| | | | | | | ------------ |
| | | | | | | 0.044 |

FLOW RATE ADJUSTMENT FOR LOOP  1 = -0.005 CFS

| | | | | | | |
|---|---|---|---|---|---|---|
| 2 | 5 | 12.0 | 2200. | 120. | 1.779 | 4.267 |
| 2 | 6 | 18.0 | 750. | 120. | -4.861 | -1.298 |
| 2 | 7 | 24.0 | 2600. | 120. | -7.411 | -2.421 |
| 2 | 2 | 24.0 | 900. | 120. | -5.416 | -0.469 |
| | | | | | | ------------ |
| | | | | | | 0.079 |

FLOW RATE ADJUSTMENT FOR LOOP  2 =  -0.014 CFS


### ITERATION NO.  4
========= === ==

| LOOP NO. | PIPE NO. | DIAMETER (IN) | LENGTH (FT) | ROUGHNESS | FLOW RATE (CFS) | HEAD LOSS (FT) |
|---|---|---|---|---|---|---|
| 1 | 1 | 18.0 | 2000. | 120. | 7.190 | 7.149 |
| 1 | 2 | 24.0 | 900. | 120. | 5.425 | 0.470 |
| 1 | 3 | 12.0 | 2800. | 120. | -2.000 | -6.742 |
| 1 | 4 | 24.0 | 1100. | 120. | -6.810 | -0.876 |
| | | | | | | ------------- |
| | | | | | | 0.002 |

```
        FLOW RATE ADJUSTMENT FOR LOOP   1 =   -0.000 CFS


            2           5          12.0        2200.        120.          1.765          4.206
            2           6          18.0         750.        120.         -4.875         -1.305
            2           7          24.0        2600.        120.         -7.425         -2.429
            2           2          24.0         900.        120.         -5.425         -0.470
                                                                                  -------------
                                                                                        0.001


        FLOW RATE ADJUSTMENT FOR LOOP   2 =   -0.000 CFS
```

**13.30** Use the computer program written for Prob. 13.28 to solve for the flow rate in each pipe of the network shown in Fig. 13-22a.

❚ It is necessary to assume an initial value of flow rate for each pipe in the network. The values assumed for this example are shown in Fig. 13-22b.

### Input

```
1 2 3 4 5 6 7 8 9 10 11 12 13 14 15 16 17 18 19 20 21 22 23 24 25 26 27 28 29 30 31 32 33 34 35 36 37 38 39 40 41 42 43 44 45 46 47 48 49 50 51 52 53 54 55 56 57 58 59 60 61 62 63 64 65 66 67 68 69 70 71 72 73 74 75 76 77 78 79 80
 10   7SAMPLE PIPE NETWØRK ANALYSIS
  4  4 5 6 3 4 4
    1   0      1000.      1000.      -0.6      100.
    2   2       750.       925.       0.4      100.
    3   3       750.      1000.       1.15     100.
    4   0       750.       925.      -2.6      100.
   10   0      1000.      1000.      -2.0      120.
   11   0      1000.       925.      10.0      120.
    9   4       750.      1000.       4.0      120.
    2   1       750.       925.      -0.4      100.
    3   1       750.      1000.      -1.15     100.
    8   4       500.       650.       2.25     100.
    7   0       500.       400.      -0.75     100.
    6   0       500.       671.      -1.75     100.
    5   0       500.       350.      -2.75     100.
    9   2       750.      1000.      -4.0      120.
   12   0      1000.       800.       5.0      120.
   14   5      1000.       650.       2.0      120.
   21   6       500.       800.       0.75     120.
   22   0       500.      1000.      -2.0      120.
    8   3       500.       650.      -2.25     100.
   13   0       750.       763.       2.0      120.
   15   7       750.       400.       0.5      120.
   14   4      1000.       650.      -2.0      120.
   21   4       500.       800.      -0.75     120.
   18   7       500.       125.       0.75     120.
   19   0       500.       800.      -0.75     120.
   20   0       500.       125.      -1.75     120.
   15   5       750.       400.      -0.5      120.
   16   0       750.       125.       0.5      120.
   17   0       750.       400.      -0.5      120.
   18   6       500.       125.      -0.75     120.
```

### Output

```
SAMPLE PIPE NETWORK ANALYSIS
```

| LOOP NO. | PIPE NO. | DIAMETER (MM) | LENGTH ( M) | ROUGHNESS | FLOW RATE (CMS) | HEAD LOSS ( M) |
|----------|----------|---------------|-------------|-----------|-----------------|----------------|
| 1 | 1 | 1000.0 | 1000. | 100. | -0.532 | -0.655 |
| 1 | 2 | 750.0 | 925. | 100. | 2.537 | 44.438 |
| 1 | 3 | 750.0 | 1000. | 100. | 0.211 | 0.481 |
| 1 | 4 | 750.0 | 925. | 100. | -2.532 | -44.269 |

*continued*

| LOOP NO. | PIPE NO. | DIAMETER (MM) | LENGTH ( M) | ROUGHNESS | FLOW RATE (CMS) | HEAD LOSS ( M) |
|---|---|---|---|---|---|---|
| 2 | 10 | 1000.0 | 1000. | 120. | -4.068 | -20.247 |
| 2 | 11 | 1000.0 | 925. | 120. | 7.932 | 64.482 |
| 2 | 9 | 750.0 | 1000. | 120. | 0.152 | 0.187 |
| 2 | 2 | 750.0 | 925. | 100. | -2.537 | -44.438 |
| | | | | | | |
| 3 | 3 | 750.0 | 1000. | 100. | -0.211 | -0.481 |
| 3 | 8 | 500.0 | 650. | 100. | 1.478 | 82.694 |
| 3 | 7 | 500.0 | 400. | 100. | 0.258 | 1.998 |
| 3 | 6 | 500.0 | 671. | 100. | -0.742 | -23.884 |
| 3 | 5 | 500.0 | 350. | 100. | -1.742 | -60.446 |
| | | | | | | |
| 4 | 9 | 750.0 | 1000. | 120. | -0.152 | -0.187 |
| 4 | 12 | 1000.0 | 800. | 120. | 6.780 | 41.703 |
| 4 | 14 | 1000.0 | 650. | 120. | 3.932 | 12.357 |
| 4 | 21 | 500.0 | 800. | 120. | 0.940 | 31.420 |
| 4 | 22 | 500.0 | 1000. | 120. | -0.220 | -2.676 |
| 4 | 8 | 500.0 | 650. | 100. | -1.478 | -82.694 |
| | | | | | | |
| 5 | 13 | 750.0 | 763. | 120. | 1.848 | 14.537 |
| 5 | 15 | 750.0 | 400. | 120. | -0.942 | -2.191 |
| 5 | 14 | 1000.0 | 650. | 120. | -3.932 | -12.357 |
| | | | | | | |
| 6 | 21 | 500.0 | 800. | 120. | -0.940 | -31.420 |
| 6 | 18 | 500.0 | 125. | 120. | 1.050 | 6.025 |
| 6 | 19 | 500.0 | 800. | 120. | 0.840 | 25.500 |
| 6 | 20 | 500.0 | 125. | 120. | -0.160 | -0.186 |
| | | | | | | |
| 7 | 15 | 750.0 | 400. | 120. | 0.942 | 2.191 |
| 7 | 16 | 750.0 | 125. | 120. | 1.790 | 2.246 |
| 7 | 17 | 750.0 | 400. | 120. | 0.790 | 1.580 |
| 7 | 18 | 500.0 | 125. | 120. | -1.050 | -6.025 |

ITERATIONS WERE REQUIRED.

For illustrative purposes, the actual flow rates as determined by the computer program are shown on Fig. 13-22c.

**13.31** Prepare a compute program written in Basic to determine the flow in each pipe in a pipe network by the Hardy Cross method.

```
 10  'HARDY-CROSS LOOP BALANCING NETWORK PROGRAM,              NET.BAS
 20  'U.S. CUSTOMARY(USC) OR SI UNITS(SI) MAY BE USED.
 30  'HAZEN-WILLIAMS(HW) OR DARCY-WEISBACH(DW) EQUATION MAY BE USED FOR PIPES.
 40  'DATA ENTRY VIA READ AND DATA STATEMENTS.
 50  '*******READ DATA AND PRINT NETWORK INFORMATION***************************
 60  DEFINT I,J,K,N
 70  DIM ITYPE(1000),ELEM(500),IND(500),Q(100),H(100),S(20),IX(240)
 80  FOR J=1 TO 100:        ITYPE(J)=5:       H(J)=-1000!:   NEXT J
 90  READ TITLE$:           LPRINT:          LPRINT" ";TITLE$
100  READ TT$,KK,TOL,VNU,DEFA 'NT$=USC OR SI, KK=NO. OF ITER., VNU=KIN. VISC.
110  'TOL=TOLERANCE IN ITERATION, DEFA=DEFAULT COEF.- EITHER C OR EPS
120  IF(TT$="SI" OR TT$="si") THEN GOTO 150
130  UNITS=4.727:           G=32.174:
140  LPRINT" US CUSTOMARY UNITS SPEC.,VISCOSITY IN FT^2/S=";VNU: GOTO 170
150  UNITS=10.674:          G=9.806:
160  LPRINT" SI UNITS SPEC.,VISCOSITY IN  M^2/S=";  VNU
170  LPRINT" DESIRED TOLERANCE=";TOL;  " NO. OF ITERATIONS=";KK
180  LPRINT"   PIPE Q(CFS OR M^3/S)  L(FT OR M)      D(FT OR M)  HW C OR EPS"
190  READ NPI,TT$:          IF NPI=0 THEN 310
200  'NPI=NO. OF PIPES IN NETWORK, TT$="HW" OR "DW"
210  FOR I1=1 TO NPI:       READ I,QQ,L,D,X3:         IF X3=0! THEN X3=DEFA
220  'I=PIPE NO.,QQ=FLOW,L=LENGTH,D=DIAMETER,X3=C OR EPS
230  Q(I)=QQ:               KP=4*(I-1)+1
240  IF TT$="HW" OR TT$="hw" THEN 280
250  ITYPE(I)=2:            ELEM(KP)=L/(2!*G*D^5*.7854^2):          'DW
260  ELEM(KP+1)=1!/(.7854*D*VNU): ELEM(KP+2)=X3/(3.7*D):
270  EX=2!:                 GOTO 290
280  ITYPE(I)=1:            EX=1.852:   ELEM(KP)=UNITS*L/(X3^EX*D^4.8704)  'HW
290  EN=EX-1!:              LPRINT"  ";I;:
300  LPRINT USING"   #####.#####";Q(I);L;D;X3:          NEXT I1
310  READ NPS,TT$:          IF NPS=0 THEN 360
320  'NPS=NO. OF PSEUDO ELEMENTS, TT$="PS"
330  FOR I1=1 TO NPS:       READ I,DH:        ITYPE(I)=3:       KP=4*(I-1)+1
```

```
340 ELEM(KP)=DH:                LPRINT"    ";I;" RESERVOIR ELEV. DIFFERENCE=";DH
350 NEXT II
360 READ NPU,TT$:              IF NPU=0 THEN 480
370 'NPU=NO. OF PUMPS, TT$="PU"
380 FOR II=1 TO NPU:           READ I,QQ,DQ,H1,H2,H3,H4:          ITYPE(I)=4
390 'I=PUMP NO.,QQ=FLOW,DQ=DEL Q, H1,H2,H3,H4=EQUALLY SPACED PTS ON PUMP CURVE
400 KP=4*(I-1)+1:              Q(I)=QQ:          ELEM(KP)=H1
410 ELEM(KP+3)=(H4-3!*(H3-H2)-H1)/(6!*DQ^3)
420 ELEM(KP+2)=(H3-2!*H2+H1)/(2!*DQ^2)-3!*ELEM(KP+3)*DQ
430 ELEM(KP+1)=(H2-H1)/DQ-ELEM(KP+2)*DQ-ELEM(KP+3)*DQ^2
440 LPRINT"    ";I;" PUMP CURVE, DQ=";DQ;" H=";H1,H2,H3,H4
450 LPRINT"        COEF. IN PUMP EQ.=";ELEM(KP);ELEM(KP+1);ELEM(KP+2);ELEM(KP+3)
460 NEXT II
470 '*******READ LOOP INDEXING DATA AND BALANCE ALL LOOPS**********************
480 READ NI,TT$:               IF NI=0 THEN 820
490 'NI=NO. OF ITEMS IN VECTOR IND, TT$="IND"
500 FOR I=1 TO NI:             READ IND(I):      NEXT I:        IND(NI+1)=0
510 LPRINT " IND=";;           FOR I=1 TO NI:    LPRINT IND(I);: NEXT I
520 FOR K=1 TO KK:             DDQ=0!:           IP=1
530 I1=IND(IP):                IF I1=0 THEN 780
540 DH=0!:                     HDQ=0!
550 FOR J=1 TO I1:             I=IND(IP+J):      IF I<0 THEN 570
560 IF I=0 THEN 710 ELSE GOTO 580
570 S(J)=-1!:                  I=-I:             GOTO 590
580 S(J)=1!
590 NTY=ITYPE(I):              KP=4*(I-1)+1
600 ON NTY GOTO 610,620,680,690
610 R=ELEM(KP):                GOTO 660
620 REY=ELEM(KP+1)*ABS(Q(I)):                    IF REY<1! THEN REY=1!
630 IF REY<2000! THEN 640 ELSE GOTO 650
640 R=ELEM(KP)*64!/REY:        GOTO 660
650 R=ELEM(KP)*1.325/(LOG(ELEM(KP+2)+5.74/REY^.9))^2
660 DH=DH+S(J)*R*Q(I)*ABS(Q(I))^EN
670 HDQ=HDQ+EX*R*ABS(Q(I))^EN:                   GOTO 710
680 DH=DH+S(J)*ELEM(KP):       GOTO 710
690 DH=DH-S(J)*(ELEM(KP)+Q(I)*(ELEM(KP+1)+Q(I)*(ELEM(KP+2)+Q(I)*ELEM(KP+3))))
700 HDQ=HDQ-(ELEM(KP+1)+2!*ELEM(KP+2)*Q(I)+3!*ELEM(KP+3)*Q(I)^2)
710 NEXT J
720 IF ABS(HDQ)<.0001 THEN HDQ=1!
730 DQ=-DH/HDQ:                DDQ=DDQ+ABS(DQ)
740 FOR J=1 TO I1:             I=ABS(IND(IP+J)):  IF ITYPE(I)=3 THEN 760
750 Q(I)=Q(I)+S(J)*DQ
760 NEXT J
770 IP=IP+I1+1:                GOTO 530
780 LPRINT:                    LPRINT" ITERATION NO.";K;" SUM OF FLOW CORRECTIONS=";
790 LPRINT USING "###.####";DDQ
800 IF DDQ<TOL THEN 820
810 NEXT K
820 LPRINT" ELEMENT FLOW":     FOR I=1 TO 100:    NTY=ITYPE(I)
830 ON NTY GOTO 840,840,850,840,850
840 LPRINT"    ";I;;           LPRINT USING"   ###.####";Q(I)
850 NEXT I
855 '******* DATA FOR PATH THRU SYSTEM TO COMPUTE HGL*********************+****
860 READ NU,TT$:   IF NU=0 THEN 80  'NU=NO. OF NODES WITH GIVEN HGL,TT$=NODES
865 FOR I=1 TO NU:             READ I1,H2:       H(I1)=H2:       NEXT I
870 READ NI,TT$:               IF NI=0 THEN 80  'NI=NO. OF ITEMS IN PATH,TT$="IX"
875 FOR I=1 TO NI:             READ IX(I):       NEXT I:        IX(NI+2)=0
880 LPRINT" IX=";;             FOR I=1 TO NI:    LPRINT IX(I);; NEXT I: LPRINT
885 IP=1
890 FOR J=1 TO 238 STEP 2:    IF J=1 THEN K=IX(IP)
895 I=IX(IP+J):                N=IX(IP+J+1)
900 IF I<1 THEN SS=-1!:        I=-I:             GOTO 910
905 IF I=0 THEN 955 ELSE SS=1!
910 NTY=ITYPE(I):              KP=4*(I-1)+1
915 ON NTY GOTO 920,925,945,950,955
920 R=ELEM(KP):                GOTO 940
925 REY=ELEM(KP+1)*ABS(Q(I)):                    IF REY<1! THEN REY=1!
930 IF REY<2000! THEN R=ELEM(KP)*64!/REY:        GOTO 940
935 IF REY>=2000! THEN R=ELEM(KP)*1.325/(LOG(ELEM(KP+2)+5.74/REY^.9))^2
940 H(N)=H(K)-SS*R*Q(I)*ABS(Q(I))^EN:            GOTO 955
945 H(N)=H(K)-SS*ELEM(KP):                        GOTO 955
950 H(N)=H(K)+SS*(ELEM(KP)+Q(I)*(ELEM(KP+1)+Q(I)*(ELEM(KP+2)+Q(I)*ELEM(KP+3))))
955 IF IX(J+IP+3)=0 THEN 975
960 IF IX(J+IP+2)=0 THEN 970
965 K=N:                       NEXT J
970 IP=IP+J+3:                 GOTO 890
975 LPRINT" JUNCTION HEAD"
980 FOR N=1 TO 100:            IF H(N)=-1000! THEN 990
985 LPRINT"    ";N;;           LPRINT USING" ###.###";H(N)
990 NEXT N
995 GOTO 80
```

(a) Pipe network

**Fig. 13-22**

(b) Initial flow rate estimates

**Fig. 13-22** (*continued*)

(c) Adjusted flow rates after 16 iterations

**Fig. 13-22** (*continued*)

**13.32** Use the computer program written for Prob. 13.31 to solve the network problem displayed in Fig. 13-23. The pump data are as follows:

| $Q$, m³/s | 0 | 0.03 | 0.06 | 0.09 |
|---|---|---|---|---|
| $H$, m | 30 | 29 | 26 | 20 |

*/*

**Input**

```
DATA EXAMPLE
DATA SI,30,.001,.000001,100.
DATA 5,HW
DATA 1,.12,600.,0.3,.0
DATA 2,.03,300.,.15,.0
DATA 3,.0,500.,.6,.0
DATA 4,.03,400.,.3,.0
DATA 5,.03,300.,.3,.0
DATA 2,PS
DATA 6,15.0
DATA 7,18.0
DATA 1,PU
DATA 8,.06,.03,30.,29.,26.,20.
DATA 16,IND
DATA 3,2,1,-3,3,4,-5,3,3,6,-4,-1,3,5,7,8
DATA 1,NODES
DATA 5,117.
DATA 9,IX
DATA 5,8,4,2,2,1,1,4,3
```

**Output**

```
    EXAMPLE
SI UNITS SPEC.,VISCOSITY IN  M^2/S= .000001
DESIRED TOLERANCE= .001  NO. OF ITERATIONS= 30
   PIPE Q(CFS OR M^3/S)  L(FT OR M)      D(FT OR M)  HW C OR EPS
    1        0.12000      600.00000       0.30000    100.00000
    2        0.03000      300.00000       0.15000    100.00000
    3        0.00000      500.00000       0.60000    100.00000
    4        0.03000      400.00000       0.30000    100.00000
    5        0.03000      300.00000       0.30000    100.00000
    6  RESERVOIR ELEV. DIFFERENCE= 15
    7  RESERVOIR ELEV. DIFFERENCE= 18
    8  PUMP CURVE, DQ= .03 H= 30          29          26          20
       COEF. IN PUMP EQ.= 30 -11.11112 -555.5555 -6172.841
IND= 3  2  1 -3  3  4 -5  3  3  6 -4 -1  3  5  7  8
ITERATION NO. 1  SUM OF FLOW CORRECTIONS=  0.1385

ITERATION NO. 2  SUM OF FLOW CORRECTIONS=  0.1040

ITERATION NO. 3  SUM OF FLOW CORRECTIONS=  0.0372

ITERATION NO. 4  SUM OF FLOW CORRECTIONS=  0.0034

ITERATION NO. 5  SUM OF FLOW CORRECTIONS=  0.0006
ELEMENT   FLOW
    1      0.143
    2     -0.034
    3      0.027
    4      0.080
    5      0.094
    8      0.087
IX= 5  8  4  2  2  1  1  4  3
JUNCTION HEAD
    1   137.811
    2   150.044
    3   135.044
    4   137.797
    5   117.000
```

**Fig. 13–23**

# CHAPTER 14
# Flow in Open Channels

**14.1** Water flows in a rectangular, concrete, open channel that is 12.0 m wide at a depth of 2.5 m. The channel slope is 0.0028. Find the water velocity and the flow rate.

$$v = (1.0/n)(R^{2/3})(s^{1/2}) \qquad n = 0.013 \qquad \text{(from Table A-13)}$$
$$R = A/p_w = (12.0)(2.5)/(2.5 + 12.0 + 2.5) = 1.765 \text{ m}$$
$$v = (1.0/0.013)(1.765)^{2/3}(0.0028)^{1/2} = 5.945 \text{ m/s} \qquad Q = Av = [(12.0)(2.5)](5.945) = 178 \text{ m}^3/\text{s}$$

**14.2** Water flows in the symmetrical trapezoidal channel lined with asphalt shown in Fig. 14-1. The channel bottom drops 0.1 ft vertically for every 100 ft of length. What are the water velocity and flow rate?

$$v = (1.486/n)(R^{2/3})(s^{1/2}) \qquad n = 0.015 \qquad \text{(from Table A-13)}$$
$$R = A/p_w \qquad A = (16.0)(4.5) + (2)\{(4.5)[(3)(4.5)]/2\} = 132.8 \text{ ft}^2$$
$$p_w = 16.0 + 2\sqrt{(4.5)^2 + [(3)(4.5)]^2} = 44.46 \text{ ft} \qquad R = 132.8/44.46 = 2.987 \text{ ft}$$
$$s = 0.1/100 = 0.00100 \qquad v = (1.486/0.015)(2.987)^{2/3}(0.00100)^{1/2} = 6.498 \text{ ft/s}$$
$$Q = Av = (132.8)(6.498) = 863 \text{ ft}^3/\text{s}$$

**Fig. 14-1**

**14.3** Water is to flow at a rate of 30 m³/s in the concrete channel shown in Fig. 14-2. Find the required vertical drop of the channel bottom per kilometer of length.

$$A = (3.6)(2.0) + (4.0 - 2.0)[(1.6 + 3.6)/2] = 12.40 \text{ m}^2 \qquad v = (1.0/n)(R^{2/3})(s^{1/2}) = Q/A = 30/12.40 = 2.419 \text{ m/s}$$
$$p_w = 3.6 + 2.0 + \sqrt{(4.0 - 2.0)^2 + (3.6 - 1.6)^2} + 1.6 = 10.03 \text{ m} \qquad R = A/p_w = 12.40/10.03 = 1.236 \text{ m}$$
$$2.419 = (1.0/0.013)(1.236)^{2/3}(s)^{1/2} \qquad s = 0.000746$$

This slope represents a drop of the channel bottom of 0.000746 m per meter of length, or 0.746 m per kilometer of length.

**Fig. 14-2**

**14.4** Water flows in the triangular steel channel shown in Fig. 14-3 at a velocity of 2.9 ft/s. Find the depth of flow if the channel slope is 0.0015.

$$v = (1.486/n)(R^{2/3})(s^{1/2}) \qquad R = A/p_w = 2\{[(d)(d \tan 27.5°)/2]/(2d/\cos 27.5°)\} = 0.2309d$$

$$2.9 = (1.486/0.014)(0.2309d)^{2/3}(0.0015)^{1/2} \qquad d = 2.57 \text{ ft}$$

**Fig. 14-3**

**14.5** After a flood had passed an observation station on a river, an engineer visited the site and, by locating flood marks, performing appropriate surveying, and doing necessary computations, determined that the cross-sectional area, wetted perimeter, and water-surface slope at the time of the peak flooding were 2960 m², 341 m, and 0.00076, respectively. The engineer also noted that the channel bottom was "earth with grass and weeds" ($n = 0.030$). Estimate the peak flood discharge.

$$v = (1.0/n)(R^{2/3})(s^{1/2}) = (1.0/0.030)(2960/341)^{2/3}(0.00076)^{1/2} = 3.881 \text{ m/s}$$

$$Q = Av = (2960)(3.881) = 11\,490 \text{ m}^3/\text{s}$$

**14.6** A rectangular, concrete channel 50 ft wide is to carry water at a flow rate of 800 cfs. The channel slope is 0.00025. Find the depth of flow.

$$v = (1.486/n)(R^{2/3})(s^{1/2}) = Q/A = 800/50d = 16.00/d \qquad R = A/p_w = 50d/(50 + 2d)$$

$$16.00/d = (1.486/0.013)[50d/(50 + 2d)]^{2/3}(0.00025)^{1/2}$$

This equation is not readily solvable, but a trial-and-error solution (not shown here) reveals that $d = 3.92$ ft.

**14.7** Prepare a computer program that will determine the depth of flow of water in a rectangular channel (as in Prob. 14.6).

```
C       THIS PROGRAM DETERMINES THE DEPTH OF FLOW FOR OPEN CHANNEL FLOW
C       IN RECTANGULAR SECTIONS.  IT CAN BE USED FOR PROBLEMS IN BOTH THE
C       ENGLISH SYSTEM OF UNITS AND THE INTERNATIONAL SYSTEM OF UNITS.
C
C       INPUT DATA MUST BE SET UP AS FOLLOWS.
C
C       CARD 1    COLUMN 1        ENTER 0 (ZERO) OR BLANK IF ENGLISH SYSTEM
C                                 OF UNITS IS TO BE USED.  ENTER 1 (ONE) IF
C                                 INTERNATIONAL SYSTEM OF UNITS IS TO BE
C                                 USED.
C                 COLUMNS 2-79    ENTER TITLE, DATE, AND OTHER INFORMATION,
C                                 IF DESIRED.
C       CARD 2    COLUMNS 1-10    ENTER NUMBER INCLUDING DECIMAL GIVING
C                                 WIDTH OF CHANNEL (IN FEET OR METERS).
C                 COLUMNS 11-20   ENTER NUMBER INCLUDING DECIMAL GIVING
C                                 FLOW RATE (IN CUBIC FEET PER SECOND OR
C                                 CUBIC METERS PER SECOND).
C                 COLUMNS 21-30   ENTER NUMBER INCLUDING DECIMAL GIVING
C                                 SLOPE.
C                 COLUMNS 31-40   ENTER NUMBER INCLUDING DECIMAL GIVING
C                                 MANNING N-VALUE.
C
C       MULTIPLE DATA SETS FOR SOLVING ANY NUMBER OF PROBLEMS MAY BE
C       INCLUDED FOR PROCESSING.
C
        DIMENSION TITLE(13)
        REAL N
        INTEGER UNITS
1       READ(5,100,END=2)UNITS,TITLE
```

```
100 FORMAT(I1,13A6)
    WRITE(6,105)TITLE
105 FORMAT('1',13A6,////)
    COEFF=1.486
    IF(UNITS.EQ.1)COEFF=1.0
    READ(5,101)W,Q,S,N
101 FORMAT(4F10.0)
    D=0.001
    TRY1=COEFF/N*(W*D/(W+2.0*D))**(2.0/3.0)*SQRT(S)-Q/W/D
104 D=D+0.001
    TRY2=COEFF/N*(W*D/(W+2.0*D))**(2.0/3.0)*SQRT(S)-Q/W/D
    IF(TRY1*TRY2)102,102,103
103 TRY1=TRY2
    GO TO 104
102 D=D-0.0005
    IF(UNITS.EQ.0)WRITE(6,106)W,Q,S,N,D
106 FORMAT(1X,'GIVEN DATA FOR AN OPEN CHANNEL FLOW IN A RECTANGULAR SE
   *CTION',//5X,'WIDTH =',F7.1,' FT',//5X,'FLOW RATE =',F7.1,' CU FT/S
   *',//5X,'SLOPE =',F10.7,//5X,'MANNING N-VALUE =',F6.3,////1X,'THE D
   *EPTH OF FLOW WILL BE',F7.2,' FT')
    IF(UNITS.EQ.1)WRITE(6,107)W,Q,S,N,D
107 FORMAT(1X,'GIVEN DATA FOR AN OPEN CHANNEL FLOW IN A RECTANGULAR SE
   *CTION',//5X,'WIDTH =',F7.1,' M ',//5X,'FLOW RATE =',F7.1,' CU M/S
   *',//5X,'SLOPE =',F10.7,//5X,'MANNING N-VALUE =',F6.3,////1X,'THE D
   *EPTH OF FLOW WILL BE',F7.2,' M')
    GO TO 1
2   STOP
    END
```

**14.8**  Solve Prob. 14.6 using the computer program of Prob. 14.7.

### Input

```
1 2 3 4 5 6 7 8 9 10 11 12 13 14 15 16 17 18 19 20 21 22 23 24 25 26 27 28 29 30 31 32 33 34 35 36 37 38 39 40 41 42 43 44 45 46 47 48 49 50 51 52 53 54 55 56 57 58 59 60 61 62 63 64 65 66 67 68 69 70 71 72 73 74 75 76 77 78 79 80
0SAMPLE ANALYSIS ØF ØPEN CHANNEL FLØW IN A RECTANGULAR SECTIØN
50.0        800.0       0.00025     0.013
```

### Output

```
SAMPLE ANALYSIS OF OPEN CHANNEL FLOW IN A RECTANGULAR SECTION


GIVEN DATA FOR AN OPEN CHANNEL FLOW IN A RECTANGULAR SECTION

     WIDTH =    50.0 FT

     FLOW RATE =   800.0 CU FT/S

     SLOPE = 0.0002500

     MANNING N-VALUE = 0.013




THE DEPTH OF FLOW WILL BE   3.92 FT
```

**14.9**  A rectangular channel ($n = 0.016$) 20 m wide is to carry water at a flow rate of 30 m$^3$/s at a slope of 0.00032. Find the depth of flow using the computer program of Prob. 14.7.

### Input

```
1 2 3 4 5 6 7 8 9 10 11 12 13 14 15 16 17 18 19 20 21 22 23 24 25 26 27 28 29 30 31 32 33 34 35 36 37 38 39 40 41 42 43 44 45 46 47 48 49 50 51 52 53 54 55 56 57 58 59 60 61 62 63 64 65 66 67 68 69 70 71 72 73 74 75 76 77 78 79 80
1SAMPLE ANALYSIS ØF ØPEN CHANNEL FLØW IN A RECTANGULAR SECTIØN
20.0        30.0        0.00032     0.016
```

**Output**

SAMPLE ANALYSIS OF OPEN CHANNEL FLOW IN A RECTANGULAR SECTION


GIVEN DATA FOR AN OPEN CHANNEL FLOW IN A RECTANGULAR SECTION

    WIDTH =   20.0 M

    FLOW RATE =   30.0 CU M/S

    SLOPE = 0.0003200

    MANNING N-VALUE = 0.016



THE DEPTH OF FLOW WILL BE  1.25 M

**14.10** A corrugated metal pipe of 500 mm diameter flows half-full at a slope of 0.0050 (see Fig. 14-4). What is the flow rate for this condition?

$$v = (1.0/n)(R^{2/3})(s^{1/2}) = (1.0/0.024)[(\tfrac{500}{1000})/4]^{2/3}(0.0050)^{1/2} = 0.7366 \text{ m/s}$$
$$Q = Av = \{[(\pi)(\tfrac{500}{1000})^2/4]/2\}(0.7366) = 0.0723 \text{ m}^3/\text{s}$$

(*Note:* The hydraulic radius for both a circular cross section and a semicircular cross section is one-fourth the diameter.)

500 mm

250 mm

**Fig. 14-4**

**14.11** A 24-in-diameter cast iron pipe on a $\frac{1}{400}$ slope carries water at a depth of 5.6 in, as shown in Fig. 14-5a. What is the flow rate?

$v = (1.486/n)(R^{2/3})(s^{1/2})$, $R = A/p_w$. The applicable area in this problem is the shaded area $(AECDA)$ in Fig. 14-5b: $AB = BC = 12$ in   (both are radii), $BE = 12 - 5.6 = 6.4$ in. Therefore, $AE = EC = \sqrt{(12)^2 - (6.4)^2} = 10.15$ in and $\angle ABE = \angle EBC = \arccos (6.4/12) = 57.77°$, $(\text{Area})_{ABCDA} = [(\pi)(24)^2/4][(2)(57.77°)/360°] = 145.19 \text{ in}^2$, $(\text{Area})_{ABEA} = (\text{Area})_{BCEB} = (6.4)(10.15)/2 = 32.48 \text{ in}^2$, $(\text{Area})_{AECDA} = (\text{Area})_{ABCDA} - (2)(\text{Area})_{ABEA} = 145.19 - (2)(32.48) = 80.23 \text{ in}^2$. The applicable wetted perimeter in this problem is the arc distance $ADC$ in Fig. 14-5: $p_w = ADC = (\pi)(24)[(2)(57.77°)/360°] = 24.20$ in. $R = 80.23/24.20 = 3.315$ in, or 0.2763 ft, $v = (1.486/0.012)(0.2763)^{2/3}(\frac{1}{400})^{1/2} = 2.627$ ft/s, $Q = Av = (80.23/144)(2.627) = 1.46 \text{ ft}^3/\text{s}$.

24 in

5.6 in

(a)

24 in

12 in

5.6 in

(b)

**Fig. 14-5**

**14.12** A 500-mm-diameter concrete pipe on a $\frac{1}{500}$ slope is to carry water at a flow rate of 0.040 m³/s. Find the depth of flow. See Fig. 14-6a.

▌ $v = (1.0/n)(R^{2/3})(s^{1/2}) = Q/A = 0.040/A$  $\qquad 0.040/A = (1.0/0.013)(R^{2/3})(\frac{1}{500})^{1/2}$  $\qquad AR^{2/3} = 0.01163$

Since $R = A/p_w$,

$$A^{5/3}/p_w^{2/3} = 0.01163 \tag{1}$$

Equation (1) contains two unknowns, $A$ and $p_w$; however, both unknowns can be expressed in terms of the unknown depth of flow, $d$. The applicable area in this problem is the shaded area (AECDA) in Fig. 14-6b: $AB = BC = 0.25$ m  (both are radii), $BE = 0.25 - d$. Therefore, $AE = CE = \sqrt{(0.25)^2 - (0.25 - d)^2}$, $\measuredangle ABE = \measuredangle EBC = \arccos[(0.25 - d)/0.25]$,

$$(\text{Area})_{ABCDA} = \left[\frac{(\pi)(0.50)^2}{4}\right]\left\{\frac{(2)\arccos[(0.25 - d)/0.25]}{360°}\right\} = (0.001091)\left(\arccos\frac{0.25 - d}{0.25}\right)$$

$$(\text{Area})_{ABEA} = (\text{Area})_{BCEB} = \frac{(0.25 - d)\sqrt{(0.25)^2 - (0.25 - d)^2}}{2}$$

$$(\text{Area})_{AECDA} = (\text{Area})_{ABCDA} - 2(\text{Area})_{ABEA} = (0.001091)\left(\arccos\frac{0.25 - d}{0.25}\right) - (2)\left[\frac{(0.25 - d)\sqrt{(0.25)^2 - (0.25 - d)^2}}{2}\right]$$

$$p_w = ADC = (\pi)(0.50)\left\{\frac{(2)\arccos[(0.25 - d)/0.25]}{360°}\right\} = (0.008727)\left(\arccos\frac{0.25 - d}{0.25}\right)$$

Therefore, substituting into Eq. (1),

$$\frac{[(0.001091)\{\arccos[(0.25 - d)/0.25]\} - (0.25 - d)\sqrt{(0.25)^2 - (0.25 - d)^2}]^{5/3}}{[(0.008727)\{\arccos[(0.25 - d)/0.25]\}]^{2/3}} = 0.01163 \ \frac{\text{m}^3}{\text{s}}$$

This equation is not readily solvable, but a trial-and-error solution (not shown here) reveals that $d = 0.166$ m or 166 mm.

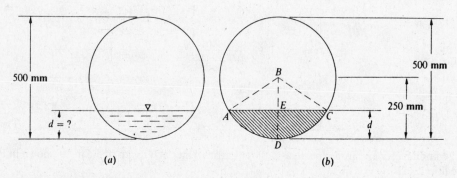

(a) $\qquad\qquad\qquad\qquad\qquad$ (b) $\qquad\qquad\qquad$ **Fig. 14-6**

**14.13** Solve Prob. 14.11 utilizing Fig. A-18.

▌ $s = \frac{1}{400} = 0.0025$ ft/ft. From Fig. A-15, $Q_{\text{full}} = 11.4$ ft³/s and $v_{\text{full}} = 3.6$ ft/s. These values of $Q_{\text{full}}$ and $v_{\text{full}}$ must be adjusted for an $n$ value of 0.012 for the given cast iron pipe (Fig. A-15 is based on an $n$ value of 0.013): $(Q_{\text{full}})_{n=0.012}/11.4 = 0.013/0.012$, $(Q_{\text{full}})_{n=0.012} = 12.4$ ft³/s, $(v_{\text{full}})_{n=0.012}/3.6 = 0.013/0.012$, $(v_{\text{full}})_{n=0.012} = 3.9$ ft/s, $d/d_{\text{full}} = 5.6/24 = 0.23$, or 23 percent. Enter the ordinate of Fig. A-18 with a value of $d/d_{\text{full}}$ of 23 percent, move horizontally to the line marked "discharge," and then vertically downward to the abscissa to read $Q/Q_{\text{full}} = 12$ percent. In a similar manner using the "velocity" line, read $v/v_{\text{full}} = 63$ percent. Therefore, $Q = (0.12)(12.4) = 1.5$ ft³/s, $v = (0.63)(3.9) = 2.5$ ft/s.

**14.14** Solve Prob. 14.12 utilizing Fig. A-18.

▌ $s = \frac{1}{500} = 0.00200$. From Fig. A-16, $Q_{\text{full}} = 0.169$ m³/s: $Q/Q_{\text{full}} = 0.040/0.169 = 0.24$, or 24 percent. From Fig. A-18, $d/d_{\text{full}} = 34$ percent; $d = (0.34)(500) = 170$ mm.

**14.15** A 30-in-diameter concrete storm sewer pipe must carry a flow rate of 9.0 cfs at a minimum velocity of 2.5 ft/s. Find the required slope and water depth.

▌ $\qquad\qquad A = Q/v = 9.0/2.5 = 3.600$ ft²  $\qquad A_{\text{full}} = [(\pi)(\frac{30}{12})^2/4] = 4.909$ ft²

$$A/A_{\text{full}} = 3.600/4.909 = 0.73 \quad \text{or} \quad 73 \text{ percent}$$

From Fig. A-18, $d/d_{full} = 69$ percent and $R/R_{full} = 116$ percent.

$$d = (0.69)(30) = 20.7 \text{ in} \qquad v = (1.486/n)(R^{2/3})(s^{1/2})$$

$$2.5 = (1.486/0.013)\{(1.16)[(\tfrac{30}{12})/4]\}^{2/3}(s^{1/2}) \qquad s = 0.000734$$

**14.16** A concrete pipe must carry water at a slope of 0.0075, at a velocity of 0.76 m/s, and at a depth of flow equal to one-tenth its diameter. What is the required pipe diameter?

▌ $$v = (1.0/n)(R^{2/3})(s^{1/2}) \qquad 0.76 = (1.0/0.013)(R^{2/3})(0.0075)^{1/2} \qquad R = 0.03853 \text{ ft}$$

From Fig. A-18 with $d/d_{full} = 10$ percent, $R/R_{full} = 25$ percent.

$$0.03853/R_{full} = 0.25 \qquad R_{full} = 0.1541 \text{ m} \qquad 0.1541 = d/4 \qquad d = 0.616 \text{ m} \quad \text{or} \quad 616 \text{ mm}$$

**14.17** Prepare a computer program that will determine either the depth of flow or the flow rate for open channel flow in circular sections.

```
▌ C     THIS PROGRAM DETERMINES EITHER THE DEPTH OF FLOW OR THE FLOW RATE
  C     FOR OPEN CHANNEL FLOW IN CIRCULAR SECTIONS.  IT CAN BE USED FOR
  C     PROBLEMS IN BOTH THE ENGLISH SYSTEM OF UNITS AND THE INTERNATIONAL
  C     SYSTEM OF UNITS.
  C
  C     INPUT DATA MUST BE SET UP AS FOLLOWS.
  C
  C     CARD 1    COLUMN 1         ENTER 0 (ZERO) OR BLANK IF ENGLISH SYSTEM
  C                                OF UNITS IS TO BE USED.  ENTER 1 (ONE) IF
  C                                INTERNATIONAL SYSTEM OF UNITS IS TO BE
  C                                USED.
  C               COLUMNS 2-79     ENTER TITLE, DATE, AND OTHER INFORMATION,
  C                                IF DESIRED.
  C     CARD 2    COLUMNS 1-10     ENTER NUMBER INCLUDING DECIMAL GIVING
  C                                DIAMETER OF CHANNEL (IN INCHES OR MILLI-
  C                                METERS).
  C               COLUMNS 11-20    ENTER NUMBER INCLUDING DECIMAL GIVING
  C                                DEPTH OF FLOW (IN INCHES OR MILLIMETERS).
  C               COLUMNS 21-30    ENTER NUMBER INCLUDING DECIMAL GIVING
  C                                SLOPE.
  C               COLUMNS 31-40    ENTER NUMBER INCLUDING DECIMAL GIVING
  C                                MANNING N-VALUE.
  C               COLUMNS 41-50    ENTER NUMBER INCLUDING DECIMAL GIVING
  C                                FLOW RATE (IN CUBIC FEET PER SECOND OR
  C                                CUBIC METERS PER SECOND).
  C
  C ******************************************************************
  C *                                                                *
  C *    NOTE WELL....EITHER THE DEPTH OF FLOW (COLUMNS 11-20) OR THE *
  C *    FLOW RATE (COLUMNS 41-50), WHICHEVER ONE IS TO BE DETERMINED BY *
  C *    THIS PROGRAM, SHOULD BE LEFT BLANK ON CARD 2.                *
  C *                                                                *
  C ******************************************************************
  C
  C     MULTIPLE DATA SETS FOR SOLVING ANY NUMBER OF PROBLEMS MAY BE
  C     INCLUDED FOR PROCESSING.
  C
        DIMENSION TITLE(13)
        COMMON D,R,D1,DIAM,PI,FACTOR,AREA,WP
        REAL N
        INTEGER UNITS
        PI=3.14159265
      1   READ(5,100,END=2)UNITS,TITLE
    100 FORMAT(I1,13A6)
        WRITE(6,105)TITLE
    105 FORMAT('1',13A6,////)
        COEFF=1.486
        FACTOR=12.0
        IF(UNITS.EQ.1)COEFF=1.0
        IF(UNITS.EQ.1)FACTOR=1000.0
        READ(5,101)DIAM,D,S,N,Q
    101 FORMAT(5F10.0)
        R=DIAM/2.0
        IF(Q.GT.0.0001)GO TO 102
        D1=D
```

```
        CALL AREAWP
        HR=AREA/WP

        Q=AREA*COEFF/N*HR**(2.0/3.0)*SQRT(S)
        IF(UNITS.EQ.0)WRITE(6,103)DIAM,D,S,N,Q
103 FORMAT(1X,'GIVEN DATA FOR AN OPEN CHANNEL FLOW IN A CIRCULAR SECTI
    *ON'//5X,'DIAMETER =',F7.2,' IN',//5X,'DEPTH OF FLOW =',F7.2,' IN',
    *//5X,'SLOPE =',F10.7,//5X,'MANNING N-VALUE =',F6.3,////1X,'THE FLO
    *W RATE WILL BE',F8.3,' CU FT/S')
        IF(UNITS.EQ.1)WRITE(6,104)DIAM,D,S,N,Q
104 FORMAT(1X,'GIVEN DATA FOR AN OPEN CHANNEL FLOW IN A CIRCULAR SECTI
    *ON'//5X,'DIAMETER =',F7.1,' MM',//5X,'DEPTH OF FLOW =',F7.1,' MM',
    *//5X,'SLOPE =',F10.7,//5X,'MANNING N-VALUE =',F6.3,////1X,'THE FLO
    *W RATE WILL BE',F8.3,' CU M/S')
        GO TO 1
102 AWP=Q*N/COEFF/SQRT(S)
        D=0.01
        D1=D
        CALL AREAWP
        TRY1=AREA**(5.0/3.0)/WP**(2.0/3.0)-AWP
108 D=D+0.01
        IF(D.GT.DIAM)GO TO 112
        D1=D
        CALL AREAWP
        TRY2=AREA**(5.0/3.0)/WP**(2.0/3.0)-AWP
        IF(TRY1*TRY2)106,106,107
107 TRY1=TRY2
        GO TO 108
106 D=D-0.005
        IF(UNITS.EQ.0)WRITE(6,109)DIAM,Q,S,N,D
109 FORMAT(1X,'GIVEN DATA FOR AN OPEN CHANNEL FLOW IN A CIRCULAR SECTI
    *ON',//5X,'DIAMETER =',F7.2,' IN',//5X,'FLOW RATE =',F8.3,' CU FT/S
    *',//5X,'SLOPE =',F13.7,//5X,'MANNING N-VALUE =',F6.3,////1X,'THE D
    *EPTH OF FLOW WILL BE',F7.2,' IN').
        IF(UNITS.EQ.1)WRITE(6,110)DIAM,Q,S,N,D
110 FORMAT(1X,'GIVEN DATA FOR AN OPEN CHANNEL FLOW IN A CIRCULAR SECTI
    *ON',//5X,'DIAMETER =',F7.1,' MM',//5X,'FLOW RATE =',F8.3,' CU M/S
    *',//5X,'SLOPE =',F13.7,//5X,'MANNING N-VALUE =',F6.3,////1X,'THE D
    *EPTH OF FLOW WILL BE',F7.1,' MM')
        GO TO 1
112 WRITE(6,116)
116 FORMAT(1X,'THIS CIRCULAR CONDUIT CANNOT CARRY THIS GREAT A FLOW AS
    * OPEN CHANNEL FLOW.')
        GO TO 1
2       STOP
        END
        SUBROUTINE AREAWP
        COMMON D,R,D1,DIAM,PI,FACTOR,AREA,WP
        IF(D.GT.R)D1=DIAM-D
        ABCDA=DIAM**2/4.0*ARCOS((R-D1)/R)
        ABEA=(R-D1)*SQRT(R**2-(R-D1)**2)/2.0
        AREA=ABCDA-2.0*ABEA
        WP=DIAM*ARCOS((R-D1)/R)
        IF(D.GT.R)AREA=PI*DIAM**2/4.0-AREA
        IF(D.GT.R)WP=PI*DIAM-WP
        AREA=AREA/FACTOR**2
        WP=WP/FACTOR
        RETURN
        END
```

**14.18** Solve Prob. 14.11 utilizing the computer program of Prob. 14.17.

**❚ Input**

```
1 2 3 4 5 6 7 8 9 10 11 12 13 14 15 16 17 18 19 20 21 22 23 24 25 26 27 28 29 30 31 32 33 34 35 36 37 38 39 40 41 42 43 44 45 46 47 48 49 50 51 52 53 54 55 56 57 58 59 60 61 62 63 64 65 66 67 68 69 70 71 72 73 74 75 76 77 78 79 80
0SAMPLE ANALYSIS ØF AN ØPEN CHANNEL FLØW IN A CIRCULAR SECTIØN
24.0      5.6      0.0025     0.012
```

**Output**

SAMPLE ANALYSIS OF AN OPEN CHANNEL FLOW IN A CIRCULAR SECTION


GIVEN DATA FOR AN OPEN CHANNEL FLOW IN A CIRCULAR SECTION

    DIAMETER = 24.00 IN

    DEPTH OF FLOW = 5.60 IN

    SLOPE = 0.0025000

    MANNING N-VALUE = 0.012


THE FLOW RATE WILL BE 1.463 CU FT/S

**14.19**    Solve Prob. 14.12 utilizing the computer program of Prob. 14.17.

    **▌ Input**

1 2 3 4 5 6 7 8 9 10 11 12 13 14 15 16 17 18 19 20 21 22 23 24 25 26 27 28 29 30 31 32 33 34 35 36 37 38 39 40 41 42 43 44 45 46 47 48 49 50 51 52 53 54 55 56 57 58 59 60 61 62 63 64 65 66 67 68 69 70 71 72 73 74 75 76 77 78 79 80
1SAMPLE ANALYSIS ØF AN ØPEN CHANNEL FLØW IN A CIRCULAR SECTIØN
500.                0.002     0.013     0.040

**Output**

SAMPLE ANALYSIS OF AN OPEN CHANNEL FLOW IN A CIRCULAR SECTION


GIVEN DATA FOR AN OPEN CHANNEL FLOW IN A CIRCULAR SECTION

    DIAMETER = 500.0 MM

    FLOW RATE = 0.040 CU M/S

    SLOPE = 0.0020000

    MANNING N-VALUE = 0.013


THE DEPTH OF FLOW WILL BE 165.7 MM

**14.20**    An open channel is to be designed to carry $1.0 \text{ m}^3/\text{s}$ at a slope of 0.0065. The channel material has an $n$ value of 0.011. Find the most efficient cross section for a semicircular section.

    **▌**
$$v = (1.0/n)(R^{2/3})(s^{1/2}) \qquad Q/A = (1.0/n)(A/p_w)^{2/3}(s^{1/2})$$
$$A^{5/3}/p_w^{2/3} = Qn/s^{1/2} = (1.0)(0.011)/0.0065^{1/2} = 0.1364$$
$$A = \pi d^2/8 \qquad p_w = \pi d/2 \qquad (\pi d^2/8)^{5/3}/(\pi d/2)^{2/3} = 0.1364 \qquad d = 0.951 \text{ m} \quad \text{or} \quad 951 \text{ mm}$$

($d$ is the diameter of the semicircular section; the depth of flow would, of course, be half of $d$.)

**14.21**    Find the most efficient cross section for Prob. 14.20 for a rectangular section.

    **▌** $A^{5/3}/p_w^{2/3} = 0.1364$   (from Prob. 14.20). The most efficient rectangular section has a width equal to twice its depth. Letting $d$ = depth, $A = (d)(2d) = 2d^2$, $p_w = d + 2d + d = 4d$, $(2d^2)^{5/3}/(4d)^{2/3} = 0.1364$, $d = 0.434$ m, or 434 mm; width $= 2d = (2)(434) = 868$ mm.

**14.22**    Find the most efficient cross section for Prob. 14.20 for a triangular section.

    **▌** $A^{5/3}/p_w^{2/3} = 0.1364$  (from Prob. 14.20). The most efficient triangular section has a 90° angle and 1:1 side slopes (see Fig. 14-7). $A = (\frac{1}{2})(d\sqrt{2})(d\sqrt{2}) = d^2$, $p_w = (2)(d\sqrt{2}) = 2.828d$, $(d^2)^{5/3}/(2.828d)^{2/3} = 0.1364$, $d = 0.614$ m, or 614 mm; sides $= d\sqrt{2} = (614)(\sqrt{2}) = 868$ mm.

**Fig. 14-7**

**14.23**  Find the most efficient cross section for Prob. 14.20 for a trapezoidal section.

▮  $A^{5/3}/p^{2/3} = 0.1364$   (from Prob. 14.20). The most efficient trapezoidal section is half a regular hexagon (see Fig. 14-8). $A = (1.155d)(d) + (2)[(d)(d \tan 30°)/2] = 1.732d^2$, $p_w = (3)(1.155d) = 3.465d$, $(1.732d^2)^{5/3}/(3.465d)^{2/3} = 0.1364$, $d = 0.459$ m; sides and bottom: each $= 1.155d = (1.155)(0.459) = 0.530$ m.

**Fig. 14-8**

**14.24**  For the same conditions given in Prob. 14.2, determine the status of flow (i.e., is it critical, subcritical, or supercritical?).

▮  $N_F = v/\sqrt{gd_m}$   $v = 6.498$ ft/s (from Prob. 14.2)   $d_m = A/B$   $A = 132.8$ ft$^2$ (from Prob. 14.2)
$$B = (3)(4.5) + 16.0 + (3)(4.5) = 43.0 \text{ ft} \qquad d_m = 132.8/43.0 = 3.088 \text{ ft}$$
$$N_F = 6.498/\sqrt{(32.2)(3.088)} = 0.652$$

Since $N_F < 1.0$, the flow is subcritical.

**14.25**  The triangular channel ($n = 0.012$) shown in Fig. 14-9 is to carry water at a flow rate of 10 m$^3$/s. Find the critical depth, critical velocity, and critical slope of the channel.

▮  $B/A^3 = g/Q^2$    $B = 6d_c$    $A = 2[(d_c)(3d_c)/2] = 3d_c^2$    $6d_c/(3d_c^2)^3 = 9.807/10^2$    $d_c = 1.178$ m
$$v_c = Q/A = 10/[(3)(1.178)^2] = 2.402 \text{ m/s} \qquad s_c = \{nv_c/[(1.0)(R_c^{2/3})]\}^2 \qquad R = A/p_w$$
$$R_c = [(3)(1.178)^2]/[(2)(\sqrt{10})(1.178)] = 0.5588 \text{ m} \qquad s_c = \{(0.012)(2.402)/[(1.0)(0.5588)^{2/3}]\}^2 = 0.00181$$

**Fig. 14-9**

**14.26**  The semicircular channel ($n = 0.010$) shown in Fig. 14-10 is to carry water at a depth of 1.0 ft. Find the velocity, slope, and discharge at the critical stage.

▮        $d_m = A/B = [(\frac{1}{2})(\pi)(2.0)^2/4]/2.0 = 0.7854 \text{ ft} \qquad v_c = \sqrt{gd_m} = \sqrt{(32.2)(0.7854)} = 5.029 \text{ ft/s}$
$$s_c = \{nv_c/[(1.486)(R_c^{2/3})]\}^2 = \{(0.010)(5.029)/[(1.486)(2.0/4)^{2/3}]\}^2 = 0.00289$$
$$Q = Av_c = [(\frac{1}{2})(\pi)(2.0)^2/4](5.029) = 7.90 \text{ ft}^3/\text{s}$$

**14.27**  A flow rate of 2.1 m$^3$/s is to be carried in an open channel at a velocity of 1.3 m/s. Determine the dimensions of the channel cross section and required slope if the cross section is rectangular with depth equal to one-half the width. Use $n = 0.020$.

**Fig. 14-10**

$Q = Av$. Let $B$ = channel width. Then channel depth = $B/2$; $2.1 = [(B)(B/2)](1.3)$, $B = 1.797$ m. Hence, the required width is 1.797 m and depth is 1.797/2, or 0.898 m.

$$R = A/p_w = (1.797)(0.898)/[1.797 + (2)(0.898)] = 0.4491 \text{ m}$$
$$s = \{nv/[(1.0)(R^{2/3})]\}^2 = \{(0.020)(1.3)/[(1.0)(0.4491)^{2/3}]\}^2 = 0.00197$$

**14.28** Repeat Prob. 14.27 if the depth must be equal to twice the width. Compare answers with Prob. 14.27.

$Q = Av$. Let $B$ = channel width. Then channel depth = $2B$; $2.1 = [(B)(2B)](1.3)$, $B = 0.899$ m. Hence, the required width is 0.899 m and depth is (2)(0.899), or 1.798 m.

$$R = A/p_w = (1.798)(0.899)/[0.899 + (2)(1.798)] = 0.3596 \text{ m}$$
$$s = \{nv/[(1.0)(R^{2/3})]\}^2 = \{(0.020)(1.3)/[(1.0)(0.3596)^{2/3}]\}^2 = 0.00264$$

The channel area is the same (neglecting round-off errors) but a steeper slope is required for the narrower channel.

**14.29** Repeat Prob. 14.27 if the channel cross section is semicircular.

$$Q = Av \qquad 2.1 = [(\tfrac{1}{2})(\pi d^2/4)](1.3) \qquad d = 2.028 \text{ m} \qquad r = 2.028/2 = 1.014 \text{ m}$$
$$s = \{nv/[(1.0)(R^{2/3})]\}^2 = \{(0.020)(1.3)/[(1.0)(2.028/4)^{2/3}]\}^2 = 0.00167$$

**14.30** Repeat Prob. 14.27 if the channel cross section is trapezoidal, with depth equal to the width of the channel bottom and side slopes of 1:1.

$Q = Av$. Let depth and channel bottom width = $B$ (see Fig. 14-11). Then surface width = $3B$; $A = (3B)(B) - (2)[(\tfrac{1}{2})(B)(B)] = 2B^2$, $2.1 = (2B^2)(1.3)$, $B = 0.899$ m.

$$R = A/p_w = (2)(0.899)^2/[0.899 + (2)(0.899)\sqrt{2}] = 0.4696 \text{ m}$$
$$s = \{nv/[(1.0)(R^{2/3})]\}^2 = \{(0.020)(1.3)/[(1.0)(0.4696)^{2/3}]\}^2 = 0.00185$$

**Fig. 14-11**

**14.31** For each of the channel cross sections shown in Fig. 14-12, compute the area, wetted perimeter, and hydraulic radius.

**(a)** $A = \tfrac{1}{2}[(\pi)(4.0)^2/4] = 6.283 \text{ m}^2$  $\qquad p_w = \tfrac{1}{2}[(\pi)(4.0)] = 6.283 \text{ m}$  $\qquad R = A/p_w = 6.283/6.283 = 1.000 \text{ m}$

**(b)** $A = (5.0)(2.5) = 12.50 \text{ m}^2$  $\qquad p_w = 2.5 + 5.0 + 2.5 = 10.00 \text{ m}$  $\qquad R = 12.50/10.00 = 1.250 \text{ m}$

**(c)** $A = (5.0)(1.2) + (2)[(\tfrac{1}{2})(1.2)(1.2)] = 7.440 \text{ m}^2$  $\qquad p_w = 5.0 + (2)[(1.2)(\sqrt{2})] = 8.394 \text{ m}$

$$R = 7.440/8.394 = 0.886 \text{ m}$$

(a)          (b)

(c)                    **Fig. 14-12**

**14.32** Water is to flow in a rectangular flume at a rate of 1.42 m³/s and at a slope of 0.0028. Determine the dimensions of the channel cross section if width must be equal to twice the depth. Use $n = 0.017$.

▌ $Q = (A)(1.0/n)(R^{2/3})(s^{1/2})$. Let $B$ = channel width and $B/2$ = channel depth; $1.42 = [(B)(B/2)][(1.0/0.017)][(B)(B/2)/(B/2 + B + B/2)]^{2/3}(0.0028)^{1/2}$, $B = 1.366$ m. Hence, required channel width = 1.366 m and depth = 1.366/2, or 0.683 m.

**14.33** Rework Prob. 14.32, assuming width must be equal to the depth. Note which solution gives the smaller (and therefore more efficient) cross section.

▌ $Q = (A)(1.0/n)(R^{2/3})(s^{1/2})$. Let $B$ = channel width and depth; $1.42 = [(B)(B)][(1.0/0.017)][(B)(B)/(B + B + B)]^{2/3}(0.0028)^{1/2}$, $B = 0.981$ m. Hence, required channel width and depth are each 0.981 m. $A = (0.981)(0.981) = 0.962$ m². For Prob. 14.32, $A = (1.366)(0.683) = 0.933$ m². The cross section of Prob. 14.32 has the smaller cross-sectional area.

**14.34** A rectangular channel ($n = 0.011$) 18 m wide is to carry water at a flow rate of 35 cfs. The slope of the channel is 0.00078. Determine the depth of flow.

▌ $\qquad Q = (A)(1.0/n)(R^{2/3})(s^{1/2}) \qquad 35 = (18d)(1.0/0.011)[18d/(18 + 2d)]^{2/3}(0.00078)^{1/2}$
$\qquad\qquad d = 0.885$ m $\qquad$ (by trial and error)

**14.35** The trapezoidal channel shown in Fig. 14-13 is laid on a slope of 0.00191. The channel must carry 60 cfs. Determine the depth of flow. Use $n = 0.015$.

▌ $\qquad\qquad Q = (A)(1.486/n)(R^{2/3})(s^{1/2}) \qquad A = 4.0d + (2)[(d)(d)/2] = 4.0d + d^2$
$p_w = 4.0 + (2)(d\sqrt{2}) = 4.0 + 2d\sqrt{2} \qquad 60 = (4.0d + d^2)(1.486/0.015)[(4.0d + d^2)/(4.0 + 2d\sqrt{2})]^{2/3}(0.00191)^{1/2}$
$\qquad\qquad d = 2.00$ ft $\qquad$ (by trial and error)

**Fig. 14-13**

**14.36** A 36-in-diameter concrete pipe on a 0.0015 slope carries water at a depth of 26 in. Determine the flow rate for this pipe.

▌ See Fig. 14-14.

$$Q = (A)(1.486/n)(R^{2/3})(s^{1/2}) \qquad OE = 26 - \tfrac{36}{2} = 8 \text{ in} \quad \text{or} \quad 0.6667 \text{ ft} \qquad \sphericalangle COE = \arccos[8/(\tfrac{36}{2})] = 63.61°$$

$$EC = \sqrt{[(\tfrac{36}{2})/12]^2 - 0.6667^2} = 1.344 \text{ ft}$$

$$A = \{[360 - (2)(63.61)]/360\}[(\pi)(\tfrac{36}{12})^2/4] + (2)[(\tfrac{1}{2})(1.344)(0.6667)] = 5.467 \text{ ft}^2$$

$$p_w = \{[360 - (2)(63.61)]/360\}[(\pi)(\tfrac{36}{12})] = 6.094 \text{ ft}$$

$$Q = (5.467)(1.486/0.013)(5.467/6.094)^{2/3}(0.0015)^{1/2} = 22.5 \text{ ft}^3/\text{s}$$

**Fig. 14-14**

**14.37** Rework Prob. 14.36 using Fig. A-18.

▌ From Fig. A-15, $Q_{full} = 25.8 \text{ ft}^3/\text{s}$; $d/d_{full} = \tfrac{26}{36} = 0.722$, or 72.2 percent. From Fig. A-18, $Q/Q_{full} = 87.5$ percent; $Q = (0.875)(25.8) = 22.6 \text{ ft}^3/\text{s}$.

**14.38** A sewer pipe, for which $n = 0.014$, is laid on a slope of 0.00018 and is to carry 2.76 $m^3/s$ when the pipe flows at 80 percent of full depth. Determine the required diameter of pipe.

▌ See Fig. 14-15.

$$Q = (A)(1.0/n)(R^{2/3})(s^{1/2}) \qquad OE = 0.80D - D/2 = 0.3000D \qquad \alpha = \arccos[0.30D/(D/2)] = 53.13°$$

$$CE = (0.3000D)(\tan 53.13°) = 0.4000D$$

$$A = \{[360 - (2)(53.13)]/360\}(\pi D^2/4) + 2[(\tfrac{1}{2})(0.3000D)(0.4000D)] = 0.6736D^2$$

$$p_w = \{[360 - (2)(53.13)]/360\}(\pi D) = 2.214D$$

$$2.76 = (0.6736D^2)(1.0/0.014)(0.6736D^2/2.214D)^{2/3}(0.00018)^{1/2} \qquad D = 2.32 \text{ m}$$

**Fig. 14-15**

**14.39** Rework Prob. 14.38 using Fig. A-18.

▌ $D/D_{full} = 0.80$. From Fig. A-18, $Q/Q_{full} = 0.96$, $A/A_{full} = 0.84$, and $R/R_{full} = 1.21$: $Q_{full} = 2.76/0.96 = 2.88 \text{ m}^3/\text{s}$. Figure A-16 cannot be used for $Q_{full} > 1.0 \text{ m}^3/\text{s}$: $Q = (A)(1.0/n)(R^{2/3})(s^{1/2})$, $A = (0.84)(\pi D^2/4) = 0.6597D^2$, $R = (1.21)(D/4) = 0.3025D$, $2.76 = (0.6597D^2)(1.0/0.014)(0.3025D)^{2/3}(0.00018)^{1/2}$, $D = 2.34 \text{ m}$.

**14.40** A 72-in-diameter vitrified sewer pipe ($n = 0.014$) is laid on a slope of 0.00025 and carries wastewater at a flow rate of 50 cfs. What is the depth of flow?

▌ From Fig. A-15, $Q_{full} = 67 \text{ ft}^3/\text{s}$: $[(Q_{full})_{n=0.014}]/67 = 0.013/0.014$, $(Q_{full})_{n=0.014} = 62.2 \text{ ft}^3/\text{s}$; $Q/Q_{full} = 50/62.2 = 0.804$, or 80.4 percent. From Fig. A-18, $D/D_{full} = 69$ percent, $D = (0.69)(72) = 49.7$ in.

**14.41** A 1.0-m-diameter pipe must carry a discharge of 0.40 m³/s at a velocity of 0.80 m/s. Determine the slope and the depth of water.

$$A = Q/v = 0.40/0.80 = 0.5000 \text{ m}^2 \qquad A_{\text{full}} = (\pi)(1.0)^2/4 = 0.7854 \text{ m}^2$$
$$A/A_{\text{full}} = 0.5000/0.7854 = 0.64 \quad \text{or} \quad 64 \text{ percent}$$

From Fig. A-18, $D/D_{\text{full}} = 0.63$ and $R/R_{\text{full}} = 1.12$.

$$D = (0.63)(1.0) = 0.630 \text{ m} \qquad v = (1.0/n)(R^{2/3})(s^{1/2}) \qquad R_{\text{full}} = 1.0/4 = 0.2500 \text{ m}$$
$$R = (1.12)(0.2500) = 0.2800 \text{ m} \qquad 0.80 = (1.0/0.013)(0.2800)^{2/3}(s^{1/2}) \qquad s = 0.000590$$

**14.42** The trapezoidal channel of Fig. 14-16 is to carry 500 cfs of water. The maximum allowable velocity of flow is 3.0 fps to avoid scouring. Determine the depth of flow, $d$, and the width of the channel bottom, $B$, if the hydraulic radius of the channel is one-half the depth of flow. Also, determine the slope of the channel bottom. Use $n = 0.025$.

$$R = d/2 = A/p_w \qquad A = Bd + 2[(\tfrac{1}{2})(1.5d)(d)] = Bd + 1.5d^2 \qquad p_w = B + 2\sqrt{d^2 + (1.5d)^2} = B + 3.606d$$
$$d/2 = (Bd + 1.5d^2)/(B + 3.606d) \qquad B + 3.606d = 2B + 3.0d \qquad B = 0.606d \qquad A = Q/v$$
$$Bd + 1.5d^2 = 500/3.0$$

Substituting $B = 0.606d$, $(0.606d)(d) + 1.5d^2 = 166.7$, $d = 8.90$ ft; $B = (0.606)(8.90) = 5.39$ ft.

$$v = (1.486/n)(R^{2/3})(s^{1/2}) \qquad R = 8.90/2 = 4.45 \text{ ft} \qquad 3.0 = (1.486/0.025)(4.45)^{2/3}(s^{1/2}) \qquad s = 0.000348$$

**Fig. 14-16**

**14.43** An open channel to be made of concrete is to be designed to carry 1.5 m³/s at a slope of 0.00085. Find the most efficient cross section for a semicircular section.

$$v = (1.0/n)(R^{2/3})(s^{1/2}) \qquad Q/A = (1.0/n)(A/p_w)^{2/3}(s^{1/2})$$
$$A^{5/3}/p_w^{2/3} = Qn/s^{1/2} = (1.5)(0.013)/0.00085^{1/2} = 0.6688 \qquad A = \pi d^2/8$$
$$p_w = \pi d/2 \qquad (\pi d^2/8)^{5/3}/(\pi d/2)^{2/3} = 0.6688 \qquad d = 1.727 \text{ m}$$

($d$ is the diameter of the semicircular section; the depth of flow would, of course, be half of $d$.)

**14.44** Find the most efficient cross section for Prob. 14.43 for a rectangular section.

$A^{5/3}/p_w^{2/3} = 0.6688$ (from Prob. 14.43). The most efficient rectangular section has a width equal to twice its depth. Letting $d = $ depth, $A = (d)(2d) = 2d^2$, $p_w = d + 2d + d = 4d$, $(2d^2)^{5/3}/(4d)^{2/3} = 0.6688$, $d = 0.789$ m; width $= 2d = (2)(0.789) = 1.578$ m.

**14.45** Find the most efficient cross section for Prob. 14.43 for a triangular section.

$A^{5/3}/p_w^{2/3} = 0.6688$ (from Prob. 14.43). The most efficient triangular section has a 90° angle and 1:1 side slopes (see Fig. 14-7): $A = (\tfrac{1}{2})(d\sqrt{2})(d\sqrt{2}) = d^2$, $p_w = (2)(d\sqrt{2}) = 2.828d$, $(d^2)^{5/3}/(2.828d)^{2/3} = 0.6688$, $d = 1.115$ m; sides $= d\sqrt{2} = (1.115)(\sqrt{2}) = 1.577$ m.

**14.46** Find the most efficient cross section for Prob. 14.43 for a trapezoidal section.

$A^{5/3}/p_w^{2/3} = 0.6688$ (from Prob. 14.43). The most efficient trapezoidal section is half a regular hexagon (see Fig. 14-8): $A = (1.155d)(d) + 2[(d)(d \tan 30°)/2] = 1.732d^2$, $p_w = (3)(1.155d) = 3.465d$, $(1.732d^2)^{5/3}/(3.465d)^{2/3} = 0.6688$, $d = 0.832$ m. Sides and bottom: each $= 1.155d = (1.155)(0.832) = 0.961$ m.

**14.47** For the conditions given in Prob. 14.32, determine whether the flow is critical, subcritical, or supercritical.

$$v = Q/A = 1.42/[(1.366)(0.683)] = 1.522 \text{ m/s} \qquad N_F = v/\sqrt{gd_m} = 1.522/\sqrt{(9.807)(0.683)} = 0.588$$

Since $N_F < 1.0$, the flow is subcritical.

**14.48** A rectangular channel with a width of 3.0 m and an $n$ value of 0.014 is to carry water at a flow rate of 13.4 m³/s. Determine the critical depth, velocity, and channel slope.

▮  $d_c = [(Q/B)^2/g]^{1/3} = [(13.4/3.0)^2/9.807]^{1/3} = 1.267$ m     $v_c = Q/A = 13.4/[(1.267)(3.0)] = 3.525$ m/s

$$R = A/p_w = (1.267)(3.0)/(1.267 + 3.0 + 1.267) = 0.6868 \text{ m}$$

$$s_c = \{nv_c/[(1.0)(R^{2/3})]\}^2 = \{(0.014)(3.525)/[(1.0)(0.6868)^{2/3}]\}^2 = 0.00402$$

**14.49** The semicircular channel ($n = 0.013$) shown in Fig. 14-17 is to carry water while flowing full (i.e., at a depth of 1.5 ft). Determine the velocity, slope, and discharge when flow is critical.

▮  $d_m = A/B = \frac{1}{2}[(\pi)(3.0)^2/4]/3.0 = 1.178$ ft     $v_c = \sqrt{gd_m} = \sqrt{(32.2)(1.178)} = 6.159$ ft/s

$$s_c = \{nv_c/[(1.486)(R^{2/3})]\}^2 = \{(0.013)(6.159)/[(1.486)(3.0/4)^{2/3}]\}^2 = 0.00426$$

$$Q = Av = \{\frac{1}{2}[(\pi)(3.0)^2/4]\}(6.159) = 21.8 \text{ ft}^3/\text{s}$$

**Fig. 14-17**

**14.50** Determine the dimensions of the most economical trapezoidal brick-lined ($n = 0.016$) channel to carry 200 m³/s with a slope of 0.0004.

▮  The most economical trapezoidal channel has a cross section as shown in Fig. 14-8 and $R = d/2$ and $A = \sqrt{3}d^2$; $Q = (A)(1.0/n)(R^{2/3})(s^{1/2})$, $200 = (\sqrt{3}d^2)(1.0/0.016)(d/2)^{2/3}(0.0004)^{1/2}$, $d = 6.491$ m. Bottom width = $(1.155)(6.491) = 7.497$ m.

**14.51** Determine the discharge for a trapezoidal channel with a bottom width of 8 ft and side slopes 1:1. The depth is 6 ft, and the slope of the bottom is 0.0009. The channel has a finished concrete lining ($n = 0.012$).

▮     $A = (8)(6) + (2)[(\frac{1}{2})(6)(6)] = 84.00 \text{ ft}^2$     $p_w = 8 + (2)[(6)(\sqrt{2})] = 24.97$ ft

$$Q = (A)(1.486/n)(R^{2/3})(s^{1/2}) = (84.00)(1.486/0.012)(84.00/24.97)^{2/3}(0.0009)^{1/2} = 701 \text{ ft}^3/\text{s}$$

**14.52** What depth is required for 4-m³/s flow in a rectangular planed-wood ($n = 0.012$) channel 2 m wide with a bottom slope of 0.002?

▮     $Q = (A)(1.0/n)(R^{2/3})(s^{1/2})$     $4 = (2d)(1.0/0.012)[2d/(2 + 2d)]^{2/3}(0.002)^{1/2}$

$$d^{5/3}/(2 + 2d)^{2/3} = 0.3381 \qquad d = 0.888 \text{ m} \qquad \text{(by trial and error)}$$

**14.53** A developer has been required by environmental regulatory authorities to line an open channel to prevent erosion. The channel is trapezoidal in cross section and has a slope of 0.0009. The bottom width is 10 ft and side slopes are 2:1 (horizontal to vertical). If he uses rubble ($\gamma_s = 135$ lb/ft³) for the lining, what is the minimum $D_{50}$ of the rubble that can be used? The design flow is 1000 cfs. Assume the shear that rubble can withstand is described by $\tau = (0.040)(\gamma_s - \gamma)(D_{50})$ (lb/ft²), in which $\gamma_s$ is the unit weight of rock and $D_{50}$ is the average rock diameter in feet.

▮  A Manning $n$ of 0.03 is appropriate for rubble.

$$Q = (A)(1.486/n)(R^{2/3})(s^{1/2})$$

$$1000 = [(d)(10 + 2d)](1.486/0.03)\{(d)(10 + 2d)/[10 + (2)(\sqrt{5})(d)]\}^{2/3}(0.0009)^{1/2}$$

$$[(d)(10 + 2d)]^{5/3}/[10 + (2)(\sqrt{5})(d)]^{2/3} = 672.9 \qquad d = 8.63 \text{ ft} \qquad \text{(by trial and error)} \qquad \tau_0 = \gamma Rs$$

$$R = 8.63[10 + (2)(8.63)]/[10 + (2)(\sqrt{5})(8.63)] = 4.841 \text{ ft} \qquad \tau_0 = (62.4)(4.841)(0.0009) = 0.2719 \text{ lb/ft}^2$$

To find the $D_{50}$ size for incipient movement $\tau = \tau_0$ and $0.2719 = (0.040)(135 - 62.4)(D_{50})$, $D_{50} = 0.0936$ ft.

**14.54** A metal-lined rectangular sluiceway is to carry 1.0 m³/s at a slope of 0.010. Determine the minimum area of galvanized iron ($n = 0.011$) needed per meter of length. Neglect freeboard.

▮ For minimum area, $D = B/2$ and $R = B/4$.

$$Q = (A)(1.0/n)(R^{2/3})(s^{1/2}) \qquad 1.0 = [(B)(B/2)](1.0/0.011)(B/4)^{2/3}(0.010)^{1/2} \qquad B = 0.8015 \text{ m}$$
$$D = 0.8015/2 = 0.4008 \text{ m} \qquad A_{\text{metal}}/L = 0.4008 + 0.8015 + 0.4008 = 1.6 \text{ m}^2/\text{m}$$

**14.55** The sides of a trapezoidal channel are inclined 63.4° to the vertical; the channel is to carry 18 m³/s with a bottom slope of 0.0009. Determine the bottom width, depth, and velocity for the best hydraulic section ($n = 0.026$).

▮
$$Q = (A)(1.0/n)(R^{2/3})(s^{1/2}) \qquad \text{and} \qquad m = \tan 63.4° = 2.000$$
$$p_w = 4d\sqrt{1+m^2} - 2md = (4)(d)\sqrt{1+(2.000)^2} - (2)(2.000)(d) = 4.944d = B + 2\sqrt{5}d$$
$$B = 0.4719d \qquad A = Bd + 2d^2 = (0.4719d)(d) + 2d^2 = 2.472d^2$$
$$18 = (2.472d^2)(1.0/0.026)(2.472d^2/4.944d)^{2/3}(0.0009)^{1/2} \qquad d = 2.373 \text{ m}$$
$$B = (0.4719)(2.373) = 1.120 \text{ m} \qquad v = Q/A = 18/[(2.472)(2.373)^2] = 1.293 \text{ m/s}$$

**14.56** A semicircular corrugated-metal ($n = 0.025$) channel must transport 2.4 m³/s a distance of 1000 m with a head loss of 2 m. Compute the required radius.

▮
$$Q = (A)(1.0/n)(R^{2/3})(s^{1/2}) \qquad 2.4 = (\pi r^2/2)(1.0/0.025)(r/2)^{2/3}(\tfrac{2}{1000})^{1/2} \qquad r = 1.121 \text{ m}$$

**14.57** Determine the best hydraulic trapezoidal section to convey 86 m³/s with a bottom slope of 0.002. The lining is finished concrete ($n = 0.012$).

▮
$$Q = (A)(1.0/n)(R^{2/3})(s^{1/2}) \qquad 86 = (A)(1.0/0.012)(A/p_w)^{2/3}(0.002)^{1/2} \qquad A^{5/3}/p_w^{2/3} = 23.08$$

The best trapezoidal section is half a regular hexagon (see Fig. 14-8) for which $A = 1.732d^2$ and $p_w = 3.465d$ (from Prob. 14.23). $(1.732d^2)^{5/3}/(3.465d)^{2/3} = 23.08$, $d = 3.141$ m. Sides and bottom: each $= 1.155d = (1.155)(3.141) = 3.628$ m.

**14.58** Calculate the discharge in steady flow through the channel and floodway of Fig. 14-18; take $s = 0.0010$ and $y = 2.438$ m.

▮
$$Q = (A)(1.0/n)(R^{2/3})(s^{1/2})$$
$$A_1 = (12)(5 + 2.438) + (2)(5 + 2.438)(5 + 2.438)/2 - (2.438)(2.438)/2 = 141.6 \text{ m}^2$$
$$(p_w)_1 = \sqrt{(5 + 2.438)^2 + (5 + 2.438)^2} + 12 + \sqrt{5^2 + 5^2} = 29.59 \text{ m}$$
$$Q_1 = (141.6)(1.0/0.025)(141.6/29.59)^{2/3}(0.0010)^{1/2} = 508.6 \text{ m}^3/\text{s}$$
$$A_2 = (120)(2.438) + (2.438)(2.438)/2 = 295.5 \text{ m}^2$$
$$(p_w)_2 = 120 + \sqrt{2.438^2 + 2.438^2} = 123.4 \text{ m} \qquad Q_2 = (295.5)(1.0/0.040)(295.5/123.4)^{2/3}(0.0010)^{1/2} = 418.1 \text{ m}^3/\text{s}$$
$$Q = Q_1 + Q_2 = 508.6 + 418.1 = 926.7 \text{ m}^3/\text{s}$$

**Fig. 14-18**

**14.59** For 25 000 cfs through the section of Fig. 14-18, find the depth of flow in the floodway (i.e., evaluate $y$) in feet when the slope of the energy grade line is 0.0004.

▮
$$Q = (A)(1.0/n)(R^{2/3})(s^{1/2}) \qquad A_1 = (12)(5 + y) + (2)(5 + y)(5 + y)/2 - (y)(y)/2 = y^2/2 + 22y + 85$$
$$(p_w)_1 = \sqrt{(5 + y)^2 + (5 + y)^2} + 12 + \sqrt{y^2 + y^2} = (5)(\sqrt{2}) + 12 + (2y)(\sqrt{2})$$
$$A_2 = 120y + (y)(y)/2 = 120y + y^2/2$$
$$(p_w)_2 = 120 + \sqrt{y^2 + y^2} = 120 + (y)(\sqrt{2}) \qquad Q = (25\,000)(0.3048)^3 = 707.9 \text{ m}^3/\text{s}$$
$$707.9 = (y^2/2 + 22y + 85)(1.0/0.025)\{(y^2/2 + 22y + 85)/[(5)(\sqrt{2}) + 12 + (2y)(\sqrt{2})]\}^{2/3}(0.0004)^{1/2}$$
$$+ (120y + y^2/2)(1.0/0.040)\{(120y + y^2/2)/[120 + (y)(\sqrt{2})]\}^{2/3}(0.0004)^{1/2}$$
$$y = 2.79 \text{ m} \quad \text{or} \quad 9.15 \text{ ft} \qquad \text{(by trial and error)}$$

**14.60** Find the critical depth for flow at 1.6 m³/s per meter of width.

$$y_c = (q^2/g)^{1/3} = (1.6^2/9.807)^{1/3} = 0.639 \text{ m}$$

**14.61** Compute the critical depth for flow at 0.4 m³/s through the cross section of Fig. 14-19.

$$Q^2T/gA^3 = 1 \qquad T = (2)(y \tan 60°/2) = 1.155y$$
$$A = (y)(y \tan 60°/2) = 0.5774y^2 \qquad (0.4)^2(1.155y)/[(9.807)(0.5774y^2)^3] = 1 \qquad y = 0.628 \text{ m}$$

**Fig. 14-19**

**14.62** Determine the critical depth for flow at 8.49 m³/s through a trapezoidal channel with bottom 2.5 m wide and with 45° sides.

$$Q^2T/gA^3 = 1 \qquad T = 2.5 + 2y \qquad A = By + my^2 = 2.5y + (1)(y)^2$$
$$(8.49)^2(2.5 + 2y)/[(9.807)(2.5y + y^2)^3] = 1 \qquad y = 0.928 \text{ m} \qquad \text{(by trial and error)}$$

**14.63** Design a transition from a trapezoidal section, 8 ft bottom width and side slopes 1 on 1, depth 4 ft, to a rectangular section, 6 ft wide and 6 ft deep, for a flow of 250 cfs. The transition is to be 20 ft long, and the loss is one-tenth the difference between velocity heads. Show the bottom profile, and do not make any sudden changes in cross-sectional area.

$A_1 = 8 \times 4 + 4^2 = 48 \text{ ft}^2$, $A_2 = 36 \text{ ft}^2$, loss $= 0.1[(v_2^2/2g) - (v_1^2/2g)]$, and $y_1 + (v_1^2/2g) + z_1 = y_2 + (v_2^2/2g) + z_2 +$ loss. Assume a linear change in area, $b$, and $T$: $b = 8 - 2(x/L)$ and $T = 16 - 10(x/L)$. Hence: $A = (b + T)(y/2) = 48 - 12(x/L)$ and $y = 2\{[4 - (x/L)]/[2 - (x/L)]\}$.

| $x/L$ | $A$ | $v^2/2g$ | loss | EGL | $y$ | $z$ |
|-------|-----|----------|------|-----|-----|-----|
| 0 | 48 | 0.421 | | 4.421 | 4.0 | 0 |
| | | | 0.008 | | | |
| 0.333 | 44 | 0.501 | | 4.413 | 4.4 | −0.488 |
| | | | 0.011 | | | |
| 0.667 | 40 | 0.607 | | 4.402 | 5.001 | −1.206 |
| | | | 0.014 | | | |
| 1.0 | 36 | 0.749 | | 4.388 | 6.0 | −2.361 |

The profile is shown in Fig. 14-20.

**Fig. 14-20**

**14.64** In a transition from a rectangular channel (2.5 m wide, 2 m deep) to a trapezoidal channel (bottom width 4 m, side slopes 2 on 1, depth 1.3 m) the energy loss is equal to 0.4 times the difference between velocity heads. The discharge is 5.8 m³/s. Determine the difference between elevations of channel bottoms.

$$0 + v_1^2/2g + y_1 = \Delta z + v_2^2/2g + y_2 + (0.4)(v_1^2/2g - v_2^2/2g) \qquad A_1 = (2.5)(2) = 5.00 \text{ m}^2$$

$$A_2 = (4)(1.3) + (2)\{[(1.3)(2)](1.3)/2\} = 8.580 \text{ m}^2 \qquad v_1 = Q/A_1 = 5.8/5.00 = 1.160 \text{ m/s}$$

$$v_2 = 5.8/8.580 = 0.6760 \text{ m/s}$$

$$v_1^2/2g = 1.160^2/[(2)(9.807)] = 0.06860 \text{ m} \qquad v_2^2/2g = 0.6760^2/[(2)(9.807)] = 0.02330 \text{ m}$$

$$0 + 0.06860 + 2 = \Delta z + 0.02330 + 1.3 + (0.4)(0.06860 - 0.02330) \qquad \Delta z = 0.727 \text{ m}$$

**14.65** A dam gate (Fig. 14-21) admits water to a horizontal canal. Considering the pressure distribution hydrostatic at section $O$, compute the depth at $O$ and the discharge per meter of width when $y = 1.0$ m.

$$d_0 = C_c y = (0.85)(1.0) = 0.85 \text{ m} \qquad p_1/\gamma + v_1^2/2g + z_1 = p_2/\gamma + v_2^2/2g + z_2 + h_L$$

$$0 + 0 + 6 = 0 + v_t^2/[(2)(9.807)] + 0.85 + 0 \qquad v_t = 10.05 \text{ m/s} \qquad v_a = C_v v_t = (0.95)(10.05) = 9.548 \text{ m/s}$$

$$Q = Av = [(0.85)(1)](9.548) = 8.12 \text{ m}^3/\text{s per meter of width}$$

**Fig. 14-21**

**14.66** A discharge of 4.5 m³/s occurs in a rectangular channel 1.83 m wide with $s = 0.002$ and $n = 0.012$. Find the normal depth for uniform flow and determine the critical depth. Is the flow subcritical or supercritical?

$$Q = (A)(1.0/n)(R^{2/3})(s^{1/2}) \qquad 4.5 = (1.83y_n)(1.0/0.012)[1.83y_n/(y_n + 1.83 + y_n)]^{2/3}(0.002)^{1/2}$$

$$y_n = 1.060 \text{ m} \qquad \text{(by trial and error)}$$

$$Q^2/g = A^3/B \qquad 4.5^2/9.807 = (1.83y_c)^3/1.83 \qquad y_c = 0.851 \text{ m}$$

Since $y_c < y_n$, the flow is subcritical.

**14.67** Determine the rate of a 17-in-deep flow in a 48-in circular corrugated-metal ($n = 0.022$) pipe on a slope of 0.003.

$$Q = (A)(1.486/n)(R^{2/3})(s^{1/2})$$

$$Q_{full} = [(\pi)(\tfrac{48}{12})^2/4](1.486/0.022)[(\tfrac{48}{12})/4]^{2/3}(0.003)^{1/2} = 46.49 \text{ ft}^3/\text{s} \qquad d/d_{full} = \tfrac{17}{48} = 0.35 \quad \text{or} \quad 35 \text{ percent}$$

From Fig. A-18, $Q/Q_{full} = 25$ percent, $Q = (0.25)(46.49) = 11.6 \text{ ft}^3/\text{s}$.

**14.68** At what depth will 9.0 ft³/s flow in a 3-ft diameter concrete pipe on a slope of 0.004?

$$Q = (A)(1.486/n)(R^{2/3})(s^{1/2}) \qquad Q_{full} = [(\pi)(3)^2/4](1.486/0.013)[3/4]^{2/3}(0.004)^{1/2} = 42.18 \text{ ft}^3/\text{s}$$

$$Q/Q_{full} = 9.0/42.18 = 0.21 \quad \text{or} \quad 21\%$$

From Fig. A-18, $d/d_{full} = 32\%$, $d = (0.32)(3) = 0.96$ ft.

**14.69** Find the flow rate in a 400-mm-diameter concrete pipe on a slope of 0.004, if the depth is 85 mm?

$$Q = (A)(1.0/n)(R^{2/3})(s^{1/2}) \qquad Q_{full} = [(\pi)(0.400)^2/4](1.0/0.013)[0.400/4]^{2/3}(0.004)^{1/2} = 0.1317 \text{ m}^3/\text{s}$$

$$d/d_{full} = 85/400 = 0.21 \quad \text{or} \quad 21\%$$

From Fig. A-18, $Q/Q_{full} = 10\%$, $Q = (0.10)(0.1317) = 0.0132 \text{ m}^3/\text{s}$.

**14.70** Repeat Prob. 14.68 if the pipe is corrugated metal ($n = 0.024$).

$$Q = (A)(1.486/n)(R^{2/3})(s^{1/2}) \qquad Q_{full} = [(\pi)(3)^2/4](1.486/0.024)[3/4]^{2/3}(0.004)^{1/2} = 22.85 \text{ ft}^3/\text{s}$$

$$Q/Q_{full} = 9.0/22.85 = 0.39 \quad \text{or} \quad 39\%$$

From Fig. A-18, $d/d_{full} = 43\%$, $d = (0.43)(3) = 1.29$ ft.

**14.71** Find the smallest downslope of 3-ft-diameter cast iron piping ($n = 0.015$) that will yield an 18 ft³/s, 2.4-ft-deep flow.

$d/d_{full} = 2.4/30.8$. From Fig. A-18, $Q/Q_{full} = 96\%$, $Q_{full} = 18/0.96 = 18.75 \text{ ft}^3/\text{s}$, $Q = (A)(1.486/n)(R^{2/3})(s^{1/2})$, $18.75 = [(\pi)(3)^2/4](1.486/0.015)[3/4]^{2/3}(s)^{1/2}$, $s = 0.00105$.

**14.72** Find the slope of a 10-ft-wide rectangular channel having $n = 0.012$ that makes critical flow occur at a depth of 4.0 ft.

$$Q = \sqrt{A^3 g/B} = \sqrt{[(4)(10)]^3(32.2)/10} = 454.0 \text{ ft}^3/\text{s} = (A)(1.486/n)(R^{2/3})(s^{1/2})$$

$$454.0 = [(4)(10)](1.486/0.012)[(4)(10)/(4.0 + 10 + 4.0)]^{2/3}(s^{1/2}) \qquad s = 0.00290$$

**14.73** Is the flow of Prob. 14.68 subcritical or supercritical?

For critical flow, $Q^2/g = A^3/B$. For a 3-ft pipe transporting at a depth of 0.96 ft, $B \approx 2.8$ ft and $A \approx 1.94 \text{ ft}^2$. Because $9.0^2/32.2 = 2.5 < 1.94^3/2.8 = 2.6$, the flow is barely subcritical.

**14.74** Water flows steadily at 16.0 ft³/s in a triangular sluice with side slopes 1 on 1; the bottom has slope 0.0039. At a certain section, the depth of flow is 2.00 ft. Characterize the flow at this section as subcritical or supercritical.

$Q^2/g = A^3/B$. At critical flow, $A = (2)[(y_c)(y_c)/2] = y_c^2$, $16.0^2/32.2 = (y_c^2)^3/(2y_c)$, $y_c = 1.74$ ft. Since $y_c < 2.00$ ft, the flow at this section is subcritical.

**14.75** Water flows at 8.5 m³/s in a 3.0-m-wide open channel of rectangular cross section. The bottom slopes up, rising 2 mm per meter in the direction of flow. If the water depth decreases from 2.10 m to 1.65 m in a 155 m length of channel, determine Manning's $n$.

▌ $$z_1 + y_1 + v_1^2/2g = z_2 + y_2 + v_2^2/2g + h_L$$

$$v_1 = Q/A_1 = 8.5/[(2.10)(3.0)] = 1.349 \text{ m/s} \qquad v_2 = 8.5/[(1.65)(3.0)] = 1.717 \text{ m/s}$$

$$0 + 2.10 + 1.349^2/[(2)(9.807)] = (0.002)(155) + 1.65 + 1.717^2/[(2)(9.807)] + h_L \qquad h_L = 0.0825 \text{ m}$$

$$v = (1.0/n)(R^{2/3})(s^{1/2}) = v_{\text{avg}} = (1.349 + 1.717)/2 = 1.533 \text{ m/s} \qquad R = A/p_w$$

$$R_1 = (2.10)(3.0)/(2.10 + 3.0 + 2.10) = 0.8750 \text{ m} \qquad R_2 = (1.65)(3.0)/(1.65 + 3.0 + 1.65) = 0.7857 \text{ m}$$

$$R_{\text{avg}} = (0.8750 + 0.7857)/2 = 0.8304 \text{ m} \qquad 1.533 = (1.0/n)(0.8304)^{2/3}(0.0825/155)^{1/2} \qquad n = 0.013$$

**14.76** A flow of 1000 ft³/s occurs in a long 10-ft-wide rectangular channel of constant bottom slope. The Manning equation yields 7.0 ft as the normal depth of flow for this flow rate. Will the depth of flow increase, decrease, or remain the same as one proceeds downstream from a point where the depth is 3.0 ft?

▌ $$y_c = (q^2/g)^{1/3} = [(\tfrac{1000}{10})^2/32.2]^{1/3} = 6.77 \text{ ft}$$

which is between 3.0 ft and 7.0 ft. Hence, the depth of flow will increase.

**14.77** Determine the cross section of greatest hydraulic efficiency for the trapezoidal channel of Fig. 14-22, if the discharge is 10.0 m³/s and the channel slope is 0.0005. Use $n = 0.020$.

▌ $R = A/p_w = (1.5y + x)(y)/\{x + (2)[\sqrt{y^2 + (1.5y)^2}]\}$. For greatest hydraulic efficiency, $R = y/2$. Hence, $(1.5y + x)(y)/[x + (2)\sqrt{y^2 + (1.5y)^2}] = y/2$, $x = 0.606y$.

$$Q = (A)(1.0/n)(R^{2/3})(s^{1/2}) \qquad A = (1.5y + x)(y) = (1.5y + 0.606y)(y) = 2.106y^2$$

$$10.0 = (2.106y^2)(1.0/0.020)(y/2)^{2/3}(0.0005)^{1/2} \qquad y = 2.05 \text{ m} \qquad x = (0.606)(2.05) = 1.24 \text{ m}$$

**Fig. 14-22**

**14.78** Specify a canal with $n = 0.0182$ that will carry 500 ft³/s of water a distance of 1 mile, if the total drop must not exceed 53 ft.

▌ Use a trapezoidal channel with side slopes of 1:1, and assume $v_{\text{max}} = 4.0$ ft/s.

$$s_{\text{max}} = 53/5280 = 0.0100 \qquad A_{\text{min}} = Q/v_{\text{max}} = 500/4.0 = 125.0 \text{ ft}^2 \qquad Q = (A)(1.486/n)(R^{2/3})(s^{1/2})$$

Try $b = 8$ ft and $y = 8$ ft: $A = (8)(8 + 8) = 128.0$ ft², $p_w = 8 + (2)(\sqrt{8^2 + 8^2}) = 30.63$ ft, $500 = (128.0)(1.486/0.0182)(128.0/30.63)^{2/3}(s)^{1/2}$, $s = 0.000340$. Because $s < s_{\text{max}}$ this design is acceptable.

**14.79** Given an open channel with a parabolic cross section ($x = 1.0$ m and $y = 1.0$ m in Fig. 14-23) on a slope of 0.02 with $n = 0.015$, find the normal depth and the critical depth for a flow rate of 2.0 m³/s.

▌ $Q = (A)(1.0/n)(R^{2/3})(s^{1/2})$. Equation of parabola: $x^2 = y$.

$$A = (2)(\tfrac{2}{3})(xy) = (\tfrac{4}{3})(y^{1/2}y) = 4y^{3/2}/3$$
$$p_w = (2x)[1 + (\tfrac{2}{3})(y/x)^2 - (\tfrac{2}{5})(y/x)^4 + \cdots] = (2y^{1/2})[1 + 2y/3 - 2y^2/5 + \cdots]$$

Try $y_n = 0.5$ m: $A = (4)(0.5)^{3/2}/3 = 0.4714$ m², $p_w = (2)(0.5)^{1/2}[1 + (2)(0.5)/3 - (2)(0.5)^2/5 + \cdots] = 1.744$ m, $Q = (0.4714)(1.0/0.015)(0.4714/1.744)^{2/3}(0.02)^{1/2} = 1.86$ m³/s. Since this value of $Q$ (1.86 m³/s) is slightly less than the given value (2.0 m³/s), try a slightly higher value of $y_n$, say 0.52 m: $A = (4)(0.52)^{3/2}/3 = 0.5000$ m², $p_w = (2)(0.52)^{1/2}[1 + (2)(0.52)/3 - (2)(0.52)^2/5 + \cdots] = 1.786$ m, $Q = (0.5000)(1.0/0.015)(0.5000/1.786)^{2/3}(0.02)^{1/2} = 2.02$ m³/s. An additional iteration (not shown) indicates an appropriate value of $y_n$ of 0.518 m. Critical depth occurs when $Q^2/g = A^3/B$: $2.0^2/9.807 = (4y_c^{3/2}/3)^3/[(2)(\sqrt{y_c})]$, $y_c = 0.766$ m.

**Fig. 14-23**

**14.80** Consider the frictionless open-channel flow of Fig. 14-24a, in which the water depth $y_\infty$ is 2 ft and the volume flow rate per unit width $\dot{Q}/w$ is 5 ft²/s. Discuss what happens as the height of the obstacle on the channel floor is increased.

    **∎** We first find out whether the flow upstream of the obstacle is subcritical or supercritical by calculating the critical depth $y_c$, using $y_c = [(\dot{Q}/w)^2/g]^{1/3} = [(5)^2/32.17]^{1/3} = 0.9194$ ft. The upstream Froude number $F_\infty$ is, then, $F_\infty = (y_c/y_\infty)^{3/2} = (0.9194/2)^{3/2} = 0.3117$. $F_\infty$ is less than 1, showing that the upstream flow is subcritical.

    Next we calculate the height of an obstacle which results in critical flow. The specific head of the flow upstream of the obstacle is $H_\infty = (\dot{Q}/w)^2/(2gy_\infty^2) + y_\infty = (5)^2/[(2)(32.17)(2)^2] + 2 = 2.097$ ft. The specific head of the flow at the critical point is the minimum specific head for that $\dot{Q}/w$: $H_{min} = \frac{3}{2}y_c = (\frac{3}{2})(0.9194) = 1.379$ ft. Then the height of an obstacle resulting in critical flow $h_c$ is $h_c = H_\infty - H_{min} = 2.097 - 1.379 = 0.718$ ft.

    Let us calculate the surface profile of a flow over an obstacle which is not high enough to cause critical flow, say an obstacle 0.5 ft high. Since the flow is subcritical, the water surface must dip or depress over the obstacle since $dy/dx$ is opposite in sign to $dh/dx$. The water surface passes through a minimum depth right over the crest of the obstacle. The specific head of the flow over the crest is $H_{crest} = H_\infty - h_{crest} = 2.097 - 0.5 = 1.597$ ft. A dimensionless specific head of $H_{crest}/y_c = 1.597/0.9194 = 1.737$ corresponds to a dimensionless depth for subcritical flow of $y_{crest}/y_c = 1.52$. The depth of the flow over the crest is $y_{crest} = (1.52)(0.9194) = 1.4$ ft. The Froude number of the flow over the crest is $F_{crest} = (y_c/y_{crest})^{3/2} = (1/1.52)^{3/2} = 0.534$.

    The local surface depth all along the obstacle is calculated in exactly the same way using the local height $h(x)$ of the obstacle. Figure 14-24a shows the water surface and Froude-number variation for this flow over a circular obstacle. The flow is symmetrical about the crest of the obstacle. Obstacles of the same crest height but different shape change the shape of the water-surface depression but not its minimum depth.

    Suppose the height of the crest of the obstacle is increased to precisely $h_c = 0.718$ ft, the height at which critical flow occurs. The flow decreases in depth from $y_\infty$ upstream of the obstacle to $y_c$ at the crest with a corresponding increase in Froude number from $N_F = 0.3117$ to 1. Downstream of the crest there are two possible surface profiles. The first possibility is that the flow remains subcritical and its depth increases from $y_c$ back to $y_\infty$ as the Froude number decreases from 1 to $N_F = 0.3117$. This is the limiting case of subcritical flow. The flow is symmetrical about the crest. The second possibility is that the flow passes through the critical point and its Froude number continues to increase while its depth continues to decrease. The flow becomes supercritical. Downstream of the obstacle the flow has a new depth corresponding to the supercritical branch for the same specific head as upstream. Although the specific head is the same on either side of the obstacle, there is a different distribution between the kinetic and potential energy. Upstream of the obstacle the flow is subcritical, and most of the energy is potential, whereas downstream of the obstacle the flow is supercritical and a larger proportion of the energy is kinetic. The transition from subcritical to supercritical flow downstream of the critical point depends on whether the conditions downstream are favorable for maintaining supercritical flow.

    Let us calculate the depth downstream of the obstacle for supercritical flow $y_{\infty, sup}$. The specific head is the same as upstream, $H_\infty = 2.097$ ft or $H/y_c = 2.281$. $y/y_c = 0.535$ is on the supercritical branch for this value of $H/y_c$. The downstream depth is then $y_{\infty, sup} = 0.535(0.9194) = 0.492$ ft. The Froude number there is $(N)_{\infty, sup} = (y_c/y_{\infty, sup})^{3/2} = (N)(1/0.535)^{3/2} = 2.56$.

    The surface elevation undergoes its largest changes when the Froude number is in the vicinity of 1. At low subcritical Froude numbers $1 - N_F^2$ is nearly 1, and the decrease in surface elevation is about the same as the increase in obstacle elevation. As the Froude number in supercritical flow increases, $1 - N_F^2$ becomes increasingly more negative, causing the surface elevation to change less and less with changes in obstacle elevation.

    Figure 14-24b shows the surface profile and Froude-number variation for subcritical and supercritical flow downstream of the crest.

    What happens if the obstacle height is increased above $h_c$, the obstacle height which results in critical flow? The flow can no longer take place with the same values of $y_\infty$ and $\dot{Q}/w$ as before. There must be an adjustment in either $y_\infty$ or $\dot{Q}/w$ or both to raise $H_\infty$ to at least the value which results in critical flow at the crest. This is the minimum specific head required to sustain the flow over the obstacle.

**14.81** A large reservoir 5 m deep has a rectangular sluice gate 1 m wide. How does the volume flow rate change as the sluice gate is raised (Fig. 14-25)?

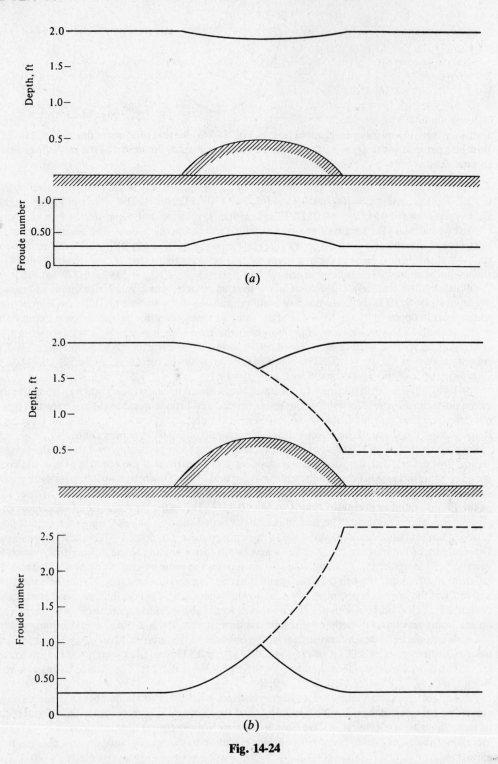

**Fig. 14-24**

▌ Apply the ideal Bernouilli equation in head form at the water surface and far enough up and downstream of the gate for the streamlines to be parallel: $(p_{atm}/\rho g) + (V_\infty^2/2g) + y_\infty = (p_{atm}/\rho g) + (V^2/2g) + y$. In terms of specific head $H_\infty = H$. This is a constant-specific-head flow. The sluice gate is not an obstacle which changes the specific head of the flow; it simply changes the distribution of energy between kinetic and potential.

Now we examine $H_\infty$, $H_\infty = (V_\infty^2/2g) + y_\infty$. Assume that the reservoir is so large that $y_\infty$ upstream of the gate does not change with changes in sluice-gate opening. Furthermore, $V_\infty$ is very small. Then, $H_\infty \approx y_\infty$ and is the same for all sluice-gate openings. Thus, the outflow has the same specific head regardless of the height of the sluice-gate opening.

The flow rate is then given by $\dot{Q} = wy\sqrt{2g(y_\infty - y)}$. The critical depth $y_c$ changes with the flow rate and consequently with $y$: $y_c = [(\dot{Q}/w)^2/g]^{1/3} = [2y^2(y_\infty - y)]^{1/3}$. The corresponding Froude number is $N_F = (y_c/y)^{3/2} = \sqrt{2[(y_\infty/y) - 1]}$.

What happens as the sluice gate is raised? First, consider a small opening which results in a $y$ of, say, 0.5 m. The flow rate is $\dot{Q} = (1)(0.5)\sqrt{2(9.8)(5 - 0.5)} = 4.696$ m³/s and the corresponding Froude number is $N_F = \sqrt{2[(5/0.5) - 1]} = 4.243$. The flow is supercritical.

As the gate is raised further, $y$ increases, causing the Froude number to decrease and the flow rate to increase. Finally, at a gate height which gives $y = \frac{2}{3}y_\infty = (\frac{2}{3})(5) = 3.333$ m, The Froude number is 1 and the maximum flow rate is passed: $\dot{Q}_{max} = w\sqrt{\frac{8}{27}gy_\infty^3} = (1)\sqrt{\frac{8}{27}(9.8)(5)^3} = 19.05$ m³/s.

If the gate is raised farther, the flow becomes subcritical, the Froude number passes through 1 and decreases, and the flow rate starts to decrease.

Fig. 14-25

**14.82** Reconsider the flow leaving the sluice gate in Prob. 14-81 in which $y_\infty = 5$ m. The flow can be made subcritical by placing a step downstream of the gate. Suppose the flow leaving the sluice gate is $y_1 = 0.5$ m deep. Estimate the height of the step $h$ required to cause a subcritical flow $y_3 = 2$ m deep over the step.

❚ The problem is solved by considering the flow in two parts (Fig. 14-26). First, a hydraulic jump occurs in front of the step, resulting in subcritical flow and a dissipation of mechanical energy. Then the level of the flow coming over the step decreases in a frictionless flow.

First we calculate the characteristics of the hydraulic jump. From Prob. 13.2 $(N_F)_1 = 4.243$. The depth ratio across the hydraulic jump is calculated from $y_2/y_1 = \frac{1}{2}(\sqrt{1 + 8(N_F)_1^2} - 1) = \frac{1}{2}[\sqrt{1 + (8)(4.243)^2} - 1] = 5.521$. The mechanical-energy dissipation across the hydraulic jump is $(h_f)_{1\to2}/y_1 = [(y_2/y_1) - 1]^3/(4y_2/y_1) = (5.521 - 1)^3/(4)(5.521) = 4.184$, $(h_f)_{1\to2} = 2.092$ m. The frictionless flow over the step is described by $H_2 = H_3 + h$. Now, $h_{t_2} = h_{t_1} - (h_f)_{1\to2}$, or $H_2 = H_1 - (h_f)_{1\to2}$. $H_1 = H_\infty = y_\infty = 5$ m. Then $H_2 = H_1 - (h_f)_{1\to2} = 5 - 2.092 = 2.908$ m, $H_3 = [(\dot{Q}/w)^2/2gy_3^2] + y_3 = \{(4.696)^2/[2(9.8)(2)^2]\} + 2 = 2.281$ m. Finally, $h = H_2 - H_3 = 2.908 - 2.281 = 0.627$ m.

Fig. 14-26

**14.83** A long channel with a rectangular cross section and an unfinished concrete surface ($n = 0.017$) is 35 ft wide and has a constant slope of 0.5°. What is the water depth when the channel carries 3500 cfs? Is the channel slope mild or steep?

❚ $Q = (A)(1.486/n)(R^{2/3})(s^{1/2})$      $3500 = (35d)(1.486/0.017)[35d/(35 + 2d)]^{2/3}(\tan 0.5°)^{1/2}$

$d = 4.97$ ft      (by trial and error)      $d_c = (q^2/g)^{1/3} = [(\frac{3500}{35})^2/32.2]^{1/3} = 6.77$ ft

Since $d < d_c$, flow is supercritical and the channel slope is steep.

**14.84** Find the depth for uniform flow in Fig. 14-27 when the flow rate is 225 cfs if $s = 0.0006$ and $n$ is assumed to be 0.016. Compute the corresponding value of $\epsilon$.

$$Q = (A)(1.486/n)(R^{2/3})(s^{1/2}) \qquad A = (y_0)(10 + 2y_0) \qquad p_w = 10 + (2)(\sqrt{5})(y_0)$$

$$225 = [(y_0)(10 + 2y_0)](1.486/0.016)\{[(y_0)(10 + 2y_0)]/[10 + (2)(\sqrt{5})(y_0)]\}^{2/3}(0.0006)^{1/2}$$

$$y_0 = 3.41 \text{ ft} \qquad \text{(by trial and error)} \qquad n = 0.093f^{1/2}R^{1/6}$$

$$R = 3.41[10 + (2)(3.41)]/[10 + (2)(\sqrt{5})(3.41)] = 2.272 \text{ ft} \qquad 0.016 = (0.093)(f^{1/2})(2.272)^{1/6} \qquad f = 0.0225$$

$$1/\sqrt{f} = 2 \log (14.8R/\epsilon) \qquad 1/\sqrt{0.0225} = 2 \log [(14.8)(2.272)/\epsilon] \qquad \epsilon = 0.0156 \text{ ft}$$

**Fig. 14-27**

**14.85** In Fig. 14-28, water flows uniformly at a steady rate of 14.0 cfs in a very long triangular flume which has side slopes of 1:1. The bottom of this flume is on a slope of 0.006, and $n = 0.012$. (*a*) Is the flow subcritical or supercritical? (*b*) Find the relation between $v_c^2/2g$ and $y_c$ for this channel.

**(a)** $\qquad Q = (A)(1.486/n)(R^{2/3})(s^{1/2}) \qquad A = (y)(2y)/2 = y^2 \qquad p_w = (2)(\sqrt{2})(y) = 2.828y$

$$14.0 = (y^2)(1.486/0.012)(y^2/2.828y)^{2/3}(0.006)^{1/2} \qquad y = 1.49 \text{ ft} \qquad \text{(by trial and error)}$$

$$Q^2/g = A^3/B \qquad 14.0^2/32.2 = (y_c^2)^3/2y_c \qquad y_c = 1.65 \text{ ft}$$

Since $y < y_c$, flow is supercritical.

**(b)** $\qquad\qquad\qquad v_c^2/g = A_c/B_c = y_c^2/2y_c = y_c/2 \qquad y_c^2/2g = y_c/4$

Consequently, we see that the relation between $v^2/2g$ and $y$ for critical-flow conditions depends on the geometry of the flow section. If the vertex angle of the triangle had been different, the relation would have been different.

**Fig. 14-28**

**14.86** In Fig. 14-29, uniform flow of water occurs at 27 cfs in a 4-ft-wide rectangular flume at a depth of 2.00 ft. (*a*) Is the flow subcritical or supercritical? (*b*) If a hump of height $\Delta t = 0.30$ ft is placed in the bottom of the flume, calculate the water depth on the hump. Neglect head loss in flow over the hump. (*c*) If the hump height is raised to $\Delta z = 0.60$ ft, what then are the water depths upstream and downstream of the hump? Once again neglect head loss over the hump.

**(a)** First find critical depth: $y_c = (q^2/g)^{1/3} = [(\frac{27}{4})^2/32.2]^{1/3} = 1.12$ ft. Since the normal depth (2.00 ft) is greater than the critical depth, the flow is subcritical and the channel slope is mild.

**(b)** Find the critical hump height. Write the energy equation between sections 1 and 2, assume critical flow on the hump and apply continuity.

$$2.00 + (V_1^2/2g) = (\Delta z)_{\text{crit}} + 1.12 + (V_2^2/2g) \tag{1}$$

$$V_2 = 27/(4 \times 1.12) = 6.03 \text{ fps} \tag{2}$$

$$V_1 = 27/(4 \times 2) = 3.38 \text{ fps} \tag{3}$$

Substituging (2) and (3) in (1) gives $(\Delta z)_{\text{crit}} = 0.49$ ft. Thus the minimum-height hump that will produce critical depth on the hump is 0.49 ft.

Since the actual hump height, $\Delta z = 0.30$ ft, is less than the critical hump height, 0.49 ft, critical flow does not occur on the hump and there is no damming action.

Let us now find the depth $y_2$ on the hump:

Energy: $\qquad\qquad\qquad\qquad 2.00 + (V_1^2/2g) = 0.30 + y_2 + (V_2^2/2g) \tag{4}$

Continuity: $\qquad\qquad\qquad\qquad (4 \times 2)V_1 = 4y_2V_2 = 27 \text{ cfs} \tag{5}$

Eliminating $V_1$ and $V_2$ from Eqs. (4) and (5) gives three roots for $y_2$; $y_2 = 1.60$ ft, 0.82 ft, or a negative answer that has no physical meaning. Since the hump height is less than $(\Delta z)_{crit}$, the flow on the hump must be subcritical (that is, $y_2 > y_c$). Hence $y_2 = 1.60$ ft and the drop in the water surface on the hump = $2.00 - (0.30 + 1.60) = 0.10$ ft.

(c) In this case the hump height $\Delta z = 0.60$ ft which is greater than the critical hump height. Hence critical depth ($y_c = 1.12$ ft) will occur on the hump. Writing the energy equation for this case, we have

$$y_1 + (V_1^2/2g) = 0.60 + 1.12 - (V_2^2/2g) \qquad (6)$$

From continuity,
$$(4 \times y_1)V_1 = 27 \text{ cfs} \qquad (7)$$

and, for critical flow,
$$V_2^2/2g = \tfrac{1}{2}y_2 = 0.56 \text{ ft} \qquad (8)$$

Combining Eqs. (6), (7), and (8) gives $y_1 + [(\tfrac{27}{4})^2/(2g)y_1^2] = 2.28$ from which $y_1 = 2.12$ ft, 0.66 ft, or a negative answer which has no physical meaning. In this case, damming action occurs and the depth $y$, upstream of the hump, is 2.12 ft. On the hump the depth passes through critical depth of 1.12 ft and just downstream of the hump the depth will be 0.66 ft. The depth will then increase in the downstream direction following an $M_2$ water-surface profile until a hydraulic jump occurs to return the depth to the normal uniform flow depth of 2.00 ft.

**General case**

**Fig. 14-29**

**14.87** For the channel of Prob. 14.84, compute the "open-channel Reynolds number" assuming that water at 50 °F is flowing. Refer to Fig. A-5 to verify whether or not the flow is wholly rough. Determine $\epsilon$ from Fig. A-5 and compare it with the value computed in Prob. 14.84.

▌ $N_R = Rv/\nu$. From Prob. 14.84, $Q = 225$ cfs, $R = 2.272$ ft, $f = 0.0225$, and $A = 3.41[10 + (2)(3.41)] = 57.36$ ft$^2$; $v = Q/A = 225/57.36 = 3.923$ ft/s.

Open channel:
$$N_R = (2.272)(3.923)/(1.40 \times 10^{-5}) = 6.37 \times 10^5$$

Equivalent pipe:
$$N_R = (4)(6.37 \times 10^5) = 2.55 \times 10^6$$

From Fig. A-5 with $N_R = 2.55 \times 10^6$ and $f = 0.0225$, $\epsilon/D = 0.0018$ and the flow is wholly rough: $\epsilon = 0.0018D = (0.0018)(4R) = 0.0018[(4)(2.272)] = 0.0164$ ft. This value of $\epsilon$ (0.0164) is close to the value of 0.0156 computed in Prob. 14.84.

**14.88** On the assumption that Fig. A-5 applies also to open channels, find the rate of discharge of water at 60 °F in a 100-in-diameter smooth concrete pipe flowing half full ($R = D/4$), if the pipe is laid on a grade of 1.8 ft/mile.

▌ $\epsilon/D = 0.001/(\tfrac{100}{12}) = 0.00012$. Try turbulent flow with $f = 0.0135$:

$$v = \sqrt{(8g/f)(Rs)} \qquad R = D/4 = (\tfrac{100}{12})/4 = 2.083 \text{ ft} \qquad s = \tfrac{1.8}{5280} = 0.0003409$$
$$v = \sqrt{[(8)(32.2)/0.0135][(2.083)(0.0003409)]} = 3.681 \text{ ft/s}$$
$$N_R = Dv/\nu = [(4)(2.083)](3.681)/(1.21 \times 10^{-5}) = 2.53 \times 10^6$$

From Fig. A-5, $f = 0.013$. Try $f = 0.013$:
$$v = \sqrt{[(8)(32.2)/0.013][(2.083)(0.0003409)]} = 3.751 \text{ ft/s} \qquad N_R = [(4)(2.083)](3.751)/(1.21 \times 10^{-5}) = 2.58 \times 10^6$$

$$f = 0.013 \quad \text{(O.K.)} \qquad Q = Av = [(\tfrac{1}{2})(\pi)(\tfrac{100}{12})^2/4](3.751) = 102 \text{ ft}^3/\text{s}$$

**14.89** For the channel of Prob. 14.84, compute the flow rate for depth 8 ft.

▎ $Q = (A)(1.486/n)(R^{2/3})(s^{1/2})$. For $y_0 = 8$ ft, $A = 8[10 + (2)(8)] = 208$ ft$^2$, $p_w = 10 + (2)(\sqrt{5})(8) = 45.78$ ft, $Q = (208)(1.486/0.016)(208/45.78)^{2/3}(0.0006)^{1/2} = 1298$ ft$^3$/s.

**14.90** Figure 14-30 shows a cross section of a canal designed to carry 1590 cfs. The canal is lined with concrete ($n = 0.014$). Find the grade of the canal and the drop in elevation per mile.

▎
$$Q = (A)(1.486/n)(R^{2/3})(s^{1/2}) \qquad A = (10.2)(50.6 + 20)/2 = 360.1 \text{ ft}^2$$

$$p_w = 18.39 + 20 + 18.39 = 56.78 \text{ ft}$$

$$1590 = (360.1)(1.486/0.014)(360.1/56.78)^{2/3}(s)^{1/2} \qquad s = 0.000147$$

$$\text{Drop in elevation} = (0.000147)(5280) = 0.776 \text{ ft/mile}$$

**Fig. 14-30**

**14.91** If the flow in the canal of Prob. 14.90 were halved, all other data, including the slope, being the same, what would be the depth of water?

▎
$$Q = (A)(1.486/n)(R^{2/3})(s^{1/2}) \qquad A = y[20 + (2)(1.5)(y) + 20]/2 = 20y + 1.5y^2$$

$$p_w = 20 + (2)[\sqrt{y^2 + (1.5y)^2}] = 20 + 3.606y$$

$$795 = (20y + 1.5y^2)(1.486/0.014)[(20y + 1.5y^2)/(20 + 3.606y)]^{2/3}(0.000147)^{1/2}$$

$$y = 7.10 \text{ ft} \qquad \text{(by trial and error)}$$

**14.92** Evaluate $\epsilon$ for Prob. 14.90.

▎ $1/\sqrt{f} = 2 \log(14.8R/\epsilon)$, $n = 0.093f^{1/2}R^{1/6}$. Therefore,

$$2 \log(14.8R/\epsilon) = 0.093R^{1/6}/n \qquad R = 360.1/56.78 = 6.342 \text{ ft} \qquad \text{(from Prob. 14.90)}$$

$$2 \log[(14.8)(6.342)/\epsilon] = (0.093)(6.342)^{1/6}/0.014 \qquad \epsilon = 0.00284 \text{ ft}$$

**14.93** Find the capacity of the canal of Prob. 14.90, assuming the grade to be 1.5 ft/mile.

▎ $Q = (A)(1.486/n)(R^{2/3})(s^{1/2})$. From Prob. 14.90, $A = 360.1$ ft$^2$, $n = 0.014$, $R = 360.1/56.78 = 6.342$ ft, $Q = (360.1)(1.486/0.014)(6.342)^{2/3}(1.5/5280)^{1/2} = 2207$ ft$^3$/s.

**14.94** Water flows uniformly in a 2-m-wide rectangular channel at a depth of 0.5 m. The channel slope is 0.0025 and $n = 0.015$. Find the flow rate.

▎
$$Q = (A)(1.0/n)(R^{2/3})(s^{1/2}) \qquad R = (2)(0.5)/(0.5 + 2 + 0.5) = 0.3333 \text{ m}$$

$$Q = [(2)(0.5)](1.0/0.015)(0.3333)^{2/3}(0.0025)^{1/2} = 1.60 \text{ m}^3/\text{s}$$

**14.95** At what depth will 4 m$^3$/s of water flow in a 3-m-wide rectangular channel if $n = 0.016$ and $s = 0.0009$.

▎
$$Q = (A)(1.0/n)(R^{2/3})(s^{1/2}) \qquad 4 = (3d)(1.0/0.016)[3d/(d + 3 + d)]^{2/3}(0.0009)^{1/2}$$

$$d = 1.00 \text{ m} \qquad \text{(by trial and error)}$$

**14.96** The water cross-sectional area in Fig. 14-31 measures 191 ft$^2$, and the wetted perimeter is 39.1 ft. If the flow is 1580 ft$^3$/s and $n = 0.012$, find the slope.

$I$ $\quad Q = (A)(1.486/n)(R^{2/3})(s^{1/2}) \quad 1580 = (191)(1.486/0.012)(191/39.1)^{2/3}(s)^{1/2} \quad s = 0.000538$

16'-0"

16'-0"

13'-2"

12'-4½"

**Fig. 14-31**

**14.97** Rework Prob. 14.96 for a completely filled conduit.

$I$ $\quad Q = (A)(1.486/n)(R^{2/3})(s^{1/2}) \quad 1580 = [(\pi)(16)^2/4](1.486/0.012)(\tfrac{16}{4})^{2/3}(s)^{1/2} \quad s = 0.000634$

**14.98** A 30-in-diameter pipe is known to have a Manning's $n$ of 0.01800. Calculate Manning's $n$ for a 96-in-diameter pipe that has the same $\epsilon$-value as a 30-in pipe for which $n = 0.01800$.

$I$ $\quad 1/\sqrt{f} = 2\log(14.8R/\epsilon)$, $n = 0.093f^{1/2}R^{1/6}$. Therefore, $2\log(14.8R/\epsilon) = 0.093R^{1/6}/n$. For 30-in pipe: $R = (\tfrac{30}{12})/4 = 0.6250$ ft, $2\log[(14.8)(0.6250)/\epsilon] = (0.093)(0.6250)^{1/6}/0.01800$, $\epsilon = 0.03780$ ft. For 96-in pipe: $R = (\tfrac{96}{12})/4 = 2.000$ ft, $2\log[(14.8)(2.000)/0.03780] = (0.093)(2.000)^{1/6}/n$, $n = 0.01804$.

**14.99** The dimensions indicated in Fig. 14-32 pertain to a flow of 30 cfs with $n = 0.018$. Compute the required slope.

$I$ $\quad Q = (A)(1.0/n)(R^{2/3})(s^{1/2}) \quad\quad A = (2.5)(5) + (2.5-2)[(25-5)/2]/2 = 15.00 \text{ ft}^2$

$$p_w = 2.5 + 5 + 2 + \sqrt{(2.5-2)^2 + [(25-5)/2]^2} = 19.51 \text{ ft}$$

$$30 = (15.00)(1.0/0.018)(15.00/19.51)^{2/3}(s)^{1/2} \quad s = 0.00184$$

$w = 25$ ft

$d = 2.5$ ft

$a = 2$ ft

$b = 5$ ft

**Fig. 14-32**

**14.100** Water flows at 10 ft/s in a rectangular trough 6 ft wide for which $n = 0.013$. Find the slope needed for a water depth of 3 ft.

$I$ $\quad v = (1.486/n)(R^{2/3})(s^{1/2}) \quad 10 = (1.486/0.013)[(6)(3)/(3+6+3)]^{2/3}(s)^{1/2} \quad s = 0.00446$

**14.101** In Fig. 14-33, area $A_1$ is 100 ft by 2 ft, $A_2$ is 30 ft by 10 ft, and $A_3$ is 200 ft by 3 ft. Compute the flow rate if $s = 0.0018$, $n_1 = n_3 = 0.03$, and $n_2 = 0.020$.

$I$ $\quad\quad\quad Q = (A)(1.486/n)(R^{2/3})(s^{1/2})$

$$Q_1 = [(100)(2)](1.486/0.03)[(100)(2)/(2+100)]^{2/3}(0.0018)^{1/2} = 658 \text{ ft}^3/\text{s}$$

$$Q_2 = [(30)(10)](1.486/0.02)\{(30)(10)/[(10-2)+30+(10-3)]\}^{2/3}(0.0018)^{1/2} = 3350 \text{ ft}^3/\text{s}$$

$$Q_3 = [(200)(3)](1.486/0.03)[(200)(3)/(3+200)]^{2/3}(0.0018)^{1/2} = 2597 \text{ ft}^3/\text{s}$$

$$Q = Q_1 + Q_2 + Q_3 = 658 + 3350 + 2597 = 6605 \text{ ft}^3/\text{s}$$

**Fig. 14-33**

**14.102** In Prob. 14.99 the parameters are changed as follows: $a = 3$ ft, $b = 6$ ft, $d = 5$ ft, $w = 36$ ft; $n = 0.020$. Given the slope $s = 0.0015$, determine the rate of flow.

▌ $Q = (A)(1.486/n)(R^{2/3})(s^{1/2})$. For depth = 5 ft: $A_1 = (6)(5) = 30.00$ ft², $A_2 = (5 - 3)[(36 - 6)]/2 = 30.00$ ft², $(p_w)_1 = 5 + 6 + 3 = 14.00$ ft, $(p_w)_2 = \sqrt{(5 - 3)^2 + [(36 - 6)]^2} = 30.07$ ft, $Q = (30.00)(1.486/0.020)(30.00/14.00)^{2/3}(0.0015)^{1/2} + (30.00)(1.486/0.020)(30.00/30.07)^{2/3}(0.0015)^{1/2} = 230$ ft³/s.

**14.103** Determine the depth below the surface of clear water at which the velocity $(u)$ as given by the von Karman equation is equal to the mean velocity $(v)$.

▌ $u = v + (1/K)(\sqrt{gy_0 s})[1 + 2.3 \log (y/y_0)]$. Where $u = v$, $v = v + (1/K)(\sqrt{gy_0 s})[1 + 2.3 \log (y/y_0)]$, $y/y_0 = 0.367$. Hence, the velocity $(u)$ is equal to the mean velocity $(v)$ when $y/y_0 = 0.367$, or at a depth below the surface of $1 - 0.367$, or 0.633 times the channel depth.

**14.104** Figure 14-34 is the longitudinal section of a very wide channel. The fluid is clear water $(K = 0.40)$. Given $a = 2.50$ ft and $n = 0.020$, find $b$.

▌ $v = v + (1/K)(\sqrt{gy_0 s})[1 + 2.3 \log (y/y_0)]$. Working with a 1-ft width of channel,

$$v = (1.486/0.020)[(2 + 4 + 3)(1)/1]^{2/3}(s)^{1/2} \qquad s = 0.000009676 v^2 \qquad u_a^2/2g = 2.80 \text{ ft}$$

$$u_a = \sqrt{(2)(32.2)(2.50)} = 12.69 \text{ ft/s} \qquad u = v + (1/K)(\sqrt{gy_0 s})[1 + 2.3 \log (y/y_0)]$$

$$12.69 = v + (1/0.40)[\sqrt{(32.2)(2 + 4 + 3)(0.000009676 v^2)}][1 + 2.3 \log (2 + 4)/(2 + 4 + 3)] \qquad v = 11.76 \text{ ft/s}$$

$$u_b = 11.76 + (1/0.40)[\sqrt{(32.2)(2 + 4 + 3)(0.000009676)(12.45^2)}]\{1 + 2.3 \log [2/(4 + 3 + 2)]\} = 10.98 \text{ ft/s}$$

$$b = 10.98^2/[(2)(32.2)] = 1.87 \text{ ft}$$

**Fig. 14-34**

**14.105** Water flows uniformly in a very wide rectangular channel at a depth of 1.5 m ($s = 0.006$ and $n = 0.015$). Calculate the velocities at two-thirds and at full depth.

▌ $v = (1.0/n)(R^{2/3})(s^{1/2})$. For a very wide channel, $R = y_0 = 1.5$ m: $v = (1.0/0.015)(1.5)^{2/3}(0.006)^{1/2} = 6.767$ m/s, $u = v + (1/K)(\sqrt{gy_0 s})[1 + 2.3 \log (y/y_0)]$. At $y = 1.0$ m, $u = 6.767 + (1/0.40)[\sqrt{(9.807)(1.5)(0.006)}][1 + 2.3 \log (1.0/1.5)] = 7.21$ m/s. At $y = 1.5$ m, $u = 6.767 + (1/0.40)[\sqrt{(9.807)(1.5)(0.006)}][1 + 2.3 \log (1.5/1.5)] = 7.51$ m/s.

**14.106** For a rectangular channel cross section of area 5 m², tabulate hydraulic radius versus width over the range of widths from 1 m to 5 m.

▮ $R = A/p_w = A/(w + 2d) = Aw(w^2 + 2A)$; see following table.

| $w$, m | $R$, m |
|---|---|
| 1 | 0.455 |
| 2 | 0.715 |
| 3 | 0.790 |
| 4 | 0.769 |
| 5 | 0.715 |

**14.107** Set up a general expression for the wetted perimeter $p_w$ of a trapezoidal channel in terms of the cross-sectional area $A$, depth $y$, and angle of side slope $\phi$. Then differentiate $p_w$ with respect to $y$ with $A$ and $\phi$ held constant. From this, prove that $R = y/2$ for the section of greatest hydraulic efficiency (i.e., smallest $p_w$ for a given $A$).

▮ Let $B =$ bottom width.

$$A = By + (y)(y \tan \phi) = By + y^2 \tan \phi \qquad B = A/y - y \tan \phi$$

$$p_w = B + 2y \sec \phi = A/y - y \tan \phi + 2y \sec \phi$$

$$dp_w/dy = -A/y^2 - \tan \phi + 2 \sec \phi = -(By + y^2 \tan \phi)/y^2 - \tan \phi + 2 \sec \phi$$

Setting $dp_w/dy = 0$, $(By + y^2 \tan \phi)/y^2 = 2 \sec \phi - \tan \phi$, $B = 2y \sec \phi - 2y \tan \phi = (2y)(\sec \phi - \tan \phi)$.

$$R = \frac{A}{p_w} = \frac{By + y^2 \tan \phi}{B + 2y \sec \phi} = \frac{(2y)(\sec \phi - \tan \phi)(y) + y^2 \tan \phi}{(2y)(\sec \phi - \tan \phi) + 2y \sec \phi} = \frac{y}{2}$$

**14.108** Prove that the most efficient triangular section is the one with a 90° vertex angle.

▮ See Fig. 14-35.

$$A = a^2 \sin \phi \cos \phi \qquad p_w = 2a \qquad R = A/p_w$$

$$R = (a^2 \sin \phi \cos \phi)/2a = (a/2)(\sin \phi \cos \phi) \qquad dR/d\phi = (a/2)(\cos^2 \phi - \sin^2 \phi) = 0$$

Hence, $\cos \phi = \sin \phi$; or $\phi = 45°$ and the vertex angle $= (2)(45)$, or 90°.

**Fig. 14-35**

**14.109** A canal cut in smooth earth ($n = 0.03123$) must transport 9.0 m³/s of water at a depth of at most 1.5 m (see Fig. 14-36). If the side slopes are 2:1 and the channel slope is 0.0004, what must be the width at the bottom?

▮ $Q = (A)(1.0/n)(R^{2/3})(s^{1/2})$      $A = 1.5[b + b + (2)(1.5) + (2)(1.5)]/2 = 1.5b + 4.5$

$$p_w = b + (2)\sqrt{1.5^2 + [(2)(1.5)]^2} = b + 6.708$$

$$9.0 = (1.5b + 4.5)(1.0/0.03123)[(1.5b + 4.5)/(b + 6.708)]^{2/3}(0.0004)^{1/2}$$

$$b = 6.00 \text{ m} \qquad \text{(by trial and error)}$$

**Fig. 14-36**

**14.110**  Refer to Fig. 14-36. If the discharge in the canal ($n = 0.03123$) is to be 6 m³/s at a depth of 1.5 m and if the velocity is not to exceed 1.0 m/s, what must be the width at the bottom and the slope?

▮      $A = Q/v = 6/(1.0) = 6.00 \text{ m}^2$      $(b)(1.5) + (2)[(2)(1.5)(1.5)]/2 = 6.00$      $b = 1.0 \text{ m}$

$$v = (1.0/n)(R^{2/3})(s^{1/2}) \qquad p_w = 1.00 + (2)\sqrt{1.5^2 + [(2)(1.5)]^2} = 7.708 \text{ m}$$

$$\tfrac{60}{60} = (1.0/0.03123)(6.00/7.708)^{2/3}(s)^{1/2} \qquad s = 0.001362$$

**14.111**  A rectangular flume of smooth wood ($n = 0.012$) slopes 1 ft per 1000 ft. (*a*) Compute the rate of discharge if the width is 4 ft and the depth of water is 2 ft. (*b*) What would be the rate of discharge if the width were 2 ft and the depth of water 4 ft? (*c*) Which of the two forms would have the greater capacity and which would require less lumber?

▮                    $$Q = (A)(1.486/n)(R^{2/3})(s^{1/2})$$

(*a*)          $Q = [(4)(2)](1.468/0.012)[(4)(2)/(2 + 4 + 2)]^{2/3}(\tfrac{1}{1000})^{1/2} = 31.3 \text{ ft}^3/\text{s}$

(*b*)          $Q = [(4)(2)](1.468/0.012)[(4)(2)/(4 + 2 + 4)]^{2/3}(\tfrac{1}{1000})^{1/2} = 27.0 \text{ ft}^3/\text{s}$

(*c*)          Lumber ratio $= (2 + 4 + 2)/(4 + 2 + 4) = 0.80$          Flow ratio $= 31.3/27.0 = 1.16$

Hence, the first design provides 16 percent more flow capacity while requiring only 80 percent as much lumber.

**14.112**  What diameter of semicircular channel will provide the same capacity as a rectangular channel of width 6 m and depth 3 m? Assume $s$ and $n$ are the same for both channels.

▮  $Q = (A)(1.486/n)(R^{2/3})(s^{1/2})$. Since $Q_s = Q_r$ and $(1.486/n)(s^{1/2})$ is constant, $A_s R_s^{2/3} = A_r R_r^{2/3}$.

$$[(\pi d^2/4)/2](d/4)^{2/3} = [(6)(3)][(6)(3)/(3 + 6 + 3)]^{2/3} \qquad d = 6.57 \text{ m}$$

**14.113**  Consider steady flow of water in a circular concrete pipe ($n = 0.016$) of diameter 10 ft at a depth of 4 ft. Using Fig. A-18, determine the flow rate and the average velocity of flow ($s = 0.0004$).

▮      $Q = (A)(1.486/n)(R^{2/3})(s^{1/2})$      $Q_{\text{full}} = [(\pi)(10)^2/4](1.486/0.016)(\tfrac{10}{4})^{2/3}(0.0004)^{1/2} = 269 \text{ ft}^3/\text{s}$

$$v_{\text{full}} = Q_{\text{full}}/A_{\text{full}} = 269/[(\pi)(10)^2/4] = 3.43 \text{ ft/s} \qquad d/d_{\text{full}} = \tfrac{4}{10} = 0.40 \quad \text{or} \quad 40 \text{ percent}$$

From Fig. A-18, $Q/Q_{\text{full}} = 32$ percent and $v/v_{\text{full}} = 88$ percent: $Q = (0.32)(269) = 86.1 \text{ ft}^3/\text{s}$, $v = (0.88)(3.43) = 3.02 \text{ ft/s}$.

**14.114**  At what depth will water flow at 0.25 m³/s in a 1.0-m-diameter concrete pipe ($n = 0.012$) on a slope of 0.0004?

▮                    $$Q = (A)(1.0/n)(R^{2/3})(s^{1/2})$$

$$Q_{\text{full}} = [(\pi)(1.0)^2/4](1.0/0.012)[1.0/4]^{2/3}(0.0004)^{1/2} = 0.519 \text{ m}^3/\text{s}$$

$Q/Q_{\text{full}} = 0.25/0.519 = 0.48$, or 48 percent. From Fig. A-18, $d/d_{\text{full}} = 51$ percent: $d = (0.51)(1.0) = 0.51 \text{ m}$.

**14.115** Evaluate the friction factor $f$ for laminar flow in terms of the Reynolds number, and compare with the equation for pipe flow. (*Note:* Recall that for a wide channel the hydraulic radius is approximately equal to the depth.)

$\blacksquare$ $\quad N_R = (4R)(v)/v = 4y_0 v/v \qquad q = (g/v)(y_0^3/3)(s) \qquad v = Q/A = q/y_0 = gsy_0^2/3v = [(8g/f)(R)(s)]^{1/2}$

Therefore, $[(8g/f)(R)(s)] = vgsy_0^2/3v$. With $R = y_0$, $f = 24v/y_0 v = 96v/4y_0 v = 96/N_R$. This compares with $f = 64/N_R$ for pipe flow.

**14.116** A viscous fluid ($v = 0.0015$ ft$^2$/s) flows down a flat plate 10 ft wide. Find the maximum rate of discharge for laminar flow, assuming a critical Reynolds number of 500.

$\blacksquare$ $\quad N_R = Rv/v = y_0 q/vy_0 = q/v \qquad q = (500)(0.0015) = 0.7500$ ft$^3$/s per ft $\qquad Q = (10)(0.7500) = 7.50$ ft$^3$/s

**14.117** At what rate will water at 60 °F flow in a wide rectangular channel on a slope of 0.00018 if the depth is 0.01 ft?

$\blacksquare$ $\quad$ Assuming laminar flow, $q = (g/v)(y_0^3/3)(s) = [32.2/(1.21 \times 10^{-5})](0.01^3/3)(0.00018) = 1.60 \times 10^{-4}$ ft$^3$/s/ft, $N_R = Rv/v = y_0 v/v = (y_0)(q/y_0)/v = q/v = 1.60 \times 10^{-4}/(1.21 \times 10^{-5}) = 13$. Since $N_R < 500$, the assumption of laminar flow is justified.

**14.118** At what rate will water at 15 °C flow in a wide, smooth, rectangular channel on a slope of 0.0002, if the depth is 8.0 mm?

$\blacksquare$ $\quad$ Assuming laminar flow, $q = (g/v)(y_0^3/3)(s) = [9.807/(1.16 \times 10^{-6})][(0.008)^3/3](0.0002) = 2.89 \times 10^{-4}$ (m$^3$/s)/m. Checking the assumption: $N_R = Rv/v = y_0 v/v = (y_0)(q/y_0)/v = q/v = 2.89 \times 10^{-4}/(1.16 \times 10^{-6}) = 249$ (laminar).

**14.119** Water flows with a velocity of 4 fps and at a depth of 2 ft in a wide rectangular channel. Is the flow subcritical or supercritical? Find the alternate depth for the same discharge and specific energy.

$\blacksquare$ $\qquad\qquad v^2/2g = 4^2/[(2)(32.2)] = 0.2484$ ft $\qquad y/2 = \frac{2}{2} = 1.000$ ft

Since $0.2484 < 1.000$, the flow is subcritical.

$$E = y + (1/2g)(q^2/y^2) = 2 + 0.2484 = 2.2484 \text{ ft} \qquad q = (4)(2) = 8.000 \text{ (ft}^2\text{/s)/ft}$$

$$2.2484 = y + \{1/[(2)(32.2)]\}(8.00^2/y^2) \qquad 2.2484y^2 - y^3 - 0.9938 = 0$$

Since $y = 2$ is one known solution, divide by $(y - 2)$ to yield $y^2 - 0.2484y - 0.4968 = 0$, $y = [-(-0.2484) \pm \sqrt{(-0.2484)^2 - (4)(1)(-0.4968)}]/[(2)(1)] = 0.840$ ft.

**14.120** Water flows down a wide rectangular channel of concrete ($n = 0.014$) laid on a slope of 2.4 mm/m. Find the depth and rate of flow for critical conditions.

$\blacksquare$ $\qquad v = (1.0/n)(R^{2/3})(s^{1/2}) \qquad v_c = \sqrt{gy_c} \qquad \sqrt{(9.807)(y_c)} = (1.0/0.014)(y_c)^{2/3}(0.0024)^{1/2}$

$\qquad y_c = 0.514$ m $\qquad q = y_c v_c = (y_c)(\sqrt{gy_c}) = (0.514)[\sqrt{(9.807)(0.514)}] = 1.15$ (m$^3$/s)/m

**14.121** Water flows at 15 ft/s in a rectangular channel at a depth of 2 ft. Find the critical depth for (**a**) this specific energy, (**b**) this rate of discharge.

$\blacksquare$ (**a**) $\qquad E = y + v^2/2g = 2 + 15^2/[(2)(32.2)] = 5.494$ ft $\qquad y_c = (\frac{2}{3})(E) = (\frac{2}{3})(5.494) = 3.66$ ft

$\quad$ (**b**) $\qquad q = yv = (2)(15) = 30.00$ ft$^3$/s/ft $\qquad y_c = (q^2/g)^{1/3} = (30.00^2/32.2)^{1/3} = 3.03$ ft

**14.122** A flow of 120 ft$^3$/s is carried in a rectangular channel 10 ft wide at a depth of 1.5 ft, the channel is made of smooth concrete ($n = 0.013$). Find (**a**) the necessary slope, (**b**) the roughness coefficient needed to produce uniform critical flow for the given rate of discharge on this slope.

**▮ (a)**  $Q = (A)(1.486/n)(R^{2/3})(s^{1/2})$     $120 = [(1.5)(10)][1.486/0.013][(1.5)(10)/(1.5 + 10 + 1.5)]^{2/3}(s)^{1/2}$

$$s = 0.00405$$

**(b)**     $y_c = (q^2/g)^{1/3}$     $y_c = [(\frac{120}{10})^2/32.2]^{1/3} = 1.648 \text{ ft}$

$$v_c = \sqrt{gy_c} = \sqrt{(32.2)(1.648)} = 7.285 \text{ ft/s}$$

$$7.285 = (1.486/n)[(1.648)(10)/(1.648 + 10 + 1.648)]^{2/3}(0.00405)^{1/2} \quad n = 0.0150$$

**14.123**  A rectangular channel 4 m wide shows a wavy surface at depth 2.5 m. Estimate the rate of discharge.

**▮** The wavy surface indicates a near-critical depth; hence $y_c \approx 2.5 \text{ m}$. $q_{max} = \sqrt{gy_c^3} = \sqrt{(9.807)(2.5)^3} = 12.38 \text{ m}^3/\text{s/m}$, $Q = (4)(12.38) = 49.5 \text{ m}^3/\text{s}$.

**14.124**  Probing for oil, geologists drive a small pipe vertically down into the bed of a fast-running stream. The upstream "wake" of the pipe has a spread $\theta = 120°$. Estimate the stream velocity.

**▮**     $\sin(\theta/2) = c/v$     $c = \sqrt{gy} = \sqrt{(32.2)(2)} = 8.025 \text{ ft/s}$     $\sin 60° = 8.025/v$     $v = 9.3 \text{ ft/s}$

**14.125**  A speedboat in shallow water lifts a 1-ft wave (height above undisturbed surface), which travels at 9 mph. Find the approximate depth of the water.

**▮**     $c = \sqrt{(g)(y + \Delta y)[(y + \Delta y/2)/y]} = (9)(5280)/3600 = 13.2 \text{ ft/s}$     $13.2 = \sqrt{(32.2)(y + 1)[(y + \frac{1}{2})/y]}$

$$174.2 = 32.2y + 48.30 + 16.1/y \qquad 32.2y^2 - 125.9y + 16.1 = 0$$

The larger root is

$$y = [-(-125.9) + \sqrt{(-125.9)^2 - (4)(32.2)(16.1)}]/[(2)(32.2)] = 3.78 \text{ ft}$$

**14.126**  A rectangular channel 10 ft wide carries a flow of 180 cfs. Find the critical depth and critical velocity for this flow.

**▮**     $y_c = (q^2/g)^{1/3} = [(\frac{180}{10})^2/32.2]^{1/3} = 2.16 \text{ ft}$     $v_c = \sqrt{gy_c} = \sqrt{(32.2)(2.16)} = 8.34 \text{ ft/s}$

**14.127**  Water flows at 600 ft³/min in a isosceles right-triangular flume for which $n = 0.012$. Find the critical depth and critical slope.

**▮** $Q^2/g = A^3/B$. If $y = $ depth, $A = y^2$ and $B = 2y$.

$$(600/60)^2/32.2 = (y^2)^3/2y \quad y = y_c = 1.44 \text{ ft} \qquad Q = (A)(1.486/n)(R^{2/3})(s^{1/2})$$

$$R = A/p_w = 1.44^2/[(2)(\sqrt{1.44^2 + 1.44^2})] = 0.5091 \text{ ft} \qquad 10 = (1.44^2)(1.486/0.012)(0.5091)^{2/3}(s)^{1/2}$$

$$s = s_c = 0.00373$$

**14.128**  A trapezoidal canal with side slopes of 2:1 has a bottom width of 4 m and carries a flow of 23 m³/s. Calculate the critical depth.

**▮** $Q^2/g = A^3/B$. If $y = $ depth, $A = 4y + 2y^2$ and $B = 4 + 4y$.

$$23^2/9.807 = (4y + 2y^2)^3/(4 + 4y) \qquad y = y_c = 1.22 \text{ m} \qquad \text{(by trial and error)}$$

**14.129**  Find the specific energy at depth 3 ft for flow of 100 ft³/s through a 10-ft-diameter tunnel.

**▮** $E = y + v^2/2g$. For $y = 3$ ft:

$$y/y_{full} = \frac{3}{10} = 0.30 \quad \text{or} \quad 30 \text{ percent} \quad \text{From Fig. A-18} \quad A/A_{full} = 25 \text{ percent} \quad A = [(\pi)(10)^2/4](0.25) = 19.6 \text{ ft}^2$$

$$v = Q/A = 100/19.6 = 5.10 \text{ ft/s} \qquad E = 3 + 5.10^2/[(2)(32.2)] = 3.40 \text{ ft}$$

**14.130** A circular conduit flowing half full carries 500 ft³/s at velocity 10 ft³/s. If $n = 0.13$, will the flow be subcritical or supercritical?

**∎**
$$A = Q/v = (\pi d^2/4)/2 \qquad \tfrac{500}{10} = (\pi d^2/4)/2 \qquad d = 11.28 \text{ ft}$$

$$A = [(\pi)(11.28)^2/4]/2 = 49.97 \text{ ft}^2 \qquad N_F = v/\sqrt{gy} = v/\sqrt{(g)(A/B)} = 10/\sqrt{(32.2)(49.97/11.28)} = 0.837$$

Since $N_F < 1.0$, the flow is subcritical.

**14.131** Figure 14-37 describes the cross section of an open channel for which $s_0 = 0.02$ and $n = 0.015$. The sketch is drawn to the scale shown. When the flow rate is 100 cfs, find (**a**) the depth for uniform flow and (**b**) the critical depth.

**∎** The cross-sectional area is found by planimetry and the wetted perimeter by use of dividers.

(**a**)

| $y_0$, ft | $A$, ft | $p_w$, ft | $R$, ft | $Q = (A)(1.486/n)(R^{2/3})(s^{1/2})$, cfs |
|---|---|---|---|---|
| 1 | 1.30 | 3.29 | 0.395 | 10 |
| 2 | 4.90 | 6.12 | 0.801 | 59 |
| 3 | 10.48 | 8.96 | 1.170 | 163 |

A plot of $Q$ versus $y_0$ (not shown) indicates that $y_0 = 2.50$ ft for $Q = 100$ cfs.

(**b**)

| $y_c$, ft | $A$, ft² | $B(x_L + x_R)$, ft | $Q = (gA^3/B)^{1/2}$, cfs |
|---|---|---|---|
| 1 | 1.30 | 2.60 | 5 |
| 2 | 4.90 | 4.60 | 29 |
| 3 | 10.48 | 6.55 | 75 |
| 4 | 18.30 | 9.10 | 147 |

A plot of $Q$ versus $y_c$ (not shown) indicates that $y_c = 3.35$ ft for $Q = 100$ cfs.

**14.132** Refer to Fig. 14-37 and replace feet dimensions with meters. Let the slope be 0.007 with $n = 0.015$. When the flow rate is 50 m³/s, find (**a**) the depth for uniform flow and (**b**) the critical depth.

| $y$ ft | $x_L$ ft | $x_R$ ft |
|---|---|---|
| 0 | 0 | 0 |
| 1 | 1.5 | 1.1 |
| 2 | 2.6 | 2.0 |
| 3 | 3.25 | 3.3 |
| 4 | 3.6 | 5.5 |
| 5 | 3.85 | 8.2 |

**Fig. 14-37**

**∎** The cross-sectional area is found by planimetry and the wetted perimeter by use of dividers.

(**a**)

| $y_0$, m | $A$, m² | $p_w$, m | $R$, m | $Q = (A)(1.0/n)(R^{2/3})(s^{1/2})$, m³/s |
|---|---|---|---|---|
| 1 | 1.30 | 3.29 | 0.395 | 4 |
| 2 | 4.90 | 6.12 | 0.801 | 24 |
| 3 | 10.48 | 8.96 | 1.170 | 65 |

A plot of $Q$ versus $y_0$ (not shown) indicates that $y_0 = 2.70$ m for $Q = 50$ m³/s.

(b)

| $y_c$, m | $A$, m² | $B(x_L + x_R)$, m | $Q = (gA^3/B)^{1/2}$, m³/s |
|---|---|---|---|
| 1 | 1.30 | 2.60 | 3 |
| 2 | 4.90 | 4.60 | 16 |
| 3 | 10.48 | 6.55 | 42 |
| 4 | 18.30 | 9.10 | 81 |

A plot of $Q$ versus $y_c$ (not shown) indicates that $y_c = 3.25$ m for $Q = 50$ m³/s.

**14.133** A rectangular channel 2 m wide carries 2.2 m³/s of water in subcritical uniform flow at a depth of 1.0 m. What is the lowest transverse hump in the bottom such that $y_c$ is attained at the peak?

$$(\Delta z)_c = E_0 - E_{min} \qquad E = y + v^2/2g \qquad v = Q/A = 2.2/[(2)(1.0)] = 1.10 \text{ m/s}$$

$$E_0 = 1.0 + 1.10^2/[(2)(9.807)] = 1.0617 \text{ m}$$

$$E_{min} = (\tfrac{3}{2})(y_c) \qquad y_c = (q^2/g)^{1/3} = [(\tfrac{2.2}{2})^2/9.807]^{1/3} = 0.4978 \text{ m}$$

$$E_{min} = (\tfrac{3}{2})(0.4978) = 0.7467 \text{ m} \qquad (\Delta z)_c = 1.0617 - 0.7467 = 0.315 \text{ m}$$

**14.134** Rework Prob. 14.86 ($y_0 = 2.0$ ft) for the case where the flow rate is 50 cfs.

(a) 
$$q = Q/B = \tfrac{50}{4} = 12.50 \text{ (ft}^3\text{/s)/ft} \qquad y_c = (q^2/g)^{1/3} = (12.50^2/32.2)^{1/3} = 1.693 \text{ ft}$$

Since $y_0 > y_c$, the flow is subcritical (and slope is mild).

(b) Install hump with $\Delta z = 0.30$ ft. First find critical hump height:

$$E = y + (\tfrac{1}{2}g)(q^2/y^2) \qquad E_0 = 2.00 + \{1/[(2)(32.2)]\}(12.50^2/2.00^2) = 2.607 \text{ ft}$$

$$E_{min} = (\tfrac{3}{2})(y_c) = (\tfrac{3}{2})(1.693) = 2.540 \text{ ft} \qquad E_0 = \Delta z_{crit} + E_{min} \qquad 2.607 = \Delta z_{crit} + 2.540 \qquad \Delta z_{crit} = 0.067 \text{ ft}$$

Since $\Delta z > \Delta z_{crit}$, $y_c$ occurs on the hump and damming action occurs. Hence, water depth on hump = 1.693 ft.

(c) Increase hump height to 0.60 ft. Still, $\Delta z > \Delta z_{crit}$, so $y_c$ occurs on the hump. Let $y$ = upstream or downstream depth, with specific energy $E$.

$$E = \Delta z + E_{min} = y + (1/2g)(q^2/y^2) \qquad \Delta z + E_{min} = y + (1/2g)(q^2/y^2)$$

$$0.60 + 2.540 = y + \{1/[(2)(32.2)]\}(12.50^2/y^2) \qquad y = 2.84 \text{ ft} \quad \text{or} \quad 1.09 \text{ ft} \qquad \text{(by trial and error)}$$

This hump causes damming action, and the depths just upstream and downstream of the hump are 2.84 ft (subcritical) and 1.09 ft (supercritical), respectively.

**14.135** A flow of 2.0 m³/s is carried in a rectangular channel 1.8 m wide at a depth of 1.0 m. Will critical depth occur at a section where (a) a frictionless hump 15 cm high is installed across the bed? (b) a frictionless sidewall reduces the channel width to 1.3 m? (c) the hump and the sidewall constriction are installed together?

(a) 
$$\Delta z_{crit} = E_0 - E_{min} \qquad E = y + (1/2g)(q^2/y^2) \qquad q = 2.0/1.8 = 1.111 \text{ (m}^3\text{/s)/m}$$

$$E_0 = 1.0 + \{1/[(2)(9.807)]\}(1.111^2/1.0^2) = 1.063 \text{ m} \qquad E_{min} = (\tfrac{3}{2})(y_c) \qquad y_c = (q^2/g)^{1/3}$$

$$y_c = (1.111^2/9.807)^{1/3} = 0.5011 \text{ m} \qquad E_{min} = (\tfrac{3}{2})(0.5011) = 0.7517 \text{ m}$$

$$\Delta z_{crit} = 1.063 - 0.7517 = 0.3113 \text{ m}$$

Since $\Delta z = 0.15 < 0.3113 = \Delta z_{crit}$, $y_c$ does not occur at the hump.

**(b)**     $q = 2.0/1.3 = 1.538$ m     $y_c = (1.538^2/9.807)^{1/3} = 0.6225$ m     $E_{min} = (\frac{3}{2})(0.6225) = 0.9338$ m

Since $E_{min} = 0.9338 < 1.063 = E_0$, $y_c$ does not occur at the constriction.

**(c)** $\Delta z_{crit} = E_0 - E_{min}$. With constriction, $E_{min} = 0.9338$ m; hence, $\Delta z_{crit} = 1.063 - 0.9338 = 0.1292$ m. Since $\Delta z = 0.15 > 0.1292 = \Delta z_{crit}$, $y_c$ does occur at the hump with constriction.

**14.136**  Rework Prob. 14.86 ($y_0 = 2.0$ ft) for the case where the flow rate is 16 cfs.

▌ **(a)**     $q = Q/B = \frac{16}{4} = 4.000$ ft$^2$/s/ft     $y_c = (q^2/g)^{1/3} = (4.000^2/32.2)^{1/3} = 0.7921$ ft

Since $y_0 > y_c$, the flow is subcritical (and slope is mild).

**(b)** Install hump with $\Delta z = 0.30$ ft. First find critical hump height.

$$E = y + (\tfrac{1}{2}g)(q^2/y^2)     E_0 = 2.00 + \{1/[(2)(32.2)]\}(4.000^2/2.00^2) = 2.062 \text{ ft}$$

$$E_{min} = (\tfrac{3}{2})(y_c) = (\tfrac{3}{2})(0.7921) = 1.188 \text{ ft}     E_0 = \Delta z_{crit} + E_{min}     2.062 = \Delta z_{crit} + 1.188     \Delta z_{crit} = 0.874 \text{ ft}$$

Since $\Delta z < \Delta z_{crit}$, $y_c$ does not occur on the hump and damming action does not result.

$$E_h = E_0 - \Delta z     E = y + (1/2g)(q^2/y^2) = y_h + \{1/[(2)(32.2)]\}(4.000^2/y_h^2)$$

$$y_h + \{1/[(2)(32.2)]\}(4.000^2/y_h^2) = 2.062 - 0.30$$

By trial and error, $y_h = 1.67$ ft (subcritical) and 0.432 ft (supercritical). The flow does not pass through $y_c$, so it cannot become supercritical. Therefore, the water depth on the 0.30-ft hump is 1.67 ft.

**(c)** Increase hump height to 0.60 ft. Still, $\Delta z < \Delta z_{crit}$, so $y_c$ does not occur on the hump and damming action does not result. $y_h + \{1/[(2)(32.2)]\}(4.000^2/y_h^2) = 2.062 - 0.60$. By trial and error $y_h = 1.32$ ft (subcritical) and 0.511 ft (supercritical). The flow does not pass through $y_c$, so it cannot become supercritical. Therefore, the water depth on the 0.60-ft hump is 1.32 ft.

**14.137**  A 4-ft-wide rectangular ditch carries 40 ft$^3$/s of water at a depth of 2.80 ft. A man standing in the middle of the ditch presents a 1-ft width to the stream. Find the local change in the water-surface elevation.

▌ $v_0 = Q/A = 40/[(2.80)(4)] = 3.571$ ft/s. At the man.

$$v = 40/[(4-1)(y)] = 13.33/y     E = y + v^2/2g = y + (13.33/y)^2/[(2)(32.2)] = y + 2.759/y^2$$

$$E_0 = 2.80 + 3.571^2/[(2)(32.2)] = 2.998 \text{ ft}     y + 2.759/y^2 = 2.998$$

By trial and error, $y = 2.59$ ft or 1.26 ft. Since locally supercritical conditions are impossible, $y = 2.59$ ft. Change in water depth $= 2.80 - 2.59 = 0.21$ ft drop.

**14.138**  A rectangular channel 10 ft wide carries 20 ft$^3$/s in uniform flow at depth 1.0 ft. Find the local change in water-surface elevation caused by an obstruction 0.20 ft high on the floor of the channel.

▌ $y_c = (q^2/g)^{1/3} = [(\tfrac{20}{10})^2/32.2]^{1/3} = 0.499$ ft. Since $y_c < 1.0$ ft, the flow is subcritical.

$$E = y + v^2/2g     v = Q/A = 20/[(1.0)(10)] = 2.000 \text{ ft/s}     E_1 = 1.0 + 2.000^2/[(2)(32.2)] = 1.062 \text{ ft}$$

$$E_2 = E_1 - \Delta z = 1.062 - 0.20 = 0.862 \text{ ft}     E = y + (1/2g)(q^2/y^2)$$

$$E_2 = y_2 + \{1/[(2)(32.2)]\}[(\tfrac{20}{10})^2/y_2^2]     0.862 = y_2 + 0.06211/y_2^2     y_2 = 0.752 \text{ ft}     \text{(by trial and error)}$$

Change in water-surface elevation $= 1.00 - (0.20 + 0.752) = 0.048$ ft (drop).

**14.139**  Rework Prob. 14.138 for a flow at depth 0.31 ft.

▌ Since $y_c = 0.499$ ft $> 0.31$ ft, the flow is supercritical.

$$E = y + v^2/2g     v = Q/A = 20/[(0.31)(10)] = 6.452 \text{ ft/s}     E_1 = 0.31 + 6.452^2/[(2)(32.2)] = 0.9564 \text{ ft}$$

$$E_2 = E_1 - \Delta z = 0.9564 - 0.20 = 0.7564 \text{ ft}     E = y + (1/2g)(q^2/y^2)     E_2 = y_2 + \{1/[(2)(32.2)]\}[(\tfrac{20}{10})^2/y_2^2]$$

$$0.7564 = y_2 + 0.06211/y_2^2     y_2 = 0.45 \text{ ft}     \text{(by trial and error)}$$

Change in water-surface elevation $= (0.20 + 0.45) - 0.31 = 0.34$ ft (rise).

**14.140** Repeat Prob. 14.137 for flow depth 0.90 ft.

**▮** $E = y + v^2/2g$     $v_0 = Q/A = 40/[(4)(0.90)] = 11.11$ ft/s     $E_0 = 0.90 + 11.11^2/[(2)(32.2)] = 2.817$ ft

As in Prob. 14.137, $v = 13.33/y$.

$$E = y + (13.33/y)^2/[(2)(32.2)] = y + 2.759/y^2 \qquad 2.817 = y + 2.759/y^2$$
$$y = 1.39 \text{ ft} \quad \text{(by trial and error)}$$

Change in water-surface elevation = $1.39 - 0.90 = 0.49$ ft (rise).

**14.141** If 1.4 m³/s of water flows uniformly in a channel of width 1.8 m at a depth of 0.75 m, what is the change in water-surface elevation at a section contracted to a 1.2 m width with a 6-cm depression in the bottom?

**▮**     $y_c = (q^2/g)^{1/3}$     $(y_c)_2 = [(1.4/1.2)^2/9.807]^{1/3} = 0.5177$ m     $E_2 = E_1 + \Delta z$     $E = y + v^2/2g$

$v_1 = Q/A = 1.4/[(0.75)(1.8)] = 1.037$ m/s     $E_1 = 0.75 + 1.037^2/[(2)(9.807)] = 0.8048$ m

$v_2 = 1.4/(1.2y_2) = 1.167/y_2$

$E_2 = y_2 + (1.167/y_2)^2/[(2)(9.807)] = y_2 + 0.06943/y_2^2$     $y_2 + 0.06943/y_2^2 = 0.8048 + \frac{6}{100}$

$y_2 = 0.737$ m (subcritical) or 0.377 m (supercritical)   (by trial and error). $y_2$ cannot be less than $(y_c)_2$; hence, $y_2 = 0.737$ m. Change in water-surface elevation = $0.75 - (0.737 - \frac{6}{100}) = 0.073$ m (drop).

**14.142** A finished-concrete channel 8 ft wide has a slope of 0.5° and a water depth of 4 ft. Predict the uniform flow rate by (a) Manning's formula with $n = 0.012$ and (b) the friction-factor analysis with $\epsilon = 0.0032$ ft.

**▮** (a)                    $R = A/p_w = (8)(4)/(4 + 8 + 4) = 2.000$ ft

$$Q = (A)(1.486/n)(R^{2/3})(s^{1/2}) = [(8)(4)](1.486/0.012)(2.000)^{2/3}(\tan 0.5°)^{1/2} = 588 \text{ ft}^3/\text{s}$$

(b)          $C = (8g/f)^{1/2}$     $1/f^{1/2} = 2.0 \log (3.7D_h/\epsilon)$     $D_h = 4R = (4)(2.000) = 8.000$ ft

$1/f^{1/2} = (2.0)\log [(3.7)(8.000)/(0.0032)]$     $f = 0.01589$     $C = [(8)(32.2)/0.01589]^{1/2} = 127.3$

$$Q = CA(Rs)^{1/2} = (127.3)[(4)(8)][(2.000)(\tan 0.5°)]^{1/2} = 538 \text{ ft}^3/\text{s}$$

**14.143** The asphalt-lined trapezoidal channel in Fig. 14-38 carries 300 cfs of water under uniform flow conditions when $s = 0.0015$. What is the normal depth? Use $n = 0.016$.

**▮**     $Q = (A)(1.486/n)(R^{2/3})(s^{1/2})$     $A = (\frac{1}{2})(6 + b_0)(y_n) = 6y_n + y_n^2 \cot 50°$     $p_w = 6 + 2y_n \csc 50°$

$$300 = (6y_n + y_n^2 \cot 50°)(1.486/0.016) [(6y_n + y_n^2 \cot 50°)/(6 + 2y_n \csc 50°)]^{2/3}(0.0015)^{1/2}$$
$$y_n = 4.58 \text{ ft} \quad \text{(by trial and error)}$$

**Fig. 14-38**

**14.144** What are the best dimensions for a rectangular brick ($n = 0.015$) channel designed to carry 5 m³/s of water in uniform flow with $s = 0.001$?

**▮** The best dimensions are for width ($b$) equal to twice the depth ($y$): $Q = (A)(1.0/n)(R^{2/3})(s^{1/2})$,
$5 = [(y)(2y)](1.0/0.015)[(y)(2y)/(y + 2y + y)]^{2/3}(0.001)^{1/2}$, $y = 1.268$ m, $b = (2)(1.268) = 2.536$ m.

**14.145** A wide rectangular clean-earth channel ($n = 0.022$) has a flow rate $q$ of 50 cfs/ft (a) What is the critical depth? (b) What type of flow exists if $y = 3$ ft? (c) What is the critical slope?

**▮** (a)                              $y_c = (q^2/g)^{1/3} = (50^2/32.2)^{1/3} = 4.27$ ft

(b) For $y < y_c$, the flow will be supercritical.

(c)                    $s_c = gn^2/(2.208y_c^{1/3}) = (32.2)(0.022)^2/[(2.208)(4.27)^{1/3}] = 0.00435$

**14.146**  The 50° triangular channel in Fig. 14-39 has a flow rate $Q$ of 16 m³/s. Compute (a) $y_c$, (b) $v_c$, and (c) $s_c$ if $n = 0.018$.

**I**   (a)   $$gA^3 = b_0 Q^2 \qquad A = y^2 \cot 50° \qquad b_0 = 2y \cot 50°$$
$$(9.807)(y^2 \cot 50°)^3 = (2y \cot 50°)(16^2) \qquad y = y_c = 2.37 \text{ m}$$

   (b)   $$v_c = Q/A = 16/[(2.37)^2(\cot 50)] = 3.39 \text{ m/s}$$

   (c)   $$R = A/p_w = y^2 \cot 50°/(2y \csc 50°) = (2.37)^2(\cot 50°)/[(2)(2.37)(\csc 50°)] = 0.7617 \text{ m}$$
$$v = (1.0/n)(R^{2/3})(s^{1/2}) \qquad 3.39 = (1.0/0.018)(0.7617)^{2/3}(s)^{1/2} \qquad s = s_c = 0.00535$$

Fig. 14-39

**14.147**  The formula for shallow-water wave-propagation speed, $c_0 = \sqrt{gy}$, is independent of the physical properties of the liquid, i.e., density, viscosity, or surface tension. Does this mean that waves propagate at the same speed in water, mercury, gasoline, and glycerin? Explain.

**I**   $c_0 = \sqrt{gy}$ is correct for any fluid except for viscosity and surface tension (very small wave) effects. It would be accurate for water, mercury, and gasoline but inaccurate for glycerin (too viscous).

**14.148**  A 10-cm-high wave travels over (shallow) water of depth 1.2 m. Compute the wave speed $c$ and the velocity $\delta v$ induced by the wave.

**I**   $$c = \sqrt{gy(1 + \delta y/y)[1 + (\tfrac{1}{2})(\delta y/y)]} = \sqrt{(9.79)(1.2)(1 + 1/12)[1 + (\tfrac{1}{2})(1/12)]} = 2.60 \text{ m/s}$$
$$\delta v = \frac{(c)(\delta y/y)}{1 + (\delta y/y)} = \frac{(2.60)(1/12)}{13/12} = 0.20 \text{ m/s}$$

**14.149**  Water flows rapidly in a flat wide channel 0.4 m deep. Pebbles dropped successively in the water at the same spot create two circular ripples which are shown from above in Fig. 14-40. Determine the current speed $V$.

**I**   The centers of the circles move at current $V$; hence,
$$(x_0)(c_0) = 4V \qquad \text{(smaller circle)}$$
$$(x_0 + 4 + 6 + 9)(c_0) = 9V \qquad \text{(larger circle)}$$

Subtracting these equations, $19c_0 = 5V$, $V = 19c_0/5$, $c_0 = \sqrt{gy} = \sqrt{(9.807)(0.4)} = 1.981$ m/s, $V = (19)(1.981)/5 = 7.53$ m/s.

Fig. 14-40

**14.150**  Rework Prob. 14.149 if the channel depth is 0.6 m and Fig. 14-41 applies.

**I**   The centers of the circles move at current $V$. If the pebble drop site is at distance $x_0$ ahead of the small circle,
$$(x_0)(c_0) = 3V \qquad \text{(smaller circle)}$$
$$(x_0 + 4)(c_0) = 9V \qquad \text{(larger circle)}$$

Subtracting these equations, $4c_0 = 6V$, $V = 2c_0/3$, $c_0 = \sqrt{gy} = \sqrt{(9.807)(0.6)} = 2.426$ m/s, $V = (2)(2.426)/3 = 1.62$ m/s.

**Fig. 14-41**

**14.151** Consider flow in a wide channel over a bump, as in Fig. 14-42. One can estimate the water-depth change or *transition* with frictionless flow. Use continuity and the Bernoulli equation to show that $dy/dx = -(dh/dx)/(1 - V^2/gy)$. Explain under what conditions the surface might rise above its upstream position $y_0$.

▮ From the Bernoulli equation, $p_0/\gamma + V_0^2/2g + y_0 = p_1/\gamma + V_1^2/2g + y_1 + h$, $0 + V_0^2/2g + y_0 = 0 + V_1^2/2g + y_1 + h$,

$$V \, dV/g + dy + dh = 0 \tag{1}$$

From continuity, $V_0 y_0 = V_1 y_1 =$ constant, $V \, dy + y \, dV = 0$, $dV = -V \, dy/y$. Substituting this value of $dV$ into Eq. (1), $(V)(-V \, dy/y)/g + dy + dh = 0$, $dy/dx = -(dh/dx)/(1 - V^2/gy)$. If $dh/dx > 0$ (a bump) and $V^2 < gy$ (subcritical Froude number), $dy/dx$ will be negative and the water level will drop across the bump. If $V^2 > gy$ (supercritical Froude number), the water level will rise.

**Fig. 14-42**

**14.152** In Fig. 14-42, the flow rate is $1.2 \text{ m}^3/\text{s}$ per meter of channel width. Compute the velocity and water depth at the top of the bulge ($h_{\max} = 0.1 \text{ m}$). Characterize the flow there.

▮ $\quad p_0/\gamma + V_0^2/2g + y_0 = p_1/\gamma + V_1^2/2g + y_1 + h \qquad 0 + V_0^2/2g + y_0 = 0 + V_1^2/2g + y_1 + h_{\max}$

$\quad q = (V_0)(0.85) = V_1 y_1 = 1.2 \qquad V_0 = 1.412 \text{ m/s} \qquad y_1 = 1.2/V_1$

$\quad 1.412^2/[(2)(9.807)] + 0.85 = V_1^2/[(2)(9.807)] + 1.2/V_1 + 0.1 \qquad 0.05098 V_1^3 - 0.8516 V_1 + 1.2 = 0$

$\quad V_1 = 1.7 \text{ m/s} \quad \text{(by trial and error)} \qquad y_1 = 1.2/1.7 = 0.706 \text{ m}$

$\quad (N_F)_1 = V_1/\sqrt{gy_1} = 1.7/\sqrt{(9.807)(0.706)} = 0.646 \quad \text{(subcritical)}.$

**14.153** Rework Prob. 14.152 for a flow rate three times as large.

▮ $\quad p_0/\gamma + V_0^2/2g + y_0 = p_1/\gamma + V_1^2/2g + y_1 + h \qquad 0 + V_0^2/2g + y_0 = 0 + V_1^2/2g + y_1 + h$

$\quad q = (V_0)(0.85) = V_1 y_1 = 3.6 \qquad V_0 = 4.24 \text{ m/s} \qquad y_1 = 3.6/V_1$

$\quad 4.24^2/[(2)(9.807)] + 0.85 = V_1^2/[(2)(9.807)] + 3.6/V_1 + 0.1 \qquad 0.05098 V_1^3 - 1.666 V_1 + 3.6 = 0$

$\quad V_1 = 3.6 \text{ m/s} \quad \text{(by trial and error)} \qquad y_1 = 3.6/3.6 = 1.00 \text{ m}$

$\quad (N_F)_1 = V_1/\sqrt{gy_1} = 3.6/\sqrt{(9.807)(1.00)} = 1.15 \quad \text{(supercritical)}.$

**14.154** Given the flow of a channel of large width $b$ under a sluice gate, as shown in Fig. 14-43 and assuming frictionless steady flow with negligible upstream kinetic energy, derive a formula for the dimensionless ratio $Q^2/y_1^3 b^2 g$ as a function of the ratio $y_2/y_1$.

∎ From the Bernoulli equation, $y_1 + V_1^2/2g = y_2 + V_2^2/2g$. Assuming $V_1^2/2g$ to be negligible, $y_1 = y_2 + V_2^2/2g$. From continuity: $V_1 y_1 = V_2 y_2 = Q/b$, $V_2 = Q/by_2$, $y_1 = y_2 + (Q/by_2)^2/2g$. Multiplying by $y_2^3/y_1^3$, $y_2^3/y_1^3 = y_2^3/y_1^3 + Q^2/2gb^2y_1^3$, $Q^2/2gb^2y_1^3 = (y_2/y_1)^2 - (y_2/y_1)^3$.

**Fig. 14-43**

**14.155** With reference to Prob. 14.154, take $y_1 = 1.00$ m and $y_2 = 0.60$ m, $b = 10$ m. Compute flow rate $Q$ if the upstream kinetic energy is (a) neglected and (b) considered.

∎ (a) If $V_1^2/2g$ is neglected, the equation developed in Prob. 14.154 applies: $Q^2/2gb^2y_1^3 = (y_2/y_1)^2 - (y_2/y_1)^3$,
$Q^2/[(2)(9.807)(10)^2(1.00)^3] = [(0.60)/(1.00)]^2 - [(0.60)/(1.00)]^3$, $Q = 16.8 \text{ m}^3/\text{s}$.

(b) From the Bernoulli equation,

$$y_1 + V_1^2/2g = y_2 + V_2^2/2g \qquad V_1 y_1 = V_2 y_2 \qquad (V_1)(1.00) = (V_2)(0.60) \qquad V_2 = 1.667V_1$$

$$1.00 + V_1^2/[(2)(9.807)] = 0.60 + (1.667V_1)^2/[(2)(9.807)] \qquad V_1 = 2.10 \text{ m/s}$$

$$Q = Av = [(10)(1.00)](2.10) = 21.0 \text{ m}^3/\text{s}$$

**14.156** For laminar = sheet draining, the flow may become turbulent if $N_R$ exceeds 500. If $s = 0.0018$, what is the maximum sheet thickness $y_0$ to ensure laminar flow of water at 20 °C?

∎ $$N_R = gy_0^3 s/3v^2 \qquad 500 = (9.807)(y_0^3)(0.0018)/[(3)(1.02 \times 10^{-6})^2] \qquad y_0 = 4.45 \text{ mm}$$

**14.157** A rectangular channel 3 m wide contains water 2 m deep. If the slope is 0.5°, compute the discharge for uniform flow by the Manning formula ($n = 0.014$).

∎ $$Q = (A)(1.0/n)(R^{2/3})(s^{1/2}) = [(3)(2)](1.0/0.014)[(3)(2)/(2+3+2)]^{2/3}(\tan 0.5°)^{1/2} = 36.13 \text{ m}^3/\text{s}$$

**14.158** Solve Prob. 14.157 by the Moody formula ($\epsilon = 0.0024$ m).

∎ $$C = \sqrt{8g/f} \qquad 1/f^{1/2} = 2\log(3.7D_h/\epsilon) \qquad D_h = 4R = (4)[(3)(2)/(2+3+2)] = 3.429 \text{ m}$$

$$1/f^{1/2} = 2\log[(3.7)(3.429)/(0.0024)] \qquad f = 0.01804 \qquad C = \sqrt{(8)(9.807)/0.01804} = 65.95$$

$$Q = CA(Rs)^{1/2} = (65.95)[(3)(2)]\{[(3)(2)/(2+3+2)](\tan 0.5°)\}^{1/2} = 34.22 \text{ m}^3/\text{s}$$

**14.159** A large metal trough ($n = 0.014$) slopes at 1:500, its section corresponds to Fig. 14-35 with $\phi = 30°$. Find the water depth for uniform flow at 222.2 L/s.

∎ $$Q = (A)(1.0/n)(R^{2/3})(s^{1/2}) \qquad A = (y)[y\tan 30°] = 0.5774y^2 \qquad p_w = 2y/\cos 30° = 2.309y$$

$$R = A/p_w = 0.5774y^2/2.309y = 0.2501y \qquad 0.2222 = (0.5774y^2)(1.0/0.014)(0.2501y)^{2/3}(\tfrac{1}{500})^{1/2} \qquad y = 0.639 \text{ m}$$

**14.160** A trapezoidal channel similar to that of Fig. 14-38 has a bottom width of 10 ft, a side angle of 45°, and a depth of 3 ft. Compute the uniform flow discharge for a clean earth channel ($n = 0.021$) with $s = 0.0004$.

∎ $$A = (10)(3) + (3)[3\tan(90° - 45°)] = 39 \text{ ft}^2 \qquad p_w = 10 + (2)[3/\cos(90° - 45°)] = 18.49 \text{ ft}$$

$$Q = (A)(1.486/n)(R^{2/3})(s^{1/2}) = (39)(1.486/0.021)(39/18.49)^{2/3}(0.0004)^{1/2} = 90.8 \text{ ft}^3/\text{s}$$

**14.161** For the channel of Prob. 14.160, compute the normal depth if the flow rate is 170 cfs.

∎ $$Q = (A)(1.486/n)(R^{2/3})(s^{1/2}) \qquad A = 10y + (y)[y\tan(90° - 45°)] = 10y + y^2$$

$$p_w = 10 + (2)[y/\cos(90° - 45°)] = 10 + 2.828y$$

$$170 = (10y + y^2)(1.486/0.021)[(10y + y^2)/(10 + 2.828y)]^{2/3}(0.0004)^{1/2}$$

$$y = 5.63 \text{ ft} \qquad \text{(by trial and error)}$$

**14.162** Use Moody's formula to express Manning's $n$ as a function of channel size $R$.

▌ Combining $C = \sqrt{8g/f} = (1.486/n)(R^{1/6})$, $1/f^{1/2} = 2 \log (3.7D_h/\epsilon)$, and $D_h = 4R$, we obtain $n = R^{1/6}/(\alpha + \beta \log R)$, for suitable constants $\alpha$ and $\beta$.

**14.163** A 7-ft-diameter metal ($n = 0.020$) conduit is flowing half-full on a slope of $1:880$. Estimate the discharge.

▌ $\quad Q = (A)(1.486/n)(R^{2/3})(s^{1/2}) = [(\tfrac{1}{2})(\pi)(7)^2/4](1.486/0.020)(7/4)^{2/3}(1/880)^{1/2} = 70.0 \text{ ft}^3/\text{s}$

**14.164** A trapezoidal conduit similar to that of Fig. 14-38 has a bottom width of 6 m, a side angle of 35°, and carries 60 m³/s of water at a depth of 4 m. If $n = 0.014$, find the slope.

▌ $\quad Q = (A)(1.0/n)(R^{2/3})(s^{1/2}) \qquad A = (6)(4) + 4[4 \tan (90° - 35°)] = 46.85 \text{ m}^2$

$p_w = 6 + (2)[4/\cos (90° - 35°)] = 19.95 \text{ m} \qquad 60 = (46.85)(1.0/0.014)(46.85/19.95)^{2/3}(s)^{1/2} \qquad s = 0.000103$

**14.165** For the conduit of Prob. 14.164, determine the water depth given a slope of 0.000512 and a discharge of 40 m³/s.

▌ $\quad Q = (A)(1.0/n)(R^{2/3})(s^{1/2}) \qquad A = 6y + y[y \tan (90° - 35°)] = 6y + 1.428y^2$

$$p_w = 6 + 2[y/\cos (90° - 35°)] = 6 + 3.487y$$

$$40 = (6y + 1.428y^2)(1.0/0.014)[(6y + 1.428y^2)/(6 + 3.487y)]^{2/3}(0.000512)^{1/2}$$

$$y = 2.14 \text{ m} \qquad \text{(by trial and error)}$$

**14.166** Uniform water flow in a wide brick channel ($n = 0.015$) of slope 0.02° moves over a 10-cm bump as in Fig. 14-44. A slight depression in water surface results. If the minimum water depth over the bump is 50 cm, compute the velocity over the bump and the flow rate per meter of width.

▌ $\quad v = (1.0/n)(R^{2/3})(s^{1/2}) \qquad v_1 = (1.0/0.015)(y_1)^{2/3}(\tan 0.02°)^{1/2} = 1.246 y_1^{2/3}$

From continuity, $v_1 y_1 = v_2 y_2$, $(1.246 y_1^{2/3})(y_1) = (v_2)(\tfrac{50}{100})$, $v_2 = 2.492 y_1^{5/3}$. Applying the Bernoulli equation and neglecting bottom slope,

$$y_1 + v_1^2/2g = y_2 + v_2^2/2g + h_{max} \qquad y_1 + (1.246 y_1^{2/3})^2/[(2)(9.807)] = \tfrac{50}{100} + (2.492 y_1^{5/3})^2/[(2)(9.807)] + \tfrac{10}{100}$$

$$y_1 = 0.623 \text{ m} \qquad \text{(by trial and error)} \qquad v_1 = (1.246)(0.623)^{2/3} = 0.9089 \text{ m/s}$$

$$v_2 = (2.492)(0.623)^{5/3} = 1.13 \text{ m/s} \qquad q = v_1 y_1 = (0.9089)(0.623) = 0.566 \text{ (m}^3/\text{s)/m}$$

Fig. 14-44

**14.167** A tar-coated ($n = 0.014$) triangular channel has 45° sides and a slope of 0.0004. Compute the normal depth for a discharge of 140 ft³/s.

▌ $\quad Q = (A)(1.486/n)(R^{2/3})(s^{1/2}) \qquad A = (y)(y \cot 45°) = 1.000y^2$

$$p_w = (2y)(\csc 45°) = 2.828y \qquad R = A/p_w = 1.000y^2/2.828y = 0.3536y$$

$$140 = (1.000y^2)(1.486/0.014)(0.3536y)^{2/3}(0.0004)^{1/2} \qquad y = 6.24 \text{ ft}$$

**14.168** A brick ($n = 0.015$) rectangular channel with $s = 0.002$ is designed to carry 230 cfs of water in uniform flow. There is an argument over whether the channel width should be 4 ft or 8 ft. Which design needs fewer bricks? By what percentage?

▌ $Q = (A)(1.486/n)(R^{2/3})(s^{1/2})$. For 4-ft width: $230 = (4y)(1.486/0.015)[4y/(y + 4 + y)]^{2/3}(0.002)^{1/2}$, $y = 9.31$ ft (by trial and error), $p_w = 9.31 + 4 + 9.31 = 22.62$ ft. For 8-ft width: $230 = (8y)(1.486/0.015)[8y/(y + 8 + y)]^{2/3}(0.002)^{1/2}$, $y = 4.07$ ft (by trial and error), $p_w = 4.07 + 8 + 4.07 = 16.14$ ft. Hence, the 8-ft-width design needs fewer bricks by $(22.62 - 16.14)/22.62 = 0.286$, or 28.6 percent.

**14.169** In the T-shaped channel of Fig. 14-45, the two arms have rougher walls than does the trunk. Approximate the discharge, if $y_1 = 20$ ft, $y_2 = 5$ ft, $b_1 = 50$ ft, $b_2 = 100$ ft, $n_1 = 0.018$, $n_2 = 0.036$, with a slope of 0.0004.

▌
$$Q = (A)(1.486/n)(R^{2/3})(s^{1/2})$$
$$Q_1 = [(50)(20 + 5)](1.486/0.018)[(50)(20 + 5)/(20 + 50 + 20)]^{2/3}(0.0004)^{1/2} = 11\,925 \text{ ft}^3/\text{s}$$
$$Q_2 = [(100)(5)](1.486/0.036)[(100)(5)/(100 + 5)]^{2/3}(0.0004)^{1/2} = 1168 \text{ ft}^3/\text{s}$$
$$Q = 11\,925 + (2)(1168) = 14\,261 \text{ ft}^3/\text{s}$$

**Fig. 14-45**

**14.170** Repeat Prob. 14.169 for the following parameters: $y_1 = b_1 = 9$ m, $n_1 = 0.018$, $y_2 = 1$ m, $b_2 = 100$ m, $n_2 = 0.036$. The angle of downslope is 0.1°.

▌
$$Q = (A)(1.0/n)(R^{2/3})(s^{1/2})$$
$$Q_1 = [(9)(9 + 1)](1.0/0.018)[(9)(9 + 1)/(9 + 9 + 9)]^{2/3}(\tan 0.1°)^{1/2} = 466 \text{ m}^3/\text{s}$$
$$Q_2 = [(100)(1)](1.0/0.036)[(100)(1)/(100 + 1)]^{2/3}(\tan 0.1°)^{1/2} = 115 \text{ m}^3/\text{s}$$
$$Q = 466 + (2)(115) = 696 \text{ m}^3/\text{s}$$

**14.171** A 1-m-diameter clay tile ($n = 0.014$) sewer pipe runs half-full on a slope of 0.004. Compute the flow rate by the Manning formula.

▌ $Q = (A)(1.0/n)(R^{2/3})(s^{1/2}) = [(\frac{1}{2})(\pi)(1)^2/4](1.0/0.014)(\frac{1}{4})^{2/3}(0.004)^{1/2} = 0.704 \text{ m}^3/\text{s}$

**14.172** Solve Prob. 14.171 by the Moody formula, with $\epsilon = 0.0026$ m.

▌
$$1/f^{1/2} = 2 \log (3.7D/\epsilon) = 2 \log [(3.7)(1)/0.0026] \qquad f = 0.02514$$
$$C = \sqrt{8g/f} = \sqrt{(8)(9.807)/0.02514} = 55.86$$
$$Q = CA(Rs)^{1/2} = 55.86[(\tfrac{1}{2})(\pi)(1)^2/4][(\tfrac{1}{4})(0.004)]^{1/2} = 0.694 \text{ m}^3/\text{s}$$

**14.173** Four of the sewer pipes from Prob. 14.171 empty into a single finished-cement ($n = 0.012$) pipe, also sloping at 0.004°. If the large pipe is also to run at half-full, what should its diameter be?

▌ $Q = (A)(1.0/n)(R^{2/3})(s^{1/2})$ $\qquad$ $(4)(0.704) = [(\tfrac{1}{2})(\pi)(D)^2/4](1.0/0.012)(D/4)^{2/3}(0.004)^{1/2}$ $\qquad$ $D = 1.59$ m

**14.174** For the circular channel of Fig. 14-46, if $n = 0.016$, $D = 3$ m, and the slope is 0.2°, find the normal depth for a discharge of 20 m³/s.

▌
$$Q = (A)(1.0/n)(R^{2/3})(s^{1/2}) \qquad y = (D/2)(1 + \sin \theta) = (\tfrac{3}{2})(1 + \sin \theta)$$
$$A = (\tfrac{1}{2})(D/2)^2(\pi + 2\theta + \sin 2\theta) = (\tfrac{1}{2})(\tfrac{3}{2})^2(\pi + 2\theta + \sin 2\theta) = (1.125)(\pi + 2\theta + \sin 2\theta)$$
$$p_w = (D/2)(\pi + 2\theta) = (\tfrac{3}{2})(\pi + 2\theta)$$
$$20 = [(1.125)(\pi + 2\theta + \sin 2\theta)](1.0/0.016)[(1.125)(\pi + 2\theta + \sin 2\theta)/(\tfrac{3}{2})(\pi + 2\theta)]^{2/3}(\tan 0.2°)^{1/2}$$
$$\theta = 31.57° \quad \text{(by trial and error)} \qquad y = (\tfrac{3}{2})(1 + \sin 31.57°) = 2.29 \text{ m}$$

**Fig. 14-46**

**14.175** For fixed $n$ and $s$, find the diameter of a semicircular channel that will have the same discharge as a rectangular channel with $b = 4$ m and $y = 2$ m.

$$Q = (A)(1.0/n)(R^{2/3})(s^{1/2}) \qquad Q_1 = Q_2$$

$$[(4)(2)][(1.0/n)][(4)(2)/(2 + 4 + 2)]^{2/3}(s)^{1/2} = [(\tfrac{1}{2})(\pi)(D)^2/4](1.0/n)(D/4)^{2/3}(s)^{1/2} \qquad D = 4.38 \text{ m}$$

**14.176** A trapezoidal channel has $n = 0.020$ and $s = 0.0004$ and is made in the shape of a half-hexagon for maximum efficiency (see Fig. 14-47). For what length of side will the channel carry 250 cfs of water?

$$Q = (A)(1.486/n)(R^{2/3})(s^{1/2}) \qquad A = (\tfrac{3}{2})(b^2)(\sin 60°) = 1.299b^2$$

$$250 = (1.299b^2)(1.486/0.020)(1.299b^2/3b)^{2/3}(0.0004)^{1/2} \qquad b = 7.64 \text{ ft}$$

**Fig. 14-47**

**14.177** Compute the discharge of a semicircular channel of the same area, slope, and $n$-value as the channel of Prob. 14.176.

$$A = (1.299)(7.64)^2 = 75.82 \text{ ft}^2 \qquad \text{(from Prob. 14.176)} \qquad 75.82 = (\tfrac{1}{2})(\pi)(d)^2/4 \qquad d = 13.90 \text{ ft}$$

$$Q = (A)(1.486/n)(R^{2/3})(s^{1/2}) = 75.82(1.486/0.020)(13.9/4)^{2/3}(0.0004)^{1/2} = 258 \text{ ft}^3/\text{s}$$

**14.178** Find the optimal dimensions of a wooden ($n = 0.012$) rectangular channel that will carry $4.0$ m³/s at $s = 0.0008$.

For an optimum, width = twice depth $(w = 2y)$: $Q = (A)(1.0/n)(R^{2/3})(s^{1/2})$, $4.0 = [(y)(2y)](1.0/0.012)[(y)(2y)/(y + 2y + y)]^{2/3}(0.0008)^{1/2}$, $y = 1.118$ m, $w = (2)(1.118) = 2.236$ m.

**14.179** How deep should an asphalt ($n = 0.015$) trapezoidal channel, with sides sloping at 45°, be to carry 5 m³/s at $s = 0.0006$?

In an optimal design, $R = y/2$. $Q = (A)(1.0/n)(R^{2/3})(s^{1/2})$, $A = (y^2)[(2)(1 + \cot 45°)^{1/2} - \cot 45°] = 1.828y^2$, $5 = (1.828y^2)(1.0/0.015)(y/2)^{2/3}(0.0006)^{1/2}$, $y = 1.44$ m.

**14.180** Compare the channel of Prob. 14.179, with a semicircular channel of the same slope and area.

$$A = (1.828)(1.44)^2 = 3.79 \text{ m}^2 \qquad \text{(from Prob. 14.179)} \qquad 3.79 = (\tfrac{1}{2})(\pi)(D)^2/4 \qquad D = 3.107 \text{ m}$$

$$Q = (A)(1.0/n)(R^{2/3})(s^{1/2}) = (3.79)(1.0/0.015)(3.107/4)^{2/3}(0.0006)^{1/2} = 5.23 \text{ m}^3/\text{s}$$

The semicircular channel carries $(5.23 - 5)/5 = 0.046$, or 4.6 percent more than the trapezoidal channel.

**14.181**  If the side angles of the trapezoidal channel of Prob. 14.176 are reduced to $\theta = 20°$ and if the bottom flat width is 8 ft, find the normal depth of the new channel.

**❚**  $Q = (A)(1.486/n)(R^{2/3})(s^{1/2})$        $A = by + y^2 \cot 20° = 8y + 2.747y^2$        $p_w = b + 2y \csc 20° = 8 + 5.848y$

$$250 = (8y + 2.747y^2)(1.486/0.020)[(8y + 2.747y^2)/(8 + 5.848y)]^{2/3}(0.0004)^{1/2}$$

$$y = 4.40 \text{ ft} \qquad \text{(by trial and error)}$$

**14.182**  A 16-ft-deep, clean-earth river ($n = 0.030$) has a flow rate $q = 140$ cfs/ft. Calculate the Froude number of the river.

**❚**        $y_c = (q^2/g)^{1/3} = (140^2/32.2)^{1/3} = 8.47 \text{ ft}$        $v = 140/16 = 8.750 \text{ ft/s}$

$$N_F = v/\sqrt{gy_c} = 8.75/\sqrt{(32.2)(8.47)} = 0.530$$

**14.183**  Find the critical slope in Prob. 14.182 by the Moody method, using $\epsilon = 0.78$ ft.

**❚**        $s_c = f/8$        $1/f^{1/2} = 2 \log (3.7D/\epsilon)$

Using data from Prob. 14.182, $D = 4R = 4y_c = (4)(8.47) = 33.88$ ft, $1/f^{1/2} = 2 \log [(3.7)(33.88)/0.78]$, $f = 0.05137$, $s_c = 0.05137/8 = 0.00642$.

**14.184**  Are the normal flows corresponding to the two widths in Prob. 14.168 subcritical or supercritical?

**❚**  $y_c = (Q^2/b^2g)^{1/3}$. For $w = 4$ ft, $y_c = \{230^2/[(4)^2(32.2)]\}^{1/3} = 4.68$ ft $< 9.31$ ft; so the flow is subcritical. For $w = 8$ ft, $y_c = \{230^2/[(8)^2(32.2)]\}^{1/3} = 2.95$ ft $< 4.07$ ft; so the flow is subcritical.

**14.185**  In a 250-mm-deep test channel, a (stationary) ship model sets up the bow wave shown in Fig. 14-48. Determine the flow velocity in the channel.

**❚**        $\sin \theta = 1/N_F = \sqrt{gh}/v$        $\sin 30° = \sqrt{(9.807)(0.250)}/v$        $v = 3.13$ m/s

**Fig. 14-48**

**14.186**  Suppose that the wave of Fig. 14-48 is seen instead on the surface of water flowing half-full in a circular channel of diameter 1.0 m. What is the flow rate if the surface is finished cement?

**❚**  Flow is supercritical.

$$v = N_F v_c \qquad N_F = \csc 30° = 2.00 \qquad A_c = (\tfrac{1}{2})(\pi)(1.0)^2/4 = 0.3927 \text{ m}^2$$

$$v_c = (gA_c/b_0)^{1/2} = [(9.807)(0.3927)/1.0]^{1/2} = 1.96 \text{ m/s} \qquad v = (2.00)(1.96) = 3.92 \text{ m/s}$$

$$Q = Av = (0.3927)(3.92) = 1.539 \text{ m}^3/\text{s}$$

**14.187**  Suppose that the wave of Fig. 14-48 is seen instead on the surface of water flowing full in a half-hexagon of side length 300 mm. What is the flow rate if the sides are of planed wood?

**❚**  Flow is supercritical.

$$v = N_F v_c \qquad N_F = \csc 30° = 2.00 \qquad v_c = (gA_c/b_0)^{1/2}$$

$$b_0 = 0.300 + (2)(0.300)(\cos 60°) = 0.600 \text{ m} \qquad y = (0.300)(\sin 60°) = 0.260 \text{ m}$$

$$A_c = (\tfrac{1}{2})(0.600 + 0.300)(0.260) = 0.1169 \text{ m}^2 \qquad v_c = [(9.807)(0.1169)/0.600]^{1/2} = 1.38 \text{ m/s}$$

$$v = (2.00)(1.38) = 2.76 \text{ m/s} \qquad Q = Av = (0.1169)(2.76) = 0.323 \text{ m}^3/\text{s}$$

**14.188** Find the critical depth and the critical slope for the conduit of Prob. 14.164.

$\blacksquare$ $\qquad A = (b_0 Q^2/g)^{1/3}$ $\qquad b_0 = 6 + 2y_c \cot 35° = 6 + 2.856 y_c$ $\qquad A = (y_c)(6 + y_c \cot 35°) = 6y_c + 1.428 y_c^2$

$6y_c + 1.428 y_c^2 = [(6 + 2.856 y_c)(50^2)/9.807]^{1/3}$ $\qquad y_c = 1.86$ m $\qquad$ (by trial and error)

$v_c = (1.0/n)(R^{2/3})(s^{1/2})$ $\qquad A = (6)(1.86) + (1.428)(1.86)^2 = 16.10$ m$^2$ $\qquad v_c = Q/A = 60/16.10 = 3.727$ m/s

$p_w = 6 + (2)(1.86)/\sin 35° = 12.49$ m $\qquad 3.727 = (1.0/0.014)(16.10/12.49)^{2/3}(s_c)^{1/2}$ $\qquad s_c = 0.00194$

**14.189** For the river of Prob. 14.182, find the depth $y_2$ at which the specific energy is equal to that at the bottom, $y_1 = 16$ ft.

$\blacksquare$ $\qquad E = y + v^2/2g$ $\qquad v_1 = 8.750$ ft/s $\qquad$ (from Prob. 14.182)

$E_1 = 16 + 8.750^2/[(2)(32.2)] = 17.19$ ft $\qquad E_2 = E_1 = 17.19 = y_2 + (140/y_2)^2/[(2)(32.2)]$

$17.19 y_2^2 - y_2^3 - 304.3 = 0$ $\qquad y_2 = 5.00$ ft $\qquad$ (by trial and error)

**14.190** Determine the critical slope of a clay-tile ($n = 0.014$) triangular channel with sides sloping at 60° and carrying 12 m$^3$/s.

$\blacksquare$ $\qquad gA_c^3 = b_0 Q^2$ $\qquad A_c = y_c^2 \cot 60° = 0.5774 y_c^2$ $\qquad b_0 = 2y_c \cot 60° = 1.155 y_c$

$(9.807)(0.5774 y_c^2)^3 = (1.155 y_c)(12)^2$ $\qquad y_c = 2.45$ m

$v_c = Q/A_c = 12/[(0.5774)(2.45)^2] = 3.46$ m/s

$v_c = (1.0/n)(R^{2/3})(s^{1/2})$

$3.46 = (1.0/0.014)\{(0.5774)(2.45)^2/[(2)(2.45)/\sin 60°]\}^{2/3}(s_c)^{1/2}$ $\qquad s_c = 0.00451$

**14.191** A triangular duct flows partly full as shown in Fig. 14-49. If the critical depth is 0.6 m and $n = 0.016$, compute (a) the critical flow rate and (b) the critical slope.

$\blacksquare$ (a) $\qquad b_0 = \{[(\sqrt{3}/2) - 0.6]/(\sqrt{3}/2)\}(1) = 0.3072$ m $\qquad A = (\frac{1}{2})(0.3072 + 1)(0.6) = 0.3922$ m$^2$

$Q = A\sqrt{gA/b_0} = (0.3922)\sqrt{(9.807)(0.3922)/0.3072} = 1.39$ m$^3$/s

(b) $\qquad Q = (A)(1.0/n)(R^{2/3})(s^{1/2})$ $\qquad p_w = 1 + (2)[0.6/\sin 60°] = 2.386$ m

$1.39 = (0.3922)(1.0/0.016)(0.3922/2.386)^{2/3}(s)^{1/2}$ $\qquad s = 0.0357$

1 m $\qquad$ 1 m

1 m $\qquad$ **Fig. 14-49**

**14.192** For the triangular duct of Prob. 14.191, if the critical flow rate is 1.0 m$^3$/s, compute (a) the critical depth and (b) the critical slope.

$\blacksquare$ (a) $\qquad Q = A\sqrt{gA/b_0}$ $\qquad b_0 = [((\sqrt{3}/2) - y_c)/(\sqrt{3}/2)](1) = 1 - 1.155 y_c$

$A = \frac{1}{2}[(1 - 1.155 y_c) + 1](y_c) = y_c - 0.5775 y_c^2$

$1.0 = (y_c - 0.5775 y_c^2)\sqrt{(9.807)(y_c - 0.5775 y_c^2)/(1 - 1.155 y_c)}$ $\qquad y_c = 0.493$ m $\qquad$ (by trial and error)

(b) $\qquad Q = (A)(1.0/n)(R^{2/3})(s^{1/2})$ $\qquad p_w = 1 + (2)[0.493/\sin 60°] = 2.139$ m

$A = 0.493 - (0.5775)(0.493)^2 = 0.3526$ m$^2$ $\qquad 1.0 = (0.3526)(1.0/0.016)(0.3526/2.139)^{2/3}(s)^{1/2}$

$s = 0.228$

**14.193** A rectangular channel 2 m wide and 1 m deep has critical slope of 0.009. Estimate the surface roughness height.

$\blacksquare$ $\qquad s_c = gn^2 p_w/(1.0b_0 R^{1/3})$ $\qquad R = A/p_w = (2)(1)/(1 + 2 + 1) = 0.5000$ m

$0.009 = (9.807)(n)^2(1 + 2 + 1)/[(1.0)(2)(0.5000)^{1/3}]$ $\qquad n = 0.0191$

$n = 0.0382\epsilon^{1/6}$ $\qquad 0.0191 = 0.0382\epsilon^{1/6}$ $\qquad \epsilon = 0.0156$ m $= 15.6$ mm

**14.194** For a water flow in a rectangular channel 10 m wide, the critical slope is 0.003491. Find the critical depth and discharge, given $n = 0.014$.

$\blacksquare$ $\quad s_c = gn^2 p_w/(1.0 b_0 R^{1/3})$ $\qquad 0.003491 = (9.807)(0.014)^2(10 + 2y_c)/\{(1.0)(10)[10y_c/(10 + 2y_c)]^{1/3}\}$

$y_c = 0.196$ m $\quad$ (by trial and error) $\qquad Q_c = (gA^3/b_0)^{1/3} = \{(9.807)[(10)(0.196)]^3/10\}^{1/3} = 1.95$ m³/s

**14.195** A circular aluminium ($n = 0.020$) channel 8 ft in diameter has Froude number 0.5 in uniform half-full flow. Compute the channel slope.

$\blacksquare$ $\qquad v_c = (gA/b_0)^{1/2} = \{(32.2)[(\frac{1}{2})(\pi)(8)^2/4]/8\}^{1/2} = 10.06$ ft/s

$v = (1.486/n)(R^{2/3})(s^{1/2}) = v_c/2 = 10.06/2 = 5.03$ ft/s

$5.03 = (1.486/0.020)(\frac{8}{4})^{2/3}(s)^{1/2} \qquad s = 0.00182$

**14.196** Water is in steady flow through the finished-concrete, semicircular channel shown in Fig. 14-50. If the bed slope is 0.0016, what is the flow rate?

$\blacksquare$ $\qquad A = (\frac{1}{2})(\pi)(10)^2 + (3)(10 + 10) = 217.1$ ft² $\qquad R_h = A/p_w = 217.1/[(\pi)(10) + 3 + 3] = 5.802$ ft

$Q = (A)(1.486/n)(R_h^{2/3})(s_0^{1/2}) = (217.1)(1.486/0.012)(5.802)^{2/3}(0.0016)^{1/2} = 3472$ ft³/s

**Fig. 14-50**

**14.197** In a planed-wood ($n = 0.012$) rectangular channel of width 4 m, water is flowing at the rate of 20 m³/s. If the slope of the channel is 0.0012, what is the depth of uniform flow?

$\blacksquare$ $\qquad Q = (A)(1.0/n)(R^{2/3})(s^{1/2}) \qquad 20 = (4D)(1.0/0.012)[4D/(D + 4 + D)]^{2/3}(0.0012)^{1/2}$

$D = 1.80$ m $\quad$ (by trial and error)

**14.198** In a planed-wood rectangular channel of width 4 m, water is flowing at a rate of 5 m³/s. The slope of the channel is 0.0001. The roughness coefficient is 0.5 mm. What is the height $h$ of the cross section of flow for normal, steady flow? The water has a temperature of 10 °C.

$\blacksquare$ $\qquad v = Q/A = (8gsR/f)^{1/2} \qquad 5/4y = \{(8)(9.807)(0.0001)[4y/(y + 4 + y)]/f\}^{1/2}$

Like problems in Chap. 9, this requires a trial-and-error solution whereby a value of the friction factor ($f$) is guessed and subsequent computations and work with Fig. A-5 are done to find the right combination of parameter values to satisfy the problem. Inasmuch as a number of trial-and-error problems of this type were presented in detail in Chap. 9, we shall save time and space in problems of this type in this chapter by "guessing" the correct value of $f$ on the first try! Guess $f = 0.0132$.

$5/4y = \{(8)(9.807)(0.0001)[4y/(y + 4 + y)]/0.0132\}^{1/2} \qquad y = 1.69$ m $\quad$ (by trial and error)

$v = Q/A = 5/[(4)(1.69)] = 0.7396$ m/s $\qquad R = (4)(1.69)/(1.69 + 4 + 1.69) = 0.9160$ m

$N_R = D_h v/v = (4R)(v)/v = (4)(0.9160)(0.7396)/(1.30 \times 10^{-6}) = 2.08 \times 10^6$

$\epsilon/D_h = \epsilon/4R = (0.5/1000)/[(4)(0.9160)] = 0.00014$

From Fig. A-5, $f = 0.0132$ (O.K.). Therefore, $y = 1.69$ m.

**14.199** Do Prob. 14.196 using friction-factor formulas with $\epsilon = 0.0032$ ft. Assume a temperature of 60 °F.

$\blacksquare$ $v^*\epsilon/v = (\sqrt{gRs})(\epsilon)/v$. Using values from Prob. 14.196, $v^*\epsilon/v = [\sqrt{(32.2)(5.802)(0.0016)}](0.0032)/(1.21 \times 10^{-5}) = 145$. Since $v^*\epsilon/v > 100$, we are in the fully rough zone and $1/f^{1/2} = 2.16 - 2\log(\epsilon/R) = 2.16 -$

2 log $(0.0032/5.802)$, $f = 0.01328$, $v = (8gsR/f)^{1/2} = \{(8)(32.2)(0.0016)(5.802)/0.01328\}^{1/2} = 13.42$ ft/s, $Q = Av = (217.1)(13.42) = 2913$ ft$^3$/s.

**14.200** A wide rectangular section has flow $q = 80$ cfs/ft. Find the critical depth and the critical slope, if $\epsilon = 0.0035$ ft.

**▮**
$$y_c = (q^2/g)^{1/3} = (80^2/32.2)^{1/3} = 5.84 \text{ ft}$$

$$s_c = f/8 \qquad N_R = D_h v/v \qquad D_h = 4R = (4)(5.84) = 23.36 \text{ ft}$$

$$N_R = (23.36)\{80/[(5)(1)]\}/(1.21 \times 10^{-5}) = 3.09 \times 10^7$$

$$\epsilon/D_h = 0.0035/23.36 = 0.000150$$

From Fig. A-5, $f = 0.0128$. $s_c = 0.0128/8 = 0.0016$.

**14.201** For steady laminar flow in a thin sheet over a flat surface (see Fig. 14-51), $V_z = [(\gamma \sin \theta)/\mu][(3q\mu/\gamma \sin \theta)^{1/3}y - y^2/2]$, where $q$ is the volumetric flow per unit width. Find the thickness $t$ of such a flow of water at 5 °C for $\theta = 25°$ and $q = 18$ (L/min)/m.

**▮** When $y = t$, $dV_z/dy = 0$. $dV_z/dy = 0 = [(\gamma \sin \theta)/\mu][(3q\mu/\gamma \sin \theta)^{1/3} - t]$, $t = (3q\mu/\gamma \sin \theta)^{1/3}$:

$$t = \left\{ \frac{(3)[(18 \times 10^{-3})/60](1.52 \times 10^{-3})}{(9.79 \times 10^{-3})(\sin 25°)} \right\}^{1/3} = 0.692 \text{ mm}$$

**Fig. 14-51**

**14.202** A film of oil ($v = 1.664 \times 10^{-4}$ ft$^2$/s) of thickness 0.0022 ft moves at uniform speed down an inclined surface having an angle $\theta = 25°$ (see Fig. 14-51). Compute the volume flow per unit width.

**▮** Assuming laminar flow,

$$t = (3q\mu/\gamma \sin \theta)^{1/3} = (3qv/g \sin \theta)^{1/3} \qquad \text{(from Prob. 14.201)}$$

$$0.0022 = \{(3)(q)(1.664 \times 10^{-4})/[(32.17)(\sin 25°)]\}^{1/3} \qquad q = 2.9 \times 10^{-4} \text{ cfs/ft}$$

**14.203** Check the assumption of laminar flow in (*a*) Prob. 14.201, (*b*) Prob. 14.202.

**▮** (*a*)
$$N_R = q/v = (3 \times 10^{-4})/(1.52 \times 10^{-6}) \approx 2 < 500 \quad \text{(O.K.)}.$$

(*b*)
$$N_R = q/v = (2.9 \times 10^{-4})/(1.664 \times 10^{-4}) \approx 2 < 500 \quad \text{(O.K.)}.$$

**14.204** Water at 5 °C is flowing 1 m deep in a finished-concrete rectangular channel of width 8 m with a slope of 0.001. Find the volume of flow.

**▮**
$$Q = (A)(1.0/n)(R^{2/3})(s^{1/2}) = [(8)(1)](1.0/0.012)[(8)(1)/(1 + 8 + 1)]^{2/3}(0.001)^{1/2} = 18.2 \text{ m}^3/\text{s}$$

**14.205** A wide rectangular channel dug from clean earth ($n = 0.024$) is to conduct a flow of 5 m$^3$/s per meter of width. The slope of the bed is 0.0016. What would be the depth of flow for normal flow?

**▮**
$$s = (n/1.0)^2(q^2/y_N^{10/3}) \qquad 0.0016 = (0.024/1.0)^2(5^2/y_N^{10/3}) \qquad y_N = 1.93 \text{ m}$$

**14.206**   The channel of Fig. 14-52 has $n = 0.020$ and carries 8 ft$^3$/s of water through a total drop of 10 ft. How long should the channel be for normal flow?

$$Q = (A)(1.486/n)(R^{2/3})(s^{1/2}) \qquad A = (2)(2 \tan 45°) = 4.000 \text{ ft}^2$$

$$R = A/p_w = 4.000/[(2)(2/\cos 45°)] = 0.7071 \text{ ft}$$

$$8 = (4.000)(1.486/0.020)(0.7071)^{2/3}[10/L]^{1/2} \qquad L = 8694 \text{ ft}$$

Fig. 14-52

**14.207**   Find the depth of flow in a 2-m-wide rectangular channel ($n = 0.016$) that carries 5 m$^3$/s of water 1 km with a head loss of 7.5 m.

$$s = 7.5/1000 = 0.0075 \qquad Q = (A)(1.0/n)(R^{2/3})(s^{1/2})$$

$$5 = (2y_N)(1.0/0.016)[2y_N/(y_N + 2 + y_N)]^{2/3}(0.0075)^{1/2} \qquad y_N = 0.795 \text{ m} \qquad \text{(by trial and error)}$$

**14.208**   The channel of Fig. 14-53 has a slope of 0.0018 and carries 57 m$^3$/s. Determine the flow depth, if $n = 0.015$.

$$Q = (A)(1.0/n)(R^{2/3})(s^{1/2}) \qquad A = 10y_N - (\tfrac{10}{2})[(\tfrac{10}{2})(\tan 10°)] = 10y_N - 4.408$$

$$p_w = 2[y_N - (\tfrac{10}{2})(\tan 10°) + (\tfrac{10}{2})/\cos 10°] = 2y_N + 8.391 \qquad R = A/p_w = (10y_N - 4.408)/(2y_N + 8.391)$$

$$57 = (10y_N - 4.408)(1.0/0.015)[(10y_N - 4.408)/(2y_N + 8.391)]^{2/3}(0.0018)^{1/2}$$

$$y_N = 2.11 \text{ m} \qquad \text{(by trial and error)}$$

Fig. 14-53

**14.209**   Shown in Fig. 14-54 is a partially filled pipe. If Manning's $n$ is 0.020, what slope is necessary for a steady flow of 60 cfs?

$$Q = (A)(1.486/n)(R^{2/3})(s^{1/2}) \qquad \theta = \arcsin \tfrac{2}{4} = 30.0°$$

$$A = \{[180° + (2)(30.0°)]/360°\}[(\pi)(8)^2/4] + (2)(4 \cos 30.0°) = 40.44 \text{ ft}^2$$

$$p_w = \{[180° + (2)(30.0°)]/360°\}[(\pi)(8)] = 16.76 \text{ ft} \qquad R = A/p_w = 40.44/16.76 = 2.413 \text{ ft}$$

$$60 = (40.44)(1.486/0.020)(2.413)^{2/3}(s)^{1/2} \qquad s = 0.0001232$$

$D = 8$ ft    Fig. 14-54

**14.210**   Find the flow in Fig. 14-55 if the slope of the channel is 0.0004 and $n = 0.024$. Side slopes are all 1:1.

∎ $A = (3)(3)/2 + (3)(20) + (5)(3 + 3 + 5)/2 + (8)(3 + 5) + (5)(3 + 3 + 5)/2 + (3)(25) + (3)(3)/2 = 263.0 \text{ m}^2$

$p_w = \sqrt{3^2 + 3^2} + 20 + \sqrt{5^2 + 5^2} + 8 + \sqrt{5^2 + 5^2} + 25 + \sqrt{3^2 + 3^2} = 75.63 \text{ m}$ $\qquad R = A/p_w = 263.0/75.63 = 3.477 \text{ m}$

$$Q = (A)(1.0/n)(R^{2/3})(s^{1/2}) = (263.0)(1.0/0.024)(3.477)^{2/3}(0.0004)^{1/2} = 503 \text{ m}^3/\text{s}$$

**Fig. 14-55**

**14.211** The channel in Prob. 14.196 is to be replaced by a rectangular channel of width 16 ft. What is the ratio of cost of the concrete allowing 2 ft of freeboard (distance above the free surface) for the walls of the channels?

∎ From Prob. 14.196, $Q = 3472 \text{ ft}^3/\text{s}$, $s = 0.0016$, and $n = 0.012$.

$$Q = (A)(1.486/n)(R^{2/3})(s^{1/2}) \qquad 3472 = (16y)(1.486/0.012)[16y/(y + 16 + y)]^{2/3}(0.0016)^{1/2}$$

$$y = 14.65 \text{ ft} \qquad \text{(by trial and error)} \qquad (p_w)_1 = 16 + (2)(14.65) + (2)(2) = 49.30 \text{ ft}$$

For the channel of Prob. 14.196, $(p_w)_2 = (\pi)(10) + 3 + 3 + (2)(2) = 41.42 \text{ ft}$. Ratio of cost $= (p_w)_1/(p_w)_2 = 49.30/41.42 = 1.19$.

**14.212** Repeat Prob. 14.204 using the friction-factor approach. Check your assumption of rough-zone flow.

∎ $R = A/p_w = (1)(8)/(1 + 8 + 1) = 0.8 \text{ m} \qquad 1/f^{1/2} = 2.16 - 2 \log (\epsilon/R) = 2.16 - 2 \log (0.001/0.8)$

$$f = 0.01576 \qquad v = (8gsR/f)^{1/2} = [(8)(9.807)(0.001)(0.8)/0.01576]^{1/2} = 1.996 \text{ m/s}$$

$$Q = Av = [(1)(8)](1.996) = 16.0 \text{ m}^3/\text{s}$$

$$N_R = D_h v/\nu = (4R)(v)/\nu = [(4)(0.8)](1.996)/(1.52 \times 10^{-6}) = 4.20 \times 10^6$$

The values $f = 0.01576$ and $N_R = 4.20 \times 10^6$ fall in in the rough zone of Fig. A-5.

**14.213** For a rectangular channel with a flow of 20 m³/s at a velocity of 5 m/s, what should width $b$ and depth $y$ be for the best hydraulic section?

∎ $A = Q/v = \frac{20}{5} = 4.000 \text{ m}^2$. For a rectangular section, the best hydraulic section has a width equal to twice the depth. Hence, $(b)(b/2) = 4.000$, $b = 2.828 \text{ m}$, $y = 2.828/2 = 1.414 \text{ m}$.

**14.214** For steady flow at 6 m³/s on a slope of 0.0018, what is the width of the rectangular channel of least wetted perimeter? Take $n = 0.0015$.

∎ $Q = (A)(1.0/n)(R^{2/3})(s^{1/2})$. The best hydraulic section has a width $b$ equal to twice the depth. Hence, $A = (b)(b/2) = b^2/2$, $R = A/p_w = (b^2/2)/(b/2 + b + b/2) = b/4$, $6 = (b^2/2)(1.0/0.015)(b/4)^{2/3}(0.0018)^{1/2}$, $b = 2.43 \text{ m}$.

**14.215** A stream has a speed of 12 ft/s and is 2 ft deep. Find the angular spread of the wave set up by a thin stationary obstruction.

∎ $$\sin \theta/2 = \sqrt{gy}/v = \sqrt{(32.2)(2)}/12 \qquad \theta = 84°$$

**14.216** A small boat is moving in shallow still water where the depth is 2 m; its bow wave makes an angle of 60° with the line of motion (see Fig. 14-56). Compute the speed of the boat.

∎ $$\sin \alpha = \sqrt{gy}/v \qquad \sin 60° = \sqrt{(9.807)(2)}/v \qquad v = 5.11 \text{ m/s}$$

**14.217** A stone is dropped into a pond; the ripples have an amplitude of about 1 in and travel at 4 ft/s. About how deep is the pond?

**Fig. 14-56**

$\blacksquare$    $(g)(y + \Delta y/2)(y + \Delta y) = c^2 y$    $32.2[y + (\frac{1}{12})/2](y + \frac{1}{12}) = (4)^2 y$    $y^2 - 0.3719y + 0.003472 = 0$

Choosing the larger root,

$$y = [-(-0.3719) + \sqrt{(-0.3719)^2 - (4)(1)(0.003472)}]/[(2)(1)] = 0.362 \text{ ft}$$

**14.218**  Two fishermen are stationed a distance $D$ apart along a straight running stream (depth $y$, speed $v$). At a certain moment, a small gravity wave is initiated midway between them. How much later than the downstream man does the upstream man feel the wave?

$\blacksquare$  With $c = \sqrt{gy}$ as the wave velocity,

$$\Delta t = \frac{D/2}{v - c} - \frac{D/2}{v + c} = \frac{Dc}{v^2 - c^2} = \frac{D\sqrt{gy}}{v^2 - gy}$$

**14.219**  A wide rectangular channel excavated from clean earth has a flow of 4 $(m^3/s)/m$. Determine the minimum specific energy.

$\blacksquare$        $y_c = (q^2/g)^{1/3} = (4^2/9.807)^{1/3} = 1.177 \text{ m}$        $E_{min} = (\frac{3}{2})(y_c) = (\frac{3}{2})(1.177) = 1.766 \text{ m}$

**14.220**  Find the critical slope in Prob. 14.219, assuming water at 5 °C ($\epsilon = 0.047$ m).

$\blacksquare$        $s_c = f/8$        $D_h = 4R = 4y_c = (4)(1.177) = 4.708 \text{ m}$        $v = q/y_c = 4/1.177 = 3.398 \text{ m/s}$

$N_R = D_h v/\nu = (4.708)(3.398)/(1.52 \times 10^{-6}) = 1.05 \times 10^7$        $\epsilon/D_h = 0.047/4.708 = 0.010$

From Fig. A-5, $f = 0.038$ and so $s_c = 0.038/8 = 0.00475$.

**14.221**  Find the flow in a wide rectangular channel for which the critical depth is 3 m.

$\blacksquare$                $y_c = (q^2/g)^{1/3}$        $3 = (q^2/9.807)^{1/3}$        $q = 16.27 \ (m^3/s)/m$

**14.222**  By what factor does the Froude number of a wide rectangular channel increase when the depth is decreased by a factor of 3/4, keeping the mass flow fixed?

$\blacksquare$  Since $N_F \propto y^{-3/2}$,

$$\text{factor} = \left(\frac{4}{3}\right)^{3/2} = \frac{8}{3\sqrt{3}} = 1.54$$

**14.223**  At a section in the rectangular channel shown in Fig. 14-57, the average velocity is 28.28 fps. Is the flow tranquil?

$\blacksquare$        $q = (Av)/b = [(10)(50)](28.28)/50 = 282.8 \text{ cfs/ft}$        $y_c = (q^2/g)^{1/3} = (282.8^2/32.2)^{1/3} = 13.54 \text{ ft}$

Since $y_c > 10$ ft, flow is not tranquil but shooting.

**Fig. 14-57**

**14.224**  How deep is a flow of 5 (m³/s)/m in a wide rectangular channel having a critical slope of 0.002? The fluid is water at 5 °C.

❚  Assume rough-flow zone:

$$v = (8gsR/f)^{1/2} \qquad s_c = f/8 \qquad 0.002 = f/8 \qquad f = 0.016$$
$$5/[(1)(y)] = [(8)(9.807)(0.002)(y)/0.016]^{1/2} \qquad y = 1.366 \text{ m}$$

**14.225**  Compute the minimum specific energy in an 8-ft-wide channel through which 150 ft³/s of water is flowing.

❚  $$gA^3 = bQ^2 \qquad (32.2)(8y_c)^3 = (8)(150)^2 \qquad y_c = 2.218 \text{ ft} \qquad E_{min} = A_c/2b + y_c$$
$$E_{min} = (8)(2.218)/[(2)(8)] + 2.218 = 3.327 \text{ ft}$$

**14.226**  Find the critical slope of a rectangular finished-concrete channel of width 4 m that carries 3 m³/s of water at 5 °C ($\epsilon = 0001$ m).

❚  $$gA^3 = bQ^2 \qquad (9.807)(4y_c)^3 = (4)(3)^2 \qquad y_c = 0.386 \text{ m} \qquad s_c = fp_w/8b \qquad \epsilon/D_h = \epsilon/4R$$
$$R = A/p_w = (4)(0.386)/(0.386 + 4 + 0.386) = 0.3236 \text{ m}$$
$$\epsilon/D_h = 0.001/[(4)(0.3236)] = 0.00077$$
$$N_R = D_h v/\nu = (4R)(Q/A)/\nu = (4)(0.3236)\{3/[(4)(0.386)]\}/(1.52 \times 10^{-6}) = 1.65 \times 10^6$$

From Fig. A-5, $f = 0.0186$.  $s_c = (0.0186)(0.386 + 4 + 0.386)/[(8)(4)] = 0.00277$.

**14.227**  What is the critical depth for a right triangular cross section for a flow of 4 m³/s?

❚  See Fig. 14-58.
$$gA^3 = bQ^2 \qquad b = (2)(y_c \tan 45°) = 2.0y_c \qquad A = (y_c)(y_c \tan 45°) = 1.0y_c^2$$
$$(9.807)(1.0y_c^2)^3 = (2.0y_c)(4)^2 \qquad y_c = 1.267 \text{ m}$$

**Fig. 14-58**

**14.228**  What is the critical depth of a trapezoidal cross section for a flow of 10 m³/s? The width at the base is 3 m and angle $\theta$ at the sides is 60°.

❚  $$gA^3 = bQ^2 \qquad b = 3 + (2)(y_c/\tan 60°) = 3 + 1.155y_c$$
$$A = 3y_c + (y_c)(y_c/\tan 60°) = 3y_c + 0.5774y_c^2$$
$$(9.807)(3y_c + 0.5774y_c^2)^3 = (3 + 1.155y_c)(10)^2 \qquad y_c = 0.976 \text{ m} \qquad \text{(by trial and error)}$$

**14.229**  Shown in Fig. 14-59 is a partially filled pipe discharging 450 cfs. What is the critical depth?

❚  $$gA^3 = bQ^2 \qquad b = (2)(4 \cos \theta) = 8 \cos \theta$$
$$A = [(\pi)(8^2/4)]/2 + [(\pi)(8^2/4)](2\theta/360) + (\tfrac{8}{2})(\cos \theta)(\tfrac{8}{2})(\sin \theta)$$
$$= 25.13 + 0.2793\theta + (16)(\cos \theta)(\sin \theta)$$
$$(32.2)[25.13 + 0.2793\theta + (16)(\cos \theta)(\sin \theta)]^3 = (8 \cos \theta)(450)^2$$
$$\theta = 20.5° \qquad \text{(by trial and error)}$$
$$y_c = \tfrac{8}{2} + (\tfrac{8}{2})(\sin 20.5°) = 5.40 \text{ ft}$$

Fig. 14-59

**14.230** Calculate the critical depth of the parabolic channel of Fig. 14-60, if the flow is $4 \mathrm{~m}^3/\mathrm{s}$.

$$gA^3 = bQ^2$$

$$A_c = (2)\int_0^{y_c} x \, dy = (2)\int_0^{y_c} (y/2)^{1/2} \, dy = 1.414[(\tfrac{2}{3})(y)^{3/2}]_0^{y_c}$$

$$= 1.414[(\tfrac{2}{3})(y_c)^{3/2}] = 0.9427 y_c^{3/2}$$

$$b = 2x = (2)(\sqrt{y_c/2}) = 1.414 y_c^{1/2}$$

$$(9.807)(0.9427 y_c^{3/2})^3 = (1.414 y_c^{1/2})(4)^2 \qquad y_c = 1.288 \mathrm{~m}$$

$y = 2x_2$   $dA = x \, dy$

$ds$   $dy$   $y_c$

Fig. 14-60

**14.231** In Prob. 14.230, what is the critical slope for normal flow? The friction factor is 0.014.

$s_c = (f)(p_w)_c/8b$. To evaluate $(p_w)_c$, consider a differential length along the wetted perimeter ($ds$ in Fig. 14-60): $ds = \sqrt{dx^2 + dy^2} = \sqrt{1 + (dy/dx)^2} \, dx$. From Prob. 14.230, when $y = y_c$, $x = \sqrt{y_c/2} = \sqrt{1.288/2} = 0.8025 \mathrm{~m}$.

$$(p_w)_c = \int ds = (2)\int_0^{0.7470} \sqrt{1 + (dy/dx)^2} \, dx \qquad y = 2x^2 \qquad dy/dx = 4x$$

$$(p_w)_c = (2)\int_0^{0.7470} \sqrt{1 + (4x)^2} \, dx = (8)\int_0^{0.8025} \sqrt{(\tfrac{1}{16} + x^2)} \, dx$$

$$= (\tfrac{8}{2})[(x)(\sqrt{\tfrac{1}{16} + x^2}) + \tfrac{1}{16}\{\log(x + \sqrt{\tfrac{1}{16} + x^2})\}]_0^{0.8025} = 2.711 \mathrm{~m}$$

$$b = (2)(0.8025) = 1.605 \mathrm{~m} \qquad s_c = (0.014)(2.711)/[(8)(1.605)] = 0.00296$$

**14.232** At a section in the triangular channel of Fig. 14-61 the average velocity is 8 ft/s. Is the flow tranquil or shooting?

$$\tan(45°/2) = (b/2)/10 \qquad b = 8.284 \mathrm{~ft} \qquad A = 10b/2 = (10)(8.282)/2 = 41.41 \mathrm{~ft}^2$$

$$Q = Av = (41.41)(8) = 331.3 \mathrm{~ft}^3/\mathrm{s} \qquad gA_c^3 = b_c Q^2$$

$$\tan(45°/2) = (b_c/2)/y_c \qquad b_c = 0.8284 y_c$$

$$A_c = y_c b_c/2 = (y_c)(0.8284 y_c)/2 = 0.4142 y_c^2$$

$$(32.2)(0.4142 y_c^2)^3 = (0.8284 y_c)(331.3)^2 \qquad y_c = 8.31 \mathrm{~ft} < 10 \mathrm{~ft} = y$$

The flow is tranquil.

**14.233** Water is moving in Fig. 14-62 with a velocity of 1 ft/s and a depth of 3 ft. It approaches a smooth rise in the channel bed of 1 ft. What should the estimated depth be after the rise? The channel is rectangular.

$$E = q^2/2y^2 g + y \qquad Q = Av = (3b)(1) = 3b \qquad q = Q/b = 3b/b = 3 \mathrm{~cfs/ft}$$

$$E_1 = 3^2/[(2)(3)^2(32.2)] + 3 = 3.016 \mathrm{~ft}$$

**Fig. 14-61**

Assuming no losses, $E_1 = E_2 + 1$, $3.016 = E_2 + 1$, $E_2 = 2.016$ ft; $y_c = (q^2/g)^{1/3} = (3^2/32.2)^{1/3} = 0.654$ ft. With surface elevation increasing, $E$ must decrease. We must have one value of $y$ which must be for tranquil flow. Hence, $2.016 = 3^2/[(2)(y_2)^2(32.2)] + y_2$, $y_2 = 1.98$ ft   (by trial and error).

**Fig. 14-62**

**14.234** In Fig. 14-63, a flow of 0.2 cfs flows over the rectangular channel of width 3 ft. If there is a smooth drop of 2 in, what is the elevation of the free surface above the bed of the channel after the drop? The velocity before the drop is 0.3 fps.

▌ $$Q = Av \qquad 0.2 = (3y_1)(0.3) \qquad y_1 = 0.2222 \text{ ft}$$
$$y_c = (q^2/g)^{1/3} = [(0.2/3)^2/32.2]^{1/3} = 0.05168 \text{ ft}$$

Since $y_c < y_1$, we have tranquil flow: $E = q^2/2y^2g + y$, $E_1 = (0.2/3)^2/[(2)(0.2222)^2(32.2)] + 0.2222 = 0.2236$ ft. Assuming no losses, $E_2 = E_1 + \frac{2}{12} = 0.2236 + \frac{2}{12} = 0.3903$ ft. We must have one depth downstream greater than $y_1$: $0.3903 = (0.2/3)^2/[(2)(y_2)^2(32.2)] + y_2$, $y_2 = 0.390$ ft   (by trial and error).

0.3 ft/s

2 in

**Fig. 14-63**

**14.235** Using the Powell equation, what quantity of liquid will flow in a smooth rectangular channel 2 ft wide, on a slope of 0.10 if the depth is 1.00 ft? Use $v = 0.00042$ ft²/s.

▌ $C = -42 \log(C/N_R + \epsilon/R)$. For smooth channels, $\epsilon/R$ is small and can be neglected; hence,

$$C = 42 \log(N_R/C) \qquad v = C\sqrt{Rs} \qquad N_R = 4Rv/v = 4RC\sqrt{Rs}/v$$
$$N_R/C = (4)[(2)(1)/(1 + 2 + 1)]^{3/2}(0.010)^{1/2}/0.00042 = 336.7$$
$$C = (42)(\log 336.7) = 106.1$$
$$Q = CA\sqrt{Rs} = (106.1)[(2)(1)]\sqrt{[(2)(1)/(1 + 2 + 1)](0.010)} = 15.0 \text{ ft}^3/\text{s}$$

**14.236** Determine $C$ by the Powell equation for a 2 ft by 1 ft rectangular channel, if $v = 5.50$ fps, $\epsilon/R = 0.0020$, and $v = 0.00042$ ft²/s.

▌ $$N_R = 4Rv/v = (4)[(2)(1)/(1 + 2 + 1)](5.50)/0.00042 = 26\,190$$
$$C = -42 \log(C/N_R + \epsilon/R) = (-42)[\log(C/26\,190 + 0.0020)] \qquad C = 95 \qquad \text{(by trial and error)}$$

**14.237** Show a correlation between roughness factor $f$ and roughness factor $n$.

▌ Taking the Manning formula as a basis of correlation, $C = \sqrt{8g/f} = 1.486R^{1/6}/n$, $1/\sqrt{f} = 1.486R^{1/6}/(n\sqrt{8g})$, $f = 8gn^2/(2.208R^{1/3})$.

**14.238** What is the average shear stress at the sides and bottom of a rectangular flume 12 ft wide, flowing 4 ft deep, and laid on a slope of 1.60 ft per 1000 ft?

▌ $$\tau_0 = \gamma Rs = 62.4[(4)(12)/(4 + 12 + 4)](1.60/1000) = 0.240 \text{ lb/ft}^2$$

**14.239** What flow can be expected in a 4-ft-wide rectangular, concrete-lined channel laid on a slope of 4 ft in 10 000 ft, if the water flows 2 ft deep. Solve using the Manning formula with $n = 0.015$.

■ $Q = (A)(1.486/n)(R^{2/3})(s^{1/2}) = [(4)(2)](1.486/0.015)[(4)(2)/(2 + 4 + 2]^{2/3}(4/10\,000)^{1/2} = 15.9 \text{ ft}^3/\text{s}$

**14.240** Solve Prob. 14.239 using Kutter's $C$.

■ $Q = AC\sqrt{Rs}$      $s = 4/10\,000 = 0.0004$      $R = A/p_w = (4)(2)/(2 + 4 + 2) = 1.000 \text{ ft}$

From Table A-15, with $s = 0.0004$, $R = 1.000$ ft, and $n = 0.015$, $C = 98$. $Q = [(4)(2)](98)\sqrt{(1.000)(0.0004)} = 15.7 \text{ ft}^3/\text{s}$.

**14.241** In a hydraulics laboratory, a flow of 14.56 cfs was measured from a rectangular channel flowing 4 ft wide and 2 ft deep. If the slope of the channel was 0.00040, what is the roughness factor for the lining of the channel? Solve using Kutter's formula.

■ $Q = AC\sqrt{Rs}$    $14.56 = [(4)(2)](C)\sqrt{[(4)(2)/(2 + 4 + 2)](0.00040)}$    $C = 91$

From Table A-15, by interpolation, $n = 0.016$.

**14.242** Solve Prob. 14.241 using Manning's formula.

■ $$Q = (A)(1.486/n)(R^{2/3})(s^{1/2})$$
$$14.56 = [(4)(2)](1.486/n)[(4)(2)/(2 + 4 + 2)]^{2/3}(0.00040)^{1/2}    n = 0.0163$$

**14.243** On what slope should a 24-in vitrified sewer pipe be laid in order that 6.00 cfs will flow when the sewer is half-full? Use $n = 0.013$.

■ $Q = (A)(1.486/n)(R^{2/3})$    $6.00 = [(\frac{1}{2})(\pi)(\frac{24}{12})^2/4](1.486/0.013)[(\frac{24}{12})/4]^{2/3}(s)^{1/2}$    $s = 0.00281$

**14.244** What would the slope be in Prob. 14.243 if the sewer flows full?

■ $Q = (A)(1.486/n)(R^{2/3})(s^{1/2})$    $6.00 = [(\pi)(\frac{24}{12})^2/4](1.486/0.013)[(\frac{24}{12})/4]^{2/3}(s)^{1/2}$    $s = 0.000703$

**14.245** A trapezoidal channel, bottom width 20 ft and side slopes 1 to 1, flows 4 ft deep on a slope of 0.0009. For a value of $n = 0.025$, what is the uniform discharge?

■ $$Q = (A)(1.486/n)(R^{2/3})(s^{1/2})    A = (20)(4) + (4)(4) = 96.00 \text{ ft}^2$$
$$p_w = 20 + (2)(4)(\sqrt{2}) = 31.31 \text{ ft}$$
$$Q = (96.00)(1.486/0.025)(96.00/31.31)^{2/3}(0.0009)^{1/2} = 361 \text{ ft}^3/\text{s}$$

**14.246** Two concrete pipes ($C = 100$) must carry the flow from an open channel of half-square section 6 ft wide and 3 ft deep ($C = 120$). The slope of both structures is 0.00090. (a) Determine the diameter of the pipes. (b) Find the depth of water in the rectangular channel after it has become stabilized, if the slope is changed to 0.00160, using $C = 120$.

■ (a)        $Q_{\text{channel}} = Q_{\text{pipes}}$    $(AC\sqrt{Rs})_{\text{channel}} = (2)(AC\sqrt{Rs})_{\text{pipe}}$
$[(3)(6)](120)\sqrt{[(3)(6)/(3 + 6 + 3)](0.00090)} = (2)[(\pi)(D)^2/4](100)\sqrt{(D/4)(0.00090)}$    $D = 4.08 \text{ ft}$
   (b)        $Q = [(3)(6)](120)\sqrt{[(3)(6)/(3 + 6 + 3)](0.00090)} = 79.36 \text{ ft}^3/\text{s}$
$79.36 = [(y)(6)](120)\sqrt{[(y)(6)/(y + 6 + y)](0.001600)}$    $y = 2.39 \text{ ft}$    (by trial and error)

**14.247** How deep will water flow at the rate of 240 cfs in a rectangular channel 20 ft wide, laid on a slope of 0.00010? Use $n = 0.0149$.

■ $Q = (A)(1.486/n)(R^{2/3})(s^{1/2})$    $240 = (20y)(1.486/0.0149)[20y/(y + 20 + y)]^{2/3}(0.00010)^{1/2}$
$y = 5.27 \text{ ft}$    (by trial and error)

**14.248** How wide must a rectangular channel be constructed in order to carry 500 cfs at a depth of 6 ft on a slope of 0.00040? Use $n = 0.010$.

■ $Q = (A)(1.486/n)(R^{2/3})(s^{1/2})$    $500 = (6b)(1.486/0.010)[6b/(6 + b + 6)]^{2/3}(0.00040)^{1/2}$
$b = 13.1 \text{ ft}$    (by trial and error)

**14.249** A rectangular channel carries 200 cfs. Find the critical depth and the critical velocity for (a) a width of 12 ft and (b) a width of 9 ft. (c) What slope will produce the critical velocity in (a) if $n = 0.020$?

▌ (a)      $y_c = (q^2/g)^{1/3} = [(\frac{200}{12})^2/32.2]^{1/3} = 2.05$ ft      $v_c = \sqrt{gy_c} = \sqrt{(32.2)(2.05)} = 8.13$ ft/s

     (b)      $y_c = [(\frac{200}{9})^2/32.2]^{1/3} = 2.48$ ft      $v_c = \sqrt{(32.2)(2.48)} = 8.94$ ft/s

     (c)      $v = (1.486/n)(R^{2/3})(s^{1/2})$      $8.13 = (1.486/0.020)[(12)(2.05)/(2.05 + 12 + 2.05)]^{2/3}(s)^{1/2}$      $s = 0.00680$

**14.250** A trapezoidal channel with side slopes of 2 horizontal to 1 vertical is to carry a flow of 590 cfs. For a bottom width of 12 ft, calculate the (a) critical depth and (b) critical velocity.

▌ (a)      $Q^2/g = A^3/b$      $b = 12 + 4y$      $A = 12y + 2y^2$

     $590^2/32.2 = (12y + 2y^2)^3/(12 + 4y)$      $y = y_c = 3.46$ ft      (by trial and error)

     (b)      $v_c = \sqrt{gA/b} = \sqrt{(32.2)[(12)(3.46) + (2)(3.46)^2]/[12 + (4)(3.46)]} = 9.03$ ft/s

**14.251** A trapezoidal channel has a bottom width of 20 ft, side slopes of 1 to 1, and flows at a depth of 3.00 ft. For $n = 0.015$, and a discharge of 360 cfs, calculate (a) the normal slope, (b) the critical slope and critical depth for 360 cfs, and (c) the critical slope at the normal depth of 3.00 ft.

▌ (a)      $Q = (A)(1.486/n)(R^{2/3})(s^{1/2})$      $A = (20)(3) + (3)(3) = 69.00$ ft$^2$      $p_w = 20 + (2)(\sqrt{3^2 + 3^2}) = 28.49$ ft

     $360 = (69.00)(1.486/0.015)(69.00/28.49)^{2/3}(s)^{1/2}$      $s = 0.000853$

     (b)      $v_c = \sqrt{gA/b} = \sqrt{(32.2)(20y_c + y_c^2)/(20 + 2y_c)}$

Also,

     $v_c = Q/A_c = 360/(20y_c + y_c^2)$      $\sqrt{(32.2)(20y_c + y_c^2)/(20 + 2y_c)} = 360/(20y_c + y_c^2)$

     $y_c = 2.08$ ft      (by trial and error)

$360 = [(20)(2.08) + 2.08^2](1.486/0.015)\{[(20)(2.08) + 2.08^2]/[20 + (2)(\sqrt{2.08^2 + 2.08^2})]\}^{2/3}(s_c)^{1/2}$      $s_c = 0.00291$

     (c)      $v_c = \sqrt{(32.2)[(20)(3.00) + 3.00^2]/[20 + (2)(3.00)]} = 9.24$ ft/s      $v_c = (1.486/n)(R^{2/3})(s^{1/2})$

     $9.24 = (1.486/0.015)\{[(20)(3.00) + 3.00^2]/[20 + (2)(\sqrt{3.00^2 + 3.00^2})]\}^{2/3}(s_c)^{1/2}$      $s_c = 0.00267$

**14.252** A rectangular channel, 30 ft wide, carries 270 cfs when flowing 3.00 ft deep. (a) What is the specific energy? (b) Is the flow subcritical or supercritical?

▌ (a)      $v = Q/A = 270/[(3.00)(30)] = 3.00$ ft/s      $E = v^2/2g + y = 3.000^2/[(2)(32.2)] + 3.00 = 3.14$ ft

     (b)      $y_c = (q^2/g)^{1/3} = [(\frac{270}{30})^2/32.2]^{1/3} = 1.36$ ft

Since $[y_c = 1.36] < [y = 3.00]$, the flow is subcritical.

**14.253** A trapezoidal channel has a bottom width of 20 ft and side slopes of 2 horizontal to 1 vertical. When the depth of water is 3.50 ft, the flow is 370 cfs. (a) What is the specific energy? (b) Is the flow subcritical or supercritical?

▌ (a)      $A = (20)(3.50) + (3.50)[(2)(3.50)] = 94.50$ ft$^2$      $v = Q/A = 370/94.50 = 3.915$ ft/s

     $E = v^2/2g + y = 3.915^2/[(2)(32.2)] + 3.50 = 3.74$ ft

     (b) $Q^2/g = A_c^3/b$      $370^2/32.2 = \{20y_c + (y_c)[(2)(y_c)]\}^3/[20 + (2)(2y_c)]$      $y_c = 2.05$ ft      (by trial and error)

Since $[y_c = 2.05] < [y = 3.50$ ft$]$, the flow is subcritical.

**14.254** The discharge through a rectangular channel ($n = 0.012$) 15 ft wide is 400 cfs when the slope is 1 ft in 100 ft. Is the flow subcritical or supercritical?

▌      $y_c = (q^2/g)^{1/3} = [(\frac{400}{15})^2/32.2]^{1/3} = 2.81$ ft      $Q = (A)(1.486/n)(R^{2/3})(s^{1/2})$

     $400 = [(15)(2.81)](1.486/0.012)[(15)(2.81)/(2.81 + 15 + 2.81)]^{2/3}(s_c)^{1/2}$      $s_c = 0.00226$

Since $[s_c = 0.00226] < [s = \frac{1}{100}]$, the flow is supercritical.

**14.255** A rectangular channel, 10 ft wide, carries 400 cfs. (a) Tabulate depth of flow against specific energy for depths from 1 ft to 8 ft. (b) Determine the minimum specific energy. (c) What type of flow exists when the depth is 2 ft and when it is 8 ft? (d) For $C = 100$, what slopes are necessary to maintain the depths in (c)?

▌ **(a)** $E = v^2/2g + y = (Q/A)^2/2g + y = (Q/10y)^2/2g + y$. For $y = 1$ ft, $E = \{400/[(10)(1)]\}^2/[(2)(32.2)] + 1 = 25.8$ ft. For succeeding depths,

| y, ft | E, ft |
|-------|-------|
| 1 | 25.8 |
| 2 | 8.21 |
| 3 | 5.76 |
| 4 | 5.55 |
| 5 | 5.99 |
| 6 | 6.69 |
| 7 | 7.51 |
| 8 | 8.39 |

**(b)** $y_c = (q^2/g)^{1/3} = [(\frac{400}{10})^2/32.2]^{1/3} = 3.676$ ft    $E_{min} = \{400/[(10)(3.676)]\}^2/[(2)(32.2)] + 3.676 = 5.51$ ft

**(c)** Since $[y = 2] < [y_c = 3.676]$, the flow is supercritical for a 2-ft depth. Since $[y = 8] > [y_c = 3.676]$, the flow is subcritical for an 8-ft depth.

**(d)** $Q = CA\sqrt{Rs}$. For a 2-ft depth: $400 = (100)[(2)(10)]\sqrt{[(2)(10)/(2 + 10 + 2)](s)}$, $s = 0.0280$. For an 8-ft depth: $400 = (100)[(8)(10)]\sqrt{[(8)(10)/(8 + 10 + 8)](s)}$, $s = 0.000812$.

**14.256** A rectangular flume ($n = 0.012$) is laid on a slope of 0.0036 and carries 580 cfs. For critical-flow conditions, what width is required?

▌ $y_c = (q^2/g)^{1/3} = [(Q/b)^2/g]^{1/3}$. Try $b = 8.0$ ft:

$$y_c = [(580/8.0)^2/32.2]^{1/3} = 5.465 \text{ ft}$$

$Q = (A)(1.486/n)(R^{2/3})(s^{1/2}) = [(5.465)(8.0)](1.486/0.012)$

$$\times [(5.465)(8.0)/(5.465 + 8.0 + 5.465)]^{2/3}(0.0036)^{1/2} = 568 \text{ ft}^3/\text{s}$$

Try $b = 8.5$ ft:

$$y_c = [(580/8.5)^2/32.2]^{1/3} = 5.249 \text{ ft}$$

$$Q = [(5.249)(8.5)](1.486/0.012)[(5.249)(8.5)/(5.249 + 8.5 + 5.249)]^{2/3}(0.0036)^{1/2} = 586 \text{ ft}^3/\text{s}$$

Try $b = 8.33$ ft:

$$y_c = [(580/8.33)^2/32.2]^{1/3} = 5.320 \text{ ft}$$

$$Q = [(5.320)(8.33)](1.486/0.012)[(5.320)(8.33)/(5.320 + 8.33 + 5.320)]^{2/3}(0.0036)^{1/2} = 580 \text{ ft}^3/\text{s}$$

Hence, $b = 8.33$ ft.

**14.257** For a constant specific energy of 6.60 ft, what maximum flow may occur in a rectangular channel 10.0 ft wide?

▌    $y_c = (\frac{2}{3})(E) = (\frac{2}{3})(6.60) = 4.40$ ft    $v_c = \sqrt{gy_c} = \sqrt{(32.2)(4.40)} = 11.90$ ft/s

$$Q_{max} = A_c v_c = [(10.0)(4.40)](11.90) = 524 \text{ ft}^3/\text{s}$$

**14.258** A rectangular channel, 20 ft wide, $n = 0.025$, flows 5 ft deep on a slope of 14.7 ft in 10 000 ft. A suppressed weir $C$, 2.45 ft high, is built across the channel ($m = 3.45$). Taking the elevation of the bottom of the channel just upstream from the weir to be 100.00 ft, estimate (using one reach) the elevation of the water surface at a point $A$, 1000 ft upstream from the weir. See Fig. 14-64.

▌ $Q = (A)(1.486/n)(R^{2/3})(s^{1/2}) = [(20)(5)](1.486/0.025)[(20)(5)/(5 + 20 + 5)]^{2/3}(14.7/10\ 000)^{1/2} = 509 \text{ ft}^3/\text{s}$

Calculate the new elevation of the water surface at $B$ (before dropdown). Note that the flow is nonuniform since the depths, velocities, and areas are not constant after the weir is installed. Estimate a depth of 6 ft just upstream from the weir (i.e., at $B$). $v_{approach} = Q/A = 509/[(20)(6)] = 4.24$ ft. The applicable weir formula is $Q = mb[(H + v^2/2g)^{3/2} - (v^2/2g)^{3/2}]$.

$$509 = (3.45)(20)\left\{\left[H + \frac{4.24^2}{(2)(32.2)}\right]^{3/2} - \left[\frac{4.24^2}{(2)(32.2)}\right]^{3/2}\right\} \qquad H = 3.56 \text{ ft}$$

$$y_B = 3.56 + 2.45 = 6.01 \text{ ft} \qquad \text{(estimated depth of 6 ft O.K.)}$$

The new elevation at $A$ must lie between $101.47 + 5 = 106.47$ ft and $101.47 + 6 = 107.47$ ft. Try an elevation of 106.90 ft:

$$(A_A)_{new} = (20)(106.90 - 101.47) = 108.6 \text{ ft}^2 \qquad (v_A)_{new} = Q/A = 509/108.6 = 4.69 \text{ ft/s}$$

$$v_{mean} = (4.24 + 4.69)/2 = 4.46 \text{ ft/s} \qquad y_A = (106.90 - 101.47) = 5.43 \text{ ft}$$

$$R_{mean} = \frac{\frac{1}{2}[(6)(20) + 108.6]}{\frac{1}{2}[(6 + 20 + 6) + (5.43 + 20 + 5.43)]} = 3.637 \text{ ft}$$

$$h_L = (vn/1.486R^{2/3})^2(L) = \{(4.46)(0.025)/[(1.486)(3.637)^{2/3}]\}^2(1000) = 1.01 \text{ ft}$$

Check by the Bernoulli equation: $v_A^2/2g + z_A = v_B^2/2g + z_B + h_L$, $4.69^2/[(2)(32.2)] + 106.90 = 4.24^2/[(2)(32.2)] + 106.00 + 1.01$, $107.24 = 107.29$ (approximately). Further refinement is not necessary. Hence, use an elevation of 106.90 ft at $A$.

Fig. 14-64

**14.259** Develop a formula for the length–energy–slope relationship for nonuniform flow problems similar to Prob. 14.258.

▌ Energy at $1 -$ head loss = energy at 2, $(z_1 + y_1 + v_1^2/2g) - h_L = (z_2 + y_2 + v_2^2/2g)$. Let $s =$ slope of the energy line and $s_0 =$ slope of the channel bottom: $s = h_L/L$, $h_L = sL$, $s_0 = (z_1 - z_2)/L$, $z_1 - z_2 = s_0L$. Therefore, $s_0L + (y_1 - y_2) + (v_1^2/2g - v_2^2/2g) = sL$, $L = [(y_1 + v_1^2/2g) - (y_2 + v_2^2/2g)]/(s - s_0) = (E_1 - E_2)/(s - s_0)$.

**14.260** A rectangular flume ($n = 0.013$) is 6 ft wide and carries 66 cfs of water. At a certain section $F$, the depth is 3.20 ft. If the slope of the channel bed is constant at 0.000400, determine the distance from $F$ where the depth is 2.70 ft. (Use one reach.)

▌ Assume the depth is upstream from $F$. Let subscript 2 refer to point $F$ and subscript 1 to the other point.

$$L = [(y_1 + v_1^2/2g) - (y_2 + v_2^2/2g)]/(s - s_0) \qquad A_1 = (6)(2.70) = 16.20 \text{ ft}^2$$

$$v_1 = Q/A_1 = 66/16.20 = 4.074 \text{ ft/s}$$

$$R = A/p_w \qquad R_1 = 16.20/(2.70 + 6 + 2.70) = 1.421 \text{ ft}$$

$$A_2 = (6)(3.20) = 19.20 \text{ ft}^2 \qquad v_2 = Q/A_2 = 66/19.20 = 3.438 \text{ ft/s}$$

$$R_2 = 19.20/(3.20 + 6 + 3.20) = 1.548 \text{ ft} \qquad v_{mean} = (4.074 + 3.438)/2 = 3.756 \text{ ft/s}$$

$$R_{mean} = (1.421 + 1.548)/2 = 1.484 \text{ ft}$$

$$v = (1.486/n)(R^{2/3})(s^{1/2}) \qquad 3.756 = (1.486/0.013)(1.484)^{2/3}(s)^{1/2} \qquad s = 0.000638$$

$$L = \frac{\{2.70 + 4.074^2/[(2)(32.2)]\} - \{3.20 + 3.438^2/[(2)(32.2)]\}}{0.000638 - 0.000400} = -1789 \text{ ft}$$

The minus sign signifies that the section with the 2.70-ft depth is downstream from $F$, not upstream as assumed.

**14.261** A rectangular channel, 40 ft wide, carries 900 cfs of water. The slope of the channel is 0.00283. At section 1 the depth is 4.50 ft and at section 2, 300 ft downstream, the depth is 5.00 ft. What is the average value of roughness factor $n$?

▌
$$L = [(y_1 + v_1^2/2g) - (y_2 + v_2^2/2g)]/(s - s_0) \qquad A_1 = (40)(4.50) = 180.0 \text{ ft}^2$$

$$v_1 = Q/A_1 = 900/180.0 = 5.000 \text{ ft/s}$$

$$R = A/p_w \qquad R_1 = 180.0/(4.50 + 40 + 4.50) = 3.673 \text{ ft}$$

$$A_2 = (40)(5.00) = 200.0 \text{ ft}^2 \qquad v_2 = Q/A = 900/200.0 = 4.500 \text{ ft/s}$$

$$R_2 = 200.0/(5.00 + 40 + 5.00) = 4.000 \text{ ft} \qquad v_{mean} = (5.000 + 4.500)/2 = 4.750 \text{ ft/s}$$

$$R_{mean} = (3.673 + 4.000)/2 = 3.836 \text{ ft}$$

$$300 = \frac{\{4.50 + 5.000^2/[(2)(32.2)]\} - \{5.00 + 4.500^2/[(2)(32.2)]\}}{s - 0.00283} \qquad s = 0.001409$$

$$v = (1.486/n)(R^{2/3})(s^{1/2}) \qquad 4.750 = (1.486/n)(3.836)^{2/3}(0.001409)^{1/2} \qquad n = 0.0288$$

**14.262** A rectangular channel, 20 ft wide, has a slope of 1 ft per 1000 ft. The depth at section 1 is 8.50 ft and at section 2, 2000 ft downstream, the depth is 10.25 ft. If $n = 0.011$, determine the probable flow.

**∎** $\qquad E = y + v^2/2g + z \qquad E_1 = 8.50 + v_1^2/2g + (2000)(\frac{1}{1000}) \qquad E_2 = 10.25 + v_2^2/2g + 0$

$$s = \text{head loss}/L = [(10.50 - 10.25) + (v_1^2/2g - v_2^2/2g)]/2000$$

$$Q = (A)(1.486/n)(R^{2/3})(s^{1/2})$$

$$A_1 = (20)(8.50) = 170.0 \text{ ft}^2 \qquad R = A/p_w$$

$$R_1 = 170.0/(8.50 + 20 + 8.50) = 4.595 \text{ ft} \qquad A_2 = (20)(10.25) = 205.0 \text{ ft}^2$$

$$R_2 = 205.0/(10.25 + 20 + 10.25) = 5.062 \text{ ft} \qquad A_{mean} = (170.0 + 205.0)/2 = 187.5 \text{ ft}^2$$

$$R_{mean} = (4.595 + 5.062)/2 = 4.828 \text{ ft}$$

Assume $s = 0.000144$: $Q = (187.5)(1.486/0.011)(4.828)^{2/3}(0.000144)^{1/2} = 868.3 \text{ ft}^3/\text{s}$. Check on $s$:

$$v_1 = Q/A_1 = 868.3/170.0 = 5.108 \text{ ft/s} \qquad v_1^2/2g = 5.108^2/[(2)(32.2)] = 0.4052 \text{ ft}$$

$$v_2 = 868.3/205.0 = 4.236 \text{ ft/s}$$

$$v_2^2/2g = 4.236^2/[(2)(32.2)] = 0.2786 \text{ ft} \qquad s = [(10.50 - 10.25) + (0.4052 - 0.2786)]/2000 = 0.000188$$

This value of $s$ (0.000188) does not equal the assumed value (0.000144); hence, try $s = 0.000210$:
$Q = (187.5)(1.486/0.011)(4.828)^{2/3}(0.000210)^{1/2} = 1049 \text{ ft}^3/\text{s}$. Check on $s$:

$$v_1 = Q/A_1 = 1049/170.0 = 6.171 \text{ ft/s} \qquad v_1^2/2g = 6.171^2/[(2)(32.2)] = 0.5913 \text{ ft}$$

$$v_2 = 1049/205.0 = 5.117 \text{ ft/s}$$

$$v_2^2/2g = 5.117^2/[(2)(32.2)] = 0.4066 \text{ ft} \qquad s = [(10.50 - 10.25) + (0.5913 - 0.4066)]/2000 = 0.000217$$

This is close to the assumed value of $s$ of 0.000210; hence, approximate $Q = 1050 \text{ ft}^3/\text{s}$.

**14.263** A reservoir feeds a rectangular channel, 15 ft wide, $n = 0.015$, as shown in Fig. 14-65. At the entrance, the depth of water is 6.22 ft above the channel bottom. The flume is 800 ft long and drops 0.72 ft in this length. The depth behind a weir at the discharge end of the channel is 4.12 ft. Determine, using one reach, the capacity of the channel assuming the loss at the entrance to be $0.25v_1^2/2g$.

**∎** $\qquad p_A/\gamma + v_A^2/2g + z_A = p_1/\gamma + v_1^2/2g + z_1 + h_m \qquad 0 + 0 + 6.22 = 0 + v_1^2/2g + y_1 + 0.25v_1^2/2g$

$$L = [(v_2^2/2g + y_2) - (v_1^2/2g + y_1)]/(s_0 - s) \qquad s = (nv/1.486R^{2/3})^2$$

Solve these equations by successive trials until $L$ approximates or equals 800 ft. Try $y_1 = 5.0$ ft:

$$6.22 = v_1^2/[(2)(32.2)] + 5.0 + 0.25v_1^2/[(2)(32.2)] \qquad v_1 = 7.928 \text{ ft/s}$$

$$q = y_1 v_1 = (5.0)(7.928) = 39.64 \text{ ft}^3/\text{s/ft} \qquad v_2 = q/y_2 = 39.64/4.12 = 9.621 \text{ ft/s}$$

$$v_{mean} = (7.928 + 9.621)/2 = 8.774 \text{ ft/s}$$

$$R_1 = (15)(5.0)/(5.0 + 15 + 5.0) = 3.000 \text{ ft} \qquad R_2 = (15)(4.12)/(4.12 + 15 + 4.12) = 2.659 \text{ ft}$$

$$R_{mean} = (3.000 + 2.659)/2 = 2.830 \text{ ft}$$

$$s = \{(0.015)(8.774)/[(1.486)(2.830)^{2/3}]\}^2 = 0.001960$$

$$L = \frac{\{(9.621)^2/[(2)(32.2)] + 4.12\} - \{(7.928)^2/[(2)(32.2)] + 5.0\}}{0.72/800 - 0.001960} = 395 \text{ ft}$$

Since $L$ is not equal to 800 ft, try $y_1 = 5.21$ ft:

$$6.22 = v_1^2/[(2)(32.2)] + 5.21 + 0.25v_1^2/[(2)(32.2)] \qquad v_1 = 7.214 \text{ ft/s}$$

$$q = y_1 v_1 = (5.21)(7.214) = 37.58 \text{ ft}^3/\text{s/ft}$$

$$v_2 = q/y_2 = 37.58/4.12 = 9.121 \text{ ft/s} \qquad v_{mean} = (7.214 + 9.121)/2 = 8.168 \text{ ft/s}$$

$$R_1 = (15)(5.21)/(5.21 + 15 + 5.21) = 3.074 \text{ ft}$$

$$R_2 = (15)(4.12)/(4.12 + 15 + 4.12) = 2.659 \text{ ft} \qquad R_{\text{mean}} = (3.074 + 2.659)/2 = 2.866 \text{ ft}$$

$$s = \{(0.015)(8.168)/[(1.486)(2.866)^{2/3}]\}^2 = 0.001670$$

$$L = \frac{\{(9.121)^2/[(2)(32.2)] + 4.12\} - \{(7.214)^2/[(2)(32.2)] + 5.21\}}{0.72/800 - 0.001670} = 787 \text{ ft}$$

$L$ is not exactly equal to 800 ft, but additional computations (not shown here) show that 5.21 ft is the best value of $y_1$ to the nearest hundredth of a foot. Hence, $Q = (37.58)(15) = 564 \text{ ft}^3/\text{s}$.

**Fig. 14-65**

**14.264** A rectangular concrete channel 15 ft wide carries water as shown in Fig. 14-66. The channel bed slope is 0.0010. Find the theoretical rate of flow in the channel.

**I**

$$v_1^2/2g + d_1 + s_0L = v_2^2/2g + d_2 + sL \qquad v_1 = Q/[(15)(5.1)] = 0.01307Q$$

$$v_2 = Q/[(15)(3.9)] = 0.01709Q$$

$$s = (nv_m/1.486R_m^{2/3})^2 \qquad v_m = (0.01307Q + 0.01709Q)/2 = 0.01508Q \qquad R = A/p_w$$

$$R_1 = (15)(5.1)/(5.1 + 15 + 5.1) = 3.036 \text{ ft} \qquad R_2 = (15)(3.9)/(3.9 + 15 + 3.9) = 2.566 \text{ ft}$$

$$R_m = (3.036 + 2.566)/2 = 2.801 \text{ ft}$$

$$s = \{(0.013)(0.01508Q)/[(1.486)(2.801)^{2/3}]\}^2 = 4.408 \times 10^{-9}Q^2$$

$$(0.01307Q)^2/[(2)(32.2)] + 5.1 + (0.0010)(1000)$$

$$= (0.01709Q)^2/[(2)(32.2)] + 3.9 + (4.408 \times 10^{-9}Q^2)(1000) \qquad Q = 591 \text{ ft}^3/\text{s}$$

Water surface

$d_1 = 5.1$ ft

$d_2 = 3.9$ ft

1000 ft

**Fig. 14-66**

**14.265** Water flowing at the normal depth in a rectangular concrete channel that is 12.0 m wide encounters an obstruction, as shown in Fig. 14-67, causing the water level to rise above the normal depth at the obstruction and for some distance upstream. The water discharge is 126 m³/s and the channel bottom slope is 0.00086. The depth of water just upstream from the obstruction ($d_0$) is 4.55 m. Find the distance upstream to the point where the water surface is at the normal depth.

**I**

$$d_c = (q^2/g)^{1/3} = [(126/12.0)^2/9.807]^{1/3} = 2.24 \text{ m} \qquad Q = (A)(1.0/n)(R^{2/3})(s^{1/2})$$

$$126 = (12.0d)(1.0/0.013)[12.0d/(d + 12.0 + d)]^{2/3}(0.00086)^{1/2}$$

$$2.256[12.0d/(d + 12.0 + d)]^{2/3} - 10.5/d = 0$$

$$d = 2.95 \text{ m} \qquad \text{(by trial and error)}$$

Since $d > d_c$, flow is subcritical, and computations should proceed upstream. The problem now is to determine the distance from the point where the depth is 4.55 m to the point upstream where the depth is 2.95 m. This will be done in ten equal depth increments of 0.16 m. The computations are given in the table below.

| (1)<br>$d$, m | (2)<br>$v$, m/s<br>$\dfrac{126}{12.0 \times (1)}$ | (3)<br>$v_m$, m/s | (4)<br>$v^2/2g$, m<br>$\dfrac{(2)^2}{2 \times 9.807}$ | (5)<br>$R$, m<br>$\dfrac{12.0 \times (1)}{12.0 + 2 \times (1)}$ | (6)<br>$R_m$, m | (7)<br>$s$<br>$\left[\dfrac{0.013 \times (3)^2}{(6)^{2/3}}\right]^2$ | (8)<br>$L$, m<br>$\dfrac{[(4) + (1)]_2 - [(4) + (1)]_1}{0.00086 - (7)}$ |
|---|---|---|---|---|---|---|---|
| 4.55 | 2.308 |       | 0.2716 | 2.588 |       |           |        |
|      |       | 2.350 |        |       | 2.562 | 0.0002662 | −236   |
| 4.39 | 2.392 |       | 0.2917 | 2.535 |       |           |        |
|      |       | 2.437 |        |       | 2.508 | 0.0002946 | −243   |
| 4.23 | 2.482 |       | 0.3141 | 2.481 |       |           |        |
|      |       | 2.531 |        |       | 2.453 | 0.0003272 | −253   |
| 4.07 | 2.580 |       | 0.3394 | 2.425 |       |           |        |
|      |       | 2.633 |        |       | 2.396 | 0.0003654 | −266   |
| 3.91 | 2.685 |       | 0.3676 | 2.367 |       |           |        |
|      |       | 2.743 |        |       | 2.338 | 0.0004098 | −284   |
| 3.75 | 2.800 |       | 0.3997 | 2.308 |       |           |        |
|      |       | 2.863 |        |       | 2.277 | 0.0004626 | −311   |
| 3.59 | 2.925 |       | 0.4362 | 2.246 |       |           |        |
|      |       | 2.993 |        |       | 2.214 | 0.0005246 | −353   |
| 3.43 | 3.061 |       | 0.4777 | 2.182 |       |           |        |
|      |       | 3.136 |        |       | 2.150 | 0.0005989 | −429   |
| 3.27 | 3.211 |       | 0.5257 | 2.117 |       |           |        |
|      |       | 3.294 |        |       | 2.083 | 0.0006893 | −613   |
| 3.11 | 3.376 |       | 0.5811 | 2.048 |       |           |        |
|      |       | 3.468 |        |       | 2.013 | 0.0007997 | −1580  |
| 2.95 | 3.559 |       | 0.6458 | 1.978 |       |           |        |
|      |       |       |        |       |       |           | −4568 m |

Hence, the answer to the problem is 4568 m.

Upper end of
the channel where
the flow is uniform

Water surface

Fig. 14-67

**14.266** Water flows in a rectangular concrete channel that is 5.0 ft wide, as shown in Fig. 14-68a, at a discharge of 16.5 cfs. Find the water-surface profile through the channel.

▮ $d_c = (q^2/g)^{1/3} = [(16.5/5.0)^2/32.2]^{1/3} = 0.70$ ft. In segment $AB$,

$$Q = (A)(1.486/n)(R^{2/3})(s^{1/2})$$

$$16.5 = (5.0d)(1.486/0.013)[5.0d/(d + 5.0 + d)]^{2/3}(0.00040)^{1/2}$$

$$2.286[5.0d/(d + 5.0 + d)]^{2/3} - 3.300/d = 0 \qquad d = 1.50 \text{ ft} \qquad \text{(by trial and error)}$$

Since $d > d_c$, the flow in segment $AB$ is subcritical.

In segment $BC$,

$$16.5 = (5.0d)(1.486/0.013)[5.0d/(d + 5.0 + d)]^{2/3}(0.025)^{1/2}$$

$$18.07[5.0d/(d + 5.0 + d)]^{2/3} - 3.300/d = 0$$

$$d = 0.38 \text{ ft} \qquad \text{(by trial and error)}$$

Since $d < d_c$, the flow in segment $BC$ is supercritical.

(a)

(b)

(c)

Fig. 14-68

Figure 14-68b shows the location of the critical depth (the dashed line), which is constant throughout, along with the normal depths of flow in segment AB and segment BC. Obviously, the water-surface profile cannot drop instantaneously at point B from the normal depth in segment AB (1.50 ft) to the normal depth in segment BC (0.38 ft); there must be a transition zone on both sides of point B as shown by the dashed line in Fig. 14-68b.

As a matter of fact, as the flow changes from subcritical to supercritical in going from segment AB to segment BC, it passes through the critical state at point B. Hence, the depth of flow at point B will be 0.70 ft. The problem now becomes one of determining the flow profile from the critical depth of 0.70 ft at point B upstream to the point where the normal depth of 1.50 ft is reached and downstream to the point where the normal depth of 0.38 ft is reached. These computations are carried out in the tables below. Each profile will be analyzed using five equal depth increments.

**For Segment AB**

| (1) d, ft | (2) $v$, ft/s $\dfrac{16.5}{5.0 \times (1)}$ | (3) $v_m$, ft/s | (4) $v^2/2g$, ft $\dfrac{(2)^2}{2 \times 32.2}$ | (5) $R$, ft $\dfrac{5.0 \times (1)}{5.0 + 2 \times (1)}$ | (6) $R_m$, ft | (7) $s$ $\left[\dfrac{0.013 \times (3)}{1.486 \times (6)^{2/3}}\right]^2$ | (8) $L$, ft $\dfrac{[(4)+(1)]_2 - [(4)+(1)]_1}{0.00040 - (7)}$ |
|---|---|---|---|---|---|---|---|
| 0.70 | 4.714 | | 0.3451 | 0.5469 | | | |
| | | 4.276 | | | 0.5934 | 0.0028063 | −18 |
| 0.86 | 3.837 | | 0.2286 | 0.6399 | | | |
| | | 3.536 | | | 0.6822 | 0.0015934 | −79 |
| 1.02 | 3.235 | | 0.1625 | 0.7244 | | | |
| | | 3.016 | | | 0.7630 | 0.0009985 | −199 |
| 1.18 | 2.797 | | 0.1215 | 0.8016 | | | |
| | | 2.630 | | | 0.8370 | 0.0006711 | −489 |
| 1.34 | 2.463 | | 0.0942 | 0.8724 | | | |
| | | 2.332 | | | 0.9050 | 0.0004755 | −1868 |
| 1.50 | 2.200 | | 0.0752 | 0.9375 | | | |

**For Segment BC**

| (1) d, ft | (2) $v$, ft/s $\dfrac{16.5}{5.0 \times (1)}$ | (3) $v_m$, ft/s | (4) $v^2/2g$, ft $\dfrac{(2)^2}{2 \times 32.2}$ | (5) $R$, ft $\dfrac{5.0 \times (1)}{5.0 + 2 \times (1)}$ | (6) $R_m$, ft | (7) $s$ $\left[\dfrac{0.013 \times (3)}{1.486 \times (6)^{2/3}}\right]^2$ | (8) $L$, ft $\dfrac{[(4)+(1)]_2 - [(4)+(1)]_1}{0.025 - (7)}$ |
|---|---|---|---|---|---|---|---|
| 0.70 | 4.714 | | 0.3451 | 0.5469 | | | |
| | | 4.935 | | | 0.5283 | 0.004364 | 0.4 |
| 0.64 | 5.156 | | 0.4128 | 0.5096 | | | |
| | | 5.473 | | | 0.4869 | 0.005985 | 2.0 |
| 0.57 | 5.789 | | 0.5204 | 0.4642 | | | |
| | | 6.130 | | | 0.4439 | 0.008493 | 4.2 |
| 0.51 | 6.471 | | 0.6502 | 0.4236 | | | |
| | | 6.986 | | | 0.3989 | 0.012720 | 12.5 |
| 0.44 | 7.500 | | 0.8734 | 0.3741 | | | |
| | | 8.092 | | | 0.3520 | 0.020164 | 49.1 |
| 0.38 | 8.684 | | 1.1710 | 0.3299 | | | |

Based on the values computed above, the water-surface profile is illustrated in Fig. 14-68c.

**14.267** Prepare a computer program to solve nonuniform flow problems like those of Probs. 14.265 and 14.266. The program should handle data in both the English Gravitational Unit System and the International System of Units.

```
C     THIS PROGRAM DETERMINES THE FLOW PROFILE FOR A NON-UNIFORM FLOW
C     IN A RECTANGULAR OPEN CHANNEL.  IT CAN BE USED FOR PROBLEMS IN
C     BOTH THE ENGLISH SYSTEM OF UNITS AND THE INTERNATIONAL SYSTEM OF
C     UNITS.
C
C     THE PROGRAM CONSIDERS TWO ADJACENT CHANNEL SEGMENTS.  CONSIDER
C     ONE SEGMENT TO GO FROM POINT "A" TO POINT "B" AND THE OTHER
C     SEGMENT TO GO FROM POINT "B" TO POINT "C" IN THE DOWNSTREAM
C     DIRECTION.  THE PROGRAM COMPUTES THE FLOW PROFILE IN SEGMENT AB
C     IN THE UPSTREAM DIRECTION (I.E., FROM "B" TO "A") BASED ON SUB-
C     CRITICAL FLOW IN SEGMENT AB AND THE FLOW PROFILE IN SEGMENT BC IN
C     THE DOWNSTREAM DIRECTION (I.E., FROM "B" TO "C") BASED ON SUPER-
C     CRITICAL FLOW IN SEGMENT BC.  THE PROGRAM CAN BE USED EITHER FOR
C     TWO SEGMENTS AS DESCRIBED ABOVE OR FOR A SINGLE SEGMENT WITH
C     COMPUTATIONS TO PROCEED UPSTREAM (SEGMENT AB) OR DOWNSTREAM (SEG-
C     MENT BC), AS DESIRED.  IN THE CASE OF A SINGLE SEGMENT, ENTER DATA
C     FOR SEGMENT AB IF FLOW IS SUBCRITICAL AND FOR SEGMENT BC IF FLOW
C     IS SUPERCRITICAL.
C
C     THIS PROGRAM IS BASED ON CONSTANT CHANNEL WIDTH, CONSTANT FLOW
C     RATE, AND CONSTANT MANNING N-VALUE THROUGHOUT AND ON A SEPARATE
C     CONSTANT CHANNEL SLOPE IN EACH OF SEGMENTS AB AND BC.
C
C     INPUT DATA MUST BE SET UP AS FOLLOWS.
C
C     CARD 1    COLUMN 1          ENTER 0 (ZERO) OR BLANK IF THE ENGLISH
C                                 SYSTEM OF UNITS IS TO BE USED.  ENTER 1
C                                 (ONE) IF THE INTERNATIONAL SYSTEM OF
C                                 UNITS IS TO BE USED.
C               COLUMN 2          ENTER 1 (ONE) IF ONLY A SINGLE UPSTREAM
C                                 COMPUTATION IS DESIRED.  (IN THIS CASE,
C                                 ENTER DEPTHS AT POINTS A AND B AND SLOPE
C                                 IN SEGMENT AB.  LEAVE DEPTH AT POINT C
C                                 AND SLOPE IN SEGMENT BC BLANK.)  ENTER 2
C                                 (TWO) IF ONLY A SINGLE DOWNSTREAM COM-
C                                 PUTATION IS DESIRED.  (IN THIS CASE, ENTER
C                                 DEPTHS AT POINTS B AND C AND SLOPE IN
C                                 SEGMENT BC.  LEAVE DEPTH AT POINT A AND
C                                 SLOPE IN SEGMENT AB BLANK.)  ENTER 3
C                                 (THREE) IF COMPUTATIONS FOR BOTH SEGMENTS
C                                 ARE DESIRED. (IN THIS CASE, ENTER DEPTHS
C                                 AT POINTS A, B, AND C AND SLOPES IN SEG-
C                                 MENTS AB AND BC.)
C               COLUMNS 3-5       ENTER INTEGER NUMBER (RIGHT ADJUSTED)
C                                 GIVING NUMBER OF LENGTH INCREMENTS TO BE
C                                 USED IN COMPUTING THE FLOW PROFILE IN
C                                 EACH SEGMENT.
C               COLUMNS 6-80      ENTER TITLE, DATE, AND OTHER INFORMATION,
C                                 IF DESIRED.
C     CARD 2    COLUMNS 1-10      ENTER NUMBER INCLUDING DECIMAL GIVING
C                                 WIDTH OF RECTANGULAR CHANNEL (IN FEET OR
C                                 METERS).
C               COLUMNS 11-20     ENTER NUMBER INCLUDING DECIMAL GIVING
C                                 FLOW RATE (IN CUBIC FEET PER SECOND OR
C                                 CUBIC METERS PER SECOND).
C               COLUMNS 21-30     ENTER NUMBER INCLUDING DECIMAL GIVING
C                                 DEPTH AT POINT A (IN FEET OR METERS).
C                                 THIS VALUE MAY BE LEFT BLANK, IN WHICH
C                                 CASE THE "NORMAL DEPTH" IN SEGMENT AB
C                                 WILL AUTOMATICALLY BE USED.
C               COLUMNS 31-40     ENTER NUMBER INCLUDING DECIMAL GIVING
C                                 DEPTH AT POINT B (IN FEET OR METERS).
C                                 THIS VALUE MAY BE LEFT BLANK, IN WHICH
C                                 CASE THE "CRITICAL DEPTH" WILL AUTO-
C                                 MATICALLY BE USED.
C               COLUMNS 41-50     ENTER NUMBER INCLUDING DECIMAL GIVING
C                                 DEPTH AT POINT C (IN FEET OR METERS).
C                                 THIS VALUE MAY BE LEFT BLANK, IN WHICH
C                                 CASE THE "NORMAL DEPTH" IN SEGMENT BC
C                                 WILL AUTOMATICALLY BE USED.
C               COLUMNS 51-60     ENTER NUMBER INCLUDING DECIMAL GIVING
C                                 CHANNEL SLOPE IN SEGMENT AB.
C               COLUMNS 61-70     ENTER NUMBER INCLUDING DECIMAL GIVING
```

```
C                           CHANNEL SLOPE IN SEGMENT BC.
C               COLUMNS 71-80   ENTER NUMBER INCLUDING DECIMAL GIVING
C                               MANNING N-VALUE.
C
C       MULTIPLE DATA SETS FOR SOLVING ANY NUMBER OF PROBLEMS MAY BE
C       INCLUDED FOR PROCESSING.
C
        COMMON DMID,DUP,NSEG,Q,W,N,COEFF,G,SUP
        DIMENSION TITLE(13)
        REAL N
        INTEGER UNITS,CODE
1       READ(5,100,END=2)UNITS,CODE,NSEG,TITLE
  100   FORMAT(2I1,I3,12A6,A3)
        WRITE(6,106)TITLE
  106   FORMAT('1',12A6,A3,////)
        READ(5,101)W,Q,DUP,DMID,DDOWN,SUP,SDOWN,N
  101   FORMAT(8F10.0)
        COEFF=1.486
        IF(UNITS.EQ.1)COEFF=1.0
        G=32.2
        IF(UNITS.EQ.1)G=9.807
        IF(DMID.LT.0.0001)DMID=((Q/W)**2/G)**(1.0/3.0)
        IF(CODE.EQ.2)GO TO 110
        IF(DUP.LT.0.0001)DUP=DNORM(W,Q,SUP,N,COEFF)
        IF(CODE.EQ.1)GO TO 109
  110   IF(DDOWN.LT.0.0001)DDOWN=DNORM(W,Q,SDOWN,N,COEFF)
        IF(CODE.EQ.2)GO TO 105
  109   X1=' UPST'
        X2='REAM '
        IF(UNITS.EQ.0)WRITE(6,107)X1,X2,W,Q,DUP,DMID,SUP,N
  107   FORMAT(1X,'GIVEN DATA FOR ' ,2A5,' FLOW PROFILE FOR A RECTANGULAR
       *OPEN CHANNEL',//5X,'WIDTH OF CHANNEL =',F7.1,' FT',6X,'FLOW RATE O
       *F WATER =',F7.1,' CU FT/S',//5X,'DEPTH OF WATER AT UPSTREAM END OF S
       * SEGMENT =',F7.2,' FT',//5X,'DEPTH OF WATER AT DOWNSTREAM END OF S
       *EGMENT =',F7.2,' FT',//5X,'SLOPE =',F10.7,20X,'MANNING N-VALUE =',
       *F6.3,///1X,'THE FLOW PROFILE WITHIN THE SEGMENT IS GIVEN IN THE TA
       *BLE BELOW'///4X,'DEPTH (FT)',10X,'VELOCITY (FT/S)',10X,'LENGTH OF
       *SUBSEGMENT (FT)')
        IF(UNITS.EQ.1)WRITE(6,108)X1,X2,W,Q,DUP,DMID,SUP,N
  108   FORMAT(1X,'GIVEN DATA FOR ' ,2A5,' FLOW PROFILE FOR A RECTANGULAR
       *OPEN CHANNEL',//5X,'WIDTH OF CHANNEL =',F7.1,' M',6X,'FLOW RATE OF
       * WATER =',F7.1,' CU M/S',//5X,'DEPTH OF WATER AT UPSTREAM END OF S
       *EGMENT =',F7.2,' M',//5X,'DEPTH OF WATER AT DOWNSTREAM END OF SEGM
       *ENT =',F7.2,' M',//5X,'SLOPE =',F10.7,20X,'MANNING N-VALUE =',
       *F6.3,///1X,'THE FLOW PROFILE WITHIN THE SEGMENT IS GIVEN IN THE TA
       *BLE BELOW'///4X,'DEPTH (M)',10X,'VELOCITY (M/S)',10X,'LENGTH OF SU
       *BSEGMENT (M)')
        CALL LENGTH
        IF(CODE.EQ.1)GO TO 1
  105   X1='DOWNS'
        X2='TREAM'
        IF(UNITS.EQ.0)WRITE(6,107)X1,X2,W,Q,DMID,DDOWN,SDOWN,N
        IF(UNITS.EQ.1)WRITE(6,108)X1,X2,W,Q,DMID,DDOWN,SDOWN,N
        DUP=DDOWN
        SUP=SDOWN
        CALL LENGTH
        GO TO 1
2       STOP
        END
        FUNCTION DNORM(W,Q,S,N,COEFF)
        D=0.001
        TRY1=COEFF/N*(W*D/(W+2.0*D))**(2.0/3.0)*SQRT(S)-Q/W/D
  104   D=D+0.001
        TRY2=COEFF/N*(W*D/(W+2.0*D))**(2.0/3.0)*SQRT(S)-Q/W/D
        IF(TRY1*TRY2)102,102,103
  103   TRY1=TRY2
        GO TO 104
  102   DNORM=D-0.0005
        RETURN
        END
        SUBROUTINE LENGTH
        COMMON DMID,DUP,NSEG,Q,W,N,COEFF,G,SUP
        DINC=(DMID-DUP)/FLOAT(NSEG)
```

```
            D1=DMID
            TOTAL=0.0
            DO 102  J=1,NSEG
            D2=D1-DINC
            V1=Q/W/D1
            V2=Q/W/D2
            VMEAN=(V1+V2)/2.0
            V2G1=V1**2/2.0/G
            V2G2=V2**2/2.0/G
            HR1=W*D1/(W+2.0*D1)
            HR2=W*D2/(W+2.0*D2)
            HRMEAN=(HR1+HR2)/2.0
            SMEAN=(N*VMEAN/COEFF/HRMEAN**(2.0/3.0))**2
            SEGL=(V2G2+D2-(V2G1+D1))/(SUP-SMEAN)
            TOTAL=TOTAL+SEGL
            WRITE(6,103)D1,V1,SEGL
      103   FORMAT(1X,F10.2,11X,F12.3,/,55X,F8.1)
      102   D1=D2
            WRITE(6,103)D2,V2
            WRITE(6,104)TOTAL
      104   FORMAT(41X,'TOTAL LENGTH =',F8.1,//)
            RETURN
            END
```

**14.268**  Solve Prob. 14.265 using the computer program developed in Prob. 14.267.

**❚ Input**

```
1 2 3 4 5 6 7 8 9 10 11 12 13 14 15 16 17 18 19 20 21 22 23 24 25 26 27 28 29 30 31 32 33 34 35 36 37 38 39 40 41 42 43 44 45 46 47 48 49 50 51 52 53 54 55 56 57 58 59 60 61 62 63 64 65 66 67 68 69 70 71 72 73 74 75 76 77 78 79 80
11 10SAMPLE ANALYSIS ØF PRØFILE IN NØN-UNIFØRM FLØW
12.0       126.0              4.55           0.00086          0.013
```

**Output**

SAMPLE ANALYSIS OF PROFILE IN NON-UNIFORM FLOW

GIVEN DATA FOR  UPSTREAM  FLOW PROFILE FOR A RECTANGULAR OPEN CHANNEL

WIDTH OF CHANNEL =   12.0 M     FLOW RATE OF WATER =  126.0 CU M/S

DEPTH OF WATER AT UPSTREAM END OF SEGMENT =   2.95 M

DEPTH OF WATER AT DOWNSTREAM END OF SEGMENT =   4.55 M

SLOPE = 0.0008600              MANNING N-VALUE = 0.013

THE FLOW PROFILE WITHIN THE SEGMENT IS GIVEN IN THE TABLE BELOW

| DEPTH (M) | VELOCITY (M/S) | LENGTH OF SUBSEGMENT (M) |
|-----------|----------------|--------------------------|
| 4.55      | 2.308          |                          |
|           |                | -235.2                   |
| 4.39      | 2.392          |                          |
|           |                | -242.8                   |
| 4.23      | 2.482          |                          |
|           |                | -252.6                   |
| 4.07      | 2.579          |                          |
|           |                | -265.6                   |
| 3.91      | 2.685          |                          |
|           |                | -283.6                   |
| 3.75      | 2.799          |                          |
|           |                | -309.9                   |
| 3.59      | 2.924          |                          |
|           |                | -351.8                   |
| 3.43      | 3.060          |                          |
|           |                | -427.9                   |
| 3.27      | 3.209          |                          |
|           |                | -607.7                   |
| 3.11      | 3.374          |                          |
|           |                | -1530.9                  |
| 2.95      | 3.556          |                          |

TOTAL LENGTH = -4508.0

**14.269**  Solve Prob. 14.266 using the computer program developed in Prob. 14.267.

**❙ Input**

```
1 2 3 4 5 6 7 8 9 10111213141516171819202122232425262728293031323334353637383940414243444546474849505152535455565758596061626364656667686970717273747576777879 80
03  5SAMPLE ANALYSIS ØF PRØFILE IN NØN-UNIFØRM FLØW
5.0       16.5                                        0.00040   0.025     0.013
```

**Output**

SAMPLE ANALYSIS OF PROFILE IN NON-UNIFORM FLOW

GIVEN DATA FOR  UPSTREAM  FLOW PROFILE FOR A RECTANGULAR OPEN CHANNEL

    WIDTH OF CHANNEL =    5.0 FT     FLOW RATE OF WATER =   16.5 CU FT/S

    DEPTH OF WATER AT UPSTREAM END OF SEGMENT =   1.50 FT

    DEPTH OF WATER AT DOWNSTREAM END OF SEGMENT =   0.70 FT

    SLOPE = 0.0004000                  MANNING N-VALUE = 0.013

THE FLOW PROFILE WITHIN THE SEGMENT IS GIVEN IN THE TABLE BELOW

| DEPTH (FT) | VELOCITY (FT/S) | LENGTH OF SUBSEGMENT (FT) |
|---|---|---|
| 0.70 | 4.736 | |
| | | -17.6 |
| 0.86 | 3.845 | |
| | | -78.9 |
| 1.02 | 3.236 | |
| | | -201.2 |
| 1.18 | 2.793 | |
| | | -501.2 |
| 1.34 | 2.457 | |
| | | -1986.4 |
| 1.50 | 2.193 | |

                TOTAL LENGTH =  -2785.2

GIVEN DATA FOR DOWNSTREAM FLOW PROFILE FOR A RECTANGULAR OPEN CHANNEL

    WIDTH OF CHANNEL =    5.0 FT     FLOW RATE OF WATER =   16.5 CU FT/S

    DEPTH OF WATER AT UPSTREAM END OF SEGMENT =   0.70 FT

    DEPTH OF WATER AT DOWNSTREAM END OF SEGMENT =   0.38 FT

    SLOPE = 0.0250000                  MANNING N-VALUE = 0.013

THE FLOW PROFILE WITHIN THE SEGMENT IS GIVEN IN THE TABLE BELOW

| DEPTH (FT) | VELOCITY (FT/S) | LENGTH OF SUBSEGMENT (FT) |
|---|---|---|
| 0.70 | 4.736 | |
| | | 0.5 |
| 0.63 | 5.208 | |
| | | 1.9 |
| 0.57 | 5.783 | |
| | | 4.5 |
| 0.51 | 6.501 | |
| | | 11.0 |
| 0.44 | 7.423 | |
| | | 46.1 |
| 0.38 | 8.650 | |

                TOTAL LENGTH =   63.9

**14.270** A rectangular channel 12.0 m wide is laid on a slope of 0.0028. The depth of flow at one section is 1.50 m, while the depth of flow at another section 500 ft downstream is 1.80 m. Determine the probable rate of flow, if $n = 0.026$.

**I**

$$Q = A_m v_m = (A_m)(1.0/n)R_m^{2/3}s^{1/2} \qquad A_m = [(12.0)(1.50) + (12.0)(1.80)]/2 = 19.8 \text{ m}^2$$

$$R_m = [(12.0)(1.50)/(1.50 + 12.0 + 1.50) + (12.0)(1.80)/(1.80 + 12.0 + 1.80)]/2 = 1.29 \text{ m}$$

$$Q = (19.8)(1.0/0.026)(1.29)^{2/3}s^{1/2} = 902.44s^{1/2} \qquad (1)$$

$$v_1^2/2g + d_1 + s_0 L = v_2^2/2g + d_2 + sL \qquad v_1 = Q/A_1 = Q/[(12.0)(1.50)] = Q/18.0$$

$$v_2 = Q/A_2 = Q/[(12.0)(1.80)] = Q/21.6$$

$$(Q/18.0)^2/[(2)(9.807)] + 1.50 + (0.0028)(500) = (Q/21.6)^2/[(2)(9.807)] + 1.80 + 500s \qquad (2)$$

Substituting (1) into (2) gives $(902.44s^{1/2}/18.0)^2/[(2)(9.807)] + 1.50 + 1.40 = (902.44s^{1/2}/21.6)^2[(2)(9.807)] + 1.80 + 500s$, $460.8s = 1.10$, $s = 0.002387$; $Q = (902.44)(0.002387)^{1/2} = 44.1 \text{ m}^3/\text{s}$.

**14.271** Solve Prob. 14.265 using five equal depth increments to determine the distance upstream to the point where the water surface is at the normal depth. Compare the answer with the one obtained in Prob. 14.265.

**I** Using data found in Prob. 14.265,

| $d$ | $v$ | $v_m$ | $v^2/2g$ | $R$ | $R_m$ | $s_g$ | $L, \text{m}$ |
|-----|-----|-------|----------|-----|-------|-------|---------------|
| 4.55 | 2.308 | | 0.2716 | 2.588 | | | |
| | | 2.395 | | | 2.535 | 0.0002805 | −479 |
| 4.23 | 2.482 | | 0.3141 | 2.481 | | | |
| | | 2.584 | | | 2.424 | 0.0003465 | −519 |
| 3.91 | 2.685 | | 0.3676 | 2.367 | | | |
| | | 2.805 | | | 2.307 | 0.0004362 | −593 |
| 3.59 | 2.925 | | 0.4362 | 2.246 | | | |
| | | 3.068 | | | 2.182 | 0.0005621 | −774 |
| 3.27 | 3.211 | | 0.5257 | 2.117 | | | |
| | | 3.385 | | | 2.048 | 0.0007446 | −1732 |
| 2.95 | 3.559 | | 0.6458 | 1.978 | | | |
| | | | | | | | −4097 |

The answer using five increments is $(4568 − 4097)/4568 = 0.103$, or 10.3 percent smaller than that using ten increments.

**14.272** Water flows in a rectangular concrete channel similar to the one depicted in Fig. 14-68a. If the channel width is 3.0 ft and the discharge is 12.0 cfs, determine the water-surface profile throughout the channel shown.

**I**

$$d_c = [(Q/B)^2/g]^{1/3} = [(12.0/3.0)^2/32.2]^{1/3} = 0.79 \text{ ft} \qquad Q = Av = (A)(1.486/n)(R^{2/3})(s^{1/2})$$

$$12.0 = (3.0d_{AB})(1.486/0.013)[3.0d_{AB}/(3.0 + 2d_{AB})]^{2/3}(0.00052)^{1/2}$$

By trial and error, $d_{AB} = 1.76$ ft. $12.0 = (3.0d_{BC})(1.486/0.013)[3.0d_{BC}/(3.0 + 2d_{BC})]^{2/3}(0.030)^{1/2}$. By trial and error, $d_{BC} = 0.42$ ft. Hence, the water-surface profile changes from the normal depth in segment $AB$ (1.76 ft) to the critical depth at $B$ (0.79 ft) to the normal depth in segment $BC$ (0.42 ft).

**For Segment AB (B to A)**

| d | v | $v_m$ | $v^2/2g$ | R | $R_m$ | $s_g$ | L, ft |
|---|---|---|---|---|---|---|---|
| 0.79 | 5.063 | | 0.398 | 0.517 | | | |
| | | 4.572 | | | 0.555 | 0.003508 | −17 |
| 0.98 | 4.082 | | 0.259 | 0.593 | | | |
| | | 3.736 | | | 0.626 | 0.001995 | −81 |
| 1.18 | 3.390 | | 0.178 | 0.660 | | | |
| | | 3.155 | | | 0.688 | 0.001254 | −196 |
| 1.37 | 2.920 | | 0.132 | 0.716 | | | |
| | | 2.734 | | | 0.742 | 0.0008516 | −510 |
| 1.57 | 2.548 | | 0.101 | 0.767 | | | |
| | | 2.410 | | | 0.788 | 0.0006107 | −1863 |
| 1.76 | 2.273 | | 0.080 | 0.810 | | | |

**For Segment BC (B to C)**

| d | v | $v_m$ | $v^2/2g$ | R | $R_m$ | $s_g$ | L, ft |
|---|---|---|---|---|---|---|---|
| 0.79 | 5.063 | | 0.398 | 0.517 | | | |
| | | 5.310 | | | 0.502 | 0.005409 | 0.4 |
| 0.72 | 5.556 | | 0.479 | 0.486 | | | |
| | | 5.903 | | | 0.468 | 0.007340 | 2.1 |
| 0.64 | 6.250 | | 0.607 | 0.449 | | | |
| | | 6.634 | | | 0.431 | 0.010346 | 4.5 |
| 0.57 | 7.018 | | 0.765 | 0.413 | | | |
| | | 7.591 | | | 0.391 | 0.015425 | 13.0 |
| 0.49 | 8.163 | | 1.035 | 0.369 | | | |
| | | 8.844 | | | 0.349 | 0.024362 | 53.7 |
| 0.42 | 9.524 | | 1.408 | 0.328 | | | |

**14.273**   In Fig. 14-69, 400 cfs flows through the transition. The rectangular section is 8 ft wide and $y_1 = 8$ ft. The trapezoidal section is 6 ft wide at the bottom with side slopes 1:1, and $y_2 = 7.5$ ft. Determine the rise $z$ in the bottom through the transition.

$$v_1^2/2g + y_1 = v_2^2/2g + y_2 + z + E_1 \qquad v_1 = Q/A_1 = 400/[(8)(8)] = 6.250 \text{ ft/s}$$

$$v_2 = 400/[(6)(7.5) + (7.5)(7.5)] = 3.951 \text{ ft/s}$$

$$E_1 = (0.3)(v_1^2/2g - v_2^2/2g) = (0.3)\{6.250^2/[(2)(32.2)] - 3.951^2/[(2)(32.2)]\} = 0.109 \text{ ft}$$

$$6.250^2/[(2)(32.2)] + 8 = 3.951^2/[(2)(32.2)] + 7.5 + z + 0.109 \qquad z = 0.755 \text{ ft}$$

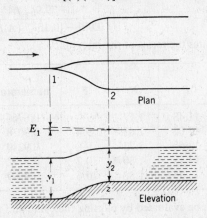

Plan

$E_1$

$y_2$

$y_1$

$z$

Elevation     **Fig. 14-69**

**14.274**   In a critical-depth meter 2 m wide with $z = 0.3$ m, the depth $y_1$ is measured to be 0.75 m. Find the discharge.

$q = (0.517)(g^{1/2})[y_1 - z + (0.55/g)(q^2/y_1^2)]^{3/2}$. Initially, neglect the last term in the equation above. $q = (0.517)(9.807)^{1/2}(0.75 - 0.3)^{3/2} = 0.489 \text{ (m}^3\text{/s)/m}$. Now, try $q = 0.500 \text{ (m}^3\text{/s)/m}$ in the whole equation:

$q = (0.517)(9.807)^{1/2}[0.75 - 0.3 + (0.55/9.807)(0.500^2/0.75^2)]^{3/2} = 0.530$ (m³/s)/m. Try $q = 0.536$ (m³/s)/m: $q = (0.517)(9.807)^{1/2}[0.75 - 0.3 + (0.55/9.807)(0.536^2/0.75^2)]^{3/2} = 0.536$ (m³/s)/m, $Q = (0.536)(2) = 1.07$ m³/s.

**14.275** At section 1 of a canal, the cross section is trapezoidal with bottom width $b_1 = 10$ m, depth $y_1 = 7$ m, and side slopes of 2 horizontal to 1 vertical. At section 2, 200 m downstream, the bottom is 0.08 m higher than at section 1, $b_2 = 15$ m, and side slopes are 3 horizontal to 1 vertical ($Q = 200$ m³/s and $n = 0.035$). Determine the depth of water at section 2.

■ $$\Delta L = [(v_1^2 - v_2^2)/2g + y_1 - y_2]/(s - s_0) \qquad s = (nQ/1.0AR^{2/3})^2$$

Try $y = 6.92$ m:

$$A_1 = (10)(7) + (7)[(2)(7)] = 168.0 \text{ m}^2 \qquad A_2 = (15)(6.92) + (6.92)[(3)(6.92)] = 247.5 \text{ m}^2$$
$$A_{avg} = (168.0 + 247.5)/2 = 207.8 \text{ m}^2 \qquad R = A/p_w \qquad (p_w)_1 = 10 + (2)(7)(\sqrt{2^2 + 1}) = 41.30 \text{ m}$$
$$(p_w)_2 = 15 + (2)(6.92)(\sqrt{3^2 + 1}) = 58.77 \text{ m} \qquad (p_w)_{avg} = (41.30 + 58.77)/2 = 50.04 \text{ m}$$
$$R = 207.8/50.04 = 4.153 \text{ m}$$
$$s = \{(0.035)(200)/[(1.0)(207.8)(4.153)^{2/3}]\}^2 = 0.0001700$$
$$v_1 = Q/A_1 = 200/168.0 = 1.190 \text{ m/s} \qquad v_2 = 200/247.5 = 0.8081 \text{ m/s}$$
$$\Delta L = \frac{(1.190^2 - 0.8081^2)/[(2)(9.807)] + 7 - 6.92}{0.0001700 - (-0.08/200)} = 209 \text{ m}$$

This value of $\Delta L$ (209 m) is not equal to the given value (200 m). However, reworking the problem with $y_2 = 6.93$ m (not shown) yields a value of $\Delta L$ of 191 m. Thus $y_2$ must be approximately 6.925 m.

**14.276** A trapezoidal channel, $b = 3$ m, side slopes of 1:1, $n = 0.014$, $s_0 = 0.001$, carries 28 m³/s. If the depth is 3 m at section 1, determine the water-surface profile for the next 700 m downstream.

■ To determine whether the depth increases or decreases, the slope of the energy grade line must be computed.
$$A = (3)(3) + (3)(3) = 18.00 \text{ m}^2 \qquad R = A/p_w = 18.00/[3 + (2)(3)(\sqrt{1^2 + 1^2})] = 1.567 \text{ m}$$
$$s = (nQ/1.0AR^{2/3})^2 = \{(0.014)(28)/[(1.0)(18.00)(1.567)^{2/3}]\}^2 = 0.0002606$$
$$(Q^2/gA_c^3)(T_c) = \{28^2/[(9.807)(18.00)^3]\}(9) = 0.1234$$

Since $(Q^2/gA_c^3)(T_c) < 1.0$, the depth is above critical. With the depth greater than critical and the energy grade line less steep than the bottom of the channel, the specific energy is increasing. When the specific energy increases above critical, the depth of flow increases, $y$ is then positive.

$$L = \int_{y_1}^{y_2} \frac{1 - Q^2T/gA^3}{s_0 - (nQ/1.0AR^{2/3})^2} \, dy$$

$$L = \int_3^y \frac{1 - (28^2)(T)/[(9.807)(A)^3]}{0.001 - \{(0.014)(28)/[(1.0)(AR^{2/3})]\}^2} \, dy$$

$$L = \int_3^y \frac{1 - 79.94T/A^3}{0.001 - [0.1537/(A^2R^{4/3})]} \, dy$$

The following table evaluates the terms of the integrand:

| $y$ | $A$ | $P$ | $R$ | $T$ | numerator | $10^6 \times$ denominator | $F(y)$ | $L$ |
|---|---|---|---|---|---|---|---|---|
| 3 | 18 | 11.48 | 1.57 | 9 | 0.8766 | 739 | 1185 | 0 |
| 3.2 | 19.84 | 12.05 | 1.65 | 9.4 | 0.9038 | 799 | 1131 | 231.6 |
| 3.4 | 21.76 | 12.62 | 1.72 | 9.8 | 0.9240 | 843 | 1096 | 454.3 |
| 3.6 | 23.76 | 13.18 | 1.80 | 10.2 | 0.9392 | 876 | 1072 | 671.1 |
| 3.8 | 25.84 | 13.75 | 1.88 | 10.6 | 0.9509 | 901 | 1056 | 883.9 |

The integral $\int F(y) \, dy$ can be evaluated by plotting the curve and taking the area under it between $y = 3$ and the following values of $y$. As $F(y)$ does not vary greatly in this example, the average of $F(y)$ can be used for each reach (the trapezoidal rule); and when it is multiplied by $\Delta y$, the length of reach is obtained. Between $y = 3$ and $y = 3.2$, $[(1185 + 1131)/2](0.2) = 231.6$. Between $y = 3.2$ and $y = 3.4$, $[(1131 + 1096)/2](0.2) = 222.7$ and so on. Five points on it are known, so the water surface can be plotted.

**14.277** After contracting below a sluice gate, water flows onto a wide horizontal floor with a velocity of 15 m/s and a depth of 0.7 m. Find the equation for the water-surface profile ($n = 0.015$).

$$x = -(\tfrac{3}{13})(1.0/nq)^2(y^{13/3} - y_1^{13/3}) + (3/4g)(1.0/n)^2(y^{4/3} - y_1^{4/3})$$

$$q = (0.7)(15) = 10.5 \ (\text{m}^3/\text{s})/\text{m}$$

$$x = -(\tfrac{3}{13})\{1.0/[(0.015)(10.5)]\}^2(y^{13/3} - 0.7^{13/3}) + \{3/[(4)(9.807)]\}(1.0/0.015)^2(y^{4/3} - 0.7^{4/3})$$

$$= -9.303y^{13/3} + 339.9y^{4/3} - 209.3$$

$$y_c = (q^2/g)^{1/3} = (10.5^2/9.807)^{1/3} = 2.240 \ \text{m}$$

The depth must increase downstream, since the specific energy decreases, and the depth must move toward the critical value for less specific energy. The equation does not hold near the critical depth because of vertical accelerations that have been neglected in the derivation of gradually varied flow. If the channel is long enough for critical depth to be attained before the end of the channel, the high-velocity flow downstream from the gate may be drowned or a jump may occur. The water-surface calculation for the subcritical flow must begin with critical depth at the downstream end of the channel.

**14.278** Prepare a computer program in BASIC to calculate the steady gradually varied water-surface profile in any prismatic rectangular, symmetric trapezoidal, or triangular channel.

```
10 REM B:PROFILES          WATER SURFACE PROFILES--STEADY STATE
20 ' WATER SURFACE PROFILE IN RECT, TRAPEZOIDAL, OR TRIANGULAR CHANNEL.
30 ' XL=LENGTH, B=BOT WIDTH, Z=SIDE SLOPE, RN=MANNING N,S0=BOT SLOPE,Q=FLOW.
40 ' YCONT=CONTROL DEPTH. IF YCONT=0 IN DATA, YCONT IS SET EQUAL TO YC.
50 ' IN SUBCRITICAL FLOW, CONTROL IS DOWNSTREAM AND DISTANCES ARE MEASURED
60 ' IN THE UPSTREAM DIRECTION
70 ' IN SUPERCRITICAL FLOW, CONTROL IS U.S. AND DISTANCES ARE MEASURED D.S.
80 LPRINT CHR$(14);"STEADY-STATE WATER-SURFACE PROFILES"
90 LPRINT CHR$(14);"    DATE=";DATE$;" TIME=";TIME$
100 DEF FNAREA(YY)=YY*(B+Z*YY): DEF FNPER(YY)=B+2!*YY*SQR(1!+Z^2)
110 DEF FNYCRIT(YY)=1!-Q^2*(B+2!*Z*YY)/(G*FNAREA(YY)^3)
120 DEF FNYNORM(YY)=1!-Q^2*CON/(FNAREA(YY)^3.333*FNPER(YY)^1.333)
130 DEFINT I:    DEF FNDL(YY)=FNYCRIT(YY)/(FNYNORM(YY)*S0)
140 DEF FNFPM(YY)=GAM*(YY^2*(.5*B+Z*YY/3!)+Q^2/(G*FNAREA(YY)))
150 ISI$="SI":    DEF FNENERGY(YY)=YY+Q^2/(2!*G*FNAREA(YY)^2)
160 READ IUNIT$,XL,B,Z,RN,S0,Q,YCONT: DATA "SI",200.,2.5,.8,.012,.025,25.,.0
170 IF IUNIT$=ISI$ THEN 190
180 GAM=62.4:G=32.2:CON=(RN/1.486)^2/S0:LPRINT" USC UNITS":LPRINT: GOTO 200
190 GAM=9802!:    G=9.806001:    CON=RN^2/S0:    LPRINT"SI UNITS":LPRINT
200 LPRINT" CHANNEL LENGTH=";XL;" DISCHARGE=";Q;" B=";B;" Z=";Z;"RN=";RN;
    " S0=";S0
210 ' DETERMINATION OF CRITICAL AND NORMAL DEPTHS
220 NN=30:   DN=0!:   UP=30!: YC=15!: FOR I= 1 TO 15: IF FNYCRIT(YC)=0! THEN 250
230 IF FNYCRIT(YC)<0! THEN DN=YC ELSE UP=YC
240 YC=.5*(UP+DN):   NEXT I
250 IF YCONT=0! THEN YCONT=YC
260 IF S0<=0! THEN 320
270 UP=40!:   DN=0!:   YN=20!:   FOR I= 1 TO 15
280 X=FNYNORM(YN):   IF X<0! THEN DN=YN:   GOTO 300
290 IF X>0! THEN UP=YN ELSE GOTO 310
300 YN=.5*(UP+DN):   ' NEXT I
310 LPRINT:   LPRINT" NORMAL DEPTH=";YN;" CRITICAL DEPTH=";YC:   GOTO 330
320 YN=3!*YC:   LPRINT:   LPRINT" CRITICAL DEPTH=";YC
330 IF YN<YC THEN 410
340 ' MILD,ADVERSE,OR HORIZONTAL CHANNEL YN>YC
350 IF YCONT<YC THEN 390
360 ' SUBCRITICAL FLOW, YCONT>=YC
370 SIGN=-1!:   DY=(YCONT-YN)*.998/NN:   LPRINT:
    LPRINT"CONTROL IS DOWNSTREAM,DEPTH=";YCONT:   GOTO 460
380 ' SUPERCRITICAL FLOW
390 SIGN=1!:   DY=(YC-YCONT)/NN:   LPRINT:
    LPRINT"CONTROL IS UPSTREAM, DEPTH=";YCONT:   GOTO 460
400 ' STEEP CHANNEL, YN<YC
410 IF YCONT<=YC THEN 450
420 ' SUBCRITICAL FLOW, YCONT>YC
430 SIGN=-1!:   DY=(YCONT-YC)/NN:   LPRINT:
    LPRINT" CONTROL IS DOWNSTREAM, DEPTH=";YCONT:   GOTO 460
440 ' SUPERCRITICAL FLOW, YCONT<=YC
450 SIGN=1!:   NN=2*NN:   DY=(YN-YCONT)*.998/NN:   LPRINT:
    LPRINT" CONTROL IS UPSTREAM, DEPTH=";YCONT
460 SL=0!:   Y=YCONT:   E=FNENERGY(Y):   FM=FNFPM(Y):   LPRINT
470 LPRINT"    DISTANCE    DEPTH    ENERGY    F+M":   GOSUB 550
480 ' WATER-SURFACE PROFILE CALCULATION
490 FOR I=1 TO NN STEP 2:   Y2=YCONT+SIGN*DY*(I+1)
500 DX=DY*(FNDL(Y)+FNDL(Y2)+4!*FNDL(YCONT+SIGN*I*DY))/3!
510 SL=SL+DX:   IF SL>XL THEN 540
520 Y=Y2:   E=FNENERGY(Y):   FM=FNFPM(Y):   IF (I=NN-1) AND (SL<0!) THEN SL=XL
530 GOSUB 550:   NEXT I:   GOTO 160
540 Y=Y2-SIGN*2!*DY*(SL-XL)/DX:   E=FNENERGY(Y):   FM=FNFPM(Y):   SL=XL:
    GOSUB 550:   GOTO 160
550 LPRINT SPC(5) USING"   ####.#";SL;:   LPRINT USING"    ###.###";Y;E;:
    LPRINT USING"   ########.";FM:   RETURN
560 DATA "SI",600.,2.5,.8,.012,.0002,25.,.907
570 DATA "SI",600.,2.5,.8,.012,.0002,25.,2.
```

Input data include the specification of the system of units (SI or USC) in the first columns of the data, followed by the channel dimensions, discharge, and water-surface control depth. If the control depth is set to zero in data, it is automatically assumed to be the critical depth in the program. For subcritical flow the control is downstream, and distances are measured in the upstream direction. For supercritical flow the control depth is upstream, and distances are measured in the downstream direction.

The program begins with several line functions to compute the various variables and functions in the problem. After the necessary data input, critical depth is computed, followed by the normal-depth calculation if normal depth exists. The bisection method is used in these calculations. The type of profile is then categorized, and finally the water-surface profile, specific energy, and $F + M$ are calculated and printed. Simpson's rule is used in the integration for the water-surface profile.

The program begins with several line functions to compute the various variables and functions in the problem. After the necessary data input, critical depth is computed, followed by the normal-depth calculation if normal.

**14.279** A trapezoidal channel, $B = 2.5$ m, side slope $= 0.8$, has two bottom slopes. The upstream portion is 200 m long, $S_0 = 0.025$, and the downstream portion, 600 m long, $S_0 = 0.0002$, $n = 0.012$. A discharge of 25 m$^3$/s enters at critical depth from a reservoir at the upstream end, and at the downstream end of the system the water depth is 2 m. Determine the water-surface profiles throughout the system, including jump location, using the computer program of Prob. 14.278.

**▮** Three separate sets of data are included in the program and are needed to obtain the results used to plot the solution as shown in Fig. 14-70. The first set for the steep upstream channel has a control depth equal to zero since it will be automatically assumed critical depth in the program. The second set is for the supercritical flow in the mild channel. It begins at a control depth equal to the end depth from the upstream channel and computes the water surface downstream to the critical depth. The third set of data uses the 2-m downstream depth as the control depth and computes in the upstream direction. Computer output from the last two data sets are given below. The jump is located by finding the position of equal $F + M$ from the output of the last two data sets.

```
SI UNITS

CHANNEL LENGTH= 600  DISCHARGE= 25  B= 2.5  Z= .8 RN= .012  SO= .0002

NORMAL DEPTH= 3.190308  CRITICAL DEPTH= 1.780243

CONTROL IS UPSTREAM, DEPTH= .9070001
```

| DISTANCE | DEPTH | ENERGY | F+M |
|---|---|---|---|
| 0.0 | 0.907 | 4.630 | 225573 |
| 25.8 | 0.965 | 4.160 | 211573 |
| 51.2 | 1.023 | 3.786 | 199573 |
| 76.0 | 1.082 | 3.487 | 189272 |
| 100.1 | 1.140 | 3.247 | 180431 |
| 123.3 | 1.198 | 3.054 | 172859 |
| 145.5 | 1.256 | 2.900 | 166400 |
| 166.5 | 1.315 | 2.777 | 160926 |
| 186.0 | 1.373 | 2.679 | 156334 |
| 204.0 | 1.431 | 2.603 | 152534 |
| 220.0 | 1.489 | 2.544 | 149455 |
| 233.9 | 1.547 | 2.500 | 147035 |
| 245.4 | 1.606 | 2.469 | 145221 |
| 254.1 | 1.664 | 2.448 | 143970 |
| 259.6 | 1.722 | 2.437 | 143242 |
| 261.5 | 1.780 | 2.433 | 143007 |

```
SI UNITS

CHANNEL LENGTH= 600  DISCHARGE= 25  B= 2.5  Z= .8 RN= .012  SO= .0002

NORMAL DEPTH= 3.190308  CRITICAL DEPTH= 1.780243

CONTROL IS DOWNSTREAM, DEPTH= 2
```

| DISTANCE | DEPTH | ENERGY | F+M |
|---|---|---|---|
| 0.0 | 2.000 | 2.474 | 146109 |
| 33.0 | 2.079 | 2.504 | 148634 |
| 80.5 | 2.158 | 2.541 | 151844 |
| 145.4 | 2.238 | 2.583 | 155711 |
| 231.3 | 2.317 | 2.630 | 160211 |
| 343.0 | 2.396 | 2.681 | 165326 |
| 486.6 | 2.475 | 2.734 | 171040 |
| 600.0 | 2.524 | 2.769 | 174857 |

$Y_c = 1.780$ m

Energy grade line

$Y_n = 0.856$ m
$Q = 25$ m³/s

2.46 m

2.0 m

0.907 m    1.22 m    $Y_n = 3.190$ m

**Fig. 14-70**

**14.280** A rectangular channel 3 m wide and 2 m deep, discharging 18 m³/s, suddenly has the discharge reduced to 12 m³/s at the downstream end. Compute the height and speed of the surge wave.

**❚**
$$(v_1 + c)(y_1) = (v_2 + c)(y_2) \qquad (\gamma/2)(y_1^2 - y_2^2) = (\gamma/g)(y_1)(v_1 + c)(v_2 + c - v_1 - c)$$
$$v_1 = Q/A_1 = 18/[(3)(2)] = 3 \text{ m/s} \qquad v_2 = 12/(3y_2) \qquad v_2 y_2 = 4 \text{ m}^2/\text{s}$$
$$(3 + c)(2) = 4 + cy_2 \qquad 6 = 4 + (c)(y_2 - 2)$$
$$(9.79/2)(2^2 - y_2^2) = (9.79/9.807)(2)(3 + c)(v_2 + c - 3 - c)$$

Eliminating $c$ and $v_2$ gives $y_2^2 - 4 = (4/9.807)[2/(y_2 - 2) + 3](3 - 4/y_2)$, $y_2 = 2.75$ m   (by trial and error), $v_2 = 4/2.75 = 1.455$ m/s. The height of the surge wave is $2.75 - 2 = 0.75$ m, and the speed of the wave is $c = 2/(y_2 - 2) = 2/(2.75 - 2) = 2.67$ m/s.

**14.281** In Fig. 14-71 find the Froude number of the undisturbed flow such that the depth $y_1$ at the gate is just zero when the gate is suddenly closed. For $v_0 = 20$ ft/s, find the liquid-surface elevation.

**❚** It is required that $v_1 = 0$ when $y_1 = 0$ at $x = 0$ for any time after $t = 0$.
$$v = v_0 - (2)(\sqrt{g})(\sqrt{y_0} - \sqrt{y}) \qquad 0 = v_0 - (2)(\sqrt{g})(\sqrt{y_0} - \sqrt{0})$$
$$v_0 = (2)(\sqrt{gy_0}) \qquad N_F = v_0/\sqrt{gy_0} = 2$$
$$x = [v_0 - (2)(\sqrt{gy_0}) + (3)(\sqrt{gy})](t) \qquad y_0 = v_0^2/4g = 20^2/[(4)(32.2)] = 3.106 \text{ ft}$$
$$x = \{20 - (2)[\sqrt{(32.2)(3.106)}] + (3)(\sqrt{32.2y})\}(t) = 17.02ty^{1/2}$$

The liquid surface is a parabola with vertex at the origin and surface concave upward.

$y_1$    $V_1$    $V$    $c$    $y$    $y_0$    $v_0$

**Fig. 14-71**

**14.282** In Fig. 14-71 the gate is partially closed at the instant $t = 0$ so that the discharge is reduced by 50 percent ($v_0 = 6$ m/s, $y_0 = 3$ m). Find $v_1$, $y_1$, and the surface profile.

**❚** The new discharge is
$$q = (6)(3)/2 = 9 = v_1 y_1 \qquad v = v_0 - (2)(\sqrt{g})(\sqrt{y_0} - \sqrt{y})$$
$$v_1 = 6 - (2)(\sqrt{9.807})(\sqrt{3} - \sqrt{y_1})$$
$$v_1 = 4.25 \text{ m/s} \qquad \text{and} \qquad y_1 = 2.11 \text{ m} \qquad \text{(by trial and error)}$$
$$x = [v_0 - (2)(\sqrt{gy_0}) + (3)(\sqrt{gy})](t) = \{6 - (2)[\sqrt{(9.807)(3)}] + (3)[\sqrt{(9.807)(y)}]\}(t)$$
$$= [(9.39)(\sqrt{y}) - 4.85](t)$$

This holds for the range of values of $y$ between 2.12 m and 3 m.

**14.283** A discharge of 160 cfs occurs in a rectangular open channel 6 ft wide with $s_0 = 0.002$ and $n = 0.012$. If the channel ends in a free outfall, calculate the depth at the brink, $y_n$, and $y_c$. Determine the shape of the water-surface profile for a distance of 100 ft upstream from the brink.

**❚**
$$Q = (A)(1.486/n)(R^{2/3})(s^{1/2}) \qquad 160 = (6y_n)(1.486/0.012)[6y_n/(y_n + 6 + y_n)]^{2/3}(0.002)^{1/2}$$
$$y_n = 3.5 \text{ ft} \qquad \text{(by trial and error)}$$
$$Q^2/g = A^3/B \qquad 160^2/32.2 = (6y_c)^3/6 \qquad y_c = 2.81 \text{ ft}$$

Since $y_n > y_c$, the flow is subcritical and the water-surface profile is $M_2$ (see Fig. A-19). The depth at the outfall is approximately $0.7y_c = (0.7)(2.81) = 2.0$ ft. Critical depth occurs at about $4y_c = (4)(2.81) = 11$ ft upstream from the brink. Computations for the water-surface profile are given below.

| $y$, ft | $A$, ft² | $B + 2y$, ft | $R$, ft | $V$, ft/s | $\frac{V^2}{2g}$, ft | $y + \frac{V^2}{2g}$, ft | $\Delta\left(y + \frac{V^2}{2g}\right)$, ft | $V_{avg}$, ft/s | $R_{avg}$, ft | $S$ | $S - S_0$ | $x$, ft | $\Sigma x$,* ft |
|---|---|---|---|---|---|---|---|---|---|---|---|---|---|
| 2.81 | 16.86 | 11.62 | 1.451 | 9.49 | 1.398 | 4.208 | | | | | | | |
| | | | | | | | 0.005 | 9.34 | 1.463 | 0.00341 | 0.00141 | 4 | 4 |
| 2.90 | 17.40 | 11.80 | 1.475 | 9.20 | 1.313 | 4.213 | | | | | | | |
| | | | | | | | 0.014 | 9.04 | 1.488 | 0.00312 | 0.00112 | 12 | 16 |
| 3.00 | 18.00 | 12.00 | 1.500 | 8.89 | 1.227 | 4.227 | | | | | | | |
| | | | | | | | 0.022 | 8.74 | 1.512 | 0.00284 | 0.00084 | 26 | 42 |
| 3.10 | 18.60 | 12.20 | 1.525 | 8.60 | 1.149 | 4.249 | | | | | | | |
| | | | | | | | 0.029 | 8.47 | 1.536 | 0.00262 | 0.00062 | 47 | 89 |
| 3.20 | 19.20 | 12.40 | 1.548 | 8.33 | 1.078 | 4.278 | | | | | | | |

* Summation $x$ is measured from the point of critical depth 11 ft upstream from the brink.

The water-surface profile is sketched in Fig. 14-72.

**Fig. 14-72**

**14.284** Examine the flow conditions in a 10-ft-wide open rectangular channel of rubble masonry with $n = 0.017$ when the flow rate is 400 cfs. The channel slope is 0.020, and an ogee weir 5.0 ft high with $C_w = 3.8$ is located in the downstream end of the channel (see Fig. 14-73).

**❚**
$$Q = (A)(1.486/n)(R^{2/3})(s^{1/2}) \qquad 400 = (10y_n)(1.486/0.017)[10y_n/(y_n + 10 + y_n)]^{2/3}(0.020)^{1/2}$$
$$y_n = 2.36 \text{ ft} \quad \text{(by trial and error)} \qquad y_c = (q^2/g)^{1/3} \qquad y_c = [(\tfrac{400}{10})^2/32.2]^{1/3} = 3.68 \text{ ft}$$

Since $y_n < y_c$, the flow is supercritical. The head required at the weir to discharge the given flow is found from the equation $Q = (C_w L)(h + V_0^2/2g)^{3/2}$:

$$V_0 = Q/A = 400/[(5 + h)(10)] = 400/(50 + 10h) \qquad 400 = (3.8)(10)\{h + [400/(50 + 10h)]^2/[(2)(32.2)]\}^{3/2}$$
$$h = 4.53 \text{ ft} \qquad \text{(by trial and error)}$$

Hence, the depth of water just upstream of the weir is $5 + 4.53 = 9.53$ ft, which is greater than $y_c$. The flow at this point is subcritical, and a hydraulic jump must occur upstream. The depth $y_2$ after the jump is found from $y_2 = -y_1/2 + (y_1^2/4 + 2V_1^2 y_1/g)^{1/2}$, $V_1 = Q/A_1 = 400/[(2.36)(10)] = 16.95$ ft/s, $y_2 = -2.36/2 + [2.36^2/4 + (2)(16.95)^2(2.36)/32.2]^{1/2} = 5.42$ ft. The distance from the weir to the jump is determined by the equation $x = [(y_A + V_A^2/2g) - (y_B + V_B^2/2g)]/(s - s_0)$:

$$V_A = Q/A = 400/[(5.42)(10)] = 7.380 \text{ ft/s} \qquad V_B = 400/[(9.53)(10)] = 4.197 \text{ ft/s}$$
$$s = (nV/1.486R^{2/3})^2 \qquad V_m = (7.380 + 4.197)/2 = 5.788 \text{ ft/s}$$
$$(p_w)_A = 5.42 + 10 + 5.42 = 20.84 \text{ ft} \qquad (p_w)_B = 9.53 + 10 + 9.53 = 29.06 \text{ ft}$$
$$R_m = [(5.42)(10)/20.84 + (9.53)(10)/29.06]/2 = 2.940 \text{ ft}$$
$$s = \{(0.017)(5.788)/[(1.486)(2.940)^{2/3}]\}^2 = 0.001041$$
$$x = \{5.42 + 7.380^2/[(2)(32.2)] - 9.53 - 4.197^2/[(2)(32.2)]\}/(0.001041 - 0.020) = 187 \text{ ft}$$

Fig. 14-73

**14.285** A rectangular channel is 8 ft wide, has an 0.008 slope, discharge of 150 cfs, and $n = 0.014$. Find $y_n$ and $y_c$. If the actual depth of flow is 5 ft, what type of profile exists?

$$Q = (A)(1.486/n)(R^{2/3})(s^{1/2}) \qquad 150 = (8y_n)(1.486/0.014)[8y_n/(y_n + 8 + y_n)]^{2/3}(0.008)^{1/2}$$

$$y_n = 1.74 \text{ ft} \quad \text{(by trial and error)} \qquad y_c = (q^2/g)^{1/3} = [(\tfrac{150}{8})^2/32.2]^{1/3} = 2.22 \text{ ft}$$

From Fig. A-19, this is an $S_1$ water-surface profile.

**14.286** A rectangular channel is 3 m wide and ends in a free outfall. If the discharge is 10 m³/s, slope is 0.0025, and $n = 0.016$, find $y_n$, $y_c$, and the water-surface profile for a distance of 150 m upstream from the outfall.

$$Q = (A)(1.0/n)(R^{2/3})(s^{1/2}) \qquad 10 = (3y_n)(1.0/0.016)[3y_n/(y_n + 3 + y_n)]^{2/3}(0.0025)^{1/2}$$

$$y_n = 1.34 \text{ m} \quad \text{(by trial and error)} \qquad y_c = (q^2/g)^{1/3} = [(\tfrac{10}{3})^2/9.807]^{1/3} = 1.04 \text{ m}$$

The depth at the outfall is approximately $0.7y_c = (0.7)(1.04) = 0.73$ m. Critical depth occurs at about $4y_c = (4)(1.04) = 4$ m upstream from the brink. Computations for the water-surface profile are given below $[s = (nv/1.0R^{2/3})^2, \ x = E/(s - s_0)]$.

| $y_n$, m | $v$, m/s | $v^2/2g$, m | $E$, m | $v_m$, m/s | $R_m$, m | $s$ | $\sum x$, m | $x$, m |
|---|---|---|---|---|---|---|---|---|
| 1.04 | 3.205 | 0.524 | 1.546 | | | | | |
| | | | | 3.06 | 0.630 | 0.00443 | 6 | 6 |
| 1.14 | 2.924 | 0.436 | 1.576 | | | | | |
| | | | | 2.80 | 0.662 | 0.00346 | 33 | 39 |
| 1.24 | 2.688 | 0.368 | 1.608 | | | | | |
| | | | | 2.58 | 0.694 | 0.00276 | 185 | 224 |
| 1.34 | 2.488 | 0.316 | 1.656 | | | | | |

**14.287** A rectangular drainage channel is 15 ft wide and is to carry 500 cfs. The channel is lined with rubble masonry ($n = 0.017$) and has a bottom slope of 0.0015. It discharges into a stream which may reach a stage 10 ft above the channel bottom during floods. Calculate $y_n$, $y_c$, and the distance from the channel outlet to the point where normal depth would occur under this condition. Use one step.

$$Q = (A)(1.486/n)(R^{2/3})(s^{1/2}) \qquad 500 = (15y_n)(1.486/0.017)[15y_n/(y_n + 15 + y_n)]^{2/3}(0.0015)^{1/2}$$

$$y_n = 4.81 \text{ ft} \quad \text{(by trial and error)} \qquad y_c = (q^2/g)^{1/3} = [(\tfrac{500}{15})^2/32.2]^{1/3} = 3.26 \text{ ft}$$

$$\Delta x = [(y_1 + V_1^2/2g) - (y_2 + V_2^2/2g)]/(s - s_0) \qquad V_1 = Q/A = 500/[(4.81)(15)] = 6.930 \text{ ft/s}$$

$$V_2 = 500/[(10)(15)] = 3.333 \text{ ft/s}$$

$$s = (nV/1.486R^{2/3})^2 \qquad V_m = (6.930 + 3.333)/2 = 5.132 \text{ ft/s}$$

$$(p_w)_1 = 4.81 + 15 + 4.81 = 24.62 \text{ ft} \qquad (p_w)_2 = 10 + 15 + 10 = 35.00 \text{ ft}$$

$$R_m = [(4.81)(15)/24.62 + (10)(15)/35.00]/2 = 3.608 \text{ ft}$$

$$s = \{(0.017)(5.132)/[(1.486)(3.608)^{2/3}]\}^2 = 0.0006229$$

$$\Delta x = \{4.81 + 6.930^2/[(2)(32.2)] - 10 - 3.333^2/[(2)(32.2)]\}/(0.0006229 - 0.0015) = 5264 \text{ ft}$$

**14.288** Solve Prob. 14.287 using three steps.

▮ Using data from Prob. 14.287, $s = (nv/1.486R^{2/3})^2$, $x = \Delta E/(s - s_0)$.

| $y_n$, ft | $V$, ft/s | $V^2/2g$, ft | $E$, ft | $V_m$, ft/s | $R_m$, ft | $s$ | $x$, ft |
|---|---|---|---|---|---|---|---|
| 10 | 3.33 | 0.17 | 10.17 | | | | |
| | | | | 3.75 | 4.09 | 0.000285 | 1560 |
| 8 | 4.17 | 0.27 | 8.27 | | | | |
| | | | | 4.86 | 3.60 | 0.000559 | 1900 |
| 6 | 5.55 | 0.48 | 6.48 | | | | |
| | | | | 6.28 | 3.12 | 0.000112 | 2420 |
| 4.8 | 7.00 | 0.76 | 5.56 | | | | |
| | | | | | | | 5880 |

**14.289** A trapezoidal channel with a bottom width of 10 ft and side slopes of 2 horizontal to 1 vertical has a horizontal curve with a radius of 100 ft without superelevation. If the discharge is 800 cfs and the water surface at the inside of the curve is 5 ft above the channel bottom, find the water-surface elevation at the outside of the curve. Assume the flow is subcritical.

▮ See Fig. 14-74. $y_2 - y_1 = v^2 B/gr$, $x - 5 = \{800/[(x+5)(10+x) - 25 - x^2]\}^2(10 + 10 + 2x)/[(32.2)(100)]$, $x = 5.53$ ft (by trial and error).

Fig. 14-74

**14.290** A rectangular flume of planed timber ($n = 0.011$) is 1.5 m wide and carries 1.7 m³/s of water. The bed slope is 0.0005, and at section 1 the depth is 0.9 m. Find the distance to section 2, where the depth is 0.75 m. Is 2 upstream or downstream of 1?

▮ $\Delta x = (y_1 + v_1^2/2g - y_2 - v_2^2/2g)/(s - s_0)$. Assume the 0.75 m depth is downstream.

$$v_1 = Q/A_1 = 1.7/[(0.9)(1.5)] = 1.259 \text{ m/s}$$

$$v_2 = Q/A_2 = 1.7/[(0.75)(1.5)] = 1.511 \text{ m/s} \qquad s = (nv/1.0R^{2/3})^2$$

$$v_m = (1.259 + 1.511)/2 = 1.385 \text{ m/s} \qquad (p_w)_1 = 0.9 + 1.5 + 0.9 = 3.30 \text{ m}$$

$$(p_w)_2 = 0.75 + 1.5 + 0.75 = 3.00 \text{ m}$$

$$R_m = [(0.9)(1.5)/3.30 + (0.75)(1.5)/3.00]/2 = 0.3920 \text{ m}$$

$$s = \{(0.011)(1.385)/[(1.0)(0.3920)^{2/3}]\}^2 = 0.000809$$

$$\Delta x = \{0.9 + 1.259^2/[(2)(9.807)] - 0.75 - 1.511^2/[(2)(9.807)]\}/(0.000809 - 0.0005) = 370 \text{ m}$$

Since $\Delta x$ is positive, the 0.75 m depth is downstream, as assumed.

**14.291** The flume of Prob. 14.290 still carries 1.7 m³/s, but now the depth varies from 1.2 m at one section to 0.9 m at a section 200 m downstream. Find the new bed slope.

▮ $\qquad \Delta x = (y_1 + v_1^2/2g - y_2 - v_2^2/2g)/(s - s_0) \qquad v_1 = Q/A_1 = 1.7/[(1.2)(1.5)] = 0.9444 \text{ m/s}$

$$v_2 = Q/A_2 = 1.7/[(0.9)(1.5)] = 1.259 \text{ m/s} \qquad s = (nv/1.0R^{2/3})^2$$

$$v_m = (0.9444 + 1.259)/2 = 1.102 \text{ m/s}$$

$$(p_w)_1 = 1.2 + 1.5 + 1.2 = 3.90 \text{ m} \qquad (p_w)_2 = 0.9 + 1.5 + 0.9 = 3.30 \text{ m}$$

$$R_m = [(1.2)(1.5)/3.90 + (0.9)(1.5)/3.30]/2 = 0.4353 \text{ m}$$

$$s = \{(0.012)(1.102)/[(1.0)(0.4353)^{2/3}]\}^2 = 0.0005301$$

$$200 = \{1.2 + 0.9444^2/[(2)(9.807)] - 0.9 - 1.259^2/[(2)(9.807)]\}/(0.0005301 - s_0) \qquad s_0 = -0.000793$$

See Fig. 14-75.

**Fig. 14-75**

**14.292** The slope of the flume in Prob. 14.290 is now increased to 0.01. For the same flow as before, find the depth 35 m downstream from a section where the flow is 0.40 m deep.

▌ $y_c = (q^2/g)^{1/3} = [(1.7/1.5)^2/9.807]^{1/3} = 0.508$ m. Since $y < y_c$, the flow is supercritical.

$$\Delta x = (y_1 + v_1^2/2g - y_2 - v_2^2/2g)/(s - s_0) \qquad v_1 = Q/A_1 = 1.7/[(0.40)(1.5)] = 2.833 \text{ m/s}$$

$$v_2 = Q/A_2 = 1.7/(1.5y_2) = 1.133/y_2 \qquad s = (nv/1.0R^{2/3})^2$$

$$v_m = (2.833 + 1.133/y_2)/2 = 1.417 + 0.5665/y_2$$

$$(p_w)_1 = 0.4 + 1.5 + 0.4 = 2.30 \text{ m} \qquad (p_w)_2 = y_2 + 1.5 + y_2 = 1.5 + 2y_2$$

$$R_m = [(0.4)(1.5)/2.30 + 1.5y_2/(1.5 + 2y_2)]/2 = 0.2609 + 0.75y_2/(1.5 + 2y_2)$$

$$s = \left\{ \frac{(0.011)(1.417 + 0.5665/y_2)}{1.0[0.2609 + 0.75y_2/(1.5 + 2y_2)]^{2/3}} \right\}^2$$

$$35 = \{0.40 + 2.833^2/[(2)(9.807)] - y_2 - (1.133/y_2)^2/[(2)(9.807)]\}/(s - 0.01)$$

$$y_2 = 0.31 \text{ m} \qquad \text{(by trial and error)}$$

**14.293** A rectangular flume 12 in wide and 30 ft long yielded the following test results: with still water, $z_1 - z_2 = 0.010$ ft; with a measured flow of 0.20 cfs, $y_1 = 0.400$ ft, $y_2 = 0.405$ ft. Evaluate the roughness coefficient $n$.

▌ $Q = (A)(1.486/n)(R^{2/3})(s^{1/2})$ \qquad $A_1 = (0.400)(\frac{12}{12}) = 0.400$ ft$^2$ \qquad $A_2 = (0.405)(\frac{12}{12}) = 0.405$ ft$^2$

$$A_m = (0.400 + 0.405)/2 = 0.4025 \text{ ft}^2 \qquad (p_w)_1 = 0.40 + \tfrac{12}{12} + 0.40 = 1.800 \text{ ft}$$

$$(p_w)_2 = 0.405 + \tfrac{12}{12} + 0.405 = 1.810 \text{ ft}$$

$$R_m = (0.40/1.80 + 0.405/1.81)/2 = 0.2230 \text{ ft} \qquad x = (y_1 + v_1^2/2g - y_2 - v_2^2/2g)/(s - s_0)$$

$$v_1 = Q/A_1 = 0.20/0.40 = 0.500 \text{ ft/s} \qquad v_2 = Q/A_2 = 0.2/0.405 = 0.4938 \text{ ft/s}$$

$$30 = \{0.40 + 0.500^2/[(2)(32.2)] - 0.405 - 0.4938^2/[(2)(32.2)]\}/(s - 0.01/30) \qquad s = 0.0001699$$

$$0.20 = (0.4025)(1.486/n)(0.2230)^{2/3}(0.0001699)^{1/2} \qquad n = 0.0143$$

**14.294** A rectangular flume 10 ft wide is built of planed timber ($n = 0.012$) on a bed slope of 0.2 ft per 1000 ft ending in a free overfall. If the measured depth at the fall is 1.82 ft, find (a) the rate of flow and (b) the distance upstream from the fall to where the depth is 4 ft.

▌ **(a)** \qquad $y_{brink} = 0.72y_c$ \qquad $1.82 = 0.72y_c$ \qquad $y_c = 2.528$ ft

$$y_c = (q^2/g)^{1/3} \qquad 2.528 = (q^2/32.2)^{1/3} \qquad q = 22.81 \text{ cfs/ft} \qquad Q = (22.81)(10) = 228 \text{ ft}^3/\text{s}$$

**(b)** Between $y_{brink}$ and $y_c$ the flow is rapidly varying. So Manning's equation is not valid there and cannot be used there to determine $Q$. $s = (nV/1.486R^{2/3})^2$, $\Delta x = (y_1 + V_1^2/2g - y_2 - V_2^2/2g)/(s - s_0)$.

| $y$, ft | $V$, fps | $V^2/2g$, ft | $E$, ft | $A$, $\text{ft}^2$ | $P$, ft | $R$, ft | $R_m$, ft | $V_m$, fps | $s$ | $\Delta x$, ft |
|---|---|---|---|---|---|---|---|---|---|---|
| 4.0 | 5.70 | 0.505 | 4.505 | 40 | 18.0 | 2.22 | | | | |
| | | | | | | | 2.12 | 6.20 | 0.000915 | 568 |
| 3.4 | 6.71 | 0.698 | 4.095 | 34 | 16.8 | 2.02 | | | | |
| | | | | | | | 1.949 | 7.15 | 0.001363 | 173 |
| 3.0 | 7.60 | 0.897 | 3.897 | 30 | 16.0 | 1.875 | | | | |
| | | | | | | | 1.814 | 8.02 | 0.001888 | 53 |
| 2.7 | 8.44 | 1.108 | 3.808 | 27 | 15.4 | 1.753 | | | | |
| | | | | | | | 1.717 | 8.73 | 0.00240 | 8 |
| 2.53 | 9.01 | 1.261 | 3.791 | 25.3 | 15.1 | 1.680 | | | | |
| $y_b$ | | | | | | | | | | $4y_c = 10$ |
| | | | | | | | | | | $\sum (\Delta x) = 812$ |

**14.295** For a wide rectangular channel dredged in earth ($n = 0.030$), of slope 0.002273 and carrying 100 $(\text{ft}^3/\text{s})/\text{ft}$, find the water depth 1.5 mi upstream of a location (2) where the depth is 30 ft.

$$\Delta x = (y_1 + v_1^2/2g - y_2 - v_2^2/2g)/(s - s_0) \qquad v_1 = q/y_1 = 100/y_1$$
$$v_2 = q/y_2 = 100/30 = 3.333 \text{ ft/s}$$
$$v_m = (100/y_1 + 3.333)/2 = 50.0/y_1 + 1.667 \qquad s = (nv/1.486R^{2/3})^2$$
$$R_m = (y_1 + 30)/2 = y_1/2 + 15.00$$
$$s = \{(0.030)(50.0/y_1 + 1.667)/[(1.486)(y_1/2 + 15.00)^{2/3}]\}^2$$
$$(1.5)(5280) = \{y_1 + (100/y_1)^2/[(2)(32.2)] - 30 - 3.333^2/[(2)(32.2)]\}/(s - 0.002273)$$
$$y_1 = 11.9 \text{ ft} \qquad \text{(by trial and error)}$$

**14.296** The slope of a stream of rectangular cross section is 0.00022, the width is 150 ft, and the value of the Chezy $C$ is 80 $\text{ft}^{1/2}/\text{s}$. Find the depth for a uniform flow of 128.45 cfs per foot of width of the stream.

$$v_0 = q/y_0 = 88.55/y_0 \qquad v_0 = C\sqrt{Rs} = (80.0)\{\sqrt{[150y_0/(y_0 + 150 + y_0)](0.00022)}\}$$
$$128.45/y_0 = 80\{\sqrt{[150y_0/(y_0 + 150 + y_0)](0.00022)}\} \qquad y_0 = 25.0 \text{ ft} \qquad \text{(by trial and error)}$$

**14.297** If the stream of Prob. 14.296 is dammed, find the distance between a section where the increase in depth is 5 ft and one where the increase is 1 ft. Use reaches with 1-ft depth increments.

$$v = C\sqrt{Rs} \qquad \Delta x = (y_1 + v_1^2/2g - y_2 - v_2^2/2g)/(s - s_0)$$

| $y$, ft | $A$ (150y), $\text{ft}^2$ | $p$, ft | $R$ (A/p), ft | $v$ (q/y), fps | $v^2/2g$, ft | $E$, ft | $\Delta E$ ($E_1 - E_2$), ft | $v_m$, fps | $R_m$, ft | $s$ | $\Delta x$, ft |
|---|---|---|---|---|---|---|---|---|---|---|---|
| 30 | 4500 | 210 | 21.43 | 4.28 | 0.284 | 30.284 | | | | | |
| | | | | | | | -0.979 | 4.36 | 21.17 | 0.0001403 | 12 284 |
| 29 | 4350 | 208 | 20.91 | 4.43 | 0.305 | 29.305 | | | | | |
| | | | | | | | -0.978 | 4.51 | 20.65 | 0.0001539 | 14 796 |
| 28 | 4200 | 206 | 20.39 | 4.59 | 0.327 | 28.327 | | | | | |
| | | | | | | | -0.975 | 4.68 | 20.12 | 0.0001701 | 19 539 |
| 27 | 4050 | 204 | 19.85 | 4.76 | 0.352 | 27.352 | | | | | |
| | | | | | | | -0.973 | 4.85 | 19.58 | 0.0001877 | 30 124 |
| 26 | 3900 | 202 | 19.31 | 4.94 | 0.379 | 26.379 | | | | | |
| | | | | | | | | | | | $\sum (\Delta x) = 76\,743$ |

**14.298** A portion of an outfall sewer is approximately a circular conduit 5 ft in diameter and with a slope of 1 ft in 1100 ft. It is of brick, for which $n = 0.013$. What would be its maximum capacity for uniform flow?

$$Q = (A)(1.486/n)(R^{2/3})(s^{1/2}) \qquad Q_{\text{full}} = [(\pi)(5)^2/4](1.486/0.013)(\tfrac{5}{4})^{2/3}(\tfrac{1}{1100})^{1/2} = 78.53 \text{ ft}^3/\text{s}$$

From Fig. A-18, $(Q_{\max}/Q_{\text{full}}) = 1.06$. $Q_{\max} = (78.53)(1.06) = 83.2 \text{ ft}^3/\text{s}$.

**14.299** If the outfall sewer in Prob. 14.298 discharges 120 cfs with a depth at the end of 2.90 ft, how far back from the end must it become a pressure conduit? Proceeding from the mouth upstream, find by tabular solution the length of sewer that is not flowing full. Use three reaches with equal depth increments.

$$s = (nV/1.486R^{2/3})^2 \qquad \Delta x = (y_1 + v_1^2/2g - y_2 - v_2^2/2g)/(s - s_0)$$

| $y$, ft | $\theta$, deg | $A$, ft$^2$ | $P$, ft | $R$ $(A/P)$, ft | $V$ $(Q/A)$, fps | $v^2/2g$, ft | $E$, ft | $\Delta E$ $(E_1 - E_2)$, ft | $V_m$, fps | $R_m$, ft | $S$ | $\Delta x$, ft |
|---|---|---|---|---|---|---|---|---|---|---|---|---|
| 2.9 | 99.2 | 11.81 | 8.66 | 1.364 | 10.16 | 1.603 | 4.503 | | | | | |
| | | | | | | | | 0.073 | 9.05 | 1.429 | 0.00387 | 24.7 |
| 3.6 | 116.1 | 15.13 | 10.13 | 1.494 | 7.93 | 0.976 | 4.576 | | | | | |
| | | | | | | | | 0.417 | 7.31 | 1.504 | 0.00236 | 287 |
| 4.3 | 136.1 | 17.96 | 11.87 | 1.513 | 6.68 | 0.693 | 4.993 | | | | | |
| | | | | | | | | 0.587 | 6.40 | 1.382 | 0.00202 | 527 |
| 5.0 | 180 | 19.63 | 15.71 | 1.250 | 6.11 | 0.580 | 5.580 | | | | | |
| | | | | | | | | | | | | $\sum (\Delta x) = 839$ |

The sewer must become a pressure conduit 839 ft back from the end.

**14.300** For the channel of Prob. 14.131, find the separation of a section where the depth is 2.0 ft and one where the depth is 3.0 ft.

▮ Assume the 2.0 ft depth is upstream. Then $y_1 = 2.0$ ft and $y_2 = 3.0$ ft. From Prob. 14.131, $Q = 100$ cfs, $s = 0.02$, and $n = 0.015$.

$$\Delta x = (y_1 + v_1^2/2g - y_2 - v_2^2/2g)/(s - s_0) \qquad v_1 = Q/A_1 = 100/4.90 = 20.41 \text{ ft/s}$$
$$v_2 = Q/A_2 = 100/10.48 = 9.542 \text{ ft/s}$$
$$s = (nv/1.486R^{2/3})^2 \qquad v_m = (20.41 + 9.542)/2 = 14.98 \text{ ft/s} \qquad R = A/p_w$$
$$R_1 = 4.90/6.12 = 0.801 \text{ ft}$$
$$R_2 = 10.48/8.96 = 1.170 \text{ ft} \qquad R_m = (0.801 + 1.170)/2 = 1.9855 \text{ ft}$$
$$s = \{(0.015)(14.98)/[(1.486)(0.9855)^{2/3}]\}^2 = 0.02331$$
$$\Delta x = \{2.0 + 20.41^2/[(2)(32.2)] - 3.0 - 9.542^2/[(2)(32.2)]\}/(0.02331 - 0.02) = 1225 \text{ ft}$$

Since $\Delta x$ is positive, the assumption that the 2.0 ft depth was upstream is correct.

**14.301** Sketch the water-surface profile in a long rectangular channel ($n = 0.014$), if the channel is 10 ft wide; the flow rate is 340 cfs; and there is an abrupt change in slope from 0.0016 to 0.0150.

▮ See Fig. 14-76. $y_c = (q^2/g)^{1/3} = [(340/10)^2/32.2]^{1/3} = 3.30$ ft, $Q = (A)(1.486/n)(R^{2/3})(s^{1/2})$. In upstream segment, $340 = (10y_n)(1.486/0.014)[10y_n/(y_n + 10 + y_n)]^{2/3}(0.0016)^{1/2}$, $y_n = 4.50$ ft (by trial and error). In downstream segment, $340 = (10y_n)(1.486/0.014)[10y_n/(y_n + 10 + y_n)]^{2/3}(0.0150)^{1/2}$, $y_n = 2.04$ ft (by trial and error). Thus, the flow is subcritical before the break and supercritical after the break.

Fig. 14-76

**14.302** Repeat Prob. 14.301 for the case where the flow rate is 130 cfs.

▮ $y_c = (q^2/g)^{1/3} = [(130/10)^2/32.2]^{1/3} = 1.74$ ft, $Q = (A)(1.486/n)(R^{2/3})(s^{1/2})$. In upstream segment, $130 = (10y_n)(1.486/0.014)[10y_n/(y_n + 10 + y_n)]^{2/3}(0.0016)^{1/2}$, $y_n = 2.28$ ft (by trial and error). In downstream segment,

$130 = (10y_n)(1.486/0.014)[10y_n/(y_n + 10 + y_n)]^{2/3}(0.0150)^{1/2}$, $y_n = 1.08$ ft (by trial and error). Thus, the flow is subcritical before the break and supercritical after the break, so that Fig. 14-76 still applies.

**14.303** Repeat Prob. 14.302 if the slope abruptly changes from 0.0016 to 0.0006. Estimate the distance upstream from the break to the point where normal depth occurs.

▮ $y_c = 1.74$ ft (from Prob. 14.302). In upstream segment, $y_n = 2.28$ ft (from Prob. 14.302). In downstream segment, $130 = (10y_n)(1.486/0.014)[10y_n/(y_n + 10 + y_n)]^{2/3}(0.0006)^{1/2}$, $y_n = 3.20$ ft (by trial and error). Thus, the flow is subcritical before and after the break. A water surface on a gentle slope cannot deviate from $y_n$ in the upstream direction. Therefore, the depth at the break is 3.20 ft. See Fig. 14-77.

$$\Delta x = (y_1 + v_1^2/2g - y_2 - v_2^2/2g)/(s - s_0) \qquad v_1 = Q/A_1 = 130/[(2.28)(10)] = 5.702 \text{ ft/s}$$
$$v_2 = Q/A_2 = 130/[(3.20)(10)] = 4.063 \text{ ft/s} \qquad s = (nv/1.486R^{2/3})^2$$
$$v_m = (5.702 + 4.063)/2 = 4.883 \text{ ft/s}$$
$$(p_w)_1 = 2.28 + 10 + 2.28 = 14.56 \text{ ft} \qquad (p_w)_2 = 3.20 + 10 + 3.20 = 16.40 \text{ ft/s}$$
$$R_m = [(2.28)(10)/14.56 + (3.20)(10)/16.40]/2 = 1.759 \text{ ft}$$
$$s = \{(0.014)(4.883)/[(1.486)(1.759)^{2/3}]\}^2 = 0.0009967$$
$$\Delta x = \{2.28 + 5.702^2/[(2)(32.2)] - 3.20 - 4.063^2/[(2)(32.2)]\}/(0.0009967 - 0.0016) = 1113 \text{ ft}$$

Fig. 14-77

**14.304** In a 6-ft-wide rectangular channel ($s = 0.002$, $n = 0.014$), water flows at 240 cfs. A temporary dam increases the depth to 9.1 ft. Examine the water-surface profile upstream from the dam.

▮ $y_c = (q^2/g)^{1/3} = [(240/6)^2/32.2]^{1/3} = 3.68$ ft, $Q = (A)(1.486/n)(R^{2/3})(s^{1/2})$. In upstream segment, $240 = (6y_n)(1.486/0.014)[6y_n/(y_n + 6 + y_n)]^{2/3}(0.002)^{1/2}$, $y_n = 5.44$ ft (by trial and error). Since $y_n > y_c$, the channel slope is gentle. Also, since 9.1 ft $= y > y_n > y_c$, upstream of the dam is an $M_1$ profile (see Fig. A-19) with the depth gradually decreasing to the normal depth (5.44 ft). Critical depth (3.68 ft) occurs on the dam.

**14.305** Solve Prob. 14.304 if the channel slope is 0.0004.

▮ $y_c = 3.68$ ft (from Prob. 14.304), $Q = (A)(1.486/n)(R^{2/3})(s^{1/2})$. In upstream segment, $240 = (6y_n)(1.486/0.014)[6y_n/(y_n + 6 + y_n)]^{2/3}(0.0004)^{1/2}$, $y_n = 10.7$ ft (by trial and error). Since $y_n > y_c$, the channel slope is gentle. But insertion of a dam cannot lower the water surface below $y_n$ to 9.1 ft, so this situation is impossible.

**14.306** Solve Prob. 14.304 if the channel slope is 0.0010.

▮ $y_c = 3.68$ ft (from Prob. 14.304), $Q = (A)(1.486/n)(R^{2/3})(s^{1/2})$. In upstream segment, $240 = (6y_n)(1.486/0.014)[6y_n/(y_n + 6 + y_n)]^{2/3}(0.0010)^{1/2}$, $y_n = 7.22$ ft (by trial and error). The conclusions follow Prob. 14.304.

**14.307** Solve Prob. 14.304 if the channel slope is 0.006.

▮ $y_c = 3.68$ ft (from Prob. 14.304), $Q = (A)(1.486/n)(R^{2/3})(s^{1/2})$. In upstream segment, $240 = (6y_n)(1.486/0.014)[6y_n/(y_n + 6 + y_n)]^{2/3}(0.006)^{1/2}$, $y_n = 3.53$ ft (by trial and error). Since $y_n < y_c$, the channel slope is steep. Also, since 9.1 ft $= y > y_c > y_n$, upstream of the dam is an $S_1$ profile (see Fig. A-19), preceded by a hydraulic jump. Upstream of the jump is straight supercritical uniform flow with depth of 3.53 ft.

**14.308** A rectangular channel changes in width from 3 ft to 5 ft, as shown in Fig. 14-78. Measurements indicate that $y_1 = 2.00$ ft and $Q = 60$ cfs. Determine the depth $y_2$, neglecting head loss.

**▮**
$$E_1 = E_2 + h_L \qquad E = y + v^2/2g \qquad v_1 = Q/A_1 = 60/[(3)(2.00)] = 10.00 \text{ ft/s}$$
$$E_1 = 2.0 + 10.00^2/[(2)(32.2)] = 3.553 \text{ ft} \qquad v_2 = Q/A_2 = 60/(5y_2) = 12.0/y_2$$
$$E_2 = y_2 + (12.0/y_2)^2/[(2)(32.2)] = y_2 + 2.236/y_2^2$$
$$h_L = 0 \qquad 3.553 = y_2 + 2.236/y_2^2 + 0 \qquad y_2 = 3.36 \text{ ft} \qquad \text{(by trial and error)}$$

**Fig. 14-78**

**14.309** Solve Prob. 14.308, assuming $h_L = (0.3)(v_1^2/2g - v_2^2/2g)$.

**▮** Using data from Prob. 14.308, $h_L = 0.3\{10.00^2/[(2)(32.2)] - (12/y_2)^2/[(2)(32.2)]\} = 0.4658 - 0.6708/y_2^2$, $3.553 = y_2 + 2.236/y_2^2 + 0.4658 - 0.6708/y_2^2$, $y_2 = 2.90$ ft (by trial and error).

**14.310** Consider a rectangular flume 4.5 m wide, built of unplaned planks ($n = 0.014$), leading from a reservoir in which the water surface is maintained constant at a height of 1.8 m above the bed of the flume at entrance (see Fig. 14-79). The flume is on a slope of 0.001. The depth 300 m downstream from the head end of the flume is 1.20 m. Assuming an entrance loss of $0.2V_1^2/2g$, find the flow rate for the given conditions.

**▮** For a first approximate answer we shall consider the entire flume as one reach. The equations to be satisfied are

Energy at entrance:
$$y_1 + (1.2V_1^2/2g) = 1.80 \tag{1}$$

Energy equation for the entire reach:
$$y_1 + (V_1^2/2g) = 1.20 + (V_2^2/2g) + (S - 0.001)L \tag{2}$$

where $S$ is given by
$$S = (nV_m/R_m^{2/3})^2 \tag{3}$$

The procedure is to make successive trials of the upstream depth $y_1$. This determines corresponding values of $V_1$, $q$, $V_2$, $V_m$, $R_m$, and $S$. The trials are repeated until the value of $\Delta x$ from Eq. (2) is close to 300 m. The solution is conveniently set in tabular form as follows:

| Trial $y_1$, m | $V_1$, Eq. (1), m/s | $q = y_1 V_1$, m³/s/m | $V_2 = q/1.20$, m/s | $V_m$, m/s | $R_{h_1}$, m | $R_{h_2}$, m | $R_{h_m}$, m | $S$, Eq. (3) | $\Delta x$, Eq. (2), m |
|---|---|---|---|---|---|---|---|---|---|
| 1.50 | 2.22 | 3.33 | 2.78 | 2.50 | 0.90 | 0.78 | 0.89 | 0.00143 | 358 |
| 1.48 | 2.29 | 3.39 | 2.82 | 2.56 | 0.89 | 0.78 | 0.835 | 0.00163 | 226 |

Thus $y_1 \approx 1.49$ m and the flow rate $Q = qB \approx 3.36 \times 4.5 \approx 15.1$ m³/s. The accuracy of the result, of course, depends on one's ability to select the correct value for Manning's $n$. If $n$ was assumed to be 0.015, for example, rather than 0.014, the result would have been quite different. Also, a more accurate result can be obtained by dividing the flume into reaches in which the depth change is about 10 percent of the depth.

**Fig. 14-79**

**14.311** A rectangular flume of planed timber ($n = 0.012$) 20 ft wide, 1000 ft long, with horizontal bed leads from a reservoir in which the still-water surface is 10 ft above the flume bed. Assume that the depth of the downstream end of the flume is fixed at 8 ft by some control section downstream. Allowing an entrance loss of 0.2 times the velocity head, find the flow rate in the flume using one reach.

▌ Energy at entrance:

$$y_1 + 1.2V_1^2/2g = 10 \tag{1}$$

Energy equation for the entire reach:

$$y_1 + V_1^2/2g = 8 + V_2^2/2g + (S - 0)(L) \tag{2}$$

where $S$ is given by

$$S = (nV_m/1.486R_m^{2/3})^2 \tag{3}$$

This problem involves a trial-and-error solution exactly like that of Prob. 14.310. After successive trials (not shown), the solution is reached with $y_1 = 8.82$ ft and $Q = qB = (70.25)(20) = 1405$ ft³/s.

**14.312** Find the flow rate in the flume of Prob. 14.311 if it ends in a free fall, all other conditions remaining the same. The critical depth may be supposed to occur at about $6y_c$ back from the fall. Thus the length of the reach is $1000 - 6y_c$, and $y_2 = y_c$.

▌ Take the flume in one reach with $y_1$ and $V_1$ at the upstream end and $y_c$ and $V_c$ at the downstream end at a distance $L = (1000 - 6y_c)$ ft. Then, energy at entrance:

$$y_1 + 1.2V_1^2/2g = 10 \tag{1}$$

Energy equation for the entire reach:

$$y_1 + V_1^2/2g = y_c + V_c^2/2g + (S - 0)(L) \tag{2}$$

where $S$ is given by

$$S = (nV_m/1.486R_m^{2/3})^2 \tag{3}$$

First trial: Assume $y_1 = 7$ ft.

$$7 + 1.2V_1^2/[(2)(32.2)] = 10 \qquad V_1 = 12.69 \text{ ft/s} \qquad q = V_1 y_1 = (12.69)(7) = 88.83 \text{ cfs/ft}$$
$$y_c = (q^2/g)^{1/3} = (88.83^2/32.2)^{1/3} = 6.26 \text{ ft}$$
$$V_c = q/y_c = 88.83/6.26 = 14.19 \text{ ft/s} \qquad V_m = (12.69 + 14.19)/2 = 13.44 \text{ ft/s}$$
$$R_m = [(20)(7)/(7 + 20 + 7) + (20)(6.26)/(6.26 + 20 + 6.26)]/2 = 3.984 \text{ ft}$$
$$S = \{(0.012)(13.44)/[(1.486)(3.984)^{2/3}]\}^2 = 0.001865$$
$$7 + 12.69^2/[(2)(32.2)] = 6.26 + 14.19^2/[(2)(32.2)] + 0.001865L \qquad L = 61.1 \text{ ft}$$

But $L = 1000 - (6)(6.26) = 962$ ft. Since the computed value of $L$ (61.1 ft) is not equal to the actual length of approximately 962 ft, the assumed value of $y_1$ of 7 ft is incorrect. As a second trial, $y_1$ should be increased. After successive trials (not shown), a value of $y_1 = 8.46$ ft is found to be appropriate. Then, $8.46 + 1.2V_1^2/[(2)(32.2)] = 10$, $V_1 = 9.091$ ft/s, $q = (9.091)(8.46) = 76.91$ cfs/ft, $Q = (76.91)(20) = 1538$ ft³/s.

**14.313** An 8-ft-wide rectangular channel has bottom slope $s_0 = 0.0015$ and roughness $n = 0.013$; it carries 250 ft³/s of water. Compute the separation between sections where depths are 4.00 ft and 4.20 ft.

$$\Delta x = (y_1 + v_1^2/2g - y_2 - v_2^2/2g)/(s - s_0) \qquad s = (n v_{avg}/1.486 R_{avg}^{2/3})^2$$

The calculations in the following table give the separation as 512 ft.

| $y,$ ft | $A,$ ft$^2$ | $p$ $(8+2y),$ ft | $R_h,$ ft | $v,$ fps | $v^2/2g,$ ft | $y + (v^2/2g)$ | Numerator $\Delta[y + (v^2/2g)]$ | $v_{avg},$ fps | $R_{avg},$ ft | $s$ | Denominator $s - s_0$ | $\Delta x,$ ft |
|---|---|---|---|---|---|---|---|---|---|---|---|---|
| 4.00 | 32.00 | 16.00 | 2.00 | 7.81 | 0.947 | 4.947 | | | | | | |
| | | | | | | | 0.055 | 7.72 | 2.013 | 0.00179 | 0.00029 | 190 |
| 4.10 | 32.80 | 16.20 | 2.025 | 7.62 | 0.902 | 5.002 | | | | | | |
| | | | | | | | 0.058 | 7.53 | 2.037 | 0.00168 | 0.00018 | 322 |
| 4.20 | 33.60 | 16.40 | 2.049 | 7.44 | 0.860 | 5.060 | | | | | | |
| | | | | | | | | | | | | $\sum (\Delta x) = 512$ |

**14.314** Assume the channel of Prob. 14.145 has a bottom slope of 0.0048. If $y_0 = 3$ ft at $x = 0$, how far along the channel $x = L$ does it take the depth to rise to $y_L = 4$ ft (see Fig. 14-80). Use increments of $y = 0.2$ ft and Manning's formula. Is the 4-ft-depth position upstream or downstream in Fig. 14-80?

▌$y_c = 4.27$ ft   (from Prob. 14.145). Since $[y = 3 \text{ ft}] < [y_c = 4.27 \text{ ft}]$, the flow is supercritical; and the given channel slope (0.0048) is greater than $s_c$ of 0.00435 (from Prob. 14.145). Therefore, we must be on an $S$ profile. $Q = (A)(1.486/n)(R^{2/3})(s^{1/2})$. Since the channel is "wide", use $R = y_n$: $50 = [(1)(y_n)](1.486/0.022)(y_n)^{2/3}(0.0048)^{1/2}$, $y_n = 4.14$ ft. Thus $y_0$ and $y_L$ are less than $y_n$, which is less than $y_c$ and we *must* be on an $S_3$ curve. For the numerical solution, we tabulate $y = 3.0$ to $4.0$ in intervals of 0.2, computing six values of $V = q/y$, $E = y + (V^2/2g)$, and $S = n^2 V^2/2.208 y^{4/3}$, from which $S_{av}$ and $\Delta x$ follow. The slope $S_0 = 0.0048$ is constant.

| $y,$ ft | $V = 50/y$ | $E = y + (V^2/2g)$ | $S$ | $S_{av}$ | $\Delta x$ | $x = \sum \Delta x$ |
|---|---|---|---|---|---|---|
| 3.0 | 16.67 | 7.313 | 0.01407 | | | 0.0 |
| | | | | 0.01271 | 40.7 | |
| 3.2 | 15.62 | 6.991 | 0.01135 | | | 40.7 |
| | | | | 0.01031 | 42.3 | |
| 3.4 | 14.71 | 6.758 | 0.00927 | | | 83.0 |
| | | | | 0.00847 | 44.4 | |
| 3.6 | 13.89 | 6.595 | 0.00766 | | | 127.4 |
| | | | | 0.00703 | 48.0 | |
| 3.8 | 13.16 | 6.488 | 0.00640 | | | 175.4 |
| | | | | 0.00590 | 56.4 | |
| 4.0 | 12.50 | 6.426 | 0.00539 | | | $L = 231.8$ ft |

For this depth increment of 0.2 ft, gradually-varied-flow theory predicts that a length of about 232 ft is required for the depth to rise from 3 to 4 ft in this supercritical flow. Using 10 increments by reducing $\Delta y$ to 0.1 ft would give an estimate of $L = 235.2$ ft (calculations not shown). It should be clear that the $y = 4$ ft position is downstream.

Fig. 14-80

**14.315** If a rectangular channel 10 ft wide, with $n = 0.014$ and $s_0 = 0.0009$, carries $Q = 300$ cfs, is the slope steep or mild?

$$Q = (A)(1.486/n)(R^{2/3})(s^{1/2}) \qquad 300 = (10y_n)(1.486/0.014)[10y_n/(y_n + 10 + y_n)]^{2/3}(0.0009)^{1/2}$$

$$y_n = 5.09 \text{ ft} \qquad \text{(by trial and error)} \qquad y_c = (q^2/g)^{1/3} = [(\tfrac{300}{10})^2/32.2]^{1/3} = 3.03 \text{ ft}$$

Since $y_c < y_n$, the slope is mild.

**14.316** Repeat Prob. 14.315 if the slope is 0.006.

$$Q = (A)(1.486/n)(R^{2/3})(s^{1/2}) \qquad 300 = (10y_n)(1.486/0.014)[10y_n/(y_n + 10 + y_n)]^{2/3}(0.006)^{1/2}$$

$$y_n = 2.57 \text{ ft} \qquad \text{(by trial and error)} \qquad y_c = (q^2/g)^{1/3} = [(\tfrac{300}{10})^2/32.2]^{1/3} = 3.03 \text{ ft}$$

Since $y_c > y_n$, the slope is steep.

**14.317** Water flows in a wide clean-earth ($n = 0.020$) channel at $s_0 = 0.006$ and $q = 10$ (m³/s)/m. Give the ranges of depths corresponding to flow on a type 1, 2, or 3 curve.

$$Q/B = (A/B)(1.0/n)(R^{2/3})(s^{1/2}) \qquad 10 = (y_n)(1.0/0.020)(y_n)^{2/3}(0.006)^{1/2}$$

$$y_n = 1.77 \text{ m} \qquad y_c = (q^2/g)^{1/3} = (10^2/9.807)^{1/3} = 2.17 \text{ m}$$

Since $y_c > y_n$, the slope is steep. Flow will be $S_1$ for $y > 2.17$ m (see Fig. A-19); $S_2$ for $1.77 \text{ m} < y < 2.17$ m; and $S_3$ for $y < 1.77$ m.

**14.318** Repeat Prob. 14.317 if the channel slope is 0.0014.

$$Q/B = (A/B)(1.0/n)(R^{2/3})(s^{1/2}) \qquad 10 = (y_n)(1.0/0.020)(y_n)^{2/3}(0.0014)^{1/2}$$

$$y_n = 2.73 \text{ m} \qquad y_c = (q^2/g)^{1/3} = (10^2/9.807)^{1/3} = 2.17 \text{ m}$$

Since $y_c < y_n$, the slope is mild. Flow will be $M_1$ for $y > 2.73$ m (see Fig. A-19); $M_2$ for $2.17 \text{ m} < y < 2.73$ m; and $M_3$ for $y < 2.17$ m.

**14.319** Water flows at 130 cfs in an isosceles right-triangular sluice with $n = 0.015$ and $s_0 = 0.010$. For what ranges of depths will the flow be a type 1, 2, or 3 curve?

$$Q = (A)(1.486/n)(R^{2/3})(s^{1/2}) \qquad A = y_n y_n = y_n^2 \qquad p_w = 2y_n/(\cos 45°)$$

$$R = A/p_w = y_n^2/[2y_n/(\cos 45°)] = 0.3536 y_n$$

$$130 = (y_n^2)(1.486/0.015)(0.3536 y_n)^{2/3}(0.010)^{1/2} \qquad y_n = 3.41 \text{ ft}$$

$$A_c = (bQ^2/g)^{1/3} \qquad y_c^2 = [(2y_c)(130)^2/32.2]^{1/3} \qquad y_c = 4.02 \text{ ft}$$

Since $y_c > y_n$, the slope is steep. Flow will be $S_1$ for $y > 4.02$ ft (see Fig. A-19); $S_2$ for $3.41 \text{ ft} < y < 4.02$ ft; and $S_3$ for $y < 3.41$ ft.

**14.320** A riveted-steel ($n = 0.014$) duct is 2 m in diameter and laid on a slope of 0.000758. The duct is half-full of water flowing at 2.5 m³/s. Specify the type of the flow curve.

$Q = (A)(1.0/n)(R^{2/3})(s^{1/2})$, $Q_{full} = [(\pi)(2)^2/4](1.0/0.014)(\tfrac{2}{4})^{2/3}(0.000758)^{1/2} = 3.89$ m³/s, $Q/Q_{full} = 2.5/3.89 = 0.64$. From Fig. A-18, $D/D_{full} = 58$ percent. $y_n = (0.58)(2) = 1.16$ m, $A_c = (bQ^2/g)^{1/3}$. Try $y_c = 0.70$ m: $y_c/y_{full} = 0.70/2 = 0.35$. From Fig. A-18, $A_c/A_{full} = 30\%$.

$$A_c = 0.30[(\pi)(2)^2/4] = 0.9425 \text{ m}^2 \qquad b = 2[\sqrt{1^2 - (1 - y_c)^2}] \quad \text{(see Fig. 14-81)}$$

$$b = 2[\sqrt{1^2 - (1 - 0.70)^2}] = 1.908 \text{ m} \qquad 0.9425 = (1.908 Q^2/9.807)^{1/3} \qquad Q = 2.07 \text{ m}^3/\text{s}$$

Since the computed value of $Q$ (2.07 m³/s) is not equal to the given value (2.5 m³/s), try $y_c = 0.75$ m: $y_c/y_{full} = 0.75/2 = 0.375$. From Fig. A-18, $A_c/A_{full} = 34\%$.

$$A_c = 0.34[(\pi)(2)^2/4] = 1.068 \text{ m}^2 \qquad b = 2[\sqrt{1^2 - (1 - 0.75)^2}] = 1.936 \text{ m}$$

$$1.068 = (1.936 Q^2/9.807)^{1/3} \qquad Q = 2.48 \text{ m}^3/\text{s}$$

This value of $Q$ is close enough to the given value of 2.5 m³/s so that $y_c$ can be taken as 0.75 m. Since $y_c = 0.75$ m $< 1.16$ m $= y_n$, the slope is mild, and an $M_2$ curve (Fig. A-19) is appropriate.

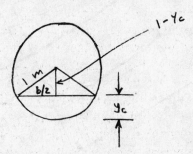

**Fig. 14-81**

**14.321** Consider the gradual change from the profile beginning at point $a$ in Fig. 14-82$a$, on a mild slope to a mild but steeper slope downstream. Sketch and label the curve $y(x)$ expected.

$\blacksquare$ See Fig. 14-82$b$.

$(a)$ $(b)$

**Fig. 14-82**

**14.322** Consider the wide-channel flow in Fig. 14-83$a$, which changes from a mild to a steep slope. Beginning at point $a$, sketch the water-surface profile $y(x)$ which is expected for gradually varied flow.

$\blacksquare$ See Fig. 14-83$b$.

$(a)$ $(b)$

**Fig. 14-83**

**14.323** In Fig. 14-84$a$ the channel slope changes from steep to less steep. Beginning at point $a$, sketch and label the expected surface curve $y(x)$ from gradually-varied-flow theory.

$\blacksquare$ See Fig. 14-84$b$.

**Fig. 14-84$a$**

**Fig. 14-84b** (*continued*)

**14.324** Repeat Prob. 14.314 to find the length $L$ using only two depth increments of 0.5 ft each. Is the accuracy sufficient?

▮ Using data from Prob. 14.314,

| $y$ | $V$ | $E$ | $S$ | $S_{av}$ | $\Delta x$ |
|-----|-----|-----|-----|----------|-----------|
| 3.0 | 16.67 | 7.313 | 0.01407 | | |
| | | | | 0.01125 | 100 |
| 3.5 | 14.29 | 6.669 | 0.00842 | | |
| | | | | 0.00691 | 115 |
| 4.0 | 12.50 | 6.426 | 0.00539 | | $L = 215$ ft |

Using only two depth increments gives $(231.8 - 215)/231.8 = 0.07$, or 7 percent smaller length than that using five increments.

**14.325** A wide-channel flow of $q = 4$ (m³/s)/m, with $n = 0.016$, climbs an adverse slope $s_0 = -0.0009$. Using distance increments of 125 mm, approximate the distance required for the depth to change from 2.5 m to 2.0 m.

▮ $y_c = (q^2/g)^{1/3} = (4^2/9.807)^{1/3} = 1.18$ m. Hence we have an $A_2$ profile (see Fig. A-19).

| $y$ | $v = 4/y$ | $E$ | $S = n^2 v^2/y^{4/3}$ | $s_{avg}$ | $\Delta x = \Delta E/(s_0 - s_{avg})$ |
|-----|-----------|-----|----------------------|-----------|---------------------------------------|
| 2.5 | 1.600 | 2.6305 | 0.0001931 | | |
| | | | | 0.0002111 | 99.8 |
| 2.375 | 1.684 | 2.5196 | 0.0002291 | | |
| | | | | 0.0002518 | 94.2 |
| 2.25 | 1.778 | 2.4111 | 0.0002745 | | |
| | | | | 0.0003032 | 87.7 |
| 2.125 | 1.882 | 2.3056 | 0.0003319 | | |
| | | | | 0.0003692 | 80.1 |
| 2.0 | 2.000 | 2.2039 | 0.0004064 | | $\sum (\Delta x) = L \approx 362$ m |

**14.326** Figure 14-85 illustrates "free overfall" or "dropdown," where a channel flow accelerates down a slope and falls freely over a sharp edge. As shown, the flow becomes critical just before the overfall. Between $y_c$ and the edge the flow fluctuates rapidly and does not obey gradually-varied-flow theory. If the flow rate is $q = 1.1$ (m³/s)/m and the surface is unfinished concrete ($n = 0.014$), estimate the water depth 300 m upstream.

▮ $$q = (1.0/n)(y)(y^{2/3})(s^{1/2}) \qquad 1.1 = (1.0/0.014)(y_n)(y_n)^{2/3}(0.001047°)^{1/2} \qquad y_n = 0.64 \text{ m}$$
$$y_c = (q^2/g)^{1/3} = (1.1^2/9.807)^{1/3} = 0.50 \text{ m}$$

Hence we have an $M_2$ profile (see Fig. A-19). Use $\Delta y = 0.02$ m.

| $y$ | $v = 1.1/y$ | $E$ | $s$ | $s_{av}$ | $\Delta x$ |
|---|---|---|---|---|---|
| 0.50 | 2.200 | 0.74669 | 0.0023904 | | |
| | | | | 0.0022439 | 1.2 |
| 0.52 | 2.115 | 0.74808 | 0.0020975 | | |
| | | | | 0.0019735 | 3.7 |
| 0.54 | 2.037 | 0.75149 | 0.0018495 | | |
| | | | | 0.0017439 | 7.4 |
| 0.56 | 1.964 | 0.75666 | 0.0016384 | | |
| | | | | 0.0015480 | 13.3 |
| 0.58 | 1.897 | 0.76333 | 0.0014575 | | |
| | | | | 0.0013796 | 24.0 |
| 0.60 | 1.833 | 0.77131 | 0.0013018 | | |
| | | | | 0.0012344 | 48.0 |
| 0.62 | 1.774 | 0.78044 | 0.0011670 | | |
| | | | | 0.0011367 | 55.2 |
| 0.63 | 1.746 | 0.78538 | 0.0011064 | | |
| | | | | 0.0010920 | 57.3 |
| 0.635 | 1.732 | 0.78795 | 0.0010776 | | |
| | | | | 0.0010692 | 70.9 |
| 0.638 | 1.724 | 0.78951 | 0.0010608 | | |
| | | | | 0.0010594 | 21.6 |
| 0.6385 | 1.723 | 0.78977 | 0.0010580 | | |
| | | | | | $\sum (\Delta x) \approx 300$ m |

Fig. 14-85

**14.327** The clean-earth ($n = 0.020$) channel in Fig. 14-86 is 6 m wide and laid on a slope of 0.005236. Water flows at 30 m³/s in the channel and enters a reservoir so that the channel depth is 3 m just before the entry. Assuming gradually varied flow, calculate the distance $L$.

**I** $\qquad Q = (A)(1.0/n)(R^{2/3})(s^{1/2}) \qquad 30 = (6y_n)(1.0/0.022)[6y_n/(y_n + 6 + y_n)]^{2/3}(0.005236)^{1/2}$

$\qquad y_n = 1.52$ m $\quad$ (by trial and error) $\qquad y_c = (q^2/g)^{1/3} = [(30/6)^2/9.807]^{1/3} = 1.37$ m

Hence we have an $M_1$ profile (see Fig. A-19). Use $\Delta y = 0.25$ m.

| $y$ | $v = 30/6y$ | $E$ | $R_h$ | $s = n^2v^2/R_h^{4/3}$ | $s_{av}$ | $\Delta x$ |
|---|---|---|---|---|---|---|
| 3.0 | 1.667 | 3.1416 | 1.500 | 0.0006474 | | |
| | | | | | 0.0007321 | 50 |
| 2.75 | 1.818 | 2.9185 | 1.435 | 0.0008168 | | |
| | | | | | 0.0009373 | 50 |
| 2.50 | 2.000 | 2.7039 | 1.364 | 0.0010577 | | |
| | | | | | 0.001235 | 51 |
| 2.25 | 2.222 | 2.5017 | 1.286 | 0.0014122 | | |
| | | | | | 0.0016864 | 52 |
| 2.0 | 2.500 | 2.3186 | 1.200 | 0.0019605 | | |
| | | | | | $L = \sum (\Delta x) \approx 203$ m | |

**Fig. 14-86**

**14.328** The channel of Fig. 14-87 is sometimes called a *venturi flume,* because measurements of $y_1$ and $y_2$ can be used to meter the flow rate. Suppose that $b_1 = 4.5$ m, $b_2 = 3$ m, $y_1 = 2$ m, and $y_2 = 1.4$ m. Assuming no losses, compute the flow rate $Q$.

**∎**
$$y_1 + v_1^2/2g = y_2 + v_2^2/2g \qquad Q = y_1 b_1 v_1 = y_2 b_2 v_2$$

Solve these equations simultaneously: $v_2 = \{(2g)(y_1 - y_2)/[1 - (y_2 b_2/y_1 b_1)^2]\}^{1/2}$, $Q = y_2 b_2 v_2 = \{(2g)(y_1 - y_2)/[(1/b_2^2 y_2^2) - (1/b_1^2 y_1^2)]\}^{1/2} = \{(2)(9.807)(2 - 1.4)/[1/(3^2)(1.4)^2 - 1/(4.5)^2(2)^2]\}^{1/2} = 16.3$ m³/s.

Top view

Side view

**Fig. 14-87**

**14.329** In Prob. 14.328, holding $y_1$, $b_1$, and $b_2$ fixed, find $y_2$ for critical exit flow $v_2$.

**∎** $y_c = [(Q/b)^2/g]^{1/3}$. $y_2$ must equal $y_c$. $Q = \{(2g)(y_1 - y_2)/[(1/b_2^2 y_2^2) - (1/b_1^2 y_1^2)]\}^{1/2}$ (from Prob. 14.328). Try $y_2 = 1.5$ m: $Q = \{(2)(9.807)(2 - 1.5)/[1/(3^2)(1.5)^2 - 1/(4.5)^2(2)^2]\}^{1/2} = 16.27$ m³/s, $y_c = [(16.27/3)^2/9.807]^{1/3} = 1.44$ m. Since $y_c = 1.44$ m is not equal to the assumed $y_2 = 1.5$ m, try $y_2 = 1.45$ m: $Q = \{(2)(9.807)(2 - 1.45)/[1/(3^2)(1.45)^2 - 1/(4.5)^2(2)^2]\}^{1/2} = 16.32$ m³/s, $y_c = [(16.32/3)^2/9.807]^{1/3} = 1.45$ m (O.K.). Hence critical flow of 16.32 m³/s will occur at $y_2 = 1.45$ m.

**14.330** In Prob. 14.328, holding $y_2$, $b_1$, and $b_2$ fixed, find $y_1$ for a flow rate of 12.2 m³/s.

**∎**
$$Q = \{(2g)(y_1 - y_2)/[(1/b_2^2 y_2^2) - (1/b_1^2 y_1^2)]\}^{1/2} \qquad \text{(from Prob. 14.328)}$$
$$12.2 = \{(2)(9.807)(y_1 - 1.4)/[(1/(3^2)(1.4)^2 - 1/(4.5)^2(y_1)^2]\}^{1/2} \qquad y_1 = 1.70 \text{ m} \qquad \text{(by trial and error)}$$

**14.331** In the venturi flume of Fig. 14-87, let $b_1 = 5$ ft, $b_2 = 4$ ft, $y_1 = 3$ ft, and $Q = 75.9$ cfs. Compute $y_2$, assuming no losses.

**∎**
$$Q = \{(2g)(y_1 - y_2)/[(1/b_2^2 y_2^2) - (1/b_1^2 y_1^2)]\}^{1/2} \qquad \text{(from Prob. 14.328)}$$
$$75.9 = \{(2)(32.2)(2 - y_2)/[(1/(3^2)(y_2)^2 - 1/(4)^2(2)^2]\}^{1/2} \qquad y_2 = 2.00 \text{ ft} \qquad \text{(by trial and error)}$$

**14.332** A flow of 35 m³/s flows along a trapezoidal concrete channel where (see Fig. 14-88) the base $a$ is 4 m and $\beta$ is 45°. If at section 1, the depth of the flow is 3 m, what is the water-surface profile up to a distance 600 m downstream. The channel is finished concrete and has a constant slope $S_0$ of 0.001.

**∎**
$$\Delta L = ([1 - (Q^2 b/gA^3)]/\{S_0 - (n/\kappa)^2[Q^2/(R_H^{4/3}A^2)]\}) \, \Delta y$$

We start with $y_1 = 3$ m. At this section we know that

$$A_1 = (3)(4) + \tfrac{1}{2}(3)(3)(2) = 21 \text{ m}^2$$
$$(R_H)_1 = A_1/(p_w)_1 = 21/[4 + (2)(3/0.707)] = 1.6818 \text{ m} \tag{1}$$
$$b_1 = 4 + (2)(3) = 10 \text{ m}$$

We take $n = 0.012$ and $\kappa = 1.00$ and let $y_2 = 3.1$ m. Now we compute $A_2$, $(R_H)_2$, and $b_2$.*

$$A_2 = (3.1)(4) + \tfrac{1}{2}(3.1)(3.1)(2) = 22.01 \text{ m}^2$$
$$(R_H)_2 = 22.01/[4 + 2(3.1/0.707)] = 1.7236 \text{ m} \tag{2}$$
$$b_2 = 4 + (2)(3.1) = 10.20 \text{ m}$$

In the first interval, the average values of $A$, $R_H$, and $b$ are

$$(A_{1-2})_{av} = 21.505 \text{ m}^2$$
$$[(R_H)_{1-2}]_{av} = 1.7027 \text{ m} \tag{3}$$
$$(b_{1-2})_{av} = 10.10 \text{ m}$$

$$(\Delta L)_{1-2} = \left\{ \frac{1 - (35^2)(10.10)/[(9.81)(21.505^3)]}{0.01 - (0.012/1)^2(35^2)/[(1.7027^{4/3})(21.505^2)]} \right\} (0.1) = 107.4 \text{ m}$$

We thus have two points on the free-surface profile. Next we compute $A_3$, $R_{H_3}$, and $b_3$ for $y = 3.2$ m. Using Eqs. (2) we now find the average values of these quantities in the interval 2–3. For instance, $(A_{2-3})_{av} = \tfrac{1}{2}\{(22.01) + [(3.2)(4) + (3.2^2)]\} = 22.525 \text{ m}^2$. We then proceed as indicated for the first interval.

The following table gives the results using six sections so that $L_{total} \approx 600$ m.

| section | $y$, m | $\Delta y$, m | $\Delta L$, m | $L_{total}$, m |
|---------|--------|---------------|---------------|----------------|
| 1 | 3.0 | 0.1 | 0 | 0 |
| 2 | 3.1 | 0.1 | 107.4 | 107.4 |
| 3 | 3.2 | 0.1 | 106.9 | 214.3 |
| 4 | 3.3 | 0.1 | 105.6 | 319.9 |
| 5 | 3.4 | 0.1 | 104.7 | 424.6 |
| 6 | 3.5 | 0.1 | 104.1 | 528.7 |
| 7 | 3.6 | 0.1 | 103.8 | 632.5 |

We can now plot $y$ versus $L$, that is, the second and last columns, starting with $y = 3$ for $L = 0$ m and going on to $y = 3.6$ m for $L = 632.5$ m. A smooth curve through these parts gives the approximate desired profile.

**Fig. 14-88**

**14.333** Water enters a rectangular channel which is 3 ft wide at an average velocity of 0.8 ft/s and a depth of 3 in. The channel is laid on a slope of 0.003499. If Manning's $n$ is 0.012, how far must the water travel before its depth becomes 4 in? Use four intermediary depths ($\Delta y = 0.2$ in) in the calculation.

■
$$\Delta x = ([1 - (Q^2 b/gA^3)]/\{s_0 - (n/\kappa)^2[Q^2/(R_h^{4/3}A^2)]\}) \Delta y$$
$$s_0 = 0.003499 \qquad R_h = by/(b + 2y) = y$$

Point 1:

$$v_1 = 0.8 \text{ ft/s} \qquad (f_R)_1 = v\sqrt{b}/\sqrt{gA} = (0.8)(\sqrt{3})/\sqrt{(g)(3)(\tfrac{1}{4})} = 0.282 \qquad \Delta y = 0.2/12 \text{ ft}$$

Point 2:

$$v_2 = (y_0/y_1)v_1 = (3/3.2)(0.8) = 0.750 \text{ ft/s} \qquad (f_R)_2 = (0.750)(\sqrt{3})/\sqrt{(g)(3)(3.2/12)} = 0.256$$

---

* We are assuming that the depth $y$ relative to the channel is increasing. If we get a positive result for $\Delta L$, we know that our assumption is correct. If $\Delta L$ is negative, then the depth must be decreasing along the channel flow. Use $y_2$ less than 3 m in that case.

1–2:

$$(v_{1-2})_{av} = (0.8 + 0.750)/2 = 0.775 \text{ ft/s} \qquad (f_{R,1-2})_{av} = (0.282 + 0.256)/2 = 0.269$$

$$(\Delta x)_{1-2} = \frac{1 - 0.269^2}{0.00349 - (0.012/1.486)^2[0.775^2/(3.1/12)^{4/3}]}\left(\frac{0.2}{12}\right) = 4.75 \text{ ft}$$

Point 3:

$$v_3 = (3/3.4)(0.8) = 0.706 \text{ ft/s} \qquad (f_R)_3 = 0.706/[g(3.4/12)]^{1/2} = 0.234$$

2–3:

$$(v_{2-3})_{av} = (0.750 + 0.706)/2 = 0.728 \text{ ft/s} \qquad (f_{R,2-3})_{av} = (0.256 + 0.234)/2 = 0.245$$

$$(\Delta x)_{2-3} = \frac{1 - 0.245^2}{0.00349 - (0.012/1.486)^2[0.728^2/(3.3/12)^{4/3}]}\left(\frac{0.2}{12}\right) = 4.75 \text{ ft}$$

Point 4:

$$v_4 = (3/3.6)(0.8) = 0.667 \text{ ft/s} \qquad (f_R)_4 = 0.667/\sqrt{(g)(3.6/12)} = 0.215$$

3–4:

$$(v_{3-4})_{av} = (0.706 + 0.667)/2 = 0.686 \text{ ft/s} \qquad (f_{R,3-4})_{av} = (0.234 + 0.215)/2 = 0.224$$

$$(\Delta x)_{3-4} = \frac{1 - 0.224^2}{0.00349 - (0.012/1.486)^2[0.686^2/(3.5/12)^{4/3}]}\left(\frac{0.2}{12}\right) = 4.75 \text{ ft}$$

Point 5:

$$v_5 = (3/3.8)(0.8) = 0.632 \text{ ft/s} \qquad (f_R)_5 = 0.632/\sqrt{(g)(3.8/12)} = 0.198$$

4–5:

$$(v_{4-5})_{av} = (0.667 + 0.632)/2 = 0.650 \text{ ft/s} \qquad (f_{R,4-5})_{av} = (0.215 + 0.198)/2 = 0.206$$

$$(\Delta x)_{4-5} = \frac{1 - 0.206^2}{0.00349 - (0.012/1.486)^2[0.650^2/(3.7/12)^{4/3}]}\left(\frac{0.2}{12}\right) = 4.75 \text{ ft}$$

Point 6:

$$v_6 = (\tfrac{3}{4})(0.8) = 0.600 \text{ ft/s} \qquad (f_R)_6 = 0.600/[g(4.0/12)]^{1/2} = 0.183$$

5–6:

$$(v_{5-6})_{av} = (0.632 + 0.600)/2 = 0.616 \text{ ft/s} \qquad (f_{R,5-6})_{av} = (0.198 + 0.183)/2 = 0.190$$

$$(\Delta x)_{5-6} = \frac{1 - 0.190^2}{0.00349 - (0.012/1.486)^2[0.616^2/(3.9/12)^{4/3}]}\left(\frac{0.2}{12}\right) = 4.75 \text{ ft}$$

$$x = \sum (\Delta x) = 23.75 \text{ ft}$$

**14.334** In Prob. 14.332, compute the distance $L$ in one calculation where the free surface has a depth of 3.6 m. Do not average.

❚ 
$$\Delta L = \left([1 - (Q^2 b/gA^3)]/\{S_0 - (n/\kappa)^2[Q^2/(R_H^{4/3} A^2)]\}\right)\Delta y$$

$$A = (3)(4) + (3)(3) = 21.00 \text{ m}^2 \qquad p_w = 4 + (2)(\sqrt{3^2 + 3^2}) = 12.49 \text{ m}$$

$$R_H = 21.00/12.49 = 1.681 \text{ m} \qquad b = 3 + 4 + 3 = 10 \text{ m}$$

$$\Delta L = \left\{\frac{1 - (35)^2(10)/(9.807)(21.00)^3}{0.001 - (0.012/1)^2[35^2/(1.681)^{4/3}(21.00)^2]}\right\}(3.6 - 3) = 649 \text{ m}$$

**14.335** Do Prob. 14.334 using a linear average in the calculations. What is the percent error in not averaging?

❚ 
$$\Delta L = \left([1 - (Q^2 b/gA^3)]/\{S_0 - (n/\kappa)^2[Q^2/(R_H^{4/3} A^2)]\}\right)\Delta y$$

At the section where $y = 3.6 \text{ m}$,

$$A = (3.6)(4) + (3.6)(3.6) = 27.36 \text{ m}^2 \qquad p_w = 4 + (2)(\sqrt{3.6^2 + 3.6^2}) = 14.18 \text{ m}$$

$$R_H = 27.36/14.18 = 1.929 \text{ m} \qquad b = 3.6 + 4 + 3.6 = 11.20 \text{ m}$$

Using corresponding values where $y = 3$ m from Prob. 14.334,

$$A_{av} = (21.00 + 27.36)/2 = 24.18 \text{ m}^2 \qquad (R_H)_{av} = (1.681 + 1.929)/2 = 1.805 \text{ m}$$

$$b_{av} = (10 + 11.20)/2 = 10.60 \text{ m}$$

$$\Delta L = \left\{ \frac{1 - (35)^2(10.60)/(9.807)(24.18)^3}{0.001 - (0.012/1)^2[35^2/(1.805)^{4/3}(24.18)^2]} \right\}(3.6 - 3) = 630 \text{ m}$$

Error $= (649 - 630)/630 = 0.03$, or 3 percent.

**14.336** A wide sluice, made of finished concrete and having slope $s_0 = 0.0004$, is fed from a large reservoir through a sharp-edged gate (Fig. 14-89). The coefficient of contraction $C_c$ is 0.82 and the coefficient of friction $C_f = 0.86$. How far from the vena contracta does the water increase depth by 30 mm? Make a one-stage calculation with linear averages.

**❚**
$$v_1^2/2g + y_1 = v_2^2/2g + y_2 \qquad y_2 = (0.30)(0.82) = 0.2460 \text{ m}$$

$$0 + 12 = v_2^2/[(2)(9.807)] + 0.2460 \qquad v_2 = 15.18 \text{ m/s}$$

$$q_{\text{theoretical}} = y_2 v_2 = (0.2460)(15.18) = 3.734 \text{ (m}^3/\text{s)/m} \qquad q_{\text{actual}} = (0.86)(3.734) = 3.21 \text{ (m}^3/\text{s)/m}$$

$$\Delta x = ([1 - (Q^2 b/gA^3)]/\{s_0 - (n/\kappa)^2[Q^2/(R_h^{4/3}A^2)]\}) \, \Delta y$$

$$\Delta y_{\text{downstream}} = 0.246 + 0.03 = 0.276 \text{ m}$$

$$\Delta x = \left\{ \frac{1 - (3.21)^2/(9.807)(0.276)^3}{0.0004 - (0.012/1)^2[3.21^2/(0.276)^{4/3}(0.276)^2]} \right\}(0.030) = 13.60 \text{ m}$$

**Fig. 14-89**

**14.337** Water is moving at 5 m/s in a very wide horizontal channel at a depth of 1 m. If $n = 0.024$, find analytically the distance $x$ downstream to the section of depth 1.1 m.

**❚**
$$x = (\kappa/n)^2 \left[ \frac{3}{4g}(y^{4/3} - y_1^{4/3}) - \frac{3}{13q^2}(y^{13/3} - y_1^{13/3}) \right]$$

$$= \left(\frac{1}{0.024}\right)^2 \left\{ \left[ \frac{3}{(4)(9.807)} \right](1.1^{4/3} - 1^{4/3}) - \left[ \frac{3}{(13)(5)^2} \right](1.1^{13/3} - 1^{13/3}) \right\} = 9.80 \text{ m}$$

**14.338** In a wide rectangular earth channel obstructed with stones ($n = 0.032$), the depth of flow increases from 1 m to 1.1 m over a 10 m distance. Compute the volumetric flow per unit width.

**❚**
$$x = \left(\frac{\kappa}{n}\right)^2 \left[ \frac{3}{4g}(y^{4/3} - y_1^{4/3}) - \frac{3}{13q^2}(y^{13/3} - y_1^{13/3}) \right]$$

$$10 = \left(\frac{1}{0.032}\right)^2 \left\{ \left[ \frac{3}{(4)(9.807)} \right](1.1^{4/3} - 1^{4/3}) - \left(\frac{3}{13q^2}\right)(1.1^{13/3} - 1^{13/3}) \right\}$$

$$q = 30.96 \text{ (m}^3/\text{s)/m}$$

**14.339** In Prob. 14.196, for a flow of 4900 cfs, calculate the distance downstream over which the depth increases from 13 ft to 14 ft. Make a one-step calculation without averaging.

▮
$$\Delta x = ([1 - (Q^2 b / g A^3)]/\{s_0 - (n/\kappa)^2[Q^2/(R_h^{4/3} A^2)]\}) \, \Delta y$$

From Prob. 14.196, $A = 217.1$ ft$^2$, $R_h = 5.802$ ft, $n = 0.012$, $s_0 = 0.0016$, $b = 20$ ft, and $\Delta y = 1$ ft.

$$\Delta x = \left\{ \frac{1 - (4900)^2(20)/(32.2)(217.1)^3}{0.0016 - (0.012/1.486)^2[4900^2/(5.802)^{4/3}(217.1)^2]} \right\}(1) = 288 \text{ ft}$$

**14.340** Repeat Prob. 14.339, using a linear average over the interval.

▮
$$\Delta x = ([1 - (Q^2 b / g A^3)]/\{s_0 - (n/\kappa)^2[Q^2/(R_h^{4/3} A^2)]\}) \, \Delta y$$

At $y = 14$ ft, $A = (\frac{1}{2})(\pi)(10)^2 + (4)(10 + 10) = 237.1$ ft$^2$ and $R_h = 237.1/[(\pi)(10) + 4 + 4] = 6.015$ ft. Therefore,

$$A_{av} = (217.1 + 237.1)/2 = 227.1 \text{ ft}^2 \qquad (R_h)_{av} = (5.802 + 6.015)/2 = 5.908 \text{ ft}$$

$$\Delta x = \left\{ \frac{1 - (4900)^2(20)/(32.2)(227.1)^3}{0.0016 - (0.012/1.486)^2[4900^2/(5.908)^{4/3}(227.1)^2]} \right\}(1) = 220 \text{ ft}$$

**14.341** Water flows at a rate of 500 ft$^3$/s through a rectangular section 10.0 ft wide from a "steep" slope to a "mild" slope creating a hydraulic jump, in the manner illustrated in Fig. 14-90. The upstream depth of flow ($d_1$) is 3.1 ft. Find the (a) downstream depth, (b) energy (head) loss in the hydraulic jump, and (c) upstream and downstream velocities.

▮ (a)   $d_2 = (d_1/2)(\sqrt{1 + 8q^2/g d_1^3} - 1) = (3.1/2)\{\sqrt{1 + (8)(500/10.0)^2/[(32.2)(3.1)^3]} - 1\} = 5.7$ ft

(b)   $E_j = (d_2 - d_1)^3/4 d_1 d_2 = (5.7 - 3.1)^3/[(4)(3.1)(5.7)] = 0.25$ ft of water

(c)   $v_1 = Q/A_1 = 500/[(3.1)(10.0)] = 16.1$ ft/s     $v_2 = 500/[(5.7)(10.0)] = 8.8$ ft/s

**Fig. 14-90**

**14.342** Rework Prob. 14.341 for a flow rate of 20.0 m$^3$/s, a channel width of 4.0 m, and an upstream depth of flow of 1.20 m.

▮ (a)   $d_2 = (d_1/2)(\sqrt{1 + 8q^2/g d_1^3} - 1) = (1.20/2)\{\sqrt{1 + (8)(20.0/4.0)^2/[(9.807)(1.20)^3]} - 1\} = 1.55$ m

(b)   $E_j = (d_2 - d_1)^3/4 d_1 d_2 = (1.55 - 1.20)^3/[(4)(1.20)(1.55)] = 0.006$ m of water

(c)   $v_1 = Q/A_1 = 20.0/[(1.20)(4.0)] = 4.17$ m/s     $v_2 = 20.0/[(1.55)(4.0)] = 3.23$ m/s

**14.343** Water flows over a concrete spillway into a rectangular channel 9.0 m wide through a hydraulic jump, in the manner illustrated in Fig. 14-90. The depths before and after the jump are 1.55 m and 3.08 m, respectively. Find the rate of flow in the channel.

$\blacksquare$  $\qquad q = \{[(d_1 + d_2)/2](gd_1d_2)\}^{1/2} = \{[(1.55 + 3.08)/2][(9.807)(1.55)(3.08)]\}^{1/2} = 10.41 \; (m^3/s)/m$

$$Q = (10.41)(9.0) = 93.7 \; m^3/s$$

**14.344** Rework Prob. 14.343 if the channel width is 20 ft and the depths of flow before and after the jump are 4.5 ft and 8.2 ft, respectively.

$\blacksquare$  $\qquad q = \{[(d_1 + d_2)/2](gd_1d_2)\}^{1/2} = \{[(4.5 + 8.2)/2][(32.2)(4.5)(8.2)]\}^{1/2} = 86.86 \; (ft^3/s)/ft$

$$Q = (86.86)(20) = 1737 \; ft^3/s$$

**14.345** A hydraulic jump occurs downstream from a 15-m-wide sluice gate. The depth is 1.5 m, and the velocity is 20 m/s. Determine (*a*) the Froude number and the Froude number corresponding to the conjugate depth, (*b*) the depth and velocity after the jump, and (*c*) the power dissipated by the jump.

$\blacksquare$ (*a*)  $\qquad N_F = v/\sqrt{gy} \qquad (N_F)_1 = 20/\sqrt{(9.807)(1.5)} = 5.21$

$$(N_F)_2 = (2)(\sqrt{2})(N_F)_1/[\sqrt{1 + (8)(N_F)_1^2} - 1]^{3/2} = (2)(\sqrt{2})(5.21)/[\sqrt{1 + (8)(5.21)^2} - 1]^{3/2} = 0.288$$

(*b*)  $\qquad (N_F)_2 = v_2/\sqrt{gy} \qquad v_2 y_2 = v_1 y_1 = (1.5)(20) = 30.00 \; m^2/s$

$$v_2^2 = (N_F)_2^2 gy_2 = (N_F)_2^2 (g)(30.00/v_2) = (0.288)^2 (9.807)(30.00/v_2)$$

$$v_2 = 2.90 \; m/s \qquad y_2 = 30.00/2.90 = 10.34 \; m$$

(*c*)  $\qquad E_j = (y_2 - y_1)^3/4y_1 y_2 = (10.34 - 1.5)^3/[(4)(1.5)(10.34)] = 11.13 \; m \; of \; water$

$\qquad P = \gamma Q E_j \qquad Q = Av = [(15)(1.5)](20) = 450 \; m^3/s \qquad P = (9.79)(450)(11.13) = 49.0 \times 10^3 \; kW$

**14.346** Water flows at a rate of 16 m³/s at half critical depth in a trapezoidal channel with $b = 4$ m and $m = 0.4$ (see Fig. 14-91), before a hydraulic jump occurs. Prepare and execute a computer program to find the height after the jump and the energy loss in kilowatts.

$\blacksquare$
```
10 REM B:EX124                  JUMP IN A TRAPEZOIDAL CHANNEL
20 DEFINT I:    DEF FNC(DY)=Q^2*(B+2!*M*DY)-G*(DY*(B+M*DY))^3  '
30 DEF FNFM(DY)=.5*B*DY^2+M*DY^3/3!+Q^2/(G*DY*(B+M*DY))        '
40 READ G,Q,B,M,GAM:    DATA 9.806,16.,4.,.4,9806.
50 YMAX=16!:    YMIN=0!:    LPRINT:    LPRINT"G,Q,B,M,GAM=";G;Q;B;M;GAM
60 FOR I=1 TO 15:    YC=.5*(YMAX+YMIN)
70 IF FNC(YC)>0! THEN YMIN=YC ELSE YMAX=YC
80 PRINT YMAX;YMIN;YC
90 NEXT I:    LPRINT"Y1,YC=";.5*YC;YC
100 Y1=.5*YC:    YMIN=YC:    YMAX=3!*YC:    FM=FNFM(Y1)
110 FOR I=1 TO 15:    Y2=.5*(YMAX+YMIN)
120 IF FNFM(Y2)-FM>0! THEN YMAX=Y2 ELSE YMIN=Y2
130 PRINT"YMAX,YMIN,Y2=";YMAX;YMIN;Y2    NEXT I
140 A1=Y1*(B+M*Y1):    A2=Y2*(B+M*Y2):    V1=Q/A1:    V2=Q/A2
150 LOSS=(V1^2-V2^2)/(2!*G)+Y1-Y2:    POWER=GAM*Q*LOSS/1000!
160 LPRINT"Y1,Y2,V1,V2,LOSS,POWER=";
170 LPRINT USING"###.### ";Y1;Y2;V1;V2;LOSS;POWER

G,Q,B,M,GAM= 9.806001 16 4 .4 9806
Y1,YC= .5661621 1.132324
Y1,Y2,V1,V2,LOSS,POWER= 0.566 1.973 6.687 1.693 0.726 113.970
```

**Fig. 14-91**

**14.347** A rectangular channel 10 ft wide carries 100 cfs in uniform flow at a depth of 1.67 ft. Suppose that an obstruction such as a submerged weir is placed across the channel, rising to a height of 6 in above the bottom. (*a*) Will this obstruction cause a hydraulic jump to form upstream? Why? (*b*) Find the water depth over the obstruction, and classify the surface profile, if possible, to be found upstream of the weir.

$\blacksquare$ (*a*) $y_c = (q^2/g)^{1/3} = [(\frac{100}{10})^2/32.2]^{1/3} = 1.459$ ft. Since $[y_0 = 1.67 \; ft] > [y_c = 1.459 \; ft]$, the slope is mild, flow is subcritical, and no jump can occur.

(*b*)  $\qquad E = y + (1/2g)(q^2/y^2) \qquad E_0 = 1.67 + \{1/[(2)(32.2)]\}[(\frac{100}{10})^2/1.67^2] = 2.227$ ft

$$E_{min} = (\tfrac{3}{2})(y_c) = (\tfrac{3}{2})(1.459) = 2.188 \; ft \qquad \Delta z_{crit} = E_0 - E_{min} = 2.227 - 2.188 = 0.039 \; ft$$

Since $[\Delta z = 0.5 \; ft] > [\Delta z_{crit} = 0.039 \; ft]$, $y_c$ occurs over the obstruction. Hence, the water depth over the obstruction = 1.459 ft. Let $y_1$ = water depth just upstream of the obstruction. Assuming no energy loss over the obstruction, $E_1 = E_{min} + \Delta z$, or $y_1 + \{1/[(2)(32.2)]\}[(\frac{100}{10})^2/y_1^2] = 2.188 + 0.5 = 2.688$, $y_1 = 2.42$ ft or 0.943 ft (by trial and error). There is no mild profile that enables normal flow to become supercritical downstream, so $y_1 = 0.943$ ft is not possible here. Thus $y_1 = 2.42$ ft. Upstream of $y_1$, the water surface tends asymptotically to $y_0 = 1.67$ ft.

**14.348** Suppose that the slope and roughness of the channel in Prob. 14.347 are such that uniform flow of 120 cfs occurs at 1.00 ft. Consider an obstruction rising 3 in above the bottom of the channel. Will a hydraulic jump form upstream? As in Prob. 14.347, classify the surface profile found just upstream from the obstruction.

▮ Since $y_0 = 1.000$ ft $< 1.459$ ft $= y_c$, the slope is steep and for $y = y_0$, the flow is supercritical.

$$E = y + (1/2g)(q^2/y^2) \qquad E_0 = 1.000 + \{1/[(2)(32.2)]\}[(\frac{120}{10})^2/1.00^2] = 3.236 \text{ ft}$$

$$E_{min} = 2.188 \text{ ft} \qquad \text{(from Prob. 14.347)} \qquad \Delta z_{crit} = E_0 - E_{min} = 3.236 - 2.188 = 1.048 \text{ ft}$$

Since $\Delta z = 0.250$ ft $< 1.048$ ft $= \Delta z_{crit}$, the obstruction is not sufficiently high to produce critical flow. Hence, upstream the flow is straight supercritical uniform flow, and a hydraulic jump cannot form.

**14.349** In a rectangular channel 4 m wide with a flow of 7.53 m³/s the depth is 0.3 m. If a hydraulic jump occurs, find the depth downstream of it. Calculate the loss of specific energy through the jump.

▮ $$y_2 = (y_1/2)(\sqrt{1 + 8q^2/gy_1^3} - 1) = (0.3/2)\{\sqrt{1 + (8)(7.53/4)^2/[(9.807)(0.3)^3]} - 1\} = 1.410 \text{ m}$$

$$E_j = (y_2 - y_1)^3/4y_1 y_2 = (1.410 - 0.3)^3/[(4)(0.3)(1.410)] = 0.808 \text{ m}$$

**14.350** In a horizontal rectangular channel 8 ft wide the depths before and after a hydraulic jump are 0.82 ft and 3.16 ft. Find the rate of flow and the specific energy loss.

▮ $$q = \{[(d_1 + d_2)/2](gd_1d_2)\}^{1/2} = \{[(0.82 + 3.16)/2][(32.2)(0.82)(3.16)]\}^{1/2} = 18.18 \text{ (ft}^3/\text{s)/m}$$

$$Q = (18.18)(8) = 145 \text{ ft}^3/\text{s} \qquad E_j = (d_2 - d_1)^3/4d_1d_2 = (3.16 - 0.82)^3/[(4)(0.82)(3.16)] = 1.24 \text{ ft}$$

**14.351** A tidal *bore* (progressive hydraulic jump) rises 13 ft above the normal low-tide river depth of 8 ft. The bore travels upstream at 16 mph (relative to the bank). Find the velocity of the undisturbed river. Is this subcritical or supercritical flow?

▮ $$q = \{[(d_1 + d_2)/2](gd_1d_2)\}^{1/2} = \{[(13 + 8 + 8)/2][(32.2)(13 + 8)(8)]\}^{1/2} = 280.1 \text{ cfs/ft}$$

$$v_1 = q/d_1 = 280.1/8 = 35.01 \text{ ft/s} \qquad \text{(relative to the bore)} \qquad v_{jump} = (16)(\tfrac{5280}{3600}) = 23.47 \text{ ft/s}$$

The bore is moving upstream at a velocity of 23.47 ft/s while the river is moving into the bore at 35.01 ft/s. Therefore, the river is moving downstream at $35.01 - 23.47 = 11.54$ ft/s; $v^2/2g = 11.54^2/[(2)(32.2)] = 2.07$ ft. Since $v^2/2g = 2.07$ ft $< 4$ ft $= y/2$, the flow is subcritical.

**14.352** A hydraulic jump occurs in a triangular flume having side slopes 1:1. The flow rate is 17.72 cfs and the depth before the jump is 1.0 ft. Find the specific-energy loss in the jump.

▮ $Q^2/gA_1 + (h_c)_1 A_1 = Q^2/gA_2 + (h_c)_2 A_2$. For the triangular section, $h_c = y/3$, $A = y^2$, and $y_1 = 1.0$ ft.

$$17.72^2/[(32.2)(1.0)^2] + (1.0/3)(1.0)^2 = 17.72^2/[(32.2)(y_2)^2] + (y_2/3)(y_2)^2$$

$$y_2 = 3.00 \text{ ft} \qquad \text{(by trial and error)}$$

$$v_1 = Q/A_1 = 17.72/1.0^2 = 17.72 \text{ ft/s} \qquad v_2 = 17.72/3.00^2 = 1.97 \text{ ft/s}$$

$$E_j = (y_1 + v_1^2/2g) - (y_2 + v_2^2/2g) = 1.0 + 17.72^2/[(2)(32.2)] - 3.00 - 1.97^2/[(2)(32.2)] = 2.815 \text{ ft}$$

**14.353** A hydraulic jump occurs in a 4-m-wide rectangular channel carrying 5 m³/s on a slope of 0.004. The depth after the jump is 1.2 m. Find (**a**) the depth before the jump, (**b**) the losses of energy and power in the jump.

▮ (**a**) $$y_1 = (y_2/2)(\sqrt{1 + 8q^2/gy_2^3} - 1) = (1.2/2)\{\sqrt{1 + (8)(\tfrac{5}{4})^2/[(9.807)(1.2)^3]} - 1\} = 0.191 \text{ m}$$

(**b**) $$E_j = (y_2 - y_1)^3/4y_1 y_2 = (1.2 - 0.191)^3/[(4)(0.191)(1.2)] = 1.120 \text{ m}$$

$$P = Q\gamma E_j = (5)(9.79)(1.120) = 54.8 \text{ kW}$$

**14.354** Analyze the water-surface profile in a long rectangular channel with concrete lining ($n = 0.013$). The channel is 10 ft wide, the flow rate is 400 cfs, and there is an abrupt change in channel slope from 0.0150 to 0.0016. Find also the horsepower loss in the jump.

**▌** $$Q = (1.486/n)(A)(R^{2/3})(s^{1/2}) \qquad 400 = (1.486/0.013)(10y_{0_1})[10y_{0_1}/(10 + 2y_{0_1})]^{2/3}(0.015)^{1/2}$$

By trial, $y_{0_1} = 2.17$ ft (normal depth on upper slope). Using a similar procedure, the normal depth $y_{0_2}$ on the lower slope is found to be 4.80 ft. $y_c = (q^2/g)^{1/3} = [(\frac{400}{10})^2/32.2]^{1/3} = 3.68$ ft. Thus flow is supercritical ($y_{0_1} < y_c$) before break in slope and subcritical ($y_{0_2} > y_c$) after break, so a hydraulic jump must occur. $y_2 = (y_1/2)[-1 + \sqrt{1 + (8q^2/gy_1^3)}]$, $y_2' = (2.17/2)(-1 + \{1 + [8(40)^2/32.2(2.17)^3]\}^{1/2}) = 5.75$ ft. Therefore the depth conjugate to the upper-slope normal depth of 2.17 ft is 5.75 ft. This jump cannot occur because the normal depth $y_{0_2}$ on the lower slope is less than 5.75 ft. $y_1' = (4.80/2)(-1 + \{1 + [8(40)^2/32.2(4.8)^3]\}^{1/2}) = 2.76$ ft. The lower conjugate depth of 2.76 ft will occur downstream of the break in slope. The location of the jump (i.e., its distance below the break in slope) may be found by $\Delta x = (E_1 - E_2)/(S - S_0)$, $E_1 = 2.17 + [(400/21.7)^2/64.4] = 7.45$ ft, $E_2 = 2.76 + [(400/27.6)^2/64.4] = 6.02$ ft, $V_m = (\frac{1}{2})[(400/21.7) + (400/27.6)] = 16.46$ fps, $R_m = (\frac{1}{2})[(21.7/14.34) + (27.6/15.52)] = 1.645$ ft, $S = (nv/1.486R^{2/3})^2 = [(0.013)(16.45)/1.49(1.645)^{2/3}]^2 = 0.01060$, $\Delta x = (7.45 - 6.02)/(0.0106 - 0.0016) = 160$ ft. Thus depth on the upper slope is 2.17 ft; downstream of the break the depth increases gradually ($M_3$ curve) to 2.76 ft over a distance of approximately 160 ft; then a hydraulic jump occurs from a depth of 2.76 ft to 4.80 ft; downstream of the jump the depth remains constant (i.e., normal) at 4.80 ft. HP loss = $\gamma Q h_{L_j}/550$ where $h_{L_j} = \Delta E$.

Before jump: $E_1' = 2.76 + [(400/27.6)^2/64.4] = 6.02$ ft. After jump: $E_{0_2} = 4.80 + [(400/48.0)^2/64.4] = 5.88$ ft. Hence HP loss = $62.4(400)(6.02 - 5.88)/550 = 6.35$.

**14.355** A very wide rectangular channel with bed slope $S_0 = 0.0003$ and roughness $n = 0.020$ carries a steady flow of 50 cfs/ft of width. If a sluice gate is so adjusted as to produce a minimum depth of 1.5 ft in the channel, determine whether a hydraulic jump will occur downstream, and if so, find (using one reach) the distance from the gate to the jump.

**▌** $$q = (y)(1.486/n)(y^{2/3})(s^{1/2}) \qquad 50 = (y_0)(1.486/0.020)(y_0)^{2/3}(0.0003)^{1/2} \qquad y_0 = 8.99 \text{ ft}$$

$$y_c = (q^2/g)^{1/3} = (50^2/32.2)^{1/3} = 4.27 \text{ ft}$$

Since $y_c < y_0$, the slope is mild. Since $y = 1.5$ ft caused by the gate $< y_c < y_0$, downstream of the sluice gate is an $M_3$ profile, which must be followed by a hydraulic jump to enable the flow to return to normal depth.

$$y_1 = (y_2/2)(\sqrt{1 + 8q^2/gy_2^3} - 1) = (8.99/2)\{\sqrt{1 + (8)(50)^2/[(32.2)(8.99)^3]} - 1\} = 1.63 \text{ ft}$$

$$\Delta x = (E_1 - E_2)/(S - S_0) \qquad E = y + v^2/2g$$

For $y = 1.5$ ft, $v = q/y = 50/1.5 = 33.33$ ft/s, $E = 1.5 + 33.33^2/[(2)(32.2)] = 18.75$ ft. For $y = 1.63$ ft, $v = q/y = 50/1.63 = 30.67$ ft/s, $E = 1.63 + 30.67^2/[(2)(32.2)] = 16.24$ ft.

$$v_m = (33.33 + 30.67)/2 = 32.00 \text{ ft/s} \qquad R_m = (1.5 + 1.63)/2 = 1.565 \text{ ft}$$

$$S = (nv/1.486R^{2/3})^2 = \{(0.020)(32.00)/[(1.486)(1.565)^{2/3}]\}^2 = 0.1021$$

$$\Delta x = (18.75 - 16.24)/(0.1021 - 0.0003) = 24.7 \text{ ft}$$

**14.356** A rectangular channel 8 ft wide carries 280 cfs in uniform flow at a depth of 5 ft. Around a 40-ft-radius bend, how much higher should the outside wall be than the inside wall?

**▌** $v = Q/A = 280/[(5)(8)] = 7.000$ ft/s, $N_F = v/\sqrt{gy} = 7.000/\sqrt{(32.2)(5)} = 0.552$. Since $N_F < 1.0$, the flow is subcritical. $\Delta y = v^2B/gr = (7.000)^2(10)/[(32.2)(40)] = 0.380$ ft.

**14.357** Repeat Prob. 14.356 if normal depth is 2 ft.

**▌** $v = Q/A = 280/[(2)(8)] = 17.5$ ft/s, $N_F = v/\sqrt{gy} = 17.5/\sqrt{(32.2)(2)} = 2.18$. Since $N_F > 1.0$, the flow is supercritical. $\Delta y = v^2B/gr = (17.5)^2(8)/[(32.2)(40)] = 1.90$ ft. Because of (supercritical) wave action, maximum water depth at inside wall = $y_0$, and maximum water depth at outside wall = $(y_0 + \Delta y)$. Therefore, the required difference in wall elevations = 1.90 ft.

**14.358** Rework Prob. 14.356 for these data: $b = 4$ m, $Q = 8$ m$^3$/s, $y = 2.0$ m, $r = 30$ m.

**▌** $v = Q/A = 8/[(4)(2.0)] = 1.000$ m/s, $N_F = v/\sqrt{gy} = 1.000/\sqrt{(9.807)(2.0)} = 0.226$. Since $N_F < 1.0$, the flow is subcritical. $\Delta y = v^2B/gr = (1.000)^2(4)/[(9.807)(30)] = 0.0136$ m.

**14.359** Find the maximum volumetric flow in a 3-ft by 3-ft concrete box culvert ($n = 0.013$) with a rounded entrance ($k_e = 0.05$, $C_d = 0.95$) if the culvert slope is 0.0065, the length 100 ft, and the headwater level 5 ft above the culvert invert? Assume free outlet conditions. Neglect headwater and tailwater velocity heads.

**I** Headwater/$d = \frac{5}{3} = 1.7$. Since $1.7 > 1.2$, conditions are those of Fig. 14-92$b$ or 14-92$c$. Assume case $b$.

$$R = A/p_w = (3)(3)/(3 + 3 + 3 + 3) = 0.750 \text{ ft}$$
$$(h_L)_{1-3} = (y_1 - y_3) + (z_1 - z_3) = y_1 - y_3 + s_0 L$$
$$\Delta h = (k_e + 29n^2 L/R^{4/3} + 1)(v^2/2g) \qquad y_1 - y_3 + s_0 L = (k_e + 29n^2 L/R^{4/3} + 1)(v^2/2g)$$
$$5 - 3 + (0.0065)(100) = [0.05 + (29)(0.013)^2(100)/0.75^{4/3} + 1]\{v^2/[(2)(32.2)]\} \qquad v = 9.82 \text{ ft/s}$$
$$Q = Av = [(3)(3)](9.82) = 88.38 \text{ ft}^3/\text{s} = (A)(1.486/n)(R^{2/3})(s^{1/2})$$

Now find the depth $y_0$ which occurs with normal uniform flow at this flow rate: $88.38 = (3y_0)(1.486/0.013)[3y_0/(y_0 + 3 + y_0)]^{2/3}(0.0065)^{1/2}$, $y_0 = 3.16$ ft (by trial and error). Since $y_0 > d$, the culvert flows full. Free discharge at the outlet is given; therefore, the preceding assumption and computations are valid, and $Q = 88.38 \text{ ft}^3/\text{s}$.

(a) Submerged outlet

(b) Normal depth > barrel height

(c) Entrance control, normal depth < barrel height          **Fig. 14-92**

**14.360** Solve Prob. 14.359 given that the tailwater elevation is 1 ft above the top of the box at the outlet.

**I** Headwater/$d = \frac{5}{3} = 1.7$. Since $1.7 > 1.2$, conditions are those of Fig. 14-92$a$ with $y_2 = 4$ ft.

$$R = A/p_w = (3)(3)/(3 + 3 + 3 + 3) = 0.75 \text{ ft}$$
$$(h_L)_{1-3} = (y_1 - y_3) + (z_1 - z_3) = y_1 - y_3 + s_0 L$$
$$\Delta h = (k_e + 29n^2 L/R^{4/3} + 1)(v^2/2g) \qquad y_1 - y_3 + s_0 L = (k_e + 29n^2 L/R^{4/3} + 1)(v^2/g)$$
$$5 - 4 + (0.0065)(100) = [0.05 + (29)(0.013)^2(100)/0.75^{4/3} + 1]\{v^2/[(2)(32.2)]\} \qquad v = 7.75 \text{ ft/s}$$
$$Q = Av = [(3)(3)](7.75) = 69.8 \text{ ft}^3/\text{s}$$

**14.361** Solve Prob. 14.359 given that the tailwater elevation is 2 ft above the top of the box at the outlet.

**I** Headwater/$d = \frac{6}{4} = 1.5$. Since $1.5 > 1.2$, conditions are those of Fig. 14-92$a$ with $y_2 = 6$ ft.

$$R = A/p_w = (4)(4)/(4 + 4 + 4 + 4) = 1.000 \text{ ft}$$
$$(h_L)_{1-3} = (y_1 - y_3) + (z_1 - z_3) = y_1 - y_3 + s_0 L$$
$$\Delta h = (k_e + 29n^2 L/R^{4/3} + 1)(v^2/2g) \qquad y_1 - y_3 + s_0 L = (k_e + 29n^2 L/R^{4/3} + 1)(v^2/2g)$$
$$5 - 5 + (0.0065)(100) = [0.05 + (29)(0.013)^2(100)/0.75^{4/3} + 1]\{v^2/[(2)(32.2)]\} \qquad v = 4.864 \text{ ft/s}$$
$$Q = Av = [(3)(3)](4.864) = 43.8 \text{ ft}^3/\text{s}$$

**14.362** Repeat Prob. 14.359 if the culvert slope is 0.043.

▮ Headwater/$d = \frac{5}{3} = 1.7$. Since $1.7 > 1.2$, conditions are those of Fig. 14-92b or 14-92c. Assume case b.

$$R = A/p_w = (3)(3)/(3 + 3 + 3 + 3) = 0.75\,\text{ft}$$
$$(h_L)_{1-3} = (y_1 - y_3) + (z_1 - z_3) = y_1 - y_3 + s_0 L$$
$$\Delta h = (k_e + 29n^2 L/R^{4/3} + 1)(v^2/2g) \qquad y_1 - y_3 + s_0 L = (k_e + 29n^2 L/R^{4/3} + 1)(v^2/2g)$$
$$5 - 3 + (0.043)(100) = [0.05 + (29)(0.013)^2(100)/0.75^{4/3} + 1]\{v^2/[(2)(32.2)]\} \qquad v = 15.14\,\text{ft/s}$$
$$Q = Av = [(3)(3)](15.14) = 136\,\text{ft}^3/\text{s} = (A)(1.486/n)(R^{2/3})(s^{1/2})$$

Now find the depth $y_0$ which occurs with normal uniform flow at this flow rate: $136 = (3y_0)(1.486/0.013)[3y_0/(y_0 + 3 + y_0)]^{2/3}(0.043)^{1/2}$, $y_0 = 2.1\,\text{ft}$  (by trial and error). Since $y_0 < d$, the situation is actually case c; therefore, $Q = C_d A\sqrt{2gh} = (0.95)[(3)(3)]\sqrt{(2)(32.2)(3)} = 119\,\text{ft}^3/\text{s}$.

**14.363** Repeat Prob. 14.359 if the culvert slope is 0.08.

▮ Headwater/$d = \frac{5}{3} = 1.7$. Since $1.7 > 1.2$, conditions are those of Fig. 14-92b or 14-92c. Assume case b.

$$R = A/p_w = (3)(3)/(3 + 3 + 3 + 3) = 0.75\,\text{ft}$$
$$(h_L)_{1-3} = (y_1 - y_3) + (z_1 - z_3) = y_1 - y_3 + s_0 L$$
$$\Delta h = (k_e + 29n^2 L/R^{4/3} + 1)(v^2/2g) \qquad y_1 - y_3 + s_0 L = (k_e + 29n^2 L/R^{4/3} + 1)(v^2/2g)$$
$$5 - 3 + (0.08)(100) = [0.05 + (29)(0.013)^2(100)/0.75^{4/3} + 1]\{v^2/[(2)(32.2)]\} \qquad v = 19.08\,\text{ft/s}$$
$$Q = Av = [(3)(3)](19.08) = 172\,\text{ft}^3/\text{s} = (A)(1.486/n)(R^{2/3})(s^{1/2})$$

Now find the depth $y_0$ which occurs with normal uniform flow at this flow rate: $172 = (3y_0)(1.486/0.013)[3y_0/(y_0 + 3 + y_0)]^{2/3}(0.08)^{1/2}$, $y_0 = 1.98\,\text{ft}$  (by trial and error). Since $y_0 < d$, our assumption was wrong and we have, as in Prob. 14.362, $Q = C_d A\sqrt{2gh} = (0.95)[(3)(3)]\sqrt{(2)(32.2)(3)} = 119\,\text{ft}^3/\text{s}$.

**14.364** A culvert under a road must carry $4.3\,\text{m}^3/\text{s}$. The culvert length will be 30 m and the slope will be 0.003. If the maximum permissible headwater level is 3.6 m above the culvert invert, what size corrugated-pipe culvert ($n = 0.025$) would you select? The outlet will discharge freely. Neglect velocity of approach. Assume square-edged entrance with $k_e = 0.5$ and $C_d = 0.65$.

▮ Assume $d < 3.0$ m, so that [headwater/$d$] > [3.6/3.0 = 1.2] and the conditions are those of Fig. 14-92b or 14-92c. Assume case b.

$$(h_L)_{1-3} = (y_1 - y_3) + (z_1 - z_3) = y_1 - y_3 + s_0 L \qquad \Delta h = (k_e + 19.62n^2 L/R^{4/3} + 1)(v^2/2g)$$
$$y_1 - y_3 + s_0 L = (k_e + 19.62n^2 L/R^{4/3} + 1)(v^2/2g) \qquad v = Q/A = 4.3/(\pi d^2/4) = 5.475/d^2$$
$$3.6 - d + (0.003)(30) = [0.5 + (19.62)(0.025)^2(30)/(d/4)^{4/3} + 1]\{(5.475/d^2)^2/[(2)(9.807)]\}$$
$$d = 1.20\,\text{m} \qquad \text{(by trial and error)} \qquad Q = (A)(1.0/n)(R^{2/3})(s^{1/2})$$

Now find the diameter $d_0$ which just flows full with normal uniform flow: $4.3 = (\pi d_0^2/4)(1.0/0.025)(d_0/4)^{2/3}(0.003)^{1/2}$, $d_0 = 1.99\,\text{m}$. Since $d_0 > d$, the culvert flows full, and free discharge at the outlet is given. Therefore, the above assumptions and analysis are valid, and $d = 1.20\,\text{m}$.

**14.365** Repeat Prob. 14.364 for a culvert length of 100 m.

▮ Assume $d < 3.0$ m, so that [headwater/$d$] > [3.6/3.0 = 1.2] and the conditions are those of Fig. 14-92b or 14-92c. Assume case b.

$$(h_L)_{1-3} = (y_1 - y_3) + (z_1 - z_3) = y_1 - y_3 + s_0 L \qquad \Delta h = (k_e + 19.62n^2 L/R^{4/3} + 1)(v^2/2g)$$
$$y_1 - y_3 + s_0 L = (k_e + 19.62n^2 L/R^{4/3} + 1)(v^2/2g) \qquad v = Q/A = 4.3/(\pi d^2/4) = 5.475/d^2$$
$$3.6 - d + (0.003)(100) = [0.5 + (19.62)(0.025)^2(100)/(d/4)^{4/3} + 1]\{(5.475/d^2)^2/[(2)(9.807)]\}$$
$$d = 1.41\,\text{m} \qquad \text{(by trial and error)} \qquad Q = (A)(1.0/n)(R^{2/3})(s^{1/2})$$

Now find the diameter $d_0$ which just flows full with normal uniform flow: $4.3 = (\pi d_0^2/4)(1.0/0.025)(d_0/4)^{2/3}(0.003)^{1/2}$, $d_0 = 1.99\,\text{m}$. Since $d_0 > d$, the culvert flows full, and free discharge at the outlet is given. Therefore, the above assumptions and analysis are valid, and $d = 1.41$ m. Use standard diameter 1.50 m.

**14.366** Water flows in a wide channel at $q = 18$ (ft³/s)/ft, $y_1 = 1$ ft, and then undergoes a hydraulic jump. Compute $y_2$, $v_2$, $(N_F)_2$, $E_j$, and the power dissipated per unit width.

$$N_F = v/\sqrt{gy} \qquad v_1 = q/y_1 = \tfrac{18}{1} = 18.00 \text{ ft/s}$$
$$(N_F)_1 = 18.00/\sqrt{(32.2)(1)} = 3.17 \qquad E = y + v^2/2g$$
$$E_1 = 1 + 18.00^2/[(2)(32.2)] = 5.03 \text{ ft}$$
$$y_2 = (y_1/2)(\sqrt{1 + 8q^2/gy_1^3} - 1) = (\tfrac{1}{2})\{\sqrt{1 + (8)(18)^2/[(32.2)(1)^3]} - 1\} = 4.01 \text{ ft}$$
$$v_2 = q/y_2 = 18/4.01 = 4.49 \text{ ft/s} \qquad (N_F)_2 = 4.49/\sqrt{(32.2)(4.01)} = 0.395$$
$$E_j = (y_2 - y_1)^3/4y_1 y_2 = (4.01 - 1)^3/[(4)(4.01)(1)] = 1.70 \text{ ft}$$
$$p_{\text{dissipated}} = q\gamma E_j = (18)(62.4)(1.7) = 1909 \text{ (ft-lb/s)/ft} = 3.47 \text{ hp/ft}$$

**14.367** A wide-channel flow at depth 60 cm passes through a hydraulic jump and emerges at a depth of 3.0 m. Compute the velocities on either side of the jump and the critical depth of flow.

$$2y_2/y_1 = -1 + [1 + 8(N_F)_1^2]^{1/2} \qquad (2)(3.0)/(0.60) = -1 + [1 + 8(N_F)_1^2]^{1/2} \qquad (N_F)_1 = 3.87$$
$$N_F = v/\sqrt{gy} \qquad 3.87 = v_1/\sqrt{(9.807)(0.60)} \qquad v_1 = 9.39 \text{ m/s}$$
$$v_1 y_1 = v_2 y_2 \qquad (9.39)(0.60) = (v_2)(3.0) \qquad v_2 = 1.88 \text{ m/s}$$
$$q = yv = (3.0)(1.88) = 5.64 \text{ (m}^3\text{/s)/m} \qquad y_c = (q^2/g)^{1/3} = (5.64^2/9.807)^{1/3} = 1.48 \text{ m}$$

**14.368** The flow downstream of a hydraulic jump in a rectangular channel is 8 m deep and has a velocity of 3.6 m/s. Obtain the velocity and depth upstream of the jump.

$$2y_1/y_2 = -1 + [1 + 8(N_F)_2^2]^{1/2} \qquad N_F = v/\sqrt{gy} \qquad (N_F)_2 = 3.6/\sqrt{(9.807)(8)} = 0.406$$
$$2y_1/8 = -1 + [1 + (8)(0.406)^2]^{1/2} \qquad y_1 = 2.09 \text{ m} \qquad y_1 v_1 = y_2 v_2$$
$$2.09v_1 = (8)(3.6) \qquad v_1 = 13.78 \text{ m/s}$$

**14.369** Water in a horizontal channel accelerates smoothly over a bump and then undergoes a hydraulic jump, as in Fig. 14-93. If $y_1 = 1$ m and $y_3 = 30$ cm, estimate $v_1$, $v_3$, and $y_4$. Neglect friction.

$$E_1 = E_3 \qquad y_1 + v_1^2/2g = y_3 + v_3^2/2g \qquad 1 + v_1^2/[(2)(9.807)] = 0.30 + v_3^2/[(2)(9.807)]$$
$$y_1 v_1 = y_3 v_3 \qquad v_1 = y_3 v_3/y_1 = (0.30)(v_3)/1 = 0.3000v_3$$
$$1 + (0.3000v_3)^2/[(2)(9.807)] = 0.30 + v_3^2/[(2)(9.807)]$$
$$v_3 = 3.88 \text{ m/s} \qquad v_1 = (0.3000)(3.88) = 1.16 \text{ m/s}$$
$$2y_4/y_3 = -1 + [1 + 8(N_F)_3^2]^{1/2} \qquad N_F = v/\sqrt{gy}$$
$$(N_F)_3 = 3.88/\sqrt{(9.807)(0.30)} = 2.26 \qquad 2y_4/(0.30) = -1 + [1 + (8)(2.26)^2]^{1/2} \qquad y_4 = 82 \text{ cm}$$

**Fig. 14-93**

**14.370** Evaluate the bump height $h$ in Prob. 14.369.

$$y_1 + v_1^2/2g = y_2 + v_2^2/2g + h$$
$$1 + 1.16^2/[(2)(9.807)] = y_2 + v_2^2/[(2)(9.807)] + h$$
$$y_1 v_1 = y_2 v_2 \qquad (1)(1.16) = y_2 v_2 \qquad v_2 = 1.16/y_2$$

Since flow over the bump is critical,

$$v_2 = v_c = \sqrt{gy_2} \qquad 1.16/y_2 = \sqrt{9.807y_2} \qquad y_2 = 0.516 \text{ m} \qquad v_2 = 1.16/0.516 = 2.248 \text{ m/s}$$

$$1 + 1.16^2/[(2)(9.807)] = 0.516 + 2.248^2/[(2)(9.807)] + h \qquad h = 29.5 \text{ m}$$

**14.371** Repeat Prob. 14.369 if $y_1 = 3$ ft and $y_3 = 1$ ft.

$$E_1 = E_3 \qquad y_1 + v_1^2/2g = y_3 + v_3^2/2g \qquad 3 + v_1^2/[(2)(32.2)] = 1 + v_3^2/[(2)(32.2)]$$

$$y_1 v_1 = y_3 v_3 \qquad v_1 = y_3 v_3/y_1 = (1)(v_3)/3 = 0.3333v_3$$

$$3 + (0.3333v_3)^2/[(2)(32.2)] = 1 + v_3^2/[(2)(32.2)]$$

$$v_3 = 12.04 \text{ ft/s} \qquad v_1 = (0.3333)(12.04) = 4.01 \text{ ft/s} \qquad 2y_4/y_3 = -1 + [1 + 8(N_F)_3^2]^{1/2}$$

$$N_F = v/\sqrt{gy} \qquad (N_F)_3 = 12.04/\sqrt{(32.2)(1)} = 2.12$$

$$2y_4/(1) = -1 + [1 + (8)(2.12)^2]^{1/2} \qquad y_4 = 2.54 \text{ ft}$$

**14.372** Evaluate the bump height $h$ in Prob. 14.371.

$$y_1 + v_1^2/2g = y_2 + v_2^2/2g + h$$

$$3 + 4.01^2/[(2)(32.2)] = y_2 + v_2^2/[(2)(32.2)] + h$$

$$y_1 v_1 = y_2 v_2 \qquad (3)(4.01) = y_2 v_2 \qquad v_2 = 12.03/y_2$$

Since flow over the bump is critical,

$$v_2 = v_c = \sqrt{gy_2} \qquad 12.03/y_2 = \sqrt{32.2y_2} \qquad y_2 = 1.65 \text{ ft} \qquad v_2 = 12.03/1.65 = 7.29 \text{ ft/s}$$

$$3 + 4.01^2/[(2)(32.2)] = 1.65 + 7.29^2/[(2)(32.2)] + h \qquad h = 0.775 \text{ ft} = 9.3 \text{ in.}$$

**14.373** Find the power loss in a 100-ft-wide hydraulic jump from depth 3 ft to depth 12 ft.

$$E_j = (y_2 - y_1)^3/4y_1 y_2 = (12 - 3)^3/[(4)(12)(3)] = 5.063 \text{ ft} \qquad P = Q\gamma E_j \qquad N_F = v/\sqrt{gy}$$

$$2y_2/y_1 = -1 + [1 + 8(N_F)_1^2]^{1/2} \qquad (2)(12)/3 = -1 + [1 + (8)(N_F)_1^2]^{1/2} \qquad (N_F)_1 = 3.16$$

$$3.16 = v_1/\sqrt{(32.2)(3)} \qquad v_1 = 31.06 \text{ ft/s} \qquad Q = A_1 v_1 = [(100)(3)](31.06) = 9318 \text{ ft}^3/\text{s}$$

$$P = (9318)(62.4)(5.063) = 2.944 \times 10^6 \text{ ft-lb/s} = 5353 \text{ hp}$$

**14.374** At a certain section of a rectangular channel 10 ft wide, the depth of flow is 2 ft and the flow rate is 400 ft$^3$/s. If a hydraulic jump occurs, will it be upstream or downstream of this section?

$$v_c = \sqrt{gy} = \sqrt{(32.2)(2)} = 8.02 \text{ ft/s} \qquad v = Q/A = 400/[(10)(2)] = 20.00 \text{ ft/s}$$

Since $v > v_c$, the flow is supercritical, and a hydraulic jump must occur downstream.

**14.375** A still canal is 2 m deep and the water behind a bore (Prob. 14.351) is 4 m deep. Obtain the propagation speed of the bore.

$$N_F = v/\sqrt{gy} \qquad 2y_2/y_1 = -1 + [1 + 8(N_F)_1^2]^{1/2} \qquad (2)(4)/2 = -1 + [1 + (8)(N_F)_1^2]^{1/2}$$

$$(N_F)_1 = 1.73 \qquad 1.73 = v_1/\sqrt{(9.807)(2)} \qquad v_1 = c_{\text{bore}} = 7.66 \text{ m/s}$$

**14.376** In Prob. 14.375 suppose that the water upstream flows toward the bore at 4 m/s ground speed. What is the ground speed of the bore?

$$v_{\text{bore/ground}} = v_{\text{bore/water}} + v_{\text{water/ground}} = -7.66 + 4 = -3.66 \text{ m/s}.$$

**14.377** Consider the flow under the sluice gate of Fig. 14-94. If $y_1 = 9$ ft and all losses are neglected except the dissipation in the jump, calculate $y_2$ and $y_3$ and the percentage dissipation. The channel is horizontal and wide.

$$y_1 + V_1^2/2g = y_2 + V_2^2/2g \qquad 10 + 2^2/[(2)(32.2)] = y_2 + V_2^2/[(2)(32.2)]$$

$$V_1 y_1 = V_2 y_2 \qquad (2.2)(9) = V_2 y_2 \qquad y_2 = 19.8/V_2$$

$$9 + 2.2^2/[(2)(32.2)] = 19.8/V_2 + V_2^2/[(2)(32.2)] \qquad V_2 = 23.0 \text{ ft/s} \qquad \text{(by trial and error)}$$

$$y_2 = 19.8/23.0 = 0.861 \text{ ft} \qquad 2y_3/y_2 = -1 + [1 + 8(N_F)_2^2]^{1/2} \qquad N_F = V/\sqrt{gy}$$

$$(N_F)_2 = 23.0/\sqrt{(32.2)(0.861)} = 4.37 \qquad 2y_3/0.861 = -1 + [1 + (8)(4.37)^2]^{1/2} \qquad y_3 = 4.91 \text{ ft}$$

$$E_j = (y_3 - y_2)^3/4y_3 y_2 = (4.91 - 0.861)^3/[(4)(4.91)(0.861)] = 3.93 \text{ ft}$$

$$E_2 = y_2 + V_2^2/2g = 0.861 + 23^2/[(2)(32.2)] = 9.08 \text{ ft}$$

$$\text{Percentage dissipation} = (3.93/9.08)(100\%) = 43.3\%$$

**Fig. 14-94**

**14.378** If bottom friction is included in the sluice-gate flow of Fig. 14-94, the depths $y_1$, $y_2$, and $y_3$ will vary with $x$. What type of solution curves will we have in regions 1, 2, and 3?

❚ See Fig. 14-95.

**Fig. 14-95**

**14.379** A 10-cm-high bump in a wide horizontal water channel creates a hydraulic jump just upstream and the flow pattern in Fig. 14-96. Neglecting losses except in the jump, for the case $y_3 = 30$ cm, estimate $V_4$, $y_4$, $V_1$, and $y_1$.

❚ Since $y_4 < y_3$, assume flow over the hump is critical.

$$V = \sqrt{gy} \qquad V_3 = \sqrt{(9.807)(0.30)} = 1.715 \text{ m}$$

$$E_2 = E_3 = E_4 = y_3 + V_3^2/2g + h = E_4 = 0.30 + 1.715^2/[(2)(9.807)] + 0.10 = E_4$$

$$E_2 = E_3 = E_4 = 0.5500 \text{ m} \qquad E_2 = y_2 + V_2^2/2g = 0.5500 \text{ m}$$

$$q = V_2 y_2 = V_3 y_3 = V_4 y_4 = (1.715)(0.30) = 0.5145 \text{ m}^2/\text{s} \qquad V_2 = 0.5145/y_2$$

$$y_2 + (0.5145/y_2)^2/[(2)(9.807)] = 0.5500$$

$$y_2 = 0.495 \text{ m} \qquad \text{(by trial and error)} \qquad V_2 = 0.5145/0.495 = 1.039 \text{ m/s}$$

$$N_F = V/\sqrt{gy} \qquad (N_F)_2 = 1.039/\sqrt{(9.807)(0.495)} = 0.472 \qquad \text{(subcritical)}$$

$$2y_1/y_2 = -1 + [1 + 8(N_F)_2^2]^{1/2} \qquad 2y_1/0.495 = -1 + [1 + (8)(0.472)^2]^{1/2} \qquad y_1 = 0.165 \text{ m}$$

$$V_1 = q/y_1 = 0.5145/0.165 = 3.12 \text{ m/s} \qquad E_3 = E_4 = 0.5500 \text{ m} \qquad y_4 + V_4^2/2g = 0.5500 \text{ m}$$

$$V_4 y_4 = 0.5145 \text{ m}^2/\text{s} \qquad V_4 = 0.5145/y_4 \qquad y_4 + (0.5145/y_4)^2/[(2)(9.807)] = 0.5500 \text{ m}$$

$$y_4 = 0.195 \text{ m} \qquad \text{(by trial and error)} \qquad V_4 = 0.5145/0.195 = 2.64 \text{ m/s}$$

Bump: $h = 10$ cm    **Fig. 14-96**

**14.380** Water flows in a wide channel at $q = 10 \text{ (m}^3/\text{s)/m}$ and $y_1 = 1.25$ m. If the flow undergoes a hydraulic jump, compute (a) $y_2$, (b) $V_2$, (c) $(N_F)_2$, (d) $E_j$, (e) the percentage dissipation, (f) the power dissipated per unit width, and (g) the temperature rise due to dissipation if $c_p = 4200 \text{ J/(kg-K)}$.

**▮ (a)**   $V_1 = q/y_1 = 10/1.25 = 8.00 \text{ m/s}$    $N_F = V/\sqrt{gy}$    $(N_F)_1 = 8.00/\sqrt{(9.807)(1.25)} = 2.285$

$2y_2/y_1 = -1 + [1 + 8(N_F)_1^2]^{1/2}$    $2y_2/1.25 = -1 + [1 + (8)(2.285)^2]^{1/2}$    $y_2 = 3.46 \text{ m}$

**(b)**    $V_1 y_1 = V_2 y_2$    $(8.00)(1.25) = (V_2)(3.46)$    $V_2 = 2.89 \text{ m/s}$

**(c)**    $(N_F)_2 = 2.89/\sqrt{(9.807)(3.46)} = 0.496$

**(d)**    $E_j = (y_2 - y_1)^3/4y_2 y_1 = (3.46 - 1.25)^3/[(4)(3.46)(1.25)] = 0.624 \text{ m}$

**(e)**    $E_1 = y_1 + v_1^2/2g = 1.25 + 8.00^2/[(2)(9.807)] = 4.51 \text{ m}$

Percentage dissipation $= E_j/E_1 = 0.624/4.51 = 0.138$   or   13.8 percent

**(f)**    $P = q\gamma E_j$    $P_{\text{dissipated}} = (10)(9.79)(0.624) = 61.1 \text{ kW/m}$

**(g)**    $P = \dot{m}c_p \, \Delta T = \rho q c_p \, \Delta T$    $(61.1)(1000) = (1000)(10)(4200)(\Delta T)$    $\Delta T = 0.0015 \text{ K}$

**14.381**   Water flows over a spillway into a sluice 10 m wide. Before the jump the water has a depth of 1 m and a velocity of 18 m/s. Determine the Froude number before the jump and the depth of flow after the jump.

**▮**    $N_F = V/\sqrt{gy}$    $(N_F)_1 = 18/\sqrt{(9.807)(1)} = 5.75$

$y_2 = [-y_1 + \sqrt{y_1^2 + (8Q^2/gb^2)(1/y_1)}]/2$    $Q = A_1 V_1 = [(10)(1)](18) = 180 \text{ ft}^3/\text{s}$

$y_2 = \{-1 + \sqrt{1^2 + [(8)(180)^2/(9.807)(10)^2](\tfrac{1}{1})}\}/2 = 7.64 \text{ m}$

**14.382**   A rectangular channel has a width of 5 ft and a flow of 10 cfs; the depth in front of a hydraulic jump is 3 in. Determine the specific energy (in ft-lb/slug) behind the jump.

**▮**    $y_2 = [-y_1 + \sqrt{y_1^2 + (8Q^2/gb^2)(1/y_1)}]/2$

$= \{-\tfrac{3}{12} + \sqrt{(\tfrac{3}{12})^2 + [(8)(10)^2/(32.2)(5)^2][1/(\tfrac{3}{12})]}\}/2 = 0.880 \text{ ft}$

$\text{KE} = v^2/2$    $v_2 = Q/A_2 = 10/[(0.880)(5)] = 2.27 \text{ ft/s}$    $(\text{KE})_2 = 2.27^2/2 = 2.58 \text{ ft-lb/slug}$

**14.383**   Water in a rectangular, finished-concrete channel overflows a dam, as shown in Fig. 14-97, and goes into a stilling basin at which there is a hydraulic jump. If the channel is of width 6 m, has slope 0.003, and carries 18 m$^3$/s, calculate the depth y at O.

**▮**    $\Delta x = ([1 - (Q^2 b/gA^3)]/\{s_0 - (n/\kappa)^2[Q^2/(R_h^{4/3}A^2)]\}) \, \Delta y$

$A = (5)(6) = 30.00 \text{ m}^2$    $R_h = A/p_w = 30.00/(5 + 6 + 5) = 1.875 \text{ m}$

$60 = \left\{ \dfrac{1 - (18^2)(6)/(9.807)(30.00)^3}{0.003 - (0.012/1.0)^2[18^2/(1.875)^{4/3}(30.00)^2]} \right\}(\Delta y)$

$\Delta y = 0.180 \text{ m}$    $y = 5 + 0.180 = 5.180 \text{ m}$

$S_0 = 0.003$    Fig. 14-97

**14.384**   Does linear averaging significantly improve the solution of Prob. 14.383?

**▮**    $\Delta x = ([1 - (Q^2 b/gA^3)]/\{s_0 - (n/\kappa)^2[Q^2/(R_h^{4/3}A^2)]\}) \, \Delta y$

At $O$, $A = (5.180)(6) = 31.08$ m$^2$ and $R_h = A/p_w = 31.08/(5.180 + 6 + 5.180) = 1.900$ m; therefore,

$$A_{av} = (30.00 + 31.08)/2 = 30.54 \text{ m}^2 \qquad (R_h)_{av} = (1.875 + 1.900)/2 = 1.888 \text{ m}$$

$$60 = \left\{ \frac{1 - (18^2)(6)/(9.807)(30.54)^3}{0.003 - (0.012/1.0)^2[18^2/(1.888)^{4/3}(30.54)^2]} \right\}(\Delta y)$$

$$\Delta y = 0.180 \text{ m}$$

**14.385** Determine the depth $y_2$ at the hydraulic jump of Prob. 14.383.

**I** Neglecting friction at the dam,

$$E_1 = E_0 \qquad E = y + q^2/2y^2g \qquad E_0 = 5.180 + (\tfrac{18}{6})^2/[(2)(5.180)^2(9.807)] = 5.197 \text{ m}$$

$$E_1 = y_1 + (\tfrac{18}{6})^2/[(2y_1^2)(9.807)] = 5.197 \text{ m} \qquad y_1 = 0.306 \text{ m} \qquad \text{(by trial and error)}$$

$$y_2 = [-y_1 + \sqrt{y_1^2 + (8Q^2/gb^2)(1/y_1)}]/2$$

$$= \{-0.306 + \sqrt{0.306^2 + [(8)(18)^2/(9.807)(6)^2](1/0.306)}\}/2 = 2.301 \text{ m}$$

**14.386** Water flows in a 4-m-wide rectangular channel at Froude number $\sqrt{10}$; the depth of flow is 1 m. If the water undergoes a hydraulic jump, what is the Froude number downstream of the jump?

**I**
$$N_F = v/\sqrt{gy} \qquad \sqrt{10} = v/\sqrt{(9.807)(1)} \qquad v = 9.903 \text{ m/s}$$

$$Q = Av = [(4)(1)](9.903) = 39.61 \text{ m}^3/\text{s}$$

$$y_2 = [-y_1 + \sqrt{y_1^2 + (8Q^2/gb^2)(1/y_1)}]/2$$

$$= \{-1 + \sqrt{1^2 + [(8)(39.61)^2/(9.807)(4)^2](1/1)}\}/2 = 4.000 \text{ m}$$

$$v_2 = Q/A_2 = 39.61/[(4.000)(4)] = 2.476 \text{ m/s} \qquad (N_F)_2 = 2.476/\sqrt{(9.807)(4.000)} = 0.395$$

**14.387** Water in a 10-m-wide rectangular channel experiences a jump in depth from 2 m to 6 m. Find the Froude numbers on either side of the jump.

**I**
$$y_2 = [-y_1 + \sqrt{y_1^2 + (8Q^2/gb^2)(1/y_1)}]/2 \qquad 6 = \{-2 + \sqrt{2^2 + [8Q^2/(9.807)(10)^2](\tfrac{1}{2})}\}/2$$

$$Q = 217 \text{ m}^3/\text{s}$$

$$N_F = v/\sqrt{gy} \qquad v_1 = Q/A_1 = 217/[(10)(2)] = 10.85 \text{ m/s} \qquad (N_F)_1 = 10.85/\sqrt{(9.807)(2)} = 2.45$$

$$v_2 = Q/A_2 = 217/[(10)(6)] = 3.62 \text{ m/s} \qquad (N_F)_2 = 3.62/\sqrt{(9.807)(6)} = 0.472$$

**14.388** Water flows in a rectangular concrete channel and undergoes a hydraulic jump such that 60 percent of its mechanical energy is to be dissipated. If the volume flow rate is 100 m$^3$/s and the width of the channel is 5 m, what must the Froude number be just before the jump? Set up the proper equations but do not actually solve.

**I**
$$N_F = v/\sqrt{gy} \qquad E = y + v^2/2g \qquad v_1 = Q/A_1 = 100/(5y_1) = 20.00/y_1$$

$$E_1 = y_1 + (20.00/y_1)^2/[(2)(9.807)] = y_1 + 20.39/y_1^2 \qquad E_j = [y_2^3 - y_1^3 + y_1 y_2(y_1 - y_2)]/4y_1 y_2$$

$$0.60E_1 = (0.60)(y_1 + 20.39/y_1^2) = [y_2^3 - y_1^3 + y_1 y_2(y_1 - y_2)]/4y_1 y_2 \qquad (1)$$

$$y_2 = [-y_1 + \sqrt{y_1^2 + (8Q^2/gb^2)(1/y_1)}]/2 = \{-y_1 + \sqrt{y_1^2 + [(8)(100)^2/(9.807)(5)^2](1/y_1)}\}/2 \qquad (2)$$

Solve Eqs. (1) and (2) to find $y_1$ and $y_2$. Then, $v_1 = Q/A_1 = 100/5y_1$, $(F_R)_1 = (100/5y_1)/\sqrt{(9.807)(y_1)} = 6.386y^{-3/2}$.

**14.389** Water is flowing from a spillway into a stilling basin, as shown in Fig. 14-98. The elevation $y_A$ ahead of the spillway is 8 m. The width of the rectangular channel is 10 m. If the stilling basin dissipates half of the mechanical energy, what is the volume flow rate? Set up three simultaneous equations for $y_1$, $y_2$, and $Q$.

$$E = y + v^2/2g \qquad E_A = E_1$$

$$8 + Q^2/[(80)^2(2g)] = y_1 + Q^2/[(10y_1)^2(2g)] \qquad (1)$$

$$0.50E_1 = E_j$$

$$0.50\{8 + Q^2/[(80)^2(2g)]\} = [y_2^3 - y_1^3 + y_1 y_2(y_1 - y_2)]/4y_1 y_2 \qquad (2)$$

$$y_2 = [-y_1 + \sqrt{y_1^2 + (8Q^2/gb^2)(1/y_1)}]/2 = [-y_1 + \sqrt{y_1^2 + [8Q^2/(9.807)(10)^2](1/y_1)}]/2 \qquad (3)$$

$y_A = 8\,\text{m}$

**Fig. 14-98**

**14.390** Water having a steady known volumetric flow rate $Q$ is moving in a rectangular channel at supercritical speed up an adverse slope. It undergoes a hydraulic jump as shown in Fig. 14-99. If we know $y_B$ and $y_A$ at positions $L_1$ and $L_2$ apart, how do we approximately locate the position of the hydraulic jump, i.e., how do we get $L_1$ and $L_2$? The channel has a known value of $n$. Explain the simplest method. The width is $b$.

■ Use the equation $\Delta L = ([1 - (Q^2 b/gA^3)]/\{S_0 - (n/\kappa)^2[Q^2/(R_H^{4/3}A^2)]\})\,\Delta y$ and guess at a value $L_1$ and solve for $(\Delta y)_A$. Now compute $y_1 = y_A + (\Delta y)_A$ before the jump. Again go to the equation above and take $L_2$ and again solve for $(\Delta y)_B$ in the flow upstream of the jump: $y_2 = y_B - (\Delta y)_B$. Now insert $y_1$ and $y_2$ into the jump equation $y_2 = [-y_1 + \sqrt{y_1^2 + (8Q^2/gb^2)(1/y_1)}]/2$ to see whether this equation is satisfied for the given volumetric flow $Q$. If not, go back and choose a different value of $L_1$ proceeding in this way until the jump equation is satisfied.

$y_A$ $\qquad$ $y_B$ $\qquad$ $y_{cr}$

$L_1$ $\qquad$ $L_2$

**Fig. 14-99**

**14.391** The distance $L_1$ in Fig. 14-99 is 200 m and $y_A$ is half the critical value. The channel is finished concrete with an adverse slope of $-0.005$. What is $y_B$ at a distance $L_2 = 100$ m from the hydraulic jump? Solve in the simplest manner to get an approximate result. The flow rate is 200 m³/s and the width of the channel is 7 m.

■
$$y_c = (q^2/g)^{1/3} = [(\tfrac{200}{7})^2/9.807]^{1/3} = 4.366 \text{ m} \qquad y_A = 4.366/2 = 2.183 \text{ m}$$

$$\Delta L = ([1 - (Q^2 b/gA^3)]/\{S_0 - (n/\kappa)^2[Q^2/(R_H^{4/3}A^2)]\})\,\Delta y$$

$$A_1 = (7)(2.183) = 15.28 \text{ m}^2 \qquad R_H = A/p_w \qquad (R_H)_1 = 15.28/(2.183 + 7 + 2.183) = 1.344 \text{ m}$$

$$200 = \left\{ \frac{1 - (200)^2(7)/[(9.807)(15.28)^3]}{-0.005 - (0.012/1)^2[200^2/(1.344)^{4/3}(15.28)^2]} \right\}(\Delta y)_A \qquad (\Delta y)_A = 0.618 \text{ m}$$

$$y_1 = 2.183 + 0.618 = 2.801 \text{ m}$$

$$y_2 = [-y_1 + \sqrt{y_1^2 + (8Q^2/gb^2)(1/y_1)}]/2 = \{-2.801 + \sqrt{2.801^2 + [(8)(200)^2/(9.807)(7)^2](1/2.801)}\}/2 = 6.435 \text{ m}$$

$$A_2 = (7)(6.435) = 45.04 \text{ m}^2 \qquad (R_H)_2 = 45.04/(6.435 + 7 + 6.435) = 2.267 \text{ m}$$

$$100 = \left\{ \frac{1 - (200)^2(7)/[(9.807)(45.04)^3]}{-0.005 - (0.012/1)^2[200^2/(2.267)^{4/3}(45.04)^2]} \right\}(\Delta y)_B \qquad (\Delta y)_B = -0.866 \text{ m}$$

$$(\Delta y)_B = y_B - y_2 \qquad -0.866 = y_B - 6.435 \qquad y_B = 5.569 \text{ m}$$

**14.392** For a rectangular channel, develop an expression for the relation between the depths before and after a hydraulic jump. Refer to Fig. 14-100.

■ For the free body between sections 1 and 2, considering a unit width of channel and unit flow $q$, $P_1 = \gamma \bar{h} A = \gamma(\tfrac{1}{2}y_1)y_1 = \tfrac{1}{2}\gamma y_1^2$ and, similarly, $P_2 = \tfrac{1}{2}\gamma y_2^2$. From the principle of impulse and momentum, $\Delta P_x\, dt = \Delta$ linear momentum $= (W/g)(\Delta V_x)$, $\tfrac{1}{2}\gamma(y_2^2 - y_1^2)\, dt = (\gamma q\, dt/g)(V_1 - V_2)$.

Since $V_2 y_2 = V_1 y_1$ and $V_1 = q/y_1$, the above equation becomes

$$q^2/g = \tfrac{1}{2} y_1 y_2 (y_1 + y_2) \qquad (1)$$

Since $q^2/g = y_c^3$,

$$y_c^3 = \tfrac{1}{2} y_1 y_2 (y_1 + y_2) \qquad (2)$$

Supercritical Flow ——▶|1   Transition   2|—— Subcritical Flow   **Fig. 14-100**

**14.393** A rectangular channel, 20 ft wide, carries 400 cfs and discharges onto a 20-ft-wide apron with no slope with a mean velocity of 20 fps. What is the height of the hydraulic jump? What energy is lost in the jump?

$\blacksquare$  $q^2/g = y_1 y_2 (y_1 + y_2)/2$   $y_1 = q/v_1 = (\tfrac{400}{20})/20 = 1.00$ ft   $(\tfrac{400}{20})^2/32.2 = (1.00 y_2)(1.00 + y_2)/2$

$0.500 y_2^2 + 0.500 y_2 - 12.42 = 0$   $y_2 = 4.51$ ft   Height of jump $= 4.51 - 1.00 = 3.51$ ft

$$E_j = (y_2 - y_1)^3/4 y_1 y_2 = (4.51 - 1.00)^3/[(4)(4.51)(1.00)] = 2.40 \text{ ft}$$

$$\text{Energy lost} = Q\gamma E_j = (400)(62.4)(2.40) = 59\,900 \text{ ft-lb/s}$$

**14.394** A rectangular channel, 16 ft wide, carries a flow of 192 cfs. The depth of water on the downstream side of a hydraulic jump is 4.20 ft. What is the upstream depth? What is the loss of head?

$\blacksquare$  $q^2/g = y_1 y_2 (y_1 + y_2)/2$   $(\tfrac{196}{16})^2/32.2 = (y_1)(4.20)(y_1 + 4.20)/2$

$2.10 y_1^2 + 8.82 y_1 - 4.472 = 0$   $y_1 = 0.457$ ft

$$E_j = (y_2 - y_1)^3/4 y_1 y_2 = (4.20 - 0.457)^3/[(4)(0.457)(4.20)] = 6.83 \text{ ft}$$

**14.395** After flowing over the concrete spillway of a dam, 9000 cfs then passes over a level concrete apron ($n = 0.013$). The velocity of the water at the bottom of the spillway is 42.0 ft/s and the width of the apron is 180 ft. Conditions will produce a hydraulic jump, the depth in the channel below the apron being 10.0 ft. In order that the jump be contained on the apron, (**a**) how long should the apron be built? (**b**) How much energy is lost from the foot of the spillway to the downstream side of the jump?

$\blacksquare$  Refer to Fig. 14-101.

(**a**)   $q^2/g = y_1 y_2 (y_1 + y_2)/2$   $(\tfrac{9000}{180})^2/32.2 = 10 y_2 (10 + y_2)/2$

$5 y_2^2 + 50 y_2 - 77.64 = 0$   $y_2 = 1.37$ ft

$y_1 = q/V_1 = (\tfrac{9000}{180})/42.0 = 1.19$ ft   $L = [(V_2^2/2g + y_2) - (V_1^2/2g + y_1)]/(S_0 - S)$

$V_2 = q/y_2 = (\tfrac{9000}{180})/1.37 = 36.50$ ft/s   $S = (nV/1.486 R^{2/3})^2$

$V_{av} = (42.0 + 36.40)/2 = 39.20$ ft/s

$R = A/p_w$   $R_1 = (180)(1.19)/(1.19 + 180 + 1.19) = 1.174$ ft

$R_2 = (180)(1.37)/(1.37 + 180 + 1.37) = 1.349$ ft

$R_{av} = (1.174 + 1.349)/2 = 1.262$ ft   $S = \{(0.013)(39.20)/[(1.486)(1.262)^{2/3}]\}^2 = 0.08623$

$L = \{36.50^2/[(2)(32.2)] + 1.37 - 42.0^2/[(2)(32.2)] - 1.19\}/(0 - 0.08623) = 75.7$ ft

The length of the jump $L_3$ from $B$ to $C$ is from $4.3 y_3$ to $5.2 y_3$. Assuming the conservative value of $5.0 y_3$, $L_3 = (5.0)(10.0) = 50$ ft. Hence, $L_{ABC} = 76 + 50 = 126$ ft (approximately).

(**b**)   $E = y + V^2/2g$   $E_A = 1.19 + 42.0^2/[(2)(32.2)] = 28.58$ ft

$V_3 = Q/A_3 = 9000/[(10.0)(180)] = 5.00$ ft/s

$E_C = 10.0 + 5.00^2/[(2)(32.2)] = 10.39$ ft   $E_{lost} = 28.58 - 10.39 = 18.19$ ft

$P = Q\gamma E$   $P_{lost} = (9000)(62.4)(18.19) = 10.22 \times 10^6$ ft-lb/s

**Fig. 14-101**

**14.396** Determine the elevation of the spillway apron if $q = 50$ cfs/ft, $h = 9$ ft, $D = 63$ ft, and the spillway crest is at elevation 200.0 ft.

▮

$$(\pi)_1(\pi_3 - \pi_2)(\pi_3 + 1)^{1/2} + 0.353 = \sqrt{\tfrac{1}{8} + (2.828)(\pi_1)(\pi_3 + 1)^{3/2}}$$

$$\pi_1 = g^{1/2}h^{3/2}/q = (32.2)^{1/2}(9)^{3/2}/50 = 3.06$$

$$\pi_2 = D/h = \tfrac{63}{9} = 7.00 \qquad \pi_3 = d/h = d/9$$

$$(3.06)(d/9 - 7.00)(d/9 + 1)^{1/2} + 0.353 = \sqrt{\tfrac{1}{8} + (2.828)(3.06)(d/9 + 1)^{3/2}}$$

$d = 77.9$ ft    (by trial and error)    Elevation of spillway apron $= 200.0 - 77.9 = 122.1$ ft

**14.397** Establish the equation for flow over a broad-crested weir assuming no lost head. See Fig. 14-102.

▮ At the section where critical flow occurs, $q = V_c y_c$. But $y_c = V_c^2/g = \tfrac{2}{3}E$, and $V_c = \sqrt{g(\tfrac{2}{3}E_c)}$. Hence the theoretical value of flow $q$ becomes $q = \sqrt{g(\tfrac{2}{3}E_c)} \times \tfrac{2}{3}E_c = 3.09E_c^{3/2}$. However, the value of $E_c$ is difficult to measure accurately, because the critical depth is difficult to locate. The practical equation becomes $q = CH^{3/2} \approx 3H^{3/2}$. The weir should be calibrated in place to obtain accurate results.

**Fig. 14-102**

**14.398** Develop an expression for a critical-flow meter (Fig. 14-103).

▮ An excellent method of measuring flow in open channels is by means of a constriction. The measurement of the critical depth is not required. The depth $y_1$ is measured a short distance upstream from the constriction. The raised floor should be about $3y_c$ long and of such height as to have the critical velocity occur on it.

For a rectangular channel of constant width, the Bernoulli equation is applied between sections 1 and 2, in which the lost head in accelerated flow is taken as one-tenth of the difference in velocity heads, i.e., $y_1 + (V_1^2/2g) - (\tfrac{1}{10})[(V_c^2/2g) - (V_1^2/2g)] = [y_c + (V_c^2/2g) + z]$, which neglects the slight drop in the channel bed between 1 and 2. Recognizing that $E_c = y_c + V_c^2/2g$, we rearrange as follows: $[y_1 + (1.10V_1^2/2g)] = z + 1.0E_c + (\tfrac{1}{10})(\tfrac{1}{3}E_c)$, $[y_1 - z + (1.10V_1^2/2g)] = 1.033E_c = (1.033)(\tfrac{3}{2}\sqrt[3]{q^2/g})$, or

$$q = (2.94)(y_1 - z + 1.10V_1^2/2g)^{3/2} \qquad (1)$$

Since $q = V_1 y_1$, $\qquad\qquad q = (2.94)(y_1 - z + 0.0171q^2/y_1^2)^{3/2} \qquad (2)$

**14.399** Consider a rectangular channel 10 ft wide with the critical-flow meter of Fig. 14-103 having dimension $z = 1.10$ ft. If the measured depth $y_1$ is 2.42 ft, what is the discharge?

▮ $q = (2.94)(y_1 - z + 0.0171q^2/y_1^2)^{3/2} = (2.94)(2.42 - 1.10 + 0.0171q^2/2.42^2)^{3/2}$. As a first approximation, neglect the last term involving $q$. $q = (2.94)(2.42 - 1.10)^{3/2} = 4.46$ cfs/ft. Try $q = 4.80$ cfs/ft: $q = 2.94[2.42 - 1.10 + (0.0171)(4.80)^2/2.42^2]^{3/2} = 4.80$ cfs/ft  (O.K.), $Q = (4.80)(10) = 48.0$ ft³/s.

**Fig. 14-103**

**14.400** In a 100-ft-wide rectangular channel, the depth upstream from a hydraulic jump is 4 ft. What flow rate is required in order for the downstream depth to be (*a*) twice the upstream depth and (*b*) ten times the upstream depth?

$$q = \{[(d_1 + d_2)/2](g d_1 d_2)\}^{1/2}$$

(*a*)     $d_2 = (2)(4) = 8\text{ ft}$     $q = \{[(4 + 8)/2][(32.2)(4)(8)]\}^{1/2} = 78.63\text{ cfs/ft}$

$$Q = (100)(78.63) = 7863\text{ ft}^3/\text{s}$$

(*b*)     $d_2 = (10)(4) = 40\text{ ft}$     $q = \{[(4 + 40)/2][(32.2)(4)(40)]\}^{1/2} = 336.7\text{ cfs/ft}$

$$Q = (100)(336.7) = 33\,670\text{ ft}^3/\text{s}$$

**15.1** If the channel width for the stream described by Fig. 15-1 is $b = 28$ ft at stage $y = 3.5$ ft, determine the local velocity of a small-amplitude wave.

▌ $u = (1/b)(dQ/dy)$. From Fig. 15-1, $dQ/dy = 300$ cfs/ft when $y = 3.5$ ft. $u = (1/28)(300) = 10.7$ ft/s.

**Fig. 15-1**

**15.2** Assume the stage-discharge relation $Q = (k)(y - a)^b$, where, in SI units, $a = 0.059$, $b = 2.55$, and $k = 10.30$. Compute the velocity of a monoclinal wave of small height at stage 1.5 m, where the channel is 19 m wide.

▌ $u = (1/w)(dQ/dy)$    $Q = (10.30)(y - 0.059)^{2.55}$    $dQ/dy = (2.55)(10.30)(y - 0.059)^{1.55}$

For $y = 1.5$ m, $dQ/dy = (2.55)(10.30)(1.5 - 0.059)^{1.55} = 46.27$ (m³/s)/m, $u = (1/19)(46.27) = 2.44$ m/s.

**15.3** Repeat Prob. 15.1 if $b = 8$ m at $y = 0.8$ m.

▌ $u = (1/w)(dQ/dy)$. From Fig. 15-1, $dQ/dy = 20$ (m³/s)/m when $y = 0.8$ m. $u = (1/8)(20) = 2.50$ m/s.

**15.4** Given the hydrographs tabulated below [$I$ = inflow, $O$ = outflow; M = midnight, N = noon], find the 12-hourly storage values for the reach. Ignore local inflow.

| day | hour | $I$, m³/s | $O$, m³/s | day | hour | $I$, m³/s | $O$, m³/s |
|-----|------|-----------|-----------|-----|------|-----------|-----------|
| 1 | M | 36 | 58 | 7 | N | 196 | 341 |
| 2 | N | 43 | 46 | | M | 153 | 272 |
| | M | 121 | 42 | 8 | N | 128 | 219 |
| 3 | N | 348 | 64 | | M | 101 | 180 |
| | M | 575 | 149 | 9 | N | 86 | 153 |
| 4 | N | 719 | 329 | | M | 71 | 124 |
| | M | 741 | 536 | 10 | N | 60 | 104 |
| 5 | N | 615 | 678 | | M | 55 | 88 |
| | M | 440 | 681 | 11 | N | 46 | 73 |
| 6 | N | 328 | 560 | | M | 41 | 62 |
| | M | 255 | 439 | 12 | N | 37 | 52 |

▌$[(I_1 + I_2)/2]t - [(O_1 + O_2)/2]t = S_2 - S_1$. Taking minimum storage equal to zero, and noting that $t = 0.5$ day $= 0.432 \times 10^5$ s, computations are carried out as in the following table. The last two columns are for use in Prob. 15.8.

| day | hour | $I$, m³/s | $O$, m³/s | $\bar{I} = (I_1+I_2)/2$ | $\bar{O} = (O_1+O_2)/2$ | $\Delta S/t = \bar{I} - \bar{O}$ | $S/t = \sum (\bar{I} - \bar{O})$ | $S$, $10^5$, m³ | $xI + (1-x)O$ $x = 0.2$ | $xI + (1-x)O$ $x = 0.3$ |
|---|---|---|---|---|---|---|---|---|---|---|
| 1 | M | 36 | 58 | | | | 13 | 6 | 54 | 51 |
| | | | | 39 | 52 | −13 | | | | |
| 2 | N | 43 | 46 | | | | 0 | 0 | 45 | 45 |
| | | | | 82 | 44 | 38 | | | | |
| | M | 121 | 42 | | | | 38 | 16 | 58 | 66 |
| | | | | 234 | 53 | 181 | | | | |
| 3 | N | 348 | 64 | | | | 219 | 95 | 121 | 149 |
| | | | | 462 | 106 | 356 | | | | |
| | M | 575 | 149 | | | | 575 | 248 | 234 | 277 |
| | | | | 647 | 239 | 408 | | | | |
| 4 | N | 719 | 329 | | | | 983 | 425 | 407 | 446 |
| | | | | 730 | 432 | 298 | | | | |
| | M | 741 | 536 | | | | 1281 | 553 | 577 | 598 |
| | | | | 678 | 607 | 71 | | | | |
| 5 | N | 615 | 678 | | | | 1352 | 584 | 665 | 659 |
| | | | | 528 | 679 | −151 | | | | |
| | M | 440 | 681 | | | | 1201 | 519 | 633 | 609 |
| | | | | 384 | 620 | −236 | | | | |
| 6 | N | 328 | 560 | | | | 965 | 417 | 514 | 490 |
| | | | | 292 | 499 | −207 | | | | |
| | M | 255 | 439 | | | | 758 | 327 | 402 | 384 |
| | | | | 226 | 390 | −164 | | | | |
| 7 | N | 196 | 341 | | | | 594 | 257 | 312 | 298 |
| | | | | 174 | 306 | −132 | | | | |
| | M | 153 | 272 | | | | 462 | 200 | 248 | 236 |
| | | | | 140 | 245 | −105 | | | | |
| 8 | N | 128 | 219 | | | | 357 | 154 | 201 | 192 |
| | | | | 114 | 199 | −85 | | | | |
| | M | 101 | 180 | | | | 272 | 118 | 164 | 156 |
| | | | | 94 | 166 | −72 | | | | |
| 9 | N | 86 | 153 | | | | 200 | 86 | 140 | 133 |
| | | | | 78 | 138 | −60 | | | | |
| | M | 71 | 124 | | | | 140 | 60 | 113 | 108 |
| | | | | 66 | 114 | −48 | | | | |
| 10 | N | 60 | 104 | | | | 92 | 40 | 95 | 91 |
| | | | | 58 | 96 | −38 | | | | |
| | M | 55 | 88 | | | | 54 | 23 | 81 | 78 |
| | | | | 50 | 80 | −30 | | | | |
| 11 | N | 46 | 73 | | | | 24 | 10 | 68 | 65 |
| | | | | 44 | 68 | −24 | | | | |
| | M | 41 | 62 | | | | 0 | 0 | 58 | 56 |

**15.5**  Given the following inflow data to a reservoir and the reservoir relationships of Fig. 15-2, route the flood through the reservoir. The initial outflow from the reservoir is 1.7 m³/s, and the initial value of $2S/t + O$ is 9.0 m³/s.

| day | hour | $I$, m³/s |
|---|---|---|
| 1 | Noon | 2.0 |
| | Midnight | 5.2 |
| 2 | Noon | 10.1 |
| | Midnight | 12.2 |
| 3 | Noon | 8.5 |
| | Midnight | 4.7 |
| 4 | Noon | 2.3 |

**Fig. 15-2**

$$I_1 + I_2 + (2S_1/t - O_1) = 2S_2/t + O_2 \qquad 2S_1/t - O_1 = 2S_2/t + O_2 - 2O_2$$

$$2S_1/t - O_1 = 9.0 - (2)(1.7) = 5.6 \, \text{m}^3/\text{s} \qquad 2.0 + 5.2 + 5.6 = 2S_2/t + O_2 \qquad 2S_2/t + O_2 = 12.8 \, \text{m}^3/\text{s}$$

From Fig. 15-2, $O_2 = 2.3 \, \text{m}^3/\text{s}$. Succeeding computations are carried out in the same manner with the results as tabulated below.

| day | hour | $I$, m³/s | $(2S/t) - O$, m³/s | $(2S/t) + O$, m³/s | $O$, m³/s |
|-----|------|-----------|--------------------|--------------------|-----------|
| 1 | Noon | 2.0 | 5.6 | 9.0 | 1.7 |
|   | Midnight | 5.2 | 8.2 | 12.8 | 2.3 |
| 2 | Noon | 10.1 | 14.9 | 23.5 | 4.3 |
|   | Midnight | 12.2 | 16.2 | 37.2 | 10.5 |
| 3 | Noon | 8.5 | 16.3 | 36.9 | 10.3 |
|   | Midnight | 4.7 | 16.3 | 29.5 | 6.6 |
| 4 | Noon | 2.3 |  | 23.3 | 4.2 |

**15.6** A reservoir has an area of 295 acres at spillway level, and its banks are vertical for several feet above spillway level. The spillway is 16 ft long and has a coefficient of 3.75. If the inflow data of Prob. 15.4, with the units changed to cfs, apply to this reservoir, compute the maximum pool level and maximum discharge to be expected if the reservoir is initially at the spillway level at midnight on the first.

**❚** The discharge over the spillway is ($C = 3.75$ and $L = 16$ ft): $Q = CLy^{3/2} = (3.75)(16)y^{3/2} = 60y^{3/2}$ cfs. The storage is equal to the reservoir area multiplied by the height of water above the spillway crest: $S = 295y$ acre-ft $= 149y$ cfs-day. From these equations and $t = 0.5$ day, we obtain:

| $y$, ft | $Q$, cfs | $S$, cfs-day | $(2S/t) + O$, cfs |
|---------|----------|--------------|-------------------|
| 1 | 60 | 149 | 656 |
| 2 | 170 | 298 | 1362 |
| 3 | 312 | 447 | 2100 |
| 4 | 480 | 596 | 2864 |
| 5 | 671 | 745 | 3651 |

Fig. 15-3

These data provide the basis for deriving the routing curves of Fig. 15-3. The inflow hydrograph of Prob. 15.4 is routed as follows: $I_1 + I_2 + [(2S_1/t) - O_1] = (2S_2/t) + O_2$.

| day | hour | $I$, cfs | $(2S_1/t) - O$, cfs | $(2S_2/t) + O$, cfs | $O$, cfs |
|-----|------|----------|---------------------|---------------------|----------|
| 1 | M | 36 | 0 | 0 | 0 |
| 2 | N | 43 | 71 | 79 | 4 |
|   | M | 121 | 207 | 235 | 14 |
| 3 | N | 348 | 562 | 676 | 57 |
|   | M | 575 | 1131 | 1485 | 177 |
| 4 | N | 719 | 1715 | 2425 | 355 |
|   | M | 741 | 2145 | 3175 | 515 |
| 5 | N | 615 | 2321 | 3501 | 590 |
|   | M | 440 | 2256 | 3376 | 560 |
| 6 | N | 328 |  | 3024 | 480 |

The peak flow is approximately 590 cfs and the maximum pool elevation is $y = (590/60)^{0.6667} = 4.59$ ft.

**15.7** Tabulated below are the elevation-storage and elevation-discharge data for a small reservoir. The inflow hydrograph of Prob. 15.4, in units of cfs, applies to this reservoir. Assuming the pool elevation to be 875 ft at midnight on the first, find the maximum pool elevation and peak outflow rate.

| elevation, ft | storage, acre-ft | discharge, cfs | elevation, ft | storage, acre-ft | discharge, cfs |
|---------------|------------------|----------------|---------------|------------------|----------------|
| 862 | 0 | 0 | 882 | 1220 | 101 |
| 865 | 41 | 0 | 884 | 1633 | 232 |
| 870 | 202 | 0 | 886 | 2270 | 395 |
| 875 | 502 | 0 | 888 | 3152 | 600 |
| 880 | 1003 | 0 |  |  |  |

❚ From the data provided, and noting that $t = 0.5$ day and 1 acre-foot $= 0.504$ cfs-day:

| elevation, ft | $Q$, cfs | $S$, cfs-day | $(2S/t) + O$ |
|:---:|:---:|:---:|:---:|
| 862 | 0 | 0 | 0 |
| 865 | 0 | 21 | 84 |
| 870 | 0 | 102 | 408 |
| 875 | 0 | 253 | 1012 |
| 880 | 0 | 506 | 2024 |
| 882 | 101 | 615 | 2561 |
| 884 | 232 | 823 | 3524 |
| 886 | 395 | 1144 | 4971 |
| 888 | 600 | 1589 | 6956 |

These data provide the basis for deriving the routing curves of Fig. 15-4. The inflow hydrograph of Prob. 15.4 is routed as follows: $I_1 + I_2 + [(2S_1/t) - O_1] = (2S_2/t) + O_2$.

Fig. 15-4

| day | hour | $I$, cfs | $(2S_1/t) - O$, cfs | $(2S_2/t) + O$, cfs | $O$, cfs |
|:---:|:---:|:---:|:---:|:---:|:---:|
| 1 | M | 36 | 1010 | 1010 | 0 |
| 2 | N | 43 | 1089 | 1089 | 0 |
|   | M | 121 | 1253 | 1253 | 0 |
| 3 | N | 348 | 1722 | 1722 | 0 |
|   | M | 575 | 2419 | 2645 | 113 |
| 4 | N | 719 | 3207 | 3713 | 253 |
|   | M | 741 | 3941 | 4667 | 363 |
| 5 | N | 615 | 4439 | 5297 | 429 |
|   | M | 440 | 4594 | 5494 | 450 |
| 6 | N | 328 | 4488 | 5362 | 437 |
|   | M | 255 |      | 5071 | 405 |

The peak flow is approximately 450 cfs and the maximum pool elevation is 886.6 ft.

**15.8**  The Muskingum method of flood routing involves a time parameter $K$ and a weighting factor $x$. Find these quantities for the flood of Prob. 15.4 (flows in $m^3/s$).

▌ Columns 4, 10, and 11 of the solution tableau of Prob. 15.4 give the weighted flows corresponding to $x = 0$, 0.2, and 0.3; these flows are graphed in Fig. 15-5. It will be seen that increasing the value of $x$ from 0.2 to 0.3 reverses the "loop" in the lower range of flow. Even so, a slight additional increase in $x$ might provide a slightly better fit, without materially affecting $K$. Take $K = 26.7$ h, $x = 0.3$, and $t = 12$ h.

**15.9**  For a flood with $K = 25.9$ h and $x = 0.25$, determine the Muskingum routing coefficients for $t = 12$ h and $t = 24$ h.

▌
$$c_0 = (Kx - 0.5t)/(K - Kx + 0.5t) \qquad c_1 = (Kx + 0.5t)/(K - Kx + 0.5t)$$
$$c_2 = (K - Kx - 0.5t)/(K - Kx + 0.5t)$$

For $t = 12$ h, with $K = 25.9$ h and $x = 0.25$,

$$c_0 = -\frac{(25.9)(0.25) - (0.5)(12)}{25.9 - (25.9)(0.25) + (0.5)(12)} = -0.019 \qquad c_1 = \frac{(25.9)(0.25) + (0.5)(12)}{25.9 - (25.9)(0.25) + (0.5)(12)} = 0.491$$

$$c_2 = \frac{25.9 - (25.9)(0.25) - (0.5)(12)}{25.9 - (25.9)(0.25) + (0.5)(12)} = 0.528$$

For $t = 24$ h, with $K = 25.9$ h and $x = 0.25$,

$$c_0 = -\frac{(25.9)(0.25) - (0.5)(24)}{25.9 - (25.9)(0.25) + (0.5)(24)} = 0.176 \qquad c_1 = \frac{(25.9)(0.25) + (0.5)(24)}{25.9 - (25.9)(0.25) + (0.5)(24)} = 0.588$$

$$c_2 = \frac{25.9 - (25.9)(0.25) - (0.5)(24)}{25.9 - (25.9)(0.25) + (0.5)(24)} = 0.236$$

The 24-h routing period is to be preferred, since negative coefficients lead to instability when flow changes rapidly.

**15.10**  For a river reach with $K = 26$ h and $x = 0.22$, determine the Muskingum routing coefficients for $t = 12$ h.

▌
$$c_0 = (Kx - 0.5t)/(K - Kx + 0.5t) \qquad c_1 = (Kx + 0.5t)/(K - Kx + 0.5t)$$
$$c_2 = (K - Kx - 0.5t)/(K - Kx + 0.5t)$$

$$c_0 = -\frac{(26)(0.22) - (0.5)(12)}{26 - (26)(0.22) + (0.5)(12)} = 0.011 \qquad c_1 = \frac{(26)(0.22) + (0.5)(12)}{26 - (26)(0.22) + (0.5)(12)} = 0.446$$

$$c_2 = \frac{26 - (26)(0.22) - (0.5)(12)}{26 - (26)(0.22) + (0.5)(12)} = 0.543$$

**15.11**  The outflow hydrograph of Prob. 15.4 becomes the inflow hydrograph for a reach. Using the routing coefficients determined in Prob. 15.10, find the peak reach outflow by the Muskingum method.

▌ $O_2 = c_0 I_2 + c_1 I_1 + c_2 O_1$. The routing computations are as follows:

| day | hour | $I$, $m^3/s$ | $0.011 I_2$, $m^3/s$ | $0.446 I_1$, $m^3/s$ | $0.543 O_1$, $n^3/s$ | $O$, $m^3/s$ |
|-----|------|------|------|------|------|------|
| 1 | M | 58 | | | | 60 (est.) |
| 2 | N | 46 | 1 | 26 | 33 | 60 |
|   | M | 42 | 0 | 21 | 33 | 54 |
| 3 | N | 64 | 1 | 19 | 29 | 49 |
|   | M | 149 | 2 | 29 | 27 | 58 |
| 4 | N | 329 | 4 | 66 | 31 | 101 |
|   | M | 536 | 6 | 147 | 55 | 208 |
| 5 | N | 678 | 7 | 239 | 113 | 359 |
|   | M | 681 | 7 | 302 | 195 | 504 |
| 6 | N | 560 | 6 | 304 | 274 | 584 |
|   | M | 439 | 5 | 250 | 317 | 572 |
| 7 | N | 341 | 4 | 196 | 311 | 511 |
|   | M | 272 | 3 | 152 | 277 | 432 |

The routed peak flow is 584 $m^3/s$.

Fig. 15-5

**15.12** For the inflow hydrograph given below for a reach, route the flood through the reach by the Muskingum method. Take $K = 11$ h, $t = 6$ h, and $x = 0.13$; hence, $c_0 = 0.124$, $c_1 = 0.353$, and $c_2 = 0.523$.

| day | hour | $I$, cfs |
|-----|------|----------|
| 1 | 6 a.m. | 10 |
| | Noon | 30 |
| | 6 p.m. | 68 |
| | Midnight | 50 |
| 2 | 6 a.m. | 40 |
| | Noon | 31 |
| | 6 p.m. | 23 |

Assume an initial outflow of 10 cfs.

**∎** $$O_2 = c_0 I_2 + c_1 I_1 + c_2 O_1$$

| day | hour | $I$ | $c_0 I_2$ | $c_1 I_1$ | $c_2 O_1$ | $O$ |
|-----|------|-----|-----------|-----------|-----------|-----|
| 1 | 6 a.m. | 10 | — | — | — | 10 |
| | Noon | 30 | 3.7 | 3.5 | 5.2 | 12.4 |
| | 6 p.m. | 68 | 8.4 | 10.6 | 6.5 | 25.5 |
| | Midnight | 50 | 6.2 | 24.0 | 13.3 | 43.5 |
| 2 | 6 a.m. | 40 | 5.0 | 17.7 | 22.7 | 45.4 |
| | Noon | 31 | 3.8 | 14.1 | 23.7 | 41.6 |
| | 6 p.m. | 23 | 2.9 | 10.9 | 21.8 | 35.6 |

All tabular values are in cubic feet per second.

**15.13** Refer to Fig. 15-6. At time $t_1$, an infinitesimal gravity wave is initiated at the dam (station $x_1$). Obtain an analytical expression for the time $t$ at which the wavefront passes through station $x$. Assume still water and a constant bed slope.

**∎** $u = dx/dt = \sqrt{gy}$. But the depth equation is

$$\frac{y - y_1}{x - x_1} = \frac{y_2 - y_1}{x_2 - x_1} \equiv s_0 \qquad (1)$$

so that $dx/dt = (1/s_0)(dy/dt)$ and

$$\frac{1}{s_0}\frac{dy}{dt} = \sqrt{gy}$$

$$\int_{t_1}^{t} dt = \frac{1}{s_0\sqrt{g}} \int_{y_1}^{y} \frac{dy}{\sqrt{y}} \qquad (2)$$

$$t - t_1 = \frac{2}{s_0\sqrt{g}}(\sqrt{y} - \sqrt{y_1})$$

Finally, substitute for $y$ in (2) from (1), obtaining

$$t - t_1 = \frac{2}{s_0\sqrt{g}}[\sqrt{y_1 + s_0(x - x_1)} - \sqrt{y_1}] \qquad (3)$$

Fig. 15-6

**15.14**  Represent the wave of Prob. 15.13 in the $xt$ plane, for the following assignment of parameters: $t_1 = 0$ s, $x_1 = 0$ ft, $y_1 = 55$ ft, $x_2 = 16\,000$ ft, $y_2 = 22$ ft.

▮ Using $s_0 = (22 - 55)/(6000 - 0) = -0.0055$ and $g = 32.2$ ft/s² in (3) of Prob. 15.13, we generate Table 15-1 and the rising curve of Fig. 15-7.

**TABLE 15-1**

| $x$, ft | $t$, s | $x$, ft | $t$, s |
|---|---|---|---|
| 0 | 0 | 8 000 | 207 |
| 1 000 | 24 | 9 000 | 236 |
| 2 000 | 48 | 10 000 | 265 |
| 3 000 | 73 | 11 000 | 296 |
| 4 000 | 99 | 12 000 | 327 |
| 5 000 | 125 | 13 000 | 360 |
| 6 000 | 152 | 14 000 | 394 |
| 7 000 | 179 | 15 000 | 429 |
|  |  | 16 000 | 466 |

**15.15**  Refer to Probs. 15.13 and 15.14. Suppose that a similar wave is initiated at $x_2$ ($= 16\,000$ ft) at time $t_1$ ($= 0$ s) and propagates toward the dam. Superimpose this second wave on Fig. 15-7. Where and when will the two waves meet?

▮ Making the appropriate replacements in (3) of Prob. 15.13, we obtain for the second wavefront

$$t - t_1 = \frac{2}{s^*\sqrt{g}}[\sqrt{y_2 + s^*(x_2 - x)} - \sqrt{y_2}] \qquad \text{where} \qquad s^* = -s_0$$

This equation generates Table 15-2 and the falling curve of Fig. 15-7. The two curves intersect at approximately $x = 8900$ ft, $t = 234$ s.

**TABLE 15-2**

| $x$, ft | $t$, s | $x$, ft | $t$, s |
|---|---|---|---|
| 0 | 466 | 8 000 | 259 |
| 1 000 | 442 | 9 000 | 230 |
| 2 000 | 417 | 10 000 | 200 |
| 3 000 | 392 | 11 000 | 170 |
| 4 000 | 367 | 12 000 | 138 |
| 5 000 | 341 | 13 000 | 106 |
| 6 000 | 314 | 14 000 | 72 |
| 7 000 | 287 | 15 000 | 37 |
|  |  | 16 000 | 0 |

**15.16**  A uniform rectangular channel 4 m wide with $n = 0.014$ and slope of 0.0003 is flowing at a depth of 1.7 m. A sudden gate opening increases the depth to 2.0 m. Calculate the propagation speed $c$ and the ground velocity $v$ of the resulting abrupt wave.

▮
$$c = \sqrt{(g)(J_2 - J_1)/[(A_1)(1 - A_1/A_2)]} \qquad J = Ad/2 \qquad A_1 = (4)(1.7) = 6.80 \text{ m}^2$$
$$J_2 = (6.80)(1.7/2) = 5.780 \text{ m}^3$$
$$A_2 = (4)(2.0) = 8.00 \text{ m}^2 \qquad J_1 = (8.00)(2.0/2) = 8.000 \text{ m}^3$$
$$c = \sqrt{(9.807)(8.000 - 5.780)/[(6.80)(1 - 6.80/8.00)]} = 4.62 \text{ m/s}$$
$$u = v_1 + c \qquad v = (1.0/n)(R^{2/3})(s^{1/2}) \qquad v_1 = (1.0/0.014)[6.80/(1.7 + 4 + 1.7)]^{2/3}(0.0003)^{1/2} = 1.17 \text{ m/s}$$
$$u = 1.17 + 4.62 = 5.79 \text{ m/s}$$

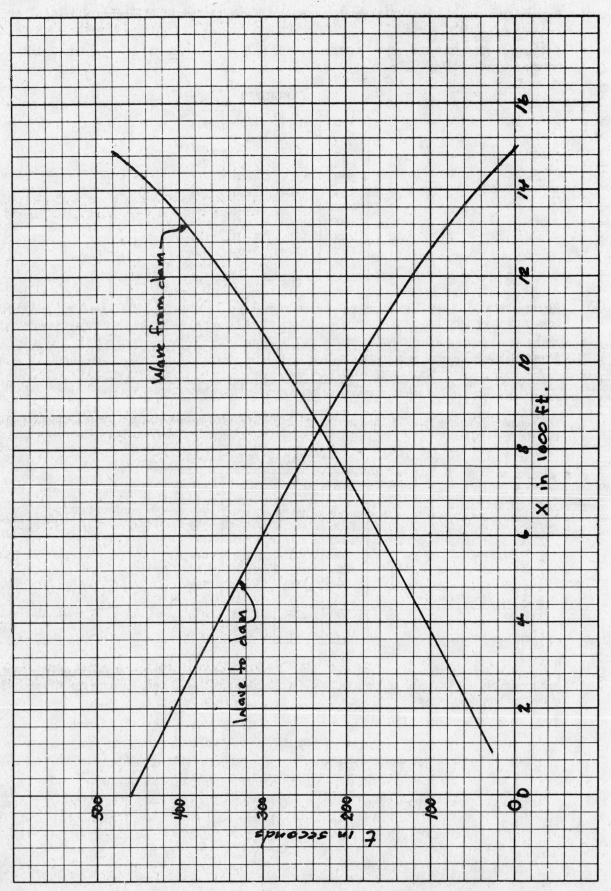

Wave from dam

Wave to dam

X in 1000 Ft.

T in seconds

500
400
300
200
100
0 0

2  4  6  8  10  12  14  16

Fig. 15-7

# CHAPTER 16
# Flow of Compressible Fluids

**16.1** For air at a temperature of 30 °C and pressure of 470 kPa abs, find the specific weight of the air.

$$\gamma = p/RT = 470/[(29.3)(30 + 273)] = 0.0529 \text{ kN/m}^3$$

**16.2** Find the mass density of helium at a temperature of 39 °F and a pressure of 26.9 psig when the atmospheric pressure is 14.9 psig.

$$\rho = p/RT = (26.9 + 14.9)(144)/[(12\,420)(39 + 460)] = 0.000971 \text{ slug/ft}^3$$

**16.3** Methane at 22 °C flows through a pipeline at a velocity of 416.4 m/s. Is the flow subsonic, sonic, supersonic, or hypersonic?

$N_M = v/\sqrt{kgRT} = 416.4/\sqrt{(1.32)(9.807)(52.9)(22 + 273)} = 0.93$. Since $N_M < 1.0$, the flow is subsonic.

**16.4** Chlorine gas at 51 °F flows through a pipeline. Find the velocity at which the flow will be sonic.

$$N_M = v/\sqrt{kgRT} \qquad 1.0 = v/\sqrt{(1.34)(32.2)(21.8)(51 + 460)} \qquad v = 693 \text{ ft/s}$$

**16.5** Compute the change in internal energy and the change in enthalpy of 15 kg of air if its temperature is raised from 20 to 30 °C. The initial pressure is 95 kPa abs.

$\Delta i = c_v(T_2 - T_1) = (0.716)(30 - 20) = 7.16 \text{ kJ/kg}$ Change in internal energy $= (\Delta i)(15) = 107.4 \text{ kJ}$

$\Delta h = c_p(T_2 - T_1) = 1.003(10) = 10.03 \text{ kJ/kg}$ Change in enthalpy $= (\Delta h)(15) = 150.0 \text{ kJ}$

**16.6** Suppose that 15 kg of air ($T_1 = 20$ °C) of Prob. 16.5 were compressed isentropically to 40 percent of its original volume. Find the final temperature and pressure, the work required, and the changes in internal energy and enthalpy.

The following relations apply: $pv = RT$ and $pv^k = $ constant, where $k = 1.40$ for air. $pv^k = pv(v^k/v) = (RT/v)v^k = RTv^{k-1} = $ constant. Since $R = $ constant, $Tv^{k-1} = $ constant, and $T_2 = T_1(v_1/v_2)^{k-1} = (273 + 20)(1.0/0.4)^{1.40-1} = 422 \text{ K} = 149 \text{ °C}$, $pv/T = R = $ constant, $p_1 = 95$ kPa abs (from Prob. 16.5), $p_1v_1/T_1 = p_2v_2/T_2$ and $v_2 = 0.4v_1$, $95v_1/293 = [p_2(0.4v_1)/422]$, $p_2 = 342$ kPa abs. Since this is an isentropic process, the work required is equal to the change in internal energy. This can be confirmed by computing the values of the pressure and corresponding volumes occupied by the gas during the isentropic process, plotting a pressure-vs.-volume curve, and finding the area under the curve and thereby determining the work done on the fluid. Thus the work required is

$$\int_{s_1}^{s_2} F \, ds = \int_{s_1}^{s_2} \left(\frac{F}{A}\right) A \, ds = \int_{\text{vol 1}}^{\text{vol 2}} p \, d(\text{vol})$$

or $\Delta i = c_v(T_2 - T_1) = 0.716(422 - 293) = 92.4 \text{ kJ/kg}$, $(\Delta i)(15 \text{ kg}) = 1385 \text{ kJ} = $ work required = change in internal energy, $\Delta h = c_p(T_2 - T_1) = 1.003(129) = 129.4 \text{ kJ/kg}$, $(\Delta h)(15 \text{ kg}) = 1941 \text{ kJ} = $ change in enthalpy.

**16.7** Using the data of Prob. 16.6 compute $\Delta(p/\rho)$ and thus show that $\Delta h = \Delta i + \Delta(p/\rho)$.

From Prob. 16.6, $\Delta h = 129.4 \text{ kJ/kg}$, $\Delta i = 92.4 \text{ kJ/kg}$, $R = 0.287 \text{ kJ/kg} \cdot \text{K}$; $p_1/\rho_1 = RT_1 = (0.287)(293) = 84.1 \text{ kJ/kg}$; $p_2/\rho_2 = RT_2 = (0.287)(422) = 121.1 \text{ kJ/kg}$; $\Delta(p/\rho) = p_2/\rho_2 - p_1/\rho_1 = 37.0 \text{ kJ/kg}$. $\Delta i + \Delta(p/\rho) = 92 + 37 = 129.4$ which checks accurately with the value of $\Delta h$.

**16.8** Using the data of Prob. 16.6, determine the work done in compressing the air by finding the area under a pressure-vs.-volume curve. Compute and tabulate volumes and pressures using volume increments which are 10 percent of the original volume.

$R = 0.287 \text{ kJ/kg} \cdot \text{K}$, $k = 1.4$. Initially, given: $T_1 = 273 + 20 = 293 \text{ K}$; $p_1 = 95 \text{ kPa}$; $v_1 = RT_1/p_1 = (0.287)(293)/95 = 0.885 \text{ m}^3/\text{kg}$. After compression, from Prob. 16.6 $T = T_1(v_1/v)^{1-k} = (293)(v_1/v)^{0.4}$; $v = v_1/(v_1/v)$; $V = mv = 15v$; $p = RT/v$.

| $v_1/v$ | $T$, K | $v$, m³/kg | $V$, m³ | $p$, kPa |
|---|---|---|---|---|
| $\frac{100}{100}$ | 293 | 0.885 | 13.28 | 95.0 |
| $\frac{100}{90}$ | 306 | 0.797 | 11.95 | 110.1 |
| $\frac{100}{80}$ | 320 | 0.708 | 10.62 | 129.9 |
| $\frac{100}{70}$ | 338 | 0.619 | 9.29 | 156.6 |
| $\frac{100}{60}$ | 359 | 0.531 | 7.97 | 194.1 |
| $\frac{100}{50}$ | 387 | 0.443 | 6.64 | 251.0 |
| $\frac{100}{40}$ | 423 | 0.354 | 5.31 | 343.0 |

These data are plotted in Fig. 16-1. Measure or calculate area $A$: Work done in compression $= A = 1.395$ MJ (by calculation).

**Fig. 16-1**

**16.9**  Compute the change in enthalpy of 500 lb of oxygen if its temperature is increased from 125 °F to 160 °F.

▮    $c_p = 5437$ ft-lb/(slug-°R)    $\Delta h = c_p(T_2 - T_1) = (5437)(160 - 125) = 190\,000$ ft-lb/slug

Five hundred pounds is equivalent to 15.53 slugs ($m = W/g$). Thus change in enthalpy $= (190\,000)(15.53) = 2.95 \times 10^6$ ft-lb.

**16.10**  Suppose the 500 lb of oxygen of Prob. 16.9 was compressed isentropically to 75 percent of its original volume. Find the final temperature and pressure, the work required, and the change in enthalpy. Assume $T_1 = 120$ °F and $p_1 = 200$ psia.

▮  $pv^k = $ constant and $pv = RT$. Eliminating $p$: $(RT/v)V^k = $ constant $= RTv^{k-1}$. Since $R = $ constant also, $Tv^{k-1} = $ constant; $k = 1.4$. From which: $T_2 = T_1(v_1/v_2)^{k-1} = (120 + 460)(1/0.75)^{0.4} = 650.7$ °R $= 190.7$ °F. $p_1v_1/T_1 = p_2v_2/T_2$; so $200v_1/580 = p_2(0.75v_1)/651$; $p_2 = 299$ psia; $c_v = 3883$, $c_p = 5437$ ft²/(s²-°R). Work required $= $ change of internal energy (as isentropic). Work required $= (\Delta i)(W) = c_v(\Delta T)(500/32.2)$ where $\Delta T = 190.7 - 120 = 70.7$ °F. Work required $= (3883)(70.7)(500/32.2) = 4.26 \times 10^6$ ft-lb; increase in enthalpy $= (5437)(70.7)(500/32.2) = 5.97 \times 10^6$ ft-lb.

**16.11**  Determine the sonic velocity in air at sea level and at 10 000, 20 000, and 30 000 ft. Assume standard atmosphere (Table A-7).

▮  $c = (kRT)^{1/2}$. In all cases $k = 1.4$ and $R = 1715$ ft-lb/(slug-°R). From Table A-7, temperatures $= 59$ °F, 23.4 °F, $-12.3$ °F, and $-47.8$ °F. Substitute for $T = (460 + $ temperature$)$ and compute the resulting $c$'s.

| elevation ft | temperature °F | $c$, fps |
|---|---|---|
| 0 | 59 | 1 116 |
| 10 000 | 23.4 | 1 077 |
| 20 000 | −12.3 | 1 037 |
| 30 000 | −47.8 | 995 |

**16.12**  Repeat Prob. 16.11 for sea level, 6 km, and 10 km.

▮  $c = (kRT)^{1/2}$, $k = 1.4$, and $R = 0.287$ kJ/kg · K $= 287$ m²/s² · K; $g = 9.81$ m/s². From Table A-8: temperatures $= 15.0$, $-24.0$, and $-49.9$ °C. Substitute for $T = (273 + $ temp. °C$)$ and compute the resulting $c$'s.

| elevation, m | temperature, °C | $c$, m/s |
|---|---|---|
| 0 | 15.0 | 340 |
| 6 000 | −24.0 | 316 |
| 10 000 | −49.9 | 299 |

**16.13**  Find the stagnation pressure and temperature in nitrogen flowing at 600 fps if the pressure and temperature in the undisturbed flow field are 100 psia and 200 °F, respectively.

$$c = \sqrt{kRT} = \sqrt{(1.40)(1774)(460 + 200)} = 1280 \text{ ft/s} \qquad N_M = v/c = \tfrac{600}{1280} = 0.469$$

$$\rho_0 = kp_0/c_0^2 = (1.40)[(100)(144)]/1280^2 = 0.0123 \text{ slug/ft}^3$$

$$p_s = p_0 + (\rho_0)(v_0^2/2)[1 + (N_M)_0^2/4 + \cdots] = (100)(144) + (0.0123)(600^2/2)(1 + 0.469^2/4)$$

$$= 16\,736 \text{ lb/ft}^2 \quad \text{or} \quad 116 \text{ lb/in}^2$$

$$(c_p/g)(T_1) + V_1^2/2g = (c_p/g)(T_s)$$

$$(6210/32.2)(460 + 200) + 600^2/[(2)(32.3)] = (6210/32.2)(T_s) \qquad T_s = 689\,°R \quad \text{or} \quad 229\,°F$$

**16.14**  Find the stagnation pressure and temperature in air flowing at 90 fps if the pressure and temperature in the undisturbed flow field are 14.7 psia and 60 °F, respectively.

$$c = \sqrt{kRT} = \sqrt{(1.40)(1716)(460 + 60)} = 1118 \text{ ft/s} \qquad N_M = v/c \qquad (N_M)_0 = \tfrac{90}{1118} = 0.0805$$

$$(c_p/g)(T_0) + V_0^2/2g = (c_p/g)(T_s)$$

$$(6000/32.2)(460 + 60) + 90^2/[(2)(32.2)] = (6000/32.2)(T_s) \qquad T_s = 521\,°R \quad \text{or} \quad 61\,°F$$

$$p_s = p_0 + (\rho_0)(v_0^2/2)[1 + (N_M)_0^2/4 + \cdots] \qquad \rho_0 = p_0/RT_0 = (14.7)(144)/[(1716)(460 + 60)] = 0.00237 \text{ slug/ft}^3$$

$$p_s = 14.7 + [(0.00237)(90^2/2)/144](1 + 0.0805^2/4) = 14.77 \text{ psia}$$

**16.15**  Air flows past an object at 500 fps. Determine the stagnation pressures and temperatures in the standard atmosphere at elevations of sea level, 5000 ft, and 30 000 ft.

From Table A-7: at sea level, 59 °F, $\rho = 0.00238$ slug/ft³; at 5000 ft, 41.2 °F, $\rho = 0.00205$ slug/ft³; at 30 000, −47.8 °F, $\rho = 0.00089$ slug/ft³.

$$c = \sqrt{kRT} \qquad [R = 1716 \text{ lb-ft/(slug-°R)}] \qquad N_M = v_0/c \qquad (\text{where } v_0 = 500 \text{ fps})$$

$$T_s = T_0 + (V_0^2/2c_p) \qquad [\text{where } c_p = 6000 \text{ lb-ft/(slug-°R)}]$$

$$p_s = p_0 + [(\rho_0)(v_0^2/2)/144][1 + (N_M)_0^2/4 + \cdots]$$

| elevation, ft | $p_0$, psi | $c$, fps | $N_M$ | $T_s$, °F | $p_s$, psi |
|---|---|---|---|---|---|
| 0 | 14.70 | 1 116 | 0.448 | 79.8 | 16.87 |
| 5 000 | 12.24 | 1 097 | 0.456 | 62.0 | 14.11 |
| 30 000 | 4.37 | 995 | 0.503 | −27.0 | 5.21 |

**16.16**  Repeat Prob. 16.15 for an air speed of 180 m/s and elevation 2 km.

From Table A-8: at 2000 m, 2 °C, $p = 79.50$ kPa, $\rho = 1.007$ kg/m³.

$$c = \sqrt{kRT} = \sqrt{(1.40)(287)(2 + 273)} = 332 \text{ m/s} \qquad N_M = v_0/c = \tfrac{180}{332} = 0.542$$

$$T_s = T_0 + (v_0^2/2c_p) = 2 + 180^2/[(2)(1003)] = 18\,°C$$

$$p_s = p_0 + (\rho_0)(v_0^2/2)[1 + (N_M)_0^2/4 + \cdots] = 79.50 + (1.007)(180^2/2)(1 + 0.542^2/4)/(1000) = 97.0 \text{ kPa abs}$$

**16.17**  Air at 250 psia is moving at 550 fps in a high-pressure wind tunnel at a temperature of 100 °F. Find the stagnation pressure and temperature. Note the magnitude of the sonic velocity for the 250-psia 100 °F air.

❚ $\qquad c = \sqrt{kRT} = \sqrt{(1.40)(1716)(460 + 100)} = 1160 \text{ ft/s}$ $\qquad N_M = v/c = \frac{550}{1160} = 0.474$

Since $[v_0 = 550 \text{ ft/s}] < [c = 1160 \text{ ft/s}]$,

$$p_s/p_0 = [1 + (v_0/c_0)^2(k-1)/2]^{k/(k-1)} \qquad p_s/250 = [1 + (\tfrac{550}{1160})^2(1.40-1)/2]^{1.40/(1.40-1)}$$

$$p_s = 292 \text{ psia} \qquad c_p T_0 + v_0^2/2g = c_p T_s \qquad (6000/32.2)(460+100) + 550^2/[(2)(32.2)] = (6000/32.2)(T_s)$$

$$T_s = 585 \,°\text{R} \quad \text{or} \quad 125 \,°\text{F}$$

**16.18** Show that the equation $p_s = p_0 + (\rho_0)(v_0^2/2)[1 + (N_M)_0^2/4 + \cdots]$ results from the binomial expansion of the equation $p_s/p_0 = [1 + (v_0/c)^2(k-1)/2]^{k/(k-1)}$.

❚ $p_s/p_0 = \{1 + (v_0/c)^2[(k-1)/2]\}^{k/(k-1)}$; substitute $k = \rho_0 c_0^2/p_0$:

$$\frac{p_s}{p_0} = \left[1 + N_M^2\left(\frac{\rho_0 c_0^2 - p_0}{2p_0}\right)\right]^{\rho_0 c_0^2/(\rho_0 c_0^2 - p_0)}$$

Expanding by the binomial theorem: $(a + b)^n = a^n + (na^{n-1})(b) + [n(n-1)/2!]a^{n-1}b^2 + \cdots$ and then simplifying leads to $p_s = p_0 + (\rho_0 v_0^2/2)[1 + (N_M^2/4) + \cdots]$   Q.E.D.

**16.19** Compute the value of $R$ from the values of $k$ and $c_p$ for air.

❚ $\qquad R = [(k-1)/k](c_p) = [(1.40-1)/1.40](6000) = 1714 \text{ ft-lb/(slug-}°\text{R)}$

**16.20** Compute the enthalpy change in 5 kg of oxygen when the initial conditions are $p_1 = 130 \text{ kPa abs}$ and $T_1 = 10 \,°\text{C}$, and the final conditions are $p_2 = 500 \text{ kPa abs}$ and $T_2 = 95 \,°\text{C}$ $[c_p = 0.917 \text{ kJ/kg} \cdot \text{K}]$.

❚ $\qquad h_2 = h_1 = (c_p)(T_2 - T_1) \qquad H_2 - H_1 = (5)(c_p)(T_2 - T_1) = (5)(0.917)(95 - 10) = 390 \text{ kJ}$

**16.21** A cylinder containing 2 kg nitrogen at 0.14 MPa abs and 5 °C is compressed isentropically to 0.30 MPa abs. Find the final temperature and the work required $[c_v = 0.741 \text{ kJ/kg} \cdot \text{K}]$.

❚ $T_2 = (T_1)(p_2/p_1)^{(k-1)/k} = (273 + 5)(0.30/0.14)^{(1.40-1)/1.40} = 346 \text{ K}$, or 73 °C. From the principle of conservation of energy, the work done on the gas must equal its increase in internal energy, since there is no heat transfer in an isentropic process; that is, $u_2 - u_1 = (c_v)(T_2 - T_1) = $ work per kilogram: Work $= (2)(0.741)(73 - 5) = 101 \text{ kJ}$.

**16.22** If 3.0 slugs of air are involved in a reversible polytropic process in which the initial conditions $p_1 = 12 \text{ psia}$ and $T_1 = 60 \,°\text{F}$ change to $p_2 = 20 \text{ psia}$ and volume $V = 1011 \text{ ft}^3$, determine the **(a)** formula for the process and **(b)** work done on the air.

❚ **(a)** $\rho_1 = p_1/RT_1 = (12)(144)/(53.3)(32.17)(460 + 60) = 0.00194 \text{ slug/ft}^3$. $R$ was converted to foot-pounds per slug and degree Rankine by multiplying by 32.17. Also, $\rho_2 = \frac{3}{1011} = 0.002967 \text{ slug/ft}^3$; $p_1/\rho_1^n = p_2/\rho_2^n$, $n = [\ln(p_2/p_1)]/[\ln(\rho_2/\rho_1)] = [\ln(\tfrac{20}{12})]/[\ln(0.002967/0.00194)] = 1.20$; hence $p/\rho^{1.2} = \text{const}$ describes the polytropic process.

  **(b)** Work of expansion is

$$W = \int_{V_1}^{V_2} p \, dV$$

This is the work done by the gas on its surroundings. Since $p_1 V_1^n = p_2 V_2^n = pV^n$, by substituting into the integral,

$$W = p_1 V_1^n \int_{V_1}^{V_2} \frac{dV}{V^n} = \frac{p_2 V_2 - p_1 V_1}{1 - n} = \frac{mR}{1 - n}(T_2 - T_1)$$

if $m$ is the mass of gas. $V_2 = 1011 \text{ ft}^3$ and $V_1 = V_2(p_2/p_1)^{1/n} = (1011)(\tfrac{20}{12})^{1/1.2} = 1547 \text{ ft}^3$. Then $W = [(20)(144)(1011) - (12)(144)(1548)]/(1 - 1.2) = -1\,184\,000 \text{ ft-lb}$. Hence, the work done on the gas is $1\,184\,000 \text{ ft-lb}$.

**16.23** For Prob. 16.22, find the **(a)** amount of heat transfer and **(b)** entropy change.

❚ **(a)** From the first law of thermodynamics the heat added minus the work done by the gas must equal the increase in internal energy; that is, $Q_H - W = U_2 - U_1 = c_v m(T_2 - T_1)$. First $T_2 = p_2/\rho_2 R = (20)(144)/(0.002965)(53.3)(32.17) = 566 \,°\text{R}$. Then $Q_H = -(1\,184\,000/778) + (0.17)(32.17)[3(566 - 520)] = -761 \text{ Btu}$ and 761 Btu was transferred from the mass of air.

**(b)**　$s_2 - s_1 = c_v \ln [(p_2/p_1)(\rho_1/\rho_2)^k]$　　$s_2 - s_1 = (0.171) \ln [(\frac{20}{12})(0.00194/0.002967)^{1.4}] = -0.01436$ Btu/(lbm-°R)

and　　　　　　　　　　　$S_2 - S_1 = -0.01436(3)(32.17) = -1.386$ Btu/°R

A rough check on the heat transfer can be made using an average temperature $T = (520 + 566)/2 = 543$ and by remembering that the losses are zero in a reversible process: $Q_H = T(S_2 - S_1) = (543)(-1.386) = -753$ Btu.

**16.24**　Carbon tetrachloride has a bulk modulus of elasticity of 1.124 GPa and a density of 1593 kg/m³. What is the speed of sound in the medium?

　$\blacksquare$　　　　　　　　$c = \sqrt{K/\rho} = \sqrt{(1.124 \times 10^9)/1593} = 840$ m/s

**16.25**　What is the speed of sound in dry air at sea level when $T = 68$ °F and in the stratosphere when $T = -67$ °F?

　$\blacksquare$　$c = \sqrt{kRT}$. At sea level, $c = \sqrt{(1.40)(1716)(460 + 68)} = 1126$ ft/s. In the stratosphere, $c = \sqrt{(1.40)(1716)(460 - 67)} = 972$ ft/s.

**16.26**　If 4 kg of a perfect gas, molecular weight 36, had its temperature increased 2 K when 6.4 kJ of work was done on it in an insulated constant-volume chamber, determine $c_v$ and $c_p$.

　$\blacksquare$　　　　　$c_v = \Delta u/\Delta t = 6.4/(4)(2) = 0.8$ kJ/kg · K　　　$c_p = c_v + R$

　　　　　　　　　　　$R = 8.312/M = 8.312/36 = 0.231$ kJ/kg · K

　　　　　　　　　　　　　$c_p = 0.8 + 0.231 = 1.03$ kJ/kg · K

**16.27**　A gas of molecular weight 46 has $c_p = 1.558$ kJ/kg · K. What is $c_v$ for this gas?

　$\blacksquare$　　　　$R = 8.312/46 = 0.181$ kJ/kg · K　　　$c_v = c_p - R = 1.558 - 0.181 = 1.377$ kJ/kg · K

**16.28**　Calculate the specific heat ratio $k$ for Probs. 16.26 and 16.27.

　$\blacksquare$　$k = c_p/c_v$. For Prob. 16.26: $k = 1.03/0.8 = 1.288$. For Prob. 16.27: $k = 1.558/1.377 = 1.131$.

**16.29**　The enthalpy of a gas is increased by 0.5 Btu/(lbm-°R) when heat is added at constant pressure and the internal energy is increased by 0.4 Btu/(lbm-°R) when the volume is maintained constant and heat is added. Calculate the molecular weight.

　$\blacksquare$　　　　$R = 1545/M = c_p - c_v$　　　$c_p = 0.5$ Btu/(lbm-°R)　　　$c_v = 0.4$ Btu/(lbm-°R)

　　　　　　　$R = (0.5 - 0.4)(778) = 77.8$ ft-lb/(lbm-°R)　　　$77.8 = 1545/M$　　　$M = 19.86$

**16.30**　Calculate the enthalpy change (in kcal) of 3 kg carbon monoxide from $p_1 = 16$ kPa abs and $T_2 = 5$ °C to $p_2 = 30$ kPa abs and $T_2 = 170$ °C.

　$\blacksquare$　　　　　　　$H = mc_p \Delta T = (3)(0.249)(170 - 5) = 123.3$ kcal

**16.31**　Calculate the entropy change in Prob. 16.30.

　$\blacksquare$　　　　$\Delta S = mc_v \ln [(T_2/T_1)^k (p_2/p_1)^{1-k}]$　　　$T_2 = 170 + 273 = 443$ K　　　$T_1 = 5 + 273 = 278$ K

　　　　　　$\Delta S = (3)(0.178)\{\ln [(\frac{443}{278})^{1.40}(\frac{30}{16})^{1-1.40}]\} = 0.215$ kcal/K

**16.32**　In an isentropic process, 1 kg of oxygen at 17 °C has its absolute pressure doubled. What is the final temperature?

　$\blacksquare$　　　　　$T_2/T_1 = (p_2/p_1)^{(k-1)/k}$　　　$T_2/(17 + 273) = (2)^{(1.40-1)/1.40}$　　　$T_2 = 354$ K　or　84 °C

**16.33**　Work out the expression for density change with temperature for a reversible polytropic process.

　$\blacksquare$　　　$p/\rho^n =$ constant　　　$p/\rho T =$ constant　　　$\rho^{n-1}/T =$ constant　　　$T_1/T_2 = (\rho_1/\rho_2)^{n-1}$

**16.34**　Hydrogen at 50 psia and 40 °F has its temperature increased to 120 °F by a reversible polytropic process with $n = 1.20$. Calculate the final pressure.

　$\blacksquare$　　　　$T_2/T_1 = (p_2/p_1)^{(n-1)/n}$　　　$(120 + 460)/(40 + 460) = (p_2/50)^{(1.20-1)/1.20}$　　　$p_2 = 122$ psia

**16.35**　A gas has a density decrease of 15 percent in a reversible polytropic process when the temperature decreases from 50 °C to 10 °C. Compute the exponent $n$ for the process.

**❚** From Prob. 16.33, $T_1/T_2 = (\rho_1/\rho_2)^{n-1}$, $(50 + 273)/(10 + 273) = (1/0.85)^{n-1}$, $\log 1.141 = (n - 1) \log 1.176$, $n = 1.81$.

**16.36** A projectile moves through water at 80 °F at 1800 fps. What is the Mach number?

**❚** $c = \sqrt{K/\rho} = \sqrt{(322\,000)(144)/1.93} = 4902$ ft/s $\qquad N_M = v/c = \frac{1800}{4902} = 0.367$.

**16.37** If an airplane travels at 1350 km/h at sea level, $p = 101$ kPa abs, $T = 25$ °C, and at the same speed in the stratosphere where $T = -55$ °C, how much greater is the Mach number in the latter case?

**❚** $(N_M)_2/(N_M)_1 = c_1/c_2 = \sqrt{T_1/T_2} = \sqrt{(273 + 25)/(273 - 55)} = 1.169$. Thus, $(N_M)_2$ is 16.9 percent greater than $(N_M)_1$.

**16.38** What is the speed of sound through hydrogen at 75 °F?

**❚** $c = \sqrt{kRT} = \sqrt{(1.40)(24\,649)(75 + 460)} = 4297$ ft/s

**16.39** Argon ($M = 39.944$, $k = 1.67$) flows through a tube such that its initial condition is $p_1 = 250$ psia and $\rho_1 = 1.16$ lbm/ft$^3$ and its final condition is $p_2 = 30$ psia and $T_2 = 265$ °F. Estimate the (a) initial temperature, (b) final density, (c) change in enthalpy, and (d) change in entropy.

**❚** (a) $\qquad R = 49\,720/39.944 = 1245$ ft$^2$/(s$^2$-°R) $\qquad \rho_1 = 1.16/32.2 = 0.03602$ slug/ft$^3$

$\qquad T_1 = p_1/R_1\rho_1 = (250)(144)/[(1245)(0.03602)] = 803$ °R $\quad$ or $\quad$ 343 °F

$\quad$ (b) $\qquad \rho_2 = p_2/RT_2 = (30)(144)/[(1245)(460 + 265)] = 0.00479$ slug/ft$^3$

$\qquad$ (c) $\qquad c_p = kR/(k - 1) = (1.67)(1245)/(1.67 - 1) = 3103$ ft$^2$/(s$^2$-°R)

$\qquad h_2 - h_1 = (c_p)(T_2 - T_1) = (3103)[(460 + 265) - 803] = -242\,000$ ft$^2$/s$^2$

$\qquad$ (d) $\qquad s_2 - s_1 = c_p \ln(T_2/T_1) - R \ln(p_2/p_1) = (3103)\{\ln[(460 + 265)/803]\}$

$\qquad\qquad\qquad\qquad\qquad - (1245)\{\ln[(30)(144)/(250)(144)]\} = 2323$ ft$^2$/(s$^2$-°R)

**16.40** Estimate the speed of sound of carbon monoxide [$k = 1.40$, $R = 0.297$ kJ/kg · K] at 200 °C.

**❚** $c = \sqrt{kRT} = \sqrt{(1.40)(297)(200 + 273)} = 443$ m/s

**16.41** A gas flows adiabatically through a duct. At section 1, $p_1 = 195$ psia, $T_1 = 480$ °F, and $V_1 = 240$ fps, while farther downstream $V_2 = 1000$ fps and $p_2 = 35$ psia. Calculate $T_2$ and $s_2 - s_1$ if the gas is air [$c_p = 6010$ ft$^2$/(s$^2$-°R)].

**❚** $\qquad c_pT_1 + v_1^2/2 = c_pT_2 + v_2^2/2 \qquad (6010)(480) + 240^2/2 = 6010T_2 + 1000^2/2 \qquad T_2 = 402$ °F

$s_2 - s_1 = c_p \ln(T_2/T_1) - R \ln(p_2/p_1) = (6010)\{\ln[(402 + 460)/(480 + 460)]\} - (1716)[\ln(\frac{35}{195})] = 2427$ ft$^2$/(s$^2$-°R)

**16.42** Rework Prob. 16.41 if the gas is argon [$c_p = 3103$ ft$^2$/(s$^2$-°R), $k = 1.67$, $R = 1245$ ft$^2$/(s$^2$-°R)].

**❚** $\qquad c_pT_1 + v_1^2/2 = c_pT_2 + v_2^2/2 \qquad (3103)(480) + 240^2/2 = 3103T_2 + 1000^2/2 \qquad T_2 = 328$ °F

$s_2 - s_1 = c_p \ln(T_2/T_1) - R \ln(p_2/p_1) = (3103)\{\ln[(328 + 460)/(480 + 460)]\} - (1245)[\ln(\frac{35}{195})] = 1591$ ft$^2$/(s$^2$-°R)

**16.43** Solve Prob. 16.41 if the gas is steam. Assume an ideal gas, with $M = 18.02$ and $k = 1.33$.

**❚** $\qquad c_pT_1 + v_1^2/2 = c_pT_2 + v_2^2/2 \qquad R = 49\,720/M = 49\,720/18.02 = 2759$ ft$^2$/(s$^2$-°R)

$\qquad\qquad c_p = kR/(k - 1) = (1.33)(2759)/(1.33 - 1) = 11\,120$ ft$^2$/(s$^2$-°R)

$\qquad\qquad (11\,120)(480) + 240^2/2 = 11\,120T_2 + 1000^2/2 \qquad T_2 = 438$ °F

$s_2 - s_1 = c_p \ln(T_2/T_1) - R \ln(p_2/p_1) = (11\,120)\{\ln[(438 + 460)/(480 + 460)]\} - (2759)[\ln(\frac{35}{195})] = 4231$ ft$^2$/(s$^2$-°R)

**16.44** Solve Prob. 16.41 if the gas is steam. Assume a real gas and use the steam tables.

**❚** $h_1 + v_1^2/2 = h_2 + v_2^2/2$. With $p_1 = 195$ psi and $T_1 = 480$ °F, $h_1 = 1258.2$ Btu/lbm and $s_1 = 1.6153$ Btu/(lbm-°F) (from the steam tables). $1258.2 + 240^2/[(2)(32.2)(778)] = h_2 + 1000^2/[(2)(32.2)(778)]$, $h_2 = 1239.4$ Btu/lbm. With $p_2 = 35$ psi and $h_2 = 1239.4$ Btu/lbm, $T_2 = 406$ °F and $s_2 = 1.7641$ Btu/(lbm-°F) (from the steam tables). $s_2 - s_1 = 1.7641 - 1.6153 = 0.149$ Btu/(lbm-°F).

**16.45** If 5 kg of oxygen in a closed tank at 240 kPa is heated from 100 °C to 300 °C, calculate the new pressure, the heat added, and the change in entropy.

**I** $\qquad p_2/p_1 = T_2/T_1 \qquad p_2/240 = (300+273)/(100+273) \qquad p_2 = 369 \text{ kPa} \qquad Q = mc_v\,\Delta T$

$c_p = kR/(k-1) = (1.40)(0.260)/(1.40-1) = 0.910 \text{ kJ/kg} \cdot \text{K} \qquad c_v = c_p/k = 0.910/1.40 = 0.650 \text{ kJ/kg} \cdot \text{K}$

$$Q = (5)(0.650)(300-100) = 650 \text{ kJ}$$

$$s_2 - s_1 = mc_v \ln(T_2/T_1) = (5)(0.650)\{\ln[(300+273)/(100+273)]\} = 1.395 \text{ kJ/K}$$

**16.46** Steam at 450 °F and 80 psia is compressed isentropically to 100 psia. What is the new temperature? Assume an ideal gas with $k = 1.33$.

**I** $\qquad T_2/T_1 = (p_2/p_1)^{(k-1)/k} \qquad T_2/(450+460) = (\tfrac{100}{80})^{(1.33-1)/1.33} \qquad T_2 = 962\,°\text{R} \quad \text{or} \quad 502\,°\text{F}$

**16.47** Solve Prob. 16.46 assuming a real gas and using the steam tables.

**I** With $p_1 = 80$ psia and $T_1 = 450\,°\text{F}$, $s_1 = 1.7078\,\text{Btu/(lbm-°F)}$ (from the steam tables). With $p_2 = 100$ psia and $s_2 = s_1 = 1.7078\,\text{Btu/(lbm-°F)}$, $T_2 = 500\,°\text{F}$ (from the steam tables).

**16.48** Carbon dioxide ($k = 1.30$, $M = 44.01$) enters a constant-area duct at 340 °F, 100 psia, and 500 fps. Farther downstream the properties are $V_2 = 1000$ fps and $T_2 = 940\,°\text{F}$. Compute (a) $p_2$, (b) the heat added between sections, (c) the entropy change between sections, and (d) the mass flow per unit area.

**I** (a) $\qquad p_2/p_1 = (T_2/T_1)(V_1/V_2) \qquad p_2/100 = [(940+460)/(340+460)](\tfrac{500}{1000}) \qquad p_2 = 87.5$ psia

(b) $\quad q = (c_p)(T_2 - T_1) + (v_2^2 - v_1^2)/2 \qquad c_p = kR/(k-1) \qquad R = 49\,720/M = 49\,720/44.01 = 1130 \text{ ft}^2/(\text{s}^2\text{-}°\text{R})$

$$c_p = (1.30)(1130)/(1.30-1) = 4897 \text{ ft}^2/(\text{s}^2\text{-}°\text{R})$$

$q = (4897)(940-340) + (1000^2 - 500^2)/2 = 3.313 \times 10^6 \text{ ft-lb/siug} = 3.313 \times 10^6/[(32.2)(778)] = 132$ Btu/lbm

(c) $\quad s_2 - s_1 = c_p \ln(T_2/T_1) - R \ln(p_2/p_1) = (4897)\{\ln[(940+460)/(340+460)]\} - (1130)[\ln(87.5/100)]$

$$= 2891 \text{ ft-lb/(slug-°R)}$$

(d) $\qquad \dot{m} = \rho_1 A_1 v_1 \qquad \rho_1 = p_1/RT_1 = (100)(144)/[(1130)(340+460)] = 0.01593 \text{ slug/ft}^3$

$$\dot{m}/A_1 = (0.01593)(500) = 7.97 \text{ (slugs/s)/ft}^2$$

**16.49** Steam enters a duct at $p_1 = 60$ psia, $T_1 = 350\,°\text{F}$, $V_1 = 200$ fps, and leaves at $p_2 = 120$ psia, $T_2 = 850\,°\text{F}$, and $V_2 = 1200$ fps. How much heat was added?

**I** $q = (h_2 - h_1) + (v_2^2 - v_1^2)/2$. From the steam tables, $h_1 = 1208.2$ Btu/lbm and $h_2 = 1453.4$ Btu/lbm. $q = (1453.4 - 1208.2) + [(1200^2 - 200^2)/2]/[(32.2)(778)] = 273$ Btu/lbm.

**16.50** The internal energy of a hypothetical perfect gas is given as $u = T^{1/2}/50 + 100$. Determine $c_v$ and $c_p$. Take $R = 50$ ft-lb/(lbm-°F).

**I** $\qquad c_v = du/dt = (\tfrac{1}{2})(T^{-1/2})/50 = 0.0100T^{-1/2} \text{ Btu/(lbm-°R)}$

$$c_p = R + c_v = \tfrac{50}{778} + 0.0100T^{-1/2} = 0.0643 + 0.0100T^{-1/2} \text{ Btu/(lbm-°R)}$$

**16.51** Air at 15 °C and 101 325 Pa is compressed to a pressure of 345 000 Pa. If the compression is adiabatic and reversible, what is the final specific volume? How much work is done per kilogram of the gas?

**I** $\qquad p_1/p_2 = [(V_s)_2/(V_s)_1]^k \qquad V_s = RT/p \qquad (V_s)_1 = (287)(15+273)/101\,325 = 0.8158 \text{ m}^3/\text{kg}$

$$101\,325/345\,000 = [(V_s)_2/(0.8158)]^{1.40} \qquad (V_s)_2 = 0.340 \text{ m}^3/\text{kg} \qquad W = \int p\,dv$$

Since $p(V_s)^k = \text{constant}$,

$$W = \int_{0.8158}^{0.340} \frac{\text{constant}}{(V_s)^k}\,dv = (\text{constant})\left[\frac{1/(V_s)^{k-1}}{1-k}\right]_{0.8158}^{0.340}$$

$$= (\text{constant})\left[\frac{(1/0.340)^{1.40-1} - (1/0.8158)^{1.40-1}}{1-1.40}\right] = (-1.137)(\text{constant})$$

$$\text{Constant} = (101\,325)(0.8158)^{1.40} = 76\,196 \qquad W = (-1.137)(76\,196) = -86\,600 \text{ J/kg}$$

**16.52** Do Prob. 16.51 for an isothermal compression.

❚ $(p_1)(V_s)_1 = (p_2)(V_s)_2$. Using data from Prob. 16.51, $(101\,325)(0.8158) = (345\,000)(V_s)_2$, $(V_s)_2 = 0.240\ \text{m}^3/\text{kg}$.

$$W = \int p\,dv = \text{constant} \int_{0.8158}^{0.240} \frac{dV_s}{V_s} = \text{constant}\,[\ln V_s]_{0.8158}^{0.240} = (\text{constant})\left(\ln \frac{0.240}{0.8158}\right) = (\text{constant})(-1.224)$$

$$\text{Constant} = (101\,325)(0.8158) = 82\,661 \qquad W = (82\,661)(-1.224) = -101\,000\ \text{J/kg}$$

**16.53** An airplane is capable of attaining a flight Mach number of 0.9. When it is flying at an altitude of 1000 ft in standard atmosphere, what is the ground speed if the air is not moving relative to the ground? What is the ground speed if the plane is at an altitude of 35 000 ft in standard atmosphere?

❚ At altitude 1000 ft, $c = 1113$ ft/s, $v_{\text{plane}} = (0.9)(1113) = 1002$ ft/s. At altitude 35 000 ft, $c = 973$ ft/s, $v_{\text{plane}} = (0.9)(973) = 876$ ft/s.

**16.54** Do the first part of Prob. 16.53 if the air is moving at 30 mph directly opposite to the direction of flight.

❚ $$v_{\text{plane}} = 1002 - (30)(5280)/3600 = 958\ \text{ft/s}$$

**16.55** What is the value of $k$ for standard atmosphere at an altitude of 30 000 ft?

❚ $$c = \sqrt{kp/\rho} \qquad 995 = \sqrt{(k)(628)/(0.374)(0.00238)} \qquad k = 1.40$$

**16.56** Suppose that a plane is moving horizontally relative to the ground at a speed of twice the velocity of sound (340.5 m/s) and that the air is moving in the opposite direction at a speed of one-fourth the velocity of sound relative to the ground. What is the Mach angle?

❚ Relative to the ground: $V_{\text{source}} = (2)(340.5) = 681.0$ m/s, $V_{\text{air}} = 340.5/4 = 85.1$ m/s. Velocity of source relative to the air is $681.0 + 85.1 = 766.1$ m/s. Hence, $N_M = 766.1/340.5 = 2.25$, $\sin \alpha = 1/2.25$, $\alpha = 26.4°$.

**16.57** Suppose that a cruise missile under test is moving horizontally at $N_m = 2$ in the atmosphere at an elevation of 300 m above the earth's surface. How long does it take for an observer to hear the disturbance from the instant when it is directly overhead? Assume standard atmosphere.

❚ See Fig. 16-2. $\sin \alpha = 1/2$, $\alpha = 30.0°$; $\tan 30° = 300/d$, $d = 519.6$ m. Speed of shell is $(2)(340) = 680$ m/s. $t = d/V = 519.6/680 = 0.764$ s.

**Fig. 16-2**

**16.58** Suppose in Prob. 16.57 that an observer in a plane is moving in the same direction as the missile at a speed of one-half the speed of sound at an elevation of 300 m above the missile. What is the time elapsed between the instant when the missile is directly below and the instant when the observer hears the sound? Neglect change of $c$ from 300 m to 600 m elevation.

❚ See Fig. 16-3. Using data from Prob. 16.57, $680t = 519.6 + \left(\frac{680}{4}\right)(t)$, $t = 1.02$ s.

**Fig. 16-3**

**16.59** A certain aircraft flies at the same Mach number regardless of its altitude. It flies 80 km/h slower at 10 km than at sea level. What is its Mach number?

❚ $c = \sqrt{kRT}$. At sea level, $c_1 = \sqrt{kRT_1} = 340.3$ m/s. At 10 000 m, $c_2 = \sqrt{kRT_2} = 299.4$ m/s, $\Delta v = (N_M)(c_2 - c_1)$, $(80)(1000)/3600 = (N_M)(340.3 - 299.4)$, $N_M = 0.543$.

**16.60** At 150 °C and 1 atm, compute the speed of sound of (a) air, (b) oxygen, (c) hydrogen, (d) steam, (e) carbon monoxide.

$$c = \sqrt{kRT}$$

(a) $c = \sqrt{(1.40)(287)(150 + 273)} = 412$ m/s  (b) $c = \sqrt{(1.40)(260)(150 + 273)} = 392$ m/s

(c) $c = \sqrt{(1.41)(4127)(150 + 273)} = 1569$ m/s  (d) $c = \sqrt{(1.33)(461)(150 + 273)} = 509$ m/s

(e) $c = \sqrt{(1.40)(297)(150 + 273)} = 419$ m/s

**16.61** Assuming that water follows the liquid equation of state $K = (n)(B + 1)(p_a)(\rho/\rho_a)^n$ with $n = 7$ and $B = 3000$, compute the bulk modulus and speed of sound at (a) 1 atm and (b) 1000 atm.

(a) At 1 atm (101 350 Pa), $\rho = \rho_a$, $K = (7)(3000 + 1)(101\ 350)(1)^7 = 2.13$ GPa, $c = \sqrt{K/\rho} = \sqrt{(2.13 \times 10^9)/998} = 1461$ m/s.

(b) At 1100 atm, $\rho/\rho_a = [(1000 + 3000)/3001]^{1/7} = 1.042$, $K = (K_{atm})(1.042)^7 = (2.13 \times 10^9)(1.042)^7 = 2.84 \times 10^9$ Pa, $c = \sqrt{(2.84 \times 10^9)/[(998)(1.042)]} = 1653$ m/s.

**16.62** The measured value of the speed of sound in water at 20 °C and 9000 atm is 2650 m/s.* Compare this with the value computed by the analysis of Prob. 16.61.

$c = \sqrt{K/\rho}$, $K = (n)(B + 1)(p_a)(\rho/\rho_a)^n$, $n = 7$, $B = 3000$, $\rho_a = 998$ kg/m³. From Prob. 16.61, $9000 = (3001)(\rho/998)^7 - 3000$, $\rho = 1216.5$ kg/m³; $K = (7)(3000 + 1)(101\ 350)(1216.5/998)^7 = 8.513$ GPa, $c = \sqrt{(8.513 \times 10^9)/1216.5} = 2645$ m/s. This is $(2650 - 2645)/2650 = 0.0019$, or 0.19 percent less than the measured value.

**16.63** Mercury at 1 atm has a bulk modulus of about 28 GPa. It has also $n = 7$ in the equation of state of Prob. 16.61. What value of $B$ in that equation best fits the measured bulk modulus? Estimate the bulk modulus and speed of sound of mercury at 2000 atm.

From Prob. 16.61, $K_a = (n)(B + 1)(p_a) = 2.8 \times 10^{10} = (7)(B + 1)(101\ 350)$, $B = 39\ 466$. At 2000 atm,

$$p/p_a = 2000 = (39\ 466 + 1)(\rho/\rho_a)^7 - 39\ 466 \qquad \rho/\rho_a = 1.007$$

$$K = (K_a)(\rho/\rho_a)^7 = (2.8 \times 10^{10})(1.007)^7 = 29.4\ \text{GPa}$$

$$\rho = 1.007\rho_a = (1.007)(13\ 550) = 13\ 645\ \text{kg/m}^3 \qquad c = \sqrt{K/\rho} = \sqrt{(2.94 \times 10^{10})/13\ 645} = 1468\ \text{m/s}$$

**16.64** Why do (a) water and mercury and (b) aluminum and steel have nearly equal speeds of sound in spite of the fact that the second material of each pair is much heavier than the first? Can this behavior be predicted from molecular theory?

In both cases, the lighter material has a proportionately smaller bulk modulus.

**16.65** An airplane flies at 360 m/s through air at $-10$ °C and 40 kPa. Is the airplane supersonic?

$c = \sqrt{kRT} = \sqrt{(1.40)(287)(273 - 10)} = 325$ m/s, $N_M = v/c = \frac{360}{325} = 1.1$. Since $N_M > 1.0$, it is supersonic.

**16.66** A weak pressure wave (sound wave) with a pressure change $\Delta p = 50$ Pa propagates through air at 20 °C and 1 atm. Estimate the (a) velocity change across the wave, (b) density change, and (c) temperature change.

(a)  $\Delta p = \rho c\ \Delta v$  $50 = (1.205)(343)(\Delta v)$  $\Delta v = 0.1210$ m/s

(b)  $\Delta \rho = (\rho + \Delta \rho)(\Delta v/c) = (1.205 + \Delta \rho)(0.1210/343) = 0.0004253$ kg/m³

(c) Approximately isentropic:

$$(T + \Delta T)/T = [(p + \Delta p)/p]^{(k-1)/k}$$

$$(273 + 20 + \Delta T)/(273 + 20) = [(101\ 350 + 50)/101\ 350]^{(1.40-1)/1.40} \qquad \Delta T = 0.041\ °C$$

**16.67** Air at 65 °F flows isothermally through a 6-in-diameter pipe. The pressure at one section is 82 psia, and that at a section 550 ft downstream is 65 psia. If the pipe surface is "smooth," find the weight flow rate of the air.

---

* A. H. Smith and A. W. Larson, *J. Chem. Phys.*, vol. 22, p. 351, 1954.

▋ $$p_1^2 - p_2^2 = (G^2 RT/g^2 A^2)[fL/D + 2\ln(p_1/p_2)] \qquad A = (\pi)(\tfrac{6}{12})^2/4 = 0.1963 \text{ ft}^2$$

Assume $f = 0.007$: $(82^2 - 65^2)(144)^2 = (G^2)(53.3)(65 + 460)/[(32.2)^2(0.1963)^2][(0.007)(550)/(\tfrac{6}{12}) + (2)(\ln\tfrac{82}{65})]$, $G = 95.2$ lb/s. This is the weight flow rate of the air if the assumed value of $f$ of 0.007 is correct. This must be checked.

$$N_R = \rho D/\mu \qquad \rho = p/RT = (82)(144)/[(1716)(65 + 460)] = 0.01311 \text{ lb-s}^2/\text{ft}^4 \qquad G = \gamma A v$$

$$\gamma = \rho g = (0.01311)(32.2) = 0.4221 \text{ lb/ft}^3 \qquad 95.2 = (0.4221)(0.1963)(v) \qquad v = 1149 \text{ ft/s}$$

$$N_R = (0.01311)(\tfrac{6}{12})(1149)/(3.78 \times 10^{-7}) = 1.99 \times 10^7$$

$$f = 0.007 \qquad \text{(from Fig. A-5, using the ``smooth pipes'' line)}$$

Since this value of $f$ is the same as the assumed value, the computed weight flow rate of 95.2 lb/s is taken as the correct value.

**16.68** Air at 18 °C flows isothermally through a 300-mm-diameter pipe at of 0.450 kN/s (weight flow). The pressure at one section is 550 kPa abs, and the pipe surface is smooth. Find the pressure at a section 200 m downstream.

▋ $$p_1^2 - p_2^2 = (G^2 RT/g^2 A^2)[fL/D + 2\ln(p_1/p_2)] \qquad A = (\pi)(0.300)^2/4 = 0.07069 \text{ m}^2$$

$$N_R = DG/gA\mu = (300 \times 10^{-3})(0.450 \times 10^3)/(9.807)(0.07069)(1.81 \times 10^{-8})$$

From Fig. A-5, $f = 0.0080$. Substituting into the equation given above and neglecting temporarily the second term inside the brackets, $550^2 - p_2^2 = \{(0.450)^2(29.3)(273 + 18)/[(9.807)^2(0.07069)^2]\}[(0.0080)(200)/0.300]$, $p_2 = 532$ kPa. Substituting this value of $p_2$ into the term that was neglected and solving for the $p_2$ on the left side of the equation, $550^2 - p_2^2 = \{(0.450)^2(29.3)(273 + 18)/[(9.807)^2(0.07069)^2]\}[(0.0080)(200)/0.300 + (2)(\ln\tfrac{550}{532})]$, $p_2 = 532$ kPa  (O.K.).

**16.69** Air at 100 °F flows isothermally through a 4-in-diameter pipe. Pressures at sections 1 and 2 are 120 psia and 80 psia, respectively. Section 2 is located 400 ft downstream from section 1. Determine the weight flow rate of the air. Assume the pipe has a smooth surface.

▋ $$p_1^2 - p_2^2 = (G^2 RT/g^2 A^2)[fL/D + 2\ln(p_1/p_2)] \qquad A = (\pi)(\tfrac{4}{12})^2/4 = 0.08727 \text{ ft}^2$$

Assume $f = 0.007$: $(120^2 - 80^2)(144)^2 = (G^2)(53.3)(100 + 460)/[(32.2)^2(0.08727)^2][(0.007)(400)/(\tfrac{4}{12}) + (2)(\ln\tfrac{120}{80})]$, $G = 69.0$ lb/s. This is the weight flow rate of the air if the assumed value of $f$ of 0.007 is correct. Further checking (not shown here but following the pattern of Prob. 16.67) indicates that the assumed value of $f$ of 0.007 is correct, and therefore $G = 69.0$ lb/s.

**16.70** Air at 85 °F flows isothermally through a 6-in-diameter pipe at a flow rate of 10 lb/s. The pipe surface is very smooth. If the pressure at one section is 70 psia, determine the pressure at a section 600 ft downstream from the first section.

▋ $$p_1^2 - p_2^2 = (G^2 RT/g^2 A^2)[fL/D + 2\ln(p_1/p_2)] \qquad A = (\pi)(\tfrac{6}{12})^2/4 = 0.1963 \text{ ft}^2 \qquad N_R = \rho Dv/\mu$$

$$\gamma = p/RT = (70)(144)/[(53.3)(460 + 85)] = 0.3470 \text{ lb/ft}^3 \qquad \rho = \gamma/g = 0.3470/32.2 = 0.01078 \text{ slug/ft}^3$$

$$G = \gamma A v \qquad 10 = (0.3470)(0.1963)(v) \qquad v = 146.8 \text{ ft/s} \qquad N_R = (0.01708)(\tfrac{6}{12})(146.8)/(3.78 \times 10^{-7}) = 2.09 \times 10^6$$

From Fig. A-5, $f = 0.0103$. Substituting into the equation given above and neglecting temporarily the second term inside the brackets, $(70^2)(144)^2 - (p_2)^2(144)^2 = (10)^2(53.3)(460 + 85)/[(32.2)^2(0.1963)^2][(0.0103)(600)/(\tfrac{6}{12})]$, $p_2 = 69.69$ lb/in$^2$. Substituting this value of $p_2$ into the term that was neglected and solving for the $p_2$ on the left side of the equation, $(70^2)(144)^2 - (p_2)^2(144)^2 = \{(10)^2(53.3)(460 + 85)/[(32.2)^2(0.1963)^2]\} \times \{(0.0103)(600)/(\tfrac{6}{12}) + (2)[\ln(70/69.69)]\}$, $p_2 = 69.69$ lb/in$^2$  (O.K.).

**16.71** Prepare a computer program to solve for either the weight flow rate or the pressure at a section downstream for isothermal flow of a compressible fluid.

```
▋    C    THIS PROGRAM DETERMINES EITHER THE FLOW RATE OR THE PRESSURE AT A
     C    SECOND (DOWNSTREAM) POINT FOR CLOSED CONDUIT, COMPRESSIBLE FLOW.
     C    IT IS BASED ON AN ISOTHERMAL FLOW ANALYSIS AND IS THEREFORE SUB-
     C    JECT TO THE RESTRICTIONS AND/OR ASSUMPTIONS IMPLICIT IN ISOTHERMAL
     C    FLOW.  IT CAN BE USED FOR PROBLEMS IN BOTH THE ENGLISH SYSTEM OF
```

```
C      UNITS AND THE INTERNATIONAL SYSTEM OF UNITS.  THE APPLICATION OF
C      THIS PROGRAM IS LIMITED TO CASES INVOLVING A SINGLE CONDUIT WITH
C      A CONSTANT DIAMETER.
C
C      INPUT DATA MUST BE SET UP AS FOLLOWS.
C
C      CARD 1    COLUMN 1          ENTER 0 (ZERO) OR BLANK IF ENGLISH SYSTEM
C                                  OF UNITS IS TO BE USED.  ENTER 1 (ONE) IF
C                                  INTERNATIONAL SYSTEM OF UNITS IS TO BE USED.
C                COLUMNS 2-79      ENTER TITLE, DATE, AND OTHER INFORMATION,
C                                  IF DESIRED.
C      CARD 2    COLUMNS 1-10      ENTER NUMBER INCLUDING DECIMAL GIVING
C                                  ABSOLUTE PRESSURE AT UPSTREAM POINT (IN
C                                  POUNDS PER SQUARE INCH OR KILO-
C                                  PASCALS).
C                COLUMNS 11-20     ENTER NUMBER INCLUDING DECIMAL GIVING
C                                  ABSOLUTE PRESSURE AT DOWNSTREAM POINT (IN
C                                  POUNDS PER SQUARE INCH OR KILO-
C                                  PASCALS).
C                COLUMNS 21-30     ENTER NUMBER INCLUDING DECIMAL GIVING GAS
C                                  CONSTANT (IN FEET PER DEGREE-RANKINE OR
C                                  METERS PER DEGREE-KELVIN).
C                COLUMNS 31-40     ENTER NUMBER INCLUDING DECIMAL GIVING
C                                  TEMPERATURE (IN DEGREES-FAHRENHEIT OR
C                                  DEGREES-CELSIUS).
C                COLUMNS 41-50     ENTER NUMBER INCLUDING DECIMAL GIVING
C                                  DIAMETER OF CONDUIT (IN INCHES OR MILLI-
C                                  METERS).
C                COLUMNS 51-60     ENTER NUMBER INCLUDING DECIMAL GIVING
C                                  LENGTH OF CONDUIT (IN FEET OR METERS).
C                COLUMNS 61-70     ENTER NUMBER INCLUDING DECIMAL GIVING
C                                  ROUGHNESS (IN FEET OR METERS).  ENTER A
C                                  VALUE OF 0 (ZERO) FOR "SMOOTH" CONDUITS.
C                COLUMNS 71-80     ENTER NUMBER INCLUDING DECIMAL GIVING
C                                  VISCOSITY OF FLUID (IN POUNDS-SECONDS
C                                  PER SQUARE FOOT OR KILONEWTONS-SECONDS PER
C                                  SQUARE METER).
C      CARD 3    COLUMNS 1-10      ENTER NUMBER INCLUDING DECIMAL GIVING
C                                  WEIGHT FLOW RATE (IN POUNDS PER
C                                  SECOND OR KILONEWTONS PER SECOND).
C                COLUMNS 11-34     ENTER TYPE OF FLUID.
C                COLUMNS 35-58     ENTER TYPE OF CONDUIT.
C
C ******************************************************************
C *                                                                *
C *    NOTE WELL....EITHER THE FLOW RATE (COLUMNS 1-10 ON CARD 3) OR *
C *    THE PRESSURE AT DOWNSTREAM POINT (COLUMNS 11-20 ON CARD 2),  *
C *    WHICHEVER ONE IS TO BE DETERMINED BY THIS PROGRAM, SHOULD BE *
C *    LEFT BLANK.                                                  *
C *                                                                *
C ******************************************************************
C
C      MULTIPLE DATA SETS FOR SOLVING ANY NUMBER OF PROBLEMS MAY BE
C      INCLUDED FOR PROCESSING.
C
       DIMENSION TITLE(13),FLUID(4),PIPE(4)
       COMMON F,ED,RN
       PI=3.14159265
       REAL L
1      READ(5,100,END=2)UNITS,TITLE
100    FORMAT(I1,13A6)
       WRITE(6,105)TITLE
105    FORMAT('1',13A6,//)
       READ(5,101)P1,P2,R,T,D,L,E,VIS,Q,FLUID,PIPE
101    FORMAT(8F10.0,/,F10.0,8A6)
       FACTOR=12.0
       IF(UNITS.EQ.1)FACTOR=1000.0
       PFACT=144.0
       IF(UNITS.EQ.1)PFACT=1.0
       TFACT=460.0
       IF(UNITS.EQ.1)TFACT=273.0
       G=32.2
       IF(UNITS.EQ.1)G=9.807
```

```
                A=PI*D**2/FACTOR**2/4.0
                SW=P1*PFACT/R/(T+TFACT)
                RHO=SW/G
                ED=E/D*FACTOR
                IF(Q.GT.0.0001) GO TO 102
                FF=0.02
            104 Q=SQRT((P1**2-P2**2)*PFACT**2*G*G*A*A/(R*(T+TFACT)*(FF*L/D*FACTOR+
               *2.0*ALOG(P1/P2))))
                V=Q/SW/A
                RN=RHO*D/FACTOR*V/VIS
                CALL ROUGH
                DIFF=ABS(F-FF)
                IF(DIFF.LT.0.0001)GO TO 103
                FF=F
                GO TO 104
            103 IF (UNITS.EQ.0)WRITE(6,106)P1,P2,D,E,R,T,L,FLUID,PIPE,Q
            106 FORMAT(1X,'GIVEN DATA FOR A COMPRESSIBLE FLOW IN A CIRCULAR CLOSED
               * CONDUIT',//5X,'PRESSURE AT POINT 1 =',F8.1,' PSI',//5X,'PRESSURE
               *AT POINT 2 =',F8.1,' PSI',//5X,'DIAMETER OF CONDUIT =',F7.2,
               *' IN',//5X,'ROUGHNESS =',F10.7,' FT',//5X,'GAS CONSTANT =',F7.1,
               *' FT/DEG-R',//5X,'TEMPERATURE =',F7.1,' DEG-F',//5X,'LENGTH OF CON
               *DUIT =',F8.1,' FT',//5X,'FLUID FLOWING IS ',4A6,//5X,'CONDUIT MATE
               *RIAL IS ',4A6,//1X,'THE FLOW RATE WILL BE',F7.1,' LB/S')
                IF (UNITS.EQ.1)WRITE(6,107)P1,P2,D,E,R,T,L,FLUID,PIPE,Q
            107 FORMAT(1X,'GIVEN DATA FOR A COMPRESSIBLE FLOW IN A CIRCULAR CLOSED
               * CONDUIT',//5X,'PRESSURE AT POINT 1 =',F8.1,' KPA',//5X,'PRESSURE
               *AT POINT 2 =',F8.1,' KPA',//5X,'DIAMETER OF CONDUIT =',F7.1,
               *' MM',//5X,'ROUGHNESS =',F10.7,' M ',//5X,'GAS CONSTANT =',F7.1,
               * ' M/DEG-K',//5X,'TEMPERATURE =',F7.1,' DEG-C',//5X,'LENGTH OF CON
               *DUIT =',F8.1,' M ',//5X,'FLUID FLOWING IS ',4A6,//5X,'CONDUIT MATE
               *RIAL IS ',4A6,//1X,'THE FLOW RATE WILL BE',F7.3,' KN/S')
                GO TO 1
            102 V=Q/SW/A
                RN=RHO*D/FACTOR*V/VIS
                CALL ROUGH
                P2=SQRT(P1**2-Q**2*R*(T+TFACT/G**2/A**2*F*L/D*FACTOR/PFACT**2)
            109 P22=SQRT(P1**2-Q**2**R*(T+TFACT)/G**2/A**2*(F*L/D*FACTOR+2.0*
               *ALOG(P1/P2))/PFACT**2)
                DIFF=ABS(P22-P2)
                IF(DIFF.LT.0.01)GO TO 108
                P2=P22
                GO TO 109
            108 IF(UNITS.EQ.0)WRITE(6,111)P1,D,E,R,T,L,Q,FLUID,PIPE,P2
            111 FORMAT(1X,'GIVEN DATA FOR A COMPRESSIBLE FLOW IN A CIRCULAR CLOSED
               * CONDUIT',//5X,'PRESSURE AT POINT 1 =',F8.1,' PSI',//5X,'DIAMETER
               *OF CONDUIT =',F7.2,' IN',//5X,'ROUGHNESS =',F10.7,' FT',//5X,'GAS
               *CONSTANT =',F7.1,' FT/DEG-R',//5X,'TEMPERATURE =',F7.1,' DEG-F',
               *//5X,'LENGTH OF CONDUIT =',F8.1,' FT',//5X,'FLOW RATE =',F7.1,
               *' LB/S ',//5X,'FLUID FLOWING IS ',4A6,//5X,'CONDUIT MATERIAL IS ',
               *4A6,//1X,'THE PRESSURE AT POINT 2 WILL BE',F8.1,' PSI')
                IF(UNITS.EQ.1)WRITE(6,110)P1,D,E,R,T,L,Q,FLUID,PIPE,P2
            110 FORMAT(1X,'GIVEN DATA FOR A COMPRESSIBLE FLOW IN A CIRCULAR CLOSED
               * CONDUIT',//5X,'PRESSURE AT POINT 1 =',F8.1,' KPA',//5X,'DIAMETER
               *OF CONDUIT =',F7.1,' MM',//5X,'ROUGHNESS =',F10.7,' M ',//5X,'GAS
               *CONSTANT =',F7.1,' M/DEG-K ',//5X,'TEMPERATURE =',F7.1,' DEG-C',
               *//5X,'LENGTH OF CONDUIT =',F8.1,' M ',//5X,'FLOW RATE =',F7.3,
               *' KN/S ',//5X,'FLUID FLOWING IS ',4A6,//5X,'CONDUIT MATERIAL IS ',
               *4A6,//1X,'THE PRESSURE AT POINT 2 WILL BE ',F8.1,'KPA')
                GO TO 1
          2     STOP
                END
                SUBROUTINE ROUGH
                COMMON F,ED,RN
                IF(RN.LE.2000.0)F=64.0/RN
                IF(RN.LE.2000.0)RETURN
                IF(RN.LT.4000.0)WRITE(6,103)
            103 FORMAT(1X,'A REYNOLDS NUMBER IS IN THE CRITICAL ZONE, FOR WHICH TH
               *E FRICTION FACTOR IS UNCERTAIN.  HENCE, PROGRAM EXECUTION WAS TERM
               *INATED.')
                IF(RN.LT.4000.0)STOP
                F=0.006
                TRY1=1.0/SQRT(F)+2.0*ALOG10(ED/3.7+2.51/RN/SQRT(F))
```

```
102 F=F+0.00001
    TRY2=1.0/SQRT(F)+2.0*ALOG10(ED/3.7+2.51/RN/SQRT(F))
    IF(TRY1*TRY2)100,100,101
101 TRY1=TRY2
    GO TO 102
100 F=F-0.000005
    RETURN
    END
```

**16.72** Solve Prob. 16.67 using the computer program of Prob. 16.71.

**❚ Input**

```
1 2 3 4 5 6 7 8 9 1011121314151617181920212223242526272829303132343536373839404142434445464748495051525354555657585960616263646566676869707172737475767778798090
0SAMPLE ANALYSIS ØF CØMPRESSIBLE FLØW
82.0        65.0        53.3        65.0        6.0        550.0                           .000000378
            AIR                              "SMØØTH"
```

**Output**

```
SAMPLE ANALYSIS OF COMPRESSIBLE FLOW

GIVEN DATA FOR A COMPRESSIBLE FLOW IN A CIRCULAR CLOSED CONDUIT

     PRESSURE AT POINT 1 =     82.0 PSI

     PRESSURE AT POINT 2 =     65.0 PSI

     DIAMETER OF CONDUIT =    6.00 IN

     ROUGHNESS = 0.0000000 FT

     GAS CONSTANT =    53.3 FT/DEG-R

     TEMPERATURE =    65.0 DEG-F

     LENGTH OF CONDUIT =   550.0 FT

     FLUID FLOWING IS AIR

     CONDUIT MATERIAL IS "SMOOTH"

THE FLOW RATE WILL BE   95.2 LB/S
```

**16.73** Solve Prob. 16.68 using the computer program of Prob. 16.71.

**❚ Input**

```
1 2 3 4 5 6 7 8 9 1011121314151617181920212223242526272829303132333435363738394041424344454647484950515253545556575859606162636465666768697071727374757677787980
1SAMPLE ANALYSIS ØF CØMPRESSIBLE FLØW
550.0                   29.3        18.0        300.0        200.0                        .000000018
0.450       AIR                              "SMØØTH"
```

**Output**

```
SAMPLE ANALYSIS OF COMPRESSIBLE FLOW

GIVEN DATA FOR A COMPRESSIBLE FLOW IN A CIRCULAR CLOSED CONDUIT

     PRESSURE AT POINT 1 =    550.0 KPA

     DIAMETER OF CONDUIT =   300.0 MM

     ROUGHNESS = 0.0000000 M

     GAS CONSTANT =    29.3 M/DEG-K

     TEMPERATURE =    18.0 DEG-C

     LENGTH OF CONDUIT =    200.0 M
```

```
FLOW RATE =  0.450 KN/S

FLUID FLOWING IS AIR

CONDUIT MATERIAL IS "SMOOTH"

THE PRESSURE AT POINT 2 WILL BE   532.1 KPA
```

**16.74**    Air flows isothermally at 65 °F through a horizontal 10- by 14-in rectangular duct at 100 lb/s. If the pressure at a section is 80 psia, find the pressure at a second section 500 ft downstream from the first. Assume the duct surface is very smooth; hence the lowest curve of Fig. A-5 may be used to determine $f$.

**∎**    $(N_M)_1^2/(N_M)_2^2 = 1 - k(N_M)_1^2\{2\ln[(N_M)_2/(N_M)_1] + f(L/D)\}$     $R_h = A/p_w = \frac{140}{48} = 2.92$ in $= 0.243$ ft
   $R_1 = DV\rho/\mu = GD/\mu g A = G(4R)/\mu g A = [100(4 \times 0.243)]/[(3.78 \times 10^{-7})(32.2)\frac{140}{144}] = 8.2 \times 10^6$

$R_2 = R_1$ since $\rho_1 V_1 = \rho_2 V_2$ and $\mu_1 = \mu_2$. From Fig. A-5, $f = 0.0083$.

   $\gamma_1 = pg/RT = [(80 \times 144)32.2]/[1715(460 + 65)] = 0.41$ lb/ft$^3$     $V_1 = G/\gamma_1 A = 100/[(0.41)\frac{140}{144}] = 250$ fps
   $c = \sqrt{kRT} = \sqrt{(1.4)(1715)(525)} = 1123$ fps     $(N_M)_1 = V_1/c = \frac{250}{1123} = 0.222$

The limiting value of $(N_M)_1$ is $1/\sqrt{1.4} = 0.845$. Therefore,

   $(0.222)^2/(0.845)^2 = 1 - 1.4(0.222)^2\{2\ln(0.845/0.222) + 0.0083[L/4(0.243)]\}$     $L = 1260$ ft

Thus the equation $p_1^2 - p_2^2 = (G^2RT/g^2A^2)[f(L/D) + 2\ln(p_1/p_2)]$ applies for all values of $L < 1260$ ft. Substituting $L = 500$ ft and neglecting the usually small logarithmic term, $[(80)(144)]^2 - p_2^2 = [(100)^2(1715)(460 + 65)/(32.2)^2(\frac{140}{144})^2]\{(0.0083)[500/4(0.243)]\}$ from which $p_2 = 67.1$ psia. Substituting this value of $p_2$ and considering the logarithmic term yields $p_2 = 66.6$ psia. Repeating the process again will give a more accurate answer.

**16.75**    Refer to Prob. 16.74. Neglecting the logarithmic term, find the pressures at sections 100, 300, and 800 ft downstream of the section where the pressure is 80 psia.

**∎**    $$p_1 - p_2 = (G^2RT/g^2A^2)[f(L/D) + 2\ln(p_1/p_2)]$$
   $$[(80)(144)]^2 - p_2^2 = [(100)^2 1715(460 + 65)/32.2^2(\tfrac{140}{144})^2]\{(0.0085)[L/4(0.243)]\}$$
   $$p_2^2 = 1.327 \times 10^8 - 80\,300L$$

At $L = 100$ ft, $p_2 = 11\,170$ psfa $= 77.5$ psia. At $L = 300$ ft, $p_2 = 10\,420$ psfa $= 72.4$ psia. At $L = 800$ ft, $p_2 = 8270$ psfa $= 57.5$ psia.
   Express in terms of $p + \gamma(V^2/2g)$:

At $L = 100$ ft, $\gamma = 0.41(77.5/80) = 0.397$ pcf.

At $L = 300$ ft, $\gamma = 0.41(72.4/80) = 0.371$ pcf.

At $L = 800$ ft, $\gamma = 0.41(57.5/80) = 0.295$ pcf.

$$\gamma_1 V_1 = \gamma_2 V_2 = \gamma_3 V_3$$

At $L = 100$ ft, $V = 250(0.41/0.397) = 258$ fps. At $L = 300$ ft, $V = 250(0.41/0.371) = 276$ fps. At $L = 800$ ft, $V = 250(0.41/0.295) = 347$ fps.

| $L$, ft | $p$, psia | $V$, fps | $V^2/2g$ | $\gamma V^2/2g$, psi | $p + \gamma V^2/2g$, psia |
|---|---|---|---|---|---|
| 0 | 80.0 | 250 | 970 | 2.76 | 82.8 |
| 100 | 77.5 | 258 | 1034 | 2.86 | 80.4 |
| 300 | 72.4 | 276 | 1183 | 3.05 | 75.5 |
| 800 | 57.5 | 347 | 1870 | 3.83 | 61.3 |

$\Delta p = 80 - 57.5 = 22.5$ psi over the first 800 ft of length. *Note:* Greater accuracy can be achieved by considering the logarithmic term.

**16.76**    Air flows isothermally through a long horizontal pipe of uniform diameter. At a section where the pressure is 90 psia, the velocity is 110 fps. Because of fluid friction the pressure at a distant point is 30 psia. (*a*) What is the increase in kinetic energy per pound of air? (b) What is the amount of thermal energy in Btu per pound of air that must be transferred in order to maintain the temperature constant? (c) Is this heat transferred to the air in the pipe or removed from it? (d) If the temperature of the air is 95 °F and the diameter of the pipe is 6 in find the total heat transferred in Btu per hour.

$\blacksquare$ $pV$ = const for isothermal flow, therefore $V_2 = p_1V_1/p_2 = 90(110)/30 = 330$ fps, $V_1^2/2g = 188$ ft-lb/lb, $V_2^2/2g = 1691$ ft-lb/lb.

(a) $V_1^2/2g + Q_B = V_2^2/2g$. Increase in KE of air $= 1691 - 188 = 1503$ ft-lb/lb.

(b) Thermal energy/lb transferred $= Q_H = \Delta KE = \frac{1503}{778} = 1.932$ Btu/lb.

(c) KE increases, therefore heat is transferred *to* the air in the pipe.

(d) 
$$\gamma_1 = gp_1/RT_1 = [(32.2)(90)(144)]/[(1715)(460+95)] = 0.438 \text{ pcf}$$
$$G = \gamma_1 V_1 A = (0.438)(110)[(\pi/4)(\tfrac{6}{12})^2] = 9.46 \text{ lb/s}$$
$$(1.932 \times 9.46) = 18.28 \text{ Btu/s} = 65\,808 \text{ Btu/h}$$

**16.77** For the case of Prob. 16.74 with a duct length of 500 ft, compute the thermal energy (heat) that must be added to the fluid to maintain isothermal conditions.

$\blacksquare$ Since the flow is isothermal, $p_1/\rho_1 = p_2/\rho_2 = RT =$ constant; $p_1 = 80$ psia and $p_2 = 66.6$ psia. Thus $\rho_1/\rho_2 = 80/66.6 = 1.20$ and $V_2/V_1 = 1.20$ since $\rho V =$ constant from continuity.

So $V_2 = 1.20(250) = 300$ fps; $V_1^2/2g + Q_H = V_2^2/2g$, $Q_H = (300)^2/[(2)(32.2)] - (250)^2/[(2)(32.2)] = 427$ ft-lb/lb of air. Since $G = 100$ lb/s, the rate at which heat must be added to the fluid is $(100)(427) = 42\,700$ ft-lb/s. *Note*: if $Q_H > 427$ ft-lb/lb of air, $T_2 > T_1$, and if $Q_H < 427$ ft-lb/lb of air, $T_2 < T_1$.

**16.78** Carbon dioxide flows isothermally at 100 °F through a horizontal 6-in-diameter pipe. At this temperature $\mu = 4.0 \times 10^{-7}$ lb-s/ft². The pressure changes from 160.0 to 150.0 psig in a 120-ft length of pipe. Determine the flow rate if the atmospheric pressure is 14.5 psia and $\epsilon$ for the pipe is 0.002 ft.

$\blacksquare$ At section 1 (inlet), $p_1 = 160.0$ psig $+ 14.5$ psi $= 174.5$ psia, $R = 1123$ ft²/(s²-°R), $\gamma = gp/RT$, $\gamma_1 = [(32.2)(174.5)(144)]/[(1123)(460+100)] = 1.287$ pcf; $\epsilon/D = 0.002/0.5 = 0.004$. Assume $N_R > 10^6$, Fig. A-5: $f = 0.0285$. At section 2, $p_2 = 150$ psig $+ 14.5$ psi $= 164.5$ psia.

$$p_1^2 - p_2^2 = (G^2RT/g^2A^2)[f(L/D) + 2\ln(p_1/p_2)]$$
$$[(174.5)(144)]^2 - [(164.5)(144)]^2 = [G^2(1123)(460+100)]/[32.2^2(0.1963)^2][0.0285(120/0.5) + 2\ln(174.5/164.5)]$$
$$G = 25.3 \text{ lb/s} \quad \text{(based on the above assumption)} \quad V_1 = G/\gamma_1 A = 25.3/[1.287(0.1963)] = 100.1 \text{ fps}$$
$$N_R = (DV\gamma/g)/\mu \simeq [(\tfrac{6}{12})100.1(1.287/32.2)]/(4.0 \times 10^{-7}) = 5.00 \times 10^6$$

Hence assumption for $N_R$ was O.K., and $G = 25.3$ lb/s.

**16.79** Methane gas is to be pumped through a 24-in-diameter welded-steel pipe connecting two compressor stations 24 mi apart. At the upstream station the pressure is not to exceed 65 psia, and at the downstream station it is to be at least 25 psia. Determine the maximum possible rate of flow (in cubic feet per day at 70 °F and 65 psia). Assume isothermal flow at 70 °F.

$\blacksquare$ For methane: $R = 3100$ ft²/(s²-°R), $\mu = 2.40 \times 10^{-7}$ lb-s/ft². Absolute viscosity is independent of pressure over a wide range of pressures but kinematic viscosity varies with pressure: $\nu_1 = \mu/\rho_1 = \mu/(p_1/RT_1) = 2.40 \times 10^{-7}/[(65)(144)/(3100)(530)] = 4.21 \times 10^{-5}$ ft²/s, $\epsilon = 0.00015$ ft; $\epsilon/D = 0.000075$. As a first trial, assume $N_R > 10^7$, in which case from Fig. A-5 $f = 0.0115$.

$$p_1^2 - p_2^2 = (G^2RT/g^2A^2)[f(L/D) + 2\ln(p_1/p_2)]$$
$$[(65)(144)]^2 - [(25)(144)]^2 = \{G^2(3100)(460+70)/32.2^2[(\pi)(\tfrac{24}{12})^2/4]\}\{(0.0115)[(24)(5280)/2] + 2\ln\tfrac{65}{25}\}$$
$$G = 25.2 \text{ lb/s}$$
$$N_R = GD/\mu gA = (25.2)(2)/(2.40 \times 10^{-7})[(32.2)(\pi)] = 2.08 \times 10^6$$

Thus, from Fig. A-5: $f = 0.0125$; initially assumed $f$ was too low. Substituting $f = 0.0125$ gives $G = 24.2$ lb/s, $\gamma_1 = gp_1/RT_1 = (32.2)(65)(144)/[(3100)(530)] = 0.1834$ lb/ft³, $Q = G/\gamma = 24.2/0.1834 = 132.0$ ft³/s $= 11.40 \times 10^6$ ft³/day at 70 °F and 65 psia.

**16.80** Air flows isothermally in a long pipe. At one section the pressure is 84 psia, the temperature is 70 °F, and the velocity is 100 fps. At a second section some distance from the first the pressure is 14 psia. Find the energy head loss due to friction, and determine the thermal energy that must have been added to or taken from the fluid between the two sections. The diameter of the pipe is constant.

**I** For isothermal flow: $\rho_1/\rho_2 = p_1/p_2 = \frac{84}{14} = 6$. But from continuity $\rho_1 A_1 V_1 = \rho_2 A_2 V_2$, thus $V_2 = 6V_1$, $V_2^2/2g + Q_H = V_2^2/2g$, $Q_H = 600^2/64.4 - 100^2/64.4 = 5430$ ft-lb/lb of air. For head loss: $dp/\gamma + d(V^2/2g) = -(2\tau\,ds/\gamma r) = -dh_L$, $\gamma = g\rho = gp/RT$; substituting for $\gamma$, $(RT/g)(dp/p) + d(V^2/2g) = -dh_L$. Integrating

$$h_L = (RT/g)\ln(p_1/p_2) + (V_1^2 - V_2^2)/2g = [1715(460 + 70)/32.2]\ln 6 + (100^2 - 600^2)/2(32.2)$$

$$= 50\,580 - 5430 = 45\,150 \text{ ft-lb/lb of air}$$

*Note:* The total energy at section 2 is greater than that at section 1 because external heat is added. Thus $h_L$ represents a degradation of mechanical energy, but not a net loss of energy, because it is converted to another form of energy, namely internal heat.

**16.81** Air is flowing at constant temperature through a 3-in-diameter horizontal pipe, $f = 0.018$. At the entrance, $V_1 = 320$ ft/s, $T = 110\,°F$, and $p_1 = 30$ psia. What is the maximum pipe length for this flow, and how much heat is transferred to the air per pound mass?

**I**
$$L_{max} = (D/f)[(1 - kN_M^2)/kN_M^2 + \ln(kN_M^2)] \qquad N_M = V/\sqrt{kRT}$$

$$(N_M)_1 = 320/\sqrt{(1.40)(1716)(110 + 460)} = 0.273$$

$$L_{max} = [(\tfrac{3}{12})/0.018]\{[1 - (1.40)(0.273)^2]/(1.40)(0.273)^2 + \ln[(1.40)(0.273)^2]\} = 87.8 \text{ ft} \qquad q_H = (V_2^2 - V_1^2)/2g$$

$$V_2 = (N_M)_2(c) \qquad (N_M)_2 = 1/\sqrt{k} = 1/\sqrt{1.40} = 0.845$$

$$V_2 = (0.845)(1180) = 997.1 \text{ ft/s} \qquad q_H = (997.1^2 - 320^2)/[(2)(32.2)(778)] = 17.8 \text{ Btu/lbm}$$

**16.82** Air at 15 °C flows through a 20-mm-diameter pipe at constant temperature. At the entrance $V_1 = 60$ m/s, and at the exit $V_2 = 90$ m/s ($f = 0.018$). What is the length of the pipe?

**I**
$$(f/D)(\Delta L) = [(1 - kN_M^2)/kN_M^2 + \ln(kN_M^2)]_1 - [(1 - kN_M^2)/k_M^2 + \ln(kN_M^2)]_2 \qquad N_M = V/\sqrt{kRT}$$

$$(N_M)_1 = 60/\sqrt{(1.40)(287)(15 + 273)} = 0.176 \qquad (N_M)_2 = 90/\sqrt{(1.40)(287)(15 + 273)} = 0.265$$

$$[0.018/0.020](\Delta L) = \{[1 - (1.40)(0.176)^2]/(1.40)(0.176)^2 + \ln[(1.40)(0.176)^2]\}$$

$$- \{[1 - (1.40)(0.265)^2]/(1.40)(0.265)^2 + \ln[(1.40)(0.265)^2]\} \qquad \Delta L = 13.4 \text{ m}$$

**16.83** If the pressure at the entrance of the pipe of Prob. 16.82 is 2.0 atm, find the pressure at the exit.

**I**
$$p^{*\prime}/p = (\sqrt{k})(N_M) \qquad p^{*\prime}/p_1 = (\sqrt{k})(N_M)_1 \qquad p^{*\prime}/p_2 = (\sqrt{k})(N_M)_2$$

Using data from Prob. 16.82,

$$p_2/p_1 = (N_M)_1/(N_M)_2 = 0.176/0.265 = 0.664 \qquad p_2 = (0.664)(2.0) = 1.328 \text{ atm}$$

**16.84** Hydrogen enters a pipe from a converging nozzle at $N_M = 1$, $p = 5$ psia, and $T = 0\,°F$. Determine, for isothermal flow, the maximum length of pipe, in diameters, and the pressure change over this length ($f = 0.018$).

**I**
$$L = (D/f)[(1 - kN_M^2)/kN_M^2 + \ln(kN_M^2)]$$

$$L/D = (1/0.018)\{[1 - (1.40)(1)^2]/(1.40)(1)^2 + \ln[(1.40)(1)^2]\} = 2.82$$

$$p^{*\prime}/p = (\sqrt{k})(N_M) \qquad p^{*\prime}/5 = (\sqrt{1.40})(1) \qquad p^{*\prime} = 5.916 \text{ psia} \qquad \Delta p = 5.916 - 5 = 0.916 \text{ psia}$$

**16.85** Oxygen flows at constant temperature of 20 °C from a pressure tank, $p = 130$ atm, through 3-mm-ID tubing to another tank where $p = 110$ atm, $f = 0.016$. Determine the mass rate of flow.

**I**
$$\dot{m} = \rho A v \qquad \rho = p/RT \qquad \rho_1 = (130)(101\,310)/[(260)(20 + 273)] = 172.9 \text{ kg/m}^3$$

$$A_1 = (\pi)(0.003)^2/4 = 0.000007069 \text{ m}^2 \qquad N_M = v/\sqrt{kRT}$$

$$(f/D)(\Delta L) = [(1 - kN_M^2)/kN_M^2 + \ln(kN_M^2)]_1 - [(1 - kN_M^2)/kN_M^2 + \ln(kN_M^2)]_2$$

$$p_2/p_1 = (N_M)_1/(N_M)_2 \qquad \text{(from Prob. 16.83)}$$

$$p_2/p_1 = \tfrac{110}{130} = 0.8462 \qquad (N_M)_2 = (N_M)_1/0.8462 = 1.182(N_M)_1$$

$$[0.016/0.003][(10)(0.3048)] = \{[1 - (1.40)(N_M)_1]^2/(1.40)(N_M)_1^2 + \ln [(1.40)(N_M)_1^2]\}$$
$$- (\{1 - (1.40)[1.182(N_M)_1]^2\}/(1.40)[1.182(N_M)_1]^2 + \ln (1.40)[1.182(N_M)_1]^2)$$
$$(N_M)_1 = 0.111 \qquad 0.111 = v_1/\sqrt{(1.40)(260)(15 + 273)} \qquad v_1 = 35.94 \text{ m/s}$$
$$\dot{m} = (172.9)(0.000007069)(35.94) = 0.0439 \text{ kg/s}$$

**16.86**  In isothermal flow of nitrogen at 80 °F, 3 lbm/s is to be transferred 100 ft from a tank where $p = 200$ psia to a tank where $p = 160$ psia. What minimum size of tubing, $f = 0.016$, is needed?

$$\blacksquare \qquad (f/D)(\Delta L) = [(1 - kN_M^2)/kN_M^2 + \ln (kN_M^2)]_1 - [(1 - kN_M^2)/k_M^2 + \ln (kN_M^2)]_2 \qquad N_M = v/c$$
$$\dot{m} = \rho_1 A_1 v_1 \qquad \rho = p/RT \qquad \rho_1 = (200)(144)/[(1773)(80 + 460)] = 0.03008 \text{ slug/ft}^3$$
$$\rho_2 = (\rho_1)(p_2/p_1) = (0.03008)(\tfrac{160}{200}) = 0.02406 \text{ slug/ft}^3 \qquad 3 = (0.03008)(32.17)(\pi D^2/4)(v_1) \qquad v_1 = 3.947/D^2$$
$$v_2 = (\tfrac{200}{160})(3.947/D^2) = 4.934/D^2 \qquad c = \sqrt{kRT} = \sqrt{(1.40)(55.2)(32.17)(80 + 460)} = 1159 \text{ ft/s}$$
$$(N_M)_1 = (3.947/D^2)/1159 = 0.003406/D^2 \qquad (N_M)_2 = (4.934/D^2)/1159 = 0.004257/D^2$$
$$(0.016/D)(100) = \{[1 - (1.40)(0.003406/D^2)^2]/(1.40)(0.003406/D^2)^2 + \ln [(1.40)(0.003406/D^2)^2]\}$$
$$- \{[1 - (1.40)(0.004257/D^2)^2]/(1.40)(0.004257/D^2)^2 + \ln [(1.40)(0.004257/D^2)^2]\}$$
$$D = 0.150 \text{ ft} \quad \text{or} \quad 1.80 \text{ in} \qquad \text{(by trial and error)}$$

**16.87**  Air enters a pipe of 1 in diameter at subsonic velocity and $p_1 = 30$ psia, $T_1 = 550$ °R. If the pipe is 10 ft long, $f = 0.025$, and the exit pressure $p_2 = 20$ psia, compute the mass flow for isothermal flow.

$$\blacksquare \qquad G^2 = (\dot{m}/A)^2 = (p_1^2 - p_2^2)/\{RT[fL/D + 2\ln (p_1/p_2)]\}$$
$$fL/D + 2\ln (p_1/p_2) = [(0.025)(10)/\tfrac{1}{12}] + 2\ln \tfrac{30}{20} = 3.81$$
$$G^2 = [30(144)]^2 - [20(144)]^2/[(1717)(550)(3.81)] = 2.88 \text{ lb}^2\text{-s}^2/\text{ft}^6 \qquad \text{or} \qquad G = 1.70 \text{ lb-s/ft}^3 = 1.70 \text{ slugs/(s-ft}^2)$$
$$A = \tfrac{1}{4}\pi D^2 = \tfrac{1}{4}\pi(\tfrac{1}{12})^2 = 0.00545 \text{ ft}^2 \qquad \text{Hence} \qquad \dot{m} = GA = (1.70)(0.00545) = 0.00926 \text{ slug/s}$$

Check the Mach number at inlet and exit:

$$a_1 = a_2 \approx 49(550)^{1/2} = 1149 \text{ ft/s} \qquad \rho_1 = p_1/RT = (30)(144)/(1717)(550) = 0.00457 \text{ slug/ft}^3$$
$$\rho_2 = p_2/RT = (20)(144)/(1717)(550) = 0.00305 \text{ slug/ft}^3$$

Then

$$V_1 = G/\rho_1 = 1.70/0.00457 = 371 \text{ ft/s} \qquad (N_M)_1 = V_1/a_1 = \tfrac{371}{1149} = 0.323 \qquad V_2 = G/\rho_2 = 557 \text{ ft/s}$$
$$(N_M)_2 = \tfrac{557}{1149} = 0.484$$

Since these are well below choking, the solution is accurate.

**16.88**  Isentropic flow of nitrogen occurs in a 2-in-diameter pipe. At one point the velocity of flow, pressure, and specific weight are 409 fps, 85 psia, and 0.655 lb/ft³, respectively. At a second point a short distance away, the pressure is 83 psia. What is the velocity at the second point?

$$\blacksquare \qquad (v_2^2 - v_1^2)/2g = (p_1/\gamma_1)[k/(k - 1)][1 - (p_2/p_1)^{(k-1)/k_1}]$$
$$(v_2^2 - 409^2)/[(2)(32.2)] = [(85)(144)/0.655][1.40/(1.40 - 1)][1 - (\tfrac{83}{85})^{(1.40-1)/1.40}] \qquad v_2 = 443 \text{ ft/s}$$

**16.89**  At one point on a streamline in an isentropic airflow, the velocity, pressure, and specific weight are 30.5 m/s, 350 kPa, and 0.028 kN/m³, respectively. At a second point on the streamline, the velocity is 150 m/s. Find the pressure at the second point.

$$\blacksquare \qquad (v_2^2 - v_1^2)/2g = (p_1/\gamma_1)[k/(k - 1)][1 - (p_2/p_1)^{(k-1)/k}]$$
$$(150^2 - 30.5^2)/[(2)(9.807)] = (350/0.028)[1.40/(1.40 - 1)][1 - (p_2/350)^{(1.40-1)/(1.40)}] \qquad p_2 = 320 \text{ kPa}$$

**16.90**  Isentropic flow of oxygen occurs in a 100-m-diameter pipe. At one point the velocity of flow, pressure, and specific weight are 125 m/s, 450 kPa, and 0.058 kN/m³, respectively. At a second point a short distance away, the pressure is 360 kPa. What is the velocity at the second point?

$$\blacksquare \qquad (v_2^2 - v_1^2)/2g = (p_1/\gamma)[k/(k - 1)][1 - (p_2/p_1)^{(k-1)/k}]$$
$$(v_2^2 - 125^2)/[(2)(9.807)] = (450/0.058)[1.40/(1.40 - 1)][1 - (\tfrac{360}{450})^{(1.40-1)/1.40}] \qquad v_2 = 220 \text{ m/s}$$

**16.91** At one point on a streamline in an isentropic airflow, the velocity, pressure, and specific weight are 80 m/s, 405 kPa, and 0.046 kN/m$^3$, respectively. At a second point on the streamline, the velocity is 165 m/s. Find the pressure at the second point.

$$(v_2^2 - v_1^2)/2g = (p_1/\gamma)[k/(k-1)][1 - (p_2/p_1)^{(k-1)/k}]$$

$$(165^2 - 80^2)/[(2)(9.807)] = (405/0.046)[1.40/(1.40-1)][1 - (p_2/405)^{(1.40-1)/1.40}] \qquad p_2 = 358 \text{ kPa}$$

**16.92** Prepare a computer program that determines the pressure either at point 1 or point 2 or the velocity at point 1 or point 2 for closed conduit, compressible, isentropic flow.

```
C     THIS PROGRAM DETERMINES EITHER THE PRESSURE AT POINT 1 OR
C     POINT 2 OR THE VELOCITY AT POINT 1 OR POINT 2 FOR CLOSED CONDUIT,
C     COMPRESSIBLE FLOW.  IT IS BASED ON AN ISENTROPIC FLOW ANALYSIS
C     AND IS THEREFORE SUBJECT TO THE RESTRICTIONS AND/OR ASSUMPTIONS
C     IMPLICIT IN ISENTROPIC FLOW.  IT CAN BE USED FOR PROBLEMS IN BOTH
C     THE ENGLISH SYSTEM OF UNITS AND THE INTERNATIONAL SYSTEM OF
C     UNITS.
C
C     INPUT DATA MUST BE SET UP AS FOLLOWS.
C
C     CARD 1    COLUMN 1          ENTER 0 (ZERO) OR BLANK IF ENGLISH SYSTEM
C                                 OF UNITS IS TO BE USED.  ENTER 1 (ONE) IF
C                                 INTERNATIONAL SYSTEM OF UNITS IS TO BE
C                                 USED.
C               COLUMNS 2-79      ENTER TITLE, DATE, AND OTHER INFORMATION,
C                                 IF DESIRED.
C     CARD 2    COLUMNS 1-10      ENTER NUMBER INCLUDING DECIMAL GIVING
C                                 ABSOLUTE PRESSURE AT UPSTREAM POINT
C                                 (POINT 1) (IN POUNDS PER SQUARE
C                                 INCH OR KILOPASCALS).
C               COLUMNS 11-20     ENTER NUMBER INCLUDING DECIMAL GIVING
C                                 ABSOLUTE PRESSURE AT DOWNSTREAM POINT
C                                 (POINT 2) (IN POUNDS PER SQUARE
C                                 INCH OR KILOPASCALS).
C               COLUMNS 21-30     ENTER NUMBER INCLUDING DECIMAL GIVING
C                                 VELOCITY AT POINT 1 (IN FEET PER SECOND
C                                 OR METERS PER SECOND).
C               COLUMNS 31-40     ENTER NUMBER INCLUDING DECIMAL GIVING
C                                 VELOCITY AT POINT 2 (IN FEET PER SECOND
C                                 OR METERS PER SECOND).
C               COLUMNS 41-50     ENTER NUMBER INCLUDING DECIMAL GIVING
C                                 SPECIFIC (OR UNIT) WEIGHT AT POINT 1 (IN
C                                 POUNDS PER CUBIC FOOT OR KILO-
C                                 NEWTONS PER CUBIC METER).
C               COLUMNS 51-60     ENTER NUMBER INCLUDING DECIMAL GIVING
C                                 SPECIFIC (OR UNIT) WEIGHT AT POINT 2 (IN
C                                 POUNDS PER CUBIC FOOT OR KILO-
C                                 NEWTONS PER CUBIC METER).
C               COLUMNS 61-70     ENTER NUMBER INCLUDING DECIMAL GIVING
C                                 SPECIFIC HEAT RATIO.
C               COLUMNS 71-80     ENTER TYPE OF FLUID.
C
C     ***************************************************************
C     *                                                             *
C     *  NOTE WELL....EITHER THE PRESSURE AT POINT 1 (COLUMNS 1-10 ON  *
C     *  CARD 2) OR AT POINT 2 (COLUMNS 11-20) OR THE VELOCITY AT POINT 1 *
C     *  (COLUMNS 21-30) OR AT POINT 2 (COLUMNS 31-40), WHICHEVER ONE IS *
C     *  TO BE DETERMINED BY THIS PROGRAM, SHOULD BE LEFT BLANK.  IF    *
C     *  EITHER THE PRESSURE OR THE VELOCITY AT POINT 1 IS BEING DETER-  *
C     *  MINED, ENTER A VALUE FOR THE SPECIFIC (OR UNIT) WEIGHT OF THE   *
C     *  FLUID AT POINT 2 (COLUMNS 51-60) AND LEAVE SPECIFIC (OR UNIT)   *
C     *  WEIGHT OF THE FLUID AT POINT 1 (COLUMNS 41-50) BLANK.  IF EITHER *
C     *  THE PRESSURE OR THE VELOCITY AT POINT 2 IS BEING DETERMINED,    *
C     *  ENTER A VALUE FOR THE SPECIFIC (OR UNIT) WEIGHT OF THE FLUID AT  *
C     *  POINT 1 (COLUMNS 41-50) AND LEAVE SPECIFIC (OR UNIT) WEIGHT OF   *
C     *  THE FLUID AT POINT 2 (COLUMNS 51-60) BLANK.                     *
C     *                                                             *
C     ***************************************************************
C
C     MULTIPLE DATA SETS FOR SOLVING ANY NUMBER OF PROBLEMS MAY BE
C     INCLUDED FOR PROCESSING.
```

```
C
      REAL K
      DIMENSION TITLE(13),FLUID(2)
1     READ(5,100,END=2)UNITS,TITLE
100   FORMAT(I1,13A6)
      WRITE(6,105)TITLE
105   FORMAT('1',13A6,//)
      READ(5,102)P1,P2,V1,V2,SW1,SW2,K,FLUID
102   FORMAT(7F10.0,2A5)
      G=32.2
      IF(UNITS.EQ.1)G=9.807
      PFACT=144.0
      IF(UNITS.EQ.1)PFACT=1.0
      IF(P1.LT.0.0001)GO TO 103
      IF(P2.LT.0.0001)GO TO 104
      IF(V1.LT.0.0001)GO TO 106
      V2=SQRT(2.0*G*P1*PFACT/SW1*K/(K-1.0)*(1.0-(P2/P1)**((K-1.0)/K))
     *+V1**2)
      IF(UNITS.EQ.0)WRITE(6,107)P1,P2,V1,SW1,FLUID,V2
107   FORMAT(1X,'GIVEN DATA FOR A COMPRESSIBLE FLOW IN A CIRCULAR CLOSED
     * CONDUIT',//5X,'PRESSURE AT POINT 1 =',F8.1,' PSI',//5X,'PRESSURE
     *AT POINT 2 =',F8.1,' PSI',//5X,'VELOCITY AT POINT 1 =',F7.1,' FT/S
     *',//5X,'SPECIFIC (OR UNIT) WEIGHT OF FLUID AT POINT 1 =',F8.3,
     * ' LB/CU FT',//5X,'FLUID FLOWING IS ',2A5,//1X,'THE VELOCITY AT PO
     *INT 2 WILL BE',F7.1,' FT/S')
      IF(UNITS.EQ.1)WRITE(6,108)P1,P2,V1,SW1,FLUID,V2
108   FORMAT(1X,'GIVEN DATA FOR A COMPRESSIBLE FLOW IN A CIRCULAR CLOSED
     * CONDUIT',//5X,'PRESSURE AT POINT 1 =',F8.1,' KPA',//5X,'PRESSURE
     *AT POINT 2 =',F8.1,' KPA',//5X,'VELOCITY AT POINT 1 =',F7.1,' M/S
     *',//5X,'SPECIFIC (OR UNIT) WEIGHT OF FLUID AT POINT 1 =',F8.3,
     *' KN/CU M ',//5X,'FLUID FLOWING IS ',2A5,//1X,'THE VELOCITY AT PO
     *INT 2 WILL BE',F7.1,' M/S')
      GO TO 1
103   P1=(((V2**2-V1**2)/2.0/G/(P2*PFACT)*SW2/K*(K-1.0)+1.0)*(P2*PFACT)
     ***((K-1.0)/K))**(K/(K-1.0)))/PFACT
      IF(UNITS.EQ.0)WRITE(6,109)P2,V1,V2,SW2,FLUID,P1
109   FORMAT(1X,'GIVEN DATA FOR A COMPRESSIBLE FLOW IN A CIRCULAR CLOSED
     * CONDUIT',//5X,'PRESSURE AT POINT 2 =',F8.1,' PSI',//5X,'VELOCITY
     *AT POINT 1 =',F7.1,' FT/S',//5X,'VELOCITY AT POINT 2 =',F7.1,' FT/
     *S',//5X,'SPECIFIC (OR UNIT) WEIGHT OF FLUID AT POINT 2 =',F8.3,
     * ' LB/CU FT',//5X,'FLUID FLOWING IS ',2A5,//1X'THE PRESSURE AT POI
     *NT 1 WILL BE ',F8.1,' PSI')
      IF(UNITS.EQ.1)WRITE(6,110)P2,V1,V2,SW2,FLUID,P1
110   FORMAT(1X,'GIVEN DATA FOR A COMPRESSIBLE FLOW IN A CIRCULAR CLOSED
     * CONDUIT',//5X,'PRESSURE AT POINT 2 =',F8.1,' KPA',//5X,'VELOCITY
     *AT POINT 1 =',F7.1,' M/S ',//5X,'VELOCITY AT POINT 2 =',F7.1,' M/S
     * ',//5X,'SPECIFIC (OR UNIT) WEIGHT OF FLUID AT POINT 2 =',F8.3,
     *' KN/CU M ',//5X,'FLUID FLOWING IS ',2A5,//1X'THE PRESSURE AT POI
     *NT 1 WILL BE ',F8.1,' KPA')
      GO TO 1
104   P2=((1.0-(V2**2-V1**2)/2.0/G/(P1*PFACT)*SW1/K*(K-1.0))*(P1*PFACT)
     ***((K-1.0)/K))**(K/(K-1.0)))/PFACT
      IF(UNITS.EQ.0)WRITE(6,111)P1,V1,V2,SW1,FLUID,P2
111   FORMAT(1X,'GIVEN DATA FOR A COMPRESSIBLE FLOW IN A CIRCULAR CLOSED
     * CONDUIT',//5X,'PRESSURE AT POINT 1 =',F8.1,' PSI',//5X,'VELOCITY
     *AT POINT 1 =',F7.1,' FT/S',//5X,'VELOCITY AT POINT 2 =',F7.1,' FT/
     *S',//5X,'SPECIFIC (OR UNIT) WEIGHT OF FLUID AT POINT 1 =',F8.3,
     * ' LB/CU FT',//5X,'FLUID FLOWING IS ',2A5,//1X'THE PRESSURE AT POI
     *NT 2 WILL BE ',F8.1,' PSI')
      IF(UNITS.EQ.1)WRITE(6,112)P1,V1,V2,SW1,FLUID,P2
112   FORMAT(1X,'GIVEN DATA FOR A COMPRESSIBLE FLOW IN A CIRCULAR CLOSED
     * CONDUIT',//5X,'PRESSURE AT POINT 1 =',F8.1,' KPA',//5X,'VELOCITY
     *AT POINT 1 =',F7.1,' M/S ',//5X,'VELOCITY AT POINT 2 =',F7.1,' M/S
     * ',//5X,'SPECIFIC (OR UNIT) WEIGHT OF FLUID AT POINT 1 =',F8.3,
     *' KN/CU M ',//5X,'FLUID FLOWING IS ',2A5,//1X'THE PRESSURE AT POI
     *NT 2 WILL BE ',F8.1,' KPA')
      GO TO 1
106   V1=SQRT(V2**2-2.0*G*P2*PFACT/SW2*K/(K-1.0)*((P1/P2)**((K-1.0)/K)
     *-1.0))
      IF(UNITS.EQ.0)WRITE(6,113)P1,P2,V2,SW2,FLUID,V1
113   FORMAT(1X,'GIVEN DATA FOR A COMPRESSIBLE FLOW IN A CIRCULAR CLOSED
     * CONDUIT',//5X,'PRESSURE AT POINT 1 =',F8.1,' PSI',//5X,'PRESSURE
     *AT POINT 2 =',F8.1,' PSI',//5X,'VELOCITY AT POINT 2 =',F7.1,' FT/S
```

```
         *',//5X,'SPECIFIC (OR UNIT) WEIGHT OF FLUID AT POINT 2 =',F8.3,
         * ' LB/CU FT',//5X,'FLUID FLOWING IS ',2A5,//1X,'THE VELOCITY AT PO
         *INT 1 WILL BE',F7.1,' FT/S')
          IF(UNITS.EQ.1)WRITE(6,114)P1,P2,V2,SW2,FLUID,V1
      114 FORMAT(1X,'GIVEN DATA FOR A COMPRESSIBLE FLOW IN A CIRCULAR CLOSED
         * CONDUIT',//5X,'PRESSURE AT POINT 1 =',F8.1,' KPA',//5X,'PRESSURE
         *AT POINT 2 =',F8.1,' KPA',//5X,'VELOCITY AT POINT 2 =',F7.1,' M/S
         *',//5X,'SPECIFIC (OR UNIT) WEIGHT OF FLUID AT POINT 2 =',F8.3,
         *' KN/CU M  ',//5X,'FLUID FLOWING IS ',2A5,//1X,'THE VELOCITY AT PO
         *INT 1 WILL BE',F7.1,' M/S')
          GO TO 1
      2   STOP
          END
```

**16.93**    Solve Prob. 16.88 using the computer program of Prob. 16.92.

**❙ Input**

```
1 2 3 4 5 6 7 8 9 10111213141516171819202122232425262728293031323334353637383940414243444546474849505152535455565758596061626364656667686970717273747576777879808182
OSAMPLE ANALYSIS ØF CØMPRESSIBLE FLØW
85.0        83.0        409.0                   0.655               1.40        NITRØGEN
```

**Output**

```
SAMPLE ANALYSIS OF COMPRESSIBLE FLOW


GIVEN DATA FOR A COMPRESSIBLE FLOW IN A CIRCULAR CLOSED CONDUIT

    PRESSURE AT POINT 1 =    85.0 PSI

    PRESSURE AT POINT 2 =    83.0 PSI

    VELOCITY AT POINT 1 =   409.0 FT/S

    SPECIFIC (OR UNIT) WEIGHT OF FLUID AT POINT 1 =   0.655 LB/CU FT

    FLUID FLOWING IS NITROGEN

THE VELOCITY AT POINT 2 WILL BE   442.5 FT/S
```

**16.94**    Solve Prob. 16.89 using the computer program of Prob. 16.92.

**❙ Input**

```
1 2 3 4 5 6 7 8 9 10111213141516171819202122232425262728293031323334353637383940414243444546474849505152535455565758596061626364656667686970717273747576777879808182
1SAMPLE ANALYSIS ØF CØMPRESSIBLE FLØW
350.0                   30.5        150.0       0.028               1.40        AIR
```

**Output**

```
SAMPLE ANALYSIS OF COMPRESSIBLE FLOW


GIVEN DATA FOR A COMPRESSIBLE FLOW IN A CIRCULAR CLOSED CONDUIT

    PRESSURE AT POINT 1 =   350.0 KPA

    VELOCITY AT POINT 1 =    30.5 M/S

    VELOCITY AT POINT 2 =   150.0 M/S

    SPECIFIC (OR UNIT) WEIGHT OF FLUID AT POINT 1 =   0.028 KN/CU M

    FLUID FLOWING IS AIR

THE PRESSURE AT POINT 2 WILL BE    320.2 KPA
```

**16.95** Air at a pressure of 200 psia and a temperature of 140 °F expands in a suitable nozzle to 20 psia. (*a*) If the flow is frictionless and adiabatic and the initial velocity is negligible, find the final velocity. (*b*) Find the final temperature at the end of the expansion.

**|** (*a*)    $V_1 \simeq 0$     $V_2^2/2g = (RT_1/g)[k/(k-1)][1 - (p_2/p_1)^{(k-1)/k}]$

$V_2^2/2g = [(1715)(600)/32.2](1.4/0.4)[1 - (\frac{20}{200})^{0.2860}]$     $V_2 = 1864$ fps

(*b*)    $V_2^2 - V_1^2 = 2c_p(T_1 - T_2)$     $1864^2 - 0 = 2(6000)(600 - T_2)$     $T_2 = 310 °R = -150 °F$

**16.96** Derive the equation for isentropic flow: $(V_2^2 - V_1^2)/2g = (p_1/\gamma_1)[k/(k-1)][1 - (p_2/p_1)^{(k-1)/k}]$ by integrating the Euler equation.

**|** Euler equation:     $dp/\gamma + (V \, dV)/g = 0$     (1)

$pv^k = \text{constant} = p/\rho^k$ because $v = 1/\rho$; so $p/\gamma^k = C$:

$p = C\gamma^k$:     $dp = Ck\gamma^{k-1} \, d\gamma$     (2)

Substitute (2) into (1) and integrate: $Ck\gamma^{k-2} \, d\gamma + (1/g)V \, dV = 0$.

$$\left[\frac{Ck\gamma^{k-1}}{k-1}\right]_{\gamma_1}^{\gamma_2} + \left[\frac{V^2}{2g}\right]_{V_1}^{V_2} = 0 \qquad \frac{V_2^2 - V_1^2}{2g} = \frac{k}{k-1}\left[\frac{p}{\gamma}\right]_2^1 = \frac{k}{k-1}\left(\frac{p_1}{\gamma_1} - \frac{p_2}{\gamma_2}\right)$$

Note that $p_1/\gamma_1 - p_2/\gamma_2 = p_1/\gamma_1[1 - (p_2/\gamma_2)/(p_1/\gamma_1)]$ where $\gamma_1 = (1/C)p_1^{1/k}$ and $\gamma_2 = (1/C)p_2^{1/k}$, $p_2\gamma_1/p_1\gamma_2 = (p_2^{1-(1/k)})/(p_1^{1-(1/k)}) = (p_2/p_1)^{(k-1)/k}$. Thus $(V_2^2 - V_1^2)/2g = (p_1/\gamma_1)[k/(k-1)][1 - (p_2/p_1)^{(k-1)/k}]$    Q.E.D.

**16.97** Carbon dioxide flows isentropically. At a point in the flow the velocity is 45 fps and the temperature is 120 °F. At a second point on the same streamline the temperature is 80 °F. What is the velocity at the second point?

**|** For carbon dioxide: $k = 1.28$; $R = 1123$ ft-lb/(slug-°R). Flow is isentropic, therefore adiabatic. $V_2^2 - V_1^2 = [2k/(k-1)](RT_1)[1 - (T_2/T_1)]$, $V_2^2 - 45^2 = [2(1.28)/0.28](1123)(460 + 120)[1 - (460 + 80)/(460 + 120)]$; $V_2 = 642$ fps. Check to see if sonic velocity is exceeded: $c = (kRT)^{1/2} = [1.28(1123)(460 + 80)]^{1/2} = 881$ fps. Since $V_2 < c$, flow is subsonic, $V_2 = 642$ fps is correct.

**16.98** Refer to Prob. 16.97. If the pressure at the first point were 20 psi, determine the pressure and temperature on the nose of a streamlined object placed in the flow at that point.

**|**    $\rho_0 = p_0/RT_0 = (20)(144)/(1123)(580) = 0.00442$ slug/ft$^3$     $c = (kRT_1)^{1/2} = [1.29(1123)(580)]^{1/2} = 917$ fps

$N_M = V/c = \frac{45}{917} = 0.0491$     $p_s = p_0 + \rho_0(V_0^2/2)(1 + N_M^2/4)(\frac{1}{144}) = 20.0$ psia     $(c_p/g)T_1 + V_1^2/2g = (c_p/g)T_2$

$(5132/32.2)(580) + (45^2/64.4) = (5132/32.2)T_s$     $T_s = 580 °R = 120 °F$

Temperature increase is negligible.

**16.99** From the equation $s_2 - s_1 = c_v \ln [(T_2/T_1)(\rho_1/\rho_2)^{k-1}]$ and the perfect gas law, derive the equation of state for isentropic flow.

**|** Since $s_2 - s_1 = c_v = 0$ for isentropic flow,

$\ln [(T_2/T_1)(\rho_1/\rho_2)^{k-1}] = 0$     $(T_2/T_1)(\rho_1/\rho_2)^{1-k} = 1$     $T_2/T_1 = (p_2/\rho_2)/(p_1/\rho_1)$    (from perfect gas law)

Eliminating $T$'s, $p_2/\rho_2^k = p_1/\rho_1^k$.

**16.100** Compute the enthalpy change per slug for helium from $T_1 = 0 °F$, $p_1 = 15$ psia to $T_2 = 140 °F$ in an isentropic process $[c_p = 1.25$ Btu/(lbm-°R)$]$.

**|**     $H = mc_p \, \Delta T = (1)[(1.25)(32.17)](140 - 0) = 5630$ Btu

**16.101** Isentropic flow of air occurs at a section of a pipe where $p = 50$ psia, $T = 100 °F$, and $V = 540$ ft/s. An object is immersed in the flow, which brings the velocity to zero. What are the temperature and pressure at the stagnation point?

**|** $V^2/2 = [kR/(k-1)](T_0 - T)$     $540^2/2 = [(1.40)(1716)/(1.40 - 1)][T_0 - (100 + 460)]$     $T_0 = 584 °R$   or   $124 °F$

$T_2/T_1 = (p_2/p_1)^{(k-1)/k}$     $584/(100 + 460) = (p_2/50)^{(1.40-1)/1.40}$     $p_2 = 57.9$ psia

**16.102** What is the Mach number for Prob. 16.101?

∎ $N_M = V/\sqrt{kRT} = 540/\sqrt{(1.40)(1716)(100 + 460)} = 0.466$

**16.103** How do the temperature and the pressure at the stagnation point in isentropic flow compare with reservoir conditions?

∎ They are the same.

**16.104** Air flows isentropically from a reservoir at 80 °C, 6 atm. Calculate the temperature, pressure, density, and velocity at a section where $N_M = 0.65$.

∎    $T_0/T = 1 + [(k-1)/2]N_M^2$    $(273 + 80)/T = 1 + [(1.40-1)/2](0.65)^2$    $T = 325$ K   or   52 °C

$p_0/p = \{1 + [(k-1)/2]N_M^2\}^{k/(k-1)}$    $(6)(101.310)/p = \{1 + [(1.40-1)/2](0.65)^2\}^{1.40/(1.40-1)}$

$p = 458$ kPa    $\rho = p/RT = 458\,000/[(287)(325)] = 4.91$ kg/m³

$c = \sqrt{kRT} = \sqrt{(1.40)(287)(325)} = 361$ m/s    $V = cN_M = (361)(0.65) = 235$ m/s

**16.105** Oxygen flows from a reservoir where $p_0 = 110$ psia, $T_0 = 100$ °F, to a 6-in-diameter section where the velocity is 500 fps. Calculate the mass rate of flow (isentropic) and the Mach number, pressure, and temperature in the 6-in section.

∎    $T = T_0 - [(k-1)/2kR]V^2 = (100 + 460) - [(1.40-1)/(2)(1.40)(1554)](500)^2 = 537$ °R   or   77 °F

$N_M = V/\sqrt{kRT} = 500/\sqrt{(1.40)(1554)(537)} = 0.463$    $\dot{m} = \rho AV$

$\rho_0 = p_0/RT_0 = (110)(144)/[(1554)(100 + 460)] = 0.01820$ slug/ft³

$\rho_0/\rho = \{1 + [(k-1)/2]N_M^2\}^{1/(k-1)}$    $0.01820/\rho = \{1 + [(1.40-1)/2](0.463)^2\}^{1/(1.40-1)}$    $\rho = 0.01639$ slug/ft³

$\dot{m} = 0.01639[(\pi)(\tfrac{6}{12})^2/4](500) = 1.61$ slugs/s

$p = \rho RT = (0.01639)(1554)(537) = 13\,677$ lb/ft²   or   95.0 lb/in²

**16.106** Air flows adiabatically through a duct. At point 1 the velocity is 800 ft/s, $T_1 = 500$ °R, and $p_1 = 25$ lb/in² abs $= 3600$ lb/ft². Compute (a) $T_0$; (b) $p_{01}$; (c) $\rho_0$; (d) $N_M$; (e) $V_{max}$; (f) $V^*$. At point 2 further downstream $V_2 = 962$ ft/s, and $p_2 = 2850$ lb/ft². (g) What is the stagnation pressure $p_{02}$?

∎ For air take $k = 1.4$, $c_p = 6010$, and $R = 1717$ (BG units). With $V_1$ and $T_1$ known, we can compute $T_0$ without using ratios.

(a)      $T_{01} = T_1 + [\tfrac{1}{2}(V_1)^2/c_p] = 500 + [\tfrac{1}{2}(800)^2/6010] = 553$ °R

Then compute $N_M$ from the known ratio $T/T_0$, using $(N_M)^2 = 5[(T_0/T) - 1]$:

(d)      $(N_M)_1^2 = 5(\tfrac{553}{500} - 1)$      $(N_M)_1 = 0.73$

The stagnation pressure follows:

(b)      $p_{01} = p_1[1 + 0.2(N_M)_1^2]^{3.5} = 3600[1 + 0.2(0.73)^2]^{3.5} = 5130$ lb/ft²

We need the density before we can compute stagnation density: $\rho_1 = p_1/RT_1 = 3600/[1717(500)] = 0.00419$ slug/ft³. Then

(c)      $\rho_0 = \rho_1[1 + 0.2(N_M)_1^2]^{2.5} = 0.00419[1 + 0.2(0.73)^2]^{2.5} = 0.00540$ slug/ft³

However, if we were clever, we could compute $\rho_0$ directly from $p_0$ and $T_0$: $\rho_0 = p_0/RT_0 = 5130/[1717(553)] = 0.00540$ slug/ft³. The maximum velocity follows:

(e)      $V_{max} = (2c_p T_0)^{1/2} = [2(6010)(553)]^{1/2} = 2580$ ft/s

and the sonic velocity from $V^* = \{[2k/(k+1)]RT_0\}^{1/2}$:

(f)      $V^* = \{[2(1.4)/(1.4+1)](1717)(553)\}^{1/2} = 1050$ ft/s

At point 2, the temperature is not given, but since we know it is adiabatic flow, $T_{02} = T_{01} = 553$ °R from solution (a). Thus we can compute $T_2$ from $T_2 = T_{02} - (\tfrac{1}{2}V_2^2/c_p) = 553 - [\tfrac{1}{2}(962)^2/6010] = 476$ °R (Trying to find $T_2$ from the Mach-number relations is a frustratingly laborious procedure.) The speed of sound $a_2$ thus equals $(kRT_2)^{1/2} \approx 49(476)^{1/2} = 1069$ ft/s, whence the Mach number $(N_M)_2 = V_2/a_2 = \tfrac{962}{1069} = 0.90$. Finally compute

(g)      $p_{02} = p_2[1 + 0.2(N_M)_2^2]^{3.5} = 28.50[1 + 0.2(0.9)^2]^{3.5} = 4820$ lb/ft²      (6 percent less than $p_{01}$)

**16.107** Air flows isentropically through a duct. At section 1 the area is 1 ft² and $V_1 = 600$ ft/s, $p_1 = 12\,000$ lb/ft², and

$T_1 = 850\,°R$. Compute (a) $T_0$, (b) $(N_M)_1$, (c) $p_0$, and (d) $A^*$. If $A_2 = 0.75\,\text{ft}^2$, compute $(N_M)_2$ and $p_2$ if $V_2$ is (e) subsonic or (f) supersonic.

**I** With $V_1$ and $T_1$ known, the energy equation gives

(a)
$$T_0 = T_1 + (V_1^2/2c_p) = 850 + [(600)^2/2(6010)] = 880\,°R$$

The sound speed $a_1 \approx 49(850)^{1/2} = 1429\,\text{ft/s}$; hence

(b)
$$(N_M)_1 = V_1/a_1 = \tfrac{600}{1429} = 0.42$$

With $(N_M)_1$ known $p_0/p_1 = [1 + 0.2(N_M)_1^2]^{3.5} = 1.129$. Hence

(c)
$$p_0 = 1.129p_1 = 1.129(12\,000) = 13\,550\,\text{lb/ft}^2$$

Similarly, from $A/A^* = (1/N_M)[(1 + 0.2N_M^2)^3/1.728]$, $A_1/A^* = [1 + 0.2(0.42)^2]^3/[1.728(0.42)] = 1.529$. Hence

(d)
$$A^* = 1/1.529 = 0.654\,\text{ft}^2$$

This throat must actually be present in the duct to expand the subsonic $(N_M)_1$ to supersonic flow downstream.

Given $A_2 = 0.75\,\text{ft}^2$, we can compute $A_2/A^* = 0.75/0.654 = 1.147$. For subsonic flow, use $N_M \approx 1 - 0.88[\ln (A/A^*)]^{0.45}$:

(e) $\quad (N_M)_2 \approx 1 - 0.88(\ln 1.147)^{0.45} = 0.64 \quad$ whence $\quad p_2 = p_0/[1 + 0.2(0.64)^2]^{3.5} = 13\,550/1.317 = 10\,300\,\text{lb/ft}^2$

Fig. 16-4

For case (e) there is *no* sonic throat; i.e., the duct area has not decreased sufficiently to create supersonic flow. It may possibly do this further downstream, as illustrated in Fig. 16-4. If, on the other hand, the flow at section 2 is supersonic, we use $N_M = 1 + 1.2[(A/A^*) - 1]^{1/2}$:

(f) $\quad (N_M)_2 \approx 1 + 1.2(1.147 - 1)^{1/2} = 1.46 \quad$ whence $\quad p_2 = 13\,550/[1 + 0.2(1.46)^2]^{3.5} = 13\,550/3.46 = 3910\,\text{lb/ft}^2$

These answers are accurate within less than 1 percent. Note that the supersonic-flow pressure level is much less than the subsonic-flow condition at section 2 for the same duct area, and a sonic throat ($A^* = 0.654\,\text{ft}^2$) *must* have occurred between sections 1 and 2.

**16.108** It is desired to expand air from $p_0 = 200\,\text{kPa}$ and $T_0 = 500\,\text{K}$ through a throat to an exit Mach number of 2.5. If the desired mass flow is 3 kg/s, compute (a) the throat area and the exit (b) pressure, (c) temperature, (d) velocity, and (e) area, assuming isentropic flow, with $k = 1.4$.

**I** (a) The throat area follows from the following equation, because the throat flow must be sonic to produce a supersonic exit: $A^* = [\dot{m}(RT_0)^{1/2}]/(0.6847p_0) = \{3.0[287(500)]^{1/2}\}/[0.6847(200\,000)] = 0.00830\,\text{m}^2 = \tfrac{1}{4}\pi D^{*2}$ or $D_{\text{throat}} = 10.3\,\text{cm}$. With the exit Mach number known, the isentropic-flow relations give pressure and temperature:

(b)
$$p_e = p_0/[1 + 0.2(2.5)^2]^{3.5} = 200/17.08 = 11.70\,\text{kPa}$$

(c)
$$T_e = T_0/[1 + 0.2(2.5)^2] = 500/2.25 = 222\,\text{K}$$

(d) The exit velocity follows from the known Mach number and temperature $V_e = (N_M)_e(kRT_e)^{1/2} = 2.5[1.4(287)(222)]^{1/2} = 747\,\text{m/s}$.

(e) The exit area follows from the known throat area and exit Mach number and $A/A^* = (1/N_M)[(1 + 0.2N_M^2)^3/1.728]$: $A_e/A^* = [1 + 0.2(2.5)^2]^3/[1.728(2.5)] = 2.64$ or $A_e = 2.64A^* = 2.64(0.0083) = 0.0219\,\text{m}^2 = \tfrac{1}{4}\pi D_e^2$: $D_e = 16.7\,\text{cm}$.

**16.109** A flow of air ($c_p = 1005$ J/kg · K, $k = 1.40$) has $V = 850$ m/s, $p = 130$ kPa, and $T = 180$ °C. Compute the pressure and temperature of the air if brought isentropically to rest.

$$T_0 = T + V^2/2c_p = (180 + 273) + 850^2/[(2)(1005)] = 812 \text{ K} \quad \text{or} \quad 539 \text{ °C}$$
$$p_0 = (p)(T_0/T)^{k/(k-1)} = 130[812/(180 + 273)]^{1.40/(1.40-1)} = 1002 \text{ kPa}$$

**16.110** Rework Prob. 16.108 for argon ($c_p = 518$ J/kg · K, $k = 1.67$).

$$T_0 = T + V^2/2c_p = (180 + 273) + 850^2/[(2)(518)] = 1150 \text{ K} \quad \text{or} \quad 877 \text{ °C}$$
$$p_0 = (p)(T_0/T)^{k/(k-1)} = (130)[1150/(180 + 273)]^{1.67/(1.67-1)} = 1326 \text{ kPa}$$

**16.111** Rework Prob. 16.108 for steam.

$h_0 = h + V^2/2$. From the steam tables, at 130 kPa and 180 °C, $h = 2834.6$ kJ/kg and $s = 7.6705$ kJ/kg · K: $h_0 = 2835 + (850^2/2)/1000 = 3196$ kJ/kg, $s_0 = 7.695$ kJ/kg · K. From the steam tables, $T_0 = 360$ °C and $p_0 = 480$ kPa.

**16.112** Compute the frontal stagnation temperature of an airplane flying at $N_M = 2.0$ at 8000 m altitude. At what Mach number would it fly to have a frontal stagnation temperature of 380 °C?

$$T = 273 - 36.9 = 236.1 \text{ K} \quad \text{(from Table A-8)}$$
$$T_0 = T\{1 + [(k-1)/2](N_M)^2\} = 236.1\{1 + [(1.40-1)/2](2.0)^2\} = 425 \text{ K} \quad \text{or} \quad 152 \text{ °C}$$
$$(380 + 273) = 236.1\{1 + [(1.40-1)/2](N_M)^2\} \qquad N_M = 2.97$$

**16.113** Air ($c_p = 1005$ J/kg · K) flows at $V = 280$ m/s, $p = 100$ kPa, and $T = 160$ °C. Compute the maximum velocity attainable by adiabatic expansion of this flow.

$$T_0 = T + V^2/2c_p = (160 + 273) + 280^2/[(2)(1005)] = 472 \text{ K} \qquad V_{max} = (2c_p T_0)^{1/2} = [(2)(1005)(472)]^{1/2} = 974 \text{ m/s}$$

**16.114** Rework Prob. 16.113 for helium ($c_p = 5224$ J/kg · K).

$$T_0 = T + V^2/2c_p = (160 + 273) + 280^2/[(2)(5224)] = 441 \text{ K} \qquad V_{max} = (2c_p T_0)^{1/2} = [(2)(5224)(441)]^{1/2} = 2147 \text{ m/s}$$

**16.115** Air expands isentropically through a duct from $p_1 = 120$ kPa and $T_1 = 110$ °C to $p_2 = 75$ kPa and $V_2 = 325$ m/s. Compute (a) $T_2$, (b) $(N_M)_2$, (c) $T_0$, (d) $p_0$, (e) $V_1$.

(a)
$$T_2 = (T_1)(p_2/p_1)^{(k-1)/k} = (110 + 273)(\tfrac{75}{120})^{(1.40-1)/1.40} = 335 \text{ K}$$

(b) $\quad N_M = V/c \qquad c = \sqrt{kRT} \qquad c_2 = \sqrt{(1.40)(287)(335)} = 367 \text{ m/s} \qquad (N_M)_2 = \tfrac{325}{367} = 0.886$

(c) $\quad (T_0)_1 = (T_0)_2 = (T_2)\{1 + [(k-1)/2](N_M)_2^2\} \qquad T_0 = 335\{1 + [(1.40-1)/2](0.886)^2\} = 388 \text{ K}$

(d)
$$(p_0)_1 = (p_0)_2 = (p_2)\{1 + [(k-1)/2](N_M)_2^2\}^{k/(k-1)}$$
$$p_0 = 75\{1 + [(1.40-1)/2](0.886)^2\}^{1.40/(1.40-1)} = 125 \text{ kPa}$$

(e) $\quad T_0 = T + V^2/2c_p \qquad 388 = (110 + 273) + V_1^2/[(2)(1005)] \qquad V_1 = 100 \text{ m/s}$

**16.116** Air flows isentropically in a duct. At section 1, $(N_M)_1 = 0.8$, $T_1 = 250$ °C, and $p_1 = 280$ kPa. At section 2, $(N_M)_2 = 3.0$. Compute (a) $(p_0)_2$, (b) $p_2$, and (c) $T_2$.

(a) $(p_0)_2 = (p_0)_1 = (p_1)\{1 + [(k-1)/2](N_M)_1^2\}^{k/(k-1)} = 280\{1 + [(1.40-1)/2](0.8)^2\}^{1.40/(1.40-1)} = 427$ kPa

(b) $\quad p_2 = (p_0)_2/\{1 + [(k-1)/2](N_M)_2^2\}^{k/(k-1)} = 427/\{1 + [(1.40-1)/2](3.0)^2\}^{1.40/(1.40-1)} = 11.6$ kPa

(c) $\quad (T_0)_2 = (T_0)_1 = (T_1)\{1 + [(k-1)/2](N_M)_1^2\} = (250 + 273)\{1 + [(1.40-1)/2](0.8)^2\} = 590$ K

$$T_2 = (T_0)_2/\{1 + [(k-1)/2](N_M)_2^2\} = 590/\{1 + [(1.40-1)/2](3.0)^2\} = 211 \text{ K}$$

**16.117** Show that for isentropic flow of a perfect gas, if a pitot-static probe measures $p_0$, $p$, and $T_0$, the gas velocity can be calculated from $V^2 = 2c_p T_0[1 - (p/p_0)^{(k-1)/k}]$. What would be a source of error if a shock wave is formed in front of the probe?

$T = (T_0)(p/p_0)^{(k-1)/k} = T_0 - V^2/2c_p$. Solving for $V^2$ gives $V^2 = 2c_p T_0[1 - (p/p_0)^{(k-1)/k}]$. If a shock wave is formed (i.e., if $V$ is supersonic), the probe measures $(p_0)_2$ inside the shock wave, which is less than stream stagnation pressure $(p_0)_1$.

**16.118** At the exit of a nozzle flow, the air temperature is 60 °C and the velocity is 355 m/s. Find the Mach number and stagnation temperature there. If the flow is adiabatic, what is the Mach number upstream, where $T = 120$ °C?

$$N_M = V/c \qquad c = \sqrt{kRT} \qquad c_2 = \sqrt{(1.40)(287)(60 + 273)} = 366 \text{ m/s}$$

$$(N_M)_2 = 355/366 = 0.97 \qquad (T_0)_2 = T_2 + V_2^2/2c_p = (60 + 273) + 355^2/[(2)(1005)] = 396 \text{ K}$$

$$T_0/T_1 = 1 + [(k-1)/2](N_M)_1^2 \qquad 396/(120 + 273) = 1 + [(1.40-1)/2](N_M)_1^2 \qquad (N_M)_1 = 0.195$$

**16.119** The compressed-air tank in Fig. 16-5 exhausts from a nozzle at 240 m/s. Assuming isentropic flow, compute the pressure in the tank and in the atmosphere. What is the exit Mach number?

**Fig. 16-5**

$$N_M = V/c \qquad c = \sqrt{kRT} \qquad T_0 = t_{\text{exit}} + V_{\text{exit}}^2/2c_p \qquad (27 + 273) = T_{\text{exit}} + 240^2/[(2)(1005)] \qquad T_{\text{exit}} = 271 \text{ K}$$

$$c_{\text{exit}} = \sqrt{(1.40)(287)(271)} = 330 \text{ m/s} \qquad (N_M)_{\text{exit}} = \tfrac{240}{330} = 0.727$$

$$p_{\text{exit}} = (p_0)/\{1 + [(k-1)/2](N_M)_{\text{exit}}^2\}^{k/(k-1)} \qquad p_0/p_{\text{exit}} = \{1 + [(1.40-1)/2](0.727)^2\}^{1.40/(1.40-1)} = 1.421$$

Since $\rho_{\text{Hg}} \gg \rho_0$, we can simply take $\rho_0 = 1.5$ kg/m³. $p_0 - p_{\text{exit}} = (\rho_{\text{Hg}} - \rho_0)gh = (13\,600 - 1.5)(9.807)(0.300) = 40$ kPa. Then $p_0/p_{\text{exit}} - 1 = 1.421 - 1 = (p_0 - p_{\text{exit}})/p_{\text{exit}} = 40/p_{\text{exit}}$, $p_{\text{exit}} = 95$ kPa, $p_0 = p_{\text{tank}} = 135$ kPa.

**16.120** Air flows adiabatically through a duct. At one section, $V_1 = 420$ fps, $T_1 = 210$ °F, and $p_1 = 35$ psia, while farther downstream $V_2 = 1000$ fps and $p_2 = 18$ psia. Compute (a) $(N_M)_2$, (b) $U_{\text{max}}$, and (c) $(p_0)_2/(p_0)_1$.

$(a)$    $(N_M)_2 = V_2/c_2 \qquad c_2 = \sqrt{kRT_2} \qquad T_2 = (T_0)_2 - V_2^2/2c_p \qquad (T_0)_2 = (T_0)_1 = T_1 + V_1^2/2c_p$

$(T_0)_2 = (210 + 460) + 420^2/[(2)(6010)] = 685$ °R     $T_2 = 685 - 1000^2/[(2)(6010)] = 602$ °R

$$c_2 = \sqrt{(1.40)(1716)(602)} = 1203 \text{ ft/s} \qquad (N_M)_2 = \tfrac{1000}{1203} = 0.831$$

$(b)$    $$U_{\text{max}} = \sqrt{2c_p T_0} = \sqrt{(2)(6010)(685)} = 2869 \text{ ft/s}$$

$(c)$    $(p_0) = (p)\{[1 + (k-1)/2](N_M)^2\}^{k/(k-1)} \qquad (N_M)_1 = V_1/\sqrt{kRT_1} = 420/\sqrt{(1.40)(1716)(210 + 460)} = 0.331$

$$(p_0)_1 = 35\{1 + [(1.40-1)/2](0.331)^2\}^{1.40/(1.40-1)} = 37.76 \text{ psia}$$

$$(p_0)_2 = 18\{1 + [(1.40-1)/2](0.831)^2\}^{1.40/(1.40-1)} = 28.31 \text{ psia}$$

$$(p_0)_2/(p_0)_1 = 28.31/37.76 = 0.75$$

**16.121** Air flows isentropically from a reservoir, where $p = 320$ kPa and $T = 490$ K, to section 1 in a duct, where $A_1 = 0.2$ m² and $V_1 = 600$ m/s. Compute (a) $T_1$, (b) $(N_M)_1$, (c) $p_1$, and (d) $\dot{m}$.

$(a)$    $$T_1 = T_0 - V_1^2/2c_p = 490 - 600^2/[(2)(1005)] = 311 \text{ K}$$

$(b)$    $N_M = V/c \qquad c = \sqrt{kRT} \qquad c_1 = \sqrt{(1.40)(287)(311)} = 353 \text{ m/s} \qquad (N_M)_1 = \tfrac{600}{353} = 1.70$

$(c)$    $$p_1 = p_0/\{1 + [(k-1)/2](N_M)_1^2\}^{k/(k-1)} = 320/\{1 + [(1.40-1)/2](1.70)^2\}^{1.40/(1.40-1)} = 64.8 \text{ kPa}$$

$(d)$    $\rho = p/RT \qquad \rho_1 = (64.8)/[(0.287)(311)] = 0.726 \text{ kg/m}^3$

$$\dot{m} = \rho_1 A_1 V_1 = (0.726)(0.2)(600) = 87.1 \text{ kg/s}$$

**16.122** Repeat Prob. 16.121 for argon ($k = 1.67$, $R = 0.208$ kJ/kg·K, $c_p = 0.519$ kJ/kg·K).

**(a)** $$T_1 = T_0 - V_1^2/2c_p = 490 - 600^2/[(2)(519)] = 143 \text{ K}$$

**(b)** $N_M = V/c$ $\quad c = \sqrt{kRT}$ $\quad c_1 = \sqrt{(1.67)(208)(143)} = 223 \text{ m/s}$ $\quad (N_M)_1 = \frac{600}{223} = 2.69$

**(c)** $p_1 = p_0/\{1 + [(k - 1)/2](N_M)_1^2\}^{k/(k-1)} = 320/\{1 + [(1.67 - 1)/2](2.69)^2\}^{1.67/(1.67-1)} = 14.9$ kPa

**(d)** $\rho = p/RT$ $\quad \rho_1 = (14.9)/[(0.208)(143)] = 0.501 \text{ kg/m}^3$

$$\dot{m} = \rho_1 A_1 V_1 = (0.501)(0.2)(600) = 60.1 \text{ kg/s}$$

**16.123** Sketch a passage which will **(a)** increase the pressure in a subsonic flow isentropically, **(b)** increase the pressure in a supersonic flow isentropically, **(c)** increase the Mach number in a supersonic flow isentropically, **(d)** decrease the Mach number in a subsonic flow isentropically.

▮ See Fig. 16-6.

**Fig. 16-6**

**16.124** Air at 28 °C flows from a large tank through a convergent nozzle, which has an exit diameter of 10 mm. Discharge is to the atmosphere where atmospheric pressure is 96.5 kPa. Air pressure inside the tank is 40.0 kPa gage. What is the weight flow rate of air through the nozzle?

▮ $$(p_2/p_1)_c = [2/(k + 1)]^{k/(k-1)} = [2/(1.40 + 1)]^{1.40/(1.40-1)} = 0.5283$$

$p_2' =$ atmospheric pressure $= 96.5$ kPa (absolute) $\qquad p_1 = 40 + 96.5 = 136.5$ kPa (absolute)

$$p_2'/p_1 = 96.5/136.5 = 0.7070$$

Since the value of the ratio $p_2'/p_1$ of 0.7070 is greater than the value of the critical pressure ratio of 0.5283, flow through the nozzle will be subsonic and

$$G = A_2\sqrt{[2gk/(k - 1)]p_1\gamma_1[(p_2/p_1)^{2/k} - (p_2/p_1)^{(k+1)/k}]} \qquad A_2 = (\pi)(0.010)^2/4 = 0.00007854 \text{ m}^2$$

$\gamma = p/RT$ $\qquad \gamma_1 = 136.5/[(29.3)(28 + 273)] = 0.01548$ kN/m³ $\qquad p_2 = p_2' = 96.5$ kN/m²

$$G = (0.00007854)\sqrt{\frac{(2)(9.807)(1.40)}{1.40 - 1}(136.5)(0.01548)\left[\left(\frac{96.5}{136.5}\right)^{2/1.40} - \left(\frac{96.5}{136.5}\right)^{(1.40+1)/1.40}\right]}$$

$$= 0.000227 \text{ kN/s} \quad \text{or} \quad 0.227 \text{ N/s}$$

**16.125** Air at 75 °F flows from a large tank through a convergent nozzle, which has an exit diameter of 1.5 in. The discharge is to the atmosphere, where atmospheric pressure is 14.0 psia. Air pressure inside the tank is 25.0 psig. What is the weight flow rate of air through the nozzle?

▮ $$(p_2/p_1)_c = 0.5283 \quad \text{(as determined for air in Prob. 16.124)}$$

$p_2' = 14.0$ psia $\qquad p_1 = 25.0 + 14.0 = 39.0$ psia $\qquad p_2'/p_1 = 14.0/39.0 = 0.3590$

Since the value of the ratio $p_2'/p_1$ of 0.3590 is less than the value of the critical pressure ratio of 0.5283, flow through the nozzle will be sonic. However, the values substituted into the equation below for $p_1$ must be the pressure that makes the ratio $p_2'/p_1$ equal to the critical pressure ratio $(p_2/p_1)_c$.

$$G = (A_2p_1/\sqrt{T_1})\sqrt{(gk/R)[2/(k + 1)]^{(k+1)/(k-1)}} \qquad A_2 = (\pi)(1.5)^2/4 = 1.767 \text{ in}^2 = 0.01227 \text{ ft}^2$$

$$p_1 = p_2'/(p_2/p_1)_c = 14.0/0.5283 = 26.5 \text{ psia} \qquad T_1 = 75° + 460 = 535 °\text{R}$$

$$G = \{(0.01227)[(26.5)(144)]/\sqrt{535}\}\sqrt{[(32.2)(1.40)/53.3][2/(1.40 + 1)]^{(1.40+1)/(1.40-1)}} = 1.08 \text{ lb/s}$$

**16.126** Air at 308 °C flows from a large tank through a convergent nozzle, which has an exit diameter of 20 mm. Discharge is to the atmosphere where atmospheric pressure is 95.0 kPa. Air pressure inside the tank is 50.0 kPa gage. What is the weight flow of air through the nozzle?

$$(p_2/p_1)_c = [2/(k+1)]^{k/(k-1)} = [2/(1.40+1)]^{1.40/(1.40-1)} = 0.5283 \qquad p_2' = 95.0 \text{ kPa}$$

$$p_1 = 50 + 95.0 = 145.0 \text{ kPa} \qquad p_2'/p_1 = 95.0/145.0 = 0.6522$$

Since $[p_2'/p_1 = 0.6552] > [(p_2/p_1)_c = 0.5283]$, the flow through the nozzle will be subsonic:

$$G = A_2\sqrt{[(2gk)/(k-1)](p_1\gamma_1)[(p_2/p_1)^{2/k} - (p_2/p_1)^{(k+1)/k_1}]}$$

$$A_2 = (\pi)(0.02)^2/4 = 0.0003142 \text{ m}^2 \qquad \gamma = p/RT$$

$$\gamma_1 = 145.0/[(29.3)(30+273)] = 0.01633 \text{ kN/m}^3 \qquad p_2 = p_2' = 95.0 \text{ kN/m}^2$$

$$G = 0.0003142\sqrt{[(2)(9.807)(1.40)/(1.40-1)](145)(0.01633)(95.0/145.0)^{2/1.40} - (95.0/145.0)^{(1.40+1)/1.40}}$$

$$= 0.00100 \text{ kN/s} = 1.00 \text{ N/s}$$

**16.127** Air at 60 °F flows from a large tank through a convergent nozzle, which has an exit diameter of 1.0 in. The discharge is to the atmosphere, where atmospheric pressure is 14.5 psia. Air pressure inside the tank is 30.0 psig. What is the weight flow rate of air through the nozzle?

**▌** From Prob. 16.126, $(p_2/p_1)_c = 0.5283$, $p_2' = 14.5$ psia, $p_1 = 30.0 + 14.5 = 44.5$ psia, $p_2'/p_1 = 14.5/44.5 = 0.3258$. Since $[p_2'/p_1 = 0.3258] < [(p_2/p_1)_c = 0.5283]$, the flow through the nozzle will be sonic.

$$G = [A_2 p_1/\sqrt{T_1}]\sqrt{(gk/R)[2/(k+1)]^{(k+1)/(k-1)}} \qquad A = (\pi)(\tfrac{1}{12})^2/4 = 0.005454 \text{ ft}^2$$

$$p_1 = p_2'/(p_2/p_1)_c = 14.5/0.5283 = 27.45 \text{ psia} \qquad T_1 = 60 + 460 = 520 \text{ °R}$$

$$G = [(0.005454)(27.45)(144)/\sqrt{520}]\sqrt{[(32.2)(1.40)/53.3][2/(1.40+1)]^{(1.40+1)/(1.40-1)}} = 0.503 \text{ lb/s}$$

**16.128** Air at 80 °F flows from a large tank through a converging nozzle of 2.0-in exit diameter. The discharge is to an atmospheric pressure of 13.5 psia. Determine the flow through the nozzle for pressures within the tank of 5, 10, 15, and 20 psig. Assume isentropic conditions. Plot $G$ as a function of $p_1$. Assume that the temperature within the tank is 80 °F in all cases.

**▌** From Prob. 16.126 the critical pressure ratio for air is $(p_2/p_1)_c = 0.528$. If the flow at the throat is subsonic, $p_2/p_1 > (p_2/p_1)_c$. Thus for subsonic flow at the throat, $p_2/p_1 > 0.528$ and $p_3 = p_2$. So $p_3/p_1 > 0.528$ and $p_1 < p_3/0.528$.

Since $p_3 = 13.5$ psia, the flow at the throat will be subsonic if $p_1 < 25.6$ psia (12.1 psig) and sonic if $p_1 > 25.6$ psia (12.1 psig).

To find the flow rate for conditions where $p_1 < 12.1$ psig (subsonic flow at throat), we use

$$G = A_2\sqrt{(2g)[k/(k-1)]p_1\gamma_1[(p_2/p_1)^{2/k} - (p_2/p_1)^{(k+1)/k}]} \tag{1}$$

Substituting the appropriate value of $p_1$ into the equation and noting that for this condition $p_2 = p_3 = 13.5$ psia, we get: for $p_1 = 5$ psig (18.5 psia); $G = 1.20$ lb/s; for $p_1 = 10$ psig (23.5 psia), $G = 1.69$ lb/s. To find the

Fig. 16-7

flow rate for conditions where $p_1 > 12.1$ psig (sonic flow at the throat) we use

$$G' = g(A_2 p_1 / \sqrt{T_1}) \sqrt{(k/R)[2/(k+1)]^{(k+1)/(k-1)}} \tag{2}$$

We get: for $p_1 = 15$ psig (28.5 psia), $G' = 2.04$ lb/s; for $p_1 = 20$ psig (33.5 psia), $G' = 2.40$ lb/s. Substituting $p_1 = 25.6$ psia in Eq. (1) for subsonic flow gives $G = 1.84$ lb/s as does Eq. (2). This is the threshold point at which the flow in the throat changes from subsonic to sonic. When $p_1 = 25.6$ psia the flow rate as found from Eq. (2) is $G'_{p=25.6\,\text{psia}} = [32.2(0.0218)(144)(25.6)]/\sqrt{540}\sqrt{(1.4/1715)(2/2.4)^{2.4/0.4}} = 1.84$ lb/s. As $p_1$ is increased beyond 12.1 psig, sonic flow prevails at the throat and the flow rate increases linearly with $(p_1)_{\text{abs}}$ as indicated by Eq. (2). The variation of the flow rate with $p_1$ is shown in Fig. 16-7. Other information concerning various aspects of this problem can be found by applying the gas law ($pv = RT$ or $\gamma = gp/RT$), the equation of state ($pv^k = $ constant or $p/\gamma^k = $ constant), continuity ($G = \gamma AV$), and the energy equation. Applying these, for example, for the case where $p_1 = 5$ psig (18.5 psia) yields $\gamma_1 = 0.093$ lb/ft$^3$, $\gamma_2 = 0.0743$ lb/ft$^3$, $V_2 = 740$ fps, $T_2 = 491$ °R, and $p_2 = 13.5$ psia. Note that in this case $p_2 = p_3$.

**16.129** Refer to Prob. 16.128. If the pressure in the tank is 5 psig, find $G$, $p_2$, and $T_2$.

❚  Air   $p_1 = 5$ psig   $p_1 = 5 + p_{\text{atmos}} = 5 + 13.5 = 18.50$ psia

$$\gamma_1 = gp_1/RT_1 = (32.2)(18.50)(144)/[(17715)(540)] = 0.0926 \text{ lb/ft}^3$$

$$p_3/p_1 = 14.5/18.5 = 0.730 > 0.528 = (p_2/p_1)_c \text{ so } G = A_2\sqrt{2g[k/(k-1)]p_1\gamma_1[(p_2/p_1)^{2/k} - (p_2/p_1)^{(k+1)/k}]}$$

$$= (\pi/4)(\tfrac{2}{12})^2\sqrt{64.4(1.4/0.4)[(18.5)(144)(0.0926)][(13.5/18.5)^{2/1.4} - (13.5/18.5)^{2.4/1.4}]}$$

$$= 1.204 \text{ lb/s}.$$

$$V_2^2/2g = (p_2/\gamma_2)[k/(k-1)][(p_1/p_2)^{(k-1)/k} - 1] \tag{1}$$

$$G = \gamma_2 A_2 V_2 \tag{2}$$

$$\gamma_2 = gp_2/RT_2 \tag{3}$$

$$p_1/\gamma_1^k = p_2/\gamma_2^k \tag{4}$$

From (2)

$$\gamma_2 = G/A_2 V_2 = 1.204/(0.0218 V_2) = 55.0/V_2 \tag{5}$$

From (4) $p_2 = p_1(\gamma_2^k/\gamma_1^k) = [(18.5)(144)][\gamma_2^k/(0.0926)^{1.4}] = (2660\gamma_2^k)/0.0357 = 74\,500\gamma_2^k$. Substitute for $\gamma_2$ from (5): $p_2 = 74\,500(55/V_2)^{1.4} = 20.4 \times 10^6/V_2^{1.4}$. From (4): $p_1/p_2 = (\gamma_1/\gamma_2)^{1.4} = [0.0926/(55/V_2)]^{1.4} = 0.0001308V_2^{1.4}$. Substitute into (1):

$$V_2^2 = (64.4)[(20.4 \times 10^6)(1.4/0.4)/(V_2^{1.4})(55/V_2)][0.0001308V_2^{1.4})^{0.4/1.4} - 1]$$

$V_2^{2.4} = (83.6 \times 10^6)[0.0777V_2^{0.4} - 1]$; by $T$ and $E$, $V_2 = 741$ fps. From (2): $\gamma_2 = G/A_2 V_2 = 1.204/(0.0218)(741) = 0.0745$ lb/ft$^3$. From (4): $p_2 = (18.5)[(0.0742)^{1.4}/(0.0926)^{1.4}] = 13.57$ psia. From (3): $T_2 = gp_2/R\gamma_2 = (32.2)(13.57)(144)/[(1715)(0.0742)] = 492$ °R.

**16.130** Air discharges from a tank through a converging-diverging nozzle with a 2.0-in-diameter throat. Within the tank the pressure is 50 psia and the temperature is 80 °F, while outside the tank the pressure is 13.5 psia. The nozzle is to operate with supersonic flow throughout its diverging section with a 13.5-psia pressure at its outlet. Find the required diameter of the nozzle outlet. Determine the flow rate and the velocities and temperatures at sections 2 and 3. Assume isentropic flow.

❚  The pressure at the throat must be such that sonic velocity will occur there. Hence $p_2 = p_c = 0.528p_1 = 26.4$ psia.

The velocity at the outlet may be found from $V_3^2/2g = (RT_1/g)[k/(k-1)][1 - (p_3/p_1)^{(k-1)/k}] = [(1715)(540)/32.2](1.4/0.4)[1 - (13.5/50)^{0.4/1.4}] = 31\,400$ ft, $V_3 = 1420$ fps. The flow rate is computed from $G = (gA_2 p_1/\sqrt{T_1})\sqrt{(k/R)[2/(k+1)]^{(k+1)/(k-1)}}$, $G' = [(32.2)(0.0218)(50)(144)/\sqrt{540}]\sqrt{(1.4/1715)(2/2.4)^{2.4/0.4}} = 3.58$ lb/s. The temperature at 3 may be determined by using $V_3^2 - V_1^2 = 2c_p(T_1 - T_2)$, $(1420)^2 = 2(6000)(540 - T_3)$, $T_3 = 372$ °R $= -88$ °F. From the perfect-gas law $p_3/\gamma_3 = RT_3/g$. $\gamma_3 = (32.2)(13.5)(144)/[(1715)(372)] = 0.098$ lb/ft$^3$. Isentropic flow between 2 and 3 may be assumed, since the shock wave does not occur within that region. Thus $p_2/\gamma_2^{1.4} = p_3/\gamma_3^{1.4}$. $26.4/\gamma_2^{1.4} = 13.5/(0.098)^{1.4}$, $\gamma_2 = 0.158$ lb/ft$^3$. The velocity at 2 may now be computed: $V_2 = G/\gamma_2 A_2 = 3.58/[(0.158)(0.0218)] = 1040$ fps. The temperature at 2 results from $(1420)^2 - (1040)^2 = 2(6000)(T_2 - 372)$, $T_2 = 450$ °R $= -10$ °F. The area at 3 is computed from $A_3 = G/\gamma_3 V_3 = 3.58/[(0.098)(1420)] = 0.0257$ ft$^2$. Finally, $D_3 = 2.17$ in, the required outlet diameter.

Check for sonic velocity at throat: $c_2 = \sqrt{kRT_2} = \sqrt{(1.4)(1715)(450)} = 1040$ fps $= V_2$. With sonic velocity at the throat, if $D_3 < 2.17$ in, there will be supersonic flow throughout the tube and a shock wave will occur in the flow field downstream of the nozzle exit. If $D_3 > 2.17$ in, with sonic flow in the throat, in order to satisfy pressure conditions, a shock wave will occur in the tube somewhere between the throat and the nozzle exit.

**16.131** Work Prob. 16.130 with all data the same except for the pressure within the tank, which is 100 psia rather than 50 psia.

> ❙ Pressure at throat: $p_2 = p_c = 0.528(100) = 52.8$ psia. Find the velocity at outlet from $V_3 = (2g)[k/(k-1)](R/g)(T)[1 - (p_2/p_1)^{(k-1)/k}] = (64.4)(1.4/0.4)(1715/32.2)(540)[1 - (13.5/100)^{0.4/1.4}]$; $V_3 = 1680$ fps. By computation similar to that of Prob. 16.130, $G' = 7.20$ lb/s (exactly twice that of Prob. 16.130), $V_3^2 = [2k/(k-1)](RT_1)[1 - (T_3/T_1)]$, $1680^2 = (2.8/0.4)(1715)(540)[1 - (T_3/540)]$; $T_3 = 305\,°R = -155\,°F$. $\gamma = pg/RT$; $\gamma_3 = (13.5)(144)(32.2)/[(1715)(305)] = 0.1197$ pcf. $p/\gamma^k = $ constant (isentropic), therefore, $5.28/\gamma_2^{1.4} = 13.5/(0.1197)^{1.4}$; $\gamma_2 = 0.317$ pcf. $V_2 = G/\gamma_2 A_2 = 7.20/[(0.317)(0.0218)] = 1042$ fps, $T_2 = 448\,°R = -12\,°F$, $A_3 = G/\gamma_3 V_3 = 7.20/[(0.1197)(1680)] = 0.0358$ ft$^2$ and $D = 2.56$ in.

**16.132** Air within a tank at 100 °F flows isentropically through a 3-in-diameter convergent nozzle into a 14.2-psia atmosphere. Find the flow rate for a tank pressure of 20 psig.

> ❙ For air $(k = 1.4)$, $(p_2/p_1)_c = 0.528$; $p_2 = 14.2$ psia. If $p_2/p_1 > 0.528$, that is, $p_1 < 14.2/0.528 = 26.9$ psia, find $G$ from

$$G = A_2\sqrt{(2g)[k/(k-1)](p_1\gamma_1)[(p_2/p_1)^{2/k} - (p_2/p_1)^{(k+1)/k}]} \tag{1}$$

If $p_1 > 26.9$ psia, find $G'$ from

$$G' = g(A_2 p_1/\sqrt{T_1})\sqrt{(k/R)[2/(k+1)]^{(k+1)/(k-1)}} \tag{2}$$

Hence, for $p_1 = 20 + 14.2 = 34.2$ psia, $G' = (32.2)[(0.0491)(34.2)(144)/\sqrt{560}]\sqrt{(1.4/1715)(2/2.4)^{2.4/0.4}} = 5.44$ lb/s.

**16.133** Air flows at 150 °F from a large tank through a 1.5-in-diameter converging nozzle. Within the tank the pressure is 85 psia. Calculate the flow rate for external pressures of 10, 30, 50, and 70 psia. Assume isentropic conditions. Assume that the temperature within the tank is 150 °F in all cases. Compute also the temperature at the nozzle outlet for each condition.

> ❙ For air $(k = 1.4)$, $(p_2/p_1)_c = 0.528$. If $p_3 > 0.528(85) = 44.9$ psia the flow is subsonic and $p_3 = p_2$. If $p_3 < 0.528(85) = 44.9$ psia the flow is sonic and $p_3 < p_2$. For sonic flows (i.e., $p_3 < 44.9$ psia), $G'$ is given by

$$G' = (g)(A_2 p_1/\sqrt{T_1})\sqrt{(k/R)[2/(k+1)]^{(k+1)/(k-1)}} \tag{1}$$
$$G' = [(32.2\pi)(1.5)^2/4][85/\sqrt{(460+150)}]\sqrt{(1.4/1715)(2/2.4)^{2.4/0.4}} = 3.24 \text{ lb/s}$$

For subsonic flows (i.e., $p_3 > 44.9$ psia), $G$ is given by

$$G = A_2\sqrt{(2g)[k/(k-1)](p_1\gamma_1)[(p_2/p_1)^{2/k} - (p_2/p_1)^{(k+1)/k}]} \tag{2}$$
$$\gamma_1 = gp_1/RT_1 = (32.2)(85)(144)/[(1715)(610)] = 0.378 \text{ pcf}$$

Substitute into Eq. (2): $p_2 = p_3$ (express in psf), $A_2 = 0.01227$ ft$^2$, $\gamma_1 = 0.378$ pcf, $p_1 = 85$ psia $= 12\,240$ psf, $k = 1.4$. For $p_2 < 44.9$ psia $= 6470$ psf, $G' = 3.35$ lb/s [Eq. (1)]. For $p_2 = p_3 = 50$ psia $= 7200$ psf, $G = 3.22$ lb/s [Eq. (2)]. For $p_2 = p_3 = 70$ psia $= 10\,000$ psf, $G = 2.49$ lb/s [Eq. (2)]. Determination of temperatures: For each case substitute the corresponding values of $p_1$ and $p_2$ into $T_2 = T_1 - (p_1 g/R\gamma_1)[1 - (p_2/p_1)^{(k-1)/k}]$. Also substitute $T_1 = 150\,°F$, $\gamma_1 = 0.378$ pcf. Results are

| $p_1$, psia | $T_1$, °F | $p_3$, psia | $p_2$, psia | $G$ or $G'$, lb/s | $T_2$, °F |
|---|---|---|---|---|---|
| 85 | 150 | 10 | 44.9 | 3.35 | 49 |
| 85 | 150 | 30 | 44.9 | 3.35 | 49 |
| 85 | 150 | 50 | 50 | 3.22 | 64 |
| 85 | 150 | 70 | 70 | 2.49 | 117 |
| 85 | 150 | 85 | 85 | 0 | 150 |

**16.134** Air enters a converging-diverging nozzle at a pressure of 100 psia and a temperature of 100 °F. Neglecting the entrance velocity and assuming a frictionless process, find the Mach number at the cross section where the pressure is 30 psia.

$$T_2 = T_1(p_2/p_1)^{(k-1)/k} \quad \text{(from } pv^k = \text{const} \quad \text{and} \quad pv = RT) \qquad T_2 = 560(\tfrac{30}{100})^{0.4/1.4} = 397\,°\text{R}$$

$$V_2^2 = [2k/(k-1)](RT_1)[1-(T_2/T_1)] = 2(1.40/0.40)(1715)(560)[1-(\tfrac{397}{560})] \qquad V_2 = 1399 \text{ fps}$$

$$c = (kRT_2)^{1/2} = [1.4(1715)397]^{1/2} = 976 \text{ fps} \qquad N_M = V/c = \tfrac{1399}{976} = 1.433$$

**16.135** Carbon dioxide within a tank at 42 psia and 90 °F discharges through a convergent nozzle into a 14.2-psia atmosphere. Find the velocity, pressure, and temperature at the nozzle outlet. Assume isentropic conditions.

$$\quad (p_2/p_1)_c = [2/(k+1)]^{k/(k-1)} \qquad k = 1.28 \qquad (p_2/p_1)_c = (2/2.28)^{1.28/0.28} = 0.549$$

If $p_3 > 0.549p_1 = 0.549(42) = 23.1$ psia, the flow is subsonic. But $p_3 = 14.2$ psia, which is $< 23.1$ psia; therefore, flow is sonic. The pressure at the nozzle outlet is 23.1 psia. $T_2 = T_1 - (p_1 g/R\gamma_1)[1-(p_2/p_1)^{(k-1)/k}]$. Noting that $p_1 g/R\gamma_1 = T_1 = (460+90)$ °R: $T_2 = 90 - 550[1-(23.1/42)^{0.28/1.28}] = 22.6$ °F, $V_2^2 = [2k/(k-1)](RT_1)[1-(T_2/T_1)] = 2(1.28/0.28)(1123)(550)[1-(\tfrac{483}{550})]$; $V_2 = 829$ fps.

**16.136** In Prob. 16.135 if the pressure and temperature within the tank had been 18 psia and 100 °F, what would have been the velocity, pressure, and temperature at the nozzle outlet? Assume isentropic conditions.

$\quad$ As for solution to Prob. 16.135: $(p_2/p_1)_c = 0.549$. If $p_3 > 0.549p_1 = 9.88$ psia, the flow is subsonic. $p_3 = 14.2$ psia $> 9.88$ psia, therefore flow *is* subsonic. With subsonic flow $p_2 = p_3 = 14.2$ psia, $T_2 = T_1 - (p_1 g/R\gamma_1)[1-(p_2/p_1)^{(k-1)/k}] = 100 - 560[1-(14.2/18)^{0.219}] = 71.7$ °F, $V_2^2 = [2k/(k-1)](RT_1)[1-(T_2/T_1)] = (2)(1.28/0.28)(1123)(560)[1-(531.7/560)]$; $V_2 = 539$ fps.

**16.137** Air discharges from a large tank through a converging-diverging nozzle. The throat diameter is 3.0 in, and the exit diameter is 4.0 in. Within the tank the air pressure and temperature are 40 psia and 150 °F, respectively. Calculate the flow rate for external pressures of 39, 38, 36, and 30 psia. Assume no friction.

$\quad \gamma_1 = gp_1/RT_1 = (32.2)(40)(144)/[(1715)(610)] = 0.1773$ pcf. If sonic flow occurs,
$G' = (g)(A_2 p_1/\sqrt{T})\sqrt{(k/R)[2/(k+1)]^{(k+1)/(k-1)}} = (32.2)[(0.0491)(40)(144)/\sqrt{460+150}]\sqrt{(1.4/1715)(2.0/2.4)^{2.4/0.4}} = 6.10$ lb/s. For subsonic flow:
$G = A_2\sqrt{(2g)[k/(k-1)](p_1\gamma_1)[(p_2/p_1)^{2/k} - (p_2/p_1)^{(k+1)/k}]} =$
$(\pi/4)(\tfrac{4}{12})^2\sqrt{(2)(32.2)(1.4/0.4)(40)(144)(0.1773)[(p_3/40)^{2/1.4} - (p_3/40)^{2.4/1.4}]}$. By $T$ and $E$ using the above equation, $p_3 = p_2 = 36.75$ psia for $G = 6.10$ lb/s.

| $p_3$, psia | $G$, lb/s | $G'$, lb/s |
|---|---|---|
| 40 | 0 | |
| 39 | 3.49 | |
| 38 | 4.87 | |
| 36.75 | 6.10 | 6.10 |
| 36 | — | 6.10 |
| 30 | — | 6.10 |

**16.138** Air is to flow through a converging-diverging nozzle at 16 lb/s. At the throat the pressure, temperature, and velocity are to be 21 psia, 120 °F, and 480 fps, respectively. At the outlet the velocity is to be 200 fps. Determine the throat diameter. Assume isentropic flow.

$$\quad G = \gamma_2 A_2 V^2 \qquad \gamma = p/RT \qquad \gamma_2 = (21)(144)/[(53.3)(120+460)] = 0.09782 \text{ lb/ft}^3$$
$$16 = (0.09782)(\pi D^2/4)(480) \qquad D = 0.6587 \text{ ft} \quad \text{or} \quad 7.90 \text{ in}$$

**16.139** Air in a tank under a pressure of 150 psia and 80 °F flows out into the atmosphere through a 2.00-in-diameter converging nozzle. (*a*) Find the flow rate. (*b*) If a diverging section with an outlet diameter of 3.00 in were attached to the converging nozzle, what then would be the flow rate? Neglect friction.

$\quad$ (*a*) $p_3/p_2 = 14.7/150 = 0.098 < 0.528$; so flow is sonic: Use $G' = (g)[A_2 p_1/\sqrt{T_1}] \times \sqrt{(k/R)[2/(k+1)]^{(k+1)/(k-1)}}$, $G' = (32.2)[(0.0218)(150)(144)/\sqrt{460+80}]\sqrt{(1.4/1715)(2/2.4)^{2.4/0.4}} = 10.8$ lb/s.
(*b*) Attaching a diverging nozzle with sonic flow (and no change in $\Delta p$) will not change the flow rate.

**16.140** Repeat Prob. 16.139 for the case where the air within the tank is at 18 psia. Assume all other data to be the same.

▮ (a) $p_3/p_1 = 14.7/18 = 0.817 > 0.528$; so flow is subsonic. $\gamma_1 = gp_1/RT_1 = (32.2)(18)(144)/[(1715)(540)] = 0.09012$ pcf, $G = A_2\sqrt{(2g)[k/(k-1)](p_1\gamma_1)[(p_2/p_1)^{2/k} - (p_2/p_1)^{(k+1)/k}]} = (\pi/4)(\frac{2}{12})^2\sqrt{2(32.2)(1.4/0.4)(18)(144)(0.09012)(0.817^{2/1.4} - 0.817^{2.4/1.4})} = 1.03$ lb/s.

(b) Attaching a diverging nozzle with subsonic flow (when the air behaves like an incompressible fluid) *will* increase the flow rate. It only changes $A_2$ in the equation above. Thus $G = 1.03(3.00/2.00)^2 = 2.32$ lb/s *providing* subsonic flow still occurs in the throat. Check this by computing $G'$, which is $G_{max}$ if $p_1$ is not increased. $G' = (g)(A_2p_1/\sqrt{T_1})\sqrt{(k/R)[2/(k+1)]^{(k+1)/(k-1)}} = (32.2)(\pi/4)(2.0)^2(18/\sqrt{460+80})\sqrt{(1.4/1715)(2/2.4)^{2.4/0.4}} = 1.296$ lb/s. Thus, with the 3.0-in diameter outlet the flow at the throat *has* become sonic, and $G = 1.296$ lb/s.

**16.141** Air discharges from a large tank through a converging-diverging nozzle with a 20-mm-diameter throat into the atmosphere. The gage pressure and temperature in the tank are 705 kPa and 42 °C; the barometric pressure is 99.5 kPa. (a) Find the nozzle-tip diameter required for $p_3$ to be equal to the atmospheric pressure. For this case, what are the flow velocity, sonic velocity, and Mach number at the nozzle exit?

▮ $p_1 = 705 + 99.5 = 804.5$ kPa abs. Assume sonic velocity at the throat:

$$V_3^2/2g = (RT/g)[k/(k-1)][1 - (p_2/p_1)^{(k-1)/k}] \qquad V_3^2/19.62 = [287(315)/9.81](1.4/0.4)[1 - (99.5/804.5)^{0.4/1.4}]$$

$$V_3 = 533 \text{ m/s}$$

$$G' = (g)(A_2p_1/\sqrt{T_1})\sqrt{(k/R)[2/(k+1)]^{(k+1)/(k-1)}}$$

$$= [9.81(\pi)(0.01^2)(804\,500)/\sqrt{273+42}]\sqrt{(1.4/287)(2/2.4)^{2.4/0.4}} = 5.65 \text{ N/s}$$

$$V_3^2 - V_1^2 = 2c_p(T_1 - T_3) \qquad 533^2 - 0 = 2(1003)(315 - T_3) \qquad T_3 = 173.4 \text{ K}$$

$$\gamma_3 = gp_3/RT_3 = (9.81)(99.5)/[(0.287)(173.4)] = 19.61 \text{ N/m}^3$$

$$A_3 = G/\gamma_3V_3 = 5.65/(19.61)(533) = 0.000541 \text{ m}^2 = 541 \text{ mm}^2$$

$$D_3 = 26.3 \text{ mm} \qquad c = [kRT]^{1/2} = [1.4(287)173.4]^{1/2} = 264 \text{ m/s} \qquad M_3 = V_3/c_3 = \tfrac{533}{264} = 2.02$$

Thus sonic velocity occurs at the throat and the shock wave is located downstream of the nozzle tip.

**16.142** Helium enters a 100-mm-ID pipe from a converging-diverging nozzle at $N_M = 1.30$, $p = 14$ kPa abs, $T = 225$ K. Determine for isothermal flow (a) the maximum length of pipe for no choking, (b) the downstream conditions, and (c) the length from the exit to the section where $N_M = 1.0$ ($f = 0.016$, $k = 1.66$).

▮ (a)
$$(f/D)L_{max} = [(1 - kN_M^2)/kN_M^2] + \ln(kN_M^2)$$

$$0.016L_{max}/0.1 = \{[1 - 1.66(1.3)^2]/(1.66)(1.3)^2\} + \ln[1.66(1.3^2)]$$

from which $L_{max} = 2.425$ m.

(b) $\qquad p^{*t}/p = \sqrt{k}N_M \qquad p^{*t} = p\sqrt{k}N_M = 14\sqrt{1.66}(1.30) = 23.45$ kPa abs

The Mach number at the exit is $1/\sqrt{1.66} = 0.776$.

$$\int_V^{V^{*t}} \frac{dV}{V} = \frac{1}{2}\int_M^{1/\sqrt{k}} \frac{dN_M^2}{N_M^2} \qquad \text{or} \qquad \frac{V^{*t}}{V} = \frac{1}{\sqrt{k}N_M}$$

At the upstream section $V = N_M\sqrt{kRT} = (1.30)\sqrt{(1.66)(2077)(225)} = 1145$ m/s and $V^{*t} = V/(\sqrt{k}N_M) = 1145/(\sqrt{1.66})(1.30) = 683.6$ m/s.

(c) Substituting into the equation in (a) for $N_M = 1$, $(0.016/0.1)L'_{max} = [(1 - 1.66)/1.66] + \ln 1.66$ or $L'_{max} = 0.683$ m. $N_M = 1$ occurs 0.683 m from the exit.

**16.143** Helium discharges from a $\frac{1}{2}$-in-diameter converging nozzle at its maximum rate for reservoir conditions of $p = 4$ atm, $T = 25$ °C. What restrictions are placed on the downstream pressure? Calculate the mass flow rate and velocity of the gas at the nozzle.

$$\dot{m}_{max} = (A^*p_0/\sqrt{T_0})\sqrt{(k/R)[2/(k+1)]^{(k+1)/(k-1)}} \qquad A^* = (\pi)[(\tfrac{1}{2})(2.54)/100]^2/4 = 0.0001267 \text{ m}^2$$

$$p_0 = (4)(101\,310) = 405\,240 \text{ Pa} \qquad T_0 = 25 + 273 = 298 \text{ K} \qquad k = 1.66 \qquad R = 2077 \text{ N-m/(kg-K)}$$

$$\dot{m}_{max} = [(0.0001267)(405\,240)/\sqrt{298}]\sqrt{(1.66/2077)[2/(1.66+1)]^{(1.66+1)/(1.66-1)}} = 0.0473 \text{ kg/s}$$

$$V^* = \dot{m}_{max}/\rho^* A^* \qquad \rho^* = \rho_0 (p^*/p_0)^{1/k} \qquad \rho_0 = p_0/RT_0 = 405\,240/[(2077)(298)] = 0.655 \text{ kg/m}^3$$
$$p^*/p_0 = [2/(k+1)]^{k/(k-1)} = [2/(1.66+1)]^{1.66/(1.66-1)} = 0.488 \qquad p^* = (0.488)(405\,240) = 197\,757 \text{ Pa}$$
$$\rho^* = 0.655(197\,757/405\,240)^{1/1.66} = 0.425 \text{ kg/m}^3$$
$$V^* = 0.0473/(0.425)(0.0001267) = 878 \text{ m/s} \qquad p_{downstream} < p^*$$

**16.144** Air in a reservoir at 285 psia, $T = 295\,°F$, flows through a 2-in-diameter throat in a converging-diverging nozzle. For $N_M = 1$ at the throat, calculate $p$, $\rho$, and $T$ there.

▮ For critical conditions at the throat, $T^* = 0.833$, $p^* = 0.528$, and $\rho^* = 0.632$.

$$T_{throat} = (0.833)(295 + 460) = 629\,°R \qquad p_{throat} = (0.528)(285) = 150 \text{ psia}$$
$$\rho = p/RT \qquad \rho_0 = (285)(144)/[(1716)(295 + 460)] = 0.0317 \text{ slug/ft}^3$$
$$\rho_{throat} = (0.632)(0.0317) = 0.0200 \text{ slug/ft}^3$$

**16.145** What must be the velocity, pressure, density, temperature, and diameter at a cross section of the nozzle of Prob. 16.144 where $N_M = 2.5$?

▮
$$T_0/T = 1 + [(k-1)/2]N_M^2 \qquad (295 + 460)/T = 1 + [(1.40 - 1)/2](2.5)^2 \qquad T = 336\,°R$$
$$p_0/p = \{1 + [(k-1)/2]N_M^2\}^{k/(k-1)} \qquad 285/p = \{1 + [(1.40-1)/2](2.5)^2\}^{1.40/(1.40-1)} \qquad p = 16.7 \text{ psia}$$
$$\rho_0/\rho = \{1 + [(k-1)/2]N_M^2\}^{1/(k-1)} \qquad 0.0317/\rho = \{1 + [(1.40-1)/2](2.5)^2\}^{1/(1.40-1)} \qquad \rho = 0.00417 \text{ slug/ft}^3$$
$$N_M = V/\sqrt{kRT} \qquad 2.5 = V/\sqrt{(1.40)(1716)(336)} \qquad V = 2246 \text{ ft/s}$$
$$A/A^* = (1/N_M)[(5 + N_M^2)/6]^3 \qquad A/[(\pi)(\tfrac{2}{12})^2/4] = (1/2.5)[(5 + 2.5^2)/6]^3 \qquad A = 0.0575 \text{ ft}^2$$
$$\pi D^2/4 = 0.0575 \qquad D = 0.271 \text{ ft} \quad \text{or} \quad 3.25 \text{ in}$$

**16.146** Nitrogen in sonic flow at a 25-mm-diameter throat section has a pressure of 60 kPa abs, $T = -15\,°C$. Determine the mass flow rate.

▮
$$\rho^* = p^*/RT^* = 60/[(0.297)(273 - 15)] = 0.7830 \text{ kg/m}^3$$
$$V^* = \sqrt{kRT^*} = \sqrt{(1.40)(297)(273 - 15)} = 327.5 \text{ m/s}$$
$$\dot{m} = \rho^* A^* V^* = 0.7830[(\pi)(\tfrac{25}{1000})^2/4](327.5) = 0.126 \text{ kg/s}$$

**16.147** What is the Mach number for Prob. 16.146 at a 40-mm-diameter section in supersonic and in subsonic flow?

▮
$$A/A^* = (1/N_M)[(5 + N_M^2)/6]^3 \qquad [(\pi)(40)^2/4]/[(\pi)(25)^2/4] = (1/N_M)[(5 + N_M^2)/6]^3$$
$$(N_M)_1 = 0.234 \qquad (N_M)_2 = 2.47$$

**16.148** What diameter throat section is needed for critical flow of 0.6 lbm/s of carbon monoxide from a reservoir where $p = 300$ psia and $T = 100\,°F$?

▮
$$\dot{m} = \rho^* A^* V^* \qquad \rho_0 = p_0/RT_0 = (300)(144)/[(55.2)(100 + 460)] = 1.398 \text{ lbm/ft}^3$$
$$\rho^* = 0.634\rho_0 = (0.634)(1.398) = 0.8863 \text{ lbm/ft}^3 \qquad T^* = 0.833 T_0 = (0.833)(100 + 460) = 466\,°R$$
$$V^* = \sqrt{kRT^*} = \sqrt{(1.40)[(55.2)(32.2)](466)} = 1077 \text{ ft/s} \qquad 0.6 = (0.8863)(A^*)(1077)$$
$$A^* = 0.0006286 \text{ ft}^2 \qquad \pi D^2/4 = 0.0006286 \qquad D = 0.02829 \text{ ft} \quad \text{or} \quad 0.34 \text{ in}$$

**16.149** A supersonic nozzle is to be designed for airflow with $N_M = 3.0$ at the exit section, which is 200 mm in diameter and has a pressure of 6 kPa abs and temperature of $-80\,°C$. Calculate throat area and reservoir conditions.

▮
$$A/A^* = (1/N_M)[(5 + N_M^2)/6]^3 \qquad A_{exit} = (\pi)(\tfrac{200}{1000})^2/4 = 0.03142 \text{ m}^2$$
$$0.03142/A^* = (1/3.0)[(5 + 3.0^2)/6]^3 \qquad A^* = 0.007420 \text{ m}^2$$
$$T_0 = (T)\{1 + [(k-1)/2]N_M^2\} = (273 - 80)\{1 + [(1.40 - 1)/2](3.0)^2\} = 540 \text{ K}$$
$$p_0 = (p)\{1 + [(k-1)/2]N_M^2\}^{k/(k-1)} = (6)\{1 + [(1.40 - 1)/2](3.0)^2\}^{1.40/(1.40-1)} = 220.4 \text{ kPa}$$
$$\rho_0 = p_0/RT_0 = 220.4/[(0.287)(540)] = 1.42 \text{ kg/m}^3$$

**16.150** In Prob. 16.149, calculate the diameter of cross section for $N_M = 1.5$, 2.0, and 2.5.

**▮** Using data from Prob. 16.149,

$$\pi(D^*)^2/4 = 0.007420 \text{ m}^2 \qquad D^* = 0.09720 \text{ m} \qquad D = (D^*)(1/\sqrt{N_M})[(5+N_M^2)/6]^{1.5}$$

$$N_M = 1.5 \qquad D = (0.09720)(1/\sqrt{1.5})[(5+1.5^2)/6]^{1.5} = 0.1054 \text{ m}$$

$$N_M = 2.0 \qquad D = (0.09720)(1/\sqrt{2.0})[(5+2.0^2)/6]^{1.5} = 0.1263 \text{ m}$$

$$N_M = 2.5 \qquad D = (0.09720)(1/\sqrt{2.5})[(5+2.5^2)/6]^{1.5} = 0.158 \text{ m}$$

**16.151** Calculate the exit velocity and the mass rate of flow of nitrogen from a reservoir where $p = 3$ atm and $T = 20\,°C$, through a converging nozzle of 60 mm diameter discharging to the atmosphere.

**▮** $T^* = 0.833T_0 = (0.833)(20 + 273) = 244 \text{ K} \qquad V^* = \sqrt{kRT^*} = \sqrt{(1.40)(297)(244)} = 319 \text{ m/s} \qquad \dot{m} = \rho^*A^*V^*$

$$\rho^* = 0.634\rho_0 \qquad \rho_0 = p_0/RT_0 = (3)(101\,350)/[(297)(20 + 273)] = 3.494 \text{ kg/m}^3$$

$$\rho^* = (0.634)(3.494) = 2.215 \text{ kg/m}^3 \qquad A^* = (\pi)(\tfrac{60}{1000})^2/4 = 0.002827 \text{ m}^2$$

$$\dot{m} = (2.215)(0.002827)(319) = 2.00 \text{ kg/s}$$

**16.152** A converging nozzle has a throat area of 1 in$^2$ and stagnation air conditions of 120 lb/in$^2$ abs and 600 °R. Compute the exit pressure and mass flow if the back pressure is (**a**) 90 lb/in$^2$ abs and (**b**) 45 lb/in$^2$ abs. Assume $k = 1.4$.

**▮** We are given $p_0 = 120 \text{ lb/in}^2$ abs $= 17\,280 \text{ lb/ft}^2$, $T_0 = 600\,°R$. Hence $\rho_0 = p_0/RT_0 = 17\,280/(1717)(600) = 0.0168 \text{ slug/ft}^3$, $p^*/p_0 = 0.5283$, or $p^* = 63.4 \text{ lb/in}^2$ abs $= 9130 \text{ lb/ft}^2$.
(**a**) Since $p_b = 90 \text{ lb/in}^2$ abs $> p^*$, the flow is subsonic throughout. The throat Mach number is found from $p_e = p_b$ and

$$N_M^2 = 5[(p_0/p)^{2/7} - 1] \qquad (N_M)_e^2 = 5[(\tfrac{120}{90})^{2/7} - 1] = 0.4283 \qquad (N_M)_e = 0.6545$$

$$T_0/T_e = 1 + 0.2(N_M)_e^2 = 1 + 0.2(0.6545)^2 = 1.0857$$

$$\rho_0/\rho_e = [1 + 0.2(N_M)_e^2]^{2.5} = [1 + (0.2)(0.6545)^2]^{2.5} = 1.2281$$

Hence $T_e = 600/1.0857 = 553\,°R$, so that $a_e \approx 49(553)^{1/2} = 1152 \text{ ft/s}$. $\rho_e = 0.0168/1.2281 = 0.0137 \text{ slug/ft}^3$. The mass flow is thus given by $\dot{m} = \rho_e A_e V_e = \rho_e A_e[(N_M)a_e] = 0.0137(\tfrac{1}{144})[0.6545(1152)] = 0.0715 \text{ slug/s}$. The exit pressure equals the back pressure: $p_e = p_b = 90 \text{ lb/in}^2$ abs.
(**b**) Since $p_b = 45 \text{ lb/in}^2$ abs $< p^*$, the throat is choked. The exit pressure is sonic: $p_e = p^* = 63.4 \text{ lb/in}^2$ abs, $\dot{m} = \dot{m}_{\max} = 0.6847p_0A_t/(RT_0)^{1/2} = 0.6847(17\,280)(\tfrac{1}{144})/[1717(600)]^{1/2} = 0.0810 \text{ slug/s}$. Any back pressure less than or equal to $p^* = 63.4 \text{ lb/in}^2$ abs would cause this same choked condition. Notice that the 50 percent increase in throat Mach number from 0.6545 to 1.0 increases the mass flux only 13 percent, from 0.0715 to 0.0810 slug/s.

**16.153** A converging-diverging nozzle has a throat area of 0.002 m$^2$ and an exit area of 0.008 m$^2$. Air stagnation conditions are $p_0 = 1000 \text{ kPa}$ and $T_0 = 500 \text{ K}$. Compute the exit pressure and mass flow for (**a**) design condition and the exit pressure and mass flow if (**b**) $p_b \approx 300 \text{ kPa}$ and (**c**) $p_b \approx 900 \text{ kPa}$. Assume $\gamma = 1.4$.

**▮** (**a**) The design condition corresponds to supersonic isentropic flow at the given area ratio $A_e/A_t = 0.008/0.002 = 4.0$. We can find the design Mach number by $(N_M)_e = [216(A_1/A^*) - 254(A/A^*)^{2/3}]^{1/5}$, $(N_M)_{e,\text{design}} \approx [216(4.0) - 254(4.0)^{2/3}]^{1/5} \approx 2.95$, $p_0/p_e = [1 + 0.2(N_M)_e^2]^{3.5}$, $p_0/p_e = [1 + 0.2(2.95)^2]^{3.5} = 34.1$ or $p_{e,\text{design}} = 1000/34.1 = 29.3 \text{ kPa}$. Since the throat is clearly sonic at design conditions, $\dot{m}_{\text{design}} = \dot{m}_{\max} = 0.6847p_0A_t/(RT_0)^{1/2} = 0.6847(10^6)(0.002)/[287(500)]^{1/2} = 3.61 \text{ kg/s}$.
(**b**) For $p_b = 300 \text{ kPa}$ we are definitely far below the subsonic isentropic condition but we may even be below condition with a normal shock in the exit, where oblique shocks occur outside the exit plane. If it is, $p_e = p_{e,\text{design}} = 29.3 \text{ kPa}$ because no shock has yet occurred. To find out, compute condition $F$ by assuming an exit normal shock with $(N_M)_1 = 2.95$, that is, the design Mach number just upstream of the shock. $p_2/p_1 = [1/(k+1)][2k(N_M)_1^2 - (k-1)] = (1/2.4)[2.8(2.95)^2 - 0.4] = 9.99$ or $p_2 = 9.99p_1 = 9.99p_{e,\text{design}} = 293 \text{ kPa}$. Since this is less than the given $p_b = 300 \text{ kPa}$, there is a normal shock just upstream of the exit plane. The exit flow is subsonic and equals the back pressure $p_e = p_b = 300 \text{ kPa}$. Also $\dot{m} = \dot{m}_{\max} = 3.61 \text{ kg/s}$. The throat is still sonic and choked at its maximum mass flow.
(**c**) Finally, for $p_b = 900 \text{ kPa}$, we compute $(N_M)_e$ and $p_e$. Again $A_e/A_t = 4.0$ for this condition, with a subsonic $(N_M)_e$: $(N_M)_e = [1 + 0.27(A/A^*)^{-2}]/(1.728)(A/A^*)$, $(N_M)_e(C) \approx [1 + 0.27/(4.0)^2]/(1.728)(4.0) = 0.147$. Then the isentropic exit-pressure ratio for this condition is $p_0/p_e = [1 + 0.2(0.147)^2]^{3.5} = 1.0152$ or $p_e = 1000/1.0152 = 985 \text{ kPa}$. The given back pressure of 900 kPa is less than this value. Thus for this case there is a normal shock just downstream of the throat and the throat is choked: $p_e = p_b = 900 \text{ kPa}$, $\dot{m} = \dot{m}_{\max} = 3.61 \text{ kg/s}$. For this large

exit-area ratio the exit pressure would have to be larger than 985 kPa to cause a subsonic flow in the throat and a mass flow less than maximum.

**16.154** Steam in a tank at 450 °F and 100 psia exhausts through a converging nozzle of throat area 0.1 in$^2$ to a 1-atm environment. Compute the initial mass flow rate. Assume an ideal gas with $k = 1.33$.

▌ It is sure to be choked since $p_0/p_e = 7$; therefore, $A_e = A^*$. $\dot{m} = \dot{m}_{max} = (k^{1/2})[2/(k + 1)]^{(k+1)/2(k-1)}(A^* p_0/\sqrt{RT_0})$, $\dot{m} = 1.33^{1/2}[2/(1.33 + 1)]^{(1.33+1)/2(1.33-1)}\{(0.1/144)[(100)(144)]/\sqrt{(2759)(450 + 460)}\} = 0.00424$ slug/s.

**16.155** Air in a tank at 700 kPa and 20 °C exhausts through a converging nozzle of throat area 65 mm$^2$ to a 1-atm environment. Compute the initial mass flow rate.

▌ It is sure to be choked since $p_0/p_e = 7$; therefore, $A_e = A^*$. $\dot{m} = \dot{m}_{max} = (k^{1/2})[2/(k + 1)]^{(k+1)/2(k-1)}(A^* p_0/\sqrt{RT_0})$, $\dot{m} = 1.40^{1/2}[2/(1.40 + 1)]^{(1.40+1)/2(1.40-1)}\{(65 \times 10^{-6})[(700)(1000)]/\sqrt{(287)(20 + 273)}\} = 0.107$ kg/s.

**16.156** At a point upstream of the throat of a converging-diverging nozzle the properties are $V_1 = 190$ m/s, $T_1 = 305$ K, and $p_1 = 120$ kPa. If the exit flow is supersonic, compute, from isentropic theory, $\dot{m}$ and $A_1$. The throat area is 35 cm$^2$.

▌ $A_1/A^* = (1 + 0.2N_M^2)^3/1.728N_M$    $c = \sqrt{kRT} = \sqrt{(1.40)(287)(305)} = 350$ m/s    $N_M = V/c = \frac{190}{350} = 0.543$

Since the exit flow is supersonic, $A^* = A_{throat} = 35$ cm$^2$.

$$A_1/35 = [1 + (0.2)(0.543)^2]^3/[(1.728)(0.543)] \qquad A_1 = 44.3 \text{ cm}^2$$
$$\rho_1 = p_1/RT = 120/[(0.287)(305)] = 1.371 \text{ kg/m}^3 \qquad \dot{m} = \rho_1 A_1 V_1 = (1.371)(44.3 \times 10^{-4})(190) = 1.15 \text{ kg/s}$$

**16.157** In transonic wind tunnel testing the small area decrease caused by model blockage can be significant. Suppose the test section Mach number is 1.10 and the section area is 1 m$^2$ for air with $T_0 = 21$ °C. Find, according to one-dimensional theory, the percentage change in test-section velocity caused by a model of cross section 0.005 m$^2$ (a 0.5 percent blockage).

▌

$$T_1 = T_0/(1 + 0.2N_M^2) = (21 + 273)/[1 + (0.2)(1.10)^2] = 237 \text{ K} \qquad c = \sqrt{kRT}$$
$$c_1 = \sqrt{(1.40)(287)(237)} = 309 \text{ m/s} \qquad V_1 = c_1(N_M)_1 = (309)(1.1) = 340 \text{ m/s} \qquad \text{(unblocked)}$$
$$A_1/A^* = (1 + 0.2N_M^2)^3/1.728N_M \qquad 1/A^* = [1 + (0.2)(1.1)^2]^3/[(1.728)(1.1)] \qquad A^* = 0.9921 \text{ m}^2$$

If $A_1$ is blocked by the 0.5 percent model, $A_1 = 0.995$ m$^2$. $A_1/A^* = 0.995/0.9921 = 1.00292$. For the blocked condition,

$$A_1/A^* = (1 + 0.2N_M^2)^3/1.728N_M \qquad 1.00292 = (1 + 0.2N_M^2)^3/1.728N_M \qquad N_M = 1.011$$
$$T_1 = (21 + 273)/[1 + (0.2)(1.011)^2] = 244 \text{ K} \qquad c_1 = \sqrt{(1.40)(287)(244)} = 313 \text{ m/s}$$
$$V_1 = (313)(1.011) = 316 \text{ m/s} \qquad \text{(blocked)}$$

Thus a 0.5 percent decrease in area causes a $(340 - 316)/340 = 0.071$, or 7.1 percent decrease in velocity for this transonic flow.

**16.158** In flow of air in a converging-diverging duct with a supersonic exit, the throat area is 9 cm$^2$ and the throat pressure is 316 kPa. Find the pressure on either side of the throat where $A = 26$ cm$^2$.

▌ Supersonic exit means $A^* = A_{throat} = 9$ cm$^2$ and $p^* = 316$ kPa.

**Subsonic side:**

$$A_1/A^* = (1 + 0.2N_M^2)^3/1.728N_M \qquad 26/9 = (1 + 0.2N_M^2)^3/1.728N_M \qquad N_M = 0.204$$
$$p_1 = p_0/(1 + 0.2N_M^2)^{k/(k-1)} \qquad p_0 = p^*/[2/(k + 1)]^{k/(k-1)} = 316/[2/(1.40 + 1)]^{1.40/(1.40-1)} = 598 \text{ kPa}$$
$$p_1 = 598/[1 + (0.2)(0.204)^2]^{1.40/(1.40-1)} = 581 \text{ kPa}$$

**Subsonic side:**

$$A_1/A^* = (1 + 0.2N_M^2)^3/1.728N_M \qquad 26/9 = (1 + 0.2N_M^2)^3/1.728N_M \qquad N_M = 2.60$$
$$p_1 = 598/[1 + (0.2)(2.60)^2]^{1.40/(1.40-1)} = 30.0 \text{ kPa}$$

**Fig. 16-8**

•

**16.159** A force $F = 2.4$ kN pushes a piston through a 12-cm-diameter insulated cylinder containing air at 21 °C, as in Fig. 16-8. The exit diameter is 4 mm and the external pressure is 1 atm. Estimate **(a)** $V_e$, **(b)** $V_p$, and **(c)** $\dot{m}_e$.

∎ **(a)** $\qquad V_e = \{[2k/(k+1)](RT)\}^{1/2} = \{[(2)(1.40)/(1.40+1)](287)(21+273)\}^{1/2} = 314$ m/s

**(b)** $\qquad\qquad V_p = (V_e)(A_e/A_p) = (314)(4/120)^2 = 0.349$ m/s

**(c)** $\quad \dot{m}_e = \dot{m}_{max} = 0.6847 p_0 A^*/(RT_0)^{1/2} \qquad p_0 = F/A_p + p_{atm} = 2.4/[\pi(0.12)^2/4] + 101.35 = 313.6$ kPa

$$A^* = (\pi)(0.004)^2/4 = 0.0000126 \text{ m}^2$$

$$\dot{m}_e = (0.6847)(313\,600)(0.0000126)/[(287)(21+273)]^{1/2} = 0.00931 \text{ kg/s}$$

**16.160** Air flows steadily and subsonically from a tank at 21 °C through a nozzle of exit area 18 cm². Compute **(a)** exit Mach number, **(b)** exit velocity, **(c)** tank pressure (assume 1 atm exit pressure).

∎ **(a)** Assume an ideal gas, $\rho_e V_e^2 = \gamma p_e (N_M)_e^2$, $\gamma p_e A_e (N_M)_e^2 = 140$, $(N_M)_e = \sqrt{140/[(1.40)(101\,350)(18 \times 10^{-4})]} = 0.740$.

**(b)** $\qquad V_e = c_e N_M \qquad T_e = T_0/(1 + 0.2 N_M^2) = (21 + 273)/[1 + (0.2)(0.740)^2] = 265$ K

$\qquad c_e = \sqrt{kRT} = \sqrt{(1.40)(287)(265)} = 326$ m/s $\qquad V_e = (326)(0.740) = 241$ m/s

**(c)** $\quad p_0 = (p_e)\{1 + [(k-1)/2](N_M)_e^2\}^{k/(k-1)} = (101.350)\{1 + [(1.40-1)/2](0.740)^2\}^{1.40/(1.40-1)} = 146$ kPa abs

**16.161** Air flows through a converging-diverging nozzle from a reservoir, where $p_0 = 390$ kPa. The throat area is 8 cm². A normal shock stands in the duct where $A = 12$ cm². Compute the pressure just downstream of this shock.

∎ $\qquad A/A^* = (1 + 0.2 N_M^2)^3/1.728 N_M = 12/8 = 1.50 \qquad 1.50 = (1 + 0.2 N_M^2)^3/1.728 N_M \qquad N_M = 1.85$

$$p_1 = p_0/(1 + 0.2 N_M^2)^{k/(k-1)} = 390/[1 + (0.2)(1.85)^2]^{1.40/(1.40-1)} = 62.9 \text{ kPa}$$

$$p_2/p_1 = [1/(k+1)][2k(N_M)_1^2 - (k-1)] = [1/(1.40+1)][(2)(1.40)(1.85)^2 - (1.40-1)] = 3.826$$

$$p_2 = (3.826)(62.9) = 241 \text{ kPa}$$

**16.162** Air in a large tank at 102 °C and 152 kPa exhausts to the atmosphere ($p_{atm} = 101.3$ kPa) through a converging nozzle with a 6-cm² throat area. Compute the exit mass flow.

∎ $\qquad\qquad \dot{m} = \rho_e A_e V_e \qquad \rho_e = \rho_0/[1 + (0.2)(N_M)_e^2]^{2.5}$

$$\rho_0 = p_0/RT_0 = 152/[(0.287)(101.3 + 273)] = 1.415 \text{ kg/m}^3$$

$$p_0/p_a = [1 + (0.2)(N_M)_e^2]^{3.5} \qquad \tfrac{152}{101.3} = [1 + (0.2)(N_M)_e^2]^{3.5} \qquad (N_M)_e = 0.784$$

$$\rho_e = 1.415/[1 + (0.2)(0.784)^2]^{2.5} = 1.059 \text{ kg/m}^3$$

$$T_e = T/[1 + 0.2(N_M)_e^2] = (102 + 273)/[1 + (0.2)(0.784)^2] = 334 \text{ K}$$

$$c_e = \sqrt{kRT_e} = \sqrt{(1.40)(287)(334)} = 366 \text{ m/s} \qquad V_e = (c_e)(N_M)_e = (366)(0.784) = 287 \text{ m/s}$$

$$\dot{m} = (1.059)(6 \times 10^{-4})(287) = 0.182 \text{ kg/s}$$

**16.163** Air flows through a converging-diverging nozzle between two reservoirs; a manometer is located at the throat, as shown in Fig. 16-9. Estimate the downstream reservoir pressure.

**Fig. 16-9**

❚ Assume critical flow in the throat: $p_t = p^* = 0.528p_0 = (0.528)(301) = 158.9$ kPa. Guessing $\rho_{exit} = 1.2$ kg/m³, we have:

$$p^* - p_e = (\rho_M - \rho_e)(gh) \qquad 158\,900 - p_e = (13\,600 - 1.2)(9.807)(0.18)$$
$$p_e = 134\,900 \text{ Pa} = 134.9 \text{ kPa} \qquad \text{(nonisentropic flow)}$$

**16.164**  Air in a tank at 121 kPa and 301 K exhausts to the atmosphere through a 5-cm²-throat converging nozzle at a rate of 0.12 kg/s. What is the atmospheric pressure? What is the maximum mass flow possible at low atmospheric pressure?

❚ $\dot{m}_{max} = 0.6847p_0A^*/(RT_0)^{1/2} = (0.6847)[(121)(1000)](5 \times 10^{-4})/[(287)(300)]^{1/2} = 0.141$ kg/s. Therefore, the given $\dot{m} = 0.12$ kg/s is less than critical and we have a subsonic exit.

$$\dot{m} = \rho_e A_e V_e = (\rho_0)(\rho_e/\rho_0)(A_e)(N_M)_e(c_0)(c_e/c_0) = (\rho_0 A_e c_0)(N_M)_e[1 + 0.2(N_M)_e^2]^{-3}$$
$$\rho_0 = p_0/RT_0 = 121/[(0.287)(301)] = 1.401 \text{ kg/m}^3 \qquad c_0 = \sqrt{kRT_0} = \sqrt{(1.40)(287)(301)} = 347.8 \text{ m/s}$$
$$0.12 = [(1.401)(5 \times 10^{-4})(347.8)](N_M)_e[1 + 0.2(N_M)_e^2]^{-3} \qquad (N_M)_e = 0.619 \qquad \text{(by trial and error)}$$
$$p_e = 121/[1 + (0.2)(0.619)^2]^{3.5} = 93.4 \text{ kPa}$$

**16.165**  A supersonic nozzle with an exit area of 10 cm² discharges air at $N_M = 2.5$, $p = 101$ kPa, and $T = 302$ K. Compute (a) $p_0$, (b) $T_0$, and (c) $\dot{m}$.

❚ (a) $$p_0 = (p)(1 + 0.2N_M^2)^{3.5} = 101[1 + (0.2)(2.5)^2]^{3.5} = 1726 \text{ kPa}$$

(b) $$T_0 = (T)(1 + 0.2N_M^2) = 302[1 + (0.2)(2.5)^2] = 680 \text{ K}$$

(c) At $(N_M)_e = 2.5$, $A_e/A^* = 2.637$, $A^* = 10/2.637 = 3.79$ cm².

$$\dot{m} = \dot{m}_{max} = 0.6847p_0A^*/(RT_0)^{1/2}$$
$$\dot{m} = 0.6847[(1726)(1000)](3.79 \times 10^{-4})/[(287)(680)]^{1/2} = 1.014 \text{ kg/s}$$

**16.166**  A perfect gas (not air) at 305 kPa expands isentropically through a supersonic nozzle with an exit area 5 times its throat area. If the exit Mach number is 3.8, what is the exit pressure of the gas?

❚ First find the specific-heat ratio:

$$A_e/A^* = [1/(N_M)_e]\{[2 + (k - 1)(N_M)_e^2]/(k + 1)\}^{(k+1)/(2)(k-1)}$$
$$5 = (1/3.8)\{[2 + (k - 1)(3.8)^2]/(k + 1)\}^{(k+1)/(2)(k-1)} \qquad k = 1.67 \qquad \text{(by trial and error)}$$

Then, $p_e = p_0/\{1 + [(k - 1)/2](N_M)_e^2\}^{k/(k-1)} = 305/\{1 + [(1.66 - 1)/2](3.8)^2\}^{1.66/(1.66-1)} = 3.72$ kPa.

**16.167**  Air with $p_0 = 300$ kPa and $T_0 = 500$ K flows through a converging–diverging nozzle with throat area of 1 cm² and exit area of 4 cm² into a receiver tank. The mass flow is 195.2 kg/h. For what range of receiver pressure is this mass flow possible?

❚ $\dot{m} = 195.2/3600 = 0.0542$ kg/s, $\dot{m}_{max} = 0.6847p_0A^*/(RT_0)^{1/2} = 0.6847[(3 \times 10^5)(1 \times 10^{-4})]/[(287)(500)]^{1/2} = 0.0542$ kg/s. Therefore the throat is choked, and $p_b$ can lie anywhere between $p_b = 0$ and $p_e$ for isentropic subsonic exit: $A_e/A^* = 4.00$, $(N_M)_e = 0.147$, $(p_e)_{max} = p/[1 + (0.2)(N_M)_e^2]^{3.5} = 300/[1 + (0.2)(0.147)^2]^{3.5} = 296$ kPa. Thus the possible exit pressure range is $0 \leq p_b \leq 296$ kPa.

**16.168**  A rocket has a thrust of 900 000 lb when operating at its design point ($p_e = p_a$). If the chamber pressure and temperature are 500 psia and 5000 °R and the gas approximates air with $k = 1.4$, compute the throat diameter of the engine.

■ $F = \rho_e A_e V_e = k p_e A_e (N_M)_e^2$     $p_0/p_e = [1 + (0.2)(N_M)_e^2]^{3.5}$     $500/14.7 = [1 + (0.2)(N_M)_e^2]^{3.5}$     $(N_M)_e = 2.95$

$\qquad$ $900\,000 = 1.4[(14.7)(144)](A_e)(2.95)^2$     $A_e = 34.90 \text{ ft}^2$     $\pi D_e^2/4 = 34.90$     $D_e = 6.67 \text{ ft}$

At $(N_M)_e = 2.95$, $A_e/A^* = [1 + (0.2)(2.95)^2]^3/[(1.728)(2.95)] = 4.038$, $A^* = A_t = 34.90/4.038 = 8.643 \text{ ft}^2$, $\pi D_t^2/4 = 8.643$, $D_t = 3.32 \text{ ft}$.

**16.169**  A nozzle for an ideal rocket is to operate at an altitude of 15.25 km, where the pressure is 11.60 kPa, and is to give a 6.67-kN thrust when the chamber pressure is 1345 kPa and the chamber temperature is 2760 °C. Determine the throat and exit areas and the exit velocity and temperature. Take $k = 1.4$ and $R = 0.355 \text{ kJ/kg} \cdot \text{K}$, and assume the exit pressure to be the ambient pressure.

■ We have $p/p_0$ for the exit, $p/p_0 = 11.60/1345 = 0.008265$. From Table A-16, we see that $M_{\text{exit}} = 3.8$, and we have an area ratio $A_{\text{exit}}/A_{\text{throat}} = 8.95$. Thus, we get for the exit temperature $T_e$: $T_e/T_0 = 0.257$, $T_e = (2760 + 273)(0.257) = 779 \text{ K} = 506 °C$. We can now determine the exit velocity: $V_e = M_e c = M_e \sqrt{kRT_e} = 3.8\sqrt{(1.4)(355)(779)} = 2364 \text{ m/s}$. To ascertain the *throat* and *exit* areas, we must consider the thrust. Using a control volume comprising the interior of the combustion chamber and nozzle and considering this control volume to be inertial we have from linear momentum considerations

$$6670 = (\rho_e V_e A_e)V_e = \rho_e A_e (2364^2) \qquad (1)$$

By the ideal gas law, $p_e = \rho_e R T_e = 11\,600/(355)(779) = 0.04195 \text{ kg/m}^3$. Hence, from Eq. (1), $A_e = 6670/(2364^2)(0.04195) = 0.02845 \text{ m}^2$ and so $A^* = 0.0285/8.95 = 0.00318 \text{ m}^2$.

**16.170**  The inlet velocity of an isentropic diffuser is 305 m/s and the undisturbed pressure and temperature are 34.5 kPa abs and 235 °C, respectively. If the pressure is increased by 30 percent at the exit of the diffuser, determine the exit velocity and temperature. Use tables.

■ Find $(N_M)$ first:

$$(N_M)_1 = V_1/\sqrt{kRT_1} = 305/\sqrt{(1.4)(287)(235 + 273)} = 0.675 \qquad p/p_0 = 0.737$$
$$p_2/p_0 = (p_2/p)(p/p_0) = (1.30)(0.737) = 0.958$$
$$(N_M)_2 = 0.249 \qquad T_2/T_0 = 0.988$$

But $T_1/T_0 = 0.916$, $T_0 = 0.508/0.916 = 555$. Hence, $T_2 = (555)(0.988) = 548 \text{ K}$. Also $V_2 = (0.249)\sqrt{(1.4)(287)(1)(548)} = 116.8 \text{ m/s}$.

**16.171**  Air is kept in a tank at pressure $p_0 = 0.689 \text{ MPa}$ abs and temperature $T_0 = 17 °C$. If one allows the air to issue out in a one-dimensional isentropic flow, what is the greatest possible flow per unit area? What is the flow per unit area at the exit of the nozzle where $p = 0.101325 \text{ MPa}$?

■ $\qquad\qquad\qquad p/p_0 = 0.147 \qquad (N_M)_e = 1.91$

For the largest $G$ we go to the throat:

$$G^* = \sqrt{k/R}(p_0/\sqrt{T_0})[2/(k + 1)]^{(k+1)/[2(k-1)]} = \sqrt{1.4/287}(0.689 \times 10^6/\sqrt{290})(2/2.4)^{2.4/0.8} = 1635 \text{ kg/m}^2 \cdot \text{s}$$

At the exit, $G = \sqrt{k/R}(p_0/\sqrt{T_0})(N_M/\{1 + [(k-1)/2]N_M^2\}^{(k+1)/[2(k-1)]}) = \sqrt{1.4/287}(0.689 \times 10^6/\sqrt{290}) \times (1.91/\{1 + [(1.4-1)/2](1.91)^2\}^{2.4/0.8}) = 1043 \text{ kg/m}^2 \cdot \text{s}$.

**16.172**  Determine the exit area and the exit velocity for isentropic flow of a perfect gas with $k = 1.4$ in the nozzle shown in Fig. 16-10. How small can we make the exit area and still have isentropic flow with the given conditions entering the nozzle? Use tables.

$p_1 = 208 \text{ kPa}$
$\rho_1 = 1.3 \text{ kg/m}^3$
$V_1 = 47.3 \text{ m/s}$
$A_1 = 0.1 \text{ m}^2$

$p_2 = 172.5 \text{ kPa}$     **Fig. 16-10**

▮ First find $(N_M)_1$: $N_M = V/c$, $c = \sqrt{kp/\rho} = \sqrt{(1.4)(208\,000)/1.3} = 473$ m/s, $(N_M)_1 = 47.3/473 = 0.1$. From Table A-16, $A/A^* = 5.82$, $A^* = 0.1/5.82 = 0.0172$ m². Also $p/p_0 = 0.993$, $p_0 = 208/0.993 = 209.47$ kPa. Now go to exit: $p/p_0 = 172.5/209.47 = 0.824$. Hence, $(N_M)_e = 0.53$ and $A_e/A^* = 1.28$, $A_e = (0.0172)(1.28) = 0.0220$ m². The smallest possible area is $A^* = 0.0172$ m².

**At $N_M = 0.1$:**

$$\rho_1/\rho_0 = 0.995 \qquad \rho_0 = 1.3/0.995 = 1.31 \text{ kg/m}^3$$

**At $(N_M)_e = 0.53$ (exit):**

$$\rho_2/\rho_0 = 0.871 \qquad \rho_2 = (0.871)(1.31) = 1.14 \text{ kg/m}^3$$

**From continuity:**

$$\rho_1 V_1 A_1 = \rho_2 V_2 A_2 \qquad (1.3)(47.3)(0.1) = (1.14)V_2(0.0220) \qquad V_2 = 245 \text{ m/s}$$

**16.173** Determine the exit area in Prob. 16.172 for a nozzle efficiency of 90 percent. Take $c_p = 1.005$ kJ/kg · K, $k = 1.4$, and $R = 0.287$ kJ/kg · K.

▮ We shall need $(T_2)_{\text{isen}}$. To compute $(T_2)_{\text{isen}}$ we use the result that $(N_M)_1 = 0.1$ from Prob. 16.172. Hence, $T_1 = (T_0)(0.998)$, $T_0 = [208/(1.3)(0.287)]0.998 = 556$ K, $T_1 = 555$ K. Since $(N_M)_2 = 0.53$ from Prob. 16.172 we have $T_2/T_0 = 0.947$, $T_2 = 527$ K, $\eta = (V_2^2/2)_{\text{act}}/[V_1^2/2 + c_p(T_1 - T_2)]_{\text{isen}}$, $0.9 = (V_2^2/2)_{\text{act}}/[47.3^2/2 + (1005)(555 - 527)]$, $(V_2)_{\text{act}} = 229$ m/s. We need $(\rho_2)_{\text{act}}$ now. For this we must evaluate $(T_2)_{\text{act}}$. If there were no "reheating" then $V_2^2/2 = 30\,259$ instead of $27\,233$. Thus we can expect an increase in exit temperature due to reheating of $\Delta T_2 = (30\,259 - 27\,233)/(1005) = 3.01$ K. Hence $(T_2)_{\text{act}} = 527 + 3 = 530$ K. Using the equation of state: $\rho_2 = 172\,500/(287)(530) = 1.134$ kg/m³. From continuity we get $(1.3)(0.1)(47.3) = (1.134)(229)(A_2)$, $A_2 = 0.0237$ m².

**16.174** A rocket has an area ratio $A_{\text{exit}}/A^*$ of 3.5 for the nozzle, and the stagnation pressure is 5200 kPa abs. Fuel burns at the rate of 46 kg/s and the stagnation temperature is 2872 °C. Calculate the throat area and exit area. Take $R = 0.355$ kJ/kg · K and $k = 1.4$.

▮ Using Table A-16 we get for $A_e/A^* = 3.5$, $(N_M)_2 = 2.80$, $T_2/T_0 = 0.389$, $p_2/p_0 = 0.037$. Hence $T_2 = (0.389)(2872 + 273) = 1223$ K, $p_2 = (0.037)(5200) = 192$ kPa. Then

$$\rho_2 = p_2/RT_2 = 192/(0.355)(1223) = 0.442 \text{ kg/m}^3$$
$$V_2 = \sqrt{kRT_2}(N_M)_2 = \sqrt{(1.4)(355)(1223)}(2.80) = 2183 \text{ m/s}$$

By continuity, $A_2 = 46/(0.442)(2183) = 0.0477$ m². Finally, $A^* = 0.0477/3.5 = 0.01363$ m².

**16.175** A nozzle expands air from a pressure $p_0 = 180$ lb/in² abs and temperature $T_0 = 100$ °F to a pressure of 18 lb/in² abs. If the mass flow $w$ is 50 lbm/s, what is the throat area and the exit area? Take $k = 1.4$ and $R = 53.3$ ft-lb/(lbm-°R).

▮ $p_e/p_0 = \frac{18}{180} = 0.1$. From Table A-16:

$$(N_M)_e = 2.16 \qquad T_e/T_0 = 0.517 \qquad T_e = (560)(0.517) = 289.5 \text{ °R}$$
$$\rho_e/\rho_0 = 0.192 \qquad A_e/A^* = 1.94$$
$$pv = RT \qquad \rho_e = p_e/RT_e = (18)(144)/[(53.3)(32.2)(289.5)] = 5.217 \times 10^{-3} \text{ slug/ft}^3$$
$$V_e = (N_M)_e c_e \qquad c_e = \sqrt{kRT_e} = \sqrt{(1.4)(53.3)(32.2)(289.5)} = 834 \text{ ft/s}$$
$$V_e = (2.16)(834) = 1801 \text{ ft/s}$$

**Continuity:**

$$(50/g) = \rho_e V_e A_e = (5.217 \times 10^{-3})(1801)(A_e) \qquad A_e = 0.1653 \text{ ft}^2$$
$$A^* = 0.1653/1.94 = 0.0852 \text{ ft}^2$$

**16.176** Determine the throat and the exit areas of an ideal rocket motor to give a static thrust of 6.67 kN at 6.1-km altitude standard atmosphere if the chamber pressure is 1.035 MPa abs and chamber temperature is 3588 K. Find the velocity at the throat. Take $k = 1.4$ and $R = 0.355$ kJ/kg · K. Assume that exit pressure is that of surroundings.

▮ From Table A-8, $p_e = 46.583$ kPa and so $p_e/p_0 = 46.583/1035 = 0.045$. Hence, from Table A-16, $(N_M)_e = 2.67$. Also $T_e/T_0 = 0.412$, $T_e = (0.412)(3588) = 1478$ K. Hence, $V_e = (2.67)\sqrt{(1.4)(355)(1478)} = 2290$ m/s, $\rho = p/RT$, $\rho_e = 46.583/(0.355)(1478) = 0.0888$ kg/m³. Now, by the momentum equation, using gage pressures,

$6670 = \rho_e V_e^2 A_e$, $A_e = 6670/[(0.0888)(2290)^2] = 0.0143 \text{ m}^2$. Also $A_e/A^* = 3.09$, $A^* = 0.0143/3.09 = 0.00463$. Finally we want $V^*$. For this we need $T^*$. From Table A-16: $T^*/T_0 = 0.833$, $T^* = (3588)(0.833) = 2988$ K. Hence, $V^* = \sqrt{(1.4)(355)(2988)} = 1220$ m/s.

**16.177** Determine the exit area in Prob. 16.176 for a nozzle efficiency of 85 percent. Take $c_p = 1.214$ kJ/kg · K.

▌ $\eta = (V_2^2/2)_{act}/[c_p(T_1 - T_2)_{isen}]$     $(V_2^2/2)_{act} = (0.85)(1214)(3588 - 1478) = 2.177 \times 10^6$     $(V_2)_{act} = 2087$ m/s

Next we get $(T_2)_{act}$ from the first law of thermodynamics:

$$(V_2^2/2)_{act} = c_p[T_0 - (T_2)_{act}] \qquad 2.177 \times 10^6 = (1214)[3588 - (T_2)_{act}] \qquad (T_2)_{act} = 1795 \text{ K}$$

$$\rho = p/RT \qquad (\rho_2)_{act} = 46.583/[(0.355)(1795)] = 0.07310 \text{ kg/m}^3$$

Finally, from momentum considerations, $6670 = (\rho V_e^2 A_e)_{act}$, $A_e = 6670/[(0.07310)(2087)^2] = 0.021 \text{ m}^2$.

**16.178** Figure 16-11 is a convergent-divergent nozzle attached to a chamber (tank 1) where the pressure is 100 lb/in² abs and the temperature is 200 °F. The area of the throat is 3 in² and $A_1$, where we happen to have a normal shock, is 4 in². Finally $A_e$ is 6 in². What is the Mach number right after the shock wave? What is the Mach number at exit? Compute the stagnation pressure and actual pressure for the jet in tank 2. What is the stagnation temperature at exit? The fluid is air.

▌ **Data:**

$$T_0 = 660 \text{ °R} \qquad p_0 = 100 \text{ psia} \qquad A^* = 3 \text{ in}^2 \qquad A_1 = 4 \text{ in}^2 \qquad A_e = 6 \text{ in}^2$$

Find $(N_M)_1$ to right of shock: Since $A/A^* = \frac{4}{3} = 1.333$, we have from Table A-16, $(N_M)_1 = 1.69$. $(N_M)_2$ from Table A-17 is 0.644. Find $N_M$ at exit: For $N_M = 0.644$ Table A-16 gives $A_1/A^* = 1.16$. Therefore, $A^* = 4/1.16 = 3.45 \text{ m}^2$. Now at exit: $A_e/A^* = 6/3.45 = 1.74$. From Table A-16: $(N_M)_e = 0.360$, $p_1/p_0 = 0.206$, $p_1 = (100)(0.206) = 20.6$ psia. From Table A-17: $p_2/p_1 = 3.165$, $p_2 = (3.165)(20.6) = 65.2$. From Table A-16: $p_2/p_0 = 0.759$, $p_0 = 652/0.759 = 86.0$ psia; $p_e/p_0 = 0.914$, $p_e = (86)(0.914) = 78.5$ psia, $T_0 = 200$ °F.

**Fig. 16-11**

**16.179** A convergent nozzle has an exit area of 10 cm². It permits flow of air to proceed from a large tank in which the pressure of the air is 138 kPa abs and the temperature is 20 °C. If the ambient pressure outside the tank is 101.325 kPa, what are the velocity of the flow on leaving the nozzle and the mass flow? Neglect friction.

▌ $$p^*/p = 0.528 \qquad p^* = (138)(0.528) = 72.8 \text{ kPa}$$

Hence fluid leaves subsonically and $p_e/p_0 = 101.325/138 = 0.734$. Hence:

$$(N_M)_e = 0.68 \qquad T_e/T_0 = 0.915 \qquad T_e = 268 \text{ K}$$

$$V_e = c(N_M)_e = \sqrt{kRT_e}(N_M)_e = \sqrt{(1.4)(287)(268)}(0.68) = 223 \text{ m/s}$$

$$\rho = p/RT \qquad \rho_e = 101.325/[(0.287)(268)] = 1.317 \text{ kg/m}^3$$

$$\dot{m} = \rho VA = (1.317)(223)(0.0010) = 0.294 \text{ kg/s}$$

**16.180** In Prob. 16.179, what is the ambient pressure that will permit the maximum flow through the nozzle? What are the maximum mass flow and temperature of the air leaving the nozzle? Neglect friction.

▮ Maximal pressure is 72.8 kPa, the critical pressure. At $N_M = 1$:

$$T^*/T_0 = 0.833 \qquad T^* = (293)(0.833) = 244 \text{ K}$$
$$V = \sqrt{kRT} \qquad V_e = \sqrt{(1.4)(287)(244)} = 313 \text{ m/s}$$
$$\rho = p/RT \qquad \rho_e = 72.8/[(0.287)(244)] = 1.040 \text{ kg/m}^3$$
$$\dot{m} = \rho VA = (1.040)(313)(1.0 \times 10^{-3}) = 0.326 \text{ kg/s}$$

**16.181** A convergent-divergent nozzle with a throat area of 13 cm$^2$ and an exit area of 20 cm$^2$ is connected to a tank wherein air is kept at a pressure of 552 kPa abs and a temperature of 15 °C. If the nozzle is operating at design conditions, what should be the ambient pressure outside and the mass flow? What is the critical pressure? Neglect friction.

▮ For design operation we can say $A_e/A^* = 1.46$. From isentropic tables: $(N_M)_e = 1.82$, $p_e/p_0 = 0.169$, $p_e = (0.169)(552) = 93.3$ kPa. Also

$$T_e/T_0 = 0.602 \qquad T_e = (0.602)(288) = 173 \text{ K}$$
$$V_e = c_e(N_M)_e = \sqrt{kRT_e}(N_M)_e = \sqrt{(1.4)(287)(173)}(1.82) = 480 \text{ m/s} \qquad \rho = p/RT$$
$$\rho_e = 93.3/[(0.287)(173)] = 1.879 \text{ kg/m}^3 \qquad \dot{m} = \rho VA = (1.879)(480)(0.0020) = 1.80 \text{ kg/s}$$

At throat $p^*/p_0 = 0.528$, $p^* = 291.5$ kPa.

**16.182** In Prob. 16.181, what is the ambient pressure at which a shock will first appear just inside the nozzle? What is the ambient pressure for the completely subsonic flow of maximum mass flow? Neglect friction.

▮ The Mach number just before the shock will be 1.82. From the normal shock tables we have just downstream of the shock $(N_M)_e = 0.612$. Also $p_e/p_1 = 3.698$. Hence $p_e = (3.698)(93.3) = 345$ kPa. For completely subsonic flow we assume isentropic flow and using $A_e/A^* = 1.46$ we have from the isentropic tables $p_e/p_0 = 0.874$. Hence $p_e = (552)(0.874) = 482.45$ kPa.

**16.183** How would you determine the approximate position and strength of the shock in Fig. 16-12?

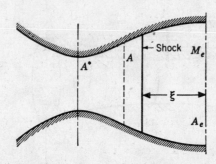

**Fig. 16-12**

▮ 1. First find $(N_M)_2$ as a function of $\xi$ using isentropic data starting from exit conditions and working in.
2. Find $(N_M)_1$ as a function also of $\xi$ using isentropic data starting from the throat and working downstream.
3. From these curves plot $(N_M)_2$ versus $(N_M)_1$.
4. Also plot $(N_M)_2$ versus $(N_M)_1$ using normal shock relations. Where they intersect gives the proper $(N_M)_1$ and $(N_M)_2$.
5. Now with either $(N_M)_1$ or $(N_M)_2$ go back to the isentropic data of step 2 or 1, respectively, to obtain $\xi$.

**16.184** Air flows through a 2-in constriction in a 3-in diameter pipeline. The pressure and temperature of the air in the pipeline are 108 psig and 105 °F, respectively, and the pressure in the constriction is 81 psig. Barometric pressure is 14.5 psia. Find the weight flow rate of the air in the pipeline.

▮ 
$$G = [A_2/\sqrt{1 - (p_2/p_1)^{2/k}(A_2/A_1)^2}]\sqrt{[2gk/(k-1)](p_1\gamma_1)[(p_2/p_1)^{2/k} - (p_2/p_1)^{(k+1)/k}]}$$
$$A_2 = (\pi)(2)^2/4 = 3.142 \text{ in}^2 = 0.02182 \text{ ft}^2 \qquad p_2 = 81 + 14.5 = 95.5 \text{ psia}$$
$$p_1 = 108 + 14.5 = 122.5 \text{ psia} \qquad A_1 = (\pi)(3)^2/4 = 7.069 \text{ in}^2 = 0.04909 \text{ ft}^2$$
$$\gamma = p/RT \qquad \gamma_1 = (122.5)(144)/[(53.3)(105 + 460)] = 0.5858 \text{ lb/ft}^3$$

$$G = \frac{0.02182}{\sqrt{1 - (95.5/122.5)^{2/1.40}(0.02182/0.04909)^2}} \sqrt{\frac{(2)(32.2)(1.40)}{1.40 - 1}(122.5)(144)(0.5858)}$$

$$\times \sqrt{\left[\left(\frac{95.5}{122.5}\right)^{2/1.40} - \left(\frac{95.5}{122.5}\right)^{(1.40+1)/1.40}\right]} = 7.87 \text{ lb/s}$$

**16.185** Air flows through a 1-in constriction in a 2-in-diameter pipeline. The pressure and temperature of the air in the pipeline are 100 psig and 102 °F, respectively, and the pressure in the constriction is 81 psig. Barometric pressure is 14.7 psia. Find the weight flow rate of the air in the pipeline.

$$\blacksquare \quad G = [A_2/\sqrt{1 - (p_2/p_1)^{2/k}(A_2/A_1)^2}]\sqrt{[(2gk)/(k-1)](p_1\gamma_1)}\sqrt{[(p_2/p_1)^{2/k} - (p_2/p_1)^{(k+1)/k}]}$$

$$A_2 = (\pi)(\tfrac{1}{12})^2/4 = 0.005454 \text{ ft}^2 \qquad p_2 = 78 + 14.7 = 92.7 \text{ psia} \qquad p_1 = 100 + 14.7 = 114.7 \text{ psia}$$

$$A_1 = (\pi)(\tfrac{2}{12})^2/4 = 0.02182 \text{ ft}^2 \qquad \gamma = p/RT \qquad \gamma_1 = (114.7)(144)/[(53.3)(102 + 460)] = 0.5514 \text{ lb/ft}^3$$

$$G = [(0.005454)/\sqrt{1 - (92.7/114.7)^{2/1.4}(0.005454/0.02182)^2}]$$

$$\times \sqrt{[(2)(32.2)(1.40)/(1.40-1)](114.7)(144)(0.5514)}$$

$$\times \sqrt{[(92.7/114.7)^{2/1.40} - (92.7/114.7)^{(1.40+1)/1.40}]} = 1.67 \text{ lb/s}$$

**16.186** A normal shock wave occurs in the flow of air where $p_1 = 10$ psia, $T_1 = 40$ °F, and $V_1 = 1400$ ft/s. Find $p_2$, $V_2$, and $T_2$.

$$\blacksquare \qquad p_2/p_1 = [2k(N_M)_1^2 - (k-1)]/(k+1) \qquad N_M = V/c$$

$$c = \sqrt{kRT} \qquad c_1 = \sqrt{(1.40)(1716)(40+460)} = 1096 \text{ ft/s}$$

$$(N_M)_1 = \tfrac{1400}{1096} = 1.28 \qquad p_2/10 = [(2)(1.40)(1.28)^2 - (1.40-1)]/(1.40+1) \qquad p_2 = 17.4 \text{ psia}$$

$$V_2/V_1 = [(k-1)(N_M)_1^2 + 2]/[(k+1)(N_M)_1^2] \qquad V_2/1400 = [(1.40-1)(1.28)^2 + 2]/[(1.40+1)(1.28)^2] \qquad V_2 = 945 \text{ ft/s}$$

$$\rho = p/RT \qquad \rho_1 V_1 = \rho_2 V_2 \qquad \rho_1 = (10)(144)/[(1716)(40+460)] = 0.001678 \text{ slug/ft}^3$$

$$(0.001678)(1400) = (\rho_2)(945) \qquad \rho_2 = 0.00249 \text{ slug/ft}^3$$

$$0.00249 = (17.4)(144)/(1716T_2) \qquad T_2 = 586 \text{ °R} \quad \text{or} \quad 126 \text{ °F}$$

**16.187** Rework Prob. 16.186 if $p_1 = 70$ Pa abs, $T_1 = 5$ °C, and $V_1 = 425$ m/s.

$$\blacksquare \quad p_2/p_1 = [2k(N_M)_1^2 - (k-1)]/(k+1) \qquad N_M = V/c \qquad c = \sqrt{kRT} \qquad c_1 = \sqrt{(1.40)(287)(5+273)} = 334 \text{ m/s}$$

$$(N_M)_1 = \tfrac{425}{334} = 1.27 \qquad p_2/70 = [(2)(1.40)(1.27)^2 - (1.40-1)]/(1.40+1) \qquad p_2 = 120 \text{ Pa}$$

$$V_2/V_1 = [(k-1)(N_M)_1^2 + 2]/[(k+1)(N_M)_1^2] \qquad V_2/425 = [(1.40-1)(1.27)^2 + 2]/[(1.40+1)(1.27)^2]$$

$$V_2 = 290 \text{ m/s}$$

$$\rho = p/RT \qquad \rho_1 V_1 = \rho_2 V_2 \qquad \rho_1 = 70/[(287)(5+273)] = 0.0008773 \text{ kg/m}^3$$

$$(0.0008773)(425) = (\rho_2)(290) \qquad \rho_2 = 0.00129 \text{ kg/m}^3 \qquad 0.00129 = 120/(287T_2) \qquad T_2 = 324 \text{ K} \quad \text{or} \quad 51 \text{ °C}$$

**16.188** The pressure, velocity, and temperature just upstream of a normal shock wave in air are 12 psia, 2300 fps, and 24 °F. Determine the pressure, velocity, and temperature just downstream of the wave.

$$\blacksquare \quad p_2 = p_1\{[2k(N_M)_1^2 - (k-1)]/(k+1)\} \qquad N_M = V/c \qquad c_1 = [kRT]^{1/2} = [1.4(1715)484]^{1/2} = 1078 \text{ fps}$$

$$(N_M)_1 = \tfrac{2300}{1078} = 2.13 \qquad p_2 = (12)\{[2(1.4)(2.13)^2 - 0.4]/2.4\} = 61.5 \text{ psia} \qquad p_1 - p_2 = \gamma_1 V_1/g(V_2 - V_1)$$

$$V_2 = V_1 + [(p_1 - p_2)/\rho_1 V_1] = 2300 + \{[(12 - 61.5)144]/[(0.0670/32.2)(2300)]\} = 811 \text{ fps}$$

$$V_2^2 - V_1^2 = [2k/(k-1)]\{[p_1/(\gamma_1/g)] - (p_2/\rho_2)\}$$

$$\gamma_1 = gp_1/RT_1 = (32.2)(12)(144)/[(1715)(484)] = 0.0670 \text{ pcf}$$

$$811^2 - 2300^2 = [2(1.4)/0.4]\{[(12)(144)/(0.0670/32.2)] - [(61.5)(144)/\rho_2]\}$$

$$\rho_2 = 0.00593 \text{ slug/ft}^3 = p_2/RT_2 \qquad T_2 = (61.5)(144)/[(1715)(0.00593)] = 871 \text{ °R} = 411 \text{ °F}$$

**16.189** Just downstream of a normal shock wave the pressure, velocity, and temperature are 52 psia, 398 fps, and 115 °F. Compute the Mach number upstream of the shock wave. Consider air and carbon dioxide.

$$\blacksquare \qquad (N_M)_2^2 = [2 + (k-1)(N_M)_1^2]/[2k(N_M)_1^2 - (k-1)]$$

**For air:**

$$N_M = V/c \qquad c = \sqrt{kRT} \qquad c_2 = \sqrt{(1.40)(1716)(115 + 460)} = 1175 \text{ ft/s}$$

$$(N_M)_2 = \tfrac{398}{1175} = 0.339 \qquad 0.339^2 = [2 + (1.40 - 1)(N_M)_1^2]/[(2)(1.40)(N_M)_1^2 - (1.40 - 1)]$$
$$(N_M)_1^2 = -26.2 \qquad \text{(impossible)}$$

**For carbon dioxide:**

$$c_2 = \sqrt{(1.28)(1123)(115 + 460)} = 909 \text{ ft/s} \qquad (N_M)_2 = \tfrac{398}{909} = 0.438$$
$$0.438^2 = [2 + (1.28 - 1)(N_M)_1^2]/[(2)(1.28)(N_M)_1^2 - (1.28 - 1)] \qquad (N_M)_1 = 3.12$$

**16.190** A schlieren photograph of a bullet shows a Mach angle of 28°. The air is at a pressure of 14 psia and 60 °F. Find the approximate speed of the bullet.

∎ $\quad N_M = V/c = 1/\sin\beta = 1/\sin 28° = 2.13 \qquad c = \sqrt{kRT} = \sqrt{(1.40)(1716)(60 + 460)} = 1118 \text{ ft/s}$
$$2.13 = V/1118 \qquad V = 2381 \text{ ft/s}$$

**16.191** If a normal shock wave occurs in the flow of helium, $p_1 = 1$ psia, $T_1 = 40$ °F, $V_1 = 4500$ ft/s, find $p_2$, $\rho_2$, $V_2$, and $T_2$.

∎ $\quad p_2 = [1/(k + 1)][2\rho_1 V_1^2 - (k - 1)p_1] \qquad \rho_1 = p_1/RT_1 = 1(144)/\{386[32.17(460 + 40)]\} = 0.0000232 \text{ slug/ft}^3$
$$p_2 = [1/(1.66 + 1)][2(0.0000232)(4500^2) - (1.66 - 1)(144)(1)] = 317 \text{ lb/ft}^2 \text{ abs} \qquad p_1 + \rho_1 V_1^2 = p_2 + \rho_1 V_1 V_2$$
$$V_2 = V_1 - [(p_2 - p_1)/\rho_1 V_1] = 4500 - [(317 - 144)(4500)(0.0000232)] = 2843 \text{ ft/s}$$
$$\rho_2 = \rho_1(V_1/V_2) = (0.0000232)(\tfrac{4500}{2843}) = 0.0000367 \text{ slug/ft}^3$$
$$T_2 = p_2/\rho_2 R = 317/\{(0.0000367)[(386)(32.17)]\} = 696 \text{ °R} \quad \text{or} \quad 236 \text{ °F}$$

**16.192** A shock wave occurs in a duct carrying air where the upstream Mach number is 2.0 and upstream temperature and pressure are 17 °C and 21 kPa abs. Calculate the Mach number, pressure, temperature, and velocity after the shock wave.

∎ For $(N_M)_1 = 2.0$, from Table A-17, $(N_M)_2 = 0.577$, $p_2/p_1 = 4.500$, and $T_2/T_1 = 1.688$.

$$p_2 = (4.500)(21) = 94.5 \text{ kPa} \qquad T_2 = (1.688)(17 + 273) = 490 \text{ K} \quad \text{or} \quad 217 \text{ °C}$$
$$c_2 = \sqrt{kRT} = \sqrt{(1.40)(287)(490)} = 444 \text{ m/s} \qquad V_2 = c_2(N_M)_2 = (444)(0.577) = 256 \text{ m/s}$$

**16.193** Show that entropy has increased across the shock wave of Prob. 16.192.

∎ $\quad \Delta s = c_v \ln[(T_2/T_1)^k(p_2/p_1)^{1-k}]$, and $[490/(17 + 273)]^{1.40}(94.5/21)^{1-1.40}$ is slightly larger than 1.

**16.194** Conditions immediately before a normal shock wave in airflow are $p_u = 6$ psia, $T_u = 139$ °F, and $V_u = 1860$ ft/s. Find $(N_M)_u$, $(N_M)_d$, $p_d$, and $T_d$, where the subscript $d$ refers to conditions just downstream from the shock wave.

∎ $\quad N_M = V/c \qquad c = \sqrt{kRT} = \sqrt{(1.40)(1716)(139 + 460)} = 1200 \text{ ft/s} \qquad (N_M)_u = \tfrac{1860}{1200} = 1.55$

From Table A-17, $(N_M)_d = 0.683$, $p_d/p_u = 2.65$, and $T_d/T_u = 1.356$. $p_d = (2.65)(6) = 15.9$ psia, $T_d = (1.356)(139 + 460) = 812$ °R, or 352 °F.

**16.195** For $A = 0.20$ ft$^2$ in Prob. 16.194, calculate the entropy increase across the shock wave in Btu/s and degrees Rankine.

∎ $\quad \Delta s = (0.171)\ln[(T_2/T_1)^k(p_2/p_1)^{1-k}] = (0.171)\ln\{[812/(139 + 460)]^{1.40}(15.9/6)^{1-1.40}\} = 0.00617 \text{ Btu/(lbm-°R)}$
$$\Delta s = \dot{m}\,\Delta s$$
$$\dot{m} = \rho_u A V_u = (p_u/RT_u)AV_u = [(6)(144)/(1716)(139 + 460)](0.20)(1860) = 0.3127 \text{ slug/s}$$
$$\Delta s = (0.3127)(32.17)(0.00617) = 0.0621 \text{ Btu/(s-°R)}$$

**16.196** An explosion in air, $\gamma = 1.4$, creates a spherical shock wave propagating radially into still air at standard conditions. At the instant shown in Fig. 16-13, the pressure just inside the shock is 200 lb/in$^2$ abs. Estimate (a) the shock speed $C$ and (b) the air velocity $V$ just inside the shock.

∎ (a) In spite of the spherical geometry the flow across the shock moves normal to the spherical wavefront; hence the normal-shock relations apply. Fixing our control volume to the moving shock, we find that the proper conditions to use are $C = V_1$, $p_1 = 14.7$ lb/in$^2$ abs, $T_1 = 520$ °R, $V = V_1 - V_2$, $p_2 = 200$ lb/in$^2$ abs. The speed of sound outside the shock is $c_1 \approx 49T_1^{1/2} = 1117$ ft/s. We can find $(N_M)_1$ from the known pressure ratio across the shock: $p_2/p_1 = 200/14.7 = 13.61 = [1/(k + 1)][2k(N_M)_1^2 - (k - 1)]$, $13.61 = (1/2.4)[2.8(N_M)_1^2 - 0.4]$ or $(N_M)_1 = 3.436$. Then, by definition of the Mach number, $C = V_1 = (N_M)_1 c_1 = 3.436(1117) = 3840$ ft/s.

$p = 14.7 \text{ lb/in}^2 \text{ abs}$
$T = 520°\text{R}$

$C$

$200 \text{ lbf/in}^2 \text{ abs}$

$V$

**Fig. 16-13**

**(b)**
$$T_2/T_1 = [2 + (k-1)(N_M)_1^2]\{[2k(N_M)_1^2 - (k-1)]/(k+1)^2(N_M)_1^2\}$$
$$= [2 + (0.4)(3.436)^2]\{[(2.8)(3.436)^2 - 0.4]/(2.4)^2(3.436)^2\} = 3.228$$
$$T_2 = 3.228 T_1 = 3.228(520) = 1679 \,°\text{R}$$

At such a high temperature we should account for non-perfect-gas effects or at least use the gas tables, but we won't. Here just estimate from the perfect-gas energy equation that $V_2^2 = 2c_p(T_1 - T_2) + V_1^2 = 2(6010)(520 - 1679) + (3840)^2 = 815\,500$ or $V_2 \approx 903$ ft/s. Notice that we did this without bothering to compute $(N_M)_2$, which equals 0.454, or $c_2 \approx 49 T_2^{1/2} = 2000$ ft/s.

Finally, the air velocity behind the shock is $V = V_1 - V_2 = 3840 - 903 \approx 2940$ ft/s. Thus a powerful explosion creates a brief but intense blast wind as it passes.

**16.197** Air from a reservoir at 22 °C and 498 kPa flows through a duct and forms a normal shock downstream of a throat of area 12 cm². If the stagnation pressure downstream of this shock exactly equals the throat pressure, what is the area where the shock wave stands?

$\blacksquare$ $A/A^* = (1/N_M)[(1 + 0.2N_M^2)^3/1.728]$ $\quad p_1^* = (p_0)_1/(1 + 0.2N_M^2)^{3.5} = 498/[1 + (0.2)(1)^2]^{3.5} = 263$ kPa $= (p_0)_2$
$$(p_0)_2/(p_0)_1 = \tfrac{263}{498} = 0.528$$

From Table A-17, $(N_M)_1 = 2.43$, $A_1/12 = (1/2.43)\{[1 + (0.2)(2.43)^2]^3/1.728\}$, $A_1 = 29.65$ cm².

**16.198** Air passes through a normal shock with upstream conditions $V_1 = 790$ m/s, $p_1 = 110$ kPa, and $T_1 = 306$ K. What are the downstream conditions $V_2$ and $p_2$?

$\blacksquare$
$$p_2/p_1 = [2k(N_M)_1^2 - (k-1)]/(k+1) \qquad N_M = V/c \qquad c = \sqrt{kRT}$$
$$c_1 = \sqrt{(1.40)(287)(306)} = 351 \text{ m/s} \qquad (N_M)_1 = \tfrac{790}{351} = 2.251$$
$$p_2/110 = [(2)(1.40)(2.251)^2 - (1.40 - 1)]/(1.40 + 1) \qquad p_2 = 632 \text{ kPa}$$
$$V_2/V_1 = [(k-1)(N_M)_1^2 + 2]/(k+1)(N_M)_1^2 \qquad V_2/790 = [(1.40 - 1)(2.251)^2 + 2]/(1.40 + 1)(2.251)^2$$
$$V_2 = 262 \text{ m/s}$$

**16.199** Repeat Prob. 16.197 if the *static* pressure downstream of the shock exactly equals the throat pressure. What is the area where the shock stands?

$\blacksquare$
$$p_2 = p_{\text{throat}} = p_1^* = 263 \text{ kPa} \qquad \text{(from Prob. 16.197)}$$
$$p_1 = (p_0)_1/(1 + 0.2N_M^2)^{3.5} = 498/(1 + 0.2N_M^2)^{3.5}$$
$$p_2/p_1 = [(k+1)(N_M)_1^2 - (k-1)]/(k+1) = [2.80(N_M)_1^2 - 0.40]/2.40$$

By trial and error, $(N_M)_1 = 2.15$, $p_1 = 50.3$ kPa, $p_2 = 263$ kPa, $A/A_1^* = 1.919$, $A_1 = (12)(1.919) = 23.0$ cm².

**16.200** An atomic explosion propagates into still air at 14.7 lb/in² abs and 515 °R. The pressure just inside the shock is 4950 lb/in² abs. Assuming $k = 1.4$, what is the speed $C$ of the shock and the velocity $V$ just inside the shock?

$\blacksquare$ $C = (N_M)_1(c_1)$ $\quad p_2/p_1 = [2k(N_M)_1^2 - (k-1)]/(k+1)$ $\quad 4950/14.7 = [(2)(1.4)(N_M)_1^2 - (1.4 - 1)]/(1.4 + 1)$
$(N_M)_1 = 17.00$ $\quad c = \sqrt{kRT}$ $\quad c_1 = \sqrt{(1.4)(1716)(515)} = 1112$ ft/s $\quad C = (17.00)(1112) = 18\,904$ ft/s $= V_1$
$V_2/V_1 = [(k-1)(N_M)_1^2 + 2]/(k+1)(N_M)_1^2$ $\quad V_2/18\,904 = [(1.4 - 1)(17.00)^2 + 2]/(1.4 + 1)(17.00)^2 = 0.1696$
$$V_2 = 3206 \text{ ft/s} \qquad V = C - V_2 = 18\,904 - 3206 = 15\,698 \text{ ft/s}$$

**16.201** The normal-shock wave from an explosion propagates at 1450 m/s into still air at 21 °C and 101 kPa. What are the pressure, velocity, and temperature just inside the shock?

$$p_2/p_1 = [2k(N_M)_1^2 - (k-1)]/(k+1) \qquad N_M = V/c \qquad c = \sqrt{kRT} = \sqrt{(1.4)(287)(21+273)} = 344 \text{ m/s}$$

$$(N_M)_1 = \tfrac{1450}{344} = 4.22 \qquad p_2/101 = [(2)(1.4)(4.22)^2 - (1.4-1)]/(1.4+1) \qquad p_2 = 208.2 \text{ kPa}$$

$$T_2/T_1 = [2 + (k-1)(N_M)_1^2]\{[(2k)(N_M)_1^2 - (k-1)]/(k+1)^2(N_M)_1^2\}$$

$$T_2/(21+273) = [2 + (1.4-1)(4.22)^2]\{[(2)(1.4)(4.22)^2 - (1.4-1)]/(1.4+1)^2(4.22)^2\}$$

$$T_2 = 1293 \text{ K} \qquad V_2/V_1 = [(k-1)(N_M)_1^2 + 2]/(k+1)(N_M)_1^2 \qquad V_2/1450 = [(1.4-1)(4.22)^2 + 2]/(1.4+1)(4.22)^2$$

$$V_2 = 310 \text{ m/s} \qquad V_{\text{inside shock}} = C - V_2 = 1450 - 310 = 1140 \text{ m/s}$$

**16.202** Air is moving at Mach number 3 in a duct and undergoes a normal shock. If the undisturbed pressure ahead of the shock is 69 kPa abs, what is the increase in pressure after the shock? What is the loss in stagnation pressure across the shock?

▌ From Table A-17, $(N_M)_2 = 0.475$, $p_2/p_1 = 10.333$, $(p_0)_2/(p_0)_1 = 0.328$. Hence $p_2 = (10.333)(69) = 712.98$ kPa. Now go to an isentropic table for $(p_0)_1$: $p_1/(p_0)_1 = 0.027$, $(p_0)_1 = 69/0.027 = 2560$ kPa. Hence $(p_0)_2 = (0.328)(2560) = 838$ kPa, and $\Delta p_0 = 2560 - 838 = 1720$ kPa.

**16.203** An airplane having a diffuser designed for subsonic flight has a normal shock attached to the edge of the diffuser when the plane is flying at a certain Mach number. If at the exit of the diffuser the Mach number is 0.3, what must the flight Mach number be for the plane, assuming isentropic diffusion behind the shock? The inlet area is 0.24 m² and the exit area is 0.38 m².

▌ From Table A-16, $A_e/A^* = 2.04$, $A^* = 0.38/2.04 = 0.186$ m². At inlet, $A_1 = 0.24$ m². Hence $A_1/A^* = 0.24/0.186 = 1.290$. From isentropic tables, $(N_M)_2 = 0.527$. This is the Mach number behind the inlet shock. From Table A-17, $(N_M)_1 = 2.360$.

**16.204** A normal shock forms ahead of the diffuser of a turbojet plane flying at Mach number 1.2. If the plane is flying at 35 000-ft altitude in standard atmosphere, what is the entering Mach number for the diffuser and the stagnation pressure.

▌ $(N_M)_1 = 1.2$. Hence from *normal shock* tables (A-17), $(N_M)_2 = 0.842$.

$$(p_0)_2/(p_0)_1 = 0.993 \tag{1}$$

Also, from *isentropic* tables (A-16) we have for $(N_M)_1 = 1.2$: $p_1/(p_0)_1 = 0.412$. Hence, $(p_0)_1 = 0.498/0.412 = 1210$ psf, where we have used the standard atmosphere table. Now using Eq. (1) we get $(p_0)_2 = (0.993)(1210) = 1200$ psf.

**16.205** Consider a supersonic flow of air through a stationary duct wherein a stationary shock is present. The Mach number ahead of the shock is 2.1 and the pressure and temperature are 101.3 kPa abs and 37 °C. What is the velocity of propagation of the shock relative to the air ahead of the shock?

▌ $\qquad (N_M)_1 = V_1/c_1 = V_1/\sqrt{kRT_1} \qquad 2.1 = V_1/\sqrt{(1.40)(287)(37+273)} \qquad V_1 = 741$ m/s

**16.206** A jet plane is diving at constant supersonic speed. There is a curved shock wave ahead of it. A static pressure gage near the nose of the plane measures 37.5 kPa abs. The ambient pressure and temperature of the atmosphere are 12 kPa and 254 K. What are the flight Mach number for the plane and its speed if one assumes that in front of the static pressure gage the shock wave is plane?

▌ Consider the reference of observation is from the plane. The ratio of the pressures across the shock wave is $p_2/p_1 = 37.5/12 = 3.125$, $(N_M)_1 = 1.68$. From Tables A-17 the plane is moving at a Mach number of 1.68. The velocity is determined next. The speed of sound in the undisturbed region ahead of the shock is $c = \sqrt{kRT} = \sqrt{(1.4)(287)(254)} = 319.5$ m/s. Therefore, $V = (1.68)(319.5) = 537$ m/s, which is also the velocity of the plane.

**16.207** A duct having a square cross section 0.300 m on a side has 25 kg/s of air flowing in it. The air, originally in a chamber where the temperature is 90 °C, has been insulated by the duct walls against heat transfer to the outside. The duct is operating in a choked condition. If the duct has a relative roughness of 0.002, determine the Mach number at a position 6 m from the exit of the duct.

▌ We may solve for $N_M$ at this position by employing $[(1/M^2) - 1] + [(k+1)/2] \ln [(k+1)M^2/2\{1 + [(k-1)/2]M^2\}] = (fk/D_H)L$. To do this, we estimate $f$ to be 0.024, from the Moody diagram (Fig. A-5). The hydraulic diameter $D_H$ is $D_H = 4A/p_w = (4)(0.3^2)/(4)(0.3) = 0.3$ m. Using $k = 1.4$, we then have $[(1/N_M^2) - 1] + 1.2 \ln [1.2N_M^2/(1 + 0.2N_M^2)] = [(0.024)(1.4)/0.3](6) = 0.672$. Solving by trial and error, we get $N_M = 0.6$.

To check our friction factor, we must compute other conditions at this section of the pipe. The temperature $T$ is determined from Table A-16 for $N_M = 0.6$. Thus $T/T_0 = 0.933$. Therefore $T = 339$ K $= 66\,°C$. We may determine the viscosity of air at that temperature to be $2.15 \times 10^{-5}$ Pa · s by making use of Fig. A-1, since viscosity does not depend greatly on the pressure. Noting that $G = \rho V = \dot{m}/A = 25/0.300^2 = 277.8$ kg/m$^2$ · s, we have $N_R = GD_H/\mu = (277.8)(0.300)/(2.15 \times 10^{-5}) = 3.88 \times 10^6$. Returning to the Moody diagram, we see that our choice of $f$ is close enough not to require further computation and the desired Mach number is 0.6.

**16.208** A constant-area duct having a circular cross-sectional area of 0.19 m$^2$ is operating in a choked condition. It is highly insulated against heat transfer, and the inside surface has a relative roughness of 0.002. At a distance 9 m from the end of the duct, what is the Mach number of a 36 kg/s airflow? The stagnation temperature is 96 °C. Perform one iteration.

❚  $$[(1/N_M^2) - 1] + [(k + 1)/2] \ln ((k + 1)N_M^2/2\{1 + [(k - 1)/2]N_M^2\}) = (fk/D_H)L$$
$$\pi D^2/4 = 0.19\ \text{m}^2 \qquad D = 0.4918\ \text{m}$$

Try $f = 0.023$:

$$[(1/N_M^2) - 1] + [(1.40 + 1)/2] \ln ((1.40 + 1)N_M^2/(2)\{1 + [(1.40 - 1)/2]N_M^2\}) = [(0.023)(1.40)/0.4918](9)$$
$$N_M = 0.61 \qquad N_R = \rho DV/\mu \qquad \dot{m} = \rho AV \qquad V = cN_M \qquad c = \sqrt{kRT}$$

From Table A-16, $T/T_0 = 0.931$:

$$T = (0.931)(96 + 273) = 344\ \text{K} \qquad c = \sqrt{(1.40)(287)(344)} = 372\ \text{m/s} \qquad V = (372)(0.61) = 227\ \text{m/s}$$
$$36 = (\rho)(0.19)(227) \qquad \rho = 0.835\ \text{kg/m}^3 \qquad N_R = (0.835)(0.4918)(227)/(2.2 \times 10^{-5}) = 4.24 \times 10^6$$

From Fig. A-5 with $\epsilon/D = 0.002$, $f = 0.023$. Since this value is the same as the guessed value of $f$, $N_M = 0.61$.

**16.209** In Prob. 16.208, determine where Mach number 0.5 is assumed, and ascertain the pressure and temperature of the flow there.

❚  From Fig. A-72, $(fk/H)L = 1.4$, $[(0.023)(1.40)/0.4918]L = 1.4$, $L = 21.4$ m. From Table A-16, $T/T_0 = 0.952$:

$$T = (0.952)(96 + 273) = 351\ \text{K} \quad \text{or} \quad 77\,°C \qquad p = (G/N_M)(RT_0/k\{1 + [(k - 1)/2]N_M^2\})^{1/2}$$
$$p = [(36/0.19)/0.5]((287)(96 + 273)/(1.40)\{1 + [(1.40 - 1)/2](0.5)^2\})^{1/2} = 101.7\ \text{kPa}$$

**16.210** A constant-area duct is operating in the choked condition. The cross section is rectangular, having sides 6 by 4 ft, and the surface has a relative roughness of 0.0001. At 22 ft from the end of the duct, the pressure is 18 lb/in$^2$ abs. If there is no heat transferred through the walls, determine the Mach number and Reynolds number at this section of air flow. The exit pressure is that of ambient pressure of the surroundings, which is 14.7 lb/in$^2$.

❚  We first compute $H$.

$$H = 4A/p_w = (4)(24)/20 = 4.8\ \text{ft} \qquad p = (G/N_M)(RT_0/k\{1 + [(k - 1)/2]N_M^2\})^{1/2}$$
$$(14.7)(144) = (G/1)\{1716T_0/(1.40)[1 + (0.4/2)(1)^2]\}^{1/2}$$
$$GT_0^{1/2} = 66.2 \tag{1}$$

At position 20 ft from end we get $(18)(144) = (G/N_M)[1716T_0/1.4(1 + 0.2N_M^2)]^{1/2}$:

$$74.0 = (G/N_M)[T_0/(1 + 0.2N_M^2)]^{1/2} \tag{2}$$

We thus have two unknowns if we consider $GT_0^{1/2}$ and $N_M$ as unknowns. Substituting for $GT_0^{1/2}$ from Eq. (1) into Eq. (2) we get $74.0 = 66.2/N_M(1 + 0.2N_M^2)^{1/2}$. Solve by trial and error to get $N_M = 0.838$. From Fig. A-72 we get $(fk/H)L = 0.08$, $f = (0.08)(4.8)/(1.4)(22) = 0.0125$. Now using $f = 0.0125$ and $\epsilon/D = 0.0001$, we get from the Moody diagram (Fig. A-5): $N_R = 3.5 \times 10^6$.

**16.211** Determine the maximum length of 50-mm-ID pipe, $f = 0.025$ for flow of air, when the Mach number at the entrance to the pipe is 0.35.

❚  $$fL_{max}/D = \tfrac{5}{7}\{[1/(N_M)_0^2 - 1]\} + \tfrac{6}{7} \ln \{6(N_M)_0^2/[(N_M)_0^2 + 5]\} \qquad k = 1.4$$
$$(0.025/0.05)L_{max} = \tfrac{5}{7}[(1/0.35^2) - 1] + \tfrac{6}{7} \ln [6(0.35^2)/(0.35^2 + 5)]$$

from which $L_{max} = 6.90$ m.

**16.212** A 4.0-in-ID pipe, $f = 0.020$, has air at 14.7 psia and at $T = 60\,°F$ flowing at the upstream end with Mach number 3.0. Determine $L_{max}$, $p^*$, $V^*$, $T^*$, and values of $p_0'$, $V_0'$, $T_0'$, and $L$ at $N_M = 2.0$.

**∎** $fL_{max}/D = \frac{5}{7}\{[1/(N_M)_0^2] - 1\} + \frac{6}{7}\ln\{6(N_M)_0^2/[(N_M)_0^2 + 5]\}$     $(0.02/0.333)L_{max} = \frac{5}{7}(\frac{1}{9} - 1) + \frac{6}{7}\ln[6(3^2)/(3^2 + 5)]$

from which $L_{max} = 8.69$ ft. If the flow originated at $N_M = 2$, the length $L_{max}$ is given by the same equation: $(0.02/0.333)L_{max} = \frac{5}{7}(\frac{1}{4} - 1) + \frac{6}{7}\ln[6(2^2)/(2^2 + 5)]$ from which $L_{max} = 5.08$ ft.

Hence, the length from the upstream section at $N_M = 3$ to the section where $N_M = 2$ is $8.69 - 5.08 = 3.61$ ft. The velocity at the entrance is $V = \sqrt{kRTN_M} = \sqrt{1.4(53.3)[32.17(460 + 60)]}(3) = 3352$ ft/s.

$$p^*/p_1 = (N_M)_0\sqrt{[(k - 1)(N_M)_0^2 + 2]/(k + 1)} \qquad V^*/V_0 = [1/(N_M)_0]\sqrt{[(k - 1)(N_M)_0^2 + 2]/(k + 1)}$$
$$T^*/T_0 = [(k - 1)(N_M)_0^2 + 2]/(k + 1) \qquad p^*/14.7 = 3\sqrt{[0.4(3^2) + 2]/2.4} = 4.583$$
$$V^*/3352 = \frac{1}{3}\sqrt{[0.4(3^2) + 2]/2.4} = 0.509 \qquad T^*/520 = [0.4(3^2) + 2]/2.4 = \frac{7}{3}$$

So $p^* = 67.4$ psia, $V^* = 1707$ ft/s, $T^* = 1213\,°R$. For $N_M = 2$ the same equations are now solved for $p_0'$, $V_0'$, and $T_0'$: $67.4/p_0' = 2\sqrt{[0.4(2^2) + 2]/2.4} = 2.45$, $1707/V_0' = \frac{1}{2}\sqrt{[0.4(2^2) + 2]/2.4} = 0.6124$, $1213/T_0' = [0.4(2^2) + 2]/2.4 = \frac{3}{2}$. So $p_0' = 27.5$ psia, $V_0' = 2787$ ft/s, and $T_0' = 809\,°R$.

**16.213** What length of 120-mm-diameter insulated duct, $f = 0.020$, is needed when oxygen enters at $N_M = 3.0$ and leaves at $N_M = 2.0$?

**∎** $$fL/D = \frac{5}{7}\{[1/(N_M)_0^2] - [1/(N_M)^2]\} + \frac{6}{7}\ln([(N_M)_0/N_M]^2\{[(N_M)^2 + 5]/[(N_M)_0^2 + 5]\})$$
$$0.020L/0.120 = \frac{5}{7}[(1/3.0^2) - (1/2.0^2)] + \frac{6}{7}\ln\{(3.0/2.0)^2[(2.0^2 + 5)/(3.0^2 + 5)]\} \qquad L = 1.303\ m$$

**16.214** Air enters an insulated pipe at $N_M = 0.3$ and leaves at $N_M = 0.6$. What portion of the duct length is required for the flow to occur at $N_M = 0.5$?

**∎** $$\frac{L_1}{L_2} = \frac{5\{[1(N_M)_0^2] - [1/(N_M)_1^2]\} + 6\ln\{[(N_M)_0/(N_M)_1]^2\{[(N_M)_1^2 + 5]/[(N_M)_0^2 + 5]\}}{5\{[1/(N_M)_0^2] - [1/(N_M)_2^2]\} + 6\ln\{[(N_M)_0/(N_M)_2]^2\{[(N_M)_2^2 + 5]/[(N_M)_0^2 + 5]\}\}}$$
$$= \frac{5[(1/0.3^2) - (1/0.5^2)] + 6\ln\{(0.3/0.5)^2[(0.5^2 + 5)/(0.3^2 + 5)]\}}{5[(1/0.3^2) - (1/0.6^2)] + 6\ln\{(0.3/0.6)^2[(0.6^2 + 5)/(0.3^2 + 5)]\}} = 0.885$$

Thus 88.5 percent of the duct length is required.

**16.215** Determine the maximum length, without choking, for the adiabatic flow of air in a 100-mm-diameter duct, $f = 0.0227$, when upstream conditions are $T = 60\,°C$, $V = 200$ m/s, and $p = 2$ atm.

**∎** $$fL_{max}/D = \frac{5}{7}\{[1/(N_M)_0^2] - 1\} + \frac{6}{7}\ln\{6(N_M)_0^2/[(N_M)_0^2 + 5]\}$$
$$N_M = V/c = V/\sqrt{kRT} \qquad (N_M)_0 = 200/\sqrt{(1.40)(287)(50 + 273)} = 0.5552$$
$$0.0227L_{max}/0.100 = \frac{5}{7}[(1/0.5552^2) - 1] + \frac{6}{7}\ln[(6)(0.5552)^2/(0.5552^2 + 5)] \qquad L_{max} = 3.077\ m$$

**16.216** What minimum size insulated duct is required to transport nitrogen 980 ft? The upstream temperature is 75 °F, and the velocity there is 210 fps ($f = 0.021$).

**∎** $$fL_{max}/D = \frac{5}{7}\{[1/(N_M)_0^2] - 1\} + \frac{6}{7}\ln\{6(N_M)_0^2/[(N_M)_0^2 + 5]\}$$
$$N_M = V/c = V/\sqrt{kRT} \qquad (N_M)_0 = 210/\sqrt{(1.40)(1776)(75 + 460)} = 0.1821$$
$$(0.021)(980)/D = \frac{5}{7}[(1/0.1821^2) - 1] + \frac{6}{7}\ln[(6)(0.1821)^2/(0.1821^2 + 5)] \qquad D = 1.140\ ft$$

**16.217** Find the upstream and downstream pressures in Prob. 16.216 for 3-lbm/s flow.

**∎** $$\rho = p/RT \qquad \dot{m} = \rho AV \qquad 3/32.2 = (\rho_u)[(\pi)(1.140)^2/4](210) \qquad \rho_u = 0.0004347\ slug/ft^3$$
$$0.0004347 = p_u/[(1776)(75 + 460)] \qquad p_u = 413.0\ psfa \quad or \quad 2.87\ psia$$
$$p^*/p_1 = (N_M)_0\sqrt{[(k - 1)(N_M)_0^2 + 2]/(k + 1)} \qquad p^*/413.0 = (0.1821)\sqrt{[(1.40 - 1)(0.1821)^2 + 2]/(1.40 + 1)}$$
$$p^* = p_d = 68.9\ psfa \quad or \quad 0.478\ psia$$

**16.218** What is the maximum mass rate of flow of air from a reservoir, $T = 20\,°C$, through 6 m of insulated 25-mm-diameter pipe, $f = 0.020$, discharging to a standard atmosphere?

**∎** $$\dot{m} = \rho_0 AV_0 \qquad \rho = p/RT \qquad p^*/p_0 = (N_M)_0\sqrt{[(k - 1)(N_M)_0^2 + 2]/(k + 1)}$$
$$fL_{max}/D = \frac{5}{7}\{1/(N_M)_0^2] - 1\} + \frac{6}{7}\ln\{6(N_M)_0^2/[(N_M)_0^2 + 5]\}$$

$$(0.020)(6)/0.025 = \tfrac{5}{7}\{[1/(N_M)_0^2] - 1\} + \tfrac{6}{7}\ln\{6(N_M)_0^2/[(N_M)_0^2 + 5]\} \qquad (N_M)_0 = 0.311$$

$$101.3/p_0 = 0.311\sqrt{[(1.40)(0.311)^2 + 2]/(1.40 + 1)} \qquad p_0 = 353\text{ kPa abs}$$

$$\rho_0 = 353/[(0.287)(20 + 273)] = 4.20\text{ kg/m}^3 \qquad V = (N_M)c = (N_M)\sqrt{kRT}$$

$$V_0 = (0.311)\sqrt{(1.40)(287)(20 + 273)} = 106.7\text{ m/s} \qquad \dot{m} = (4.20)[(\pi)(0.025)^2/4](106.7) = 0.220\text{ kg/s}$$

**16.219** Air flows subsonically in an adiabatic 1-in-diameter duct. The average friction factor is 0.024. (*a*) What length of duct is necessary to accelerate the flow from $(N_M)_1 = 0.1$ to $(N_M)_2 = 0.5$? (*b*) What additional length will accelerate it to $(N_M)_3 = 1.0$? Assume $k = 1.4$.

**┃** (*a*) $f(\Delta L/D) = (fL^*/D)_1 - (fL^*/D)_2$. From Table A-24,

$$(fL^*/D)_{N_M=0.1} = 66.9216 \qquad (fL^*/D)_{N_M=0.5} = 1.0691$$

$$(0.024)(\Delta L)/(\tfrac{1}{12}) = 66.9216 - 1.0691 \qquad \Delta L = 229\text{ ft}$$

(*b*) $\qquad f(\Delta L/D) = (fL^*/D)_{N_M=0.5} \qquad (0.024)(\Delta L)/(\tfrac{1}{12}) = 1.0691 \qquad \Delta L = 3.71\text{ ft}$

**16.220** For the duct flow of Prob. 16.219 assume that at $(N_M)_1 = 0.1$ we have $p_1 = 100\text{ lb/in}^2$ abs and $T_1 = 600\text{ °R}$. Compute at section 2 farther downstream $[(N_M)_2 = 0.5]$ (*a*) $p_2$; (*b*) $T_2$; (*c*) $V_2$; and (*d*) $(p_0)_2$.

**┃** As preliminary information we can compute $V_1$ and $(p_0)_1$ from the given information: $V_1 = (N_M)_1 c_1 = 0.1[49(600)^{1/2}] = 120\text{ ft/s}$, $(p_0)_1 = p_1[1 + \tfrac{1}{2}(k - 1)(N_M)_1^2]^{3.5} = 100[1 + 0.2(0.1)^2]^{3.5} = 100.7\text{ lb/in}^2$ abs. Now enter Table A-24, to find the following ratios:

| section | $N_M$ | $p/p^*$ | $T/T^*$ | $V/V^*$ | $p_0/p_0^*$ |
|---------|-------|---------|---------|---------|-------------|
| 1 | 0.1 | 10.9435 | 1.1976 | 0.1094 | 5.8218 |
| 2 | 0.5 | 2.1381 | 1.1429 | 0.5345 | 1.3399 |

Use these ratios to compute all properties downstream:

(*a*) $\qquad p_2 = p_1(p_2/p^*)(p^*/p_1) = 100(2.1381/10.9435) = 19.5\text{ lb/in}^2$ abs

(*b*) $\qquad T_2 = T_1(T_2/T^*)(T^*/T_1) = 600(1.1429)/1.1976 = 573\text{ °R}$

(*c*) $\qquad V_2 = V_1(V_2/V^*)(V^*/V_1) = 120(0.5345)/0.1094 = 586\text{ ft/s}$

(*d*) $\qquad (p_0)_2 = (p_0)_1[(p_0)_2/p_0^*][p_0^*/(p_0)_1] = 100.7(1.3399)/5.8218 = 23.2\text{ lb/in}^2$ abs

**16.221** Air enters a duct of $L/D = 38$ at $V_1 = 195\text{ m/s}$ and $T_1 = 302\text{ K}$. The flow at the exit is choked. What is the average friction factor in the duct for adiabatic flow?

**┃** $\qquad N_M = V/c = V/\sqrt{kRT} \qquad (N_M)_1 = 195/\sqrt{(1.40)(287)(302)} = 0.560$

From Table A-24, $fL/D = 0.6736$, $(f)(38) = 0.6736$, $f = 0.0177$.

**16.222** Air in a tank at $p_0 = 100$ psia and $T_0 = 518\text{ °R}$ flows through a converging nozzle into pipe of 1 in diameter. What will be the mass flow through the pipe if its length is (*a*) 0 ft, (*b*) 1 ft? Assume $f = 0.0409$ and the pressure outside the duct is negligibly small.

**┃** (*a*) $\qquad \dot{m} = 0.6847 p_0 A^*/(RT_0)^{1/2} \qquad$ (nozzle is choked)

$$= (0.6847)[(100)(144)][(\pi)(\tfrac{1}{12})^2/4]/[(1716)(518)]^{1/2} = 0.0570\text{ slug/s}$$

(*b*) $\qquad fL/D = (0.0409)(1)/\tfrac{1}{12} = 0.4908 \qquad$ (section 2 is choked)

From Table A-24 $\qquad N_M = 0.6000 \qquad \dot{m} = \rho_1 A_1 V_1$

$\rho = p/RT \qquad p_1 = p_0/\{1 + [(k - 1)/2](N_M)^2\}^{3.5} = 100/\{1 + [(1.40 - 1)/2](0.6000)^2\}^{3.5} = 78.4\text{ psia}$

$$T_1 = T_0/\{1 + [(k - 1)/2](N_M)^2\} = 518/\{1 + [(1.40 - 1)/2](0.6000)^2\} = 483\text{ °R}$$

$$\rho = (78.4)(144)/[(1716)(483)] = 0.01362\text{ slug/ft}^3$$

$$V = (N_M)(c) = (N_M)\sqrt{kRT} = (0.6000)\sqrt{(1.40)(1716)(483)} = 646\text{ ft/s}$$

$$\dot{m} = (0.01362)[(\pi)(\tfrac{1}{12})^2/4](646) = 0.0480\text{ slug/s}$$

**16.223** Hydrogen ($k = 1.41$ and $R = 4.124$ kJ/kg · K) enters a 6-cm-diameter pipe at $p_1 = 502$ kPa, $V_1 = 298$ m/s, and $T_1 = 21\,°C$. The friction factor is 0.025. How long is the duct if the flow is choked? What is the exit pressure?

$$fL/D = \{[1 - (N_M)_1^2]/k(N_M)_1^2\} + [(k+1)/2k] \ln \{(k+1)(N_M)_1^2/[2 + (k-1)(N_M)_1^2]\}$$
$$N_M = V/c = V/\sqrt{kRT} \qquad (N_M)_1 = 298/\sqrt{(1.41)(4124)(21 + 273)} = 0.227$$
$$0.025L/0.06 = [(1 - 0.227^2)/(1.41)(0.227^2)]$$
$$+ [(1.41 + 1)/(2)(1.41)] \ln \{(1.41 + 1)(0.227)^2/[2 + (1.41 - 1)(0.227)^2]\} \qquad L = 25.6 \text{ m}$$
$$p_1/p^* = (1/N_M)\{(k+1)/[2 + (k-1)(N_M)_2]\}^{1/2}$$
$$502/p^* = (1/0.227)\{(1.41 + 1)/[2 + (1.41 - 1)(0.227)^2]\}^{1/2} \qquad p^* = p_2 = 104 \text{ kPa}$$

**16.224** Air enters a 5-cm by 5-cm square duct at $V_1 = 900$ m/s and $T_1 = 300$ K. The friction factor is 0.018. For what length duct will the flow decelerate to exactly $N_M = 1.0$? If the duct length is 2 m, will there be a normal shock in the duct? If so, at what Mach number will it occur?

$$N_M = V/c = V/\sqrt{kRT} \qquad (N_M)_1 = 900/\sqrt{(1.40)(287)(300)} = 2.59 \qquad \text{(supersonic entrance)}$$

From Table A-24, $fL^*/D = 0.451$, $0.018L^*/0.05 = 0.451$, $L^* = 1.25$ m. If $L = 2$ m, which is larger than $L^*$, there will be a normal shock in the duct. By trial and error, the shock occurs at $N_M = 1.97$ where $fL/D = 0.296$, or $\Delta L_1 = (0.451 - 0.296)(0.05)/0.018 = 0.43$ m. On the downstream side of the shock, $N_M = 0.582$ and $fL^*/D = 0.565$, so $\Delta L_2 = (0.565)(0.05)/0.018 = 1.57$ m; $L_{\text{duct}} = \Delta L_1 + \Delta L_2 = 0.43 + 1.57 = 2.00$ m.

**16.225** Air enters a 0.5-in-diameter pipe subsonically at $p_1 = 62$ psia and $T_1 = 595\,°R$. The pipe length is 25 ft, $f = 0.020$, and the receiver pressure outside the pipe entrance is 18 psia. Compute the mass flow in the pipe, assuming isothermal flow.

$$(\dot{m}/A)^2 = (p_1^2 - p_2^2)/RT[(fL/D) + 2 \ln (p_1/p_2)]$$
$$= \{[(62)(144)]^2 - [(18)(144)]^2\}/(1716)(595)\{[(0.020)(25)/(0.5/12)] + 2 \ln \tfrac{62}{18}\} = 4.939[\text{slugs}/(\text{s-ft}^2)]^2$$
$$\dot{m}/A = 2.222 \text{ slugs}/(\text{s-ft}^2) \qquad \dot{m} = (2.222)[(\pi)(0.5/12)^2/4] = 0.00303 \text{ slug/s}$$

**16.226** Oxygen ($R = 0.260$ kJ/kg · K, $k = 1.40$, and $\mu = 2 \times 10^{-5}$ Pa · s) enters a 120-m-long smooth pipe at 250 kPa and $60\,°C$. The pipe exits into a low-pressure reservoir. The desired mass flow is 0.25 kg/s. Using the Moody chart (Fig. A-5) to compute $f$, estimate the maximum pipe diameter to transport this flow and the resulting exit pressure and temperature.

Guess $D = 0.05$ m

$$\dot{m} = \rho_1 A_1 V_1 \qquad \rho = p/RT \qquad \rho_1 = 250/[(0.260)(60 + 273)] = 2.888 \text{ kg/m}^3$$
$$0.25 = (2.888)[(\pi)(0.05)^2/4](V_1) \qquad V_1 = 44.09 \text{ m/s}$$
$$N_M = V/c = V/\sqrt{kRT} \qquad (N_M)_1 = 44.09/\sqrt{(1.40)(260)(60 + 273)} = 0.1272$$

From Table A-24, $fL^*/D = 40.39$. $N_R = \rho DV/\mu = (2.888)(0.05)(44.09)/(2 \times 10^{-5}) = 3.18 \times 10^5$. From Fig. A-5, $f = 0.0143$. $(0.0143)(120)/D = 40.39$, $D = 0.0425$ m. Additional iterations are required; they yield a value of $D = 0.0485$ m at $(N_M)_1 = 0.135$.

**16.227** Air enters a 1-in-diameter cast iron duct 19 ft long at $V_1 = 200$ fps, $p_1 = 39$ psia, and $T_1 = 521\,°R$. Compute the exit pressure and mass flow using the Moody chart (Fig. A-5) to predict the friction factor.

$$\dot{m} = \rho_1 A_1 V_1 \qquad \rho = p/RT \qquad \rho_1 = (39)(144)/[(1716)(521)] = 0.006282 \text{ slug/ft}^3$$
$$A_1 = (\pi)(\tfrac{1}{12})^2/4 = 0.005454 \text{ ft}^2 \qquad \dot{m} = (0.006282)(0.005454)(200) = 0.006852 \text{ slug/s}$$
$$N_M = V/c = V/\sqrt{kRT} \qquad (N_M)_1 = 200/\sqrt{(1.40)(1716)(521)} = 0.1788$$

From Table A-24, $(fL^*/D)_1 = 18.882$.

$$(fL^*/D)_2 = (fL^*/D)_1 - fL/D \qquad N_R = \rho DV/\mu$$
$$(N_R)_1 = (0.006282)(\tfrac{1}{12})(200)/(3.74 \times 10^{-7}) = 2.80 \times 10^5 \qquad \epsilon/D = 0.00085/(\tfrac{1}{12}) = 0.0102$$

From Fig. A-5, $f = 0.0384$.

$$(fL^*/D)_2 = 18.882 - (0.0384)(20)/(\tfrac{1}{12}) = 9.666 \qquad (N_M)_2 = 0.237$$
$$p_1/p^* = 6.112 \quad \text{and} \quad p_2/p^* = 4.601 \qquad p_2 = (4.601/6.112)(39) = 29.4 \text{ psia}$$

**16.228** Air at 320 K is to be transported through a duct 45 m long. If $f = 0.020$, what is the minimum diameter of duct that can carry the flow without choking, if $V_1 = 50$ m/s? Assume adiabatic flow.

**I**
$$N_M = V/c = V/\sqrt{kRT} \qquad (N_M)_1 = 50/\sqrt{(1.40)(287)(320)} = 0.139$$

From Fig. A-24, $(fL^*/D)_1 = 32.80$, $(0.020)(45)/D = 32.80$, $D = 0.0274$ m

**16.229** Very cold air, to be used in an air-conditioning system of a test chamber, passes through a rectangular duct of cross-sectional area $A$ ft$^2$ and of length $L$ ft. It enters the duct at a temperature of $T_1$ °F at a pressure of $p_1$ psia. It is estimated that $Q$ Btu per unit length per lbm will be transferred from the surroundings into the flow of air into the duct. If the exit temperature is to be $T_2$ °F at ambient pressure $p_2$, how much flow should there be? Set up equations only. Explain how you might go about solving the equations.

**I**.

| equations | unknowns |
|---|---|
| (1) $QL = c_p[(T_0)_2 - (T_0)_1]$ | $(T_0)_1$, $(T_0)_2$ |
| (2) $c_1 = \sqrt{kRT_1}$ | $c_1$ |
| (3) $c_2 = \sqrt{kRT_2}$ | $c_2$ |
| (4) $\dot{m} = \rho_1 V_1 A_1 = (p_1/RT_1)V_1 A_1$ | $\dot{m}, V_1$ |
| (5) $\dot{m} = (p_2/RT_2)V_2 A_2$ | $V_2$ |
| (6) $(N_M)_1 = V_1/c_1$ | $(N_M)_1$ |
| (7) $(N_M)_2 = V_2/c_2$ | $(N_M)_2$ |
| (8) $T_2/(T_0)_2 = 1/\{1 + [(k-1)/2](N_M)_2^2\}$ | |
| (9) $T_1/(T_0)_1 = 1/\{1 + [(k-1)/2](N_M)_1^2\}$ | |

We thus have nine unknowns and nine equations. One way of proceeding is as follows. Solve for $c_2$ and $c_3$ immediately from Eqs. (2) and (3). Now guess at a "$\dot{m}$." Solve for $V_2$ from Eq. (5) . Get $(N_M)_2$ from Eq. (7) and then $(T_0)_2$ from Eq. (8). From Eq. (4) get $V_1$ then go to Eq. (6) for $(N_M)_1$ and finally to Eq. (9) for $(T_0)_1$. Finally go to Eq. (1) and see if the computed $(T_0)_1$ and $(T_0)_2$ satisfy the equation. If not, select another $\dot{m}$, etc.

**16.230** Air at $V_1 = 300$ ft/s, $p = 40$ psia, $T = 60$ °F flows into a 4.0-in-diameter duct. How much heat transfer per unit mass is needed for sonic conditions at the exit? Determine pressure, temperature, and velocity at the exit and at the section where $N_M = 0.70$.

**I** $(N_M)_1 = V_1/\sqrt{kRT_1} = 300/\sqrt{1.4(53.3)(32.17)(460 + 60)} = 0.268$. The isentropic stagnation temperature at the entrance is $T_{01} = T_1\{1 + [(k-1)/2](N_M)_1^2\} = 520[1 + 0.2(0.268^2)] = 527$ °R. The isentropic stagnation temperature at the exit is $T_0^* = T_0(1 + kN_M^2)^2/\{(k+1)N_M^2[2 + (k-1)N_M^2]\} = 527[1 + 1.4(0.268^2)]^2/\{(2.4 \times 0.268^2)[2 + 0.4(0.268^2)]\} = 1827$ °R. The heat transfer per slug of air flowing is $q_H = c_p(T_0^* - T_{01}) = 0.24[32.17(1827 - 527)] = 10\,037$ Btu/slug. The pressure at the exit is $p^* = p[(1 + kN_M^2)/(k + 1)] = (40/2.4)[1 + 1.4(0.268^2)] = 18.34$ psia and the temperature is $T^* = T[(1 + kN_M^2)/(k+1)N_M]^2 = 520\{[1 + 1.4(0.268^2)]/2.4(0.268)\}^2 = 1522$ °R. At the exit, $V^* = c^* = \sqrt{kRT^*} = \sqrt{1.4(53.3)(32.17)(1522)} = 1911$ ft/s. At the section where $N_M = 0.7$, $p = p^*[(k+1)/(1 + kN_M^2)] = 18.34(2.4)/[1 + 1.4(0.7^2)] = 26.1$ psia, $T = T^*[(k+1)N_M/(1 + kN_M^2)]^2 = 1522\{2.4(0.7)/[1 + 1.4(0.7^2)]\}^2 = 1511$ °R and $V = N_M\sqrt{kRT} = 0.7\sqrt{1.4(53.3)(32.17)(1511)} = 1333$ ft/s.

**16.231** In frictionless oxygen flow through a duct, the following conditions prevail at inlet and outlet: $V_1 = 298$ fps, $T_1 = 82$ °F, $(N_M)_2 = 0.5$. Find the heat added per slug and the pressure ratio $p_1/p_2$.

**I**
$$p_1/p_2 = [1 + k(N_M)_2^2]/[1 + k(N_M)_1^2] \qquad N_M = V/c = V/\sqrt{kRT}$$
$$(N_M)_1 = 298/\sqrt{(1.40)(1554)(82 + 460)} = 0.274$$
$$p_1/p_2 = [1 + (1.40)(0.5)^2]/[1 + (1.40)(0.274)^2] = 1.22 \qquad q = c_p[(T_0)_2 - (T_0)_1]$$
$$(T_0) = T\{1 + [(k-1)/2]N_M^2\} \qquad (T_0)_1 = (82 + 460)\{1 + [(1.40 - 1)/2](0.274)^2\} = 550 \text{ °R}$$
$$T_2 = T_1([(N_M)_2/(N_M)_1]\{[1 + k(N_M)_1^2]/[1 + k(N_M)_2^2]\})^2$$
$$= (82 + 460)((0.5/0.274)\{[1 + (1.40)(0.274)^2]/[1 + (1.40)(0.5^2)]\})^2 = 1209 \text{ °R}$$
$$(T_0)_2 = (1209)\{1 + [(1.40 - 1)/2](0.5)^2\} = 1269 \text{ °R} \qquad q = (0.219)(32.17)(1269 - 550) = 5066 \text{ Btu/slug}$$

**16.232** In frictionless air flow through a 100-mm-diameter duct, 0.16 kg/s enters at $T = 0$ °C and $p = 8$ kPa abs. At what rate can heat be added without choking the flow?

❚ rate $= \dot{m} c_p [T_p^* - (T_0)_1]$     $T_0^* = T_0 (1 + k N_M^2)^2 / (N_M)^2 (k + 1)[2 + (k - 1)N_M^2]$     $T_0 = T\{1 + [(k - 1)/2]N_M^2\}$

$N_M = V/c = V/\sqrt{kRT}$     $\dot{m} = \rho A V$     $\rho = p/RT = 8/[(0.287)(0 + 273)] = 0.1021 \text{ kg/m}^3$

$0.16 = (0.1021)[(\pi)(0.100)^2/4](V)$     $V = 199.5 \text{ m/s}$     $(N_M)_1 = 199.5/\sqrt{(1.40)(287)(0 + 273)} = 0.602$

$(T_0)_1 = (0 + 273)\{1 + [(1.40 - 1)/2](0.602)^2\} = 293 \text{ K}$

$T_0^* = (293)[1 + (1.40)(0.602)^2]^2/(0.602)^2(1.40 + 1)[2 + (1.40 - 1)(0.602)^2] = 357 \text{ K}$

rate $= (0.16)(1.005)(357 - 293) = 10.3 \text{ kW}$

**16.233** Frictionless flow through a duct with heat transfer causes the Mach number to decrease from 2 to 1.70 ($k = 1.4$). Determine the temperature, velocity, pressure, and density ratios.

❚     $p_1/p_2 = [1 + k(N_M)_2^2]/[1 + k(N_M)_1^2] = [1 + (1.40)(1.70)^2]/[1 + (1.40)(2)^2] = 0.765$

$T_1/T_2 = [(N_M)_1/(N_M)_2]^2 (p_1/p_2)^2 = (2/1.70)^2 (0.765)^2 = 0.529$

$\rho_2/\rho_1 = [(N_M)_1/(N_M)_2]\sqrt{T_1/T_2} = (2/1.70)\sqrt{0.529} = 0.856$     $V_1/V_2 = \rho_2/\rho_1 = 0.856$

**16.234** In Prob. 16.233 the duct is 3 in square, $p_1 = 16$ psia, and $V_1 = 2000$ fps. Calculate the mass rate of flow for air flowing.

❚     $\dot{m} = \rho_1 A_1 V_1$     $\rho_1 = p_1/RT_1$     $RT_1 = V_1^2/kN_m^2$

$\rho_1 = p_1/(V_1^2/kN_M^2) = (16)(144)/\{2000^2/[(1.40)(2)^2]\} = 0.003226 \text{ slug/ft}^3$

$\dot{m} = (0.003226)[(\tfrac{3}{12})(\tfrac{3}{12})](2000) = 0.403 \text{ slug/s}$

**16.235** How much heat must be transferred per kilogram to cause the Mach number to increase from 2 to 2.6 in a frictionless duct carrying air? ($V_1 = 500$ m/s)

❚     $q_H = c_p[(T_0)_2 - (T_0)_1]$     $T_0 = T\{1 + [(k - 1)/2]N_M^2\}$     $T = V^2/N_M^2 kR$

$T_1 = 500^2/(2)^2(1.40)(287) = 156 \text{ K}$

$(T_0)_1 = (156)\{1 + [(1.40 - 1)/2](2)^2\} = 281 \text{ K}$     $T_1/T_2 = \{[(N_M)_1/(N_M)_2](p_1/p_2)\}^2$

$p_1/p_2 = [1 + k(N_M)_2^2]/[1 + k(N_M)_1^2] = [1 + (1.40)(2.6)^2]/[1 + (1.40)(2)^2] = 1.585$

$T_1/T_2 = [(2/2.6)(1.585)]^2 = 1.487$     $T_2 = 156/1.487 = 105 \text{ K}$

$(T_0)_2 = (105)\{1 + [(1.40 - 1)/2](2.6)^2\} = 247 \text{ K}$

$q_H = (1.005)(247 - 281) = -52.3 \text{ kJ/kg}$     (i.e., cooling)

**16.236** Oxygen at $V_1 = 530$ m/s, $p = 80$ kPa abs, and $T = -10$ °C flows in a 60-mm-diameter frictionless duct. How much heat transfer per kilogram is needed for sonic conditions at the exit?

❚     $q_H = c_p[(T_0)_2 - (T_0)_1]$     $T_0 = T\{1 + [(k - 1)/2]N_M^2\}$     $N_M = V/c = V/\sqrt{kRT}$

$(N_M)_1 = 530/\sqrt{(1.40)(260)(-10 + 273)} = 1.71$     $(T_0)_1 = (-10 + 273)\{1 + [(1.40 - 1)/2](1.71)^2\} = 417 \text{ K}$

$T_2 = T_1\{[(N_M)_2/(N_M)_1](p_2/p_1)\}^2$     $p_1/p_2 = (1 + k)/[1 + k(N_M)_1^2] = (1 + 1.40)/(1 + [(1.40)(1.71)^2]) = 0.4712$

$T_2 = (-10 + 273)[(1/1.70)(1/0.4712)]^2 = 410 \text{ K}$     $(T_0)_2 = (410)\{1 + [(1.40 - 1)/2](1)^2\} = 492 \text{ K}$

$q_H = (0.916)(492 - 417) = 68.6 \text{ kJ/kg}$

**16.237** Apply the first law of thermodynamics, $q_H + p_1/\rho_1 + gz_1 + v_1^2/2 + u_1 = w_s + p_2/\rho_2 + gz_2 + v_2^2/2 + u_2$, to isothermal flow of a perfect gas in a horizontal pipeline, and develop an expression for the heat added per slug flowing.

❚ For isothermal flow in a horizontal pipeline, $w = 0$, $z_1 = z_2$, $T_1 = T_2$, $p_1/\rho_1 = p_2/\rho_2$, $u_1 = u_2$. Hence, $q_H = (v_2^2 - v_1^2)/2$.

**16.238** A fuel mixture, assumed equivalent to air, enters a duct combustion chamber at $V_1 = 95$ m/s and $T_1 = 398$ K. Determine the heat input per unit mass that will cause the exit flow to be choked. What will be the exit Mach number and temperature if 1000 kJ/kg is added during combustion?

❚     $q_{\text{choke}} = c_p[T_0^* - (T_0)_1]$     $T_0 = T + V^2/2c_p$     $(T_0)_1 = 398 + [95^2/(2)(1005)] = 402 \text{ K}$

$N_M = V/c = V/\sqrt{kRT}$     $(N_M)_1 = 95/\sqrt{(1.40)(287)(398)} = 0.238$

From Table A-25, $(T_0)_1/T_0^* = 0.2361$. $T_0^* = 402/0.2361 = 1703$ K, $q_{\text{choke}} = (1.005)(1703 - 402) = 1308$ kJ/kg. For

$q = 1000$ kJ/kg; $(T_0)_2 = (T_0)_1 + (q/c_p) = 402 + (1000/1.005) = 1397$ K, $(T_0)_2/T_0^* = 1397/1703 = 0.8203$. From Table A-25, $(N_M)_2 = 0.601$. $T_2 = (T_0)_2/\{1 + [(k-1)/2](N_M)_2^2\} = (1398)/\{1 + [(1.40-1)/2](0.601)^2\} = 1304$ K.

**16.239** What happens to the inlet flow of Prob. 16.238 if the combustion results in 2480 kJ/kg heat addition and $(p_0)_1$ and $(T_0)_1$ remain the same? How much is the mass flow reduced?

**I**
$$(T_0)_2 = (T_0)_1 + (q/c_p)$$

Choking:

$$(T_0)_2 = T_0^* = 402 + (2480/1.005) = 2870 \text{ K}$$
$$(T_0)_1/T_0^* = 402/2870 = 0.1400$$

From Table A-25, new $(N_M)_1 = 0.1777$.

$$\dot{m}/A = \rho_1 V_1 = \rho_1 c_1 (N_M)_1 = \rho_0 c_0/\{1 + [(k-1)/2](N_M)_1^2\}^3$$

$$\frac{\dot{m} \text{ (new)}}{\dot{m} \text{ (old)}} = \frac{N_M \text{ (new)}}{N_M \text{ (old)}} \left[ \frac{1 + [(1.40-1)/2]N_M^2 \text{ (old)}}{1 + [(1.40-1)/2]N_M^2 \text{ (new)}} \right]^3 = \frac{0.1777}{0.238} \left[ \frac{1 + (0.2)(0.238)^2}{1 + (0.2)(0.1777)^2} \right]^3 = 0.750$$

The mass flow is reduced by 25%.

# CHAPTER 17
# Flow Measurement

**17.1** Water is being discharged through a 3-in-diameter pipe directly into a container that has a volume of $27.0\,\text{ft}^3$. Find the rate and velocity of flow through the pipe if the time required to fill the container is 4 min 19.6 s.

$\blacksquare$ $\qquad Q = V/t = 27.0/[(4)(60) + 19.6] = 0.104\,\text{ft}^3/\text{s}$ $\qquad v = Q/A = 0.104/[(\pi)(\tfrac{3}{12})^2/4] = 2.12\,\text{ft/s}$.

**17.2** Water is being discharged through a 6-in-diameter pipe directly into a container that has a volume of $200.0\,\text{ft}^3$. Find the rate and velocity of flow through the pipe if the time required to fill the container is 3 min 21.2 s.

$\blacksquare$ $\qquad Q = V/t = 200.0/[(3)(60) + 21.2] = 0.994\,\text{ft}^3/\text{s}$ $\qquad v = Q/A = 0.994/[(\pi)(\tfrac{6}{12})^2/4] = 5.06\,\text{ft/s}$

**17.3** A Pitot tube being used to determine the velocity of flow of water in a closed conduit indicates a difference between water levels in the Pitot tube and in the piezometer of 48 mm. What is the velocity of flow?

$\blacksquare$ $\qquad v = \sqrt{2gh} = \sqrt{(2)(9.807)(0.048)} = 0.970\,\text{m/s}$

**17.4** A Pitot tube being used to determine the velocity of flow of water in a closed conduit indicates a difference between water levels in the Pitot tube and in the piezometer of 58 mm. What is the velocity of flow?

$\blacksquare$ $\qquad v = \sqrt{2gh} = \sqrt{(2)(9.807)(0.058)} = 1.07\,\text{m/s}$

**17.5** An airplane uses a Pitot measuring device to determine its velocity. The instruments show a stagnation pressure ($p_s$) of 508 mmHg abs, a static pressure ($p_0$) of 325 mmHg abs, and a stagnation temperature of 60 °C. What is the plane's velocity?

$\blacksquare$ $\qquad N_M = \sqrt{[2/(k-1)][(p_s/p_0)^{(k-1)/k} - 1]} = \sqrt{[2/(1.40-1)][(\tfrac{508}{325})^{(1.40-1)/1.40} - 1]} = 0.825$

Since ($N_M = 0.825$) < 1.0, flow is subsonic, and

$$v = \sqrt{[2kRT/(k-1)][(p_s/p_0)^{(k-1)/k} - 1]} \qquad T = 60 + 273 = 333\,\text{K}$$
$$v = \sqrt{[(2)(1.40)(287)(333)/(1.40-1)][(\tfrac{508}{325})^{(1.40-1)/1.40} - 1]} = 302\,\text{m/s}$$

**17.6** An airplane uses a Pitot measuring device to determine its velocity. The instruments show a stagnation pressure ($p_s$) of 498 mmHg abs, a static pressure ($p_0$) of 318 mmHg abs, and a stagnation temperature of 75 °C. What is the plane's velocity?

$\blacksquare$ $\qquad N_M = \sqrt{[2/(k-1)][(p_s/p_0)^{(k-1)/k} - 1]} = \sqrt{[2/(1.40-1)][(\tfrac{498}{318})^{(1.40-1)/1.40} - 1]} = 0.827$

Since ($N_M = 0.827$) < 1.0, flow is subsonic, and

$$v = \sqrt{[2kRT/(k-1)][(p_s/p_0)^{(k-1)/k} - 1]} \qquad T = 75 + 273 = 348\,\text{K}$$
$$v = \sqrt{[(2)(1.40)(287)(348)/(1.40-1)][(\tfrac{498}{318})^{(1.40-1)/1.40} - 1]} = 309\,\text{m/s}$$

**17.7** Air ($\gamma = 0.075$ pcf) is flowing in Fig. 17-1. If $u = 13.9$ fps and $V_c = 15.7$ fps, determine the manometer reading.

**Fig. 17-1**

$$u^2/2g = (\gamma_{H_2O}/\gamma_{air} - 1)(x) \qquad 13.9^2/[(2)(32.2)] = (62.4/0.075 - 1)(x)$$
$$x = 0.00361 \text{ ft} \quad \text{or} \quad 0.0433 \text{ in}$$

(Such a reading is too small for practical use.)

**17.8** Repeat Prob. 17.7 for Fig. 17-2.

Fig. 17-2

$$V_c/2g - u^2/2g = (\gamma_{oil}/\gamma_{air} - 1)(x)$$
$$15.7^2/[(2)(32.2)] - 13.9^2/[(2)(32.2)] = [(0.86)(62.4)/0.075 - 1](x) \qquad x = 0.00116 \text{ ft} \quad \text{or} \quad 0.0139 \text{ in}$$

(As in Prob. 17.7, such a reading is too small for practical use.)

**17.9** In Fig. 17-3, pressure gage $A$ reads 10.8 psi, while pressure gage $B$ reads 11.7 psi. Find the velocity if air at 50 °F is flowing. Atmospheric pressure is 26.9 inHg. Assume $C_I = 1.0$ and neglect compressibility effects.

Fig. 17-3

$$p = (26.9/29.9)(14.7) + 10.8 = 24.03 \text{ psia} \qquad \gamma = p/RT = (24.03)(144)/[(53.3)(50 + 460)] = 0.1273 \text{ lb/ft}^3$$
$$u = C_I\sqrt{2g(p_s/\gamma - p_0/\gamma)} = (1.0)\sqrt{(2)(32.2)[(11.7)(144)/0.1273 - (10.8)(144)/0.1273]} = 256 \text{ ft/s}$$

**17.10** In Prob. 17.9, if the two pressure gages were replaced by a differential manometer containing water, what would be the reading on the manometer ($y$)?

$$\Delta p/\gamma = (\text{s.g.}_M/\text{s.g.}_F - 1)(y) \qquad \Delta p/\gamma = (11.7)(144)/0.1273 - (10.8)(144)/0.1273 = 1018 \text{ ft}$$
$$1018 = (62.4/0.1273 - 1)(y) \qquad y = 2.081 \text{ ft} \quad \text{or} \quad 25.0 \text{ in}$$

**17.11** In Fig. 17-3, assume kerosene (s.g. = 0.81) is flowing. The pressure gages at $A$ and $B$ read 74 Pa and 152 Pa, respectively. Find the velocity $u$, assuming $C_I = 1.0$.

$$\gamma = (0.81)(9.79) = 7.930 \text{ kN/m}^3 \quad \text{or} \quad 7930 \text{ N/m}^3$$
$$u = C_I\sqrt{2g(p_s/\gamma - p_0/\gamma)} = (1.00)\sqrt{(2)(9.807)(\tfrac{152}{7930} - \tfrac{74}{7930})} = 0.439 \text{ m/s}$$

**17.12** The piezometer in Fig. 17-4 is connected to a mercury manometer, and the reading is 92 mm. The velocity is 3.5 m/s. If carbon tetrachloride (s.g. = 1.59) is flowing, what is $C_I$ for the measurement?

Fig. 17-4

$$u = C_I\sqrt{2g(p_s/\gamma - p_0/\gamma)} = C_I\sqrt{2g(\Delta p/\gamma)}$$
$$\Delta p/\gamma = (\text{s.g.}_M/\text{s.g.}_F - 1)(y) = (13.6/1.59 - 1)(0.092) = 0.6949 \text{ m}$$
$$3.5 = (C_I)\sqrt{(2)(9.807)(0.6949)} \qquad C_I = 0.948$$

**17.13** In Fig. 17-4, suppose air at 90 °F is flowing. The pitometer is attached to a manometer containing a liquid (s.g. = 0.89). Find velocity $u$ versus manometer readings assuming $C_I = 0.93$. Assume the air is at standard atmospheric pressure.

▎ From Table A-4, $\rho_{air} = 0.00224$ slug/ft$^3$. $\rho_{liquid}/\rho_{air} = (0.89)(1.94)/0.00224 = 770.8$. Thus 1 ft of manometer reading is equivalent to 770.8 ft of air. $u = C_I\sqrt{2g(p_s/\gamma - p_0/\gamma)} = C_I\sqrt{2g(\Delta p/\gamma)}$. Letting $y$ = manometer reading, $u = C_I\sqrt{2g(770.8y)}$. For $y = 1$ in, $u = (0.93)\sqrt{(2)(32.2)[(770.8)(\frac{1}{12})]} = 59.8$ ft/s. For $y = 5$ in, $u = (0.93)\sqrt{(2)(32.2)[(770.8)(\frac{5}{12})]} = 134$ ft/s. For $y = 10$ in, $u = (0.93)\sqrt{(2)(32.2)[(770.8)(\frac{10}{12})]} = 189$ ft/s. For $y = 20$ in, $u = (0.93)\sqrt{(2)(32.2)[(770.8)(\frac{20}{12})]} = 267$ ft/s. (At larger manometer readings, the effect of compressibility must be considered.)

**17.14** A Pitot tube is placed in a pipe carrying water at 25 °C. The Pitot tube and a wall piezometer tube are connected to a water-mercury manometer which registers a differential of 66 mm. Assuming $C_I = 0.98$, what is the velocity approaching the tube?

▎
$$\Delta p/\gamma = (s.g._M/s.g._F - 1)(y) = (13.6/1 - 1)(0.066) = 0.8316 \text{ m}$$
$$u = C_I\sqrt{2g(p_s/\gamma - p_0/\gamma)} = C_I\sqrt{2g(\Delta p/\gamma)} = (0.98)\sqrt{(2)(9.807)(0.8316)} = 3.96 \text{ m/s}$$

**17.15** Suppose that the fluids of Prob. 17.14 are reversed so that mercury is flowing in the pipe and water is the gage fluid (with the manometer now inverted). With the same gage differential, what would be the velocity of the mercury?

▎
$$\Delta p/\gamma = (1 - s.g._M/s.g._F)(y) = (1 - 1/13.6)(0.066) = 0.06115 \text{ m}$$
$$u = C_I\sqrt{2g(p_s/\gamma - p_0/\gamma)} = C_I\sqrt{2g(\Delta p/\gamma)} = (0.98)\sqrt{(2)(9.807)(0.06115)} = 1.07 \text{ m/s}$$

**17.16** A Prandtl tube is placed on the centerline of a smooth 10-in-diameter pipe in which water at 80 °F is flowing. The reading on a differential manometer attached to this Prandtl tube is 11 in of carbon tetrachloride (s.g. = 1.59). Find the flow rate.

▎
$$V/u_{max} = 1/(1 + 1.33f^{1/2}) \qquad h = (1.59 - 1)(\tfrac{11}{12}) = 0.5408 \text{ ft}$$
$$u_{max} = \sqrt{2gh} = \sqrt{(2)(32.2)(0.5408)} = 5.901 \text{ ft/s}$$

Try $V = 5$ ft/s: $N_R = DV/\nu = (\frac{10}{12})(5)/(9.15 \times 10^{-6}) = 4.55 \times 10^5$. From Fig. A-5, $f = 0.013$. $V/5.901 = 1/[1 + (1.33)(0.013)^{1/2}]$, $V = 5.12$ ft/s. ($f$ is still 0.013 for $V = 5.12$ ft/s.) $Q = AV = [(\pi)(\frac{10}{12})^2/4](5.12) = 2.79$ ft$^3$/s.

**17.17** A Pitot-static tube for which $C_I = 0.99$ is connected to an inverted $U$-tube containing oil (s.g. = 0.83). Water is flowing. What is the velocity if the manometer reading is 3.7 in?

▎
$$\Delta p/\gamma = (1 - s.g._M/s.g._F)(y) = (1 - 0.83/1)(3.7/12) = 0.0524 \text{ ft}$$
$$u = C_I\sqrt{2g(p_s/\gamma - p_0/\gamma)} = C_I\sqrt{2g(\Delta p/\gamma)} = (0.99)(\sqrt{(2)(32.2)(0.0524)}) = 1.82 \text{ ft/s}$$

**17.18** A Pitot tube having a coefficient of 0.98 is used to measure the velocity of water at the center of a pipe, as shown in Fig. 17-5. What is the velocity?

**Fig. 17-5**

▎
$$u = C_I\sqrt{2g(p_s/\gamma - p_0/\gamma)} = (0.98)\sqrt{(2)(32.2)(18.6 - 15.5)} = 13.8 \text{ ft/s}$$

**17.19** Air flows through a duct, and the Pitot-static tube measuring the velocity is attached to a differential gage containing water. If the deflection of the gage is 4 in, calculate the air velocity, assuming the specific weight of air is constant at 0.0761 lb/ft³ and that the coefficient of the tube is 0.98.

$$(p_s/\gamma - p_0/\gamma) = (\tfrac{4}{12})(62.4)/0.0761 = 273.3 \text{ ft of air}$$
$$u = C_I\sqrt{2g(p_s/\gamma - p_0/\gamma)} = (0.98)\sqrt{(2)(32.2)(273.3)} = 130 \text{ ft/s}$$

**17.20** Carbon tetrachloride (s.g. = 1.60) flows through a pipe. The differential gage attached to the Pitot-static tube shows a 3-in deflection of mercury. Assuming $C = 1.00$, find the velocity.

$$p_s - p_0 = (\tfrac{3}{12})(13.6 - 1.60)(62.4) = 187.2 \text{ lb/ft}^2$$
$$u = C_I\sqrt{2g(p_s/\gamma - p_0/\gamma)} = (1.00)\sqrt{(2)(32.2)\{187.2/[(1.60)(62.4)]\}} = 11.0 \text{ ft/s}$$

**17.21** Water flows at a velocity of 4.65 fps. A differential gage ($C_I = 1.0$) which contains a liquid of specific gravity 1.25 is attached to the Pitot-static tube. What is the deflection of the gage fluid?

$$u = C_I\sqrt{2g(p_s/\gamma - p_0/\gamma)} = C_I\sqrt{2g(\Delta p/\gamma)} \qquad 4.65 = (1.0)\sqrt{(2)(32.2)(\Delta p/\gamma)} \qquad \Delta p/\gamma = 0.3358 \text{ ft of water}$$
$$\Delta p/\gamma = (\text{s.g.}_M/\text{s.g.}_F - 1)(y) \qquad 0.3358 = [(1.25/1) - 1](y) \qquad y = 1.34 \text{ ft}$$

**17.22** Develop the expression for measuring the flow of a gas with a Pitot tube.

The flow from $A$ to $B$ in the figure of Prob. 17.18 may be considered adiabatic and with negligible loss. Using the Bernoulli equation, $A$ to $B$, we obtain $\{[k/(k - 1)](p_A/\gamma_A) + (V_A^2/2g) + 0\} - \text{negligible loss} = [k/(k - 1)](p_A/\gamma_A)(p_B/p_A)^{(k-1)/k} + 0 + 0$ or

$$V_A^2/2g = [k/(k - 1)](p_A/\gamma_A)[(p_B/p_A)^{(k-1)/k} - 1] \qquad (1)$$

The term $p_B$ is the stagnation pressure. This expression (1) is usually rearranged, introducing the ratio of the velocity at $A$ to the acoustic velocity $c$ of the undisturbed fluid. The acoustic velocity $c = \sqrt{E/\rho} = \sqrt{kp/\rho} = \sqrt{kpg/\gamma}$. Combining with Eq. (1) above,

$$V_A^2 = [c^2/(k - 1)][(p_B/p_A)^{(k-1)/k} - 1] \qquad \text{or} \qquad p_B/p_A = \{1 + [(k - 1)/2](V_A/c)^2\}^{k/(k-1)} \qquad (2)$$

Expanding by the binomial theorem,

$$p_B/p_A = 1 + (k/2)(V_A/c)^2\{1 + \tfrac{1}{4}(V_A/c)^2 - [(k - 2)/24](V_A/c)^4 + \cdots\} \qquad (3)$$

In order to compare this expression with the formula in Prob. 17.18, multiply through by $p_A$ and replace $kp_A/c^2$ by $\rho_A$, obtaining

$$p_B = p_A + \tfrac{1}{2}\rho_A V_A^2\{1 + \tfrac{1}{4}(V_A/c)^2 - [(k - 2)/24](V_A/c)^4 + \cdots\} \qquad (4)$$

The above expressions apply to all compressible fluids for ratios of $V/c$ less than unity. For ratios over unity, shock wave and other phenomena occur, the adiabatic assumption is not sufficiently accurate, and the derivation no longer applies. The ratio $V/c$ is called the *Mach number*.

The term in braces in (4) is greater than unity and the first two terms provide sufficient accuracy. The effect of compressibility is to increase the stagnation-point pressure over that of an incompressible fluid.

**17.23** Air flowing under atmospheric conditions ($\gamma = 0.0763$ lb/ft³ at 60 °F) at a velocity of 300 fps is measured by a Pitot tube. Calculate the error in the stagnation pressure by assuming the air to be incompressible.

$$p_B = p_A + \rho V^2/2 = (14.7)(144) + (0.0763/32.2)(300)^2/2 = 2223 \text{ lb/ft}^2 \text{ abs}$$

Also:
$$p_B = p_A + \tfrac{1}{2}\rho_A V_A^2\{1 + \tfrac{1}{4}(V_A/c)^2 - [(k - 2)/24](V_A/c)^4 + \cdots\}$$
$$c = \sqrt{kgRT} = \sqrt{(1.40)(32.2)(53.3)(60 + 460)} = 1118 \text{ ft/s}$$
$$p_B = (14.7)(144) + [(0.0763/32.2)(300)^2/2]\{1 + (\tfrac{1}{4})(\tfrac{300}{1118})^2 - [(1.40 - 2)/24](\tfrac{300}{1118})^4 + \cdots\} = 2225 \text{ lb/ft}^2 \text{ abs}$$

The error in the stagnation pressure is $(2225 - 2223)/2225 = 0.0009$, or 0.09 percent.

**17.24** The difference between the stagnation pressure and the static pressure measured by a Pitot-static device is 412 lb/ft². The static pressure is 14.5 psia and the temperature in the air stream is 60 °F. What is the velocity of the air, assuming the air is (a) compressible and (b) incompressible.

**(a)** Using Eq. (2) of Prob. 17.22,

$$p_B/p_A = \{1 + [(k-1)/2](V_A/c)^2\}^{k/(k-1)}$$
$$c = \sqrt{kgRT} = \sqrt{(1.40)(32.2)(53.3)(60+460)} = 1118 \text{ ft/s}$$
$$[(14.5)(144)+412]/(14.5)(144) = \{1 + [(1.40-1)/2](V_A/1118)^2\}^{1.40/(1.40-1)} \qquad V_A = 574 \text{ ft/s}$$

**(b)**
$$\gamma = p/RT = (14.5)(144)/[(53.3)(60+460)] = 0.07534 \text{ lb/ft}^3$$
$$V = \sqrt{2g(\Delta p/\gamma)} = \sqrt{(2)(32.2)(412/0.07534)} = 593 \text{ ft/s}$$

**17.25** Air flows at 800 fps through a duct. At standard barometer, the stagnation pressure is −5.70 ft of water gage. The stagnation temperature is 145 °F. What is the static pressure in the duct?

▌ Using Eq. (2) of Prob. 17.22, $p_B/p_A = \{1 + [(k-1)/2](V_A/c)^2\}^{k/(k-1)}$. With two unknowns in the equation, assume a $V/c$ ratio of 0.72. $(-5.70+34.0)(62.4)/p_A = \{1 + [(1.40-1)/2](0.72)^2\}^{1.40/(1.40-1)}$, $p_A = 1250 \text{ lb/ft}^2$ abs. Check the assumption using the adiabatic relation $T_B/T_A = (p_B/p_A)^{(k-1)/k}$:

$$(145+460)/T_A = [(-5.70+34.0)(62.4)/1250]^{(1.40-1)/1.40} \qquad T_A = 548 \text{ °R}$$
$$c = \sqrt{kgRT} = \sqrt{(1.40)(32.2)(53.3)(548)} = 1147 \text{ ft/s}$$
$$V/c = \tfrac{800}{1147} = 0.697 \qquad (-5.70+34.0)(62.4)/p_A = \{1 + [(1.40-1)/2](0.697)^2\}^{1.40/(1.40-1)}$$
$$p_A = 1277 \text{ lb/ft}^2 \text{ abs}$$

No further refinement is necessary.

**17.26** A static tube indicates a static pressure that is 900 Pa too low when liquid is flowing at 2.1 m/s. Calculate the correction to be applied to the indicated pressure for the liquid flowing at 6 m/s.

▌ Since $\Delta h$ is a function of $V^2$, $\Delta h_2/\Delta h_1 = V_2^2/V_1^2$, $\Delta p_2/\Delta p_1 = V_2^2/V_1^2$, $\Delta p_2/900 = 6^2/2.1^2$, $\Delta p_2 = 7350 \text{ Pa}$.

**17.27** A simple Pitot tube is inserted into a small stream of flowing oil, $\gamma = 56 \text{ lb/ft}^3$, $\mu = 0.65 \text{ P}$, $\Delta h = 1.7 \text{ in}$, $h_0 = 6 \text{ in}$. What is the velocity at point 1?

▌
$$v = \sqrt{2g\,\Delta h} = \sqrt{(2)(32.2)(1.7/12)} = 3.02 \text{ ft/s}$$

**17.28** A stationary body immersed in a river has a maximum pressure of 73 kPa exerted on it at a distance of 5.9 m below the free surface. Calculate the river velocity at this depth.

▌
$$V^2/2g = p_t/\gamma - p_s/\gamma \qquad V^2/[(2)(9.807)] = 73/9.79 - 5.9 \qquad V = 5.53 \text{ m/s}$$

**17.29** For Fig. 17-6 derive the equation for velocity at point 1.

(a)     (b)     **Fig. 17-6**

▌ For Fig. 17-6a (Pitot tube and piezometer opening), $V_1^2/2g = (p_2 - p_1)/\gamma$.

**Manometer equation:**

$$p_1(S/\gamma) + kS + R'S_0 - (k+R')S = p_2(S/\gamma) \qquad (p_2-p_1)/\gamma = R'[(S_0-S)/S]$$
$$V_1^2/2g = R'[(S_0-S)/S] \qquad V_1 = \sqrt{2gR'(S_0/S-1)}$$

For Fig. 17-6b (Pitot-static tube), we analyze this system in a manner similar to that for Fig. 17-6a and show that the same relations hold, but the uncertainty in the measurement of static pressure requires a corrective coefficient $C_t$ to be applied so that $V_1 = C_t\sqrt{2gR'(S_0/S-1)}$.

**17.30** In Fig. 17-6a air is flowing ($p = 20$ psia, $T = 48$ °F) and water is in the manometer. For $R' = 0.9$ in, calculate the velocity of air.

$$V = \sqrt{2gR'(S_0/S - 1)} \qquad \rho = p/RT = (20)(144)/[(53.3)(48 + 460)] = 0.1064 \text{ lbm/ft}^3$$

$$S_0/S = 62.4/0.1064 = 586.5 \qquad V = \sqrt{(2)(32.2)(0.9/12)(586.5 - 1)} = 53.2 \text{ ft/s}$$

**17.31** In Fig. 17-6 air is flowing ($p = 102$ kPa abs and $T_1 = 9$ °C) and mercury is in the manometer. For $R' = 188$ mm, calculate the velocity at 1 for air considered incompressible.

$$V = \sqrt{2gR'(S_{Hg}/S_{air} - 1)} \qquad \rho = p/RT \qquad \rho_{air} = 102/[(0.287)(9 + 273)] = 1.260 \text{ kg/m}^3$$

$$S_{Hg}/S_{air} = (13.6)(1000)/1.260 = 10\,794 \qquad V = \sqrt{(2)(9.807)(0.188)(10\,794 - 1)} = 199 \text{ m/s}$$

**17.32** Work Prob. 17.31 for isentropic compression of air between 1 and 2.

$$V_1^2/2 = c_p T_1[(p_2/p_1)^{(k-1)/k} - 1] \qquad (p_2 - p_1)/\gamma_{air} = R'[(S_{Hg} - S_{air})/S_{air}]$$

$$p_2 - p_1 = R'\gamma_{Hg} - R'\gamma_{air} \approx R'\gamma_{Hg} \qquad p_2 - 102 = (0.188)[(13.6)(9.79)] \qquad p_2 = 127.0 \text{ kPa}$$

$$V_1^2/2 = (1005)(9 + 273)[(127.0/102)^{(1.40-1)/1.40} - 1] \qquad V_1 = 191 \text{ m/s}$$

**17.33** A Pitot-static tube directed into a 3-m/s water stream has a gage difference of 20 mm on a water-mercury differential manometer. Determine the coefficient for the tube.

$$V = C_I\sqrt{2gR'(S_0/S - 1)} \qquad 3 = C_I\sqrt{(2)(9.807)(0.020)[(13.6/1) - 1]} \qquad C_I = 1.35$$

**17.34** A Pitot-static tube, $C_I = 1.14$, has a gage difference of 15 mm on a water-mercury manometer when directed into a water stream. Calculate the velocity.

$$V = C_I\sqrt{2gR'(S_0/S - 1)} = (1.14)\sqrt{(2)(9.807)(0.015)[(13.6/1) - 1]} = 2.19 \text{ m/s}$$

**17.35** A Pitot-static tube of the Prandtl type has the following values of gage difference $R'$ for the radial distance from the center of a 3-ft-diameter pipe:

| r, ft | 0.0 | 0.3 | 0.6 | 0.9 | 1.2 | 1.48 |
|-------|------|------|------|------|------|------|
| R', in | 4.00 | 3.89 | 3.72 | 3.40 | 2.94 | 2.30 |

Water is flowing, and the manometer fluid has a specific gravity of 2.83. Calculate the discharge.

$V = C_I\sqrt{2gR'(S_0/S - 1)}$. For a Prandtl tube, $C_I = 1.0$

$$V_i = (1.0)\sqrt{(2)(32.2)(R'/12)[(2.83/1) - 1]} = 3.134\sqrt{R'} \qquad (R' \text{ in inches})$$

$$A_1 = (\pi)[(r_2 - r_1)/2]^2 \qquad A_i = (\pi)\{[(r_{i+1} + r_i)/2]^2 - [(r_i + r_{i-1})/2]^2\}$$

$$A_6 = (\pi)\{r_7^2 - [(r_5 + r_6)/2]^2\}$$

| i | 1 | 2 | 3 | 4 | 5 | 6 | 7 | |
|-----|-------|-------|-------|-------|-------|-------|-----|------|
| $r_i$ | 0 | 0.3 | 0.6 | 0.9 | 1.2 | 1.48 | 1.5 | ft |
| $V_i$ | 6.268 | 6.181 | 6.045 | 5.779 | 5.374 | 4.753 | — | ft/s |
| $A_i$ | 0.071 | 0.565 | 1.131 | 1.696 | 2.177 | 1.428 | — | ft² |

$$Q = \sum_{i=1}^{6} V_i A_i = (6.268)(0.071) + (6.181)(0.565) + (6.045)(1.131)$$

$$+ (5.779)(1.696) + (5.374)(2.177) + (4.753)(1.428) = 39.1 \text{ ft}^3/\text{s}$$

**17.36** Determine the gage difference on a water-nitrogen manometer for flow of nitrogen at 180 m/s, using a Pitot-static tube. The static pressure is 160 kPa abs, and the corresponding temperature is 35 °C. True static pressure is measured by the tube.

❚ $$p_2 = (p_1)(V_1^2/2c_pT_1 + 1)^{k/(k-1)} = (160)\{180^2/[(2)(1040)(35 + 273)] + 1\}^{1.40/(1.40-1)} = 190.2 \text{ kPa}$$
$$R' = (p_t - p_s)/\gamma = (190.2 - 160)/9.79 = 3.08 \text{ m}$$

**17.37** Measurements in an air stream indicate that the stagnation pressure is 18 psia, the static pressure is 13 psia, and the stagnation temperature is 109 °F. Determine the temperature and velocity of the air stream.

❚ $$V_1^2/2 = c_pT_2[1 - (p_1/p_2)^{(k-1)/k}] = (6000)(109 + 460)[1 - (\tfrac{13}{18})^{(1.40-1)/1.40}]$$
$$V_1 = 779 \text{ ft/s} \qquad T_1 = T_2 - V^2/2c_p = (109 + 460) - 779^2/[(2)(6000)] = 518 \text{ °R} \quad \text{or} \quad 58 \text{ °F}$$

**17.38** If 0.6 kg/s nitrogen flows through a 60-mm-diameter tube with stagnation temperature of 42 °C and undisturbed temperature of 14 °C, find the velocity and static and stagnation pressures.

❚ $$V_1 = \sqrt{2c_p(T_2 - T_1)} = \sqrt{(2)(1040)[(42 + 273) - (14 + 273)]} = 241 \text{ m/s}$$
$$p_1 = \rho_1 RT_1 = (\dot{m}/AV)RT_1$$
$$A = (\pi)(0.060)^2/4 = 0.002827 \text{ m}^2 \qquad p_1 = \{0.6/[(0.002827)(241)]\}(0.297)(14 + 273) = 75.067 \text{ kPa}$$
$$p_2 = (p_1)(T_2/T_1)^{k/(k-1)} = (75.067)[(42 + 273)/(14 + 273)]^{1.40/(1.40-1)} = 103.980 \text{ kPa}$$

**17.39** A Pitot tube is inserted into a flowing air stream at 25 °C and 120 kPa. The differential pressure reads 18 mm on an air-mercury manometer. What is the indicated air speed?

❚ $$\Delta p = (\rho_{Hg} - \rho_{air})gh = \rho_{air}V^2/2 \qquad \rho_{air} = p/RT = 120/[(0.287)(25 + 273)] = 1.40 \text{ kg/m}^3$$
$$(13\,570 - 1.40)(9.807)(0.018) = (1.40)(V^2/2) \qquad V = 58.5 \text{ m/s}$$

**17.40** For the Pitot-static pressure arrangement of Fig. 17-7, compute the **(a)** centerline velocity, **(b)** pipe volume flow, and **(c)** wall shear stress. The manometer fluid is Meriam red oil (s.g. = 0.827).

**Fig. 17-7**

❚ **(a)** $$p_0 - p = (\rho_{oil} - \rho_{air})gh = [(0.827)(998) - 1.20](9.807)(0.046) = 371.8 \text{ Pa}$$
$$V_{C_L} = \sqrt{2\Delta p/\rho_{air}} = \sqrt{(2)(371.8)/1.20} = 24.9 \text{ m/s}$$

**(b)** $V_{avg} = V_{C_L}/(1 + 1.33\sqrt{f})$. Estimate $V_{avg} = 0.85V_{C_L} = (0.85)(24.9) = 21.2$ m/s. $N_R = \rho DV/\mu = (1.20)(0.100)(21.2)/(1.80 \times 10^{-5}) = 1.41 \times 10^5$. From Fig. A-5, $f = 0.0178$.

$$V_{avg} = 24.9/(1 + 1.33\sqrt{0.0178}) = 21.1 \text{ m/s} \qquad \text{(O.K.)} \qquad Q = AV = [(\pi)(0.100)^2/4](21.1) = 0.1657 \text{ m}^3/\text{s}$$
**(c)** $$\tau = (\tfrac{1}{8})(f\rho V_{avg}^2) = (\tfrac{1}{8})(0.0178)(1.20)(21.1)^2 = 1.19 \text{ Pa}$$

**17.41** For the water flow at 20 °C of Fig. 17-8, use the manometer measurement to estimate the **(a)** centerline velocity and **(b)** volume flow in the 8-in-diameter pipe. Assume a smooth wall.

❚ **(a)** $V_{C_L} = \sqrt{2\Delta p/\rho_{H_2O}}$. Assume at first that static pressure at $A$ = static pressure at $B$.

$$\Delta p = (p_0)_B - p_B = (p_0)_B - p_A = (\rho_{Hg} - \rho_{H_2O})gh = [(13.6 - 1)(1.94)](32.2)(\tfrac{3}{12}) = 196.8 \text{ lb/ft}^2$$
$$V_{C_L} = \sqrt{(2)(196.8)/1.94} = 14.2 \text{ ft/s}$$

**(b)** $V_{avg} = V_{C_L}/(1 + 1.33\sqrt{f})$. Estimate $V_{avg} = 0.87V = (0.87)(14.2) = 12.4$ ft/s. $N_R = \rho DV/\mu = (1.94)(\tfrac{8}{12})(12.4)/(2.04 \times 10^{-5}) = 7.86 \times 10^5$. From Fig. A-5, $f = 0.0133$.

$$V_{avg} = 14.2/(1 + 1.33\sqrt{0.0133}) = 12.3 \text{ ft/s} \qquad \text{(O.K.)}$$
$$Q = AV = [(\pi)(\tfrac{8}{12})^2/4](12.3) = 4.29 \text{ ft}^3/\text{s}$$

0.5 ft

3 in

Mercury

**Fig. 17-8**

Check:

$$p_A - p_B = \rho g f(L/D)(V^2/2g) = (1.94)(32.2)(0.0133)[0.5/(\tfrac{8}{12})]\{12.3^2/[(2)(32.2)]\} = 1.5 \text{ lb/ft}^2$$

This gives a $1.5/196.8 = 0.008$, or 0.8 percent error in $\Delta p$.

**17.42** A Pitot tube placed in a flow of methane ($\rho = 0.666$ kg/m$^3$) at 20 °C and 1 atm shows a methane-water differential manometer reading of 14 mm. What is the methane velocity? What will the manometer reading be if the methane velocity is 35 m/s?

$$\Delta p = (\rho_{\text{H}_2\text{O}} - \rho_{\text{CH}_4})gh = (998 - 0.666)(9.807)(0.014) = 136.9 \text{ Pa}$$
$$V_{\mathbb{C}} = \sqrt{2\Delta p/\rho_{\text{CH}_4}} = \sqrt{(2)(136.9)/0.666} = 20.3 \text{ m/s}$$

If $V_{\mathbb{C}} = 35$ m/s, $\Delta p = (\rho_{\text{H}_2\text{O}} - \rho_{\text{CH}_4})gh = \rho V^2/2$, $(998 - 0.666)(9.807)(h) = (0.666)(35)^2/2$, $h = 0.0417$ m, or 41.7 mm.

**17.43** An airplane flying at altitude 7 km uses a Pitot stagnation probe without a static tube. The measured stagnation pressure is 44 kPa. Find the indicated airplane speed and its probable uncertainty.

$\Delta p = p_0 - p = \rho V^2/2$. From Table A-8, $p$ at 7000 m $= 41.44$ kPa and $\rho = 0.593$ kg/m$^3$. $(44 - 41.44)(1000) = (0.593)(V)^2/2$, $V = 92.9$ m/s. As for the uncertainty, assume $\rho_{\text{air}}$ and $p_{\text{air}}$ have ±3 percent error and $p_0$ has ±1 percent. The worst case would be high $\rho$ and $p$ and low $p_0$. $\Delta p = [(44)(0.99) - (41.44)(1.03)](1000) = [(0.593)(1.03)](V)^2/2$, $V = 53.6$ m/s. This is $(92.9 - 53.6)/92.9 = 0.42$, or 42 percent less. Thus, small errors in this case make for a large error in the velocity estimate.

**17.44** An engineer who took college fluid mechanics on a pass-fail basis has placed the static pressure hole far upstream of the stagnation probe, as in Fig. 17-9, thus contaminating the Pitot measurement ridiculously with pipe friction losses. If the pipe flow is air at 20 °C and 1 atm and the manometer fluid is Meriam red oil (s.g. = 0.827), estimate the air centerline velocity for the given manometer reading. Assume a smooth-walled tube.

|← 11 m →|

→ Air

170 mm

$D = 80$ mm

**Fig. 17-9**

$$(p_0)_B - p_A = (\rho_{\text{oil}} - \rho_{\text{air}})gh = [(0.827)(998) - 1.20](9.807)(-0.170) = -1374 \text{ Pa}$$

Consider friction loss.

$$p_A - p_B = (f)(L/D)(V_{\text{avg}}^2/2)(\rho)$$
$$(p_0)_B - p_B = [(p_0)_B - p_A] + (p_A - p_B) = -1374 + (f)(L/D)(V_{\text{avg}}^2/2)(\rho) = (\rho)(V_{\mathbb{C}})_B^2/2$$

Guess $f = 0.016$ and $V_{avg} = 0.85V_{\mathscr{C}}$:

$$-1374 + (0.016)[11/0.080](V_{avg}^2/2)(1.20) = (1.20)(V_{avg}/0.85)^2/2 \qquad V_{avg} = 53.0 \text{ m/s}$$
$$N_R = \rho DV/\mu = (1.20)(0.080)(53.0)/(1.81 \times 10^{-5}) = 2.81 \times 10^5$$

From Fig. A-5, $f = 0.016$ (O.K.).

$$V_{\mathscr{C}}/V_{avg} = 1 + 1.33f^{1/2} = 1 + (1.33)(0.016)^{1/2} = 1.168 \qquad V_{avg}/V_{\mathscr{C}} = 1/1.168 = 0.856 \qquad \text{(O.K.)}$$

Hence, $V_{avg} = 53.0 \text{ m/s}$ and $V_{\mathscr{C}} = (53.0)(1.168) = 61.9 \text{ m/s}$.

**17.45** The loss of head due to friction in an orifice nozzle, or tube, may be expressed as $h_L = kV^2/2g$, where $V$ is the actual velocity of the jet. Compute $k$ for the three tubes in Fig. 17-10.

$$C_v = 0.98 \qquad\qquad C_v = 0.82 \qquad\qquad C_v \approx 0.74$$
$$C_c = 1.00 \qquad\qquad C_c = 1.00 \qquad\qquad C_c = 1.00$$
(a) \qquad\qquad (b) \qquad\qquad (c) \qquad\qquad **Fig. 17-10**

∎
$$h_L = kV^2/2g = (1/C_v^2 - 1)(V^2/2g) \qquad k = 1/C_v^2 - 1$$

**(a)** $\qquad\qquad C_v = 0.98 \qquad k = (1/0.98^2) - 1 = 0.0412$

**(b)** $\qquad\qquad C_v = 0.82 \qquad k = (1/0.82^2) - 1 = 0.487$

**(c)** $\qquad\qquad C_v = 0.74 \qquad k = (1/0.74^2) - 1 = 0.826$

**17.46** If the tubes in Prob. 17.45 and Fig. 17-10 discharge water under a head of 6 ft, compute the loss of head in each case.

∎ $\quad V = C_v\sqrt{2gh} \qquad h_L = kV^2/2g = (1/C_v^2 - 1)(V^2/2g) = (1/C_v^2 - 1)(C_v\sqrt{2gh})^2/2g = (1/C_v^2 - 1)(2hC_v^2/2)$

**(a)** $\qquad C_v = 0.98 \qquad h_L = [(1/0.98^2) - 1][(2)(6)(0.98)^2/2)] = 0.238 \text{ ft}$

**(b)** $\qquad C_v = 0.82 \qquad h_L = [(1/0.82^2) - 1][(2)(6)(0.82)^2/2)] = 1.97 \text{ ft}$

**(c)** $\qquad C_v = 0.74 \qquad h_L = [(1/0.74^2) - 1][(2)(6)(0.74)^2/2)] = 2.71 \text{ ft}$

**17.47** The diverging tube shown in Fig. 17-11 discharges water when $h = 4$ ft. The area $A$ is twice area $A_0$. Neglecting all friction losses, find the **(a)** velocity at the throat and **(b)** pressure head at the throat.

**Fig. 17-11**

∎ **(a)**
$$V = \sqrt{2gh} = \sqrt{(2)(32.2)(4)} = 16.05 \text{ ft/s}$$
$$AV = A_0V_0 \qquad (2A_0)(16.05) = A_0V_0 \qquad V_0 = 32.1 \text{ ft/s}$$

**(b)** $\qquad p_0/\gamma = V^2/2g - V_0^2/2g = 16.05^2/[(2)(32.2)] - 32.1^2/[(2)(32.2)] = -12.0 \text{ ft}$

**17.48** If the barometric pressure is 14.6 psia and the water temperature is 85 °F, what is the maximum value of $h$ at which the tube will flow full, all other data being the same as in Prob. 17.47? What will happen if the value of $h$ is made greater than this?

$$(p_0)_{min} = p_v = 87.2 \text{ lb/ft}^2 \quad \text{(from Table A-1)}$$

$$p_0/\gamma + V_0^2/2g = p_a/\gamma + V^2/2g \qquad V_0 = 2V \quad \text{(from Prob. 17.47)}$$

$$87.2/62.2 + (2V)^2/2g = (14.6)(144)/62.2 + V^2/2g \qquad V^2/2g = 10.80 \text{ ft}$$

$$h_{max} = V^2/2g = 10.80 \text{ ft}$$

If $h > 10.80$ ft, the jet will spring free from the diverging tube, so that $p_0 = p_a$, and discharge will decrease.

**17.49** For a rounded entrance and tube flowing full as in the sketch for Prob. 17.47, $C_c = 1.0$ both for the throat and for the exit, and thus $C_v = C_d$ for both sections. For the throat, assume the value of $C_v$ as given in Fig. 17-10a, and assume that for the tube as a whole the discharge coefficient applied to the exit end is 0.70. If $h = 4$ ft, find the velocity and the pressure head at the throat, and compare with Prob. 17.47.

$Q = AV = C_c A_0 (C_v \sqrt{2g \, \Delta h})$. Since $A = C_c A_0$,

$$V = 0.98\sqrt{(2)(32.2)(4)} = 15.73 \text{ ft/s} \qquad V_0 = 2V \quad \text{(from Prob. 17.47)}$$

$$V_0 = (2)(15.73) = 31.46 \text{ ft/s}$$

$$p_1/\gamma + v_1^2/2g + z_1 = p_0/\gamma + v_0^2/2g + z_0 + h_L$$

$$0 + 0 + 4 = p_0/\gamma + 31.46^2/[(2)(32.2)] + 0 + (1 - 0.98^2)(4) \qquad p_0/\gamma = -11.5 \text{ ft}$$

**17.50** Suppose that the diverging tube shown in the figure for Prob. 17.47 is discharging water when $h = 2.7$ m. The area $A$ is $1.7A_0$. Neglecting all friction losses, find the **(a)** velocity and **(b)** pressure head at the throat.

$$V = \sqrt{2gh} = \sqrt{(2)(9.807)(2.7)} = 7.277 \text{ m/s}$$

**(a)** $$AV = A_0 V_0 \qquad (1.7A_0)(7.277) = A_0 V_0 \qquad V_0 = 12.4 \text{ m/s}$$

**(b)** $$p_0/\gamma = V^2/2g - V_0^2/2g = 7.277^2/[(2)(9.807)] - 12.4^2/[(2)(9.807)] = -5.14 \text{ m}$$

**17.51** If the tube of Prob. 17.50 is operating at standard atmospheric conditions at a 2500-m elevation, what would be the maximum value of $h$ at which the tube will flow full?

$p_0/\gamma + V_0^2/2g = p_{atm}/\gamma + V^2/2g$. From Table A-8, at 2500-m elevation, $p_{atm} = 74.66$ kPa and $T = 0.4$ °C. From Table A-2, at 0.4 °C, $p_v = 0.636$ kPa and $\gamma = 9.81$ kN/m³. Hence, $p_v/\gamma = 0.636/9.81 = 0.0648$ m, $V_0 = 1.7V$, $0.0648 + (1.7V)^2/2g = 74.66/9.81 + V^2/2g$, $V^2/2g = 3.99$ m. The tube will not flow full if $h > 3.99$ m.

**17.52** Find the maximum theoretical head at which the Borda tube of Fig. 17-12 will flow full if the liquid is water at 90 °F and the barometer reads 28.9 inHg. Assume $C_d = 0.72$ for the tube flowing full.

**Fig. 17-12**

$V = 0.72(2gh)^{1/2}$, $h = (V^2/2g)/0.72^2 = 1.929v^2/2g$. If $C_c = 0.52$ and $C_v = 0.98$ for the contracted throat section, $V_0^2/2g = (V^2/2g)/0.52^2 = (0.72^2/0.52^2)(h) = 1.917h$, $p_a/\gamma + V_a^2/2g + z_a = p_0/\gamma + V_0^2/2g + z_0 + h_L$. From Table A-1, at 80 °F, $p_v = 101$ lb/ft² and $\gamma = 62.1$ lb/ft³. Hence, $(p_0/\gamma)_{min} = p_v/\gamma = 101/62.1 = 1.626$ ft,

$(28.9/12)(13.6) + 0 + h = 1.626 + 1.917h + (1 - 0.98^2)(h + 33.9)$. $[(h + 33.9)$ ft is the total absolute pressure head acting on the orifice.$]$ $h = h_{max} = 31.1$ ft.

**17.53** For the 4-in-diameter short tube shown in Fig. 17-13, **(a)** what flow of water at 75 °F will occur under a head of 30 ft? **(b)** What is the pressure head at section $B$? Use $C_v = 0.82$.

**Fig. 17-13**

▮ For a standard short tube, the stream contracts at $B$ to about 0.62 of the area of the tube. The lost head from $A$ to $B$ has been measured at about 0.042 times the velocity head at $B$.

**(a)**
$$p_A/\gamma + V_A^2/2g + z_A = p_C/\gamma + V_C^2/2g + z_C + h_L$$
$$0 + 0 + 30 = 0 + V_C^2/[(2)(32.2)] + 0 + [(1/0.82^2) - 1]V_C^2/[(2)(32.2)] \qquad V_C = 36.04 \text{ ft/s}$$
$$Q = A_C V_C = [(\pi)(\tfrac{4}{12})^2/4](36.04) = 3.15 \text{ ft}^3/\text{s}$$

**(b)**
$$p_A/\gamma + V_A^2/2g + z_A = p_B/\gamma + V_B^2/2g + z_B + h_L \qquad V_B = V_C/0.62 = 36.04/0.62 = 58.13 \text{ ft/s}$$
$$h_L = 0.042V_B^2/2g = (0.042)(58.13)^2/[(2)(32.2)] = 2.204 \text{ ft}$$
$$0 + 0 + 30 = p_B/\gamma + 58.13^2/[(2)(32.2)] + 0 + 2.204 \qquad p_B/\gamma = -24.7 \text{ ft of water}$$

**17.54** In Prob. 17.53, what maximum head can be used if the tube is to flow full at the exit?

▮ As the head causing flow through the short tube is increased, the pressure head at $B$ will become less and less. For steady flow (and with the tube full at the exit), the pressure head at $B$ must not be less than the vapor pressure head for the liquid at the particular temperature. From Table A-1, for water at 75 °F, $p_v = 63.0$ lb/ft$^2$ and $\gamma = 62.3$ lb/ft$^3$; hence, $p_v/\gamma = 63.0/62.3 = 1.0$ ft abs at sea level $(-33.0$ ft gage). Applying the Bernoulli equation between $A$ and $B$ [see Prob. 17.53(**(b)**)], $C_c A V_B = A V_C = A C_v \sqrt{2gh}$, $V_B(C_v/C_c)\sqrt{2gh}$, $V_B^2/2g = (C_v/C_c)^2(h) = (0.82/0.62)^2(h) = 1.749h$, $0 + 0 + h = -33.0 + 1.749h + 0 + (0.042)(1.749h)$, $h = 40.1$ ft of water. Any head over this value will cause the stream to spring free of the sides of the tube. The tube will then function as an orifice.

**17.55** A 4-ft-diameter tank contains oil of specific gravity 0.75. A 3-in-diameter short tube is installed near the bottom of the tank $(C = 0.85)$. How long will it take to lower the level of the oil from 6 ft above the tube to 4 ft above the tube?

▮
$$t = t_2 - t_1 = (2A_T/CA_0\sqrt{2g})(h_1^{1/2} - h_2^{1/2}) \qquad A_T = (\pi)(4)^2/4 = 12.57 \text{ ft}^2$$
$$A_0 = (\pi)(\tfrac{3}{12})^2/4 = 0.04909 \text{ ft}^2$$
$$t = [(2)(12.57)/(0.85)(0.04909)\sqrt{(2)(32.2)}](6^{1/2} - 4^{1/2}) = 33.7 \text{ s}$$

**17.56** A Borda mouthpiece 60 mm in diameter has a discharge coefficient of 0.52. What is the diameter of the issuing jet?

▮
$$C_c = (D_{jet}/D)^2 \qquad 1 = 2C_d C_v = 2C_v^2 C_c \qquad 1 = (2)(0.52)(C_v) \qquad C_v = 0.9615$$
$$1 = (2)(0.9615)^2(C_c) \qquad C_c = 0.5408 \qquad 0.5408 = (D_{jet}/60)^2 \qquad D_{jet} = 44.1 \text{ mm}$$

**17.57** A reservoir of variable area is drained by a 150-mm-diameter short pipe with a valve attached. The valve is being adjusted so that the loss (in velocity heads) for the piping system is $K = 1.5 + 0.04t + 0.0001t^2$ with $t$ in seconds. The reservoir area is given by $A = 4 + 0.1y + 0.01y^2$ m$^2$ where $y$ is the elevation of the reservoir surface above the centerline of the valve. If $y = 20$ m at $t = 0$, determine $y$, $A$, $K$, and the discharge $Q$ for 300 s. Solve using a computer program.

$\blacksquare$ $Q\,dt = -A\,dy$, $y + 0 + 0 = KV^2/2g = KQ^2/2gA_0^2$, $dy = -HQ/A$, and $Q = \sqrt{C_1 y/K}$ in which $C_1 = 2gA_0^2$. $H$ is the time increment used in Runge–Kutta.

```
10  REM EXAMPLE DRAINAGE OF RESERVOIR—2ND ORDER RUNGE-KUTTA
20  DEF FNQ(YD,TD)=SQR(C1*YD/(K1+K2*TD+K3*TD^2))/(A1+A2*YD+A3*YD^2)
30  DEFINT I: READ A1,A2,A3,K1,K2,K3,Y,TMAX,PI,H,G,D
40  DATA 4.,.1,.01,1.5,.04,.0001,20!,300!,3.1416,1.,9.806001,.15
50  LPRINT: LPRINT"A1,A2,A3,K1,K2,K3=";A1;A2;A3;K1;K2;K3
60  LPRINT"Y,TMAX,PI,H,G,D=";Y;TMAX;PI;H;G;D: LPRINT
70  C1=2!*G*(.25*PI*D^2)^2: T=0!: I=0: A=A1+Y*(A2+A3*Y): Q=FNQ(Y,T)*A: K=K1
80  LPRINT"    T,s      Y,m    Q,m^3/s AREA,m^2    K": LPRINT
90  LPRINT USING"###.### ";T;Y;Q;A;K
100 IF T>= TMAX THEN STOP
110 I=I+1: U1=-H*FNQ(Y,T): U2=-H*FNQ(Y+U1,T+H): Y=Y+.5*(U1+U2): T=T+H
120 A=A1+A2*Y+A3*Y^2: Q=FNQ(Y,T)*A: K=K1+T*(K2+K3*T)
130 IF I MOD 30 = 0 THEN 90 ELSE GOTO 110
140 IF T<TMAX THEN 110 ELSE STOP

A1,A2,A3,K1,K2,K3= 4  .1  .01  1.5  .04  .0001
Y,TMAX,PI,H,G,D= 20  300  3.1416  1  9.806001  .15


    T,s     Y,m    Q,m^3/s AREA,m^2    K

   0.000  20.000  0.286   10.000    1.500
  30.000  19.265  0.206    9.638    2.790
  60.000  18.687  0.164    9.361    4.260
  90.000  18.202  0.137    9.133    5.910
 120.000  17.778  0.119    8.938    7.740
 150.000  17.401  0.105    8.768    9.750
 180.000  17.060  0.094    8.616   11.940
 210.000  16.748  0.085    8.480   14.310
 240.000  16.460  0.077    8.355   16.860
 270.000  16.191  0.071    8.241   19.590
 300.000  15.941  0.066    8.135   22.500
```

**17.58** A 2-in ISA flow nozzle is installed in a 3-in pipe carrying water at 72 °F. If a water-air manometer shows a differential of 2 in, find the flow.

$\blacksquare$ This is a trial-and-error type of solution. First assume a reasonable value of $K$. From Fig. A-20, for $D_2/D_1 = 0.67$, for the level part of the curve, $K = 1.06$. Then $Q = KA_2\sqrt{2g[(p_1/\gamma + z_1) - (p_2/\gamma + z_2)]}$, where $A_2 = (\pi/4)(2^2/144) = 0.0218\ \text{ft}^2$ and $\Delta(p/\gamma + z) = \frac{2}{12} = 0.167\ \text{ft}$. Thus $Q = (1.06)(0.0218)\sqrt{(64.4)(0.167)} = 0.0757\ \text{cfs}$. With this first determination of $Q$, $V_1 = Q/A = 0.0757/0.0492 = 1.54\ \text{fps}$. Then $D_1'' V_1 = (3)(1.54) = 4.62$. From Fig. A-20, $K = 1.04$ and $Q = (1.04/1.06)(0.0757) = 0.0743\ \text{cfs}$. No further correction is necessary.

**17.59** The velocity of water in a 6-in-diameter pipe is 12 ft/s. At the end of the pipe is a nozzle whose velocity coefficient is 0.98. If the pressure in the pipe is 10 psi, what is the velocity in the jet? What is the diameter of the jet? What is the rate of discharge? What is the head loss?

$\blacksquare$ $p_1/\gamma + V_1^2/2g = p_2/\gamma + V_2^2/2g$. Since $p_2 = p_{\text{atm}} = 0$,

$$(V_2)_{\text{ideal}} = [2g(p_1/\gamma + V_1^2/2g)]^{1/2}$$

$$V_2 = C_v(V_2)_{\text{ideal}} = (0.98)\{(2)(32.2)[(10)(144)/62.4 + 12^2/(2)(32.2)]\}^{1/2} = 39.6\ \text{ft/s}$$

$$A_{\text{jet}} = Q/V \qquad Q = AV = [(\pi)(\tfrac{6}{12})^2/4](12) = 2.356\ \text{ft}^3/\text{s} \qquad A_{\text{jet}} = 2.356/39.6 = 0.05949\ \text{ft}^2$$

$$0.05949 = \pi D_{\text{jet}}^2/4 \qquad D_{\text{jet}} = 0.2752\ \text{ft} \quad\text{or}\quad 3.30\ \text{in}$$

$$h_L = h - V^2/2g \qquad h = \text{original head} = p/\gamma + V^2/2g + z$$

$$h = (10)(144)/62.4 + 12^2/[(2)(32.2)] + 0 = 25.31\ \text{ft} \qquad h_L = 25.31 - 39.6^2/[(2)(32.2)] = 0.960\ \text{ft}$$

**17.60** A jet of water 4 in in diameter is discharged through a nozzle whose velocity coefficient is 0.96. If the pressure in the pipe is 14 psi and the pipe diameter is 9 in and if it is assumed that there is no contraction of the jet, what is the velocity at the tip of the nozzle? What is the rate of discharge?

$\blacksquare$ $V = \{C_v/[1 - C_v^2(D_2/D_1)^4]^{1/2}\}(2gp/\gamma)^{1/2} = \{0.96/[1 - (0.96)^2(\tfrac{4}{9})^4]^{1/2}\}\{(2)(32.2)[(14)(144)]/62.4)\}^{1/2} = 44.6\ \text{ft/s}$

$Q = AV = [(\pi)(\tfrac{4}{12})^2/4](44.6) = 3.89\ \text{ft}^3/\text{s}$

**17.61** The nozzle in Fig. 17-14 throws a stream of water vertically upward so that the power available in the jet at point 2 is 3.50 hp. If the pressure at the base of the nozzle, point 1, is 36.0 psi, find the **(a)** theoretical height to

Fig. 17-14

which the jet will rise, (b) coefficient of velocity, (c) head loss between 1 and 2, and (d) theoretical diameter of the jet at a point 18 ft above point 2.

$$P = Q\gamma V_2^2/2g = A_2\gamma V_2^3/2g$$

$$(3.50)(550) = [(\pi)(1.0/12)^2/4](62.4)(V_2^3)/[(2)(32.2)] \qquad V_2 = 71.42 \text{ ft/s}$$

(a) $$h_{\text{theor}} = V_2^2/2g = 71.42^2/[(2)(32.2)] = 79.2 \text{ ft}$$

(b) $$C_v = V/V_i \qquad V_i = \sqrt{2g(p_1/\gamma + V_1^2/2g + z_1 - z_2)}$$

$$V_1^2/2g = (V_2^2/2g)(D_2/D_1)^4 = \{71.42^2/[(2)(32.2)]\}(1.0/3)^4 = 0.9778 \text{ ft}$$

$$V_i = \sqrt{(2)(32.2)[(36.0)(144)/62.4 + 0.9778 + 0 - \tfrac{13}{12}]} = 73.10 \text{ ft/s}$$

$$C_v = 71.42/73.10 = 0.977$$

(c) $$h_L = (1 - C_v^2)h = (1 - 0.977^2)\{73.10^2/[(2)(32.2)]\} = 3.77 \text{ ft}$$

(d) $$V_2^2/2g = (V_3^2/2g)(D_3/D_2)^4 \qquad V_3^2/2g = V_2^2/2g - z_3$$

$$D_3 = D_2[(V_2^2/2g)/(V_3^2/2g)]^{1/4} = (1.0)\{[71.42^2/(2)(32.2)]/[71.42^2/(2)(32.2) - 18.0]\}^{1/4} = 1.07 \text{ in}$$

**17.62** Suppose a 125-mm ISA flow nozzle is used to measure the flow of water at 50 °C. What would be the reading on a mercury manometer for the flow rate 16 L/s?

$$\Delta(p/\gamma + z) = R'(S_M/S_F - 1) = R'[(13.6/1) - 1] = 12.6R'$$

$$Q = KA_2\sqrt{2g\,\Delta(p/\gamma + z)} = KA_2\sqrt{(2g)(12.6R')}$$

Let $K = 1.0$: $Q = (1.0)[(\pi)(0.125)^2/4]\sqrt{(2)(9.807)(12.6)R'} = 0.1929\sqrt{R'}$. Then

$$16 \times 10^{-3} = 0.1929\sqrt{R'} \qquad R' = 0.00688 \text{ m} = 6.88 \text{ mm}$$

Of course, one should check the Reynolds number to confirm the value of $K$ used.

**17.63** A 9-in ISA nozzle is used to measure the flow of crude oil (s.g. = 0.855) at 15 °F. If a mercury manometer shows a reading of 5.5 in, what is the flow? Assume $D_2/D_1 = 0.70$, $\nu = 0.00034 \text{ ft}^2/\text{s}$.

$$\Delta(p/\gamma + z) = R'(S_M/S_F - 1) = R'[(13.6/0.855) - 1] = 14.91R'$$

$$Q = KA_2\sqrt{2g\,\Delta(p/\gamma + z)} = KA_2\sqrt{(2g)(14.91R')}$$

Let $K = 1.08$: $Q = (1.08)[(\pi)(\tfrac{9}{12})^2/4]\sqrt{(2)(32.2)(5.5/12)(14.91)} = 10.01 \text{ ft}^3/\text{s}$. Check $N_R$ to confirm that the assumed $K$ is applicable: $V = Q/A = 10.01/[(\pi)(\tfrac{9}{12})^2/4] = 22.66 \text{ ft/s}$, $N_R = DV/\nu = (\tfrac{9}{12})(22.66)/0.00034 = 5.00 \times 10^4$. From Fig. A-20, $K = 1.06$ approximately. $Q = (1.06/1.08)(10.01) = 9.82 \text{ ft}^3/\text{s}$.

**17.64** Water flows through a 4-in pipe at the rate of 0.952 cfs and thence through a nozzle attached to the end of the pipe. The nozzle tip is 2 in in diameter and the coefficients of velocity and contraction for the nozzle are 0.950 and 0.930, respectively. What pressure head must be maintained at the base of the nozzle if atmospheric pressure surrounds the jet?

■ Apply the Bernoulli equation between base of nozzle and jet.

$$p_b/\gamma + V_b^2/2g + z_b = p_j/\gamma + V_j^2/2g + z_j + h_L \qquad A_b V_b = A_j V_j = (C_c A_{2\text{-in}})V_j = 0.952$$

$$V_b = 0.952/A_b = 0.952/[(\pi)(\tfrac{4}{12})^2/4] = 10.91 \text{ ft/s}$$

$$V_j = 0.952/(C_c A_{2\text{-in}}) = 0.952/\{(0.930)[(\pi)(\tfrac{2}{12})^2/4]\} = 46.92 \text{ ft/s}$$

$$h_L = (1/C_v^2 - 1)(V_j^2/2g)$$

$$p_b/\gamma + 10.91^2/[(2)(32.2)] + 0 = 0 + 46.92^2/[(2)(32.2)] + 0 + (1/0.950^2 - 1)\{46.92^2/[(2)(32.2)]\}$$

$$p_b/\gamma = 36.0 \text{ ft of water}$$

**17.65** A 4-in base diameter by 2-in tip diameter nozzle points downward and the pressure head at the base of the nozzle is 26.0 ft of water. The base of the nozzle is 3.0 ft above the tip and the coefficient of velocity is 0.962. Determine the horsepower in the jet of water.

■ $P_j = Q\gamma H_j$. For a nozzle, unless $C_c$ is given, it may be taken as unity. Hence, $V_j = V_t$.

$$p_b/\gamma + V_b^2/2g + z_b = p_t/\gamma + V_t^2/2g + z_t + h_L \qquad A_b V_b = A_t V_t$$

$$[(\pi)(4^2/4)](V_b) = [(\pi)(2^2/4)](V_t) \qquad V_b = V_t/4 \qquad h_L = (1/C_v^2 - 1)(V_t^2/2g)$$

$$26.0 + (V_t/4)^2/[(2)(32.2)] + 3.0 = 0 + V_t^2/[(2)(32.2)] + 0 + (1/0.962^2 - 1)\{V_t^2/[(2)(32.2)]\}$$

$$V_t = 42.83 \text{ ft/s} \qquad Q = A_t V_t = [(\pi)(\tfrac{2}{12})^2/4](42.83) = 0.9344 \text{ ft}^3/\text{s}$$

$$H_j = 0 + V_t^2/[(2)(32.2)] + 0 = 42.83^2/[(2)(32.2)] = 28.48 \text{ ft}$$

$$P_j = (0.9344)(62.4)(28.48) = 1661 \text{ ft-lb/s} = \tfrac{1661}{550} = 3.02 \text{ hp}$$

**17.66** A nozzle with a 4-in-diameter tip is installed in a 10-in pipe. Medium fuel oil at 80 °F flows through the nozzle at the rate of 3.49 cfs. Assume the calibration of the nozzle is represented by curve $\beta = 0.40$ on Fig. A-21. Calculate the differential gage reading if a liquid of specific gravity 13.6 is the gage liquid.

■
$$Q = A_4 V_4 = A_4 C\sqrt{[2g(p_A/\gamma - p_B/\gamma)]/[1 - (A_4/A_{10})^4]}$$
$$= A_4 C\sqrt{[2g(p_A/\gamma - p_B/\gamma)]/[1 - (\tfrac{4}{10})^4]} \qquad (1)$$

Figure A-21 indicates that $C$ varies with Reynolds number: $V_4 = Q/A_4 = 3.49/[\tfrac{1}{4}\pi(\tfrac{4}{12})^2] = 40.0 \text{ ft/s}$ and $N_R = (40.0)(\tfrac{4}{12})/(3.65 \times 10^{-5}) = 365\,000$. Curve for $\beta = 0.40$ gives $C = 0.993$. Thus $3.49 = \tfrac{1}{4}\pi(\tfrac{4}{12})^2(0.993)\sqrt{[2g(p_A/\gamma - p_B/\gamma)]/[1 - (\tfrac{4}{10})^4]}$ and $(p_A/\gamma - p_B/\gamma) = 24.6 \text{ ft of fuel oil}$. Differential gage principles produce, using specific gravity of the oil = 0.851, $24.6 = h[(13.6/0.851) - 1]$ and $h = 1.64 \text{ ft}$ (gage reading).

**17.67** Derive an expression for the flow of a compressible fluid through a nozzle flowmeter and a Venturi meter.

■ Since the change in velocity takes place in a very short period of time, little heat can escape and adiabatic conditions will be assumed. The Bernoulli theorem for compressible flow gives $\{[k/(k-1)](p_1/\gamma_1) + (V_1^2/2g) + z_1\} - H_L = \{[k/(k-1)](p_1/\gamma_1)(p_2/p_1)^{(k-1)/k} + (V_2^2/2g) + z_2\}$.

For a nozzle meter and for a horizontal Venturi meter, $z_1 = z_2$ and the lost head will be taken care of by means of the coefficient of discharge. Also, since $C_c = 1.00$, $W = \gamma_1 A_1 V_1 = \gamma_2 A_2 V_2$. Then upstream $V_1 = W/\gamma_1 A_1$, downstream $V_2 = W/\gamma_2 A_2$. Substituting and solving for $W$, $W^2/\gamma_2^2 A_2^2 - W^2/\gamma_1^2 A_1^2 = 2g[k/(k-1)](p_1/\gamma_1)[1 - (p_2/p_1)^{(k-1)/k}]$ or (ideal) $W = [\gamma_2 A_2/\sqrt{1 - (\gamma_2/\gamma_1)^2(A_2/A_1)^2}]\sqrt{[2gk/(k-1)](p_1/\gamma_1)[1 - (p_2/p_1)^{(k-1)/k}]}$. It may be more practical to eliminate $\gamma_2$ under the radical. Since $\gamma_2/\gamma_1 = (p_2/p_1)^{1/k}$,

$$\text{(ideal)} \quad W = \gamma_2 A_2\sqrt{[2gk/(k-1)](p_1/\gamma_1)[1 - (p_2/p_1)^{(k-1)/k}]/[1 - (A_2/A_1)^2(p_2/p_1)^{2/k}]} \qquad (2)$$

The true value of $W$ in pounds per second is obtained by multiplying the right-hand side of the equation by coefficient $C$. From Eq. (1) of Prob. 17.66, the following equation may be written: $W = \gamma Q = [\gamma A_2 C/\sqrt{1 - (A_2/A_1)^2}]\sqrt{2g(\Delta p/\gamma)}$ or $W = \gamma K A_2\sqrt{2g(\Delta p/\gamma)}$. The above equation can be expressed more generally so that it will apply both to compressible and incompressible fluids. An expansion (adiabatic) factor $Y$ is introduced and the value of $\gamma_1$ at inlet is specified. The fundamental relation is then

$$W = \gamma_1 K A_2 Y\sqrt{2g(\Delta p/\gamma_1)} \qquad (2)$$

For incompressible fluids, $Y = 1$. For compressible fluids, equate expressions (1) and (2) and solve for $Y$. By so doing,

$$Y = \sqrt{\{[1 - (A_2/A_1)^2]/[1 - (A_2/A_1)^2(p_2/p_1)^{2/k}]\}\{[k/(k-1)][1 - (p_2/p_1)^{(k-1)/k}](p_2/p_1)^{2/k}/(1 - p_2/p_1)\}}$$

This expansion factor $Y$ is a function of three dimensionless ratios. Table A-18 lists some typical values for nozzle flowmeters and for Venturi meters.

**17.68** Air at a temperature of 80 °F flows through a 4-in pipe and through a 2-in flow nozzle. The pressure differential is 0.522 ft of oil, s.g. = 0.910. The pressure upstream from the nozzle is 28.3 psi gage. How many pounds per second are flowing for a barometric reading of 14.7 psi, assuming the air has constant density.

▌ $\gamma = p/RT$, $\gamma_1 = [(28.3 + 14.7)144]/[53.3(460 + 80)] = 0.215$ lb/ft$^3$. From differential gage principles, using pressure heads in feet of air, $\Delta p/\gamma_1 = 0.522[(\gamma_{oil}/\gamma_{air}) - 1] = 0.522\{[(0.910)(62.4)/0.215] - 1\} = 137$ ft of air.

Assuming $C = 0.980$ and using equation (1) of Prob. 17.66 after multiplying by $\gamma_1$, we have $W = \gamma_1 Q = (0.215)(\frac{1}{4}\pi)(\frac{2}{12})^2(0.980)\sqrt{2g(137)}/[1 - (\frac{2}{4})^4] = 0.445$ lb/s.

To check the value of $C$, find Reynolds number and use the appropriate curve on Fig. A-21 (here $\gamma_1 = \gamma_2$ and $v = 16.9 \times 10^{-5}$ ft$^2$/s at standard atmosphere).: $V_2 = W/A_2\gamma_2 = W/(\pi d_2^2/4)\gamma_2$. Then $N_R = V_2 d_2/v = 4W/\pi d_2 v\gamma_2 = 4(0.445)/[\pi(\frac{2}{12})(16.9 \times 14.7/43.0)10^{-5}(0.215)] = 274\,000$.

From Fig. A-21, $C = 0.986$. Recalculating, $W = 0.447$ lb/s.

Further refinement in calculation is not warranted inasmuch as the Reynolds number will not be changed materially, nor will the value of $C$ read from Fig. A-21.

**17.69** Find $Q$ for flow of water in a horizontal pipe of inside diameter $D = 100$ mm using a long-radius-type flow nozzle. The value of $h$ for the manometer is 140 mm. The throat diameter of the nozzle is 60 mm. Take $\rho = 999$ kg/m$^3$ and $v = 1.12 \times 10^{-3}$ m$^2$/s.

▌
$$p_1 - p_2 = h(\gamma_{Hg} - \gamma_{H_2O}) = 0.140(13.6 - 1)(999)(9.81) = 17.300 \text{ kPa} \tag{1}$$
$$C_d = 0.99622 + 0.00059D - (6.36 + 0.13D - 0.24\beta^2)(1/\sqrt{N_R})$$
$$(C_d)_{noz} = [0.99622 + 0.00059(10.0/2.54)] - [6.36 + 0.13(10.0/2.54) - (0.24)(0.6^2)](1/\sqrt{N_R})$$
$$= 0.998 - 6.785(1/\sqrt{N_R}) \tag{2}$$
$$q_{act} = (C_d)_{noz}A_2\{2[(p_1 - p_2)/\rho]/[1 - (A_2/A_1)^2]\}^{1/2} = (C_d)_{noz}A_2\{2[(p_1 - p_2)/\rho]/(1 - \beta^4)\}^{1/2}$$
$$A_2V_2 = [0.998 - 6.785(1/\sqrt{N_R})](A_2)[2(17\,300/999)/(1 - 0.6^4)]^{1/2}$$

This equation becomes
$$V_2 = (0.998 - 6.785/\sqrt{\rho V_1 d/\mu})(6.31) = (0.998 - \{6.785/\sqrt{[(999)(V_1)(0.60)]/(1.12 \times 10^{-3})}\})(6.31) \tag{3}$$

From *continuity*,
$$V_2 = (\tfrac{100}{60})^2 V_1 \tag{4}$$

Therefore, substituting for $V_2$ in Eq. (3) using Eq. (4), $V_1 = [0.998 - (0.0293/\sqrt{V_1})](2.77)$. Solving by trial and error we get $V_1 = 2.22$ m/s. The Reynolds number is then $N_R = [(999)(0.060)(2.22)/(1.12 \times 10^{-3})] = 1.189 \times 10^5$. We are well within the range of Eq. (2), so we get $Q = [(\pi)(0.100^2)/4](2.22) = 0.01744$ m$^3$/s.

**17.70** Determine the flow through a 6-in-diameter water line that contains a 4-in-diameter flow nozzle. The mercury-water differential manometer has a gage difference of 10 in. Water temperature is 60 °F.

▌ $Q = CA_2\sqrt{2gR'(S_0/S_1 - 1)}$. From Fig. A-22, for $A_2/A_1 = (\frac{4}{6})^2 = 0.4444$, assume that the horizontal region of the curves applies. Hence, $C = 1.056$.

$$Q = (1.056)[(\pi)(\tfrac{4}{12})^2/4]\sqrt{(2)(32.2)(\tfrac{10}{12})[(13.6/1) - 1]} = 2.40 \text{ ft}^3/\text{s}$$
$$V = Q/A = 2.40/[(\pi)(\tfrac{6}{12})^2/4] = 12.22 \text{ ft/s}$$
$$N_R = \rho DV/\mu = (1.94)(\tfrac{6}{12})(12.22)/(2.35 \times 10^{-5}) = 5.04 \times 10^5$$

Figure A-22 shows the value of $C$ to be correct; therefore, the discharge is 2.40 ft$^3$/s.

**17.71** Air flows through a 100-mm-diameter ISA flow nozzle in a 150-mm-diameter pipe. $p_1 = 140$ kPa abs; $T_1 = 10$ °C; and a differential manometer with liquid of s.g. 2.80 has a gage difference of 0.9 m when connected between taps. Calculate the mass rate of flow. [MW of air = 29.]

▌ $\dot{m} = \rho Q = \rho CYA_2\sqrt{2\,\Delta p/\rho}$    $\rho = p/RT = 0.140/[(8.312/29)(10 + 273)] = 1.726$ kg/m$^3$
$$A_2/A_1 = (D_2/D_1)^2 = (\tfrac{100}{150})^2 = 0.4444$$

From Fig. A-22, $C$ is assumed to be 1.055.

$$p_2 = p_1 - \gamma\,\Delta h = 140 - (9.79)(2.80)(0.9) = 115.3 \text{ kPa}$$
$$p_2/p_1 = 115.3/140 = 0.824 \qquad D_2/D_1 = \tfrac{100}{150} = 0.667$$

From Fig. A-26, $Y = 0.875$.

$$A_2 = (\pi)(0.100)^2/4 = 0.007854 \text{ m}^2$$

$$\dot{m} = (1.726)(1.055)(0.875)(0.007854)\sqrt{(2)(140 - 115.3)(1000)/1.726} = 2.12 \text{ kg/s}$$

$$N_R = \rho DV/\mu \qquad V = \dot{m}/\rho A \qquad A_1 = (\pi)(0.150)^2/4 = 0.01767 \text{ m}^2$$

$$V_1 = 2.12/[(1.726)(0.01767)] = 69.51 \text{ m/s}$$

$$N_R = (1.726)(0.150)(69.51)/(1.77 \times 10^{-5}) = 1.02 \times 10^6$$

From Fig. A-22, assumed $C = 1.055$ is O.K.; hence, $\dot{m} = 2.12 \text{ kg/s}$.

**17.72** A 3.5-in-diameter ISA nozzle is used to measure flow of water at 50 °C in a 9-in-diameter pipe. What gage difference on a water-mercury manometer is required for 400 gpm?

$\blacksquare$
$$R' = \Delta p/[(\gamma)(S_0/S_1 - 1)] \qquad \Delta p = (\rho/2)(Q/CA_2)^2 \qquad A_2/A_1 = (3.5/9)^2 = 0.1512$$

$$V_1 = Q/A_1 = (400)(0.002228)/[(\pi)(\tfrac{9}{12})^2/4] = 2.017 \text{ ft/s}$$

$$(N_R)_1 = V_1 D_1/\nu_1 = (2.017)(\tfrac{9}{12})/(1.40 \times 10^{-5}) = 1.08 \times 10^5$$

From Fig. A-22, $C = 0.995$.

$$A_2 = (\pi)(3.5/12)^2/4 = 0.06681 \text{ ft}^2$$

$$\Delta p = (1.94/2)\{(400)(0.002228)/[(0.995)(0.06681)]\}^2 = 174.3 \text{ lb/ft}^2$$

$$R' = 174.3/\{(62.4)[(13.6/1) - 1]\} = 0.2217 \text{ ft} \quad \text{or} \quad 2.66 \text{ in}$$

**17.73** We want to meter the volume flow of water ($\rho = 1000 \text{ kg/m}^3$, $\nu = 1.02 \times 10^{-6} \text{ m}^2/\text{s}$) moving through a 200-mm-diameter pipe at an average velocity of 2.0 m/s. If the differential pressure gage selected reads accurately at $p_1 - p_2 = 50 \text{ kPa}$, what size meter should be selected for installing a long-radius flow nozzle? What would be the nonrecoverable head loss?

$\blacksquare$ Here the unknown is the $\beta$ ratio of the meter. Since the discharge coefficient is a complicated function of $\beta$, iteration will be necessary. We are given $D = 0.2 \text{ m}$ and $V_1 = 2.0 \text{ m/s}$. The pipe-approach Reynolds number is thus $N_R = V_1 D/\nu = (2.0)(0.2)/(1.02 \times 10^{-6}) = 392\,000$. The generalized formula gives the throat velocity as $V_t = V_1/\beta^2 = \alpha[2(p_1 - p_2)/\rho]^{1/2}$ where everything is known except $\alpha$ and $\beta$. Solve for $\beta^2$:

$$\beta^2 = (1/\alpha)(\rho V_1^2/2 \, \Delta p)^{1/2} \tag{1}$$

With $V_1 = 2.0 \text{ m/s}$ and $\Delta p = 50\,000 \text{ Pa}$, $\beta^2 = (1/\alpha)\{(1000)(2.0)^2/[2(50\,000)]\}^{1/2} = 0.2/\alpha$ or

$$\beta = 0.447/\alpha^{1/2} \tag{2}$$

The solution depends only on getting the proper flow coefficient. A good guess for flow-nozzle design is $\alpha = 1.0$. Iterate Eq. (2) and list the results:

| $\alpha$ | $\beta$, Eq. (2) | $C_d$, Fig. A-23 | $E = (1 - \beta^4)^{-1/2}$ | $\alpha = EC_d$ |
|---|---|---|---|---|
| 1.0 | 0.447 | 0.9895 | 1.0206 | 1.0099 |
| 1.0099 | 0.445 | 0.9895 | 1.0202 | 1.0095 |
| 1.0095 | 0.445 | | | |

Convergence is rapid to $\beta = 0.445$, $d = \beta D = 89 \text{ mm}$. The throat velocity is $2.0/(0.445)^2 = 10.1 \text{ m/s}$; the throat head is $(10.1)^2/[2(9.81)] = 5.2 \text{ m}$. From Fig. A-24 for the nozzle read $K_m \approx 0.7$. Then the nozzle loss is $h_m = 0.7(5.2 \text{ m}) = 3.6 \text{ m}$.

**17.74** Gasoline at 20 °C flows at 80 L/s through a 200-mm-diameter pipe and is metered by a 100-mm-diameter nozzle. What is the expected pressure drop across the nozzle? ($\rho = 680 \text{ kg/m}^3$; $\nu = 4.29 \times 10^{-7} \text{ m}^2/\text{s}$)

$\blacksquare$
$$Q = C_d A_t[(2 \, \Delta p/\rho)/(1 - \beta^4)]^{1/2} \qquad C_d = 0.9965 - 0.00653\beta^{1/2}[10^6/(N_R)_D]^{1/2}$$

$$\beta = d/D = 100/200 = 0.500 \qquad N_R = DV/\nu \qquad V = Q/A = 80 \times 10^{-3}/[(\pi)(0.200)^2/4] = 2.546 \text{ m/s}$$

$$(N_R)_D = (0.200)(2.546)/(4.29 \times 10^{-7}) = 1.19 \times 10^6$$

$$C_d = 0.9965 - (0.00653)(0.500)^{1/2}[10^6/(1.19 \times 10^6)]^{1/2} = 0.9923$$

$$80 \times 10^{-3} = (0.9923)[(\pi)(0.100)^2/4][(2 \, \Delta p/680)/(1 - 0.500^4)]^{1/2} \qquad \Delta p = 33.59 \text{ kPa}$$

**17.75** Ethyl alcohol at 20 °C flowing in a 100-mm-diameter pipe is metered through a 60-mm-diameter nozzle. If the measured pressure drop is 55 kPa, calculate the volume flow. ($\rho = 789$ kg/m³; $v = 1.52 \times 10^{-6}$ m²/s)

**❚** $Q = C_d A_t [(2\,\Delta p/\rho)/(1-\beta^4)]^{1/2}$. Estimate $C_d = 0.99$.

$$\beta = d/D = 60/100 = 0.600$$
$$Q = (0.99)[(\pi)(0.060)^2/4]\{[(2)(55\,000)/789]/(1-0.600^4)\}^{1/2} = 0.03543 \text{ m}^3/\text{s}$$
$$C_d = 0.9965 - 0.00653\beta^{1/2}[10^6/(N_R)_D]^{1/2} \qquad N_R = DV/v$$
$$V = Q/A = 0.03543/[(\pi)(0.100)^2/4] = 4.511 \text{ m/s}$$
$$(N_R)_D = (0.100)(4.511)/(1.52 \times 10^{-6}) = 2.97 \times 10^5$$
$$C_d = 0.9965 - (0.00653)(0.600)^{1/2}[10^6/(2.97 \times 10^5)]^{1/2} = 0.9872$$
$$Q = (0.9872)[(\pi)(0.060)^2/4]\{[(2)(55\,000)/789]/(1-0.600^4)\}^{1/2} = 0.03533 \text{ m}^3/\text{s} = 127 \text{ m}^3/\text{h}$$

**17.76** Kerosene at 20 °C flows at 23 m³/h in a 100-mm-diameter pipe. The flow is to be metered by an ISA 1932 flow nozzle so that the pressure drop is 8 kPa. What is the necessary nozzle diameter? ($\rho = 804$ kg/m³; $v = 2.39 \times 10^{-6}$ m²/s)

**❚** $Q = C_d A_t(\beta)^2[(2\,\Delta p/\rho)/(1-\beta^4)]^{1/2}$. Estimate $C_d = 0.99$.

$$23/3600 = (0.99)[(\pi)(0.100)^2/4](\beta)^2\{[(2)(8000)/804]/(1-\beta^4)\}^{1/2}$$
$$\beta = 0.426 \qquad C_d = 0.9965 - 0.00653\beta^{1/2}[10^6/(N_R)_D]^{1/2}$$
$$N_R = DV/v \qquad V = Q/A = (23/3600)/[(\pi)(0.100)^2/4] = 0.8135 \text{ m/s}$$
$$(N_R)_D = (0.100)(0.8135)/(2.39 \times 10^{-6}) = 3.40 \times 10^4$$
$$C_d = 0.9965 - (0.00653)(0.426)^{1/2}[10^6/(3.40 \times 10^4)]^{1/2} = 0.9734$$
$$23/3600 = (0.9734)[(\pi)(0.100)^2/4](\beta)^2\{[(2)(8000)/804]/(1-\beta^4)\}^{1/2}$$
$$\beta = 0.430 \qquad d_{\text{nozzle}} = \beta D = (0.430)(100) = 43 \text{ mm}$$

**17.77** Water flows from a large tank through an orifice and discharges to the atmosphere, as shown in Fig. 17-15. The coefficients of velocity and contraction are 0.96 and 0.62, respectively. Find the diameter and actual velocity in the jet and the discharge from the orifice.

10 ft    Water

2-in-diameter, sharp-edged orifice

**Fig. 17-15**

**❚** 
$$C_c = a/A \qquad 0.62 = a/[(\pi)(\tfrac{2}{12})^2/4] \qquad a = 0.01353 \text{ ft}^2$$
$$(\pi)(D_{\text{jet}})^2/4 = 0.01353 \qquad D_{\text{jet}} = 0.1313 \text{ ft} \quad \text{or} \quad 1.58 \text{ in}$$

In order to determine the actual velocity in the jet, the discharge from the orifice will be determined next:
$C = C_c C_v = (0.96)(0.62) = 0.595$, $Q = CA\sqrt{2gh} = (0.595)[(\pi)(\tfrac{2}{12})^2/4]\sqrt{(2)(32.2)(10)} = 0.329$ ft³/s, $v = Q/a = 0.329/0.01353 = 24.3$ ft/s.

**17.78** Oil discharges from a pipe through a sharp-crested, round orifice, as shown in Fig. 17-16. The coefficients of contraction and velocity are 0.62 and 0.98, respectively. Find the discharge from the orifice and the diameter and actual velocity in the jet.

**Fig. 17-16**

■

$$Q = CA\sqrt{2g(p_1/\gamma)}/\sqrt{1 - C^2(D/D_p)^4} \qquad C = C_v C_c = (0.98)(0.62) = 0.608$$

$$p_1/\gamma + 2.5 - (13.6/0.84)(7.2/12) = 0 \qquad p_1/\gamma = 7.21 \text{ ft of oil}$$

$$Q = (0.608)[(\pi)(\tfrac{3}{12})^2/4]\sqrt{(2)(32.2)(7.21)}/\sqrt{1 - (0.608)^2(\tfrac{3}{12})^4} = 0.644 \text{ ft}^3/\text{s}$$

$$C_c = a/A \qquad 0.62 = a/[(\pi)(\tfrac{3}{12})^2/4] \qquad a = 0.03043 \text{ ft}^2$$

$$(\pi)(D_{\text{jet}})^2/4 = 0.03043 \qquad D_{\text{jet}} = 0.1968 \text{ ft} \quad \text{or} \quad 2.36 \text{ in}$$

$$v = Q/a = 0.644/0.03043 = 21.2 \text{ ft/s}$$

**17.79** Oil flows through a pipe as shown in Fig. 17-17. The coefficient of discharge for the orifice in the pipe is 0.63. What is the discharge of oil in the pipe?

**Fig. 17-17**

■

$$Q = CA\sqrt{2g[(p_1 - p_2)/\gamma]}[1 + (C^2/2)(D/D_p)^4] \qquad A = (\pi)(0.100)^2/4 = 0.007854 \text{ m}^2$$

$$p_1/\gamma + y + 0.350 - (13.6/0.91)(0.350) - y = p_2/\gamma \qquad (p_1 - p_2)/\gamma = 4.881 \text{ m of oil}$$

$$Q = (0.63)(0.007854)\sqrt{(2)(9.807)(4.881)}[1 + (0.63^2/2)(\tfrac{100}{250})^4] = 0.0487 \text{ m}^3/\text{s}$$

**17.80** The tank shown in Fig. 17-18a has the form of a frustum of a cone with dimensions as shown in the figure. The bottom of the tank contains an orifice that has a coefficient of discharge of 0.62. The tank contains water to its depth of 3.5 m. Find the diameter of the orifice needed to empty the tank in 8 min.

■ $dt = A_s \, dh/CA\sqrt{2gh}$. Rearranging this equation gives

$$CA\sqrt{2gh} \, dt = A_s \, dh \tag{1}$$

In order to solve this problem, $A_s$ must be expressed as a function of the water depth, $h$. This can be accomplished as follows: By observing Fig. 17-18b, it is evident that $y/(y + 3.5) = 1.5/3$, $y = 3.5$; $x/1.5 = (3.5 + h)/(3.5 + 3.5)$,

$$x = 0.7500 + 0.2143h \tag{2}$$

$$A = (\pi)(D)^2/4 = 0.7854D^2$$

$$A_s = \pi x^2 \tag{3}$$

**Fig. 17-18**

Substituting Eq. (2) into Eq. (3) gives $A_s = (\pi)(0.7500 + 0.2143h)^2$. Therefore, substituting into Eq. (1),

$$(0.62)(0.7854D^2)\sqrt{(2)(9.807)}h^{1/2}\,dt = (\pi)(0.7500 + 0.2143h)^2\,dh$$

$$2.157D^2h^{1/2}\,dt = (1.767 + 1.010h + 0.1443h^2)\,dh$$

$$2.157D^2 = (1.767h^{-1/2} + 1.010h^{1/2} + 0.1443h^{3/2})\,dh/dt$$

$$= \left[\int_0^{3.5}(1.767h^{-1/2} + 1.010h^{1/2} + 0.1443h^{3/2})\,dh\right]\Bigg/\left(\int_0^{480}dt\right)$$

$$= [3.534h^{1/2} + 0.6733h^{3/2} + 0.05772h^{5/2}]_0^{3.5}/[t]_0^{480}$$

$$= 12.34/480$$

$$D = 0.109 \text{ m} \quad \text{or} \quad 109 \text{ mm}$$

**17.81** A rectangular tank is divided by a partition into two chambers, as shown in Fig. 17-19a. A 6-in-diameter orifice, for which $C$ is 0.65, is located near the bottom of the partition. At a certain time, the water level in chamber $A$ is 10.0 ft higher than that in chamber $B$. Find the time it will take for the water surfaces in the two chambers to be at the same level.

6-in-diameter, sharp-edged orifice
$C = 0.65$

(a)

(b)                                                                                    **Fig. 17-19**

▌ $dV = CA\sqrt{2gh}\ dt$. Let $h$ = difference in elevations of the two water levels at any time, $dh$ = change in $h$ in time $dt$, and $dV$ = volume of water flowing into chamber $B$ in time $dt$. As illustrated in Fig. 17-19b, during a time $dt$, the water level in chamber $A$ drops an amount indicated by $dh_A$ while the level in chamber $B$ rises an amount indicated by $dh_B$. During the same time, the difference in elevations of the two water levels decreases from $h_1$ to $h_2$. The volume of water leaving chamber $A$ equals the volume entering chamber $B$. Hence,

$$(6)(24)(dh_A) = (6)(8)(dh_B) \tag{1}$$
$$dh_B = 3.00\,dh_A$$

Also from Fig. 17-19b,

$$dh = dh_A + dh_B \tag{2}$$

Substituting Eq. (1) into Eq. (2) gives $dh = dh_A + 3.00\,dh_A = 4.00\,dh_A$ or $dh_A = 0.250\,dh$, $dV = (0.250\,dh)(24)(6) = 36.0\,dh$, $A = (\pi)(0.5)^2/4 = 0.1963\ \text{ft}^2$, $36.0\,dh = (0.65)(0.1963)\sqrt{(2)(32.2)}h^{1/2}\ dt$, $35.16h^{-1/2}\ dh = dt$,

$$\int_0^t dt = 35.16\int_0^{10} h^{-1/2}\ dh \qquad [t]_0^t = 35.16[2h^{1/2}]_0^{10} \qquad t = 222\ \text{s}$$

**17.82** Water flows from a large tank through an orifice and discharges to the atmosphere, as shown in Fig. 17-15, but the diameter of the orifice is 3 in and the head is 15 ft. The coefficients of velocity and contraction are 0.97 and 0.61, respectively. Find the diameter and actual velocity in the jet and the discharge from the orifice.

$$C_c = a/A \qquad 0.61 = a/[(\pi)(\tfrac{3}{12})^2/4] \qquad a = 0.02994 \text{ ft}^2$$
$$(\pi)(D_{\text{jet}})^2/4 = 0.02994 \qquad D_{\text{jet}} = 0.1952 \text{ ft} \quad \text{or} \quad 2.34 \text{ in}$$

In order to determine the actual velocity in the jet, the discharge from the orifice will be determined next:
$C = C_c C_v = (0.97)(0.61) = 0.592$; $Q = CA\sqrt{2gh} = (0.592)[(\pi)(\tfrac{3}{12})^2/4]\sqrt{(2)(32.2)(15)} = 0.903 \text{ ft}^3/\text{s}$, $v = Q/a = 0.903/0.02994 = 30.2 \text{ ft/s}$.

**17.83** Oil with a specific gravity of 0.86 discharges from a 250-mm-diameter pipe through a 100-mm-diameter sharp-edged orifice, as shown in Fig. 17-20. Coefficients of velocity and contraction are 0.98 and 0.63, respectively. Determine the discharge of oil in the pipe, diameter of the jet, and velocity of oil in the jet.

Fig. 17-20

$$Q = CA\sqrt{2g(p_1/\gamma)}/\sqrt{1 - C^2(D/D_p)^4} \qquad C = C_v C_c = (0.98)(0.63) = 0.617$$
$$p_1/\gamma + 0.760 - (13.6/0.86)(0.290) = 0 \qquad p_1/\gamma = 3.826 \text{ m of oil}$$
$$Q = (0.617)[(\pi)(0.100)^2/4]\sqrt{(2)(9.807)(3.826)}/\sqrt{1 - (0.617)^2(0.100)^4} = 0.0420 \text{ m}^3/\text{s}$$
$$C_c = a/A \qquad 0.63 = a/[(\pi)(0.100)^2/4] \qquad a = 0.004948 \text{ m}^2$$
$$(\pi)(D_{\text{jet}})^2/4 = 0.004948 \qquad D_{\text{jet}} = 0.0794 \text{ m} \quad \text{or} \quad 79.4 \text{ mm}$$
$$v = Q/a = 0.0420/0.004948 = 8.49 \text{ m/s}$$

**17.84** Oil flows in the pipe shown in Fig. 17-21. The orifice in the pipe has a coefficient of discharge of 0.64. Compute the discharge of the oil in the pipe.

Fig. 17-21

$$Q = CA\sqrt{2g(p_1 - p_2)/\gamma}[1 + (C^2/2)(D/D_p)^4] \qquad A = (\pi)(\tfrac{3}{12})^2/4 = 0.04909 \text{ ft}^2$$
$$p_1/\gamma + y + 15.2/12 - (13.6/0.86)(15.2/12) - y = p_2/\gamma \qquad (p_1 - p_2)/\gamma = 18.76 \text{ ft of oil}$$
$$Q = (0.64)(0.04909)\sqrt{(2)(32.2)(18.76)}[1 + (0.64^2/2)(\tfrac{3}{10})^4] = 1.09 \text{ ft}^3/\text{s}$$

**17.85** The water tank shown in Fig. 17-22a has the form of the frustum of a cone. The diameter of the top of the tank is 12 ft, while that at the bottom is 8 ft. The bottom of the tank contains a round, sharp-edged orifice, which has a diameter of 4 in. The discharge coefficient of the orifice is 0.60. If the tank is full at a depth of 10.0 ft, as shown in Fig. 17-22a, how long will it take to empty the tank?

Fig. 17-22

▌ $dt = A_s \, dh/(CA\sqrt{2gh})$. From Fig. 17-22b,

$$y/(y + 10) = \tfrac{8}{12} \qquad y = 20 \text{ ft} \qquad d/12 = (20 + h)/30 \qquad d = 8 + 0.4h$$

$$A = (\pi)(\tfrac{4}{12})^2/4 = 0.08727 \text{ ft}^2 \qquad A_s = \pi d^2/4 = (\pi)(8 + 0.4h)^2/4 = 50.27 + 5.027h + 0.1257h^2$$

$$dt = (50.27 + 5.027h + 0.1257h^2)\, dh/[(0.60)(0.08727)\sqrt{(2)(32.2)(h)}] = (119.6h^{-1/2} + 11.96h^{1/2} + 0.2991h^{3/2})\, dh$$

$$\int_0^t dt = \int_0^{10} (119.6h^{-1/2} + 11.96h^{1/2} + 0.2991h^{3/2})\, dh$$

$$[t]_0^t = [239.2h^{1/2} + 7.973h^{3/2} + 0.1196h^{5/2}]_0^{10} \qquad t = 1046.4 \text{ s} = 17.44 \text{ min}$$

**17.86** A rectangular water tank is divided by a partition into two chambers, as shown in Fig. 17-23a. In the bottom of the partition is a round, sharp-edged orifice with a diameter of 150 mm. The coefficient of discharge for the orifice is 0.62. At a certain instant, the water level in chamber $B$ is 2.5 m higher than it is in chamber $A$. How long will it take for the water surfaces in the two chambers to be at the same level?

(a)

(b)                                                                 **Fig. 17-23**

▌ $dV = CA\sqrt{2gh}\, dt$. From Fig. 17-23b,

$$(2)(7.5)(dh_B) = (3)(2)(dh_A) \qquad dh_A = 2.5\, dh_B$$

$$dh = dh_A + dh_B = 2.5\, dh_B + dh_B = 3.5\, dh_B \qquad dh_B = 0.2857\, dh$$

$$dV = (dh_B)(7.5)(2.0) = (0.2857\, dh)(15) = 4.286\, dh \qquad A = (\pi)(0.150)^2/4 = 0.01767 \text{ m}^2$$

$$4.286\, dh = (0.62)(0.01767)\sqrt{(2)(9.807)(h)}\, dt = 0.04852h\, dt \qquad dt = 88.33h^{-1/2}\, dh$$

$$\int_0^t dt = \int_0^{2.5} 88.33h^{-1/2}\, dh \qquad [t]_0^t = [176.66h^{1/2}]_0^{2.5} \qquad t = 279 \text{ s}$$

**17.87** A 2-in circular orifice (not standard) at the end of a 3-in-diameter pipe discharges into the atmosphere a measured flow of 0.60 cfs of water when the pressure in the pipe is 10.0 psi, as shown in Fig. 17-24. The jet velocity is determined by a Pitot tube to be 39.2 fps. Find the values of the coefficients $C_v$, $C_c$, $C_d$. Find also the head loss for inlet to throat.

**Fig. 17-24**

▮ Define the inlet as section 1 and the throat as section 2. $p_1/\gamma = 10(144/62.4) = 23.1$ ft, $V_1 = Q/A_1 = 0.60/0.0491 = 12.22$ fps, $V_1^2/2g = 2.32$ ft. Express the ideal energy equation from 1 to 2 to determine the ideal velocity at 2: $(p_1/\gamma) + (V_1^2/2g) = V_2^2/2g$, $23.1 - 2.3 = V_2^2/2g$, $(V_2)_{ideal} = 40.4$ fps, $C_v = V/V_i = 39.2/40.4 = 0.97$. Area of jet: $A_2 = Q/V = 0.60/39.2 = 0.0153$ ft$^2$, $C_c = A_2/A_0 = 0.0153/0.0218 = 0.70$. Hence, $C_d = C_c C_v = 0.68$, $h_{L_{1-2}} = [(1/C_v^2) - 1][1 - (A_2/A_1)^2](V_2^2/2g) = \{[1/(0.97)^2] - 1\}[1 - (\frac{2}{3})^4](V_2^2/2g) = 0.051(V_2^2/2g) = 0.051[(39.2)^2/64.4] = 1.22$ ft. As a check, determine the actual velocity at 2 by expressing the real energy equation from 1 to 2: $(p_1/\gamma) + (V_1^2/2y) - h_{L_{1-2}} = V_2^2/2g$, $23.1 + 2.3 - 1.2 = V_2^2/2g$, $(V_2)_{actual} = 39.5$ fps, which is a good check. A better check would result if all numbers were carried out to more places.

**17.88** Water issues from a circular orifice under a head of 45 ft. The diameter of the orifice is 4.5 in. If the discharge is found to be 740 ft$^3$ in 3.4 min, what is the coefficient of discharge? If the diameter at the vena contracta is measured to be 3.54 in, what is the coefficient of contraction and what is the coefficient of velocity?

▮
$$Q = C_d A_0 (2gh)^{1/2} \qquad 740/[(3.4)(60)] = (C_d)[(\pi)(\tfrac{4.5}{12})^2/4][(2)(32.2)(45)]^{1/2} \qquad C_d = 0.610$$
$$C_c = A/A_0 = [(\pi)(3.54)^2/4]/[(\pi)(4.5)^2/4] = 0.619 \qquad C_v = C_d/C_c = 0.610/0.619 = 0.985$$

**17.89** A jet discharges from an orifice in a vertical plane under a head of 16 ft. The diameter of the orifice is 2.0 in, and the measured discharge is 0.42 cfs. The coordinates of the centerline of the jet are 15.39 ft horizontally from the vena contracta and 4.0 ft below the center of the orifice. Find the coefficients of discharge, velocity, and contraction.

▮
$$V = (x)(g/2z)^{1/2} = (15.39)\{32.2/[(2)(4.0)]\}^{1/2} = 30.88 \text{ ft/s} = (C_v)(2gh)^{1/2}$$
$$30.88 = (C_v)[(2)(32.2)(16)]^{1/2} \qquad C_v = 0.962$$
$$Q = C_d A_0 (2gh)^{1/2} \qquad 0.42 = (C_d)[(\pi)(2.0/12)^2/4][(2)(32.2)(16)]^{1/2} \qquad C_d = 0.600$$
$$C_v = C_d/C_c \qquad 0.962 = 0.600/C_c \qquad C_c = 0.624$$

**17.90** In Fig. 17-25, the Pitot tube in a water jet at elevation 120 ft registers a pressure of 22.7 psi. The orifice at the bottom of the large open tank has a diameter of 1.50 in. Find $C_c$ and $C_v$ of the orifice. Neglect air resistance. The flow rate is 0.39 cfs.

**Fig. 17-25**

▮
$$V_3^2/2g = p_3/\gamma = (22.7)(144)/62.4 = 52.38 \text{ ft} \qquad V_3^2/[(2)(32.2)] = 52.38 \qquad V_3 = 58.08 \text{ ft/s}$$
$$A_3 = Q/V_3 = 0.39/58.08 = 0.006715 \text{ ft}^2$$
$$V_{jet}^2/[(2)(32.2)] + 135 = (22.7)(144)/62.4 + 120 \qquad V_{jet} = 49.07 \text{ ft/s}$$
$$V_{ideal} = (2gh)^{1/2} = [(2)(32.2)(176 - 135)]^{1/2} = 51.38 \text{ ft/s}$$
$$C_v = V_{jet}/V_{ideal} = 49.07/51.38 = 0.955$$
$$A_{jet} V_{jet} = C_c A_2 V_{jet} = A_3 V_3 \qquad (C_c)[(\pi)(\tfrac{1.5}{12})^2/4](49.07) = (0.006715)(58.08) \qquad C_c = 0.648$$

**17.91** Water flows from one tank to an adjacent tank through a 60-mm sharp-edged orifice. The head of water on one side of the orifice is 2.9 m and that on the other is 0.7 m. Assuming $C_c = 0.63$ and $C_v = 0.96$, calculate the flow rate.

$$Q = C_c C_v A(2gh)^{1/2} = (0.63)(0.96)[(\pi)(0.060)^2/4][(2)(9.807)(2.9 - 0.7)]^{1/2} = 0.0112 \text{ m}^3/\text{s}$$

**17.92** Repeat Prob. 17.63 for a VDI orifice.

$$Q = KA_2\sqrt{(2g)(14.91R')} \quad \text{(from Prob. 17.63)}$$

Let $K = 0.70$ (see Fig. A-25): $Q = (0.70)[(\pi)(\frac{9}{12})^2/4]\sqrt{(2)(32.2)[(14.91)(5.5/12)]} = 6.49 \text{ ft}^3/\text{s}$. Check $N_R$ to confirm that the assumed $K$ is applicable.

$$V = Q/A = 6.49/[(\pi)(\tfrac{9}{12})^2/4] = 14.69 \text{ ft/s} \qquad N_R = DV/v = (\tfrac{9}{12})(14.69)/0.00034 = 3.24 \times 10^4$$

$$D_0/D_1 = 0.70 \quad \text{(from Prob. 17.63)}$$

From Fig. A-25, $K = 0.72$, approximately, $Q = (0.72/0.70)(6.49) = 6.68 \text{ ft}^3/\text{s}$.

**17.93** Helium, for which $k = 1.66$ and $R = 12\,420$ ft-lb/(slug-°R), is in a tank under a pressure of 60 psia and a temperature of 60 °F. It flows out through an orifice 0.6 in in diameter. For such an orifice, $C_v = 0.98$ and $C_c = 0.62$ for liquids. Find the rate of flow if the pressure into which the gas discharges is 48 psia. Assume $Y = 0.95$.

$$G = CYA_2\sqrt{2g\gamma_1\{(p_1 - p_2)/[1 - (D_2/D_1)^4]\}} \qquad C = C_c C_v = (0.62)(0.98) = 0.608$$

$$Y = 0.95 \quad \text{(from Fig. A-26)}$$

$$\gamma_1 = p_1/RT = (60)(144)/[(12\,420/32.2)(60 + 460)] = 0.04308 \text{ lb/ft}^3$$

$$G = (0.608)(0.95)[(\pi)(0.6/12)^2/4]\sqrt{(2)(32.2)(0.04308)[(60 - 48)(144)/(1 - 0)]} = 0.0785 \text{ lb/s}$$

**17.94** For the data in Prob. 17.93, find the rate of discharge if $p_2 = p_c$.

$$G = CYA_2\sqrt{2g\gamma_1\{(p_1 - p_2)/[1 - (D_2/D_1)^4]\}}$$

$$p_2 = p_c = 0.488p_1 = (0.488)(60) = 29.3 \text{ psia} \qquad Y = 0.85 \quad \text{(from Fig. A-26)}$$

Using data from Prob. 17.93, for $D_0/D_1 = 1/\infty$,

$$G = (0.608)(0.85)[(\pi)(0.6/12)^2/4]\sqrt{(2)(32.2)(0.04308)[(60 - 29.3)(144)/(1 - 0)]} = 0.112 \text{ lb/s}.$$

**17.95** Air is in a tank under a pressure of 1350 kPa abs and a temperature of 90 °C. It flows out through an orifice having an area of 8 cm² into a space where the pressure is 660 kPa abs. Compute the mass flow, assuming $C_d = 0.60$.

$$\dot{m} = CYA_2\sqrt{2\rho_1\{(p_1 - p_2)/[1 - (D_2/D_1)^4]\}} \qquad Y = 0.85 \quad \text{(from Fig. A-26)}$$

$$\rho_1 = p_1/RT_1 = 1350/[(0.287)(90 + 273)] = 12.9 \text{ kg/m}^3$$

$$\dot{m} = (0.60)(0.85)(8 \times 10^{-4})\sqrt{(2)(12.9)[(1350 - 660)(1000)/(1 - 0)]} = 1.72 \text{ kg/s}$$

**17.96** Using the same data as in Prob. 17.95, what would be the flow if the air discharged into a space where the pressure is 125 kPa abs?

With $p_2 = 125$ kPa, $D_0/D_1 = 0$, $p_2/p_1 = 125/1350 = 0.093$, $Y = 0.73$; hence,

$$\dot{m} = (0.60)(0.73)(8 \times 10^{-4})\sqrt{(2)(12.9)[(1350 - 125)(1000)/(1 - 0)]} = 1.97 \text{ kg/s}$$

**17.97** Air in a tank at 1600 kPa abs and 70 °C flows out through a 60-mm-diameter orifice into a space where the pressure is 600 kPa abs. Compute the rate of discharge assuming $C_d = 0.60$.

$p_2/p_1 = 600/1600 = 0.375$. Since $p_2/p_1 < 0.528$, sonic velocity occurs at the throat and
$\dot{m} = C_d(A_2 p_1/\sqrt{T_1})\sqrt{(k/R)[2/(k + 1)]^{(k+1)/(k-1)}}$.

$$A_2 = (\pi)(0.060)^2/4 = 0.002827 \text{ m}^2$$

$$\dot{m} = (0.60)[(0.002827)(16 \times 10^5)/\sqrt{70 + 273}]\sqrt{(1.40/287)[2/(1.40 + 1)]^{(1.40+1)/(1.40-1)}} = 5.92 \text{ kg/s}$$

**17.98** A 4-in-diameter standard orifice discharges water under a 20.0-ft head as shown in Fig. 17-26. What is the flow?

**Fig. 17-26**

$\blacksquare$ $Q = CA_0\sqrt{2gh}$. From Table A-19, $C = 0.594$, $Q = (0.594)[(\pi)(\frac{4}{12})^2/4]\sqrt{(2)(32.2)(20.0)} = 1.86$ ft$^3$/s.

**17.99** The actual velocity in the contracted section of a jet of liquid flowing from a 2-in-diameter orifice is 28.0 ft/s under a head of 15 ft. (**a**) What is the value of the coefficient of velocity? (**b**) If the measured discharge is 0.403 cfs, determine the coefficients of contraction and discharge.

$\blacksquare$ (**a**) $\qquad\qquad V_{\text{actual}} = C_v\sqrt{2gH} \qquad 28.0 = C_v\sqrt{(2)(32.2)(15)} \qquad C_v = 0.901$

(**b**) $\qquad\qquad Q_{\text{actual}} = CA\sqrt{2gH} \qquad 0.403 = (C)[(\pi)(\frac{2}{12})^2/4]\sqrt{(2)(32.2)(15)} \qquad C = 0.594$

$$C = C_vC_c \qquad 0.594 = 0.901C_c \qquad C_c = 0.659$$

**17.100** Oil flows through a standard 1-in-diameter orifice under an 18.0-ft head at the rate of 0.111 cfs. The jet strikes a wall 5.00 ft away and 0.390 ft vertically below the centerline of the contracted section of the jet. Compute the coefficients.

$\blacksquare$ $\qquad Q = CA\sqrt{2gH} \qquad 0.111 = (C)[(\pi)(\frac{1}{12})^2/4]\sqrt{(2)(32.2)(18.0)} \qquad C = 0.598 \qquad V = C_v\sqrt{2gH}$

Letting $x$ and $y$ represent the coordinates of the jet as measured, from kinematic mechanics, $x = Vt$, $y = gt^2/2$. Eliminating $t$ gives

$$x^2 = (2V^2/g)(y) \qquad 5.00^2 = (2V^2/32.2)(0.390) \qquad V = V_{\text{actual}} = 32.13 \text{ ft/s}$$

$$32.13 = C_v\sqrt{(2)(32.2)(18.0)} \qquad C_v = 0.944 \qquad C = C_cC_v \qquad 0.598 = (C_c)(0.944) \qquad C_c = 0.633$$

**17.101** The tank in Prob. 17.98 is closed and the air space above the water is under pressure, causing the flow to increase to 2.65 cfs. Find the pressure in the air space.

$\blacksquare$ $Q = CA_0\sqrt{2gH}$. Table A-19 indicates that $C$ does not change appreciably at the range of head under consideration. Using $C = 0.593$, $2.65 = (0.593)[(\pi)(\frac{4}{12})^2/4]\sqrt{(2)(32.2)(20 + p/\gamma)}$, $p/\gamma = 20.72$ ft of water, $p' = \gamma h = (62.4)(20.72)/144 = 8.98$ lb/in$^2$.

**17.102** Oil of specific gravity 0.720 flows through a 3-in-diameter orifice whose coefficients of velocity and contraction are 0.950 and 0.650, respectively. What must be the reading of gage $A$ in Fig. 17-27 in order for the power in the jet $C$ to be 8.00 hp?

$\blacksquare$ $\qquad p_B/\gamma + V_B^2/2g + z_B = p_C/\gamma + V_C^2/2g + z_C + h_L \qquad P_{\text{jet}} = \gamma QH_{\text{jet}} = (\gamma)(C_cA_0V_{\text{jet}})(0 + V_{\text{jet}}^2/2g + 0)$

$$(8.00)(550) = [(0.720)(62.4)](0.650)[(\pi)(\frac{3}{12})^2/4](V_{\text{jet}})\{V_{\text{jet}}^2/[(2)(32.2)]\}$$

$$V_{\text{jet}} = V_C = 58.25 \text{ ft/s} \qquad h_L = (1/C_v - 1)(V_{\text{jet}}^2/2g)$$

$$p_A/[(0.720)(62.4)] + 0 + 9.0 = 0 + 58.25^2/[(2)(32.2)] + 0 + (1/0.950 - 1)\{58.25^2/[(2)(32.2)]\}$$

$$p_A = 2087 \text{ lb/ft}^2 \quad \text{or} \quad 14.5 \text{ lb/in}^2$$

**Fig. 17-27**

**17.103** Water at 100 °F flows at the rate of 0.525 cfs through a 4-in diameter orifice used in an 8-in pipe. What is the difference in pressure head between the upstream section and the contracted section (vena contracta section)?

▌ In Fig. A-27, it is observed that $C'$ varies with the Reynolds number. Note that the Reynolds number must be calculated for the orifice cross section, not for the contracted section of the jet nor for the pipe section. This value is $N_R = V_o D_o / v = (4Q/\pi D_o^2)D_o/v = 4Q/v\pi D_o = 4(0.525)/[\pi(0.00000739)(\frac{4}{12})] = 272\,000$. Figure A-27 for $\beta = 0.500$ gives $C' = 0.604$.

Applying the Bernoulli theorem, pipe section to jet section, produces the general equation for incompressible fluids, as follows: $[(p_8/\gamma) + (V_8^2/2g) + 0] - [(1/C_v^2) - 1](V_{jet}^2/2g) = [(p_{jet}/\gamma) + (V_{jet}^2/2g) + 0]$ and $Q = A_8 V_8 = (C_c A_4)V_{jet}$.

Substituting for $V_8$ in terms of $V_{jet}$ and solving, $V_{jet}^2/2g = C_v^2\{(p_8/\gamma - p_{jet}/\gamma)/[1 - C^2(A_4/A_8)^2]\}$ or $V_{jet} = C_v\sqrt{2g(p_8/\gamma - p_{jet}/\gamma)/[1 - C^2(D_4/D_8)^4]}$. Then $Q = A_{jet}V_{jet} = (C_c A_4)(C_v)\sqrt{2g(p_8/\gamma - p_{jet}/\gamma)/[1 - C^2(D_4/D_8)^4]} = CA_4\sqrt{2g(p_8/\gamma - p_{jet}/\gamma)/[1 - C^2(D_4/D_8)^4]}$.

More conveniently, for an orifice with velocity of approach and a contracted jet, the equation can be written

$$Q = [C'A_4/\sqrt{1 - (D_4/D_8)^4}]\sqrt{2g(\Delta p/\gamma)} \tag{1}$$

or

$$Q = KA_4\sqrt{2g(\Delta p/\gamma)} \tag{2}$$

where $K$ is called the flow coefficient. The meter coefficient $C'$ may be determined experimentally for a given ratio of diameter of orifice to diameter of pipe, or the flow coefficient $K$ may be preferred.

Proceeding with the solution by substituting in Eq. (1),

$$0.525 = \frac{(0.604)[(\pi)(\frac{4}{12})^2/4]}{\sqrt{1 - (\frac{1}{2})^4}}\sqrt{(2)(32.2)\left(\frac{\Delta p}{\gamma}\right)} \qquad \frac{\Delta p}{\gamma} = 1.44 \text{ ft of water}$$

**17.104** For the pipe orifice in Prob. 17.103, what pressure difference in pounds per square inch would cause the same quantity of turpentine at 68 °F (s.g. = 0.862, $v = 0.0000186$ ft$^2$/s) to flow?

▌
$$Q = [C'A_4/\sqrt{1 - (D_4/D_8)^4}]\sqrt{2g(\Delta p/\gamma)}$$

$$V = Q/A = 0.525/[(\pi)(\tfrac{4}{12})^2/4] = 6.016 \text{ ft/s} \qquad N_R = DV/v = (\tfrac{4}{12})(6.016)/0.0000186 = 1.08 \times 10^5$$

From Fig. A-27, for $\beta = 0.500$, $C' = 0.607$.

$$0.525 = \frac{(0.607)[(\pi)(\frac{4}{12})^2/4]}{\sqrt{1 - (\frac{1}{2})^4}}\sqrt{(2)(32.2)\left[\frac{\Delta p}{(0.862)(62.4)}\right]}$$

$$p = 76.9 \text{ lb/ft}^2 \quad \text{or} \quad 0.534 \text{ lb/in}^2$$

**17.105** Determine the flow of water at 70 °F through a 6-in orifice installed in a 10-in pipeline if the pressure head differential for vena-contracta taps is 3.62 ft of water.

▌
$$Q = [C'A_6/\sqrt{1 - (D_6/D_{10})^4}]\sqrt{2g(\Delta p/\gamma)}$$

The value of $C'$ cannot be found inasmuch as the Reynolds number cannot be computed. Referring to Fig. A-27, for $\beta = 0.600$, a value of $C'$ will be assumed at 0.610. Using this assumed value,

$$Q = \frac{(0.610)[(\pi)(\frac{6}{12})^2/4]}{\sqrt{1 - (0.600)^4}}\sqrt{(2)(32.2)(3.62)} = 1.96 \text{ ft}^3/\text{s}$$

$$V = Q/A = 1.96/[(\pi)(\tfrac{6}{12})^2/4] = 9.982 \text{ ft/s} \qquad N_R = DV/v = (\tfrac{6}{12})(9.982)/0.00001059 = 4.71 \times 10^5$$

From Fig. A-27, for $\beta = 0.600$, $C' = 0.609$.

$$Q = \frac{(0.609)[(\pi)(\frac{6}{12})^2/4]}{\sqrt{1-(0.600)^4}} \sqrt{(2)(32.2)(3.62)} = 1.96 \text{ ft}^3/\text{s}$$

**17.106** Carbon dioxide discharges through a $\frac{1}{2}$-in hole in the wall of a tank in which the pressure is 110 psig and the temperature is 68 °F. What is the velocity in the jet (standard atmosphere)?

$$\gamma_1 = p_1/RT_1 = (110 + 14.7)(144)/[(35.1)(68 + 460)] = 0.9689 \text{ lb/ft}^3$$
$$(p_2/p_1)_{\text{critical}} = [2/(k+1)]^{k/(k-1)} = [2/(1.30+1)]^{1.30/(1.30-1)} = 0.546$$
$$p_{\text{atm}}/p_{\text{tank}} = 14.7/(110+14.7) = 0.118$$

Since the latter ratio is less than the critical pressure ratio, the pressure of the escaping gas is 0.546 times $p_1$; hence,

$$p_2 = (0.546)(110 + 14.7) = 68.1 \text{ psia} \qquad V_2 = c_2 = \sqrt{kgRT_2} \qquad T_2/T_1 = (p_2/p_1)^{(k-1)/k}$$
$$T_2/(68 + 460) = 0.546^{(1.30-1)/1.30} \qquad T_2 = 459 \text{ °R} \qquad V_2 = \sqrt{(1.30)(32.2)(35.1)(459)} = 821 \text{ ft/s}$$

**17.107** Nitrogen flows through a duct in which changes in cross section occur. At a particular cross section the velocity is 1200 ft/s, the pressure is 12.0 psia, and the temperature is 90 °F. Assuming no friction losses and adiabatic conditions, (a) what is the velocity at a section where the pressure is 18.0 psia and (b) what is the Mach number at this section?

(a)
$$V_2^2/2g - V_1^2/2g = [k/(k-1)](p_1/\gamma_1)[1 - (p_2/p_1)^{(k-1)/k}]$$
$$\gamma_1 = p_1/RT_1 = (12.0)(144)/[(55.1)(90+460)] = 0.05702 \text{ lb/ft}^3$$
$$\{V_2^2/[(2)(32.2)]\} - \{1200^2/[(2)(32.2)]\} = [1.40/(1.40-1)][(12.0)(144)/0.05702]$$
$$\times [1 - (18.0/12.0)^{(1.40-1)/1.40}] \qquad V_2 = 775 \text{ ft/s}$$

(b)
$$N_M = V_2/c_2 = V_2/\sqrt{kgRT} \qquad T_2/T_1 = (p_2/p_1)^{(k-1)/k}$$
$$T_2/(90+460) = (18.0/12.0)^{(1.40-1)/1.40} \qquad T_2 = 618 \text{ °R}$$
$$N_M = 775/\sqrt{(1.40)(32.2)(55.1)(618)} = 0.626$$

**17.108** Establish the formula to determine the time to lower the liquid level in a tank of constant cross section by means of an orifice. Refer to Fig. 17-28.

**Fig. 17-28**

▮ Inasmuch as the head is changing with time, we know that $\partial V/\partial t \neq 0$, i.e., we do not have steady flow. This means that the energy equation should be amended to include an acceleration term, which complicates the solution materially. As long as the head does not change too rapidly, no appreciable error will be introduced by assuming steady flow, thus neglecting the acceleration-head term.

**Case A:** With *no inflow* taking place, the instantaneous flow will be $Q = CA_0\sqrt{2gh}$ cfs. In time interval $dt$, the small volume $dV$ discharged will be $Q\,dt$. In the same time interval, the head will decrease $dh$ ft and the volume discharged will be the area of the tank $A_T$ times $dh$. Equating these values, $(CA_0\sqrt{2gh})\,dt = -A_T\,dh$ where the negative sign signifies that $h$ decreases as $t$ increases. Solving for $t$ yields

$$t = \int_{t_1}^{t_2} dt = (-A_T/CA_0\sqrt{2g}) \int_{h_1}^{h_2} h^{-1/2}\,dh$$

or

$$t = t_2 - t_1 = (2A_T/CA_0\sqrt{2g})(h_1^{1/2} - h_2^{1/2}) \qquad (1)$$

In using this expression, an average value of coefficient of discharge $C$ may be used without producing significant error in the result. As $h_2$ approaches zero, a vortex will form and the orifice will cease to flow full. However, using $h_2 = 0$ will not produce serious error in most cases.

Equation $(1)$ can be rewritten by multiplying and dividing by $(h_1^{1/2} + h_2^{1/2})$. There results

$$t = t_2 - t_1 = A_T(h_1 - h_2)/[\tfrac{1}{2}(CA_0\sqrt{2gh_1} + CA_0\sqrt{2gh_2})] \qquad (2)$$

Noting that the volume discharged in time $(t_2 - t_1)$ is $A_T(h_1 - h_2)$, this equation simplifies to

$$t = t_2 - t_1 = \text{volume discharged}/[\tfrac{1}{2}(Q_1 + Q_2)] = \text{volume discharged/average flow } Q \qquad (3)$$

**Case B.** With a constant rate of inflow less than the flow through the orifice taking place,

$$-A_T\, dh = (Q_{\text{out}} - Q_{\text{in}})\, dt \qquad \text{and} \qquad t = t_2 - t_1 = \int_{h_1}^{h_2} \frac{-A_T\, dh}{Q_{\text{out}} - Q_{\text{in}}}$$

Should $Q_{\text{in}}$ exceed $Q_{\text{out}}$, the head would increase, as would be expected.

**17.109** The initial head on an orifice was 9 ft and when the flow was terminated the head was measured at 4 ft. Under what constant head $H$ would the same orifice discharge the same volume of water in the same time interval? Assume coefficient $C$ is constant.

❚ Volume under falling head = volume under constant head

$$\tfrac{1}{2}CA_0\sqrt{2g}\,(h_1^{1/2} + h_2^{1/2})(t) = CA_0\sqrt{2gH}\,(t)$$

$$\tfrac{1}{2}CA_0\sqrt{2g}\,(9^{1/2} + 4^{1/2})(t) = CA_0\sqrt{2g}\,\sqrt{H}\,(t) \qquad H = 6.25 \text{ ft}$$

**17.110** A 75-mm-diameter orifice under a head of 4.88 m discharges 8900 N water in 32.6 s. The trajectory was determined by measuring $x_0 = 4.76$ m for a drop of 1.22 m. Determine $C_v$, $C_c$, $C_d$, the head loss per unit weight, and the power loss.

❚ The theoretical velocity $V_{2t}$ is $V_{2t} = \sqrt{2gH} = \sqrt{2(9.806)(4.88)} = 9.783$ m/s. The actual velocity is determined from the trajectory. The time to drop 1.22 m is $t = \sqrt{2y_0/g} = \sqrt{2(1.22)/9.806} = 0.499$ s and the velocity is expressed by $x_0 = V_{2a}t$, $V_{2a} = 4.76/0.499 = 9.539$ m/s. Then $C_v = V_{2a}/V_{2t} = 9.539/9.783 = 0.975$. The actual discharge $Q_a$ is $Q_a = (8900/32.6)/9806 = 0.0278$ m³/s. $C_d = Q_a/(A_0\sqrt{2gH}) = 0.0278/[\pi(0.0375^2)\sqrt{2(9.806)(4.88)}] = 0.643$, $C_c = C_d/C_v = 0.643/0.975 = 0.659$. The head loss is loss $= H(1 - C_v^2) = 4.88(1 - 0.975^2) = 0.241$ m. The power loss is $Q\gamma(\text{loss}) = 0.0278(9806)(0.241) = 65.7$ W.

**17.111** A tank has a horizontal cross-sectional area of 2 m² at the elevation of the orifice, and the area varies linearly with elevation so that it is 1 m² at a horizontal cross section 3 m above the orifice. For a 100-mm-diameter orifice, $C_d = 0.65$, compute the time to lower the surface from 2.5 m to 1 m above the orifice.

❚
$$t = -\frac{1}{C_d A_0 \sqrt{2g}} \int_{y_1}^{y_2} A_r y^{-1/2}\, dy \qquad A_r = 2 - \frac{y}{3}$$

$$t = -\frac{1}{(0.65)[(\pi)(\frac{100}{1000})^2/4]\sqrt{(2)(9.807)}} \int_{2.5}^{1} \left(2 - \frac{y}{3}\right) y^{-1/2}\, dy = -44.23[4y^{1/2} - \tfrac{2}{9}y^{3/2}]_{2.5}^{1} = 73.8 \text{ s}$$

**17.112** Determine the equation for trajectory of a jet discharging horizontally from a small orifice with head of 8 m and velocity coefficient of 0.96. Neglect air resistance.

❚
$$x = V_2 t \qquad y = gt^2/2 \qquad V_2 = C_v\sqrt{2gH} \qquad t = x/V_2$$

$$y = (g)(x/V_2)^2/2 = (g)[x/(C_v\sqrt{2gH})]^2/2 = (9.807)\{x/[0.96\sqrt{(2)(9.807)(8)}]\}^2/2 = 0.0339x^2$$

**17.113** An orifice of area 25 cm² in a vertical plate has a head of 1.13 m of oil, s.g. = 0.89. It discharges 6192 N of oil in 88.8 s. Trajectory measurements yield $x_0 = 2.40$ m, $y_0 = 1.37$ m. Determine $C_v$, $C_c$, and $C_d$.

❚ Letting subscript $a$ denote "actual" and $t$, "theoretical,"

$$Q_a = W/\gamma = (6192/88.8)/[(0.89)(9790)] = 0.008003 \text{ m}^3/\text{s}$$

$$(V_2)_a = x\sqrt{g/2y} = (2.40)\sqrt{(9.807)/[(2)(1.37)]} = 4.541 \text{ m/s}$$

$$(V_2)_t = \sqrt{2gH} = \sqrt{(2)(9.807)(1.13)} = 4.708 \text{ m/s} \qquad C_v = (V_2)_a/(V_2)_t = 4.541/4.708 = 0.965$$

$$C_d = Q_a/[A(V_2)_t] = 0.008003/[(25 \times 10^{-4})(4.708)] = 0.680 \qquad C_c = C_d/C_v = 0.680/0.965 = 0.705$$

**17.114** Calculate $Y$, the maximum rise of a jet from an inclined plate (Fig. 17-29), in terms of $H$ and $\alpha$. Neglect losses.

**Fig. 17-29**

$\blacksquare$
$$y = -gt^2/2 + (V \cos \alpha)(t) \qquad x = Vt \sin \alpha \qquad V = \sqrt{2gH}$$

Substituting for $V$ and $t$,

$$y = -x^2/(4H \sin^2 \alpha) + x \cot \alpha \qquad dy/dx = -x/(2H \sin^2 \alpha) + \cot \alpha = 0$$
$$x = 2H \sin \alpha \cos \alpha \qquad y_{\max} = H \cos^2 \alpha$$

**17.115** In Fig. 17-29, for $\alpha = 45°$, $Y = 0.48H$. Neglecting air resistance of the jet, find $C_v$ for the orifice.

$\blacksquare$ From Prob. 17.114 but with $V = C_v\sqrt{2gH}$, $Y = C_v^2 H \cos^2 \alpha$, $C_v = \sqrt{Y/H}/\cos \alpha = \sqrt{0.48H/H}/\cos 45° = 0.980$.

**17.116** Show that the locus of maximum points of the jet of Fig. 17-29 is given by $X^2 = 4Y(H - Y)$ when losses are neglected.

$\blacksquare$ From Prob. 17.114, maximum point coordinates are given by $X = 2H \sin \alpha \cos \alpha$ and $Y = H \cos^2 \alpha$. Thus, $\cos \alpha = \sqrt{Y/H}$ and $\sin \alpha = \sqrt{1 - \cos^2 \alpha} = \sqrt{1 - Y/H}$. Hence, $X = 2H\sqrt{Y/H}\sqrt{1 - Y/H}$, $X^2 = 4H^2[Y/H - (Y/H)^2] = 4Y(H - Y)$.

**17.117** A 4-in-diameter orifice discharges 135 ft³ of liquid, s.g. = 1.10, in 95.5 s under a 10-ft head. The velocity at the vena contract is determined by a Pitot-static tube with coefficient 1.0. The manometer liquid is acetylene tetrabromide, s.g. 2.96, and the gage difference is $R' = 3.62$ ft. Determine $C_v$, $C_c$, and $C_d$.

$\blacksquare$
$$V_a = C\sqrt{2gR'(S_0/S - 1)} = (1.0)\sqrt{(2)(32.2)(3.62)[(2.96/1.10) - 1]} = 19.85 \text{ ft/s}$$
$$V_t = \sqrt{2gH} = \sqrt{(2)(32.2)(10)} = 25.38 \text{ ft/s}$$
$$C_v = V_a/V_t = 19.85/25.38 = 0.782 \qquad Q_a = V/t = 135/95.5 = 1.41 \text{ ft}^3/\text{s}$$
$$C_d = Q_a/AV_t = 1.41/\{[(\pi)(\tfrac{4}{12})^2/4](25.38)\} = 0.637 \qquad C_c = C_d/C_v = 0.637/0.782 = 0.815.$$

**17.118** A 115-mm-diameter orifice discharges 60.0 L/s of water under a head of 2.70 m. A flat plate held normal to the jet downstream from the vena contracta requires a force of 423 N to resist impact of the jet. Find $C_d$, $C_v$, and $C_c$.

$\blacksquare$
$$F = \rho(V_2)_a Q_a \qquad (V_2)_a = F/\rho Q_a = 423/(1000)(60.0 \times 10^{-3}) = 7.050 \text{ m/s}$$
$$C_v = (V_2)_a/\sqrt{2gH} = 7.050/\sqrt{(2)(9.807)(2.70)} = 0.969$$
$$C_d = Q_a/(A\sqrt{2gH}) = (60.0 \times 10^{-3})/\{[(\pi)(0.115)^2/4]\sqrt{(2)(9.807)(2.70)}\} = 0.794$$
$$C_c = C_d/C_v = 0.794/0.969 = 0.819$$

**17.119** Compute the discharge from the tank shown in Fig. 17-30.

**Fig. 17-30**

$\blacksquare$
$$Q = C_d A\sqrt{2g(H + p_a/\gamma)} = (0.76)[(\pi)(0.090)^2/4]\sqrt{(2)(9.807)\{3 + 24/[(0.90)(9.79)]\}} = 0.0512 \text{ m}^3/\text{s}$$

**17.120** For $C_v = 0.95$ in Fig. 17-30, calculate the head loss.

❚ $$h_L = (H + p_a/\gamma)(1 - C_v^2) = \{3 + 24/[(0.90)(9.79)]\}(1 - 0.95^2) = 0.558 \text{ m}$$

**17.121** Calculate the discharge through the orifice of Fig. 17-31.

**Fig. 17-31**

❚ $$H' = (3 + 50/9.79) - (2 + 15/9.79) = 4.575 \text{ m}$$
$$Q = C_d A\sqrt{2gH'} = (0.86)[(\pi)(0.140)^2/4]\sqrt{(2)(9.807)(4.575)} = 0.125 \text{ m}^3/\text{s}$$

**17.122** For $C_v = 0.92$ in Fig. 17-31, determine the power loss.

❚ $H' = 4.575$ m     (from Prob. 17.121)     $h_L = H'(1 - C_v^2) = (4.575)(1 - 0.92^2) = 0.703$ m

$Q = 0.125$ m³/s     (from Prob. 17.121)     power loss $= Q\gamma h_L = (0.125)[(9.79)(1000)](0.703) = 860$ W

**17.123** A 3-in-diameter orifice discharges 0.826 cfs of liquid under a head of 9.5 ft. The diameter of the jet at the vena contracta is found by calipering to be 2.65 in. Calculate $C_v$, $C_d$, and $C_c$.

❚ $$V_2 = \sqrt{2gH} = \sqrt{(2)(32.2)(9.5)} = 24.73 \text{ ft/s} \qquad A_0 = (\pi)(\tfrac{3}{12})^2/4 = 0.04909 \text{ ft}^2$$
$$C_d = Q_a/V_2A_0 = 0.826/[(24.73)(0.04909)] = 0.680$$
$$C_c = A_2/A_0 = [(\pi)(2.65/12)^2/4]/[(\pi)(\tfrac{3}{12})^2/4] = 0.780 \qquad C_v = C_d/C_c = 0.680/0.780 = 0.872$$

**17.124** A 90-mm-diameter orifice, $C_d = 0.80$, is placed in the bottom of a vertical tank that has a diameter of 2.0 m. How long does it take to draw the surface down from 4 to 3.5 m?

❚ $$t = (2A_r/C_d A_0\sqrt{2g})(y_1^{1/2} - y_2^{1/2}) \qquad A_r = (\pi)(2.0)^2/4 = 3.142 \text{ m}^2$$
$$A_0 = (\pi)(0.090)^2/4 = 0.006362 \text{ m}^2$$
$$t = \{(2)(3.142)/[(0.80)(0.006362)\sqrt{(2)(9.807)}]\}(4^{1/2} - 3.5^{1/2}) = 36.0 \text{ s}$$

**17.125** Select the size of orifice that permits a tank of horizontal cross section 2.0 m² to have the liquid surface drawn down at the rate of 140 mm/s for 3.50-m head on the orifice ($C_d = 0.65$).

❚ $$Q = (A_r)(\Delta y/\Delta t) = (2.0)(140 \times 10^{-3}) = 0.2800 \text{ m}^3/\text{s}$$
$$A_0 = Q/(C_d\sqrt{2gH}) = 0.2800/[0.65\sqrt{(2)(9.807)(3.50)}] = 0.05199 \text{ m}^2 = 51\,990 \text{ mm}^2$$
$$51\,990 = \pi D^2/4 \qquad D = 257 \text{ mm}$$

**17.126** A 3-in-diameter orifice in the side of a 5-ft-diameter tank draws the surface down from 7 ft to 5 ft above the orifice in 53.0 s. Calculate the discharge coefficient.

❚ $$t = [2A_r/(C_d A_0\sqrt{2g})](y_1^{1/2} - y_2^{1/2}) \qquad A_r = (\pi)(5)^2/4 = 19.63 \text{ ft}^2$$
$$A_0 = (\pi)(\tfrac{3}{12})^2/4 = 0.04909 \text{ ft}^2$$
$$53.0 = \{(2)(19.63)/[(C_d)(0.04909)\sqrt{(2)(32.2)}]\}(7^{1/2} - 5^{1/2}) \qquad C_d = 0.770$$

**17.127** Select a reservoir of such size and shape that the liquid surface drops 1.1 m/min over a 4-m distance for flow through a 90-mm-diameter orifice ($C_d = 0.75$).

$$dy/dt = -Q/A_r = -(C_d A_0 \sqrt{2gy})/A_r$$

$$A_r = C_d A_0 \sqrt{2g}(\Delta t/\Delta y)y^{1/2} = (0.75)[(\pi)(0.090)^2/4]\sqrt{(2)(9.807)}(60/1.1)y^{1/2} = 1.153y^{1/2}$$

$$\pi D^2/4 = 1.153y^{1/2} \qquad D = (1.212 \text{ m}^{3/4})y^{1/4}$$

**17.128** In Fig. 17-32 the truncated cone has an angle $\theta = 60°$. How long does it take to draw the liquid surface down from $y = 5$ m to $y = 2$ m?

1 m diam
80 mm diam
$C_d = 0.80$
$\theta$

**Fig. 17-32**

$$t = -\frac{1}{C_d A_0 \sqrt{2g}} \int_{y_1}^{y_2} \frac{A_r}{\sqrt{y}} dy \qquad A_r = \pi \left(r_0 + y \tan \frac{\theta}{2}\right)^2$$

$$t = \frac{\pi}{C_d A_0 \sqrt{2g}} \int_{y_2}^{y_1} \frac{[r_0 + y \tan(\theta/2)]^2}{\sqrt{y}} dy = \frac{\pi}{C_d A_0 \sqrt{2g}} \left[ r_0^2(2\sqrt{y}) + 2r_0(\tfrac{2}{3}y^{3/2}) \tan \frac{\theta}{2} + \tfrac{2}{5}y^{5/2} \tan^2 \frac{\theta}{2} \right]_{y_2}^{y_1}$$

$$A_0 = (\pi)(0.080)^2/4 = 0.005026 \text{ m}^2$$

$$t = -\frac{\pi}{(0.80)(0.005026)\sqrt{(2)(9.807)}} [(\tfrac{1}{2})^2(2\sqrt{y}) + (2)(\tfrac{1}{2})(\tfrac{2}{3})y^{3/2} \tan \tfrac{60}{2} + \tfrac{2}{5}y^{5/2} \tan^2 \tfrac{60}{2}]_5^2 = 1822 \text{ s} \quad \text{or} \quad 30.4 \text{ min}$$

**17.129** Calculate the dimensions of a tank such that the surface velocity varies inversely as the distance from the centerline of an orifice draining the tank. When the head is 400 mm, the velocity of fall of the surface is 40 mm/s, orifice diameter is 15 mm, $C_d = 0.64$.

$$dy/dt = -Q/A_r = k/y \qquad k = y(dy/dt) = (0.400)(-0.040) = -0.01600 \qquad -C_d A_0 \sqrt{2gy}/(\pi D^2/4) = k/y$$

$$D = y^{3/4} \sqrt{\frac{C_d A_0 \sqrt{2g}}{\pi/4}} \left(-\frac{1}{k}\right) = y^{3/4} \sqrt{\frac{(0.64)[(\pi)(0.015)^2/4]\sqrt{(2)(9.807)}}{\pi/4}} \left(-\frac{1}{-0.01600}\right)$$

$$= (1.578 \text{ m}^{1/4})y^{3/4}$$

**17.130** Determine the time required to raise the right-hand surface of Fig. 17-33 by 3 ft.

H = difference between surface elevations

7 ft
$y_2$
4 in diam
$C_d = 0.92$
$y_1$
$A_R = 20 \text{ ft}^2$
$A_R = 25 \text{ ft}^2$

**Fig. 17-33**

$$A_1 y_1 + A_2 y_2 = C \qquad A_1 \, dy_1 + A_2 \, dy_2 = 0 \qquad H = y_1 - y_2$$

$$dH = dy_1 - dy_2 = -(1 + A_2/A_1) \, dy_2$$

$$dy_2 = -dH/(1 + A_2/A_1) \qquad Q \, dt = C_d A_0 \sqrt{2gH} \, dt = A_2 \, dy_2 = -[A_2 \, dH/(1 + A_2/A_1)]$$

Hence,

$$t = \frac{A_2}{C_d A_0 \sqrt{2g}(1 + A_2/A_1)} \int_{H_1}^{H_2} H^{-1/2} \, dH = \frac{2A_2(\sqrt{H_1} - \sqrt{H_2})}{C_d A_0 \sqrt{2g}(1 + A_2/A_1)} \qquad H_1 = 7 \text{ ft}$$

Rise of the right-hand surface by 3 ft will create a fall of the left-hand surface by $(3)(\frac{25}{20}) = 3.75$ ft. Thus, $H_2 = 7 - 3 - 3.75 = 0.25$ ft. $t = (2)(25)(\sqrt{7} - \sqrt{0.25})/\{(0.92)[(\pi)(\frac{4}{12})^2/4]\sqrt{(2)(32.2)}(1 + \frac{25}{20})\} = 74.0$ s.

**17.131** How long does it take to raise the water surface of Fig. 17-34 by 1 m? The left-hand surface is a large reservoir of constant water-surface elevation.

**Fig. 17-34**

∎ As in Prob. 17.130, $t = [2A_2(\sqrt{H_1} - \sqrt{H_2})]/[C_d A_0 \sqrt{2g}(1 + A_2/A_1)]$ where $A_1 = \infty$, $A_0 = (\pi)(0.200)^2/4 = 0.03142$ m$^2$, $t = [(2)(\pi 6^2/4)(\sqrt{4} - \sqrt{4-1})]/[(0.86)(0.03142)\sqrt{(2)(9.807)}(1 + 0)] = 126.6$ s.

**17.132** Determine the discharge in a 350-mm-diameter line with a 180-mm-diameter VDI orifice for water at 20 °C when the gage difference is 320 mm on an acetylene tetrabromide (s.g. = 2.94, $v = 1.01 \times 10^{-6}$ m$^2$/s)–water differential manometer.

∎ $$Q = CA_0\sqrt{2gR'(S_0/S_1 - 1)} \qquad A_0/A_1 = (180/350)^2 = 0.264$$

From Fig. A-28, assume $C = 0.632$ (for high $N_R$).

$$Q = (0.632)[(\pi)(0.180)^2/4]\sqrt{(2)(9.807)(0.320)[(2.94/1) - 1]} = 0.0561 \text{ m}^3/\text{s}$$
$$N_R = DV/v = 4Q/\pi Dv = (4)(0.0561)/[(\pi)(0.350)(1.01 \times 10^{-6})] = 2.02 \times 10^5$$

From Fig. A-28, $C = 0.632$ as assumed is O.K.

**17.133** A 15-mm-diameter VDI orifice is installed in a 30-mm-diameter pipe carrying nitrogen at $p_1 = 10$ atm, $T_1 = 30$ °C. For a pressure drop of 120 kPa across the orifice, calculate the mass flow rate.

∎ $$\dot{m} = CYA_0\sqrt{2\rho_1 \Delta p} \qquad A_0/A_1 = (15/30)^2 = 0.2500$$

From Fig. A-28, assume $C = 0.63$.

$$k = 1.4 \qquad D_0/D_1 = 0.5 \qquad p_2/p_1 = [(10)(101.3) - 120]/[(10)(101.3)] = 0.882$$

From Fig. A-26, $Y = 0.97$.

$$\rho = p/RT = (10)(101.3)/[(0.297)(30 + 273)] = 11.26 \text{ kg/m}^3$$
$$\dot{m} = (0.63)(0.97)[(\pi)(0.015)^2/4]\sqrt{(2)(11.26)[(120)(1000)]} = 0.1775 \text{ kg/s} \qquad N_R = \rho DV/\mu$$
$$V = \dot{m}/\rho A = 0.1775/\{(11.26)[(\pi)(0.030)^2/4]\} = 22.30 \text{ m/s}$$
$$N_R = (11.26)(0.030)(22.30)/(2.15 \times 10^{-5}) = 3.50 \times 10^5$$

From Fig. A-28, $C = 0.63$; hence, $\dot{m} = 0.1775$ kg/s.

**17.134** Air at 1 atm, $T = 29$ °C, flows through a 1.1-m-square duct that contains a 400-mm-diameter square-edged orifice. With a head loss of 55 mmH$_2$O across the orifice, compute the mass flow.

∎ $$\dot{m} = CYA_0\sqrt{2\rho \Delta p} \qquad A_0/A_1 = [(\pi)(0.400)^2/4]/[(1.1)(1.1)] = 0.104$$

From Fig. A-28, assume $C = 0.605$ (for high $N_R$).

$$p_2 = p_1 - \gamma h = 101\,300 - (9790)(0.055) = 100\,762 \text{ Pa} \qquad p_2/p_1 = 100\,762/101\,300 = 0.995$$
$$D_0/D_1 = (0.400)/[(1.1)(1.1)] = 0.331 \qquad k = 1.4$$

From Fig. A-26, $Y = 0.995$.

$$\rho = p/RT = 101.3/[(0.287)(29 + 273)] = 1.170 \text{ kg/m}^3$$

$$\dot{m} = (0.605)(0.995)[(\pi)(0.400)^2/4]\sqrt{(2)(1.170)(9790)(0.055)} = 2.685 \text{ kg/s}$$

$$N_R = \rho D_h V/\mu = (1.170)[(4)(1.1)]\{2.29/[(1.1)(1.1)]\}/(1.915 \times 10^{-5}) = 5.09 \times 10^5$$

From Fig. A-28, $C = 0.605$, as assumed.

**17.135** A 4-in-diameter VDI orifice is installed in an 8-in-diameter oil line, $\mu = 6$ cP, $\gamma = 54$ lb/ft$^3$. An oil-air differential manometer is used. For a gage difference of 18 in determine the flow rate in gallons per minute.

$\blacksquare$
$$Q = CA_0\sqrt{2gR'} \qquad A_0/A_1 = (\tfrac{4}{8})^2 = 0.250$$

From Fig. A-28, assume $C = 0.625$ (for high $N_R$).

$$Q = (0.625)[(\pi)(\tfrac{4}{12})^2/4]\sqrt{(2)(32.2)(\tfrac{18}{12})} = 0.536 \text{ ft}^3/\text{s}$$

$$N_R = \rho DV/\mu = 4Q\rho/\pi D\mu = (4)(0.536)(54/32.2)/\{(\pi)(\tfrac{8}{12})[(\tfrac{6}{100})/479]\} = 1.37 \times 10^4$$

From Fig. A-28, $C = 0.63$ (O.K.) $Q = 0.536/0.002228 = 241$ gpm.

**17.136** Solve Prob. 17.73 for an orifice with $D : \tfrac{1}{2}D$ taps (instead of a long-radius flow nozzle).

$\blacksquare$ A good initial guess for an orifice is $\alpha \approx 0.62$. From Eq. (2) of Prob. 17.73 compute $\beta \approx 0.447/(0.62)^{1/2} = 0.568$. From Fig. A-29 compute the discharge coefficient $C_d = 0.6064$ for $\beta = 0.568$ and $N_R = 392\,000$. Then $E = [1 - (0.568)^4]^{-1/2} = 1.0565$ and $\alpha = C_d E = 0.6407$. Iterate Eq. (2) again: $\beta = 0.447/(0.6407)^{1/2} = 0.558$, $C_d = 0.6061$, and $\alpha = 0.6378$. Stop. We have converged satisfactorily to a design value: $\beta \approx 0.56$, $d = \beta D = 112$ mm. The throat velocity is $V_t = V_1/\beta^2 = 2.0/(0.56)^2 = 6.38$ m/s. The throat head is $V_t^2/2g = (6.38)^2/[2(9.81)] = 2.07$ m. From Fig. A-24 for the orifice at $\beta = 0.56$, estimate $K_m \approx 1.7$. Then the nonrecoverable loss of the orifice will be $h_m = K_m(V_t^2/2g) \approx 1.7(2.07) = 3.5$ m.

**17.137** Water at 20 °C in a 15-cm-diameter pipe flows through a 7-cm-diameter thin-plate orifice with $D : \tfrac{1}{2}D$ taps ($\beta = 0.5$). If the measured pressure drop is 73 kPa, find the volume flow.

$\blacksquare$ $Q = C_d A_t\sqrt{2\,\Delta p/[(\rho)(1 - \beta^4)]}$. Guess $C_d = 0.61$:

$$Q = (0.61)[(\pi)(0.07)^2/4]\sqrt{(2)(73)(1000)/[(998)(1 - 0.5^4)]} = 0.0293 \text{ m}^3/\text{s}.$$

$$V = Q/A = 0.0293/[(\pi)(0.15)^2/4] = 1.658 \text{ m/s}$$

$$N_R = \rho DV/\mu = (998)(0.15)(1.658)/(1.02 \times 10^{-3}) = 2.43 \times 10^5$$

From Fig. A-29, $C_d = 0.605$. $Q = (0.605)[(\pi)(0.07)^2/4]\sqrt{(2)(73)(1000)/[(998)(1 - 0.5^4)]} = 0.0291$ m$^3$/s, or 29 L/s.

**17.138** Gasoline at 20 °C flows in a 60-mm-diameter pipe at 30 m$^3$/h. If a 30-mm-diameter thin-plate orifice with corner taps is installed in the pipe, what will the measured pressure drop be? $\beta = 0.4$. Use $\rho = 680$ kg/m$^3$ and $\mu = 0.000292$ Pa $\cdot$ s.

$\blacksquare$
$$Q = C_d A_t\sqrt{2\,\Delta p/[(\rho)(1 - \beta^4)]} \qquad V = Q/A = (30/3600)/[(\pi)(0.060)^2/4] = 2.947 \text{ m/s}$$

$$N_R = \rho DV/\mu = (680)(0.060)(2.947)/0.000292 = 4.12 \times 10^5$$

For corner taps,

$$f(\beta) = 0.5959 + 0.0312\beta^{2.1} - 0.184\beta^8 = 0.5959 + (0.0312)(0.4)^{2.1} - (0.184)(0.4)^8 = 0.6003$$

$$C_d = f(\beta) + 91.71\beta^{2.5}N_R^{-0.75} = 0.6003 + (91.71)(0.4)^{2.5}(4.12 \times 10^5)^{-0.75} = 0.6009$$

$$30/3600 = (0.6009)[(\pi)(0.030)^2/4]\sqrt{2\,\Delta p/[(680)(1 - 0.4^4)]} \qquad \Delta p = 128\,000 \text{ Pa} = 128 \text{ kPa}$$

**17.139** A 2-m-diameter cylindrical tank in Fig. 17-35 is initially filled with kerosene at 20 °C to a depth of 3 m. There is a 40-mm-diameter thin-plate orifice in the bottom. When the orifice is opened, what is the initial flow rate?

$\blacksquare$ $Q = C_d A_t\sqrt{2\,\Delta p/[(\rho)(1 - \beta^4)]}$, $\Delta p$ across the orifice is $\rho gh(t)$. The orifice simulates "corner taps" as $\beta \to 0$, so $C_d \approx 0.596$. $Q_0 = C_d A_t\sqrt{2gh_0/(1 - \beta^4)} = (0.596)[(\pi)(0.040)^2/4]\sqrt{(2)(9.807)(3)/(1 - 0^4)} = 0.005745$ m$^3$/s, or 5.745 L/s.

**Fig. 17-35**

**17.140** For the conditions given in Prob. 17.139, how long will it take $h(t)$ to drop from 3 m to 1.9 m?

■ $Q(t) = -dV_{\text{tank}}/dt = -(A_0)(dh/dt)$. On the other hand, from Prob. 17.139,
$$Q(t) = (0.596)[(\pi)(0.040)^2/4]\sqrt{(2)(9.807)(h)/(1-0^4)} = 0.003317h^{1/2}$$
Then, since $A_0 = (\pi)(2)^2/4 = 3.142 \text{ m}^2$,

$$(-3.142)(dh/dt) = 0.003317h^{1/2} \qquad \int_3^{1.9} \frac{dh}{h^{1/2}} = -\left(\frac{0.003317}{3.142}\right)\int_0^t dt \qquad [2h^{1/2}]_3^{1.9} = -0.001056[t]_0^t$$
$$-0.001056t = (2)(1.9^{1/2} - 3.0^{1/2}) \qquad t = 670 \text{ s} = 11.2 \text{ min}$$

**17.141** Water flowing in a 12-in-diameter pipe passes through a Venturi meter with a 6-in throat diameter. The measured pressure head differential is 150.5 in of water. Assuming a discharge coefficient of 0.98, find the flow rate of water through the pipe.

■
$$Q = CA_2\sqrt{[2g(p_1-p_2)/\gamma]/[1-(A_2/A_1)^2]} \qquad A_2 = (\pi)(\tfrac{6}{12})^2/4 = 0.1963 \text{ ft}^2$$
$$A_1 = (\pi)(\tfrac{12}{12})^2/4 = 0.7854 \text{ ft}^2$$
$$Q = (0.98)(0.1963)\sqrt{[(2)(32.2)(150.5/12)]/[1-(0.1963/0.7854)^2]} = 5.65 \text{ ft}^3/\text{s}$$

**17.142** Oil is flowing upward through a Venturi meter as shown in Fig. 17-36. Assume a discharge coefficient of 0.984. What is the rate of flow of the oil?

**Fig. 17-36**

$$Q = CA_2\sqrt{[2g(p_1 - p_2)/\gamma]/[1 - (A_2/A_1)^2]} \qquad A_2 = (\pi)(0.200)^2/4 = 0.03142 \text{ m}^2$$
$$A_1 = (\pi)(0.500)^2/4 = 0.1963 \text{ m}^2$$
$$p_1/\gamma + y - (1.26/0.82)(1.5) - x = p_2/\gamma \qquad x + 1.5 = 0.5 + y \qquad x = y - 1.0$$
$$p_1/\gamma + y - (1.26/0.82)(1.5) - (y - 1.0) = p_2/\gamma \qquad (p_1 - p_2)/\gamma = 1.305 \text{ m of oil}$$
$$Q = (0.984)(0.03142)\sqrt{(2)(9.807)(1.305)/[1 - (0.03142/0.1963)^2]} = 0.158 \text{ m}^3/\text{s}$$

**17.143** Carbon dioxide at 20 °C flowing in a 200-mm-diameter pipe passes through a Venturi meter with a 100-mm throat diameter. Measured (gage) pressures are $p_1 = 22.0$ kPa and $p_2 = 5.2$ kPa. Atmospheric pressure is 762 mmHg, and the discharge coefficient is 0.98. Find the weight flow rate of carbon dioxide through the pipe.

$$G = C\gamma_2 A_2\sqrt{[2gk/(k-1)](p_1/\gamma_1)[1 - (p_2/p_1)^{(k-1)/k}]/[1 - (A_2/A_1)^2(p_2/p_1)^{2/k}]} \qquad \gamma = p/RT$$
$$p_2 = 5.2 + (0.762)(9.79)(13.6) = 106.7 \text{ kPa} \qquad \gamma_2 = 106.7/[(19.3)(20 + 273)] = 0.01887 \text{ kN/m}^3$$
$$p_1 = 22.0 + (0.762)(9.79)(13.6) = 123.5 \text{ kPa} \qquad \gamma_1 = 123.5/[(19.3)(20 + 273)] = 0.02184 \text{ kN/m}^3$$
$$A_2 = (\pi)(0.100)^2/4 = 0.007854 \text{ m}^2 \qquad A_1 = (\pi)(0.200)^2/4 = 0.03142 \text{ m}^2$$
$$G = (0.98)(0.01887)(0.007854)$$
$$\times \sqrt{\frac{[(2)(9.807)(1.30)/(1.30 - 1)](123.5/0.02184)[1 - (106.7/123.5)^{(1.30-1)/1.30}]}{1 - (0.007854/0.03142)^2(106.7/123.5)^{2/1.30}}} = 18.8 \text{ N/s}$$

**17.144** Water flows through a Venturi meter, as shown in Fig. 17-37. Determine the discharge coefficient of the Venturi meter if the discharge is determined to be 2.12 cfs.

**Fig. 17-37**

$$Q = CA_2\sqrt{[2g(p_1 - p_2)/\gamma]/[1 - (A_2/A_1)^2]} \qquad A_2 = (\pi)(\tfrac{4}{12})^2/4 = 0.08727 \text{ ft}^2$$
$$A_1 = (\pi)(\tfrac{12}{12})^2/4 = 0.7854 \text{ ft}^2$$
$$p_1/\gamma + y + \tfrac{9}{12} - (13.6)(\tfrac{9}{12}) - y = p_2/\gamma \qquad (p_1 - p_2)/\gamma = 9.45 \text{ ft of water}$$
$$2.12 = (C)(0.08727)\sqrt{[(2)(32.2)(9.45)]/[1 - (0.08727/0.7854)^2]} \qquad C = 0.979$$

**17.145** A Venturi meter having a throat diameter of 150 mm is installed in a horizontal 300-mm-diameter water main, as shown in Fig. 17-38. The coefficient of discharge is 0.982. Determine the difference in level of the mercury columns of the differential manometer attached to the Venturi meter if the discharge is 0.142 m³/s.

**Fig. 17-38**

❚
$$Q = CA_2\sqrt{[2g(p_1 - p_2)/\gamma]/[1 - (A_2/A_1)^2]} \qquad A_2 = (\pi)(0.150)^2/4 = 0.01767 \text{ m}^2$$
$$A_1 = (\pi)(0.300)^2/4 = 0.07069 \text{ m}^2$$
$$0.142 = (0.982)(0.01767)\sqrt{[(2)(9.807)(p_1 - p_2)/\gamma]/[1 - (0.01767/0.07069)^2]}$$
$$(p_1 - p_2)/\gamma = 3.201 \text{ m}$$
$$p_1/\gamma + x + y - 13.6y - x = p_2/\gamma \qquad (p_1 - p_2)/\gamma = 12.6y = 3.201 \qquad y = 0.254 \text{ m} = 254 \text{ mm}$$

**17.146** Determine the mass flow rate when air at 40 °C and 755 kPa abs flows through a Venturi meter, if the pressure at the throat of the meter is 455 kPa abs. The diameters at inlet and throat are 350 mm and 175 mm, respectively. Assume that $C = 0.985$.

❚
$$\dot{m} = CYA_2\sqrt{2\rho_1\{(p_1 - p_2)/[1 - (D_2/D_1)^4]\}}$$
$$p_2/p_1 = 455/755 = 0.60 \qquad D_2/D_1 = 1/2 \qquad Y = 0.74 \qquad \text{(from Fig. A-26)}$$
$$\rho = p/RT \qquad \rho_1 = 755/[(0.287)(40 + 273)] = 8.40 \text{ kg/m}^3$$
$$\dot{m} = (0.985)(0.74)[(\pi)(0.175)^2/4]\sqrt{2(8.40)\{(3 \times 10^5)/[1 - (1/2)^4]\}} = 40.7 \text{ kg/s}$$

**17.147** Find the flow rate of water at 72 °F for the Venturi tube of Fig. 17-39 if the mercury manometer reads $y = 5$ in for the case where $D_1 = 10$ in, $D_2 = 5$ in, and $\Delta z = 2.0$ ft. Assume the discharge coefficients of Fig. A-30 are applicable. In addition, find the head loss from inlet to throat. Also, determine the total head loss across the meter. Assume a diverging cone angle of 10°.

**Fig. 17-39**

❚
$$Q = \{CA_2/[1 - (D_2/D_1)^4]^{1/2}\}[2g\,\Delta(z + p/\gamma)]^{1/2}$$
$$(z_1 + p_1/\gamma) - (z_2 + p_2/\gamma) = (R')(S_M/S_F - 1) = (\tfrac{5}{12})[(13.6/1) - 1] = 5.25 \text{ ft}$$
$$A_2 = (\pi)(\tfrac{5}{12})^2/4 = 0.1364 \text{ ft}^2$$

From Fig. A-30, assume $C = 0.95$.

$$Q = \{(0.95)(0.1364)/[1 - (\tfrac{5}{10})^4]^{1/2}\}[(2)(32.2)(5.25)]^{1/2} = 2.46 \text{ ft}^3/\text{s}$$
$$V_2 = Q/A_2 = 2.46/0.1364 = 18.0 \text{ ft/s} \qquad D''V_2 = (5)(18.0) = 90.0$$

From Fig. A-30, $C = 0.985$. Since the curve is flat in this area, $C = 0.985$ is O.K.

$$Q = (0.985/0.95)(2.46) = 2.55 \text{ ft}^3/\text{s} \qquad V_2 = 2.55/0.1364 = 18.7 \text{ ft/s}$$
$$V_3 = V_2/4 = 18.7/4 = 4.68 \text{ ft/s}$$
$$(h_L)_{1-2} = (1/C^2 - 1)[1 - (D_2/D_1)^4](V_2^2/2g) = [(1/0.985^2) - 1][1 - (\tfrac{5}{10})^4]\{18.7^2/[(2)(32.2)]\} = 0.156 \text{ ft}$$
$$(h_L)_{2-3} = k'(V_2 - V_3)^2/2g$$

From Fig. A-31, $k' = 0.175$.

$$(h_L)_{2-3} = (0.175)(18.7 - 4.68)^2/[(2)(32.2)] = 0.534 \text{ ft} \qquad (h_L)_{\text{total}} = 0.156 + 0.534 = 0.690 \text{ ft}$$

**17.148** Repeat Prob. 17.147 for the case of a horizontal Venturi tube (i.e., $\Delta z = 0$) with all other data the same.

❚ The answers are the same as for Prob. 17.147, since $R'(S_M/S_F - 1) = \Delta(p/\gamma + z)$.

**17.149** Refer to Prob. 17.147. What would be the flow rate if the manometer fluid was carbon tetrachloride (s.g. = 1.59) with all other data remaining the same?

$$Q = \{CA_2/[1 - (D_2/D_1)^4]^{1/2}\}[2g\,\Delta(z + p/\gamma)]^{1/2}$$

Assume $C = 0.95$ (see Fig. A-30).

$$A_2 = 0.1364\,\text{ft}^2 \quad \text{(from Prob. 17.147)} \quad \Delta(p/\gamma + z) = (\tfrac{5}{12})[(1.59/1) - 1] = 0.2458\,\text{ft}$$
$$Q = \{(0.95)(0.1364)/[1 - (\tfrac{5}{10})^4]^{1/2}\}[(2)(32.2)(0.2458)]^{1/2} = 0.532\,\text{ft}^3/\text{s}$$
$$V_2 = Q/A_2 = 0.532/0.1364 = 3.90\,\text{ft/s} \qquad D''V_2 = (5)(3.90) = 19.50$$

From Fig. A-30, $C = 0.985$.

$$Q = (0.985/0.95)(0.532) = 0.552\,\text{ft}^3/\text{s}$$

**17.150** Find the flow rate of water for the Venturi tube shown in Fig. 17-39 if $D_1 = 70$ cm, $D_2 = 35$ cm, $\Delta z = 186$ cm, and $y = 13$ cm of mercury.

$$Q = \{CA_2/[1 - (D_2/D_1)^4]^{1/2}\}\sqrt{2gy(S_0/S_1 - 1)}. \text{ Try } C = 0.985.$$
$$Q = \{(0.985)[(\pi)(0.35)^2/4]/[1 - (\tfrac{35}{70})^4]^{1/2}\}\sqrt{(2)(9.807)(0.13)[(13.6/1) - 1]} = 0.555\,\text{m}^3/\text{s}$$
$$V_2 = Q/A_2 = 0.555/[(\pi)(0.35)^2/4] = 5.77\,\text{m/s} \qquad N_R = DV/\nu = (0.35)(5.77)/10^{-6} = 2.02 \times 10^6$$

From Fig. A-30, $C = 0.985$. Hence, $Q = 0.555\,\text{m}^3/\text{s}$.

**17.151** In Fig. 17-39 suppose $D_1 = 4$ in, $D_2 = 2$ in, and $\Delta z = 8$ in. Oil (s.g. = 0.89) with a kinematic viscosity of $0.0005\,\text{ft}^2/\text{s}$ is flowing. Determine the manometer reading $y$ if mercury is used as the manometer fluid. The rate of flow is 0.12 cfs.

$$Q = \{CA_2/[1 - (D_2/D_1)^4]^{1/2}\}\sqrt{2gy(S_0/S - 1)}. \text{ Try } C = 0.96:$$
$$0.12 = \{(0.96)[(\pi)(\tfrac{2}{12})^2/4]/[1 - (\tfrac{2}{4})^4]^{1/2}\}\sqrt{(2)(32.2)(y)[(13.6/0.89) - 1]}$$
$$y = 0.03346\,\text{ft} \quad \text{or} \quad 0.40\,\text{in} \qquad V = Q/A = 0.12/[(\pi)(\tfrac{2}{12})^2/4] = 5.50\,\text{ft/s}$$
$$N_R = DV/\nu = (\tfrac{2}{12})(5.50)/0.0005 = 1833$$

From Fig. A-30, $C = 0.905$. $y = (0.905/0.96)^2(0.40) = 0.36\,\text{in}$.

**17.152** Assume that air at 50 °F and 90 psia flows through a Venturi tube and that the pressure at the throat is 54 psia. The inlet area is $0.55\,\text{ft}^2$, and the throat area is $0.10\,\text{ft}^2$. The tube coefficient is 0.98, and $k$ is 1.4. Find the rate of discharge.

$$G = CA_t\sqrt{2g[k/(k - 1)]p_1\gamma_1(p_2/p_1)^{2/k}\{[1 - (p_2/p_1)^{(k-1)/k}]/[1 - (A_t/A)^2(p_2/p_1)^{2/k}]\}}$$
$$p_1 = (90)(144) = 12\,960\,\text{lb/ft}^2 \qquad \gamma = p/RT$$
$$\gamma_1 = 12\,960/[(53.3)(50 + 460)] = 0.4768\,\text{lb/ft}^3$$
$$G = (0.98)(0.10)\sqrt{(2)(32.2)\left(\frac{1.4}{1.4 - 1}\right)(12\,960)(0.4768)(\tfrac{54}{90})^{2/1.4}\left[\frac{1 - (\tfrac{54}{90})^{(1.4-1)/1.4}}{1 - (0.10/0.55)^2(\tfrac{54}{90})^{2/1.4}}\right]} = 29.8\,\text{lb/s}$$

**17.153** Solve Prob. 17.152 by evaluating $Y$ from Fig. A-26 and using it to find the rate of discharge.

$$G = CYA_t\sqrt{2g\gamma_1\{(p_1 - p_2)/[1 - (D_2/D_1)^4]\}}$$

For $D_2/D_1 = \sqrt{\tfrac{0.10}{0.55}} = 0.426$ and $p_2/p_1 = \tfrac{54}{90} = 0.6$, from Fig. A-26, $Y = 0.745$.

$$G = (0.98)(0.745)(0.10)\sqrt{(2)(32.2)(0.4768)\{[(90 - 54)(144)]/(1 - 0.426^4)\}} = 29.6\,\text{lb/s}$$

**17.154** What would be the value of $Y$ and the rate of discharge for a square-edged orifice for the same data as in Prob. 17.152, assuming $C = 0.60$?

$$G = CYA_t\sqrt{2g\gamma_1\{(p_1 - p_2)/[1 - (D_0/D_1)^4]\}}$$

With $D_0/D_1 = 0.426$ and $p_2/p_1 = 0.6$, $Y = 0.875$ (from Fig. A-26).

$$G = (0.60)(0.875)(0.10)\sqrt{(2)(32.2)(0.4768)[(90 - 54)(144)/(1 - 0.426^4)]} = 21.3\,\text{lb/s}$$

**17.155** What is the value of the throat velocity in Prob. 17.152?

❚ $\quad V^2/2 = C_pT_1[1 - (p_2/p_1)^{(k-1)/k}] = (6000)(50 + 460)[1 - (\frac{54}{90})^{(1.40-1)/1.40}] \qquad V = 912 \text{ ft/s}$

**17.156** Air flows through a 20-cm by 10-cm Venturi meter. At the inlet the air temperature is 30 °C and the pressure is 150 kPa. Determine the mass flow if a mercury manometer reads 27 cm. Assume an atmospheric pressure of 101.3 kPa abs.

❚ $\quad \dot{m} = CYA_t\sqrt{2\rho_1\{(p_1 - p_2)/[1 - (D_0/D_1)^4]\}}$. From Fig. A-30, $C = 0.985$.

$$\rho = p/RT \qquad \rho_1 = (150 + 101.3)/[(0.287)(30 + 273)] = 2.89 \text{ kg/m}^3$$

$$p_1 - p_2 = \gamma h = [(13.6)(9.79)](0.27) = 35.9 \text{ kPa}$$

$$p_1 = 150 + 101.3 = 251.3 \text{ kPa} \qquad p_2 = 251.3 - 35.9 = 215.4 \text{ kPa}$$

For $D_2/D_1 = 0.5$ and $p_2/p_1 = 215.4/251.3 = 0.857$, from Fig. A-26, $Y = 0.92$. Thus

$$\dot{m} = (0.985)(0.92)[(\pi)(0.10/2)^2]\sqrt{2(2.89)[(35.9)(1000)/(1 - 0.5^4)]} = 3.34 \text{ kg/s}.$$

**17.157** Natural gas, for which $k = 1.3$ and $R = 3100$ ft-lb/(slug-°R), flows through a Venturi tube with pipe and throat diameters of 10 in and 5 in, respectively. The initial pressure of the gas is 159 psia, and its temperature is 40 °F. If the meter coefficient is 0.98, find the rate of flow for a throat pressure of 109 psia.

❚ $\quad G = CA_t\sqrt{2g[k/(k-1)]p_1\gamma_1(p_2/p_1)^{2/k}\{[1 - (p_2/p_1)^{(k-1)/k}]/[1 - (A_2/A_1)^2(p_2/p_1)^{2/k}]\}}$

$$A_t = (\pi)(\tfrac{5}{12})^2/4 = 0.1364 \text{ ft}^2 \qquad \gamma = p/RT$$

$$\gamma_1 = (159)(144)/[(3100/32.2)(40 + 460)] = 0.4756 \text{ lb/ft}^3$$

$G = (0.98)(0.1364)$

$\qquad \times \sqrt{(2)(32.2)[1.3/(1.3-1)][(159)(144)](0.4756)(\frac{109}{159})^{2/1.3}\{[1 - (\frac{109}{159})^{(1.3-1)/1.3}]/[1 - (5^2/10^2)^2(\frac{109}{159})^{2/1.3}]\}}$

$$= 51.2 \text{ lb/s}$$

**17.158** Referring to Fig. 17-40, assume that liquid flows from $A$ to $C$ at the rate of 222 L/s and that the friction loss between $A$ and $B$ is negligible but that between $B$ and $C$ it is $0.1V_B^2/2g$. Find the pressure heads at $A$ and $C$.

**Fig. 17-40**

❚

$$p_A/\gamma + V_A^2/2g + z_A = p_B/\gamma + V_B^2/2g + z_B + h_L$$

$$V_A = Q/A_A = 0.222/[(\pi)(0.36)^2/4] = 2.18 \text{ m/s} \qquad V_B = 0.222/[(\pi)(0.12)^2/4] = 19.6 \text{ m/s}$$

$$p_A/\gamma + 2.18^2/[(2)(9.807)] + 0 = 0.68 + 19.6^2/[(2)(9.807)] + 0 + 0 \qquad p_A/\gamma = 20.0 \text{ m}$$

$$p_B/\gamma + V_B^2/2g + z_B = p_C/\gamma + V_C^2/2g + z_C + h_L$$

$$0.68 + 19.6^2/[(2)(9.807)] + 0 = p_C/\gamma + 2.18^2/[(2)(9.807)] + 0 + 0 + (0.1)\{19.6^2/[(2)(9.807)]\}$$

$$p_C/\gamma = 18.1 \text{ m}$$

**17.159** Water flows through a 12-in by 6-in Venturi meter at the rate of 1.49 cfs and the differential gage is deflected 3.50 ft, as shown in Fig. 17-41. The specific gravity of the gage liquid is 1.25. Determine the coefficient of the meter.

❚ Applying the Bernoulli equation, $A$ to $B$, ideal case, yields $(p_A/\gamma + V_{12}^2/2g + 0)$ — no lost head $= (p_B/\gamma + V_6^2/2g + 0)$ and $V_{12}^2 = (A_6/A_{12})^2V_6^2$. Solving, $V_6 = \sqrt{2g(p_A/\gamma - p_B/\gamma)/[1 - (A_6/A_{12})^2]}$ (no lost head).

The true velocity (and hence the true value of flow $Q$) will be obtained by multiplying the ideal value by the coefficient $C$ of the meter. Thus

$$Q = A_6V_6 = A_6C\sqrt{2g(p_A/\gamma - p_B/\gamma)/[1 - (A_6/A_{12})^2]} \qquad (1)$$

**Fig. 17-41**

To obtain the differential pressure head above, the principles of the differential gage must be used. $p_C = p_{C'}$, $(p_A/\gamma - z) = p_B/\gamma - (z + 3.50) + 1.25(3.50)$ or $p_A/\gamma - p_B/\gamma = 0.875$ ft.
Substituting in (1), $1.49 = \frac{1}{4}\pi(\frac{1}{2})^2 C\sqrt{2g(0.875)/(1 - \frac{1}{16})}$ and $C = 0.978$ (use 0.98).

**17.160** Water flows upward through a vertical 12-in by 6-in Venturi meter whose coefficient is 0.980. The differential gage deflection is 3.88 ft of liquid of specific gravity 1.25, as shown in Fig. 17-42. Determine the flow in cfs.

**Fig. 17-42**

❚ Reference to the Bernoulli equation in Prob. 17.159 indicates that, for this problem, $z_A = 0$ and $z_B = 1.50$ ft.
Then $Q = CA_6\sqrt{2g[(p_A/\gamma - p_B/\gamma) - 1.50]/[1 - (\frac{1}{2})^4]}$.
Using the principles of the differential gage to obtain $\Delta p/\gamma$,

$$p_C/\gamma = p_D/\gamma \quad \text{(ft of water units)}$$
$$p_A/\gamma + (n + 3.88) = p_B/\gamma + m + 1.25(3.88)$$
$$(p_A/\gamma - p_B/\gamma) - (m - n) = 3.88(1.25 - 1.00)$$
$$(p_A/\gamma - p_B/\gamma) - 1.50 = 0.97 \text{ ft of water}$$

Substituting in the equation for flow, $Q = 0.980(\frac{1}{4}\pi)(\frac{1}{2})^2\sqrt{2g(0.97)/(1 - \frac{1}{16})} = 1.57$ cfs.

**17.161** An 8-in by 4-in Venturi meter is used to measure the flow of carbon dioxide at 68 °F. The deflection of the water column in the differential gage is 71.8 in and the barometer reads 30.0 in of mercury. For a pressure at entrance of 18.0 psi abs, calculate the weight flow.

❚ The absolute pressure at entrance $= p_1 = (18.0)(144) = 2590$ psf abs and the specific weight $\gamma_1$ of the carbon dioxide is $\gamma_1 = 2590/[34.9(460 + 68)] = 0.1405$ lb/ft³. The pressure difference $= (71.8/12)(62.4 - 0.141) = 372$ psf and hence the absolute pressure at the throat $= p_2 = 2590 - 372 = 2220$ psf abs.
To obtain the specific weight $\gamma_2$, we use $p_2/p_1 = \frac{2220}{2590} = 0.858$ and $\gamma_2/\gamma_1 = (0.858)^{1/k}$. Thus $\gamma_2 = (0.1405)(0.858)^{1/1.30} = 0.1250$ lb/ft³. $W = \gamma_1 K A_2 Y\sqrt{2g(\Delta p/\gamma_1)}$ in lb/s. Using $k = 1.30$, $d_2/d_1 = 0.50$ and $p_2/p_1 =$

0.858, $Y$ (Table A-18) = 0.909 by interpolation. Assuming $C = 0.985$, from Fig. A-32, and noting that $K = 1.032$, we have $W = (0.1405)[(1.032)(0.985)][\frac{1}{4}\pi(\frac{4}{12})^2](0.909)\sqrt{2g(372/0.1405)} = 4.67$ lb/s.

To check the assumed value of $C$, determine the Reynolds number and use the appropriate curve on Fig. A-32. $N_R = 4W/\pi d_2 \nu \gamma_2 = 4(4.67)/\{\pi(\frac{4}{12})[(9.1)(14.7)/18.0](10^{-5})](0.1250)\} = 1.92 \times 10^6$.

From Fig. A-32, $C = 0.984$. Recalculating, $W = 4.66$ lb/s.

**17.162** A 100-mm-diameter pipe has a 50-mm-diameter, long constriction. If the mercury manometer indicates a column-height difference of 200 mm, estimate the flow rate of water passed by the pipe.

▮ We assume that the streamlines are parallel at locations 1, 2, and 3 (Fig. 17-43). Applying the Bernoulli equation with mechanical-energy loss for the sudden contraction between points 1 and 2 gives $p_1/\rho g + \bar{V}_1^2/2g - K_{SC}(\bar{V}_2^2/2g) = p_2/\rho g + \bar{V}_2^2/2g$. The Bernoulli equation with mechanical-energy loss for the sudden enlargement between points 2 and 3 gives $p_2/\rho g + \bar{V}_2^2/2g - K_{SE}(\bar{V}_2^2/2g) = p_3/\rho g + \bar{V}_3^2/2g$. The steady incompressible continuity equation gives $\bar{V}_1 = \bar{V}_3 = (d/D)^2\bar{V}_2$. The pressure change indicated by the manometer is $p_1 - p_3 = (p_1 - p_2) + (p_2 - p_3)$. Using the Bernoulli equation to eliminate the pressure differences $p_1 - p_2$ and $p_2 - p_3$ and continuity to eliminate $\bar{V}_1$ and $\bar{V}_3$, we get $p_1 - p_3 = (K_{SC} + K_{SE})\frac{1}{2}\rho\bar{V}_2^2$. But $\dot{Q} = (\pi d^2/4)\bar{V}_2$ and $p_1 - p_3 = \rho g(\text{s.g.}_{Hg} - 1)h$. Then $\dot{Q} = (\pi d^2/4)\sqrt{2gh(\text{s.g.}_{Hg} - 1)/(K_{SC} + K_{SE})}$. From Fig. A-33, $K_{SC} = 0.478$ and from $K_{SE} = (1 - A_1/A_2)^2$, $K_{SE} = (1 - 0.25)^2 = 0.5625$. Then $\dot{Q} = (\pi/4)(0.050)^2\sqrt{2(9.80)(0.200)(13.6 - 1)/(0.478 + 0.5625)}$ 0.0135 m³/s.

**Fig. 17-43**

**17.163** Find $q$ for flow of water in a pipe of inside diameter $D$ of 100 mm using a Venturi meter with a machined convergent section. The throat diameter $d$ of the Venturi meter is 60 mm. The Venturi meter is in a pipe section having an inclination $\theta$ (see Fig. 17-44) of 45°. The distance from 1 to 2 in the meter is 120 mm and the value of $h$ for the manometer is 140 mm. Take $\rho = 999$ kg/m³ and $\mu = 1.12 \times 10^{-3}$ Pa · s.

**Fig. 17-44**

∎ We note that $(C_d)_{\text{ven}}$ for this case is 0.995. We need next the value of $p_1 - p_2$. This is easily determined as follows from manometry: $p_1 + \gamma_{\text{H}_2\text{O}}(\delta + l + h) = p_2 + l\gamma_{\text{H}_2\text{O}} + h\gamma_{\text{Hg}}$, $p_1 - p_2 = h(\gamma_{\text{Hg}} - \gamma_{\text{H}_2\text{O}}) - \gamma_{\text{H}_2\text{O}}\delta = (0.140)(13.6 - 1)(999)(9.81) - (999)(9.81)(0.120)(0.707) = 16\,456$ Pa. We now compute $V_2$ in $q_{\text{act}} = (C_d)_{\text{ven}}A_2(2\{[(p_1 - p_2)/\rho] + (z_1 - z_2)g\}/[1 - (A_2/A_1)^2])^{1/2}$ by deleting $A_2$ in the formula. Thus, $V_2 = 0.995(2\{[16\,456/999] + (0.120)(0.707)(9.81)\}/[1 - (\frac{60}{100})^4]\})^{1/2}$ where we used diameters squared in place of areas. We get from the last equation and continuity $V_2 = 6.27$ m/s, $V_1 = (0.6^2)(6.27) = 2.26$ m/s. We must next compute $N_R$ for the pipe: $N_R = \rho V_1 D/\mu = [(999)(2.26)(0.100)/(1.12 \times 10^{-3})] = 2.02 \times 10^5$. We see that the Reynolds number is within the proper range for the $(C_d)_{\text{ven}}$ we have used, and so $q = (V_2)(A_2) = (6.27)[(\pi)(0.060^2)/4] = 0.01773$ m³/s.

**17.164** Consider in Prob. 17.163 that we have a flow of air at a temperature of 40 °C and a pressure of 200 kPa abs. The pipe as before is 100 mm in inside diameter, but the diameter in the Venturi is now 20 mm. The height $h$ is 140 mmHg. What is the mass flow rate?

∎ We can use the same $C_d$ as in incompressible flow, which is 0.995. We next determine $Y$, so we can use $\dot{m} = \rho_1 C_d A_3\{2[(p_1 - p_2)/\rho_1]/[1 - (A_2/A_1)^2]\}^{1/2}Y$. We need $p_2$ for this purpose. Thus, $p_2 = p_1 - \Delta p = 200\,000 - (0.140)(9806)(13.6) = 181.3$ kPa. The ratio $p_2/p_1$ is then $p_2/p_1 = 181.3/200 = 0.907$. Noting that $A_2/A_1 = (d/D)^2 = (0.020/0.100)^2 = 0.04$, we go to Fig. A-34 to get $Y$: $Y = 0.95$. Finally, we need $\rho_1$. Using the equation of state of a perfect gas, $\rho_1 = p_1/RT_1 = 200\,000/[(287)(313)] = 2.2264$ kg/m³, $\dot{m} = \rho_1 C_d A_2\{2[(p_1 - p_2)/\rho_1]/[1 - (A_2/A_1)^2]\}^{1/2}Y = (2.2264)(0.995)[(\pi)(0.020^2)/4]\{2[(200\,000 - 181\,300)/2.2264]/[1 - (0.04^2)]\}^{1/2}(0.95) = 0.0858$ kg/s.

**17.165** A 3-m by 1.5-m Venturi meter carries water at 25 °C. A water-air differential manometer has a gage difference of 72 mm. What is the discharge?

∎ $Q = C_v A_2\sqrt{2g(\Delta p/\gamma)/[1 - (D_2/D_1)^4]}$. Assume $N_R > 3 \times 10^5$. From Fig. A-35, $C_v = 0.983$.

$$Q = (0.983)[(\pi)(1.5)^2/4]\sqrt{(2)(9.807)(0.072)/[1 - (\tfrac{1.5}{3})^4]} = 2.13 \text{ m}^3/\text{s}$$
$$N_R = D_1 V_1/\nu \qquad V_2 = Q/A_2 = 2.13/[(\pi)(1.5)^2/4] = 1.21 \text{ m/s}$$
$$A_1 V_1 = A_2 V_2 \qquad [(\pi)(3)^2/4](V_1) = [(\pi)(1.5)^2/4](1.21) \qquad V_1 = 0.302 \text{ m/s}$$
$$N_R = (3)(0.302)/(9.02 \times 10^{-7}) = 1.00 \times 10^6$$

From Fig. A-35, $C_v = 0.983$. Hence, $Q = 2.13$ m³/s.

**17.166** What is the pressure difference between the upstream section and throat of a 140-mm by 70-mm horizontal Venturi meter carrying 46 L/s of water at 58 °C?

∎ $$Q = C_v A_2\sqrt{2g(\Delta p/\gamma)/[1 - (D_2/D_1)^4]} \qquad V_1 = Q/A_1 = (46 \times 10^{-3})/[(\pi)(0.140)^2/4] = 2.988 \text{ m/s}$$
$$N_R = V_1 D_1/\nu = (2.988)(0.140)/(5.70 \times 10^{-7}) = 7.34 \times 10^5$$

From Fig. A-35, $C_v = 0.984$.

$$46 \times 10^{-3} = (0.984)[(\pi)(0.070)^2/4]\sqrt{(2)(9.807)(\Delta p/9.79)/[1 - (\tfrac{70}{140})^4]} \qquad \Delta p = 69.0 \text{ kPa}$$

**17.167** A 14-in by 7-in Venturi meter is mounted in a vertical pipe with the flow upward. 2900 gpm of oil, s.g. = 0.82, $\mu = 0.1$ P, flows through the pipe. The throat section is 7 in above the upstream section. What is $p_1 - p_2$?

∎ $$Q = C_v A_2\sqrt{2g[-h + (p_1 - p_2)/\gamma]/[1 - (D_2/D_1)^4]}$$
$$V_1 = Q/A_1 = (2900)(0.002228)/[(\pi)(\tfrac{14}{12})^2/4] = 6.044 \text{ ft/s}$$
$$N_R = \rho D_1 V_1/\mu = [(0.82)(1.94)](\tfrac{14}{12})(6.044)/(0.1/479) = 5.37 \times 10^4$$

From Fig. A-35, $C_v = 0.962$.

$$(2900)(0.002228) = (0.962)[(\pi)(\tfrac{7}{12})^2/4]\sqrt{(2)(32.2)\{-\tfrac{7}{12} + (p_1 - p_2)/[(0.82)(62.4)]\}/[1 - (\tfrac{7}{14})^4]}$$
$$p_1 - p_2 = 500.3 \text{ lb/ft}^2 \quad \text{or} \quad 3.47 \text{ lb/in}^2$$

**17.168** Air flows through a Venturi meter in a 50-mm-diameter pipe having a throat diameter of 25 mm, $C_v = 0.97$. For $p_1 = 800$ kPa abs, $T_1 = 35$ °C, $p_2 = 520$ kPa abs, calculate the mass per second flowing.

$$\dot{m} = C_v Y A_2\sqrt{2\rho_1\,\Delta p/[1 - (D_2/D_1)^4]} \qquad p_2/p_1 = 520/800 = 0.650 \qquad D_2/D_1 = 25/50 = 1/2$$

From Fig. A-26, $Y = 0.78$.

$$\rho = p/RT \qquad \rho_1 = 800/[(0.287)(35 + 273)] = 9.050 \text{ kg/m}^3$$
$$\dot{m} = (0.97)(0.78)[(\pi)(0.025)^2/4]\sqrt{(2)(9.050)[(800 - 520)(1000)]}/[1 - (\tfrac{1}{2})^4] = 0.864 \text{ kg/s}$$

17.169 Oxygen, $p_1 = 37$ psia, $T_1 = 100\,°\text{F}$, flows through a 2-in by 1-in Venturi meter with a pressure drop of 5 psi. Find the mass per second flowing.

❚ $\dot{m} = C_v Y A_2 \sqrt{(2\rho_1\,\Delta p)/[1 - (D_2/D_1)^4]}$. Assume $C_v = 0.984$.

$$p_2/p_1 = (37 - 5)/37 = 0.865 \qquad D_2/D_1 = 1/2 = 0.500$$

From Fig. A-26, $Y = 0.91$.

$$\rho = p/RT \qquad \rho_1 = (37)(144)/[(1551)(100 + 460)] = 0.006134 \text{ slug/ft}^3$$
$$\dot{m} = (0.984)(0.91)[(\pi)(1/12)^2/4]\sqrt{(2)(0.006134)[(5)(144)]}/[1 - (1/2)^4] = 0.01499 \text{ slug/s}$$
$$V_1 = \dot{m}/\rho_1 A_1 = 0.01499/\{(0.006134)[(\pi)(\tfrac{2}{12})^2/4]\} = 112 \text{ ft/s}$$
$$N_R = \rho_1 D_1 V_1/\mu = (0.006134)(\tfrac{2}{12})(112)/(4.18 \times 10^{-7}) = 2.74 \times 10^5$$

From Fig. A-35, $C_v = 0.984$ is O.K.

17.170 Solve Prob. 17.73 for a Venturi nozzle.

❚ For the Venturi guess $\alpha \approx 1.0$ and iterate Eq. (2) of Prob. 17.73: $\alpha = 1.0$, $\beta = 0.447$, $C_d = 0.9806$. This rapidly converges to give $\beta = 0.4468$, $d = \beta D = 89$ mm. The throat velocity is $2.0/(0.4468)^2 = 10.0$ m/s, and the throat head is $(10.0)^2/[2(9.81)] = 5.12$ m. From Fig. A-24 for the Venturi estimate $K_m \approx 0.15$; hence $h_m = 0.15(5.12 \text{ m}) = 0.8 \text{ m}$.

17.171 It is planned to equip a 20-cm-diameter pipe carrying water at 20 °C with a modern Venturi nozzle. In order for international standards to be met, what is the permissible range of flow rates and nozzle diameters?

❚ **Flow range:**

$$1.5 \times 10^5 < N_R < 2 \times 10^6 \qquad 1.5 \times 10^5 < 4Q\rho/\pi\mu D < 2 \times 10^6$$
$$1.5 \times 10^5 < (4)(Q)(998)/[(\pi)(1.02 \times 10^{-3})(0.20)] < 2 \times 10^6 \qquad 0.0241 \text{ m}^3/\text{s} < Q < 0.321 \text{ m}^3/\text{s}$$

**Size range:**

$$0.316 < [\beta = d/D] < 0.775 \qquad 0.316 < d/20 < 0.775 \qquad 6.32 \text{ cm} < d < 15.50 \text{ cm}$$

17.172 For Prob. 17.171, what is the permissible range of pressure drops? Would compressibility be a problem for the highest pressure-drop condition?

❚ From Fig. A-36, $0.9236 < C_d < 0.9847$.

$$Q = (C_d \pi d^2/4)\sqrt{(2\,\Delta p)/[\rho(1 - \beta^4)]} \qquad d = \beta D = 0.20\beta$$
$$\Delta p = (1.598 \times 10^6)(1 - \beta^4)(Q/C_d\beta^2)^2$$

Smallest $\Delta p$ occurs at largest $\beta$ and smallest $Q$.

$$(\Delta p)_{\min} = (1.598 \times 10^6)(1 - 0.775^4)\{0.0241/[(0.9847)(0.775)^2]\}^2 = 1.696 \text{ kPa}$$

Largest $\Delta p$ occurs at smallest $\beta$ and largest $Q$.

$$(\Delta p)_{\max} = (1.598 \times 10^6)(1 - 0.316^4)\{0.321/[(0.9236)(0.316)^2]\}^2 = 19.2 \text{ MPa}$$

Since $(\Delta p)_{\max}$ is much smaller than the bulk modulus of water (2190 MPa), compressibility is not important.

17.173 Light oil (s.g. $= 0.92$, $v = 10^{-5}$ ft$^2$/s) flows down an 8-in vertical pipe through a 4-in Venturi nozzle as in Fig. 17-45. If the mercury manometer reads a deflection of 5 in, what is the estimated flow rate?

**Fig. 17-45**

$$Q = C_d A_t \sqrt{(2 \, \Delta p)/[\rho(1-\beta^4)]} \qquad C_d = 0.9858 - 0.196\beta^{4.5} \qquad \beta = d/D = \tfrac{4}{8} = 0.500$$
$$C_d = 0.9858 - (0.196)(0.500)^{4.5} = 0.9771$$
$$\Delta p = (\rho_M - \rho_0)gh = (\text{s.g.}_M - \text{s.g.}_0)\gamma h = (13.6 - 0.92)(62.4)(\tfrac{5}{12}) = 329.7 \, \text{lb/ft}^2$$
$$Q = (0.9771)[(\pi)(\tfrac{4}{12})^2/4]\sqrt{(2)(329.7)/\{[(0.92)(1.94)](1-0.500^4)\}} = 1.69 \, \text{ft}^3/\text{s}$$
$$N_R = 4Q/\pi \nu D = (4)(1.69)/(\pi)(10^{-5})(\tfrac{8}{12}) = 3.23 \times 10^5 \quad (\text{O.K.})$$

**17.174** A Venturi meter (Fig. 17-46) is a carefully designed constriction whose pressure difference is a measure of the flow rate in a pipe. Using Bernoulli's equation for steady incompressible flow with no losses, show that flow rate $Q$ is related to manometer reading $h$ by $Q \approx [A_2/\sqrt{1-(D_2/D_1)^4}]\sqrt{2gh(\rho_M - \rho)/\rho}$ where $\rho_M$ is the density of the manometer fluid.

**Fig. 17-46**

**Continuity:** $\rho_1 A_1 V_1 = \rho_2 A_2 V_2$. Since $\rho_1 = \rho_2$, $V_2 = V_1(D_1/D_2)^2$.

**Bernoulli (no losses):** $p_1/\rho + V_1^2/2 + gz_1 = p_2/\rho + V_2^2/2 + gz_2$. Since $gz_1 = gz_2$, $V_2 = \sqrt{2(p_1-p_2)/\{\rho[1-(D_2/D_1)^4]\}}$, $Q = A_2 V_2$.

**Manometer:** $p_1 - p_2 = (\rho_M - \rho)gh$.
Combining these, $Q \approx [A_2/\sqrt{1-(D_2/D_1)^4}]\sqrt{2gh(\rho_M - \rho)/\rho}$.

**17.175** Water flows from a reservoir over a dam that is essentially a broad-crested weir. The dam is 20 ft long. How much higher will the water surface in the reservoir be above the top of the dam when the discharge over the dam is 1000 cfs?

$$Q = CLH^{1.5} \qquad 1000 = (3.09)(120)(H^{1.5}) \qquad H = 1.94 \, \text{ft}$$

**17.176** A sharp-crested, suppressed weir is under a head of 1.65 m. The weir is 2.5 m long and has a height of 4.50 m. What is the flow rate of water over the weir?

▮ $$C = 0.611 + 0.075(H/H_w) = 0.611 + 0.075(1.65/4.50) = 0.6385$$
$$Q = C(\tfrac{2}{3})L\sqrt{2g}\,H^{1.5} = (0.6385)(\tfrac{2}{3})(2.50)\sqrt{(2)(9.807)}(1.65)^{1.5} = 9.99 \text{ m}^3/\text{s}$$

**17.177** A sharp-crested, contracted, rectangular weir is to carry water at a flow rate of 12.75 cfs. What must the weir width be in order for the head of flow to be 1.50 ft?

▮ $$Q = CL^{1.02}H^{1.47} \qquad 12.75 = (3.10)(L^{1.02})(1.50)^{1.47} \qquad L = 2.23 \text{ ft}$$

**17.178** A sharp-crested, contracted, triangular weir with a weir angle of 80° carries water under a head of 0.955 m. Find the flow of water over the weir.

▮ $$Q = C(\tfrac{8}{15})\sqrt{2g}\,[\tan(\theta/2)]H^{2.5} = (0.60)(\tfrac{8}{15})\sqrt{(2)(9.807)}\,[\tan(80°/2)](0.955)^{2.5} = 1.06 \text{ m}^3/\text{s}$$

**17.179** Water flows over a Cipolletti weir 8.0 ft long under a head of 1.2 ft. What is the flow rate of water over the weir?

▮ $$Q = CLH^{1.5} = (3.367)(8.0)(1.2)^{1.5} = 35.4 \text{ ft}^3/\text{s}$$

**17.180** For the same conditions as given in Prob. 17.178 except that the weir is submerged and the downstream water surface is 0.450 m above the top of the weir, find the flow rate over the submerged weir.

▮ $$Q = Q_1[1 - (d/h)^n]^{0.385} \qquad Q_1 = 1.06 \text{ m}^3/\text{s} \quad \text{(from Prob. 17.178)}$$
$$Q = (1.06)[1 - (0.450/0.955)^{2.5}]^{0.385} = 0.995 \text{ m}^3/\text{s}$$

**17.181** A sharp-crested, suppressed weir is under a head of 2.2 ft. The weir's length is 10.0 ft, and its height is 4.0 ft. Determine the flow rate of water over the weir.

▮ $$C = 0.611 + 0.075(H/H_w) = 0.611 + 0.075(2.2/4.0) = 0.6522$$
$$Q = C(\tfrac{2}{3})L\sqrt{2g}\,H^{1.5} = (0.6522)(\tfrac{2}{3})(10)\sqrt{(2)(32.2)}\,(2.2)^{1.5} = 114 \text{ ft}^3/\text{s}$$

**17.182** A sharp-crested, contracted, rectangular weir carries water at a flow rate of 1.81 m³/s. What must the length of the weir be in order for the head of flow to be 0.8 m?

▮ $$Q = CL^{1.02}H^{1.47} \qquad 181 = (1.69)(L^{1.02})(0.8)^{1.47} \qquad L = 1.475 \text{ m}$$

**17.183** A sharp-crested, contracted, triangular weir with a weir angle of 60° carries water under a head of 1.23 ft. Determine the flow rate of water over the weir.

▮ $$Q = C(\tfrac{8}{15})\sqrt{2g}\,[\tan(\theta/2)]H^{2.5} = (0.60)(\tfrac{8}{15})\sqrt{(2)(32.2)}\,[\tan(60°/2)](1.23)^{2.5} = 2.49 \text{ ft}^3/\text{s}$$

**17.184** The rate of flow of water over a sharp-crested, contracted, 45° triangular weir is 0.823 cfs. Determine the head of flow.

▮ $$Q = C(\tfrac{8}{15})\sqrt{2g}\,[\tan(\theta/2)]H^{2.5} \qquad 0.823 = (0.60)(\tfrac{8}{15})\sqrt{(2)(32.2)}\,[\tan(45°/2)](H)^{2.5} \qquad H = 0.902 \text{ ft}$$

**17.185** What length of a Cipolletti weir should be constructed in order that the head of flow will be 0.91 m when the rate of flow is 3.45 m³/s?

▮ $$Q = CLH^{1.5} \qquad 3.45 = (1.859)(L)(0.910)^{1.5} \qquad L = 2.14 \text{ m}$$

**17.186** Compute the rate of flow of water over a 4.5-ft-long Cipolletti weir under a head of 1.15 ft.

▮ $$Q = CLH^{1.5} = (3.367)(4.5)(1.15)^{1.5} = 18.7 \text{ ft}^3/\text{s}$$

**17.187** Water flows from a reservoir over a dam that is essentially a broad-crested weir. The dam is 45 m long, and the head of flow is 0.83 m. Compute the discharge of water over the dam.

▮ $$Q = CLH^{1.5} = (1.71)(45)(0.83)^{1.5} = 58.2 \text{ m}^3/\text{s}$$

**17.188** Given the same conditions as Prob. 17.181 except that the weir is submerged and the downstream water surface is 0.96 ft above the top of the weir, determine the flow rate of water over the weir.

▮ $$Q = Q_1[1 - (d/h)^n]^{0.385} \qquad Q_1 = 114 \text{ ft}^3/\text{s} \quad \text{(from Prob. 17.181)}$$
$$Q = (114)[1 - (0.96/2.2)^{1.5}]^{0.385} = 100 \text{ ft}^3/\text{s}$$

**17.189** Flow is occurring in a rectangular channel at a velocity of 3 fps and depth of 1.0 ft. Neglecting the effect of velocity of approach, determine the height of a sharp-crested, suppressed weir that must be installed to raise the water depth upstream of the weir to 4 ft.

$\blacksquare$ $$Q = 3.33LH^{3/2} = Av = (1.0L)(3) = 3.00L \qquad 3.00L = (3.33L)(H)^{3/2}$$
$$H = 0.93 \text{ ft} \qquad \text{Height of weir} = 4 - 0.93 = 3.07 \text{ ft}$$

**17.190** A rectangular, sharp-crested weir 1.0 m high extends across a rectangular channel which is 3.3 m wide. When the head is 0.49 m, find the rate of discharge by neglecting the velocity of approach.

$\blacksquare$ $$Q = 1.84LH^{3/2} = (1.84)(3.3)(0.49)^{3/2} = 2.08 \text{ m}^3/\text{s}$$

**17.191** Suppose the rectangular weir of Prob. 17.190 is contracted at both ends. Find the rate of discharge for a head of 0.49 m by the Francis formula.

$\blacksquare$ $$Q = 1.84LH^{3/2} = (1.84)[3.3 - (2)(0.1)(0.49)](0.49)^{3/2} = 2.02 \text{ m}^3/\text{s}$$

**17.192** Develop a table of $C_d$ versus $P/H$ with $H$ as a parameter. Use the Rehbock formula.

$\blacksquare$ $C_d = 0.605 + 1/305H + 0.08H/P$. For $P/H = 0.5$ and $H = 0.2$, $C_d = 0.605 + 1/[(305)(0.2)] + 0.08/0.5 = 0.781$. Similarly, for additional values of $P/H$ and $H$:

| | $C_d$ | | |
|---|---|---|---|
| $P/H$ | $H = 0.2$ ft | 1.0 ft | 5.0 ft |
| 0.5 | 0.781 | 0.768 | 0.765 |
| 1.0 | 0.701 | 0.688 | 0.685 |
| 2.0 | 0.661 | 0.648 | 0.645 |
| 5.0 | 0.637 | 0.624 | 0.621 |
| 10.0 | 0.629 | 0.616 | 0.613 |

**17.193** What is the rate of discharge of water over a 60° triangular weir when the head is 0.9 ft?

$\blacksquare$ $Q = C_d(\frac{8}{15})\sqrt{2g} [\tan (\theta/2)](H^{5/2})$. From Fig. A-37, $C_d = 0.582$. $Q = (0.582)(\frac{8}{15})\sqrt{(2)(32.2)} [\tan (60°/2)](0.9)^{5/2} = 1.11 \text{ ft}^3/\text{s}$.

**17.194** With the same head, what would be the increase in discharge obtained by doubling the notch angle in Prob. 17.193?

$\blacksquare$ $Q = C_d(\frac{8}{15})\sqrt{2g} [\tan (\theta/2)](H^{5/2})$. From Fig. A-37, $C_d = 0.581$. $Q = (0.581)(\frac{8}{15})\sqrt{(2)(32.2)} [\tan (120°/2)](0.9)^{5/2} = 3.31 \text{ ft}^3/\text{s}$, increase $= (3.31 - 1.11)/1.11 = 1.98$, or 198 percent.

**17.195** What would be the head for a discharge of 2.6 cfs of water over a 90° triangular weir? Use a value of $C_d$ of 0.58.

$\blacksquare$ $Q = C_d(\frac{8}{15})\sqrt{2g} [\tan (\theta/2)](H^{5/2}) \qquad 2.6 = (0.58)(\frac{8}{15})\sqrt{(2)(32.2)} [\tan (90°/2)](H^{5/2}) \qquad H = 1.02 \text{ ft}$

**17.196** For the Cipolletti weir, derive the slope ($\frac{1}{4}$:1) of the sides of the trapezoid by setting the reduction in discharge due to contraction equal to the increase in discharge due to the triangular area added.

$\blacksquare$ By the Francis formula, the decrease in flow due to two contractions is $Q = C_d(\frac{2}{3})\sqrt{2g} LH^{3/2}$, $\Delta Q = C_d(\frac{2}{3})\sqrt{2g} [(2)(0.1h)]H^{3/2}$. This deficiency is made up by the equivalent triangular weir of angle $\theta$, for which $Q = C_d(\frac{8}{15})\sqrt{2g} [\tan (\theta/2)](H^{5/2})$. Equating these, $C_d(\frac{2}{3})\sqrt{2g} [(2)(0.1H)]H^{3/2} = C_d(\frac{8}{15})\sqrt{2g} [\tan (\theta/2)](H^{5/2})$. Assuming $C_d$ is the same for both weirs, $\tan (\theta/2) = 0.2500$, or side slope is $\frac{1}{4}$:1.

**17.197** Develop in general terms an expression for the percent of error in $Q$ over a triangular weir if there is a small error in the measurement of the vertex angle. Assume there is no error in the weir coefficient. Compute the percent error in $Q$ if there is a 3° error in the measurement of the total vertex angle of a triangular weir having a total vertex angle of 60°.

▌ $$Q = C_d(\tfrac{8}{15})(2g)^{1/2}[\tan(\theta/2)]H^{5/2} \qquad \partial Q/\partial \theta = C_d(\tfrac{8}{15})(2g)^{1/2}H^{5/2}\{(\partial/\partial\theta)[\tan(\theta/2)]\}$$
$$\partial Q/Q = (\partial/\partial\theta)\{[\tan(\theta/2)]\,\partial\theta\}/\tan(\theta/2) = \sec^2(\theta/2)(\tfrac{1}{2})\,\partial\theta/[\tan(\theta/2)] = \partial\theta/[2\sin(\theta/2)\cos(\theta/2)]$$

For the case where $\partial\theta = 3°$ and $\theta = 60°$: $\theta/2 = 30°$, so that $\partial Q/Q = 3(\pi/180)/[2(0.500)0.866] = 0.0453 = 4.53$ percent.

**17.198** A broad-crested weir rises 0.4 m above the bottom of a horizontal channel. With a measured head of 0.7 m above the crest, what is the rate of discharge per unit width?

▌ $$Q = L\sqrt{g}\,(\tfrac{2}{3})^{3/2}E^{3/2} = L\sqrt{9.807}\,(\tfrac{2}{3})^{3/2}(0.7 + V_1^2/2g)^{3/2}$$
$$Q/L = q = (1.705)\{0.7 + V_1^2/[(2)(9.807)]\}^{3/2}$$

But $q = (0.7 + 0.4)V_1$: $V_1 = q/1.1$, $q = (1.705)\{0.7 + (q/1.1)^2/[(2)(9.807)]\}^{3/2}$, $q = 1.11$ m³/s/m (by trial and error).

**17.199** A broad-crested weir of height 2.50 ft in a channel 6.00 ft wide has a flow over it of 11.50 cfs. What is the water depth just upstream of the weir?

▌ $$q = 11.5/6 = 1.92 \text{ cfs/ft} \qquad y_2 = y_c = (q^2/g)^{1/3} = (1.92^2/32.2)^{1/3} = 0.4856 \text{ ft}$$
$$V_2 = q/y_2 = 1.92/0.4856 = 3.954 \text{ ft/s} \qquad y_1 + V_1^2/2g = 2.5 + 0.4856 + 3.954^2/[(2)(32.2)]$$

But $V_1 = 1.92/y_1$: $y_1 + (1.92/y_1)^2/[(2)(32.2)] = 3.228$, $y_1 = 3.22$ ft (by trial and error).

**17.200** A 90° V-notch weir is used to measure a flow rate of 0.29 cfs. Assuming $C_d$ is known precisely, compute the percentage of error in $Q$ that would result from an error of 0.01 ft in head measurement.

▌ $Q = C_d(\tfrac{8}{15})(2g)^{1/2}[\tan(\theta/2)]H^{5/2}$. For $\theta = 90°$, $C_d = 0.587$, and $Q = 0.29$ cfs: $H = 0.422$ ft. $\partial Q/\partial H = C_d(\tfrac{8}{15})(2g)^{1/2}[\tan(\theta/2)](\tfrac{5}{2})H^{3/2}$, $\partial Q/Q = \tfrac{5}{2}H^{3/2}(\partial H)/H^{5/2} = 5(\partial H)/2H = 5(0.01)/[2(0.422)] = 0.0592 = 5.92$ percent.

**17.201** Work Prob. 17.200 for a rectangular weir with end contractions having a crest length of 3 ft.

▌ $$Q = 3.33LH^{3/2} \qquad L = 3 \text{ ft} \quad \text{and} \quad Q = 0.29 \text{ cfs} \qquad H = 0.0945 \text{ ft}$$
$$\partial Q/\partial H = C_w L(\tfrac{3}{2})H^{1/2} \qquad \partial Q/Q = \tfrac{3}{2}H^{1/2}(\partial H)/H^{3/2} = 3(\partial H)/2H = 3(0.01)/[2(0.0945)] = 0.159 = 15.9 \text{ percent}$$

**17.202** Derive the theoretical formula for flow over a rectangular weir. Refer to Fig. 17-47.

(a)

(b)

**Fig. 17-47**

▌ Consider the rectangular opening in Fig. 17-47 to extend the full width $W$ of the channel ($b = W$). With the liquid surface in the dashed position, application of the Bernoulli theorem between $A$ and an elemental strip of height $dy$ in the jet produces, for ideal conditions, $(0 + V_A^2/2g + y) - $ no losses $= (0 + V_{jet}^2/2g + 0)$ where $V_A$ represents the average velocity of the particles approaching the opening.

Thus the ideal $V_{jet} = \sqrt{2g(y + V_A^2/2g)}$ and

$$\text{Ideal } dQ = dAV_{jet} = (b\,dy)V_{jet} = b\sqrt{2g}\left(y + \frac{V_A^2}{2g}\right)^{1/2}dy = b\sqrt{2g}\int_{h_1}^{h_2}\left(y + \frac{V_A^2}{2g}\right)^{1/2}dy$$

A weir exists when $h_1 = 0$. Let $H$ replace $h_2$ and introduce a coefficient of discharge $C$ to obtain the actual

flow. Then

$$Q = Cb\sqrt{2g} \int_0^H \left(y + \frac{V_A^2}{2g}\right)^{1/2} dy = \tfrac{2}{3}Cb\sqrt{2g}\left[\left(H + \frac{V_A^2}{2g}\right)^{3/2} - \left(\frac{V_A^2}{2}\right)^{3/2}\right] = mb\left[\left(H + \frac{V_A^2}{2g}\right)^{3/2} - \left(\frac{V_A^2}{2g}\right)^{3/2}\right] \quad (1)$$

*Notes:*

(1) For a fully contracted rectangular weir the end contractions cause a reduction in flow. Length $b$ is corrected to recognize this condition, and the formula becomes

$$Q = m(b - \tfrac{2}{10})[H + V_A^2/2g)^{3/2} - (V_A^2/2g)^{3/2}] \quad (2)$$

(2) For high weirs and most contracted weirs the velocity head of approach is negligible and

$$Q = m(b - \tfrac{2}{10}H)H^{3/2} \qquad \text{for contracted weirs} \quad (3)$$

or $\qquad\qquad\qquad Q = mbH^{3/2} \qquad\qquad \text{for suppressed weirs} \quad (4)$

(3) Coefficient of discharge $C$ is not constant. It embraces the many complexities not included in the derivation, such as surface tension, viscosity, density, nonuniform velocity distribution, secondary flows, and possibly others.

**17.203**  Derive the theoretical formula for flow through a triangular-notched weir. Refer to Fig. 17-48.

**Fig. 17-48**

▌ From Prob. 17.202, $V_{\text{jet}} = \sqrt{2g(y + \text{negligible } V^2/2g)}$ and ideal $dQ = dAV_{\text{jet}} = x\,dy\sqrt{2gy}$.
By similar triangles, $x/b = (H - y)/H$ and $b = 2H\tan(\theta/2)$. Then

$$\text{Actual } Q = \left(\frac{b}{H}\right)C\sqrt{2g}\int_0^H (H - y)y^{1/2}\,dy$$

Integrating and substituting,

$$Q = \tfrac{8}{15}C\sqrt{2g}\,H^{5/2}\tan(\tfrac{1}{2}\theta) \quad (1)$$

A common $V$-notch is the 90° opening. Expression (1) then becomes $Q = 4.28CH^{5/2}$ where an *average* value of $C$ is about 0.60 for heads above 1.0 ft.

**17.204**  During a test on an 8-ft suppressed weir which was 3 ft high, the head was maintained constant at 1.000 ft. In 38.0 s, 7600 gal of water were collected. Find weir factor $m$ in Eqs. (1) and (4) of Prob. 17.202.

▌ (a) Change the measured flow to cfs. $Q = 7600/[(7.48)(38)] = 26.7$ cfs.
(b) Check the velocity of approach. $V = Q/A = 26.7/[(8)(4)] = 0.834$ ft/s. Then $V^2/2g = (0.834)^2/2g = 0.018$ ft.
(c) Using (1), $Q = mb[(H + V^2/2g)^{3/2} - (V^2/2g)^{3/2}]$ or $26.7 = (m)(8)[(1.000 + 0.011)^{3/2} - (0.0108)^{3/2}]$ and $m = 3.29$. Using (4), $Q = 26.7 = mbH^{3/2} = (m)(8)(1.000)^{3/2}$ and $m = 3.34$ (about 1.8 percent higher neglecting the velocity of approach terms).

**17.205**  Determine the flow over a suppressed weir 10.0 ft long and 4.00 ft high under a head of 3.000 ft. The value of $m$ is 3.46.

▌ Since the velocity head term cannot be calculated, an approximate flow is $Q = mbH^{3/2} = 3.46(10)(3.000)^{3/2} = 178.8$ cfs. For this flow, $V = 178.8/[(10)(7)] = 2.554$ ft/s and $V^2/2g = 0.101$ ft. Using Eq. (1) of Prob. 17.202, $Q = 3.46(10)[(3.000 + 0.101)^{3/2} - (0.101)^{3/2}] = 188$ cfs.

This second calculation shows an increase of 9 cfs or about 5.0 percent over the first calculation. Further calculation will generally produce an unwarranted refinement, i.e., beyond the accuracy of the formula itself. However, to illustrate, the revised velocity of approach would be $V = 188/[(10)(7)] = 2.69$ ft/s, $V^2/2g = 0.112$ ft and $Q = 3.46(10)[(3.000 + 0.112)^{3/2} - (0.112)^{3/2}] = 189$ cfs.

**17.206** A suppressed weir, 25.0 ft long, is to discharge 375.0 cfs into a channel. The weir factor $m = 3.42$. To what height $Z$ (nearest $\frac{1}{100}$ ft) may the weir be built, if the water behind the weir must not exceed 6 ft in depth?

**∎** Velocity of approach $V = Q/A = 375.0/[(25)(6)] = 2.50$ ft/s. Then $375.0 = (3.42)(25.0)\{[H + (2.50)^2/2g]^{3/2} - [(2.50)^2/2g]^{3/2}\}$ and $H = 2.59$ ft. Height of weir is $Z = 6.00 - 2.59 = 3.41$ ft.

**17.207** A contracted weir, 4.00 ft high, is to be installed in a channel 8 ft wide. The maximum flow over the weir is 60.0 cfs when the total depth back of the weir is 7.00 ft. What length of weir should be installed if $m = 3.40$?

**∎** Velocity of approach $V = Q/A = 60.0/[(8)(7)] = 1.071$ ft/s. It appears that the velocity head is negligible in this case. $Q = m(b - \frac{2}{10}H)(H)^{3/2}$, $60.0 = 3.40[b - \frac{2}{10}(3.00)](3.00)^{3/2}$, $b = 4.00$ ft long.

**17.208** The discharge from a 6-in-diameter orifice, under a 10.0-ft head, $C = 0.600$, flows into a rectangular weir channel and over a contracted weir. The channel is 6 ft wide and, for the weir, $Z = 5.00$ ft and $b = 1.00$ ft. Determine the depth of water in the channel if $m = 3.35$.

**∎** The discharge through the orifice is $Q = CA\sqrt{2gh} = (0.600)[\frac{1}{4}\pi(\frac{1}{2})^2]\sqrt{2g(10.0)} = 2.99$ cfs. For the weir, $Q = m(b - \frac{2}{10}H)H^{3/2}$ (velocity head neglected) or $2.99 = 3.35(1.00 - 0.20H)H^{3/2}$ and $H^{3/2} - 0.20H^{5/2} = 0.893$. By successive trials, $H = 1.09$ ft; and the depth $= Z + H = 5.00 + 1.09 = 6.09$ ft.

**17.209** The discharge of water over a 45° triangular weir is 0.750 cfs. For $C = 0.580$, determine the head on the weir.

**∎**
$$Q = C(\tfrac{8}{15})\sqrt{2g}\,[\tan(\theta/2)]H^{5/2}$$
$$0.750 = (0.580)(\tfrac{8}{15})\sqrt{(2)(32.2)}\,[\tan(45°/2)]H^{5/2} \qquad H = 0.881 \text{ ft}$$

**17.210** What length of Cipolletti weir should be constructed so that the head will be 1.54 ft when the discharge is 122 cfs?

**∎**
$$Q = 3.367bH^{3/2} \qquad 122 = (3.367)(b)(1.54)^{3/2} \qquad b = 19.0 \text{ ft}$$

**17.211** Derive the expression for the time to lower the liquid level in a tank, lock, or canal, by means of a suppressed weir.

**∎**
$$Q\,dt = -A_T\,dH \qquad (mLH^{3/2})\,dt = -A_T\,dH$$
$$t = \int_{t_1}^{t_2} dt = \left(\frac{-A_T}{mL}\right)\int_{H_1}^{H_2} H^{-3/2}\,dH \qquad t = t_2 - t_1 = \left(\frac{2A_T}{mL}\right)(H_2^{-1/2} - H_1^{-1/2})$$

**17.212** A rectangular flume, 50 ft long and 10 ft wide, feeds a suppressed weir under a head of 1.000 ft. If the supply to the flume is cut off, how long will it take for the head on the weir to decrease to 4 in? Use $m = 3.33$.

**∎**
$$t = (2A_T/mL)(H_2^{-1/2} - H_1^{-1/2}) \qquad \text{(from Prob. 17.211)}$$
$$= (2)[(50)(10)]/[(3.33)(10)][(\tfrac{4}{12})^{-1/2} - 1.000^{-1/2}] = 22.0 \text{ s}$$

**17.213** The head on a sharp-crested, rectangular weir of height 2.0 ft and crest length 4.0 ft was incorrectly observed to be 0.38 ft when it was actually 0.40 ft. Determine the percentage error in the computed value of flow rate.

**∎**
$$Q = CLH^{3/2} \qquad \partial Q/\partial H = (\tfrac{3}{2})CLH^{1/2} \approx \Delta Q/\Delta H \qquad \Delta Q \approx \Delta H(\tfrac{3}{2})CLH^{1/2}$$
$$\Delta Q/Q \approx (\tfrac{3}{2})(\Delta H/H) = (\tfrac{3}{2})(0.02/0.40) = 0.075 \quad \text{or} \quad 7.5 \text{ percent}$$

**17.214** It is desired to measure a discharge which may vary from 0.6 to 2.4 cfs with a relative accuracy of at least 0.4 percent through the entire range. A stage recorder which is accurate to the nearest 0.001 ft is available. What is the maximum width of rectangular sharp-crested weir that will satisfy these conditions? Assume the weir has end contractions and $C_w = 3.5$. Neglect velocity of approach.

**∎**
$$Q = 3.5Lh^{3/2} \qquad \partial Q/\partial h = 5.25Lh^{1/2} \approx \Delta Q/\Delta h$$

Thus $\Delta Q = 5.25Lh^{1/2}(\Delta h)$. $\Delta Q/Q = (5.25/3.50)(\Delta h/h) = 1.5(\Delta h/h)$, $0.004 = 1.5(0.001/h)$, $h = 0.0015/0.004 = 0.375$ ft. $h$ must be $\geq 0.375$ ft to acquire desired accuracy: $Q = 0.6 = 3.5L(0.375)^{3/2}$, $L \leq 0.747$ ft.

**17.215** What is the maximum permissible vertex angle of a $V$-notch weir that will satisfy the conditions of Prob. 17.214?

■ $$Q = 4.28(0.58)h^{5/2}[\tan{(\theta/2)}] = 2.48h^{5/2}[\tan{(\theta/2)}]$$
$$\partial Q/\partial h = \tfrac{5}{2}(2.48)h^{3/2}[\tan{(\theta/2)}] = 6.2h^{3/2}[\tan{(\theta/2)}] \approx \Delta Q/\Delta h$$

Thus $\Delta Q = 6.2h^{3/2}[\tan{(\theta/2)}]\,\Delta h$; $\Delta Q/Q = 2.5(\Delta h/h)$, $0.004 = 2.5(0.001/h)$, $h \geq 0.625$ ft; $Q = 0.6 = 2.48h^{5/2}[\tan{(\theta/2)}] = 2.48(0.625)^{5/2}[\tan{(\theta/2)}]$, $\tan{(\theta/2)} = 0.6/[2.48(0.309)] = 0.783$, $\theta/2 \leq 38°04'$, $\theta < 76°08'$.

**17.216** With a head of 0.12 m on a 60° $V$-notch weir, what error in the measured head will produce the same percentage error in the computed flow rate as an error of 1° in the vertex angle?

■ $$Q = 2.48h^{5/2}[\tan{(\theta/2)}] \qquad \partial Q/\partial h = 6.2h^{3/2}[\tan{(\theta/2)}] \approx \partial Q/\partial h$$

Thus $\Delta Q = 6.2h^{3/2}[\tan{(\theta/2)}]\,\Delta h$; $\Delta Q/Q = 2.5(\Delta h/h)$, $\partial Q/\partial\theta = 2.48h^{5/2}[\sec^2{(\theta/2)}]\tfrac{1}{2} = 1.24h^{5/2}[\sec^2{(\theta/2)}] = \Delta Q/\Delta\theta$. Thus $\Delta Q = 1.24h^{5/2}[\sec^2{(\theta/2)}]\,\Delta\theta$; $\Delta Q/Q = [(\Delta\theta)\sec^2{(\theta/2)}]/(2\tan{\theta})$. For $\theta = 30°$, $\sec^2{(\theta/2)}/\tan{(\theta/2)} = 2.31$, $\Delta Q = 2.5(\Delta h/0.12) = 2.31(1°)/2 = (2.31/2)(1/57.3)]$, $\Delta h = 0.00097$ m $= 0.10$ cm.

**17.217** Tests on a 60°-notch weir yield the following values of head $H$ on the weir and discharge $Q$:

| $H$, ft | 0.345 | 0.356 | 0.456 | 0.537 | 0.568 | 0.594 | 0.619 | 0.635 | 0.654 | 0.665 |
|---|---|---|---|---|---|---|---|---|---|---|
| $Q$, cfs | 0.107 | 0.110 | 0.205 | 0.303 | 0.350 | 0.400 | 0.435 | 0.460 | 0.490 | 0.520 |

Use the theory of least squares to determine the constants in $Q = CH^m$ for this weir.

■ Taking the logarithm of each side of the equation, $\ln Q = \ln C + m \ln H$ or $y = B + mx$, it is noted that the best values of $B$ and $m$ are needed for a straight line through the data when plotted on log-log paper.

By the theory of least squares, the best straight line through the data points is the one yielding a minimum value of the sums of the squares of vertical displacements of each point from the line; or, from Fig. 17-49,

$$F = \sum_{i=1}^{i=n} s_i^2 = \sum [y_i - (B + mx_i)]^2$$

where $n$ is the number of experimental points. To minimize $F$, $\partial F/\partial B$ and $\partial F/\partial m$ are taken and set equal to zero, yielding two equations in the two unknowns $B$ and $m$: $\partial F/\partial B = 0 = 2\sum [y_i - (B + mx_i)](-1)$ from which

$$\sum y_i - nB - m\sum x_i = 0 \qquad (1)$$

and $\partial F/\partial m = 0 = 2\sum [y_i - (B + mx_i)](-x_i)$ or

$$\sum x_i y_i - B\sum x_i - m\sum x_i^2 = 0 \qquad (2)$$

**Fig. 17-49**

Solving Eqs. (1) and (2) for $m$ gives $m = (\sum x_i y_i/\sum x_i - \sum y_i/n)/(\sum x_i^2/\sum x_i - \sum x_i/n)$, $B = (\sum y_i - m \sum x_i)/n$. These equations are readily solved by hand calculator having the $\sum$ key, or a simple program can be written for the digital computer. The answer for the data of this problem is $m = 2.437$, $C = 1.395$.

**17.218** A rectangular, sharp-crested weir 15 m long with end contractions suppressed is 1.6 m high. Determine the discharge when the head is 300 mm.

▌ Since the height of the weir is large compared to $H$, the upstream velocity head can be omitted; hence, $Q = 1.83LH^{1.5} = (1.83)(15)(0.300)^{1.5} = 4.51$ m³/s.

**17.219** In Fig. 17-50, $L = 9$ ft, $P = 1.9$ ft, and $H = 0.90$ ft. Estimate the discharge over the weir ($C = 3.33$).

**Fig. 17-50**

▌ $Q = CL(H + \alpha V^2/2g)^{1.5}$     $V = Q/[(L)(P + H)]$     $Q = CL\{H + \alpha Q^2/[(2gL^2)(P + H)^2]\}^{1.5}$
$Q = (3.33)(9)\{0.90 + 1.4Q^2/[(2)(32.2)(9)^2(1.9 + 0.90)^2]\}^{1.5} = 26.7$ ft³/s     (by trial and error)

**17.220** A rectangular, sharp-crested weir with end contractions is 1.4 m long. How high should it be placed in a channel to maintain an upstream depth of 2.35 m for 0.40 m³/s flow?

▌ $Q = 1.83L'H^{3/2}$ where $L' = L - 0.2H$ (due to end contractions).

$0.40 = (1.83)(1.4 - 0.2H)(H)^{3/2}$     $H = 0.30$ m     $P = 2.35 - 0.30 = 2.05$ m     (height of weir)

**17.221** Determine the head on a 45° $V$-notch weir for a discharge of 225 L/s.

▌ $Q = C(\frac{8}{15})\sqrt{2g}\,[\tan{(\theta/2)}]H^{5/2}$     $\frac{225}{1000} = (0.58)(\frac{8}{15})\sqrt{(2)(9.807)}\,[\tan{(45°/2)}]H^{5/2}$     $H = 0.691$ m

**17.222** Tests on a 90° $V$-notch weir gave the following results: $H = 200$ mm, $Q = 21.3$ L/s, $H = 430$ mm, $Q = 165$ L/s. Determine the formula for the weir.

▌ Let $Q = AH^n$: $\frac{165}{1000} = A(\frac{430}{1000})^n$, $21.3/1000 = A(\frac{200}{1000})^n$. Dividing one equation by the other, $7.746 = 2.150^n$, $n = 2.67$. $\frac{165}{1000} = A(\frac{430}{1000})^{2.67}$, $A = 1.57$; $Q = 1.57H^{2.67}$.

**17.223** A broad-crested weir 1.8 m high and 9 m long has a well-rounded upstream corner. What head is required for a flow of 8.80 m³/s?

▌ $Q = 1.67LH^{3/2}$     $8.80 = (1.67)(9)H^{3/2}$     $H = 0.700$ m

**17.224** A weir in a horizontal channel is 12 ft wide and 4 ft high. The upstream water depth is 5.2 ft. Estimate the discharge if the weir is sharp-crested.

▌ $C_w = 0.611 + 0.075H/Y = 0.611 + (0.075)(5.2 - 4)/4 = 0.6335$

$q = (\frac{2}{3})(C_w)(2g)^{1/2}H^{3/2} = (\frac{2}{3})(0.6335)[(2)(32.2)]^{1/2}(5.2 - 4)^{3/2} = 4.455$ ft³/s/ft     $Q = (12)(4.455) = 53.5$ ft³/s

**17.225** Solve Prob. 17.224 if the weir is broad-crested.

▌ $C_w = 0.65/(1 + H/Y)^{1/2} = 0.65/[1 + (5.2 - 4)/4]^{1/2} = 0.5701$. Since everything else is the same as in Prob. 17.224, we simply scale down the sharp-crested result: $Q = (0.5701/0.6335)(53.5) = 48.1$ ft³/s.

**17.226** A weir in a horizontal channel is 10 m wide and 1.4 m high. If the upstream depth is 3 m, estimate the channel discharge for a sharp-crested weir.

**I** $$C_w = 0.611 + 0.075 H/Y = 0.611 + (0.075)(3-1.4)/1.4 = 0.6967$$

$$q = (\tfrac{2}{3})(C_w)(2g)^{1/2}H^{3/2} = (\tfrac{2}{3})(0.6967)[(2)(9.807)]^{1/2}(3-1.4)^{3/2} = 4.163 \,(\text{m}^3/\text{s})/\text{m} \qquad Q = (10)(4.163) = 41.6 \,\text{m}^3/\text{s}$$

**17.227** Solve Prob. 17.226 if the weir is broad-crested.

**I** $C_w = 0.65/(1+H/Y)^{1/2} = 0.65/[1+(3-1.4)/1.4]^{1/2} = 0.4440$. Since everything else is the same as in Prob. 17.226, we simply scale down the sharp-crested result: $Q = (0.4440/0.6967)(41.6) = 26.5 \,\text{m}^3/\text{s}$.

**17.228** Show that the discharge of the $V$-notch weir in Fig. 17-51 is given by $Q = C_w(\tfrac{8}{15})(2g)^{1/2}(\tan\alpha)H^{5/2}$. Experiments show that $C_w$ varies from 0.65 at $\alpha = 5°$ to 0.58 at $\alpha = 50°$.

**Fig. 17-51**

**I** Assume $V$ at any strip $= \sqrt{2gz}$, where $z$ is measured down from the top.

$$Q = \int V\,dA = \int_0^H (2gz)^{1/2}b\,dz \qquad \text{where } b = b_0\left(1-\frac{z}{H}\right)$$

$$= \int_0^H (2gz)^{1/2}\left[b_0\left(1-\frac{z}{H}\right)\right]dz = (\tfrac{4}{15})(2g)^{1/2}\left(\frac{b_0}{H}\right)(H^{5/2})$$

$$b_0/H = 2\tan\alpha \qquad Q = (\tfrac{8}{15})(2g)^{1/2}(\tan\alpha)H^{5/2}$$

A correction factor $C_w$ is inserted to account for end and edge effects and upstream kinetic energy; hence, $Q = C_w(\tfrac{8}{15})(2g)^{1/2}(\tan\alpha)H^{5/2}$.

**17.229** Wastewater flows through a Parshall flume with a throat width of 3.0 ft at a depth of flow of 2.5 ft. Find the flow rate.

**I** $$Q = 4.0Bh^{1.522B^{0.026}} = (4.0)(3.0)(2.5)^{(1.522)(3.0)^{0.026}} = 50.4 \,\text{ft}^3/\text{s}$$

**17.230** Water is to flow through a proposed Parshall flume at a maximum flow rate of 30 m³/s. The width of the Parshall flume's throat is to be 10.0 m. What should be the channel depth entering the flume?

**I** $$Q = (2.293B + 0.474)h^{1.6} \qquad 30 = [(2.293)(10.0)+0.474]h^{1.6} \qquad h = 1.17\,\text{m}$$

**17.231** Wastewater flows through a Parshall flume with a throat width of 5.0 ft at a depth of flow of 3.2 ft. Determine the rate of flow of the wastewater.

**I** $$Q = 4.0Bh^{1.522B^{0.026}} = (4.0)(5.0)(3.2)^{(1.522)(5.0)^{0.026}} = 127 \,\text{ft}^3/\text{s}$$

**17.232** Water is to flow through a Parshall flume at a maximum flow rate of 25 m³/s. The width of the throat of the Parshall flume is 8.0 m. What will be the depth of flow at maximum flow rate?

**I** $$Q = (2.293B + 0.474)h^{1.6} \qquad 25 = [(2.293)(8.0)+0.474]h^{1.6} \qquad h = 1.19\,\text{m}$$

**17.233** The stilling-well depth-measurement scale of a Parshall flume with a 3-ft throat gives a head reading that is 0.04 ft too large. Compute the percentage errors in flow rate when the observed head readings are 0.6 ft and 2.2 ft.

**I** $$Q = 4Bh^{1.522B^{0.026}} \qquad B = 3\,\text{ft} \qquad \Delta h = 0.04\,\text{ft} \qquad e = 0.04\,\text{ft}$$

$$\% \text{ Error} = \{[(h+e)/h]^{1.522B^{0.026}} - 1\}(100) \qquad 3^{0.026} = 1.0290$$

$$\% \text{ Error} = [(0.64/0.60)^{1.566} - 1] = 0.106 \quad \text{or} \quad 10.6 \text{ percent}$$

If $h = 2.2$ ft, $\%$ Error $= [(2.24/2.20)^{1.566} - 1] = 0.0286$, or 2.86 percent.

**17.234** It is desired to measure the flow in a canal which may carry between 30 cfs and 90 cfs. The flow is to be measured to an accuracy of 2 percent, and the available water-level recorder is accurate to 0.01 ft. The canal is on a very flat slope and the head loss in the measuring device should be as small as possible. What type and size of flow-measuring device would you recommend?

▮ Try an 8-ft-throat Parshall flume: $Q = 4Bh^{1.552B^{0.026}} = (4)(8)h^{1.522(8)^{0.026}} = 32h^{1.607}$, $dQ = 51.4h^{0.607}\,dh$. For $Q = 30$ cfs, $h = 0.961$ ft, $dQ/Q = (51.4)(0.961)^{0.607}(0.01)/30 = 0.0167 < 0.02$. For $Q = 90$ cfs, $h = 1.90$ ft, $dQ/Q = (51.4)(1.90)^{0.607}(0.01)/90 = 0.0084 < 0.02$. So a Parshall flume with an 8-ft throat provides the required accuracy.

**17.235** A rectangular channel 2.0 m wide contains a sluice gate which extends across the width of the channel. If the gate produces free flow when it is open 0.15 m with an upstream depth of 1.15 m, find the rate of discharge, assuming $C_d = 0.60$ and $C_c = 0.62$.

▮ $$Q = C_d Ba[2g(y_1 - y_2 + V_1^2/2g)]^{1/2} \qquad y_2 = C_c a = (0.62)(0.15) = 0.0930 \text{ m}$$

Find a trial $Q$ by neglecting the $V_1^2/2g$ term in the equation above.

$$Q = (0.60)(2.0)(0.15)[(2)(9.807)(1.15 - 0.0930 + 0)]^{1/2} = 0.820 \text{ m}^3/\text{s}$$

$$V_1 = Q/A_1 = 0.820/[(2.0)(1.15)] = 0.357 \text{ m/s} \qquad V_1^2/2g = 0.357^2/[(2)(9.807)] = 0.00650 \text{ m}$$

$$Q = (0.60)(2.0)(0.15)[(2)(9.807)(1.15 - 0.0930 + 0.00650)]^{1/2} = 0.822 \text{ m}^3/\text{s}$$

**17.236** For the sluice gate shown in Fig. 17-52, if $C_v = 0.98$, what is the flow rate? If $C_c = 0.62$, what is the height of the opening?

**Fig. 17-52**

▮ $V_2 = C_v[2g(y_1 - y_2 + V_1^2/2g)]^{1/2}$. Find a trial $V_2$ by neglecting the $V_1^2/2g$ term in the equation above.

$$V_2 = (0.98)[(2)(9.807)(2.2 - 0.9 + 0)]^{1/2} = 4.949 \text{ m/s} \qquad V_2 y_2 = V_1 y_1$$

$$(4.949)(0.9) = (V_1)(2.2) \qquad V_1 = 2.025 \text{ m/s}$$

$$V_1^2/2g = 2.025^2/[(2)(9.807)] = 0.209 \text{ m} \qquad V_2 = (0.98)[(2)(9.807)(2.2 - 0.9 + 0.209)]^{1/2} = 5.332 \text{ m/s}$$

**Second trial:**

$$(5.332)(0.9) = (V_1)(2.2) \qquad V_1 = 2.181 \text{ m/s}$$

$$V_1^2/2g = 2.181^2/[(2)(9.807)] = 0.243 \text{ m} \qquad V_2 = (0.98)[(2)(9.807)(2.2 - 0.9 + 0.243)]^{1/2} = 5.391 \text{ m/s}$$

**Third trial:**

$$(5.391)(0.9) = (V_1)(2.2) \qquad V_1 = 2.205 \text{ m/s}$$

$$V_1^2/2g = 2.205^2/[(2)(9.807)] = 0.248 \text{ m} \qquad V_2 = (0.98)[(2)(9.807)(2.2 - 0.9 + 0.248)]^{1/2} = 5.400 \text{ m/s}$$

**Fourth trial:**

$$(5.400)(0.9) = (V_1)(2.2) \qquad V_1 = 2.209 \text{ m/s} \qquad V_1^2/2g = 2.209^2/[(2)(9.807)] = 0.249 \text{ m}$$

$$V_2 = (0.98)[(2)(9.807)(2.2 - 0.9 + 0.249)]^{1/2} = 5.402 \text{ m/s} \qquad \text{(O.K.)}$$

$$q = y_2 V_2 = (0.9)(5.402) = 4.86 \text{ (m}^3/\text{s)/m} \qquad a = y_2 C_c = 0.9/0.62 = 1.45 \text{ m}$$

**17.237** A disk meter has a volumetric displacement of 30 mL for one complete oscillation. Calculate the flow for 76.5 oscillations per minute.

$$Q = (\text{displacement per oscillation})(\text{number of oscillations per unit time})$$

$$= (30)(76.5) = 2295 \text{ mL/min} = 2.295 \text{ L/min}$$

**17.238** A disk water meter with volumetric displacement of 35 mL per oscillation requires 450 oscillations per minute to pass 0.30 L/s and 3740 oscillations per minute to pass 2.19 L/s. Calculate the percent error, or slip, in the meter.

$$\text{Slip} = (Q_{\text{actual}} - Q_{\text{indicated}})/Q_{\text{actual}}$$

$$Q = (\text{displacement per oscillation})(\text{number of oscillations per unit time})$$

$Q_1 = (0.035)(450)/60 = 0.2625 \text{ L/s}$ $\quad$ $\text{Slip}_1 = (0.30 - 0.2625)/0.30 = 0.0278$ $\quad$ or $\quad$ 2.78 percent

$Q_2 = (0.035)(3740)/60 = 2.182 \text{ L/s}$ $\quad$ $\text{Slip}_2 = (2.19 - 2.182)/2.19 = 0.0037$ $\quad$ or $\quad$ 0.37 percent

**17.239** A volumetric tank 3 ft in diameter and 6 ft high was filled with oil in 17 min 22.4 s. What is the average discharge in gallons per minute?

$$Q = V/t = [(\pi)(3)^2/4](6)/[17 + 22.4/60] = 2.441 \text{ ft}^3/\text{min} = (2.441)(7.48) = 18.3 \text{ gpm}$$

**17.240** A weigh tank receives 79 N of liquid, s.g. = 0.83, in 13.9 s. What is the volume flow rate?

$$V = W/\gamma = 79/[(0.83)(9.79)(1000)] = 0.009722 \text{ m}^3$$

$$Q = V/t = 0.009722/(13.9) = 0.0007 \text{ m}^3/\text{s} = 0.7 \text{ L/s}$$

**17.241** A critical-depth meter 15 ft wide has a rise in the bottom of 1.8 ft. For an upstream depth of 3.45 ft, determine the flow through the meter.

$$q = 0.517 g^{1/2}(y_1 - z + 0.55 q^2/g y_1^2)^{3/2}$$

$$= (0.517)(32.2)^{1/2}\{3.45 - 1.8 + 0.55 q^2/[(32.2)(3.45)^2]\}^{3/2}$$

$$= (2.934)(1.65 + 0.001435 q^2)^{3/2}$$

$$q = 6.57 \text{ cfs/ft} \quad \text{(by trial and error)}$$

$$Q = (6.57)(15) = 98.6 \text{ cfs}$$

# CHAPTER 18
# Dimensional Analysis and Similitude

**18.1**    The *Brinkman number* $N_B$, often used in analysis of organic–liquid flows, is the ratio of viscous dissipation to heat conduction in a fluid. It is a dimensionless combination of viscosity $\mu$, flow velocity $V$, thermal conductivity $k$, and fluid temperature $T$. Derive the Brinkman number, using the fact that it is proportional to viscosity.

■  $N_B = f(\mu, V, k, T)$ and is proportional to $\mu$.

$$N_B = \{1\} = \mu^1 V^a k^b T^c = \{M/LT\}^1 \{L/T\}^a \{ML/T^3\theta\}^b \{\theta\}^c = M^0 L^0 T^0 \theta^0$$

$$L: \ -1 + a + b = 0 \qquad M: \ 1 + b = 0 \qquad T: \ -1 - a - 3b = 0 \qquad \theta: \ -b + c = 0$$

Hence, $a = 2$ and $b = c = -1$. $N_B = \mu V^2 / kT$.

**18.2**    The velocity of sound $c$ of a gas varies with pressure $p$ and density $\rho$. Show by dimensional reasoning that the proper form must be $c = (\text{constant})(p/\rho)^{1/2}$.

■
$$c = f(p, \rho) = (\text{constant}) p^a \rho^b \qquad \{L/T\} = \{M/LT^2\}^a \{M/L^3\}^b$$
$$M: \ 0 = a + b \qquad L: \ 1 = -a - 3b \qquad T: \ -1 = -2a$$

Hence, $a = \frac{1}{2}$ and $b = -\frac{1}{2}$. $c = (\text{constant})(p/\rho)^{1/2}$.

**18.3**    The speed of propagation $C$ of a capillary wave in deep water is known to be a function only of density $\rho$, wavelength $\lambda$, and surface tension $\sigma$. Find the proper functional relationship, completing it with a dimensionless constant. For a given density and wavelength, how does the propagation speed change if surface tension is doubled?

■
$$C = f(\rho, \lambda, \sigma) = (\text{constant}) \rho^a \lambda^b \psi^c \qquad \{L/T\} = \{M/L^3\}^a \{L\}^b \{M/T^2\}^c$$
$$M: \ 0 = a + c \qquad L: \ 1 = -3a + b \qquad T: \ -1 = -2c$$

Hence, $a = -\frac{1}{2}$, $b = -\frac{1}{2}$, $c = \frac{1}{2}$. $C = (\text{constant})(\sigma/\rho\lambda)^{1/2}$. If $\sigma$ is doubled, $C$ increases as $\sqrt{2}$, or 41 percent.

**18.4**    The excess pressure $\Delta p$ inside a bubble is known to be a function of the surface tension and the radius. By dimensional reasoning determine how the excess pressure will vary if we double (*a*) the surface tension and (*b*) the radius.

■
$$\Delta p = f(\sigma, r) = (\text{constant}) \sigma^a r^b \qquad \{F/L^2\} = \{F/L\}^a \{L\}^b$$
$$F: \ 1 = a \qquad L: \ -2 = -a + b$$

Hence, $a = 1$ and $b = -1$. $\Delta p = (\text{constant})(\sigma/r)$. (*a*) If $\sigma$ is doubled, $\Delta p$ is doubled. (*b*) If $r$ is doubled, $\Delta p$ is halved.

**18.5**    The pressure drop in a Venturi meter varies only with fluid density, velocity of approach, and the diameter ratio of the meter. Venturi meter 1 in water at 20 °C shows a 6-kPa drop when the approach velocity is 5 m/s. Meter 2, geometrically similar to 1, is used in a $\frac{1}{6}$-m$^3$/s flow of benzene ($\rho = 680$ kg/m$^3$). Find the upstream pipe diameter that will yield a 16-kPa drop in meter 2.

■
$$\Delta p = f(\rho, V, d/D) \qquad \Delta p/\rho V^2 = g(d/D) \qquad (\Delta p/\rho V^2)_1 = (\Delta p/\rho V^2)_2$$
$$(\Delta p/\rho V^2)_1 = (6)(1000)/[(998)(5)^2] = 0.2405$$
$$(\Delta p/\rho V^2)_2 = (16)(1000)/(680 V^2) = 23.53/V^2 \qquad 0.2405 = 23.53/V^2 \qquad V = 9.891 \text{ m/s}$$
$$Q = AV \qquad \tfrac{1}{6} = (\pi D_p^2/4)(9.891) \qquad D_p = 0.146 \text{ m} \quad \text{or} \quad 146 \text{ mm}$$

**18.6**    For two hydraulic machines to be homologous, they must (*a*) be geometrically similar, (*b*) have the same discharge coefficient when viewed as an orifice, $Q_1/(A_1\sqrt{2gH_1}) = Q_2/(A_2\sqrt{2gH_2})$, and (*c*) have the same ratio of peripheral speed to fluid velocity, $\omega D/(Q/A)$. Show that the scaling ratios can be expressed as $Q/ND^3 = \text{constant}$ and $H/(ND)^2 = \text{constant}$. $N$ is the rotational speed.

■ (1) From condition ($b$),

$$Q_1/(A_1\sqrt{H_1}) = Q_2/(A_2\sqrt{H_2}) \qquad (1)$$

From condition ($c$),

$$\omega_1 D_1 A_1/Q_1 = \omega_2 D_2 A_2/Q_2 \qquad (2)$$

$$(Q_1/\omega_1 D_1)(\omega_2 D_2/Q_2) = A_1/A_2$$

Geometrical similarity yields $A_1/A_2 = D_1^2/D_2^2 = (Q_1/\omega_1 D_1)(\omega_2 D_2/Q_2)$. Hence, $Q_1/\omega_1 D_1^3 = Q_2/\omega_2 D_2^3$.

(2) From Eq. (1), $Q_1/A_1 = \sqrt{H_1/H_2}Q_2/A_2$. From Eq. (2), $Q_1/A_1 = (Q_2/A_2)(\omega_1 D_1/\omega_2 D_2)$. Hence, $\omega_1 D_1/\omega_2 D_2 = \sqrt{H_1/H_2}$, $H_1/(\omega_1 D_1)^2 = H_2/(\omega_2 D_2)^2$. ($N$ is rotational speed in revolutions per minute; $\omega$ is in radians per second. They are equivalent for modeling purposes.)

**18.7** Use the scaling ratios of Prob. 18.6 to determine the head and discharge of a 1:5 model of a centrifugal pump that discharges 42 m³/min at a 33 m·head when turning at 200 rpm. The model operates at 1000 rpm.

■ From Prob. 18.6, $H_1/(\omega_1 D_1)^2 = H_2/(\omega_2 D_2)^2$ or $H_m = H_p[(\omega_m/\omega_p)(D_m/D_p)]^2 = (33)[(\frac{1000}{200})(\frac{1}{5})]^2 = 33.00$ m; $Q_1/\omega_1 D_1 = Q_2/\omega_2 D_2$ or $Q_m = Q_p(\omega_m/\omega_p)(D_m/D_p)^3 = (42)(\frac{1000}{200})(\frac{1}{5})^3 = 1.68$ m³/min.

**18.8** For an ideal liquid, express the flow $Q$ through an orifice in terms of the density of the liquid, the diameter of the orifice, and the pressure difference.

■ $Q = f(\rho, p, d)$ or $Q = K\rho^a p^b d^c$. Then, dimensionally, $F^0 L^3 T^{-1} = (F^a T^{2a} L^{-4a})(F^b L^{-2b})(L^c)$ and $0 = a + b$, $3 = -4a - 2b + c$, $-1 = 2a$, from which $a = -\frac{1}{2}$, $b = \frac{1}{2}$, $c = 2$. Substituting, $Q = K\rho^{-1/2}p^{1/2}d^2$ or ideal $Q = Kd^2\sqrt{p/\rho}$. Factor $K$ must be obtained by physical analysis and/or experimentation.

For an orifice in the side of a tank under head $h$, $p = \gamma h$. To obtain the familiar orifice formula, let $K = \sqrt{2}(\pi/4)$. Then ideal $Q = \sqrt{2}(\pi/4)d^2\sqrt{\gamma h/\rho}$. But $g = \gamma/\rho$; hence ideal $Q = \frac{1}{4}\pi d^2\sqrt{2gh}$.

**18.9** Determine the dynamic pressure exerted by a flowing incompressible fluid on an immersed object, assuming the pressure is a function of the density and the velocity.

■ $p = f(\rho, V)$ or $p = K\rho^a V^b$. Then, dimensionally, $F^1 L^{-2} T^0 = (F^a T^{2a} L^{-4a})(L^b T^{-b})$ and $1 = a$, $-2 = -4a + b$, $0 = 2a - b$, from which $a = 1$, $b = 2$. Substituting, $p = K_\rho V^2$.

**18.10** Develop the expression for lost head in a horizontal pipe for turbulent incompressible flow.

■ For any fluid, the lost head is represented by the drop in the pressure gradient and is a measure of the resistance to flow through the pipe. The resistance is a function of the diameter of the pipe, the viscosity and density of the fluid, the length of the pipe, the velocity of the fluid, and the roughness $K$ of the pipe. We may write $(p_1 - p_2) = f(d, \mu, \rho, L, V, K)$ or

$$(p_1 - p_2) = Cd^a\mu^b\rho^c L^d V^e(\epsilon/d)^f \qquad (1)$$

From experiment and observation, the exponent of the length $L$ is unity. The value of $K$ is usually expressed as a ratio of the size of the surface protuberances $\epsilon$ to the diameter $d$ of the pipe, a dimensionless number. We may now write $F^1 L^{-2} T^0 = (L^a)(F^b T^b L^{-2b})(F^c T^{2c} L^{-4c})(L^e T^{-e})(L^f/L^f)$ and $1 = b + c$, $-2 = a - 2b - 4c + 1 + e + f - f$, $0 = b + 2c - e$ from which the values of $a$, $b$, and $c$ may be determined in terms of $e$, or $c = e - 1$, $b = 2 - e$, $a = e - 3$. Substituting in (1), $(p_1 - p_2) = Cd^{e-3}\mu^{2-e}\rho^{e-1}L^1 V^e(\epsilon/d)^f$. Dividing the left side of the equation by $w$ and the right side by its equivalent $\rho g$, $(p_1 - p_2)/w = $ lost head $= C(\epsilon/d)^f L(d^{e-3}V^e\rho^{e-1}\mu^{2-e})/(\rho g)$, which becomes (introducing 2 in numerator and in denominator): lost head $= 2C(\epsilon/d)^f(L/d)(V^2/2g)(d^{e-2}V^{e-2}\rho^{e-2}/\mu^{e-2}) = K'(R_E^{e-2})(L/d)(V^2/2g) = f(L/d)(V^2/2g)$ (Darcy formula).

**18.11** To find the drag on a 1-mm-diameter sphere in slowly streaming water, a 100-mm sphere is tested in glycerin at $V = 300$ mm/s. The measured drag on the model is 1.3 N. Under similar conditions, what are the water velocity and the drag on the 1-mm sphere?

■ Water (prototype): $\qquad \mu_p = 0.001$ Pa·s $\qquad \rho_p = 999$ kg/m³

Glycerin (model): $\qquad \mu_m = 1.5$ Pa·s $\qquad \rho_m = 1263$ kg/m³

$$(N_R)_m = \rho_m V_m L_m/\mu_m = (1263)(0.300)(0.100)/1.5 = 25.3$$

$$(C_F)_m = F_m/\rho_m V_m^2 L_m^2 = 1.3/[(1263)(0.300)^2(0.100)^2] = 1.14$$

$$(N_R)_p = (N_R)_m = 25.3 = 999V_p(0.001)/0.001 \quad \text{or} \quad V_p = 0.0253 \text{ m/s}$$

$$(C_F)_p = (C_F)_m = 1.14 = F_p/[999(0.0253)^2(0.001)^2] \quad \text{or} \quad F_p = 0.731 \,\mu\text{N}$$

**18.12**   A 0.1-ft-diameter steel sphere ($\rho_s = 15.2$ slugs/ft$^3$) is dropped in water [$\rho = 1.94$ slugs/ft$^3$, $\mu = 0.000021$ slug/(ft-s)] until it reaches terminal velocity or zero acceleration. From the sphere data in Fig. A-38 compute the terminal velocity of the falling sphere in feet per second.

▎   At terminal velocity, the net weight of the sphere equals the drag; hence the drag is known in this problem: $D = W_{net} = (\rho_s - \rho)g(\pi/6)d^3 = (15.2 - 1.94)(32.2)(\pi/6)(0.1)^3 = 0.224$ lb. We can compute that portion of $C_D$ and $N_R$ which excludes the unknown velocity: $C_D = D/[\frac{1}{2}\rho(\pi/4)d^2V^2] = 0.224/[1.94(\pi/8)(0.1)^2V^2] = 29.4/V^2$, $N_R = \rho Vd/\mu = (1.94)(V)(0.1)/(2.1 \times 10^{-5}) = 9240V$. Now we will just have to guess an initial velocity $V$ to get started on the iteration.
   Guess $V = 1.0$ ft/s; then $N_R = 9240(1.0) = 9240$. From Fig. A-38 read $C_D \approx 0.38$; then $V \approx (29.4/C_D)^{1/2} = 8.8$ ft/s. Now try again with this new guess.
   Guess $V = 8.8$ ft/s, $N_R = 9240(8.8) = 81\,000$. From Fig. A-38 read $C_D \approx 0.52$, $V \approx (29.4/0.52)^{1/2} = 7.5$ ft/s. One more try will give pretty good convergence.
   Guess $V = 7.5$ ft/s, $N_R = 9240(7.5) = 69\,000$. From Fig. A-38 read $C_D \approx 0.51$, $V \approx (29.4/0.51)^{1/2} = 7.6$ ft/s. To the accuracy of the figure, $V_{term} \approx 7.6$ ft/s.

**18.13**   Repeat Prob. 18.12, using the regrouped chart, Fig. A-39.

▎   We must repeat the calculation of the net weight to establish that $D = W_{net} = 0.224$ lb. But now we can go directly to the new drag coefficient: $C'_F = D\rho/\mu^2 = (0.224)(1.194)/(2.1 \times 10^{-5})^2 = 9.85 \times 10^8$. Now enter Fig. A-39 and read $N_R = 70\,000$. Then the desired velocity is $V = \mu N_R/\rho d = (2.1 \times 10^{-5})(70\,000)/[(1.94)(0.1)] = 7.58$ ft/s.

**18.14**   A 7-cm-diameter sphere is tested in water at 20 °C and a velocity of 4 m/s and has a measured drag of 8 N. What will be the velocity and drag force of a 3-m-diameter weather balloon moving in air at 20 °C and 1 atm under similar conditions?

▎   $(N_R)_m = (N_R)_p$     $N_R = \rho DV/\mu$     $(998)(0.07)(4)/(1.02 \times 10^{-3}) = (1.20)(3)(V_b)/(1.81 \times 10^{-5})$
   $V_b = 1.38$ m/s
   $(C_F)_m = (C_F)_p$     $C_F = F/\rho V^2 D^2$     $8/[(998)(4)^2(0.07)^2] = F_b/[(1.20)(1.38)^2(3)^2]$     $F_b = 2.10$ N

**18.15**   To find the drag on an airplane whose velocity is 275 mph, could one test a 1:25 scale model in a subsonic wind tunnel at the same pressure and temperature?

▎   $(N_R)_m = (N_R)_p$     $N_R = LV/\nu$     $L_m = L_p/25$

Since $p_m = p_p$ and $T_m = T_p$, $\nu_m = \nu_p$. Therefore, $V_m = V_p(L_p/L_m) = (275)(25) = 6875$ mph. As this velocity is hypersonic, a subsonic test is impossible.

**18.16**   A 1:25 scale model of a submarine is tested at 180 ft/s in a wind tunnel using sea-level standard air. What is the prototype speed in sea water at 20 °C for dynamic similarity? If the model drag is 1.6 lb, what is the prototype drag?

▎   $(N_R)_m = (N_R)_p$     $N_R = \rho LV/\mu$     $L_m = L_p/25$
   $(0.00234)(L_p/25)(180)/(3.78 \times 10^{-7}) = (2.00)(L_p)(V_p)/(2.23 \times 10^{-5})$     $V_p = 0.497$ ft/s
   $(C_F)_m = (C_F)_p$     $C_F = F/\rho V^2 L^2$
   $1.6/[(0.00234)(180)^2(L_p/25)^2] = F_p/[(2.00)(0.497)^2(L_p)^2]$     $F_p = 6.52$ lb

**18.17**   A blimp is designed to move in air at 20 °C at 7 m/s. If a 1:20 scale model is tested in water at 20 °C, what should the water velocity be? If the measured water drag on the model is 3.09 kN, calculate the drag on the prototype blimp and the power required to propel it.

▎   $(N_R)_m = (N_R)_p$     $N_R = \rho LV/\mu$     $L_m = L_p/20$
   $(998)(L_p/20)(V_m)/(1.02 \times 10^{-3}) = (1.20)(L_p)(7)/(1.81 \times 10^{-5})$     $V_m = 9.49$ m/s
   $(C_F)_m = (C_F)_p$     $C_F = F/\rho V^2 L^2$     $3.09/[(998)(9.49)^2(L_p/20)^2] = F_p/[(1.20)(7)^2(L_p)^2]$
   $F_p = 0.809$ kN     $P_p = F_pV_p = (0.809)(7) = 5.663$ kW

**18.18**   A 1:6 scale model automobile is tested in a wind tunnel in the same air properties as the prototype. The prototype velocity is 48 km/h. For dynamically similar conditions, the model drag is 320 N. What are the drag of the prototype automobile and the power required to overcome this drag?

▮ $$(N_R)_m = (N_R)_p \qquad N_R = \rho LV/\mu \qquad L_m = L_p/6$$

$$(\rho)(L_p/6)(V_m)/\mu = (\rho)(L_p)[(48)(1000)/3600]/\mu \qquad V_m = 80.00 \text{ m/s}$$

$$(C_F)_m = (C_F)_p \qquad C_F = F/\rho V^2 L^2 \qquad 320/[(\rho)(80.00)^2(L_p/6)^2] = F_p/[(\rho)[(48)(1000)/3600]^2(L_p)^2]$$

$$F_p = 320 \text{ N} \qquad P_p = F_p V_p = (320)[(48)(1000)/3600] = 4267 \text{ W}$$

**18.19** A model airplane is built to the scale 1:20. If the prototype is to fly at 425 mph, what must be the air velocity in the wind tunnel to achieve the same Reynolds number at the same air temperature and pressure?

▮ $$(N_R)_m = (N_R)_p \qquad N_R = \rho LV/\mu \qquad L_m = L_p/20 \qquad (\rho)(L_p/20)(V_m)/\mu = (\rho)(L_p)(425)/\mu$$

$$V_m = 8500 \text{ mph} \quad \text{[impractically high]}$$

**18.20** A model airplane has dimensions that are 1/18 those of its prototype. It is desired to test it in a pressure wind tunnel at a speed the same as that of the prototype. If the air temperature is the same and the Reynolds number is the same, what must be the pressure in the wind tunnel relative to the atmospheric pressure?

▮ $$(N_R)_m = (N_R)_p \qquad N_R = LV/\nu \qquad L_m = L_p/18 \qquad (L_p/18)(V)\nu_m = (L_p)(V)\nu_p$$

$$\nu_m/\nu_p = \tfrac{1}{18} = (\mu_m/\rho_m)/(\mu_p/\rho_p) = (\mu_m/\mu_p)(\rho_p/\rho_m)$$

For $T_m = T_p$, assume $\mu_m = \mu_p$; hence, $\rho_p/\rho_m = \gamma_p/\gamma_m = \tfrac{1}{18}$, $\gamma_p/\gamma_m = (p_p/RT)/(p_m/RT) = \tfrac{1}{18}$. Therefore, $p_m = 18p_p = 18$ atm.

**18.21** What weight flow rate of air at 70 °F ($\mu = 3.82 \times 10^{-7}$ lb-s/ft$^2$) at 58 psi in a 2-in-diameter pipe will give dynamic similarity to a 275-gpm flow of water at 60 °F in a 6-in-diameter pipe?

▮ $$(N_R)_{\text{air}} = (N_R)_{\text{H}_2\text{O}} \qquad N_R = DV/\nu \qquad \nu = \mu/\rho$$

$$\rho_{\text{air}} = p/RT = (58)(144)/[(1716)(70 + 460)] = 0.009183 \text{ slug/ft}^3$$

$$\nu_{\text{air}} = (3.82 \times 10^{-7})/0.009183 = 4.160 \times 10^{-5} \text{ ft}^2/\text{s}$$

$$V_{\text{air}} = Q/[(\pi)(\tfrac{2}{12})^2/4] = 45.84Q \qquad V_{\text{H}_2\text{O}} = Q/A = (275)(0.002228)/[(\pi)(\tfrac{6}{12})^2/4] = 3.120 \text{ ft/s}$$

$$(\tfrac{2}{12})(45.84Q)/(4.160 \times 10^{-5}) = (\tfrac{6}{12})(3.120)/(1.21 \times 10^{-5}) \qquad Q = 0.7020 \text{ ft}^3/\text{s}$$

$$\gamma = \rho g \qquad \gamma_{\text{air}} = (0.009183)(32.2) = 0.2957 \text{ lb/ft}^3 \qquad G = \gamma Q = (0.2957)(0.7020) = 0.208 \text{ lb/s}$$

**18.22** A 1:40 scale model of a submarine is tested in a wind tunnel. It is desired to know the drag on the submarine when it is operating at 9 knots (15.2 fps) in ocean water at 40 °F. At what velocity should the object be tested in a wind tunnel containing air at 70 °F at atmospheric pressure?

▮ $$(N_R)_{\text{air}} = (N_R)_{\text{H}_2\text{O}} \qquad N_R = LV/\nu \qquad L_{\text{air}} = L_{\text{H}_2\text{O}}/40$$

$$(L_{\text{H}_2\text{O}}/40)(V_{\text{air}})/(1.64 \times 10^{-4}) = (L_{\text{H}_2\text{O}})(15.2)/(1.66 \times 10^{-5}) \qquad V_{\text{air}} = 6007 \text{ ft/s}$$

Since this velocity exceeds sonic velocity, the model will not operate properly to indicate prototype behavior.

**18.23** A ship 650 ft long is to operate at a speed of 22 mph in ocean water whose kinematic viscosity is 0.00001261 ft$^2$/s. What should be the kinematic viscosity of a fluid used with an 11-ft model so that both the Reynolds number and the Froude number would be the same?

▮ $$(N_R)_m = (N_R)_p \qquad N_R = LV/\nu$$

Let subscript $r$ denote ratio of prototype to model.

$$L_r V_r/\nu_r = 1 \qquad (N_F)_m = (N_F)_p \qquad N_F = V/(gL)^{1/2} \qquad V_r/(g_r L_r)^{1/2} = 1 \qquad L_r V_r/\nu_r = V_r/(g_r L_r)^{1/2}$$

$$g_r = 1.0 \qquad L_r = \tfrac{650}{11} = 59.1 \qquad (59.1)(V_r)/\nu_r = V_r/[(1.0)(59.1)]^{1/2} \qquad \nu_r = 454.3$$

Hence, $\nu_m = 0.00001261/454.3 = 0.0000000278$ ft$^2$/s [impossibly small].

**18.24** The valve coefficients $K = \Delta p/(\rho V^2/2)$ for a 600-mm-diameter valve are to be determined from tests on a geometrically similar 300-mm-diameter valve using atmospheric air at 80 °F ($\nu = 1.57 \times 10^{-5}$ m²/s). What range of air flows is required to model flows of water at 70 °F ($\nu = 9.96 \times 10^{-7}$ m²/s) between 1 m/s and 2.5 m/s?

$\blacksquare$ $\qquad N_R = DV/\nu \qquad (N_R)_{min} = (0.600)(1)/(9.96 \times 10^{-7}) = 6.02 \times 10^5$

$$(N_R)_{max} = (0.600)(2.5)/(9.6 \times 10^{-7}) = 1.51 \times 10^6$$

For air:

$$6.02 \times 10^5 = (0.300)(V_{min})/(1.57 \times 10^{-5}) \qquad V_{min} = 31.5 \text{ m/s}$$

$$1.51 \times 10^6 = (0.300)(V_{max})/(1.57 \times 10^{-5}) \qquad V_{max} = 79.0 \text{ m/s}$$

$$Q_{min} = AV_{min} = [(\pi)(0.300)^2/4](31.5) = 2.23 \text{ m}^3/\text{s} \qquad Q_{max} = AV_{max} = [(\pi)(0.300)^2/4](79.0) = 5.58 \text{ m}^3/\text{s}$$

**18.25** A model of a Venturi meter has linear dimensions 1/6 those of the prototype. The prototype operates on water at 10 °C, and the model on water at 100 °C. For a throat diameter of 700 mm and a velocity at the throat of 7 m/s, what discharge is needed through the model for similitude?

$\blacksquare$ $(N_R)_m = (N_R)_p \qquad N_R = LV/\nu \qquad (L_p/6)(V_m)/(2.94 \times 10^{-7}) = (L_p)(7)/(1.30 \times 10^{-6}) \qquad V_m = 9.498 \text{ m/s}$

$\qquad Q_m = A_m V_m \qquad D_m = D_p/6 = \frac{700}{6} = 117 \text{ mm} \qquad Q_m = [(\pi)(0.117)^2/4](9.498) = 0.102 \text{ m}^3/\text{s}$

**18.26** The losses in a Y in a 1.3-m-diameter pipe system carrying gas ($\rho = 40$ kg/m³, $\mu = 2 \times 10^{-4}$ Pa·s, $V = 28$ m/s) are to be determined by testing a model with water at 40 °C. The laboratory has a water capacity of 80 L/s. What should be the diameter of the model?

$\blacksquare$ $\qquad (N_R)_m = (N_R)_p \qquad N_R = \rho DV/\mu \qquad V_m = Q_m/A_m = (80 \times 10^{-3})/(\pi D_m^2/4) = 0.1019/D_m^2$

$$(998)(D_m)(0.1019/D_m^2)/(6.51 \times 10^{-4}) = (40)(1.3)(28)/(2 \times 10^{-4})$$

$$D_m = 0.02146 \text{ m} \quad \text{or} \quad 21 \text{ mm}$$

**18.27** A 1:4 scale model of a water pumping station pumping system is to be tested to determine overall head losses. Air at 30 °C, 1 atm, is available. For a prototype velocity of 600 mm/s in a 3-m-diameter section with water at 20 °C, determine the air velocity and quantity needed. How may losses determined from the model be converted into prototype losses?

$\blacksquare$ $\qquad (N_R)_m = (N_R)_p \qquad N_R = DV/\nu \qquad [(3)(\frac{1}{4})](V_m)/(1.60 \times 10^{-5}) = (3)(0.600)/(1.02 \times 10^{-6})$

$\qquad V_m = 37.65 \text{ m/s} \qquad Q_m = A_m V_m = \{(\pi)[(3)(\frac{1}{4})]^2/4\}(37.65) = 16.6 \text{ m}^3/\text{s}$

Losses are the same when expressed in velocity heads.

**18.28** Water at 60 °F flows at 12.0 fps in a 6-in pipe. At what velocity must medium fuel oil at 90 °F ($\nu = 3.19 \times 10^{-5}$ ft²/s) flow in a 3-in pipe for the two flows to be dynamically similar?

$\blacksquare$ $(N_R)_{H_2O} = (N_R)_{oil} \qquad N_R = DV/\nu \qquad (\frac{6}{12})(12.0)/(1.21 \times 10^{-5}) = (\frac{3}{12})(V_{oil})/(3.19 \times 10^{-5}) \qquad V_{oil} = 63.3 \text{ ft/s}$

**18.29** Air at 68 °F is to flow through a 24-in pipe at an average velocity of 6.00 fps. For dynamic similarity, what size pipe carrying water at 60 °F at 3.65 fps should be used?

$\blacksquare$ $(N_R)_{air} = (N_R)_{H_2O} \qquad N_R = DV/\nu \qquad (\frac{24}{12})(6.00)/(1.63 \times 10^{-4}) = (D/12)(3.65)/(1.21 \times 10^{-5}) \qquad D = 2.93 \text{ in}$

(Would probably use a 3-in pipe.)

**18.30** A 1:5 model of a submarine is to be tested in a towing tank containing salt water. If the submarine moves at 12.0 mph, at what velocity should the model be towed for dynamic similarity?

$\blacksquare$ $\qquad (N_R)_m = (N_R)_p \qquad N_R = LV/v \qquad (L/15)(V_m)/v = (L)(12.0)/v \qquad V_m = 180$ mph

**18.31** A model of a torpedo is tested in a towing tank at a velocity of 80.0 fps. The prototype is expected to attain a velocity of 20.0 fps in water at 60 °F. (**a**) What model scale has been used? (**b**) What would be the model speed if tested in a wind tunnel under a pressure of 20 atm and at constant temperature 80 °F?

$\blacksquare$ $\qquad\qquad\qquad\qquad (N_R)_m = (N_R)_p \qquad N_R = DV/v$

(**a**) $(L/x)(80.0)/v = (L)(20.0)/v$, $x = 4.00$. Hence, the scale model is 1:4.

(**b**) $\qquad v = \mu/\rho \qquad \rho_{air} = p/RT = (20)(14.7)(144)/[(1716)(80+460)] = 0.04569$ slug/ft$^3$

$$v_{air} = (3.85 \times 10^{-7})/0.04569 = 8.426 \times 10^{-6} \text{ ft}^2/\text{s}$$

$$(L/4)(V_m)/(8.426 \times 10^{-6}) = (L)(20.0)/(1.21 \times 10^{-5}) \qquad V_m = 55.7 \text{ ft/s}$$

**18.32** A centrifugal pump pumps medium lubricating oil at 60 °F ($v = 188 \times 10^{-5}$ ft$^2$/s) while rotating at 1200 rpm. A model pump, using air at 68 °F, is to be tested. If the diameter of the model is 3 times the diameter of the prototype, at what speed should the model run?

$\blacksquare$ $(N_R)_m = (N_R)_p$, $N_R = DV/v$. Using the peripheral speeds (which equal radius times angular velocity in radians per second) as the velocities in Reynolds number, we obtain $(3D)[(3D/2)(\omega_m)]/(1.63 \times 10^{-4}) = (D)[(D/2)(\omega_p)]/(188 \times 10^{-5})$, $\omega_p = 103.8\omega_m$. Hence, model speed $= 1200/103.8 = 11.6$ rpm.

**18.33** An airplane wing of 3-ft chord is to move at 90 mph in air. A model of 3-in chord is to be tested in a wind tunnel with air velocity at 108 mph. For air temperature of 68 °F in each case, what should be the pressure in the wind tunnel?

$\blacksquare$ $\qquad (N_R)_m = (N_R)_p \qquad N_R = LV/v \qquad (\frac{3}{12})(108)/v_m = (3)(90)/(1.63 \times 10^{-4}) \qquad v_m = 1.63 \times 10^{-5} \text{ ft}^2/\text{s}$

The pressure that produces this kinematic viscosity of air at 68 °F can be found by remembering that the absolute viscosity is not affected by pressure changes. The kinematic viscosity equals absolute viscosity divided by density. But density increases with pressure (temperature constant). $v = \mu/\rho$, $v_p/v_m = (1.63 \times 10^{-4})/(1.63 \times 10^{-5}) = 10$. Thus the density of air in the tunnel must be ten times standard (68 °F) air and the resulting pressure in the tunnel must be 10 atm.

**18.34** A dam spillway is to be tested using Froude scaling with a 1:20 model. The model flow has an average velocity of 0.7 m/s and a volume flow of 0.06 m$^3$/s. What are the velocity and flow of the prototype?

$\blacksquare$ For Froude scaling,

$$V_p/V_m = 1/\sqrt{\alpha} \qquad \alpha = L_m/L_p = \frac{1}{20} \qquad V_p/0.7 = 1/\sqrt{\frac{1}{20}} \qquad V_p = 3.13 \text{ m/s}$$

$$Q_p/Q_m = (V_p/V_m)(L_p/L_m)^2 = (1/\sqrt{\alpha})(1/\alpha)^2 = (1/\alpha)^{5/2} \qquad Q_p/0.06 = [1/(\frac{1}{20})]^{5/2} \qquad Q_p = 107 \text{ m}^3/\text{s}$$

**18.35** A prototype ship is 40 m long and designed to cruise at 10 m/s (about 19 knots). Its drag is to be simulated by a 0.8-m-long model pulled in a tow tank. For Froude scaling find (**a**) the tow speed, (**b**) the ratio of prototype to model drag, and (**c**) the ratio of prototype to model power.

$\blacksquare$ For Froude scaling,

(**a**) $\qquad V_p/V_m = 1/\sqrt{\alpha} \qquad \alpha = L_m/L_p = \frac{0.8}{40} = \frac{1}{50} \qquad 10/V_m = 1/\sqrt{\frac{1}{50}} \qquad V_m = 1.41 \text{ m/s}$

(**b**) $\qquad F_m/F_p = (\rho_m/\rho_p)(V_m/V_p)^2(L_m/L_p)^2 = (1)(\sqrt{\alpha})^2(\alpha)^2 = \alpha^3 = (\frac{1}{50})^3 = 1/125\,000$

(**c**) $\qquad P_m/P_p = (F_m/F_p)(V_m/V_p) = (\alpha)^3(\sqrt{\alpha}) = \alpha^{3.5} = (\frac{1}{50})^{3.5} = 1/883\,883$

**18.36** A ship 555 ft long is to operate at a speed of 22 mph. If a model is 8 ft long, what should be its speed to give the same Froude number? What is the value of the Froude number?

$\blacksquare$ $\qquad\qquad (N_F)_m = (N_F)_p \qquad N_F = V/(gL)^{1/2} \qquad V_m/[(g)(8)]^{1/2} = 22/[(g)(555)]^{1/2}$

$$V_m = 2.64 \text{ mph} \quad \text{or} \quad 3.87 \text{ ft/s} \qquad N_F = 3.87/[(32.2)(8)]^{1/2} = 0.241$$

**18.37** In a $1:30$ model of the flow in a spillway, the velocity at point $A$ is 0.4 m/s and the force exerted on a small area about $A$ is 0.10 N. What would be the force on the corresponding area in the prototype?

$$\blacksquare \qquad (N_F)_m = (N_F)_p \qquad N_F = V/(gL)^{1/2} \qquad 0.4/[(g)(L_p/30)]^{1/2} = V_p/(gL_p)^{1/2} \qquad V_p = 2.19 \text{ m/s}$$

$$(C_F)_m = (C_F)_p \qquad C_F = F/\rho V^2 L^2 \qquad 0.10/[(\rho)(0.4)^2(L_p/30)^2] = F_p/[(\rho)(2.19)^2(L_p)^2] \qquad F_p = 2698 \text{ N}$$

**18.38** A $1:400$ model is constructed to study tides. What length of time in the model corresponds to a day in the prototype? Suppose the model could be transported to the moon and tested there. What then would be the time relationship between the model and prototype? $g$ of earth equals six times $g$ of moon.

$$\blacksquare \qquad (N_F)_m = (N_F)_p \qquad N_F = V/(gL)^{1/2} \qquad [V/(gL)^{1/2}]_m = [V/(gL)^{1/2}]_p$$

$$V_r = V_p/V_m = (gL)_r^{1/2} \qquad T_r = L_r/V_r = L_r(gL)_r^{1/2} \qquad T_p/T_m = [L/(gL)^{1/2}]_p/[L/(gL)^{1/2}]_m$$

**On earth:**

$$1/T_m = \{L_p/[(g)(L_p)]^{1/2}\}/\{(L_p/400)/[(g)(L_p/400)]^{1/2}\} \qquad T_m = 0.05000 \text{ day} \quad \text{or} \quad 1.20 \text{ h}$$

**On the moon:**

$$1/T_m = \{L_p/[(g_p)(L_p)]^{1/2}\}/\{(L_p/400)/[(g_p/6)(L_p/400)]^{1/2}\} \qquad T_m = 0.1225 \text{ day} \quad \text{or} \quad 2.94 \text{ h}$$

**18.39** A sectional model of a spillway 2 ft high is placed in a laboratory flume of 9-in width. Under a head of 0.345 ft the flow is 0.62 cfs. What flow does this represent in the prototype if the scale model is $1:20$ and the spillway is 700 ft long?

$$\blacksquare \quad (N_F)_m = (N_F)_p \qquad N_F = V/(gL)^{1/2} \qquad (N_F)_r = (N_F)_p/(N_F)_m = [V/(gL)^{1/2}]_p/[V/(gL)^{1/2}]_m = V_r/L_r^{1/2} = 1.00$$

$$V_r/20^{1/2} = 1.00 \qquad V_r = 4.47 \qquad Q_r = A_r V_r \qquad A_r = A_p/A_m = (700)[(0.345)(20)]/[(\tfrac{9}{12})(0.345)] = 18\,667$$

$$Q_r = A_r V_r = (18\,667)(4.47) = 83\,441 \qquad Q = (83\,441)(0.62) = 51\,733 \text{ ft}^3/\text{s}$$

**18.40** The flow over a model spillway is 98 L/s per meter of width. What flow does this represent in the prototype spillway if the model scale is $1:20$?

$$\blacksquare \qquad q = Q/b = [(by)(V)]/b = yV \qquad q_r = y_r V_r \qquad y_r = L_r$$

$$V_r/L_r^{1/2} = 1.00 \qquad \text{or} \qquad V_r = L_r^{1/2} \qquad \text{(from Prob. 18.39)}$$

$$q_r = L_r L_r^{1/2} = L_r^{3/2} = 20^{3/2} = 89.44 \qquad q_p = (89.44)(98) = 8770 \text{ (L/s)/m} = 8.77 \text{ (m}^3/\text{s)/m}$$

**18.41** The velocity at a point in a model of a spillway for a dam is 2 m/s. For a geometric ratio of prototype to model of $12:1$, what is the velocity at the corresponding point in the prototype under similar conditions?

$$\blacksquare \qquad (N_F)_m = (N_F)_p \qquad N_F = V/(gL)^{1/2} \qquad 2/[(g)(L_p/12)]^{1/2} = V_p/(gL_p)^{1/2} \qquad V_p = 6.93 \text{ m/s}$$

**18.42** The wave drag on a model of a ship is 25 N at a speed of 4 m/s. For a prototype 12 times as long, what will the corresponding speed and wave drag be if the liquid is the same in each case?

$$\blacksquare \qquad (N_F)_m = (N_F)_p \qquad N_F = V/(gL)^{1/2} \qquad 4/[(g)(L_p/12)]^{1/2} = V_p/(gL_p)^{1/2} \qquad V_p = 13.86 \text{ m/s}$$

$$C_D = F/(A\rho V^2/2)$$

At the same Froude number, $C_D$ is equal for model and prototype; hence, $F_m/(A_m V_m^2/2) = F_p/(A_p V_p^2/2)$, $F_p = (F_m)(A_p/A_m)(V_p/V_m)^2 = (F_m)(L_p/L_m)^3 = (25)(12)^3 = 43\,200$ N, or 43.2 kN.

**18.43** Oil of kinematic viscosity $50 \times 10^{-5}$ ft$^2$/s is to be used in a prototype in which both viscous and gravity forces dominate. A model scale of $1:5$ is also desired. What viscosity of model liquid is ncessary to make both Froude number and Reynolds number the same in model and prototype?

$$\blacksquare \quad (N_F)_m = (N_F)_p \qquad N_F = V/(gL)^{1/2} \qquad V_r = (g_r L_r)^{1/2} = L_r^{1/2} \qquad (N_R)_m = (N_R)_p \qquad N_R = LV/\nu$$

$$V_r = \nu_r/L_r \qquad L_r^{1/2} = \nu_r/L_r \qquad \nu_r = L_r^{3/2} = (\tfrac{1}{5})^{3/2} = 0.08944$$

$$\nu_m = (0.08944)(50 \times 10^{-5}) = 4.47 \times 10^{-5} \text{ ft}^2/\text{s}$$

**18.44** A ship whose hull length is 460 ft is to travel at 25.0 fps. (*a*) Compute the Froude number. (*b*) For dynamic similarity, at what velocity should a $1:30$ model be towed through water?

**(a)**    $$N_F = V/(gL)^{1/2} = 25.0/[(32.2)(469)]^{1/2} = 0.205$$

**(b)**    $(N_F)_m = (N_F)_p$    $$V_m/[(32.2)(\tfrac{460}{30})]^{1/2} = 25.0/[(32.2)(460)]^{1/2}    V_m = 4.56 \text{ ft/s}$$

**18.45**  An airplane is designed to fly at 240 m/s at 8 km altitude. If a $1:12$ model is tested in a wind tunnel at 25 °C, find the tunnel pressure that will reproduce both the Reynolds number and the Mach number.

▌ At 8000 meters, $\rho_p = 0.526 \text{ kg/m}^3$, $T_p = -36.9$ °C, and $\mu_p = 1.53 \times 10^{-5}$ Pa · s (from Table A-8).

$$(N_M)_m = (N_M)_p    N_M = V/c    V_m/c_m = V_p/c_p    c \approx (20 \text{ m/s} \cdot \text{K}^{1/2})T^{1/2}$$
$$c_m \approx (20)(25 + 273)^{1/2} = 345.3 \text{ m/s}    c_p \approx (20)(-36.9 + 273)^{1/2} = 307.3 \text{ m/s}$$
$$V_m/345.3 = 240/307.3    V_m = 269.7 \text{ m/s}    (N_R)_m = (N_R)_p    N_R = \rho LV/\mu$$
$$(\rho_m)(L_p/12)(269.7)/(1.84 \times 10^{-5}) = (0.526)(L_p)(240)/(1.53 \times 10^{-5})    \rho_m = 6.75 \text{ kg/m}^3$$
$$p_m = \rho_m RT_m = (6.75)(287)(25 + 273) = 577\,300 \text{ Pa    or    577 kPa}$$

**18.46**  One wishes to model the flow about a missile that travels at 975 mph through the atmosphere at elevation 9000 ft. The model is to be tested in a wind tunnel at standard atmospheric conditions with air at 60 °F. What air speed in the wind tunnel is required for dynamic similarity?

▌
$$(N_M)_m = (N_M)_p    N_M = V/c = V/(p/\rho)^{1/2}$$
$$V_m/(14.7/0.00227)^{1/2} = 975/(10.54/0.001814)^{1/2}    V_m = 1029 \text{ mph    or    1510 ft/s}$$

**18.47**  A model of a supersonic aircraft is tested in a variable-density wind tunnel at 1290 fps. The air is at 90 °F with a pressure of 20 psia. At what velocity should this model be tested to maintain dynamic similarity if the air temperature is raised to 130 °F and the pressure increased to 27 psia?

▌ $N_M = V/(p/\rho)^{1/2}$, $\rho = p/RT$. At 90 °F, $\rho = (20)(144)/[(53.3)(90 + 460)] = 0.09824 \text{ slug/ft}^3$, $N_M = 1290/[(20)(144)/0.09824]^{1/2} = 7.534$. At 130 °F, $\rho = (27)(144)/[(53.3)(130 + 460)] = 0.1236 \text{ slug/ft}^3$, $N_M = V/[(27)(144)/0.1236]^{1/2} = 7.534$, $V = 1336 \text{ ft/s}$.

**18.48**  The flow about a ballistic missile which travels at 1400 fps through air at 80 °F and 14.7 psia is to be modeled in a high-speed wind tunnel with a $1:10$ model. If the air in the wind tunnel test section has a temperature of 10 °F at a pressure of 12 psia, what velocity is required in the model test section?

▌
$$(N_M)_m = (N_M)_p    N_M = V/(p/\rho)^{1/2}$$
$$\rho = p/RT    \rho_m = (12)(144)/[(53.3)(10 + 460)] = 0.06898 \text{ slug/ft}^3$$
$$\rho_p = (14.7)(144)/[(53.3)(80 + 460)] = 0.07355 \text{ slug/ft}^3$$
$$V_m/[(12)(144)/0.06898]^{1/2} = 1400/[(14.7)(144)/0.07355]^{1/2}    V_m = 1306 \text{ ft/s}$$

**18.49**  A prototype spillway has a characteristic velocity of 2 m/s and a characteristic length of 9 m. A small model is constructed using Froude scaling. What is the minimum scale ratio of the model that will ensure that its minimum Weber number is 100? Both flows use water at 30 °C.

▌    $$N_W = \rho V^2 L/\sigma    (N_W)_p = (996)(2)^2(9)/0.0712 = 503\,596$$

**Froude scaling:**
$$(N_W)_m/(N_W)_p = (\rho_m/\rho_p)(V_m/V_p)^2(L_m/L_p)/(\sigma_m/\sigma_p) = (1)(\sqrt{\alpha})^2(\alpha)(1) = \alpha^2 \geq 100/503\,596$$
$$\alpha = 0.01409    L_p/L_m \leq 1/\alpha \leq 1/0.01409 = 71:1$$

**18.50**  At low velocities (laminar flow), the volume flow $Q$ through a small-bore tube is a function only of the tube radius $R$, the fluid viscosity $\mu$, and the pressure drop per unit tube length $dp/dx$. Using the Buckingham pi theorem, find an appropriate dimensionless relationship.

▌ Write the given relation and count variables: $Q = f(R, \mu, dp/dx)$  four variables ($n = 4$). Make a list of the dimensions of these variables:

| $Q$ | $R$ | $\mu$ | $dp/dx$ |
|---|---|---|---|
| $\{L^3T^{-1}\}$ | $\{L\}$ | $\{ML^{-1}T^{-1}\}$ | $\{ML^{-2}T^{-2}\}$ |

There are three primary dimensions $(M, L, T)$, hence $j \leq 3$. By trial and error we determine that $R$, $\mu$, and $dp/dx$ cannot be combined into a pi group. Then $j = 3$, and $n - j = 4 - 3 = 1$. There is only *one* pi group, which we find by combining $Q$ in a power product with the other three: $\Pi_1 = R^a \mu^b (dp/dx)^c Q^1 = (L)^a (ML^{-1}T^{-1})^b (ML^{-2}T^{-2})^c (L^3 T^{-1}) = M^0 L^0 T^0$. Equate exponents:

Mass: $\qquad\qquad\qquad\qquad\qquad\qquad b + c \quad = 0$

Length: $\qquad\qquad\qquad\qquad\qquad a - b - 2c + 3 = 0$

Time: $\qquad\qquad\qquad\qquad\qquad\qquad - b - 2c - 1 = 0$

Solving simultaneously, we obtain $a = -4$, $b = 1$, $c = -1$. Then $\Pi_1 = R^{-4} \mu^1 (dp/dx)^{-1} Q$ or $\Pi_1 = Q\mu/[R^4(dp/dx)] = $ const. Since there is only one pi group, it must equal a dimensionless constant.

**18.51** The capillary rise $h$ of a liquid in a tube varies with tube diameter $d$, gravity $g$, fluid density $\rho$, surface tension $\sigma$, and the contact angle $\theta$. (**a**) Find a dimensionless statement of this relation. (**b**) If $h = 3$ cm in a given experiment, what will $h$ be in a similar case if diameter and surface tension are half as much, density is twice as much, and the contact angle is the same?

▮ (**a**) *Step 1* Write down the function and count variables $h = f(d, g, \rho, \sigma, \theta)$, $n = 6$ variables.

*Step 2* List the dimensions $(FLT)$:

| $h$ | $d$ | $g$ | $\rho$ | $Y$ | $\theta$ |
|:---:|:---:|:---:|:---:|:---:|:---:|
| $\{L\}$ | $\{L\}$ | $\{LT^{-1}\}$ | $\{FT^2L^{-4}\}$ | $\{FL^{-1}\}$ | None |

*Step 3* Find $j$. Several groups of three form no pi: $\sigma$, $\rho$, and $g$ or $\rho$, $g$, and $d$. Therefore $j = 3$, and we expect $n - j = 6 - 3 = 3$ dimensionless groups. One of these is obviously $\theta$, which is already dimensionless: $\Pi_3 = \theta$. If we chose carelessly to search for it using steps 4 and 5, we would still find $\Pi_3 = \theta$.

*Step 4* Select $j$ variables which do not form a pi group: $\rho$, $g$, $d$.

*Step 5* Add one additional variable in sequence to form the pi's:

Add $h$: $\qquad\qquad \Pi_1 = \rho^a g^b d^c h = (FT^2L^{-4})^a (LT^{-2})^b (L)^c (L) = F^0 L^0 T^0$

Solve for $a = b = 0$, $c = -1$. Therefore $\Pi_1 = \rho^0 g^0 d^{-1} h = h/d$. Finally add $Y$, again selecting its exponent to be 1: $\Pi_2 = \rho^a g^b d^c \sigma = (FT^2L^{-4})^a (LT^{-2})^b (L)^c (FL^{-1}) = F^0 L^0 T^0$. Solve for $a = b = -1$, $c = -2$. Therefore $\Pi_2 = \rho^{-1} g^{-1} d^{-2} \sigma = \sigma/\rho g d^2$.

*Step 6* The complete dimensionless relation for this problem is thus

$$h/d = F(\sigma/\rho g d^2, \theta) \qquad\qquad (1)$$

This is as far as dimensional analysis goes. Theory, however, establishes that $h$ is proportional to $Y$. Since $Y$ occurs only in the second parameter, we can slip it outside $(h/d)_{\text{actual}} = (\sigma/\rho g d^2) F_1(\theta)$ or $h\rho g d/\sigma = F_1(\theta)$.

(**b**) We are given $h_1$ for certain conditions $d_1$, $\sigma_1$, $\rho_1$, and $\theta_1$. If $h_1 = 3$ cm, what is $h_2$ for $d_2 = \frac{1}{2}d$, $\sigma_2 = \frac{1}{2}\sigma$, $\rho_2 = 2\rho_1$, and $\theta_2 = \theta_1$? We know the functional relation, Eq. (*1*), must still hold at condition 2: $h_2/d_2 = F(\sigma_2/\rho_2 g d_2^2, \theta_2)$. But $\sigma_2/\rho_2 g d_2^2 = \frac{1}{2}\sigma_1/2\rho_1 g(\frac{1}{2}d_1)^2 = \sigma_1/\rho_1 g d_1^2$. Therefore, functionally, $h_2/d_2 = F(\sigma_1/\rho_1 g d_1^2, \theta_1) = h_1/d_1$. We are given a condition 2 which is exactly similar to condition 1, and therefore a scaling law holds: $h_2 = h_1(d_2/d_1) = (3)(\frac{1}{2}d_1/d_1) = 1.5$ cm. If the pi groups had not been exactly the same for both conditions, we would have to know more about the functional relation $F$ to calculate $h_2$.

**18.52** Under laminar conditions, the volume flow $Q$ through a small triangular-section pore of side length $b$ and length $L$ is a function of viscosity $\mu$, pressure drop per unit length $\Delta p/L$, and $b$. Using the pi theorem, rewrite this relation in dimensionless form. How does the volume flow change if the pore size $b$ is tripled?

▮ $\qquad\qquad\qquad Q = f(\Delta p/L, \mu, b) \qquad\qquad \{L^3/T\} = \{M/L^2T^2\}\{M/LT\}\{L\}$

$n = 4$, $j = 3$ ($Q$, $\mu$, $b$ do not make a $\Pi$), $n - j = 1$ $\Pi$ expected. $\Pi_1 = Q^1 (\Delta p/L)^a (\mu)^b (b)^c = \{L^3/T\}\{M/L^2T^2\}^a \{M/LT\}^b \{L\}^c = M^0 L^0 T^0$.

$M$:
$$a + b = 0$$

$L$:
$$3 - 2a - b + c = 0$$

$T$:
$$-1 - 2a - b = 0$$

$a = -1$, $b = +1$, $c = -4$; $\Pi_1 = Q\mu/(\Delta p/L)b^4 = $ constant. If $b$ is tripled, $Q$ increases $3^4$, or 81 times.

**18.53** The power input $P$ to a centrifugal pump is assumed to be a function of volume flow $Q$, impeller diameter $D$, rotational rate $\Omega$, and the density $\rho$ and viscosity $\mu$ of the fluid. Rewrite this as a dimensionless relationship.

∎ $P_{\text{pump}} = f(Q, D, \Omega, \rho, \mu)$. $n = 6$, $j = 3$, $n - j = 6 - 3 = 3$ $\Pi$'s expected. $P/(\rho\Omega^3 D^5) = f(Q/\Omega D^3, \rho\Omega D^2/\mu)$.

**18.54** The resistance force $F$ of a surface ship is a function of its length $L$, velocity $V$, gravity $g$, and the density $\rho$ and viscosity $\mu$ of the water. Rewrite in dimensionless form.

∎ $F_{\text{ship}} = f(L, V, g, \rho, \mu)$. $m = 6$, $j = 3$, $n - j = 6 - 3 = 3$ $\Pi$'s expected. $F/(\rho V^2 L^2) = f(V^2/gL, \rho VL/\mu)$.

**18.55** The torque $M$ on an axial-flow turbine is a function of fluid density $\rho$, rotor diameter $D$, angular rotation rate $\Omega$, and volume flow $Q$. Rewrite in dimensionless form. If it is known that $M$ is proportional to $Q$ for a particular turbine, how would $M$ vary with $\Omega$ and $D$ for that turbine?

∎ $M_{\text{turbine}} = f(\rho, D, \Omega, Q)$. $n = 5$, $j = 3$, $n - j = 5 - 3 = 2$ $\Pi$'s expected. $M/(\rho\Omega^2 D^5) = f(Q/\Omega D^3)$. If $M \approx Q$, then $M \approx \Omega$ and $M \approx D^2$.

**18.56** A *weir* is an obstruction in a channel flow that can be calibrated to measure flow rate. The volume flow $Q$ varies with gravity $g$, weir width $b$, and upstream water height $H$ above the weir crest. If it is known that $Q$ is proportional to $b$, use the pi theorem to find a unique functional relationship $Q(g, b, H)$.

∎ $Q/b = f(g, H)$. $n = 3$, $j = 2(L, T)$, $n - j = 3 - 2 = 1$ $\Pi$ expected. $(Q/b)/(g^{1/2}H^{3/2}) = $ constant.

**18.57** The size of droplets produced by a liquid spray nozzle is thought to depend upon the nozzle diameter $D$, jet velocity $U$, and the properties of the liquid $\rho$, $\mu$, and $\sigma$. Rewrite this relation in dimensionless form.

∎ $d = f(D, U, \rho, \mu, \sigma)$. $n = 6$, $j = 3$, $n - j = 6 - 3 = 3$ $\Pi$'s expected. $d/D = f(\rho UD/\mu, \sigma/\rho U^2 D)$.

**18.58** Perfume (specific gravity 0.92) in a tube of 25 mm diameter has a capillary rise of 3 mm. Determine its capillary rise in a 40-mm-diameter tube.

∎ $h = f(\rho g, \sigma, D)$. $n = 4$, $j = 3$, $n - j = 4 - 3 = 1$ $\Pi$ expected. $\rho ghD/\sigma = $ constant. For a given fluid (i.e., given $\rho g$ and $\sigma$), $hD = $ constant, or $h_m D_m = h_p D_p$. Thus, $(3)(2.5) = (h_p)(4)$ or $h_p = 1.88$ mm.

**18.59** An estuary has a lunar tide of period 12.02 h and a tidal current of approximately 2.7 km/h. In a 1 : 400 scale model, what should the period and current speed be?

∎
$$T_m/T_p = \sqrt{\alpha} \qquad \alpha = L_m/L_p = \tfrac{1}{400} \qquad T_m/12.02 = \sqrt{\tfrac{1}{400}} \qquad T_m = 0.601 \text{ h} \quad \text{or} \quad 36.1 \text{ min}$$
$$V_m/V_p = \sqrt{\alpha} \qquad V_m/2.7 = \sqrt{\tfrac{1}{400}} \qquad V_m = 0.135 \text{ km/h} = 2.25 \text{ m/min}$$

**18.60** Derive the expression for the drag on a submerged torpedo. The parameters are the size of the torpedo $L$, the velocity of the torpedo $V$, the viscosity of the water $\mu$, and the density of the water $\rho$. The size of a torpedo may be represented by its diameter or its length.

∎
$$F_D = f(L, V, \rho, \mu) = (\text{constant})L^a V^b \rho^c \mu^d \qquad \{ML/T^2\} = \{L\}^a\{L/T\}^b\{M/L^3\}^c\{M/LT\}^d$$

$M$:
$$1 = c + d$$

$L$:
$$1 = a + b - 3c - d$$

$T$:
$$-2 = -b - d$$

$$c = 1 - d \qquad b = 2 - d \qquad a = 2 - d$$

$$F_D = (\text{constant})L^{2-d}V^{2-d}\rho^{1-d}\mu^d = (\text{constant})C\rho L^2 V^2(\mu/LV\rho)^d = (\text{constant})C\rho L^2 V^2(LV\rho/\mu)^{-d}$$

Since $LV\rho/\mu = N_R$, $F_D = \phi(N_R)\rho L^2 V^2$.

**18.61** Derive an expression for the drag on a surface vessel. Use the same parameters as in Prob. 18.60, and add the acceleration due to gravity $g$ to account for the effect of wave action.

**∎** $F_D = f(L, V, \rho, \mu, g)$, $f'(F_D, L, V, \rho, \mu, g) = 0$, $n = 6$, $j = 3$, $n - j = 6 - 3 = 3$ $\Pi$'s expected.

$$\Pi = \rho^a L^b V^c F_D^d \qquad \Pi_2 = \rho^a L^b V^c \mu^d \qquad \Pi_3 = \rho^a L^b V^c g^d$$
$$= F_D/L^2 V^2 \qquad\qquad = LV\rho/\mu = N_R \qquad = Lg/V^2 = 1/N_F^2$$
$$F_D = \phi(N_F, N_R)\rho L^2 V^2$$

**18.62** The discharge through a horizontal capillary tube is thought to depend upon the pressure drop per unit length, the diameter, and the viscosity. Find the form of the equation.

**∎** The quantities are listed with their dimensions:

| quantity | symbol | dimensions |
|---|---|---|
| Discharge | $Q$ | $L^3 T^{-1}$ |
| Pressure drop per length | $\Delta p/l$ | $ML^{-2}T^{-2}$ |
| Diameter | $D$ | $L$ |
| Viscosity | $\mu$ | $ML^{-1}T^{-1}$ |

Then $F(Q, \Delta p/l, D, \mu) = 0$. Three dimensions are used, and with four quantities there will be one $\Pi$ parameter: $\Pi = Q^{x_1}(\Delta p/l)^{y_1}D^{z_1}\mu$. Substituting in the dimensions gives $\Pi = (L^3 T^{-1})^{x_1}(ML^{-2}T^{-2})^{y_1}L^{z_1}ML^{-1}T^{-1} = M^0 L^0 T^0$. The exponents of each dimension must be the same on both sides of the equation. With $L$ first, $3x_1 - 2y_1 + z_1 - 1 = 0$ and similarly for $M$ and $T$: $y_1 + 1 = 0$, $-x_1 - 2y_1 - 1 = 0$; from which $x_1 = 1$, $y_1 = -1$, $z_1 = -4$, and $\Pi = Q\mu/(D^4 \Delta p/l)$. After solving for $Q$, $Q = C(\Delta p/l)(D^4/\mu)$, from which dimensional analysis yields no information about the numerical value of the dimensionless constant $C$; experiment (or analysis) shows that it is $\pi/128$.

**18.63** The discharge $Q$ of a V-notch weir is some function of the elevation $H$ of upstream liquid surface above the bottom of the notch. In addition, the discharge depends upon gravity and upon the velocity of approach $V_0$ to the weir. Determine the form of discharge equation.

**∎** A functional relation $F(Q, H, g, V_0, \phi) = 0$ is to be grouped into dimensionless parameters. $\phi$ is dimensionless; hence, it is one $\Pi$ parameter. Only two dimensions are used, $L$ and $T$. If $g$ and $H$ are the repeating variables, $\Pi_1 = H^{x_1}g^{y_1}Q = L^{x_1}(LT^{-2})^{y_1}L^3 T^{-1}$, $\Pi_2 = H^{x_2}g^{y_2}V_0 = L^{x_2}(LT^{-2})^{y_2}LT^{-1}$. Then

$$x_1 + y_1 + 3 = 0 \qquad x_2 + y_2 + 1 = 0$$
$$-2y_1 - 1 = 0 \qquad -2y_2 - 1 = 0$$

from which $x_1 = -\frac{5}{2}$, $y_1 = -\frac{1}{2}$, $x_2 = -\frac{1}{2}$, $y_2 = -\frac{1}{2}$, and $\Pi_1 = Q/(\sqrt{g}H^{5/2})$, $\Pi_2 = V_0/\sqrt{gH}$, $\Pi_3 = \phi$, or $f(Q/\sqrt{g}H^{5/2}), V_0/\sqrt{gH}, \phi) = 0$. This can be written $Q/(\sqrt{g}H^{5/2}) = f_1(V_0/\sqrt{gH}, \phi)$ in which both $f$ and $f_1$ are unknown functions. After solving for $Q$, $Q = \sqrt{g}H^{5/2}f_1(V_0/\sqrt{gH}, \phi)$. Either experiment or analysis is required to yield additional information about the function $f_1$. If $H$ and $V_0$ were selected as repeating variables in place of $g$ and $H$, $\Pi_1 = H^{x_1}V_0^{y_1}Q = L^{x_1}(LT^{-1})^{y_1}L^3 T^{-1}$, $\Pi_2 = H^{x_2}V_0^{y_2}g = L^{x_2}(LT^{-1})^{y_2}LT^{-2}$. Then

$$x_1 + y_1 + 3 = 0 \qquad x_2 + y_2 + 1 = 0$$
$$-y_1 - 1 = 0 \qquad -y_2 - 2 = 0$$

from which $x_1 = -2$, $y_1 = -1$, $x_2 = 1$, $y_2 = -2$, and $\Pi_1 = Q/H^2 V_0$, $\Pi_2 = gH/V_0^2$, $\Pi_3 = \phi$ or $f(Q/H^2 V_0, gH/V_0^2, \phi) = 0$. Since any of the $\Pi$ parameters can be inverted or raised to any power without affecting their dimensionless status, $Q = V_0 H^2 f_2(V_0/\sqrt{gH}, \phi)$. The unknown function $f_2$ has the same parameters as $f_1$, but it could not be the same function. The last form is not very useful, in general, because frequently $V_0$ may be neglected with V-notch weirs. This shows that a term of minor importance should not be selected as a repeating variable.

**18.64** The losses $\Delta p/l$ in turbulent flow through a smooth horizontal pipe depend upon velocity $V$, diameter $D$, dynamic viscosity $\mu$, and density $\rho$. Use dimensional analysis to determine the general form of the equation $F(\Delta p/l, V, D, \rho, \mu) = 0$.

▮ If $V$, $D$, and $\rho$ are repeating variables, $\Pi_1 = V^{x_1} D^{y_1} \rho^{z_1} \mu = (LT^{-1})^{x_1} L^{y_1} (ML^{-3})^{z_1} ML^{-1} T^{-1}$

$$x_1 + y_1 - 3z_1 - 1 = 0$$
$$-x_1 \qquad\quad -1 = 0$$
$$z_1 + 1 = 0$$

from which $x_1 = -1$, $y_1 = -1$, $z_1 = -1$, and $\Pi_2 = V^{x_2} D^{y_2} \rho^{z_2} (\Delta p/l) = (LT^{-1})^{x_2} L^{y_2} (ML^{-3})^{z_2} ML^{-2} T^{-2}$

$$x_2 + y_2 - 3z_2 - 2 = 0$$
$$-x_2 \qquad\quad -2 = 0$$
$$z_2 + 1 = 0$$

From which $x_2 = -2$, $y_2 = 1$, and $z_2 = -1$. Then $\Pi_1 = \mu/VD\rho$, $\Pi_2 = (\Delta p/l)/(\rho V^2/D)$, $F[VD\rho/\mu, (\Delta p/l)/(\rho V^2/D)] = 0$, since the $\Pi$ quantities can be inverted if desired. The first parameter. $VD\rho/\mu$, is the *Reynolds number* $N_R$. After solving for $\Delta p/l$ we have $\Delta p/l = f_1(N_R, \rho V^2/D)$. The usual formula is $\Delta p/l = f(N_R)(\rho V^2/2D)$ or, in terms of head loss, $\Delta h/l = f(N_R)(1/D)(V^2/2g)$.

**18.65**  A fluid-flow situation depends on the velocity $V$, the density $\rho$, several linear dimensions, $l$, $l_1$, $l_2$, pressure drop $\Delta p$, gravity $g$, viscosity $\mu$, surface tension $\sigma$, and bulk modulus of elasticity $K$. Apply dimensional analysis to these variables to find a set of $\Pi$ parameters: $F(V, \rho, l, l_1, l_2, \Delta p, g, \mu, \sigma, K) = 0$.

▮ As three dimensions are involved, three repeating variables are selected. For complete situations, $V$, $\rho$, and $l$ are generally helpful. There are seven $\Pi$ parameters: $\Pi_1 = V^{x_1} \rho^{y_1} l^{z_1} \Delta p$, $\Pi_2 = V^{x_2} \rho^{y_2} l^{z_2} g$, $\Pi_3 = V^{x_3} \rho^{y_3} l^{z_3} \mu$, $\Pi_4 = V^{x_4} \rho^{y_4} l^{z_4} \sigma$, $\Pi_5 = V^{x_5} \rho^{y_5} l^{z_5} K$, $\Pi_6 = l/l_1$, $\Pi = l/l_2$. By expanding the $\Pi$ quantities into dimensions, $\Pi_1 = (LT^{-1})^{x_1} (ML^{-3})^{y_1} L^{x_1} ML^{-1} T^{-2}$

$$x_1 - 3y_1 + z_1 - 1 = 0$$
$$-x_1 \qquad\qquad -2 = 0$$
$$y_1 \qquad\quad +1 = 0$$

from which $x_1 = -2$, $y_1 = -1$, $z_1 = 0$. $\Pi_2 = (LT^{-1})^{x_2} (ML^{-3})^{y_2} L^{z_2} LT^{-2}$

$$x_2 - 3y_2 + z_2 + 1 = 0$$
$$-x_2 \qquad\qquad -2 = 0$$
$$y_2 \qquad\qquad = 0$$

from which $x_2 = -2$, $y_2 = 0$, $z_2 = 1$. $\Pi_3 = (LT^{-1})^{x_3} (ML^{-3})^{y_3} L^{z_3} ML^{-1} T^{-1}$

$$x_3 - 3y_3 + z_3 - 1 = 0$$
$$-x_3 \qquad\qquad -1 = 0$$
$$y_3 \qquad\quad +1 = 0$$

from which $x_3 = -1$, $y_3 = -1$, $z_3 = 1$. $\Pi_4 = (LT^{-1})^{x_4} (ML^{-3})^{y_4} L^{z_4} MT^{-2}$

$$x_4 - 3y_4 + z_4 \qquad = 0$$
$$-x_4 \qquad\qquad -2 = 0$$
$$y_4 \qquad\quad +1 = 0$$

from which $x_4 = -2$, $y_4 = -1$, $z_4 = -1$. $\Pi_5 = (LT^{-1})^{x_5} (ML^{-3})^{y_5} L^{z_5} ML^{-1} T^{-2}$

$$x_5 - 3y_5 + z_5 - 1 = 0$$
$$-x_5 \qquad\qquad -2 = 0$$
$$y_5 \qquad\quad +1 = 0$$

from which $x_5 = -2$, $y_5 = -1$, $z_5 = 0$. $\Pi_1 = \Delta p/\rho V^2$, $\Pi_2 = gl/V^2$, $\Pi_3 = \mu/Vl\rho$, $\Pi_4 = \sigma/V^2 l\rho$, $\Pi_5 = K/\rho V^2$, $\Pi_6 = l/l_1$, $\Pi_7 = l/l_2$, and $f(\Delta p/\rho V^2, gl/V^2, \mu/Vl\rho, \sigma/V^2 l\rho, K/\rho V^2, l/l_1, l/l_2) = 0$. It is convenient to invert some of the parameters and to take some square roots. $f_1(\Delta p/\rho V^2, V/\sqrt{gl}, Vl\rho/\mu, V^2 l\rho/\sigma, V/\sqrt{K/\rho}, l/l_1, l_1/l_2) = 0$. The first parameter, usually written $\Delta p/(\rho V^2/2)$, is the *pressure coefficient*; the second parameter is the *Froude* number $N_F$; the third is the *Reynolds* number $N_R$; the fourth is the *Weber* number $N_W$; and the fifth is the *Mach* number $N_M$. Hence, $f_1(\Delta p/\rho V^2, N_F, N_R, N_W, N_M, l/l_1, l/l_2) = 0$. After solving for pressure drop, $\Delta p = \rho V^2 f_2(N_F, N_R, N_W, N_M, l/l_1, l/l_2)$ in which $f_1$ and $f_2$ must be determined from analysis or experiment. By selecting other repeating variables, a different set of $\Pi$ parameters could be obtained.

**18.66** The thrust due to any one of a family of geometrically similar airplane propellers is to be determined experimentally from a wind-tunnel test on a model. Use dimensional analysis to find suitable parameters for plotting test results.

▮ The thrust $F_T$ depends upon speed of rotation $\omega$, speed of advance $V_0$, diameter $D$, air viscosity $\mu$, density $\rho$, and speed of sound $c$. The function $F(F_T, V_0, D, \omega, \mu, \rho, c) = 0$ is to be arranged into four dimensionless parameters, since there are seven quantities and three dimensions. Starting first by selecting $\rho$, $\omega$, and $D$ as repeating variables, $\Pi_1 = \rho^{x_1}\omega^{y_1}D^{z_1}F_T = (ML^{-3})^{x_1}(T^{-1})^{y_1}L^{z_1}MLT^{-2}$, $\Pi_2 = \rho^{x_2}\omega^{y_2}D^{z_2}V_0 = (ML^{-3})^{x_2}(T^{-1})^{y_2}L^{z_2}LT^{-1}$, $\Pi_3 = \rho^{x_3}\omega^{y_3}D^{z_3}\mu = (ML^{-3})^{x_3}(T^{-1})^{y_3}L^{z_3}ML^{-1}T^{-1}$, $\Pi_4 = \rho^{x_4}\omega^{y_4}D^{z_4}c = (ML^{-3})^{x_4}(T^{-1})^{y_4}L^{z_4}LT^{-1}$. Writing the simultaneous equations in $x_1$, $y_1$, $z_1$, etc., as before and solving them gives $\Pi_1 = F_T/\rho\omega^2D^2$, $\Pi_2 = V_0/\omega D$, $\Pi_3 = \mu/\rho\omega D^2$, $\Pi_4 = c/\omega D$. Solving for the thrust parameter leads to $F_T/\rho\omega^2D^4 = f_1(V_0/\omega D, \rho\omega D^2/\mu, c/\omega D)$. Since the parameters can be recombined to obtain other forms, the second term is replaced by the product of the first and second terms, $VD\rho/\mu$, and the third term is replaced by the first term divided by the third term, $V_0/c$; thus $F_T/\rho\omega^2D^4 = f_2(V_0/\omega D, V_0D\rho/\mu, V_0/c)$. Of the dimensionless parameters, the first is probably of the most importance since it relates speed of advance to speed of rotation. The second parameter is a Reynolds number and accounts for viscous effects. The last parameter, speed of advance divided by speed of sound, is a Mach number, which would be important for speeds near or higher than the speed of sound. Reynolds effects are usually small, so that a plot of $F_T/\rho\omega^4$ against $V_0/\omega D$ should be most informative.

**18.67** The variation $\Delta p$ of pressure in static liquids is known to depend upon specific weight $\gamma$ and elevation difference $\Delta z$. By dimensional reasoning determine the form of the hydrostatic law of variation of pressure.

▮ 
$$p = f(\gamma, \Delta z) = (\text{constant})\gamma^a(\Delta z)^b \qquad \{M/LT^2\} = \{M/L^2T^2\}^a\{L\}^b$$

$M$: $\qquad\qquad\qquad\qquad\qquad a \qquad\quad 1$

$L$: $\qquad\qquad\qquad\qquad -2a + b = -1$

$T$: $\qquad\qquad\qquad\qquad -2a \qquad = -2$

$$a = 1 \qquad b = 1 \qquad \Delta p = (\text{constant})\gamma\Delta z$$

**18.68** When viscous and surface-tension effects are neglected, the velocity $V$ of efflux of liquid from a reservoir is thought to depend upon the pressure drop $\Delta p$ of the liquid and its density $\rho$. Determine the form of expression for $V$.

▮ 
$$V = f(\Delta p, \rho) \qquad (\Delta p)^a\rho^b = (\text{constant})V \qquad \{M/LT^2\}^a\{M/L^3\}^b = L/T$$

$M$: $\qquad\qquad\qquad\qquad a + b = 1$

$L$: $\qquad\qquad\qquad\qquad -a - 3b = 1$

$T$: $\qquad\qquad\qquad\qquad -2a \qquad = -1$

$$a = \tfrac{1}{2} \qquad b = -\tfrac{1}{2} \qquad V = (\text{constant})\sqrt{\Delta p/\rho}$$

**18.69** The buoyant force $F_b$ on a body is thought to depend upon its volume submerged $V$ and the gravitational body force acting on the fluid. Determine the form of the buoyant-force equation.

▮ 
$$F_b = f(V, \rho g) \qquad V^a(\rho g)^b = (\text{constant})F_b \qquad \{L^3\}^a\{(M/L^3)(L/T^2)\}^b = \{ML/T^2\}$$

$M$: $\qquad\qquad\qquad\qquad\qquad b = 1$

$L$: $\qquad\qquad\qquad\qquad 3a - 2b = 1$

$T$: $\qquad\qquad\qquad\qquad\quad -2b = -2$

$$a = 1 \qquad b = 1 \qquad F_b = (\text{constant})V\rho g$$

**18.70** In a fluid rotated as a solid about a vertical axis with angular velocity $\omega$, the pressure rise $p$ in a radial direction depends upon speed $\omega$, radius $r$, and fluid density $\rho$. Obtain the form of equation for $p$.

▮ 
$$\Delta p = f(\omega, r, \rho) \qquad \omega^a r^b \rho^c = (\text{constant})\Delta p \qquad \{1/T\}^a\{L\}^b\{M/L^3\}^c = \{M/LT^2\}$$

$M$: $\qquad\qquad\qquad\qquad\qquad\qquad c = 1$

$L$: $\qquad\qquad\qquad\qquad\quad b - 3c = -1$

$T$: $\qquad\qquad\qquad\qquad -a \qquad\qquad = -2$

$$a = 2 \qquad b = 2 \qquad c = 1 \qquad \Delta p = (\text{constant})\omega^2r^2\rho$$

**18.71** The Mach number $N_M$ for flow of a perfect gas in a pipe depends upon the specific-heat ratio $k$ (dimensionless), the pressure $p$, the density $\rho$, and the velocity $V$. Obtain by dimensional reasoning the form of the Mach number expression.

▌ $N_M = f(k, \rho, p, V)$, $f'(N_M, k, \rho, p, V) = 0$; $n = 5$, $j = 3$, $n - j = 5 - 3 = 2$ $\Pi$'s expected.

$$\Pi_1 = M \qquad \Pi_2 = k \qquad p^a \rho^b V = \Pi_3 \qquad \{M^a/L^a\}\{M^b/T^{2a}L^{3b}\}\{L/T\} = 1$$

$M$: $\qquad\qquad\qquad\qquad\qquad\qquad\qquad a + b \quad = 0$

$L$: $\qquad\qquad\qquad\qquad\qquad\qquad -a - 3b + 1 = 0$

$T$: $\qquad\qquad\qquad\qquad\qquad\qquad\quad -2a \quad - 1 = 0$

$$a = -\tfrac{1}{2} \qquad b = \tfrac{1}{2} \qquad N_M = f(V/\sqrt{p/\rho}, k)$$

**18.72** The moment exerted on a submarine by its rudder is to be studied with a $1:25$ scale model in a water tunnel. If the torque measured on the water model is $7\,\text{N} \cdot \text{m}$ for a tunnel velocity of $17\,\text{m/s}$, what are the corresponding torque and speed for the prototype?

▌ $\qquad V_p/V_m = L_m/L_p \qquad V_p/17 = \tfrac{1}{25} \qquad V_p = 0.68\,\text{m/s} \qquad T = (\text{force})(\text{arm}) \propto (\rho V^2 A)(L)$

$T_r = \rho_r V_r^2 A_r L_r = (1)(1/L_r)^2 L_r^2 L_r = L_r$, $\qquad T_p/T_m = L_p/L_m \qquad T_p/7 = \tfrac{25}{1} \qquad T_p = 175\,\text{N} \cdot \text{m}$

**18.73** A rotary machine is to mix paint $[\rho = 789\,\text{kg/m}^3, \mu = 1.20 \times 10^{-3}\,\text{Pa} \cdot \text{s}]$. Tests with a $1:5$ scale model in SAE30 oil $[\rho = 917\,\text{kg/m}^2, \mu = 0.29\,\text{Pa} \cdot \text{s}]$ indicate best mixing at $\omega = 1850\,\text{rpm}$. What should the rotation speed of the prototype be?

▌ $(N_R)_m = (N_R)_p$. Use a characteristic velocity $\omega D$, so that $N_R = (\rho)(\omega D)(D)/\mu = \rho \omega D^2/\mu$. Then

$$(917)(1850)(D_p/5)^2/0.29 = (789)(\omega_p)(D_p)^2/(1.20 \times 10^{-3}), \quad \omega_p = 0.356\,\text{rpm}.$$

**18.74** A one-fourteenth-scale model of a parachute has a drag of $400\,\text{lb}$ when tested at $23\,\text{ft/s}$ in a water tunnel. If Reynolds number effects are negligible, estimate the terminal fall velocity at $5000\,\text{ft}$ standard altitude of a parachutist using the prototype if chute and chutist together weight $175\,\text{lb}$. Neglect the drag coefficient of the woman.

▌ $(C_D)_m = (C_D)_p$ (if Reynolds number effects are negligible), $C_D = F/\rho V^2 D^2$.

$$\rho_p = 0.002048\,\text{slug/ft}^3 \qquad \text{(from Table A-7)} \qquad D_m = D_p/14$$

$$400/[(1.94)(23)^2(D_p/14)^2] = 175/[(0.002048)(V_p)^2(D_p)^2] \qquad V_p = 33.4\,\text{ft/s}$$

**18.75** A one-tenth-scale model of a weir has a measured flow rate of $2.5\,\text{cfs}$ when the upstream water height is $H = 6.9\,\text{in}$. Use the results of Prob. 18.56 to predict the prototype flow rate when $H = 3.1\,\text{ft}$.

▌ $(Q/b)/(g^{1/2}H^{3/2}) = \text{constant}$ (from Prob. 18.56) $\qquad [(Q/b)/(g^{1/2}H^{3/2})]_m = [(Q/b)/(g^{1/2}H^{3/2})]_p$

$b_m = b_p/10 \qquad [2.5/(b_p/10)]/[(32.2)^{1/2}(6.9/12)^{3/2}] = (Q_p/b_p)/[(32.2)^{1/2}(3.1)^{3/2}] \qquad Q_p = 313\,\text{ft}^3/\text{s}$

**18.76** For model and prototype, show that, when gravity and inertia are the only influences, the ratio of flows $Q$ is equal to the ratio of the length dimension to the five-halves power.

▌ $Q_m/Q_p = (L_m^3/T_m)/(L_p^3/T_p) = L_r^3/T_r$. The time ratio must be established for the conditions influencing the flow. Expressions can be written for the gravitation and inertia forces, as follows:

Gravity: $\qquad\qquad\qquad F_m/F_p = W_m/W_p = (\gamma_m/\gamma_p)(L_m^3/L_p^3) = \gamma_r L_r^3$

Inertia $\qquad\qquad\qquad F_m/F_p = M_m a_m/M_p a_p = (\rho_m/\rho_p)(L_m^3/L_p^3)(L_r/T_r^2) = \rho_r L_r^3(L_r/T_r^2)$

Equating the force ratios, $\gamma_r L_r^3 = \rho_r L_r^3(L_r/T_r^2)$, which, when solved for the time ratio, yields

$$T_r^2 = L_r(\rho_r/\gamma_r) = L_r/g_r \qquad\qquad\qquad (1)$$

Recognizing that the value of $g_r$ is unity, substitution in the flow ratio expression gives

$$Q_r = Q_m/Q_p = L_r^3/L_r^{1/2} = L_r^{5/2} \qquad\qquad\qquad (2)$$

**18.77** A spillway model is to be built to a scale of $1:25$ across a flume which is $2\,\text{ft}$ wide. The prototype is $37.5\,\text{ft}$ high and the maximum head expected is $5.0\,\text{ft}$. What height of model and what head on the model should be used? If the flow over the model at 0.20 ft head is 0.70 cfs, what flow per foot of prototype may be expected;

$$L_m/L_p = \tfrac{1}{25} \qquad \text{Height of model} = (37.5)(\tfrac{1}{25}) = 1.50 \text{ ft}$$
$$\text{Head on model} = (5.0)(\tfrac{1}{25}) = 0.20 \text{ ft} \qquad Q_r = L_r^{5/2} \qquad \text{(from Prob. 18.76)}$$
$$Q_m/Q_p = (L_m/L_p)^{5/2} \qquad 0.070/Q_p = (\tfrac{1}{25})^{5/2} \qquad Q_p = 2188 \text{ ft}^3/\text{s}$$
$$b_m/b_p = \tfrac{1}{25} \qquad 2/b_p = \tfrac{1}{25} \qquad b_p = 50.0 \text{ ft} \qquad q_p = Q_p/b_p = 2188/50.0 = 43.8 \text{ ft}^3/\text{s}$$

**18.78** For the model described in Prob. 18.77, if the model shows a measured hydraulic jump of 1.0 in, how high is the jump in the prototype? If the energy dissipated in the model at the hydraulic jump is 0.15 hp, what would be the energy dissipation in the prototype?

▌ Since $L_m/L_p = \tfrac{1}{25}$, height of jump $= (1.0)(25) = 25.0$ in, or 2.08 ft, $P_r = F_r L_r / T_r = (\gamma_r L_r^3)(L_r)/\sqrt{L_r/g_r}$. Since $g_r = \gamma_r = 1$, $P_r = L_r^{7/2}$, $0.15/P_p = (\tfrac{1}{25})^{7/2}$, $P_r = 11\,720$ hp.

**18.79** A model of a reservoir is drained in 4 min by opening the sluice gate. The model scale is $1:225$. How long should it take to empty the prototype?

▌ Since gravity is the dominant force, $Q_r = L_r^{5/2}$ (from Prob. 18-76), $Q_r = Q_m/Q_p = (L_m^3/t_m)/(L_p^3/t_p)$, $L_r^{5/2} = (L_r^3)(t_p/t_m)$, $t_p = t_m/L_r^{1/2} = (4)/(\tfrac{1}{225})^{1/2} = 60.0$ min.

**18.80** A rectangular pier in a river is 4 ft wide by 12 ft long and the average depth of water is 9.0 ft. A model is built to a scale of $1:16$. The velocity of flow of 2.50 fps is maintained in the model and the force acting on the model is 0.90 lb. What are the values of velocity in and force on the prototype?

▌ Since the gravity forces predominate, $V_m/V_p = \sqrt{L_r}$, $2.50/V_p = \sqrt{\tfrac{1}{16}}$, $V_p = 10.0$ ft/s. $F_m/F_p = \gamma_r L_r^3$, $0.90/F_p = (1.0)(\tfrac{1}{16})^3$, $F_p = 3686$ lb.

**18.81** If a standing wave in the model of Prob. 18.80 is 0.16 ft high, what height of wave should be expected at the nose of the pier? What is the coefficient of drag resistance?

▌
$$V_m/V_p = \sqrt{L_m}/\sqrt{L_p} = \sqrt{h_m}/\sqrt{h_p} \qquad 2.50/10.0 = \sqrt{0.16}/\sqrt{h_p} \qquad h_p = 2.56 \text{ ft}$$
$$F_D = C_D \rho A V^2/2 \qquad 3686 = (C_D)(1.94)[(4)(9)](10.0)^2/2 \qquad C_D = 1.06$$

**18.82** The measured resistance in fresh water of an 8-ft ship model moving at 6.50 fps was 9.60 lb. (**a**) What would be the velocity of the 128-ft prototype? (**b**) What force would be required to drive the prototype at this speed in salt water?

▌ (**a**) Since gravity forces predominate, $V_m/V_p = \sqrt{L_r}$, $6.50/V_p = \sqrt{\tfrac{8}{128}}$, $V_p = 26.0$ ft/s.

(**b**) $\qquad F_m/F_p = \gamma_r L_r^3 \qquad 9.60/F_p = (62.4/64.2)(\tfrac{1}{16})^3 \qquad F_p = 40\,460$ lb

**19.1** The surface of a frictionless fluid column 2.18 m long is dropping at 2 m/s when $z = 0.5$ m (measured upwards from the reference level). Find (a) the maximum value of $z$, (b) the maximum speed.

■ (a) $z = Z \cos \omega t$, $dz/dt = -\omega Z \sin \omega t$, where $\omega = \sqrt{2g/L} = \sqrt{2(9.807)/2.18} = 3.00$ rad/s. Then, if $t_1$ is the time when $z = 0.5$ and $dz/dt = 2$, $0.5 = Z \cos 3.00 t_1$, $-2 = -3.00 Z \sin 3.00 t_1$. From these,

$$\sin^2 3.00 t_1 + \cos^2 3.00 t_1 = 1 = \left(\frac{2}{3Z}\right)^2 + \left(\frac{1}{2Z}\right)^2$$

whence $Z^2 = 25/36$, or $Z = 5/6 = 0.833$ m.
(b) $\omega Z = (3.00)(0.833) = 2.50$ m/s.

**19.2** A 1.0-in-diameter U-tube contains oil, $v = 1 \times 10^{-4}$ ft²/s, with a total column length of 120 in. Applying air pressure to one of the tubes makes the gage difference 16 in. By quickly releasing the air pressure, the oil column is free to oscillate. Find the maximum velocity, the maximum Reynolds number, and the equation for position of one meniscus $z$, in terms of time.

■ The assumption is made that the flow is laminar, and the Reynolds number will be computed on this basis. The constants $m$ and $n$ are $m = 16v/D^2 = (16 \times 10^{-4})/(\frac{1}{12})^2 = 0.2302$, $n = \sqrt{(16v/D^2)^2 - (2g/L)} = \sqrt{0.2302^2 - [2(32.2)/10]} = \sqrt{-6.387} = i2.527$, or $n' = 2.527$. The liquid will oscillate above and below $z = 0$. The oscillation starts from the maximum position, that is, $Z = 0.667$ ft. The velocity (fictitious) when $z = 0$ at time $t_0$ before the maximum is determined to be $V_0 = Z\sqrt{2g/L} \exp[(m/n') \tan^{-1}(n'/m)] = 0.667\sqrt{64.4/10} \exp[(0.2302/2.527) \tan^{-1}(2.527/0.2302)] = 1.935$ ft/s and $\tan n' t_0 = n'/m$, $t_0 = (1/2.527) \tan^{-1}(2.527/0.2302) = 0.586$ s; $z = (V_0/n') e^{-mt} \sin(n't) = 0.766 \exp[-0.2302(t + 0.586)] \sin 2.527(t + 0.586)$ in which $z = Z$ at $t = 0$. The maximum velocity (actual) occurs for $t > 0$. Differentiating with respect to $t$ to obtain the expression for velocity, $V = dz/dt = -0.1763 \exp[-0.2302(t + 0.586)] \sin 2.527(t + 0.586) + 1.935 \exp[-0.2302(t + 0.586)] \cos 2.527(t + 0.586)$. Differentiating again with respect to $t$ and equating to zero to obtain maximum $V$ produces $\tan 2.527(t + 0.586) = -0.1837$. The solution in the second quadrant should produce the desired maximum, $t = 0.584$ s. Substituting this time into the expression for $V$ produces $V = -1.48$ ft/s. The corresponding Reynolds number is $VD/v = 1.48(\frac{1}{12} \times 10^4) = 1234$; hence the assumption of laminar resistance is justified.

**19.3** A U-tube consisting of 500-mm-diameter pipe with $f = 0.03$ has a maximum oscillation (Fig. 19-1) of $z_m = 6$ m. Find the minimum position of the surface and the following maximum.

■ $[1 + (fz_m/D)]e^{-fz_m/D} = [1 + (fz_{m+1}/D)]e^{-fz_{m+1}/D}$, $\{1 + [0.03(6)/0.5]\}e^{0.03(6)0.5} = (1 + 0.06z_{m+1})e^{-0.06z_{m+1}}$ or $(1 + 0.06z_{m+1})e^{-0.06z_{m+1}} = 0.9488$, which is satisfied by $z_{m+1} = -4.84$ m. Using $z_m = 4.84$ m in the previous equation, $(1 + 0.06z_{m+1})e^{-0.06z_{m+1}} = [1 + 0.06(4.84)]e^{-0.06(4.84)} = 0.9651$, which is satisfied by $z_{m+1} = -4.05$ m. Hence, the minimum water surface is $z = -4.84$ m and the next maximum is $z = 4.05$ m.

Fig. 19-1

**19.4** The pipe of Prob. 19.3 is 1000 m long. By numerical integration of the equation

$$\frac{d^2z}{dt^2} + \frac{f}{2D}\frac{dz}{dt}\left|\frac{dz}{dt}\right| + \frac{2g}{L}z = 0$$

using the third-order Runge–Kutta method [see Prob. 19.22], determine by a computer program the time to the first minimum $z$ and the next maximum $z$ and check the results of Prob. 19.3.

```
10 REM              EXAMPLE          TURBULENT OSCILLATION OF A U-TUBE.
20 DEFINT I:   DEF FNZ(ZZ,YY)=C1*YY*ABS(YY)+C2*ZZ
30 READ D,F,L,ZM,G,H,IPR:    DATA .5,.03,1000.,6.,9.806,.5,4
40 LPRINT:    LPRINT"D,F,L,ZM,G,H, IPR=";D;F;L;ZM;G;H; IPR
50 C1=-.5*F*H/D: C2=-2!*G*H/L: Y=0!: Z=6!                    ' Y IS DZ/DT
60 OZ=6!:    I=0:      EPS=.001
70 LPRINT:       LPRINT"  T,sec        Z,m        V,m/s":     LPRINT
80 I=I+1:   OOZ=OZ:    OZ=Z      ' OOZ and OZ used to determine max and min Z
90 U11=FNZ(Z,Y):                    U12=H*Y
100  U21=FNZ(Z+.3333*U12,Y+.3333*U11):    U22=H*(Y+.3333*U11)
110 U31=FNZ(Z+.6667*U22,Y+.6667*U21):   U32=H*(Y+.6667*U21)
120 Y=Y+.25*U11+.75*U31:    Z=Z+.25*U12+.75*U32:    T=T+H
130 IF (OZ<OOZ-EPS) AND (OZ<Z-EPS) THEN LPRINT"ZMIN=";OZ;"TIME=";T-H
140 IF I MOD IPR=0 THEN LPRINT USING"###.###  ";T;Z;Y
150 IF (OZ>OOZ+EPS) AND (OZ>Z+EPS) THEN LPRINT"ZM=";OZ;"TIME=";T-H: STOP
    ELSE GOTO 80

D,F,L,ZM,G,H,IPR= .5  .03  1000  6  9.806001  .5  4

   T,sec     Z,m     V,m/s

   2.000    5.767    -0.231
   4.000    5.091    -0.438
   6.000    4.042    -0.602
   8.000    2.721    -0.709
  10.000    1.247    -0.755
  12.000   -0.257    -0.740
  14.000   -1.677    -0.671
  16.000   -2.912    -0.558
  18.000   -3.884    -0.409
  20.000   -4.533    -0.236
  22.000   -4.821    -0.050
ZMIN=-4.834092 TIME= 22.5
  24.000   -4.732     0.138
  26.000   -4.278     0.313
  28.000   -3.502     0.457
  30.000   -2.478     0.559
  32.000   -1.298     0.613
  34.000   -0.061     0.616
  36.000    1.136     0.574
  38.000    2.206     0.490
  40.000    3.075     0.374
  42.000    3.688     0.235
  44.000    4.006     0.082
ZM= 4.048564 TIME= 45
```

**19.5** In Fig. 19-2 a valve is opened suddenly in the pipeline when $z_1 = 40$ ft. $L = 2000$ ft, $A_1 = 200$ ft$^2$, $A_2 = 300$ ft$^2$, $D = 3.0$ ft, $f = 0.024$, and minor losses are $3.50V^2/2g$. Determine the subsequent maximum negative and positive surges in the reservoir $A_1$.

▌ The equivalent length of minor losses is $KD/f = 3.5(3)/0.024 = 438$ ft. Then $L_e = 2000 + 438 = 2438$ and $z_m = z_1 A_1/A = 40(200/(2.25\pi)) = 1132$ ft. The corresponding $\phi$ is $\phi = f(L_e/L)(z_m/D) = 0.024(\frac{2438}{2000})(\frac{1132}{3}) = 11.04$ and $F(\phi) = (1 + \phi)e^{-\phi} = (1 + 11.04)e^{-11.04} = 0.000193$, which is satisfied by $\phi \approx -1.0$. Then $F(\phi) = (1 + 1)e^{-1} = 0.736 = (1 + \phi)e^{-\phi}$ which is satisfied by $\phi = -0.593$. The values of $z_m$ are, for $\phi = -1$. $z_m = \phi LD/fL_e = -1(2000)(3)/[0.024(2438)] = -102.6$ and, for $\phi = 0.593$, $z_m = 0.593(2000)(3)/[0.024(2438)] = 60.9$. The corresponding values of $z_1$ are $z_1 = z_m(A/A_1) = -102.6(2.25\pi/200) = -3.63$ ft and $z_1 = 60.9(2.25\pi/200) = 2.15$ ft.

**Fig. 19-2**

**19.6** In Fig. 19-3 the minor losses are $16V^2/2g$, $f = 0.030$, $L = 3000$ m, $D = 2.4$ m, and $H = 20$ m. Determine the time, after the sudden opening of a valve, for velocity to attain nine-tenths the final velocity.

$$H = f(L_e/D)(V_0^2/2g) \qquad L_e = 3000 + [16(2.4)/0.03] = 4280 \text{ m}$$
$$V_0 = \sqrt{2gHD/fL_e} = \sqrt{19.612(20)(2.4)/[0.030(4280)]} = 2.708 \text{ m/s}$$

Substituting $V = 0.9V_0$ into $t = (LV_0/2gH) \ln[(V_0 + V)/(V_0 - V)]$, $t = [3000(2.708)/19.612(20)] \ln(1.90/0.10) = 60.98$ s.

**Fig. 19-3**

**19.7** In Fig. 19-4 the water surface changes according to the equation $H_{P_A} = H_0 + \Delta H \sin \omega t$ while the right end of the pipe contains a small orifice. The frequency of the waves is set at the natural period of the pipe, $4L/a$, which yields an $\omega$ of $2\pi/(4L/a)$. Determine the resulting motion of fluid in the pipe and the head fluctuations.

One reach is used. The upstream section is solved by $H_{P_i} = C_M + B_M Q_{P_i}$ together with the head, which is a known function of time. The downstream boundary condition solves the orifice equation together with $Q_{P_B}^2 = C_1 H_{P_A}$, $H_{P_B} = C_P - B_P Q_{P_B}$. The program and results are shown below.

```
10 REM .                      TRANSIENT IN SINGLE-REACH PIPE.
20 DEFINT I:     READ F,L,D,A,DH,HO,IPR,PI,G,TM,CD,DORIF
30 DATA .018,600.,.5,1200.,3.,40.,1,3.1416,9.806,12.,.6,.03
40 AO=.25#PI#DORIF^2:    LPRINT"CD,DORIF,AO=";CD;DORIF;AO:   C1=(CD#AO)^2#2!#G
50 LPRINT"F,L,D,A,DH=";F;L;D;A;DH:     LPRINT"HO,IPR,PI,G,TM=";HO;IPR;PI;G;TM
60 PER=4!#L/A:     OM=2!#PI/PER:     AR=.25#PI#D^2
70 B=A/(G#AR):     R=.5#F#L/(G#D#AR^2):     QO=SQR(HO/(R+1!/C1))
80 T=0!:     DT=L/A:     I=0:     HA=HO:     HB=HO-R#QO^2:     QA=QO:     QB=QO
90 LPRINT"     T,s      HA,m      QA,l/s     HB,m      QB,l/s"
100 LPRINT USING"####.####  ";T;HA;1000!#QA;HB;1000!#QB
110 T=T+DT:     I=I+1:     IF T>TM THEN STOP
120 HPA=HO+DH#SIN(OM#T):     CM=HB-B#QB:     BM=B+R#ABS(QB):     QPA=(HPA-CM)/BM
130 CP=HA+B#QA:     BP=B+R#ABS(QA):     QPB=-.5#BP#C1+SQR((.5#BP#C1)^2+CP#C1)
140 HPB=CP-BP#QPB:     HA=HPA:     HB=HPB:     QA=QPA:     QB=QPB
150 IF I MOD IPR=0 THEN 100 ELSE GOTO 110
CD,DORIF,AO= .6  .03  7.0686E-04
F,L,D,A,DH= .018  600  .5  1200  3
HO,IPR,PI,G,TM= 40  1  3.1416  9.806001  12
```

| T,s | HA,m | QA,l/s | HB,m | QB,l/s |
|---|---|---|---|---|
| 0.0000 | 40.0000 | 11.8783 | 39.9960 | 11.8783 |
| 0.5000 | 43.0000 | 16.6892 | 39.9960 | 11.8783 |
| 1.0000 | 40.0000 | 11.8782 | 45.4994 | 12.6692 |
| 1.5000 | 37.0000 | -0.9676 | 39.9959 | 11.8783 |
| 2.0000 | 40.0000 | 11.8784 | 29.9865 | 10.2851 |
| 2.5000 | 43.0000 | 31.1506 | 39.9961 | 11.8783 |
| 3.0000 | 39.9999 | 11.8780 | 53.8149 | 13.7783 |
| 3.5000 | 37.0000 | -13.1930 | 39.9958 | 11.8783 |
| 4.0000 | 40.0001 | 11.8787 | 23.1428 | 9.0355 |
| 4.5000 | 43.0000 | 40.8796 | 39.9963 | 11.8783 |
| 5.0000 | 39.9999 | 11.8776 | 59.4364 | 14.4801 |
| 5.5000 | 37.0000 | -21.5050 | 39.9955 | 11.8782 |
| 6.0000 | 40.0001 | 11.8792 | 18.5504 | 8.0895 |
| 6.5000 | 43.0000 | 47.3015 | 39.9966 | 11.8784 |
| 7.0000 | 39.9998 | 11.8771 | 63.1574 | 14.9265 |
| 7.5000 | 37.0000 | -27.0247 | 39.9952 | 11.8782 |
| 8.0000 | 40.0002 | 11.8798 | 15.5371 | 7.4034 |
| 8.5000 | 43.0000 | 51.4503 | 39.9970 | 11.8784 |
| 9.0000 | 39.9998 | 11.8765 | 65.5652 | 15.2084 |
| 9.5000 | 37.0000 | -30.6034 | 39.9948 | 11.8781 |
| 10.0000 | 40.0002 | 11.8804 | 13.6031 | 6.9273 |
| 10.5000 | 43.0000 | 54.0776 | 39.9974 | 11.8785 |
| 11.0000 | 39.9998 | 11.8759 | 67.0916 | 15.3844 |
| 11.5000 | 37.0000 | -32.8746 | 39.9944 | 11.8781 |
| 12.0000 | 40.0003 | 11.8810 | 12.3853 | 6.6099 |

$DH = 3$ m

$H_0 = 40$ m

$f = 0.018$, $L = 600$ m, $D = 500$ mm

$A$

$a = 1200$ m/s

$B$

$D_0 = 30$ mm
$C_d = 0.6$

**Fig. 19-4**

**19.8** Develop the necessary boundary-condition equations for the pump in Fig. 19-5. The pump is to be started with a linear speed rise to $N_R$ in $t_0$ s. A check valve exists in the discharge pipe. The initial no-flow steady-state head on the downstream side of the check valve is $H_C$. For a steady flow of $Q_0$ there is a loss of $\Delta H_0$ across the open check valve. Assume that the check valve opens when the pump has developed enough head to exceed $H_C$.

▮ The equation for the hydraulic grade line downstream from the pump and check valve (after the check valve is open) is $H_P = C_1N^2 + C_2NQ_P + C_3Q_P^2 - Q_P^2(\Delta H_0/Q_0^2)$, $Q_P = \{(B_M - C_2N)/[2(C_3 - \Delta H_0/Q_0^2)]\}\{1 - [1 + 4(C_3 - \Delta H_0/Q_0^2)^2(C_M - C_1N^2)/(B_M - C_2N)^2]^{1/2}\}$. The equations for the boundary condition are $B_M = (a/gA) + R|Q_2|$, $C_M = H_2 - BQ_2$,

$$N = \begin{cases} N_R(t/t_0) & t \le t_0 \\ N_R & t > t_0 \end{cases}$$

If $C_1N^2 > H_C$, $Q_P$ is defined by the above equation for the quadratic, and $H_P = C_M + B_MQ_P$. If $C_1N^2 < H_C$, $Q_P = 0$ and $H_P = H_C$.

$\Delta H_0$

Final hydraulic gradeline

Initial gradient

$H_c$

$Q$

Reservoir

Check valve

$C^-$

$t$

$x$

**Fig. 19-5**

**19.9** The system of Fig. 19-6 initially has a valve opening $C_DA = 0.06$ m². At intervals of 5 s, $C_DA$ takes on the values 0.03, 0.01, 0.003, 0.001, 0.0005, 0.0002, 0.0, and remains closed. Using a computer program, calculate the transients of the system for 40 s after the valve starts to close.

▮ The data for this problem and some of the output are given below. The number of reaches was selected as $N = 4$. Hydraulic-grade-line elevations are given in meters and discharges in cubic meters per second.

```
10 REM B:WH   WATERHAMMER PROGRAM IN IBMPC BASICA FOR UPSTREAM RESERVOIR,
20 ' SINGLE PIPE, AND DOWNSTREAM VALVE. DARCY-WEISBACH FRICTION. F GIVEN.
30 ' A=WAVE SPEED, CVA=PRODUCT OF DISCHARGE COEF. AND AREA OF VALVE
40 ' OPENING, WITH VALUES GIVEN FOR INTERVALS DCV.
50 LPRINT "     BASIC WATERHAMMER PROGRAM"
60 LPRINT "     DATE=";DATE$;"   TIME=";TIME$
65 CLEAR:   DIM CVA(11):   DEFINT I,J,K
70 READ F,L,A,N,D,HRES,G,JPR,DCV,TMAX
80 DATA .022,4800.,1200.,4,2.,100.,9.806,1,5.,40.
90 LPRINT:   LPRINT"F,L,A,N,D=";F;L;A;N;D
100 LPRINT"HRES,G,JPR,DCV,TMAX=";HRES;G;JPR;DCV;TMAX
105 LPRINT"CVA=";
110 FOR I=1 TO 11:   READ CVA(I):   NEXT I:   LPRINT
115 FOR I=1 TO 11:   LPRINT CVA(I);:   NEXT I:   LPRINT
120 DATA .06,.03,.01,.003,.001,.0005,.0002,.0,.0,.0,.0
```

```
130 AR=.7854*D^2:   B=A/(G*AR):    NS=N+1:   DT=L/(A*N):   CV=CVA(1)
140 HP(1)=HRES:   J=0:  T=0!:   Q0=SQR(HRES/(F*L/(2!*G*D*AR^2)+1'/(2!*G*CV^2)))
150 H0=(Q0/CV)^2/(2!*G):   R=(HRES-H0)/(Q0^2*N)
160 FOR I=1 TO NS:  Q(I)=Q0:   H(I)=HRES-(I-1)*R*Q0^2:   NEXT I
170 LPRINT"     PIEZOMETER HEADS AND FLOWS ALONG THE PIPE"
180 LPRINT"    TIME    CV   X/L=    .0      .25      .5      .75      1."
190 LPRINT USING"###.###";T;:   LPRINT USING" #.#####";CV;:   LPRINT" H=";
200 LPRINT USING"  ###.##";H(1);H(2);H(3);H(4);H(5)
210 LPRINT SPC(17);"Q=";:   LPRINT USING"  ###.##";Q(1);Q(2);Q(3);Q(4);Q(5)
220 T=T+DT:   J=J+1:   K=FIX(T/DCV)+1:   IF T>TMAX THEN STOP
230 CV=CVA(K)+(T-(K-1)*DCV)*(CVA(K+1)-CVA(K))/DCV
240 REM DOWNSTREAM BOUNDARY CONDITION
250 CP=H(N)+B*Q(N):    BP=B+R*ABS(Q(N))
260 QP(NS)=-G*BP*CV^2+SQR((G*BP*CV^2)^2+2!*G*CV^2*CP):    HP(NS)=CP-BP*QP(NS)
270 REM UPSTREAM BOUNDARY CONDITION
280 QP(1)=(HP(1)-H(2)+B*Q(2))/(B+R*ABS(Q(2)))
290 REM INTERIOR SECTIONS
300 FOR I=2 TO N:   CP=H(I-1)+B*Q(I-1):   BP=B+R*ABS(Q(I-1))
310 CM=H(I+1)-B*Q(I+1):   BM=B+R*ABS(Q(I+1))
320 QP(I)=(CP-CM)/(BP+BM):   HP(I)=CP-BP*QP(I):   NEXT I
330 FOR I=1 TO NS:   H(I)=HP(I):   Q(I)=QP(I):   NEXT I
340 IF J MOD JPR=0 THEN 190 ELSE 220
```

```
F,L,A,N,D= .022  4800  1200  4  2
HRES,G,JPR,DCV,TMAX= 100  9.806001  1  5  36
CVA=
 .06  .03  .01  .003  .001  .0005  .0002  0  0  0  0
     PIEZOMETER HEADS AND FLOWS ALONG THE PIPE
   TIME    CV   X/L=    .0      .25      .5      .75      1.
  0.000 0.06000  H=  100.00    99.53    99.06    98.58    98.11
                 Q=    2.63     2.63     2.63     2.63     2.63
  1.000 0.05400  H=  100.00    99.53    99.06    98.58   105.14
                 Q=    2.63     2.63     2.63     2.63     2.45
  2.000 0.04800  H=  100.00    99.53    99.06   105.59   112.77
                 Q=    2.63     2.63     2.45     2.45     2.26
  3.000 0.04200  H=  100.00    99.53   106.03   113.18   121.07
                 Q=    2.63     2.63     2.45     2.26     2.05
  4.000 0.03600  H=  100.00   106.47   113.59   121.45   130.07
                 Q=    2.63     2.45     2.26     2.05     1.82
  5.000 0.03000  H=  100.00   114.00   121.83   130.41   139.85
                 Q=    2.28     2.26     2.05     1.82     1.57
  6.000 0.02600  H=  100.00   115.33   130.76   140.17   146.84
                 Q=    1.90     1.88     1.82     1.57     1.40
  7.000 0.02200  H=  100.00   116.76   133.63   147.14   154.25
                 Q=    1.48     1.46     1.40     1.40     1.21
  8.000 0.01800  H=  100.00   118.29   133.13   147.70   162.04
                 Q=    1.03     1.01     1.04     1.04     1.01
  9.000 0.01400  H=  100.00   116.37   132.36   148.02   157.80
                 Q=    0.54     0.61     0.64     0.65     0.78
 10.000 0.01000  H=  100.00   114.08   131.27   142.47   152.18
                 Q=    0.19     0.18     0.22     0.38     0.55
 11.000 0.00860  H=  100.00   114.90   124.19   135.44   139.90
                 Q=   -0.19    -0.20    -0.08     0.12     0.45
 12.000 0.00720  H=  100.00   110.11   119.07   121.62   126.02
                 Q=   -0.58    -0.45    -0.30    -0.02     0.36
 13.000 0.00580  H=  100.00   104.18   107.56   109.66   110.38
                 Q=   -0.70    -0.69    -0.38    -0.06     0.27
 14.000 0.00440  H=  100.00    97.46    94.78    96.32    99.66
                 Q=   -0.79    -0.64    -0.44    -0.09     0.19
 15.000 0.00300  H=  100.00    90.62    86.23    84.79    87.91
                 Q=   -0.57    -0.55    -0.35    -0.19     0.12
 16.000 0.00260  H=  100.00    88.78    80.63    77.83    73.65
                 Q=   -0.31    -0.28    -0.29    -0.13     0.10
 17.000 0.00220  H=  100.00    90.01    80.38    69.49    69.43
                 Q=    0.00    -0.05    -0.07    -0.01     0.08
 18.000 0.00180  H=  100.00    91.60    78.88    71.98    66.65
                 Q=    0.20     0.22     0.23     0.15     0.07
 30.000 0.00020  H=  100.00    96.69    93.19    93.90    94.64
                 Q=   -0.87    -0.72    -0.53    -0.21     0.01
 31.000 0.00016  H=  100.00    90.08    85.27    83.34    85.51
                 Q=   -0.64    -0.61    -0.43    -0.28     0.01
 32.000 0.00012  H=  100.00    88.59    80.24    76.88    72.20
                 Q=   -0.36    -0.34    -0.36    -0.21     0.00
 33.000 0.00008  H=  100.00    90.16    80.20    69.10    68.40
                 Q=   -0.05    -0.11    -0.13    -0.08     0.00
 34.000 0.00004  H=  100.00    91.61    79.03    71.72    66.12
                 Q=    0.15     0.16     0.18     0.09     0.00
 35.000 0.00000  H=  100.00    88.87    83.13    76.04    75.15
                 Q=    0.38     0.43     0.38     0.26     0.00
 36.000 0.00000  H=  100.00    91.52    85.88    86.56    86.02
                 Q=    0.72     0.60     0.51     0.29     0.00
```

**Fig. 19-6**

**19.10** Determine the period of oscillation of a U-tube containing 333 mL of water. The cross-sectional area is 270 mm². Neglect friction.

∎
$$L = V/A = (333 \times 10^{-6})/(270 \times 10^{-6}) = 1.235 \text{ m}$$
$$T = 2\pi\sqrt{L/2g} = (2)(\pi)\sqrt{1.235/[(2)(9.807)]} = 1.58 \text{ s}$$

**19.11** A fluid-filled U-tube is oscillating with maximum displacement from an equilibrium position of 0.1 m. The total column length is 0.9 m. Determine the maximum fluid velocity and the period of oscillation. Neglect friction.

∎
$$z = Z \cos \sqrt{2g/L}\, t = (0.1) \cos \sqrt{2(9.807)/0.9}\, t = 0.100 \cos 4.668t \qquad dz/dt = -0.4668 \sin 4.668t$$

Maximum velocity = 0.4668 m/s; $T = 2\pi/4.668 = 1.35$ s.

**19.12** A U-tube contains liquid oscillating with a velocity of 1.8 m/s at the instant the menisci are at the same elevation. Find the time to the instant the menisci are next at the same elevation and determine the velocity then. Take $v = 1 \times 10^{-5}$ m²/s, $D = 8$ mm, and $L = 800$ mm.

∎
$$N_R = DV/v = (0.008)(1.8)/(1 \times 10^{-5}) = 1440 \qquad \text{(laminar flow)}$$
$$m = 16v/D^2 = (16)(1 \times 10^{-5})/(0.008)^2 = 2.500$$
$$n = \sqrt{(16v/D^2)^2 - 2g/L} = \sqrt{2.500^2 - (2)(9.807)/(0.800)} = 4.274i$$
$$n' = 4.274 \qquad z = (V_0/n')(e^{-mt}) \sin(n't)$$

At next $z = 0$, $4.274t = \pi$. $t = 0.735$ s, $V = dz/dt = V_0 e^{-mt}[\cos(n't) - (m/n') \sin(n't)]$. When $n't = \pi$, $V = (1.8)[e^{-(2.500)(0.735)}][\cos \pi - (m/n') \sin \pi] = 0.287$ m/s.

**19.13** A 9-ft-diameter horizontal tunnel has 9-ft-diameter vertical shafts spaced 1.1 mi apart. When valves are closed isolating this reach of tunnel, the water surges to a depth of 48 ft in one shaft when it is 18 ft in the other shaft. For $f = 0.020$, find the height of the next two surges.

∎
$$z_m = (48 - 18)/2 = 15.00 \text{ ft} \qquad [1 + (fz_m/D)]e^{-fz_m/D} = [1 + (fz_{m+1}/D)]e^{-fz_{m+1}/D}$$
$$fz_m/D = (0.020)(15.00)/9 = 0.0333$$
$$(1 + 0.0333)e^{-0.0333} = [1 + (0.020z_{m+1}/9)]e^{-0.020z_{m+1}/9} \qquad z_{m+1} = 14.99 \text{ ft}$$

Repeat, using $z_m = 14.99$ ft.

$$fz_m/D = (0.020)(14.99)/9 = 0.0333$$
$$(1 + 0.0333)e^{-0.0333} = [1 + (0.020z_{m+1}/9)]e^{-0.020z_{m+1}/9} \qquad z_{m+1} = 14.98 \text{ ft}$$

Hence, $z_1 = 14.99$ ft and $z_2 = 14.98$ ft.

**19.14** Two standpipes 5 m in diameter are connected by 965 m of 2.0-m-diameter pipe; $f = 0.020$ and minor losses are 4.2 velocity heads. One reservoir level is 9.6 m above the other one when a valve is rapidly opened in the pipeline. Find the maximum fluctuation in water level in the standpipe.

∎
$$L_e = L + KD/f = 965 + (4.2)(2.0)/0.020 = 1385 \text{ m} \qquad z_m = z_1(A_1/A) = \left(\tfrac{9.6}{2}\right)(5/2.0)^2 = 30.00 \text{ m}$$
$$\phi = (f)(L_e/L)(z_m/D) = (0.020)(\tfrac{1385}{965})(30.00/2.0) = 0.4306$$

$F(\phi) = (1 + \phi)e^{-\phi} = 0.954$, which is satisfied by $\phi = -0.275$.

$$z_m = \phi LD/fL_e = (-0.275)(965)(2.0)/[(0.020)(1385)] = -19.16 \text{ m}$$
$$z_1 = z_m(A/A_1) = (-19.16)(2.0/5)^2 = -3.07 \text{ m}$$
$$\text{Maximum fluctuation} = \tfrac{9.6}{2} - (-3.07) = 7.87 \text{ m}$$

**19.15** A valve is quickly opened in a pipe 1270 m long, $D = 0.7$ m, with a 0.4-m-diameter nozzle on the downstream end. Minor losses are $3V^2/2g$, with $V$ the velocity in the pipe, $f = 0.024$, and $H = 8$ m. Find the time to attain 95 percent of the steady-state discharge.

**I** $\quad t = (LV/2gH) \ln [(V_0 + V)/(V_0 - V)] \qquad H = (fL_e/D)(V^2/2g) + V_1^2/2g \qquad (V_1 = \text{nozzle velocity})$

$$L_e = L + KD/f = 1270 + (3)(0.7)/0.024 = 1358 \text{ m} \qquad V_1/V = (0.7/0.4)^2 \qquad V_1 = 3.062V$$

$$8 = [(0.024)(1358)/0.7]\{V^2/[(2)(9.807)]\} + (3.062V)^2/[(2)(9.807)] \qquad V = 1.675 \text{ m/s}$$

$$t = \{(1270)(1.675)/[(2)(9.807)(8)]\} \ln (1.95/0.05) = 49.7 \text{ s}$$

**19.16** A globe valve ($K = 10$) at the end of a pipe 2400 ft long is rapidly opened. $D = 3.5$ ft, $f = 0.019$, minor losses are $2V^2/2g$, and $H = 85$ ft. How long does it take for the discharge to attain 85 percent of its steady-state value?

$$t = (LV/2gH) \ln [(V_0 + V)/(V_0 - V)] \qquad V = \sqrt{2gHD/fL_e}$$

$$L_e = L + KD/f = 2400 + (10 + 2)(3.5)/0.019 = 4611 \text{ ft}$$

$$V = \sqrt{(2)(32.2)(85)(3.5)/[(0.019)(4611)]} = 14.79 \text{ ft/s}$$

$$t = \{(2400)(14.79)/[(2)(32.2)(85)]\} \ln (1.85/0.15) = 16.3 \text{ s}$$

**19.17** Benzene ($K = 150\,000$ psi, s.g. $= 0.88$) flows through 1-in-ID steel ($E = 30\,000\,000$ psi) tubing with $\frac{3}{16}$-in wall thickness. Determine the speed of a pressure wave.

**I** $\qquad a = \sqrt{(K/\rho)/[1 + (K/E)(D/e)]}$

$$= \sqrt{\{[(150\,000)(144)]/[(0.88)(1.94)]\}/\{1 + (150\,000/30\,000\,000)[(1)/(\frac{3}{16})]\}} = 3511 \text{ ft/s}$$

**19.18** A valve is closed in 4.5 s at the downstream end of a 3200-m pipeline carrying water at 2.7 m/s ($a = 1000$ m/s). What is the peak pressure developed by the closure?

**I** $\quad 2L/a = (2)(3200)/1000 = 6.40$ s. Since $[2L/a = 6.40 \text{ s}] > 4$ s, this is rapid closure. Then, $h = aV_0/g = (1000)(2.7)/9.807 = 275.3$ m, $p = \gamma h = (9.79)(275.3) = 2695$ kPa.

**19.19** Determine the length of pipe in Prob. 19.18 subject to the peak pressure.

**I** $\qquad x = L - at_c/2 = 3200 - (1000)(4.5)/2 = 950$ m

**19.20** A valve is closed at the downstream end of a pipeline in such a manner that only one-fourth of the line is subjected to maximum pressure. During what proportion of the time $2L/a$ is it closed?

**I** $\qquad x = L - at_c/2 \qquad L/4 = L - at_c/2 \qquad 3L/4 = at_c/2 \qquad t_c/(2L/a) = \frac{3}{4}$

**19.21** A 10-mm-diameter U-tube contains oil, $v = 5$ m$^2$/s, with a total column length of 2 m. If the initial half-amplitude of displacement is 250 mm, prepare a computer program to find the first 10 maximum and minimum displacements and the times they occur.

```
I  10 REM                          LAMINAR OSCILLATION OF A U-TUBE.
   20 ' TO=ATN(N'/M)/N'
   30 ' VO=Z*N'/(EXP(-M*TO)*SIN(N'*TO))
   40 ' The angle THO is THO=N'*TO, and the
   50 ' general angle is TH=THO+I*PI, with T=TH/N'-TO.
   60 ' Z=VO/N'*EXP(-M*TH/N')*SIN(TH)
   70 DEFINT I: DEF FNZ(DT,DTH)=C1*EXP(-M*DT)*SIN(DTH)
   80 READ D,NU,L,G,DZ,PI: DATA .01,5E-6,2.,9.806,.25,3.1416
   90 LPRINT: LPRINT"D,NU,L,G,DZ,PI=";D;NU;L;G;DZ;PI: LPRINT
  100 M=16!*NU/D^2:CC=2!*G/L-M^2:IF CC>0! THEN NP=SQR(CC) ELSE
      LPRINT"NON-OSCILLATING CASE":STOP
  110 TO=ATN(NP/M)/NP:VO=NP*DZ*EXP(M*TO)/SIN(NP*TO):THO=NP*TO
  120 LPRINT"M,NP,TO,VO,THO=";M;NP;TO;VO;THO: LPRINT
  130 C1=VO/NP: FOR I=0 TO 10:TH=THO+I*PI:T=TH/NP:Z=FNZ(T,TH)
  140 LPRINT"T,Z,TH=";:LPRINT USING " ####.##### ";T-TO;Z;TH
  150 NEXT I
  160 LPRINT: LPRINT"MAXIMUM REYNOLD'S NUMBER,V*D/NU=";VO*D/NU
```

```
     D,NU,L,G,DZ,PI= .01   .000005   2   9.806001   .25   3.1416

     M,NP,TO,VO,THO= .8   3.02754   .4335071   1.107393   1.31246
```

```
T,Z,TH=      0.00000      0.25000      1.31246
T,Z,TH=      1.03767     -0.10900      4.45406
T,Z,TH=      2.07535      0.04752      7.59566
T,Z,TH=      3.11302     -0.02072     10.73726
T,Z,TH=      4.15070      0.00903     13.87886
T,Z,TH=      5.18837     -0.00394     17.02046
T,Z,TH=      6.22604      0.00172     20.16206
T,Z,TH=      7.26372     -0.00075     23.30366
T,Z,TH=      8.30139      0.00033     26.44526
T,Z,TH=      9.33907     -0.00014     29.58686
T,Z,TH=     10.37674      0.00006     32.72846

MAXIMUM REYNOLD'S NUMBER,V*D/NU= 2214.786
```

**19.22** An equation for dealing with turbulent resistance is given by

$$\frac{d^2z}{dt^2} + \frac{f}{2D}\frac{dz}{dt}\left|\frac{dz}{dt}\right| + \frac{2g}{L} = 0$$

Put this equation in suitable form for solution by the third-order Runge-Kutta method.

▌ Let $y = dz/dt$, then $dy/dt = d^2z/dt^2$, and $dy/dt = -(f/2D)y\,|y| - (2g/L)z = F_1(y, z, t)$, $dz/dt = y = F_2(y, z, t)$. The two equations are solved simultaneously, from known initial conditions $y_n$, $z_n$, $t_n$: $u_{11} = hF_1(y_n, z_n, t_n) = h[-(f/2D)y_n\,|y_n| - (2g/L)z_n]$, $u_{12} = hF_2(y_n, z_n, t_n) = hy_n$, $u_{21} = hF_1[y_n + (u_{11}/3), z_n + (u_{12}/3), t_n + (h/3)] = h\{-(f/2D)[y_n + (u_{11}/3)]\,|y_n + (u_{11}/3)| - (2g/L)[z_n + (u_{12}/3)]\}$, $u_{22} = hF_2[y_n + (u_{11}/3), z_n + (u_{12}/3), t_n + (h/3)] = h[y_n + (u_{11}/3)]$, $u_{31} = hF_1(y_n + \frac{2}{3}u_{21}, z_n + \frac{2}{3}u_{22}, t_n + \frac{2}{3}h) = h[-(f/2D)(y_n + \frac{2}{3}u_{21})\,|y_n + \frac{2}{3}u_{21}| - (2g/L)(z_n + \frac{2}{3}u_{22})]$, $u_{32} = hF_2(y_n + \frac{2}{3}u_{21}, z_n + \frac{2}{3}u_{22}, t_n + \frac{2}{3}h) = h(y_n + \frac{2}{3}u_{21})$, $y_{n+1} = y_n + (u_{11}/4) + \frac{3}{4}u_{31}$, $z_{n+1} = z_n + (u_{12}/4) + \frac{3}{4}u_{32}$, $t_{n+1} = t_n + h$. The equations for simultaneous solution have been written for a general case as well as for the specific case of solution of the given equation.

**19.23** Prepare a program to carry out the solution of Prob. 19.22 and apply it to the following case: $t = 0$, $z = 12$ ft, $V_0 = 0$, $L = 400$ ft, $f = 0.017$, $d = 2$ ft, $dt = 0.1$ s, and $t_{max} = 30$ s.

▌

```
10 REM                    TURBULENT OSCILLATION OF A U-TUBE
20 DEFINT I: DEF FNN(DY,DZ)=H*(-C1*DY*ABS(DY)-C2*DZ): DEF FNM(DY)=H*DY
30 READ Z,V,L,F,D,DT,TMAX,IPR,G: DATA 12.,,0,400.,,017,2.5,.1,30.,20,32.2
40 LPRINT: LPRINT"Z,V,L,F,D=";Z;V;L;F;D: LPRINT"DT,TMAX,IPR,G=";DT,TMAX,IPR,G
50 II=FIX(TMAX/DT)+1: H=DT: C1=.5*F/D: C2=2!*G/L: Y=0!: T=0!
60 LPRINT: LPRINT"      T            V            Z"
70 FOR I=0 TO II: IF I MOD IPR=0 THEN LPRINT USING "   ###.###  ";T;Y;Z
80 U11=FNN(Y,Z):U12=FNM(Y)
90 U21=FNN(Y+U11/3!,Z+U12/3!):U22=FNM(Y+U11/3!)
100 U31=FNN(Y+2!*U21/3,Z+2!*U22/3!):U32=FNM(Y+2!*U21/3!)
110 Y=Y+.25*U11+.75*U31:Z=Z+.25*U12+.75*U32:T=T+H
120 NEXT I
```

```
Z,V,L,F,D= 12    0    400    .017    2.5
DT,TMAX,IPR,G= .1            30            20            32.2
```

| T | V | Z |
|---|---|---|
| 0.000 | 0.000 | 12.000 |
| 2.000 | -3.434 | 8.354 |
| 4.000 | -4.681 | -0.240 |
| 6.000 | -3.083 | -8.433 |
| 8.000 | 0.308 | -11.355 |
| 10.000 | 3.459 | -7.356 |
| 12.000 | 4.418 | 0.982 |
| 14.000 | 2.702 | 8.495 |
| 16.000 | -0.585 | 10.723 |
| 18.000 | -3.466 | 6.425 |
| 20.000 | -4.162 | -1.647 |
| 22.000 | -2.345 | -8.513 |
| 24.000 | 0.834 | -10.104 |
| 26.000 | 3.458 | -5.552 |
| 28.000 | 3.911 | 2.244 |
| 30.000 | 2.011 | 8.492 |

**19.24** A surge tank of area $AR$ is to be used with a hydroelectric power plant. The turbines may change from full flow to zero flow in $TC$ seconds, and similarly accept full flow in $TC$ seconds. Assuming linear changes in flow to the turbines, determine the elevation of maximum and minimum water surfaces in the surge tank. The maximum flow is $QM$; elevation of reservoir is $Z$. Prepare a program using the Runge–Kutta second-order integration of the equations. Specifically, let $Z = 100$ m, $QM = 20$ m$^3$/s, $AR = 75$ m$^2$, and a pressure pipe $L = 10$ km, $D = 3$ m, $F = 0.016$, with $TC = 4$ s and $DT = H = 1$ s.

■

```
10 REM                    SIMPLE SURGE TANK.
20 ' Continuity equation: DY/DT=(QIN-QOUT)/AR
30 ' Eq.Motion: GAM*AR*(Z-Y-R*Q*ABS(Q))=GAM*AR*(L/G)*DQ/DT
40 ' or DQ/DT=G*AP*(Z-Y-R*Q*ABS(Q))/L: in which Y=elev of water surface in
50 ' surge tank; AP=area of pressure pipe; R=head loss in pressure pipe
60 ' per unit discharge.
70 DEFINT I: DEF FNQ(DY,DQ)=C1*(Z-DY-R*DQ*ABS(DQ))
80 READ F,L,D,G,Z,AR,QM,TC,H,TM,IPR,ICLOSE
90 DATA .016,10000.,3.,9.806,100.,75.,20.,4.,1.,900.,60,0
100 LPRINT: LPRINT"F,L,D,G,Z=";F;L;D;G;Z: LPRINT
110 LPRINT"AR,QM,TC,H,TM,IPR,ICLOSE=";AR;QM;TC;H;TM;IPR;ICLOSE: LPRINT
120 AP=.7854*D^2: R=F*L/(2!*G*.7854^2*D^5): C1=G*AP*H/L: C2=H/AR: II=FIX(TM/H)+1

130 T=0!: IF ICLOSE=1 THEN Y=Z-R*QM^2: QT=QM: Q=QM ELSE Y=Z: QT=0!: Q=0!
140 LPRINT"    TIME        DISCHARGE        Y            QT"
150 YMAX=Z: YMIN=Z: LPRINT USING"  ###.###   ";T;Q;Y;QT: IF ICLOSE=0 THEN 270
160 FOR I=0 TO II:T=I*H: IF T<=TC THEN QT=QM*(1!-T/TC) ELSE QT=0!
170 U11=FNQ(Y,Q):U12=C2*(Q-QT):TT=T+H
180 IF TT<=TC THEN QT=QM*(1!-TT/TC) ELSE QT=0!
190 U21=FNQ(Y+U12,Q+U11):U22=C2*(Q+U11-QT)
200 Q=Q+.5*(U11+U21):Y=Y+.5*(U12+U22):T=T+H
210 IF Y>YMAX THEN YMAX=Y:TMAX=T
220 IF Y<YMIN THEN YMIN=Y:TMIN=T
230 IF (I+1) MOD IPR=0 THEN LPRINT USING" ####.###   ";T;Q;Y;QT
240 NEXT I
250 GOTO 360
260 END
270 FOR I=1 TO II:T=I*H: IF T<=TC THEN QT=QM*T/TC ELSE QT=QM
280 U11=FNQ(Y,Q):U12=C2*(Q-QT):TT=T+H
290 IF TT<=TC THEN QT=QM*TT/TC ELSE QT=QM
300 U21=FNQ(Y+U12,Q+U11):U22=C2*(Q+U11-QT)
310 Q=Q+.5*(U11+U21):Y=Y+.5*(U12+U22):T=T+H
320 IF Y>YMAX THEN YMAX=Y:TMAX=T
330 IF Y<YMIN THEN YMIN=Y:TMIN=T
340 IF (I+1) MOD IPR=0 THEN LPRINT USING" ####.###   ";T;Q;Y;QT
350 NEXT I
360 LPRINT"YMAX,TMAX=";:LPRINT USING"  ####.##   ";YMAX;TMAX
370 LPRINT"YMIN,TMIN=";:LPRINT USING"  ####.##   ";YMIN;TMIN
```

F,L,D,G,Z= .016  10000  3  9.806001  100

AR,QM,TC,H,TM,IPR,ICLOSE= 75  20  4  1  900  60  0

| TIME | DISCHARGE | Y | QT |
|---|---|---|---|
| 0.000 | 0.000 | 100.000 | 0.000 |
| 60.000 | 2.978 | 85.346 | 20.000 |
| 120.000 | 10.451 | 74.568 | 20.000 |
| 180.000 | 17.524 | 69.934 | 20.000 |
| 240.000 | 21.355 | 69.706 | 20.000 |
| 300.000 | 22.561 | 71.399 | 20.000 |
| 360.000 | 22.479 | 73.466 | 20.000 |
| 420.000 | 21.944 | 75.248 | 20.000 |
| 480.000 | 21.346 | 76.561 | 20.000 |
| 540.000 | 20.839 | 77.427 | 20.000 |
| 600.000 | 20.466 | 77.940 | 20.000 |
| 660.000 | 20.220 | 78.206 | 20.000 |
| 720.000 | 20.075 | 78.318 | 20.000 |
| 780.000 | 20.000 | 78.345 | 20.000 |
| 840.000 | 19.970 | 78.331 | 20.000 |
| 900.000 | 19.964 | 78.303 | 20.000 |
| YMAX,TMAX= | 100.00 | 0.00 | |
| YMIN,TMIN= | 69.44 | 212.00 | |

**19.25** The open wedge-shaped tank in Fig. 19-7 has a length of 15 ft perpendicular to the sketch. It is drained with a 3-in-diameter pipe of length 10 ft whose discharge end is at elevation zero. The coefficient of loss at the pipe entrance is 0.50, the total of the bend loss coefficient is 0.20, and $f$ for the pipe is 0.018. Find the time required to lower the water surface in the tank from elevation 8 ft to 5 ft. Neglect the possible change of $f$ with $N_R$, and assume that the acceleration effects in the pipe are negligible.

▋ Energy equation from water surface to jet at discharge: $z - [0.5 + 0.2 + 0.018(10/0.25)](V^2/2g) = V^2/2g$, $z - 1.42(V^2/2g) = V^2/2g$, $V = 5.16z^{1/2}$; $Q_0 = AV = (\pi/4)(0.25)^2 5.16z^{1/2} = 0.253z^{1/2}$. The area of the water surface may be expressed as $A_s = 15b = 15Kz$. At the top of the tank, $A_s = 15(6) = 15K(10)$, $K = 0.6$. Thus $A_s = 15(0.6)z = 9z$.

$$t = \int_{z_1}^{z_2} \frac{A_s \, dz}{Q_i - Q_0} = \int_8^5 \frac{9z \, dz}{0 - 0.253z^{1/2}} = -\frac{9}{0.253} \int_8^5 z^{1/2} \, dz = -35.5[\tfrac{2}{3}z^{3/2}]_8^5 = 271 \text{ s}$$

Note that if the pipe had discharged at an elevation other than zero, the integral would have been different, because the head on the pipe would then have been $z + h$, where $h$ is the vertical distance of the discharge end of the pipe below ($h$ positive) or above ($h$ negative) point $A$ of the figure.

Fig. 19-7

**19.26** Two large water reservoirs are connected to one another with a 10-cm-diameter pipe ($f = 0.02$) of length 15 m. The water-surface elevation difference between the reservoirs is 2.0 m. A valve in the pipe, initially closed, is suddenly opened. Determine the times required for the flow to reach $\frac{1}{4}$, $\frac{1}{2}$, and $\frac{3}{4}$ of the steady-state flow rate. Assume the water-surface elevations remain constant. Repeat for pipe lengths of 150 m and 1500 m with all other data remaining the same. In the first case $L/D = 15/0.10 = 150$, hence minor losses are significant. Assume square-edged entrance.

▋ Square-edge entrance: $k_L = 0.5$, $k = k_L + f(L/D) = 0.5 + 0.02(15/0.10) = 3.5$. For steady flow: $V_0 = \sqrt{2gH/(1+k)} = \sqrt{2(9.81)_2/(1+3.5)} = 2.95$ m/s. For unsteady flow use $t = \{L/[(1+k)V_0]\} \ln [(V_0 + V)/(V_0 - V)] = [15/4.5(2.95)] \ln [(2.95 + V)/(2.95 - V)] = 1.129 \ln [(2.95 + V)/(2.95 - V)]$. For $Q = \frac{1}{4}Q_0$ substitute $V = \frac{1}{4}V_0$, etc.:

| $Q$ | $V$, m/s | $\dfrac{2.95 + V}{2.95 - V}$ | ln | $t$, s |
|---|---|---|---|---|
| $0.25Q_0$ | 0.74 | 1.667 | 0.511 | 0.577 |
| $0.50Q_0$ | 1.48 | 3.00 | 1.099 | 1.240 |
| $0.75Q_0$ | 2.21 | 7.00 | 1.946 | 2.197 |

For the other two lengths the results are as follows:

| $Q$ | $L = 150$ m | $L = 1500$ m |
|---|---|---|
| $0.25Q_0$ | 2.18 s | 7.04 s |
| $0.50Q_0$ | 4.69 s | 15.15 s |
| $0.75Q_0$ | 8.30 s | 26.84 s |

**19.27** In Fig. 19-8 the elasticity and dimensions of the pipe are such that the celerity of the pressure wave is 3200 fps. Suppose the pipe has a length of 2000 ft and a diameter of 4 ft. The flow rate is initially 30 cfs. Water is flowing.

Find (a) the water-hammer pressure for instantaneous valve closure; (b) the approximate water-hammer pressure at the valve if it is closed in 4.0 s; (c) the water-hammer pressure at the valve if it is manipulated so that the flow rate drops almost instantly from 30 to 10 cfs; (d) the maximum water-hammer pressure at a point in the pipe 300 ft from the reservoir if a 1.0-s valve closure reduces the flow rate from 10 cfs to zero.

$$V = Q/A = 30/(\pi 2^2) = 2.39 \text{ fps}$$

(a) $$p_h = \rho c_P V = 1.94(3200)(2.39) = 14\,840\,\text{lb/ft}^2 = 103.0\,\text{psi}$$

(b) $$p'_h \approx (\tfrac{4000}{3200}/4.0)p_h = (1.25/4.00)(103.0) = 32.2\,\text{psi}$$

(c) For this case of partial closure $\Delta p_h = -\rho c_P(\Delta V)$. $\Delta V = (10 - 30)/\pi 2^2 = -1.592$ fps,
$\Delta p_h = -1.94(3200)(-1.592) = 9880\,\text{lb/ft}^2 = 68.6\,\text{psi}$.

(d) If $2x_0/c_P = 1.0$ s, $x_0 = 1600$ ft, so that full water-hammer pressure will be developed in the pipe only in the region that is farther than 1600 ft from the reservoir.
For this cases, at valve, $p_h = 1.94(3200)(2.39/3) = 4940\,\text{lb/ft}^2 = 34.3\,\text{psi}$. At point 300 ft from reservoir:
$p_h = 34.3(\tfrac{300}{1600}) = 6.43\,\text{psi}$.

**Fig. 19-8**

**19.28** Suppose a ship lock has vertical sides and that water enters or discharges through a conduit area $A$ such that $Q = C_d A\sqrt{2gz}$, where $z$ is the variable difference in level between the water surface in the lock and that outside. Prove that for the water level in the lock to change from $z_1$ to $z_2$ the time is $t = [2A_s/(C_d A\sqrt{2g})](z_1^{1/2} - z_2^{1/2})$. (*Note:* If the lock is being filled, the signs must be reversed.)

$$t = \int_{z_1}^{z_2} \frac{A_s\,dz}{Q_i - Q_o} \qquad Q_i = 0 \qquad Q_o = C_d A(2gz)^{1/2}$$

thus

$$t = \int_{z_1}^{z_2} \frac{A_s\,dz}{-C_d A(2gz)^{1/2}} = \frac{A_s}{C_d A(2g)^{1/2}} \int_{z_1}^{z_2} \frac{dz}{-z^{1/2}} = \frac{A_s}{C_d A(2g)^{1/2}}[-2z^{1/2}]_{z_1}^{z_2} = \frac{2A_s}{C_d A(2g)^{1/2}}(z_1^{1/2} - z_2^{1/2})$$

**19.29** Suppose the lock of Prob. 19.28 is 285 ft long by 100 ft wide, and water enters through a conduit for which the discharge coefficient is 0.52. If the water surface in the lock is initially 29 ft below the level of the surface of the water upstream, how large must the conduit be if the lock is to be filled in 6 min?

$$t = [2A_s/(C_d A\sqrt{2g})](z_1^{1/2} - z_2^{1/2}) \qquad (6)(60) = \{(2)[(285)(100)]/[(0.52)(A)\sqrt{(2)(32.2)}]\}(29^{1/2} - 0^{1/2})$$
$$A = 204\,\text{ft}^2$$

**19.30** (a) Suppose a reservoir has vertical sides and that initially there is a steady flow into it such that the height of the surface above the level of a spillway $(Q = C_W L H^{3/2})$ is $z_1$. If the inflow is suddenly cut off, prove that the time required for the water level to fall from $z_1$ to $z_2$ is $t = (2A_s/C_W L)(1/\sqrt{z_2} - 1/\sqrt{z_1})$. (*Note:* $z = H$.) (b) How long will it take theoretically for the outflow to cease entirely? What factors make this theoretical answer unrealistic?

(a)
$$t = \int_{z_1}^{z_2} \frac{A_s\,dz}{Q_i - Q_o} \qquad Q_i = 0 \qquad Q_o = C_W L z^{3/2}$$

$$= \int_{z_1}^{z_2} \frac{A_s\,dz}{-C_W L z^{3/2}} = \frac{A_s}{C_W L}\left[\frac{2}{z^{1/2}}\right]_{z_1}^{z_2} = \frac{2A_s}{C_W L}\left[\frac{1}{z_2^{1/2}} - \frac{1}{z_1^{1/2}}\right]$$

(b) As $z_2 \to 0$, $t \to \infty$, so theoretically the outflow would never cease. Factors which make this theoretical answer unrealistic include surface tension, surface ripples, evaporation, and leakage/seepage losses.

**19.31** The crest of the overflow of a reservoir is 82 ft long, and the value of $C_W$ is 3.45. For the range of levels considered here the area of the water surface is constant and is 600 000 ft$^2$. Initially, there is a flow into the reservoir at such a rate that the height of the water surface above that of the spillway crest is stabilized at 4 ft, and then the inflow is suddenly diverted. Find the length of time for the water surface to fall to a height of 2 ft above the level of the spillway.

❚

$$t = \int_{z_1}^{z_2} \frac{A_s\, dz}{Q_i - Q_o} \qquad Q_i = 0 \qquad Q_o = C_W L z^{3/2}$$

$$= \int_{z_1}^{z_2} \frac{A_s\, dz}{-C_W L z^{3/2}} = \frac{2A_s}{C_W L}\left[\frac{1}{z_2^{1/2}} - \frac{1}{z_1^{1/2}}\right]$$

$$= \frac{2(600\,000)}{3.45(82)}\left[\frac{1}{2^{1/2}} - \frac{1}{4^{1/2}}\right] = 879\text{ s} = 14.6\text{ min}$$

**19.32** The crest of the overflow spillway of a reservoir is 46 ft long, and the value of $C_W$ is 3.50. The area of the water surface is assumed constant at 700 000 ft$^2$ for the range of heights considered here. Initially, the water surface is 4 ft below the level of the spillway crest. If suddenly there is turned into this reservoir a flow of 560 cfs, what will be the height of water in the reservoir for equilibrium? How long a time will be required for this height to be reached? How long a time will be required for the water surface to reach a height of 1 ft above the level of the spillway? (*Note*: This last can be solved by integration after substituting $x^3$ for $z^{3/2}$ and consulting integral tables. However, it will be easier to solve it graphically either by plotting and actually measuring the area under the curve or by computing the latter by some method, such as Simpson's rule.)

❚ $Q = C_W L z^{3/2}$, $Q_o = 3.50(46)z^{3/2} = 161z^{3/2}$. For equilibrium $Q_i = Q_o = 560$ cfs, from which $z = (\frac{560}{161})^{2/3} = 2.30$ ft. The time required to reach this equilibrium height is theoretically infinite, but surface ripples and minor irregularities in flow cause practical equilibrium to be reached in a finite time. However, this does point out that true equilibrium is not quickly obtained. This should be borne in mind in experimental work. The time for the water level to reach the spillway crest $= 4(700\,000)/560 = 5000$ s. The time for water level to rise from crest to 2 ft above it is

$$t = \int_{z_1}^{z_2} \frac{A_s\, dz}{Q_i - Q_o} = 700\,000 \int_0^1 \frac{dz}{560 - 161z^{3/2}}$$

Letting $z^{3/2} = x^3$, $(\frac{3}{2})z^{1/2}\, dz = 3x^2\, dx$ and $dz = (3x^2\, dx)/[(\frac{3}{2})z^{1/2}] = (3x^2\, dx)/[(\frac{3}{2})x] = 2x\, dx$.

$$t = 2(700\,000)\int_0^{1^{1/2}} \frac{x\, dx}{560 - 161x^3} = 1\,400\,000 \frac{(-\frac{560}{161})^{1/3}}{3(560)}\left[\frac{1}{2}\ln\frac{[x + (-\frac{560}{161})^{1/3}]^2}{(-\frac{560}{161})^{2/3} - (-\frac{560}{161})^{1/3}x + x^2} + 3^{1/2}\tan^{-1}\frac{2x - (-\frac{560}{161})^{1/3}}{(-\frac{560}{161})^{1/3}3^{1/2}}\right]_0^{1^{1/2}}$$

$$= 1\,400\,000\left(\frac{-1.515}{1680}\right)\left[\frac{1}{2}\ln\frac{(x - 1.515)^2}{2.30 + 1.515x + x^2} + 3^{1/2}\tan^{-1}\frac{2x + 1.515}{-1.515(3)^{1/2}}\right]_0^{1^{1/2}} = 3863\text{ s}$$

Total time $= 5000 + 3863 = 8863$ s

**19.33** Water enters a reservoir at such a rate that the height of water above the level of the spillway crest is 2.5 ft. The spillway ($Q = C_W L H^{3/2}$) is 80 ft long, and the value of $C_W$ is 3.45. The area of the water surface for various water levels is as follows:

| z, ft | A_s, ft$^2$ |
|-------|-------------|
| 2.50  | 730 000     |
| 2.00  | 620 000     |
| 1.50  | 490 000     |
| 1.25  | 435 000     |
| 1.00  | 380 000     |

If the inflow is suddenly reduced to 150 cfs, what will be the height of water for equilibrium? How long will it take, theoretically, for equilibrium to be attained? How long will it take for the level to drop from 2.5 ft to 1 ft above that of the spillway?

█ $Q_o = 276z^{3/2}$; when $Q_o = Q_i = 150$ cfs, $z = 0.666$ ft and the time for equilibrium is theoretically infinite.

$$t = \int_{z_1}^{z_2} \frac{A_s\, dz}{Q_i - Q_o} = \int \frac{A_s\, dz}{276z^{3/2} - 150} \qquad (z \text{ gets smaller})$$

| $z$, ft | $\Delta z$, ft | $A_s$, ft$^2$ | $\bar{A}_s$, ft$^2$ | $\bar{A}_s(\Delta z)$, ft$^3$ | $Q_o = 276z^{3/2}$, cfs | $\bar{Q}_o$, cfs | $\bar{Q}_o - 150$, cfs | $\dfrac{\bar{A}_s(\Delta z)}{\bar{Q}_o - 150}$ |
|---|---|---|---|---|---|---|---|---|
| 2.5 | | 730 000 | | | 1091 | | | |
| | 0.5 | | 675 000 | 337 500 | | 936 | 786 | 429 |
| 2.0 | | 620 000 | | | 781 | | | |
| | 0.5 | | 555 000 | 277 500 | | 644 | 494 | 562 |
| 1.5 | | 490 000 | | | 507 | | | |
| | 0.25 | | 463 000 | 115 750 | | 446 | 296 | 391 |
| 1.25 | | 435 000 | | | 386 | | | |
| | 0.25 | | 408 000 | 102 000 | | 331 | 181 | 564 |
| 1.00 | | 380 000 | | | 276 | | | |
| | | | | | | | | $\Sigma = 1946$ s |

Total time for $z$ to change from 2.5 ft to 1.0 ft is approximately 1946 s = 32.4 min.

**19.34** Work Prob. 19.33 using the same numbers but changing feet to meters, square feet to square meters, and cubic feet per second to cubic meters per second.

█ This problem is solved in the same manner as Prob. 19.33. $Q_o = 1.84(3.45/3.33)LH^{3/2} = 1.906(80)z^{3/2} = 152.5z^{3/2}$. When $Q_o = Q_i = 150$ m$^3$/s, $z = (150/152.5)^{2/3} = 0.989$ m and the time for equilibrium is theoretically infinite. Columns 2 to 5 are omitted from the following solution table since they are numerically the same as those in Prob. 19.33.

| $z$, m | $Q_o = 152.5z^{3/2}$, m$^3$/s | $\bar{Q}_o$, m$^3$/s | $\bar{Q}_o - 150$, m$^3$/s | $\dfrac{A_s(\Delta z)}{\bar{Q}_o - 150}$ |
|---|---|---|---|---|
| 2.5 | 603 | | | |
| | | 517 | 367 | 920 |
| 2.0 | 431 | | | |
| | | 356 | 206 | 1347 |
| 1.5 | 280 | | | |
| | | 246 | 96 | 1206 |
| 1.25 | 213 | | | |
| | | 182 | 32 | 3188 |
| 1.00 | 152 | | | |
| | | | | $\Sigma = 6661$ s |

Total time for $z$ to change from 2.5 m to 1.0 m is approximately 6661 s = 111.0 min.

**19.35** Figure 19-9 shows a tank whose shape is the frustum of a cone with a 1.5-ft$^2$ orifice in the bottom. Assume $C_d = 0.62$. If the water level outside the tank is constant at section 2, how long will it take the water level in the tank to drop from section 1 to section 2? (*Note:* Diameter of tank $= Ky$, and $y = z + h_2$, where $z$ is the variable distance between surface levels.)

█ $D = ky$; $k = (39.6 - 18)/34 = 0.635$ $A_s = (\pi/9.92)y^2$; $y = z + h_2 = z + 28.35$; $Q_o = C_d A(2gz)^{1/2} = 0.62(1.5)(2gz)^{1/2}$.

$$t = \int_{z_1}^{z_2} \frac{A_s\, dz}{Q_i - Q_o} = \int_{34}^{0} \frac{(\pi/9.92)(z + 28.35)^2\, dz}{-(0.62)1.5(2gz)^{1/2}} = 830 \text{ s}$$

**Fig. 19-9**

**19.36** If in Fig. 19-9 the water surface outside the tank is constant at section 1 and the tank is initially empty, how long will it take for the water level in the tank to rise from section 2 to section 1? Assume a 1.5-ft² orifice with $C_d = 0.62$.

▌ $z = h_1 = y$; $y = h_1 - z = 62.35 - z$; $Q_o = C_d A(2gz)^{1/2} = 0.62(1.5)(2gz)^{1/2}$.

$$t = \int_{z_1}^{z_2} \frac{A_s \, dz}{Q_i - Q_o} = \int_{34}^{0} \frac{(\pi/9.92)(62.35 - z)^2(-dz)}{(0.62)1.5(2gz)^{1/2}} = 1339 \text{ s}$$

**19.37** A 1-in-diameter smooth brass pipe 900 ft long drains an open cylindrical tank which contains oil having $\rho = 1.8$ slugs/ft³, $\mu = 0.0006$ lb-s/ft². The pipe discharges at elevation 80 ft. Find the time required for the oil level to drop from elevation 100 ft to elevation 88 ft if the tank is 5 ft in diameter.

▌ Check if flow is always laminar. When $V = V_{max}$, the elevation of oil in the tank is at 100 ft, then $100 - f(L/D)(V^2/2g) = 80$; i.e., $20 = f(L/D)(V^2/2g)$. Assuming laminar flow: $f = 64/N_R$, $N_R = DV\rho/\mu$. Substituting: $20 = (64\mu/DV\rho)(L/D)(V^2/2g)$, $20 = \{64(0.0006)/[(\frac{1}{12})V(1.8)]\}(900/\frac{1}{12})\{V^2/[2(32.2)]\} = 42.9V$, $V = V_{max} = 0.466$ fps; $N_{Rmax} = (\frac{1}{12})(0.466)(1.8)/0.0006 = 116.5$, $N_{Rmax} < 2000$, so the flow is always laminar. In general, from the above: elev $- 80 = 42.9V$, $V = (\text{elev} - 80)/42.9 = h/42.9$ (defining $h = \text{elev} - 80$); $Q = AV = 0.00545(h/42.9)$; $A_s \, dz = Q_i \, dt - Q_o \, dt$, $Q \, dt = -A \, dh$, $dt = -6.25\pi \, dh/[0.00545(h/42.9)] = -154\,558 \, dh/h$.

$$\int_0^t dt = -154\,558 \int_{20}^{8} \frac{dh}{h} = -154\,558[\ln h]_{20}^{8} \qquad t = -154\,558(2.08 - 3.00) = 141\,620 \text{ s} = 39.3 \text{ h}$$

**19.38** A large reservoir is being drained with a pipe system as shown in Fig. 19-10. Initially, when the pump is rotating at 190 rpm, the flow rate is 6.0 cfs. If the pump speed is increased instantaneously to 240 rpm, determine the flow rate as a function of time. Assume that the head $h_p$ developed by the pump is proportional to the square of the rotative speed; that is, $h_p \propto n^2$.

▌ Initial condition neglecting minor losses: $z - h_{L_f} + h_p = V^2/2g$, $37 - 0.022(\frac{2900}{1})V^2/2g + h_p = V^2/2g$, where $V = Q/A = 6.0/0.785 = 7.64$ fps, from which $h_p = 21.7$ ft; $h_p' = 21.7(\frac{240}{190})^2 = 34.6$ ft.

New conditions: $37 - 63.8(V^2/2g) + 34.6 = V^2/2g + (L/g)(dV/dt)$. Eventually $dV/dt = 0$, $V^2/2g = 71.6/64.8 = 1.105$ ft, $V = 8.44$ fps when steady flow is eventually achieved.

In general: $71.6 - 64.8(V^2/2g) = (2900/32.2)(dV/dt)$, $dt = 89.6 \, dV/(71.2 - V^2)$. Integrate the left side from zero to $t$; integrate the right side from 7.64 to $V_t$. Select various $V_t$'s and find the corresponding $t$'s:

$$\int dt = \frac{89.6}{16.88}\left[\ln \frac{8.44 + V}{8.44 - V}\right]_{7.64}^{V_t} = 5.31 \ln \frac{8.44 + V_t}{8.44 - V_t} - 15.93$$

| $V_t$, fps | $t$, s | $Q$, cfs |
|------------|--------|----------|
| 7.64 | 0 | 6.00 |
| 7.90 | 2.176 | 6.20 |
| 8.10 | 4.70 | 6.36 |
| 8.25 | 7.84 | 6.48 |
| 8.40 | 16.16 | 6.60 |
| 8.42 | 19.84 | 6.61 |
| 8.44 | ∞ | 6.63 |

**Fig. 19-10**

**19.39** (*a*) Repeat Prob. 19.38 with all data the same except use a 24-in-diameter pipe rather than a 12-in pipe. (*b*) Repeat also for the case of a 10-inch-diameter pipe.

▌ (*a*) **24-in pipe:** Initial conditions: $37 - 0.022(\frac{2900}{2})(V^2/2g) + h_p = V^2/2g$, $V = 6.0/3.142 = 1.91$ fps; $h_p = 32.9(V^2/2g) - 37 = -35.1$ ft. Thus a pump is not required; actually a head of 35.1 ft is available to drive a turbine when the flow rate is 6.0 cfs.

(*b*) **10-in pipe:** Initial conditions: $37 - 0.022(2900/0.833)(V^2/2g) + h_p = V^2/2g$ where $V = Q/A = 6.0/[\pi(\frac{5}{12})^2] = 11.00$ fps, $h_p = 77.6(V^2/2g) - 37 = 108.8$ ft; $h_p' = 108.8(\frac{240}{190})^2 = 173.6$ ft.

New conditions: $37 - 76.6(V^2/2g) + 173.6 = V^2/2g + (L/g)(dV/dt)$. Eventually $dV/dt = 0$, $V^2/2g = \frac{210.6}{77.6} = 2.71$ ft, $V = 13.21$ fps when steady flow is eventually achieved.

In general: $210.6 - 77.6(V^2/2g) = (2900/32.2)(dV/dt)$, $dt = 74.7 \, dV/(174.8 - V^2)$.

$$\int_0^t dt = t = \int_{11.00}^{V_t} \frac{74.7 \, dV}{(174.8 - V^2)} = \frac{74.7}{26.42}\left[\ln\frac{13.21 + V}{13.21 - V}\right]_{11.00}^{V_t} = 2.83\ln\frac{13.21 + V_t}{13.21 - V_t} - 6.77$$

| $V_t$, fps | $t$, s | $Q$, cfs |
|---|---|---|
| 11.00 | 0 | 6.00 |
| 12.00 | 1.82 | 6.54 |
| 13.00 | 6.90 | 7.09 |
| 13.20 | 15.53 | 7.20 |
| 13.21 | ∞ | 7.20 |

**19.40** Attached to the tank in Fig. 19-11 is a flexible 30-mm-diameter hose ($f = 0.016$) of length 72 m. The tank is hoisted in such a manner that $h = 6 + 0.8t$, where $h$ is the head in meters and $t$ is the time in seconds. Estimate the flow at $t = 9$ s.

▌ Find steady-state velocity for $t = 9$ s; then $h = 6 + 0.8(9) = 13.2$ m, $13.2 - 0.016(72/0.030)(V^2/2g) = V^2/2g$; $V = 2.56$ m/s at steady state. Now find the effect of changing head.

Given $h = 6 + 0.8t$, therefore $\partial h/\partial t = 0.8$ m/s, $h - 38.4(V^2/2g) = V^2/2g + (L/g)(dV/dt)$, $\partial h/\partial t = 39.4(2V/2g)(\partial h/\partial t) + (L/g)(\partial^2 V/\partial t^2)$. Neglect the last term (of higher order): $0.8 = 39.4(V/g)(\partial V/\partial t)$; $\partial V/\partial t = 0.8g/(39.4V) = 0.0778$ m/s per second. Finally, $13.2 = 39.4[V^2/2(9.81)] + (72/9.81)(0.0778)$; $39.4V = 2.51$ m/s; $Q = 0.001774$ m³/s, or 1.77 L/s.

**Fig. 19-11**

**19.41** Repeat Prob. 19.40 for a 100-mm-diameter hose, all other data remaining the same.

▌ Find steady-state velocity for $t = 9$ s: $13.2 - 0.016(72/0.100)(V^2/2g) = V^2/2g$; $V = 4.55$ m/s. Now find effect of changing head: $h = 12.52(V^2/2g) + (L/g)(dV/dt)$. Differentiating with respect to $t$ and neglecting the higher-order term: $\partial h/\partial t = 0.8 = 12.52(2V/2g)(dV/dt)$; $\partial V/\partial t = 0.8g/(12.52V) = 0.1377$ m/s per second.

Hence,

$$13.2 = 12.52[V^2/2(9.81)] + (72/9.81)(0.1377) \qquad V = 4.37 \text{ m/s} \qquad Q = 0.0343 \text{ m}^3/\text{s} = 34.3 \text{ L/s}$$

**19.42** In Prob. 19.40, suppose $h$ was changed instantaneously from 6 m to 13.2 m. Under these conditions, find the flow rate at $t = 9$ s.

❚ Initial condition where $h = 6$ m: $6 - 0.016(72/0.030)(V^2/2g) = V^2/2g$; $V = 1.728$ m/s (initially). After head is changed to 13.2 m: $13.2 = 39.4(V^2/2g) + (L/g)(dV/dt)$. Eventually $dV/dt = 0$, $V = 2.56$ m/s, $13.2 = [39.4/(2)(9.81)]V^2 + (72/9.81)(dV/dt)$; $13.2 = 2.008V^2 + 7.34(dV/dt)$; $6.57 = V^2 + 3.66(dV/dt)$; $dt = 3.66\,dV/(2.56^2 - V^2)$.

$$\int_0^t dt = \int_{1.728}^V \frac{3.66\,dV}{2.56^2 - V^2} = \frac{3.66}{5.12}\left[\ln\frac{2.56 + V}{2.56 - V}\right]_{1.728}^V \qquad t = 0.715\ln\frac{2.56 + V}{2.56 - V} - 1.172$$

| $V$, m/s | $t$, s |
|----------|--------|
| 1.728 | 0 |
| 2.1 | 0.484 |
| 2.4 | 1.283 |
| 2.5 | 2.00 |
| 2.56 | ∞ |

For all practical purposes $V = 2.00$ m/s at $t = 9$ s when $Q = (0.000707)2.00 = 0.00141$ m$^3$/s $= 1.41$ L/s.

**19.43** A 6-in-diameter pipe of length 3600 ft drains a reservoir. The elevation difference between the reservoir water surface and the pipe outlet is 95 ft. The pipe entrance is flush. Initially there is no flow since there is a plug at the outlet. The plug is then removed. Tabulate $Q$ versus $t$, assuming $f = 0.022$.

❚ $L/D = 3600/0.50 > 1000$, neglect minor losses, i.e., $k_L = 0$, $k = k_L + fL/D = 0 + 0.022(3600)/(\frac{6}{12}) = 158$, $V_0^2 = 2gH/(1 + k) = 2(32.2)95/(1 + 158)$; steady $V_0 = 6.20$ fps.

$$t = [L/(1 + k)V_0]\ln[(V_0 + V)/(V_0 - V)] = [3600/159(6.20)]\ln[(V_0 + V)/(V_0 - V)]$$

$$= 3.65\ln[(6.20 + V)/(6.20 - V)]$$

| $V$, fps | $t$, s | $Q$, cfs |
|----------|--------|----------|
| 0 | 0 | 0 |
| 1 | 1.19 | 0.196 |
| 3 | 3.85 | 0.589 |
| 5 | 8.15 | 0.982 |
| 5.6 | 10.87 | 1.100 |
| 6.20 | ∞ | 1.217 |

**19.44** Repeat Prob. 19.43 for the case where the pipe length is 360 ft rather than 3600 ft.

❚ $L/D = 360/0.50 = 900 < 1000$, minor losses are *not* negligible; $k_L = 0.5$ (flush entrance). $k = k_L + fL/D = 0.5 + 0.022(360)/(\frac{6}{12}) = 16.34$, $V_0^2 = 2gH/(1 + k) = 2(32.2)95/(1 + 16.34)$; steady $V_0 = 18.78$ fps.

$$t = [L/(1 + k)V_0]\ln[(V_0 + V)/(V_0 - V)] = [360/17.34(18.78)]\ln[(V_0 + V)/(V_0 - V)]$$

$$= 1.105\ln[(18.78 + V)/(18.78 - V)]$$

| $V$, fps | $t$, s | $Q$, cfs |
|----------|--------|----------|
| 0 | 0 | 0 |
| 5 | 0.603 | 0.982 |
| 10 | 1.312 | 1.963 |
| 15 | 2.420 | 2.945 |
| 18 | 4.258 | 3.534 |
| 18.78 | ∞ | 3.687 |

**19.45** A 10-cm-diameter pipe of length 400 m drains a reservoir. The elevation difference between the reservoir water surface and the pipe outlet is 50 m. Initially there is no flow because a valve at the pipe outlet is closed. The valve is suddenly opened; tabulate $Q$ versus $t$, given $f = 0.028$.

■ $L/D = 400/0.10 > 1000$, neglect minor losses, i.e., $k_L = 0$, $k = k_L + fL/D = 0 + 0.028(400)/0.10 = 112$, $V_p^2 = 2gH/(1 + k) = 2(9.81)50/(1 + 112)$; steady-flow velocity $V_0 = 2.95$ m/s.

$$t = [L/(1 + k)V_0] \ln [(V_0 + V)/(V_0 - V)] = [400/112(2.95)] \ln [(V_0 + V)/(V_0 - V)]$$
$$= 1.211 \ln [(2.95 + V)/(2.95 - V)]$$

| $V$, m/s | $t$, s | $Q$, L/s |
|---|---|---|
| 0 | 0 | 0 |
| 1.0 | 0.855 | 7.85 |
| 2.0 | 1.999 | 15.7 |
| 2.3 | 2.53 | 18.1 |
| 2.6 | 3.35 | 20.4 |
| 2.8 | 4.42 | 22.0 |
| 2.9 | 5.77 | 22.8 |
| 2.95 | ∞ | 23.2 |

**19.46** A large water reservoir is drained by a pipeline that consists of 225 ft of 4-in-diameter pipe ($f = 0.028$) followed by 490 ft of 12-in-diameter pipe ($f = 0.022$). The point of discharge is 125 ft below the elevation of the reservoir water surface. A valve at the discharge end of the pipe is initially closed. It is then quickly opened. Derive an equation applicable to this situation, and tabulate flow rate versus time. Neglect minor losses.

■ $h = f_1(L_1/D_1)(V_1^2/2g) - f_2(L_2/D_2)(V_2^2/2g) = V_2^2/2g + (L_1/g)(dV_1/dt) + (L_2/g)(dV_2/dt)$. By continuity: $V_1 = V_2(D_2/D_1)^2$; $dV_1 = dV_2(D_2/D_1)^2$. Eliminate $V_1$: $h - [f_1(L_1/D_1)(D_2/D_1)^4 + f_2(L_2/D_2) + 1](V_2^2/2g) = [L_1(D_2/D_1)^2 dV_2]/(g\ dt) + (L_2/g)(dV_2/dt)$. For steady-state condition when $dV_2/dt = 0$ and $V_2 = V_{20}$. Substituting known values yields steady-flow $V_{20} = 2.28$ fps. In general, rearranging: $dt = \{2[L_1(D_2/D_1)^2 + L_2]/[1 + f_1(L_1/D_1)(D_2/D_1)^4 + f_2(L_2/D_2)]\}[dV/(V_{20}^2 - V^2)]$; integrating

$$t = \{[L_1(D_2/D_1)^2 + L_2]/[1 + f_1(L_1/D_1)(D_2/D_1)^4 + f_2(L_2/D_2)]\}(1/V_{20}) \ln [(V_{20} + V_2)/(V_{20} - V_2)]$$
$$= 0.7150 \ln [(2.28 + V_2)/(2.28 - V_2)]$$

| $V_1$, fps | $V_2$, fps | $t$, s | $Q$, cfs |
|---|---|---|---|
| 0 | 0 | 0 | 0 |
| 11.25 | 1.25 | 0.88 | 0.98 |
| 13.50 | 1.5 | 1.13 | 1.18 |
| 15.75 | 1.75 | 1.45 | 1.37 |
| 18.00 | 2.0 | 1.95 | 1.57 |
| 19.80 | 2.20 | 2.88 | 1.73 |
| 20.52 | 2.28 | ∞ | 1.79 |

**19.47** A 12-in-diameter pipe ($f = 0.022$) of length 360 ft is connected to a reservoir. Entrance losses are negligible. At the discharge end of the pipe is a nozzle that produces a 4.8-in-diameter jet. The elevation difference between the jet and the water surface in the reservoir is 52 ft. The nozzle has a coefficient of velocity of 0.95. Initially, there is a tight-fitting plug in the nozzle, which is then removed. For this situation derive an equation and tabulate flow rate versus time. Assume that the liquid level in the reservoir does not drop.

■ $h - k(V^2/2g) - [(1/C_V^2) - 1][1 - (D_j/D)^4](V_j^2/2g) = (V_j^2/2g) + (L/g)(dV/dt)$. But $V_j = V(A/A_j) = V(D/D_j)^2$, so $h = C(V^2/2g) + (L/g)(dV/dt)$, where $C = k + (D/D_j)^4\{1 + [(1/C_V^2) - 1][1 - (D_j/D)^4]\}$. For steady-state conditions when $dV/dt = 0$ and $V = V_0$: $h = CV_0^2/2g$, or $V_0 = (2gh/C)^{1/2}$. Eliminating $h$ between the above two equations and rearranging: $dt = (2L/C)[dV/(V_0^2 - V^2)]$ and integrating $t = (L/CV_0) \ln [(V_0 + V)/(V_0 - V)]$. Given $k_L = 0$: $k = 0 + 0.022(360)/(\frac{12}{12}) = 7.92$. Given $C_V = 0.95$ and $D/D_j = \frac{12}{4.8}$: $C = 50.4$, $V_0 = [2(32.2)52/50.4]^{1/2} = 8.15$ fps, $t = [360/50.4(8.15)] \ln [(V_0 + V)/(V_0 - V)] = 0.876 \ln [(8.15 + V)/(8.15 - V)]$.

| $V_J$, fps | $V$, fps | $t$, s | $Q$, cfs |
|---|---|---|---|
| 0 | 0 | 0 | 0 |
| 12.5 | 2 | 0.439 | 1.57 |
| 25.0 | 4 | 0.941 | 3.14 |
| 37.5 | 6 | 1.65 | 4.71 |
| 50.0 | 8 | 4.10 | 6.28 |
| 50.6 | 8.1 | 5.07 | 6.36 |
| 50.9 | 8.15 | $\infty$ | 6.40 |

**19.48** An open tank containing oil ($s = 0.85$, $\mu = 0.0005$ lb-s/ft$^2$) is connected to a 3-in-diameter smooth pipe of length 3800 ft. The elevation drop from the liquid surface in the tank to the point of discharge is 19 ft. A valve on the discharge end of the pipe, initially closed, is then opened. Tabulate the ensuing flow rate versus time.

**∎** $L/D = 3800/(\frac{3}{12}) = 15\,200 > 1000$, neglect minor losses. Assume laminar flow, neglect velocity head: $h_L = 19 = 32(\mu/\gamma)(L/D^2)V$, $V = 19(0.85)(62.4)(0.25)^2/[32(0.0005)(3800)] = 1.036$ fps, $N_R = DV\rho/\mu = 0.25(1.036)(0.85)(1.94)/0.0005 = 854$. $N_R < 2000$, therefore flow *is* laminar, assumption was O.K. Assuming the velocity head is negligible: $h - 32(\mu/\gamma)(L/D^2)V = (L/g)(dV/dt)$, or $h - \alpha V = \beta(dV/dt)$, where $\alpha = 32(\mu/\gamma)(L/D^2) = 32(0.0005)3800/[(0.85)(62.4)(0.25)^2] = 18.3$ and $\beta = L/g = 3800/32.2 = 118$.

$$dt = \frac{\beta\, dV}{h - \alpha V} = \frac{dV}{(h/\beta) - (\alpha/\beta)V}$$

$$t = \frac{-1}{\alpha/\beta}\ln\left[\left(\frac{h}{\beta}\right) - \left(\frac{\alpha}{\beta}\right)V\right]_0^V = -6.45\ln[0.1610 - 0.155V]_0^V = -6.45\ln\left(\frac{0.1610 - 0.155V}{0.1610}\right) = -6.45\ln(1 - 0.963V)$$

| $V$, fps | $t$, s | $Q$, cfs |
|---|---|---|
| 0 | 0 | 0 |
| 0.20 | 1.38 | 0.0098 |
| 0.40 | 3.14 | 0.0196 |
| 0.60 | 5.56 | 0.0295 |
| 0.80 | 9.49 | 0.0393 |
| 1.00 | 21.26 | 0.0491 |
| 1.036 | $\infty$ | 0.0509 |

Check neglect of velocity head: $(V^2/2g)/h_L = (V^2/2g)/\alpha V = V/(2g\alpha) = V/1179$. As $V$ varies from 0 to 1.036, $(V^2/2g)/h_L$ varies from 0 to 0.000877; therefore $V^2/2g < 0.022$ percent of $h_L$, and velocity head *is* negligible.

**19.49** An open tank containing oil ($s = 0.82$, $v = 0.002$ m$^2$/s) is connected to a 125-mm-diameter pipe of length 470 m. The oil surface in the tank is 3.5 m above the pipe outlet. A valve at the end of the pipe, initially closed, is suddenly opened. Plot the ensuing flow rate as a function of time.

**∎** $L/D = 470/0.125 = 3760 > 1000$, neglect minor losses. Assuming laminar flow for $h_L$, $H - 32v(L/gD^2)V = (L/g)(dV/dt)$; $3.5 - 32(0.002)[470/9.81(0.125)^2]V = (470/9.81)(dV/dt)$, $3.5 - 196V = 47.9\,dV/dt$; $dt = (47.9\,dV)/(3.5 - 196V) = (13.69\,dV)/(1 - 56.00V)$. For steady-state conditions when $dV/dt = 0$ and $V = V_0$: $3.5 - 196V_0 = 0$; steady-flow velocity $V_0 = 0.01786$ m/s. Integrating:

$$t = -(13.69/56.00)[\ln(1 - 56.00V)]_0^V = -0.2445\ln(1 - 56.00V)$$

| $V$, m/s | $t$, s | $Q$, L/s |
|---|---|---|
| 0.001 | 0.014 | 0.0123 |
| 0.002 | 0.029 | 0.0245 |
| 0.005 | 0.080 | 0.0614 |
| 0.010 | 0.201 | 0.123 |
| 0.015 | 0.448 | 0.184 |
| 0.017 | 0.742 | 0.209 |
| 0.01786 | $\infty$ | 0.219 |

$(N_R)_{max} = (0.125)0.01786/0.002 = 1.12$. $N_R < 2000$, so flow *is* laminar (as assumed) and answers are valid.

**19.50** Find the celerity of a pressure wave in benzene (s.g. = 0.90, $E_v$ = 150 000 psi) contained in a 9-in-diameter steel pipe having a wall thickness of 0.300 in.

$$c = (E_v/\rho)^{1/2} = [(150\,000)(144)/(0.9)(1.94)]^{1/2} = 3520 \text{ fps}$$

$$c_p = c/\sqrt{1 + (D/t)(E_v/E)} = 3520/\{1 + (9/0.300)[150\,000/(30 \times 10^6)]\}^{1/2} = 3282 \text{ fps}$$

**19.51** (a) What is the celerity of a pressure wave in a 4.5-ft-diameter water pipe with 0.4-in steel walls? (b) If the pipe is 3750 ft long, what is the time required for a pressure wave to make the round trip from the valve?

(a) $c = 4720$ fps in water

$$c_p = c/\sqrt{1 + (D/t)(E_v/E)} = 4720/\{1 + [(4.5)(12)/0.4][300\,000/(30 \times 10^6)]\}^{1/2} = 3079 \text{ fps}$$

(b) $$T_r = 2L/c_p = 2(3750)/3079 = 2.44 \text{ s}$$

**19.52** In Prob. 19.51, (a) if the initial water velocity is 7 fps, what will be the rise in pressure at the valve if the time of closure is less than the time of a round trip? (b) If the valve is closed at such a rate that the velocity in the pipe decreases uniformly with respect to time and closure is completed in a time $t_c = 5L/c_P$, approximately what will be the pressure head at the valve when the first pressure unloading wave reaches the valve?

(a) $$p = (\gamma/g)c_p V = (62.4/32.2)(3079)7 = 41\,767 \text{ psf} = 669 \text{ ft}$$

(b) If velocity changes uniformly from 7.0 fps to 0 in time $5L/c_p$, the change in $V$ in $t' = T_r = 2L/c_p$ is $(\frac{2}{5})7.0 = 2.80$ fps. $\Delta p'/\gamma = c_p\,\Delta V/g = 3079(2.80)/32.2 = 268$ ft.

**19.53** For the situation depicted in Prob. 19.27, find the water-hammer pressure at the valve if a flow of 75 cfs is reduced to 20 cfs in 3.2 s. Under these conditions what would be the maximum water-hammer pressures at points 600 ft and 1600 ft from the reservoir?

$\Delta p'_h = (T_r/t_c)(\Delta p_h)$, $V_1 = 75/4\pi = 5.97$ fps, $V_2 = 20/4\pi = 1.592$ fps, $\Delta V = 4.38$ fps over 3.2 s; $\Delta p_h = \rho c_p(\Delta V) = 1.94(3200)(4.38) = 27\,200$ psf for instantaneous drop in flow rate. $T_r = 2L/c_p = \frac{4000}{3200} = 1.250$ s. At valve: $\Delta p'_h \simeq (1.25/3.2)(27\,200) = 10\,625$ psf = 73.8 psi.
600 ft from reservoir: $\Delta p'_h = (\frac{600}{2000})73.8 = 22.1$ psi.
1600 ft from reservoir: $\Delta p'_h = (\frac{1600}{2000})73.8 = 59.0$ psi.

**19.54** Water is flowing through a 35-cm-diameter welded steel pipe ($f = 0.014$) of length 2300 m that drains a reservoir under a head of 50 m. The pipe has a thickness of 9 mm. If a valve at the end of the pipe is closed in 9 s, approximately what water-hammer pressure will be developed?

$$h_f = (f)(L/D)(V^2/2g) \qquad 50 = 0.014(2300/0.35)V^2/2g \qquad V = 3.26 \text{ m/s}$$

$$p'_h = 2LV\rho/t_c = [2(2300)3.26(1000)/9] = 1.667 \text{ MPa}$$

**19.55** In Fig. 19-12, the total length of pipe is 9800 ft, its diameter is 30 in, and its thickness is $\frac{5}{8}$ in. Assume $E = 30\,000\,000$ psi and $E_v = 300\,000$ psi. If the initial velocity for steady flow is 8 fps and the valve at $G$ is partially closed so as to reduce the flow to half of the initial velocity in 3 s, find (a) the maximum pressure rise from the water hammer; (b) the location of the point of maximum total pressure.

(a) $c = 4720$ ft/s $\qquad c_p = c/\sqrt{1 + (D/t)(E_V/E)} = 4720/[1 + (30/\frac{5}{8})(300\,000/30\,000\,000)]^{1/2} = 3880 \text{ fps}$

$$\Delta p = (\gamma/g)c_p(\Delta V) = (62.4/32.2)(\tfrac{3880}{144})4 = 209 \text{ psi} \quad \text{or} \quad 482 \text{ ft}$$

(b) $x_0 = c_p t_c/2 = 3880(3)/2 = 5820$ ft. Thus water-hammer pressure rise, at $B$: $\Delta p = 209(\frac{1800}{5820}) = 64.6$ psi; at $E$: $\Delta p = 209(100 \text{ percent}) = 209$ psi. Total pressure = static pressure plus water hammer. Total pressure at $B$: $p/\gamma = 700 + 64.6(144/62.4) = 849$ ft; at $E$: $p/\gamma = 100 + 209(144/62.4) = 582$ ft. Maximum total pressure occurs at point $B$.

**19.56** Refer to Fig. 19-12, but take all the dimensions given in feet to be in meters instead. This 9.8-km-long pipe has a diameter of 1.4 m and a wall thickness of 30 mm. Assume $E = 200$ GPa and $E_v = 2$ GPa. The initial steady-flow velocity is 5 m/s. The valve at $G$ is then partially closed so as to reduce the velocity to 1 m/s in 9 s. Find (a) the maximum pressure rise due to water hammer, and (b) the location of the point of maximum total pressure.

**Fig. 19-12**

**(a)**     $c_p = 1440$ m/s     $c_p = c/\sqrt{1 + (D/t)(E_v/E)} = 1440/[1 + (\frac{1400}{30})(\frac{2}{200})]^{1/2} = 1189$ m/s

$$\Delta p = \rho c_p \, \Delta V = 1000(1189)4 = 4.756 \text{ MPa}$$

**(b)** $x_0 = c_p t_c/2 = 1189(9)/2 = 5350$ m. Thus water-hammer pressure rise, at $B$: $\Delta p = 4.756(\frac{1800}{5350}) = 1.600$ MPa, or 163.4 m; at $E$: $\Delta p = 4.756$ MPa or 486 m. Total pressure = static pressure plus water-hammer pressure. Total pressure at $B$: $p/\gamma = 700$ m + 163.4 m = 863.4 m; at $E$: $p/\gamma = 100$ m + 486 m = 586 m. Maximum total pressure occurs at point $B$.

**19.57** Derive the equation $V^2 = (2gAD^2/LA_s f^2)[1 - (fA_s/AD)z] - Ce^{-(fA_s/AD)z}$, which expresses the relationship between velocity in a pipe and water-surface level in a tank over an interval from valve closure to the top of the first surge.

▌ Write $dV/dt$ as $(dV/dz)(dz/dt)$ in $0 - f(L/D)(V^2/2g) = z + (L/g)(dV/dt)$ and substitute for $dz/dt$ from $AV = A_s \, (dz/dt)$; $0 - (fL/D)(V^2/2g) = z + (L/g)(dV/dz)(AV/A_s)$ or $z + (fL/2gd)V^2 + (LA/2gA_s)[(2V \, dV)/dz] = 0$. Define $K = fL/2gD$; $M = (L/2g)(A/A_s)$; and $V^2 = x$ so that $2V \, dV = dx$. This gives $z + kx + (M \, dx)/dz = 0$ or $dx/dz + (K/M)x + (1/M)z = 0$. Now define $\alpha = K/M$, $\beta = 1/M$; then $dx/dz + \alpha Z + \beta z = 0$. Multiply by $e^{\alpha z}$: $(dx/dz)e^{\alpha z} + \alpha x e^{\alpha z} + \beta z e^{\alpha} = 0$, i.e., $d/dz(xe^{\alpha z}) + e^{\alpha z}\beta z = 0$ or $d(xe^{\alpha z}) + e^{\alpha z}\beta z \, dz = 0$. Integrate: $\int d(xe^{\alpha z}) + \int e^{\alpha z}\beta z \, dz = 0$, $xe^{\alpha z} + \beta(e^{\alpha z}/\alpha^2)(\alpha z - 1) + C = 0$ or $x = (\beta/\alpha^2)(1 - \alpha z) - Ce^{-\alpha z}$. But $\beta/\alpha^2 = M/K^2 = 2gAD^2/LA_s f^2$, $\alpha = K/M = fA_s/AD$, and $x = V^2$; therefore $V^2 = (2gAD^2/LA_s f^2)[1 - (fA_s/AD)z] - Ce^{-(fA_s/AD)z}$.

**19.58** A 36-in steel pipe 3300 ft long supplies water to a small power plant. What height would be required for a simple surge tank 7 ft in diameter situated 59 ft upstream from the valve at a point where the centerline of the pipe is 100 ft below the water surface in the reservoir if the tank is to protect against instantaneous closure of a valve at the plant? The valve is 130 ft below reservoir level, and the discharge is 140 cfs. Take $f = 0.014$. The surge tank is not to overflow. Neglect all velocity heads and minor losses; in the surge tank (only) neglect fluid friction and inertial effects.

▌     $V^2 = (2gAD^2/LA_s f^2)[1 - (fA_s/AD)z] - Ce^{-(fA_s/AD)z}$

$$= [2(32.2)/3241](\tfrac{3}{7})^2(3/0.014)^2[1 - (0.014/3)(\tfrac{7}{3})^2 z] - Ce^{-(0.014/3)(7/3)^2 z} = 168(1 - 0.025z) - Ce^{-0.025z}$$

First find the constant of integration $C$, from initial (steady) conditions, when $Q_1 = 140$ cfs, $V_1 = Q_1/A = 140/[(\pi/4)3^2] = 19.8$ fps, and $-z_1 = (fL/D)(V_1^2/2g) = (0.014)(\frac{3241}{3})19.8^2/[(2)(32.2)] = 92.1$ ft. Substituting these values $19.8^2 = 168[1 - 0.025(-92.1)] - Ce^{-0.025(-92.1)}$, from which $C = 16.3$. For the first surge, $z = z_{max}$ when $V = 0$, so $168(1 - 0.025z_{max}) - 16.3e^{-0.025z_{max}}$. [*Note:* The last (3rd) term is always negative. For a positive middle (2nd) term we need $0.025z_{max} < 1$, i.e., $z_{max} < 1/0.025 = 40$.] By trial and error: $z_{max} = 37.5$ ft. Required height of surge tank = $(100 + 37.5) = 137.5$ ft.

**19.59** Repeat Prob. 19.58 for the case where the surge tank is to have a diameter of 11 ft.

▌ Following the procedure of Prob. 19.58: $V^2 = 67.87(1 - 0.0627z) - Ce^{-0.0627z}$. From initial (steady) conditions: $z_1 = -92.1$ ft, $C = 0.211$. (Note the large change in $C$ from Prob. 19.58.) For the first surge, we need $0.0627z_{max} < 1$, i.e., $z_{max} < 15.95$ ft. By trial and error: $z_{max} = 15.93$ ft. Required height of surge tank = $(100 + 15.93) = 115.9$ ft.

**19.60** Using the data of Prob. 19.58, find the diameter of a surge tank that will produce a surge requiring a tank height of 145 ft.

▌ As in Prob. 19.58 for initial (steady) conditions: $V_1 = 19.8$ fps, $z_1 = -92.1$ ft. Substituting these and other known values into $V^2 = (2gAD^2/LA_s f^2)[1 - (fA_s/AD)z] - Ce^{-(fA_s/AD)z}$, while writing $A/A_s$ as $(D/D_s)^2$: $19.8^2 = 8212/D_s^2 + 392 - Ce^{0.0478D_s^2}$, that is,

$$0 = 8212/D_s^2 - Ce^{0.0478D_s^2} \tag{1}$$

When $z = z_{max} = 145 - 100 = 40$ ft, $V = 0$; substituting these and other known values

$$0 = 8212/D_s^2 - 168 - Ce^{-0.0207D_s^2} \tag{2}$$

Eliminating the unknown $C$ between Eqs. (1) and (2): $1 - 0.0205D_s^2 = e^{-0.0695D_s^2}$. By trial and error: $D_s = 6.85$ ft.

**19.61** A 0.95-m-diameter steel pipe ($f = 0.015$) of length 1100 m carries drinking water from a reservoir. What height would be required for a simple surge tank 3.2 m in diameter situated 11 m upstream from the valve at a point where the centerline of the pipe is 38 m below the water surface in the reservoir? Assume instantaneous closure of the valve. The valve is 48 m below reservoir level and the flow is 3.4 m³/s. The surge tank is not to overflow. Neglect all velocity heads and minor losses; in the surge tank (only) neglect fluid friction and inertial effects.

**▋** $V^2 = (2gAD^2/LA_sf^2)[1 - (fA_s/AD)z] - Ce^{-(fA_s/AD)z}$

$= [2(9.81)/1089](0.95/3.2)^2(0.95/0.015)^2[1 - (0.015/0.95)(3.2/0.95)^2z] - Ce^{-(0.015/0.95)(3.2/0.95)^2z}$

$= 6.37(1 - 0.1792z) - Ce^{-0.1792z}$

First, find the constant of integration $C$ from initial (steady) conditions when $Q_1 = 3.4$ m³/s, $V_1 = Q_1/A = 3.4/[(\pi/4)0.95^2] = 4.80$ m/s, and $-z_1 = (fL/D)(V_1^2/2g) = (0.015)(1089/0.95)4.80^2/[(2)(9.81)] = 20.19$ m. Substituting these values, $4.80^2 = 6.37[1 - 0.1792(-20.19)] - Ce^{-0.1792(-20.19)}$, from which $C = 0.1712$. For the first surge, $z = z_{max}$ when $V = 0$, so $0 = 6.37(1 - 0.1792z_{max}) - 0.1712e^{-0.1792z_{max}}$. [*Note:* The last (3rd) term is always negative. For a positive middle (2nd) term we need $0.1792z_{max} < 1$, i.e., $z_{max} < 1./0.1792 = 5.58$ m.] By trial and error: $z_{max} = 5.52$ m. Required height of surge tank $= 38 + 5.52 = 43.52$ m.

**19.62** A 1.5-m-diameter sluice ($f = 0.018$) 980 m long carries water from a dam. When the outlet valve is closed instantaneously, water rises in a 2-m-diameter surge tank immediately adjacent to the outlet valve. Determine the maximum allowable initial discharge $Q$ so that the resulting surge will not rise more than 9 m above the reservoir water surface. Neglect all velocity heads and minor losses; in the surge tank (only) neglect fluid friction and inertial effects.

**▋** $V^2 = (2gAD^2/LA_sf^2)[1 - (fA_s/AD)z] - Ce^{-(fA_s/AD)z}$

$= [2(9.81)/980](1.5/2)^2(1.5/0.018)^2[1 - (0.018/1.5)(2/1.5)^2] - Ce^{-(0.018/1.5)(2/1.5)^2z}$

$= 78.2(1 - 0.0213z) - Ce^{-0.0213z}$

For the first surge, $z = z_{max} = 9$ m when $V = 0$, so $0 = 78.2[1 - 0.0213(9)] - Ce^{-0.0213(9)}$; from which $C = 76.6$. For the initial conditions, when $V =$ steady-flow velocity $= V_1$, $z_1 = -h_L = -(fL/D)(V_1^2/2g)$. Substitute this expression for $z_1$ for $V_1^2$, and simplify: $V_1^2 = B[1 + (V_1^2/B)] - Ce^{V_1^2/B}$, where $B = 78.2$, $B = Ce^{V_1^2/B}$ or $\ln(B/C) = V_1^2/B$, $V_1 = [B \ln(B/C)]^{1/2} = [78.2 \ln(78.2/76.6)]^{1/2} = 1.271$ m/s. So $Q = AV = (\pi/4)(1.5)^2(1.271) = 2.25$ m³/s.

# CHAPTER 20
# Pumps and Fans

**20.1** A pump delivers 0.300 m³/s against a head of 200 m with a rotative speed of 2000 rpm. Find the specific speed.

$$N_s = 51.64 N Q^{0.5}/H^{0.75} = (51.64)(2000)(0.300)^{0.5}/200^{0.75} = 1064$$

**20.2** A pump delivers 0.019 m³/s against a head of 16.76 m with a rotative speed of 1750 rpm. Find the specific speed.

$$N_s = 51.64 N Q^{0.5}/H^{0.75} = (51.64)(1750)(0.019)^{0.5}/16.76^{0.75} = 1504$$

**20.3** A radial-flow pump must deliver 2000 gpm against a head of 950 ft. Find the minimum practical rotative speed.

$N_s = N Q^{0.5}/H^{0.75}$. From Fig. A-40, it is apparent that the minimum practical specific speed for a radial-flow pump is about 500. Hence, $500 = (N)(2000)^{0.5}/950^{0.75}$, $N = 1913$ rpm.

**20.4** A radial-flow pump must deliver 300 gpm against a head of 30 ft. Find the operating rotative speed.

$N_s = N Q^{0.5}/H^{0.75}$. From Fig. A-40, $N_s = 2500$ for maximum efficiency of 92.5 percent for a radial-flow pump. Hence, $2500 = (N)(300)^{0.5}/30^{0.75}$, $N = 1850$ rpm.

**20.5** An axial-flow pump is to be operated at a rotative speed of 2500 rpm against a head of 400 m. What flow rate will be delivered by the most efficient pump?

From Fig. A-40, it is apparent that the most efficient axial-flow pump has a specific speed of around 12 500. $N_s = 51.64 N Q^{0.5}/H^{0.75}$, $12\,500 = (51.64)(2500)(Q)^{0.5}/400^{0.75}$, $Q = 75.0$ m³/s.

**20.6** A mixed-flow pump is to be operated at a rotative speed of 1500 rpm against a head of 15 m at maximum pump efficiency. Determine the flow rate the pump will deliver.

From Fig. A-40, it is apparent that the most efficient mixed-flow pump has a specific speed of around 6500. $N_s = 51.64 N Q^{0.5}/H^{0.75}$, $6500 = (51.64)(1500)(Q)^{0.5}/15^{0.75}$, $Q = 0.409$ m³/s.

**20.7** A radial-flow pump operating at maximum efficiency is to deliver 260 gpm against a head of 129 ft at a rotative speed of 2100 rpm. Find the required number of stages (i.e., impellers).

From Fig. A-40, it is apparent that maximum efficiency for a radial-flow pump is about 93 percent at a specific speed of 2500. $N_s = N Q^{0.5}/H^{0.75}$, $2500 = (2100)(260)^{0.5}/H^{0.75}$, $H = 32.29$ ft. Since the given head is 129 ft, a total of 129/32.29, or 4 stages will be needed.

**20.8** A radial-flow pump operating at maximum efficiency is to deliver 400 gpm against a head of 191 ft at a rotative speed of 1920 rpm. Find the required number of stages (i.e., impellers).

From Fig. A-40, it is apparent that maximum efficiency for a radial-flow pump is about 93 percent at a specific speed of 2500. $N_s = N Q^{0.5}/H^{0.75}$, $2500 = (1920)(400)^{0.5}/H^{0.75}$, $H = 38.18$ ft; number of stages = 191/38.18 = 5.

**20.9** The value of $(\text{NPSH})_{\min}$ for a pump is given by the manufacturer as 20 ft. Water is being pumped from a reservoir at a rate of 25 cfs. The water level in the reservoir is 6.0 ft below the pump. Atmospheric pressure is 14.7 psia and water temperature is 40 °F. If the total head loss in the suction pipe is 4.0 ft, is the pump safe from cavitation effects?

$$\text{NPSH} = p_A/\gamma - z_s - h_L - p_v/\gamma = (14.7)(144)/62.4 - 6.0 - 4.0 - 18.5/62.4 = 23.6 \text{ ft}$$

Since $[\text{NPSH} = 23.6] > [(\text{NPSH})_{\min} = 20]$, cavitation should not be a problem.

**20.10** The $(\text{NPSH})_{\min}$ for a pump is given by the manufacturer as 7.0 m. This pump is being used to pump water from a reservoir at a rate of 0.2832 m³/s. The water level in the reservoir is 1.280 m below the pump. Atmospheric pressure is 98.62 kN/m² and water temperature is 20 °C. Assume total head loss in the suction pipe is 1.158 m of water. Determine whether or not the pump is safe from cavitation effects.

**20.11** A commercial pump is operating at 2150 rpm and delivers 1800 gpm against a head of 340 ft. Find the approximate efficiency of the pump.

$$N_s = NQ^{0.5}/H^{0.75} = (2150)(1800)^{0.5}/340^{0.75} = 1152$$

From Fig. A-41, with $N_s = 1152$ and $Q = 1800$ gpm, the approximate efficiency of the pump is determined to be 82 percent.

**20.12** A pump operating at 1600 rpm delivers 0.189 m³/s against a head of 47.03 m. Determine the approximate efficiency of the pump.

$$N_s = 51.64NQ^{0.5}/H^{0.75} = (51.64)(1600)(0.189)^{0.5}/47.03^{0.75} = 2000$$

From Fig. A-41, with $N_s = 2000$ and $Q = 0.189$ m³/s, the approximate efficiency of the pump is determined to be 87 percent.

**20.13** A centrifugal pump with a 700-mm-diameter impeller runs at 1800 rpm. The water enters without whirl, and $\alpha_2 = 60°$. The actual head produced by the pump is 17 m. Find its hydraulic efficiency when $V_2 = 6$ m/s.

The theoretical head is $H = (u_2 V_2 \cos \alpha_2)/g = [(\frac{1800}{60})(2\pi)(\frac{700}{1000})/2](6)(\cos 60°)/9.807 = 20.18$ m, $e_h = 17/20.18 = 0.842$, or 84.2 percent.

**20.14** A centrifugal water pump has an impeller (Fig. 20-1a) with $r_2 = 12$ in, $r_1 = 4$ in, $\beta_1 = 20°$, $\beta_2 = 10°$. The impeller is 2 in wide at $r = r_1$ and $\frac{3}{4}$ in wide at $r = r_2$. For 1800 rpm, neglecting losses and vane thickness, determine (a) the discharge for shockless entrance when $\alpha_1 = 90°$, (b) $\alpha_2$ and the theoretical head $H$, (c) the horsepower required, and (d) the pressure rise through the impeller.

(a) The peripheral speeds are $u_1 = \frac{1800}{60}(2\pi)(\frac{1}{3}) = 62.8$ ft/s, $u_2 = 3u_1 = 188.5$ ft/s. The vector diagrams are shown in Fig. 20-1b. With $u_1$ and the angles $\alpha_1$, $\beta_1$ known, the entrance diagram is determined, $V_1 = u_1 \tan 20° = 22.85$ ft/s; hence, $Q = 22.85(\pi)(\frac{2}{3})(\frac{2}{12}) = 7.97$ cfs.

(b) At the exit the radial velocity $V_{r2}$ is $V_{r2} = 7.97(12)/[2\pi(0.75)] = 20.3$ ft/s. By drawing $u_2$ (Fig. 20-1b) and a parallel line distance $V_{r2}$ from it, the vector triangle is determined when $\beta_2$ is laid off. Thus, $v_{u2} = 20.3 \cot 10° = 115$ ft/s, $V_{u2} = 188.5 - 115 = 73.5$ ft/s, $\alpha_2 = \tan^{-1}(20.3/73.5) = 15°26'$, $V_2 = 20.3 \csc 15°26' = 76.2$ ft/s; $H = (u_2 V_2 \cos \alpha_2)/g = u_2 V_{u2}/g = 188.5(73.5)/32.2 = 430$ ft.

(c) $$\text{Power} = Q\gamma H/550 = [7.97(62.4)(430)]/550 = 338 \text{ hp}$$

(d) By applying the energy equation from entrance to exit of the impeller, including the energy $H$ added (elevation change across impeller is neglected), $H + (V_1^2/2g) + (p_1/\gamma) = (V_2^2/2g) + (p_2/\gamma)$ and $(p_2 - p_1)/\gamma = 430 + (22.85^2/64.4) - (76.2^2/64.4) = 348$ ft or $p_2 - p_1 = 348(0.433) = 151$ psi.

(a)  **Fig. 20-1**

(b)

**Fig. 20-1** (*continued*)

**20.15** Tests of a pump model indicate a $\sigma_c$ of 0.10. A homologous unit to be installed at a location where $p_a = 90$ kPa and $p_v = 3.5$ kPa is to pump water against a head of 25 m. The head loss from suction reservoir to pump impeller is 0.35 N·m/N. What is the maximum permissible suction head?

$$\sigma' = V_e^2/2gH = (p_a - p_v - \gamma z_s + h_l)/\gamma H$$

$$z_s = [(p_a - p_v)/\gamma] - \alpha' H + h_l = [(90\,000 - 3500)/9806] - 0.10(25) + 0.35 = 6.67 \text{ m}$$

**20.16** A prototype test of a mixed-flow pump with a 72-in-diameter discharge opening, operating at 225 rpm, resulted in the following characteristics:

| $H$, ft | $Q$, cfs | $e$, % | $H$, ft | $Q$, cfs | $e$, % | $H$, ft | $Q$, cfs | $e$, % |
|---|---|---|---|---|---|---|---|---|
| 60 | 200 | 69 | 47.5 | 330 | 87.3 | 35 | 411 | 82 |
| 57.5 | 228 | 75 | 45 | 345 | 88 | 32.5 | 425 | 79 |
| 55 | 256 | 80 | 42.5 | 362 | 87.4 | 30 | 438 | 75 |
| 52.5 | 280 | 83.7 | 40 | 382 | 86.3 | 27.5 | 449 | 71 |
| 50 | 303 | 86 | 37.5 | 396 | 84.4 | 25 | 459 | 66.5 |

What size and synchronous speed (60 Hz) of homologous pump should be used to produce 200 cfs at 60-ft head when operating at highest efficiency? Find the characteristic curves for this case.

❙ Subscript 1 refers to the 72-in pump. For best efficiency $H_1 = 45$, $Q_1 = 345$, $e = 88$ percent. $H/N^2D^2 = H_1/N_1^2D_1^2$, $Q/ND^3 = Q/N_1D_1^3$, or $60/N^2D^2 = 45/[225^2(72^2)]$, $200/ND^3 = 345/[225(72^3)]$. After solving for $N$ and $D$, $N = 366.7$ rpm, $D = 51.0$ in. The nearest synchronous speed (3600 divided by number of pairs of poles) is 360 rpm. To maintain the desired head of 60 ft, a new $D$ is necessary. When its size is computed, $D = \sqrt{\frac{60}{45}\left(\frac{225}{360}\right)}(72) = 52$ in, the discharge at best efficiency is $Q = Q_1ND^3/N_1D_1^3 = 345\left(\frac{360}{225}\right)\left(\frac{52}{72}\right)^3 = 208$ cfs, which is slightly more capacity than required. With $N = 360$ and $D = 52$, equations for transforming the corresponding values of $H$ and $Q$ for any efficiency can be obtained: $H = H_1(ND/N_1D_1)^2 = H_1[\left(\frac{360}{225}\right)\left(\frac{52}{72}\right)]^2 = 1.335H_1$ and $Q = Q_1(ND^3/N_1D_1^3) = Q_1\left(\frac{360}{225}\right)\left(\frac{52}{72}\right)^3 = 0.603Q_1$. The characteristics of the new pump are

| $H$, ft | $Q$, cfs | $e$, % | $H$, ft | $Q$, cfs | $e$, % | $H$, ft | $Q$, cfs | $e$, % |
|---|---|---|---|---|---|---|---|---|
| 80 | 121 | 69 | 63.5 | 200 | 87.3 | 46.7 | 248 | 82 |
| 76.7 | 138 | 75 | 60 | 208 | 88 | 43.4 | 257 | 79 |
| 73.4 | 155 | 80 | 56.7 | 219 | 87.4 | 40 | 264 | 75 |
| 70 | 169 | 83.7 | 53.5 | 231 | 86.3 | 36.7 | 271 | 71 |
| 66.7 | 183 | 86 | 50 | 239 | 84.4 | 33.4 | 277 | 66.5 |

The efficiency of the 52-in pump might be a fraction of a percent less than that of the 72-in pump, as the hydraulic radii of flow passages are smaller, so that the Reynolds number would be less.

**20.17** Develop a program for calculating homologous pump characteristics and apply it to Prob. 20.16.

```
10 REM                    HOMOLOGOUS CHARACTERISTICS
20 DEFINT I: DIM H1(20),Q1(20),E(20),H(20),Q(20)
30 FOR I=1 TO 15: READ H1(I): NEXT I
40 DATA 60.,57.5,55.,52.5,50.,47.5,45.,42.5,40.,37.5,35.,32.5,30.,27.5,25.
50 LPRINT:LPRINT"H1=";: FOR I=1 TO 15:LPRINT USING"##.# ";H1(I);: NEXT I:LPRINT
60 FOR I=1 TO 15: READ Q1(I): NEXT I
70 DATA 200.,228.,256.,280.,303.,330.,345.,362.,382.,396.,411.,425.,438.,449.,45
9.
80 LPRINT:LPRINT"Q=";: FOR I=1 TO 15: LPRINT USING"###.# ";Q1(I);:NEXT I:LPRINT
90 FOR I=1 TO 15: READ E(I): NEXT I
100 DATA 69.,75.,80.,83.7,86.,87.3,88.,87.4,86.3,84.4,82.,79.,75.,71.,66.5
110 LPRINT:LPRINT"E=";: FOR I=1 TO 15: LPRINT USING"##.# ";E(I);: NEXT I:LPRINT
120 EE=0!: FOR I=1 TO 15: IF E(I)>EE THEN II=I: EE=E(I): NEXT I
```

```
130 LPRINT"II,E(I)=";II;EE
140 READ HH,QQ,NSYN,D1,N1:DATA 60.,200.,3600.,72.,225.
150 LPRINT"HH,QQ,NSYN,D1,N1=";HH;QQ;NSYN;D1;N1
160 D=((QQ/Q1(II))^2*H1(II)/HH)^.25*D1: N=N1*SQR(HH/H1(II))*D1/D
170 LPRINT"D,N=";D;N: I=FIX(NSYN/N)
180 NN1=FIX(NSYN/I): NN2=FIX(NSYN/(I+1)): LPRINT"N,NN1,NN2=";N;NN1;NN2
190 IF (NN1-N)<(N-NN2) THEN N=NN1 ELSE N=NN2
200 D=D1*N1*SQR(HH/H1(II))/N: QQ=(D/D1)^3*N*Q1(II)/N1: LPRINT"D,N,QQ=";D;N;QQ
210 C1=(N*D/(N1*D1))^2: C2=(D/D1)^3*N/N1
220 FOR I=1 TO 15: H(I)=C1*H1(I): Q(I)=C2*Q1(I): NEXT I
230 LPRINT: LPRINT"H=";: FOR I=1 TO 15:LPRINT USING" ###.#";H(I);:NEXT I:LPRINT
240 LPRINT"Q=";: FOR I=1 TO 15: LPRINT USING" ###.#";Q(I);: NEXT I: LPRINT

H1=60.0 57.5 55.0 52.5 50.0 47.5 45.0 42.5 40.0 37.5 35.0 32.5 30.0 27.5 25.0

Q=200.0 228.0 256.0 280.0 303.0 330.0 345.0 362.0 382.0 396.0 411.0 425.0 438.0
449.0 459.0

E=69.0 75.0 80.0 83.7 86.0 87.3 88.0 87.4 86.3 84.4 82.0 79.0 75.0 71.0 66.5
II,E(I)= 7  88
HH,QQ,NSYN,D1,N1= 60 200  3600  72  225
D,N= 51.01563  366.6749
N,NN1,NN2= 366.6749  400  360
D,N,QQ= 51.96153  360  207.4853

H= 80.0 76.7  73.3  70.0  66.7  63.3  60.0  56.7  53.3  50.0  46.7  43.3  40.0
  36.7  33.3
Q= 120.3 137.1 154.0 168.4 182.2 198.5 207.5 217.7 229.7 238.2 247.2 255.6 263.4
  270.0 276.0
```

**20.18** Develop the characteristic curve for a homologous pump of the series of Prob. 20.16 for 21-in-diameter discharge and 1600 rpm.

$$Q = Q_1(N/N_1)(D/D_1)^3 = Q_1\frac{1600}{225}(1.75/6)^3 = 0.176Q_1$$

$$H = H_1(ND/N_1D_1)^2 = H_1[(1600)(1.75)/(225)(6)]^2 = 4.30H_1$$

| $H_1$ | $H$ | $Q_1$ | $Q$ |
|-------|------|-------|------|
| 60 | 258 | 200 | 35.2 |
| 55 | 236 | 256 | 45.1 |
| 50 | 215 | 303 | 53.3 |
| 45 | 194 | 345 | 60.7 |
| 40 | 172 | 382 | 67.2 |
| 35 | 150 | 411 | 72.3 |
| 30 | 129 | 438 | 77.1 |
| 25 | 108 | 459 | 80.8 |

See curve plotted in Fig. 20-2.

Fig. 20-2

**20.19** Determine the size and synchronous speed of a pump homologous to the 72-in pump of Prob. 20.16 that will produce 2.5 m³/s at 90-m head when operating at highest efficiency.

$$Q/ND^3 = Q_1/N_1D_1^3 \qquad Q_1 = 2.5/0.3048^3 = 88.3 \text{ cfs} \qquad H_1 = 90/0.3048 = 295.3 \text{ ft}$$

$$N_1D_1^3 = (Q_1/Q)ND^3 = (88.3/345)(225)(\tfrac{72}{12})^3 = 12\,439 \qquad H/N^2D^2 = H_1/N_1^2D_1^2$$

$$N_1^2D_1^2 = (H_1/H)N^2D^2 = (295.3/45)(225)^2(\tfrac{72}{12})^2 = 11\,959\,650$$

$D_1 = 1.897$ ft, or 22.76 in, and $N_1 = 1822$ rpm. Use $N = 1800$ rpm (one pair of poles). Then, $D_1 = \sqrt{H_1/H(N/N_1)}D = \sqrt{295.3/45}(\frac{225}{1800})(6) = 1.92$ ft.

**20.20** Sketch the head versus discharge curve for a centrifugal pump (Fig. 20-1a) having $r_1 = 55$ mm, $r_2 = 110$ mm, $b_1 = 27.5$ mm, $b_2 = 22$ mm, $N = 1300$ rpm, and $\beta_2 = 33°$.

▌ $H = u_2^2/g - [(u_2 Q \cot \beta_2)/(2\pi r_2 b_2 g)]$ where $u_2 = N(2\pi/60)r_2 = 1300(2\pi/60)(0.110) = 14.975$ m/s, $H = (14.975^2/9.807) - \{[(14.975 \cot 33°)Q]/[2\pi(0.110)(0.022)(9.807)]\} = 22.87 - 173.94Q$. See Fig. 20-3.

Fig. 20-3

**20.21** A centrifugal pump (see Fig. 20-1a) has an impeller with dimensions $r_1 = 75$ mm, $r_2 = 150$ mm, $b_1 = 50$ mm, $b_2 = 30$ mm, $\beta_1 = \beta_2 = 30°$. For a discharge of 70 L/s and shockless entry to vanes, compute (a) the speed, (b) the head, (c) the torque, and (d) the power. Neglect losses ($\alpha_1 = 90°$).

▌
$$V_1 = V_{r1} = Q/2\pi r_1 b_1 = (70 \times 10^{-3})/[(2)(\pi)(0.075)(0.050)] = 2.971 \text{ m/s}$$
$$u_1 = V_1/(\tan \beta_1) = 2.971/(\tan 30°) = 5.146 \text{ m/s}$$

(a) $\qquad N = (u_1/r_1)(60/2\pi) = [5.146/(0.075)](60/2\pi) = 655.2$ rpm

(b) $V_{r2} = Q/2\pi r_2 b_2 = (r_1 b_1/r_2 b_2)V_1 = [(75)(50)/(150)(30)](2.971) = 2.476$ m/s $\qquad u_2/u_1 = r_2/r_1 = 150/75 = 2$

$u_2 = 2u_1 = (2)(5.146) = 10.292$ m/s $\qquad V_{u2} = u_2 - V_{r2} \cot \beta_2 = 10.292 - 2.476 \cot 30° = 6.003$ m/s

$$H = u_2 V_{u2}/g = (10.292)(6.003)/9.807 = 6.30 \text{ m}$$

(c) $\qquad T = \rho Q(r_2 V_{u2}) = (1000)(70 \times 10^{-3})(0.150)(6.003) = 63.0 \text{ N} \cdot \text{m}$

(d) $\qquad P = T\omega = T(u_1/r_1) = (63.0)[5.146/(0.075)] = 4323 \text{ W}$

**20.22** A centrifugal pump with impeller dimensions $r_1 = 2$ in, $r_2 = 5$ in, $b_1 = 3$ in, $b_2 = 1.5$ in, $\beta_2 = 60°$ is to pump 6 cfs at a 64-ft head. Determine (a) $\beta_1$, (b) the speed, (c) the pressure rise across the impeller. Neglect losses and assume no shock at the entrance ($\alpha_1 = 90°$).

▌
$$V_{r1} = Q/2\pi r_1 b_1 = 6/[(2)(\pi)(\tfrac{2}{12})(\tfrac{3}{12})] = 22.92 \text{ ft/s}$$
$$V_{2r} = Q/2\pi r_2 b_2 = (r_1 b_1/r_2 b_2)V_{r1} = [(2)(3)/(5)(1.5)](22.92) = 18.34 \text{ ft/s}$$
$$H = u_2 V_{u2}/g = u_2(u_1 - v_{r2} \cot \beta_2) \qquad u_2^2 - u_2 V_{r2} \cot \beta_2 - gH = 0$$
$$u_2 = (V_{r2} \cot \beta_2 + \sqrt{V_{r2}^2 \cot^2 \beta_2 + 4gH})/2 = [18.34 \cot 60° + \sqrt{18.34^2 \cot^2 60° + (4)(32.2)(64)}]/2 = 51.0 \text{ ft/s}$$
$$u_1 = (r_1/r_2)u_2 = (\tfrac{2}{5})(51.0) = 20.4 \text{ ft/s}$$

(a) $\qquad \tan \beta_1 = V_{r1}/u_1 = 22.92/20.4 = 1.124 \qquad \beta_1 = 48.33°$

(b) $\qquad N = u_2/r_2 = [51.0/(\tfrac{5}{12})](60/2\pi) = 1169$ rpm

(c) $\quad V_1 = V_{r1}/(\sin \beta_1) = 22.92/(\sin 48.33°) = 30.68$ ft/s $\qquad V_2 = V_{r2}/(\sin \beta_2) = 18.34/(\sin 60°) = 21.18$ ft/s

$\Delta p = (\rho/2)(u_2^2 - u_1^2 - V_2^2 + V_1^2) = (1.94/2)(51.0^2 - 20.4^2 - 21.18^2 + 30.68^2) = 2597$ lb/ft$^2$ $\quad$ or $\quad 18.0$ lb/in$^2$

**20.23** Evaluate $r_1$, $r_2$, $\beta_1$, $\beta_2$, $b_1$, and $b_2$ of a centrifugal impeller that will take 36 L/s of water from a 100-mm-diameter suction line and increase its energy head by 18 m ($N = 1440$ rpm; $\alpha_1 = 90°$). Neglect losses.

■ Select sample values $r_1 = 50$ mm, $r_2 = 100$ mm, $b_1 = 40$ mm, $b_2 = 25$ mm. Then find $\beta_1$ and $\beta_2$.

$$V_1 = Q/2\pi r_1 b_1 = (36 \times 10^{-3})/[(2)(\pi)(0.050)(0.040)] = 2.865 \text{ m/s}$$

$$u_1 = \omega r_1 = [(2)(\pi)(1440)/60](0.050) = 7.540 \text{ m/s} \qquad u_2 = (r_2/r_1)u_1 = (\tfrac{100}{50})(7.540) = 15.08 \text{ m/s}$$

Since $\alpha_1 = 90°$, $\tan \beta_1 = V_1/u_1 = 2.865/7.540 = 0.37997$, $\beta_1 = 20.8°$.

$$V_{u2} = Hg/u_2 = (18)(9.807)/15.08 = 11.71 \text{ m/s} \qquad V_{r2} = V_{r1}(r_1 b_1/r_2 b_2) = (2.865)[(50)(40)/(100)(25)] = 2.292 \text{ m/s}$$

$$\tan \beta_2 = V_{r2}/(u_2 - V_{u2}) = 2.292/(15.08 - 14.04) = 2.20385 \qquad \beta_2 = 65.6°$$

*Note:* It is best to choose $r_1 = 50$ mm (100-mm inlet diameter). Then $r_2$, $b_2$, and $b_1$ can be arbitrarily selected within reasonable limits.

**20.24** A mercury–water differential manometer, $R' = 740$ mm, is connected from the 100-mm-diameter suction pipe to the 80-mm-diameter discharge pipe of a pump, as shown in Fig. 20-4. The centerline of the suction pipe is 330 mm below the discharge pipe. For $Q = 3.42$ m³/min of water, calculate the head developed by the pump.

■ $V_1 = Q/A_1 = (3.42/60)/[(\pi)(0.100)^2/4] = 7.257 \text{ m/s}$ $\qquad V_2 = V_1(D_1^2/D_2^2) = (7.257)(100^2/80^2) = 11.34 \text{ m/s}$

The energy equation yields $p_1/\gamma + V_1^2/2g + H_p = p_2/\gamma + V_2^2/2g + h$, $H_p = \Delta p/\gamma + (V_2^2 - V_1^2)/2g + h$. The manometer equation yields $\Delta p/\gamma = R'(\text{s.g.} - 1) - h = (0.740)(13.6 - 1) - 0.330 = 8.99$ m of water, $H_p = 8.99 + (11.34^2 - 7.257^2)/[(2)(9.807)] + 0.330 = 13.19$ m.

**Fig. 20-4**

**20.25** An air blower is to be designed to produce pressure of 90 mmH₂O when operating at 3200 rpm; $r_2 = 1.1r_1$; $\beta_2 = \beta_1$; width of impeller is 100 mm; $\alpha_1 = 90°$. Find $r_1$. Assume a temperature of 30 °C.

■ $Q = 2\pi r_1 b_1 V_1 = 2\pi r_2 b_2 V_2$. Since $r_2 = 1.1r_1$ and $b_1 = b_2$,

$$V_{r2} = V_{r1}/1.1 = V_1/1.1 \qquad u_2 = (r_2/r_1)u_1 = 1.1u_1$$

$$h_{\text{air}} = H_{\text{H}_2\text{O}}(\gamma_{\text{H}_2\text{O}}/\gamma_{\text{air}}) = (0.090)[(9.77)(1000)/11.4] = 77.13 \text{ m of air}$$

$$V_{u2} = gh_{\text{air}}/u_2 = (9.807)(77.13)/1.1u_1 = 687.6/u_1 \qquad u_2 = V_{u2} + V_{r2} \cot \beta$$

Since $\beta_1 = \beta_2 = \beta$ and $u_1 = V_1 \cot \beta$, $1.1u_1 = 687.6/u_1 + (V_1 \cot \beta)/1.1 = 687.6/u_1 + u_1/1.1$, $u_1 = 60.0$ m/s; $r_1 = u_1/\omega = 60.0/[(3200)(2\pi)/60] = 0.179$ m, or 179 mm.

**20.26** In Prob. 20.25, calculate the discharge for $\beta_1 = 32°$.

■ $$V_1 = u_1 \tan \beta_1 = 60.0 \tan 32° = 37.5 \text{ m/s}$$

$$Q = 2\pi r_1 b_1 V_1 = (2)(\pi)(0.179)(0.100)(37.5) = 4.218 \text{ m}^3/\text{s} \quad \text{or} \quad 253 \text{ m}^3/\text{min}$$

**20.27** What is the cavitation index at a point in flowing water where $T = 20$ °C, $p = 16$ kPa, and the velocity is 13 m/s.

■ $$\sigma' = (p - p_v)/(\rho V^2/2) = (16 - 2.34)(1000)/[(998)(13)^2/2] = 0.162$$

**20.28** Two reservoirs $A$ and $B$ are connected with a long pipe which has characteristics such that the head loss through the pipe is expressible as $h_L = 20Q^2$, where $h_L$ is in feet and $Q$ is the flow rate in hundreds of gpm. The water-surface elevation in reservoir $B$ is 35 ft above that in reservoir $A$. Two identical pumps are available for use to pump the water from $A$ to $B$. The characteristic curve of the pump when operating at 1800 rpm is given in the table on page 616.

At the optimum point of operation the pump delivers 200 gpm at a head of 75 ft. Determine the specific speed $N_s$ of the pump and find the rate of flow under the following conditions: (*a*) a single pump operating at 1800 rpm; (*b*) two pumps in series, each operating at 1800 rpm; (*c*) two pumps in parallel, each operating at 1800 rpm.

■ The head-capacity curves for the pumping alternatives are plotted in Fig. 20-5 and so is the $h_L$ versus $Q$ curve for the pipe system. In this case $h = \Delta z + h_L = 35 + 20Q^2$. The answers are found at the points of

| operation at 1800 rpm | |
|---|---|
| head, ft | flow rate, gpm |
| 100 | 0 |
| 90 | 110 |
| 80 | 180 |
| 60 | 250 |
| 40 | 300 |
| 20 | 340 |

intersection of the curves. They are as follows: (a) single pump, 156 gpm; (b) two pumps in series, 224 gpm; (c) two pumps in parallel, 170 gpm.

If $\Delta z$ had been greater than 100 ft, neither the single pump nor the two pumps in parallel would have delivered any water. If $\Delta z$ had been −20 ft (i.e., with the water-surface elevation in reservoir $B$ 20 ft below that in $A$), the flows would have been (a) 212 gpm; (b) 258 gpm; and (c) 232 gpm.

**Fig. 20-5**

**20.29** The diameter of the discharge pipe of a pump is 8 in, and that of the intake pipe is 10 in. The pressure gage at discharge reads 32 psi, and the vacuum gage at intake reads 12 inHg. If $Q = 4.0$ ft³/s of water and the brake horsepower is 49.0, find the efficiency. The intake and the discharge are at the same elevation.

$$V_s = Q/A_s = 4.0/[(\pi)(\tfrac{10}{12})^2/4] = 7.334 \text{ ft/s} \qquad V_p = 4.0/[(\pi)(\tfrac{8}{12})^2/4] = 11.46 \text{ ft/s}$$

$$\text{vacuum pressure} = -(12/29.9)(14.7) = -5.900 \text{ lb/ft}^2$$

$$H_p = p_d/\gamma + V_d^2/2g + z_d - (p_s/\gamma + V_s^2/2g + z_s) = (32)(144)/62.4 + 11.46^2/[(2)(32.2)] + 0$$
$$- \{(-5.900)(144)/62.4 + 7.334^2/[(2)(32.2)] + 0\} = 88.67 \text{ ft}$$

$$P = Q\gamma H_p = (4.0)(62.4)(88.67)/550 = 40.24 \text{ hp} \qquad \eta = 40.24/49.0 = 0.821 \quad \text{or} \quad 82.1\%$$

**20.30** A centrifugal pump with an impeller diameter of 2.94 in delivers 260 gpm of water at a head of 820 ft with an efficiency of 62 percent at 21 000 rpm. Compute the peripheral velocity $u$, specific speed $N_s$, and peripheral-velocity factor $\phi$.

$$u = \omega r = [(21\,000)(2\pi)/60][(2.94/2)/12] = 269 \text{ ft/s} \qquad N_s = NQ^{0.5}/H^{0.75} = (21\,000)(260)^{0.5}/820^{0.75} = 2210$$
$$u = \phi\sqrt{2gh} \qquad 269 = (\phi)\sqrt{(2)(32.2)(820)} \qquad \phi = 1.171$$

**20.31** Select the specific speed of the pump or pumps required to lift 5835 gpm of water 350 ft through 9400 ft of 2.5-ft-diameter pipe ($f = 0.022$). The pump rotative speed is to be 1700 rpm. Consider the following cases: single pump, two pumps in series, two pumps in parallel.

$$Q = (5835)(0.002228) = 13 \text{ ft}^3/\text{s} \qquad V = Q/A = 13/[(\pi)(2.5)^2/4] = 2.648 \text{ ft/s}$$

$$h_f = (f)(L/D)(V^2/2g) = (0.022)(9400/2.5)\{2.648^2/[(2)(32.2)]\} = 9.0 \text{ ft}$$

$$h_p = 350 + 9.0 = 359.0 \text{ ft} \qquad N_s = NQ^{0.5}/H^{0.75}$$

| | gpm per pump | $h_p$ per pump | $N_s$ |
|---|---|---|---|
| Single pump | 5835 | 359.0 | 1575 |
| Two pumps in series | 5835 | 179.5 | 2648 |
| Two pumps in parallel | 2918 | 359.0 | 1113 |

**20.32** The pump of Fig. 20-6a is used to lift water from one reservoir to another through a 9-in-diameter pipe ($f = 0.022$), 692 ft long. The difference $\Delta z$ in surface elevation between the reservoirs fluctuates from 20 ft to 110 ft. Plot delivery rate $Q$ versus pump head $h_p$. Plot also the corresponding efficiencies. The pump is operated at a constant speed of 1450 rpm. Neglect minor losses.

▌ With $Q$ in gpm,

$$V = 0.002228Q/A = 0.002228Q/[(\pi)(\tfrac{9}{12})^2/4] = 0.0050Q$$

$$h_p = \Delta z + (f)(L/D)(V^2/2g) = \Delta z + (0.022)[692/(\tfrac{9}{12})]\{(0.0050Q)^2/[(2)(32.2)]\} = \Delta z + (8.0 \times 10^{-6})Q^2$$

Plot the pump characteristic curve and pipe system curves for a variety of $\Delta z$'s.

| $\Delta z$, ft | $Q$, gpm | efficiency, % |
|---|---|---|
| 20 | 2500 | 70 |
| 40 | 2320 | 76 |
| 60 | 2100 | 81 |
| 80 | 1860 | 83 |
| 100 | 1350 | 77 |
| 110 | 940 | 66 |

See plotted curves in Fig. 20-6b.

(a)

Fig. 20-6

(b)                                    **Fig. 20-6**   (*continued*)

**20.33**  Repeat Prob. 20.32 for the same pump operating at 1250 rpm. Assume efficiency pattern and values remain the same.

▌ The pump characteristic curve must be transformed using $Q \propto N$ and $h \propto N^2$. Thus point $A$ in Fig. 20-6b transforms to $A'$, where $Q' = (2000)(\frac{1250}{1450}) = 1724$ gpm and $h'_p = (100)(\frac{1250}{1450})^2 = 74.3$ ft, etc.

| $\Delta z$, ft | $Q$, gpm | efficiency, % |
|---|---|---|
| 20 | 2000 | 73 |
| 40 | 1750 | 82 |
| 60 | 1350 | 84 |
| 80 | 550 | 58 |
| 82.3 | (shutoff) | 0 |

The transformed pump characteristic curve is shown in Fig. 20-6b. (The transformed efficiency curve is not shown in the figure.)

**20.34**  Repeat Prob. 20.32 for a homologous pump whose diameter is 21% smaller. Assume efficiency pattern and values remain the same.

▌ In this case the pump characteristic curve is transformed using $Q \propto D^3$ and $h \propto D^2$. Thus point $A$ in Fig. 20-6b transforms as follows: $Q = (2000)(0.79)^3 = 986$ gpm, $h = (100)(0.79)^2 = 62.4$ ft.

| $\Delta z$, ft | $Q$, gpm | efficiency, % |
|---|---|---|
| 20 | 1400 | 60 |
| 40 | 1200 | 76 |
| 60 | 800 | 81 |
| 76.8 | (shutoff) | 0 |

The transformed pump characteristic curve is shown in Fig. 20-7.

**20.35**  Repeat Prob. 20.32 if the pump diameter and rotation speed are reduced by 21% and 13.8%, respectively. Assume efficiency pattern and values remain the same.

▌ In this case the pump characteristic curve is transformed using $Q \propto ND^3$ and $h \propto N^2D^2$. Thus point $A$ in Fig. 20-6b for the solution of Prob. 20.32 transforms as follows: $Q = (2000)(0.862)(0.79)^3 = 850$ gpm, $h = (100)(0.862)^2(0.79)^2 = 46.4$ ft. Transforming the pump characteristic curve according to these rules and following

**Fig. 20-7**

the procedure as outlined in Prob. 20.34 gives

| $\Delta z$, ft | $Q$, gpm | efficiency, % |
|:---:|:---:|:---:|
| 20 | 1080 | 69 |
| 40 | 700 | 83 |
| 52.6 | Shutoff | 0 |

**20.36** A pump with a critical value of $\sigma_c$ of 0.10 is to pump against a head of 450 ft. The barometric pressure is 14.4 psia, and the vapor pressure of the water is 0.6 psia. Assume the friction losses in the intake piping are 6 ft. Find the maximum allowable height of the pump relative to the water surface at intake.

$$(z_s)_{max} = (p_0)_{abs}/\gamma - p_v/\gamma - \sigma_c h - h_L \qquad p_0/\gamma = (14.4)(144)/62.4 = 33.2 \text{ ft}$$

$$p_v/\gamma = (0.6)(144)/62.4 = 1.385 \text{ ft} \qquad (z_s)_{max} = 33.2 - 1.385 - (0.10)(450) - 6 = -19.2 \text{ ft} \qquad \text{(submerged)}$$

**20.37** A pump draws water at 212 °F from a boiler with a friction loss of 3 ft in the intake pipe. Barometric pressure is 29 in of mercury, $\sigma_c$ for the pump is 0.10, and the total pumping head is 300 ft. Locate the pump intake relative to the water surface in the boiler.

$$(z_s)_{max} = (p_0)_{abs}/\gamma - p_v/\gamma - \sigma_c h - h_L \qquad p_0 = (\tfrac{29}{12})(847.3/144) = 14.2 \text{ psia} \qquad p_0/\gamma = (14.2)(144)/59.8 = 34.2 \text{ ft}$$

$$p_v/\gamma = (14.7)(144)/59.8 = 35.4 \text{ ft} \qquad (z_s)_{max} = 34.2 - 35.4 - (0.10)(300) - 3 = -34.2 \text{ ft}$$

Thus the water surface must be 34.2 ft above the pump suction.

**20.38** Water is to be pumped at a head of 142 ft, the water temperature being 100 °F and the barometric pressure being 14.3 psia. At intake, the pressure is a vacuum of 15 inHg and the velocity is 13 fps. What are the values of NPSH and $\sigma$?

$$\text{NPSH} = (p_s)_{abs}/\gamma + V_s^2/2g - p_v/\gamma \qquad (p_s)_{abs} = p_0/\gamma + p_s/\gamma = (14.3)(144)/62.0 + (-\tfrac{15}{12})(847.3/62.0) = 16.13 \text{ ft}$$

$$V_s^2/2g = 13^2/[(2)(32.2)] = 2.62 \text{ ft} \qquad p_v/\gamma = 135/62.0 = 2.18 \text{ ft}$$

$$\text{NPSH} = 16.13 + 2.62 - 2.18 = 16.57 \text{ ft} \qquad \sigma = 16.57/142 = 0.117$$

**20.39** A pump discharges 7700 gpm of water at 140 °F at a head of 250 ft, the barometric pressure being 13.8 psia. Determine the reading on a pressure gage at the suction flange when cavitation is just starting. Assume the suction pipe diameter equals 2.1 ft and neglect the effects of prerotation. Take $\sigma_c = 0.085$.

$$\sigma = [(p_s)_{abs}/\gamma + V_s^2/2g - p_v/\gamma]/h. \text{ Let } p = \text{gage pressure at suction flange.}$$

$$(p_s)_{abs} = p_{atm} + p = 13.8 + p \qquad V_s = Q/A_s = (7700)(0.002228)/[(\pi)(2.1)^2/4] = 4.953 \text{ ft/s}$$

$$p_v/\gamma = 416/61.4 = 6.775 \text{ ft} \qquad 0.085 = \{(13.8 + p)(144)/61.4 + 4.953^2/[(2)(32.2)] - 6.775\}/250$$

$$p = -2.013 \text{ psia} = +2.013 \text{ psia vacuum}$$

**20.40** If the maximum efficiency of the pump of Prob. 20.32 is 81%, approximately what would be the efficiency of the pump of Prob. 20.34?

❚ $(1 - \eta_1)/(1 - \eta_2) = (D_2/D_1)^{1/5}$ $(1 - 0.81)/(1 - \eta_2) = (0.79/1.0)^{1/5}$ $\eta_2 = 0.801$ or 80.1%

**20.41** Figure 20-8 shows the dimensions and angles of the diffuser vanes of a centrifugal pump. The vane passages are 0.75 in wide perpendicular to the plane of the figure. If the impeller delivers water at the rate of 3 cfs under ideal and frictionless conditions, what is the rise in pressure through the diffuser?

❚ $V_1^2/2g = V_2^2/2g + \Delta p/\gamma$ $V = Q/A$ $A_1 = (2\pi)(\frac{6.5}{12})(\sin 15°)(0.75/12) = 0.05505$ ft²

$V_1 = 3/0.05505 = 54.50$ ft/s $A_2 = (2\pi)(9/12)(\sin 28°)(0.75/12) = 0.1382$ ft²

$V_2 = 3/0.1382 = 21.71$ ft/s

$54.50^2/[(2)(32.2)] = 21.71^2/[(2)(32.2)] + \Delta p/\gamma$ $\Delta p/\gamma = 38.80$ ft $\Delta p = (62.4)(38.80)/144 = 16.8$ lb/in²

**Fig. 20-8**

**20.42** Two pumps whose characteristics are given in Prob. 20.28 are to be used in parallel. They must develop a head $h = 35 + 20Q^2$ as in Prob. 20.28. One pump is to be operated at 1800 rpm. The speed of the other pump is to be gradually reduced until it no longer delivers water. At approximately what speed will this happen?

❚ From Prob. 20.28, with a single pump operating at 1800 rpm, the flow rate is 156 gpm and the head is 83.8 ft. If we add a pump in parallel to this pump with a shutoff head of 83.8 ft, the second pump will convey no flow. Thus, $h/h' = (n/n')^2$, $83.8/100 = (n/1800)^2$, $n = 1648$ rpm.

**20.43** A pump is installed to deliver water from a reservoir of surface elevation zero to another of elevation 290 ft. The 10-in-diameter suction pipe ($f = 0.022$) is 90 ft long, and the 8-in-diameter discharge pipe ($f = 0.028$) is 4900 ft long. The pump characteristic at 1100 rpm is defined by $h_p = 365 - 20Q^2$ where $h_p$, the pump head, is in feet and $Q$ is in cubic feet per second. Compute the rate at which this pump will deliver water under these conditions assuming the setting is low enough to avoid cavitation.

❚ $h_p = 365 - 20Q^2$. Also, since $h_f = (f)(L/D)(V^2/2g)$,

$$h_p = 290 + (0.022)[90/(\tfrac{10}{12})]V_{10}^2/2g + (0.028)(4900)(\tfrac{8}{12})V_8^2/2g \quad V = Q/A$$

$$V_{10} = Q/[(\pi)(\tfrac{10}{12})^2/4] = 1.833Q \quad V_8 = Q/[(\pi)(\tfrac{8}{12})^2/4] = 2.865Q$$

$$h_p = 290 + (0.022)[90/(\tfrac{10}{12})]\{(1.833Q)^2/[(2)(32.2)]\}$$

$$+ (0.028)[4900/(\tfrac{8}{12})]\{(2.865Q)^2/[(2)(32.2)]\} = 290 + 26.35Q^2$$

Equating the two expressions for $h_p$ gives $365 - 20Q^2 = 290 + 26.35Q^2$, $Q = 1.27$ ft³/s.

**20.44** Repeat Prob. 20.43 for the case of two such pumps in parallel.

❚ From Prob. 20.43, $h_p$ per pump $= 290 + 26.35Q^2 = 365 - 20(Q/2)^2$, or $Q = 1.55$ ft³/s.

**20.45** A centrifugal pump with a 14-in-diameter impeller is rated at 690 gpm against a head of 91 ft when rotating at 1800 rpm. What would be the rating of a pump of identical geometry with a 9-in impeller? Assume pump efficiencies and rotative speeds are identical.

▌ $$h \propto D^2 \qquad h = (91)(\tfrac{9}{12})^2 = 51.2 \text{ ft} \qquad Q \propto D^3 \qquad Q = (690)(\tfrac{9}{12})^3 = 291 \text{ gpm}$$

**20.46** The following are data for a centrifugal water pump: $r_1 = 4$ in, $r_2 = 7$ in, $\beta = 30°$, $\beta_2 = 20°$, speed $= 1440$ rpm. If $b_1 = b_2 = 1.75$ in, estimate (a) design-point discharge; (b) water horsepower; (c) head.

▌ (a) The angular velocity is $\omega = 2\pi(\tfrac{1440}{60}) = 150.8$ rad/s. Thus the tip speeds are $u_1 = \omega r_1 = 150.8(\tfrac{4}{12}) = 50.3$ ft/s, and $u_2 = \omega r_2 = 150.8(\tfrac{7}{12}) = 88.0$ ft/s. From the inlet-velocity diagram, Fig. 20-9a, with $\alpha_1 = 90°$ for design point, we compute $V_{n1} = u_1 \tan 30° = 29.0$ ft/s, whence the discharge is $Q = 2\pi r_1 b_1 V_{n1} = 2\pi(\tfrac{4}{12})(1.75/12)(29.0) = 8.86$ ft$^3$/s = 3980 gal/min.

(b) The outlet radial velocity follows from $Q$: $V_{n2} = Q/(2\pi r_2 b_2) = 8.86/[2\pi(\tfrac{7}{12})(1.75/12)] = 16.6$ ft/s. This enables us to construct the outlet-velocity diagram as in Fig. 20-9b, given $\beta_2 = 20°$. The tangential component is $V_{t2} = u_2 - V_{n2}\cot\beta_2 = 88.0 - 16.6\cot 20° = 42.4$ ft/s, $\alpha_2 = \tan^{-1}(16.6/42.4) = 21.4°$. The power is then computed with $V_{t1} = 0$ at the design point: $P_w = \rho Q u_2 V_{t2} = [(1.94)(8.8)(88.0)(42.4)]/550 = 117$ hp.

(c) The head is estimated as $H \approx P_w/\rho g Q = (117)(550)/[(62.4)(8.86)] = 116$ ft.

(a)  (b)  **Fig. 20-9**

**20.47** The 32-in pump of Fig. 20-10 is to pump 24 000 gal/min of water at 1170 rpm from a reservoir whose surface is at 14.7 lb/in$^2$ abs. If head loss from reservoir to pump inlet is 6 ft, where should the pump inlet be placed to avoid cavitation for water at (a) 60 °F, $p_v = 0.26$ lb/in$^2$ abs, s.g. = 1.0 and (b) 200 °F, $p_v = 11.52$ lb/in$^2$ abs, s.g. = 0.9635?

▌ For either case read from Fig. 20-10 at 24 000 gal/min that the required NPSH is 40 ft.

(a) For this case $\rho g = 62.4$ lb/ft$^3$. NPSH $\leq [(p_a - p_v)/\rho g] - Z_i - h_{fi}$ or 40 ft $\leq [(14.7 - 0.26)(144)/62.4] - Z_i - 6.0$ or $Z_i \leq 27.3 - 40 = -12.7$ ft. The pump must be placed at least 12.7 ft *below* the reservoir surface to avoid cavitation.

(From Ingersoll-Rand Corporation, Cameron Pump Division)  **Fig. 20-10**

**(b)** For this case $\rho g = 62.4(0.9635) = 60.1$ lb/ft$^3$. 40 ft $\leq [(14.7 - 11.52)(144)/60.1] - Z_i - 6.0$ or $Z_i \leq 1.6 - 40 = 38.4$ ft. The pump must now be placed at least 38.4 ft below the reservoir surface. These are unusually stringent conditions because a large, high-discharge pump requires a large NPSH.

**20.48** A pump from the family of Fig. 20-11 has $D = 21$ in and $n = 1500$ rpm. Estimate **(a)** discharge, **(b)** head, **(c)** pressure rise, and **(d)** brake horsepower of this pump for water at 60 °F and best efficiency.

**▌ (a)** In BG units take $D = \frac{21}{12} = 1.75$ ft and $n = \frac{1500}{60} = 25$ r/s. At 60 °F, $\rho$ of water is 1.94 slugs/ft$^3$. The BEP parameters are known from Fig. 20-11. The BEP discharge is thus $Q^* = C_Q \cdot nD^3 = 0.115(25)(1.75)^3 = 15.4$ ft$^3$/s = 6920 gal/min.

**(b)** Similarly, the BEP head is $H^* = C_H \cdot n^2 D^2/g = 5.0(25)^2(1.75)^2/32.2 = 297$ ft water.

**(c)** Since we are not given elevation or velocity-head changes across the pump, we neglect them and estimate $\Delta p \approx \rho g H = 1.94(32.2)(297) = 18\,600$ lb/ft$^2$ = 129 lb/in$^2$.

**(d)** Finally, the BEP power is $P^* = C_p \cdot \rho n^3 D^5 = 0.65(1.94)(25)^3(1.75)^5/550 = 588$ hp.

**Fig. 20-11**

**20.49** We want to build a pump from the family of Fig. 20-11, which delivers 3000 gal/min water at 1200 rpm at best efficiency. Estimate **(a)** the impeller diameter, **(b)** the maximum discharge, **(c)** the shutoff head, and **(d)** the NPSH at best efficiency.

**▌ (a)** 3000 gal/min = 6.68 ft$^3$/s and 1200 rpm = 20 r/s. At BEP we have $Q^* = C_Q \cdot nD^3 = 6.68$ ft$^3$/s = $(0.115)(20)D^3$ or $D = [6.68/0.115(20)]^{1/3} = 1.43$ ft = 17.1 in.

**(b)** The max $Q$ is related to $Q^*$ by a ratio of capacity coefficients: $Q_{max} = Q^* C_{Q \cdot max}/C_{Q \cdot} \approx 3000(0.23)/0.115 = 6000$ gal/min.

**(c)** From Fig. 20-11 we estimated the shutoff-head coefficient to be 6.0. Thus $H(0) \approx C_H(0)n^2D^2/g = 6.0(20)^2(1.43)^2/32.2 = 152$ ft.

**(d)** Finally, the NPSH at BEP is approximately NPSH$^* = C_{HS} \cdot n^2 D^2/g = 0.37(20)^2(1.43)^2/32.2 = 9.4$ ft. Since this is a small pump, it will be less efficient than the pumps in Fig. 20-11, probably about 82 percent maximum.

**20.50** We want to use a centrifugal pump from the family of Fig. 20-11 to deliver 100 000 gal/min of water at 60 °F with a head of 25 ft. What should be **(a)** the pump size and speed and **(b)** brake horsepower assuming operation at best efficiency?

**▌ (a)** Enter the known head and discharge into the BEP parameters: $H^* = 25$ ft $= C_H \cdot n^2 D^2/g = 5.0n^2D^2/32.2$, $Q^* = 100\,000$ gal/min = 222.8 ft$^3$/s = $C_Q \cdot nD^3 = 0.115nD^3$. The two unknowns are $n$ and $D$. Solve simultaneously for $D = 12.4$ ft, $n = 1.03$ r/s = 62 rpm.

**(b)** The most efficient horsepower is, then, bhp* $\approx C_p \cdot \rho n^3 D^5 = 0.65(1.94)(1.03)^3(12.4)^5/550 = 734$ hp.

The solution to this problem is mathematically correct but results in a grotesque pump: an impeller more than 12 ft in diameter, rotating so slowly one can visualize oxen walking in a circle turning the shaft.

**20.51** We want to use the 32-in pump of Fig. 20-10 at 1170 rpm to pump water at 60 °F from one reservoir to another 120 ft higher through 1500 ft of 16-in-ID pipe with friction factor $f = 0.030$. **(a)** What will the operating point and efficiency be? **(b)** To what speed should the pump be changed to operate the BEP?

**❙** **(a)** For reservoirs the initial and final velocities are zero; thus the system head is $H_s = z_2 - z_1 + (V^2/2g)(fL/D) = 120 + (V^2/2g)[0.030(1500)/\frac{16}{12}]$. From continuity in the pipe, $V = Q/A = Q/[\frac{1}{4}\pi(\frac{16}{12})^2]$, and so we substitute for $V$ above to get

$$H_s = 120 + 0.269Q^2 \qquad Q \text{ in ft}^3/\text{s} \tag{1}$$

Since Fig. 20-10 uses thousands of gallons per minute for the abscissa, we convert $Q$ in Eq. (1) to this unit:

$$H_s = 120 + 1.335Q^2 \qquad Q \text{ in } 10^3 \text{ gal/min} \tag{2}$$

We can plot Eq. (2) on Fig. 20-10 and see where it intersects the 32-in pump-head curve, as in Fig. 20-12. A graphical solution gives approximately $H \approx 430$ ft, $Q \approx 15\,000$ gal/min. The efficiency is about 82 percent, slightly off design.

An analytic solution is possible if we fit the pump-head curve to a parabola, which is very accurate:

$$H_{\text{pump}} \approx 490 - 0.26Q^2 \qquad Q \text{ in } 10^3 \text{ gal/min} \tag{3}$$

Equations (2) and (3) must match at the operating point: $490 - 0.26Q^2 = 120 + 1.335Q^2$ or $Q^2 = (490 - 120)/(0.26 + 1.335) = 232$, $Q = 15.2 \times 10^3$ gal/min = 15 200 gal/min; $H = 490 - 0.26(15.2)^2 = 430$ ft.

**(b)** To move the operating point to BEP, we change $n$, which changes both $Q \propto n$ and $H \propto n^2$. From Fig. 20-10, at BEP, $H^* \approx 386$ ft; thus for any $n$, $H^* = 386(n/1170)^2$. Also read $Q^* \approx 20 \times 10^3$ gal/min; thus for any $n$, $Q^* = 20(n/1170)$. Match $H^*$ to the system characteristics, Eq. (2), $H^* = 386(n/1170)^2 \approx 120 + 1.335[20(n/1170)]^2$, which gives $n^2 < 0$. Thus it is impossible to operate at maximum efficiency with this particular system and pump.

**Fig. 20-12**

**20.52** Investigate extending Prob. 20.51 by using two 32-in pumps in parallel to deliver more flow. Is this efficient?

**❙** Since the pumps are identical, each delivers $\frac{1}{2}Q$ at the same 1170 rpm speed. The system curve is the same, and the balance-of-head relation becomes $H = 490 - 0.26(\frac{1}{2}Q)^2 = 120 + 1.335Q^2$ or $Q^2 = (490 - 120)/(1.335 + 0.065)$, $Q = 16\,300$ gal/min. This is only 7 percent more than a single pump. Each pump delivers $\frac{1}{2}Q = 8150$ gal/min, for which the efficiency is only 60 percent. The total brake horsepower required is 3200, whereas a single pump used only 2000 bhp.

**20.53** Suppose the elevation change in Prob. 20.51 is raised from 120 to 500 ft, greater than a single 32-in pump can supply. Investigate using 32-in pumps in series at 1170 rpm.

**❙** Since the pumps are identical, the total head is twice as much and the constant 120 in the system-head curve is replaced by 500. The balance of heads becomes $H = 2(490 - 0.26Q^2) = 500 + 1.335Q^2$ or $Q^2 = (980 - 500)/(1.335 + 0.52)$, $Q = 16.1 \times 10^3$ gal/min. The operating head is $500 + 1.335(16.1)^2 = 846$ ft, or 97 percent more than a single pump in Prob. 20.50. Each pump is operating at $16.1 \times 10^3$ gal/min, which from Fig. 20-10 is 83 percent efficient, a pretty good match to the system. To pump at this operating point requires 4100 bhp, or about 2050 bhp for each pump.

**20.54**  A piston positive-displacement pump (PDP) has a 6-in diameter and a 2.5-in stroke. Its crankshaft rotates at 300 rpm. Calculate its output at 100 percent volumetric efficiency.

$$V_{\text{displaced}} = [(\pi)(\tfrac{6}{12})^2/4](\tfrac{2.5}{12}) = 0.04091 \text{ ft}^3 \qquad Q = (0.04091)(300) = 12.27 \text{ ft}^3/\text{min}$$

**20.55**  If the PDP of Prob. 20.54 delivers water against a total head of 24 ft, what horsepower is required at 72% efficiency?

$$P = Q\gamma E_p = (12.27/60)(62.4)(24) = 306.3 \text{ ft-lb/s} \qquad P_{\text{required}} = (306.3/550)/0.72 = 0.77 \text{ hp}$$

**20.56**  A pump delivers 1.6 m³/min of water at 20 °C against a pressure rise of 287 kPa. Kinetic- and potential-energy changes are negligible. If the driving motor supplies 10 kW, what is the overall efficiency?

$$P = Q\gamma E_p = \Delta p\, Q = (287)(1.6/60) = 7.65 \text{ kN} \cdot \text{m/s} = 7.65 \text{ kW} \qquad \eta = 7.65/10 = 0.765$$

**20.57**  A 26-hp pump delivers 475 gpm of gasoline ($\gamma = 42.5 \text{ lb/ft}^3$) at 20 °C with 78% efficiency. What head and pressure rise result across the pump?

$$P_{\text{required}} = Q\gamma E_p/\eta \qquad (26)(550) = [(475)(0.002228)](42.5)(E_p)/0.78 \qquad E_p = 248 \text{ ft}$$
$$\Delta p = \gamma E_p = (42.5)(248) = 10\,540 \text{ lb/ft}^2 \quad \text{or} \quad 73.2 \text{ lb/in}^2$$

**20.58**  A sump pump turning at 1600 rpm provides a discharge of 9 gpm against a head of 18 ft. Compute the minimum power input.

$N_s = NQ^{0.5}/H^{0.75} = (1600)(9)^{0.5}/18^{0.75} = 549$. From Fig. A-41, $\eta_{\max} = 0.32$. $P_{\min} = Q\gamma H/\eta_{\max} = [(9)(0.002228)](62.4)(18)/0.32 = 70.4$ ft-lb/s, or 0.128 hp.

**20.59**  A model pump delivering water at 180 °F at 900 gpm and 2500 rpm begins to cavitate when the inlet pressure and velocity are 13 psia and 22 fps. Find the required NPSH of a prototype which is 4 times larger and runs at 1100 rpm.

$$\text{NPSH} = (p_i - p_v)/\gamma + V_i^2/2g \qquad \text{NPSH}_{\text{model}} = [(13)(144) - 1086]/60.6 + 22^2/[(2)(32.2)] = 20.49 \text{ ft}$$
$$\text{NPSH}_{\text{proto}} = (\text{NPSH}_{\text{model}})(N_p/N_m)^2(D_p/D_m)^2 = (20.49)(\tfrac{1100}{2500})^2(4)^2 = 63.5 \text{ ft.}$$

**20.60**  Determine the specific speed of the pump whose operating characteristics are shown in Fig. 20–13. If this pump were operated at 1200 rpm, what head and discharge would be developed at rated capacity, and what power would be required?

$N_s = NQ^{0.5}/H^{0.75}$. At its rated capacity of 700 gpm, this pump develops 120 ft of head when operating at 1450 rpm. Thus $N_s = (1450)(700)^{0.5}/120^{0.75} = 1058$. If the efficiency remains constant with the speed change, at 1200 rpm, $Q = (\tfrac{1200}{1450})(700) = 579$ gpm, $H = (\tfrac{1200}{1450})^2(120) = 82.2$ ft, $P = (\tfrac{1200}{1450})^3(26) = 14.7$ hp.

**Fig. 20-13**

**20.61**  A mixed-flow centrifugal pump ($N_s = 400$, $\sigma_c = 0.30$) is to develop 50 ft of head. Find the maximum permissible suction lift on the pump if the pump is at sea level and the water is at 80 °F.

❚ Neglecting head loss between reservoir and pump, $z_s \approx (p_{atm}/\gamma - p_v/\gamma) - \text{NPSH}$, $\text{NPSH} = \sigma_c h = (0.30)(50) = 15.0$ ft, $z_s \approx [(14.7)(144)/62.4 - 73.5/62.4] - 15.0 = 17.7$ ft.

**20.62** A centrifugal pump (efficiency 79%) driven by an electric motor (efficiency 88%) lifts water through a total height of 135 ft from reservoir to discharge. Pumping is through 977 ft of 3-in-diameter pipe at the rate of 290 gpm. If $f = 0.025$ and electricity costs \$7 per megawatt-hour, find the cost of pumping one million gallons.

❚ $$Q = (290)(0.002228) = 0.6461 \text{ ft}^3/\text{s} \qquad V = Q/A = 0.6461/[(\pi)(\tfrac{3}{12})^2/4] = 13.16 \text{ ft/s}$$

$$h_f = (f)(L/D)(V^2/2g) = (0.025)[977/(\tfrac{3}{12})]\{13.16^2/[(2)(32.2)]\} = 262.7 \text{ ft} \qquad E_p = 135 + 262.7 = 397.7 \text{ ft}$$

$$P_{H_2O} = Q\gamma E_p = (0.6461)(62.4)(397.7) = 16\,034 \text{ ft-lb/s} = 21.74 \text{ kW}$$

$$P_{motor} = 21.74/[(0.79)(0.88)] = 31.27 \text{ kW} = 0.03127 \text{ MW}$$

$$\text{Cost to pump 1 000 000 gallons} = \left(\frac{10^6 \text{ gal}}{290 \times 60 \text{ gal/h}}\right)(0.03127 \text{ MW})\left(\frac{\$7}{1 \text{ MWh}}\right) = \$12.59$$

**20.63** A model of a mixed-flow water pump has been tested in the laboratory to give performance curves as is shown in Fig. 20-14. What would be the total head $\Delta H$ delivered by a prototype pump with an impeller size of 1.2 m operating at a speed of 1750 rpm and delivering 1.300 $m^3/s$ flow? What is its mechanical efficiency? Consider that dynamic similarity can be achieved between model flow and prototype flow.

❚ We first compute the flow coefficient for the prototype flow: $(Q/ND^3)_p = 1.300/\{[(1750)(2\pi)/60](1.2^3)\} = 0.00411$. Examining Fig. 20-14 we see that the efficiency for this operating point is about 75 percent for the prototype. Also note that the head coefficient is about 0.17. Therefore we have for the head $\Delta H$ of the prototype $(g\,\Delta H)/N^2D^2 = 0.17$, $\Delta H = (0.17/9.81)[1750(2\pi/60)]^2(1.2^2) = 838$ m.

**Fig. 20-14**

**20.64** A centrifugal blower compresses 1.4 lbm/s of air at a temperature of 60 °F and a pressure of 13.00 psia to a pressure of 14.50 psia. The impeller runs at 3500 rpm. The exit area of the blower is 0.14 $ft^2$ and the inlet area is 0.4 $ft^2$. How much power is needed to run the blower at 80% efficiency?

❚ $$(dW_p/dt)_{theo} = \{[(V_2)^2_{theo} - V_1^2]/2 + c_p T_1[(p_2/p_1)^{(k-1)/k} - 1]\}\dot{m}$$

Evaluate the two velocities:

$$p_1 = \rho_1 R T_1 \qquad \rho_1 = (13)(144)/[(53.3)(g)(520)] = 0.002098 \text{ slug/ft}^3 \qquad \rho_1/\rho_2 = (p_1/p_2)^{1/k}$$

$$\rho_2 = (0.002098)(14.5/13)^{1/1.4} = 0.002268 \text{ slug/ft}^3$$

$$\dot{m} = (1.4/g) = \rho_2(V_2)_{theo}A_2 \qquad 1.4/g = (0.002268)(V_2)_{theo}(0.14) \qquad (V_2)_{theo} = 136.93 \text{ ft/s}$$

Also $\dot{m} = 1.4/g = \rho_1 V_1 A_1 = (0.002098)(V_1)(0.4)$, $V_1 = 51.81$ ft/s. Consequently,

$$(dW_p/dt)_{theo} = \{[(136.93^2 - 51.81^2)/2] + (0.24)(778)(g)(520)[(14.5/13)^{0.4/1.4} - 1]\}(1.4/g)/550 = 8.47 \text{ hp}$$

$$(dW_p/dt)_{act} = 8.47/0.80 = 10.59 \text{ hp}$$

**20.65** In Prob. 20.64, if the impeller has an outside radius of 9 in and is 1.6 in wide, what should the blade angle $\beta_2$ be at exit?

▌

$$(dW_p/dt)_{theo} = (\omega r_2 \dot{m})[(U_t)_2 + (V_r)_2 \cot \beta_2]$$

$$(8.47)(550) = [(3500)(2\pi)/(60)](\tfrac{9}{12})(1.4/g)\{[(3500)(2\pi)/60](\tfrac{9}{12}) + (V_r)_2 \cot \beta_2\} \qquad (1)$$

Writing the continuity equation as $\dot{m} = \rho_2 (V_r)_2 (2\pi r_2)(b)$ and taking $\rho_2$ from Prob. 20.64, we find

$$1.4/g = (0.002268)(V_r)_2(2\pi)(\tfrac{9}{12})(1.6/12) \qquad (V_r)_2 = 30.51 \text{ ft/s}$$

Now Eq. (1) is solved to give $\beta_2 = 14.86°$.

**20.66** An air compressor has a mass flow of 0.7 kg/s, taking the fluid from a pressure of 100 kPa at the inlet to 101.67 kPa at the outlet. The inlet temperature is 15 °C. The exit angle $\beta_2$ is 160 °. The inlet area is 8000 mm$^2$ and the exit area is 7000 mm$^2$. Determine the speed of the compressor, if the diameter of the rotor is 175 mm, the width of the impeller is 80 mm, and the compressor efficiency is $\eta = 0.80$. Base your calculation on actual, not theoretical, torque.

▌

$$dW_p/dt = \{[(V_2)_{theo}^2 - V_1^2]/2 + c_p T_1[(p_2/p_1)^{(k-1)/k} - 1]\}\dot{m}/\eta$$

From the continuity equation, $\rho_2(V_2)_{theo}A_2 = 0.7$ kg/s, $V_2 = 0.7/[(1.2244)(7000 \times 10^{-6})] = 81.7$ m/s; hence,

$$dW_p/dt = \{[(81.7^2 - 71.5^2)/2] + (1003)(288)[(101.67/100)^{0.4/1.4} - 1]\}(0.7)/0.80 = 1883 \text{ N} \cdot \text{m/s}$$

Now, $dW_p/dt = T_s\omega = r_2\dot{m}[(U_t)_2 + (V_r)_2 \cot \beta_2]\omega$, or

$$1883 = (0.175/2)(0.7)[(\omega)(0.175/2) + (V_r)_2 \cot 160°]\omega \qquad (1)$$

By the continuity equation, $(1.2244)(V_{r_2})[(\pi)(0.175)](0.08) = 0.7$, or $(V_r)_2 = 13.00$ m/s. Then (1) gives:
$$\omega^2 - 408.2\omega - 3.513 \times 10^5 = 0, \quad \omega = [408.2 \pm \sqrt{408.2^2 + (4)(3.518 \times 10^5)(1)}]/(2)(1) = 831 \text{ rad/s} = 7935 \text{ rpm}.$$

**20.67** A performance chart is shown in Fig. 20-15 for a 3000-rpm blower from a bulletin of the Buffalo Forge Co. For a 21-in wheel having a flow of 3000 ft$^3$/min, what is the efficiency and the required torque? The exit diameter is 8 in and the entrance diameter is 16 in. At the entrance $p = 14.7$ lb/in$^2$ and $T = 70$ °F. The capacity is given at the exit pressure. *Note:* The upper curves apply to the left ordinate and the lower curves apply to the right ordinate.

▌ From the chart:

$$\begin{cases} hp = 21 \\ (\Delta p)_{static} = 24.4 \text{ inH}_2\text{O} = (24.4)(62.4)/12 = 126.9 \text{ psf} \end{cases}$$

$$p_2 = (14.7)(144) + 126.9 = 2244 \text{ psf} \qquad \dot{m} = \rho_2 Q = [(0.002378)(\tfrac{2216}{2117})^{1/k}](\tfrac{3000}{60}) = 0.1228 \text{ slug/s}$$

$$V_2 = Q/A_2 = (\tfrac{3000}{60})/[\pi(\tfrac{8}{12})^2/4] = 143.2 \text{ ft/s}$$

Now get $V_1$.

$$\rho_1 V_1 A_1 = \dot{m} = 0.1228 \text{ slug/s} \qquad (0.002378)(V_1)(\pi)(\tfrac{16}{12})^2/4 = 0.1228 \qquad V_1 = 36.98 \text{ ft/s}$$

$$dW_p/dm = [(V_2^2/2) - (V_1^2/2)] + c_p T_1[(p_2/p_1)^{(k-1)/k} - 1]$$

$$dW_p/dt = (dW_p/dm)\dot{m} = (0.1228)\{[(143.2^2/2) - (36.98^2/2)]$$

$$+ (0.24)(778)(g)(460 + 70)[(\tfrac{2216}{2117})^{(1.4-1)/1.4} - 1]\}/550 = 11.49 \text{ hp}$$

$$e = [(21 - 11.49)/21](100) = 45.3\% \qquad T = (11.49)(550)/[3000(2\pi/60)] = 20.12 \text{ ft-lb}$$

**20.68** A centrifugal pump in a horizontal pipe drives 213 gal/min of water from an entrance pressure $p_1 = 13$ psia. Pump data are: inside radius of impeller is 3 in; outside radius of impeller is 0.8 ft; pump outputs 25 hp running at 1650 rpm; pump efficiency is 80%; width $b$ of the vanes is 2 in. Determine the exit pressure $p_2$. Take $\rho = 1.938$ slug/ft$^3$ for water.

**Fig. 20-15** (*Courtesy Buffalo Forge Company. Reprinted with permission.*)

▌ First, find the blade angle from $T_s = (\text{theoretical power})/\omega = r_2\dot{m}[(U_t)_2 + (V_r)_2 \cot \beta_2]$:

$$(25/0.80)(550)/[(1650)(2\pi)/60]\} = (0.8)[(213)(231)/(1728)(60)](1.938)([(1650)(2\pi)/60](0.8)$$
$$+ \{[(213)(231)/(1728)(60)]/[(2\pi)(0.8)(\tfrac{2}{12})]\} \cot \beta_2)$$

$$\beta_2 = 169.4°$$

Next compute the exit velocity:

$(V_2)_{\text{rel}} \sin(\pi - \beta_2) = (V_r)_2 \qquad (V_2)_{\text{rel}} = \{[(213)(231)/(1728)(60)]/[2\pi(0.8)(\tfrac{2}{12})]\}/[\sin(180° - 169.4°)] = 3.079 \text{ ft/s}$

and, from the law of cosines,

$$V_2^2 = (U_t)_2^2 + (V_{\text{rel}})_2^2 - 2(U_t)(V_{\text{rel}})_2 \cos(\pi - \beta_2)$$
$$= \{[(1650)(2\pi)/60](0.8)\}^2 + 3.079^2 - (2)\{[(1650)(2\pi)/60](0.8)\}(3.079 \cos 10.6°) = 18\,280 \text{ ft}^2/\text{s}^2$$

Since the entrance velocity is $V_1 = [(213)(231)/(1728)(60)]/[(2\pi)(\tfrac{3}{12})(\tfrac{2}{12})] = 1.813 \text{ ft/s}$, we can now write

$$\text{theoretical power} = \left(\frac{25}{0.80}\right)(550) = \dot{m}\left[\frac{\Delta(V^2)}{2} + \frac{\Delta p}{\rho}\right] = (\tfrac{213}{60})(\tfrac{231}{1728})(1.938)\left[\frac{18\,280 - 1.813^2}{2} + \frac{\Delta p}{1.938}\right]$$

Solving, $\Delta p = 18\,507 \text{ psf} = 129 \text{ psi}$ and $p_2 = 13 + 129 = 142 \text{ psi}$.

**20.69** A centrifugal pump moves 9.0 L/s of water from a pressure of 90 kPa to a pressure of 520 kPa. The section inlet has a diameter of 50 mm and the discharge has a diameter of 40 mm. What power is required if there is an efficiency of 65 percent and the discharge pipe is 0.3 m above the inlet pipe?

$$(dW_p/dt)_{\text{theo}} = \dot{m}(\{[(V_2^2)_{\text{theo}} - V_1^2]/2\} + (p_2 - p_1)/\rho + g(z_2 - z_1)) \qquad \dot{m} = (0.0090)(1000) = 9 \text{ kg/s}$$

$$V_1 = Q/A_1 = 0.0090/[(\pi)(0.050)^2/4] = 4.584 \text{ m/s} \qquad V_2 = Q/A_2 = 0.0090/[(\pi)(0.040)^2/4] = 7.162 \text{ m/s}$$

$$dW_p/dt = (9)\{[(7.162^2 - 4.584^2)/2] + [(520 - 90)/1000] \times 10^3 + 9.81(0.3)\} = 4033 \text{ W}$$

$$P = 4033/0.65 = 6.20 \text{ kW}$$

**20.70** Shown in Fig. 20-16 are performance curves as found in a Buffalo Forge Co. bulletin for multistage centrifugal pumps. The number designations $2 \times 3$, $4 \times 4$, etc., give the exit and the inlet diameters, respectively. For a $2 \times 3$ (four-stage) pump operating at 25 hp with a flow $Q$ of 80 gal/min of water, compute the efficiency. What is the total required torque on the four rotors?

▮ $\Delta H(\text{output}) = 540 \text{ ft} \qquad \Delta H(\text{input}) = [(\text{hp})(550)/Q\gamma] = (25)(550)/\{[(80)(231)/(1728)(60)]62.4\} = 1236 \text{ ft}$

$$\eta = \frac{540}{1236}(100) = 44 \text{ percent} \qquad \text{Torque} = (25)(550)/[(3500)(2\pi)/60] = 37.5 \text{ ft-lb}$$

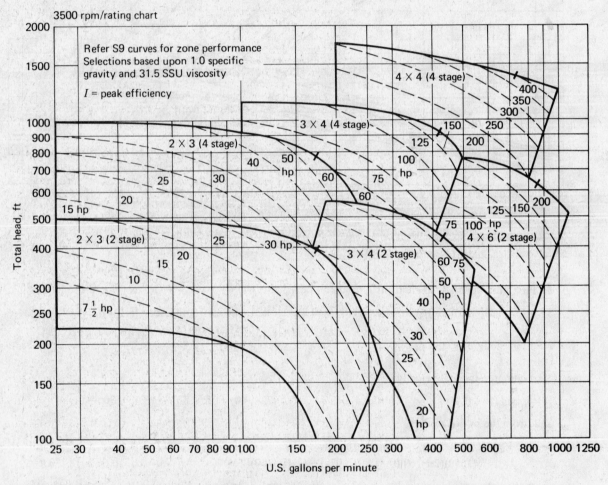

**Fig. 20-16** (*Courtesy Buffalo Forge Company. Reprinted with permission.*)

**20.71** Using Fig. 20-16, select a centrifugal pump to move 120 gal/min of water from tank $A$ to tank $B$ in Fig. 20-17. The inlet and outlet pipes are steel. Neglect head loss for the short inlet piping. Water is at 60 °F.

▮ Use the first law for pump and 2-in outlet pipe.

$$(1/g)(dW_p/dm) + (V_1^2/2g) + (p_1/\gamma) + 0 = (V_2^2/2g) + (p_2/\gamma) + 390 + [(h_L)/g] \qquad (1)$$

For $p_1$ (gage pressure) apply Bernoulli's equation between free surface of tank $A$ and pump inlet:

$$(V_{fs}^2/2g) + (p_{atm}/\gamma) + 1 = (V_1^2/2g) + (p_1/\gamma) \tag{2}$$

$$V_1 = [(120)(231)/(1728)(60)]/[(\pi)(3^2)/(4)(144)] = 5.447 \text{ ft/s} \qquad p_1/\gamma = 1 - (5.447^2/2g) = 0.539 \text{ ft}$$

$$V_2 = [(120)(231)/(1728)(60)]/[(\pi)(2^2)/(4)(144)] = 12.25 \text{ ft/s} \qquad N_R = (12.25)(\tfrac{2}{12})/(1.217 \times 10^{-5}) = 1.398 \times 10^5$$

$$\epsilon/D = 0.00015/(\tfrac{2}{12}) = 0.0009 \qquad f = 0.021 \qquad h_L = f(L/d)(V^2/2) = (0.021)(390/\tfrac{1}{6})(12.25^2/2) = 3687 \text{ ft}$$

Now substitute in Eq. (1):

$$(1/g)(dW_p/dm) + (5.447^2/2g) + 0.539 = (12.25^2/2g) + [(\gamma)(20)/\gamma] + 390 + (3687/g) = 527 \text{ ft}$$

Use $2 \times 3$(four-stage) 30-hp pump.

**Fig. 20-17**

**20.72** A pump impeller, 12 in in diameter, discharges 5.25 cfs when running at 1200 rpm. The blade angle $\beta_2$ is 160° and the exit area $A_2$ is 0.25 ft$^2$. Assuming losses of $2.8(v_2^2/2g)$ and $0.38(V_2^2/2g)$, compute the efficiency of the pump (exit area $A_2$ is measured normal to $v_2$).

**I** The absolute and relative velocities at exit must be calculated first. Velocities $u_2$ and $v_2$ are $u_2 = r_2\omega = (\tfrac{6}{12})(2\pi)(\tfrac{1200}{60}) = 62.8 \text{ ft/s}$, $v_2 = Q/A_2 = 5.25/0.25 = 21.0 \text{ ft/s}$.

From the vector diagram shown in Fig. 20-18, the value of the absolute velocity at exit is $V_2 = 43.7 \text{ ft/s}$. Head furnished by impeller, $H' = (u_2^2/2g) - (v_2^2/2g) + (V_2^2/2g) = [(62.8)^2/2g] - [(21.0)^2/2g] + [(43.7)^2/2g] = 84.0 \text{ ft}$. Head delivered to water, $H = H' - \text{losses} = 84.0 - \{2.8[(21.0)^2/2g] + 0.38[(43.7)^2/2g]\} = 53.6 \text{ ft}$.

Efficiency $e = H/H' = 53.6/84.0 = 63.8$ percent. The value of $H'$ might have been calculated by means of the commonly used expression $H' = (u_2/g)(u_2 + v_2 \cos \beta_2) = (62.8/g)[62.8 + 21.0(-0.940)] = 84.0 \text{ ft}$.

**Fig. 20-18**

**20.73** A 6-in pump delivers 1300 gpm against a head of 90 ft when rotating at 1750 rpm. The head-discharge and efficiency curves are shown in Fig. 20-19. For a geometrically similar 8-in pump running at 1450 rpm and delivering 1800 gpm, determine (a) the probable head developed by the 8-in pump. (b) Assuming a similar efficiency curve for the 8-in pump, what power would be required to drive the pump at the 1800-gpm rate?

**I** (a) The homologous pumps will have identical characteristics at corresponding flows. Choose several rates of flow for the 6-in pump and read off the corresponding heads. Calculate the values of $Q$ and $H$ so that a curve for the 8-in pump can be plotted. One such calculation is detailed below and a table of values established by similar determinations.

Using the given 1300 gpm and the 90-ft head, we obtain from the speed relation $H_8 = (D_8/D_6)^2(N_8/N_6)^2H_6 = (\frac{8}{6})^2(\frac{1450}{1750})^2H_6 = 1.22H_6 = 1.22(90) = 109.8$ ft. From the flow relation, $Q/D^3N =$ constant, we obtain $Q_8 = (D_8/D_6)^3(N_8/N_6)Q_6 = (\frac{8}{6})^3(\frac{1450}{1750})Q_6 = 1.964Q_6 = 1.964(1300) = 2550$ gpm.

Additional values, which have been plotted as the dashed line in Fig. 20-19, are as follows:

| For 6-in pump at 1750 rpm | | | For 8-in pump at 1450 rpm | | |
|---|---|---|---|---|---|
| $Q$, gpm | $H$, ft | Efficiency, % | $Q$, gpm | $H$, ft | Efficiency, % |
| 0 | 124 | 0 | 0 | 151.6 | 0 |
| 500 | 119 | 54 | 980 | 145.5 | 54 |
| 800 | 112 | 64 | 1570 | 134.5 | 64 |
| 1000 | 104 | 68 | 1960 | 127.0 | 68 |
| 1300 | 90 | 70 | 2550 | 110.0 | 70 |
| 1600 | 66 | 67 | 3140 | 80.6 | 67 |

From the head-discharge curve, for $Q = 1800$ gpm the head is 130 ft.

**(b)** The efficiency of the 8-in pump would probably be somewhat higher than that of the 6-in pump at comparable rates of flow. For this case, the assumption is that the efficiency curves are the same at comparable rates of flow. The table above lists the values for the flows indicated. The figure gives the efficiency curve for the 8-in pump and, for the 1800 gpm flow, the value is 67 percent. Then $P = \gamma QH/550e = 62.4[1800/(60)(7.48)](130)/[550(0.67)] = 88.3$ horsepower required.

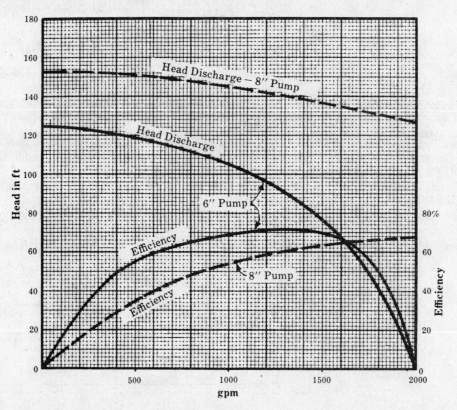

**Fig. 20-19**

**20.74** In order to predict the behavior of a small oil pump, tests are to be made on a model using air. The oil pump is to be driven by a $\frac{1}{20}$-hp motor at 1800 rpm and a $\frac{1}{4}$-hp motor is available to drive the air pump at 600 rpm. Using the s.g. of oil at 0.912 and the density of air constant at 0.00238 slug/ft³, what size model should be built?

**▌** Using the power relation, we obtain prototype $[P/(\rho D^5N^3)] =$ model$[P/(\rho D^5N^3)]$. Then $\frac{1}{20}/[0.912(62.4/32.2)D_p^5(1800)^3] = \frac{1}{4}/[0.00238D_m^5(600)^3]$ and $D_m/D_p = \frac{10}{1}$. The model should be 10 times as large as the oil pump.

**20.75** What is the power ratio of a pump and its $\frac{1}{5}$ scale model if the ratio of the heads is 4 to 1?

▮ For geometrically similar pumps, $P/(D^2H^{3/2})$ for pump = $P/(D^2H^{3/2})$ for model. Then $P_p/[(5D)^2(4H)^{3/2}] = P_m/(D^2H^{3/2})$ and $P_p = 25(4)^{3/2}P_m = 200P_m$.

**20.76** Determine the thrust coefficient of a propeller which is 4 in in diameter, revolves at 1800 rpm, and develops a thrust of 2.50 lb in fresh water.

▮ Thrust coefficient $= F/(\rho N^2 D^4) = 2.50/[1.94(\frac{1800}{60})^2(\frac{4}{12})^4] = 0.116$. The coefficient is dimensionless when $F$ is in pounds, $N$ in revolutions/per second, and $D$ in feet.

**20.77** The power and thrust coefficients of an 8-ft-diameter propeller moving forward at 100 ft/s at a rotational speed of 2400 rpm are 0.068 and 0.095, respectively. **(a)** Determine the power requirement and thrust in air ($\rho = 0.00237$ slug/ft$^3$). **(b)** If the advance-diameter ratio for maximum efficiency is 0.70, what is the air speed for the maximum efficiency?

▮ **(a)** Power $P = C_P\rho N^3 D^5$ in ft-lb/s $= [0.068(0.00237)(\frac{2400}{60})^3(8)^5]/550 = 615$ horsepower.

Thrust $F = C_F\rho N^2 D^4$ in lb $= 0.095(0.00237)(\frac{2400}{60})^2(8)^4 = 1476$ lb.

**(b)** Since $V/ND = 0.70$, $V = 0.70(\frac{2400}{60})(8) = 224$ ft/s.

**20.78** An airplane flies at 180 mph in still air, $\gamma = 0.0750$ lb/ft$^3$. The propeller is 5.5 ft in diameter and the velocity of the air through the propeller is 320 ft/s. Determine **(a)** the slipstream velocity, **(b)** the thrust, **(c)** the horsepower input, **(d)** the horsepower output, **(e)** the efficiency, and **(f)** the pressure difference across the propeller.

▮ **(a)** $V = \frac{1}{2}(V_1 + V_4)$ $\quad 320 = \frac{1}{2}[180(\frac{5280}{3600}) + V_4]$ $\quad V_4 = 376$ ft/s $\quad$ (relative to fuselage)

**(b)** $\quad$ Thrust $F = (\gamma/g)Q(V_4 - V_1) = (0.0750/32.2)[\frac{1}{4}\pi(5.5)^2(320)](376 - 264) = 1983$ lb

**(c)** $\quad$ Power input $P_i = FV/550 = 1983(320)/550 = 1154$ hp

**(d)** $\quad$ Power output $P_0 = FV_1/550 = 1983(264)/550 = 952$ hp

**(e)** $\quad$ Efficiency $e = \frac{952}{1154} = 82.5$ percent or $e = 2V_1/(V_4 + V_1) = 2(264)/(376 + 264) = 82.5$ percent.

**(f)** Pressure difference $=$ (thrust $F$)/[area $(\frac{1}{4}\pi D^2)$] $= 1983/[\frac{1}{4}\pi(5.5)^2] = 83.5$ psf or pressure difference $= (0.0750/32.2)\{[(376)^2 - (264)^2]/2\} = 83.5$ psf.

**20.79** Water is being pumped from the lower reservoir to the upper reservoir, as shown in Fig. 20-20. Find the pump horsepower required, assuming 75 percent pump efficiency. Neglect minor losses.

▮ $p_1/\gamma + V_1^2/2g + z_1 + E_p = p_2/\gamma + V_2^2/2g + z_2 + h_L$. From Fig. A-13, with $Q = 23$ cfs and $D = 24$ in, $h_1 = 0.0076$ ft/ft.

$$h_L = h_f = (0.0076)(3600) = 27.36 \text{ ft} \qquad 0 + 0 + 97.5 + E_p = 0 + 0 + 132.0 + 27.36$$

$$E_p = 61.86 \text{ ft} \qquad P = Q\gamma E_p/550 = (23)(62.4)(61.86)/550 = 161 \text{ hp}$$

Fig. 20-20

**20.80** Water is being pumped from a reservoir to the top of a hill, where it is discharged, as shown in Fig. 20-21. The pump, which is 70 percent efficient, is rated at 150 kW. Find the flow rate at which water is being discharged from the pipe. Neglect minor losses.

**I**  $p_1/\gamma + V_1^2/2g + z_1 + E_p = p_2/\gamma + V_2^2/2g + z_2 + h_L$   $E_p = P/Q\gamma = (150)(0.70)/[(Q)(9.79)] = 10.725/Q$

$$V = Q/A \quad V_2 = Q/[(\pi)(0.500)^2/4] = 5.093Q \quad h_L = h_f = (f)(L/D)(V^2/2g)$$

Assume $f = 0.018$. $h_L = (0.018)[975/0.500]\{(5.093Q)^2/[(2)(9.807)]\} = 46.42Q^2$, $0 + 0 + 111.0 + 10.725/Q = 0 + (5.093Q)^2/[(2)(9.807)] + 150.2 + 46.42Q^2$, $Q = 0.254$ m³/s (by trial and error). This solution was based on the assumed value of $f$ of 0.018. However, the value of $f$ is dependent on the Reynolds number and relative roughness. Therefore a new value of $f$ should be determined based on the computed value of $Q$ of 0.254 m³/s.

$$\epsilon/D = 0.00030/0.500 = 0.00060 \quad V = Q/A = 0.254/[(\pi)(0.500)^2/4] = 1.294 \text{ m/s}$$

$$N_R = DV/v = (0.500)(1.294)/(1.02 \times 10^{-6}) = 6.34 \times 10^5$$

From Fig. A-5, $f = 0.018$. Hence, the computed flow rate of 0.254 m³/s is correct.

Fig. 20-21

**20.81** Oil with a specific gravity of 0.87 is being pumped from a lower reservoir to an elevated tank as shown in Fig. 20-22. The pump in the system is 78 percent efficient and is rated at 185 kW. Determine the flow rate of the oil in the pipe if the total head loss from point 1 to point 2 is 12 m of oil.

**I**  $p_1/\gamma + v_1^2/2g + z_1 + E_p = p_2/\gamma + v_2^2/2g + z_2 + h_L$   $p_1/\gamma = p_2/\gamma = v_1^2/2g = 0$   $z_1 = 150$ m

$p = Q\gamma E_p$  $(0.78)(185) = (Q)[(0.87)(9.79)](E_p)$   $E_p = 16.94/Q$

$$v_2^2/2g = (Q/A)^2/2g = \{Q/[(\pi)(0.160)^2/4]\}^2/[(2)(9.807)] = 126.12Q^2$$

$z_2 = 200$ m   $h_L = 12$ m   $0 + 0 + 150 + 16.94/Q = 0 + 126.12Q^2 + 200 + 12$   $126.12Q^2 - 16.94/Q + 62 = 0$

By trial and error solution, $Q = 0.244$ m³/s.

Fig. 20-22

**20.82** For the conditions given in Prob. 20-80, find the flow rate at which water is being discharged from the pipe using the computer program written for Prob. 9.295.

**/ INPUT**

1 2 3 4 5 6 7 8 9 10 11 12 13 14 15 16 17 18 19 20 21 22 23 24 25 26 27 28 29 30 31 32 33 34 35 36 37 38 39 40 41 42 43 44 45 46 47 48 49 50 51 52 53 54 55 56 57 58 59 60 61 62 63 64 65 66 67 68 69 70 71 72 73 74 75 76 77 78 79 80
1SAMPLE ANALYSIS ØF INCØMPRESSIBLE FLØW
                         1.0        111.0      150.2     105.0
          500.0     975.0     .00000102 .00030     9.79
WATER                    (RØUGHNESS = 0.300 MM)

SAMPLE ANALYSIS OF INCOMPRESSIBLE FLOW

**/ OUTPUT**

GIVEN DATA FOR A CIRCULAR CLOSED CONDUIT CARRYING INCOMPRESSIBLE FLOW

PRESSURE AT POINT 1 =   0.0 KPA

PRESSURE AT POINT 2 =   0.0 KPA

ELEVATION AT POINT 1 =  111.0 M

ELEVATION AT POINT 2 =  150.2 M

ACTUAL ENERGY ADDED BETWEEN POINTS 1 AND 2 =105.0 KW

ACTUAL ENERGY REMOVED BETWEEN POINTS 1 AND 2 =  0.0 KW

MINOR HEAD LOSSES BETWEEN POINTS 1 AND 2 =  0.0 M

DIAMETER OF CONDUIT =  500.0 MM

LENGTH OF CONDUIT =   975.0 M

FLUID FLOWING IS WATER

CONDUIT MATERIAL IS (ROUGHNESS = 0.300 MM)

THE FLOW RATE WILL BE  0.254 CU M/S

VELOCITY AT POINT 1 =  0.00 M/S

VELOCITY AT POINT 2 =  1.29 M/S

**20.83** A pump draws water from a reservoir and then discharges it to an elevated tank, as shown in Fig. 20-23. The pipe's ends are squared-cornered (i.e., sharp-edged). There are two 90° bends as shown. The pipe roughness ($\epsilon$) is 0.0084 ft and $C$ is 120. If the rate of flow is 8.0 ft$^3$/s and the efficiency of the pump is 75 percent, determine the required horsepower of the pump. Consider both friction and minor losses.

**/**
$$p_1/\gamma + v_1^2/2g + z_1 + E_p = p_2/\gamma + v_2^2/2g + z_2 + h_L \qquad p_1/\gamma = p_2/\gamma = v_1^2/2g = v_2^2/2g = 0$$
$$z_1 = 10 \text{ ft} \qquad z_2 = 120 \text{ ft} \qquad h_L = h_f + h_m$$

For the 12-in pipe with $Q = 8$ cfs, $(h_1)_{12} = 0.032$ ft/ft. For the 10-in pipe with $Q = 8$ cfs, $(h_1)_{10} = 0.074$ ft/ft.

$$h_f = (0.032)(40) + (0.074)(200) = 16.08 \text{ ft} \qquad h_m = h_c + 2h_b + h_e \qquad h_c = K_c v_{10}^2/2g \qquad K_c = 0.45$$
$$v_{12} = Q/A = 8.0/[(\pi)(1.0)^2/4] = 10.19 \text{ ft/s} \qquad h_c = (0.45)(10.19)^2/[(2)(32.2)] = 0.73 \text{ ft}$$
$$h_b = K_b v_{10}^2/2g \qquad v_{10} = Q/A = 8.0/[(\pi)(0.833)^2/4] = 14.68 \text{ ft/s}$$

With $\epsilon/D = 0.0084/(\frac{10}{12}) = 0.101$ and $R/D = \frac{20}{10} = 2.0$, $K_b = 0.6$.

$$h_b = (0.6)(14.68)^2/[(2)(32.2)] = 2.01 \text{ ft} \qquad h_e = K_e v_{10}^2/2g \qquad K_e = 1.0$$
$$h_e = (1.0)(14.68)^2/[(2)(32.2)] = 3.35 \text{ ft} \qquad h_m = 0.73 + (2)(2.01) + 3.35 = 8.10 \text{ ft}$$
$$h_L = h_f + h_m = 16.08 + 8.10 = 24.18 \text{ ft} \qquad 0 + 0 + 10 + E_p = 0 + 0 + 120 + 24.18 \qquad E_p = 134.18 \text{ ft}$$
$$P = Q\gamma E_p = (8.0)(62.4)(134.18) = 66\,983 \text{ ft-lb/s} \qquad \text{Required horsepower} = 66\,983/[(550)(0.75)] = 162 \text{ hp}$$

**Fig. 20-23**

**20.84** Oil with a specific gravity of 0.86 is being pumped from a reservoir as shown in Fig. 20-24. The pressures at points 1 and 2 are −4.0 psi and 43.0 psi, respectively. The rate of flow in the pipe is 0.50 ft³/s. The pump is rated at 8 hp. Determine the efficiency of the pump. Neglect energy losses in the system.

■ $p_1/\gamma + v_1^2/2g + z_1 + E_p = p_2/\gamma + v_2^2/2g + z_2 + h_L$ $\quad p_1/\gamma = (-4)(144)/[(0.86)(62.4)] = -10.73$ ft

$$v_1^2/2g = v_2^2/2g \quad z_1 = h_L = 0 \quad p_2/\gamma = (43)(144)/[(0.86)(62.4)] = 115.38 \text{ ft}$$

$$z_2 = 3 \text{ ft} \quad -10.73 + E_p = 115.38 + 3 \quad E_p = 129.11 \text{ ft}$$

$$P = Q\gamma E_p = (0.50)[(0.86)(62.4)](129.11) = 3464 \text{ ft-lb/s} = \tfrac{3464}{550} = 6.30 \text{ hp}$$

Efficiency of pump = (6.30/8)(100) = 78.8 percent

**Fig. 20-24**

**20.85** A pump circulates water at the rate of 1900 gpm in a closed circuit holding 9000 gal. The total pump head is 290 ft and the pump efficiency is 0.89. Assuming the bearing friction to be negligible and the system to be adiabatic, find the rate of temperature rise in the water. [For water, $c = 25\,000$ ft-lb/(slug-F°).]

■ By the first law $mc(\Delta T/\Delta t) = Q\gamma H/\eta$, or

$$\frac{\Delta T}{\Delta t} = \frac{Q\gamma H}{mc\eta} = \frac{[(1900)(0.002228](62.4)(290)}{[(9000/7.48)(1.94)](25\,000)(0.89)} = 0.00147 \text{ F°/s}$$

or 5.3 F°/h.

**20.86** The pump in Fig. 20-25 delivers water at 3 ft³/s to a machine at section 2, which is 20 ft higher than the reservoir surface. The losses between 1 and 2 are given by $h_f = KV_2^2/2g$, where $K = 7.5$ is a dimensionless loss coefficient. Find the power required to drive the pump if it is 80 percent efficient.

**▮** The flow is steady except for the slow decrease in reservoir depth, which we neglect, taking $V_1 \approx 0$. We can compute $V_2$ from the given flow rate and diameter: $V_2 = Q/A_2 = 3/[(\pi/4)(\frac{3}{12})^2] = 61.1$ ft/s. Because of the solid walls and one-dimensional ports, the viscous work is zero. $(p_1/\rho g) + (V_1^2/2g) + z_1 = (p_2/\rho g) + (V_2^2/2g) + z_2 + h_s + h_f$. With $V_1 = z_1 = 0$ and $h_f = KV_2^2/2g$, we can solve for

$$h_s = [(p_1 - p_2)/\rho g] - z_2 - (1 + K)(V_2^2/2g) \tag{1}$$

The pressures must be in pounds per square foot for consistent units, with $\rho g = 62.4$ lb/ft³ for water. Introducing numerical values, we have

$$h_s = [(14.7 - 10)(144)/62.4] - 20 - [(1 + 7.5)(61.1)^2/2(32.2)] = -502 \text{ ft} \tag{2}$$

The pump head is negative, indicating work done on the fluid. The power delivered is $P = \dot{m}w_s = \rho Q g h_s = (1.94)(3)(32.2)(-502) = -94\,100$ ft-lb/s, hp = 94 100/550 = 171 hp. The input power thus required to drive the 80 percent efficient pump is $P_{\text{input}} = P/\text{efficiency} = 171/0.8 = 214$ hp.

$p_1 = 14.7$ lb/in² abs

Machine

$z_1 = 0$

$D_2 = 3$ in
$z_2 = 20$ ft
$p_2 = 10$ lb/in²

Water

Pump

$h_s$ (negative)

**Fig. 20-25**

**20.87** For the pump of Fig. 20-26, total friction head loss is 6 m. If the pump delivers 40 kW of power to the water, determine the exit velocity $V_e$ and the flow rate $Q$.

**▮** $p_1/\gamma + V_1^2/2g + z_1 + E_p = p_2/\gamma + V_2^2/2g + z_2 + h_L \qquad 0 + 0 + 0 + E_p = 0 + V_e^2/[(2)(9.807)] + 3 + 6$

$$E_p = 0.05098V_e^2 + 9$$

Also, $P = Q\gamma E_p$, $Q = AV = [(\pi)(0.050)^2/4](V_e) = 0.001964V_e$, $(40)(1000) = (0.001964V_e)[(9.79)(1000)](E_p)$, $E_p = 2080/V_e$. Equating expressions for $E_p$, $0.05098V_e^2 + 9 = 2080/V_e$, $V_e = 32.7$ m/s (by trial and error); $Q = (0.001964)(32.7) = 0.0642$ m³/s, or 64.2 L/s.

$D = 120$ mm $\qquad D_e = 50$ mm

$V_e$

Pump

3 m

5 m

Water

**Fig. 20-26**

$D = 3$ in

Pump

130 ft/s

10 ft

6 ft

$D = 9$ in

**Fig. 20-27**

**20.88** A fireboat pump draws sea water (s.g. = 1.025) from a 9-in submerged pipe and discharges it as 130 fps through a 3-in nozzle, as in Fig. 20-27. Total head loss is 9 ft. If the pump is 72 percent efficient, how much horsepower is required to drive it?

**▮** $\qquad P = Q\gamma E_p/\eta \qquad Q = AV = [(\pi)(\frac{3}{12})^2/4](130) = 6.381$ ft³/s

$p_1/\gamma + V_1^2/2g + z_1 + E_p = p_2/\gamma + V_2^2/2g + z_2 + h_L \qquad 0 + 0 + 0 + E_p = 0 + 130^2/[(2)(32.2)] + 10 + 9$

$E_p = 281.4$ ft $\qquad P = \{(6.381)[(1.025)(62.4)](281.4)/550\}/0.72 = 290$ hp

**20.89** The pump in Fig. 20-28 discharges water at 30 L/s. Neglecting losses and elevation changes, what power is delivered to the water by the pump?

❚ $p_1/\gamma + V_1^2/2g + z_1 + E_p = p_2/\gamma + V_2^2/2g + z_2 + h_L$     $V_1 = Q/A_1 = (30 \times 10^{-3})/[(\pi)(0.10)^2/4] = 3.820$ m/s

$$V_2 = 0.03/[(\pi)(0.04)^2/4] = 23.87 \text{ m/s}$$

$$125/9.79 + 3.820^2/[(2)(9.807)] + 0 + E_p = 409/9.79 + 23.87^2/[(2)(9.807)] + 0 + 0$$

$$E_p = 57.31 \text{ m} \qquad P = Q\gamma E_p = (30 \times 10^{-3})(9.79)(57.31) = 16.8 \text{ kW}$$

**Fig. 20-28**

**20.90** What is the flow for the system shown in Fig. 20-29a? The pump has the characteristics shown in Fig. 20-29b. What is the power required?

❚ Express the first law for thermodynamics for pipe interior as shown:

$$(p_1/\rho) + (V_1^2/2) + gz_1 + gH = (p_2/\rho) + (V_2^2/2) + gz_2 + h \tag{1}$$

Bernoulii in left tank: Use gage pressures, $[(35)(144)/1.94] + 36g = (p_1/\rho) + (V^2/2) + 0$:

$$p_1/\rho = 3757 - V^2/2 \tag{2}$$

Hydrostatics in right-hand tank:

$$p_2/\rho = (1/\rho)[(10)(144) + 11.3(62.4)] = 1106 \tag{3}$$

Substitute Eqs. (2) and (3) into Eq. (1):

$$[3757 - (V^2/2)] + gH = 1106 + (g)(45) + f(1345/2.00)(V^2/2) + 2.3(V^2/2) \tag{4}$$

Assume $Q_1 = 80$ cfs. Then $V_1 = 80/[(\pi/4)(2.00)^2] = 25.5$ ft/s, $(N_R)_1 = [(25.5)(2.00)/(0.1217 \times 10^{-4})] = 4.19 \times 10^6$, $f_1 = 0.0118$. Solve for $H_1$ in Eq. (4): $H_1 = 76.1$ ft. The point (1) for $H_1$ and $Q_1$ is above $Q$ curve. We take as a second estimate $Q_2 = 70$, $V_2 = 70/[(\pi/4)(2)^2] = 22.28$ ft/s, $(N_R)_2 = (22.28)(2)/(0.1217 \times 10^{-4}) = 3.66 \times 10^6$, $f_2 = 0.0118$, $H_2 = 49.3$ ft. New point is just below $Q$ line. Third estimate can now be easily made: $Q_3 = 71$ cfs, $V_3 = 71/[(\pi/4)(2^2)] = 22.60$ ft/s, $(N_R)_3 = (22.60)(2)/(0.1217 \times 10^{-4}) = 3.71 \times 10^6$, $f_3 = 0.0118$, $H_3 = 51.8$ ft. We are close enough to the intersection so that we can say: $Q = 71$ cfs. The power needed for this operation is then Power $= g(51.8)(1.938)(71)/[(0.73)(550)] = 572$ hp.

Fig. 20-29

(a)

(b)

# CHAPTER 21
# Turbines

**21.1** A radial-flow turbine has the following dimensions: $r_1 = 1.6$ ft, $r_2 = 1.0$ ft, and $\beta_1 = 80°$. The width of the flow passage between the two sides of the turbine is 0.8 ft. At 300 rpm the flow rate through the turbine is 120 cfs. Find **(a)** the blade angle $\beta_2$ such that the water exits from the turbine in the radial direction, **(b)** the torque exerted by the water on the runner and the horsepower thus developed, **(c)** the head utilized by the runner and the power resulting therefrom. Assume that water enters and leaves the blades smoothly. Assume the blades are so thin that they do not occupy any of the available flow area.

▌ **(a)** At the outer periphery ($r_1 = 1.6$ ft): $u_1 = \omega r_1 = (2\pi/60)(300)(1.6) = 50.3$ fps. From continuity $Q = 120 = 2\pi r_1(z)V_{r_1} = 2\pi(1.6)(0.8)(V_{r_1})$, $V_{r_1} = 120/8.04 = 14.92$ fps $= v_1 \sin \beta_1 = v_1 \sin 80°$, $v_1 = 14.92/(\sin 80°) = 15.15$ fps, $v_1 \cos 80° = (15.15)(0.1736) = 2.63$ fps, $V_1 \cos \alpha_1 = u_1 + v_2 \cos \beta_1 = 50.3 + 2.6 = 52.9$ fps. (See Fig. 21-1a.) At the inner periphery ($r_2 = 1.0$ ft): $u_2 = \omega r_2 = (2\pi/60)(300)(1.0) = 31.4$ fps, $V_{r_2} = V_{r_1}(r_1/r_2) = 14.92(1.6/1.0) = 23.9$ fps. Because the water exits from the turbine in the radial direction, $\alpha_2 = 90°$ and $V_2 = V_{r_2}$. $\tan(180° - \beta_2) = V_2/u_2 = 23.9/31.4 = 0.760$. Thus $\beta_2 = 180° - 37.2° = 142.8°$, the required blade angle. (See Fig. 21-1b.)
**(b)** $T = \rho Q(r_1 V_1 \cos \alpha_1 - r_2 V_2 \cos \alpha_2) = 1.94(120)[(1.6)(52.9 - 0)] = 19\,700$ ft-lb and power $= T\omega = 19700(31.4) = 619\,000$ ft-lb/s $= 1125$ hp.
**(c)** $h'' = (u_1 V_1 \cos \alpha_1 - u_2 V_2 \cos \alpha_2)/g = [50.3(52.9) - 0]/32.2 = 82.6$ ft and power $= \gamma Q h'' = (62.4)(120)(82.6) = 619\,000$ ft-lb/s $= 1125$ hp.

Further calculations indicate that the absolute velocity of the water changed from $V_1 = 55.0$ fps at entry to $V_2 = 23.9$ fps at exit while the velocity of the water relative to the blades changed from $v_1 = 15.15$ fps at entry to $v_2 = 39.5$ fps at exit.

Fig. 21-1

**21.2** A 6-in-diameter pipe ($f = 0.020$) of length 1000 ft delivers water from a reservoir with a water-surface elevation of 500 ft to a nozzle at elevation 300 ft. The jet from the nozzle is used to drive a small impulse turbine. If the head loss through the nozzle can be expressed as $0.04\,V_j^2/2g$, find the jet diameter that will result in maximum power in the jet. Neglect the head loss at entrance to the pipe from the reservoir. Evaluate the power in the jet.

▌ Energy equation: $500 \times 0.02(1000/0.5)(V_p^2/2g) - 0.04(V_j^2/2g) = 300 + (V_j^2/2g)$. If we define the pipe diameter and velocity as $D_p$ and $V_p$ and the jet diameter and velocity as $D_j$ and $V_j$, from continuity we get $A_p V_p = A_j V_j$, $D_p^2 V_p = D_j^2 V_j$.
Since the pipe diameter $D_p = 0.50$ ft, $0.25 V_p = D_j^2 V_j$ and $V_p = 4D_j^2 V_j$. Substituting this expression for $V_p$ in the energy equation gives $200 - (V_j^2/2g)(1.04 + 640D_j^4)$.
Assuming different values for $D_j$, we can compute corresponding values of $V_j$ and $Q$, and then the jet power can be computed using $p_{jet} = [\gamma Q(V_j^2/2g)]/550$. The results are shown in the table on page 639.
Thus a 2-in-diameter jet is the optimum; it will have about 30 hp.
An alternative procedure for solving this problem is to set up an algebraic expression for the power of the jet, $P_{jet}$, as a function of the jet diameter, $D_j$, and differentiate $P_{jet}$ with respect to $D_j$ and equate to zero to find the value of $D_J$ for which $P_{jet}$ is a maximum.

| $D_j$, in | $D_j$, ft | $V_j$, fps | $A_j$, ft$^2$ | $Q = A_j V_j$, cfs | $P_{jet}$, hp |
|---|---|---|---|---|---|
| 1.0 | 0.083 | 111 | 0.0054 | 0.60 | 12.8 |
| 1.5 | 0.125 | 105 | 0.0122 | 1.28 | 24.2 |
| 2.0 | 0.167 | 91 | 0.0218 | 2.00 | 29.8 |
| 2.5 | 0.208 | 76 | 0.0338 | 2.57 | 26.2 |
| 3.0 | 0.250 | 60 | 0.0491 | 2.94 | 18.8 |
| 4.0 | 0.333 | 38 | 0.0873 | 3.29 | 8.4 |
| 6.0 | 0.500 | 18 | 0.197 | 3.49 | 1.9 |

**21.3** A turbine is to operate at 400 rpm under a net head of 1320 ft. If a single 6-in-diameter water jet is used, find the specific speed of this machine assuming $C_v = 0.98$, $\phi = 0.45$, and $\eta = 0.85$. Find also the required pitch diameter of the wheel.

$$V = C_v\sqrt{2gh} = 0.98\sqrt{(64.4)(1320)} = 286 \text{ fps} \qquad Q = AV = (0.196)(286) = 56.0 \text{ cfs}$$
$$\text{bhp} = \eta(\gamma Qh/550) = 0.85[(62.4)(56)(1320)/550] = 7130$$
$$N_s = N\sqrt{\text{bhp}}/h^{5/4} = 400\sqrt{7130}/(1320)^{5/4} = 4.25 \qquad u = \phi\sqrt{2gh} = 0.45\sqrt{(64.4)(1320)} = 131.2 \text{ fps}$$
$$N = 400 \text{ rpm} = u(60)/\pi D = (131.2)(60)/\pi D \qquad D = 6.26 \text{ ft} = 75.2 \text{ in}$$

**21.4** In lieu of the single impulse wheel of Prob. 21.3, suppose that three identical single-nozzle wheels are to be used, operating under the same head of 1320 ft. The total flow rate is to be 56.0 cfs. Determine the required specific speed of these turbines, their pitch diameter, the jet diameter, and the operating speed. Once again, assume $C_v = 0.98$, $\phi = 0.45$, and $\eta = 0.85$.

$$Q = \tfrac{56}{3} = 18.7 \text{ cfs} \qquad \text{bhp} = \tfrac{7130}{3} = 2377 \qquad N_s = N\sqrt{2377}/(1320^{5/4}) = N(0.00613)$$
$$N = 131.2(60)/\pi D = 2056/D$$

From the two preceding expressions it is apparent that the required $N_s$ depends on the operating speed, as does $D$. Hence there are a number of possible answers. If we let the operating speed be 400 rpm (18-pole generator for 60-cycle electricity), $N_s = 400(0.00613) = 2.45$, $D = \tfrac{2506}{400} = 6.26$ ft = 75.2 in. Thus three $N_s = 2.45$ wheels of pitch diameter 75 in operating at 400 rpm would suffice. These wheels would have relatively small buckets. At such a low specific speed the optimum efficiency of impulse wheels is usually less than 0.85. An alternative solution, for example, would be to use an operating speed of 600 rpm (12-pole generator for 60-cycle electricity). For this case, $N_s = (600)(0.00613) = 3.68$, $D = \tfrac{2506}{600} = 4.18$ ft = 50.1 in. Thus three $N_s = 3.68$ wheels, of pitch diameter 50.1 in, operating at 600 rpm would suffice.

Let us now determine the required jet diameter for these 50.1-in-diameter wheels operating at 600 rpm: $Q = AV = (\pi D_j^2/4)(286) = 18.7$ cfs from which $D_j = 0.289$ ft = 3.46 in.

**21.5** A two-nozzle single-wheel turbine, with $C_v$, $\phi$, and $\eta$ as in Prob. 21.3, is designed to operate under a head of 1300 ft with a total flow of 56 cfs. Express the specific speed $N_s$ and pitch diameter $D$ as functions of the rotation speed $N$.

$$V_j = C_v(2gh)^{1/2} = (0.98)[(2)(32.2)(1300)]^{1/2} = 283.6 \text{ fps} \qquad Q_j = A_jV_j = (A_j)(283.6) = \tfrac{56}{2}$$
$$A_j = 0.09873 \text{ ft}^2 \qquad \pi D_j^2/4 = 0.09800 \qquad D_j = 0.3546 \text{ ft} \quad \text{or} \quad 4.25 \text{ in}$$
$$bp_j = (\eta)(\gamma Q_j h)/550 = (0.85)(62.4)(28.0)(1300)/550 = 3510 \text{ bhp per jet}$$
$$N_s = N(\text{bhp})^{0.5}/h^{5/4} = (N)(3510)^{0.5}/1300^{5/4} = 0.007590N$$
$$D = 60\phi(2gh)^{1/2}/\pi N = (60)(0.45)[(2)(32.2)(1300)]^{1/2}/(\pi N) = 2487/N$$

**21.6** A series of vanes is acted on by a 4-in water jet having a velocity of 102 fps, $\alpha_1 = \beta_1 = 0°$. In terms of the vane velocity $v$, find the blade angle $\beta_2$ that makes the force acting on the vanes in the direction of the jet equal 210 lb. Neglect friction.

From the velocity vector diagram of Fig. 21-2,

$$\Delta V_x = (u + v_2\cos\beta_2) - V_1 \qquad Q = AV_1 = [(\pi)(\tfrac{4}{12})^2/4](102) = 8.901 \text{ ft}^3/\text{s}$$
$$F_x = \rho Q(-\Delta V_x) = -\rho Q[(u + v_2\cos\beta_2) - V_1] \qquad 210 = -(1.94)(8.901)[(u + v_2\cos\beta_2) - 102]$$
$$u + v_2\cos\beta_2 = 89.8 \qquad v_2 = v_1 = 102 - u \qquad u + (102 - u)\cos\beta_2 = 89.8$$
$$\beta_2 = \arccos\frac{89.8 - u}{102 - u}$$

**Fig. 21-2**

**21.7** Under a net head of 1 ft, the Pelton wheel for Fig. 21-3a discharges 0.288 cfs at full nozzle opening and the maximum power is 0.026 bhp for a value of $\phi = 0.465$. The corresponding brake torque is 3.70 ft-lb as shown in Fig. 21-3b. Assuming that the similarity laws apply precisely, determine the discharge, torque, power, and rotative speed of this wheel when it operates under a head of 1600 ft.

▌ On Fig. 21-3a, $h_1 = 1.0$ ft, $\phi = 0.465$, $Q_1 = 0.288$ cfs, $T_1 = 3.70$ ft-lb, $\text{bhp}_1 = 0.026$, and $r = 1$ ft.

$$u_1 = (\phi)(2gh_1)^{1/2} = (0.465)[(2)(32.2)(1.0)]^{1/2} = 3.732 \text{ fps} \qquad \omega_1 = u_1/r = 3.732/1 = 3.732 \text{ rad/s}$$

From the similarity laws, $Q \propto h^{1/2}$, $T \propto h$, $\omega \propto h^{1/2}$, and $\text{bhp} \propto h^{3/2}$. At $h_2 = 1600$ ft,

$$Q_2 = (0.288)(1600)^{1/2} = 11.5 \text{ ft}^3/\text{s} \qquad T_2 = (3.70)(1600) = 5920 \text{ ft-lb}$$

$$\omega_2 = (3.73)(1600)^{1/2} = 149.2 \text{ rad/s} \quad \text{or} \quad 1425 \text{ rpm} \qquad \text{bhp}_2 = (0.026)(1600)^{3/2} = 1664$$

**21.8** A wheel and nozzle similar to that of Fig. 21-3a and b with a pitch diameter of 12 ft is used under a net head of 1500 ft. What are the torque, power, and rotative speed at point of best efficiency for full nozzle opening?

▌ In Fig. 21-3a, $\phi_e = 0.46$. In Fig. 21-3b, for $\phi = 0.46$, nozzle fully open, $T = 3.76$ ft-lb.

$$u = (\phi)(2gh)^{1/2} = (0.46)[(2)(32.2)(1)]^{1/2} = 3.691 \text{ fps} \qquad \omega = u/r = 3.691/1 = 3.691 \text{ rad/s}$$

$$\text{bhp} = Tu/550 = (3.76)(3.691)/550 = 0.02523 \qquad N = 60u/\pi D = (60)(3.691)/[(\pi)(2)] = 35.25 \text{ rpm}$$

For a 12-ft-diameter wheel and 1500 ft head, $T = (3.76)(\frac{12}{2})^3(1500) = 1.22 \times 10^6$ ft-lb, $P = (0.02523)(\frac{12}{2})^2(1500)^{3/2} = 52\,770$ hp, $N = (35.25)(1500)^{1/2}/6 = 228$ rpm.

**Fig. 21-3(a)**

Relation between power and speed at constant head with maximum nozzle opening. (From tests made by F. G. Switzer and R. L. Daugherty.)

**Fig. 21-3(b)**

Relation between torque and speed at constant head. (From tests made by F. G. Switzer and R. L. Daugherty.)

**21.9** Find the torque and power transferred to the buckets of an impulse wheel with $\alpha_1 = 0°$, $\beta_2 = 160°$, $k = 0.44$, $\phi = 0.46$, $C_v = 0.98$, a jet diameter of 10 in, and a pitch diameter of 10 ft. Find also, the hydraulic efficiency, and, expressed as a percentage of the total head, find the head loss in bucket friction, the energy head loss at discharge, and the head loss in the nozzle.

▮ For operation under the purely artificial value of 1-ft net head, $u = \phi\sqrt{2g} = 8.02\phi = 3.68$ fps, $V_1 = 0.98\sqrt{2g} = 7.86$ fps, and $\gamma Q = \gamma AV = (62.4)(0.545)(7.86) = 267$ lb/s. Hence, the expression for the torque exerted on the wheel by the water is $T = (r)(F_u) = 5(267/32.2)[1 + (0.940/1.2)](8.02)(C_v - \phi)$. With $C_v = 0.98$ and $\phi = 0.46$, $T = 309$ ft-lb. The power transferred from the water to the buckets is $F_u u = T\omega = T(u/r) = 309[(0.46)(8.02)/5] = 228$ ft-lb/s. The power input is $\gamma Qh = 267(1) = 267$ ft-lb/s. The hydraulic efficiency is $\frac{228}{267} = 0.85$. Note that the answers given here are for $h = 1$ ft. For other values of $h$, the flow rate, torque, and power can be found by adjusting the values for $h = 1$ ft according to similarity laws.

Under 1-ft head, $v_1 = V_1 - u = 7.86 - 3.68 = 4.18$ fps and $v_2 = v_1/\sqrt{1+k} = 4.18/\sqrt{1.44} = 3.48$ fps. Hence the head loss in bucket friction is $0.44(3.48)^2/2g = 0.083$ ft, or 8.3 percent. $V_2 \cos\alpha_2 = u + v_2 \cos\beta_2 = 3.68 + 3.48 \cos 160° = 3.68 - 3.27 = 0.41$ fps, $V_2 \sin\alpha_2 = v_2 \sin\beta_2 = 3.48 \sin 160° = 1.19$ fps. Hence $\cot\alpha_2 = 0.41/1.19 = 0.344$ or $\alpha_2 = 71°$, and $V_2 = 1.19/0.945 = 1.26$ fps from which the energy head loss at discharge is $1.26^2/2g = 0.025$ ft, or 2.5 percent. The head loss in the nozzle is approximately $(1/C_v^2 - 1)V_1^2/2g$, about 4.0 percent; so the total hydraulic loss is $8.3 + 2.5 + 4.0 \approx 14.8$ percent, which gives a close check on the computed hydraulic efficiency of 85 percent.

**21.10** A nozzle having a velocity coefficient of 0.98 discharges a jet 6 in in diameter under a head of 900 ft. The wheel diameter is 8 ft, $\beta_2 = 170°$, $\alpha_1 = 0°$, $k = 0.5$, and the mechanical efficiency of the wheel is 96%. Calculate the hydraulic efficiency and the gross efficiency, assuming $\phi = 0.45$.

▮ Nozzle efficiency $= 0.98^2 = 0.9604$, $\eta_h = (2)[1 - (\cos\beta_2/\sqrt{1+k})](C_v - \phi)(\phi) = (2)[1 - (\cos 170°/\sqrt{1+0.5})](0.98 - 0.45)(0.45) = 0.861$. As $C_v$ is included, $\eta_h$ is here the hydraulic efficiency of both the wheel and the nozzle. $\eta = \eta_h\eta_m = (0.861)(0.96) = 0.830$.

**21.11** A 24-in Pelton wheel is tested under a head of 65.5 ft. For a certain nozzle setting, the net brake load at 276 rpm is 40 lb at a brake arm of 5.0 ft, and the discharge is 1.90 cfs. What are the brake horsepower and efficiency under these conditions?

$$T_2 = dW = (5.0)(40) = 200 \text{ ft-lb} \qquad \omega = 2\pi n/60 = (2)(\pi)(276)/60 = 28.9 \text{ rad/s}$$
$$bp = T_2\omega = (200)(28.9) = 5780 \text{ ft-lb/s} \quad \text{or} \quad 10.5 \text{ hp}$$
$$P_w = Q\gamma H = (1.90)(62.4)(65.5) = 7766 \text{ ft-lb/s} \quad \text{or} \quad 14.1 \text{ hp} \qquad \eta = 10.5/14.1 = 74.5\%$$

**21.12** Supposing the bearing-friction and windage losses in Prob. 21.11 to be 0.2 hp, calculate the mechanical efficiency and the hydraulic efficiency.

▌ Bearing-friction and windage loss $= 0.2/10.5 = 0.019$, or 1.9 percent.

$$\eta_m = 100 - 1.9 = 98.1 \text{ percent} \qquad \eta_h = \eta/(\eta_v\eta_m) \qquad \eta_v = 1$$
$$\eta_{h_a} = 0.745/[(1)(0.981)] = 0.759 \quad \text{or} \quad 75.9 \text{ percent}$$

**21.13** Find the approximate hydraulic efficiency of an impulse wheel for which the nozzle velocity coefficient is 0.96 and the bucket angle is 165°, if $\phi = 0.46$ and $k = 0.1$.

$$\eta_h = (2)[1 - (\cos\beta_2/\sqrt{1+k})](C_v - \phi)(\phi) = (2)[1 - (\cos 165°/\sqrt{1+0.1})](0.96 - 0.46)(0.46)$$
$$= 0.884 \quad \text{or} \quad 88.4 \text{ percent}$$

**21.14** A double-overhung impulse-turbine installation is to develop 20 000 hp at 275 rpm under a net head of 1100 ft. Determine $N_s$, wheel-pitch diameter, and approximate jet diameter.

$$N_s = NP^{0.5}/H^{5/4} = (275)(20\,000/2)^{0.5}/1100^{5/4} = 4.34$$
$$D = 153.2\phi\sqrt{h}/N = (153.2)(0.45)(\sqrt{1100})/275 = 8.31 \text{ ft}$$
$$D_j = D/12 = 8.31/12 = 0.6925 \text{ ft} \quad \text{or} \quad 8.31 \text{ in}$$

**21.15** Repeat Prob. 21.14 for a single wheel with a single nozzle.

▌ $N_s = (275)(20\,000/1)^{0.5}/1100^{5/4} = 6.14$. Noting the computation of $D$ in Prob. 21.14, $D$ is unchanged; hence, $D_j = 8.31$ in.

**21.16** Repeat Prob. 21.14 for a single wheel with four nozzles.

▌ $N_s = (275)(20\,000/4)^{0.5}/1100^{5/4} = 3.07$. Noting the computation of $D$ in Prob. 21.14, $D$ is unchanged; hence, $D_j = 8.98$ in.

**21.17** An impulse turbine ($N_s = 5$) develops 105 800 hp under a head of 2050 ft. For 60-Hz electricity calculate the turbine speed, wheel diameter, and number of poles in the generator. Assume $\phi = 0.45$.

$$N_s = NP^{0.5}/H^{5/4} \qquad 5 = (N)(105\,800)^{0.5}/2050^{5/4} \qquad N = 212 \text{ rpm} = 7200/N_{\text{poles}}$$
$$N_{\text{poles}} = 7200/212 \approx 34 \qquad D = 153.2\phi\sqrt{h}/N = (153.2)(0.45)(\sqrt{2050})/212 = 14.72 \text{ ft}$$

**21.18** Rework Prob. 21.17 for a six-nozzle unit using the same $N_s$, bhp, and head.

$$N_s = NP^{0.5}/H^{5/4} \qquad 5 = (N)(105\,800/6)^{0.5}/2050^{5/4} \qquad N = 519 \text{ rpm} = 7200/N_{\text{poles}}$$
$$N_{\text{poles}} = 7200/519 \approx 14 \qquad D = 153.2\phi\sqrt{h}/N = (153)(0.45)(\sqrt{2050})/514 = 6.06 \text{ ft}$$

**21.19** Making suitable assumptions, estimate the wheel diameter and the rotative speed of a turbine that will develop 14 900 bhp under a head of 995 ft.

▌ $N_s = NP^{0.5}/H^{5/4}$. Best efficiency occurs at $N_s < 4.5$. Assume $N_s = 4$: $4 = (N)(14\,900)^{0.5}/995^{5/4}$, $N = 183$ rpm, $D = 153.2\phi\sqrt{h}/N$. Assume synchronous speed $= 180$ rpm and $\phi = 0.46$: $D = (153.2)(0.46)(\sqrt{995})/180 = 12.35$ ft.

**21.20** An 18-in-diameter model turbine is operated at 600 rpm under a total head of 55.0 ft; the brake horsepower and flow are measured as 330 and 61 cfs. Calculate (a) the specific speed, (b) the value of $\phi$, (c) the efficiency, and (d) the shaft torque.

(a) 
$$N_s = NP^{0.5}/H^{5/4} = (600)(330)^{0.5}/55.0^{5/4} = 72.8$$

(b) 
$$\phi = u/(2gh)^{1/2} = (2\pi N/60)r/(2gH)^{1/2} = [(2)(\pi)(600)/60][9/12]/[(2)(32.2)(55.0)]^{1/2} = 0.792$$

**(c)**     $\eta = P/\gamma QH = (330)(550)/[(62.4)(61)(55.0)] = 0.867$   or   86.7 percent

**(d)**             $T = P/\omega = (330)(550)/[(2\pi)(600)/60] = 2889$ ft-lb

**21.21**   Refer to Prob. 21.20. **(a)** What would be the efficiency of a 12-ft-diameter prototype turbine? **(b)** If the prototype is to operate at 140 rpm under a net head of 202 ft, find its horsepower output and flow rate.

▌ **(a)**   $(1 - \eta_2)/(1 - \eta_1) = (D_1/D_2)^{1/5}$     $(1 - \eta_2)/(1 - 0.867) = (1.5/12)^{1/5}$     $\eta_2 = 0.912$   or   91.2 percent

**(b)**             $N_s = NP^{0.5}/H^{5/4}$     $72.8 = 140P^{0.5}/202^{5/4}$     $P = 156\ 800$ bhp

$P = Q\gamma H/550$     $156\ 800 = (Q)(62.4)(202)(0.912)/550$     $Q = 7502$ ft$^3$/s

**21.22**   A single turbine with $N_s \approx 25$ is to develop 4180 hp under a head of 245 ft. What rotative speed would occur with 50-Hz electricity? How many poles should the generator have?

▌                 $N_s = NP^{0.5}/H^{5/4}$     $25 = (N)(4180)^{0.5}/245^{5/4}$     $N = 375$ rpm

For 50 cycles, $N_{poles} = 6000/N = \frac{6000}{375} = 16$.

**21.23**   The Francis turbine for which the test curves are shown in Figs. 21-4a and b has a 27-in-diameter runner and a maximum efficiency of 86% when discharging 39 cfs and developing 560 bhp at 600 rpm under a net head of 142.0 ft. Compute $N_s$, $\phi_e$, and $C_r$, assuming $B/D = 0.15$.

▌                 $N_s = NP^{0.5}/H^{5/4} = (600)(560)^{0.5}/142.0^{5/4} = 29.0$     $ND = 153.2\phi\sqrt{H}$

$(600)(\tfrac{27}{12}) = (153.2)(\phi_e)\sqrt{142}$     $\phi_e = 0.740$

$C_r = N_s^2/[(63\ 800)(\phi_e^2)(B/D)(\eta)] = (29.0)^2/[(63\ 800)(0.740)^2(0.15)(0.86)] = 0.1866$     $V_r = Q/A_c$

$A_c = 0.95\pi(B/D)D^2 = (0.95)(\pi)(0.15)(\tfrac{27}{12})^2 = 2.266$ ft$^2$     $V_r = 39/2.266 = 17.21$ fps

$C_r = V_r/(2gH)^{1/2} = 17.21/[(2)(32.2)(142.0)]^{1/2} = 0.1800$     (a good check)

**21.24**   If a turbine homologous to that in Prob. 21.23 were made with a runner diameter of 130 in, what would be its probable efficiency under the same head?

▌         $(1 - \eta_2)/(1 - \eta_1) = (D_1/D_2)^{1/5}$     $(1 - \eta_2)/(1 - 0.86) = (27/130)^{1/5}$     $\eta = 0.898 = 89.8\%$

**21.25**   The turbine of Prob. 21.23 has a horizontal shaft, and at the time of the test the centerline of the shaft was 12.70 ft above the surface of the water in the tailrace. The discharge edge of the runner at its highest point is 0.85 ft above the centerline of the shaft. Evaluate the cavitation factor $\sigma$, assuming a barometric pressure of 14.6 psia and a vapor pressure of 0.5 psia.

▌         $\sigma = (p_{atm}/\gamma - p_v/\gamma - z_{max})/h$     $p_{atm}/\gamma = (14.6)(144)/62.4 = 33.69$ ft

$p_v/\gamma = (0.5)(144)/62.4 = 1.15$ ft     $z_{max} = 12.70 + 0.85 = 13.55$ ft

$\sigma = (33.69 - 1.15 - 13.55)/142 = 0.134$

**21.26**   If the critical value of the cavitation factor for the turbine of Prob. 21.25 is 0.06, compute the maximum allowable height of the centerline of the shaft above the tailwater surface.

▌             $z_{max} = p_{atm}/\gamma - p_v/\gamma - \sigma_c h = 33.69 - 1.15 - (0.06)(142.0) = 24.02$ ft

Height to centerline of shaft $= 24.02 - 0.85 = 23.17$ ft

**21.27**   Rework Prob. 21.26, if the total head on the turbine is 420 ft.

▌             $z_{max} = p_{atm}/\gamma - p_v/\gamma - \sigma_c h = 33.69 - 1.15 - (0.06)(420) = 7.34$ ft

Height to centerline of shaft $= 7.34 - 0.85 = 6.49$ ft

**21.28**   Find the maximum permissible head under which an axial-flow turbine ($N_s = 160$) can operate if it is set 6 ft below the tailwater. The installation is at elevation 3150 ft where $p_{atm} = 14.15$ psia, and the water temperature is 65 °F. Assume $\sigma_c = 0.90$.

▌         $z_{max} = p_{atm}/\gamma - p_v/\gamma - \sigma_c h$     $-6 = (14.15)(144)/62.4 - 44.4/62.4 - 0.90h$     $h = 42.2$ ft

**21.29**   In Prob. 21.23, the Francis turbine has frictional losses of 2.8 hp. Assuming that the leakage is 1% of the measured discharge and that $\alpha_2 = 90°$, find the values of $\eta_h$, $\alpha_1$, and $\beta_1$.

**Fig. 21-4(a).** Francis turbine operating at variable speed under constant head with constant gate opening. Runner diameter = 27 in, head = 140.5 ft.

**Fig. 21-4(b).** Francis turbine operating at constant speed and variable gate opening. Runner diameter = 27 in.

$$\eta_v = 99\% = 0.99 \qquad \eta_m = bp/(bp + fp) = 560/(560 + 2.8) = 0.995$$

$$\eta_h = 0.86/[(0.99)(0.995)] = 0.873 \qquad C_1 = \cos \alpha_1 = 0.873/[(2)(0.74)] = 0.5899 \qquad \eta_h = 2\phi_e C_1 \cos \alpha_1$$

$$\tan \alpha_1 = (C_1 \sin \alpha_1)/(C_1 \cos \alpha_1) = C_r/(C_1 \cos \alpha_1) = 0.1866/0.5899 = 0.3163 \qquad \alpha_1 = 17.55°$$

$$\cot \beta_1 = (C_1 \cos \alpha_1 - \phi)/(C_1 \sin \alpha_1) = (C_1 \cos \alpha_1 - \phi)/C_r = (0.5899 - 0.74)/0.1866 = -0.80439 \qquad \beta_1 = 141.2°$$

**21.30** A dam turbine has runner diameter 16′ 6″. The height of the guide vanes is 34.4 in. The turbine is rated at 150 000 hp under a head of 331 ft at 120 rpm. At this power the efficiency is 88 percent and the absolute velocity of the water entering the runner is 77.2 fps. Compute the rated specific speed, $\phi$, and $C_1$ for this full-gate opening.

$$N_s = NP^{0.5}/H^{5/4} = (120)(150\,000)^{0.5}/331^{5/4} = 32.9 \qquad \phi = u_1/(2gH)^{1/2}$$

$$u_1 = r\omega = r(2\pi N/60) = [16.5/2][(2)(\pi)(120)/60] = 103.7 \text{ ft/s}$$

$$\phi = 103.7/[(2)(32.2)(331)]^{1/2} = 0.710 \qquad C_1 = V_1/(2gh)^{1/2} = 77.2/[(2)(32.2)(331)]^{1/2} = 0.529$$

**21.31**    For the data of Prob. 21.30, find $C_r$, $\alpha_1$, and $\beta_1$.

▮    $C_r = V_r/(2gh)^{1/2}$        $V_r = Q/A_c$        $A_c = 0.95\pi BD = (0.95)(\pi)(34.4/12)(16.5) = 141.2 \text{ ft}^2$

$$Q = \frac{\text{bhp}}{\gamma H \eta} = \frac{(150\,000)(550)}{(62.4)(331)(0.88)} = 4539 \text{ cfs}    V_r = 4539/141.2 = 32.15 \text{ fps}$$

$C_r = 32.15/[(2)(32.2)(331)]^{1/2} = 0.220$        $\sin \alpha_1 = C_r/C_1 = 0.220/0.529 = 0.415879$        $\alpha_1 = 24.6°$

$\tan \beta_1 = (C_1 \sin \alpha_1)/(\phi - C_1 \cos \alpha_1) = (0.529)(\sin 24.6°)/[0.710 - (0.529)(\cos 24.6°)] = 0.96157$        $\beta_1 = 43.9°$

**21.32**    How many identical turbines, operating at 139.0 rpm and 91% efficiency, are needed to exploit a head of 1200 ft and a flow of 1660 ft³/s?

▮    bhp $= \eta Q \gamma H_p/550 = (0.91)(1660)(62.4)(1200)/550 = 206\,000$. From Fig. A-42, $N_s \leq 5.4$. Let $x =$ number of identical turbines. $N_s \geq NP^{0.5}/H_p^{5/4}$, $5.4 = (139.0)(206\,000/x)^{0.5}/1200^{5/4}$, $x = 2.7$; use three turbines.

**21.33**    A Francis turbine, working a total head of 102 ft, is to produce 25 000 hp at maximum efficiency. What speed and diameter should this turbine have?

▮    For a Francis turbine, $N_s = 20$ to 80; try $N_s = 50$: $N_s = NP^{0.5}/H^{5/4}$, $50 = N(25\,000)^{0.5}/102^{5/4}$, $N = 103$ rpm; $D = 153.2\phi\sqrt{H}/N$. From Fig. A-43 for $N_s = 50$, $\phi_e = 0.78$. $D = (153.2)(0.78)\sqrt{102}/103 = 11.70$ ft.

**21.34**    Solve Prob. 21.33 for a propeller turbine.

▮    For a propeller turbine, $N_s = 100$ to 250; try $N_s = 120$: $N_s = NP^{0.5}/H^{5/4}$, $120 = N(25\,000)^{0.5}/102^{5/4}$, $N = 246$ rpm; $D = 153.2\phi\sqrt{H}/N$. From Fig. A-43 for $N_s = 120$, $\phi_e = 1.63$. $D = (153.2)(1.63)\sqrt{102}/246 = 10.25$ ft.

**21.35**    How may poles should a 50-Hz generator have, if it is connected to a turbine operating under a design head of 3000 ft with a flow of 82 cfs? Assume turbine efficiencies as given in Fig. A-44 and avoid cavitation.

▮    $N_s = NP^{0.5}/H^{5/4}$. From Fig. A-42, we must use $N_s \leq 3$; use $N_s = 3$.

$$P = \eta Q \gamma H/550 = (0.84)(82)(62.4)(3000)/550 = 23\,444 \text{ hp}$$

$$3 = (N)(23\,444)^{0.5}/3000^{5/4}    N = 435 \text{ rpm}    N_{\text{poles}} = \frac{(50)(120)}{435} = 13.8$$

Use a 12- or 14-pole generator.

**21.36**    A turbine is to be installed where the net available head is 186 ft, and the available flow will average 885 cfs. Specify type, operating speed, and number of generator poles for 60-Hz electricity, if the turbine is to have the highest tolerable specific speed without cavitation. Assume the turbine is set 5 ft above tailwater and is 91% efficient. Approximately what size of runner is required?

▮        $N_s = NP^{0.5}/H^{5/4}$        bhp $= \eta Q \gamma H/550 = (0.91)(885)(62.4)(186)/550 = 16\,995$

From Fig. A-42 for $H = 186$ ft and draft head $= +5$ ft, highest tolerable $N_s = 50$. $50 = (N)(16\,995)^{0.5}/186^{5/4}$, $N = 263$ rpm. With 60-cycle electricity, a 28-pole generator will operate at $\frac{7200}{28} = 257$ rpm. So use a 28-pole generator operating at 257 rpm. $N_s = (257)(16\,995)^{0.5}/186^{5/4} = 48.8$. Use a Francis turbine with $N_s = 48.8$. $D = 153.3\phi\sqrt{H}/N = (153.2)(0.78)(\sqrt{186})/257 = 6.34$ ft.

**21.37**    Repeat Prob. 21.36 for the following data: head, 176 ft; flow, 1000 cfs; 60-Hz power; draft head, 10 ft; efficiency, 90%.

▮    $N_s = NP^{0.5}/H^{5/4}$. From Fig. A-42 for $H = 176$ ft and draft head $= +10$ ft, $N_s \leq 48$.

bhp $= \eta Q \gamma H/550 = (0.90)(1000)(62.4)(176)/550 = 17\,971$        $48 = (N)(17\,971)^{0.5}/176^{5/4}$        $N = 230$ rpm

$N_{\text{poles}} = 7200/N = \frac{7200}{230} = 31$; use 32 poles. $32 = 7200/N$, $N = 225$ rpm; $N_s = (225)(17\,971)^{0.5}/176^{5/4} = 47.1$; use a Francis turbine. From Fig. A-43 for $N_s = 47.3$, $\phi_e = 0.77$. $D = 153.2\phi\sqrt{H}/N = (153.2)(0.77)(\sqrt{176})/225 = 6.96$ ft.

**21.38**    For the data of Prob. 21.37, select a set of identical turbines to be operated in parallel. Specify the speed and size of the units.

$$N_s = NP^{0.5}/H^{5/4} \qquad N_s \le 48 \qquad \text{bhp} = 17\,971/2 = 8986$$
$$48 = (N)(8986)^{0.5}/176^{5/4} \qquad N = 325 \qquad N_{poles} = 7200/N$$

$N_{poles} = \frac{7200}{325} = 22.2$; use 22 poles. $22 = 7200/N$, $N = 327$; $N_s = (327)(8986)^{0.5}/176^{5/4} = 48.4$. From Fig. A-43 for $N_s = 48.4$, $\phi_e = 0.75$. $D = 153.2\phi\sqrt{h}/N = (153.2)(0.75)(\sqrt{176})/327 = 4.66$ ft.

**21.39** Water enters a rotating wheel with a relative velocity of 198 fps; $r_1 = 4.0$ ft, $r_2 = 3.0$ ft, and $N = 425$ rpm. There is no pressure drop in flow over the vanes. Assume $k = 0.2$. Find the relative velocity at discharge.

▮ With no pressure drop, $(V_1^2 - u_1^2)/2g - (V_2^2 - u_2^2)/2g = kV_2^2/2g$, $u = \pi DN/60$.

$$u_1 = (\pi)[(2)(4)](425)/60 = 178.0 \text{ ft} \qquad u_2 = (\pi)[(2)(3)](425)/60 = 133.5 \text{ ft}$$
$$(198^2 - 178.0^2)/[(2)(32.2)] - (V_2^2 - 133.5^2)/[(2)(32.2)] = 0.2V_2^2/[(2)(32.2)] \qquad V_2 = 145 \text{ fps}$$

**21.40** Water enters a rotating wheel in such manner that the passages are completely filled. If $Q = 406$ cfs, $a_1 = 10$ ft$^2$, $a_2 = 8$ ft$^2$, $r_1 = 1.5$ ft, $r_2 = 1.0$ ft, $N = 538$ rpm, and $k = 0.2$, find the drop in pressure head between entrance and exit.

▮
$$(p_1/\gamma - p_2/\gamma) + (V_1^2 - u_1^2)/2g - (V_2^2 - u_2^2)/2g = kV_2^2/2g \qquad V_1 = Q/a_1 = \frac{400}{10} = 40.6 \text{ fps}$$
$$V_2 = \frac{406}{8} = 50.8 \text{ fps} \qquad u = \pi DN/60$$
$$u_1 = (\pi)[(2)(1.5)](538)/60 = 84.51 \text{ fps} \qquad u_2 = (\pi)[(2)(1.0)](538)/60 = 56.34 \text{ fps}$$
$$(p_1/\gamma - p_2/\gamma) + (40.6^2 - 84.51^2)/[(2)(32.2)] - (50.8^2 - 56.34^2)/[(2)(32.2)] = (0.2)(50.8^2)/[(2)(32.2)]$$
$$p_1/\gamma - p_2/\gamma = 103 \text{ ft}$$

**21.41** It is desired to develop 300 000 hp under a head of 48 ft and to operate at 60 rpm. How many Francis turbines with a specific speed of 80 are required?

▮
$$N_s = NP^{0.5}/H^{5/4} \qquad 80 = 60P^{0.5}/49^{5/4} \qquad P = 29\,879 \text{ bhp}$$

Number of units required $= 300\,000/29\,879 = 10.0$.

**21.42** Air flows steadily, as shown in Fig. 21-5, through a turbine which produces 700 hp. For the inlet and exit conditions shown, estimate (a) the exit velocity $V_2$ and (b) the heat transferred $\dot{Q}$.

▮ (a) The inlet and exit densities can be computed from the perfect-gas law: $\rho_1 = p_1/RT_1 = 150(144)/[1715(460 + 300)] = 0.0166$ slug/ft$^3$, $\rho_2 = p_2/RT_2 = 40(144)/[1715(460 + 35)] = 0.00679$ slug/ft$^3$. The mass flow is determined by the inlet conditions $\dot{m} = \rho_1 A_1 V_1 = (0.0166)(\pi/4)(\frac{6}{12})^2(100) = 0.326$ slug/s. Knowing mass flow, we compute the exit velocity

$$\dot{m} = 0.326 = \rho_2 A_2 V_2 = (0.00679)(\pi/4)(\tfrac{6}{12})^2 V_2 \qquad \text{or} \qquad V_2 = 245 \text{ fps}$$

(b) The steady-flow energy equation applies with $\dot{W}_v = 0$, $z_1 = z_2$, and $\hat{h} = c_p T$:

$$\dot{Q} = \dot{W}_s + \dot{m}(c_p T_2 + \tfrac{1}{2}V_2^2 - c_p T_1 - \tfrac{1}{2}V_1^2)$$
$$\dot{Q} = 700(550) + 0.326[6003(495) + \tfrac{1}{2}(245)^2 - 6003(760) - \tfrac{1}{2}(100)^2] = -125\,000 \text{ ft-lb/s}$$

The negative sign indicates that this heat transfer is a *loss* from the control volume.

Fig. 21-5

**21.43** A hydroelectric power plant (Fig. 21-6) takes in 30 m³/s of water through its turbine and discharges it at $V_2 = 2$ m/s at atmospheric pressure. The head loss in the turbine and penstock system is $h_f = 20$ m. Estimate the power extracted by the turbine.

▮ $(p_1/\rho g) + (V_1^2/2g) + z_1 = (p_2/\rho g) + (V_2^2/2g) + z_2 + h_s$. We neglect viscous work and heat transfer and take section 1 at the reservoir surface, where $V_1 \approx 0$ and $p_1 = p_a$. Section 2 is at the turbine outlet.

$$(p_a/\rho g) + \tfrac{1}{2}(0)^2 + 100 = (p_a/\rho g) + [\tfrac{1}{2}(2)^2/9.81] + 0 + h_f + h_s \qquad (1)$$

With $h_f$ given as 20 m, we solve for $h_s = 100 - 20 - 0.2 = 79.8$ m; then

$$w_s = h_s g = (79.8)(9.81) = 783 \text{ m}^2/\text{s}^2 = 783 \text{ J/kg} \qquad (2)$$

The result is positive, as expected from our sign convention that work done by the fluid on a turbine is positive. The power extracted is $P = \dot{m}w_s = \rho Q w_s = (1000)(30)(783) = 23.5 \times 10^6$ J/s = 23.5 MW.

Turbine　　　　　　**Fig. 21-6**

**21.44** It is proposed to build a dam in a river where the flow rate is 10 m³/s and a 32-m drop in elevation can be achieved for flow through a turbine. If the turbine is 82 percent efficient, what is the maximum power that can be achieved?

▮ $$P = \eta \gamma Q H = (0.82)(9.79)(10)(32) = 2570 \text{ kW}$$

**21.45** Steam enters a turbine at 400 psia, 600 °F, and 10 fps and is discharged at 100 fps and 20 psia saturated conditions. The mass flux is 2.0 lb/s, and the heat loss from the turbine is 6 Btu per pound of steam. Head losses and elevation changes are negligible. How much horsepower does the turbine develop?

▮ $Q - W_s - W_v = \dot{m}(h_2 + V_2^2/2 - h_1 - V_1^2/2)$. 1 Btu/lbm = 25 000 ft-lb/slug; hence, $Q = (-6)(25\,000)(2.0/32.2) = -9332$ ft-lb/s. From the steam tables:

At 400 psia, 600 °F:　　　　　　　　$h_1 = 1306.6$ Btu/lbm

At 20 psia saturated:　　　　　　　　$h_2 = 1156.4$ Btu/lbm

$$-9332 - W_s - 0 = (2.0/32.2)[(1156.4)(25\,040) + 100^2/2 - (1306.6)(25\,000) - 10^2/2]$$

$$W_s = 224\,337 \text{ ft-lb/s} = 224\,337/550 = 408 \text{ hp}$$

**21.46** Investigate the possibility of using a Pelton wheel similar to Fig. A-45 to deliver 30 000 bhp from a net head of 1200 ft.

▮ From Fig. A-46, the most efficient Pelton wheel occurs at about $N_s \approx 4.5 = (N)(30\,000)^{1/2}/(1200)^{1.25}$ or $N = 183$ rpm. From Fig. A-45 the best operating point is $\phi \approx 0.47 = \pi D(3.06)/[2(32.2)(1200)]^{1/2}$ or $D = 13.6$ ft. This Pelton wheel is perhaps a little slow and a trifle large. You could reduce $D$ and increase $N$ by increasing $N_s$ to, say, 6 or 7 and accepting the slight reduction in efficiency. Or you could use a double-hung, two-wheel configuration, each delivering 15 000 bhp, which changes $D$ and $N$ by the factor $2^{1/2}$.

Double wheel:　　　　　　$N = (183)2^{1/2} = 259$ rpm　　　　$D = 136/2^{1/2} = 9.6$ ft

**21.47** For the data given in Prob. 21.46, investigate the possibility of using the Francis-turbine family of Fig. A-47.

❚ The Francis wheel of Fig. A-47 must have $N_s = 29 = (N)(30\,000)^{1/2}/(1200)^{1.25}$ or $N = 1183$ rpm. Then the optimum power coefficient is $C_{P*} = 2.70 = P/\rho N^3 D^5 = 30\,000(550)/[(1.94)(\frac{1183}{60})^3 D^5]$ or $D = 3.33$ ft = 40 in. This is a faster speed than normal practice, and the casing would have to withstand 1200 ft of water or about 520 lb/in$^2$ internal pressure, but the 40-in size is extremely attractive. Francis turbines are now being operated at heads up to 1500 ft.

**21.48** A lawn sprinkler can be used as a simple turbine. As shown in Fig. 21-7$a$, the flow enters normal to the paper in the center and splits evenly into $Q/2$ and $V_{rel}$ leaving each nozzle. The arms rotate at angular velocity $\omega$ and do work on a shaft. Draw the velocity diagram for this turbine. Neglecting friction, find an expression for the power delivered to the shaft. Find the rotation rate for which the power is a maximum.

❚ The velocity diagram for this turbine is shown in Fig. 21-7$b$.

$$P = \omega T = \rho Q(u_2 V_{t2} - u_1 V_{t1}) = \rho Q u (W - u) = \rho Q \omega R (V_{rel} - \omega R) \qquad dP/du = \rho Q (W - 2u) = 0$$

$$u = W/2 \quad \text{or} \quad \omega R = V_{rel}/2 \qquad P_{max} = \rho Q (\omega R)^2 = \rho Q u^2$$

$(a)$                                $(b)$       **Fig. 21-7**

**21.49** Apply the result of Prob. 21.48 to a nozzle flow of water, $Q = 5$ L/s, where the exit diameter is 1 cm.

❚ $\qquad P_{max} = \rho Q u^2 \qquad u = V_{rel}/2 \qquad V_{rel} = (Q/2)/A_{exit} = [(5 \times 10^{-3})/2]/[(\pi)(0.01)^2/4] = 31.83$ m/s

$\qquad u = 31.83/2 = 15.92$ m/s $\qquad P_{max} = (998)(5 \times 10^{-3})(15.92)^2 = 1265$ W

**21.50** Show that if the net head $H$ is varied for a turbine operating at a given valve opening and efficiency, the speed will vary as $H^{1/2}$ and the output power as $H^{3/2}$.

❚ At constant efficiency, $gH/N^2 D^2$ is constant; and, since $g$ and $D$ are constant, $H \propto N^2$. Hence, $N \propto H^{1/2}$. Also at constant efficiency, $P/\rho N^3 D^5$ is constant; and, since $\rho$ and $D$ are constant, $P \propto N^3 \propto (H^{1/2})^3$. Hence, $P \propto H^{3/2}$.

**21.51** Turbines are to be installed where the net head is 398 ft and the flow rate is 250 000 gal/min. Discuss the type, number, and size of turbine which might be selected if the generator selected is (**a**) 48-pole, 60 cycle ($n = 150$ rpm); and (**b**) 8-pole ($n = 900$ rpm). Why are at least two turbines desirable from a planning point of view?

❚ Assume $\eta = 90$ percent. Select two turbines so that one will still be available for power if the other is shut down.

(**a**) $\qquad\qquad P = \eta \gamma QH = (0.90)(62.4)[(250\,000)(0.002228)](398)/550 = 22\,640$ bhp

$\qquad\qquad\qquad N_s = NP^{0.5}/H^{5/4} = (150)(22\,640/2)^{0.5}/398^{5/4} = 9.0$

Select two impulse turbines. $\phi = \pi ND/(2gH)^{1/2}$. Estimate $\phi = 0.47$. $0.47 = (\pi)(\frac{150}{60})(D)/[(2)(32.2)(398)]^{1/2}$, $D = 9.60$ ft.

(**b**) $N_s = (900)(22\,640/2)^{0.5}/398^{5/4} = 53.9$. Select two Francis turbines. $C_p^* = P/\rho N^3 D^5$. From Fig. A-47, $C_p^* = 2.6$. $2.6 = (22\,640/2)(550)/[(1.94)(\frac{900}{60})^3(D^5)]$, $D = 3.26$ ft.

**21.52** What type of turbine develops 54 000 bhp at 80 rpm under a head of 90 ft? Estimate the flow rate.

❚ $\qquad N_s = NP^{0.5}/H^{5/4} = (80)(54\,000)^{0.5}/90^{5/4} = 67.1$ $\qquad$ (Francis turbine) $\qquad P = \eta \gamma QH$

From Fig. A-48, $\eta = 93$ percent: $(54\,000)(550) = (0.93)(62.4)(Q)(90)$, $Q = 5687$ ft$^3$/s.

**21.53** What type of turbine delivers 25 000 bhp at 500 rpm under a net head of 5340 ft? Estimate the impeller diameter.

▮ $\quad N_s = NP^{0.5}/H^{5/4} = (500)(25\,000)^{0.5}/5340^{5/4} = 1.73$ $\quad$ (impulse turbine) $\quad\quad \phi = \pi ND/(2gH)^{1/2}$.

Estimate $\phi = 0.47$. $0.47 = (\pi)(\frac{500}{60})(D)/[(2)(32.2)(5340)]^{1/2}$, $D = 10.5$ ft.

**21.54** A Pelton wheel of 12-ft pitch diameter operates under a net head of 1990 ft. Estimate the speed, power output, and flow rate for best efficiency if the nozzle exit diameter is 4 in.

▮ $\quad V_{jet} = C_v\sqrt{2gH}$. Take $C_v = 0.94$: $V_{jet} = (0.94)\sqrt{(2)(32.2)(1990)} = 337$ ft/s. $\phi = \pi ND/(2gH)^{1/2}$. Estimate $\phi = 0.47$: $0.47 = (\pi)(N/60)(12)/[(2)(32.2)(1990)]^{1/2}$, $N = 268$ rpm.

$$Q = AV = [(\pi)(\tfrac{4}{12})^2/4](337) = 29.4 \text{ ft}^3/\text{s} \quad\quad P_w = \rho Qu(V_{jet} - u)(1 - \cos\beta)$$

Assume $\beta = 165°$.

$$u = V_{jet}/2 = \tfrac{337}{2} = 168.5 \text{ fps} \quad\quad P_w = (1.94)(29.4)(168.5)(337 - 168.5)(1 - \cos 165°)/550 = 5788 \text{ hp}$$
$$N_s = NP^{0.5}/H^{5/4} = (268)(5788)^{0.5}/1990^{5/4} = 1.53$$

Hence, $\eta = 0.75$. $P = (0.75)(5788) = 4341$ hp.

**21.55** It is planned to use the Francis-turbine family of Fig. A-47 at an installation with a head of 602 ft and a flow rate of 198 ft³/s. What are the proper impeller diameter and the optimum speed and power produced?

▮ $\quad N_s = NP^{0.5}/H^{5/4}$, $P^* = \eta\gamma QH$. Estimate $\eta = 87$ percent.

$$P^* = (0.87)(62.4)(198)(602)/550 = 11\,765 \text{ hp} \quad\quad N_s = 29 \quad\quad 29 = (N)(11\,765)^{0.5}/602^{5/4} \quad\quad N = 797 \text{ rpm}$$

$$C_H^* = gH/N^2D^2 = 9.03 \quad\quad 9.03 = (32.2)(602)/[(\tfrac{797}{60})^2(D)^2] \quad\quad D = 3.49 \text{ ft}$$

**21.56** A giant wind-tunnel fan has rotor blades 125 ft in diameter. For a 20 mph wind (the optimum output), compute (a) the power generated, (b) the rotor speed, and (c) the velocity $V_2$ behind the rotor.

▮ $\quad$ (a) $(C_p)_{max} = P/(\rho A V_1^3/2)$. From Fig. A-49, $(C_p)_{max} = 0.46$.

$$A = [(\pi)(125)^2/4] = 12\,272 \text{ ft}^2 \quad\quad V_1 = (20)(5280)/3600 = 29.33 \text{ fps}$$
$$0.46 = P/[(0.00233)(12\,272)(29.33)^3/2] \quad\quad P = 165\,934 \text{ ft-lb/s} = (165\,934)(0.3048)(4.448)/1000 = 225 \text{ kW}$$

(b) At $P_{max}$, $\omega r/V_1 = 5.5$. $(\omega)(\tfrac{125}{2})/29.33 = 5.5$, $\omega = 2.581$ rad/s, or 24.6 rpm.

(c) From ideal theory, $V_2 = V_1/3$, $V_2 = 29.33/3 = 9.78$ fps.

**21.57** A hydroelectric plant (Fig. 21-8) has a difference in elevation from head water to tail water of $H = 50$ m and a flow $Q = 5$ m³/s of water through the turbine. The turbine shaft rotates at 180 rpm, and the torque in the shaft is measured to be $T = 1.16 \times 10^5$ N · m. Output of the generator is 2100 kW. Determine (a) the reversible power for the system, (b) the irreversibility, or losses, in the system, and (c) the losses and the efficiency in the turbine and in the generator.

**Fig. 21-8**

▮ (a) The water's potential energy per unit weight is $H = 50$ J/N. Hence, for perfect conversion the reversible power is $\gamma QH = (9806)(5)(50) = 2\,451\,500$ J/s $= 2451.5$ kW.

(b) The irreversibility, or lost power, in the system is the difference between the power into and out of the system, or $2451.5 - 2100 = 351.5$ kW.

(c) The rate of work by the turbine is the product of the shaft torque and the rotational speed: $T\omega = (1.16 \times 10^5)[180(2\pi)/60] = 2186.5$ kW. The irreversibility through the turbine is then $2451.5 - 2186.5 = 265.0$ kW, or, when expressed as lost work per unit weight of fluid flowing, $(265.0)(\frac{1000}{1})(\frac{1}{9806})(\frac{1}{5}) = 5.4$ J/N. The generator power loss is $2186.5 - 2100 = 86.5$ kW, or $86.5(1000)/[9806(5)] = 1.76$ J/N. Efficiency of the turbine $\eta_t$ is $\eta_t = 100[(50 - 5.4)/50] = 89.2$ percent and efficiency of the generator $\eta_g$ is $\eta_g = 100[(50 - 5.4 - 1.76)/(50 - 5.4)] = 96.1$ percent.

**21.58** A turbine has a velocity of 6 m/s at the entrance to the draft tube and a velocity of 1.2 m/s at the exit. For friction losses of 0.1 m and a tailwater 5 m below the entrance to the draft tube, find the pressure head at the entrance.

▮ $$p_1/\gamma = -z_s - V_1^2/2g + \text{losses} = -5 - 6^2/[(2)(9.807)] + 1.2^2/[(2)(9.807)] + 0.1 = -6.66 \text{ m}$$

**21.59** The wicket gates of the propeller turbine of Fig. 21-9 are turned so that the flow makes an angle of 45° with a radial line at section 1, where the speed is 4.005 m/s. Determine the magnitude of tangential velocity component $V_u$ over section 2.

▮ Since no torque is exerted on the flow between sections 1 and 2, the moment of momentum is constant and the motion follows the free-vortex law $V_u r = $ constant. At section 1, $V_{u1} = 4.005 \cos 45° = 2.832$ m/s, $V_{u1}r_1 = (2.832)(1.5/2) = 2.124$ m²/s. Across section 2, $V_{u2} = 2.124/r$. At the hub, $V_u = 2.124/(0.45/2) = 9.44$ m/s. At the outer edge, $V_u = 2.124/(1.2/2) = 3.54$ m/s.

**Fig. 21-9**

**21.60** Assuming uniform axial velocity over section 2 of Fig. 21-9 using the data of Prob. 21.59, determine the angle of the leading edge of the propeller at $r = 0.225$, 0.45, and 0.6 m for a propeller speed of 240 rpm.

▮ At $r = 0.225$ m, $u = \frac{240}{60}(2\pi)(0.225) = 5.66$ m/s, $V_u = 9.44$ m/s. At $r = 0.45$ m, $u = \frac{240}{60}(2\pi)(0.45) = 11.3$ m/s, $V_u = 4.72$ m/s. At $r = 0.6$ m, $u = \frac{240}{60}(2\pi)(0.6) = 15.06$ m/s, $V_u = 3.54$ m/s. The discharge through the turbine is, from section 1, $Q = (0.6)(1.5)(\pi)(4.005)(\cos 45°) = 8.01$ m³/s. Hence, the axial velocity at section 2 is $V_a = 8.01/[\pi(0.6^2 - 0.225^2)] = 8.24$ m/s. Figure 21-10 shows the initial vane angle for the three positions.

Fig. 21-10

**21.61** A Pelton wheel is to be selected to drive a generator at 600 rpm. The water jet is 75 mm in diameter and has a velocity of 100 m/s. With the blade angle at 170°, the ratio of vane speed to initial jet speed at 0.47, and neglecting losses, determine (a) diameter of wheel to centerline of buckets (vanes), (b) power developed, and (c) kinetic energy per newton remaining in the fluid.

**▌** (a) The peripheral speed of the wheel is $u = 0.47(100) = 47$ m/s. Then $\frac{600}{60}[2\pi(D/2)] = 47$ m/s or $D = 1.496$ m.

(b) The power is $P = \rho Q u V_r(1 - \cos\theta) = (1000)(\pi/4)(0.075)^2(100)(47)(100 - 47)[1 - (-0.9848)] = 2.184$ MW.

(c) From Fig. 21.11, the absolute-velocity components leaving the vane are $V_x = (100 - 47)(-0.9848) + 47 = -5.2$ m/s, $V_y = (100 - 47)(0.1736) = 9.2$ m/s. The kinetic energy remaining in the jet is $(5.2^2 + 9.2^2)/[2(9.807)] = 5.69$ J/N.

(a)     (b)

(c)

Fig. 21-11

**21.62** A small impulse wheel is to be used to drive a generator for 60-Hz power. The head is 100 m, and the discharge is 40 L/s. Determine the diameter of the wheel at the centerline of the buckets and the speed of the wheel ($C_v = 0.98$). Assume efficiency of 80 percent.

**▌** The power is $P = \gamma Q H_a \eta = 9806(0.040)(100)(0.80) = 31.38$ kW. By taking a trial value of $N_s$ of 15, $N = N_s H_a^{5/4}/\sqrt{P} = 15(100^{5/4})/\sqrt{31.38} = 847$ rpm. For 60-Hz power the speed must be 3600 divided by the number of pairs of poles in the generator. For five pairs of poles the speed would be $\frac{3600}{5} = 720$ rpm, and for four pairs of poles it would be $\frac{3600}{4} = 900$ rpm. The closer speed 900 is selected. Then $N_s = N\sqrt{P}/H_a^{5/4} = 900\sqrt{31.38}/100^{5/4} = 15.94$. For $N_s = 15.94$, take $\phi = 0.448$, $u = \phi\sqrt{2gH_a} = 0.448\sqrt{2(9.807)(100)} = 19.84$ m/s and $\omega = \frac{900}{60}(2\pi) = 94.25$ rad/s. The peripheral speed $u$ and $D$ and $\omega$ are related: $u = \omega D/2$, $D = 2u/\omega = 2(19.84)/94.25 = 421$ mm. The diameter $d$ of the jet is obtained from the jet velocity $V_2$; thus $V_2 = C_v\sqrt{2gH_a} = 0.98\sqrt{2(9.807)(100)} = 43.4$ m/s, $a = Q/V_2 = (40 \times 10^{-3})/43.4 = 0.000922$ m$^2$ = 922 mm$^2$, $d = \sqrt{4a/\pi} = \sqrt{922/0.7854} = 34.3$ mm, where $a$ is the area of jet. Hence, the diameter ratio $D/d$ is $D/d = 421/34.3 = 12.27$. The desired diameter ratio for the best efficiency is $D/d = 206/N_s = 206/15.94 = 12.92$ so the ratio $D/d$ is satisfactory. The wheel diameter is 421 mm, and the speed is 900 rpm.

**21.63** If 24 m$^3$/s of water flowing through the fixed vanes of a turbine has a tangential component of 2 m/s at a radius of 1.25 m and the impeller, turning at 240 rpm, discharges in an axial direction, what torque is exerted on the impeller? Neglecting losses, what is the head on the turbine?

❚ $$T = \rho Q[(rV_t)_{in} - (rV_t)_{out}] = \rho Q(r_1 V_{u1} - r_2 V_{u2}) = (1000)(24)[(1.25)(2) - 0] = 60 \text{ kN} \cdot \text{m}$$

$$T\omega = Q\gamma H \quad (60\,000)[(\tfrac{240}{60})(2\pi)] = (24)[(9.79)(1000)](H) \quad H = 6.42 \text{ m}$$

**21.64** A turbine model test with 250-mm-diameter impeller showed an efficiency of 90 percent. What efficiency could be expected from a 1.5-m-diameter impeller?

❚ $$\eta_2 = 1 - (1 - \eta_1)(D_1/D_2)^{1/4} = 1 - (1 - 0.90)[0.250/1.5]^{1/4} = 0.936 \quad \text{or} \quad 93.6\%$$

**21.65** A tangential turbine develops 7200 hp at 200 rpm under a head of 790 ft at an efficiency of 82 percent. (*a*) If the speed factor is 0.46 compute the wheel diameter, the flow, the unit speed, unit power, unit flow, and specific speed. (*b*) For this turbine, what would be the speed, power, and flow under a head of 529 ft? (*c*) For a turbine having the same design, what size of wheel should be used to develop 3800 hp under a 600-ft head and what would be its speed and rate of discharge? Assume no change in efficiency.

❚ (*a*) Since $\phi = D_1 N/(1840\sqrt{H})$, $D_1 = 1840\sqrt{790}(0.46)/200 = 119$ in. From horsepower output = $\gamma QH/550$,
$Q = (7200)(550)/[(62.4)(790)(0.82)] = 98.0$ cfs.

$$N_u = ND_1/\sqrt{H} = (200)(119)/\sqrt{790} = 847 \text{ rpm} \qquad P_u = P/D_1^2 H^{3/2} = 7200/[(119)^2(790)^{3/2}] = 0.0000229 \text{ hp}$$

$$Q_u = Q/D_1^2\sqrt{H} = 98.0/[(119)^2\sqrt{790}] = 0.000246 \text{ cfs} \qquad N_s = N\sqrt{P}/H^{5/4} = 200\sqrt{7200}/(790)^{5/4} = 4.05 \text{ rpm}$$

(*b*) $$\text{Speed } N = N_u\sqrt{H}/D_1 = 847\sqrt{529}/119 = 164 \text{ rpm}$$

$$\text{Power } P = P_u D_1^2 H^{3/2} = 0.0000229(119)^2(529)^{3/2} = 3946 \text{ hp}$$

$$\text{Flow } Q = Q_u D_1^2\sqrt{H} = 0.000246(119)^2\sqrt{529} = 80.1 \text{ cfs}$$

The above three quantities might have been obtained by noting that, for the same turbine ($D_1$ unchanged), the speed varies as $H^{1/2}$, the power varies as $H^{3/2}$, and $Q$ varies as $H^{1/2}$. Thus $N = 200\sqrt{\frac{529}{790}} = 164$ rpm, $P = 7200(\frac{529}{790})^{3/2} = 3945$ hp, $Q = 98.0\sqrt{\frac{529}{790}} = 80.2$ cfs.

(*c*) From $P = P_u D_1^2 H^{3/2}$ we obtain $3800 = 0.0000229(D_1)^2(600)^{3/2}$ from which $D_1^2 = 11\,290$ and $D_1 = 106$ in. $N = N_u\sqrt{H}/D_1 = 847\sqrt{600}/106 = 196$ rpm, $Q = Q_u D_1^2\sqrt{H} = 0.000246(11\,290)\sqrt{600} = 68.0$ cfs.

**21.66** An impulse wheel at best produces 125 hp under a head of 210 ft. By what percent should the speed be increased for a 290-ft head? Assuming equal efficiencies, what power would result?

❚ For the same wheel, the speed is proportional to the square root of the head. Thus $N_1/\sqrt{H_1} = N_2/\sqrt{H_2}$, $N_2 = N_1\sqrt{H_2/H_1} = N_1\sqrt{\frac{290}{210}} = 1.175N_1$. Thus, the speed should be increased 17.5 percent.

$$N_s = NP^{0.5}/H^{5/4} \qquad N_1 P_1^{0.5}/H_1^{5/4} = N_2 P_2^{0.5}/H_2^{5/4} \qquad (N_1)(125)^{0.5}/210^{5/4} = (1.175N_1)(P_2)^{0.5}/290^{5/4} \qquad P_2 = 203 \text{ hp}$$

**21.67** The reaction turbines at the Hoover Dam installation have a rated capacity of 115 000 hp at 180 rpm under a head of 487 ft. The diameter of each turbine is 11 ft and the discharge is 2350 cfs. Evaluate the speed factor, the unit speed, unit discharge and unit power, and the specific speed.

❚ $$\phi = D_1 N/1840\sqrt{H} = (11)(12)(180)/(1840\sqrt{487}) = 0.585 \qquad N_u = D_1 N/\sqrt{H} = (11)(12)(180)/\sqrt{487} = 1077 \text{ rpm}$$

$$Q_u = Q/D_1^2\sqrt{H} = 2350/[(132)^2\sqrt{487}] = 0.00611 \text{ cfs}$$

$$P_u = \text{hp}/D_1^2 H^{3/2} = 115\,000/[(132)^2(487)^{3/2}] = 0.000614 \text{ hp} \qquad N_s = N_u\sqrt{P_u} = 1077\sqrt{0.000614} = 26.7$$

**21.68** A single-stage impulse turbine receives from the nozzles 6 lbm/ft³ of combustion products moving at 6000 ft/s. The radius $R$ to the nozzles is 2 ft. Find the most efficient speed $\omega$ of the turbine, given $\alpha_1 = 20°$? If the turbine runs at 0.72 of this speed, how much power does it develop?

❚ $$U_t = R\omega = \tfrac{1}{2}V_1 \cos \alpha_1 \qquad (2)(\omega) = \tfrac{1}{2}(6000) \cos 20° \qquad \omega = 1410 \text{ rad/s} = 13\,460 \text{ rpm}$$

$$\text{Actual speed} = (0.72)(1410) = 1015 \text{ rad/s}$$

$$dW_s/dt = 2\rho QU_t(V_1 \cos \alpha_1 - U_t) = (2)(6/g)(2)(1015)[6000 \cos 20° - (2)(1015)]/550 = 4963 \text{ hp}$$

**21.69** Three nozzles expand combustion products to a speed of 1800 m/s; the nozzles are at a distance $R = 0.6$ m from the axis of rotation. The density of the fluid as it reaches the turbine blades is 0.5 kg/m³. If the turbine is to rotate at its most efficient speed, 12 000 rpm, and if the power desired is 5900 hp, determine the exit areas of the nozzles.

$$U_t = \tfrac{1}{2}V_1 \cos \alpha \qquad (0.6)[(12\,000)(2\pi)/60] = \tfrac{1}{2}(1800) \cos \alpha_1 \qquad \alpha_1 = 33.1°$$

$$dW_s/dt = 2\rho Q U_t (V_1 \cos \alpha_1 - U_t)$$

$$(5900)(745.7) = (2)(0.5)(Q)(0.6)[(12\,000)(2\pi)/60]\{1800 \cos 33.1° - [(0.6)(12\,000)(2\pi)/60]\}$$

$$Q = 7.74 \text{ m}^3/\text{s} \qquad 3(A)(V_a) = 7.74 \qquad 3(A)(1800 \sin 33.1°) = 7.74$$

$$A = 2.625 \times 10^{-3} \text{ m}^2 = 2625 \text{ mm}^2$$

**21.70** The effective head $H$ for an axial-flow hydraulic turbine is 92 m and the volumetric flow $Q$ is 21 m³/s. The radius $R_o$ of the guide vanes is 1.6 m and the height $b$ is 1 m. The exit flow is axial only and the guide vanes are at an angle $\alpha$ of 11.0°. Calculate the rotor angular velocity $\omega$.

$g(\Delta H) = (U_t V_t)_2 - (U_t V_t)_1$. At any position $r$:

$$(9.81)(-92) = -(\omega)(r)(V_t)_1 \tag{1}$$

From continuity: $Q = (2)(\pi)(R_0)(b)(V_0) \sin \alpha$, $21 = (2\pi)(1.6)(1)(V_0) \sin 11°$, $V_0 = 10.95$ m/s. At any distance $r$:

$$V_t = (V_o R_o \cos \alpha)/r = [(10.95)(1.6) \cos 11°]/r = 17.20/r \tag{2}$$

Replace $(V_t)_1$ in Eq. (1) using Eq. (2): $(9.81)(92) = \omega r(17.20/r)$, $\omega = 52.5$ rad/s $= 501$ rpm.

**21.71** The first stage of an axial-flow gas turbine has a hub-to-tip diameter ratio of 0.85 with the tip having a diameter of 4 ft. Air $[R = 53.3$ ft-lbf/slug-°R] flows through the machine at 52 lbm/s. At the first stage, the pressure is 4 atm and the temperature is 1800 °R. If $\alpha_1 = 38°$ and $\beta_2 = 55°$ for the first stage and it is to develop 25 hp, at what speed should the rotor turn?

Find $V_a$.

$$Q = V_a(\pi/4)\{4^2 - [(0.85)(4)]^2\} \tag{1}$$

Also we can say $Q = (52/g)/\rho$, $\rho = p/RT = (14.7)(144)(4)/[(53.3)(32.2)(1800)] = 0.002741$ slug/ft³, $Q = (50/g)/0.002741 = 566.5$ cfs. Go back to Eq. (1): $566.5 = V_a(\pi/4)\{4^2 - [(0.85)(4)]^2\}$, $V_a = 162.45$ ft/s. Compute $\Delta H$ for the first stage.

$$-\Delta H(Q\rho g) = 25(550) \qquad \Delta H = -(25)(550)/[(566.2)(0.002741)(32.2)] = -275 \text{ ft}$$

$$\Delta H g/U_t^2 = [1 - (V_a/U_t)(\cot \alpha_1 + \cot \beta_2)] \qquad U_t^2 - (162.45)(\cot 38° + \cot 55°)U_t + (275)(32.2) = 0$$

$$U_t^2 - 322 U_t + 8855 = 0 \qquad U_t = [322 \pm \sqrt{322^2 - (4)(8855)(1)}]/(2)(1) \qquad (U_t)_1 = 292 \text{ ft/s} \qquad (U_t)_2 = 30.5 \text{ fps}$$

$$U_t = R_{av}\omega \qquad 292 = (2 - 0.15)(\omega_1) \qquad \omega_1 = 157.8 \text{ rad/s} = 1507 \text{ rpm} \qquad 30.5 = (2 - 0.15)(\omega_2)$$

$$\omega_2 = 16.49 \text{ rad/s} = 157 \text{ rpm}$$

**21.72** Water flows from an upper reservoir to a lower one while passing through a turbine, as shown in Fig. 21-12. Find the power generated by the turbine. Neglect minor losses.

$p_1/\gamma + V_1^2/2g + z_1 - E_t = p_2/\gamma + V_2^2/2g + z_2 + h_L$. From Fig. A-14, with $Q = 0.15$ m³/s and $D = 250$ mm, $h_1 = 0.037$ m/m.

$$h_L = h_f = (0.037)(100) = 3.70 \text{ m} \qquad 0 + 0 + 197.3 - E_t = 0 + 0 + 50.0 + 3.70$$

$$E_t = 143.6 \text{ m} \qquad P = Q\gamma E_t = (0.15)(9.79)(143.6) = 211 \text{ kW}$$

**Fig. 21-12**

**21.73** Water is being discharged from a reservoir through a turbine, as shown in Fig. 21-13. What water-surface elevation is required in the reservoir in order for the turbine to generate 100 hp? Neglect minor losses.

▌ $p_1/\gamma + V_1^2/2g + z_1 - E_t = p_2/\gamma + V_2^2/2g + z_2 + h_L$    $P = Q\gamma E_t$    $100 = (20)(62.4)(E_t)/550$    $E_t = 44.07$ ft

$$V_2 = Q/A_2 = 20/[(\pi)(\tfrac{36}{12})^2/4] = 2.829 \text{ ft/s}$$

From Fig. A-13, with $Q = 20$ ft³/s and $D = 36$ in, $h_1 = 0.00084$ ft/ft.

$h_L = h_f = (0.00084)(270) = 0.23$ ft    $0 + 0 + z_1 - 44.07 = 0 + 2.829^2/[(2)(32.2)] + 100 + 0.23$

$$z_1 = 144.4 \text{ ft}$$

**Fig. 21-13**

**21.74** The flow rate of water through the turbine shown in Fig. 21-14 is 0.20 m³/s, and the pressures at points 1 and 2 are 150 and −35.0 kPa, respectively. Determine the power generated by the turbine.

▌    $p_1/\gamma + v_1^2/2g + z_1 - E_t = p_2/\gamma + v_2^2/2g + z_2 + h_L$    $p_1/\gamma = 150/9.79 = 15.32$ m

$$v_1 = Q/A_1 = 0.20/[(\pi)(0.25)^2/4] = 4.07 \text{ m/s}$$

$v_1^2/2g = (4.07)^2/[(2)(9.807)] = 0.84$ m    $z_1 = 1.00$ m    $p_2/\gamma = -35/9.79 = -3.58$ m

$v_2 = Q/A_2 = 0.20/[(\pi)(0.50)^2/4] = 1.02$ m/s    $v_2^2/2g = (1.02)^2/[(2)(9.807)] = 0.05$ m    $z_2 = 0$    $h_L = 0$

$15.32 + 0.84 + 1.0 - E_t = -3.58 + 0.05 + 0 + 0$    $E_t = 20.69$ m    $P = Q\gamma E_t = (0.20)(9.79)(20.69) = 40.5$ kW

250-**mm**-diameter pipe

Turbine

1.0 **m**

500-**mm**-diameter pipe

Fig. 21-14

**21.75** Water flows from reservoir $A$ to a turbine and then discharges into tailrace $B$, as shown in Fig. 21-15. The loss of head from $A$ to point 1 is six times the velocity head in the pipe, and the loss of head from point 2 to $B$ is 0.5 times the velocity head in the pipe. If the flow rate in the pipe is 30.0 ft$^3$/s and the diameter of the pipe is 24 in, calculate the horsepower being generated by the turbine.

$$p_A/\gamma + v_A^2/2g + z_A - E_t = p_B/\gamma + v_B^2/2g + z_B + h_L \qquad p_A/\gamma = p_B/\gamma = v_A^2/2g = v_B^2/2g = 0$$

$$v = Q/A = 30/[(\pi)(2.0)^2/4] = 9.55 \text{ ft/s}$$

$$v^2/2g = (9.55)^2/[(2)(32.2)] = 1.416 \text{ ft} \qquad h_L \text{ from } A \text{ to } 1 = (6)(1.416) = 8.50 \text{ ft}$$

$$h_L \text{ from } 2 \text{ to } B = (0.5)(1.416) = 0.71 \text{ ft}$$

$$0 + 0 + 210 - E_t = 0 + 0 + 0 + 8.50 + 0.71 \qquad E_t = 200.79 \text{ ft}$$

$$P = Q\gamma E_t = (30)(62.4)(200.79) = 375\,879 \text{ ft-lb/s} = 375\,879/550 = 683 \text{ hp}$$

Elev. 210 ft

Water

Turbine

Elev. 10 ft

24-**in**-diameter pipe

Elev. 0 ft

Fig. 21-15

**21.76** Water is being discharged from a reservoir through a turbine, as shown in Fig. 21-16. Determine the elevation of the water surface in the reservoir that would be required in order for the turbine to generate 56 kW of power. Neglect minor losses. Assume $C = 120$ for all pipes.

$$p_1/\gamma + v_1^2/2g + z_1 - E_t = p_2/\gamma + v_2^2/2g + z_2 + h_L \qquad p_1/\gamma = p_2/\gamma = v_1^2/2g = 0 \qquad p = Q\gamma E_t$$

$$56 = (0.28)(9.79)(E_t) \qquad E_t = 20.43 \text{ m} \qquad v_2^2/2g = (Q/A)^2/2g = \{0.28/[(\pi)(0.60)^2/4]\}^2/[(2)(9.807)] = 0.05 \text{ m}$$

$$z_2 = 50 \text{ m}$$

With $Q = 0.28$ m$^3$/s and $D = 300$ m, $(h_1)_{300} = 0.050$ m/m. With $Q = 0.28$ m$^3$/s and $D = 600$ mm, $(h_1)_{600} = 0.0017$ m/m.

$$h_L = h_f = (0.050)(50) + (0.0017)(20) = 2.53 \text{ m} \qquad 0 + 0 + z_1 - 20.43 = 0 + 0.05 + 50 + 2.53 \qquad z_1 = 73.0 \text{ m}$$

Fig. 21-16

# CHAPTER 22

# Hydraulic and Energy Grade Lines

**22.1** Plot the hydraulic grade line and the energy grade line for the pipe shown in Fig. 22-1a. The turbine develops 45 kW; the water is at 5 °C.

**❚** First compute $a$.

$$(V_1^2/2) + (p_1/\rho) + 0 - (dW_s/dm) = (V_2^2/2) + (p_2/\rho) + 0 + h_L \tag{1}$$

$$dW_s/dm = (dW_s/dt)(dt/dm) = (45\,000)/[V_2(\pi/4)(0.3^2)(1000)] = 637/V_2 \tag{2}$$

$$h = f_1(90/0.5)(V_1^2/2) + f_2(90/0.3)(V_2^2/2) + 0.05(V_1^2/2) + K(V_1^2/2) \tag{3}$$

Let $f_1 = 0.023$, $f_2 = 0.0258$: $K = 0.5[1 - (0.3/0.5)^2]/(0.3/0.5)^4 = 2.47$. Substitute the above results into Eq. (1) noting that $V_1 = (0.3/0.5)^2 V_2 = 0.36 V_2$,

$$[(0.36V_2)^2/2] + (p_1/\rho) - (637/V_2) = (V_2^2/2) + (0.023)(90/0.5)[(0.36V_2)^2/2] + (0.0258)(90/0.3)(V_2^2/2)$$
$$+ [(0.05)(0.36V_2)^2/2] + [2.47(0.36V_2)^2/2] \tag{4}$$

But, by Bernoulli's equation, $90g = (p_1/\rho) + [(0.36V_2)^2/2] + 0$. Substitute this expression for $(p_1/\rho)$ in Eq. (4) to find: $4.80V_2^3 - 90gV_2 + 637 = 0$. Solve by trial and error: $V_2 = 13.18$ m/s, $V_1 = 4.74$ m/s.

**At A:**

$$(H_{\text{Hyd}})_a = 90 - (0.05)(4.74^2/2g) - (0.023)(90/0.5)(4.74^2/2g) \qquad (H_{\text{Hyd}})_a = 85.2 \text{ m}$$

After contraction:

$$(H_{\text{Hyd}})_{a'} = -(2.47)(4.74^2/2g) + 85.2 = 82.4 \text{ m}$$

**At B:**

$$(H_{\text{Hyd}})_b = 82.4 - (0.0258)(45/0.3)(13.18^2/2g) \qquad (H_{\text{Hyd}})_b = 48.1 \text{ m}$$

After turbine $(b')$:

$$(H_{\text{Hyd}})_{b'} = 48.1 - [637/(13.18)(g)] = 43.2 \text{ m}$$

**At C:**

$$(H_{\text{Hyd}})_c = 43.2 - (0.0258)(45/0.3)(13.18^2/2g) = 8.9 \text{ m}$$

This just equals $V_2^2/2g$, the energy head at exit, as it must. The graphs are shown in Fig. 22-1b.

*(a)*

*(b)*

Fig. 22-1

**22.2**  At a point $A$ in a horizontal 12-in pipe ($f = 0.020$) the pressure head is 200 ft. At a distance of 200 ft from $A$ the 12-in pipe reduces suddenly to a 6-in pipe. At a distance of 100 ft from this sudden reduction the 6-in pipe ($f = 0.015$) suddenly enlarges to a 12-in pipe and point $F$ is 100 ft beyond this change in size. For a velocity of 8.025 ft/s in the 12-in pipes, draw the energy and hydraulic grade lines. Refer to Fig. 22-2.

❚ The velocity heads are $V_{12}^2/2g = (8.025)^2/2g = 1.00$ ft and $V_6^2/2g = 16.0$ ft. The energy line drops in the direction of flow by the amount of the lost head. The hydraulic grade line (gradient) is below the energy line by the amount of the velocity head at any cross section. Note (in Fig. 22-2) that the *hydraulic grade line* can rise where a change (enlargement) in size occurs.

Tabulating the results to the nearest 0.1 ft,

| | lost head in feet | | elevation energy line | $V^2/2g$ | elevation of hyd. gradient |
|---|---|---|---|---|---|
| **at** | **from** | **calculated** | | | |
| $A$ | (Elevation 0.0) | | 201.0 | 1.0 | 200.0 |
| $B$ | $A$ to $B$ | $(0.020)(\frac{200}{1})(1) = 4.0$ | 197.0 | 1.0 | 196.0 |
| $C$ | $B$ to $C$ | $(K_c)(16) = (0.37)(16) = 5.9$ | 191.1 | 16.0 | 175.1 |
| $D$ | $C$ to $D$ | $(0.015)(100/\frac{1}{2})(16) = 48.0$ | 143.1 | 16.0 | 127.1 |
| $E$ | $D$ to $E$ | $\dfrac{(V_6 - V_{12})^2}{2g} = \dfrac{(32.1 - 8.0)^2}{64.4} = 9.0$ | 134.1 | 1.0 | 133.1 |
| $F$ | $E$ to $F$ | $(0.020)(\frac{100}{1})(1) = 2.0$ | 132.1 | 1.0 | 131.1 |

**Fig. 22-2**

**22.3**  The elevations of the energy line and hydraulic grade line at point $G$ are 44.0 and 42.0 ft, respectively. For the system shown in Fig. 22-3, calculate (**a**) the power extracted between $G$ and $H$ if the energy line at $H$ is at elevation 4.0 and (**b**) the pressure heads at $E$ and $F$ which are at elevation 20.0. (**c**) Draw, to the nearest 0.1 ft, the energy and hydraulic grade lines, assuming $K$ for valve $CD$ is 0.40 and $f = 0.010$ for the 6-in pipes.

❚ The flow must be from the reservoir since the energy line at $G$ is below the reservoir level. $GH$ is a turbine. Before the power extracted can be calculated, flow $Q$ and the head extracted must be obtained.

(**a**) At $G$, $V_{12}^2/2g = 2.0$ (the difference between the elevations of the energy and hydraulic grade lines). Also $V_6^2/2g = (16)(2.0) = 32.0$ and $V_{24}^2/2g = \frac{1}{16}(2.0) = 0.13$ ft. To obtain $Q$, $V_{12} = 11.34$ ft/s and $Q = \frac{1}{4}\pi(1)^2(11.34) = 8.91$ cfs. h.p. $= \gamma QH_T/550 = 62.4(8.91)(44.0 - 4.0)/550 = 40.4$ extracted.

(**b**) $F$ to $G$, datum zero

$$(\text{Energy at } F) - 0.030(\tfrac{100}{1})(2.0) = (\text{energy at } G = 44.0) \qquad \text{Energy at } F = 44.0 + 6.0 = 50.0 \text{ ft}$$

$E$ to $F$, datum zero:

$$(\text{Energy at } E) - (45.4 - 11.3)^2/2g = (\text{energy at } F = 50.0) \qquad \text{Energy at } E = 50.0 + 18.0 = 68.0 \text{ ft}$$
$$z + V^2/2g$$

The pressure head at $E = 68.0 - (20 + 32) = 16.0$ ft water. The pressure head at $F = 50.0 - (20 + 2) = 28.0$ ft water.

(**c**) Working back from $E$:

Drop in energy line $DE = 0.010(25/\frac{1}{2})(32.0) = 16.0$ ft    Drop in energy line $CD = 0.40(32.0) = 12.8$ ft

Drop in energy line $BC =$ same as $DE = 16.0$ ft    Drop in energy line $AB = 0.50(32.0) = 16.0$ ft

(Elevation at $D - 16.0$) = elevation at $E$ of 68.0    Elev. $D = 84.0$

(Elevation at $C - 12.8$) = elevation at $D$ of 84.0    Elev. $C = 96.8$

(Elevation at $B - 16.0$) = elevation at $C$ of 96.8    Elev. $B = 112.8$

(Elevation at $A - 16.0$) = elevation at $B$ of 112.8    Elev. $A = 128.8$

The hydraulic grade line is $V^2/2g$ below the energy line: 32.0 ft in the 6-in, 2.0 ft in the 12-in and 0.13 ft in the 24-in. The values are shown in Fig. 22-3.

**Fig. 22-3**

**22.4**  The head extracted by turbine $CR$ in Fig. 22-4 is 200 ft and the pressure at $T$ is 72.7 psi. For losses of $2.0(V_{24}^2/2g)$ between $W$ and $R$ and $3.0(V_{12}^2/2g)$ between $C$ and $T$, determine (*a*) how much water is flowing and (*b*) the pressure head at $R$. Draw the energy line.

❚  Because the energy line at $T$ is at elevation $\{250.0 + [(72.7)(44)/62.4] + (V_{12}^2/2g)\}$ and well above the elevation at $W$, the water flows into the reservoir $W$.

(*a*) Using $T$ to $W$, datum zero,

$$\overset{\text{at }T}{\left[\frac{(72.7)(144)}{62.4} + \frac{V_{12}^2}{2g} + 250\right]} - \overset{T\text{ to }C\quad R\text{ to }W}{\left(3.0\frac{V_{12}^2}{2g} + 2.0\frac{V_{24}^2}{2g}\right)} \overset{H_T}{- 200} = \overset{\text{at }W}{(0 + \text{negl} + 150)}$$

Substituting $V_{24}^2 = \frac{1}{16}V_{12}^2$ and solving, $V_{12}^2/2g = 32.0$ ft or $V_{12} \doteq 45.4$ ft/s. Then $Q = \frac{1}{4}\pi(1)^2(45.4) = 35.7$ cfs.

(*b*) Using $R$ to $W$, datum $R$, $[(p_R/\gamma) + \frac{1}{16}(32.0) + 0] - 2(\frac{1}{16})(32.0) = (0 + \text{negl} + 50)$ and $p_R/\gamma = 52.0$ ft. The reader may check this pressure head by applying the Bernoulli equation between $T$ and $R$.

To plot the energy line in the figure, evaluate the energy at the four sections indicated:

Elevation of energy line at $T = 168.0 + 32.0 + 250.0 = 45.0$

Elevation of energy line at $C = 450.0 - 3(32.0) = 354.0$

Elevation of energy line at $R = 354.0 - 200.0 = 154.0$

Elevation of energy line at $W = 154.0 - 2(\frac{1}{16})(32) = 150.0$

**Fig. 22-4**

**22.5** A pipeline with a pump leads to a nozzle as shown in Fig. 22-5. Find the flow rate when the pump develops a head of 80 ft. Assume that the head loss in the 6-in-diameter pipe may be expressed by $h_L = 5V_6^2/2g$, while the head loss in the 4-in-diameter pipe is $h_L = 12V_4^2/2g$. Sketch the energy line and hydraulic grade line, and find the pressure head at the suction of the pump.

▌ Select the datum as the elevation of the water surface in the reservoir. Note from continuity that $V_6 = (\frac{3}{6})^2 V_3 = 0.25V_3$ and $V_4 = (\frac{3}{4})^2 V_3 = 0.563V_3$, where $V_3$ is the jet velocity. Writing an energy equation from the surface of the reservoir to the jet, $[z_1 + (p_1/\gamma) + (V_1^2/2g)] - h_{L_6} + h_p - h_{L_4} = z_3 + (p_3/\gamma) + (V_3^2/2g)$, $0 + 0 + 0 - 5(V_6^2/2g) + 80 - 12(V_4^2/2g) = 10 + 0 + (V_3^2/2g)$. Express all velocities in terms of $V_3$: $-[5(0.25V_3)^2/2g] + 80 - 12[(0.563V_3)^2/2g] = 10 + (V_3^2/2g)$, $V_3 = 29.7$ fps; $Q = A_3V_3 = (\pi/4)(\frac{3}{12})^2(29.7) = 1.45$ cfs. Head loss in suction pipe: $h_L = 5(V_6^2/2g) = [5(0.25V_3)^2/2g] = 0.312V_3^2/2g = 4.3$ ft. Head loss in discharge pipe: $h_L = 12(V_4^2/2g) = 12(0.563V_3)^2/2g = 52.1$ ft, $V_3^2/2g = 13.7$ ft, $V_4^2/2g = 4.3$ ft, $V_6^2/2g = 0.86$ ft $\approx 0.9$ ft. The energy line and hydraulic grade line are drawn on the figure to scale. Inspection of the figure shows that the pressure head on the suction side of the pump is $p_B/\gamma = 14.8$ ft. Likewise, the pressure head at any point in the pipe may be found if the figure is to scale.

**Fig. 22-5**

**22.6** A pump lifts water at the rate of 180 cfs to a height of 375 ft, at a pipe friction loss of 27 ft. Find the power required if the pump efficiency is 88%. Sketch the energy grade line and the hydraulic grade line.

▌ Water hp $= Q\gamma(\Delta z + h_L)/550 = (180)(62.4)(375 + 27)/550 = 8210$     Input power $= 8210/0.88 = 9330$

The energy and hydraulic grade lines are sketched in Fig. 22-6.

**Fig. 22-6**

**22.7** The diameters of the suction and discharge pipes of a pump are 6 in and 4 in, respectively. The discharge pressure is read by a gage at a point 5 ft above the centerline of the pump, and the suction pressure is read by a gage 2 ft below the centerline. If the pressure gage reads 20 psi and the suction gage reads a vacuum of 10 inHg when a fluid of specific gravity 0.79 is pumped at the rate of 2.0 cfs, find the power delivered to the fluid. Sketch the energy grade line and the hydraulic grade line.

I $\quad p_1/\gamma + z_1 + V_1^2/2g + E_p = p_2/\gamma + z_2 + V_2^2/2g + h_L \qquad V_1 = Q/A_1 = 2.0/[(\pi)(\frac{6}{12})^2/4] = 10.19 \text{ fps}$

$$V_2 = 2.0/[(\pi)(\frac{4}{12})^2/4] = 22.92 \text{ fps} \qquad h_L = 0$$

$$-(\tfrac{10}{12})(13.6/0.79) - 2 + 10.19^2/[(2)(32.2)] + E_p = (20)(144)/[(0.79)(62.4)] + 5 + 22.92^2/[(2)(32.2)] + 0$$

$$E_p = 86.31 \text{ ft}$$

$$P = Q\gamma E_p = (2.0)[(0.79)(62.4)](86.31)/550 = 15.5 \text{ hp}$$

The energy and hydraulic grade lines are sketched in Fig. 22-7.

**Fig. 22-7**

**22.8**  Determine the elevation of hydraulic and energy grade lines at points $A$, $B$, $C$, $D$, and $E$ of Fig. 22-8 ($z = 10$ ft).

I  Solving for the velocity head is accomplished by applying the energy equation from the reservoir to $E$,
$10 + 60 + 0 + 0 = (V_E^2/2g) + 10 + 0 + \frac{1}{2}(V^2/2g) + 0.020(200/0.50)(V^2/2g) + 10(V^2/2g) + 0.10(V_E^2/2g)$. From the continuity equation, $V_E = 4V$. After simplifying, $60 = (V^2/2g)[16 + \frac{1}{2} + 8 + 10 + 16(0.1)] = 36.1(V^2/2g)$ and $V^2/2g = 1.66$ ft. Applying the energy equation for the portion from the reservoir to $A$ gives $70 + 0 + 0 = (V^2/2g) + (p/\gamma) + z + 0.5(V^2/2g)$. Hence, the hydraulic grade line at $A$ is

$$\left.\frac{p}{\gamma} + z\right|_A = 70 - 1.5\frac{V^2}{2g} = 70 - 1.5(1.66) = 67.51 \text{ ft}$$

The energy grade line for $A$ is $(V^2/2g) + z + (p/\gamma) = 67.51 + 1.66 = 69.17$ ft. For $B$, $70 + 0 + 0 = (V^2/2g) + (p/\gamma) + z + 0.5(V^2/2g) + 0.02(80/0.5)(V^2/2g)$ and

$$\left.\frac{p}{\gamma} + z\right|_B = 70 - (1.5 + 3.2)(1.66) = 62.19 \text{ ft}$$

The energy grade line is at $62.19 + 1.66 = 63.85$ ft.

Across the valve the hydraulic grade line drops by $10V^2/2g$, or 16.6 ft. Hence, at $C$ the energy and hydraulic grade lines are at 47.25 ft and 45.59 ft, respectively.

At point $D$, $70 = (V^2/2g) + (p/\gamma) + z + [10.5 + 0.02(200/0.50)](V^2/2g)$ and

$$\left.\frac{p}{\gamma} + z\right|_D = 70 - 19.5(1.66) = 37.6 \text{ ft}$$

with the energy grade line at $37.6 + 1.66 = 39.26$ ft.

At point $E$ the hydraulic grade line is 10 ft, and the energy grade line is $z + (V_E^2/2g) = 10 + 16(V^2/2g) = 10 + 16(1.66) = 36.6$ ft.

**Fig. 22-8**

**22.9** A pump with a shaft input of 7.5 kW and an efficiency of 70 percent is connected in a water line carrying 0.1 m³/s. The pump has a 150-mm-diameter suction line and a 120-mm-diameter discharge line. The suction line enters the pump 1 m below the discharge line. For a suction pressure of 70 kN/m², calculate the pressure at the discharge flange and the rise in the hydraulic grade line across the pump.

▌ The energy added per unit weight of water, $E$, is given by $Q\gamma E = 7500(0.70)$, or $E = 7500(0.7)/[0.1(9806)] = 5.354$ m. Applying the energy equation from suction flange to discharge flange gives $(V_s^2/2g) + (p_s/\gamma) + 0 + 5.354 = (V_d^2/2g) + (p_d/\gamma) + 1$ in which the subscripts $s$ and $d$ refer to the suction and discharge conditions, respectively. From the continuity equation, $V_s = 0.1(4)/(0.15^2\pi) = 5.66$ m/s, $V_d = 0.1(4)/(0.12^2\pi) = 8.84$ m/s. Solving for $p_d$ gives $p_d/\gamma = [5.66^2/2(9.806)] + (70\,000/9806) + 5.354 - [8.84^2/2(9.806)] - 1 = 9.141$ m and $p_d = 89.6$ kPa. The rise in hydraulic grade line is $[(p_d/\gamma) + 1] - (p_s/\gamma) = 9.141 + 1 - (70\,000/9806) = 3.002$ m.

**22.10** Sketch the hydraulic and energy grade lines for Fig. 22-9a ($H = 10$ m).

▌ $$p_1/\gamma + V_1^2/2g + z_1 = p_2/\gamma + V_2^2/2g + z_2 + h_L \qquad h_L = h_f + h_m \qquad h_f = (f)(L/D)(V^2/2g)$$

$(\epsilon/D)_1 = 0.000046/0.150 = 0.000307$; try $f_1 = 0.015$. $(\epsilon/D)_2 = 0.000046/0.300 = 0.000153$; try $f_2 = 0.015$.

$$h_f = (0.015)[25/0.150]\{V_1^2/[(2)(9.807)]\} + (0.015)[(20+8)/0.300]\{(V_1/4)^2/[(2)(9.807)]\} = 0.1319V_1^2$$

$$h_m = KV^2/2g = [1 + (1 - \tfrac{1}{4})^2 + 3.5/16 + \tfrac{1}{16}]\{V_1^2/[(2)(9.807)]\} = 0.09400V_1^2$$

$$h_L = 0.1319V_1^2 + 0.09400V_1^2 = 0.2259V_1^2 \qquad 0 + 0 + 10 = 0 + 0 + 0 + 0.2259V_1^2 \qquad V_1 = 6.653 \text{ m/s}$$

$V_1^2/2g = 6.653^2/[(2)(9.807)] = 2.257$ m $\qquad V_2 = 6.653/4 = 1.663$ m/s $\qquad V_2^2/2g = 1.663^2/[(2)(9.807)] = 0.141$ m

Friction losses are $(f_1)(L/D)(v_1^2/2g) = 5.642$ m, $(f_2)(L/D)(v_2^2/2g) = 0.197$ m. Minor losses are $(V_1 - V_2)^2/2g = 1.270$ m, $3.5V_2^2/2g = 0.494$ m. The hydraulic and energy grade lines are sketched in Fig. 22-9b. [Further checking (not shown) of the Reynolds number indicates that the assumed values of $f_1$ and $f_2$ are acceptable.]

(a)

(b)

**Fig. 22-9**

**22.11** Determine the slope of the hydraulic grade line for flow of atmospheric air through a rectangular 21-in by 7-in galavanized iron conduit ($V = 25$ fps).

∎ $h_f = (f)(L/D)(V^2/2g)$. Slope of the hydraulic grade line $= h_f/L = (f/D)(V^2/2g)$.

$$\epsilon = 0.0005 \text{ ft} \qquad R_h = (\tfrac{21}{12})(\tfrac{7}{12})/4 = 0.2552 \text{ ft} \qquad D = 4R_h = (4)(0.2552) = 1.021 \text{ ft}$$
$$\epsilon/D = 0.0005/1.021 = 0.000490 \qquad N_R = DV/\nu = (1.021)(25)/(1.69 \times 10^{-4}) = 1.51 \times 10^5$$

From Fig. A-5, $f = 0.0195$. Slope of the hydraulic grade line $= (0.0195/1.021)\{25^2/[(2)(32.2)]\} = 0.185 \text{ ft/ft}$.

**22.12** What size square conduit is needed to convey 375 L/s of water at 15 °C with a slope of the hydraulic grade line of 0.002? ($\epsilon = 1$ mm)

∎ Let $d$ = side of conduit.

$$R_h = d^2/4d = d/4 \qquad D = 4R_h = 4d/4 = d \qquad V = Q/A = (375 \times 10^{-3})/d^2$$
$$N_R = DV/\nu = (d)(0.375/d^2)/(1.16 \times 10^{-6}) = 323\,276/d \qquad \epsilon/D = 0.001/d \qquad h_f = (f)(L/D)(V^2/2g)$$
$$h_f/L = 0.002 = (f/d)\{(0.375/d^2)^2/[(2)(9.807)]\} \qquad d^5 = 3.585f$$

Assume $f = 0.022$.

$$d^5 = (3.585)(0.022) \qquad d = 0.602 \text{ m} \qquad N_R = 323\,276/0.602 = 537\,000$$

From Fig. A-5, $f = 0.023$. $d^5 = (3.585)(0.023)$, $d = 0.607$ m.

**22.13** Neglecting minor losses other than the valve, sketch the hydraulic grade line for Fig. 9-54. The globe valve has a loss coefficient of $K = 4.5$.

∎ $\qquad p_1/\gamma + V_1^2/2g + z_1 = p_2/\gamma + V_2^2/2g + z_2 + h_L \qquad h_L = (f)(L/D)(V^2/2g) + KV^2/2g$

Try $f = 0.013$.

$$h_L = (0.013)[(100 + 70 + 105)/(\tfrac{8}{12})]\{V^2/[(2)(32.2)]\} + (4.5)\{V^2/[(2)(32.2)]\} = 0.1531V^2$$
$$0 + 0 + 12 = 0 + 0 + 0 + 0.1531V^2 \qquad V = 8.853 \text{ fps}$$
$$12 - (0.013)[(100 + 70)/(\tfrac{8}{12})]\{8.853^2/[(2)(32.2)]\} = 7.97 \text{ ft} \qquad 4.5V^2/2g = (4.5)\{8.853^2/[(2)(32.2)]\} = 5.48 \text{ ft}$$

The hydraulic grade line is sketched in Fig. 22-10. [Further checking (not shown) of the Reynolds number indicates that the assumed value of $f$ is acceptable.]

**Fig. 22-10**

# CHAPTER 23
# Forces Developed by Fluids in Motion

**23.1** A hose and nozzle discharge a horizontal water jet against a nearby vertical plate, as shown in Fig. 23-1. The flow rate of water is 0.025 m³/s, and the diameter of the nozzle tip is 30 mm. Find the horizontal force necessary to hold the plate in place.

**▌** $F = \rho Q(v_2 - v_1)$. The net external force acting on the fluid ($F$ in the equation) is the horizontal force necessary to hold the plate in place (i.e., $R$ in Fig. 23-1). Assume it acts toward the left, as shown in the figure, and this direction is taken to be positive. $v_1 = Q/A_1 = 0.025/[(\pi)(0.030)^2/4] = 35.37$ m/s, $v_2 = 0$, $R = (1000)(0.025)[0 - (-35.37)] = 884$ N (leftward).

Fig. 23-1

**23.2** Water flows from a large tank through an orifice of 3 in diameter and against a block, as shown in Fig. 23-2. The water jet strikes the block at the vena contracta. The block weighs 50 lb, and the coefficient of the friction between block and floor is 0.57. The orifice's coefficient of discharge ($C$) is 0.60, and its coefficient of contraction ($C_c$) is 0.62. What is the minimum height to which water must rise in the tank ($y$ in Fig. 23-2) in order to start the block moving to the right?

**▌** $F = \rho Q(v_2 - v_1)$. The force caused by the water striking the block must equal (or slightly exceed) the friction force between block and floor ($F_f$ in Fig. 23-2). In other words, the net external force acting on the fluid ($F$ in the equation above) is $F_f$ when the block begins to move. Hence, $F = F_f = (0.57)(50) = 28.5$ lb, $-28.5 = (1.94)(Q)(0 - v_1)$, $Qv_1 = 14.69$. But $V_1 = Q/a$, where $a$ is the area of the jet at its vena contracta.

$$a = (0.62)[(\pi)(\tfrac{3}{12})^2/4] = 0.03043 \text{ ft}^2 \qquad v_1 = Q/0.03043 = 32.86Q \qquad (Q)(32.86Q) = 14.69$$

$$Q = 0.6686 \text{ ft}^3/\text{s} = CA\sqrt{2gh}$$

$$0.6686 = (0.60)[(\pi)(\tfrac{3}{12})^2/4]\sqrt{(2)(32.2)(y - 1.0)} \qquad y = 9.00 \text{ ft}$$

Fig. 23-2

**23.3** A hose and nozzle discharge a horizontal water jet against a nearby vertical plate, as shown in Fig. 23-1. The flow rate of water is 0.043 m³/s, and the diameter of the nozzle tip is 50 mm. Find the horizontal force necessary to hold the plate in place.

∎ $$F = \rho Q(v_2 - v_1) \qquad v_1 = Q/A_1 = 0.042/[(\pi)(0.050)^2/4] = 21.39 \text{ m/s}$$

$$v_2 = 0 \qquad R = (1000)(0.042)[0 - (-21.39)] = 898 \text{ N} \qquad \text{(leftward)}$$

**23.4** In Fig. 23-3 a small ingot and platform rest on a steady water jet. If the total weight supported is 825 N, what is the jet velocity?

∎ $$F = \rho Q(v_2 - v_1) \qquad Q = Av_1 = [(\pi)(0.060)^2/4](v_1) = 0.002827 v_1$$

$$825 = (1000)(0.002827 v_1)(v_1) \qquad v_1 = 17.1 \text{ m/s}$$

Water jet
$D_0 = 60$ mm

**Fig. 23-3**

**23.5** The horizontal nozzle in Fig. 23-4 has $D_1 = 10$ in and $D_2 = 6$ in. The inlet pressure $p_1 = 60$ psia, and the exit velocity $V_2 = 85$ fps. Compute the tensile force in the flange bolts. Assume incompressible steady flow.

∎ $$\Sigma F = \rho Q(V_2 - V_1) \qquad Q = AV = [(\pi)(\tfrac{6}{12})^2/4](85) = 16.69 \text{ ft}^3/\text{s} \qquad V_1 = (\tfrac{6}{10})^2(V_2) = (\tfrac{6}{10})^2(85) = 30.60 \text{ fps}$$

$$(60 - 15)[(\pi)(10)^2/4] - F_{\text{bolts}} = (1.94)(16.69)(85 - 30.60) \qquad F_{\text{bolts}} = 1773 \text{ lb}$$

$p_a = 15$ lb/in² abs

Open jet

Water

②

①

**Fig. 23-4**

**23.6** The tank in Fig. 23-5 weighs 920 N empty and contains 1.1 m³ of water at 20 °C. The entrance and exit pipes are identical, $D_1 = D_2 = 60$ mm, and both carry 80 L/s. Determine the scale reading $W$.

∎ $$\Sigma F_y = W - W_{\text{H}_2\text{O}} - W_{\text{tank}} = -\rho Q[(v_y)_2 - (v_y)_1] \qquad (v_y)_1 = Q/A = 80 \times 10^{-3}/[(\pi)(0.060)^2/4] = 28.29 \text{ m/s}$$

$$W - (1.1)[(9.79)(1000)] - 920 = -(1000)(80 \times 10^{-3})(0 - 28.29) \qquad W = 13.952 \text{ kN}$$

①

②

W

Water

Scale

**Fig. 23-5**

**23.7** Garbage is discharged onto a scow, as in Fig. 23-6; the discharge velocity is 4 fps and the weight flux is 700 lb/s. Calculate the tension in the mooring line.

▌ The $x$-equilibrium of the scow is expressed by

$$T = F_{x\text{ scow}} = -F_{x\text{ garbage}} = -\dot{m}(v_2 - v_1) = -(700/g)(0 - 4\cos 35°) = +71.3 \text{ lb}$$

**Fig. 23-6**

**23.8** The water jet in Fig. 23-7, moving at 45 ft/s, is split so that one-third of the water moves toward $A$. Calculate the magnitude and direction of the force on the stationary splitter. Assume ideal flow in a horizontal plane.

▌
$$F = \rho Q(v_2 - v_1) \qquad -F_x = \rho Q_A[(v_x)_2 - (v_x)_1] + \rho Q_B[(v_x)_2 - (v_x)_1]$$
$$F_y = \rho Q_A[(v_y)_2 - (v_y)_1] + \rho Q_B[(v_y)_2 - (v_y)_1]$$

$$Q = AV = (0.15)(45) = 6.75 \text{ ft}^3/\text{s} \qquad Q_A = 6.75/3 = 2.25 \text{ ft}^3/\text{s} \qquad Q_B = 6.75 - 2.25 = 4.50 \text{ ft}^3/\text{s}$$

$$-F_x = (1.94)(2.25)(-45\cos 60° - 45) + (1.94)(4.50)(45\cos 60° - 45) = -491 \text{ lb}$$

$$F_y = (1.94)(2.25)(-45\cos 30° - 0) + (1.94)(4.50)(45\cos 30° - 0) = 170 \text{ lb}$$

$$F_r = \sqrt{491^2 + 170^2} = 520 \text{ lb} \qquad \tan \alpha = \tfrac{170}{491} = 0.34623 \qquad \alpha = 19.1°$$

**Fig. 23-7**

**23.9** Rework Prob. 23.8, if the splitter is moving to the right at 8 fps.

▌
$$F = \rho Q(v_2 - v_1) \qquad -F_x = \rho Q_A[(v_x)_2 - (v_x)_1] + \rho Q_B[(v_x)_2 - (v_x)_1]$$
$$F_y = \rho Q_A[(v_y)_2 - (v_y)_1] + \rho Q_B[(v_y)_2 - (v_y)_1]$$

$$V = 45 - 8 = 37 \text{ ft/s} \qquad Q = AV = (0.15)(37) = 5.55 \text{ ft}^3/\text{s} \qquad Q_A = 5.55/3 = 1.85 \text{ ft}^3/\text{s}$$

$$Q_B = 5.55 - 1.85 = 3.70 \text{ ft}^3/\text{s}$$

$$-F_x = (1.94)(1.85)(-37\cos 60° - 37) + (1.94)(3.70)(37\cos 60° - 37) = -332 \text{ lb}$$

$$F_y = (1.94)(1.85)(-37\cos 30° - 0) + (1.94)(3.70)(37\cos 30° - 0) = 115 \text{ lb}$$

$$F_r = \sqrt{332^2 + 115^2} = 351 \text{ lb} \qquad \tan \alpha = \tfrac{115}{332} = 0.34639 \qquad \alpha = 19.1°$$

**23.10** A locomotive tender running at 18 mph scoops up water from a trough between the rails, as shown in Fig. 23-8. The scoop delivers water at a point 9 ft above its original level and in the direction of motion. The area of the stream of water at entrance is 45 in². The water is everywhere under atmospheric pressure. **(a)** Find the ground velocity of the water as it leaves the scoop and the reaction force acting on the tender. How slow can the train move and still deliver water to the same point?

▌
$$F_u = \rho A_1 V_1 (\Delta V_u) \qquad u_1 = u_2 = -(18)(5280)/3600 = -26.40 \text{ ft/s} \qquad v_1 = 26.40 \text{ fps}$$

Relative to the moving car,

$$p_1/\gamma + v_1^2/2g + z_1 = p_2/\gamma + v_2^2/2g + z_2 \qquad 0 + 26.40^2/[(2)(32.2)] + 0 = 0 + v_2^2/[(2)(32.2)] + 9 \qquad v_2 = -10.83 \text{ fps}$$

$$V_2 = -26.40 - 10.83 = -37.23 \text{ fps} \qquad F_u = (1.94)(\tfrac{45}{144})(26.40)(37.23) = 596 \text{ lb} \quad \text{·}$$

At minimum speed, $V^2/2g = 9$ ft, since $v_2 = 0$.

$$V^2/[(2)(32.2)] = 9 \qquad V = 24.07 \text{ fps} = 16.4 \text{ mph}$$

**Fig. 23-8**

**23.11**  Referring to Fig. 23-9, a 2-in-diameter stream of water strikes a 4-ft-square door which is at an angle of 30° with the stream's direction. The velocity of the water in the stream is 60.0 ft/s and the jet strikes the door at its center of gravity. Neglecting friction, what normal force applied at the edge of the door will maintain equilibrium?

▮ The force exerted by the door on the water will be normal to the door (no friction). Hence, since no forces act in the $W$ direction in the figure, there will be no change in momentum in that direction. Thus, using $W$ components, initial momentum $\pm 0 =$ final momentum, $+M(V \cos 30°) = +M_1 V_1 - M_2 V_2$, $(\gamma/g)(A_{jet}V)(V \cos 30°) = (\gamma/g)(A_1 V_1)V_1 - (\gamma/g)(A_2 V_2)V_2$. But $V = V_1 = V_2$ (friction neglected). Then $A_{jet} \cos 30° = A_1 - A_2$ and, from the equation of continuity, $A_{jet} = A_1 + A_2$. Solving, $A_1 = A_{jet}(1 + \cos 30°)/2 = A_{jet}(0.933)$ and $A_2 = A_{jet}(1 - \cos 30°)/2 = A_{jet}(0.067)$.

The stream divides as indicated and the momentum equation produces, for the $X$ direction, $[(62.4/32.2)(\frac{1}{4}\pi)(\frac{1}{6})^2 60](60) - F_x(1) = [(62.4/32.2)(\frac{1}{4}\pi)(\frac{1}{6})^2(0.933)(60)](52.0) + [(62.4/32.2)(\frac{1}{4}\pi)(\frac{1}{6})^2(0.067)(60)](-52.0)$ and $F_x = 38.0$ lb.

Similarly, in the $Y$ direction, $M(0) + F_y(1) = [(62.4/32.2)(0.0218)(0.933)(60)](30.0) + [(62.4/32.3)(0.0218)(0.067)(60)](-30.0)$ and $F_y = 65.9$ lb.

For the door as the free body, $\Sigma M_{hinge} = 0$ and $+38.0(1) + 65.9(2)(0.866) - P(4) = 0$ or $P = 38.0$ lb.

**Fig. 23-9**

**23.12**  The force exerted by a 1-in-diameter stream of water against a flat plate held normal to the stream's axis is 145 lb. What is the flow of water?

▮ $F = \rho Q(v_2 - v_1)$ $\qquad v_1 = Q/A = Q/[(\pi)(\frac{1}{12})^2/4] = 183.3Q$ $\qquad -145 = (1.94)(Q)(0 - 183.3Q)$ $\qquad Q = 0.639$ ft³/s

**23.13**  Water flow in open channels can be controlled and measured with the sluice gate in Fig. 23-10. At a moderate distance upstream and downstream of the gate, sections 1 and 2, the flow is uniform and the pressure is hydrostatic. Derive an expression for the force $F$ required to hold the gate as a function of $\rho$, $V_1$, $g$, $h_1$, and $h_2$. Neglect bottom friction in your analysis.

■ $\Sigma F_x = \dot{m}(V_2 - V_1)$    $-F + \gamma h_1(h_1 b)/2 - \gamma h_2(h_2 b)/2 = \rho h_1 b V_1[(h_1/h_2)(V_1) - V_1]$

$$F = \gamma b(h_1^2 - h_2^2)/2 - \rho h_1 b V_1^2(h_1/h_2 - 1)$$

Fig. 23-10

**23.14** Compute the force per unit width required to hold the gate in Fig. 23-10 if $h_1 = 4$ m, $h_2 = 0.7$ m, and $V_1 = 1.25$ m/s.

■    $F/b = \gamma(h_1^2 - h_2^2)/2 - \rho h_1 V_1^2(h_1/h_2 - 1)$

    $= (9.79)(4^2 - 0.7^2)/2 - (9.98)(4)(1.25)^2(4/0.7 - 1) = 46.5$ kN/m

**23.15** Find the thrust developed when water is pumped through a 300-mm-diameter pipe in the bow of a boat at $V = 2.0$ m/s and emitted through a 200-mm-diameter pipe in the stern.

■    $Q = A_1 V_1 = [(\pi)(0.300)^2/4](2.0) = 0.1414$ m³/s    $V_2 = (2.0)(\frac{300}{200})^2 = 4.50$ m/s

    $F = \rho Q(V_2 - V_1) = (988)(0.1414)(4.50 - 2.0) = 349$ N

**23.16** A jet of water flowing freely in the atmosphere is deflected by a curved vane, as shown in Fig. 23-11a. If the water jet has a diameter of 1.5 in and a velocity of 25.5 fps, what is the force required to hold the vane in place?

■ Forces acting on the water jet are shown in Fig. 23-11b.

$$F_x = \rho Q[(v_x)_2 - (v_x)_1]    Q = AV = [(\pi)(1.5/12)^2/4](25.5) = 0.3129 \text{ ft}^3/\text{s}$$

$R_x = (1.94)(0.3129)[0 - (-25.5)] = 15.48$ lb    (leftward)    $F_y = \rho Q[(v_y)_2 - (v_y)_1]$

$R_y = (1.94)(0.3129)(25.5 - 0) = 15.48$ lb    (upward)    $R = \sqrt{15.48^2 + 15.48^2} = 21.49$ lb

With equal $x$ and $y$ components, the direction of $R$ is at a 45° angle (i.e., $\alpha = 45°$ in Fig. 23-11b).

(a)    Fig. 23-11    (b)

**23.17** A jet of water having a velocity of 10 m/s and a flow rate of 60 L/s is deflected through a right angle, as shown in Fig. 23-12. Determine the reaction on the chute. Neglect the friction of the chute.

■    $F_x = \rho Q[(v_x)_2 - (v_x)_1]$    $R_x = (1000)(60 \times 10^{-3})[-10 \sin 30° - (-10 \cos 30°)] = 219.6$ N    (leftward)

$F_y = \rho Q[(v_y)_2 - (v_y)_1]$    $R_y = (1000)(60 \times 10^{-3})[10 \sin 60° - (-10 \cos 60°)] = 819.6$ N    (upward)

$$R = \sqrt{219.6^2 + 819.6^2} = 849 \text{ N}    \alpha = \arctan(819.6/219.6) = 75.0°$$

**Fig. 23-12**

**23.18** For the cart of Fig. 23-13, compute the force on the wheels caused by deflecting the jet and the compression of the spring, if its stiffness is 1.6 kN/m.

**I** The force on the jet is upward and to the left. The upward component is supplied by the ground, acting on the wheels:
$$F_y = \rho Q[(v_y)_2 - (V_y)_1] \qquad Q = Av = [(\pi)(0.040)^2/4](20) = 0.02513 \text{ m}^3/\text{s}$$
$$F_y = (998)(0.02513)(20 \sin 45° - 0) = 355 \text{ N}$$

The leftward component is supplied by the compressed spring:
$$F_x = \rho Q[(v_x)_2 - (V_x)_1] = (998)(0.02513)(20 \cos 45° - 20) = -147 \text{ N}$$

Then, by Hooke's law, $F_x = -k \, \Delta x$, $-147 = (1600)(\Delta x)$, $\Delta x = 92$ mm.

**Fig. 23-13**

**23.19** In Fig. 23-14 the vane turns the water jet completely around. Find an expression for the maximum jet velocity $V_0$ if the maximum possible support force is $F_0$.

**I** $\quad F = \rho Q(u_2 - u_1) \qquad F_0 = (\rho_0)(A_0 V_0)(-V_0 - V_0) = (\rho_0)[(\pi D_0^2/4)(V_0)](-V_0 - V_0) \qquad V_0 = \sqrt{2F_0/(\rho_0 D_0^2 \pi)}$

**Fig. 23-14**

**23.20** The water tank in Fig. 23-15 stands on a frictionless cart and feeds a jet of diameter 3 cm and velocity 10 m/s, which is deflected 50° by a vane. Compute the tension in the supporting cable.

**I** Contrast Prob. 23.18; here the force on the water is upward and *rightward*.
$$F_x = \rho Q[(v_x)_2 - (v_x)_1] \qquad Q = Av = [(\pi)(0.03)^2/4](10) = 0.007069 \text{ m}^3/\text{s}$$
$$T = (998)(0.007069)(10 \cos 50° - 0) = 45.3 \text{ N}$$

**23.21** Water exits to the atmosphere ($p_{atm} = 101$ kPa) through a split nozzle as shown in Fig. 23-16. Duct areas are $A_1 = 200$ cm$^2$ and $A_2 = A_3 = 90$ cm$^2$. The flow rate is $Q_2 = Q_3 = 200$ m$^3$/h, and inlet pressure $p_1 = 165$ kPa abs. Compute the force on the flange bolts at section 1.

**Fig. 23-15**

$$\Sigma F_x = \rho Q[(v_x)_2 - (v_x)_1] \qquad -F_{\text{bolts}} + p_1 A_1 = 2\rho Q_2(-v_2 \cos 30°) - \rho Q_1 v_1$$

$$v_2 = Q_2/A_2 = (\tfrac{200}{3600})/(90 \times 10^{-4}) = 6.173 \text{ m/s} \qquad v_1 = (2)(\tfrac{200}{3600})/(200 \times 10^{-4}) = 5.556 \text{ m/s}$$

$$-F_{\text{bolts}} + [(165 - 101)(1000)](200 \times 10^{-4}) = (2)(998)(\tfrac{200}{3600})(-5.556 \cos 30°) - (998)[(2)(\tfrac{200}{3600})](5.556)$$

$$F_{\text{bolts}} = 2430 \text{ N}$$

**Fig. 23-16**

**23.22** A pump in a water tank directs a water jet at 25 ft/s and 0.3 ft³/s against a vane, as in Fig. 23-17. Compute the force $F$ to hold the cart stationary if the jet follows path $A$.

▮ If the water follows path $A$, no momentum flux is created; thus, $F = 0$.

**Fig. 23-17**

**23.23** Compute the force $F$ in Prob. 23.22 if the fluid jet follows path $B$. The tank holds 50 ft³ of water at this instant.

▮ $F = \rho Q(v_2 - v_1)$. From "frictionless particle" concept, jet $B$ will have constant horizontal velocity $v = v_{\text{jet}} \cos (180 - 120)°$. $F = (1.94)(0.3)[25 \cos (180 - 120)° - 0] = 7.28 \text{ lb}$.

**23.24** The cart in Fig. 23-18 moves at constant velocity $V_0 = 25$ m/s and takes on water with a scoop 1.7 m wide which dips $h = 25$ mm into a pond. Find the mass flow into the cart and the force needed to propel the cart.

▮ $$\dot m = \rho A V_0 = (998)[(1.7)(0.025)](25) = 1060 \text{ kg/s}$$

The cart is gaining momentum at the rate $(1060 \text{ kg/s})(25 \text{ m/s}) = 26\,500 \text{ (kg} \cdot \text{m/s)/s} = 26.5 \text{ kN}$, which, by Newton's law, must be the force on the cart.

**Fig. 23-18**

**23.25** A racing car, coasting, is to be decelerated by a scoop, as shown in Fig. 23-19. If the car and scoop weigh 2000 lb and $h = 3$ in, find the deceleration when $V = 540$ ft/s. Assume that the scoop has a width $b = 1$ ft into the paper.

▌ Force on car $= \rho bhV(1 - \cos 60°) = Ma$, whence ($\rho = 1.94$ slug/ft³)

$$a = \frac{(1.94)(1)(3/12)(540)(1 - 0.5)}{(2000/32.2)} = 1138 \text{ ft/s}^2$$

**Fig. 23-19**

**23.26** A vane of turning angle $(180° - \theta)$ is mounted on a water tank, as shown in Fig. 23-20. It is struck by a 3-in-diameter, 40-fps water jet, which turns and falls off into the water tank without spilling. What is the force $F$ required to hold the water tank stationary if $\theta = 90°$?

▌ $\Sigma F_x = \rho Q(v_2 - v_1)$     $Q = Av = [(\pi)(\frac{3}{12})^2/4](40) = 1.963 \text{ ft}^3/\text{s}$     $F = (1.94)(1.963)[0 - (-40)] = 152 \text{ lb}$

**Fig. 23-20**

**23.27** In Fig. 23-21, air of density 1.2 kg/m³ flows in the 30-cm duct at 12 m/s; it is choked by a 90° cone in the exit. Estimate the force of the airflow on the cone.

▌           $\Sigma F_x = \dot{m}_2 u_2 - \dot{m}_1 u_1$     $\dot{m}_1 = \rho A_1 V_1 = (1.2)[(\pi)(0.30)^2/4](12) = 1.018 \text{ kg/s}$

$A_2 = (\pi)[0.40 + (0.01)(\cos 45°)](0.01) = 0.01279 \text{ m}^2$     $\dot{m}_2 = \rho A_2 V_2 = (1.2)(0.01279)(V_2) = 1.018$

$V_2 = 66.33 \text{ m/s}$     $F = (1.018)(66.33 \cos 45°) - (1.018)(12) = 35.5 \text{ N}$     (rightward)

**23.28** For the same situation and conditions as given in Prob. 23.16 except that the vane itself is moving to the right with a velocity of 10.0 ft/s, find the force exerted by the water on the vane.

**Fig. 23-21**

▌ $F_x = \rho Q_r V_r (\cos \beta + 1)$ ⠀⠀ $V_r = 25.5 - 10.0 = 15.5$ ft/s ⠀⠀ $Q_r = A_r V_r = [(\pi)(1.5/12)^2/4](15.5) = 0.1902$ ft$^3$/s

$R_x = (1.94)(0.1902)(15.5)(\cos 90° + 1) = 5.719$ lb ⠀⠀ $F_y = \rho Q_r V_r (\sin \beta)$

$R_y = (1.94)(0.1092)(15.5)(\sin 90°) = 5.719$ lb ⠀⠀ $R = \sqrt{5.719^2 + 5.719^2} = 8.09$ lb

As in Prob. 23.16, $R$ acts at an angle $\beta$ of 45°.

**23.29** Solve Prob. 23.28 if the "angle of the vane" is 30° (i.e., the 90° angle in Fig. 23-11b is 30° instead).

▌ $F_x = \rho Q_r V_r (\cos \beta + 1)$. Using data from Prob. 23.28, $R_x = (1.94)(0.1902)(15.5)(\cos 30° + 1) = 10.67$ lb, $F_y = Q_r V_r (\sin \beta)$, $R_y = (1.94)(0.1902)(15.5)(\sin 30°) = 2.86$ lb, $R = \sqrt{10.67^2 + 2.86^2} = 11.0$ lb. The direction can be determined by referring to Fig. 23.22: $\alpha = \arctan (2.86/10.67) = 15.0°$.

**Fig. 23-22**

**23.30** A jet of water flowing freely in the atmosphere is deflected by a curved, 90° vane (see Fig. 23-11a). The vane is part of a turbine and is moving to the right with a velocity of 20 ft/s. The water jet has a diameter of 2 in and a velocity of 60 ft/s. Determine the force exerted by the water on the vane.

▌ ⠀⠀ $F_x = \rho Q_r V_r (\cos \beta + 1)$ ⠀⠀ $V_r = 60 - 20 = 40$ ft/s ⠀⠀ $Q_r = A_r V_r = [(\pi)(\frac{2}{12})^2/4](40) = 0.8727$ ft$^3$/s

$R_x = (1.94)(0.8728)(40)(\cos 90° + 1) = 67.72$ lb ⠀⠀ $F_y = Q_r V_r (\sin \beta)$

$R_y = (1.94)(0.8728)(40)(\sin 90°) = 67.72$ lb ⠀⠀ $R = \sqrt{67.72^2 + 67.72^2} = 95.8$ lb

Since $R_x$ and $R_y$ are equal, $R$ acts at an angle of 45° with the horizontal.

**23.31** Solve Prob. 23.30 if the "angle of the vane" is 60° (i.e., the 90° angle in Fig. 23-11b is 30° instead).

▌ $F_x = \rho Q_r V_r (\cos \beta + 1)$. Using data from Prob. 23.30, $R_x = (1.94)(0.8727)(40)(\cos 60° + 1) = 101.6$ lb, $F_y = \rho Q_r V_r (\sin \beta)$, $R_y = (1.94)(0.8728)(40)(\sin 60°) = 58.65$ lb, $R = \sqrt{101.6^2 + 58.65^2} = 117.3$ lb. The direction can be determined by referring to Fig. 23-22: $\alpha = \arctan (58.65/101.6) = 30.0°$.

**23.32** As shown in Fig. 23-23, a vane is curved so that the water jet is entirely reversed in direction. The vane is moving to the right at a velocity of 9.0 m/s, and the velocity of the water jet is 21.0 m/s. If the diameter of the water jet is 50 mm, find the force exerted by the vane on the water.

▌ $F_x = \rho Q_r V_r (\cos \beta + 1)$ ⠀⠀ $V_r = 21.0 - 9.0 = 12.0$ m/s ⠀⠀ $Q_r = A_r V_r = [(\pi)(0.050)^2/4](12.0) = 0.0236$ m$^3$/s

$R_x = (1000)(0.0236)(12.0)(\cos 0° + 1) = 566$ N ⠀⠀ $F_y = \rho Q_r V_r \sin \beta$ ⠀⠀ $R_y = (1000)(0.0236)(12.0)(\sin 0°) = 0$

Hence, $R = 566$ N ⠀ (acting horizontally).

**23.33** A 2-in-diameter water jet with a velocity of 100 fps impinges on a single vane moving in the same direction (thus $F_x = F_u$) at a velocity of 60 fps. If $\beta_2 = 150°$ and friction losses over the vane are such that $v_2 = 0.9v_1$, compute the force exerted by the water on the vane. See Fig. 23-24a.

**Fig. 23-23**

▌ The velocity vector diagrams at entrance and exit to the vane are shown in Fig. 23-24b. Since $v_2 = (0.9)(40) = 36$ fps;

$$V_2 \sin \alpha_2 = v_2 \sin \beta_2 = (36)(0.5) = 18 \text{ fps} \qquad (1)$$

$$V_2 \cos \alpha_2 = u + v_2 \cos \beta_2 = 60 + 36(-0.866) = 28.8 \text{ fps} \qquad (2)$$

Solving (1) and (2) simultaneously yields $V_2 = 34$ fps, $\alpha_2 = 32°$. Hence $-F_x = \rho Q'(V_2 \cos \alpha_2 - V_1) = 1.94(0.0218)(100 - 60)(28.8 - 100) = -120.4$ lb. So $F_x = 120.4$ lb. The force of vane on water is to the left as assumed; hence force of water on vane is 120.4 lb to the right. $-F_y = \rho Q'(V_2 \sin \alpha_2 - 0) = 1.94(0.0218)(40)(-18) = -30.5$ lb. Thus $F_y = 30.5$ lb in the direction shown. The force of water on the vane is equal and opposite and thus 30.5 lb upward. If the blade were one of a series of blades, $-F_x = \rho Q(V_2 \cos \alpha_2 - V_1) = 1.94(0.0218)(100)(28.8 - 100) = -301$ lb.

**Fig. 23-24**

23.34 A jet of water with an area of 4 in$^2$ and a velocity of 175 fps strikes a single vane which reverses it through 180° without friction loss. Find the force exerted if the vane moves **(a)** in the same direction as the jet with a velocity of 65 fps; **(b)** in a direction opposite to that of the jet with a velocity of 65 fps.

▌
$$F = \rho Q(v_2 - v_1)$$

**(a)** $v_1 = 175 - 65 = 110$ ft/s $\quad Q = Av = (\frac{4}{144})(110) = 3.056$ ft$^3$/s $\quad F = (1.94)(3.056)[110 - (-110)] = 1304$ lb

**(b)** $v_1 = 175 - (-65) = 240$ ft/s $\quad Q = (\frac{4}{144})(240) = 6.667$ ft$^3$/s $\quad F = (1.94)(6.667)[240 - (-240)] = 6208$ lb

23.35 The pipe bend shown in Fig. 23-25 is in a horizontal plane. A fluid of weight density 8.615 kN/m$^3$ enters the bend with a velocity of 3.50 m/s and a pressure of 280 kPa. Neglecting any energy losses in the bend, find the force required to hold the bend in place.

▌
$$\Sigma F_x = \rho Q[(v_x)_2 - (v_x)_1] \qquad p_1 A_1 + p_2 A_2 \cos 65° - R_x = \rho Q(-v_2 \cos 65° - v_1)$$

$$A_1 = (\pi)(0.200)^2/4 = 0.03142 \text{ m}^2$$

$p_2$ can be determined using the Bernoulli equation.

$$p_1/\gamma + v_1^2/2g + z_1 = p_2/\gamma + v_2^2/2g + z_2 \qquad A_2 v_2 = A_1 v_1$$

$$A_2 = (\pi)(0.080)^2/4 = 0.005027 \text{ m}^2 \qquad 0.005027 v_2 = (0.03142)(3.50) \qquad v_2 = 21.88 \text{ m/s}$$

$$280/8.615 + 3.50^2/[(2)(9.807)] + 0 = p_2/8.615 + 21.88^2/[(2)(9.807)] + 0$$

$$p_2 = 75.10 \text{ kPa} \qquad Q = A_1 v_1 = (0.03142)(3.50) = 0.1100 \text{ m}^3/\text{s}$$

$$[(280)(1000)](0.03142) + [(75.10)(1000)](0.005027)(\cos 65°) - R_x$$

$$= [8615/9.807](0.1100)[(-21.88)(\cos 65°) - 3.50]$$

$$R_x = 10\ 190 \text{ N} \quad \text{or} \quad 10.19 \text{ kN} \qquad \Sigma F_y = \rho Q[(v_y)_2 - (v_y)_1] \qquad -p_2 A_2 \sin 65° + R_y = \rho Q(v_2 \sin 65° - 0)$$

$$-[(75.10)(1000)](0.005027)(\sin 65°) + R_y = [8615/9.807](0.1100)[(21.88)(\sin 65°) - 0]$$

$$R_y = 2262 \text{ N} \quad \text{or} \quad 2.26 \text{ kN} \qquad R = \sqrt{10.19^2 + 2.26^2} = 10.4 \text{ kN}$$

$$\alpha = \arctan(2.26/10.19) = 12.5° \qquad \text{(see Fig. 23-25}b\text{)}$$

(a)

(b)      Fig. 23-25

**23.36** The pipe bend shown in Fig. 23-26a is in a horizontal plane. Oil with a specific gravity of 0.86 enters the reducing bend at section A with a velocity of 3.2 m/s and a pressure of 150 kPa. Determine the force required to hold the bend in place. Neglect any energy loss in the bend.

❚ External forces acting on the oil within the bend are shown in Fig. 23-26b.

$$A_1 v_1 = A_2 v_2 \qquad A_1 = \pi(0.150)^2/4 = 0.01767 \text{ m}^2 \qquad A_2 = \pi(0.100)^2/4 = 0.007854 \text{ m}^2$$

$$(0.01767)(3.2) = 0.007854 v_2 \qquad v_2 = 7.199 \text{ m/s} \qquad p_1/\gamma + v_1^2/2g + z_1 = p_2/\gamma + v_2^2/2g + z_2$$

$$150/[(0.86)(9.79)] + (3.2)^2/[(2)(9.807)] + 0 = p_2/[(0.86)(9.79)] + (7.199)^2/[(2)(9.807)] + 0 \qquad p_2 = 132.1 \text{ kPa}$$

$$F_x = \rho Q(v_{x2} - v_{x1}) \qquad p_1 A_1 - p_2 A_2 \cos 30° - R_x = \rho Q(v_{x2} \cos 30° - v_{x1})$$

$$Q = Av = (0.01767)(3.2) = 0.05654 \text{ m}^3/\text{s}$$

$$(150\,000)(0.01767) - (132\,100)(0.007854)\cos 30° - R_x = (0.86)(1000)(0.05654)(7.199\cos 30° - 3.2)$$

$$R_x = 1604 \text{ N}$$

$$F_y = \rho Q(v_{y2} - v_{y1}) \qquad -p_2 A_2 \sin 30° + R_y = \rho Q(v_{y2}\sin 30° - 0)$$

$$(-132\,100)(0.007854)\sin 30° + R_y = (0.86)(1000)(0.05654)(7.199\sin 30°) \qquad R_y = 694 \text{ N}$$

$$R = \sqrt{(1604)^2 + (694)^2} = 1748 \text{ N} \qquad \alpha = \arctan\tfrac{694}{1604} = 23.4° \qquad \text{(see Fig. 23-26b)}$$

Fig. 23-26

**23.37** Find the magnitude and direction of the resultant force on the compressor shown in Fig. 23-27. Air ($\gamma = 0.075$ lb/ft$^3$) enters at $A$ through a 4-ft$^2$ area at a velocity of 12 fps. Air is discharged at $B$ through a 3-ft$^2$ area at a velocity of 14 fps.

$$F_x = \rho Q[(V_2)_x - (V_1)_x] \qquad Q = AV = (4)(12) = 48.0 \text{ ft}^3/\text{s}$$

$$F_x = (0.075/32.2)(48.0)(14 - 12\cos 50°) = 0.703 \text{ lb} \qquad \text{(rightward)}$$

$$F_y = \rho Q[(V_2)_y - (V_1)_y] = (0.075/32.2)(48.0)[0 - (-12\sin 50°)] = 1.028 \text{ lb} \qquad \text{(upward)}$$

Force of fluid on compressor is 0.703 lb leftward and 1.028 lb downward. $F = \sqrt{0.703^2 + 1.028^2} = 1.245$ lb, $\alpha = \arctan(1.028/0.703) = 55.6°$ to the horizontal.

**23.38** Repeat Prob. 23.37 for the case where a gas ($\gamma = 12.1$ N/m$^3$) enters at $A$ through a 75-cm-diameter pipe at 3.5 m/s and leaves at $B$ through a 60-cm-diameter pipe at 4.5 m/s.

Fig. 23-27

$$F_x = \rho Q[(V_2)_x - (V_1)_x] \qquad Q = AV = [(\pi)(0.75)^2/4](3.5) = 1.546 \text{ m}^3/\text{s}$$

$$F_x = (12.1/9.807)(1.546)(4.5 - 3.5 \cos 50°) = 4.29 \text{ N} \qquad (\text{rightward})$$

$$F_y = \rho Q[(V_2)_y - (V_1)_y] = (12.1/9.807)(1.546)[0 - (-3.5 \sin 50°)] = 5.11 \text{ N} \qquad (\text{upward})$$

Force of fluid on compressor is 4.29 N leftward and 5.11 N downward. $F = \sqrt{4.29^2 + 5.11^2} = 6.67$ N, $\alpha = \arctan(5.11/4.29) = 50.0°$ to the horizontal.

**23.39** In Fig. 23-28a is shown a curved pipe section of length 40 ft that is attached to the straight pipe section as shown. Determine the resultant force on the curved pipe, and find the horizontal component of the jet reaction. All significant data are given in the figure. Assume an ideal liquid with $\gamma = 55 \text{ lb/ft}^3$.

**I** The energy equation between sections 1 and 3 gives $[(30)(144)/55] + 35 = 20 + (V_3^2/2g)$ and $V_3 = 77.6$ fps (jet velocity); $Q = A_3 V_3 = 3.81$ cfs, $V_2 = Q/A_2 = 43.6$ fps. The energy equation between section 2 and 3 gives $10 + [p_2(144)/55] + [(43.6)^2/64.4] = 20 + [(77.6)^2/2g]$, and $p_2 = 28.3$ psi. The free-body diagram of the forces acting on the liquid contained in the curved pipe is shown in Fig. 23-28b. $p_2 A_2 - p_3 A_3 \cos 20° - (F_{N/L})_x = \rho Q(V_3 \cos 20° - V_2)$ where $(F_{N/L})_x$ represents the force of the curved pipe on the liquid in the $x$ direction. Since section 3 is a jet in contact with the atmosphere, $p_3 = 0$. Thus $28.3[(\pi/4)(4^2)] - (F_{N/L})_x = [(1.94)(55/62.4)](3.81)[(77.6)(0.94) - 43.6]$, $356 - (F_{N/L})_x = 191$, $(F_{N/L})_x = +165$ lb where the plus sign indicates that the assumed direction is correct. In the $y$ direction the $p_2 A_2$ force has no component. Estimating the weight of liquid $W$ as 150 lb, $(F_{N/L})_y = \rho Q[(77.6)(0.342) - 0] + 150 = +323$ lb. The resultant force of liquid on the curved pipe is equal and opposite to that of the curved pipe on liquid. The resultant force of liquid on the curved pipe is $[(165)^2 + (323)^2]^{1/2} = 363$ lb downward and to the right at an angle of 62°56′ with the horizontal.

The horizontal jet reaction is best found by taking a free-body diagram of the liquid in the system as shown in Fig. 23-28c: $(F_{S/L})_x = \rho Q(V_3 \cos 20° - 0) = 475$ lb where $(F_{S/L})_x$ respresents the force of the system on the liquid in the $x$ direction. $(F_{S/L})_x$ is equivalent to the integrated effect of the $x$ components of the pressure vectors shown in Fig. 23-28c. Equal and opposite to $(F_{S/L})_x$ is the force of the liquid on the system, i.e., the jet reaction. Hence the horizontal jet reaction is a 475-lb force to the left. Thus there is a 165 lb force to the right tending to separate the curved pipe section from the straight pipe section, while at the same time there is a 475-lb force tending to move the entire system to the left.

(a)

(b)

(c)

Fig. 23-28

**23.40** A 24-in pipe carrying 31.4 cfs of oil (s.g. = 0.85) has a 90° bend in a horizontal plane. The loss of head in the bend is 3.50 ft of oil and the pressure at entrance is 42.5 psi. Determine the resultant force exerted by the oil on the bend.

❚ Referring to Fig. 23-29, the free-body diagram indicates the static and dynamic forces acting on the mass of oil in the bend. These forces are calculated as follows:

1. $$P_1 = p_1 A = (42.5)(\tfrac{1}{4}\pi)(24)^2 = 19\,200\text{ lb}$$

2. $P_2 = p_2 A$, where $p_2 = p_1 -$ loss in psi, from the Bernoulli equation since $z_1 = z_2$ and $V_1 = V_2$. Then $P_2 = [42.5 - (0.85)(62.4)(3.50/144)](\tfrac{1}{4}\pi)(24)^2 = 18\,600$ lb.

3. Using the impulse-momentum principle, and knowing that $V_1 = V_2 = Q/A = 10.0$ ft/s, $MV_{x_1} + \Sigma$ (forces in $X$ direction)$(1) = MV_{x_2}$, $19\,200 - F_x = [(0.85)(62.4)(31.4/32.2)](0 - 10.0) = -517$ lb, and $F_x = +19\,720$ lb, to the left on the oil.

4. Similarly, for $t = 1$ s, $MV_{y_1} + \Sigma$ (forces in $Y$ direction)$(1) = MV_{y_2}$, $F_y - 18\,600 = [(0.85)(62.4)(31.4/32.2)](10.0 - 0) = +517$ lb, and $F_y = +19\,120$ lb, upward on the oil.

On the pipe bend, the resultant force $R$ acts to the right and downward and is $R = \sqrt{(19\,720)^2 + (19\,120)^2} = 27\,470$ lb at $\theta_x = \tan^{-1}(19\,120/19\,720) = 44.1°$.

**Fig. 23-29**

**23.41** The 24-in pipe of Prob. 23.40 is connected to a 12-in pipe by a standard *reducer* fitting. For the same flow of 31.4 cfs of oil, and a pressure of 40.0 psi, what force is exerted by the oil on the reducer, neglecting any lost head?

❚ Since $V_1 = 10.0$ fps, $V_2 = (\tfrac{2}{1})^2(10.0) = 40.0$ fps. Also, the Bernoulli equation between section 1 at entrance and section 2 at exit yields $\{(p_1/\gamma) + [(10)^2/2g] + 0\} -$ negligible lost head $= \{(p_2/\gamma) + [(40)^2/2g] + 0\}$. Solving, $p_2/\gamma = [(40.0)(144)/(0.85)(62.4)] + (100/2g) - (1600/2g) = 85.3$ ft of oil and $p_2' = 31.4$ psi.

Figure 23-30 represents the forces acting on the mass of oil in the reducer: $P_1 = p_1 A_1 = (40.0)(\tfrac{1}{4}\pi)(24)^2 = 18\,100$ lb (to the right), $P_2 = p_2 A_2 = (31.4)(\tfrac{1}{4}\pi)(12)^2 = 3550$ lb (to the left).

In the $X$ direction the momentum of the oil is changed. Then $MV_{x_1} + \Sigma$ (forces in $X$ direction)$(1) = MV_{x_2}$, $(18\,100 - 3550 - F_x)1 = [(0.85)(62.4)(31.4/32.2)](40.0 - 10.0)$, and $F_x = 13\,000$ lb acting to the left on the oil.

The forces in the $Y$ direction will balance each other and $F_y = 0$. Hence the force exerted by the oil on the reducer is 13 000 lb to the right.

**Fig. 23-30**

**23.42** A 45° reducing bend, 24-in diameter upstream, 12-in diameter downstream, has water flowing through it at the rate of 15.7 cfs under a pressure of 21.0 psi. Neglecting any loss in the bend, calculate the force exerted by the water on the reducing bend.

❚ $V_1 = 15.7/A_1 = 5.00$ ft/s and $V_2 = 20.0$ ft/s. The Bernoulli equation, section 1 to section 2, produces $\{[(21.0)(144)/62.4] + (25/2g) + 0\} -$ negligible lost head $= [(p_2/\gamma) + (400/2g) + 0]$ from which $p_2/\gamma = 42.6$ ft and $p_2' = 18.5$ psi.

In Fig. 23-31 is shown the mass of water acted upon by static and dynamic forces: $P_1 = p_1 A_1 = (21.0)(\tfrac{1}{4}\pi)(24)^2 = 9500$ lb, $P_2 = p_2 A_2 = (18.5)(\tfrac{1}{4}\pi)(12)^2 = 2090$ lb, $P_{2_x} = P_{2_y} = (2090)(0.707) = 1478$ lb.

In the $X$ direction, $MV_{x_1} + \Sigma$ (forces in $X$ direction)$(1) = MV_{x_2}$, $(9500 - 1478 - F_x)1 = [(62.4)(15.7/32.2)][(20.0)(0.707) - 5.00]$ and $F_x = 7740$ lb to the left.

In the $Y$ direction, $(+F_y - 1478)1 = [(62.4)(15.7/32.2)][(20.0)(0.707) - 0]$ and $F_y = 1910$ lb upward.

The force exerted by the water on the reducing bend is $F = \sqrt{(7740)^2 + (1910)^2} = 7970$ lb to the right and downward, at an angle $\theta_x = \tan^{-1} \frac{1910}{7740} = 13°52'$.

Fig. 23-31

**23.43** Determine the reaction of a jet flowing through an orifice on the containing tank.

▎ In Fig. 23-32 a mass of liquid $ABCD$ is taken as a free body. The only horizontal forces acting are $F_1$ and $F_2$, which change the momentum of the water. $(F_1 - F_2)(1) = M(V_2 - V_1)$, where $V_1$ can be considered negligible. Reaction $F = F_1 - F_2 = (\gamma Q/g)V_2 = (\gamma A_2 V_2/g)V_2$. But $A_2 = c_c A_0$ and $V_2 = c_v \sqrt{2gh}$. Hence $F = [\gamma(c_c A_0)/g]c_v^2(2gh) = (c c_v)\gamma A_0(2h)$ (to the right on the liquid).

1. For average values of $c = 0.60$ and $c_v = 0.98$, the reaction $F = 1.176\gamma h A_0$. Hence the force acting to the left on the tank is about 18 percent more than the static force on a plug which would just fill the orifice.

2. For ideal flow (no friction, no contraction), $F = 2(\gamma h A_0)$. This force is equal to twice the force on a plug which would just fill the orifice.

3. For a nozzle ($c_c = 1.00$), the reaction $F = c_v^2 \gamma A(2h)$ where $h$ would be the effective head causing the flow.

Fig. 23-32

**23.44** The jets from a garden sprinkler are 1-in in diameter and are normal to the 2-ft radius. If the pressure at the base of the nozzles is 50 psi, what force must be applied on each sprinkler pipe, 1 ft from the center of rotation, to maintain equilibrium? (Use $c_v = 0.80$ and $c_c = 1.00$.)

▎ The reaction of the sprinkler jet may be calculated from the momentum principle. Inasmuch as the force which causes a change in momentum in the $X$ direction acts along the $X$ axis, no torque is exerted. We are interested, therefore, in the change in momentum in the $Y$ direction. But the initial momentum in the $Y$ direction is zero. The jet velocity $V_Y = c_v \sqrt{2gh} = 0.80\sqrt{2g[(50)(2.31) + \text{negligible velocity head}]} = 69.0$ ft/s. Thus $F_Y \, dt = M(V_Y) = [(62.4/32.2)(\frac{1}{4}\pi)(\frac{1}{12})^2(69.0) \, dt](-69.0)$ or $F_Y = -50.3$ lb downward on the water. Hence the force of the jet on the sprinkler is $+50.3$ lb upward. Then $\Sigma M_0 = 0$, $F(1) - 2(50.3) = 0$, $F = 100.6$ lb for equilibrium.

Fig. 23-33

**23.45** A rocket engine burns 5.0 kg of fuel and oxidizer per second, and combustion gases exit the rocket at a velocity of 550 m/s relative to the rocket, at a pressure approximately equal to the ambient air pressure. Find the thrust produced by the rocket engine.

▎

$$F = \rho Q v_2 = M v = (5.0)(550) = 2750 \text{ N}$$

**23.46** A jet craft traveling at a velocity of 590 fps takes in air at a rate of 1.55 slugs/s. The air:fuel ratio is 25:1, and the exhaust velocity relative to the jet is 1950 fps. Find the thrust produced by the jet engine.

**▌** $$F = \rho_a Q_a(v_2 - v_1) + \rho_f Q_f v_2 \qquad \rho_a Q_a = M_a = 1.55 \text{ slugs/s}$$

Since the air:fuel ratio is 25:1, $\rho_f Q_f = 1.55/25 = 0.0620$ slug/s, $F = (1.55)(1950 - 590) + (0.0620)(1950) = 2229$ slug-ft/s². Since slug = lb-s²/ft, $F = 2229$ lb.

**23.47** A rocket engine burns 8.5 kg of fuel and oxidizer per second, and combustion gases exit the rocket at a velocity of 565 m/s relative to the rocket, at a pressure approximately equal to the surrounding atmospheric pressure. Find the thrust produced by the rocket engine.

**▌** $$F = \rho Q v_2 = Mv = (8.5)(565) = 4802 \text{ N}$$

**23.48** A jet craft traveling at a velocity of 700 fps takes in air at a rate of 1.68 slugs/s. The air:fuel ratio is 20:1, and the exhaust velocity relative to the jet is 2105 fps. Find the thrust produced by the jet engine.

**▌** $$F = \rho_a Q_a(v_2 - v_1) + \rho_f Q_f v_2 \qquad \rho_a Q_a = M_a = 1.68 \text{ slugs/s}$$

Since the air:fuel ratio is 20:1, $\rho_f Q_f = 1.68/20 = 0.0840$ slug/s, $F = (1.68)(2105 - 700) + (0.0840)(2105) = 2537$ slug-ft/s². Since slug = lb-s²/ft, $F = 2537$ lb.

**23.49** A jet engine is being tested in the laboratory. The engine consumes 50 lb of air per second and 0.5 lb of fuel per second. If the exit velocity of the gases is 1500 fps, what is the thrust?

**▌** $$F = W_{\text{exit}} V_2/g - W_1 V_1/g = (50 + 0.5)(1500)/32.2 - (50)(0)/32.2 = 2352 \text{ lb}$$

**23.50** A jet engine operates at 600 fps and consumes air at the rate of 50.0 lb/s. At what velocity should the air be discharged in order to develop a thrust of 1500 lb?

**▌** $$F = 1500 = (50.0/32.2)(V_{\text{exit}} - 600) \qquad V_{\text{exit}} = 1566 \text{ fps}$$

**23.51** A turbojet engine is tested in the laboratory under conditions simulating an altitude where atmospheric pressure is 785.3 lb/ft² abs, temperature is 429.5 °R, and specific weight is 0.0343 lb/ft³. If the exit area of the engine is 1.50 ft² and the exit pressure is atmospheric, what is the Mach number if the gross thrust is 1470 lb? Use $k = 1.33$.

**▌** $$F = W_{\text{exit}} V_{\text{exit}}/g = (\gamma A_{\text{exit}} V_{\text{exit}})(V_{\text{exit}})/g \qquad 1470 = (0.0343)(1.50)(V_{\text{exit}})^2/32.2 \qquad V_{\text{exit}} = 959 \text{ fps}$$
$$N_M = V_{\text{exit}}/\sqrt{kgRT} = 959/\sqrt{(1.33)(32.2)(53.3)(429.5)} = 0.969$$

**23.52** A rocket device burns its propellant at a rate of 15.2 lb/s. The exhaust gases leave the rocket at a relative velocity of 3220 fps and at atmospheric pressure. The exhaust nozzle has an area of 50.0 in² and the gross weight of the rocket is 500 lb. At the given instant, 2500 hp is developed by the rocket engine. What is the rocket velocity?

**▌** $$P = FV_{\text{rocket}} \qquad F = W_{\text{exit}} V_{\text{exit}}/g = (15.2)(3220)/32.2 = 1520 \text{ lb}$$
$$(2500)(550) = 1520 V_{\text{rocket}} \qquad V_{\text{rocket}} = 905 \text{ fps}$$

**23.53** A rocket motor is operating steadily, as shown in Fig. 23-34. The exhaust products approximate a perfect gas of molecular weight 26. Calculate the exit velocity.

**▌** $$\dot{m}_{\text{exit}} = \rho_{\text{exit}} A_{\text{exit}} V_{\text{exit}} = 0.6 + 0.12 = 0.72 \text{ slug/s} \qquad \rho = p/RT \qquad R = 49\,750/26 = 1913 \text{ ft-lb/slug-°R}$$
$$\rho_{\text{exit}} = (15)(144)/[(1913)(1000 + 460)] = 0.0007734 \text{ slug/ft}^3 \qquad A_{\text{exit}} = (\pi)(6/12)^2/4 = 0.1964 \text{ ft}^2$$
$$0.72 = (0.0007734)(0.1964)(V_{\text{exit}}) \qquad V_{\text{exit}} = 4740 \text{ fps}$$

**23.54** In contrast to the liquid rocket in Fig. 23-34, the solid-propellant rocket in Fig. 23-35 is self-contained and has no entrance ducts. Using a control-volume analysis for the conditions shown in Fig. 23-35, compute the rate of mass loss of the propellant, assuming that the exit gas has molecular weight 31.

**▌** $$\dot{m}_{\text{propellant}} = \rho_{\text{exit}} A_{\text{exit}} V_{\text{exit}} \qquad \rho = p/RT \qquad R = \tfrac{8313}{31} = 268 \text{ J/kg·K}$$
$$\rho_{\text{exit}} = (80)(1000)/[(268)(725)] = 0.4117 \text{ kg/m}^3 \qquad A_{\text{exit}} = (\pi)(0.200)^2/4 = 0.03142 \text{ m}^2$$
$$m_{\text{propellant}} = (0.4117)(0.03142)(1100) = 14.2 \text{ kg/s}$$

Fig. 23-34

Fig. 23-35

**23.55** For the rocket engine of Prob. 23.53, compute the thrust if atmospheric pressure is 15 psia.

❚
$$F = \dot{m}_{exit}V_{exit} = (0.72)(4740) = 3413 \text{ lb}.$$

**23.56** A rocket ship is moving at 8.5 km/s in outer space. It is desired to slow the rocket down to 8.4 km/s by firing a retrorocket forward. The retrorocket burns 8 kg/s of fuel and oxidizer at an exhaust velocity of 1.3 km/s relative to the rocket. If the initial mass of the rocket ship is 1550 kg, how long should the retrorocket burn, and how much fuel will be burned?

❚
$$\int_{V_0}^{V_{final}} dV = \dot{m}V_{exit}\int_0^{t_{burn}} \frac{dt}{m_0 - \dot{m}t} \qquad V_{final} - V_0 = V_{exit}\ln\left(1 - \frac{\dot{m}t_{burn}}{m_0}\right) \qquad 8.4 - 8.5 = 1.3\ln\left(1 - \frac{8t_{burn}}{1550}\right)$$

$$t_{burn} = 14.3 \text{ s} \qquad m_{burned} = (8)(14.3) = 114 \text{ kg}$$

**23.57** Determine the magnitude and direction of the resultant force exerted on the double nozzle of Fig. 23-36. The axes of the pipes and both nozzles lie in a horizontal plane. Both nozzle jets have a water velocity of 12 m/s.

❚ **Continuity:**

$$A_1V_1 = A_2V_2 + A_3V_3 \qquad [(\pi)(0.150)^2/4](V_1) = [(\pi)(0.100)^2/4](12) + [(\pi)(0.075)^2/4](12)$$

$$V_1 = 8.33 \text{ m/s} \qquad Q_1 = A_1V_1 = [(\pi)(0.150)^2/4](8.33) = 0.147 \text{ m}^3/\text{s}$$

$$Q_2 = [(\pi)(0.100)^2/4](12) = 0.094 \text{ m}^3/\text{s} \qquad Q_3 = [(\pi)(0.075)^2/4](12) = 0.053 \text{ m}^3/\text{s}$$

**Energy equation:**

$$(p_1/\gamma) + [8.33^2/2(9.81)] = 0 + [12^2/2(9.81)] \qquad p_1/\gamma = 3.80 \text{ m} \qquad p_1 = 37.3 \text{ kPa} \qquad p_1A_1 = 0.659 \text{ kN}$$

$$\sum F_x = p_1A_1 - (F_{N/L})_x = (\rho Q_2 V_{2_x} + \rho Q_3 V_{3_x}) - \rho Q_1 V_{1_x} \qquad \rho = 10^3 \text{ kg/m}^3$$

$$V_{2_x} = V_2\cos 15° = 12(0.966) = 11.6 \text{ m/s} \qquad V_{3_x} = V_3\cos 30° = 12(0.866) = 10.4 \text{ m/s} \qquad V_{1_x} = V_1 = 8.33 \text{ m/s}$$

$$0.659 - (F_{N/F})_x = 10^3(0.094)(11.6) + 10^3(0.053)(10.4) - 10^3(0.147)8.33 = 0.417 \text{ kN}$$

$$(F_{N/L})_x = 0.659 - 0.417 = 0.242 \text{ kN}$$

$$\sum F_y = (F_{N/L})_y = (\rho Q_2 V_{2_y} + \rho Q_3 V_{3_y}) - \rho Q_1 V_{1_y} \qquad V_{2_y} = V_2\sin 15° = 12(0.259) = 3.1 \text{ m/s}$$

$$V_{3_y} = -V_3\sin 30° = -12(0.50) = -6.0 \text{ m/s} \qquad V_{1_y} = 0$$

$$(F_{N/L})_y = 10^3(0.094)(3.1) + 10^3(0.053)(-6.0) - 10^3(1.47)(0) = -0.027 \text{ kN}$$

Free-body diagram

**Fig. 23-36**

The minus sign indicates that the assumed direction of $(F_{N/L})_y$ was wrong. Therefore $(F_{N/L})_y$ acts in the negative $y$ direction. Equal and opposite to $F_{N/L}$ is $F_{L/N}$: $(F_{L/N})_x = 0.242$ kN (in positive $x$ direction), $(F_{L/N})_y = 0.027$ kN (in positive $y$ direction).

**23.58** In Fig. 23-37 suppose that $\theta = 30°$, $V_1 = 100$ fps, and the stream is a jet of water with an initial diameter of 2 in. Assume friction losses such that $V_2 = 95$ fps. Find the resultant force on the blade. Assume that flow occurs in a horizontal plane.

**❙** This problem is best solved by taking a free-body diagram of the element of fluid in contact with the blade. The forces acting on this element are as shown in the sketch. The forces $(F_{B/W})_x$ and $(F_{B/W})_y$ represent the components of force of blade on water in the $x$ and $y$ directions. These forces include shear stresses tangential to the blade as well as pressure forces normal to the blade. $A = 0.0218$ ft$^2$, $-(F_{B/W})_x = \rho Q(V_{2_x} - V_{1_x}) = 1.94(0.0218)(100)[(0.866)(95) - 100] = 4.23(-17.7) = -74.9$ lb. Hence, $(F_{B/W})_x = +74.9$ lb. The plus sign indicates that the assumed direction of $(F_{B/W})_x$ was correct. $+(F_{B/W})_y = \rho Q(V_{2_y} - V_{1_y}) = 4.23[(0.50)(95) - 0] = 201$ lb. Thus the force of the blade on the fluid is the resultant of a 74.9 lb component to the left and a 201-lb component upward in the $y$ direction. Equal and opposite to this is the force of the fluid on the blade (downward and to the right). The resultant force is 215 lb at an angle of 69.6° below the horizontal.

**Fig. 23-37**

**23.59** On the right end of a 9-in diameter pipe is a nozzle which discharges a 3-in-diameter water jet into the air. The pressure in the pipe is 60 psi, and the pipe velocity is 9.9 fps. Compute the axial force on the nozzle and the head loss in the nozzle.

**❙**
$$\sum F = \rho Q(v_2 - v_1) \qquad Q = A_1 v_1 = [(\pi)(\tfrac{9}{12})^2/4](9.9) = 4.374 \text{ ft}^3/\text{s} \qquad v_2 = (\tfrac{9}{3})^2(9.9) = 89.1 \text{ fps}$$

$$-F + (60)(144)[(\pi)(\tfrac{9}{12})^2/4] = (1.94)(4.374)(89.1 - 9.9) \qquad F = 3145 \text{ lb}$$

$$h_L = p_1/\gamma + v_1^2/2g - v_2^2/2g = (60)(144)/62.4 + 9.9^2/[(2)(32.2)] - 89.1^2/[(2)(32.2)] = 16.7 \text{ ft}$$

**23.60** A jet of steam hits the blades of a steam turbine at 3700 fps (absolute speed); the exit speed is 2500 fps. $\alpha_1 = 20°$, $\alpha_2 = 150°$, $u_1 = u_2 = 475$ fps, and $r_1 = r_2 = 0.6$ ft. Find the torque exerted on the rotor and the power delivered to it, if $G = 0.5$ lb/s.

$T = \rho Q(r_1 V_1 \cos \alpha_1 - r_2 V_2 \cos \alpha_2) = (0.5/32.2)[(0.6)(3700)(\cos 20°) - (0.6)(2500)(\cos 150°)] = 52.6$ ft-lb

$$P = FV = T\omega = (52.6)(475/0.6)/550 = 75.7 \text{ hp}$$

**23.61** When a turbine runner is held fixed, the discharge under a head of 44 ft is 30 cfs. $\alpha_1 = 35°$, $\beta_2 = 155°$, $r_1 = 0.75$ ft, $r_2 = 0.45$ ft, $A_1 = 0.850$ ft$^2$, $A_2 = 0.900$ ft$^2$. Find the zero-speed torque? Neglect shock loss.

$$T = \rho Q(r_1 V_1 \cos \alpha_1 - r_2 V_2 \cos \beta_2) \qquad \rho Q = (30)(62.4)/32.2 = 58.14 \text{ slugs/s}$$

$$V_1 = Q/A_1 = 30/0.850 = 35.29 \text{ fps} \qquad v_2 = 30/0.900 = 33.33 \text{ fps}$$

For stopped runner, $V_2 = v_2$ and $\alpha_2 = \beta_2 = 155°$. $T = (58.14)[(0.75)(35.29)(\cos 35°) - (0.45)(33.33)(\cos 155°)] = 1274$ ft-lb.

**23.62** A 24-in-diameter fan drives air ($\gamma = 0.076$ lb/ft$^3$) at a rate of 1.90 lb/s. Find the horsepower required to drive the fan. Neglect losses.

$$\gamma Q = \gamma A(\Delta V/2) \qquad 1.90 = (0.076)[(\pi)(\tfrac{24}{12})^2/4](\Delta V/2) \qquad \Delta V = 15.92 \text{ fps}$$

$$F = \rho Q(\Delta V) = (1.90/32.2)(15.92) = 0.939 \text{ lb} \qquad P = FV = F(\Delta V/2) = (0.939)(15.92/2)/550 = 0.0136 \text{ hp}$$

**23.63** A fan sucks air ($\rho = 0.0022$ slug/ft$^3$) into a building through a 20-in-diameter duct. If the pressure difference across the two sides of the fan is 3.9 inH$_2$O, determine the flow rate of the air. What thrust must the fan support be designed to withstand?

$$F = A(\Delta p) = [(\pi)(\tfrac{20}{12})^2/4](62.4)(3.9/12) = 44.2 \text{ lb} \qquad Q = A(\Delta V/2)$$

$$F = \rho Q(\Delta V) = \rho[A(\Delta V/2)](\Delta V) \qquad 44.2 = (0.0022)\{[(\pi)(\tfrac{20}{12})^2/4]/2\}(\Delta V)^2$$

$$\Delta V = 135.7 \text{ fps/s} \qquad Q = [(\pi)(\tfrac{20}{12})^2/4](135.7/2) = 148 \text{ ft}^3/\text{s}$$

**23.64** Find the thrust and efficiency of two 6.5-ft-diameter propellers through which flows a total of 20 000 cfs of air (0.072 lb/ft$^3$). The propellers are attached to an airplane moving at 150 mph through still air. Find also the pressure rise across the propellers and the horsepower input to each propeller. Neglect eddy losses.

Velocity of air relative to airplane is $V_1 = 150$ mph $= (150)(44)/30 = 220$ fps. Velocity of air through the actuating disk is $V = V_1 + (\Delta V/2) = Q/A = (20\,000/2/[(\pi/4)(6.5)^2] = 301$ fps. Thus $\Delta V = 2(301 - 220) = 162$ fps, $F_T = \rho Q\,\Delta V = (0.072/32.2)(20\,000)(162) = 7240$ lb (total thrust of both propellers), $\eta = 1/[1 + \Delta V/2V_1] = 1/(1 + \tfrac{162}{440}) = 0.73 = 73$ percent. $F_T$ on one propeller $= \tfrac{7240}{2} = 3620$ lb. But $F_T = (\Delta p)(A)$, thus $3620 = \Delta p(\pi/4)(6.5)^2$, $\Delta p = 109$ psf $= 0.758$ psi, hp/propeller $= \gamma Q(\Delta p/\gamma)/550 = Q(\Delta p)/550 = 10\,000(109)/550 = 1980$. Check: hp/propeller $= F_T(V_1 + \Delta V/2)/550 = 3620(301)/550 = 1980$.

**23.65** This water passage (Fig. 23-38) is 10 ft wide normal to the figure. Determine the horizontal force acting on the shaded structure. Assume ideal flow.

In free-surface flow such as this where the streamlines are parallel, the water surface is coincident with the hydraulic grade line. Writing an energy equation from the upstream section to the downstream section,

$$6 + (V_1^2/2g) = 3 + (V_2^2/2g) \tag{1}$$

From continuity,

$$6(10)V_1 = 3(10)V_2 \tag{2}$$

Substituting Eq. (2) into Eq. (1) yields $V_1 = 8.02$ fps, $V_2 = 16.04$ fps, $Q = A_1 V_1 = A_2 V_2 = 481$ cfs. Next take a free-body diagram of the element of water shown in the figure $F_1 - F_2 - F_x = \rho Q(V_2 - V_1)$, where $F_x$ represents the force of the structure on the water in the horizontal direction. $F_1 = \gamma h_{c_1} A_1$ and $F_2 = \gamma h_{c_2} A_2$. Hence $62.4(3)(10)(6) - 62.4(1.5)(10)(3) - F_x = 1.94(481)(16.04 - 8.02)$ and $F_x = 940$ lb. The positive sign means that the assumed direction is correct. Hence the force of the water on the structure is equal and opposite, namely, 940 lb to the right.

The momentum principle will not permit one to obtain the vertical component of the force of the water on the shaded structure because the pressure distribution along the bottom of the channel is unknown. The pressure distribution along the boundary of the structure and along the bottom of the channel can be estimated by sketching a flow net and applying Bernoulli's principle. The horizontal and vertical components of the force can be found by computing the integrated effect of the pressure-distribution diagram.

Fig. 23-38

**23.66** Flow occurs over a spillway, as shown in Fig. 23-39. Determine the horizontal force on the spillway per foot of width. Assume ideal flow.

▮ $\sum F = \rho Q(v_2 - v_1)$. Energy consideration: $y_1 + v_1^2/2g = y_2 + v_2^2/2g$, $5 + v_1^2/[(2)(32.2)] = 0.9 + v_2^2/[(2)(32.2)]$. Continuity consideration: $y_1 v_1 = y_2 v_2$, $5v_1 = 0.9v_2$, $v_1 = 0.1800v_2$.

$5 + (0.1800v_2)^2/[(2)(32.2)] = 0.9 + v_2^2/[(2)(32.2)]$ $\qquad v_2 = 16.52$ fps $\qquad v_1 = (0.1800)(16.52) = 2.973$ fps

$(62.4)(5)(\frac{5}{2}) - (62.4)(0.9)(0.9/2) - F_x = (1.94)[(5)(2.973)](16.52 - 2.973)$ $\qquad F_x = 364$ lb/ft $\qquad$ (rightward)

Fig. 23-39

# CHAPTER 24
# Dynamic Drag and Lift

**24.1**    A pitcher throws a baseball through air at 50 °F at a velocity of 92 mph. The diameter of the baseball is 2.82 in. Find the drag force on the baseball.

▐    $F_D = C_D\rho(v^2/2)A$    $v = (92)(\frac{5280}{3600}) = 134.9$ ft/s    $N_R = Dv/v = (2.82/12)(134.9)/(1.52 \times 10^{-4}) = 2.09 \times 10^5$

From Fig. A-50, $C_D = 0.42$. $F_D = (0.42)(0.00242)(134.9^2/2)[(\pi)(2.82/12)^2/4] = 0.401$ lb.

**24.2**    The pole of Fig. 24-1 is a cylinder 100 mm in diameter. Wind is blowing against it with a velocity of 15 m/s, and air temperature is 30 °C. Find the bending moment about the base (at ground level). Neglect end effects.

▐              $F_D = C_D\rho(v^2/2)A$       $N_R = Dv/v = (0.100)(15)/(1.60 \times 10^{-5}) = 9.38 \times 10^4$

From Fig. A-51, $C_D = 1.3$. $F_D = (1.3)(1.16)(15^2/2)[(17)(0.100)] = 288$ N. Wind acts uniformly along the pole; hence, the resultant drag force acts at the pole's midpoint as shown in Fig. 24-1. The bending moment about the base is, therefore, $M_{base} = (288)(8.5) = 2448$ N · m.

**Fig. 24-1**

**24.3**    A flat plate, 0.1 m by 0.1 m, moves at a velocity of 5.0 m/s perpendicular to the plate. Find the drag force on the plate if it is moving through (**a**) still air ($\rho = 1.200$ kg/m³) and (**b**) still water ($\rho = 1000$ kg/m³).

▐    $F_D = C_D\rho(v^2/2)A$. From Fig. A-52 with $x/y = 0.1/0.1 = 1$, $C_D = 1.1$.

(**a**)                            $F_D = (1.1)(1.200)(5.0^2/2)[(0.1)(0.1)] = 0.165$ N

(**b**)                            $F_D = (1.1)(1000)(5.0^2/2)[(0.1)(0.1)] = 138$ N

**24.4**    A smooth, flat plate 8.0 ft wide and 80.0 ft long is being towed lengthwise (parallel to the plate) through still water at 70 °F at a velocity of 17.0 fps. What is the drag force on one side of the plate (i.e., the skin-friction drag)?

▐              $F_D = C_D\rho(v^2/2)A$    $N_R = Dv/v = (80.0)(17.0)/(1.05 \times 10^{-5}) = 1.30 \times 10^8$

From Fig. A-53, $C_D = 0.0020$. $F_D = (0.0020)(1.93)(17.0^2/2)[(80.0)(8.0)] = 357$ lb.

**24.5**    A basketball is thrown through air at 50 °F at a velocity of 27 mph. The diameter of the basketball is 9.3 in. Determine the drag force on the basketball.

▐    $F_D = C_D\rho(v^2/2)A$       $v = (27)(\frac{5280}{3600}) = 39.60$ ft/s    $N_R = Dv/v = (9.3/12)(39.60)/(1.52 \times 10^{-4}) = 2.02 \times 10^5$

From Fig. A-50, $C_D = 0.41$. $F_D = (0.41)(0.00242)(39.60^2/2)[(\pi)(9.3/12)^2/4] = 0.367$ lb.

**24.6**    A silo with a height of 10 m above ground has the shape of a cylinder 3.5 m in diameter. Wind is blowing against the silo at a velocity of 8.3 km/h when air temperature is 30 °C. Determine the bending moment about the base of the silo.

▮ $F_D = C_D\rho(v^2/2)A$     $v = (8.3)(\frac{1000}{3600}) = 2.306$ m/s     $N_R = Dv/v = (3.5)(2.306)/(1.60 \times 10^{-5}) = 5.04 \times 10^5$

From Fig. A-51, $C_D = 0.35$. $F_D = (0.35)(1.16)(2.306^2/2)[(3.5)(10)] = 37.8$ N, $M_{base} = (37.8)(\frac{10}{2}) = 189$ N-m.

**24.7**    A flat plate, 0.5 m by 2.0 m, moves at a velocity of 3.5 m/s perpendicular to the plate. Find the drag force on the plate if it is moving through (**a**) still air ($\rho = 1.200$ kg/m³) and (**b**) still water ($\rho = 1000$ kg/m³).

▮ $F_D = C_D\rho(v^2/2)A$. From Fig. A-52 with $x/y = 2.0/0.5 = 4.0$, $C_D = 1.2$.

(**a**)                      $F_D = (1.2)(1.200)(3.5^2/2)[(0.5)(2.0)] = 8.82$ N

(**b**)                      $F_D = (1.2)(1000)(3.5^2/2)[(0.5)(2.0)] = 7350$ N

**24.8**    Wind with a velocity of 35 mph is blowing perpendicularly against a 14-ft by 10-ft highway sign. If the mass density of the air is 0.00234 slug/ft³, compute the total drag force on the sign.

▮ $F_D = C_D\rho(v^2/2)A$. From Fig. A-52 with $x/y = 14/10 = 1.4$, $C_D = 1.12$. $v = (35)(\frac{5280}{3600}) = 51.33$ ft/s, $F_D = (1.12)(0.00234)(51.33^2/2)[(14)(10)] = 483$ lb.

**24.9**    A highway sign is illustrated in Fig. 24-2. Assuming the sign can be analyzed as a flat plate, determine the wind velocity required to overturn the sign if it was designed to resist a total moment of 6000 lb-ft at its base. Use a mass density of air of 0.00234 slug/ft³.

▮ $F_D = C_D\rho(v^2/2)A$. From Fig. A-52 with $x/y = \frac{15}{9} = 1.67$, $C_D = 1.16$.

$$F_D = (1.16)(0.00234)(v^2/2)[(15)(9)] = 0.1832v^2 \qquad \sum M_{base} = 0$$

$$6000 - (6 + \tfrac{9}{2})(0.1832v^2) = 0 \qquad v = 55.85 \text{ ft/s} \quad \text{or} \quad 38.1 \text{ mph}$$

Fig. 24-2

**24.10**    A smooth, flat plate 2.0 m wide and 10.0 m long is being towed lengthwise through still water at 20 °C at a velocity of 5.0 m/s. What is the drag force on one side of the plate (i.e., the skin-friction drag)?

▮           $F_D = C_D\rho(v^2/2)A$       $N_R = Dv/v = (10.0)(5.0)/(1.02 \times 10^{-6}) = 4.90 \times 10^7$

From Fig. A-53, $C_D = 0.0023$. $F_D = (0.0023)(998)(5.0^2/2)[(10.0)(2.0)] = 574$ N.

**24.11**    Find the friction drag on one side of a smooth flat plate 6 in wide and 18 in long, placed longitudinally in a stream of crude oil (s.g. $= 0.925$) at 60 °F flowing with undistributed velocity of 2 fps.

▮ From Fig. A-2, $v = 0.0010$ ft²/s. Then, at $x = L$, $N_R = LU/v = (1.5)(2)/0.0010 = 3000$, well within the laminar range; that is, $N_R < 500\,000$. $C_f = 1.328/\sqrt{N_R} = 1.328\sqrt{3000} = 0.0242$, $F_f = C_f\rho(V^2/2)BL = (0.0242)(0.925)(1.94)(2^2)(6)(18)/[(2)144] = 0.065$ lb. Find the thickness of the boundary layer and the shear stress at the trailing edge of the plate. $\delta/x = 4.91/\sqrt{N_R} = 4.91/\sqrt{3000} = 0.0896$, $\delta = (0.0896)(1.5) = 0.1344$ ft $= 1.61$ in. At $x = L$, $\tau_0 = 0.332(v\rho U/L)\sqrt{N_R} = (0.332)[(0.0010)(0.925)(1.94)(2)/1.5]\sqrt{3000} = 0.0435$ lb/ft².

**24.12** (*a*) Find the frictional drag on the top and sides of a box-shaped moving van 8 ft wide, 10 ft high, and 35 ft long, traveling at 60 mph through air ($\gamma = 0.0725$ lb/ft³) at 50 °F. Assume that the vehicle has a rounded nose so that the flow does not separate from the top and sides (see Fig. A-54*b*). Assume also that even though the top and sides of the van are relatively smooth there is enough roughness so that for all practical purposes a turbulent boundary layer starts immediately at the leading edge. (*b*) Find the thickness of the boundary layer and the shear stress of the trailing edge.

▮ (*a*) From Fig. A-2, for air at 50 °F, $v = 0.00015$ ft²/s. Then $N_R = LU/v = (35)(88)/0.00015 = 20\,530\,000$. As $N_R > 10^7$, $C_f = 0.455/(\log N_R)^{2.58} = 0.455/(7.31)^{2.58} = 0.00269$, $F_f = C_f\rho(V^2/2)BL =$ $(0.00269)(0.0725/32.2)[(88)^2/2](10 + 8 + 10)35 = 23.0$ lb.

(*b*) $$\delta/x = 0.377/N_R^{1/5} \qquad \delta_{35} = (35)(0.377)/[(205.3)^{1/5} \times 10] = 0.455 \text{ ft}$$
$$\tau_0 = 0.0587\rho(U^2/2)(v/Ux)^{1/5} \qquad (\tau_0)_{35} = (0.0587)(0.0725/32.2)(88^2/2)[0.00015/(88)(35)]^{1/5} = 0.0176 \text{ lb/ft}^2$$

**24.13** A small submarine, which may be supposed to approximate a cylinder 10 ft in diameter and 50 ft long, travels submerged at 3 knots (5.06 fps) in sea water at 40 °F. Find the friction drag assuming no separation from the sides.

▮ $$F_f = C_f\rho(V^2/2)A \qquad N_F = DV/v$$

Kinematic viscosity of sea water ≈ kinematic viscosity of fresh water = 0.0000166 ft²/s. Then $N_F = (50)(5.06)/0.0000166 = 1.52 \times 10^7$. From Fig. A-55, $C_f = 0.0028$ and $F_f = (0.0028)(64/32.2)[(5.06)^2/2](\pi)(10)(50) = 112$ lb. Find the value of the critical roughness for a point 1 ft from the nose of the submarine.

At $x = 1$ ft, $(N_F) = (5.06)(1)/0.0000166 = 305\,000$, $e_c = 26(v/V)(N_R)_x^{1/4} = [(26)(0.0000166)/5.06](305\,000)^{1/4} = 0.0020$ ft. Find the height of roughness at the midsection of the submarine which would class the surface as truly rough.

At $x = 25$ ft, $(N_R)_x = (5.05)(25)/0.0000166 = 7.61 \times 10^6$, $\tau_0/\rho = 0.0587(V^2/2)(v/Vx)^{1/5} = (0.0587)[(5.06)^2/2][1/(76.2 \times 10^5)^{1/5}] = 0.0316$ ft²/s², $\delta_t = 60v/\sqrt{\tau_0/\rho} = (60)(0.0000166)/\sqrt{0.0316} = 0.0056$ ft.

**24.14** Using the data of Prob. 24.12 determine the total drag exerted by the air on the van. Assume that $C_D \approx 0.45$ (see Fig. A-54).

▮ $$F_D = C_D\rho(V^2/2)A = 0.45(0.0725/32.2)[(88)^2/2](8)(10) = 314 \text{ lb}$$

Thus the pressure drag = 314 − 23 = 291 lb; in this case the pressure drag is responsible for about 93 percent of the total drag while the friction drag comprises only 7 percent of the total.

**24.15** Find the "free-fall" velocity of an 8.5-in-diameter sphere (bowling ball) weighing 16 lb when falling through the following fluids under the action of gravity: through the standard atmosphere at sea level; through the standard atmosphere at 10 000-ft elevation; through water at 60 °F; through crude oil (s.g. = 0.925) at 60 °F.

▮ When first released the sphere will accelerate (Fig. 24-3*a*) because the forces acting on it are out of balance. This acceleration results in a buildup of velocity which causes an increase in the drag force. After a while the drag force will increase to the point where the forces acting on the sphere are in balance, as indicated in Fig. 24-3*b*. When that point is reached the sphere will attain a constant or terminal (free-fall) velocity. Thus for free-fall conditions, $\sum F_x = W - F_B - F_D = (\text{mass})(\text{acceleration}) = 0$ where $W$ is the weight, $F_B$ the buoyant force, and $F_D$ the drag force. The buoyant force is equal to the unit weight of the fluid multiplied by the volume ($\pi D^3/6 = 0.186$ ft³) of the sphere. The given data are approximately as follows:

| fluid | $\gamma$, lb/ft³ | $\rho$, slugs/ft³ | $v = \mu/\rho$ ft²/s | $F_B$, lb |
|---|---|---|---|---|
| Air (sea level) | 0.0765 | 0.00238 | $1.57 \times 10^{-4}$ | 0.0142 |
| Air (10 000 ft) | 0.0564 | 0.00176 | $2.01 \times 10^{-4}$ | 0.0105 |
| Water, 60 °F | 62.4 | 1.94 | $1.22 \times 10^{-5}$ | 11.6 |
| Oil, 60 °F | 57.7 | 1.79 | 0.001 | 10.7 |

The detailed analysis for the sphere falling through the standard sea-level atmosphere is as follows: $16 - 0.0142 - C_D\rho(V^2/2)A = 0$ where $\rho = 0.00238$ slug/ft³ and $A = \pi(8.5/12)^2/4 = 0.394$ ft² or $15.986 = C_D(0.00238)(V^2/2)(0.394) = 0.00047C_DV^2$. A trial-and-error solution is required. Let $C_D = 0.2$, then $V = 412$ fps. $N_R = DV/v = (8.5/12)412/(1.57 \times 10^{-4}) = 1.86 \times 10^6$. The values of $C_D$ and $N_R$ check, hence $C_D = 0.2$ and $V = 412$ fps.

Following a similar procedure for the other three fluids gives the following free-fall velocities:

| fluid | $C_D$ | $N_R$ | $V_{fall}$ |
|---|---|---|---|
| Standard atmosphere at 10 000 ft | 0.20 | $1.69 \times 10^6$ | 480 fps |
| Water at 60 °F | 0.19 | 453 000 | 7.4 fps* |
| Crude oil (s.g. = 0.925) at 60 °F | 0.39 | 4 390 | 6.2 fps |

* In this instance the Reynolds number is 453 000 which, for the case of a sphere, generally indicates a turbulent boundary layer (Fig. A-56). This is very close to the point where the boundary layer changes from laminar to turbulent. If the water had been at a somewhat lower temperature and, hence, more viscous, a laminar boundary layer might have been present, in which case the free-fall velocity would have been only about 5.2 fps.

Fig. 24-3

**24.16** Find the critical Reynolds number (indicating transition from laminar to turbulent boundary layer) for a circular pipe, given the value 480 000 for a flat plate.

▌ For the plate, $\delta/x = 4.91/\sqrt{(N_R)_x} = 4.91/\sqrt{480\,000} = 0.00709$; therefore, $N_R = 480\,000 = Ux/\nu = U\delta/0.00709\nu$. Imagine the plate rolled up to form a circular pipe of radius $R$: the critical conditions become $U \to U_{centerline} = 2V$ ($V$ = cross-sectional average velocity) and $\delta = R = D/2$. Hence, for unchanged $\nu$, $480\,000 = (2V)(D/2)/0.00709\nu$ or $DV/\nu = (N_R)_{pipe} = (480\,000)(0.00709) = 3403$.

**24.17** Refer to the data of Prob. 24.12. Find the shear stress on the sides of the van at 4, 12, and 20 ft back from the front edge of the sides.

▌ $\tau_0 = 0.0587\rho(U^2/2)(\nu/Ux)^{1/5}$. At $x = 4$ ft, $\tau_0 = (0.0587)(0.0725/32.2)(88^2/2)\{0.00015/[(88)(4)]\}^{1/5} = 0.0272$ lb/ft². At $x = 12$ ft, $\tau_0 = (0.0587)(0.0725/32.2)(88^2/2)\{0.00015/[(88)(12)]\}^{1/5} = 0.0219$ lb/ft². At $x = 20$ ft, $\tau_0 = (0.0587)(0.0725/32.2)(88^2/2)\{0.00015/[(88)(20)]\}^{1/5} = 0.0197$ lb/ft².

**24.18** If the wind velocity 6 ft above a soccer field is 18 fps, approximately what are the wind velocities 0.4 ft and 0.8 ft above the field?

▌ $u/U = (y/\delta)^{1/7}$. At 0.4 ft, $u/18 = (0.4/6)^{1/7}$, $u = 12.2$ fps. At 0.8 ft, $u/18 = (0.8/6)^{1/7}$, $u = 13.5$ fps.

**24.19** A metal ball of diameter 1.2 ft and weight 99 lb is dropped into the ocean. Determine the maximum velocity the ball will achieve if, for seawater, $\rho = 2.0$ slugs/ft³ and $\mu = 3.3 \times 10^{-5}$ lb-s/ft².

▌
$$F_D = W - F_b = C_D\rho(V^2/2)A \qquad W - (\gamma)(\pi D^3/6) = C_D\rho(V^2/2)(\pi D^2/4)$$
$$99 - [(62.4)(2.0/1.94)](\pi)(1.2)^3/6 = C_D(2.0)(V^2/2)(\pi)(1.2)^2/4 \qquad 99 - 58.20 = 1.131C_DV^2 \qquad C_DV^2 = 36.07$$
$$N_R = DV\rho/\mu = (1.2)(V)(2.0)/(3.3 \times 10^{-5}) = 72\,727V$$

Try $C_D = 0.4$; $0.4V^2 = 36.07$, $V = 9.496$ ft/s, $N_R = (72\,727)(9.496) = 6.91 \times 10^5$. From Fig. A-56, $C_D = 0.20$. Try $C_D = 0.20$: $0.20V^2 = 36.07$, $V = 13.43$ ft/s, $N_R = (72\,727)(13.43) = 9.77 \times 10^5$, $C_D = 0.20$ (O.K.). Hence, $V = 13.43$ fps.

**24.20** Two 5-ft-diameter bodies—the airship hull of Fig. A-56 and a 4:3 oblate ellipsoid—move at 80.7 ft/s through air at STP. For each body, find the power required to overcome air resistance.

▮ $\qquad P = F_D V \qquad F_D = C_D \rho (V^2/2) A \qquad N_R = DV/\nu = (5)(80.7)/0.0001564 = 2.58 \times 10^6$

Assuming from Fig. A-56 that $C_D$ remains constant for $N_R > 10^6$, $C_D$ for airship hull = 0.04 and $C_D$ for 1:0.75 ellipsoid = 0.20. Thus, $(F_D)_{\text{hull}} = (0.04)(0.002378)(80.7^2/2)[(\pi)(5)^2/4] = 6.08$ lb, $P_{\text{hull}} = (6.08)(80.7)/550 = 0.892$ hp, and $(F_D)_{\text{ellip.}} = (0.20)(0.002378)(80.7^2/2)[(\pi)(5)^2/4] = 30.4$ lb, $(P)_{\text{ellip.}} = (30.4)(80.7)/550 = 4.46$ hp.

**24.21** The drag coefficient for a hemispherical shell with the concave side upstream is approximately 1.33 if $N_R > 1000$. Design a parachute that will land a 100-kg Marine as though he had jumped (parachuteless) off a 3.0-m wall. Assume standard air at sea level and neglect the parachute's weight.

▮ In jump from 3.0-m wall,

$$V = \sqrt{2gH} = \sqrt{(2)(9.80)(3.0)} = 7.67 \text{ m/s} \qquad F_D = C_D \rho (V^2/2) A$$

$$(100)(9.80) = (1.33)(1.225)(7.67^2/2)(\pi D^2/4) \qquad D = 5.10 \text{ m}$$

$$N_R = DV/\nu = (5.10)(7.67)/(1.455 \times 10^{-5}) = 2.69 \times 10^6 \qquad \text{(O.K.)}$$

**24.22** A football, 6.78 in in diameter and 14.5 oz in weight, can be modeled by the prolate ellipsoid of Fig. A-56. Find the resistance when the ball is passed through still air (14.7 psia and 40 °F) at a velocity of 35 fps. Neglect spin.

▮ $\qquad F_D = C_D \rho (V^2/2) A \qquad N_R = DV/\nu = (6.78/12)(35)/(1.46 \times 10^{-4}) = 1.35 \times 10^5$

From Fig. A-56, $C_D = 0.065$, $F_D = (0.065)(0.00247)(35^2/2)[(\pi)(6.78/12)^2/4] = 0.0247$ lb.

**24.23** Repeat Prob. 24.22, if the air temperature is 100 °F.

▮ $F_D = C_D \rho (V^2/2) A$. At $T = 100$ °F, $\rho = 0.00220$ slug/ft³ (from Table A-4). $F_D = (0.065)(0.00220)(35^2/2)[(\pi)(6.78/12)^2/4] = 0.0220$ lb. [The air is rarer at the higher temperature.]

**24.24** At sea level, a 3.2-m-diameter spherical balloon travels at 22 m/s. At a 3 km altitude, what speed will give the same drag force? Let $C_D$ be the same at both altitudes.

▮ $F_D = C_D \rho (V^2/2) A$, whereby $V \propto 1/\sqrt{\rho}$. Hence

$$V_3 = \sqrt{\frac{\rho_0}{\rho_3}} V_0 = \sqrt{\frac{1.225}{0.958}} (22) = 24.9 \text{ m/s}$$

**24.25** Find the drag force on a radio antenna 0.33 in in diameter extended to 5 ft in length on an automobile traveling through 80 °F air at 90 mph.

▮ $\qquad F_D = C_D \rho (V^2/2) A \qquad N_R = DV/\nu = (0.33/12)[\tfrac{90}{60}(88)]/(1.69 \times 10^{-4}) = 2.15 \times 10^4$

From Fig. A-57, $C_D = 1.1$.

$$F_D = (1.1)(0.00228)\{[\tfrac{90}{60}(88)]^2/2\}[(0.33/12)(5)] = 3.00 \text{ lb}$$

**24.26** A 75-mph (110-fps) wind blows across a 0.100-in-diameter wire at standard sea level ($\nu = 1.57 \times 10^{-4}$ ft²/s). What note (approximately) does the wire "sing"?

▮ $\qquad\qquad\qquad f \approx 0.20(V/D)(1 - 20/N_R) \qquad N_R = DV/\nu$

$$N_R = (0.100/12)(110)/(1.57 \times 10^{-4}) = 5839 \qquad f = (0.20)[110/(0.100/12)](1 - \tfrac{20}{5839}) = 2631 \text{ Hz}$$

On the piano, this would be an E (third octave above middle C).

**24.27** Determine the deceleration experienced by the blunt-nosed projectile of Fig. A-58 when it is moving horizontally at 950 mph (1393 fps). Assume standard sea-level atmosphere. The projectile has a diameter of 20 in and it weighs 550 lb.

■ $F_D = C_D \rho(V^2/2)A$. At sea level, $c = 1116$ fps. $V/c = \frac{1393}{1116} = 1.25$. From Fig. A-58, $C_D = 1.22$.

$$F_D = (1.22)(0.002377)(1393^2/2)[(\pi)(\tfrac{20}{12})^2/4] = 6138 \text{ lb}$$

$$F_D = -a(W/g) \qquad 6138 = -a(550/32.2) \qquad a = -359 \text{ ft/s}^2$$

**24.28** Solve Prob. 24.27 if the projectile is moving upward at an angle of 35° with the horizontal.

■ In this case, $F_D - W \sin\theta = -a(W/g)$, $6138 - 550 \sin 35° = -a(550/32.2)$, $a = -341$ ft/s².

**24.29** A sharp flat plate with $L = 1$ m and $b = 3$ m is immersed parallel to a stream of velocity 2 m/s. Find the drag on one side of the plate, and at the trailing edge find the thicknesses $\delta$, $\delta^*$, and $\theta$ for air ($\rho = 1.23$ kg/m³ and $v = 1.46 \times 10^{-5}$ m²/s).

■ The airflow Reynolds number is $VL/v = (2.0)(1.0)/(1.46 \times 10^{-5}) = 137\,000$. Since this is less than $3 \times 10^6$, we assume that the boundary layer is laminar and the drag coefficient is $C_D = 1.328/N_R^{1/2} = 1.328/(137\,000)^{1/2} = 0.00359$. Thus the drag on one side in the airflow is $D = C_D \frac{1}{2}\rho U^2 bL = 0.00359(\frac{1}{2})(1.23)(2.0)^2(3.0)(1.0) = 0.0265$ N. The boundary-layer thickness at the end of the plate is $\delta/L = 5.0/N_R^{1/2} = 5.0/(137\,000)^{1/2} = 0.0135$ or $\delta = 0.0135(1.0) = 0.0135$ m $= 13.5$ mm. We find the other two thicknesses by proportion: $\delta^* = (1.721/5.0)\delta = 4.65$ mm, $\theta = \delta^*/2.59 = 1.80$ mm.

**24.30** Solve Prob. 24.29 for water ($\rho = 1000$ kg/m³ and $v = 1.02 \times 10^{-6}$ m²/s).

■ The water Reynolds number is $N_R = 2.0(1.0)/(1.02 \times 10^{-6}) = 1.96 \times 10^6$. This is rather close to the critical value of $3 \times 10^6$, so that a rough surface or noisy free stream might trigger transition to turbulence; but let us assume that the flow is laminar. The water drag coefficient is $C_D = 1.328/(1.96 \times 10^6)^{1/2} = 0.000949$ and $D = 0.000949(\frac{1}{2})(1000)(2.0)^2(3.0)(1.0) = 5.69$ N. The drag is 215 times more for water in spite of the higher Reynolds number and lower drag coefficient because water is 57 times more viscous and 813 times denser than air.

The boundary-layer thickness is given by $\delta/L = 5.0/(1.96 \times 10^6)^{1/2} = 0.00357$ or $\delta = 0.00357(1000$ mm$) = 3.57$ mm. By ratioing down we have $\delta^* = (1.721/5.0)\delta = 1.23$ mm, $\theta = \delta^*/2.59 = 0.47$ mm. The water layer is 3.8 times thinner than the air layer, which reflects the square root of the 14.3 ratio of air to water kinematic viscosity.

**24.31** A hydrofoil 1.2 ft long and 6 ft wide is placed in a water flow of 40 fps, with $\rho = 1.99$ slugs/ft³ and $v = 0.000011$ ft²/s. Estimate the boundary-layer thickness at the end of the plate.

■ The Reynolds number is $N_R = UL/v = (40)(1.2)/0.000011 = 4.36 \times 10^6$. Thus the trailing-edge flow is certainly turbulent. The maximum boundary-layer thickness would occur for turbulent flow starting at the leading edge. $\delta/L = 0.16/N_F^{1/7} = 0.16/(4.36 \times 10^6)^{1/7} = 0.018$ or $\delta = 0.018(1.2$ ft$) = 0.0216$ ft.

**24.32** For the data in Prob. 24.31, estimate the friction drag for turbulent smooth-wall flow from the leading edge.

■ This is 7.5 times thicker than a fully laminar boundary layer at the same Reynolds number. For fully turbulent smooth-wall flow, the drag coefficient on one side of the plate is $C_D = 0.031/N_F^{1/7} = 0.031/(4.36 \times 10^6)^{1/7} = 0.00349$. Then the drag on both sides of the foil is approximately $D = 2C_D(\frac{1}{2}\rho U^2)bL = 2(0.00349)(\frac{1}{2})(1.99)(40)^2(6.0)(1.2) = 80.0$ lb.

**24.33** Solve Prob. 24.32 for laminar-turbulent flow with $(N_R)_{trans} = 5 \times 10^5$.

■ With a laminar leading edge and $(N_R)_{trans} = 5 \times 10^5$, $C_D = (0.031/N_R^{1/7}) - (1440/N_R) = 0.00349 - [1440/(4.36 \times 10^6)] = 0.00316$. The drag can be recomputed for this lower drag coefficient: $D = 2C_D(\frac{1}{2}\rho U^2)bL = 72.4$ lb.

**24.34** Solve Prob. 24.32 for turbulent rough-wall flow with $\epsilon = 0.0004$ ft.

■ For the rough wall, we calculate $L/\epsilon = 1.2/0.0004 = 3000$. From Fig. A-59 at $N_R = 4.36 \times 10^6$, this condition is just inside the fully rough regime. $C_D \approx [1.89 + 1.62 \log(L/\epsilon)]^{-2.5} = (1.89 + 1.62 \log 3000)^{-2.5} = 0.00644$ and the drag estimate is $D = 2C_D(\frac{1}{2}\rho U^2)bL = 148$ lb. This small roughness nearly doubles the drag. It is probable that the total hydrofoil drag is still another factor of 2 larger because of trailing-edge flow-separation effects.

**24.35** A square 6-in piling is acted on by a water flow of 5 ft/s 20 ft deep, as shown in Fig. 24-4. Estimate the maximum bending exerted by the flow on the bottom of the piling.

❚ Assume sea water with $\rho = 1.99$ slugs/ft$^3$ and kinematic viscosity $\nu = 0.000011$ ft$^2$/s. With piling width of 0.5 ft, we have $N_R = VD/\nu = (5)(0.5)/0.000011 = 2.3 \times 10^5$. This is the range where Table A-20 applies. The worst case occurs when the flow strikes the flat side of the piling, $C_D \approx 2.1$. The frontal area is $A = Lh = (20)(0.5) = 10$ ft$^2$. The drag is estimated by $F = C_D(\frac{1}{2}\rho V^2 A) \approx 2.1(\frac{1}{2})(1.99)(5)^2(10) = 522$ lb. If the flow is uniform, the center of this force should be at approximately middepth. Therefore the bottom bending moment is $M_0 \approx FL/2 = 522(\frac{20}{2}) = 5220$ ft-lb. According to the flexure formula from strength of materials, the bending stress at the bottom would be $S = M_0 y/I = (5220)(0.25)/[\frac{1}{12}(0.5)^4] = 251\,000$ lb/ft$^2 = 1740$ lb/in$^2$ to be multiplied, of course, by the stress-concentration factor due to the built-in end conditions.

**Fig. 24-4**

**24.36** A high-speed car with $m = 2000$ kg, $C_D = 0.3$, and $A = 1$ m$^2$, deploys a 2-m parachute to slow down from an initial velocity of 100 m/s (Fig. 24-5). Assuming constant $C_D$, brakes free, and no rolling resistance, calculate the distance and velocity of the car after 1, 10, 100, and 1000 s. For air, assume $\rho = 1.2$ kg/m$^3$ and neglect interference between the wake of the car and the parachute.

❚ Newton's law applied in the direction of motion gives $F_x = m(dV/dt) = -F_c - F_p = -\frac{1}{2}\rho V^2(C_{Dc}A_c + C_{Dp}A_p)$ where subscript $c$ is the car and subscript $p$ the parachute. This is of the form $dV/dt = -(K/m)V^2$, $K = \sum C_D A(\rho/2)$. Separate the variables and integrate

$$\int_{V_0}^{V} \frac{dV}{V^2} = -\frac{K}{m}\int_0^t dt$$

or $V_0^{-1} - V^{-1} = -(K/m)t$. Rearrange and solve for the velocity $V$: $V = V_0/[1 + (K/m)V_0 t]$, $K = (C_{Dc}A_c + C_{Dp}A_p)\rho/2$. We can integrate this to find the distance traveled; $S = (V_0/\alpha)\ln(1 + \alpha t)$, $\alpha = (K/m)V_0$. Now work out some numbers. From Table A-21, $C_{Dp} \approx 1.2$; hence $C_{Dp}A_c + C_{Dp}A_p = 0.3(1) + 1.2(\pi/4)(2)^2 = 4.07$ m$^2$. Then $(K/m)V_0 = \frac{1}{2}(4.07)(1.2)(100)/2000 = 0.122$ s$^{-1} = \alpha$. Now make a table of the results for $V$ and $S$

| $t$, s | 1 | 10 | 100 | 1000 |
|---|---|---|---|---|
| $V$, m/s | 89 | 45 | 7.6 | 0.8 |
| $S$, m | 94 | 654 | 2100 | 3940 |

**Fig. 24-5**

**24.37** A thin flat plate, 0.5 m by 1.0 m, is immersed in a stream of glycerin at 20 °C and a velocity of 6 m/s. Compute the viscous drag if the stream is parallel to the short side.

❚ $\quad\quad F_D = C_D\rho(V^2/2)A \quad\quad N_R = \rho DV/\mu = (1258)(0.5)(6)/1.49 = 2533 \quad$ (laminar)

$C_D = 1.328/\sqrt{N_R} = 1.328/\sqrt{2533} = 0.0264 \quad\quad (F_D)_{\text{total}} = (2)(0.0264)(1258)(6^2/2)[(1.0)(0.5)] = 598$ N

**24.38** Solve Prob. 24.37 if the stream is parallel to the long side.

$$F_D = C_D\rho(V^2/2)A \qquad N_R = \rho DV/\mu = (1258)(1.0)(6)/1.49 = 5066$$
$$C_D = 1.328/\sqrt{N_R} = 1.328/\sqrt{5066} = 0.0187 \qquad (F_D)_{total} = (2)(0.0187)(1258)(6^2/2)[(1.0)(0.5)] = 423 \text{ N}$$

**24.39** A ship is 180 m long and has a wetted area of 8000 m². Estimate the power expended against drag when the ship moves at 3.86 m/s in sea water at 20 °C.

$$P = F_D V \qquad F_D = C_D\rho(V^2/2)A \qquad C_D = 0.031/N_R^{1/7}$$
$$N_R = \rho DV/\mu = (1028)(180)(3.86)/(1.07 \times 10^{-3}) = 6.68 \times 10^8$$
$$C_D = 0.031/(6.68 \times 10^8)^{1/7} = 0.00170 \qquad F_D = (0.00170)(1028)(3.86^2/2)(8000) = 104\,154 \text{ N}$$
$$P = (104\,154)(3.86) = 0.402 \text{ MW}$$

**24.40** A blimp has the form of an ellipsoid of revolution, $L = 225$ ft long and $D = 45$ ft in diameter. Estimate its skin-friction drag when the blimp moves at 55 mph through still air at 68 °F and 12 psia ($\rho = 0.001906$ slug/ft³; $\mu = 3.76 \times 10^{-7}$ slug/ft-s).

$$F_D = C_D\rho(V^2/2)A \qquad C_D = 0.031/N_R^{1/7}$$
$$N_R = \rho LV/\mu = (0.001906)(225)(\tfrac{55}{60} \times 88)/(3.76 \times 10^{-7}) = 9.20 \times 10^7 \qquad C_D = 0.031/(9.20 \times 10^7)^{1/7} = 0.00226$$

By calculus, $A = (\pi LD/2)[\lambda + (\sin^{-1}\sqrt{1-\lambda^2})/\sqrt{1-\lambda^2}]$, where $\lambda = D/L = 1/5$. For an estimate, neglect $\lambda^2$:

$$A \approx \frac{\pi(225)(45)}{2}\left[\frac{1}{5} + \frac{\pi}{2}\right] \approx 28\,000 \text{ ft}^2 \qquad F_D \approx (0.00226)(0.001906)\frac{(\tfrac{55}{60} \times 88)^2}{2}(28\,000) = 392 \text{ lb}$$

**24.41** A hydrofoil 750 mm long and 3.2 m wide skims at 15 m/s. Using flat-plate theory, estimate the drag if $(N_R)_{trans} = 5 \times 10^5$ for a smooth bottom.

$$F_D = C_D\rho(V^2/2)A \qquad C_D = 0.031/N_R^{1/7} - 1440/N_R$$
$$N_R = \rho DV/\mu = (998)(0.750)(15)/(1.02 \times 10^{-3}) = 1.10 \times 10^7 \qquad (\text{turbulent})$$
$$C_D = 0.031/(1.10 \times 10^7)^{1/7} - 1440/(1.10 \times 10^7) = 0.00293$$
$$(F_D)_{total} = (2)(0.00293)(998)(15^2/2)[(0.750)(3.2)] = 1579 \text{ N}$$

**24.42** Solve Prob. 24.41 for a rough bottom ($\epsilon = 0.1$ mm).

$$F_D = C_D\rho(V^2/2)A \qquad L/\epsilon = 750/0.1 = 7500 \qquad (\text{fully rough})$$
$$C_D = [1.89 + 1.62 \log(L/\epsilon)]^{-2.5} = (1.89 + 1.62 \log 7500)^{-2.5} = 0.00525$$
$$(F_D)_{total} = (2)(0.00525)(998)(15^2/2)[(0.750)(3.2)] = 2829 \text{ N}$$

**24.43** An airplane travels at 230 m/s at an altitude of 9 km. It has a smooth wing measuring 6.5 m by 50 m (transverse dimension). Estimate the power required to overcome friction drag.

$$P = F_D V \qquad F_D = C_D\rho(V^2/2)A \qquad C_D = 0.031/N_R^{1/7}$$
$$N_R = \rho LV/\mu = (0.470)(6.5)(230)/(1.50 \times 10^{-5}) = 4.68 \times 10^7 \qquad C_D = 0.031/(4.68 \times 10^7)^{1/7} = 0.00249$$
$$(F_D)_{total} = (2)(0.00249)(0.470)(230^2/2)[(6.5)(50)] = 0.02012 \text{ MN}$$
$$P = (0.02012)(230) = 4.628 \text{ MW}$$

**24.44** If, in Prob. 24.43, the wing requires 14 MW to overcome friction, estimate the wing roughness $\epsilon$.

$$P = F_D V \qquad 14 \times 10^6 = (F_D)(230) \qquad F_D = 60\,870 \text{ N} = C_D\rho(V^2/2)A$$
$$60\,870/2 = (C_D)(0.470)(230^2/2)[(6.5)(50)] \qquad C_D = 0.00753$$

From Fig. A-59, $L/\epsilon = 1750$. $\epsilon = \frac{6.5}{1750} = 0.00371$ m, or 3.71 mm.

**24.45** A flat barge 12 m wide and 35 m long moves at 2 knots (1.029 m/s) in sea water at 20 °C. Estimate the friction drag if the bottom of the barge has roughness height 4 mm.

$$F_D = C_D\rho(V^2/2)A \qquad L/\epsilon = 35/0.004 = 8750.$$

From Fig. A-59, $C_D = 0.0052$; $F_D = (0.0052)(1028)(1.029^2/2)[(12)(35)] = 1189 \text{ N}$.

**24.46** A shark 60 cm in diameter and 5.5 m long swims at 24.69 m/s in sea water at 20 °C. Estimate the power required to overcome friction drag if $(N_R)_{crit} = 5 \times 10^5$ and $\epsilon = 0.5$ mm.

▮ $F_D = C_D \rho (V^2/2) A$     $N_R = \rho D V/\mu = (1028)(5.5)(24.69)/(1.07 \times 10^{-3}) = 1.30 \times 10^8$     (roughness prevails)

$$L/\epsilon = 5.5/5 \times 10^{-4} = 11\,000$$

From Fig. A-59, $C_D = 0.00490$. $F_D = (0.00490)(1028)(24.69^2/2)[(\pi)(0.60)(5.5)] = 15.92$ kN.

**24.47** A smokestack 2.2 m in diameter and 42 m high is subjected to a wind of 22 m/s. Estimate the drag force and the bending moment about the base ($\rho_{air} = 1.2255$ kg/m$^3$ and $\mu_{air} = 1.78 \times 10^{-5}$ Pa · s).

▮     $F_D = C_D \rho (V^2/2) A$     $N_R = \rho D V/\mu = (1.2255)(2.2)(22)/(1.78 \times 10^{-5}) = 3.33 \times 10^6$

Turbulent: $C_D \approx 0.3$. $F_D = (0.3)(1.2255)(22^2/2)[(2.2)(42)] = 8221$ N, $M_{base} = (8221)(\frac{42}{2}) = 173$ kN · m.

**24.48** A log 2.2 m in diameter and 22 m long is towed at a constant speed of 3 m/s in fresh water at 20 °C. Find the horsepower required if the log is towed lengthwise.

▮     $F_D = C_D \rho (V^2/2) A$     $N_R = \rho D V/\mu = (998)(22)(3)/(1.02 \times 10^{-3}) = 6.46 \times 10^7$

From Table A-21, $C_D \approx 1.05$ at $L/D \approx 10$. $F_D = (1.05)(998)(3^2/2)[(\pi)(2.2)^2/4] = 17.925$ kN. $P = (17.925)(3) = 53.78$ kW $= 72$ hp.

**24.49** A seine consists of 1-mm-diameter cord knotted into 20-mm squares. Estimate the drag of 1 m$^2$ of such a net when towed normal to its plane at 3.5 m/s in sea water.

▮     $F_D = C_D \rho (V^2/2) A$     $N_R = \rho D V/\mu = (1028)(0.001)(3.5)/(1.07 \times 10^{-3}) = 3.36 \times 10^3$

From Fig. A-60a, $C_D \approx 1.0$. For a single 20-mm side: $F_D = (1.0)(1028)(3.5^2/2)[(0.020)(0.001)] = 0.126$ N. Since 1 m$^2$ of net contains 5000 sides, $F_D = (0.126)(5000) = 630$ N.

**24.50** A filter may be idealized as an array of cylindrical fibers normal to the flow, as in Fig. 24-6a. Assuming that the fibers are uniformly distributed and have drag coefficients given by Fig. A-61a derive an approximate expression for the pressure drop $\Delta p$ through a filter of thickness $L$.

▮ Let $N =$ number of fibers per unit area of filter, $D =$ fiber diameter, $b =$ filter width, $H =$ filter height, and $L =$ filter length (see Fig. 24-6b). Each fiber has drag of $F_D = C_D \rho (V^2/2) Db$. Force balance on filter: $\Delta p H b = \sum F_{fibers} = NHLC_D \rho (V^2/2) Db$, $\Delta p = NLC_D \rho (V^2/2) D$.

**Fig. 24-6**

**24.51** Apply Prob. 24.50 to a filter consisting of 0.25-mm-diameter fibers packed 9/mm$^2$ section in the plane of Fig. 24-6a. For sea-level standard air flowing at 1.5 m/s, estimate the pressure drop through a filter 35 mm thick.

▮                  $\Delta p = NLC_D \rho (V^2/2) D$     $N = 9 \times 10^6/m^2$

$$N_R = \rho D V/\mu = (1.2255)(0.25 \times 10^{-3})(1.5)/(1.78 \times 10^{-5}) = 25.8$$

From Fig. A-61a, $C_D \approx 1.0$. $\Delta p = (9 \times 10^6)(0.035)(1.0)(1.2255)(1.5^2/2)(0.25 \times 10^{-3}) = 109$ Pa.

**24.52** A 1-$\mu$m-diameter volcanic dust particle has a specific weight of about 26.2 kN/m$^3$. Approximately how long will it take the particle to reach the earth from an altitude of 8 km?

■ By *Stokes' law*, the velocity of slow fall, $v$, satisfies $F_D = W_{net} = 3\pi\mu Dv$. Assume average values of $\gamma_{air} = 8.6$ N/m$^3$ and $\mu_{air} = 1.7 \times 10^{-5}$ Pa · s.

$$W = \gamma V \qquad W_{net} = (26\,200 - 8.6)(\pi D^3/6) = 13\,714 D^3$$
$$13\,714 D^3 = (3)(\pi)(1.7 \times 10^{-5})(D)(u) \qquad u = 8.56 \times 10^7 D^2 = 8.56 \times 10^{-5} \text{ m/s}$$
$$t = (8 \times 10^3)/(8.56 \times 10^{-5}) = 9.35 \times 10^7 \text{ s} \quad \text{or} \quad 1082 \text{ days}$$

**24.53** A radar installation is approximated by a 16-m-diameter sphere atop a 1.25-m-diameter, 22-m-tall pole. Estimate the bending moment at the base during hurricane winds of 100 mph. Use $\rho_{air} = 1.2255$ kg/m$^3$.

■ $F = C_D \rho (V^2/2) A$. Estimate $C_D = 0.2$ for sphere and 0.3 for rod; 100 mph = 45 m/s.

$$F_{sphere} = (0.2)(1.2255)(45^2/2)[(\pi)(16^2)/4] = 49.90 \text{ kN} \qquad F_{rod} = (0.3)(1.2255)(45^2/2)[(1.25)(22)] = 10.24 \text{ kN}$$
$$M_{base} = (49.90)(22 + \tfrac{16}{2}) + (10.24)(\tfrac{22}{2}) = 1600 \text{ kN} \cdot \text{m}$$

**24.54** A boxcar measures 10 ft high, 40 ft long, and 6 ft wide; it stands 3 ft above tracks 4.8 ft apart and has gross weight 45 000 lb. What hurricane wind would topple the boxcar?

■ $F_D = C_D \rho (V^2/2) A$. From Table A-21, $C_D = 1.2$; and the area is that of one side of the boxcar; hence, $F_D = (1.2)(0.00237)(V^2/2)[(10)(40)] = 0.5688 V^2$, $M_{overturning} = (0.5688 V^2)(3 + \tfrac{10}{2}) = 4.550 V^2$, $M_{righting} = (45\,000)(4.8/2) = 108\,000$ lb-ft, $4.550 V^2 = 108\,000$, $V = 154.1$ ft/s, or 105 mph.

**24.55** A long copper wire, 0.5 in in diameter, is stretched taut and is exposed to a wind of velocity 90.0 fps normal to the wire. Compute the drag force per foot of length.

■ $$F_D = C_D \rho (V^2/2) A \qquad N_R = DV/\nu = (0.5/12)(90.0)/(1.6 \times 10^{-4}) = 2.34 \times 10^4$$

From Fig. A-62, $C_D = 1.30$. $F_D = (1.30)(0.00233)(90.0^2/2)[(0.5/12)(1)] = 0.511$ lb per foot.

**24.56** Consider the area on one side of a moving van to be 600 ft$^2$. Determine the resultant force acting on the side of the van when the wind is blowing at 10 mph normal to the area (a) when the van is at rest and (b) when the van is moving at 30 mph normal to the direction of the wind. In (a) use $C_D = 1.30$, and in (b) use $C_D = 0.25$ and $C_L = 0.60$. ($\rho = 0.00237$ slug/ft$^3$)

■ (a) The force acting normal to the area = $C_D (\rho/2) A V^2$. Then Resultant force = $1.30(0.00237/2)(600)[(10)(\tfrac{5280}{3600})]^2 = 199$ lb normal to area.

(b) It will be necessary to calculate the relative velocity of the wind with respect to the van. From kinetic mechanics, $V_{wind} = V_{wind/van} \leftrightarrow V_{van}$.

Figure 24-7 indicates this vector relationship, i.e., $OB = OA \leftrightarrow AB = 30.0 \leftrightarrow V_{w/v}$. Thus the relative velocity = $\sqrt{(30)^2 + (10)^2} = 31.6$ mph to the right and upward at an angle $\theta = \tan^{-1} \tfrac{10}{30} = 18.4°$.

The component of the resultant force normal to the relative velocity of wind with respect to van is Lift force = $C_L (\rho/2) A V^2 = 0.60(0.00237/2)(600)[(31.6)(\tfrac{5280}{3600})]^2 = 916$ lb normal to the relative velocity. The component of the resultant force parallel to the relative motion of wind to van is Drag force = $C_D (\rho/2) A V^2 = 0.25(0.00237/2)(600)[(31.6)(\tfrac{5280}{3600})]^2 = 382$ lb parallel to the relative velocity.

Referring to Fig. 24-7b, the resultant force = $\sqrt{(916)^2 + (382)^2} = 992$ lb at an angle $\alpha = \tan^{-1} \tfrac{916}{382} = 67.4°$. Hence the angle with the longitudinal axis ($X$ axis) is $18.4° + 67.4° = 85.8°$.

**24.57** A man weighing 170 lb is descending from an airplane using an 18-ft-diameter parachute. Assuming a drag coefficient of 1.00 and neglecting the weight of the parachute, what maximum terminal velocity will be attained?

■ The forces on the parachute are the weight down and the drag force up. $F_D = C_D \rho (V^2/2) A$, $170 = (1.00)(0.00237)(V^2/2)[(\pi)(18)^2/4]$, $V = 23.7$ ft/s.

**24.58** A 1-in-diameter sphere of lead weighing 710 lb/ft$^3$ is moving downward in an oil at a constant velocity of 1.17 fps. Calculate the absolute viscosity of the oil if its specific gravity is 0.93.

■ $$F_D = (\gamma_{lead} - \gamma_{oil})(V_{lead}) = C_D \rho (V^2/2) A$$
$$[710 - (0.93)(62.4)][(\pi)(\tfrac{1}{12})^3/6] = C_D[(1.94)(0.93)](1.17^2/2)[(\pi)(\tfrac{1}{12})^2/4] \qquad C_D = 29.3$$

From Fig. A-62, $N_R = 0.85$ for $C_D = 29.3$. $N_R = \rho DV/\mu$, $0.85 = [(1.94)(0.93)](\tfrac{1}{12})(1.17)/\mu$, $\mu = 0.207$ lb-s/ft$^2$.

(a)          (b)          **Fig. 24-7**

**24.59**  A sphere, 0.5 in in diameter, rises in oil at the maximum velocity of 0.12 fps. What is the specific weight of the sphere if the density of the oil is 1.78 slugs/ft$^3$ and the absolute viscosity is 0.000710 lb-s/ft$^2$?

$\blacksquare$    $F_D = F_b - W = C_D \rho (V^2/2)(A)$    $N_R = \rho D V / \mu = (1.78)(0.5/12)(0.12)/0.000710 = 12.5$

From Fig. A-62, $C_D = 3.9$ for $N_R = 12.5$. $[(\pi)(0.5/12)^3/6][(1.78/1.94)(62.4)] - W = (3.9)(1.78)(0.12^2/2)[(\pi)(0.5/12)^2/4]$, $W = 0.002100$ lb; $\gamma = 0.002100/[(\pi)(0.5/12)^3/6] = 55.4$ lb/ft$^3$.

**24.60**  Measurements on a smooth sphere, 6 in in diameter, in an air stream (68 °F) gave a force for equilibrium equal to 0.250 lb. At what velocity was the air moving?

$\blacksquare$  Total drag = $C_D \rho A V^2/2$, where $C_D$ = overall drag coefficient. Since neither Reynolds number nor $C_D$ can be found directly, assume $C_D = 1.00$. Then $0.250 = C_D(0.00233)\frac{1}{4}\pi(\frac{1}{2})^2(V^2/2)$, $V^2 = 1093/C_D$, $V = 33.1$ ft/s. Calculate $N_R = Vd/\nu = 33.1(\frac{1}{2})/(16.0 \times 10^{-5}) = 103\,000$. From Fig. A-62, $C_D = 0.59$ (for spheres). Then $V^2 = 1093/0.59 = 1853$, $V = 43.0$ ft/s. Anticipating result, use $V = 44.0$ ft. Recalculate $N_R = Vd/\nu = 44.0(\frac{1}{2})/(16.0 \times 10^{-5}) = 137\,500$. From Fig. A-62, $C_D = 0.56$. Then $V^2 = 1093/0.56 = 1952$, $V = 44.2$ ft/s (satisfactory accuracy).

**24.61**  What should be the diameter of a sphere (s.g. = 2.50) in order that its freely falling velocity at 60 °F attains the acoustic velocity?

$\blacksquare$  $F_D = W = C_D \rho (V^2/2)A$. From Fig. A-63, $C_D = 0.80$. $V = c = \sqrt{kgRT}$. For air at 60 °F, $c = \sqrt{(1.40)(32.2)(53.3)(460 + 60)} = 1118$ ft/s, $[(2.50)(62.4)](\pi d^3/6) = (0.80)(0.00237)(1118^2/2)(\pi d^2/4)$, $d = 11.4$ ft.

**24.62**  The fixed keel of a Columbia 22 sailboat is about 38 in long (see Fig. 24-8). Moving in Lake Ontario at a speed of 3 knots, what is the skin drag from the keel? The water is at 40 °F. Solve this problem using rectangular plate of length 38 in and width 24.5 in, which is the average width of the keel. Transition takes place at a Reynolds number of $10^6$.

$\blacksquare$  Compute the plate Reynolds number for the rectangular model of the keel.

$$N_R = Vl/\nu = [(3)(1.689)](24.5/12)/(1.664 \times 10^{-5}) = 6.22 \times 10^5 \qquad (1)$$

We thus have a laminar boundary layer. We get the plate coefficient of drag using

$$C_f = 1.328/\sqrt{N_R} = 1.328/\sqrt{6.22 \times 10^5} = 1.684 \times 10^{-3} \qquad (2)$$

Next we get the skin drag, realizing that there are two sides to be considered. Thus,

$$D = 2[(C_f)(\tfrac{1}{2}\rho V_0^2)(A)] = 2\{(1.684 \times 10^{-3})(\tfrac{1}{2})(1.940)[(3)(1.689)]^2[(38)(24.5)/144]\} = 0.542 \text{ lb} \qquad (3)$$

**24.63**  Solve Prob. 24.62 using the actual dimensions of the keel as shown in Fig. 24-8. Compare answers and comment on the result.

$\blacksquare$  For this purpose, consult Fig. 24-9 in which a keel with an infinitesimal slice $dz$ is depicted. The length $l(z)$ of the slice is

$$l = \tfrac{15}{12} + (z/38)(19) = 1.250 + 0.5z \text{ ft} \qquad (4)$$

with $z$ in feet. Now let us first see if we have transition anywhere on the keel. Looking at the uppermost

**Fig. 24-8**

portion, we have for $(N_R)_{max}$

$$(N_R)_{max} = Vl_{max}/\nu = [(3)(1.689)](\tfrac{34}{12})/(1.664 \times 10^{-5}) = 8.63 \times 10^5 \tag{5}$$

We have only a laminar boundary layer. For an infinitesimal plate of length $l = 1.250 + 0.5z$ ft, $C_f = 1.328/(Vl/\nu)^{1/2} = 1.328\{[(3)(1.689)/(1.664 \times 10^{-5})](1.250 + 0.5z)\}^{-1/2}$. Now for the drag $D$,

$$D = 2\int_0^{38/12} C_f(\tfrac{1}{2}\rho V_0^2)(1.250 + 0.5z)(dz)$$

$$= 2\int_0^{38/12} 1.238\left[\frac{(3)(1.689)}{1.664 \times 10^{-5}}(1.250 + 0.5z)\right]^{-1/2}(\tfrac{1}{2})(1.940)[(3)(1.689)]^2(1.250 + 0.5z)\,dz$$

$$= 0.1199\int_0^{3.17}(1.250 + 0.5z)^{1/2}\,dz$$

Let $1.250 + 0.5z = \eta$; therefore $0.5\,dz = d\eta$, $dz = 2\,d\eta$. Hence,

$$D = \frac{0.1199}{0.5}\int_{1.250}^{2.833} \eta^{1/2}\,d\eta = \frac{0.1199}{0.5}\eta^{3/2}(\tfrac{2}{3})\Big|_{1.250}^{2.833} = 0.539\text{ lb}$$

The averaging process in Prob. 24.62 gave a very good result when compared with the result above.

$$l = \tfrac{15}{12} + \left[z/(\tfrac{38}{12})\right](\tfrac{19}{12})\text{ ft} = 1.250 + 0.5z\text{ ft}$$
$$z\text{ measured in feet}$$

**Fig. 24-9**

**24.64** The United States at one time in the thirties had three large dirigibles—the Los Angeles, the Graf Zeppelin, and the Akron. Two of them were destroyed by accidents. The largest was the Akron, having a length of 785 ft and a maximum diameter of 132 ft. Its maximum speed was 84 mph. The useful lift was 182 000 lb.

Moving at top speed, estimate the power needed to overcome skin friction, which is a significant part of the drag. Disregard effects of protrusion from engine cowlings, cabin region, etc. Assume that the surface is smooth. Take the critical Reynolds number to be 500 000. Consider the Akron at 10 000-ft standard atmosphere.

▌ We can make a reasonable estimate of the skin drag by "unwrapping" the outer surface of the Akron to form a flat plate. As a first step, we wish to calculate the plate Reynolds number. For this we find from the standard atmosphere table that $\rho = (0.7385)(0.002378) = 0.001756$ slug/ft$^3$, $T = 23.3\,°$F. From the viscosity curves in Fig. A-1, we then find that $\mu = 3.7 \times 10^{-7}$ lb-s/ft$^2$. We can now compute $N_R = \rho dV/\mu$:

$$N_R = (0.001756)[(84)(\tfrac{5280}{3600})](785)/(3.7 \times 10^{-7}) = 4.59 \times 10^8 \tag{1}$$

We will use the Prandtl–Schlichting skin-friction formula with $A = 1700$:

$$C_f = [0.455/(\log N_R)^{2.58}] - (A/N_R) = \{0.455/[\log (4.59 \times 10^8)]^{2.58}\} - [1700/(4.59 \times 10^8)] = 1.730 \times 10^{-3} \tag{2}$$

Now going to $D = C_f(\tfrac{1}{2}\rho V_0^2)(bL)$, we have for the drag $D$ owing to skin friction using the maximum diameter of 132 ft:

$$D = (1.730 \times 10^{-3})(\tfrac{1}{2})(0.001756)[84(\tfrac{5280}{3600})]^2(785)[(\pi)(132)] = 7505\text{ lb} \tag{3}$$

The power needed then is

$$\text{Power} = (7505)(84)(\tfrac{5280}{3600})/550 = 1681\text{ hp} \tag{4}$$

**24.65** In Prob. 24.64, we computed the skin drag for the dirigible Akron using smooth flat-plate theory. Let us now evaluate the admissible roughness $\epsilon_{adm}$ for the results Prob. 24.64 to be valid. Then using a roughness of 0.05 in, recompute the skin-friction drag. What is the power needed to overcome this drag at the top speed of 84 mph?

▌ We can immediately solve for the admissible roughness and hence the largest roughness that will still give hydraulically smooth flow. Using the plate Reynolds number from Prob. 24.64, we have

$$\epsilon_{adm} = L(100/N_R) = (785)[100/(4.59 \times 10^8)] = 1.710 \times 10^{-4}\text{ ft} = 0.00205\text{ in} \tag{1}$$

For the dirigible, the actual roughness coefficient is 0.05 in. Now $L/\epsilon$ is $785/(0.05/12) = 1.884 \times 10^5$, so that on consulting Fig. A-64 we are clearly in the rough zone. We accordingly use

$$C_f = [1.89 + 1.62 \log (L/\epsilon)]^{-2.5} = \{1.89 + 1.62 \log [785/(0.05/12)]\}^{-2.5} = 0.002843 \tag{2}$$

For the skin drag, we have, using 0.001756 slug/ft$^3$ for $\rho$, $D = (0.002843)(\tfrac{1}{2})(0.001756)[(84)(\tfrac{5280}{3600})]^2(785)(\pi)(132) = 12\,333$ lb. The power needed to overcome skin friction is Power $= (12\,333)(84)(\tfrac{5280}{3600})/550 = 2763$ hp.

**24.66** In Prob. 24.62, consider that the surface is rough with $\epsilon = 0.009$ in. Also, consider that transition takes place at $(N_R)_{crit} = 3.2 \times 10^6$. Calculate the skin drag on the keel as accurately as you can.

▌ From Prob. 24.62, we see that the plate Reynolds number ranges from $3.81 \times 10^5$ at the bottom of the keel to $8.63 \times 10^5$ at the top of the keel. The ratio $L/\epsilon$ ranges from $1.667 \times 10^3$ at the bottom to $3.778 \times 10^3$ at the top. Considering Fig. A-64, we see that at both extremes we are in the transition zone of flow. We accordingly will use $C_f = [0.031/(N_R)^{1/7}] - (A/N_R)$ for $C_f$. We will furthermore use strips of width $dz$ (see Fig. 24-9) and of length $(1.250 + 0.5z)$ ft, with $z$ measured in feet. We have for $C_f$ for such a strip

$$C_f = \frac{0.031}{\{[(3)(1.689)(1.250 + 0.5z)]/(1.664 \times 10^{-5})\}^{1/7}} - \frac{1050}{\{[(3)(1.689)(1.250 + 0.5z)]/(1.664 \times 10^{-5})\}}$$
$$= 0.00510(1.250 + 0.5z)^{-1/7} - 0.00345(1.250 + 0.5z)^{-1}$$

The drag $D$ then becomes

$$D = 2\int_0^{38/12} C_f(\tfrac{1}{2}\rho V_0^2)(1.250 + 0.5z)\,dz$$
$$= 2\int_0^{38/12} [(0.00510)(1.250 + 0.5z)^{-1/7} - (0.00345)(1.250 + 0.5z)^{-1}](\tfrac{1}{2})(1.490)[(3)(1.689)]^2(1.250 + 0.5z)\,dz$$
$$= \int_0^{3.17} [0.254(1.250 + 0.5z)^{6/7} - (0.1718)\,dz$$

Integrating, we get

$$D = (0.254)(1.250 + 0.5z)^{13/7}\left(\frac{7}{13}\right)\left(\frac{1}{0.5}\right)\Big|_0^{3.17} - (0.1718)(3.17) = 0.936\text{ lb}$$

We see that we get almost a doubling of the drag. In racing sailboats, every ounce of drag counts, so sailors carefully smooth the wetted surfaces of their boats.

**24.67**  We will consider the dirigible Akron once again for drag (see Prob. 24.64). This time we will use an ellipsoidal body of revolution from Table A-22 to represent the dirigible. Estimate the pressure drag.

**I**  The coefficient of drag should correspond to an ellipsoid whose $L/D_{max} = \frac{785}{132} = 5.95$. For turbulent flow, we estimate $C_D$ using simple interpolation from the 4:1 to the 8:1 ellipsoids:

$$C_D = 0.06 + [(5.95 - 4)/4](0.13 - 0.06) = 0.094 \tag{1}$$

Now the total drag is next computed using

$$D = C_D A(\rho V_0^2/2) = (0.094)[(\pi)(132^2)/4](\tfrac{1}{2})(0.001756)[84(\tfrac{5280}{3600})]^2 = 17\,143 \text{ lb} \tag{2}$$

We can now estimate the pressure drag on the Akron using the drag from Prob. 24.64. Thus considering hydraulically smooth flow, $D_{press} = 17\,143 - 7505 = 9638$ lb. The total power needed to move the Akron should be $P = (17\,143)[(84)(\tfrac{5280}{3600})]/550 = 3840$ hp. The Akron actually had 8450-hp diesel engines.

**24.68**  A well-streamlined car can have a drag coefficient as low as 0.45 compared with old-fashioned cars whose drag coefficient could be as high as 0.9 (see Fig. A-65). Consider a car with a drag coefficient of 0.45 moving at the speed of 100 km/h. The frontal area of the car is 2 m² and its mass with driver is 1300 kg. How far must it move to halve its speed if the engine is disengaged and the vehicle rolls freely along a straight flat road with negligible wind present. Neglect the rotational effects of the wheels but consider that the coefficient of rolling resistance, $a$, of the tires is 0.50 mm. The air temperature is 20 °C. The tire diameter is 450 mm.

**I**  You will recall from your statics course that the resistance to rolling $F_R$ is given as

$$F_R = Pa/r \tag{1}$$

where $P$ = normal force on wheel, $a$ = coefficient of rolling resistance, $r$ = radius of wheel.

For this problem, we have for the total effect of the four wheels $F_R = [(1300)(9.81)](0.50 \times 10^{-3})/0.225 = 28.34$ N. The drag force from the air is next computed:

$$D = C_D(\tfrac{1}{2}\rho V^2)(A) = (0.45)(\tfrac{1}{2})(\rho)(V^2)(2) \tag{2}$$

For $\rho$ we have $\rho = p/RT = 101\,325/[(287)(293)] = 1.205$ kg/m³. Hence we have for $D$

$$D = 0.542V^2 \tag{3}$$

We next express Newton's law for the car: $M(dV/dt) = F$.

$$(1300)(dV/dt) = -0.542V^2 - 28.34 \tag{4}$$

In order to separate variables with the distance $x$ appearing, we may say, using the chain rule,

$$dV/dt = (dV/dx)(dx/dt) = V(dV/dx) \tag{5}$$

Now Eq. (4) becomes $1300V(dV/dx) = -(0.542V^2 + 28.34)$. Separating variables, we get

$$(1300V\,dV)/(0.542V^2 + 28.34) = -dx \tag{6}$$

Let $\eta = 0.542V^2 + 28.34$:

$$d\eta = 1.084V\,dV \tag{7}$$

Making the above substitution of variable, we get $(1300/1.084)(d\eta/\eta) = -dx$. Integrating, $1.199 \times 10^3 \ln \eta = -x + C$:

$$1.199 \times 10^3 \ln (0.542V^2 + 28.34) = -x + C \tag{8}$$

When $x = 0$, $V = 100(\tfrac{1000}{3600}) = 27.78$ m/s. We can determine $C$: $1.199 \times 10^3 \ln [(0.542)(27.78)^2 + 28.34] = C$, $C = 1.199 \times 10^3 \ln 446.6$. Equation (8) can now be written as

$$\ln [(0.542V^2 + 28.34)/446.6] = -x/(1.199 \times 10^3) \tag{9}$$

Let $V = 50(\tfrac{1000}{3600}) = 13.89$ m/s. We get for $x$

$$\ln \{[(0.542)(13.89)^2 + 28.34]/446.6\} = -x/(1.199 \times 10^3) \qquad x = 1453 \text{ m} \tag{10}$$

The vehicle moves a distance of 1453 m. Actually, by including the rotational energy of the wheels, the vehicle would move even farther.

**24.69**  A jet aircraft discharges solid particles of matter 10 $\mu$m in diameter, s.g. = 2.5, at an altitude of 11 km. Assume the viscosity $\mu$ of air, in Pa · s, to be expressed by $\mu = 1.78 \times 10^{-5} - 3.06 \times 10^{-10}y$, where $y$ is altitude, in m.

Estimate the time for these particles to reach sea level. Neglect air currents and wind effects.

▌ By Stokes' law (Prob. 24.52), $u = (D^2/18\mu)(\gamma_s - \gamma)$. Writing $u = -dy/dt$ and recognizing the unit weight of air to be much smaller than the unit weight of the solid particles, one has $-dy/dt \approx (D^2/18)(\gamma_s/\mu)$:

$$\int_0^T dt = \int_{11\,000}^0 18(1.78 \times 10^{-5} - 3.06 \times 10^{-10}y)\left[\frac{1}{(10 \times 10^{-6})^2}\right]\left[\frac{1}{2.5(9806)}\right] dy$$

$$T = (1/86\,400)[1.78y - (3.06 \times 10^{-5}y^2)/2]_0^{11\,000}(73.45) = 15.07 \text{ days}$$

**24.70** How many 32-m-diameter parachutes ($C_D = 1.2$) should be attached to a 100-kN load to ensure a terminal speed of 10 m/s, if air pressure and temperature are 100 kPa and 10 °C?

▌ For one parachute,

$$F_D = C_D\rho(V^2/2)A \qquad \rho = p/RT = 100\,000/[(287)(10 + 273)] = 1.231 \text{ kg/m}^3$$

$$F_D = (1.2)(1.231)(10^2/2)[(\pi)(32)^2/4] = 59.4 \text{ kN}$$

Thus, two parachutes should be sufficient.

**24.71** An 85-cm cubical box is tied to the roof of a car. Estimate the additional power needed to travel at 75 km/h ($C_D = 1.1$ and $\rho_{air} = 1.2$ kg/m³).

▌
$$P = F_D V \qquad F_D = C_D\rho(V^2/2)A$$
$$V = (75)(1000)/3600 = 20.83 \text{ m/s} \qquad F_D = (1.1)(1.2)(20.83^2/2)[(0.85)(0.85)] = 206.9 \text{ N}$$
$$P = (206.9)(20.83) = 4310 \text{ W} = 5.78 \text{ hp}$$

**24.72** A semitubular cylinder of 8-in radius with concave side upstream is submerged in water flowing 2 fps. Calculate the drag for a cylinder 20 ft long.

▌ $F_D = C_D\rho(V^2/2)A$. From Table A-23, $C_D = 2.3$. $F_D = (2.3)(1.94)(2^2/2)[(20)(8 + 8)/12] = 238$ lb.

**24.73** A fully loaded, small aircraft weighing 5000 lb with a wing area (projected chord area) of 350 ft² is to take off at a horizontal velocity of 100 mph (146.7 fps). What is the necessary angle of attack (i.e., angle the wings make with the horizontal)? Assume the wings have the characteristics of the airfoil of Fig. A-66. Use $\rho_{air} = 0.00234$ slug/ft³.

▌
$$F_L = C_L\rho(v^2/2)A \qquad 5000 = C_L(0.00234)(146.7^2/2)(350) \qquad C_L = 0.567$$
From Fig. A-66, $\alpha = 6.0°$.

**24.74** An aircraft weighing 1000.0 kN when empty has a wing area of 226 m². It is to take off at a velocity of 300 km/h and a 20° angle of attack. Assume $\rho_{air} = 1.20$ kg/m³. Also, assume the wing has the characteristics of the airfoil in Fig. A-66. What is the allowable weight of cargo?

▌ $F_L = C_L\rho(v^2/2)A$. From Fig. A-66, $C_L = 1.42$. $v = (300)(1000)/3600 = 83.33$ m/s, $F_L = (1.42)(1.20)(83.33^2/2)(226) = 1\,337\,000$ N, or 1337 kN. This value (1337 kN) represents the total weight that can be lifted. Since the aircraft weighs 1000.0 kN when empty, it can carry, in theory, $1337 - 1000 = 337$ kN of cargo.

**24.75** A fully loaded aircraft weighing 10 000 lb with a wing area of 450 ft² is to take off at a horizontal velocity of 125 mph (183.3 fps). What is the necessary angle of attack? Assume the wings have the characteristics of the airfoil of Fig. A-66. Use $\rho_{air} = 0.00234$ slug/ft³.

▌
$$F_L = C_L\rho(v^2/2)A \qquad 10\,000 = C_L(0.00234)(183.3^2/2)(450) \qquad C_L = 0.565$$
From Fig. A-66, $\alpha = 6.0°$.

**24.76** An aircraft weighing 65.2 kN when empty has a wing area of 62.3 m². It is to take off at a velocity of 250 km/h and a 5.0° angle of attack. Assume $\rho_{air} = 1.20$ kg/m³. Also, assume the wing has the characteristics of the airfoil in Fig. A-66. What is the allowable weight of cargo?

▌ $F_L = C_L\rho(v^2/2)A$. From, Fig. A-66, $C_L = 0.50$. $v = (250)(1000)/3600 = 69.44$ m/s, $F_L = (0.50)(1.20)(69.44^2/2)(62.3) = 90\,000$ N, or 90.1 kN; $W_{cargo} = 90.1 - 65.2 = 24.9$ kN.

**24.77** A 3-ft by 4-ft plate moves at 44 ft/s in still air at an angle of 12° with the horizontal. Using a coefficient of drag $C_D = 0.17$ and a coefficient of lift $C_L = 0.72$, determine (a) the resultant force exerted by the air on the plate, (b) the frictional force and (c) the horsepower required to keep the plate moving. (Use $\gamma = 0.0752$ lb/ft³.)

▮ (a)   Drag force $= C_D(\gamma/g)A(V^2/2)$   $F_D = 0.17(0.752/32.2)(12)[(44)^2/2] = 4.61$ lb

   Lift force $= C_L(\gamma/g)A(V^2/2)$   $F_L = 0.72(0.0752/32.2)(12)[(44)^2/2] = 19.5$ lb

   Referring to Fig. 24-10, the resultant of the drag and lift components is $R = \sqrt{(4.61)^2 + (19.5)^2} = 20.0$ lb acting on the plate at $\theta_x = \tan^{-1}(19.5/4.61) = 76°42'$.

(b) The resultant force might also have been resolved into a normal component and a frictional component (shown dotted in the figure). From the vector triangle, frictional component $= R\cos(\theta_x + 12°) = 20.0(0.0227) = 0.45$ lb.

(c) Horsepower = (force in direction of motion)(velocity)/550 = (4.61)(44)/550 = 0.369

Fig. 24-10

**24.78** If an airplane weighs 4000 lb and has a wing area of 300 ft², what *angle of attack* must the wings make with the horizontal at a speed of 100 mph? Assume the coefficient of lift varies linearly from 0.35 at 0° to 0.80 at 6° and use $\gamma = 0.0752$ lb/ft³ for air.

▮ For equilibrium in the vertical direction, $\sum Y = 0$. Hence, lift − weight = 0, or weight $= C_L \gamma A(V^2/2g)$, $4000 = C_L(0.0752)(300)\{[(100)(\frac{5280}{3600})]^2/2g\}$, $C_L = 0.53$. By interpolation between 0° and 6°, angle of attack = 2.4°.

**24.79** What wing area is required to support a 5000-lb plane when flying at an angle of attack of 5° at 88 ft/s? Use coefficients given in Prob. 24.78.

▮ From given data, $C_L = 0.725$ for 5° angle by interpolation. As in Prob. 24.78, weight = lift, $5000 = 0.725(0.0752/32.2)A(88)^2/2$, $A = 763$ ft².

**24.80** A kite weighs 2.50 lb and has an area of 8.00 ft². The tension in the kite string is 6.60 lb when the string makes an angle of 45° with the horizontal. For a wind of 20 mph, what are the coefficients of lift and drag if the kite assumes an angle of 8° with the horizontal? Consider the kite essentially a flat plate and $\gamma_{air} = 0.0752$ lb/ft³.

▮ Figure 24-11 indicates the forces acting on the kite taken as a free body. The components of the tension are 4.66 lb each.

Fig. 24-11

From $\sum X = 0$, drag $= 4.66$ lb. From $\sum Y = 0$, lift $= 4.66 + 2.50 = 7.16$ lb. Drag force $= C_D \rho A V^2/2$, $4.66 = C_D(0.0752/32.2)(8.00)[(20)(\frac{5280}{3600})]^2/2$, $C_D = 0.58$. Lift force $= C_L \rho A V^2/2$, $7.16 = C_L(0.0752/32.2)(8.00)[(20)(\frac{5280}{3600})]^2/2$, $C_L = 0.89$.

**24.81** If the minimum landing speed ($V_{\min}$) of the Mustang racer is 1.3 times the stall speed, what is the minimum landing speed for this plane with flaps at 40°? The Mustang has a wing area of 233 ft² and a weight of 9500 lb. Take $\rho$ to be 0.002378 slug/ft³ and $(C_L)_{\max}$ as 1.65.

$\blacksquare$
$$V_{\text{stall}} = \sqrt{2W/[(\rho)(C_L)_{\max}A]} = \sqrt{(2)(9500)/[(0.002378)(1.65)(233)]} = 144.2\ \text{ft/s} \quad \text{or} \quad 98.3\ \text{mph}$$
$$V_{\min} = (1.3)(98.3) = 128\ \text{mph}$$

**24.82** When a wing of 2-m chord moves through still air ($\gamma = 11.3$ N/m³) at 138 km/h, the mean velocity along the top surface is 160 km/h (44.45 m/s) and that along the bottom surface is 130 km/h (36.11 m/s). Estimate the lift per unit length of span.

$\blacksquare$
$$\text{Lift} = (\Delta p)(c) \qquad p_{\text{top}}/\gamma + v_{\text{top}}^2/2g = p_{\text{bottom}}/\gamma + v_{\text{bottom}}^2/2g$$
$$p_{\text{top}}/11.3 + 44.45^2/[(2)(9.807)] = p_{\text{bottom}}/11.3 + 36.11^2/[(2)(9.807)]$$
$$p_{\text{bottom}} - p_{\text{top}} = \Delta p = 387.1\ \text{N/m}^2 \qquad \text{Lift} = (387.1)(2) = 774\ \text{N/m}$$

**24.83** For a rectangular Clark Y airfoil of 6-ft chord by 36-ft span, find the value of the friction coefficient $\eta$ if the angle of attack $\alpha = 5.4°$ when the wing is moving at 300 fps through standard atmosphere at altitude 10 000 ft. Find the weight which the wing will carry and the horsepower required to drive it.

$\blacksquare$ From Fig. A-67, with $\alpha = 5.4°$, $C_L = 0.8$, $C_D = 0.047$. From Fig. A-68, for $B/c = 6$, $\tau = 0.175$. $\alpha_i$(radians) $= [C_L/\pi(B/c)](1 + \tau) = [0.8/\pi(\frac{36}{6})](1 + 0.175) = 0.0499$ rad $= 2.86°$. From Fig. A-69, $\alpha_0 = \alpha - \alpha_i = 5.40 - 2.86 = 2.54°$ and since the angle of zero lift is $-5.6°$, $\alpha_0' = 2.54 + 5.6 = 8.14° = 0.1421$ rad, $\eta = C_L/2\pi\alpha_0' = 0.8/[(2\pi)(0.1421)] = 0.896$. The wing will support a weight equal to the lift force, $F_L = C_L\rho(V^2/2)(B)(c)$.
   From Table A-7, at 10 000 ft, $\rho = 0.001756$ slug/ft³. $F_L = (0.8)(0.001756)[(300)^2/2](36)(6) = 13\,650$ lb while $F_D = (0.047/0.8)(13\,650) = 802$ lb, horsepower required $= (802)(300)/550 = 437$ hp.

**24.84** A cylinder 4 ft in diameter and 25 ft long rotates at 90 rpm with its axis perpendicular to an air stream with a wind velocity of 120 fps. The specific weight of the air is 0.0765 lb/ft³. Assuming no slip between the cylinder and the circulatory flow, find (a) the value of the circulation; (b) the transverse or lift force; and (c) the position of the stagnation points.

$\blacksquare$ (a) Peripheral velocity $v_t = 2\pi Rn/60 = (2\pi)(2)(\frac{90}{60}) = 18.85$ fps, $\Gamma = 2\pi R v_t = (2\pi)(2)(18.85) = 237$ ft²/s.

(b)
$$F_L = \rho BU\Gamma = (0.0765/32.2)(25)(120)(237) = 1690\ \text{lb}$$

(c)
$$\sin \theta_s = -\Gamma/4\pi RU = -237/(4\pi 2)(120) = -0.0786$$

Therefore $\theta_x = 184.5°$, $355.5°$. Actually, the real circulation produced by surface drag of the rotating cylinder would be only about one-half of that obtained for the no-slip assumption.

**24.85** A wing with a 22-m span and 64-m² planform area moves horizontally through the standard atmosphere at 8 km, with a velocity of 760 km/h. If the wing supports 280 kN, find (a) the required value of the lift coefficient, (b) the downwash velocity, assuming semielliptical distribution of lift over the span, and (c) the induced drag.

$\blacksquare$ From Table A-8, $\rho = 0.526$ kg/m³; $V = (760)(1000)/3600 = 211.1$ m/s.

(a)
$$F_L = C_L\rho(V^2/2)A \qquad 280\,000 = C_L(0.526)(211.1^2/2)(64) \qquad C_L = 0.373$$

(b)
$$V_i/V = C_L/[\pi(B^2/A)] \qquad V_i/211.1 = 0.373/[(\pi)(22^2/64)] \qquad V_i = 3.31\ \text{m/s}$$

(c)
$$(C_D)_i = C_L^2/[\pi(B^2/A)] = 0.373^2/[(\pi)(22^2/64)] = 0.00586$$
$$(F_D)_i = (C_D)_i(F_L/C_L) = (0.00586)(280\,000/0.373) = 4399\ \text{N}$$

**24.86**  A glider, of gross weight 350 lb, has a 3-ft-chord by 18-ft-span wing of the Clark Y section. Assuming that its characteristics are the same as those for the larger wing of the same aspect ratio shown in Fig. A-67, find the angle of glide through standard air at 2000 ft which will produce the greatest horizontal range.

▌ From Fig. A-67 for maximum $C_L/C_D$: $\alpha = -0.1°$, $C_L = 0.4$, and $C_D = 0.019$. For maximum range, glide angle $\beta$ is minimum. $\beta = \arctan(F_D/F_L) = \arctan(0.019/0.4) = 2.7°$.

**24.87**  A boat fitted with a hydrofoil weighs 4000 lb. At a velocity of 60 fps, what size hydrofoil is needed to support the boat? Use the lift characteristics in Fig. A-70 at an angle of attack of 2°.

▌ $F_L = C_L \rho(V^2/2)A$. From Fig. A-70, $C_L = 0.35$. $4000 = (0.35)(1.94)(60^2/2)A$, $A = 3.27$ ft².

**24.88**  At what angle of attack should a 128-kN airplane be flown at a speed of 225 km/h, if the planform area is 75 m²? What power is required to overcome wing drag? Assume a temperature of 30 °C and flaps at zero degrees.

▌
$$F_L = C_L(\tfrac{1}{2})\rho V^2 A \qquad 128\,000 = (C_L)(\tfrac{1}{2})(\rho)(225/3.6)^2(75) \qquad \rho = p/RT$$

$$\rho = 101\,325/[(287)(303)] = 1.165 \text{ kg/m}^3 \qquad C_L = 0.751 \qquad \alpha = 8° \qquad C_D = 0.04$$

$$F_D = (0.04)(\tfrac{1}{2})(1.165)(225/3.6)^2(75) = 6.825 \text{ kN} \qquad P = (6.825)(225/3.6) = 426 \text{ kW} = 570 \text{ hp}$$

**24.89**  The takeoff speed of the aircraft of Prob. 24.88 is about 1.3 times the stall speed. Calculate the takeoff distance for a constant thrust of 24 kN and a rolling resistance of 1.8 kN. Assume that the flaps are at 40°, the air is at 30 °C, the overall coefficient of drag for the plane is 0.20, and the frontal area is 45 m².

▌ The stall speed occurs at the condition of $(C_L)_{max}$ which here is 1.64.

$$\rho = 101\,325/[(287)(303)] = 1.165 \text{ kg/m}^3$$

$$V_{stall} = \sqrt{(2)W/[(\rho)(C_L)_{max}A]} = \sqrt{(2)(128\,000)/[(1.165)(1.64)(75)]} = 42.3 \text{ m/s}$$

$V$ at takeoff is $(1.3)(42.3) = 55.0$ m/s. By Newton's second law,

$$24\,000 - 1800 - (0.2)(\rho V^2/2)(45) = (128\,000/g)V(dV/dx)$$

$$7400 - 1.748V^2 = 4354V(dV/dx) \qquad (4354)(V\,dV)/(7400 - 1.748V^2) = dx$$

Let $7400 - 1.748V^2 = \eta$, $-3.50V\,dV = d\eta$, $V\,dV = (d\eta/3.50)$.

$$\int_{7400}^{2112}(4354)\frac{(-d\eta/3.50)}{\eta} = \int_0^{L_1}dx \qquad -\frac{4354}{3.50}\ln\eta\Big]_{7400}^{2112} = L_1 \qquad L_1 = 1560 \text{ m}$$

**24.90**  An airplane with a planform wing area of 233 ft² is flying at an angle of attack of 3° at a speed of 200 mph. The air is at 40 °F, and the flaps are at zero degrees. Compute the weight of the plane.

▌
$$C_L = 0.35 \qquad L = C_L(\tfrac{1}{2})\rho V^2 A = W = (0.35)(\tfrac{1}{2})(\rho)(V_0^2)(A)$$

$$\rho = (14.7)(144)/[(53.3)(g)(500)] = 0.00247 \text{ slug/ft}^3$$

$$W = (0.35)(\tfrac{1}{2})(0.00247)[(200)\tfrac{5280}{3600}]^2(233) = 8666 \text{ lb} = 4.33 \text{ ton}$$

**24.91**  A boat is fitted with hydrofoils having a total planform area of 1.2 m². At 11 knots, the stall speed, the coefficient of lift is 1.6 and the coefficient of drag is 0.6. Find the maximum weight and the minimum power corresponding to this stall speed.

▌
$$W = F_L = C_L(\tfrac{1}{2})\rho V^2 A = (1.6)(\tfrac{1}{2})(1000)[(11)(0.5144)]^2(1.2) = 30.737 \text{ kN}$$

$$F_D = C_D(\tfrac{1}{2})\rho V^2 A = (0.6)(\tfrac{1}{2})(1000)[(11)(0.5144)]^2(1.2) = 11.526 \text{ kN}$$

$$P = F_D V = (11.526)[(11)(0.5144)] = 65.2 \text{ kW}$$

**24.92**  The 9-in-wide, 4.5-ft-long flat blades of a ceiling fan are attached to an 8-in-diameter core; the blades' angle of attack is 15°. At what speed of rotation will the core experience zero vertical force? Each blade weighs 1.6 lb, and the room temperature is 40 °F.

▎From Fig. A-71, $C_L = 0.79$. Determine $\omega$ by equating the lift on a single blade to its weight.

$$F_L = C_L \int \left(\tfrac{1}{2}\rho V^2\right) dA = (0.79) \int_{0.333}^{4.833} \left(\tfrac{1}{2}\rho V_0^2\right)\left(\frac{9\,dr}{12}\right) \qquad \rho = \frac{(14.7)(144)}{(53.3)(g)(500)} = 0.002467 \text{ slug/ft}^3$$

$$F_L = 1.6 = (0.79)(\tfrac{1}{2})(0.002467)(\tfrac{9}{12}) \int_{0.333}^{4.833} (r\omega)^2 \, dr \qquad 1.6 = (0.000731)\omega^2 \left.\frac{r^3}{3}\right|_{0.333}^{4.833} = 0.000244\omega^2(4.833^3 - 0.333^3)$$

$$\omega = 7.623 \text{ rad/s} = (7.623/2\pi)(60) = 72.8 \text{ rpm}$$

**25.1** Assuming $\rho$ to be constant, do these flows satisfy continuity? (*a*) $u = -2y$, $v = 3x$; (*b*) $u = 0$, $v = 3xy$; (*c*) $u = 2x$, $v = -2y$. Continuity for incompressible fluids is satisfied if $\partial u / \partial y + \partial v / \partial y = 0$.

**∎** (*a*)          $[\partial(-2y)/\partial x] + [\partial(3x)/\partial y] = 0 + 0 = 0$          (continuity is satisfied)

   (*b*)          $[\partial(0)/\partial x] + [\partial(3xy)/\partial y] = 0 + 3x \neq 0$          (continuity is not satisfied)

   (*c*)          $[\partial(2x)/\partial x] + [\partial(-2y)/\partial y] = 2 - 2 = 0$          (continuity is satisfied)

*Note:* If (*b*) did indeed describe a flow field, the fluid must be compressible.

**25.2** Determine whether these flows are rotational or irrotational. (*a*) $u = -2y$, $v = 3x$; (*b*) $u = 0$, $v = 3xy$; (*c*) $u = 2x$, $v = -2y$.

**∎** If $(\partial v / \partial x) - (\partial u / \partial y) = 0$   (flow is irrotational):

   (*a*)          $[\partial(3x)/\partial x] - [\partial(-2y)/\partial y] = 3 + 2 \neq 0$     (flow is rotational)

   (*b*)          $[\partial(3xy)/\partial x] - [\partial(0)/\partial y] = 3y - 0 \neq 0$     (flow is rotational)

   (*c*)          $[\partial(-2y)/\partial x] - [\partial(2x)/\partial y] = 0 - 0 = 0$     (flow is irrotational)

**25.3** Check these flows for continuity and determine the vorticity of each: (*a*) $v_t = 6r$, $v_r = 0$; (*b*) $v_t = 0$, $v_r = -5/r$.

**∎** Applying the equations $(v_r/r) + (\partial v_r / \partial r) + (\partial v_t / r \, \partial \theta) = 0$, $\xi = (\partial v_t / \partial r) + (v_t / r) - (\partial v_r / r \, \partial \theta)$.

   (*a*)          $(0/r) + [\partial(0)/\partial r] + [\partial(6r)/r \, \partial \theta] = 0$          (continuity is satisfied)

   $\xi = [\partial(6r)/\partial r] + (6r/r) - [\partial(0)/r \, \partial \theta] = 6 + 6 - 0 = 12$          (flow is rotational)

   (*b*)     $-[(5/r)/r] + [\partial(-5r^{-1})/\partial r] + [\partial(0)/r \, \partial \theta] = -(5/r^2) + (5/r^2) + 0 = 0$          (continuity is satisfied)

   $\xi = \dfrac{\partial(0)}{\partial r} + \dfrac{0}{r} - \dfrac{\partial(-5/r)}{r \, \partial \theta} = 0$          (flow is irrotational)

**25.4** A flow is defined by $u = 2x$ and $v = -2y$. Find the stream function and potential function for this flow and plot the flow net.

**∎** Check continuity: $(\partial u / \partial x) + (\partial v / \partial y) = 2 - 2 = 0$. Hence continuity is satisfied and it is possible for a stream function to exist: $d\psi = -v \, dx + u \, dy = 2y \, dx + 2x \, dy$, $\psi = 2xy + C_1$. Check to see if the flow is irrotational: $(\partial v / \partial x) - (\partial u / \partial y) = 0 - 0 = 0$. Hence the flow is irrotational and a potential function exists: $d\phi = -u \, dx - v \, dy = -2x \, dx + 2y \, dy$, $\phi = -(x^2 - y^2) + C_2$. The location of lines of equal $\psi$ can be found by substituting values of $\psi$ into the expression $\psi = 2xy$. Thus for $\psi = 60$, $x = 30/y$. This line is plotted (in the upper right-hand quadrant) in Fig. 25-1. In a similar fashion lines of equal potential can be plotted. For example, for $\phi = 60$ we have $-(x^2 - y^2) = 60$ and $x = \pm\sqrt{y^2 - 60}$. This line is also plotted on the figure. The flow net depicts *flow in a corner*. Mathematically the net will plot symmetrically in all four quadrants.

**25.5** Given a two-dimensional flow defined by $u = 4 + 3x$, $v = 0$. If this flow satisfies continuity, what can be said about the density of the fluid?

**∎** $\partial(\rho u)/\partial x = 0$, or $\rho v = c(y)$, or $\rho = c(y)/(4 + 3x)$. Therefore, for each fixed $y$, the density $\rho$ must decrease as $x$ increases.

**25.6** The flow of an incompressible fluid is defined by $u = 3$, $v = 9x$. Does a stream function exist for this flow? If so, determine the expression for the stream function.

**∎** To find if a stream function exists, check for continuity. $\partial u / \partial x = 0$, $\partial v / \partial y = 0$, $w = 0$. Thus, $(\partial u / \partial x) + (\partial v / \partial y) + (\partial w / \partial z) = 0$ (continuity) is satisfied and a stream function exists.

$$d\psi = -v \, dx + u \, dy = -9x \, dx + 3 \, dy \qquad \psi = -4.5x^2 + 3y$$

Flow net                    **Fig. 25-1**

**25.7**   Plot the streamlines in the upper right-hand quadrant for the flow defined by $\psi = 1.6x^2 + y^2$ and determine the value of the velocity at $x = 3$, $y = 4$.

▌ $u = \partial\psi/\partial y = 2y$, $v = -\partial\psi/\partial x = -3.2x$. At $(3, 4)$ $u = 8$, $v = -9.6$, $V = \sqrt{8^2 + 9.6^2} = 12.50$. To plot streamlines, rewrite as $y = (\psi - 1.6x^2)^{1/2}$. Assume $\psi$, compute $y$'s for different $x$'s. See Fig. 25-2.

**Fig. 25-2**

**25.8**   The components of the velocities of a certain flow system are $u = -(Q/2\pi)[x/(x^2 + y^2)] + By + C$, $v = -A[y/(x^2 + y^2)] + Dx + E$. **(a)** Calculate a value of $A$ consistent with continuous flow. **(b)** Sketch the streamlines for this flow system, assuming $B = C = D = E = 0$.

▌ **(a)** To satisy continuity 2-D flow: $(\partial u/\partial x) + (\partial v/\partial y) = 0$.

$$\partial u/\partial x = -(Q/2\pi)\{[(x^2 + y^2) - x(2x)]/(x^2 + y^2)^2\} = -(Q/2\pi)[(y^2 - x^2)/(x^2 + y^2)^2]$$
$$\partial v/\partial y = -A\{[(x^2 + y^2) - y(2y)]/(x^2 + y^2)^2\} = -A[(x^2 - y^2)/(x^2 + y^2)^2]$$

Substituting, $-Q/2\pi(y^2 - x^2) - A(x^2 - y^2) = 0$; thus $A = Q/2\pi$.

**(b)**   $\qquad\qquad u = -(Q/2\pi)[x/(x^2 + y^2)] \qquad v = -(Q/2\pi)[y/(x^2 + y^2)]$

$$d\psi = -v\,dx + u\,dy = (Q/2\pi)[(y\,dx)/(x^2 + y^2)] - (Q/2\pi)[(x\,dy)/(x^2 + y^2)]$$

Integrating, $\psi = (Q/2\pi)[\tan^{-1}(x/y) - \tan^{-1}(y/x)]$. See Fig. 25-3.

**Fig. 25-3**

**25.9**  A flow field is described by $\psi = x^2 - 2y$. Describe the streamlines for $\psi = 0$, 1, and 2. Derive an expression for the velocity at any point in the flow field and determine the vorticity of the flow.

■ From the equation for $\psi$, the flow field is a family of parabolas symmetric about the $y$ axis. The streamline $\psi = 0$ passes through the origin. $u = \partial u/\partial y = -2$, $v = -\partial u/\partial x = -2x$. Thus $V = (u^2 + v^2)^{1/2} = (4 + 4x^2)^{1/2}$. Vorticity: $\xi = (\partial v/\partial x) - (\partial u/\partial y) = -2$. (Since $\xi \neq 0$, the flow is rotational.)

**25.10**  Locate the stagnation point of the incompressible flow due to a source of strength $9\pi$ located at $(3, 0)$ and a source of strength $18\pi$ located at $(-4, 0)$.

■ It is obvious that this flow has a single stagnation point, located between the sources on the $x$-axis (where $v$ is automatically zero). $\psi = (q_1\theta_1/2\pi) + (q_2\theta_2/2\pi) = (9\pi/2\pi) \arctan[y/(x-3)] + (18\pi/2\pi) \arctan[y/(x+4)]$, $u = \partial\psi/\partial y = 4.5\{(x-3)/[(x-3)^2 + y^2]\} + 9\{(x+4)/[(x+4)^2 + y^2]\}$. If $y = 0$, $u = [4.5/(x-3)] + [9/(x+4)]$. At stagnation point, $u = 0 = [4.5/(x-3)] + [9/(x+4)]$; thus $x = \frac{2}{3}$.

**25.11**  A source discharging 18 m³/s per m is located at the origin and a uniform flow at 2 m/s in the +x-direction is superimposed on the source flow. Write the overall stream function in polar and rectangular coordinates

■

$$\psi_1 = 18(\theta/2\pi) = (18/2\pi)\arctan(y/x) \qquad \psi_2 = 2y = 2r\sin\theta$$

$$\psi_{\text{polar}} = (9\theta/\pi) + 2r\sin\theta \qquad \psi_{\text{rect}} = (9/\pi)\arctan(y/x) + 2y$$

**25.12**  Refer to Prob. 25.11. Find the difference in pressure head between point $A(-9\,\text{m}, 0)$ and point $B(0, 1.33\,\text{m})$.

■ $u = \partial\psi_{\text{rect}}/\partial y = (9/\pi)[x/(x^2 + y^2)] + 2$; $v = -\partial\psi_{\text{rect}}/\partial x = (9/\pi)[y/(x^2 + y^2)]$. At point $A$ $(-9\,\text{m}, 0)$: $u = 1.68\,\text{m/s}$, $v = 0$; $V_A = 1.68\,\text{m/s}$. At point $B$ $(0, 1.33\,\text{m})$: $u = 2.0\,\text{m/s}$, $v = 2.15\,\text{m/s}$; $V_B = \sqrt{2.0^2 + 2.15^2} = 2.94\,\text{m/s}$. From Bernoulli's theorem: $p_A/\gamma - p_B/\gamma = V_B^2/2g - V_A^2/2g = (2.94^2 - 1.68^2)/[2(9.81)] = 0.297\,\text{m}$.

**25.13**  (a) Does the flow $u = x^2 + 3x - 4y$, $v = -2xy - 3y$ satisfy continuity? (b) Compute the vorticity. (c) Locate all stagnation points. (d) Find the equation of the stream function and graph it in the first quadrant.

■ (a) $(\partial u/\partial x) + (\partial v/\partial y) = 2x + 3 - 2x - 3 = 0$; continuity is satisfied.
(b) $\xi = (\partial v/\partial x) - (\partial u/\partial y) = -2y - 4 \neq 0$; flow is rotational.
(c) Stagnation points occur where both $u = 0$ and $v = 0$. $v = 0 = -2xy - 3y = -2y(x + \frac{3}{2})$; true when $y = 0$ or $x = -\frac{3}{2}$, $u = 0 = x^2 + 3x - 4y$. If $x = -1$, $y = -\frac{1}{2}$. If $y = 0$, $x = 0$ or $-3$. Thus there are three stagnation points: $(0, 0)$, $(-3, 0)$, and $(-\frac{3}{2}, -\frac{1}{2})$.

(d)
$$u = \partial\psi/\partial y \qquad \psi = \int u\,\partial y = \int (x^2 + 3x - 4y)\,\partial y = x^2y + 3xy - 2y^2 + C(x)$$

$$v = -2xy - 3y = -\partial\psi/\partial x = -2xy - 3y - C'(x) \qquad C(x) = \text{constant} = 0$$

So $\psi = x^2y + 3xy - 2y^2$. Notice in the graph, Fig. 25-4, the branching of $\psi = 0$ at the origin (a stagnation point).

**25.14**  Is the function $\psi = 4x - 3y$ the stream function of a two-dimensional potential flow?

■ Yes: $\partial^2\psi/\partial x^2 + \partial^2\psi/\partial y^2 = 0 + 0 = 0$.

**Fig. 25-4**

**25.15** An ideal fluid flows in a two-dimensional 90° bend. The inner and outer radii of the bend are 0.5 and 1.5 ft. Sketch the flow net and estimate the velocity at the inner and outer walls of the bend if the velocity in the 1.0-ft-wide straight section is 9 fps. Develop an analytic expression for the stream function, noting that $v_t = -\partial\psi/\partial r$ and $v_r = \partial\psi/r\,\partial\theta$. Determine the inner and outer velocities accurately.

▌ Velocity estimates: $V = -\partial\phi/\partial s$, $-\Delta\phi \approx V\,\Delta s \approx 9\,\Delta s_0$, $V \approx 9\,\Delta s_0/\Delta s$. From Fig. 25-5, $V_{out} \approx 9(\frac{5}{8}) = 5.6$ fps, $V_{in} \approx 9(5/2.5) = 18$ fps.

Accurate velocity determinations: $v_t(\Delta r) = $ constant, $\Delta r$ is proportional to $r$. $v_t(kr) = $ constant, therefore $v_t = C/r$. Also $v_r = 0$. $\psi = \psi(r, \theta)$, so by chain rule of differentiation $d\psi = (\partial\psi/r\,\partial\theta)r\,d\theta + (\partial\psi/\partial r)\,dr = v_r r\,d\theta - v_t\,dr = 0 - v_t\,dr$. If $\psi_{in} = 0$, $\psi_{out} = \psi_{in} + dq = 0 + (1)(9) = 9$ cfs/ft. Integrating

$$\int_0^{10} d\psi = -\int_{0.5}^{1.5} v_t\,dr = -C\int_{0.5}^{1.5}\frac{dr}{r} \qquad \psi = 9 = -C[\ln r]_{0.5}^{1.5} = -C\ln\frac{1.5}{0.5} = -C\ln 3.0$$

and $C = -8.19$. Thus the analytic expression is $\psi = 8.19\ln r$; $v_t = -8.19/r$, $(v_t)_{r=0.5} = 16.4$ fps, $(v_t)_{r=1.5} = 5.5$ fps.

$V_0 = 9$ fps

Fig. 25-5

**25.16** A 4-ft-diameter cylindrical drum is fixed in an open channel 5 ft wide. Water flows at the rate of 60 cfs beneath the drum, as shown in Fig. 25-6a. Sketch the flow net, and infer values of the pressure at points $A$, $C$, $D$, and $B$. Sketch the pressure distribution, and by numerical integration estimate the horizontal thrust on the cylinder.

(a)

$p/\gamma$

0.140 A

C
0.891

D
0.317    B
        0

(b)

Horizontal Components of p/γ
(projection)

A  0.140 × 1.000 = 0.140

C  0.891 × 0.866 = 0.772

D  0.317 × 0.500 = 0.158

B  0

(c)

Fig. 25-6

❚ Sketch streamlines by eye and use continuity to determine velocities. High degree of accuracy is not to be expected. See Fig. 25-6b. At $A$, $p/\gamma = p_s/\gamma = V^2/2g = 3^2/2g = 0.140$ ft. At $B$, in contact with atmosphere ($p_{atm} = 0$ psig): $p/\gamma = 0$. At $C$, from flow net, $V \approx 4$ fps: $4 + 3^2/2g = (4 - 1.0) + p/\gamma + 4^2/2g$; $p/\gamma = 0.891$ ft. At $D$, from flow net, $V \approx 10$ fps: $4 + 3^2/2g = (4 - 1.73) + p/\gamma + 10^2/2g$; $p/\gamma = 0.317$ ft. See Fig. 25-6c. Area under projected $p/\gamma$ diagram $\approx$ avg horizontal component = area/2.2 = 0.36 ft = 0.16 psi, horizontal thrust $\approx$ 0.16(2.2)(12)(5)(12) = 253 lb.

**25.17** For the two-dimensional flow of a frictionless incompressible fluid against a flat plate normal to the initial velocity, the stream function is given by $\psi = -2axy$, while its *conjugate function*, the velocity potential, is $\phi = a(x^2 - y^2)$ where $a$ is a constant and the flow is symmetrical about the $yz$ plane. By direct differentiation demonstrate that these functions satisfy the equation $(\partial^2\phi/\partial x^2) + (\partial^2\phi/\partial y^2) = 0$. Using a scale of 1 in = 1 unit of distance, plot the streamlines given by $\psi = \pm 2a$, $\pm 4a$, $\pm 6a$, $\pm 8a$, and the equipotential lines given by $\phi = 0$, $\pm 2a$, $\pm 4a$, $\pm 6a$, $\pm 8a$. Observe that this flow net also gives the ideal flow around an inside square corner.

❚ Given: $\psi = -2axy$; $\phi = a(x^2 - y^2)$. Differentiating: $\partial\psi/\partial y = -2ax$; $\partial\phi/\partial x = 2ax$; $-\partial\psi/\partial x = 2ay$; $\partial\phi/\partial y = a(-2y)$; $\partial^2\phi/\partial x^2 = 2a$; $\partial^2\phi/\partial y^2 = -2a$; thus the equation indicated is satisfied. See Fig. 25-7 for required plot.

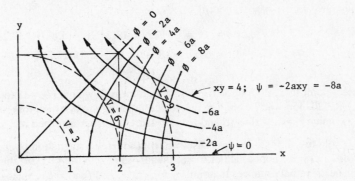

Fig. 25-7

**25.18** (*Equivalent Source Method*). The flow around the body of Fig. 25-8a may be considered as that due to the sum of two velocity potentials, $\phi_1 = -Ux$, representing an undisturbed flow of velocity $U$ in the $x$ direction, and $\phi_2 = -S \ln r$, representing the radial flow from a source located inside the body behind the stagnation point. To relate $U$ and $S$, it is observed that the total flow $2\pi S$ from the source (which is hydrodynamically equivalent to the body itself) must be equal to the flow of the main stream which is not passing through the body of width $b$, or $2\pi S = Ub$. This gives $\phi_2 = -(Ub/2\pi) \ln r$. The velocity at any point of the field is given by the vector sum of the components $U$ and $v_r = -\partial\phi_2/\partial r$. (**a**) Prove that the distance from source to stagnation point is as given in Fig. 25-8a.

There follows an ingenious method of plotting the boundary of such a streamlined body, as shown in Fig. 25-8b. Suppose that the streamlines in the undisturbed flow are spaced a distance $a$ apart, where $b/2a = n$, an integer. Next divide the upper half of the source into $n$ radial sectors, each of angle $\alpha$, that is, $n\alpha = \pi$. Then the undisturbed flow between the $x$ axis and the first streamline is associated with the source flow in the first sector from the stagnation point. Thus the intersection of the first streamline with the first line must be a point on the boundary of the body, through which there can be no flow. Similarly, the intersection of the horizontal line at $2a$ with the radial line at $2\alpha$ forms another point, and so on. Further streamlines can be plotted by connecting successive intersections of the original horizonal lines with the radial lines, recognizing that the same flow must exist between any adjacent pair of streamlines. Thus the intersection of a horizontal line $ea$ above the axis with a radial line at $f\alpha$ from the stagnation point must lie on a streamline which is $(e - f)a$ distant from the axis in the undisturbed region, where $e$ and $f$ are integers. (**b**) Given $U = 18$ fps and $b = 34$ ft, determine the velocity vector at point $E$ of Fig. 25-8b.

❚ (**a**) Given: source potential, $\phi_2 = -(Ub/2\pi) \ln r$; $v_r = -\partial\phi_2/\partial r = Ub/2\pi r$. Setting $v_r = U$ for stagnation condition (on tip of nose) and solving for $r$, the desired distance behind the nose, $r = b/2\pi$.

(**b**) For $b = 34$ ft, distance to stagnation point = $-34/2\pi = -5.41$ ft. Since $b/2a = n = 6 = 34/2a =$ (the number of radial sectors from half of source), streamline spacing = $a = \frac{34}{12} = 2.83$ ft. From Fig. 25-8b: At $E$, $r = 3a = 3(2.83) = 8.49$ ft and $v_r = Ub/2\pi r = (18)(34)/[2\pi(8.49)] = 11.47$ fps. By superposition, $V = (U^2 + v_r^2)^{1/2} = (18^2 + 11.47^2)^{1/2} = 21.3$ fps. Direction relative to axis is $\tan^{-1}(11.47/18) = 32.5°$.

Fig. 25-8

**25.19** Find the flow velocity at point $F$ of Fig. 25-8$b$.

    ❚ At $F$ ($\alpha = 30°$), $r = a/\sin 30° = 2.83/0.5 = 5.66$ ft.

$$v_r = \partial \phi_2/\partial r = Ub/2\pi r = 18(34)/[2\pi(5.66)] = 17.21 \text{ fps}$$
$$V^2 = U^2 + v_r^2 - 2Uv_r \cos \alpha = 18^2 + 17.21^2 - 2(18)17.21 \cos 30° \qquad V = 9.14 \text{ fps}$$

**25.20** Superimpose a point source ($Q = 95$ cfs/ft) on a rectilinear flow field ($U = 18$ fps). Plot the body contour at $\theta = 30, 60, 90, 120, 150, 180°$. Compute the velocities along the body contour at these points, as well as the pressures, assuming $\rho = 1.94$ slug/ft$^3$ and zero pressure in the undisturbed rectilinear flow field.

    ❚ Given: $U = 18$ fps, $v_r = 95/2\pi r$ fps. Distance from source to stagnation point: $U = v_r$, $r = 5.28/2\pi = 0.840$ ft. Refer to Prob. 25.18; for upper half let $n = 6$, thus $\theta = 30°$. $q = Ua = 95/2n$; thus $a = 95/(2nU) = 0.440$ ft. See Fig. 25-9. Scale off radial distances to points $a$, $b$, $c$, $d$, $e$, and $f$; $v_x = U - v_r \cos \theta$, $v_y = v_r \sin \alpha$. Vector addition: $V = (v_x^2 + v_y^2)^{1/2}$.

| point | $r$, ft | $\theta$ | $v_r$, fps | $v_x$, fps | $v_y$, fps | $V$, fps |
|-------|---------|----------|------------|------------|------------|----------|
| $S$ | 0.84 | 0 | 18.0 | 0 | 0 | 0 |
| $a$ | 0.85 | 30° | 17.8 | 2.6 | 8.9 | 9.3 |
| $b$ | 1.00 | 60° | 15.1 | 10.4 | 13.1 | 16.7 |
| $c$ | 1.28 | 90° | 11.8 | 18.0 | 11.8 | 21.5 |
| $d$ | 1.94 | 120° | 7.8 | 21.9 | 6.8 | 22.9 |
| $e$ | 4.20 | 150° | 3.6 | 21.1 | 1.8 | 21.2 |

$$p_0 + \rho V_0^2/2 = p + \rho V^2/2 \qquad 1.94(18^2/2) = 314 \text{ lb/ft}^2 = p + \rho V^2/2$$

Therefore, $p = 314 - \rho V^2/2$. Hence desired pressures are 314, 230, 43, $-134$, $-195$, $-122$ psf.

Fig. 25-9

# Appendixes

**Fig. A-1**

Absolute viscosity $\mu$ of fluids.

**Fig. A-2**

Kinematic viscosity $\nu$ of fluids.

**Fig. A-3**

Specific weight $\gamma$ of pure water as a function of temperature and pressure for condition where $g = 32.2 \text{ ft/s}^2$ $(9.81 \text{ m/s}^2)$.

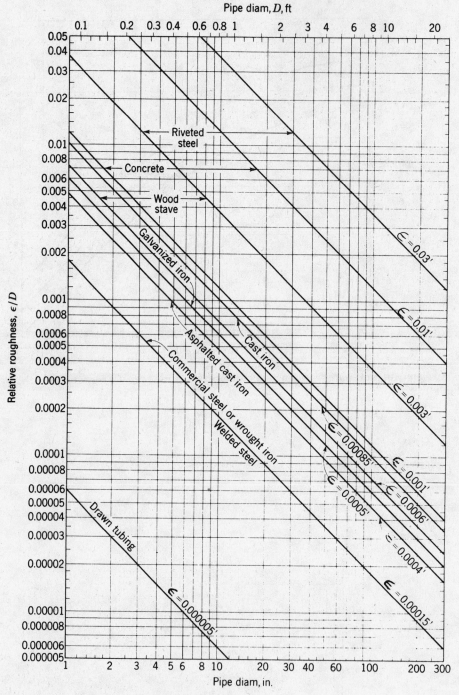

Pipe diam, $D$, ft

Relative roughness, $\epsilon/D$

Pipe diam, in.

**Fig. A-4**

Roughness factors ($\epsilon$ expressed in feet) for commercial pipes.

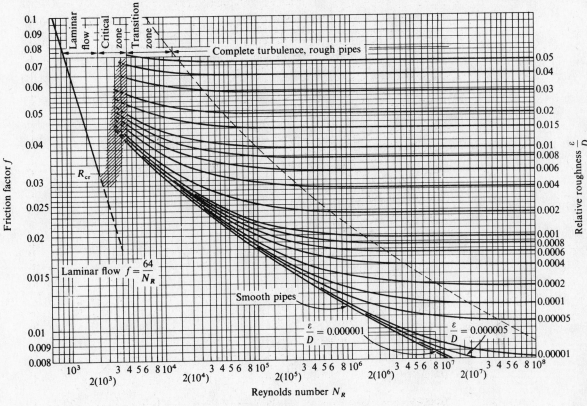

**Fig. A-5**

Moody diagram.

From Lewis F. Moody, "Friction Factors for Pipe Flows," *ASME Trans.*, vol. 66, pp. 671–684, 1944.

**Fig. A-6**

Nikuradse's sand-roughened-pipe tests.

Reentrant:
$K = 0.78$

Sharp-edged:
$K = 0.4\text{-}0.5$

Slightly
rounded:
$K = 0.2\text{-}0.25$

Well-rounded:
$K = 0.05$

*Exit losses*: $K = 1.0$ for all shapes of exit (reentrant, sharp-edged, slightly, or well-rounded)

**Fig. A-7**

Entrance and exit loss coefficients.

(a)

(b)

**Fig. A-8**

Entrance and exit loss coefficients: (*a*) reentrant inlets; (*b*) rounded and beveled inlets. Exit losses are $K \approx 1.0$ for all shapes of exit (reentrant, sharp, beveled, or rounded). (Adapted by permission from *ASHRAE Handbook of Fundamentals*, Atlanta, 1985.)

**Fig. A-9**

Sudden expansion and contraction losses. Note that the loss is based on velocity head in the small pipe.

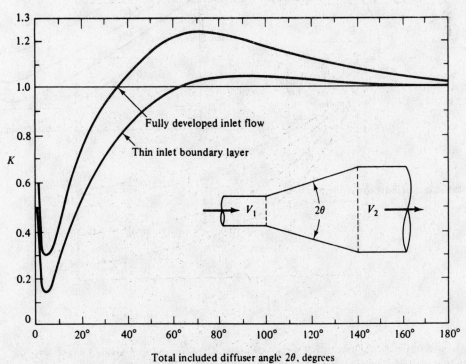

Total included diffuser angle $2\theta$, degrees

**Fig. A-10**

Flow losses in a gradual conical expansion region.

## SUDDEN AND GRADUAL CONTRACTION

If: $\theta \leqslant 45° \ldots \ldots \ldots K_2 =$ Formula 1

$45° < \theta \leqslant 180° \ldots K_2 =$ Formula 2

- **Formula 1**

$$K_2 = \frac{0.8 \left(\sin \frac{\theta}{2}\right)(1 - \beta^2)}{\beta^4} = \frac{K_1}{\beta^4}$$

- **Formula 2**

$$K_2 = \frac{0.5 \,(1 - \beta^2) \sqrt{\sin \frac{\theta}{2}}}{\beta^4} = \frac{K_1}{\beta^4}$$

$$\beta = \frac{d_1}{d_2}$$

$$\beta^2 = \left(\frac{d_1}{d_2}\right)^2 = \frac{a_1}{a_2}$$

Subscript 1 defines dimensions and coefficients with reference to the smaller diameter.

Subscript 2 refers to the larger diameter.

**Fig. A-11**

From "Flow of Fluids through Valves, Fittings, and Pipe," *Crane Co. Tech. Pap. 410,* New York, 1985.

Secondary flow pattern:

$R$

$D$ = constant

(*Note*: Resistance due to bend length must be added.)

$\frac{\epsilon}{D} = 0.01$

0.002

0.001

0.0005

0

$K$

$\frac{R}{D}$

**Fig. A-12**

Resistance coefficients for 90° bends.

**Fig. A-13**

Pipe diagram: Hazen-Williams equation ($C = 120$); English Gravitational System.

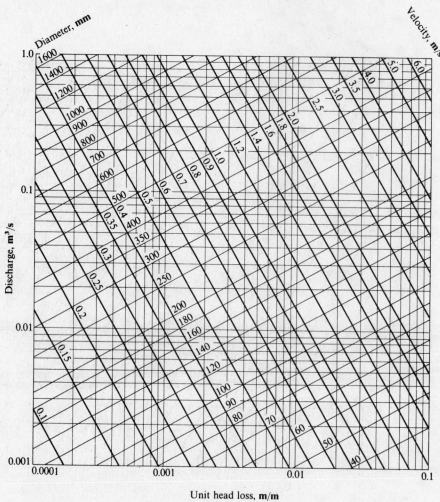

**Pipe diagram:** Hazen-Williams equation ($C = 120$); International System.

**Fig. A-15**

Pipe diagram: Manning equation ($n = 0.013$); English Gravitational System.

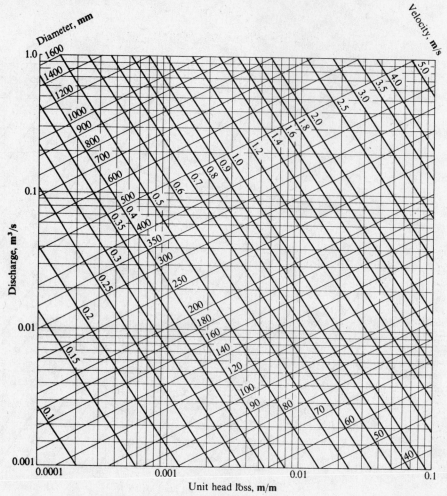

Fig. A-16

Pipe diagram: Manning equation ($n = 0.013$); International System.

# FLOW CHART

## HAZEN-WILLIAMS FORMULA, $C_1 = 100$

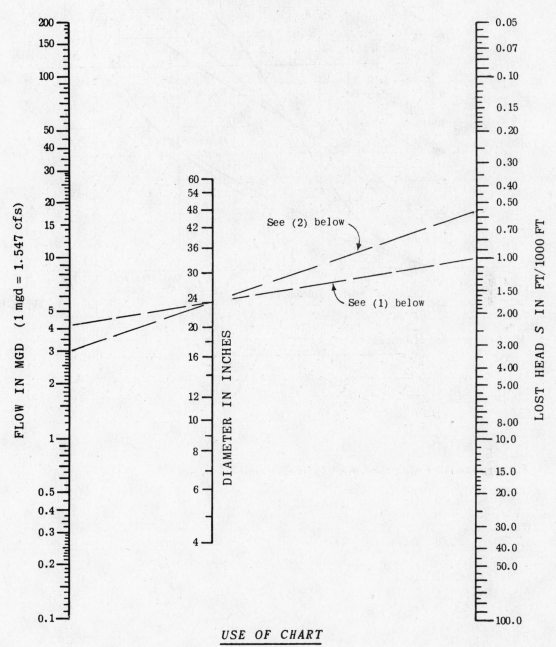

### USE OF CHART

(1) Given $D = 24''$, $S = 1.0$ ft/1000 ft, $C_1 = 120$; find flow $Q$.

Chart gives $Q_{100} = 4.2$ mgd.

For $C_1 = 120$, $Q = (120/100)(4.2) = 5.0$ mgd.

(2) Given $Q = 3.6$ mgd, $D = 24''$, $C_1 = 120$; find Lost Head.

Change $Q_{120}$ to $Q_{100}$: $Q_{100} = (100/120)(3.6) = 3.0$ mgd.

Chart gives $S = 0.55$ ft/1000 ft.

**Fig. A-17**

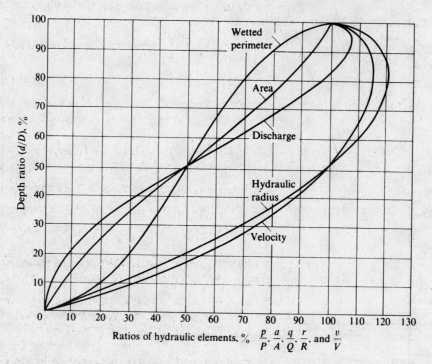

Ratios of hydraulic elements, % $\dfrac{p}{P}, \dfrac{a}{A}, \dfrac{q}{Q}, \dfrac{r}{R},$ and $\dfrac{v}{V}$

**Fig. A-18**

Hydraulic elements of a circular section.

| Channel Slope | Depth Relations | $\left(\dfrac{dy}{dL}\right)$ | Type of Profile | Symbol | Type of Flow | Form of Profile |
|---|---|---|---|---|---|---|
| Mild $0 < S < S_c$ | $y > y_N > y_c$ | $+$ | Backwater | $M_1$ | Subcritical | |
| | $y_N > y > y_c$ | $-$ | Dropdown | $M_2$ | Subcritical | |
| | $y_N > y_c > y$ | $+$ | Backwater | $M_3$ | Supercritical | |
| Horizontal $S = 0$ $y_N = \infty$ | $y > y_c$ | $-$ | Dropdown | $H_2$ | Subcritical | |
| | $y_c > y$ | $+$ | Backwater | $H_3$ | Supercritical | |
| Critical $S_N = S_c$ $y_N = y_c$ | $y > y_c = y_N$ | $+$ | Backwater | $C_1$ | Subcritical | |
| | $y_c = y = y_N$ | | Parallel to bed | $C_2$ | Uniform, Critical | |
| | $y_c = y_N > y$ | $+$ | Backwater | $C_3$ | Supercritical | |
| Steep $S > S_c > 0$ | $y > y_c > y_N$ | $+$ | Backwater | $S_1$ | Subcritical | |
| | $y_c > y > y_N$ | $-$ | Dropdown | $S_2$ | Supercritical | |
| | $y_c > y_N > y$ | $+$ | Backwater | $S_3$ | Supercritical | |
| Adverse $S < 0$ $y_N = \infty$ | $y > y_c$ | $-$ | Dropdown | $A_2$ | Subcritical | |
| | $y_c > y$ | $+$ | Backwater | $A_3$ | Supercritical | |

**Fig. A-19**

Types of profiles in open-channel flow.

Values of $(D_1''V_1)$ for water at 72°F (diameter in inches x velocity in fps)

$$Q = KA_2\sqrt{2gh}$$
$$h = \Delta(z + \frac{p}{\gamma})$$

Reynolds number of approach, $\mathbf{R} = \dfrac{D_1 V_1 \rho_1}{\mu_1}$

**Fig. A-20**

Flow coefficients for ISA nozzle. (Adapted from *ASME Flow Measurement*, 1959.)

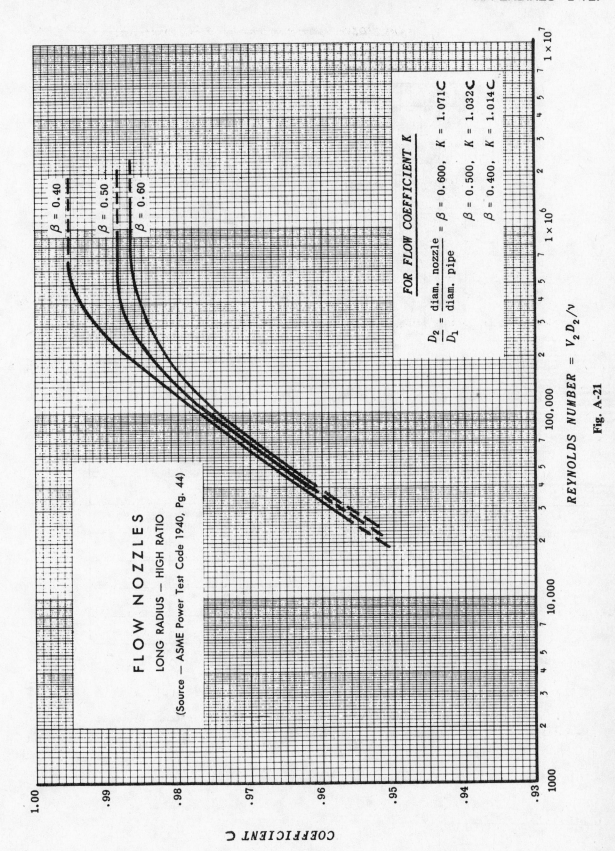

**FLOW NOZZLES**

LONG RADIUS — HIGH RATIO

(Source — ASME Power Test Code 1940, Pg. 44)

*FOR FLOW COEFFICIENT K*

$$\frac{D_2}{D_1} = \frac{\text{diam. nozzle}}{\text{diam. pipe}} = \beta = 0.600, \quad K = 1.071C$$

$$\beta = 0.500, \quad K = 1.032C$$

$$\beta = 0.400, \quad K = 1.014C$$

$\beta = 0.40$

$\beta = 0.50$

$\beta = 0.60$

COEFFICIENT C

*REYNOLDS NUMBER* $= V_2 D_2 / \nu$

**Fig. A-21**

$$C = \frac{C_v}{\sqrt{1 - \left(\frac{A_2}{A_1}\right)^2}}$$

Fig. A-22

ISA (VDI) flow nozzle and discharge coefficients. (Ref. 11 in *NACA Tech. Mem. 952.*)

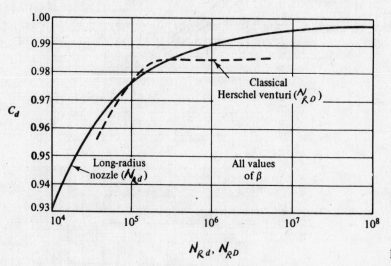

Fig. A-23

Discharge coefficient for long-radius nozzle and classical Herschel-type venturi.

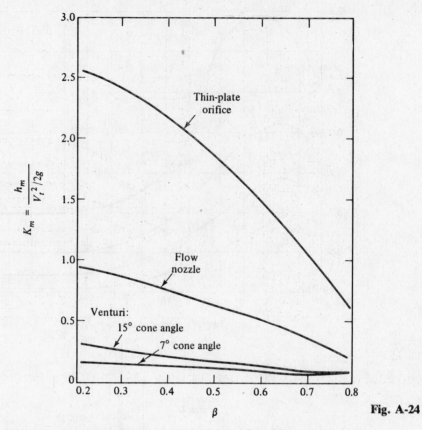

Fig. A-24

Nonrecoverable head loss in Bernoulli obstruction meters. [H. S. Bean (ed.), '*Fluid Meters: Their Theory and Application*,' 6th ed., American Society of Mechanical Engineers, New York, 1971.]

Fig. A-25

VDI orifice meter and flow coefficients for flange taps. (Adapted from *NACA Tech. Mem. 952.*)

Fig. A-26

Expansion factors.

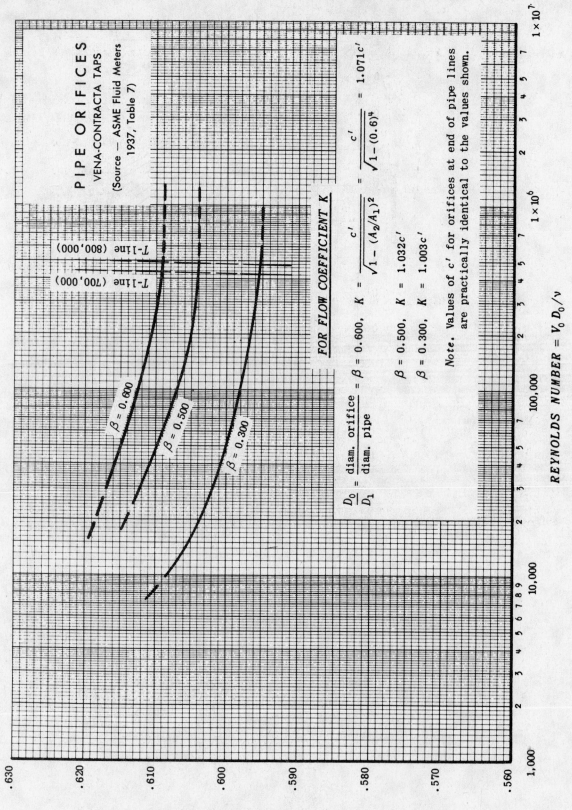

PIPE ORIFICES
VENA-CONTRACTA TAPS
(Source — ASME Fluid Meters
1937, Table 7)

$T$-line (800,000)

$T$-line (700,000)

$\beta = 0.600$

$\beta = 0.500$

$\beta = 0.300$

FOR FLOW COEFFICIENT K

$$\frac{D_0}{D_1} = \frac{\text{diam. orifice}}{\text{diam. pipe}} = \beta = 0.600, \quad K = \frac{c'}{\sqrt{1 - (A_2/A_1)^2}} = \frac{c'}{\sqrt{1 - (0.6)^4}} = 1.071c'$$

$$\beta = 0.500, \quad K = 1.032c'$$

$$\beta = 0.300, \quad K = 1.003c'$$

Note. Values of c' for orifices at end of pipe lines
are practically identical to the values shown.

REYNOLDS NUMBER = $V_0 D_0/\nu$

Fig. A-27

COEFFICIENT c'

**Fig. A-28**

VDI orifice and discharge coefficients. (Ref. 11 in *NACA Tech. Mem. 952.*)

**Fig. A-29**

Discharge coefficient for a thin-plate orifice with $D : \frac{1}{2}D$.

Values of $D_2''V_2$ for water at 72°F (diameter in inches × velocity in fps)

$$Q = CA_2 \sqrt{\frac{2gh}{1 - (D_2/D_1)^4}}$$

$D_1$ = inlet diameter
$D_2$ = throat diameter
$A_2$ = throat area

$$h = \left(z_1 + \frac{p_1}{\gamma}\right) - \left(z_2 + \frac{p_2}{\gamma}\right)$$

Reynolds number at throat, $N_R = \dfrac{D_2 V_2 \rho_2}{\mu_2}$

**Fig. A-30**

Venturi meter with conical entrance and flow coefficients for $D_2/D_1 = 0.5$.

(a)

(b)

Fig. A-31

Loss coefficient for conical diffusers.

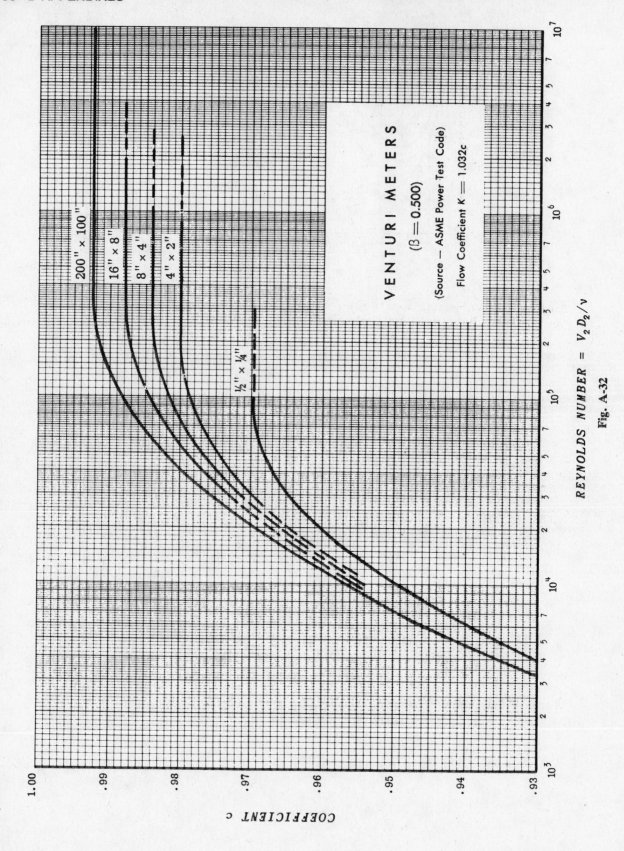

VENTURI METERS

(β = 0.500)

(Source — ASME Power Test Code)

Flow Coefficient K = 1.032c

200" × 100"

16" × 8"

8" × 4"

4" × 2"

½" × ¼"

COEFFICIENT c

REYNOLDS NUMBER = $V_2 D_2 / \nu$

Fig. A-32

Fig. A-33

Sudden-contraction loss coefficient. (Adapted from R. P. Benedict, N. A. Carlucci, and S. D. Swetz, "Flow Losses in Abrupt Enlargements and Contractions," *J. Eng. Power,* vol. 48, January 1966.)

Fig. A-34

Compressibility factor $Y$ for flow nozzles and Venturi meters. $k = 1.4$.

**Fig. A-35**

Coefficient $C_v$ for venturi meters (*Fluid Meters: Their Theory and Application*, 5th ed., American Society of Mechanical Engineers, 1959.)

**Fig. A-36**

Discharge coefficient for a venturi nozzle.

Fig. A-37

Coefficients for triangular weirs.

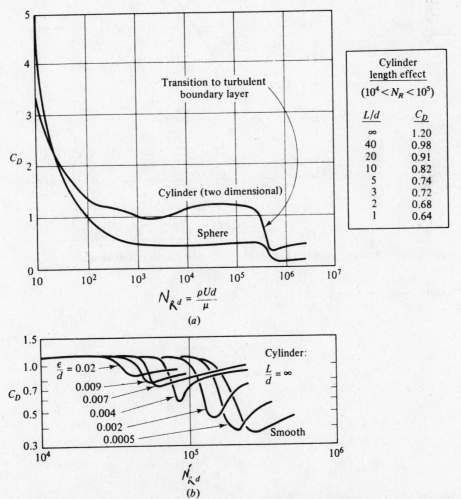

Fig. A-38

The proof of practical dimensional analysis: drag coefficients of a cylinder and sphere: (a) drag coefficient of a smooth cylinder and sphere (data from many sources); (b) increased roughness causes earlier transition to a turbulent boundary layer.

$$N_{R}^{'} = \frac{\rho V d}{\mu}$$

**Fig. A-39**

Crossplot of sphere-drag data from Fig. A-38a to isolate diameter and velocity.

$$n_s = \frac{n\sqrt{Q}}{h^{3/4}}$$

Fig. A-40

**Optimum efficiency of water pumps as a function of specific speed.**

**Fig. A-41**

**Optimum efficiency of pumps versus capacity and specific speed. (G. F. Wislicenus, *Fluid Mechanics of Turbomachinery*, 2d ed., McGraw-Hill, New York, 1965; I. J. Karassick et al., *Pump Handbook*, 2d ed., McGraw-Hill, New York, 1985.)**

**Fig. A-42**

Recommended limits of specific speed for turbines under various effective heads at sea level with water temperature at 80 °F. (After Moody.)

**Fig. A-43**

Characteristics of turbines as a function of specific speed.

$$n_s = \frac{n_e \sqrt{bhp_e}}{h^{5/4}}$$

**Fig. A-44**

Optimum values of turbine efficiency.

$$\phi = \frac{u}{(2gH)^{1/2}}$$

**Fig. A-45**

Efficiency of an impulse turbine: solid curve = ideal, $\beta = 180°$, $C_v = 1.0$; dashed curve = actual, $\beta = 160°$, $C_v = 0.94$; open circles = data, Pelton wheel, diameter = 2 ft.

**Fig. A-46**

Optimum efficiency of turbine designs.

Fig. A-47

Reaction turbines: (*a*) Francis (radial type); (*b*) Francis (mixed-flow); (*c*) propeller (axial-flow); (*d*) performance curves for a Francis turbine, $n = 600$ r/min, $D = 2.25$ ft, $N_{sp} = 29$.

Fig. A-48

Efficiency versus power level for various turbine designs at constant speed and head.

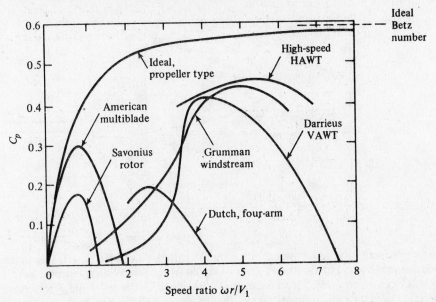

**Fig. A-49**

Estimated performance of various wind turbine designs as a function of blade-tip speed ratio. (Reprinted from *The Aeronautical Journal*, Vol. 85, No. 845, June 1981, by kind permission of The Royal Aeronautical Society.)

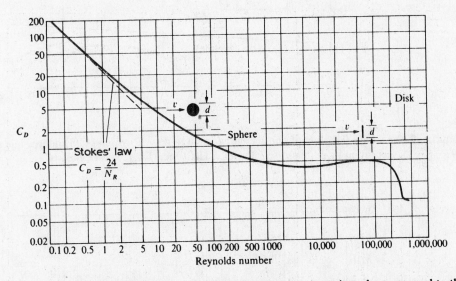

**Fig. A-50**

Drag coefficients for sphere and circular disk. Area in the drag relation is projected area normal to the stream. (Data adapted from "Das Widerstandsproblem," by F. Eisner, *Proc. 3d Intern. Congr. Appl. Mech.*, Stockholm, 1931.) Raymond C. Binder, *Fluid Mechanics*, 5e, © 1973, p. 132. Reprinted by permission of Prentice-Hall, Inc., Englewood Cliffs, N.J.

Drag coefficient for two-dimensional flow around a cylinder and a flat plate. Area in the drag relation is projected area normal to the stream. (Data adapted from "Das Widerstandsproblem," by F. Eisner, *Proc. 3d Intern. Congr. Appl. Mech.,* Stockholm, 1931.) Raymond C. Binder, *Fluid Mechanics,* 5e, © 1973, p. 134. Reprinted by permission of Prentice-Hall, Inc. Englewood Cliffs, N.J. 1973.

**Fig. A-52**

Drag coefficients for a flat plate of finite length normal to flow. Raymond C. Binder, *Fluid Mechanics,* 5e, © 1973, p. 135. Reprinted by permission of Prentice-Hall, Inc., Englewood Cliffs, N.J.

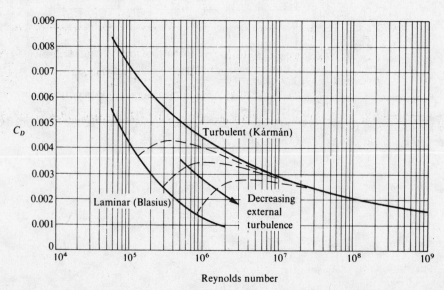

**Fig. A-53**

Skin-friction drag for smooth flat plates. T. v. Kármán, "Turbulence and Skin Friction," *J. Aeronaut. Sci.,* 1, no. 1 (January, 1934), © American Institute of Aeronautics and Astronautics, reprinted with permission.

**Fig. A-54**

Plan view of flow about a motor vehicle (delivery van). (*a*) Blunt nose with separated flow along the entire side wall and a large drag coefficient $C_D = 0.75$. (*b*) Round nose with separation at the rear of the vehicle and smaller drag coefficient $C_D = 0.45$. (Adapted from H. Schlicting, *Boundary Layer Theory*, 4th ed., McGraw-Hill Book Co., New York, N.Y., 1960, p. 34).

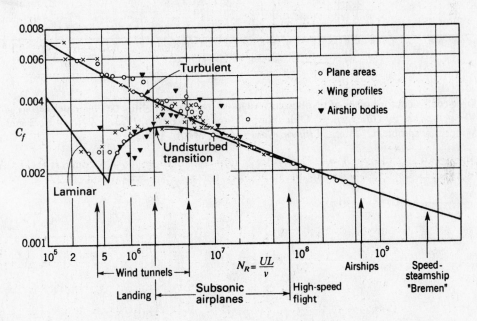

**Fig. A-55**

Drag coefficients for a smooth flat plate. (Adapted from *NACA Tech. Mem. 1218*, p. 117, 1949.)

**Fig. A-56**

Drag coefficient for bodies of revolution. (Adapted from L. Prandtl, *Ergebnisse der aerodynamischen Versuchsanstalt zu Göttingen,* R. Oldenbourg, Munich and Berlin, 1923, p. 29; and F. Eisner, "Das Widerstandsproblem," *Proc. 3d Intern. Congr. Appl. Mech.,* 1930, p. 32.)

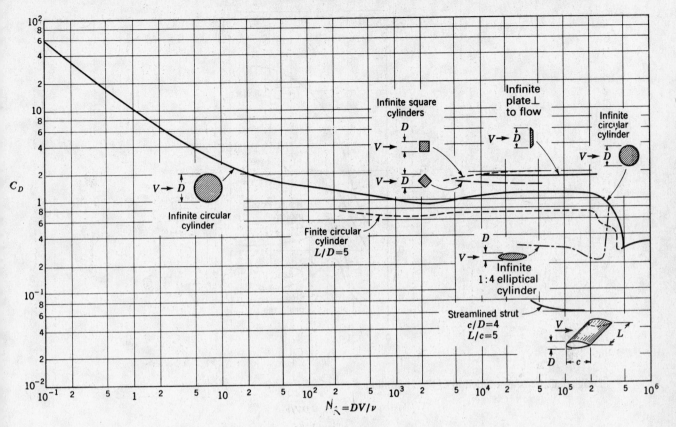

**Fig. A-57**

Drag coefficient for two-dimensional bodies. (Adapted from L. Prandtl, *Ergebnisse der aerodynamischen Versuchsanstalt zu Göttingen,* R. Oldenbourg, Munich and Berlin, 1923, p. 24; F. Eisner, "Das Widerstandsproblem," *Proc. 3d Intern. Congr. Appl. Mech.,* p. 32, 1930; A. F. Zahm, R. H. Smith, and G. C. Hill, "Point Drag and Total Drag of Navy Struts No. 1 Modified," *NACA Rept. 137,* p. 14, 1972; and W. F. Lindsey, "Drag of Cylinders of Simple Shapes," *NACA Rept. 619,* pp. 4–5, 1938.)

Fig. A-58

Drag coefficients as a function of Mach number.

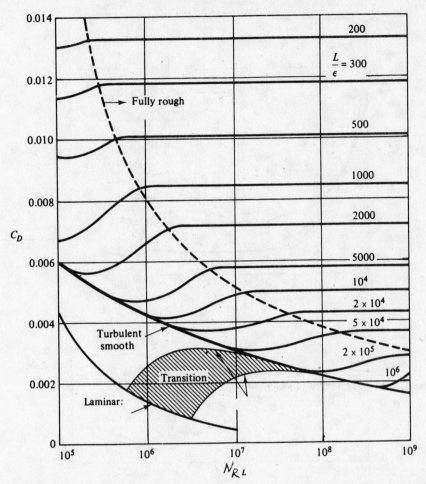

**Fig. A-59**

Drag coefficient of laminar and turbulent boundary layer on smooth and rough flat plates.

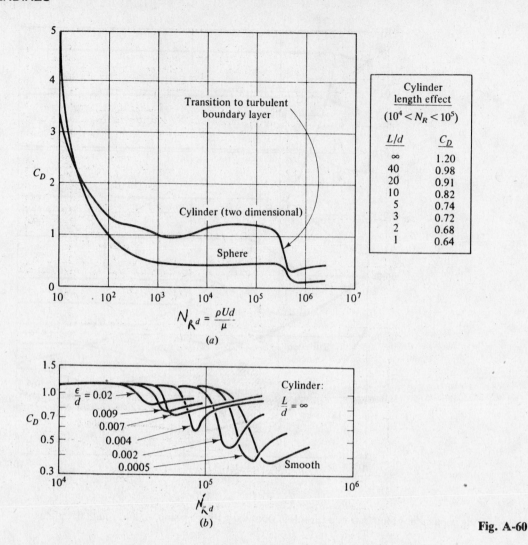

Cylinder
length effect
$(10^4 < N_R < 10^5)$

| $L/d$ | $C_D$ |
|---|---|
| $\infty$ | 1.20 |
| 40 | 0.98 |
| 20 | 0.91 |
| 10 | 0.82 |
| 5 | 0.74 |
| 3 | 0.72 |
| 2 | 0.68 |
| 1 | 0.64 |

Transition to turbulent boundary layer

Cylinder (two dimensional)

Sphere

$N_{R^d} = \dfrac{\rho U d}{\mu}$

(a)

Cylinder: $\dfrac{L}{d} = \infty$

$\dfrac{\epsilon}{d} = 0.02$
0.009
0.007
0.004
0.002
0.0005

Smooth

$N_{R^d}$

(b)

**Fig. A-60**

The proof of practical dimensional analysis: drag coefficients of a cylinder and sphere: (a) drag coefficient of a smooth cylinder and sphere (data from many sources); (b) increased roughness causes earlier transition to a turbulent boundary layer.

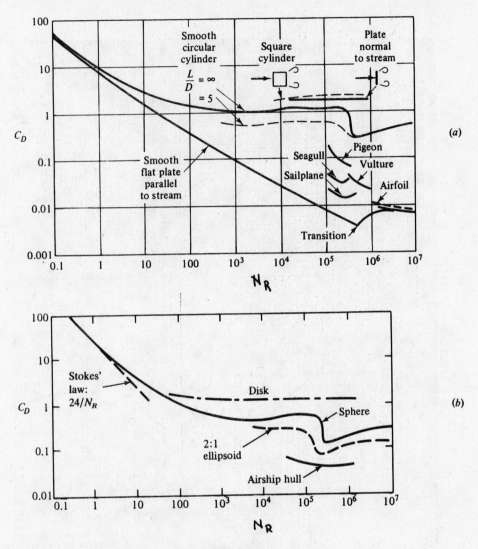

**Fig. A-61**

Drag coefficients of smooth bodies at low Mach numbers: (*a*) two-dimensional bodies; (*b*) three-dimensional bodies. Note the Reynolds-number independence of blunt bodies at high $N_R$.

COEFFICIENT OF DRAG vs $N_R$
(Use Lower Scale Unless Noted)

SPHERES (USE UPPER SCALE)

$$C_D = \frac{24}{R_g} \text{ for } R_g \leq 0.60$$

INFINITE CYLINDERS
(USE UPPER SCALE)

CIRCULAR DISK
$C_D = 1.12$

RECTANGULAR PLATES

For $x/y = 1$, $C_D = 1.16$
For $x/y = 20$, $C_D = 1.50$
For $x/y = \infty$, $C_D = 1.90$

FINITE CYLINDERS
$L/D = 5$

SPHERES

(CONTINUED AT UPPER RIGHT)

INFINITE CYLINDERS

SPHERES

COEFFICIENT OF DRAG $C_D$

REYNOLDS NUMBER $(VD/\nu)$

Fig. A-62

**MACH NUMBER**

**Fig. A-63**

Drag coefficients at supersonic velocities.

$$N_R = \frac{V_O l}{\nu}$$

**Fig. A-64**

Three zones of flow for a rough plate.

**Fig. A-65**

Time history of the aerodynamic drag of cars in comparison with streamlined bodies. (From W. H. Hucho, L. J. Janssen, and H. J. Emmelmann, *The Optimisation of Body Details—A Method For Reducing the Aerodynamic Drag of Road Vehicles,* SAE 760185, 1976.)

**Fig. A-66**

Lift and drag coefficients plotted against angle of attack for N.A.C.A. 2418 airfoil. Reynolds number = 3 060 000. (*NACA. Tech. Rept. No. 669.*)

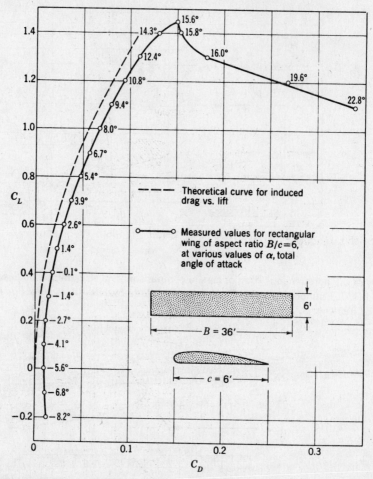

Fig. A-67

Polar diagram for rectangular Clark Y airfoil of 6-ft chord by 36-ft span. (Data from A. Silverstein, *NACA Rept. 502*, p. 15, 1934.)

Fig. A-68

Correction factors for transforming rectangular airfoils from finite to infinite aspect ratio. (From A. Silverstein, *NACA Rept. 502*, Fig. 7, 1934.)

**Fig. A-69**

Definition sketch for induced drag.

**Fig. A-70**

Typical lift and drag coefficients for an airfoil; $C_L$ and $C_D$ based on maximum projected wing area.

**Fig. A-71**

Coefficients of lift and drag for a flat plate at varying inclination $\alpha$.

**Fig. A-72**

Friction curve for a perfect gas. $k = 1.4$. Source: Mechanics of Fluids by Shames.

**TABLE A-1 Approximate physical properties of water (English Gravitational Unit System)**

| Temperature (°F) | Specific (or unit) weight, $\gamma$ (lb/ft³) | Mass density, $\rho$ (slugs/ft³) | Dynamic viscosity, $\mu$ (lb-s/ft²) | Kinematic viscosity, $\nu$ (ft²/s) | Vapor pressure, $p_v$ (lb/ft²) | Surface tension,† $\sigma$ (lb/ft) | Bulk modulus of elasticity, K (lb/in²) |
|---|---|---|---|---|---|---|---|
| 32 | 62.4 | 1.94 | $3.66 \times 10^{-5}$ | $1.89 \times 10^{-5}$ | 12.8 | 0.00518 | 293,000 |
| 40 | 62.4 | 1.94 | $3.23 \times 10^{-5}$ | $1.67 \times 10^{-5}$ | 18.5 | 0.00514 | 294,000 |
| 50 | 62.4 | 1.94 | $2.72 \times 10^{-5}$ | $1.40 \times 10^{-5}$ | 25.7 | 0.00509 | 305,000 |
| 60 | 62.4 | 1.94 | $2.35 \times 10^{-5}$ | $1.21 \times 10^{-5}$ | 36.5 | 0.00504 | 311,000 |
| 70 | 62.3 | 1.93 | $2.04 \times 10^{-5}$ | $1.05 \times 10^{-5}$ | 52.2 | 0.00500 | 320,000 |
| 80 | 62.2 | 1.93 | $1.77 \times 10^{-5}$ | $9.15 \times 10^{-6}$ | 73.5 | 0.00492 | 322,000 |
| 90 | 62.1 | 1.93 | $1.60 \times 10^{-5}$ | $8.29 \times 10^{-6}$ | 101 | 0.00486 | 323,000 |
| 100 | 62.0 | 1.93 | $1.42 \times 10^{-5}$ | $7.37 \times 10^{-6}$ | 135 | 0.00480 | 327,000 |
| 110 | 61.9 | 1.92 | $1.26 \times 10^{-5}$ | $6.55 \times 10^{-6}$ | 189 | 0.00473 | 331,000 |
| 120 | 61.7 | 1.92 | $1.14 \times 10^{-5}$ | $5.94 \times 10^{-6}$ | 251 | 0.00465 | 333,000 |
| 130 | 61.5 | 1.91 | $1.05 \times 10^{-5}$ | $5.49 \times 10^{-6}$ | 322 | 0.00460 | 334,000 |
| 140 | 61.4 | 1.91 | $9.60 \times 10^{-6}$ | $5.03 \times 10^{-6}$ | 416 | 0.00454 | 330,000 |
| 150 | 61.2 | 1.90 | $8.90 \times 10^{-6}$ | $4.68 \times 10^{-6}$ | 545 | 0.00447 | 328,000 |
| 160 | 61.0 | 1.90 | $8.30 \times 10^{-6}$ | $4.38 \times 10^{-6}$ | 693 | 0.00441 | 326,000 |
| 170 | 60.8 | 1.89 | $7.70 \times 10^{-6}$ | $4.07 \times 10^{-6}$ | 875 | 0.00433 | 322,000 |
| 180 | 60.6 | 1.88 | $7.23 \times 10^{-6}$ | $3.84 \times 10^{-6}$ | 1086 | 0.00426 | 318,000 |
| 190 | 60.4 | 1.88 | $6.80 \times 10^{-6}$ | $3.62 \times 10^{-6}$ | 1358 | 0.00419 | 313,000 |
| 200 | 60.1 | 1.87 | $6.25 \times 10^{-6}$ | $3.35 \times 10^{-6}$ | 1671 | 0.00412 | 308,000 |
| 210 | 59.9 | 1.86 | $5.95 \times 10^{-6}$ | $3.20 \times 10^{-6}$ | 2042 | 0.00405 | 301,000 |
| 212 | 59.8 | 1.86 | $5.89 \times 10^{-6}$ | $3.17 \times 10^{-6}$ | 2116 | 0.00404 | 300,000 |

† In contact with air.

**TABLE A-2 Approximate physical properties of water (International System of Units)**

| Temperature (°C) | Specific (or unit) weight, $\gamma$ (kN/m³) | Mass density, $\rho$ (kg/m³) | Dynamic Viscosity, $\mu$ (Pa·s) | Kinematic Viscosity, $\nu$ (m²/s) | Vapor Pressure (kPa) | Surface Tension,† $\sigma$ (N/m) | Bulk modulus of elasticity, K (GPa) |
|---|---|---|---|---|---|---|---|
| 0 | 9.81 | 1000 | $1.75 \times 10^{-3}$ | $1.75 \times 10^{-6}$ | 0.611 | 0.0756 | 2.02 |
| 10 | 9.81 | 1000 | $1.30 \times 10^{-3}$ | $1.30 \times 10^{-6}$ | 1.23 | 0.0742 | 2.10 |
| 20 | 9.79 | 998 | $1.02 \times 10^{-3}$ | $1.02 \times 10^{-6}$ | 2.34 | 0.0728 | 2.18 |
| 30 | 9.77 | 996 | $8.00 \times 10^{-4}$ | $8.03 \times 10^{-7}$ | 4.24 | 0.0712 | 2.25 |
| 40 | 9.73 | 992 | $6.51 \times 10^{-4}$ | $6.56 \times 10^{-7}$ | 7.38 | 0.0696 | 2.28 |
| 50 | 9.69 | 988 | $5.41 \times 10^{-4}$ | $5.48 \times 10^{-7}$ | 12.3 | 0.0679 | 2.29 |
| 60 | 9.65 | 984 | $4.60 \times 10^{-4}$ | $4.67 \times 10^{-7}$ | 19.9 | 0.0662 | 2.28 |
| 70 | 9.59 | 978 | $4.02 \times 10^{-4}$ | $4.11 \times 10^{-7}$ | 31.2 | 0.0644 | 2.25 |
| 80 | 9.53 | 971 | $3.50 \times 10^{-4}$ | $3.60 \times 10^{-7}$ | 47.4 | 0.0626 | 2.20 |
| 90 | 9.47 | 965 | $3.11 \times 10^{-4}$ | $3.22 \times 10^{-7}$ | 70.1 | 0.0608 | 2.14 |
| 100 | 9.40 | 958 | $2.82 \times 10^{-4}$ | $2.94 \times 10^{-7}$ | 101.3 | 0.0589 | 2.07 |

† In contact with air.

TABLE A-3 Approximate physical properties of some common liquids at 1 atmosphere pressure and 20 °C (68 °F)

| Liquid | Specific (or unit) weight, $\gamma$ | | Mass density, $\rho$ | | Specific gravity | Dynamic viscosity, $\mu$ | | Vapor pressure | | Surface tension,† $\sigma$ | |
|---|---|---|---|---|---|---|---|---|---|---|---|
| | (lb/ft³) | (kN/m³) | (slugs/ft³) | (kg/m³) | | (lb-s/ft²) | (Pa · s) | (lb/ft²) | (kPa) | (lb/ft) | (N/m) |
| Ammonia | 51.7 | 8.13 | 1.61 | 829 | 0.83 | $4.60 \times 10^{-6}$ | $2.20 \times 10^{-4}$ | 19,000 | 910 | 0.00146 | 0.0213 |
| Benzene | 54.8 | 8.62 | 1.70 | 879 | 0.88 | $1.36 \times 10^{-5}$ | $6.51 \times 10^{-4}$ | 210 | 10.1 | 0.00198 | 0.0289 |
| Carbon tetrachloride | 99.1 | 15.57 | 3.08 | 1,588 | 1.59 | $2.02 \times 10^{-5}$ | $9.67 \times 10^{-4}$ | 250 | 12.0 | 0.00185 | 0.0270 |
| Ethanol | 49.2 | 7.73 | 1.53 | 788 | 0.79 | $2.51 \times 10^{-5}$ | $1.20 \times 10^{-3}$ | 120 | 5.75 | 0.00156 | 0.0228 |
| Gasoline | 44.9 | 7.05 | 1.40 | 719 | 0.72 | $6.10 \times 10^{-6}$ | $2.92 \times 10^{-4}$ | 1,150 | 55.1 | | |
| Glycerin | 78.5 | 12.34 | 2.44 | 1,258 | 1.26 | $3.11 \times 10^{-2}$ | 1.49 | 0.0003 | 0.000014 | 0.00434 | 0.0633 |
| Kerosine | 51.1 | 8.03 | 1.59 | 819 | 0.82 | $4.00 \times 10^{-5}$ | $1.92 \times 10^{-3}$ | 65 | 3.11 | 0.00190 | 0.0277 |
| Mercury | 847.3 | 133.1 | 26.34 | 13,570 | 13.6 | $3.25 \times 10^{-5}$ | $1.56 \times 10^{-3}$ | 0.000023 | 0.0000011 | 0.0352 | 0.514 |
| Methanol | 49.2 | 7.73 | 1.53 | 788 | 0.79 | $1.25 \times 10^{-5}$ | $5.98 \times 10^{-4}$ | 280 | 13.4 | 0.00155 | 0.0226 |
| SAE 10 Oil | 54.2 | 8.52 | 1.68 | 869 | 0.87 | $1.70 \times 10^{-3}$ | $8.14 \times 10^{-2}$ | .......... | .......... | 0.00250 | 0.0365 |
| SAE 30 Oil | 55.4 | 8.71 | 1.72 | 888 | 0.89 | $9.20 \times 10^{-3}$ | $4.40 \times 10^{-1}$ | .......... | .......... | 0.00240 | 0.0350 |
| Water | 62.3 | 9.79 | 1.94 | 998 | 1.00 | $2.09 \times 10^{-5}$ | $1.02 \times 10^{-3}$ | 48 | 2.34 | 0.00500 | 0.0728 |
| Seawater | 64.2 | 10.08 | 2.00 | 1,028 | 1.03 | $2.23 \times 10^{-5}$ | $1.07 \times 10^{-3}$ | 48 | 2.34 | 0.00500 | 0.0728 |

† In contact with air.

**TABLE A-4 Approximate physical properties of air at standard atmospheric pressure (English Gravitational Unit System)**

| Temperature (°F) | Specific (or unit) weight, $\gamma$ (lb/ft³) | Mass density, $\rho$ (slugs/ft³) | Dynamic viscosity, $\mu$ (lb-s/ft²) | Kinematic viscosity, $\nu$ (ft²/s) |
|---|---|---|---|---|
| 32 | 0.0808 | 0.00251 | $3.59 \times 10^{-7}$ | $1.43 \times 10^{-4}$ |
| 40 | 0.0794 | 0.00247 | $3.62 \times 10^{-7}$ | $1.46 \times 10^{-4}$ |
| 50 | 0.0779 | 0.00242 | $3.68 \times 10^{-7}$ | $1.52 \times 10^{-4}$ |
| 60 | 0.0763 | 0.00237 | $3.74 \times 10^{-7}$ | $1.58 \times 10^{-4}$ |
| 70 | 0.0750 | 0.00233 | $3.82 \times 10^{-7}$ | $1.64 \times 10^{-4}$ |
| 80 | 0.0735 | 0.00228 | $3.85 \times 10^{-7}$ | $1.69 \times 10^{-4}$ |
| 90 | 0.0723 | 0.00224 | $3.90 \times 10^{-7}$ | $1.74 \times 10^{-4}$ |
| 100 | 0.0709 | 0.00220 | $3.96 \times 10^{-7}$ | $1.80 \times 10^{-4}$ |
| 110 | 0.0696 | 0.00218 | $4.02 \times 10^{-7}$ | $1.84 \times 10^{-4}$ |
| 120 | 0.0684 | 0.00215 | $4.07 \times 10^{-7}$ | $1.89 \times 10^{-4}$ |
| 130 | 0.0674 | 0.00210 | $4.10 \times 10^{-7}$ | $1.95 \times 10^{-4}$ |
| 140 | 0.0663 | 0.00206 | $4.14 \times 10^{-7}$ | $2.01 \times 10^{-4}$ |
| 150 | 0.0652 | 0.00202 | $4.18 \times 10^{-7}$ | $2.06 \times 10^{-4}$ |
| 160 | 0.0641 | 0.00199 | $4.22 \times 10^{-7}$ | $2.12 \times 10^{-4}$ |
| 170 | 0.0631 | 0.00196 | $4.28 \times 10^{-7}$ | $2.18 \times 10^{-4}$ |
| 180 | 0.0621 | 0.00193 | $4.34 \times 10^{-7}$ | $2.25 \times 10^{-4}$ |
| 190 | 0.0612 | 0.00190 | $4.42 \times 10^{-7}$ | $2.32 \times 10^{-4}$ |
| 200 | 0.0602 | 0.00187 | $4.49 \times 10^{-7}$ | $2.40 \times 10^{-4}$ |
| 210 | 0.0594 | 0.00184 | $4.57 \times 10^{-7}$ | $2.48 \times 10^{-4}$ |
| 212 | 0.0592 | 0.00184 | $4.58 \times 10^{-7}$ | $2.50 \times 10^{-4}$ |

**TABLE A-5 Approximate physical properties of air at standard atmospheric pressure (International System of Units)**

| Temperature (°C) | Specific (or unit) weight, $\gamma$ (N/m³) | Mass density, $\rho$ (kg/m³) | Dynamic viscosity, $\mu$ (Pa · s) | Kinematic viscosity, $\nu$ (m²/s) |
|---|---|---|---|---|
| 0 | 12.7 | 1.29 | $1.72 \times 10^{-5}$ | $1.33 \times 10^{-5}$ |
| 10 | 12.2 | 1.25 | $1.77 \times 10^{-5}$ | $1.42 \times 10^{-5}$ |
| 20 | 11.8 | 1.20 | $1.81 \times 10^{-5}$ | $1.51 \times 10^{-5}$ |
| 30 | 11.4 | 1.16 | $1.86 \times 10^{-5}$ | $1.60 \times 10^{-5}$ |
| 40 | 11.0 | 1.13 | $1.91 \times 10^{-5}$ | $1.69 \times 10^{-5}$ |
| 50 | 10.7 | 1.09 | $1.95 \times 10^{-5}$ | $1.79 \times 10^{-5}$ |
| 60 | 10.4 | 1.06 | $1.99 \times 10^{-5}$ | $1.89 \times 10^{-5}$ |
| 70 | 10.1 | 1.03 | $2.04 \times 10^{-5}$ | $1.99 \times 10^{-5}$ |
| 80 | 9.80 | 1.00 | $2.09 \times 10^{-5}$ | $2.09 \times 10^{-5}$ |
| 90 | 9.53 | 0.972 | $2.13 \times 10^{-5}$ | $2.19 \times 10^{-5}$ |
| 100 | 9.28 | 0.946 | $2.17 \times 10^{-5}$ | $2.30 \times 10^{-5}$ |

**TABLE A-6 Approximate physical properties of some common gases at 1 atmosphere pressure and 30 °C (68 °F)**

| Gas | Specific (or unit) weight, $\gamma$ | | Mass density, $\rho$ | | Dynamic viscosity ($\mu$) | | Gas constant ($R$) | | | |
|---|---|---|---|---|---|---|---|---|---|---|
| | (lb/ft³) | (N/m³) | (slugs/ft³) | (kg/m³) | (lb-s/ft²) | (kPa · s) | (ft/°R) | (m/k) | (lb-ft/slug-°R) | (J/kg · K) |
| Air | 0.0752 | 11.8 | 0.00234 | 1.20 | $3.78 \times 10^{-7}$ | $1.81 \times 10^{-8}$ | 53.3 | 29.3 | 1 716 | 287 |
| Carbon dioxide | 0.115 | 18.1 | 0.00357 | 1.84 | $3.10 \times 10^{-7}$ | $1.48 \times 10^{-8}$ | 35.1 | 19.3 | 1 130 | 189 |
| Helium | 0.0104 | 1.63 | 0.000323 | 0.166 | $4.11 \times 10^{-7}$ | $1.97 \times 10^{-8}$ | 385.7 | 212.0 | 12 420 | 2079 |
| Hydrogen | 0.00522 | 0.823 | 0.000162 | 0.0839 | $1.89 \times 10^{-7}$ | $9.05 \times 10^{-9}$ | 765.5 | 420.8 | 24 649 | 4127 |
| Methane | 0.0416 | 6.53 | 0.00129 | 0.666 | $2.80 \times 10^{-7}$ | $1.34 \times 10^{-8}$ | 96.2 | 52.9 | 3 098 | 519 |
| Nitrogen | 0.0726 | 11.4 | 0.00225 | 1.16 | $3.68 \times 10^{-7}$ | $1.76 \times 10^{-8}$ | 55.1 | 30.3 | 1 774 | 297 |
| Oxygen | 0.0830 | 13.0 | 0.00258 | 1.33 | $4.18 \times 10^{-7}$ | $2.00 \times 10^{-8}$ | 48.2 | 26.5 | 1 552 | 260 |

| Gas | Specific heat ratio ($k$) | Specific heat ($c_p$) | | Specific heat ($c_v$) | | Molecular weight ($M$) |
|---|---|---|---|---|---|---|
| | Dimensionless ratio $c_p/c_v$ | lb-ft/slug-°R | J/kg · K | lb-ft/slug-°R | J/kg · K | |
| Air | 1.40 | 6 000 | 1 003 | 4 285 | 716 | 29.00 |
| Carbon dioxide | 1.30 | 5 132 | 858 | 4 009 | 670 | 44.00 |
| Helium | 1.66 | 31 230 | 5 220 | 18 810 | 3 143 | 4.00 |
| Hydrogen | 1.41 | 86 390 | 14 450 | 61 710 | 10 330 | 2.02 |
| Methane | 1.32 | 13 400 | 2 250 | 10 300 | 1 730 | 16.00 |
| Nitrogen | 1.40 | 6 210 | 1 040 | 4 437 | 743 | 28.00 |
| Oxygen | 1.40 | 5 437 | 909 | 3 883 | 649 | 32.00 |

**TABLE A-7 The ICAO standard atmosphere in English units**

| Elevation above sea level, ft | Temp, °F | Absolute pressure, psia | Specific weight $\gamma$, lb/ft³ | Density $\rho$, slugs/ft³ | Viscosity $\mu \times 10^7$, lb·s/ft² |
|---|---|---|---|---|---|
| 0 | 59.0 | 14.70 | 0.07648 | 0.002377 | 3.737 |
| 5,000 | 41.2 | 12.24 | 0.06587 | 0.002048 | 3.637 |
| 10,000 | 23.4 | 10.11 | 0.05643 | 0.001756 | 3.534 |
| 15,000 | 5.6 | 8.30 | 0.04807 | 0.001496 | 3.430 |
| 20,000 | −12.3 | 6.76 | 0.04070 | 0.001267 | 3.325 |
| 25,000 | −30.1 | 5.46 | 0.03422 | 0.001066 | 3.217 |
| 30,000 | −47.8 | 4.37 | 0.02858 | 0.000891 | 3.107 |
| 35,000 | −65.6 | 3.47 | 0.02367 | 0.000738 | 2.995 |
| 40,000 | −69.7 | 2.73 | 0.01882 | 0.000587 | 2.969 |
| 45,000 | −69.7 | 2.15 | 0.01481 | 0.000462 | 2.969 |
| 50,000 | −69.7 | 1.69 | 0.01165 | 0.000364 | 2.969 |
| 60,000 | −69.7 | 1.05 | 0.00722 | 0.000226 | 2.969 |
| 70,000 | −69.7 | 0.65 | 0.00447 | 0.000140 | 2.969 |
| 80,000 | −69.7 | 0.40 | 0.00277 | 0.000087 | 2.969 |
| 90,000 | −57.2 | 0.25 | 0.00168 | 0.000053 | 3.048 |
| 100,000 | −40.9 | 0.16 | 0.00102 | 0.000032 | 3.150 |

**TABLE A-8 The ICAO standard atmosphere in SI units**

| Elevation above sea level, km | Temp, °C | Absolute pressure, kPa, abs | Specific weight $\gamma$, N/m$^3$ | Density $\rho$, kg/m$^3$ | Viscosity $\mu \times 10^5$, Pa·s |
|---|---|---|---|---|---|
| 0 | 15.0 | 101.33 | 12.01 | 1.225 | 1.79 |
| 2 | 2.0 | 79.50 | 9.86 | 1.007 | 1.73 |
| 4 | −4.5 | 60.12 | 8.02 | 0.909 | 1.66 |
| 6 | −24.0 | 47.22 | 6.46 | 0.660 | 1.60 |
| 8 | −36.9 | 35.65 | 5.14 | 0.526 | 1.53 |
| 10 | −49.9 | 26.50 | 4.04 | 0.414 | 1.46 |
| 12 | −56.5 | 19.40 | 3.05 | 0.312 | 1.42 |
| 14 | −56.5 | 14.20 | 2.22 | 0.228 | 1.42 |
| 16 | −56.5 | 10.35 | 1.62 | 0.166 | 1.42 |
| 18 | −56.5 | 7.57 | 1.19 | 0.122 | 1.42 |
| 20 | −56.5 | 5.53 | 0.87 | 0.089 | 1.42 |
| 25 | −51.6 | 2.64 | 0.41 | 0.042 | 1.45 |
| 30 | −40.2 | 1.20 | 0.18 | 0.018 | 1.51 |

**TABLE A-9 Typical wall roughness values for commercial conduits**

| Material (new) | Roughness ($\varepsilon$) | |
|---|---|---|
| | ft | m |
| Riveted steel | 0.003–0.03 | 0.0009–0.009 |
| Concrete | 0.001–0.01 | 0.0003–0.003 |
| Wood stave | 0.0006–0.003 | 0.0002–0.0009 |
| Cast iron | 0.00085 | 0.00026 |
| Galvanized iron | 0.0005 | 0.00015 |
| Asphalted cast iron | 0.0004 | 0.0001 |
| Commercial steel or wrought iron | 0.00015 | 0.000046 |
| Drawn brass or copper tubing | 0.000005 | 0.0000015 |
| Glass and plastic | "smooth" | "smooth" |

From: Lewis F. Moody, "Friction Factors for Pipe Flow," *ASME Trans.*, vol. 66, pp. 671–684, 1944.

**TABLE A-10 Laminar friction constants $fN_R$ for rectangular and triangular ducts**

| Rectangular | | Isosceles triangle | |
|---|---|---|---|
| $b/a$ | $fN_{RD_h}$ | $\theta$, deg | $fN_{RD_h}$ |
| 0.0 | 96.00 | 0 | 48.0 |
| 0.05 | 89.91 | 10 | 51.6 |
| 0.1 | 84.68 | 20 | 52.9 |
| 0.125 | 82.34 | 30 | 53.3 |
| 0.167 | 78.81 | 40 | 52.9 |
| 0.25 | 72.93 | 50 | 52.0 |
| 0.4 | 65.47 | 60 | 51.1 |
| 0.5 | 62.19 | 70 | 49.5 |
| 0.75 | 57.89 | 80 | 48.3 |
| 1.0 | 56.91 | 90 | 48.0 |

**TABLE A-11 Resistance coefficients $K = \dfrac{h_m}{v^2/2g}$ for open valves, elbows, and trees**

| Nominal diameter, in | Screwed | | | | Flanged | | | | |
|---|---|---|---|---|---|---|---|---|---|
| | $\frac{1}{2}$ | 1 | 2 | 4 | 1 | 2 | 4 | 8 | 20 |
| **Valves (fully open):** | | | | | | | | | |
| Globe | 14 | 8.2 | 6.9 | 5.7 | 13 | 8.5 | 6.0 | 5.8 | 5.5 |
| Gate | 0.30 | 0.24 | 0.16 | 0.11 | 0.80 | 0.35 | 0.16 | 0.07 | 0.03 |
| Swing check | 5.1 | 2.9 | 2.1 | 2.0 | 2.0 | 2.0 | 2.0 | 2.0 | 2.0 |
| Angle | 9.0 | 4.7 | 2.0 | 1.0 | 4.5 | 2.4 | 2.0 | 2.0 | 2.0 |
| **Elbows:** | | | | | | | | | |
| 45° regular | 0.39 | 0.32 | 0.30 | 0.29 | | | | | |
| 45° long radius | | | | | 0.21 | 0.20 | 0.19 | 0.16 | 0.14 |
| 90° regular | 2.0 | 1.5 | 0.95 | 0.64 | 0.50 | 0.39 | 0.30 | 0.26 | 0.21 |
| 90° long radius | 1.0 | 0.72 | 0.41 | 0.23 | 0.40 | 0.30 | 0.19 | 0.15 | 0.10 |
| 180° regular | 2.0 | 1.5 | 0.95 | 0.64 | 0.41 | 0.35 | 0.30 | 0.25 | 0.20 |
| 180° long radius | | | | | 0.40 | 0.30 | 0.21 | 0.15 | 0.10 |
| **Tees:** | | | | | | | | | |
| Line flow | 0.90 | 0.90 | 0.90 | 0.90 | 0.24 | 0.19 | 0.14 | 0.10 | 0.07 |
| Branch flow | 2.4 | 1.8 | 1.4 | 1.1 | 1.0 | 0.80 | 0.64 | 0.58 | 0.41 |

**TABLE A-12 Increased losses of partially open valves**

| | Ratio $K/K$(open condition) | |
| --- | --- | --- |
| Condition | Gate value | Globe value |
| Open | 1.0 | 1.0 |
| Closed, 25% | 3.0–5.0 | 1.5–2.0 |
| 50% | 12–22 | 2.0–3.0 |
| 75% | 70–120 | 6.0–8.0 |

**TABLE A-13 Values of $n$ in Manning's formula**
Prepared by R. E. Horton and others

| | $n$ | |
| --- | --- | --- |
| Nature of surface | Min | Max |
| Neat cement surface | 0.010 | 0.013 |
| Wood-stave pipe | 0.010 | 0.013 |
| Plank flumes, planed | 0.010 | 0.014 |
| Vitrified sewer pipe | 0.010 | 0.017 |
| Metal flumes, smooth | 0.011 | 0.015 |
| Concrete, precast | 0.011 | 0.013 |
| Cement mortar surfaces | 0.011 | 0.015 |
| Plank flumes, unplaned | 0.011 | 0.015 |
| Common-clay drainage tile | 0.011 | 0.017 |
| Concrete, monolithic | 0.012 | 0.016 |
| Brick with cement mortar | 0.012 | 0.017 |
| Cast iron—new | 0.013 | 0.017 |
| Cement rubble surfaces | 0.017 | 0.030 |
| Riveted steel | 0.017 | 0.020 |
| Corrugated metal pipe | 0.021 | 0.025 |
| Canals and ditches, smooth earth | 0.017 | 0.025 |
| Metal flumes, corrugated | 0.022 | 0.030 |
| Canals: | | |
|   Dredged in earth, smooth | 0.025 | 0.033 |
|   In rock cuts smooth | 0.025 | 0.035 |
|   Rough beds and weeds on sides | 0.025 | 0.040 |
|   Rock cuts, jagged and irregular | 0.035 | 0.045 |
| Natural streams: | | |
|   Smoothest | 0.025 | 0.033 |
|   Roughest | 0.045 | 0.060 |
|   Very weedy | 0.075 | 0.150 |

**TABLE A-14 Typical values of the Hazen-Williams Coefficient, $C$**

| | |
| --- | --- |
| Extremely smooth and straight pipes | 140 |
| New steel or cast iron | 130 |
| Wood; concrete | 120 |
| New riveted steel; vitrified | 110 |
| Old cast iron | 100 |
| Very old and corroded cast iron | 80 |

**TABLE A-15 Values of $C$ from the Kutter formula**

| Slope $S$ | $n$ | Hydraulic Radius $R$ in Feet | | | | | | | | | | | | | | |
|---|---|---|---|---|---|---|---|---|---|---|---|---|---|---|---|---|
| | | 0.2 | 0.3 | 0.4 | 0.6 | 0.8 | 1.0 | 1.5 | 2.0 | 2.5 | 3.0 | 4.0 | 6.0 | 8.0 | 10.0 | 15.0 |
| .00005 | .010 | 87 | 98 | 109 | 123 | 133 | 140 | 154 | 164 | 172 | 177 | 187 | 199 | 207 | 213 | 220 |
| | .012 | 68 | 78 | 88 | 98 | 107 | 113 | 126 | 135 | 142 | 148 | 157 | 168 | 176 | 182 | 189 |
| | .015 | 52 | 58 | 66 | 76 | 83 | 89 | 99 | 107 | 113 | 118 | 126 | 138 | 145 | 150 | 159 |
| | .017 | 43 | 50 | 57 | 65 | 72 | 77 | 86 | 93 | 98 | 103 | 112 | 122 | 129 | 134 | 142 |
| | .020 | 35 | 41 | 45 | 53 | 59 | 64 | 72 | 80 | 84 | 88 | 95 | 105 | 111 | 116 | 125 |
| | .025 | 26 | 30 | 35 | 41 | 45 | 49 | 57 | 62 | 66 | 70 | 78 | 85 | 92 | 96 | 104 |
| | .030 | 22 | 25 | 28 | 33 | 37 | 40 | 47 | 51 | 55 | 58 | 65 | 74 | 78 | 84 | 90 |
| .0001 | .010 | 98 | 108 | 118 | 131 | 140 | 147 | 158 | 167 | 173 | 178 | 186 | 196 | 202 | 206 | 212 |
| | .012 | 76 | 86 | 95 | 105 | 113 | 119 | 130 | 138 | 144 | 148 | 155 | 165 | 170 | 174 | 180 |
| | .015 | 57 | 64 | 72 | 81 | 88 | 92 | 103 | 109 | 114 | 118 | 125 | 134 | 140 | 143 | 150 |
| | .017 | 48 | 55 | 62 | 70 | 75 | 80 | 88 | 95 | 99 | 104 | 111 | 118 | 125 | 128 | 135 |
| | .020 | 38 | 45 | 50 | 57 | 63 | 67 | 75 | 81 | 85 | 88 | 95 | 102 | 107 | 111 | 118 |
| | .025 | 28 | 34 | 38 | 43 | 48 | 51 | 59 | 64 | 67 | 70 | 77 | 84 | 89 | 93 | 98 |
| | .030 | 23 | 27 | 30 | 35 | 39 | 42 | 48 | 52 | 55 | 59 | 64 | 72 | 75 | 80 | 85 |
| .0002 | .010 | 105 | 115 | 125 | 137 | 145 | 150 | 162 | 169 | 174 | 178 | 185 | 193 | 198 | 202 | 206 |
| | .012 | 83 | 92 | 100 | 110 | 117 | 123 | 133 | 139 | 144 | 148 | 154 | 162 | 167 | 170 | 175 |
| | .015 | 61 | 69 | 76 | 84 | 91 | 96 | 105 | 110 | 114 | 118 | 124 | 132 | 137 | 140 | 145 |
| | .017 | 52 | 59 | 65 | 73 | 78 | 83 | 90 | 97 | 100 | 104 | 110 | 117 | 122 | 125 | 130 |
| | .020 | 42 | 48 | 53 | 60 | 65 | 68 | 76 | 82 | 85 | 88 | 94 | 100 | 105 | 108 | 113 |
| | .025 | 30 | 35 | 40 | 45 | 50 | 54 | 60 | 65 | 68 | 70 | 76 | 83 | 86 | 90 | 95 |
| | .030 | 25 | 28 | 32 | 37 | 40 | 43 | 49 | 53 | 56 | 59 | 63 | 69 | 74 | 77 | 82 |
| .0004 | .010 | 110 | 121 | 128 | 140 | 148 | 153 | 164 | 171 | 174 | 178 | 184 | 192 | 197 | 193 | 203 |
| | .012 | 87 | 95 | 103 | 113 | 120 | 125 | 134 | 141 | 145 | 149 | 153 | 161 | 165 | 168 | 172 |
| | .015 | 64 | 73 | 78 | 87 | 93 | 98 | 106 | 112 | 115 | 118 | 123 | 130 | 134 | 137 | 142 |
| | .017 | 54 | 62 | 68 | 75 | 80 | 84 | 92 | 98 | 101 | 104 | 110 | 116 | 120 | 123 | 128 |
| | .020 | 43 | 50 | 55 | 61 | 67 | 70 | 77 | 83 | 86 | 88 | 94 | 99 | 104 | 106 | 110 |
| | .025 | 32 | 37 | 42 | 47 | 51 | 55 | 60 | 65 | 68 | 70 | 75 | 82 | 85 | 88 | 92 |
| | .030 | 26 | 30 | 33 | 38 | 41 | 44 | 50 | 54 | 57 | 59 | 63 | 68 | 73 | 75 | 80 |
| .001 | .010 | 113 | 124 | 132 | 143 | 150 | 155 | 165 | 172 | 175 | 178 | 184 | 190 | 195 | 197 | 201 |
| | .012 | 88 | 97 | 105 | 115 | 121 | 127 | 135 | 142 | 145 | 149 | 154 | 160 | 164 | 167 | 171 |
| | .015 | 66 | 75 | 80 | 88 | 94 | 98 | 107 | 112 | 116 | 119 | 123 | 130 | 133 | 135 | 141 |
| | .017 | 55 | 63 | 68 | 76 | 81 | 85 | 92 | 98 | 102 | 105 | 110 | 115 | 119 | 122 | 127 |
| | .020 | 45 | 51 | 56 | 62 | 68 | 71 | 78 | 84 | 87 | 89 | 93 | 98 | 103 | 105 | 109 |
| | .025 | 33 | 38 | 43 | 48 | 52 | 55 | 61 | 65 | 68 | 70 | 75 | 81 | 84 | 87 | 91 |
| | .030 | 27 | 30 | 34 | 38 | 42 | 45 | 50 | 54 | 57 | 59 | 63 | 68 | 72 | 74 | 78 |
| .01 | .010 | 114 | 125 | 133 | 143 | 151 | 156 | 165 | 172 | 175 | 178 | 184 | 190 | 194 | 196 | 200 |
| | .012 | 89 | 99 | 106 | 116 | 122 | 128 | 136 | 142 | 145 | 149 | 154 | 159 | 163 | 166 | 170 |
| | .015 | 67 | 76 | 81 | 89 | 95 | 99 | 107 | 113 | 116 | 119 | 123 | 129 | 133 | 135 | 140 |
| | .017 | 56 | 64 | 69 | 77 | 82 | 86 | 93 | 99 | 103 | 105 | 109 | 115 | 118 | 121 | 126 |
| | .020 | 46 | 52 | 57 | 63 | 68 | 72 | 78 | 84 | 87 | 89 | 93 | 98 | 102 | 105 | 108 |
| | .025 | 34 | 39 | 44 | 49 | 52 | 56 | 62 | 65 | 68 | 70 | 75 | 80 | 83 | 86 | 90 |
| | .030 | 27 | 31 | 35 | 39 | 43 | 45 | 51 | 55 | 58 | 59 | 63 | 67 | 71 | 73 | 77 |

## TABLE A-16 One-dimensional isentropic relations†

| $N_M$ | $A/A^*$ | $p/p_0$ | $\rho/\rho_0$ | $T/T_0$ | $N_M$ | $A/A^*$ | $p/p_0$ | $\rho/\rho_0$ | $T/T_0$ |
|---|---|---|---|---|---|---|---|---|---|
| 0.00 | ........ | 1.000 | 1.000 | 1.000 | 1.10 | 1.01 | 0.468 | 0.582 | 0.805 |
| 0.01 | 57.87 | 0.9999 | 0.9999 | 0.9999 | 1.12 | 1.01 | 0.457 | 0.571 | 0.799 |
| 0.02 | 28.94 | 0.9997 | 0.9999 | 0.9999 | 1.14 | 1.02 | 0.445 | 0.561 | 0.794 |
| 0.04 | 14.48 | 0.999 | 0.999 | 0.9996 | 1.16 | 1.02 | 0.434 | 0.551 | 0.788 |
| 0.06 | 9.67 | 0.997 | 0.998 | 0.999 | 1.18 | 1.02 | 0.423 | 0.541 | 0.782 |
| 0.08 | 7.26 | 0.996 | 0.997 | 0.999 | 1.20 | 1.03 | 0.412 | 0.531 | 0.776 |
| 0.10 | 5.82 | 0.993 | 0.995 | 0.998 | 1.22 | 1.04 | 0.402 | 0.521 | 0.771 |
| 0.12 | 4.86 | 0.990 | 0.993 | 0.997 | 1.24 | 1.04 | 0.391 | 0.512 | 0.765 |
| 0.14 | 4.18 | 0.986 | 0.990 | 0.996 | 1.26 | 1.05 | 0.381 | 0.502 | 0.759 |
| 0.16 | 3.67 | 0.982 | 0.987 | 0.995 | 1.28 | 1.06 | 0.371 | 0.492 | 0.753 |
| 0.18 | 3.28 | 0.978 | 0.984 | 0.994 | 1.30 | 1.07 | 0.361 | 0.483 | 0.747 |
| 0.20 | 2.96 | 0.973 | 0.980 | 0.992 | 1.32 | 1.08 | 0.351 | 0.474 | 0.742 |
| 0.22 | 2.71 | 0.967 | 0.976 | 0.990 | 1.34 | 1.08 | 0.342 | 0.464 | 0.736 |
| 0.24 | 2.50 | 0.961 | 0.972 | 0.989 | 1.36 | 1.09 | 0.332 | 0.455 | 0.730 |
| 0.26 | 2.32 | 0.954 | 0.967 | 0.987 | 1.38 | 1.10 | 0.323 | 0.446 | 0.724 |
| 0.28 | 2.17 | 0.947 | 0.962 | 0.985 | 1.40 | 1.11 | 0.314 | 0.437 | 0.718 |
| 0.30 | 2.04 | 0.939 | 0.956 | 0.982 | 1.42 | 1.13 | 0.305 | 0.429 | 0.713 |
| 0.32 | 1.92 | 0.932 | 0.951 | 0.980 | 1.44 | 1.14 | 0.297 | 0.420 | 0.707 |
| 0.34 | 1.82 | 0.923 | 0.944 | 0.977 | 1.46 | 1.15 | 0.289 | 0.412 | 0.701 |
| 0.36 | 1.74 | 0.914 | 0.938 | 0.975 | 1.48 | 1.16 | 0.280 | 0.403 | 0.695 |
| 0.38 | 1.66 | 0.905 | 0.931 | 0.972 | 1.50 | 1.18 | 0.272 | 0.395 | 0.690 |
| 0.40 | 1.59 | 0.896 | 0.924 | 0.969 | 1.52 | 1.19 | 0.265 | 0.387 | 0.684 |
| 0.42 | 1.53 | 0.886 | 0.917 | 0.966 | 1.54 | 1.20 | 0.257 | 0.379 | 0.678 |
| 0.44 | 1.47 | 0.876 | 0.909 | 0.963 | 1.56 | 1.22 | 0.250 | 0.371 | 0.672 |
| 0.46 | 1.42 | 0.865 | 0.902 | 0.959 | 1.58 | 1.23 | 0.242 | 0.363 | 0.667 |
| 0.48 | 1.38 | 0.854 | 0.893 | 0.956 | 1.60 | 1.25 | 0.235 | 0.356 | 0.661 |
| 0.50 | 1.34 | 0.843 | 0.885 | 0.952 | 1.62 | 1.27 | 0.228 | 0.348 | 0.656 |
| 0.52 | 1.30 | 0.832 | 0.877 | 0.949 | 1.64 | 1.28 | 0.222 | 0.341 | 0.650 |
| 0.54 | 1.27 | 0.820 | 0.868 | 0.945 | 1.66 | 1.30 | 0.215 | 0.334 | 0.645 |
| 0.56 | 1.24 | 0.808 | 0.859 | 0.941 | 1.68 | 1.32 | 0.209 | 0.327 | 0.639 |
| 0.58 | 1.21 | 0.796 | 0.850 | 0.937 | 1.70 | 1.34 | 0.203 | 0.320 | 0.634 |
| 0.60 | 1.19 | 0.784 | 0.840 | 0.933 | 1.72 | 1.36 | 0.197 | 0.313 | 0.628 |
| 0.62 | 1.17 | 0.772 | 0.831 | 0.929 | 1.74 | 1.38 | 0.191 | 0.306 | 0.623 |
| 0.64 | 1.16 | 0.759 | 0.821 | 0.924 | 1.76 | 1.40 | 0.185 | 0.300 | 0.617 |
| 0.66 | 1.13 | 0.747 | 0.812 | 0.920 | 1.78 | 1.42 | 0.179 | 0.293 | 0.612 |
| 0.68 | 1.12 | 0.734 | 0.802 | 0.915 | 1.80 | 1.44 | 0.174 | 0.287 | 0.607 |
| 0.70 | 1.09 | 0.721 | 0.792 | 0.911 | 1.82 | 1.46 | 0.169 | 0.281 | 0.602 |
| 0.72 | 1.08 | 0.708 | 0.781 | 0.906 | 1.84 | 1.48 | 0.164 | 0.275 | 0.596 |
| 0.74 | 1.07 | 0.695 | 0.771 | 0.901 | 1.86 | 1.51 | 0.159 | 0.269 | 0.591 |
| 0.76 | 1.06 | 0.682 | 0.761 | 0.896 | 1.88 | 1.53 | 0.154 | 0.263 | 0.586 |
| 0.78 | 1.05 | 0.669 | 0.750 | 0.891 | 1.90 | 1.56 | 0.149 | 0.257 | 0.581 |
| 0.80 | 1.04 | 0.656 | 0.740 | 0.886 | 1.92 | 1.58 | 0.145 | 0.251 | 0.576 |
| 0.82 | 1.03 | 0.643 | 0.729 | 0.881 | 1.94 | 1.61 | 0.140 | 0.246 | 0.571 |
| 0.84 | 1.02 | 0.630 | 0.719 | 0.876 | 1.96 | 1.63 | 0.136 | 0.240 | 0.566 |
| 0.86 | 1.02 | 0.617 | 0.708 | 0.871 | 1.98 | 1.66 | 0.132 | 0.235 | 0.561 |
| 0.88 | 1.01 | 0.604 | 0.698 | 0.865 | 2.00 | 1.69 | 0.128 | 0.230 | 0.556 |
| 0.90 | 1.01 | 0.591 | 0.687 | 0.860 | 2.02 | 1.72 | 0.124 | 0.225 | 0.551 |
| 0.92 | 1.01 | 0.578 | 0.676 | 0.855 | 2.04 | 1.75 | 0.120 | 0.220 | 0.546 |
| 0.94 | 1.00 | 0.566 | 0.666 | 0.850 | 2.06 | 1.78 | 0.116 | 0.215 | 0.541 |
| 0.96 | 1.00 | 0.553 | 0.655 | 0.844 | 2.08 | 1.81 | 0.113 | 0.210 | 0.536 |
| 0.98 | 1.00 | 0.541 | 0.645 | 0.839 | 2.10 | 1.84 | 0.109 | 0.206 | 0.531 |
| 1.00 | 1.00 | 0.528 | 0.632 | 0.833 | 2.12 | 1.87 | 0.106 | 0.201 | 0.526 |
| 1.02 | 1.00 | 0.516 | 0.623 | 0.828 | 2.14 | 1.90 | 0.103 | 0.197 | 0.522 |
| 1.04 | 1.00 | 0.504 | 0.613 | 0.822 | 2.16 | 1.94 | 0.100 | 0.192 | 0.517 |
| 1.06 | 1.00 | 0.492 | 0.602 | 0.817 | 2.18 | 1.97 | 0.097 | 0.188 | 0.513 |
| 1.08 | 1.01 | 0.480 | 0.592 | 0.810 | 2.20 | 2.01 | 0.094 | 0.184 | 0.508 |

**TABLE A-16 One-dimensional isentropic relations** (*continued*)

| $N_M$ | $A/A^*$ | $p/p_0$ | $\rho/\rho_0$ | $T/T_0$ | $N_M$ | $A/A^*$ | $p/p_0$ | $\rho/\rho_0$ | $T/T_0$ |
|------|---------|---------|---------------|---------|-------|---------|---------|---------------|---------|
| 2.22 | 2.04 | 0.091 | 0.180 | 0.504 | 2.74 | 3.31 | 0.040 | 0.101 | 0.400 |
| 2.24 | 2.08 | 0.088 | 0.176 | 0.499 | 2.76 | 3.37 | 0.039 | 0.099 | 0.396 |
| 2.26 | 2.12 | 0.085 | 0.172 | 0.495 | 2.78 | 3.43 | 0.038 | 0.097 | 0.393 |
| 2.28 | 2.15 | 0.083 | 0.168 | 0.490 | 2.80 | 3.50 | 0.037 | 0.095 | 0.389 |
|      |      |       |       |       | 2.82 | 3.57 | 0.036 | 0.093 | 0.386 |
| 2.30 | 2.19 | 0.080 | 0.165 | 0.486 | 2.84 | 3.64 | 0.035 | 0.091 | 0.383 |
| 2.32 | 2.23 | 0.078 | 0.161 | 0.482 |      |      |       |       |       |
| 2.34 | 2.27 | 0.075 | 0.157 | 0.477 | 2.86 | 3.71 | 0.034 | 0.089 | 0.379 |
| 2.36 | 2.32 | 0.073 | 0.154 | 0.473 | 2.88 | 3.78 | 0.033 | 0.087 | 0.376 |
|      |      |       |       |       | 2.90 | 3.85 | 0.032 | 0.085 | 0.373 |
| 2.38 | 2.36 | 0.071 | 0.150 | 0.469 | 2.92 | 3.92 | 0.031 | 0.083 | 0.370 |
| 2.40 | 2.40 | 0.068 | 0.147 | 0.465 |      |      |       |       |       |
| 2.42 | 2.45 | 0.066 | 0.144 | 0.461 | 2.94 | 4.00 | 0.030 | 0.081 | 0.366 |
| 2.44 | 2.49 | 0.064 | 0.141 | 0.456 | 2.96 | 4.08 | 0.029 | 0.080 | 0.363 |
|      |      |       |       |       | 2.98 | 4.15 | 0.028 | 0.078 | 0.360 |
| 2.46 | 2.54 | 0.062 | 0.138 | 0.452 | 3.00 | 4.23 | 0.027 | 0.076 | 0.357 |
| 2.48 | 2.59 | 0.060 | 0.135 | 0.448 |      |      |       |       |       |
| 2.50 | 2.64 | 0.059 | 0.132 | 0.444 | 3.10 | 4.66 | 0.023 | 0.0685 | 0.342 |
| 2.52 | 2.69 | 0.057 | 0.129 | 0.441 | 3.20 | 5.12 | 0.020 | 0.062 | 0.328 |
|      |      |       |       |       | 3.3 | 5.63 | 0.0175 | 0.0555 | 0.315 |
| 2.54 | 2.74 | 0.055 | 0.126 | 0.437 | 3.4 | 6.18 | 0.015 | 0.050 | 0.302 |
| 2.56 | 2.79 | 0.053 | 0.123 | 0.433 |      |      |       |       |       |
| 2.58 | 2.84 | 0.052 | 0.121 | 0.429 | 3.5 | 6.79 | 0.013 | 0.045 | 0.290 |
| 2.60 | 2.90 | 0.050 | 0.118 | 0.425 | 3.6 | 7.45 | 0.0114 | 0.041 | 0.278 |
| 2.62 | 2.95 | 0.049 | 0.115 | 0.421 | 3.7 | 8.17 | 0.0099 | 0.037 | 0.2675 |
| 2.64 | 3.01 | 0.047 | 0.113 | 0.418 | 3.8 | 8.95 | 0.0086 | 0.0335 | 0.257 |
| 2.66 | 3.06 | 0.046 | 0.110 | 0.414 |      |      |       |       |       |
| 2.68 | 3.12 | 0.044 | 0.108 | 0.410 | 3.9 | 9.80 | 0.0075 | 0.030 | 0.247 |
|      |      |       |       |       | 4.0 | 10.72 | 0.0066 | 0.028 | 0.238 |
| 2.70 | 3.18 | 0.043 | 0.106 | 0.407 |      |      |       |       |       |
| 2.72 | 3.24 | 0.042 | 0.103 | 0.403 |      |      |       |       |       |

†For a perfect gas with constant specific heat, $k = 1.4$

**TABLE A-17 One-dimensional normal-shock relations†**

| $N_{M1}$ | $N_{M2}$ | $\dfrac{p_2}{p_1}$ | $\dfrac{T_2}{T_1}$ | $\dfrac{(p_0)_2}{(p_0)_1}$ | $N_{M1}$ | $N_{M2}$ | $\dfrac{p_2}{p_1}$ | $\dfrac{T_2}{T_1}$ | $\dfrac{(p_0)_2}{(p_0)_1}$ |
|---|---|---|---|---|---|---|---|---|---|
| 1.00 | 1.000 | 1.000 | 1.000 | 1.000 | 2.04 | 0.571 | 4.689 | 1.720 | 0.702 |
| 1.02 | 0.980 | 1.047 | 1.013 | 1.000 | 2.06 | 0.567 | 4.784 | 1.737 | 0.693 |
| 1.04 | 0.962 | 1.095 | 1.026 | 1.000 | 2.08 | 0.564 | 4.881 | 1.754 | 0.683 |
| 1.06 | 0.944 | 1.144 | 1.039 | 1.000 | 2.10 | 0.561 | 4.978 | 1.770 | 0.674 |
| 1.08 | 0.928 | 1.194 | 1.052 | 0.999 | 2.12 | 0.558 | 5.077 | 1.787 | 0.665 |
| 1.10 | 0.912 | 1.245 | 1.065 | 0.999 | 2.14 | 0.555 | 5.176 | 1.805 | 0.656 |
| 1.12 | 0.896 | 1.297 | 1.078 | 0.998 | 2.16 | 0.553 | 5.277 | 1.822 | 0.646 |
| 1.14 | 0.882 | 1.350 | 1.090 | 0.997 | 2.18 | 0.550 | 5.378 | 1.839 | 0.637 |
| 1.16 | 0.868 | 1.403 | 1.103 | 0.996 | 2.20 | 0.547 | 5.480 | 1.857 | 0.628 |
| 1.18 | 0.855 | 1.458 | 1.115 | 0.995 | 2.22 | 0.544 | 5.583 | 1.875 | 0.619 |
| 1.20 | 0.842 | 1.513 | 1.128 | 0.993 | 2.24 | 0.542 | 5.687 | 1.892 | 0.610 |
| 1.22 | 0.830 | 1.570 | 1.140 | 0.991 | 2.26 | 0.539 | 5.792 | 1.910 | 0.601 |
| 1.24 | 0.818 | 1.627 | 1.153 | 0.988 | 2.28 | 0.537 | 5.898 | 1.929 | 0.592 |
| 1.26 | 0.807 | 1.686 | 1.166 | 0.986 | 2.30 | 0.534 | 6.005 | 1.947 | 0.583 |
| 1.28 | 0.796 | 1.745 | 1.178 | 0.983 | 2.32 | 0.532 | 6.113 | 1.965 | 0.575 |
| 1.30 | 0.786 | 1.805 | 1.191 | 0.979 | 2.34 | 0.530 | 6.222 | 1.984 | 0.566 |
| 1.32 | 0.776 | 1.866 | 1.204 | 0.976 | 2.36 | 0.527 | 6.331 | 2.003 | 0.557 |
| 1.34 | 0.766 | 1.928 | 1.216 | 0.972 | 2.38 | 0.525 | 6.442 | 2.021 | 0.549 |
| 1.36 | 0.757 | 1.991 | 1.229 | 0.968 | 2.40 | 0.523 | 6.553 | 2.040 | 0.540 |
| 1.38 | 0.748 | 2.055 | 1.242 | 0.963 | 2.42 | 0.521 | 6.666 | 2.060 | 0.532 |
| 1.40 | 0.740 | 2.120 | 1.255 | 0.958 | 2.44 | 0.519 | 6.779 | 2.079 | 0.523 |
| 1.42 | 0.731 | 2.186 | 1.268 | 0.953 | 2.46 | 0.517 | 6.894 | 2.098 | 0.515 |
| 1.44 | 0.723 | 2.253 | 1.281 | 0.948 | 2.48 | 0.515 | 7.009 | 2.118 | 0.507 |
| 1.46 | 0.716 | 2.320 | 1.294 | 0.942 | 2.50 | 0.513 | 7.125 | 2.138 | 0.499 |
| 1.48 | 0.708 | 2.389 | 1.307 | 0.936 | 2.52 | 0.511 | 7.242 | 2.157 | 0.491 |
| 1.50 | 0.701 | 2.458 | 1.320 | 0.930 | 2.54 | 0.509 | 7.360 | 2.177 | 0.483 |
| 1.52 | 0.694 | 2.529 | 1.334 | 0.923 | 2.56 | 0.507 | 7.479 | 2.198 | 0.475 |
| 1.54 | 0.687 | 2.600 | 1.347 | 0.917 | 2.58 | 0.506 | 7.599 | 2.218 | 0.468 |
| 1.56 | 0.681 | 2.673 | 1.361 | 0.910 | 2.60 | 0.504 | 7.720 | 2.238 | 0.460 |
| 1.58 | 0.675 | 2.746 | 1.374 | 0.903 | 2.62 | 0.502 | 7.842 | 2.260 | 0.453 |
| 1.60 | 0.668 | 2.820 | 1.388 | 0.895 | 2.64 | 0.500 | 7.965 | 2.280 | 0.445 |
| 1.62 | 0.663 | 2.895 | 1.402 | 0.888 | 2.66 | 0.499 | 8.088 | 2.301 | 0.438 |
| 1.64 | 0.657 | 2.971 | 1.416 | 0.880 | 2.68 | 0.497 | 8.213 | 2.322 | 0.431 |
| 1.66 | 0.651 | 3.048 | 1.430 | 0.872 | 2.70 | 0.496 | 8.338 | 2.343 | 0.424 |
| 1.68 | 0.646 | 3.126 | 1.444 | 0.864 | 2.72 | 0.494 | 8.465 | 2.364 | 0.417 |
| 1.70 | 0.641 | 3.205 | 1.458 | 0.856 | 2.74 | 0.493 | 8.592 | 2.396 | 0.410 |
| 1.72 | 0.635 | 3.285 | 1.473 | 0.847 | 2.76 | 0.491 | 8.721 | 2.407 | 0.403 |
| 1.74 | 0.631 | 3.366 | 1.487 | 0.839 | 2.78 | 0.490 | 8.850 | 2.429 | 0.396 |
| 1.76 | 0.626 | 3.447 | 1.502 | 0.830 | 2.80 | 0.488 | 8.980 | 2.451 | 0.389 |
| 1.78 | 0.621 | 3.530 | 1.517 | 0.821 | 2.82 | 0.487 | 9.111 | 2.473 | 0.383 |
| 1.80 | 0.617 | 3.613 | 1.532 | 0.813 | 2.84 | 0.485 | 9.243 | 2.496 | 0.376 |
| 1.82 | 0.612 | 3.698 | 1.547 | 0.804 | 2.86 | 0.484 | 9.376 | 2.518 | 0.370 |
| 1.84 | 0.608 | 3.783 | 1.562 | 0.795 | 2.88 | 0.483 | 9.510 | 2.541 | 0.364 |
| 1.86 | 0.604 | 3.869 | 1.577 | 0.786 | 2.90 | 0.481 | 9.645 | 2.563 | 0.358 |
| 1.88 | 0.600 | 3.957 | 1.592 | 0.777 | 2.92 | 0.480 | 9.781 | 2.586 | 0.352 |
| 1.90 | 0.596 | 4.045 | 1.608 | 0.767 | 2.94 | 0.479 | 9.918 | 2.609 | 0.346 |
| 1.92 | 0.592 | 4.134 | 1.624 | 0.758 | 2.96 | 0.478 | 10.055 | 2.632 | 0.340 |
| 1.94 | 0.588 | 4.224 | 1.639 | 0.749 | 2.98 | 0.476 | 10.194 | 2.656 | 0.334 |
| 1.96 | 0.584 | 4.315 | 1.655 | 0.740 | 3.00 | 0.475 | 10.333 | 2.679 | 0.328 |
| 1.98 | 0.581 | 4.407 | 1.671 | 0.730 | | | | | |
| 2.00 | 0.577 | 4.500 | 1.688 | 0.721 | | | | | |
| 2.02 | 0.574 | 4.594 | 1.704 | 0.711 | | | | | |

†For a perfect gas with $k = 1.4$.

**TABLE A-18 Some expansion factors $Y$ for compressible flow through flow-nozzles and venturi meters**

| $p_2/p_1$ | $k$ | Ratio of Diameters $(d_2/d_1)$ | | | | |
|---|---|---|---|---|---|---|
| | | 0.30 | 0.40 | 0.50 | 0.60 | 0.70 |
| 0.95 | 1.40 | 0.973 | 0.972 | 0.971 | 0.968 | 0.962 |
| | 1.30 | .970 | .970 | .968 | .965 | .959 |
| | 1.20 | .968 | .967 | .966 | .963 | .956 |
| 0.90 | 1.40 | 0.944 | 0.943 | 0.941 | 0.935 | 0.925 |
| | 1.30 | .940 | .939 | .936 | .931 | .918 |
| | 1.20 | .935 | .933 | .931 | .925 | .912 |
| 0.85 | 1.40 | 0.915 | 0.914 | 0.910 | 0.902 | 0.887 |
| | 1.30 | .910 | .907 | .904 | .896 | .880 |
| | 1.20 | .902 | .900 | .896 | .887 | .870 |
| 0.80 | 1.40 | 0.886 | 0.884 | 0.880 | 0.868 | 0.850 |
| | 1.30 | .876 | .873 | .869 | .857 | .839 |
| | 1.20 | .866 | .864 | .859 | .848 | .829 |
| 0.75 | 1.40 | 0.856 | 0.853 | 0.846 | 0.836 | 0.814 |
| | 1.30 | .844 | .841 | .836 | .823 | .802 |
| | 1.20 | .820 | .818 | .812 | .798 | .776 |
| 0.70 | 1.40 | 0.824 | 0.820 | 0.815 | 0.800 | 0.778 |
| | 1.30 | .812 | .808 | .802 | .788 | .763 |
| | 1.20 | .794 | .791 | .784 | .770 | .745 |

For $p_2/p_1 = 1.00$, $Y = 1.00$.

**TABLE A-19 Discharge coefficients for vertical sharp-edged circular orifices**
For Water at 60 °F Discharging into Air at Same Temperature

| Head in Feet | Orifice Diameter in Inches | | | | | |
|---|---|---|---|---|---|---|
| | 0.25 | 0.50 | 0.75 | 1.00 | 2.00 | 4.00 |
| 0.8 | 0.647 | 0.627 | 0.616 | 0.609 | 0.603 | 0.601 |
| 1.4 | .635 | .619 | .610 | .605 | .601 | .600 |
| 2.0 | .629 | .615 | .607 | .603 | .600 | .599 |
| 4.0 | .621 | .609 | .603 | .600 | .598 | .597 |
| 6.0 | .617 | .607 | .601 | .599 | .597 | .596 |
| 8.0 | .614 | .605 | .600 | .598 | .596 | .595 |
| 10.0 | .613 | .604 | .600 | .597 | .596 | .595 |
| 12.0 | .612 | .603 | .599 | .597 | .595 | .595 |
| 14.0 | .611 | .603 | .598 | .596 | .595 | .594 |
| 16.0 | .610 | .602 | .598 | .596 | .595 | .594 |
| 20.0 | .609 | .602 | .598 | .596 | .595 | .594 |
| 25.0 | .608 | .601 | .597 | .596 | .594 | .594 |
| 30.0 | .607 | .600 | .597 | .595 | .594 | .594 |
| 40.0 | .606 | .600 | .596 | .595 | .594 | .593 |
| 50.0 | .605 | .599 | .596 | .595 | .594 | .593 |
| 60.0 | .605 | .599 | .596 | .594 | .593 | .593 |

*Source:* F. W. Medaugh and G. D. Johnson, Civil Engr., July 1940, p. 424.

**TABLE A-20 Drag of two-dimensional bodies at $N_R \geq 10^4$**

| Shape | $C_D$ based on frontal area | Shape | $C_D$ based on frontal area |
|---|---|---|---|
| **Plate:** | 2.0 | **Half-cylinder:** | 1.2 |
| **Square cylinder:** | 2.1 | | 1.7 |
| | 1.6 | **Equilateral triangle:** | 1.6 |
| **Half tube:** | 1.2 | | 2.0 |
| | 2.3 | | |

| Elliptical cylinder: | | Laminar | Turbulent |
|---|---|---|---|
| 1:1 | | 1.2 | 0.3 |
| 2:1 | | 0.6 | 0.2 |
| 4:1 | | 0.35 | 0.15 |
| 8:1 | | 0.25 | 0.1 |

**TABLE A-21 Drag of three-dimensional bodies at $N_R \geq 10^4$**

| Body | Ratio | $C_D$ based on frontal area |
|------|-------|------------------------------|

Cube:

1.07

0.81

60° cone:

0.5

Disk:

1.17

Cup:

1.4

0.4

Parachute (low porosity):

1.2

Rectangular plate:

| $b/h$ | $C_D$ |
|-------|-------|
| 1 | 1.18 |
| 5 | 1.2 |
| 10 | 1.3 |
| 20 | 1.5 |
| ∞ | 2.0 |

Flat-faced cylinder:

| $L/d$ | $C_D$ |
|-------|-------|
| 0.5 | 1.15 |
| 1 | 0.90 |
| 2 | 0.85 |
| 4 | 0.87 |
| 8 | 0.99 |

Ellipsoid:

| $L/d$ | Laminar | Turbulent |
|-------|---------|-----------|
| 0.75 | 0.5 | 0.2 |
| 1 | 0.47 | 0.2 |
| 2 | 0.27 | 0.13 |
| 4 | 0.25 | 0.1 |
| 8 | 0.2 | 0.08 |

**TABLE A-22 Drag coefficients for three-dimensional bodies**

| Shape | | $C_D$ | Shape | | $C_D$ Laminar flow | Turbulent flow |
|-------|--|-------|-------|--|---------|---------|
| Disc | → ▯ | 1.17 | Sphere: → ◯ | | 0.47 | 0.27 |
| 60° cone | → ◁ | 0.49 | Ellipsoidal body of revolution: | | | |
| Cube | → ▢ | 1.05 | | | | |
| | → ◇ | 0.80 | 2:1 → | ⬭ | 0.27 | 0.06 |
| Hollow cup | → ◖ | 0.38 | 4:1 → | ⬭ | 0.20 | 0.06 |
| | → ◗ | 1.42 | 8:1 → | ⬭ | 0.25 | 0.13 |
| Solid hemisphere | → ◖ | 0.38 | | | | |
| | → ◗ | 1.17 | | | | |

**TABLE A-23 Typical drag coefficients for various cylinders in two-dimensional flow**

| Body shape | | $C_D$ | Reynolds number |
|------------|--|-------|-----------------|
| Circular cylinder | → ◯ | 1.2 | $10^4$ to $1.5 \times 10^5$ |
| Elliptical cylinder | → ⬭ 2:1 | 0.6 | $4 \times 10^4$ |
| | | 0.46 | $10^5$ |
| | → ⬭ 4:1 | 0.32 | $2.5 \times 10^4$ to $10^5$ |
| | → ⬭ 8:1 | 0.29 | $2.5 \times 10^4$ |
| | | 0.20 | $2 \times 10^5$ |
| Square cylinder | → ▢ | 2.0 | $3.5 \times 10^4$ |
| | → ◇ | 1.6 | $10^4$ to $10^5$ |
| Triangular cylinders | → ◁120° | 2.0 | $10^4$ |
| | → ◁120° | 1.72 | $10^4$ |
| | → ▷90° | 2.15 | $10^4$ |
| | → ◁90° | 1.60 | $10^4$ |
| | → ▷ 60° | 2.20 | $10^4$ |
| | → ◁60° | 1.39 | $10^4$ |
| | → ▷30 | 1.8 | $10^5$ |
| | → ◁30° | 1.0 | $10^5$ |
| Semitubular | → ) | 2.3 | $4 \times 10^4$ |
| | → ( | 1.12 | $4 \times 10^4$ |

From W. F. Lindsey, *NACA Tech. Rep. 619*, 1938.

**TABLE A-24 Adiabatic frictional flow in a constant-area duct for $\gamma = 1.4$**

| $N_M$ | $\bar{f}L^*/D$ | $p/p^*$ | $T/T^*$ | $\rho^*/\rho = V/V^*$ | $p_0/p_0^*$ |
|---|---|---|---|---|---|
| 0.0 | ∞ | ∞ | 1.2000 | 0.0 | ∞ |
| 0.02 | 1778.4500 | 54.7701 | 1.1999 | 0.0219 | 28.9421 |
| 0.04 | 440.3520 | 27.3817 | 1.1996 | 0.0438 | 14.4815 |
| 0.06 | 193.0310 | 18.2508 | 1.1991 | 0.0657 | 9.6659 |
| 0.08 | 106.7180 | 13.6843 | 1.1985 | 0.0876 | 7.2616 |
| 0.1 | 66.9216 | 10.9435 | 1.1976 | 0.1094 | 5.8218 |
| 0.12 | 45.4080 | 9.1156 | 1.1966 | 0.1313 | 4.8643 |
| 0.14 | 32.5113 | 7.8093 | 1.1953 | 0.1531 | 4.1824 |
| 0.16 | 24.1978 | 6.8291 | 1.1939 | 0.1748 | 3.6727 |
| 0.18 | 18.5427 | 6.0662 | 1.1923 | 0.1965 | 3.2779 |
| 0.2 | 14.5333 | 5.4554 | 1.1905 | 0.2182 | 2.9635 |
| 0.22 | 11.5961 | 4.9554 | 1.1885 | 0.2398 | 2.7076 |
| 0.24 | 9.3865 | 4.5383 | 1.1863 | 0.2614 | 2.4956 |
| 0.26 | 7.6876 | 4.1851 | 1.1840 | 0.2829 | 2.3173 |
| 0.28 | 6.3572 | 3.8820 | 1.1815 | 0.3043 | 2.1656 |
| 0.3 | 5.2993 | 3.6191 | 1.1788 | 0.3257 | 2.0351 |
| 0.32 | 4.4467 | 3.3887 | 1.1759 | 0.3470 | 1.9219 |
| 0.34 | 3.7520 | 3.1853 | 1.1729 | 0.3682 | 1.8229 |
| 0.36 | 3.1801 | 3.0042 | 1.1697 | 0.3893 | 1.7358 |
| 0.38 | 2.7054 | 2.8420 | 1.1663 | 0.4104 | 1.6587 |
| 0.4 | 2.3085 | 2.6958 | 1.1628 | 0.4313 | 1.5901 |
| 0.42 | 1.9744 | 2.5634 | 1.1591 | 0.4522 | 1.5289 |
| 0.44 | 1.6915 | 2.4428 | 1.1553 | 0.4729 | 1.4740 |
| 0.46 | 1.4509 | 2.3326 | 1.1513 | 0.4936 | 1.4246 |
| 0.48 | 1.2453 | 2.2313 | 1.1471 | 0.5141 | 1.3801 |
| 0.5 | 1.0691 | 2.1381 | 1.1429 | 0.5345 | 1.3398 |
| 0.52 | 0.9174 | 2.0519 | 1.1384 | 0.5548 | 1.3034 |
| 0.54 | 0.7866 | 1.9719 | 1.1339 | 0.5750 | 1.2703 |
| 0.56 | 0.6736 | 1.8975 | 1.1292 | 0.5951 | 1.2403 |
| 0.58 | 0.5757 | 1.8282 | 1.1244 | 0.6150 | 1.2130 |
| 0.6 | 0.4908 | 1.7634 | 1.1194 | 0.6348 | 1.1882 |
| 0.62 | 0.4172 | 1.7026 | 1.1143 | 0.6545 | 1.1656 |
| 0.64 | 0.3533 | 1.6456 | 1.1091 | 0.6740 | 1.1451 |
| 0.66 | 0.2979 | 1.5919 | 1.1038 | 0.6934 | 1.1265 |
| 0.68 | 0.2498 | 1.5413 | 1.0984 | 0.7127 | 1.1097 |
| 0.7 | 0.2081 | 1.4935 | 1.0929 | 0.7318 | 1.0944 |
| 0.72 | 0.1721 | 1.4482 | 1.0873 | 0.7508 | 1.0806 |
| 0.74 | 0.1411 | 1.4054 | 1.0815 | 0.7696 | 1.0681 |
| 0.76 | 0.1145 | 1.3647 | 1.0757 | 0.7883 | 1.0570 |
| 0.78 | 0.0917 | 1.3261 | 1.0698 | 0.8068 | 1.0471 |
| 0.8 | 0.0723 | 1.2893 | 1.0638 | 0.8251 | 1.0382 |
| 0.82 | 0.0559 | 1.2542 | 1.0578 | 0.8433 | 1.0305 |
| 0.84 | 0.0423 | 1.2208 | 1.0516 | 0.8614 | 1.0237 |
| 0.86 | 0.0310 | 1.1889 | 1.0454 | 0.8793 | 1.0179 |
| 0.88 | 0.0218 | 1.1583 | 1.0391 | 0.8970 | 1.0129 |

Source: Fluid Mechanics by White

**TABLE A-24** (*Cont.*)

| $N_M$ | $\bar{f}L^*/D$ | $p/p^*$ | $T/T^*$ | $\rho^*/\rho = V/V^*$ | $p_0/p_0^*$ |
|-------|------|--------|--------|--------|--------|
| 0.9 | 0.0145 | 1.1291 | 1.0327 | 0.9146 | 1.0089 |
| 0.92 | 0.0089 | 1.1011 | 1.0263 | 0.9320 | 1.0056 |
| 0.94 | 0.0048 | 1.0743 | 1.0198 | 0.9493 | 1.0031 |
| 0.96 | 0.0021 | 1.0485 | 1.0132 | 0.9663 | 1.0014 |
| 0.98 | 0.0005 | 1.0238 | 1.0066 | 0.9833 | 1.0003 |
| 1.0 | 0.0000 | 1.0000 | 1.0000 | 1.0000 | 1.0000 |
| 1.02 | 0.0005 | 0.9771 | 0.9933 | 1.0166 | 1.0003 |
| 1.04 | 0.0018 | 0.9551 | 0.9866 | 1.0330 | 1.0013 |
| 1.06 | 0.0038 | 0.9338 | 0.9798 | 1.0492 | 1.0029 |
| 1.08 | 0.0066 | 0.9133 | 0.9730 | 1.0653 | 1.0051 |
| 1.1 | 0.0099 | 0.8936 | 0.9662 | 1.0812 | 1.0079 |
| 1.12 | 0.0138 | 0.8745 | 0.9593 | 1.0970 | 1.0113 |
| 1.14 | 0.0182 | 0.8561 | 0.9524 | 1.1126 | 1.0153 |
| 1.16 | 0.0230 | 0.8383 | 0.9455 | 1.1280 | 1.0198 |
| 1.18 | 0.0281 | 0.8210 | 0.9386 | 1.1432 | 1.0248 |
| 1.2 | 0.0336 | 0.8044 | 0.9317 | 1.1583 | 1.0304 |
| 1.22 | 0.0394 | 0.7882 | 0.9247 | 1.1732 | 1.0366 |
| 1.24 | 0.0455 | 0.7726 | 0.9178 | 1.1879 | 1.0432 |
| 1.26 | 0.0517 | 0.7574 | 0.9108 | 1.2025 | 1.0504 |
| 1.28 | 0.0582 | 0.7427 | 0.9038 | 1.2169 | 1.0581 |
| 1.3 | 0.0648 | 0.7285 | 0.8969 | 1.2311 | 1.0663 |
| 1.32 | 0.0716 | 0.7147 | 0.8899 | 1.2452 | 1.0750 |
| 1.34 | 0.0785 | 0.7012 | 0.8829 | 1.2591 | 1.0842 |
| 1.36 | 0.0855 | 0.6882 | 0.8760 | 1.2729 | 1.0940 |
| 1.38 | 0.0926 | 0.6755 | 0.8690 | 1.2864 | 1.1042 |
| 1.4 | 0.0997 | 0.6632 | 0.8621 | 1.2999 | 1.1149 |
| 1.42 | 0.1069 | 0.6512 | 0.8551 | 1.3131 | 1.1262 |
| 1.44 | 0.1142 | 0.6396 | 0.8482 | 1.3262 | 1.1379 |
| 1.46 | 0.1215 | 0.6282 | 0.8413 | 1.3392 | 1.1501 |
| 1.48 | 0.1288 | 0.6172 | 0.8344 | 1.3520 | 1.1629 |
| 1.5 | 0.1361 | 0.6065 | 0.8276 | 1.3646 | 1.1762 |
| 1.52 | 0.1433 | 0.5960 | 0.8207 | 1.3770 | 1.1899 |
| 1.54 | 0.1506 | 0.5858 | 0.8139 | 1.3894 | 1.2042 |
| 1.56 | 0.1579 | 0.5759 | 0.8071 | 1.4015 | 1.2190 |
| 1.58 | 0.1651 | 0.5662 | 0.8004 | 1.4135 | 1.2344 |
| 1.6 | 0.1724 | 0.5568 | 0.7937 | 1.4254 | 1.2502 |
| 1.62 | 0.1795 | 0.5476 | 0.7869 | 1.4371 | 1.2666 |
| 1.64 | 0.1867 | 0.5386 | 0.7803 | 1.4487 | 1.2836 |
| 1.66 | 0.1938 | 0.5299 | 0.7736 | 1.4601 | 1.3010 |
| 1.68 | 0.2008 | 0.5213 | 0.7670 | 1.4713 | 1.3190 |
| 1.7 | 0.2078 | 0.5130 | 0.7605 | 1.4825 | 1.3376 |
| 1.72 | 0.2147 | 0.5048 | 0.7539 | 1.4935 | 1.3567 |
| 1.74 | 0.2216 | 0.4969 | 0.7474 | 1.5043 | 1.3764 |
| 1.76 | 0.2284 | 0.4891 | 0.7410 | 1.5150 | 1.3967 |
| 1.78 | 0.2352 | 0.4815 | 0.7345 | 1.5256 | 1.4175 |

**TABLE A-24** (*Contd.*) **Adiabatic frictional flow in a constant-area duct for** $\gamma = 1.4$

| $N_M$ | $\bar{f}L^*/D$ | $p/p^*$ | $T/T^*$ | $\rho^*/\rho = V/V^*$ | $p_0/p_0^*$ |
|------|------|------|------|------|------|
| 1.8 | 0.2419 | 0.4741 | 0.7282 | 1.5360 | 1.4390 |
| 1.82 | 0.2485 | 0.4668 | 0.7218 | 1.5463 | 1.4610 |
| 1.84 | 0.2551 | 0.4597 | 0.7155 | 1.5564 | 1.4836 |
| 1.86 | 0.2616 | 0.4528 | 0.7093 | 1.5664 | 1.5069 |
| 1.88 | 0.2680 | 0.4460 | 0.7030 | 1.5763 | 1.5308 |
| 1.9 | 0.2743 | 0.4394 | 0.6969 | 1.5861 | 1.5553 |
| 1.92 | 0.2806 | 0.4329 | 0.6907 | 1.5957 | 1.5804 |
| 1.94 | 0.2868 | 0.4265 | 0.6847 | 1.6052 | 1.6062 |
| 1.96 | 0.2929 | 0.4203 | 0.6786 | 1.6146 | 1.6326 |
| 1.98 | 0.2990 | 0.4142 | 0.6726 | 1.6239 | 1.6597 |
| 2.0 | 0.3050 | 0.4082 | 0.6667 | 1.6330 | 1.6875 |
| 2.02 | 0.3109 | 0.4024 | 0.6608 | 1.6420 | 1.7160 |
| 2.04 | 0.3168 | 0.3967 | 0.6549 | 1.6509 | 1.7451 |
| 2.06 | 0.3225 | 0.3911 | 0.6491 | 1.6597 | 1.7750 |
| 2.08 | 0.3282 | 0.3856 | 0.6433 | 1.6683 | 1.8056 |
| 2.1 | 0.3339 | 0.3802 | 0.6376 | 1.6769 | 1.8369 |
| 2.12 | 0.3394 | 0.3750 | 0.6320 | 1.6853 | 1.8690 |
| 2.14 | 0.3449 | 0.3698 | 0.6263 | 1.6936 | 1.9018 |
| 2.16 | 0.3503 | 0.3648 | 0.6208 | 1.7018 | 1.9354 |
| 2.18 | 0.3556 | 0.3598 | 0.6152 | 1.7099 | 1.9698 |
| 2.2 | 0.3609 | 0.3549 | 0.6098 | 1.7179 | 2.0050 |
| 2.22 | 0.3661 | 0.3502 | 0.6043 | 1.7258 | 2.0409 |
| 2.24 | 0.3712 | 0.3455 | 0.5989 | 1.7336 | 2.0777 |
| 2.26 | 0.3763 | 0.3409 | 0.5936 | 1.7412 | 2.1153 |
| 2.28 | 0.3813 | 0.3364 | 0.5883 | 1.7488 | 2.1538 |
| 2.3 | 0.3862 | 0.3320 | 0.5831 | 1.7563 | 2.1931 |
| 2.32 | 0.3911 | 0.3277 | 0.5779 | 1.7637 | 2.2333 |
| 2.34 | 0.3959 | 0.3234 | 0.5728 | 1.7709 | 2.2744 |
| 2.36 | 0.4006 | 0.3193 | 0.5677 | 1.7781 | 2.3164 |
| 2.38 | 0.4053 | 0.3152 | 0.5626 | 1.7852 | 2.3593 |
| 2.4 | 0.4099 | 0.3111 | 0.5576 | 1.7922 | 2.4031 |
| 2.42 | 0.4144 | 0.3072 | 0.5527 | 1.7991 | 2.4479 |
| 2.44 | 0.4189 | 0.3033 | 0.5478 | 1.8059 | 2.4936 |
| 2.46 | 0.4233 | 0.2995 | 0.5429 | 1.8126 | 2.5403 |
| 2.48 | 0.4277 | 0.2958 | 0.5381 | 1.8192 | 2.5880 |
| 2.5 | 0.4320 | 0.2921 | 0.5333 | 1.8257 | 2.6367 |
| 2.52 | 0.4362 | 0.2885 | 0.5286 | 1.8322 | 2.6865 |
| 2.54 | 0.4404 | 0.2850 | 0.5239 | 1.8386 | 2.7372 |
| 2.56 | 0.4445 | 0.2815 | 0.5193 | 1.8448 | 2.7891 |
| 2.58 | 0.4486 | 0.2781 | 0.5147 | 1.8510 | 2.8420 |
| 2.6 | 0.4526 | 0.2747 | 0.5102 | 1.8571 | 2.8960 |
| 2.62 | 0.4565 | 0.2714 | 0.5057 | 1.8632 | 2.9511 |
| 2.64 | 0.4604 | 0.2682 | 0.5013 | 1.8691 | 3.0073 |
| 2.66 | 0.4643 | 0.2650 | 0.4969 | 1.8750 | 3.0647 |
| 2.68 | 0.4681 | 0.2619 | 0.4925 | 1.8808 | 3.1233 |

**TABLE A-24** (*Cont.*)

| $N_M$ | $\bar{f}L^*/D$ | $p/p^*$ | $T/T^*$ | $\rho^*/\rho = V/V^*$ | $p_0/p_0^*$ |
|---|---|---|---|---|---|
| 2.7 | 0.4718 | 0.2588 | 0.4882 | 1.8865 | 3.1830 |
| 2.72 | 0.4755 | 0.2558 | 0.4839 | 1.8922 | 3.2440 |
| 2.74 | 0.4791 | 0.2528 | 0.4797 | 1.8978 | 3.3061 |
| 2.76 | 0.4827 | 0.2498 | 0.4755 | 1.9033 | 3.3695 |
| 2.78 | 0.4863 | 0.2470 | 0.4714 | 1.9087 | 3.4342 |
| 2.8 | 0.4898 | 0.2441 | 0.4673 | 1.9140 | 3.5001 |
| 2.82 | 0.4932 | 0.2414 | 0.4632 | 1.9193 | 3.5674 |
| 2.84 | 0.4966 | 0.2386 | 0.4592 | 1.9246 | 3.6359 |
| 2.86 | 0.5000 | 0.2359 | 0.4552 | 1.9297 | 3.7058 |
| 2.88 | 0.5033 | 0.2333 | 0.4513 | 1.9348 | 3.7771 |
| 2.9 | 0.5065 | 0.2307 | 0.4474 | 1.9398 | 3.8498 |
| 2.92 | 0.5097 | 0.2281 | 0.4436 | 1.9448 | 3.9238 |
| 2.94 | 0.5129 | 0.2256 | 0.4398 | 1.9497 | 3.9993 |
| 2.96 | 0.5160 | 0.2231 | 0.4360 | 1.9545 | 4.0763 |
| 2.98 | 0.5191 | 0.2206 | 0.4323 | 1.9593 | 4.1547 |
| 3.0 | 0.5222 | 0.2182 | 0.4286 | 1.9640 | 4.2346 |
| 3.02 | 0.5252 | 0.2158 | 0.4249 | 1.9686 | 4.3160 |
| 3.04 | 0.5281 | 0.2135 | 0.4213 | 1.9732 | 4.3989 |
| 3.06 | 0.5310 | 0.2112 | 0.4177 | 1.9777 | 4.4835 |
| 3.08 | 0.5339 | 0.2090 | 0.4142 | 1.9822 | 4.5696 |
| 3.1 | 0.5368 | 0.2067 | 0.4107 | 1.9866 | 4.6573 |
| 3.12 | 0.5396 | 0.2045 | 0.4072 | 1.9910 | 4.7467 |
| 3.14 | 0.5424 | 0.2024 | 0.4038 | 1.9953 | 4.8377 |
| 3.16 | 0.5451 | 0.2002 | 0.4004 | 1.9995 | 4.9304 |
| 3.18 | 0.5478 | 0.1981 | 0.3970 | 2.0037 | 5.0248 |
| 3.2 | 0.5504 | 0.1961 | 0.3937 | 2.0079 | 5.1210 |
| 3.22 | 0.5531 | 0.1940 | 0.3904 | 2.0120 | 5.2189 |
| 3.24 | 0.5557 | 0.1920 | 0.3872 | 2.0160 | 5.3186 |
| 3.26 | 0.5582 | 0.1901 | 0.3839 | 2.0200 | 5.4201 |
| 3.28 | 0.5607 | 0.1881 | 0.3807 | 2.0239 | 5.5234 |
| 3.3 | 0.5632 | 0.1862 | 0.3776 | 2.0278 | 5.6286 |
| 3.32 | 0.5657 | 0.1843 | 0.3745 | 2.0317 | 5.7358 |
| 3.34 | 0.5681 | 0.1825 | 0.3714 | 2.0355 | 5.8448 |
| 3.36 | 0.5705 | 0.1806 | 0.3683 | 2.0392 | 5.9558 |
| 3.38 | 0.5729 | 0.1788 | 0.3653 | 2.0429 | 6.0687 |
| 3.4 | 0.5752 | 0.1770 | 0.3623 | 2.0466 | 6.1837 |
| 3.42 | 0.5775 | 0.1753 | 0.3594 | 2.0502 | 6.3007 |
| 3.44 | 0.5798 | 0.1736 | 0.3564 | 2.0537 | 6.4198 |
| 3.46 | 0.5820 | 0.1718 | 0.3535 | 2.0573 | 6.5409 |
| 3.48 | 0.5842 | 0.1702 | 0.3507 | 2.0607 | 6.6642 |
| 3.5 | 0.5864 | 0.1685 | 0.3478 | 2.0642 | 6.7896 |
| 3.52 | 0.5886 | 0.1669 | 0.3450 | 2.0676 | 6.9172 |
| 3.54 | 0.5907 | 0.1653 | 0.3422 | 2.0709 | 7.0471 |
| 3.56 | 0.5928 | 0.1637 | 0.3395 | 2.0743 | 7.1791 |
| 3.58 | 0.5949 | 0.1621 | 0.3368 | 2.0775 | 7.3135 |

**TABLE A-24** (*Cont.*) **Adiabatic frictional flow in a constant-area duct for $\gamma = 1.4$**

| $N_M$ | $\bar{f}L^*/D$ | $p/p^*$ | $T/T^*$ | $\rho^*/\rho = V/V^*$ | $p_0/p_0^*$ |
|---|---|---|---|---|---|
| 3.6 | 0.5970 | 0.1606 | 0.3341 | 2.0808 | 7.4501 |
| 3.62 | 0.05990 | 0.1590 | 0.3314 | 2.0840 | 7.5891 |
| 3.64 | 0.6010 | 0.1575 | 0.3288 | 2.0871 | 7.7305 |
| 3.66 | 0.6030 | 0.1560 | 0.3262 | 2.0903 | 7.8742 |
| 3.68 | 0.6049 | 0.1546 | 0.3236 | 2.0933 | 8.0204 |
| 3.7 | 0.6068 | 0.1531 | 0.3210 | 2.0964 | 8.1691 |
| 3.72 | 0.6087 | 0.1517 | 0.3185 | 2.0994 | 8.3202 |
| 3.74 | 0.6106 | 0.1503 | 0.3160 | 2.1024 | 8.4739 |
| 3.76 | 0.6125 | 0.1489 | 0.3135 | 2.1053 | 8.6302 |
| 3.78 | 0.6143 | 0.1475 | 0.3111 | 2.1082 | 8.7891 |
| 3.8 | 0.6161 | 0.1462 | 0.3086 | 2.1111 | 8.9506 |
| 3.82 | 0.6179 | 0.1449 | 0.3062 | 2.1140 | 9.1148 |
| 3.84 | 0.6197 | 0.1436 | 0.3039 | 2.1168 | 9.2817 |
| 3.86 | 0.6214 | 0.1423 | 0.3015 | 2.1195 | 9.4513 |
| 3.88 | 0.6231 | 0.1410 | 0.2992 | 2.1223 | 9.6237 |
| 3.9 | 0.6248 | 0.1397 | 0.2969 | 2.1250 | 9.7990 |
| 3.92 | 0.6265 | 0.1385 | 0.2946 | 2.1277 | 9.9771 |
| 3.94 | 0.6282 | 0.1372 | 0.2923 | 2.1303 | 10.1581 |
| 3.96 | 0.6298 | 0.1360 | 0.2901 | 2.1329 | 10.3420 |
| 3.98 | 0.6315 | 0.1348 | 0.2879 | 2.1355 | 10.5289 |
| 4.0 | 0.6331 | 0.1336 | 0.2857 | 2.1381 | 10.7188 |

**TABLE A-25 Frictionless duct flow with heat transfer for $\gamma = 1.4$**

| $N_M$ | $T_0/T_0^*$ | $p/p^*$ | $T/T^*$ | $\rho^*/\rho = V/V^*$ | $p_0/p_0^*$ |
|---|---|---|---|---|---|
| 0.0 | 0.0 | 2.4000 | 0.0 | 0.0 | 1.2679 |
| 0.02 | 0.0019 | 2.3987 | 0.0023 | 0.0010 | 1.2675 |
| 0.04 | 0.0076 | 2.3946 | 0.0092 | 0.0038 | 1.2665 |
| 0.06 | 0.0171 | 2.3880 | 0.0205 | 0.0086 | 1.2647 |
| 0.08 | 0.0302 | 2.3787 | 0.0362 | 0.0152 | 1.2623 |
| 0.1 | 0.0468 | 2.3669 | 0.0560 | 0.0237 | 1.2591 |
| 0.12 | 0.0666 | 2.3526 | 0.0797 | 0.0339 | 1.2554 |
| 0.14 | 0.0895 | 2.3359 | 0.1069 | 0.0458 | 1.2510 |
| 0.16 | 0.1151 | 2.3170 | 0.1374 | 0.0593 | 1.2461 |
| 0.18 | 0.1432 | 2.2959 | 0.1708 | 0.0744 | 1.2406 |
| 0.2 | 0.1736 | 2.2727 | 0.2066 | 0.0909 | 1.2346 |
| 0.22 | 0.2057 | 2.2477 | 0.2445 | 0.1088 | 1.2281 |
| 0.24 | 0.2395 | 2.2209 | 0.2841 | 0.1279 | 1.2213 |
| 0.26 | 0.2745 | 2.1925 | 0.3250 | 0.1482 | 1.2140 |
| 0.28 | 0.3104 | 2.1626 | 0.3667 | 0.1696 | 1.2064 |
| 0.3 | 0.3469 | 2.1314 | 0.4089 | 0.1918 | 1.1985 |
| 0.32 | 0.3837 | 2.0991 | 0.4512 | 0.2149 | 1.1904 |
| 0.34 | 0.4206 | 2.0657 | 0.4933 | 0.2388 | 1.1822 |
| 0.36 | 0.4572 | 2.0314 | 0.5348 | 0.2633 | 1.1737 |
| 0.38 | 0.4935 | 1.9964 | 0.5755 | 0.2883 | 1.1652 |
| 0.4 | 0.5290 | 1.9608 | 0.6151 | 0.3137 | 1.1566 |
| 0.42 | 0.5638 | 1.9247 | 0.6535 | 0.3395 | 1.1480 |
| 0.44 | 0.5975 | 1.8882 | 0.6903 | 0.3656 | 1.1394 |
| 0.46 | 0.6301 | 1.8515 | 0.7254 | 0.3918 | 1.1308 |
| 0.48 | 0.6614 | 1.8147 | 0.7587 | 0.4181 | 1.1224 |
| 0.5 | 0.6914 | 1.7778 | 0.7901 | 0.4444 | 1.1141 |
| 0.52 | 0.7199 | 1.7409 | 0.8196 | 0.4708 | 1.1059 |
| 0.54 | 0.7470 | 1.7043 | 0.8469 | 0.4970 | 1.0979 |
| 0.56 | 0.7725 | 1.6678 | 0.8723 | 0.5230 | 1.0901 |
| 0.58 | 0.7965 | 1.6316 | 0.8955 | 0.5489 | 1.0826 |
| 0.6 | 0.8189 | 1.5957 | 0.9167 | 0.5745 | 1.0753 |
| 0.62 | 0.8398 | 1.5603 | 0.9358 | 0.5998 | 1.0682 |
| 0.64 | 0.8592 | 1.5253 | 0.9530 | 0.6248 | 1.0615 |
| 0.66 | 0.8771 | 1.4908 | 0.9682 | 0.6494 | 1.0550 |
| 0.68 | 0.8935 | 1.4569 | 0.9814 | 0.6737 | 1.0489 |
| 0.7 | 0.9085 | 1.4235 | 0.9929 | 0.6975 | 1.0431 |
| 0.72 | 0.9221 | 1.3907 | 1.0026 | 0.7209 | 1.0376 |
| 0.74 | 0.9344 | 1.3585 | 1.0106 | 0.7439 | 1.0325 |
| 0.76 | 0.9455 | 1.3270 | 1.0171 | 0.7665 | 1.0278 |
| 0.78 | 0.9553 | 1.2961 | 1.0220 | 0.7885 | 1.0234 |
| 0.8 | 0.9639 | 1.2658 | 1.0255 | 0.8101 | 1.0193 |
| 0.82 | 0.9715 | 1.2362 | 1.0276 | 0.8313 | 1.0157 |
| 0.84 | 0.9781 | 1.2073 | 1.0285 | 0.8519 | 1.0124 |
| 0.86 | 0.9836 | 1.1791 | 1.0283 | 0.8721 | 1.0095 |
| 0.88 | 0.9883 | 1.1515 | 1.0269 | 0.8918 | 1.0070 |

**TABLE A-25** (*Cont.*) **Frictionless duct flow with heat transfer for $\gamma = 1.4$**

| $N_M$ | $T_0/T_0^*$ | $p/p^*$ | $T/T^*$ | $\rho^*/\rho = V/V^*$ | $p_0/p_0^*$ |
|-------|-------------|---------|---------|------------------------|-------------|
| 0.9 | 0.9921 | 1.1246 | 1.0245 | 0.9110 | 1.0049 |
| 0.92 | 0.9951 | 1.0984 | 1.0212 | 0.9297 | 1.0031 |
| 0.94 | 0.9973 | 1.0728 | 1.0170 | 0.9480 | 1.0017 |
| 0.96 | 0.9988 | 1.0479 | 1.0121 | 0.9658 | 1.0008 |
| 0.98 | 0.9997 | 1.0236 | 1.0064 | 0.9831 | 1.0002 |
| 1.0 | 1.0000 | 1.0000 | 1.0000 | 1.0000 | 1.0000 |
| 1.02 | 0.9997 | 0.9770 | 0.9930 | 1.0164 | 1.0002 |
| 1.04 | 0.9989 | 0.9546 | 0.9855 | 1.0325 | 1.0008 |
| 1.06 | 0.9977 | 0.9327 | 0.9776 | 1.0480 | 1.0017 |
| 1.08 | 0.9960 | 0.9115 | 0.9691 | 1.0632 | 1.0031 |
| 1.1 | 0.9939 | 0.8909 | 0.9603 | 1.0780 | 1.0049 |
| 1.12 | 0.9915 | 0.8708 | 0.9512 | 1.0923 | 1.0070 |
| 1.14 | 0.9887 | 0.8512 | 0.9417 | 1.1063 | 1.0095 |
| 1.16 | 0.9856 | 0.8322 | 0.9320 | 1.1198 | 1.0124 |
| 1.18 | 0.9823 | 0.8137 | 0.9220 | 1.1330 | 1.0157 |
| 1.2 | 0.9787 | 0.7958 | 0.9118 | 1.1459 | 1.0194 |
| 1.22 | 0.9749 | 0.7783 | 0.9015 | 1.1584 | 1.0235 |
| 1.24 | 0.9709 | 0.7613 | 0.8911 | 1.1705 | 1.0279 |
| 1.26 | 0.9668 | 0.7447 | 0.8805 | 1.1823 | 1.0328 |
| 1.28 | 0.9624 | 0.7287 | 0.8699 | 1.1938 | 1.0380 |
| 1.3 | 0.9580 | 0.7130 | 0.8592 | 1.2050 | 1.0437 |
| 1.32 | 0.9534 | 0.6978 | 0.8484 | 1.2159 | 1.0497 |
| 1.34 | 0.9487 | 0.6830 | 0.8377 | 1.2264 | 1.0561 |
| 1.36 | 0.9440 | 0.6686 | 0.8269 | 1.2367 | 1.0629 |
| 1.38 | 0.9391 | 0.6546 | 0.8161 | 1.2467 | 1.0701 |
| 1.4 | 0.9343 | 0.6410 | 0.8054 | 1.2564 | 1.0777 |
| 1.42 | 0.9293 | 0.6278 | 0.7947 | 1.2659 | 1.0856 |
| 1.44 | 0.9243 | 0.6149 | 0.7840 | 1.2751 | 1.0940 |
| 1.46 | 0.9193 | 0.6024 | 0.7735 | 1.2840 | 1.1028 |
| 1.48 | 0.9143 | 0.5902 | 0.7629 | 1.2927 | 1.1120 |
| 1.5 | 0.9093 | 0.5783 | 0.7525 | 1.3012 | 1.1215 |
| 1.52 | 0.9042 | 0.5668 | 0.7422 | 1.3095 | 1.1315 |
| 1.54 | 0.8992 | 0.5555 | 0.7319 | 1.3175 | 1.1419 |
| 1.56 | 0.8942 | 0.5446 | 0.7217 | 1.3253 | 1.1527 |
| 1.58 | 0.8892 | 0.5339 | 0.7117 | 1.3329 | 1.1640 |
| 1.6 | 0.8842 | 0.5236 | 0.7017 | 1.3403 | 1.1756 |
| 1.62 | 0.8792 | 0.5135 | 0.6919 | 1.3475 | 1.1877 |
| 1.64 | 0.8743 | 0.5036 | 0.6822 | 1.3546 | 1.2002 |
| 1.66 | 0.8694 | 0.4940 | 0.6726 | 1.3614 | 1.2131 |
| 1.68 | 0.8645 | 0.4847 | 0.6631 | 1.3681 | 1.2264 |
| 1.7 | 0.8597 | 0.4756 | 0.6538 | 1.3746 | 1.2402 |
| 1.72 | 0.8549 | 0.4668 | 0.6445 | 1.3809 | 1.2545 |
| 1.74 | 0.8502 | 0.4581 | 0.6355 | 1.3870 | 1.2692 |
| 1.76 | 0.8455 | 0.4497 | 0.6265 | 1.3931 | 1.2843 |
| 1.78 | 0.8409 | 0.4415 | 0.6176 | 1.3989 | 1.2999 |

**TABLE A-25** (*Cont.*)

| $N_M$ | $T_0/T_0^*$ | $p/p^*$ | $T/T^*$ | $\rho^*/\rho = V/V^*$ | $p_0/p_0^*$ |
|---|---|---|---|---|---|
| 1.8 | 0.8363 | 0.4335 | 0.6089 | 1.4046 | 1.3159 |
| 1.82 | 0.8317 | 0.4257 | 0.6004 | 1.4102 | 1.3324 |
| 1.84 | 0.8273 | 0.4181 | 0.5919 | 1.4156 | 1.3494 |
| 1.86 | 0.8228 | 0.4107 | 0.5836 | 1.4209 | 1.3669 |
| 1.88 | 0.8185 | 0.4035 | 0.5754 | 1.4261 | 1.3849 |
| 1.9 | 0.8141 | 0.3964 | 0.5673 | 1.4311 | 1.4033 |
| 1.92 | 0.8099 | 0.3895 | 0.5594 | 1.4360 | 1.4222 |
| 1.94 | 0.8057 | 0.3828 | 0.5516 | 1.4408 | 1.4417 |
| 1.96 | 0.8015 | 0.3763 | 0.5439 | 1.4455 | 1.4616 |
| 1.98 | 0.7974 | 0.3699 | 0.5364 | 1.4501 | 1.4821 |
| 2.0 | 0.7934 | 0.3636 | 0.5289 | 1.4545 | 1.5031 |
| 2.02 | 0.7894 | 0.3575 | 0.5216 | 1.4589 | 1.5246 |
| 2.04 | 0.7855 | 0.3516 | 0.5144 | 1.4632 | 1.5467 |
| 2.06 | 0.7816 | 0.3458 | 0.5074 | 1.4673 | 1.5693 |
| 2.08 | 0.7778 | 0.3401 | 0.5004 | 1.4714 | 1.5924 |
| 2.1 | 0.7741 | 0.3345 | 0.4936 | 1.4753 | 1.6162 |
| 2.12 | 0.7704 | 0.3291 | 0.4868 | 1.4792 | 1.6404 |
| 2.14 | 0.7667 | 0.3238 | 0.4802 | 1.4830 | 1.6653 |
| 2.16 | 0.7631 | 0.3186 | 0.4737 | 1.4867 | 1.6908 |
| 2.18 | 0.7596 | 0.3136 | 0.4673 | 1.4903 | 1.7168 |
| 2.2 | 0.7561 | 0.3086 | 0.4611 | 1.4938 | 1.7434 |
| 2.22 | 0.7527 | 0.3038 | 0.4549 | 1.4973 | 1.7707 |
| 2.24 | 0.7493 | 0.2991 | 0.4488 | 1.5007 | 1.7986 |
| 2.26 | 0.7460 | 0.2945 | 0.4428 | 1.5040 | 1.8271 |
| 2.28 | 0.7428 | 0.2899 | 0.4370 | 1.5072 | 1.8562 |
| 2.3 | 0.7395 | 0.2855 | 0.4312 | 1.5104 | 1.8860 |
| 2.32 | 0.7364 | 0.2812 | 0.4256 | 1.5134 | 1.9165 |
| 2.34 | 0.7333 | 0.2769 | 0.4200 | 1.5165 | 1.9476 |
| 2.36 | 0.7302 | 0.2728 | 0.4145 | 1.5194 | 1.9794 |
| 2.38 | 0.7272 | 0.2688 | 0.4091 | 1.5223 | 2.0119 |
| 2.4 | 0.7242 | 0.2648 | 0.4038 | 1.5252 | 2.0451 |
| 2.42 | 0.7213 | 0.2609 | 0.3986 | 1.5279 | 2.0789 |
| 2.44 | 0.7184 | 0.2571 | 0.3935 | 1.5306 | 2.1136 |
| 2.46 | 0.7156 | 0.2534 | 0.3885 | 1.5333 | 2.1489 |
| 2.48 | 0.7128 | 0.2497 | 0.3836 | 1.5359 | 2.1850 |
| 2.5 | 0.7101 | 0.2462 | 0.3787 | 1.5385 | 2.2218 |
| 2.52 | 0.7074 | 0.2427 | 0.3739 | 1.5410 | 2.2594 |
| 2.54 | 0.7047 | 0.2392 | 0.3692 | 1.5434 | 2.2978 |
| 2.56 | 0.7021 | 0.2359 | 0.3646 | 1.5458 | 2.3370 |
| 2.58 | 0.6995 | 0.2326 | 0.3601 | 1.5482 | 2.3770 |
| 2.6 | 0.6970 | 0.2294 | 0.3556 | 1.5505 | 2.4177 |
| 2.62 | 0.6945 | 0.2262 | 0.3512 | 1.5527 | 2.4593 |
| 2.64 | 0.6921 | 0.2231 | 0.3469 | 1.5549 | 2.5018 |
| 2.66 | 0.6896 | 0.2201 | 0.3427 | 1.5571 | 2.5451 |
| 2.68 | 0.6873 | 0.2171 | 0.3385 | 1.5592 | 2.5892 |

**TABLE A-25** (*Cont.*) **Frictionless duct flow with heat transfer for** $\gamma = 1.4$

| $N_M$ | $T_0/T_0^*$ | $p/p^*$ | $T/T^*$ | $\rho^*/\rho = V/V^*$ | $p_0/p_0^*$ |
|---|---|---|---|---|---|
| 2.7 | 0.6849 | 0.2142 | 0.3344 | 1.5613 | 2.6343 |
| 2.72 | 0.6826 | 0.2113 | 0.3304 | 1.5634 | 2.6802 |
| 2.74 | 0.6804 | 0.2085 | 0.3264 | 1.5654 | 2.7270 |
| 2.76 | 0.6781 | 0.2058 | 0.3225 | 1.5673 | 2.7748 |
| 2.78 | 0.6760 | 0.2030 | 0.3186 | 1.5693 | 2.8235 |
| 2.8 | 0.6738 | 0.2004 | 0.3149 | 1.5711 | 2.8731 |
| 2.82 | 0.6717 | 0.1978 | 0.3111 | 1.5730 | 2.9237 |
| 2.84 | 0.6696 | 0.1953 | 0.3075 | 1.5748 | 2.9752 |
| 2.86 | 0.6675 | 0.1927 | 0.3039 | 1.5766 | 3.0278 |
| 2.88 | 0.6655 | 0.1903 | 0.3004 | 1.5784 | 3.0813 |
| 2.9 | 0.6635 | 0.1879 | 0.2969 | 1.5801 | 3.1359 |
| 2.92 | 0.6615 | 0.1855 | 0.2934 | 1.5818 | 3.1914 |
| 2.94 | 0.6596 | 0.1832 | 0.2901 | 1.5834 | 3.2481 |
| 2.96 | 0.6577 | 0.1809 | 0.2868 | 1.5851 | 3.3058 |
| 2.98 | 0.6558 | 0.1787 | 0.2835 | 1.5867 | 3.3646 |
| 3.0 | 0.6540 | 0.1765 | 0.2803 | 1.5882 | 3.4245 |
| 3.02 | 0.6522 | 0.1743 | 0.2771 | 1.5898 | 3.4854 |
| 3.04 | 0.6504 | 0.1722 | 0.2740 | 1.5913 | 3.5476 |
| 3.06 | 0.6486 | 0.1701 | 0.2709 | 1.5928 | 3.6108 |
| 3.08 | 0.6469 | 0.1681 | 0.2679 | 1.5942 | 3.6752 |
| 3.1 | 0.6452 | 0.1660 | 0.2650 | 1.5957 | 3.7408 |
| 3.12 | 0.6435 | 0.1641 | 0.2620 | 1.5971 | 3.8076 |
| 3.14 | 0.6418 | 0.1621 | 0.2592 | 1.5985 | 3.8756 |
| 3.16 | 0.6402 | 0.1602 | 0.2563 | 1.5998 | 3.9449 |
| 3.18 | 0.6386 | 0.1583 | 0.2535 | 1.6012 | 4.0154 |
| 3.2 | 0.6370 | 0.1565 | 0.2508 | 1.6025 | 4.0871 |
| 3.22 | 0.6354 | 0.1547 | 0.2481 | 1.6038 | 4.1602 |
| 3.24 | 0.6339 | 0.1529 | 0.2454 | 1.6051 | 4.2345 |
| 3.26 | 0.6324 | 0.1511 | 0.2428 | 1.6063 | 4.3101 |
| 3.28 | 0.6309 | 0.1494 | 0.2402 | 1.6076 | 4.3871 |
| 3.3 | 0.6294 | 0.1477 | 0.2377 | 1.6088 | 4.4655 |
| 3.32 | 0.6280 | 0.1461 | 0.2352 | 1.6100 | 4.5452 |
| 3.34 | 0.6265 | 0.1444 | 0.2327 | 1.6111 | 4.6263 |

**TABLE A-25** (*Cont.*)

| $N_M$ | $T_0/T_0^*$ | $p/p^*$ | $T/T^*$ | $\rho^*/\rho = V/V^*$ | $p_0/p_0^*$ |
|---|---|---|---|---|---|
| 3.36 | 0.6251 | 0.1428 | 0.2303 | 1.6123 | 4.7089 |
| 3.38 | 0.6237 | 0.1412 | 0.2279 | 1.6134 | 4.7929 |
| 3.4 | 0.6224 | 0.1397 | 0.2255 | 1.6145 | 4.8783 |
| 3.42 | 0.6210 | 0.1381 | 0.2232 | 1.6156 | 4.9652 |
| 3.44 | 0.6197 | 0.1366 | 0.2209 | 1.6167 | 5.0536 |
| 3.46 | 0.6184 | 0.1351 | 0.2186 | 1.6178 | 5.1435 |
| 3.48 | 0.6171 | 0.1337 | 0.2164 | 1.6188 | 5.2350 |
| 3.5 | 0.6158 | 0.1322 | 0.2142 | 1.6198 | 5.3280 |
| 3.52 | 0.6145 | 0.1308 | 0.2120 | 1.6208 | 5.4226 |
| 3.54 | 0.6133 | 0.1294 | 0.2099 | 1.6218 | 5.5188 |
| 3.56 | 0.6121 | 0.1280 | 0.2078 | 1.6228 | 5.6167 |
| 3.58 | 0.6109 | 0.1267 | 0.2057 | 1.6238 | 5.7162 |
| 3.6 | 0.6097 | 0.1254 | 0.2037 | 1.6247 | 5.8173 |
| 3.62 | 0.6085 | 0.1241 | 0.2017 | 1.6257 | 5.9201 |
| 3.64 | 0.6074 | 0.1228 | 0.1997 | 1.6266 | 6.0247 |
| 3.66 | 0.6062 | 0.1215 | 0.1977 | 1.6275 | 6.1310 |
| 3.68 | 0.6051 | 0.1202 | 0.1958 | 1.6284 | 6.2390 |
| 3.7 | 0.6040 | 0.1190 | 0.1939 | 1.6293 | 6.3488 |
| 3.72 | 0.6029 | 0.1178 | 0.1920 | 1.6301 | 6.4605 |
| 3.74 | 0.6018 | 0.1166 | 0.1902 | 1.6310 | 6.5739 |
| 3.76 | 0.6008 | 0.1154 | 0.1884 | 1.6318 | 6.6893 |
| 3.78 | 0.5997 | 0.1143 | 0.1866 | 1.6327 | 6.8065 |
| 3.8 | 0.5987 | 0.1131 | 0.1848 | 1.6335 | 6.9256 |
| 3.82 | 0.5977 | 0.1120 | 0.1830 | 1.6343 | 7.0466 |
| 3.84 | 0.5967 | 0.1109 | 0.1813 | 1.6351 | 7.1696 |
| 3.86 | 0.5957 | 0.1098 | 0.1796 | 1.6359 | 7.2945 |
| 3.88 | 0.5947 | 0.1087 | 0.1779 | 1.6366 | 7.4215 |
| 3.9 | 0.5937 | 0.1077 | 0.1763 | 1.6374 | 7.5505 |
| 3.92 | 0.5928 | 0.1066 | 0.1746 | 1.6381 | 7.6816 |
| 3.94 | 0.5918 | 0.1056 | 0.1730 | 1.6389 | 7.8147 |
| 3.96 | 0.5909 | 0.1046 | 0.1714 | 1.6396 | 7.9499 |
| 3.98 | 0.5900 | 0.1036 | 0.1699 | 1.6403 | 8.0873 |
| 4.0 | 0.5891 | 0.1026 | 0.1683 | 1.6410 | 8.2269 |

# INDEX

# SCHAUM'S SOLVED PROBLEMS SERIES

■ Learn the best strategies for solving tough problems in step-by-step detail
■ Prepare effectively for exams and save time in doing homework problems
■ Use the indexes to quickly locate the types of problems you need the most help solving
■ Save these books for reference in other courses and even for your professional library